# Springer Series in Operations Research and Financial Engineering

The Springer Series in Operations Research and Financial Engineering publishes monographs and textbooks on important topics in theory and practice of Operations Research, Management Science, and Financial Engineering. The Series is distinguished by high standards in content and exposition, and special attention to timely or emerging practice in industry, business, and government. Subject areas include:

Linear, integer and non-linear programming including applications; dynamic programming and stochastic control; interior point methods; multi-objective optimization; Supply chain management, including inventory control, logistics, planning and scheduling; Game theory Risk management and risk analysis, including actuarial science and insurance mathematics; Queuing models, point processes, extreme value theory, and heavy-tailed phenomena; Networked systems, including telecommunication, transportation, and many others; Quantitative finance: portfolio modeling, options, and derivative securities; Revenue management and quantitative marketing Innovative statistical applications such as detection and inference in very large and/or high dimensional data streams; Computational economics

More information about this series at https://link.springer.com/bookseries/3182

Johannes O. Royset · Roger J-B Wets

# An Optimization Primer

Springer

Johannes O. Royset
Department of Operations Research
Naval Postgraduate School
Monterey, CA, USA

Roger J-B Wets
Department of Mathematics
University of California, Davis
Davis, CA, USA

ISSN 1431-8598 ISSN 2197-1773 (electronic)
Springer Series in Operations Research and Financial Engineering
ISBN 978-3-030-76277-3 ISBN 978-3-030-76275-9 (eBook)
https://doi.org/10.1007/978-3-030-76275-9

Mathematics Subject Classification: 90-01, 90C05, 90C06, 90C15, 90C20, 90C25, 90C29, 90C30, 90C31, 90C26, 90C33, 90C34, 90C35, 90C46, 90C47, 90C51, 90C53, 90C55, 90C59, 90C90, 90B05, 90B06, 90B10, 90B18, 90B20, 90B40, 90B50, 49M05, 49M15, 49M20, 49M27, 49M29, 49M30, 49M37, 49N15, 49K35, 49K40, 49K45, 47J20, 47J22, 47J30, 49J40, 49J53, 65K05, 65K10, 65K15, 62-01, 62J02, 62J05, 62J07, 62H12, 62H30, 91-01, 91B06, 91B30, 91B50, 91B52, 91B54, 91G70, 62P05, 62P20, 62P30



This Springer imprint is published by the registered company Springer Nature Switzerland AG
The registered company address is: Gewerbestrasse 11, 6330 Cham, Switzerland

*to our families*

# Preface

The concerns with finding minima and maxima were for a long time mainly restricted to mathematical expressions stemming from physical phenomena. The preeminence of being able to do this was underlined by Leonard Euler when he wrote in 1744: "Nothing at all takes place in the universe in which some rule of maximum or minimum does not appear." The mathematical tools required were supplied by differential calculus as developed by Newton, Leibnitz and many others, not forgetting the Lagrange multiplier strategy for finding the local minima and maxima constrained by smooth equalities. This resulted in a fertile interaction between physics and mathematics, which was instrumental to their shared progress.

By the mid-twentieth century, "optimizers" were confronted by a radically new variety of problems coming from organizational and technological issues. Companies wanted to forecast the demand for their products and line up their supply chain accordingly. Airlines had to match crews as well as aircraft types to specific routes. Engineers needed to plan transportation networks that best serve urban as well as rural areas. Governments aimed to allocate resources efficiently to sectors of the economy. While the decision criteria might have been smooth, the resulting problems invariably included nonsmooth inequality constraints. Sometimes, they're simple sign restrictions on the decision variables but were also more complex collections of interwoven inequalities and equalities. It became evident that traditional differential calculus wasn't equipped to deal with problems of this type. There was even serious doubt whether there ever would be a satisfactory theory providing efficient algorithms as well as tools for analyzing the calculated solutions.

The breakthrough came with George Dantzig's simplex method in 1947, which demonstrated that one could, in fact, solve optimization problems involving inequalities. Despite its limitation to linear functions and concerns about scalability, the simplex method quickly became the main computational approach to many important planning problems and also motivated subsequent developments for handling more general inequalities. These algorithmic advances took place in parallel to a growing theoretical understanding of convex sets and functions. Convexity presented a tractable frontier for exploration beyond the linear paradigm, one that was unencumbered by the absence of smoothness in problem formulations.

Today, we refer to *Variational Analysis* as the broad field of mathematics supporting optimization, equilibrium, control and stability theory, with a focus on systems of equations and inequalities as well as functions that are neither smooth nor well defined in a traditional sense. The field has grown much beyond the convex setting and now provides the foundation for the development of optimization models and algorithms across a vast number of applications.

In the age of massive data and capable computers, automated approaches to prediction, planning and decision-making are becoming increasingly integral to nearly all human activity. Intelligent systems such as food delivery robots, military drones and driverless cars make complicated decisions autonomously. Data about the location, purchase history and interests of potential costumers guide decisions about online advertisement. *Optimization* has become the interdisciplinary field that supplies a wide range of applications with models and algorithms for making the best possible predictions and decisions.

The solution of problems by means of optimization models and algorithms certainly benefits from, and is often entirely contingent on, knowledge of variational analysis. At the same time, an appreciation of mathematical properties and proof techniques can't fully mature without insight about the needs in applications. The interplay between applications, models and algorithms on one side and mathematical theory on the other side is the central organizing principle of the book. Throughout, we integrate background material with the main narrative and thereby enable the reader to see the usefulness of abstract concepts in the context of concrete problems.

Applications require skills in building mathematical models of real-world situations. But models are most useful when they're tractable and can be "solved," i.e., there are algorithms that produce from the models numerical estimates, predictions and recommended courses of actions. The underlying theory guides the development of models and algorithms, establishes their correctness and informs us about computational efficiency and limitations. In this book, we teach how to formulate optimization models and construct the necessary algorithms, but also describe the underlying mathematical theory so the reader can rigorously justify a solution strategy and be prepared to tackle emerging challenges.

Modeling skills are built with the complications of the real world in mind; we give many examples from data analytics, engineering design, operations research, management science and economics. Foremost, a modeler needs to tackle the unavoidable fact that nearly all practical problems come with numerous intangibles and uncertainties. It's difficult to overestimate the need to deal squarely with nearly ubiquitous data uncertainty when coming to grips with applications. Slight levels of uncertainty about model parameters are often suitably handled by sensitivity analysis. When confronted with more serious uncertainty, for example, about material properties, product demand, environmental conditions, sensor input and the spread of an infectious disease, it's better to rely on stochastic models of uncertainty. This requires us to broach into statistical applications and machine learning with their problems of finding best predictions for random phenomena. It also leads to the need for addressing *decision-making under uncertainty* and the associated modeling and algorithmic challenges. How should we compare and optimize decisions when each one of them results in an uncertain future "cost?" We describe modeling approaches that balance concerns about present expenses with possible future costs as well as safety and reliability requirements. This gives rise to *stochastic optimization problems* and their specialized algorithms, which we cover in detail.

Our broad view of optimization brings us beyond traditional subjects such as linear and nonlinear programming as well as convex optimization to "complicated" nonsmooth and nonconvex functions, set-valued mappings and multi-agent problems. With this book, we aim to demystify these concepts and show that they're not only key to modeling of real-world problems, but also quite approachable for students as well as practitioners. Most significant in this regard is our treatment of subdifferentiability. The literature has several versions of directional derivatives and subgradients for nonconvex functions, a situation that has hampered the dissemination of these concepts beyond the specialists. We reduce the subject to revolve around one kind of subderivatives and one kind of subgradients, but still are able to address arbitrary functions by means of wide-reaching calculus rules.

The treatment of dual problems and optimality conditions brings forth the often overlooked fact that they stem from perturbations of the actual problem of interest as expressed by a *Rockafellian*, so named to honor R. T. Rockafellar for his development of this versatile perspective. Thus, a problem is viewed as an instance from a family of perturbed problems, which reflects the practically important fact that optimization technology is more a tool for identifying possibilities than an oracle for producing a definite answer. It also highlights the great variety of dual problems and optimality conditions one might consider, with numerous algorithmic possibilities.

We hope that our gradual introduction to set-convergence broadens the appeal of this important tool for analyzing approximating problems and implementable algorithms as well as carrying out sensitivity analysis and confirming local properties such as those pertaining to strong duality. In particular, we develop a road map for constructing *consistent approximations* with minimizers and stationary points that converge to minimizers and stationary points of the actual problem of interest. This concept is further refined and quantified via the truncated Hausdorff distance to produce error bounds and rates of convergence even in the nonconvex and nonsmooth setting.

## *How to Read the Book*

The book is designed to serve as an introduction to the field of optimization for students of science, engineering or mathematics at the undergraduate and graduate levels. It's expected that the reader has a foundation in differential calculus and linear algebra, with some exposure to real analysis being helpful but not prerequisite. We summarize basic concepts when needed, especially those from real analysis (sequences, limits, continuity, etc.). Experience with mathematical reasoning makes the proofs more accessible, but isn't necessary for a reading focused on modeling and algorithms. No knowledge of probability and statistics is prerequisite.

The book contains ten chapters, intended to be read sequentially, but the following sections can be skipped without jeopardizing (significantly) the understanding of the subsequent material: §1.I, §1.J, §2.J, §2.K, §3.I-§3.K, §5.J and §6.G-§6.I. Chapters 7–10 can largely be read independently of each other. The book also contains many examples from a variety of application areas and more than 100 exercises, which provide further flexibility in how to approach the material. A solution manual is available for instructors.

Since we advance on a broad front, a comprehensive treatment is often postponed when a subject is encountered for the first time. The goal is to ease the pressure on the reader in the early stages and to motivate the theory and algorithms with applications. Consequently, we return to the same subject several times, while gradually increasing the level of sophistication. We hope this approach appeals to a broad audience and to instructors teaching a "general-purpose" optimization course. For experts and mathematically advanced readers looking for all the properties associated with a specific concept, the organization might be frustrating. We hope the extensive index can help with zeroing in on the needed results.

Theorems, propositions, definitions, examples, exercises, convergence statements and so forth are numbered jointly without a specification of whether it's a theorem, proposition, definition, etc. Thus, "as seen in 4.2" references the second statement (theorem, proposition, etc.) in Chap. 4. Equations are numbered independently using parenthesis, i.e., "(4.2)" means the second equation in Chap. 4. Figures are also numbered independently, with "Figure 4.2" referencing the second figure in Chap. 4. Tables are numbered similarly and referenced, for example, as "Table 4.2." Sections are referenced in the text with the symbol § so that "§4.B" becomes Section B of Chap. 4. Mathematical background material is highlighted in gray-shaded paragraphs. An active use of the index should settle most questions about definitions and notation. Definitions are stated in italic font.

The extensive scope of the book makes it difficult for us to be entirely consistent with the notation. Still, Greek letters typically denote scalars, lowercase and uppercase roman letters represent vectors and matrices, respectively, uppercase roman letters also denote sets and calligraphic font specifies collections of sets and functions. Vectors to be optimized are usually denoted by $x, y, z$ and occasionally $a, b, c$. Multipliers are written as $y$ and $z$. Functions are denoted by $f, g, h$ and sometimes $\varphi$ (the Greek letter "phi") and $\psi$ (psi), vector-valued and set-valued mappings by $F, G, S, T$ and various functionals by calligraphic capital letters. Boldface letters indicate random variables, vectors and matrices. In particular, $\boldsymbol{\xi}$ (ksi) is a random vector representing uncertain parameters and $\mathbb{E}[\boldsymbol{\xi}]$ is its expected value. The use of superscripts can be confusing at first; $x^i$ probably doesn't mean $x$ to power $i$ but rather the $i$th vector in a collection of vectors. Occasionally when powers actually appear, we make this clear in the text as needed. Subscripts routinely designate components of a vectors. For example, $x_j$ is typically the $j$th component of a vector $x$. We often consider a sequence of points (or sets or functions) and these are usually indicated by the superscript $\nu$ (nu), which then runs over the natural numbers $\mathbb{N} = \{1, 2, 3, \ldots\}$. For example, the collection of vectors $x^1, x^2, x^3, \ldots$, is denoted by $\{x^\nu, \nu \in \mathbb{N}\}$. We write $\mathbb{R}$ for the real line and $\overline{\mathbb{R}}$ when the two special "numbers" $-\infty$ and $\infty$ have also been included, i.e., $\overline{\mathbb{R}} = \mathbb{R} \cup \{-\infty, \infty\}$.

### *Supporting Material*

The maturity of the field of optimization makes it impossible to cover all the important subjects in one volume and even our bibliography can't be complete. Among the many excellent texts, we mention a few books that supplement our exposition especially well. A foundation in linear algebra and convex optimization with many applications is provided by [22]. How to develop optimization models in the presence of uncertainty is described by [56]. Algorithmic details for linear and quadratic optimization problems as well as those involving smooth objective and constraint functions are laid out in [69]. First-order methods and their applications in machine learning are addressed by [10, 60]. The rich areas of variational inequalities and complementarity problems are systematically addressed in [35]. Theoretical refinements and more advanced subjects are dealt with in [105], which naturally would be the next source to consult for mathematical details after having read the present text. There, one also finds an extensive bibliography and notes about the historical development of variational analysis.

We omit a treatment of *discrete optimization*, where decision variables are constrained to integer values, and refer to [125] for a graduate text on the subject. We also barely mention network flow optimization and its numerous specialized algorithms; see [1].

In addition to standard mathematical software such as MATLAB, there are many open-source packages with optimization capabilities. A reputable list is found at `coin-or.org`. To solve an actual optimization problem, we would need a suitable *solver*, i.e., a computer implementation of an algorithm, but can also benefit greatly from a *modeling language*, which allows us to specify objective functions and constraints easily as well as manage subproblems, batch runs, tolerances and output formats. In particular, Pyomo (`pyomo.org`) is a Python-based, open-source modeling language that supports a diverse set of optimization problems including those involving uncertain parameters [48].

Monterey, California — Johannes O. Royset
Davis, California — Roger J-B Wets
March, 2021

# Acknowledgements

We've been blessed with numerous supporters. The first author has benefitted tremendously from the technical discussions and collaborations with Armen Der Kiureghian, Lucien Polak, Terry Rockafellar, Alexander Shapiro and Kevin Wood as well as colleagues and students at the Naval Postgraduate School among which Jesse Pietz and Matthew Miller stand out with contributions to sections in the book. Jerry Brown, Matt Carlyle and Rob Dell taught him valuable lessons on what it takes to address real-world problems. He's also indebted to Laurent El Ghaoui for inviting him to teach EE127/227 "Optimization Models" at Berkeley in 2016 and to Peter Glynn for facilitating a sabbatical visit at Stanford in 2019–2020. The second author is particularly thankful to Hedy Attouch, Jean-Pierre Aubin, Gabriella Salinetti, Jong-Shi Pang, Dave Woodruff and especially Terry Rockafellar, for interactions in various forms that helped him expand his optimization horizon.

Luca Bonfiglio, Terje Haukaas, Matthew Norton, Michael Teter and Daniele Venturi assisted with numerical simulations as well as data and George Lan, Andy Philpott and Terry Rockafellar clarified algorithmic aspects, examples and theoretical developments. The authors appreciate the assistance from the professional Springer team led by Donna Chernyk, which includes Suresh Kumar and his invaluable LaTeX support.

Funding from the Air Force Office of Scientific Research (Mathematical Optimization), Army Research Office, Defense Advanced Research Projects Agency (Defense Sciences Office) and the Office of Naval Research (Operations Research; Science of Autonomy) has been critical in developing this material over many years. In particular, we're grateful to Fariba Fahroo for constantly championing the role of rigorous optimization and to Marc Steinberg and Donald Wagner for seeing the importance of optimization technology in a broad range of application areas.

Above all, we've relied on the unwavering support and encouragement from our families. Their patience during this long journey—while suffering through plague and fire—isn't forgotten.

# Contents

# Chapter 1
# PRELUDE

Optimization technology helps us identify the best predictions and decisions in a wide range of applications. This is made possible by the formulation and subsequent solution of a diverse set of optimization problems. An *unconstrained optimization problem* aims to determine the minimum of a well-behaved multi-variate function without any restrictions and is the main subject of this chapter. In addition to theoretical foundations and algorithms for computing solutions, we'll cover applications in inventory control, statistical estimation and data analytics. *Constrained optimization problems*, involving also side conditions, as well as more general *variational problems* will be dealt with later on.

Fig. 1.1: Nicolas Oresme (left) and Pierre de Fermat (right).

## 1.A The Mathematical Curtain Rise

Let's begin our journey in the classical landscape: All functions to be minimized are smooth and there are no complicating constraints. The foundations were laid down in



J. O. Royset and R. J-B Wets, *An Optimization Primer*, Springer Series in Operations Research and Financial Engineering, https://doi.org/10.1007/978-3-030-76275-9_1

the middle of the last millennium by two genial mathematical dabblers, Nicolas Oresme and Pierre de Fermat (Figure 1.1). The rules they formulated provide the guidelines for building an optimization theory in the unconstrained setting and, as we'll see later, also in nonclassical extensions.

The modern theory of optimization starts in the middle of the 14th century with Nicolas Oresme (ca.1323-1382), part-time mathematician and full-time Bishop of Lisieux (France). A slight adaptation of an assertion in his treatise *Configurations of Qualities and Motions* [26] states that "near a minimum, the increment of a variable quantity becomes nonnegative." Translating Oresme's statement into our present-day language, it reads: For a function $f : \mathbb{R} \to \mathbb{R}$,

$$x^\star \in \operatorname{argmin} f \implies \vec{d}f(x^\star; w) \geq 0 \quad \forall \text{ (for all) } w \in \mathbb{R},$$

where

$$\operatorname{argmin} f = \{\bar{x} \mid f(\bar{x}) \leq f(x)\ \forall x\}$$

designates the *arg*uments *min*imizing $f$, i.e., the *minimizers* of $f$, and

$$\vec{d}f(\bar{x}; w) = \lim_{\tau \searrow 0} \frac{f(\bar{x} + \tau w) - f(\bar{x})}{\tau}$$

is the *directional derivative* of $f$ at $\bar{x}$ in direction $w$. This limiting quantity[1] is the rate of change of $f$ in the $w$-direction. A calculus class would refer to $\vec{d}f(\bar{x}; -1)$ and $\vec{d}f(\bar{x}; 1)$ as the left and right derivative of $f$ at $\bar{x}$, respectively; see Figure 1.2.

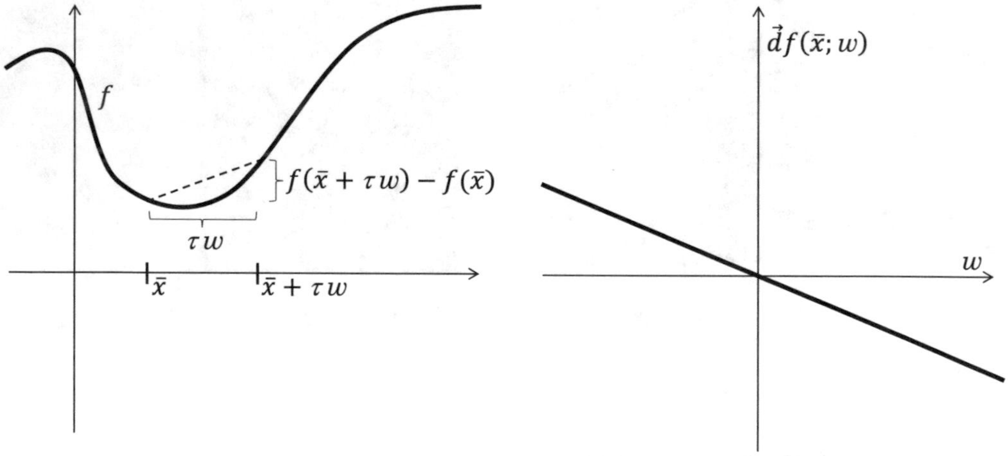


Fig. 1.2: The point $\bar{x}$ isn't a minimizer of $f$ since $\vec{d}f(\bar{x}; w) < 0$ for positive $w$.

The condition of having nonnegative directional derivatives at $\bar{x}$ doesn't exclude the possibility that $\bar{x}$ could actually be a maximizer or an inflection point. It's a *necessary* condition for $\bar{x}$ to be a minimizer but not a sufficient one. Although Oresme probably

[1] When $\tau > \alpha$ as it approaches $\alpha$, we write $\tau \searrow \alpha$; when $\tau < \alpha$, we write $\tau \nearrow \alpha$, whereas $\tau \to \alpha$ doesn't place any restrictions on the way $\tau$ approaches $\alpha$.

had only smoothly varying functions in mind, his condition for optimality reaches further since it also applies to functions with kinks. For instance, $f(x) = |x|$ has

$$\vec{d}f(0; w) = |w| \geq 0$$

and the optimality condition correctly singles out $x^\star = 0$ as a minimizer.

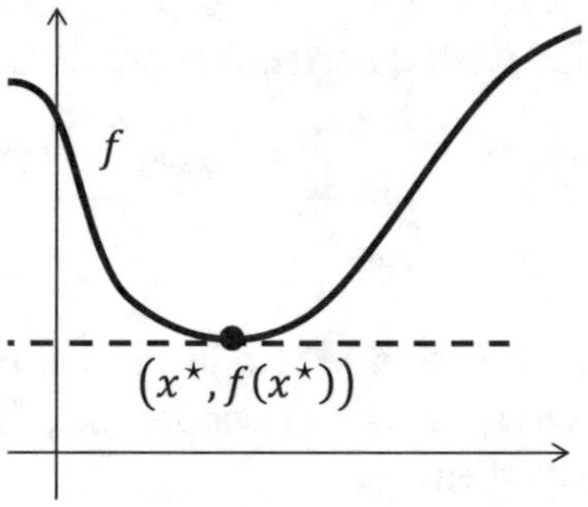


Fig. 1.3: Horizontal tangent to the graph of a function at a minimizer.

About three centuries later, Pierre de Fermat (1607-1665), another part-time mathematician and full-time councillor (lawyer) at the High Judiciary Court in Toulouse (France), while working on the long-standing tangent problem [52], observed that for $x^\star$ to be a minimizer of $f$, the tangent to the graph of $f$ at the point $(x^\star, f(x^\star))$ must be "flat;" cf. Figure 1.3. We would now express this as

$$x^\star \in \operatorname{argmin} f \quad \Longrightarrow \quad f'(x^\star) = 0,$$

where $f'(\bar{x})$ is the *derivative* of $f$ at $\bar{x}$, i.e.,

$$f'(\bar{x}) = \lim_{\substack{\tau \to 0 \\ \tau \neq 0}} \frac{f(\bar{x} + \tau) - f(\bar{x})}{\tau}.$$

Implicit in this optimality condition is the assumption that there aren't any kinks or jumps in the value of $f$ near $\bar{x}$, which then ensures that the limit exists.

Let's proceed with multi-variate functions defined on $\mathbb{R}^n$. To follow Oresme's description, we now need to consider all possible variations, not just left and right. Accordingly, with minimizers and directional derivatives naturally extended to $\mathbb{R}^n$, we obtain

| |
|---|
| the Oresme rule: $x^\star \in \operatorname{argmin} f \quad \Longrightarrow \quad \vec{d}f(x^\star; w) \geq 0 \;\; \forall w \in \mathbb{R}^n$ |

Similarly, Fermat's statement gets translated to

$$\boxed{\text{the Fermat rule:} \quad x^\star \in \operatorname{argmin} f \implies \nabla f(x^\star) = 0}$$

where 0 is an $n$-dimensional vector of zeros and

$$\nabla f(\bar{x}) = \left(\frac{\partial f(\bar{x})}{\partial x_1}, \frac{\partial f(\bar{x})}{\partial x_2}, \ldots, \frac{\partial f(\bar{x})}{\partial x_n}\right)$$

is the *gradient* of $f$ at $\bar{x}$ and the *partial derivatives* are defined by

$$\frac{\partial f(\bar{x})}{\partial x_j} = \lim_{\substack{\tau \to 0 \\ \tau \neq 0}} \frac{f(\bar{x} + \tau e^j) - f(\bar{x})}{\tau},$$

with $e^j$ being the $n$-dimensional vector with 1 in the $j$th position and 0 elsewhere[2]. In particular, $\nabla f(\bar{x}) = f'(\bar{x})$ when $n = 1$. If $f$ is *smooth*, i.e., the gradient is defined at every $x \in \mathbb{R}^n$ and varies continuously, then

$$\vec{d}f(\bar{x}; w) = \langle \nabla f(\bar{x}), w \rangle = \sum\nolimits_{j=1}^{n} \frac{\partial f(\bar{x})}{\partial x_j} w_j \quad \forall w = (w_1, \ldots, w_n) \in \mathbb{R}^n \tag{1.3}$$

and the Oresme and Fermat rules become equivalent.

Although the rules only deliver necessary conditions for optimality (and not sufficient ones), they provide us with two strategies for finding minimizers. At a current point $x \in \mathbb{R}^n$, an algorithm could

$$\text{find } w \in \mathbb{R}^n \text{ with } \vec{d}f(x; w) < 0,$$

pointing to a potential reduction in the function value when moving from $x$ to a next point in the direction of $w$. One might hope that a sequence of points generated in this manner "descends" to a minimizer of $f$. If there's no such $w$, then the algorithm stops and $f$ has nonnegative directional derivatives at the current point as requested by the Oresme rule. Alternatively, an algorithm can rely on the Fermat rule and solve the $n \times n$-system of equations

$$\frac{\partial f(x)}{\partial x_j} = 0, \quad j = 1, \ldots, n.$$

This calls attention to the close relationship between the structure of algorithms for finding minimizers and those designed for solving systems of equations.

**Exercise 1.1** (one variable). Consider the cubic function given by $f(x) = \frac{1}{3}x^3 - x + 1$ for $x \in \mathbb{R}$. Determine its derivatives and directional derivatives. Plot the values of $f(x)$ and $f'(x)$ as functions of $x$ and plot the values of $\vec{d}f(0; w)$, $\vec{d}f(1/2; w)$ and $\vec{d}f(1; w)$ as

---

[2] Superscripts often label or index quantities, such as $e^j$ for example, and shouldn't be confused with powers which we point out when needed. Subscripts routinely indicate components of a vector: $x_j \in \mathbb{R}$ is the $j$th component of $x \in \mathbb{R}^n$.

functions of $w$. Use the Oresme rule to identify points that can't be minimizers of $f$. Repeat the process but rely on the Fermat rule.

**Exercise 1.2** (two variables). Consider the quadratic function given by $f(x) = 2x_1^2 + (x_2 - 1)^2$ for $x = (x_1, x_2) \in \mathbb{R}^2$. Determine its gradients and directional derivatives. Use the Oresme rule to identify points that can't be minimizers of $f$. Repeat the process but rely on the Fermat rule.

## 1.B Data Smoothing

Given the observed values $y_1, y_2, \ldots, y_m$ of an unknown univariate function at the distinct points $x_1, x_2, \ldots, x_m$, we're interested in finding a polynomial $q : \mathbb{R} \to \mathbb{R}$ of degree $n$, i.e.,

$$q(x) = c_n x^n + \cdots + c_2 x^2 + c_1 x + c_0,$$

whose values at $x_1, \ldots, x_m$ are as close as possible to the observed values; here, of course, $x^n$ is the $n$th power of $x$. There are a number of ways to interpret this statement, but let's adopt the goal of finding the *least-squares* fit. This implies minimizing the sum of the squares of the distance between $q(x_i)$ and $y_i$. With $c = (c_0, c_1, \ldots, c_n)$ being the coefficients of the polynomial, we formulate the problem as

$$\underset{c \in \mathbb{R}^{1+n}}{\text{minimize}} \; f(c) = \sum\nolimits_{i=1}^{m} \left(c_n x_i^n + \cdots + c_2 x_i^2 + c_1 x_i + c_0 - y_i\right)^2 = \langle Dc - y, Dc - y \rangle,$$

where $D$ is the $m \times (1+n)$-matrix[3]

$$D = \begin{bmatrix} 1 & x_1 & x_1^2 & \cdots & x_1^n \\ 1 & x_2 & x_2^2 & \cdots & x_2^n \\ \vdots & \vdots & \vdots & \ddots & \vdots \\ 1 & x_m & x_m^2 & \cdots & x_m^n \end{bmatrix} \quad \text{and} \quad y = \begin{bmatrix} y_1 \\ y_2 \\ \vdots \\ y_m \end{bmatrix}.$$

The function $f$ to be minimized is referred to as the *objective function* and the vector $c$ comprises *decision variables* to be determined, typically by an algorithm. The formulation is referred to as an *optimization model* to stress its role in supporting decision-making via modeling, which involves assumptions and simplifications. A specific optimization model, especially when it takes on a canonical form and its solution emerges as the primary task, is also called an *optimization problem*.

Applying the Fermat rule, we obtain after some algebra that a minimizer $c^\star$ of $f$ must satisfy

$$\nabla f(c^\star) = 2D^\top (Dc^\star - y) = 0$$

or, equivalently, $D^\top D c^\star = D^\top y$.

[3] The attentive reader may notice that $c$ was introduced as a *row* vector, but here emerges as a *column* vector. Throughout, we view vectors as having either orientation, or both, as the meaning will be clear from the context. The notation $\langle a, b \rangle = \sum_{j=1}^{n} a_j b_j$ for the inner product avoids a commitment either way.

**Linear Algebra.** The vectors $\{v^i \in \mathbb{R}^n, i = 1, \ldots, m\}$ are *linearly independent* when $\sum_{i=1}^m \alpha_i v^i = 0$ implies that $\alpha_i = 0$ for all $i$. The *rank* of an $m \times n$-matrix $A$, denoted by rank $A$, is the number of linearly independent columns, or rows, in $A$. If $m = n$ and $A$ has *full rank* (i.e., rank $A = n$), then $A$ is *invertible* and its *inverse* $A^{-1}$ satisfies $AA^{-1} = A^{-1}A = I$, where $I$ is the *identity matrix* with ones along the diagonal and zeros elsewhere. For any $m \times n$-matrix $A$, the $n \times n$-matrix $A^\top A$ has full rank if and only if rank $A = n$.

Assuming that $m \geq 1+n$, the columns of $D$ are linearly independent because $x_1, \ldots, x_m$ are distinct. Hence, $D^\top D$ is invertible and

$$c^\star = (D^\top D)^{-1} D^\top y$$

is the unique solution to the equations derived from the Fermat rule. Instead of computing the inverse, it's usually more efficient to solve a system of equations by Gauss elimination, iterative methods (see §1.H) or factor methods such as Cholesky decomposition. Figure 1.4 illustrates a fit to $m = 20$ points by a polynomial of degree $n = 5$.

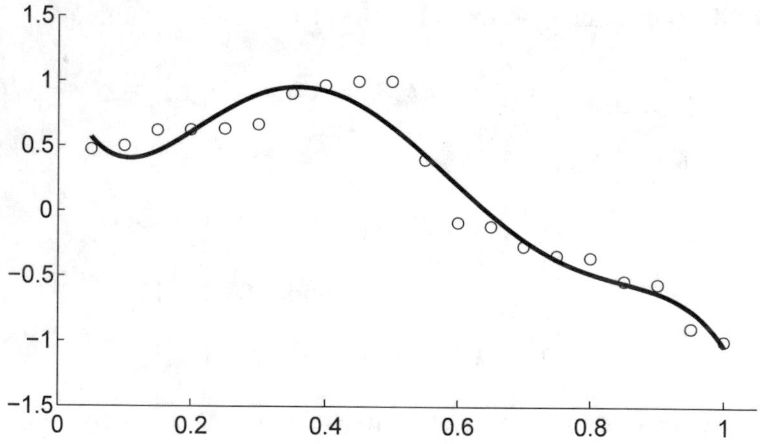


Fig. 1.4: Smoothing data (circles) with a polynomial of degree five (solid line).

**Exercise 1.3** (data smoothing). Using $x = (0, 0.1, 0.2, \ldots, 0.9, 1)$ and $y = (0.85, 0.61, 0.89, 0.46, 0.82, 0.62, 0.72, 0.18, 0.54, 0.41, 0.26)$, fit polynomials of degree $n = 1, 3, 5, \ldots, 11$ and 20. For each $n$, visualize the fit by plotting the data as well as the graph of the resulting polynomial and also report the minimum objective function value. Discuss which $n$ gives the "best" result.

## 1.C Optimization under Uncertainty

In practice, problems are rarely as clear-cut as in the preceding section. We need to select an optimization model among many alternatives that addresses the real-world problem at hand and provides valuable insight to decision-makers. Numerous considerations including the availability of data, algorithms, software and hardware as well as the urgency of the matter influence whether a particular model is suitable. A frequent issue that can't be ignored without jeopardizing the usefulness of the entire process is how to address uncertainty about data, questionable assumptions and variable conditions under which the model will be employed. Even the data smoothing problem of §1.B might come with uncertainty about locations and values of the observations.

A versatile and powerful approach for dealing with a parameter to which we can't assign a specific value, perhaps because it's intrinsically varying or represents an unknown condition, is to model the parameter as a *random variable*. This allows us to leverage the vast and sophisticated tools of probability theory and statistics and brings us to *stochastic optimization problems*.

**Example 1.4** (newsvendor problem). In a contractual agreement, a newsvendor (a firm) places an order for the daily delivery of a fixed number of newspapers (perishable items) to meet an uncertain daily demand $\boldsymbol{\xi}$, modeled as a random variable[4] with known *distribution function* $P : \mathbb{R} \to [0, 1]$. The probability that $\boldsymbol{\xi}$ takes a value no greater than $\xi$ is then $P(\xi)$. The newsvendor is charged $\gamma$ cents per paper ordered and sells each for $\delta > \gamma$ cents; unsold papers can't be returned and are worthless at the end of the day. The goal of the newsvendor is to choose an order quantity that minimizes the expected (average) loss. This is a problem from the area of *inventory management*.

**Detail.** When ordering $x$ newspapers and $\xi$ is the demand, the loss (expense minus income) turns out to be

$$\begin{cases} \gamma x - \delta x & \text{if } \xi \geq x \\ \gamma x - \delta \xi & \text{otherwise,} \end{cases}$$

with negative values implying a profit.

If $P$ is a continuous function and can be expressed by a *density function* $p : \mathbb{R} \to [0, \infty)$ such that

$$P(\xi) = \int_{-\infty}^{\xi} p(\eta)\, d\eta \quad \forall \xi,$$

then an order of $x$ newspapers would yield an expected loss of

$$\begin{aligned} f(x) &= \int_0^x (\gamma x - \delta\xi)p(\xi)\, d\xi + \int_x^\infty (\gamma - \delta)xp(\xi)\, d\xi \\ &= (\gamma - \delta)x + \delta \int_0^x (x - \xi)p(\xi)\, d\xi. \end{aligned}$$

[4] Random variables as well as other random quantities are written in boldface, such as $\boldsymbol{\xi}$, returning to regular font, such as $\xi$, for their values (outcomes).

After integration by parts followed by differentiation, or more directly relying on the Leibniz rule for differentiation under the integral, the Oresme rule predicates that a minimizer $x^\star$ of $f$ satisfies

$$\vec{d}f(x^\star; w) = \big(\gamma - \delta + \delta P(x^\star)\big)w \geq 0 \quad \forall w,$$

which holds only if $P(x^\star) = 1 - \gamma/\delta$; cf. Figure 1.5(left). In the figure, $x^\star$ is unique, but this isn't always the case because $P$ could be "flat" when its value is $1 - \gamma/\delta$.

More generally, when $P$ isn't necessarily continuous and might have jumps (discontinuities) as in Figure 1.5(right), a bit more work is required to derive $f$ and identify a minimizer. This would be the case when the random variable $\boldsymbol{\xi}$ takes on only a finite or countable number of possible values (i.e., has a discrete distribution). For all $\xi$, let

$$P(\xi_-) = \lim_{\eta \nearrow \xi} P(\eta)$$

be the value of $P$ "just" to the left of $\xi$. Since $P$ is a nondecreasing function, $P(\xi_-) \leq P(\xi)$ with equality holding when $P$ is continuous at $\xi$. Then, one can show[5] that

$$\vec{d}f(x; w) = \begin{cases} \big(\gamma - \delta + \delta P(x_-)\big)w & \text{if } w \leq 0 \\ \big(\gamma - \delta + \delta P(x)\big)w & \text{otherwise} \end{cases}$$

and a minimizer $x^\star$ must satisfy

$$P(x^\star_-) \leq 1 - \gamma/\delta \leq P(x^\star)$$

by the Oresme rule. Figure 1.5(right) illustrates this requirement and the fact that there could be a whole range of values for $\delta$ and $\gamma$ that results in the same order quantity.

In this problem the Oresme rule specifies not only a necessary but also a sufficient condition for a minimizer. Thus, the order quantities computed from the above expressions actually minimize the expected loss. A formal argument for this relies on the development in the following sections. □

Usually, we construct a stochastic optimization problem from a *deterministic version*, where uncertainty about parameters is ignored. For the newsvendor problem, we might want to minimize loss subject to a requirement that an *assumed* demand $\hat{\xi}$ is met. This results in the (trivial) model

$$\underset{x \in \mathbb{R}}{\text{minimize}} \;\; (\gamma - \delta)x \;\; \text{subject to} \;\; x = \hat{\xi},$$

which necessarily stipulates that the order quantity should be $\hat{\xi}$. However, it's overwhelmingly unlikely that the demand will exactly match $\hat{\xi}$. Since the demand isn't known at the time of the order, the actual loss has to be evaluated after the decision is

[5] This nontrivial derivation highlights the mathematical challenges associated with optimization under uncertainty. By Chap. 3, we'll have enough expertise to address them rigorously.

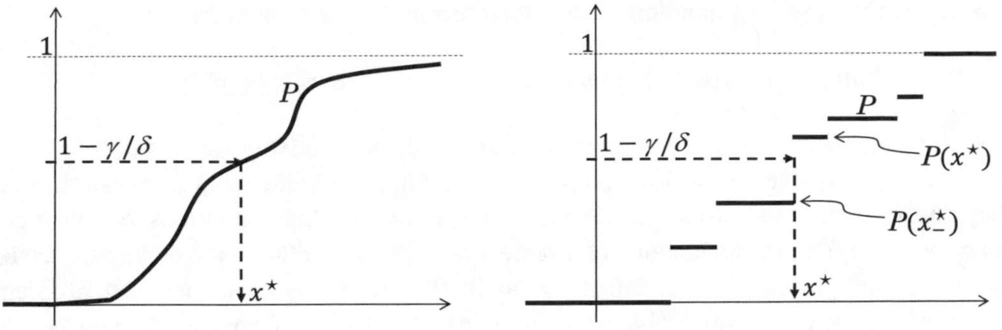


Fig. 1.5: Order quantities under continuous (left) and discrete (right) distributions.

made. Analyzing a bit more closely the decision and evaluation process, we see it follows the pattern:

$$\text{decision: } x \rightsquigarrow \text{ observation: } \xi \rightsquigarrow \text{ evaluation: } \xi - x.$$

Moreover, the expected loss $f$ in 1.4 is made up of two terms. The first one, $(\gamma - \delta)x$, corresponds to the loss accrued when ordering and selling $x$ newspapers. It's a negative value reflecting a positive margin for each sale. The second term evaluates the order quantity based on the disparity between the demand ($\xi$) and the available newspapers ($x$), which can only be measured *after* the demand is observed. The second term is the *expected recourse cost*, which can be expressed concisely as (see Figure 1.6)

$$\mathbb{E}\big[h(x - \xi)\big], \text{ with recourse cost function } h(y) = \begin{cases} 0 & \text{if } y < 0 \\ \delta y & \text{otherwise.} \end{cases}$$

It captures the fact that there's no additional cost if the demand turns out to be above the order quantity $x$. If the demand is lower, however, there's the additional cost $\delta(x - \xi)$ associated with the $x - \xi$ newspapers never sold but prematurely included in the first term.

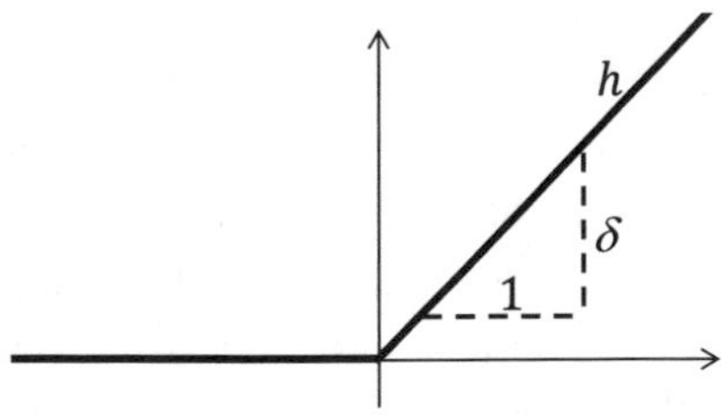


Fig. 1.6: The recourse cost function $h$ given by $h(y) = \max\{0, \delta y\}$.

Alternative expressions for the recourse cost function render its properties more tangible and foretell its extension to situations involving uncertainty about more than one parameter as well as additional, second-stage decisions in the calculation of the recourse cost. Since

$\delta > 0$, our recourse cost function admits the alternative representations

$$h(y) = \max\{0, \delta y\} = \min\{\delta y^+ \mid y^+ - y^- = y,\ y^+ \geq 0,\ y^- \geq 0\}.$$

The use of $y^+$ and $y^-$ suggests surplus and shortfall, respectively, and reflects the need for a second-stage recourse decision to be made after the value of $\boldsymbol{\xi}$ is revealed. Since they describe different cost structures in surplus versus shortfall situations, recourse cost functions are typically nonsmooth; cf. Figure 1.6. For any pair $(x, \xi)$, the recourse decision $(y^+, y^-)$ is unique and in fact rather trivial to determine. We refer to such problems as stochastic optimization problems with *simple recourse*. Of course, the newsvendor problem is only a micro-example of this class; Chap. 3 provides far-reaching extensions.

In view of this discussion, we can formulate the newsvendor problem by starting from the deterministic version and adding the expected recourse cost to obtain

$$\underset{x \in \mathbb{R}}{\text{minimize}}\ (\gamma - \delta)x + \mathbb{E}\big[h(x - \boldsymbol{\xi})\big]\,. \tag{1.4}$$

The formulation illustrates two features of stochastic optimization problems: Decisions are modeled relative to the arrival of information and minimizers are obtained in a trade-off between initial (first-stage) and recourse (second-stage) costs.

**Exercise 1.5** (newsvendor problem). With $\gamma = 3$ and $\delta = 7$, (a) find the minimizing order quantity when $\boldsymbol{\xi}$ is uniformly distributed on $[10, 20]$ and (b) write an expression for the expected loss, graph the distribution function $P$ and show that 16 is the minimizing order quantity when $\boldsymbol{\xi}$ is discretely distributed with equal probability for the values $\{10, 11, \dots, 20\}$. In both (a,b), show that $\mathbb{E}[\boldsymbol{\xi}]$, the expected demand, isn't a minimizer.

**Guide.** In (a), the density function has $p(\xi) = 0.1$ for $\xi \in [10, 20]$ and zero otherwise. In (b), if $\boldsymbol{\xi}$ has values $\xi_1, \dots, \xi_m$ with probabilities $p_1, \dots, p_m$, then the expected value of $h(\boldsymbol{\xi})$ is $\mathbb{E}[h(\boldsymbol{\xi})] = \sum_{i=1}^m p_i h(\xi_i)$. □

## 1.D Convex Analysis

In optimization, convexity plays the role of a fault line, both theoretically and practically. On the "convex side," we're on solid ground to engineer efficient numerical procedures whereas on the other side one may have to exert an exponentially growing amount of computational effort as the problem size increases or settle for algorithms that produce solutions satisfying an optimality condition derived from the Oresme or Fermat rule. In this section, we'll lay out conditions for convexity and examine the consequences.

**Definition 1.6** (convex sets and convex combinations). A set $C \subset \mathbb{R}^n$ is *convex* if it's either empty or for any $x^0, x^1 \in C$, the line segment

$$[x^0 \leftrightarrow x^1] = \big\{(1 - \lambda)x^0 + \lambda x^1 \;\big|\; \forall\, 0 \leq \lambda \leq 1\big\} \subset C.$$

For $\lambda \in [0, 1]$, a *convex combination* of $x^0$ and $x^1$ is the point $x^\lambda = (1 - \lambda)x^0 + \lambda x^1$.

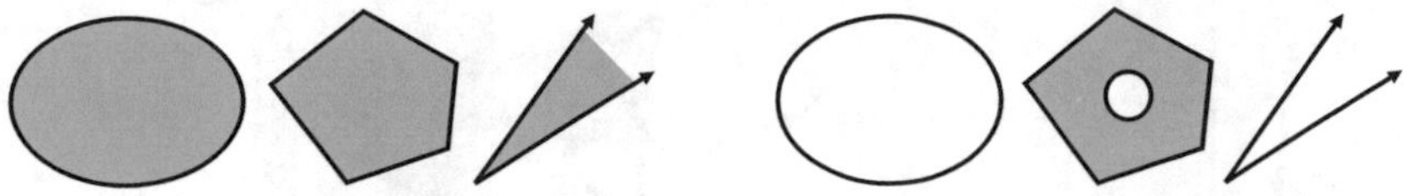

Fig. 1.7: Some convex (left) and nonconvex (right) sets in $\mathbb{R}^2$.

It's clear that $[x^0 \leftrightarrow x^1] \subset C$ if and only if $x^\lambda \in C$ for all $\lambda \in [0,1]$ so one could also define convex sets in terms of convex combinations. Balls, cubes, lines, quadrants, planes and the entire $\mathbb{R}^n$-space are examples of convex sets. Moreover, *subspaces*[6] such as the *nullspace* $\{x \in \mathbb{R}^n \mid Ax = 0\}$ of an $m \times n$-matrix $A$ are convex and the same holds for *affine sets*, i.e., displaced subspaces, for example expressed by $\{x \in \mathbb{R}^n \mid Ax = b\}$. Sets with dents or holes aren't convex; cf. Figure 1.7.

**Functions**. A function associates an *argument* (input) with a *value* (output). Nearly exclusively in this text, arguments are points in $\mathbb{R}^n$. Values are often real numbers, but to model situations in which certain decisions need to be barred and some formulae aren't valid in the traditional sense, we also allow the values $-\infty$ and $\infty$. For example, when $f(x) = 1/x$ for $x \neq 0$, one might assign $f(0) = \infty$ and thereby achieve a function that returns a value for all $x \in \mathbb{R}$. Let

$$\overline{\mathbb{R}} = \mathbb{R} \cup \{-\infty, \infty\}$$

be the real line augmented with $-\infty$ and $\infty$. We write $f : \mathbb{R}^n \to \overline{\mathbb{R}}$ for a function that returns a value $f(x) \in \overline{\mathbb{R}}$ for *every* $x \in \mathbb{R}^n$. It's *real-valued* when all its values are real numbers, which we indicate by writing $f : \mathbb{R}^n \to \mathbb{R}$.

When minimizing a function $f : \mathbb{R}^n \to \overline{\mathbb{R}}$, an argument with value $\infty$ would correspond to the worst conceivable situation; think of $f(x)$ as the cost associated with decision $x$. We would like to identify decisions that aren't of this kind. They define the *domain* of $f$:

$$\operatorname{dom} f = \big\{x \in \mathbb{R}^n \bigm| f(x) < \infty\big\}.$$

If $f$ is real-valued, then $\operatorname{dom} f = \mathbb{R}^n$. The *set of minimizers* now becomes

$$\operatorname{argmin} f = \big\{x^\star \in \operatorname{dom} f \bigm| f(x^\star) \le f(x)\ \forall x \in \mathbb{R}^n\big\}.$$

The *graph* of $f$, $\operatorname{gph} f = \big\{(x, \alpha) \in \mathbb{R}^n \times \mathbb{R} \bigm| \alpha = f(x)\big\}$, is familiar, but turns out to be less useful than the *epigraph* of $f$,

$$\operatorname{epi} f = \big\{(x, \alpha) \in \mathbb{R}^n \times \mathbb{R} \bigm| \alpha \ge f(x)\big\},$$

which often brings along a much gained perspective; see Figure 1.8.

[6] $C$ is a subspace of $\mathbb{R}^n$ if it's nonempty and $\alpha x + \beta \bar{x} \in C$ for all $x, \bar{x} \in C$ and $\alpha, \beta \in \mathbb{R}$.

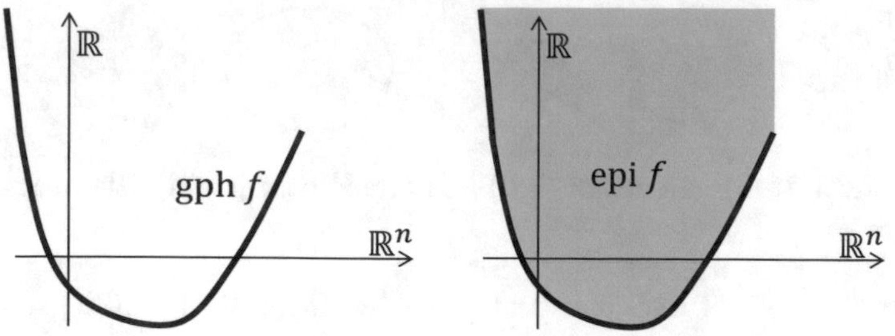


Fig. 1.8: Graph and epigraph of a function $f : \mathbb{R}^n \to \overline{\mathbb{R}}$.

**Definition 1.7** (convex functions). A function $f : \mathbb{R}^n \to \overline{\mathbb{R}}$ is *convex* if epi $f$ is a convex set.

A convex function is either linear, *affine* (i.e., linear plus a constant) or has an epigraph with just one "valley" as illustrated in Figure 1.9. Moreover, the absolute value function[7] $x \mapsto |x|$ and the exponential functions $x \mapsto e^{\alpha x}$, with $\alpha \in \mathbb{R}$, are convex functions. The sine function $x \mapsto \sin x$ serves as a typical example of nonconvexity.

**Exercise 1.8** (affine function). For $a \in \mathbb{R}^n$ and $\alpha \in \mathbb{R}$, show that the affine function given by $f(x) = \langle a, x \rangle + \alpha$ is convex using the definitions of convex sets and functions.

A main consequence of convexity is that the Oresme and Fermat rules specify conditions that are necessary *and* sufficient for a point to be a minimizer. Thus, algorithms can concentrate on computing points that satisfy one of these conditions.

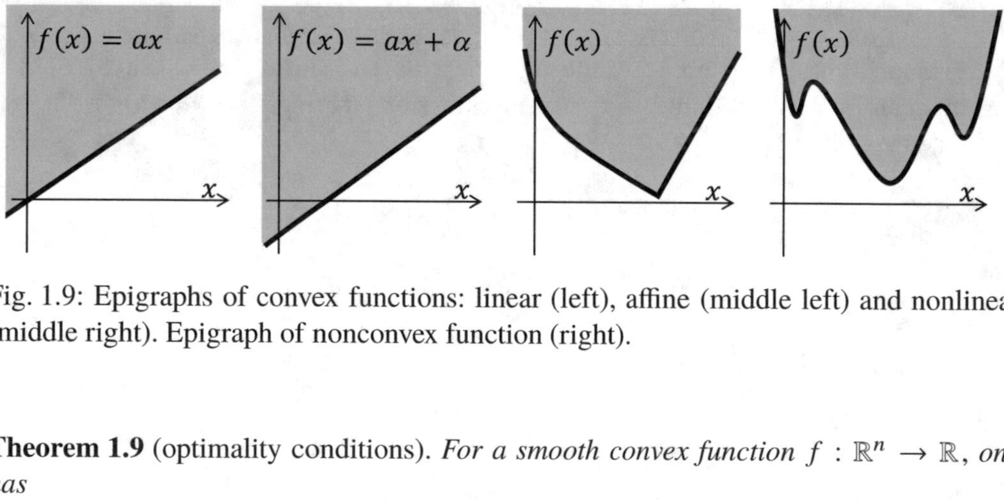


Fig. 1.9: Epigraphs of convex functions: linear (left), affine (middle left) and nonlinear (middle right). Epigraph of nonconvex function (right).

**Theorem 1.9** (optimality conditions). *For a smooth convex function* $f : \mathbb{R}^n \to \mathbb{R}$, *one has*

$$x^\star \in \operatorname{argmin} f \iff \vec{d}f(x^\star; w) \geq 0 \ \forall w \in \mathbb{R}^n \iff \nabla f(x^\star) = 0.$$

---

[7] Rather than introducing a particular name for a function, say $f$, and having to specify the formula for $f(x)$, say $e^x$, it's sometimes expedient to write $x \mapsto e^x$.

**Proof.** By (1.3), $\vec{d}f(x; w) = \langle \nabla f(x), w \rangle$, which implies that the Oresme and Fermat rules are equivalent. Next, suppose that $x^\star \in \operatorname{argmin} f$ and $w \in \mathbb{R}^n$. Then,

$$\frac{f(x^\star + \tau w) - f(x^\star)}{\tau} \geq 0 \quad \forall \tau \in (0, \infty)$$

and $\vec{d}f(x^\star; w) \geq 0$. This establishes the necessity of the conditions even without convexity (as already asserted in §1.A). For the converse, suppose that the directional derivatives are nonnegative at $x^\star$. If $x^\star$ isn't a minimizer, then there exists $w \in \mathbb{R}^n$ such that $f(x^\star + w) = f(x^\star) - \delta$ for some $\delta > 0$. Since $f$ is convex, the line segment

$$\big[(x^\star, f(x^\star)) \leftrightarrow (x^\star + w, f(x^\star + w))\big] \subset \operatorname{epi} f$$

and, in particular,

$$\big(x^\star + \tau w,\ \tau f(x^\star + w) + (1 - \tau) f(x^\star)\big) \in \operatorname{epi} f \quad \forall \tau \in (0, 1].$$

The definition of an epigraph ensures that

$$f(x^\star + \tau w) \leq \tau f(x^\star + w) + (1 - \tau) f(x^\star),$$

which yields

$$\frac{f(x^\star + \tau w) - f(x^\star)}{\tau} \leq \frac{\tau f(x^\star + w) + (1 - \tau) f(x^\star) - f(x^\star)}{\tau}$$
$$= \frac{\tau\big(f(x^\star + w) - f(x^\star)\big)}{\tau} = -\delta.$$

Thus, $\vec{d}f(x^\star; w) \leq -\delta$ and we've reached a contradiction. □

**Extended Arithmetic**. The possibilities of $-\infty$ and $\infty$ require us to extend the usual arithmetic operations. We naturally define $\infty + \alpha = \infty$ and $-\infty + \alpha = -\infty$ for $\alpha \in \mathbb{R}$; $\alpha \cdot \infty = \infty$ and $\alpha \cdot (-\infty) = -\infty$ for $\alpha \in (0, \infty)$; $\alpha \cdot \infty = -\infty$ and $\alpha \cdot (-\infty) = \infty$ for $\alpha \in (-\infty, 0)$; $\infty + \infty = \infty$; and $-\infty - \infty = -\infty$. It's convenient to let both $0 \cdot \infty$ and $0 \cdot (-\infty)$ be equal to 0. We focus on *minimizing* functions and then a value of $\infty$ represents a worst possible decision, which we would like to override any other possible benefit. Thus, we define

$$\infty + (-\infty) = \infty \quad \text{and} \quad -\infty + \infty = \infty.$$

Then, the usual associative, commutative and distributive laws apply with the exception that $\alpha(\infty - \infty) \neq (\alpha \cdot \infty - \alpha \cdot \infty)$ when $\alpha \in (-\infty, 0)$. With these adjustments, we can proceed with arithmetic calculations.

**Proposition 1.10** (convexity inequality). *A function $f : \mathbb{R}^n \to \overline{\mathbb{R}}$ is convex if and only if*

$$f\big((1-\lambda)x^0 + \lambda x^1\big) \le (1-\lambda)f(x^0) + \lambda f(x^1) \quad \forall x^0, x^1 \in \mathbb{R}^n, \lambda \in [0,1].$$

**Proof.** Suppose that $f$ is convex. Since the asserted inequality holds trivially otherwise, it suffices to consider $\lambda \in (0,1)$ and $x^0, x^1 \in \operatorname{dom} f$. By the convexity of epi $f$,

$$\big[(x^0, \alpha^0) \leftrightarrow (x^1, \alpha^1)\big] \subset \operatorname{epi} f \tag{1.5}$$

whenever $(x^0, \alpha^0), (x^1, \alpha^1) \in \operatorname{epi} f$. Hence,

$$f\big((1-\lambda)x^0 + \lambda x^1\big) \le (1-\lambda)\alpha^0 + \lambda\alpha^1.$$

This implies the convexity inequality. For the converse, suppose that $(x^0, \alpha^0), (x^1, \alpha^1) \in \operatorname{epi} f$ and $\lambda \in (0,1)$. By the convexity inequality and the definition of epigraphs, one obtains

$$f\big((1-\lambda)x^0 + \lambda x^1\big) \le (1-\lambda)f(x^0) + \lambda f(x^1) \le (1-\lambda)\alpha^0 + \lambda\alpha^1.$$

Since this holds for all $\lambda \in (0,1)$, the inclusion (1.5) is valid. Thus, $f$ is convex. □

The function $f : \mathbb{R} \to \overline{\mathbb{R}}$ defined by $f(x) = -\ln x$ for $x > 0$ and $f(x) = \infty$ otherwise is convex with $\operatorname{dom} f = (0, \infty)$; see Figure 1.10. This is clear from the convexity inequality under the rules for extended arithmetic. For example, if $x^0$ in the figure is moved to the left so that $x^0 = 0$ and $\lambda \in (0,1)$, then the convexity inequality compares a left-hand side of $f((1-\lambda)0 + \lambda x^1)$, which is finite, and a right-hand side of $(1-\lambda) \cdot \infty + \lambda f(x^1) = \infty$.

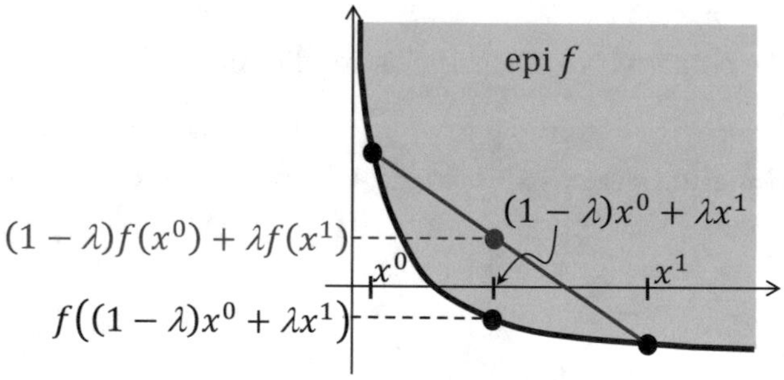


Fig. 1.10: Convexity inequality applied to the negative ln-function.

For $\alpha \in \overline{\mathbb{R}}$ and $f : \mathbb{R}^n \to \overline{\mathbb{R}}$, the *$\alpha$-level-set* of $f$ is denoted by

$$\{f \le \alpha\} = \big\{x \in \mathbb{R}^n \,\big|\, f(x) \le \alpha\big\}$$

and is illustrated in Figure 1.11. If $f$ is an objective function, then a level-set of particular interest is

$$\{f \le f(x^\star)\} = \operatorname{argmin} f, \quad \text{where} \quad x^\star \in \operatorname{argmin} f.$$

Generally, the $\alpha$-level-set of $f$ specifies the decisions that produce a "cost" of no more than $\alpha$. Another quantity that can be expressed in terms of level-sets is the domain of a function:

$$\operatorname{dom} f = \bigcup\nolimits_{\alpha\in\mathbb{R}} \{f \le \alpha\}.$$

As seen next, every level-set of $f$ is convex if $f$ is a convex function. This implies that under convexity any two minimizers can be combined to produce another minimizer. Specifically,

$$x^\star, x^{\star\star} \in \operatorname{argmin} f \quad \Longrightarrow \quad (1-\lambda)x^\star + \lambda x^{\star\star} \in \operatorname{argmin} f \quad \forall \lambda \in (0,1).$$

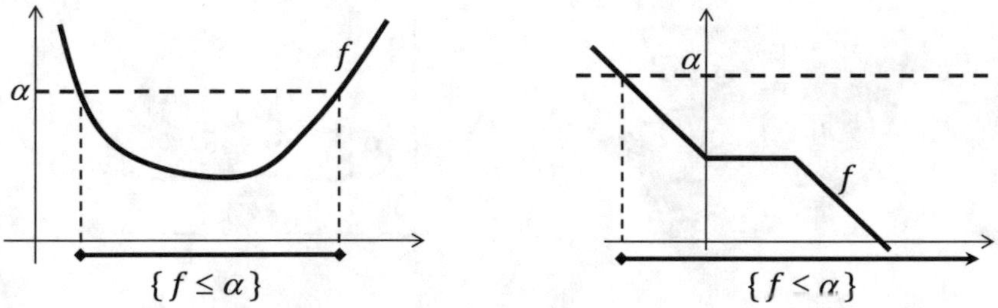


Fig. 1.11: Level-sets of convex (left) and nonconvex (right) functions.

**Proposition 1.11** (convexity of domains and level-sets). *If* $f : \mathbb{R}^n \to \overline{\mathbb{R}}$ *is a convex function, then* $\operatorname{dom} f$ *and* $\{f \le \alpha\}$ *are convex sets for all* $\alpha \in \overline{\mathbb{R}}$.

**Proof.** Let $\alpha \in \overline{\mathbb{R}}$. If $\{f \le \alpha\}$ is empty, it's convex. Otherwise, for $x^0, x^1 \in \{f \le \alpha\}$, $\lambda \in [0,1]$ and the corresponding convex combination $x^\lambda = (1-\lambda)x^0 + \lambda x^1$, one has

$$f(x^\lambda) \le (1-\lambda)f(x^0) + \lambda f(x^1) \le \alpha$$

by the convexity inequality 1.10. Thus, $x^\lambda \in \{f \le \alpha\}$ and $\{f \le \alpha\}$ is convex. A similar argument establishes that $\operatorname{dom} f$ is convex. □

Convexity of all level-sets doesn't imply convexity of $f$. For example, $f(x) = \max\{1 - x, 1\}$ for $x \le 1$ and $f(x) = -x + 2$ when $x > 1$ define a nonconvex function but $\{f \le \alpha\}$ is convex for all $\alpha$; see Figure 1.11(right). A function with all its level-sets being convex is said to be *quasi-convex*.

We next turn to rules for determining whether a set or function is convex.

**Proposition 1.12** (operations preserving convexity). *If* $\{C_\alpha \subset \mathbb{R}^n, \alpha \in \mathbb{A}\}$ *is a collection of convex sets, then the following sets are also convex:*

(a) *Intersection rule for arbitrary* $\mathbb{A}$:

$$\bigcap\nolimits_{\alpha\in\mathbb{A}} C_\alpha.$$

(b) *Product rule for* $\mathbb{A} = \{1, 2, \ldots, m\}$:

$$\prod\nolimits_{\alpha\in\mathbb{A}} C_\alpha = C_1 \times \cdots \times C_m = \big\{(x^1, \ldots, x^m) \,\big|\, x^1 \in C_1, \ldots, x^m \in C_m\big\}.$$

(c) *Linear combination rule for* $\mathbb{A} = \{1, 2, \ldots, m\}$ *and* $\lambda_1, \ldots, \lambda_m \in \mathbb{R}$:

$$\sum\nolimits_{\alpha\in\mathbb{A}} \lambda_\alpha C_\alpha = \Big\{\sum\nolimits_{\alpha\in\mathbb{A}} \lambda_\alpha x^\alpha \,\Big|\, x^1 \in C_1, \ldots, x^m \in C_m\Big\}.$$

**Proof.** These facts follow readily from the definition of convex sets. □

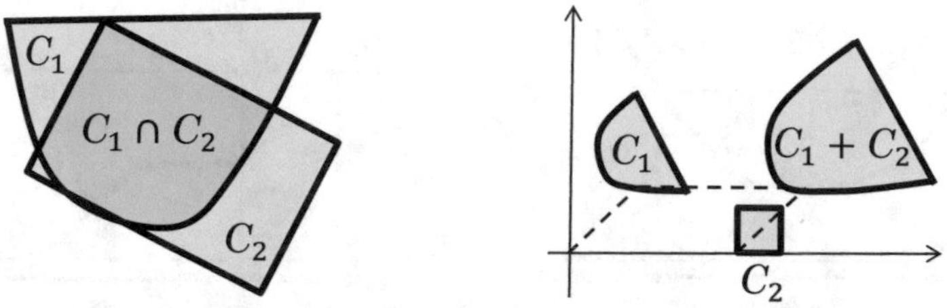


Fig. 1.12: Intersection (left) and linear combination (right) of convex sets.

Figure 1.12(left) illustrates the intersection rule. An example of a *product set* is the *box* $\prod_{j=1}^n [a_j, b_j]$ in $\mathbb{R}^n$ defined by $a_j \le b_j$ for $j = 1, \ldots, n$. Figure 1.12(right) visualizes the linear combination of two sets $\lambda_1 C_1 + \lambda_2 C_2$, i.e., all the points that can be obtained by scaling a point from each set and adding them together, but here only with $\lambda_1 = \lambda_2 = 1$.

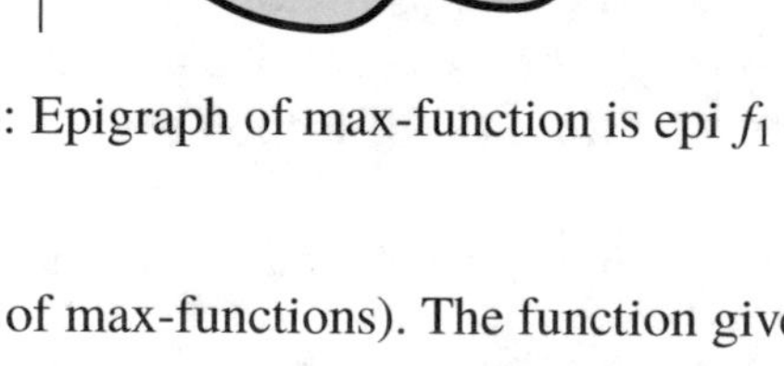

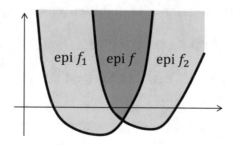

Fig. 1.13: Epigraph of max-function is epi $f_1 \cap$ epi $f_2$.

**Example 1.13** (convexity of max-functions). The function given by

$$f(x) = \max\{f_1(x), \ldots, f_m(x)\}$$

is convex provided that $f_i : \mathbb{R}^n \to \overline{\mathbb{R}}$, $i = 1, 2, \ldots, m$ are convex.

**Detail.** A function of this kind is referred to as a *max-function*. Since all epi $f_i$ are convex by definition and epi $f = \cap_{i=1}^m$ epi $f_i$ (see Figure 1.13), it follows from 1.12(a) that epi $f$ is convex. As a particular instance, the function $h$ in Figure 1.6 is convex. □

**Example 1.14** (affine mapping). A mapping $F : \mathbb{R}^n \to \mathbb{R}^m$ is *affine* when for some $m \times n$-matrix $A$ and $b \in \mathbb{R}^m$, one has $F(x) = Ax + b$ for all $x \in \mathbb{R}^n$. If $F$ is affine and $C \subset \mathbb{R}^n$ is convex, then

$$F(C) = \{F(x) \mid x \in C\}$$

is a convex set.

**Detail.** The claim follows by direct appeal to the definition of convex sets. As an illustration, for given

$$C \subset \mathbb{R}^3, \quad A = \begin{bmatrix} 1 & 0 & 0 \\ 0 & 1 & 0 \\ 0 & 0 & 0 \end{bmatrix} \quad \text{and} \quad b = \begin{bmatrix} 0 \\ 0 \\ 0 \end{bmatrix},$$

the set $\{Ax + b \mid x \in C\}$ is the projection of $C$ on the plane $\{x \in \mathbb{R}^3 \mid x_3 = 0\}$. □

**Norms.** The distance between a point $x$ and the origin in $\mathbb{R}^n$ is often taken to be the straight-line distance, but this isn't the only option. Norms are functions that tell distances between points. Specifically, a *norm* on $\mathbb{R}^n$ is a real-valued function $x \mapsto \|x\|$ such that for any $x, \bar{x} \in \mathbb{R}^n$:

$$\|\lambda x\| = |\lambda|\|x\| \quad \forall \lambda \in \mathbb{R} \qquad \text{(absolute homogeneity)}$$

$$\|x + \bar{x}\| \leq \|x\| + \|\bar{x}\| \qquad \text{(triangle inequality or subadditivity)}$$

$$\|x\| = 0 \implies x = 0 \qquad \text{(separation of points)}$$

Norms arise naturally in optimization models related to layout and geometrical considerations as well as in statistical estimation, but are also central to the notion of convergence.

**Example 1.15** (norms; closed and open balls). Any norm is convex. Specific choices include the following (see Figure 1.14):

$$\|x\|_2 = \Big(\sum_{j=1}^n x_j^2\Big)^{1/2} = \sqrt{\langle x, x\rangle} = \sqrt{x^\top x} \qquad \text{(Euclidean or } \ell^2\text{-norm)}$$

$$\|x\|_1 = \sum_{j=1}^n |x_j| \qquad \text{(Manhattan or } \ell^1\text{-norm)}$$

$$\|x\|_p = \Big(\sum_{j=1}^n |x_j|^p\Big)^{1/p}, \quad 1 \leq p < \infty \qquad (\ell^p\text{-norm})$$

$$\|x\|_\infty = \max_{j=1,\dots,n} |x_j| \qquad (\ell^\infty\text{-norm}).$$

We write $\|x\|$ whenever a statement applies to any norm.

For $\rho \in [0, \infty)$, the *closed Euclidean ball*

$$\mathbb{B}(\bar{x}, \rho) = \{x \mid \|x - \bar{x}\|_2 \leq \rho\}$$

is a convex set and the same holds for the *open Euclidean ball* $\{x \mid \|x - \bar{x}\|_2 < \rho\}$ with $\rho \in (0, \infty)$.

**Detail.** Convexity of norms follows readily from their defining properties via the convexity inequality 1.10. A brief argument based on the definitions takes care of the convexity of balls or, in the case of closed balls, one can appeal to 1.11. □

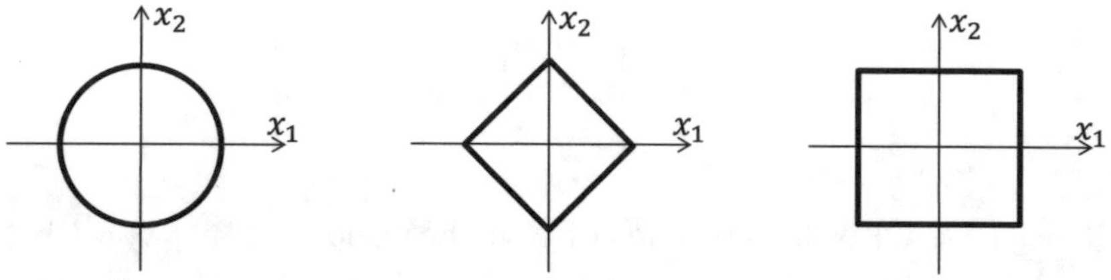


Fig. 1.14: Illustrations of points $x = (x_1, x_2)$ with $\|x\|_2 = 1$ (left), $\|x\|_1 = 1$ (middle) and $\|x\|_\infty = 1$ (right).

A slight refinement of convexity helps us to establish uniqueness of minimizers. We say that a convex function $f : \mathbb{R}^n \to \overline{\mathbb{R}}$ is *strictly convex* if

$$f\big((1-\lambda)x^0 + \lambda x^1\big) < (1-\lambda)f(x^0) + \lambda f(x^1) \;\; \forall \lambda \in (0,1), x^0 \neq x^1 \text{ with } f(x^0), f(x^1) \in \mathbb{R}.$$

Affine functions are convex, but not strictly because then the convexity inequality 1.10 holds with equality. Likewise, norms aren't strictly convex because with $x^0 = 0$ and $x^1 \neq 0$ the convexity inequality again holds with equality. The ln-function $x \mapsto -\ln x$, understood as $\infty$ when $x \leq 0$, the exponential functions $x \mapsto e^{\alpha x}$ for nonzero $\alpha \in \mathbb{R}$ and the polynomial functions $x \mapsto x^n$ for $n = 2, 4, 6, \ldots$ are all strictly convex.

**Proposition 1.16** (uniqueness of minimizer). *Suppose that $f : \mathbb{R}^n \to \overline{\mathbb{R}}$ is strictly convex and $f(x) > -\infty$ for all $x \in \mathbb{R}^n$. Then,* argmin $f$ *contains at most one point, i.e., any minimizer is unique.*

**Proof.** For the sake of contradiction, suppose that there are two distinct minimizers $x^0$ and $x^1$. By assumption, $f(x^0), f(x^1) \in \mathbb{R}$ so that the definition of strict convexity guarantees

$$f\big(\tfrac{1}{2}x^0 + \tfrac{1}{2}x^1\big) < \tfrac{1}{2}f(x^0) + \tfrac{1}{2}f(x^1) = f(x^0).$$

This contradicts the fact that $x^0$ is a minimizer. □

**Infimum and Supremum.** The lowest scalar in a set $Y \subset \overline{\mathbb{R}}$ is well understood when $Y$ contains a finite number of elements, in which case we say that $Y$ is a *finite set*. In optimization problems, however, we need to consider an uncountable number of possible scalars, say $\{f(x) \mid x \in \mathbb{R}^n\}$ for a function $f : \mathbb{R}^n \to \overline{\mathbb{R}}$. This requires some terminology. The *infimum* of a nonempty $Y \subset \overline{\mathbb{R}}$, denoted by $\inf Y$, is the *highest* scalar $\alpha \in \overline{\mathbb{R}}$ that satisfies $\alpha \leq \beta$ for all $\beta \in Y$. Moreover, $\inf \emptyset$ is defined as $\infty$. When $Y$ contains $\inf Y$, one refers to $\inf Y$ as the *minimum* of $Y$ and denotes it by $\min Y$. The distinction is necessary because neither $\mathbb{R}$ nor the interval $(0, 1)$ has a minimum, but $\inf \mathbb{R} = -\infty$ and $\inf (0, 1) = 0$. In fact, $\inf Y$ exists for any $Y \subset \overline{\mathbb{R}}$.

We're especially interested in scalars corresponding to values produced by a function $f : \mathbb{R}^n \to \overline{\mathbb{R}}$ on a set $C \subset \mathbb{R}^n$. Then, as short-hand notation,

$$\inf_C f = \inf_{x \in C} f(x) = \inf\{f(x) \mid x \in C\} \quad \text{and} \quad \inf f = \inf_{\mathbb{R}^n} f.$$

For example, $f(x) = e^x$ has $\inf f = 0$ and $g(x) = x$ has $\inf g = -\infty$. If there's $\bar{x} \in C$ such that $f(\bar{x}) = \inf_C f$, then $\inf_C f$ is the *minimum* (or *minimum value*) of $f$ on $C$ and we might replace "inf" by "min" in the expression. Moreover, the *set of minimizers*

$$\operatorname{argmin}_C f = \operatorname{argmin}_{x \in C} f(x) = \{x \in C \cap \operatorname{dom} f \mid f(x) = \inf_C f\}$$

so that any $x^\star \in \operatorname{argmin}_C f$ furnishes the minimum of $f$ on $C$ as $f(x^\star)$. We note that $\operatorname{argmin}_C f$ is empty when $C \cap \operatorname{dom} f = \emptyset$ even though $f(x) = \inf_C f = \infty$ for all $x \in C$ in that case. We wouldn't like such $x$ to be a solution of the problem of minimizing $f$ over $C$ because a "cost" of $\infty$ is intolerable.

Similarly, the *supremum* of a nonempty $Y \subset \overline{\mathbb{R}}$, denoted by $\sup Y$, is the *lowest* scalar $\alpha \in \overline{\mathbb{R}}$ that satisfies $\alpha \geq \beta$ for all $\beta \in Y$, with $\sup \emptyset$ being set to $-\infty$. When $Y$ contains $\sup Y$, one refers to $\sup Y$ as the *maximum* of $Y$ and denotes it by $\max Y$. For $f : \mathbb{R}^n \to \overline{\mathbb{R}}$ and $C \subset \mathbb{R}^n$, we write

$$\sup_C f = \sup_{x \in C} f(x) = \sup\{f(x) \mid x \in C\} \quad \text{and} \quad \sup f = \sup_{\mathbb{R}^n} f.$$

Again, when there's $x \in C$ such that $f(x) = \sup_C f$, we might replace "sup" by "max" in the expressions.

**Example 1.17** (uncertain linear system). A linear system takes as input a vector $x \in \mathbb{R}^n$ and returns an output $y = Ax$, where $A$ is an $m \times n$-matrix. We would like to choose $x$ such that $\|y\| = \|Ax\|$ is small. However, the problem is complicated by the fact that $A$ depends on an unknown parameter vector $\xi \in \Xi \subset \mathbb{R}^q$, i.e., $A = A(\xi)$. This leads to a formulation involving a supremum.

**Detail.** A conservative approach to the problem would be to consider the optimization model

$$\underset{x \in \mathbb{R}^n}{\text{minimize}} \ f(x) = \sup_{\xi \in \Xi} \big\|A(\xi)x\big\|,$$

where we attempt to find an input $x$ that guarantees the lowest possible output regardless of the value of the unknown vector $\xi \in \Xi$. The objective function $f$ is defined in terms of a supremum and could possibly take the value $\infty$ for some $x$. It's convex by (a) and (c) below. □

**Proposition 1.18** (operations preserving convexity). *Let $\{f, f_\alpha : \mathbb{R}^n \to \overline{\mathbb{R}}, \alpha \in \mathbb{A}\}$ be a collection of convex functions. Then, the following hold:*

(a) *Supremum rule for arbitrary $\mathbb{A}$:*

$$x \mapsto \sup_{\alpha \in \mathbb{A}} f_\alpha(x)$$

*is convex and strictly so if all $f_\alpha$ are strictly convex and $\mathbb{A}$ is finite.*

(b) *Sum rule for finite $\mathbb{A}$:*

$$x \mapsto \sum\nolimits_{\alpha \in \mathbb{A}} f_\alpha(x)$$

*is convex and strictly so if at least one $f_\alpha$ is strictly convex.*

(c) *Composition rule with affine function:*

$$x \mapsto f(Ax + b)$$

*is convex, where $A$ is an $n \times m$-matrix and $b \in \mathbb{R}^n$.*

(d) *Composition rule with convex nondecreasing function $\varphi : \mathbb{R} \to \overline{\mathbb{R}}$:*

$$x \mapsto \varphi\big(f(x)\big)$$

*is convex under the convention $\varphi(\infty) = \infty$ and $\varphi(-\infty) = \inf \varphi$. In particular,*

$$x \mapsto \lambda f(x)$$

*is convex when $\lambda \in [0, \infty)$ and strictly convex when $\lambda \in (0, \infty)$ and $f$ is strictly convex.*

**Proof.** Convexity in (a) holds by 1.12(a); cf. 1.13. Strict convexity is established directly from the definition. For (b), one can apply 1.10 and the definition of strict convexity. An application of 1.10 yields (c) and (d). □

**Exercise 1.19** (convexity in search problem). In planning the search for floating debris from an aircraft lost over the ocean, one would divide the area of interest into $n$ squares and then determine how much time and resources should be put toward searching each square. For simplicity, suppose that the search will be carried out by one aircraft flying at $\eta$ miles per hour, with a sensor that effectively can reach out $\omega/2$ miles, and that the debris is located in one square and is stationary. Prior knowledge stipulates the probability that the debris is in square $j$ as $p_j > 0$, with $\sum_{j=1}^n p_j = 1$. If the squares are $\alpha_1, \ldots, \alpha_n$ square miles in size and the total amount of search time available is $\tau$, then the *search problem* of determining the time allocation $x = (x_1, \ldots, x_n)$ that minimizes the probability of not finding the debris can be formulated as

$$\underset{x \in C}{\text{minimize}}\ f(x) = \sum_{j=1}^{n} p_j \exp(-\eta\omega x_j/\alpha_j),$$

with $C = \{x \in \mathbb{R}^n \,|\, \sum_{j=1}^{n} x_j = \tau,\ x_j \geq 0, j = 1, \ldots, n\}$. Show that $f$ and $C$ are convex.

**Guide.** Note that $C$ is the intersection of $n + 1$ relatively simple sets. □

**Exercise 1.20** (convexity of newsvendor problem). Show that the objective function (1.4) of the newsvendor problem is convex when $\boldsymbol{\xi}$ has possible values $\xi_1, \ldots, \xi_m$ with corresponding probabilities $p_1, \ldots, p_m$.

**Guide.** Note that $\mathbb{E}[h(\boldsymbol{\xi})] = \sum_{i=1}^{m} p_i h(\xi_i)$ and rely on 1.18. □

In the same way as a function of interest can be the supremum of some other function, it can also be given by an infimum.

**Proposition 1.21** (convexity of inf-projection). *For convex* $f : \mathbb{R}^m \times \mathbb{R}^n \to \overline{\mathbb{R}}$, *the* inf-projection $p : \mathbb{R}^m \to \overline{\mathbb{R}}$ *given by*

$$p(u) = \inf \{f(u, x) \mid x \in \mathbb{R}^n\}$$

*is a convex function.*

**Proof.** The set $C = \{(u, x, \alpha) \mid f(u, x) < \alpha < \infty\}$ is the epigraph of $f$ with its graph removed, which is still convex because the epigraph is convex. The set $D = \{(u, \alpha) \mid p(u) < \alpha < \infty\}$ is produced by an affine mapping of $C$ via the removal of the $x$-coordinate. By 1.14, $D$ is convex. Since $D$ is the epigraph of $p$ with the graph removed, epi $p$ is convex as well. □

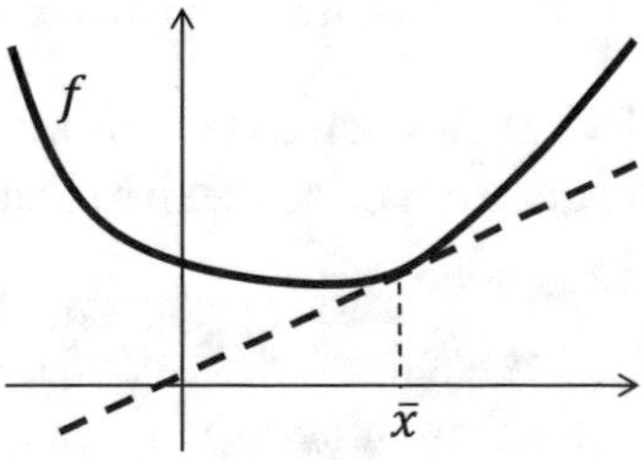


Fig. 1.15: The affine function $x \mapsto f(\bar{x}) + \langle \nabla f(\bar{x}), x - \bar{x} \rangle$ bounds $f$ from below.

The gradients of a smooth function characterize convexity in a way that also points to the possibility of constructing lower bounding approximations as illustrated in Figure 1.15.

**Proposition 1.22** (gradient inequality for convex functions). *For a smooth function* $f : \mathbb{R}^n \to \mathbb{R}$, *one has*

$$f \text{ convex} \iff f(x) \geq f(\bar{x}) + \langle \nabla f(\bar{x}), x - \bar{x} \rangle \ \ \forall x, \bar{x} \in \mathbb{R}^n.$$

**Proof.** If $f$ is convex, then

$$f\big((1-\tau)\bar{x}+\tau x\big) \le (1-\tau)f(\bar{x})+\tau f(x) \quad \forall \tau \in (0,1]$$

holds via 1.10 and, after dividing both sides by $\tau$, one obtains

$$f(x) \ge \frac{f\big(\bar{x}+\tau(x-\bar{x})\big)-(1-\tau)f(\bar{x})}{\tau} = f(\bar{x}) + \frac{f\big(\bar{x}+\tau(x-\bar{x})\big)-f(\bar{x})}{\tau}.$$

Now letting $\tau \searrow 0$, we see that

$$f(x) \ge f(\bar{x}) + \vec{d}f(\bar{x}; x-\bar{x}).$$

By (1.3), the gradient inequality holds. For the converse, the gradient inequality implies

$$f(x) = \sup_{\bar{x}\in\mathbb{R}^n} \big\{ f(\bar{x}) + \langle \nabla f(\bar{x}), x-\bar{x}\rangle \big\} \ \forall x \in \mathbb{R}^n.$$

Then, the convexity of $f$ follows by 1.18(a). □

For the quadratic function given by $f(x) = x^2$, the gradient inequality in 1.22 amounts to having $x^2 \ge \bar{x}^2 + 2\bar{x}(x-\bar{x})$ for all $x, \bar{x} \in \mathbb{R}$. Since this is equivalently written as $(x-\bar{x})^2 \ge 0$, it certainly holds.

If $f : \mathbb{R}^n \to \mathbb{R}$ has all its second-order partial derivatives at $\bar{x}$, then the *Hessian* of $f$ at $\bar{x}$ is the symmetric[8] $n \times n$-matrix

$$\nabla^2 f(\bar{x}) = \begin{bmatrix} \frac{\partial^2 f(\bar{x})}{\partial x_1 \partial x_1} & \cdots & \frac{\partial^2 f(\bar{x})}{\partial x_1 \partial x_n} \\ \vdots & \ddots & \vdots \\ \frac{\partial^2 f(\bar{x})}{\partial x_n \partial x_1} & \cdots & \frac{\partial^2 f(\bar{x})}{\partial x_n \partial x_n} \end{bmatrix}.$$

We say that $f$ is *twice smooth* if $\nabla^2 f(x)$ is defined for all $x \in \mathbb{R}^n$ and it varies continuously. For twice smooth functions, the question of convexity is addressed by linear algebra.

**Semidefinite Matrices.** An $n \times n$-matrix $Q$ is *positive semidefinite* if $\langle y, Qy\rangle \ge 0$ for all $y \in \mathbb{R}^n$. When $Q$ is symmetric, it's positive semidefinite if and only if all its eigenvalues are nonnegative. The matrix $Q$ is *positive definite* if $\langle y, Qy\rangle > 0$ for all $y \ne 0$ and, when symmetric, it's positive definite if and only if all its eigenvalues are positive.

**Proposition 1.23** (Hessian rules for convexity). *For a twice smooth function $f : \mathbb{R}^n \to \mathbb{R}$, the following hold:*

(a) $\nabla^2 f(x)$ *is positive semidefinite for all* $x \in \mathbb{R}^n \iff f$ *is convex.*
(b) $\nabla^2 f(x)$ *is positive definite for all* $x \in \mathbb{R}^n \implies f$ *is strictly convex.*

[8] The matrix $A$ is *symmetric* if $A = A^\top$.

**Proof.** We refer to [105, Theorem 2.14] for a proof. □

We observe that the quartic function given by $f(x) = x^4$ is strictly convex, but $\nabla^2 f(x) = 12x^2$ isn't positive definite for $x = 0$. Thus, the converse of (b) in the proposition fails.

**Example 1.24** (quadratic functions). A function $f : \mathbb{R}^n \to \mathbb{R}$ is *quadratic* if it's expressible as

$$f(x) = \tfrac{1}{2}\langle x, Qx\rangle + \langle c, x\rangle + \alpha$$

for some symmetric $n \times n$-matrix $Q$, $c \in \mathbb{R}^n$ and $\alpha \in \mathbb{R}$. Then,

$$\nabla f(x) = Qx + c \quad \text{and} \quad \nabla^2 f(x) = Q,$$

so $f$ is convex if and only if $Q$ is positive semidefinite by 1.23. Moreover, $f$ is strictly convex if and only if $Q$ is positive definite.

Since they're level-sets of convex functions, sets of the form

$$\left\{x \in \mathbb{R}^n \;\middle|\; \tfrac{1}{2}\langle x, Qx\rangle + \langle c, x\rangle + \alpha \leq 0\right\}$$

are convex provided that $Q$ is positive semidefinite; see 1.11. This class of sets includes closed Euclidean balls as well as general ellipsoids, paraboloids and cylinders.

**Detail.** The assertion that positive definiteness of $Q$ isn't only sufficient for strict convexity as in 1.23 but also necessary can be established as follows: If $Q$ is positive semidefinite, but not positive definite, then there's a vector $z \neq 0$ such that $\langle z, Qz\rangle = 0$, and along the line through the origin in the direction of $z$ it's impossible for $f$ to be strictly convex.

As a specific illustration, the objective function in §1.B is given by

$$f(c) = \langle Dc - y, Dc - y\rangle = \tfrac{1}{2}\langle c, 2D^\top Dc\rangle + \langle -2y^\top D, c\rangle + \|y\|_2^2.$$

Thus, $Q = 2D^\top D$ in this case, which is positive semidefinite because

$$\langle c, 2D^\top Dc\rangle = 2\langle Dc, Dc\rangle = 2\|Dc\|_2^2 \geq 0 \quad \forall c.$$

The objective function from §1.B is then convex and the solution found there is indeed a minimizer by 1.9. In fact, since $D^\top D$ is also invertible, its eigenvalues are positive, $Q$ is positive definite, the objective function is strictly convex and the minimizer is unique. □

**Exercise 1.25** (quadratic function). For the quadratic function in 1.2, check whether it's convex, strictly convex or neither. Determine all its minimizers.

The benefits of convexity aren't contingent on smoothness. (The statements in 1.9 and 1.22 are preliminary with extensions in Chap. 2.) The main consequence of convexity is that global properties of a function such as its minimizers can be ascertained from a local analysis and this fact is unrelated to smoothness.

For $f : \mathbb{R}^n \to \overline{\mathbb{R}}$, we say that $x^\star \in \operatorname{dom} f$ is a *local minimizer* of $f$ if there's $\varepsilon > 0$ such that

$$f(x^\star) \leq f(x) \quad \forall x \in \mathbb{B}(x^\star, \varepsilon).$$

A minimizer is sometimes called a *global minimizer*, for emphasis. Every minimizer is a local minimizer because then $\varepsilon$ can be taken to be any positive number; cf. Figure 1.16.

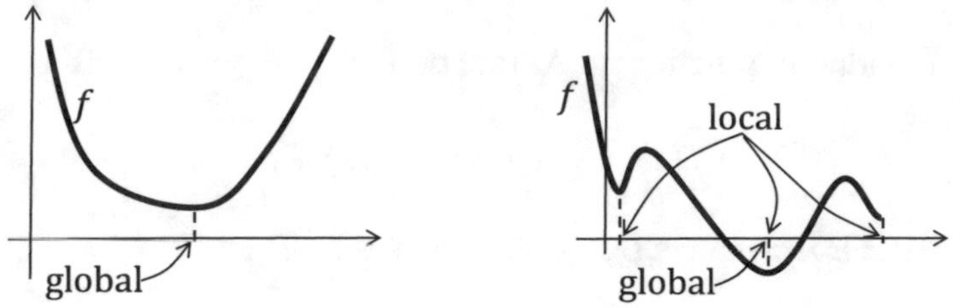


Fig. 1.16: Local and global minimizers for convex (left) and nonconvex (right) functions.

**Theorem 1.26** (local implies global minimizers). *Every local minimizer of a convex function* $f : \mathbb{R}^n \to \overline{\mathbb{R}}$ *is a (global) minimizer.*

**Proof.** Suppose $x^\star$ is a local minimizer of $f$ and, for some $\varepsilon > 0$, $f(x^\star) \leq f(x)$ for all $x \in \mathbb{B}(x^\star, \varepsilon)$. If $x^\star$ isn't a minimizer, then there exists $\bar{x}$ such that $f(\bar{x}) < f(x^\star)$. The convexity inequality 1.10 ensures that there's $\lambda > 0$, sufficiently small, such that

$$x^\lambda = \lambda x^\star + (1-\lambda)\bar{x} \in \mathbb{B}(x^\star, \varepsilon) \text{ and } f(x^\lambda) \leq \lambda f(x^\star) + (1-\lambda) f(\bar{x}) < f(x^\star).$$

However, this contradicts the fact that $x^\star$ is a local minimizer. □

Throughout, we put a firm emphasis on optimization problems for which the goal is to *minimize*, not maximize, a function, i.e., *minimization problems*. This one-sided focus has implications for definitions and concepts: $\infty$ takes precedence over $-\infty$ in extended arithmetic, the epigraph of a function is more central than its *hypograph* and convexity dominates *concavity*[9]. We'll occasionally comment on the parallel universe of *maximization problems*, but a comprehensive treatment is largely redundant as most concepts are obtained through a sign change. In particular, for $f : \mathbb{R}^n \to \overline{\mathbb{R}}$,

$$\sup f = -\inf(-f)$$

and the set of *maximizers*

$$\operatorname{argmax} f = \left\{x^\star \in \mathbb{R}^n \;\middle|\; f(x^\star) > -\infty,\ f(x^\star) \geq f(x)\ \forall x \in \mathbb{R}^n\right\} = \operatorname{argmin}(-f).$$

**Exercise 1.27** (point-to-set distance). Show that the function $f : \mathbb{R}^n \to \overline{\mathbb{R}}$ given by $f(x) = \inf_{y \in C} \|Ax - y\|$ is convex when $C \subset \mathbb{R}^m$ is a convex set and $A$ is an $m \times n$-matrix. (If $A$ is the $n \times n$ identity matrix, then $f(x)$ is the minimum distance from $x$ to $C$.) Give a counterexample to demonstrate that the convexity assumption on $C$ can't be removed in general.

[9] A function $f : \mathbb{R}^n \to \overline{\mathbb{R}}$ is concave if its hypograph, $\operatorname{hypo} f = \{(x, \alpha) \mid \alpha \leq f(x)\}$, is a convex set or, equivalently, $-f$ is convex.

**Guide.** Consider the alternative expression $f(x) = \inf_{y \in \mathbb{R}^m} g(x, y)$, with $g(x, y) = \|Ax - y\|$ if $y \in C$ and $g(x, y) = \infty$ if $y \notin C$, and apply 1.21. □

## 1.E Estimation and Classification

A common task in data analytics, learning and statistics is to predict a quantity of interest from available data. A prime example is the problem of estimating the distribution function of a random variable on the basis of observations of its values. Often, we seek a distribution function that "best" fits the data in some sense and this lands us squarely in the optimization landscape.

Suppose that we observe $m$ times the outcome of a random phenomenon such as the number of eyes obtained when rolling a six-sided die. In a real-world application, the random phenomenon might be the daily demand of a perishable product as in §1.C. Let $\{\xi_1, \ldots, \xi_m\}$ be the observed values. For example, this data could be the demand for newspapers over the previous $m$ days. We view the data as the outcome of $m$ random variables $\{\boldsymbol{\xi}_1, \ldots, \boldsymbol{\xi}_m\}$. How can we estimate the distribution function(s) of these random variables and thereby obtain a *model of uncertainty* about future values?

In general, a collection of random variables $\{\boldsymbol{\xi}_1, \ldots, \boldsymbol{\xi}_m\}$ has a distribution function defined on $\mathbb{R}^m$:

$$P_{1:m}(\eta) = \text{prob}\,\{\, \boldsymbol{\xi}_1 \leq \eta_1, \ldots, \boldsymbol{\xi}_m \leq \eta_m \,\} \quad \forall \eta = (\eta_1, \ldots, \eta_m),$$

where *prob* abbreviates "probability of." The distribution function $P_{1:m}$ is sometimes called the *joint* distribution function of the *random vector* $\boldsymbol{\xi} = (\boldsymbol{\xi}_1, \ldots, \boldsymbol{\xi}_m)$ to distinguish it from the distribution functions of the individual random variables $\boldsymbol{\xi}_1, \ldots, \boldsymbol{\xi}_m$, denoted by $P_1, \ldots, P_m$ and referred to as *marginal* distribution functions. The random variables $\{\boldsymbol{\xi}_1, \ldots, \boldsymbol{\xi}_m\}$ are *statistically independent* if

$$P_{1:m}(\eta) = \prod\nolimits_{i=1}^{m} P_i(\eta_i) \quad \forall \eta = (\eta_1, \ldots, \eta_m).$$

They're *iid* if statistically independent and identically distributed, i.e., $P_1 = P_i$ for all $i$.

Suppose that we're willing to assume that $\{\xi_1, \ldots, \xi_m\}$ is an outcome of $m$ iid random variables[10] $\{\boldsymbol{\xi}_1, \ldots, \boldsymbol{\xi}_m\}$ with common but unknown distribution function $P$. A reasonable estimate of $P$ would then be a distribution function that is most likely to yield the data actually observed; a *maximum likelihood* estimate. Suppose that $P$ is continuous and can be expressed in terms of a density function $p$. Then, we may just as well estimate the density function.

The probability that an outcome of $\boldsymbol{\xi}_i$ ends up to be $\xi_i$ or in a neighborhood, say $\pm\delta$, is

$$P(\xi_i + \delta) - P(\xi_i - \delta) = \int_{\xi_i - \delta}^{\xi_i + \delta} p(\eta)\, d\eta \approx p(\xi_i) 2\delta.$$

[10] The iid assumption needs to be examined carefully because in practice daily observations, for example, might be correlated due to underlying trends.

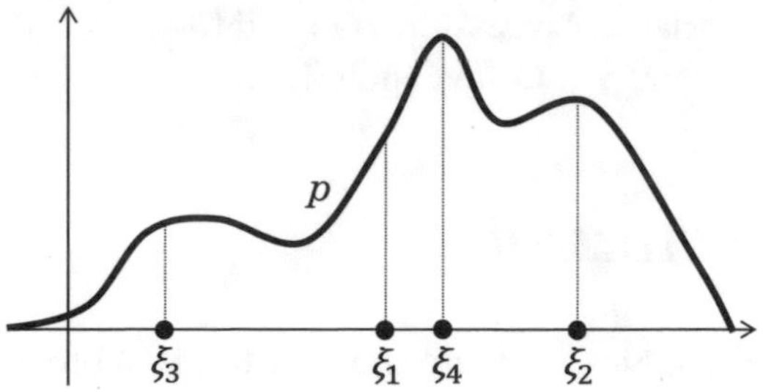


Fig. 1.17: Likelihood of observing $\xi_1, \ldots, \xi_4$ under density function $p$ is the product of the "heights" at these points.

Hence, the probability that we observe an outcome of $(\boldsymbol{\xi}_1, \ldots, \boldsymbol{\xi}_m)$ in the box $\prod_{i=1}^m [\xi_i - \delta,\ \xi_i + \delta]$ is then given by the expression $2\delta \prod_{i=1}^m p(\xi_i)$. Since $2\delta$ is a constant, maximizing this expression over some collection of density functions $\mathcal{P}$ comes down to finding $p \in \mathcal{P}$ that maximizes $p \mapsto \prod_{i=1}^m p(\xi_i)$, i.e., making the product of the "heights" in Figure 1.17 as large as possible by adjusting $p$. Equivalently, one can maximize

$$p \mapsto \ln\Big(\prod_{i=1}^m p(\xi_i)\Big)$$

because the logarithmic function is increasing. With our focus on minimization problems, we convert the problem one last time by changing the sign of the objective function and also applying the product rule for logarithms so the problem becomes

$$\underset{p \in \mathcal{P}}{\text{minimize}} \ -\frac{1}{m}\sum_{i=1}^m \ln p(\xi_i), \tag{1.6}$$

where the factor $1/m$ furnishes a normalization without changing the set of minimizers. A minimizer of the objective function in this problem is a density function that's most likely to generate the observed data $\{\xi_1, \ldots, \xi_m\}$. For example, in the context of §1.C, with the data representing demand for newspapers in the past, the density function is a model of uncertainty about future demand. The difficulty of solving the problem depends on $\mathcal{P}$, but there are many tractable cases; see the next example and §3.D.

**Example 1.28** (estimating a normal distribution). A *normal* (Gaussian) random variable $\boldsymbol{\xi}$ has a density function of the form

$$p(\xi) = \frac{1}{\sqrt{2\pi}\sigma} \exp\Big( - \frac{(\xi - \mu)^2}{2\sigma^2} \Big),$$

where

$$\mu = \mathbb{E}[\boldsymbol{\xi}] \qquad \textit{(mean, average, expected value)}$$

$$\sigma^2 = \operatorname{var}(\boldsymbol{\xi}) = \mathbb{E}\big[\, (\boldsymbol{\xi} - \mu)^2 \,\big] = \mathbb{E}[\, \boldsymbol{\xi}^2 \,] - \mu^2 \qquad \textit{(variance)}.$$

It's *standard normal* if $\mu = 0$ and $\sigma^2 = 1$.

Given the data $\{\xi_1, \ldots, \xi_m\}$ obtained from outcomes of iid normally distributed random variables with mean $\mu$ and variance $\sigma^2$, we seek to determine the maximum likelihood estimate of $\mu$ under the assumption that the variance $\sigma^2 \in (0, \infty)$ is known.

**Detail.** Since we assume that $\mathcal{P}$ in the maximum likelihood problem contains only normal density functions with a known variance, (1.6) reduces to a minimization problem over $\mu$. For such density functions

$$-\ln p(\xi_i) = f_i(\mu) = \frac{(\mu - \xi_i)^2}{2\sigma^2} + \ln \sigma + \ln \sqrt{2\pi}.$$

The function $f_i$, defined by the right-hand side above, is strictly convex by 1.24 and the same holds for the whole objective function by 1.18(b,d). The Fermat rule results in the equation

$$0 = \frac{1}{m} \sum_{i=1}^m f_i'(\mu) = \frac{1}{m} \sum_{i=1}^m \frac{\mu - \xi_i}{\sigma^2},$$

which has solution

$$\hat{\mu} = \frac{1}{m} \sum_{i=1}^m \xi_i.$$

By 1.9 and 1.16, $\hat{\mu}$ is the unique minimizer. The maximum likelihood estimate is, therefore, a normal density function with mean $\hat{\mu}$ and variance $\sigma^2$. □

**Example 1.29** (classification and logistic regression). Should an email message be classified as spam on the basis of its characteristics (wording, layout, length, signature, etc.) and those of past email messages that have been identified as spam or not? This is an example of a (statistical) *classification problem* which consists of assigning an observation to a category based on its characteristics given that we've at our disposal a collection of observations with known characteristics and category membership. The problem, also referred to as a *learning problem*, can be formulated using an optimization model.

**Detail.** We're given the observations $\{x^i \in \mathbb{R}^n, i = 1, \ldots, m\}$ and their corresponding labels $\{y_i = \pm 1, i = 1, \ldots, m\}$, which specify the categories in which the observations belong. In the email context, each $i$ corresponds to an email, the vector $x^i$ describes the characteristics of the email and $y_i$ specifies the two categories: label 1 (spam) or $-1$ (non-spam). For example, the $i$th email in the dataset might be

> Urgent! call 800 651 8721 immediately. Your complimentary 4* Las Vegas Holiday or $10,000 cash await collection

with label $y_i = 1$. The vector $x^i = (x_1^i, \ldots, x_n^i)$ could describe which words from a dictionary of $n$ words are present: $x_j^i = 1$ if the $j$th word in the dictionary is present and $x_j^i = 0$ otherwise. Typically, $n$ is large.

A maximum likelihood approach to classification, referred to as *logistic regression*, is to determine a function $h : \mathbb{R}^n \to [0, 1]$ that assigns to every $x \in \mathbb{R}^n$ an estimated probability that it corresponds to label $-1$ (non-spam). Given any $h : \mathbb{R}^n \to [0, 1]$, the probability to observe the data would then be

$$\prod_{i|y_i=-1} h(x^i) \prod_{i|y_i=1} (1 - h(x^i))$$

and the maximum likelihood estimate is a function $h$ that maximizes this quantity. The situation resembles that of (1.6) and again we need to settle on a collection of functions to optimize over. It's common to assume that

$$h(x) = \frac{1}{1 + \exp(\langle a, x \rangle + \alpha)}, \quad \text{where } a \in \mathbb{R}^n, \alpha \in \mathbb{R}.$$

Indeed, $h(x) \in (0, 1)$ and can be any such number depending on the choice of $a$ and $\alpha$. The problem then reduces to optimizing the values of $a \in \mathbb{R}^n$ and $\alpha \in \mathbb{R}$. Specifically, bringing in the logarithm, normalizing with $1/m$ and switching to minimization, we obtain the model

$$\underset{(a,\alpha)\in\mathbb{R}^{n+1}}{\text{minimize}} \; -\frac{1}{m} \ln \prod_{i=1}^{m} \frac{1}{1 + \exp(-y_i(\langle a, x^i \rangle + \alpha))} = \frac{1}{m} \sum_{i=1}^{m} \ln\left(1 + e^{-y_i(\langle a, x^i \rangle + \alpha)}\right).$$

A solution $(x^\star, \alpha^\star)$ of this problem defines a *classifier* $h^\star : \mathbb{R}^n \to [0, 1]$, with

$$h^\star(x) = \frac{1}{1 + \exp(\langle a^\star, x \rangle + \alpha^\star)}$$

furnishing the probability that $x$ corresponds to label $-1$. When each component of $x$ describes whether a particular word is included as discussed above, $a^\star$ typically has positive values for the components corresponding to words often used in spam emails and negative values for words often used in non-spam. Thus, if $x$ indeed corresponds to a spam email, then $\langle a^\star, x \rangle$ tends to be a high number and consequently $h^\star(x)$ is low: the classifier indeed predicts that $x$ is most likely spam.

In view of 1.18(b,c,d), the objective function is convex provided that $z \mapsto \ln(1 + e^z)$ is convex. The latter function is convex because its Hessian is $e^z/(1 + e^z)^2$ and one can invoke 1.23. In contrast to the previous example, it's not possible to solve this estimation problem analytically; the variables $a$ and $\alpha$ must be determined by an algorithm. □

## 1.F Gradient Descent Method

There's a symbiotic relationship between minimizing a function on $\mathbb{R}^n$ and solving $n$ equations in $n$ unknowns. In this section, we'll take the first perspective and outline a method, motivated by the Oresme rule, that leverages directional derivatives to generate points with successively lower function values. The second perspective is exploited in §1.G and leads to solutions based on the Fermat rule. Since we deal with smooth functions, the two rules are equivalent but the distinction still provides insight about the underlying algorithmic strategies.

As indicated in §1.A, an algorithm for minimizing a function $f : \mathbb{R}^n \to \mathbb{R}$ can be based on the following simple but fruitful steps: From a given $x \in \mathbb{R}^n$, first identify

$w \in \mathbb{R}^n$ such that

$$\vec{d}f(x; w) < 0;$$

second, determine $\lambda > 0$ such that

$$f(x + \lambda w) < f(x);$$

and, third, replace $x$ by $x + \lambda w$. One might hope that after many iterations of this kind, the algorithm reaches a point $x$ with

$$\vec{d}f(x; w) \geq 0 \;\; \forall w \in \mathbb{R}^n$$

as required by the Oresme rule. An algorithm that iteratively reduces the value of $f$ in this manner is a *descent method.*

We say that $w \in \mathbb{R}^n$ is a *descent direction* for $f : \mathbb{R}^n \to \mathbb{R}$ at $\bar{x} \in \mathbb{R}^n$ if there's $\bar{\lambda} > 0$ such that

$$f(\bar{x} + \lambda w) < f(\bar{x}) \quad \forall \lambda \in (0, \bar{\lambda}).$$

**Proposition 1.30** (descent direction). *When $f : \mathbb{R}^n \to \mathbb{R}$ is smooth, every $w \in \mathbb{R}^n$ with*

$$\vec{d}f(\bar{x}; w) = \langle \nabla f(\bar{x}), w \rangle < 0$$

*is a descent direction for $f$ at $\bar{x}$.*

**Proof.** Suppose that $\vec{d}f(\bar{x}; w) = -\delta$ with $\delta > 0$. Then, for some $\bar{\lambda} > 0$,

$$\frac{f(\bar{x} + \lambda w) - f(\bar{x})}{\lambda} \leq -\tfrac{1}{2}\delta \quad \forall \lambda \in (0, \bar{\lambda}).$$

The conclusion follows after multiplying each side by $\lambda$. □

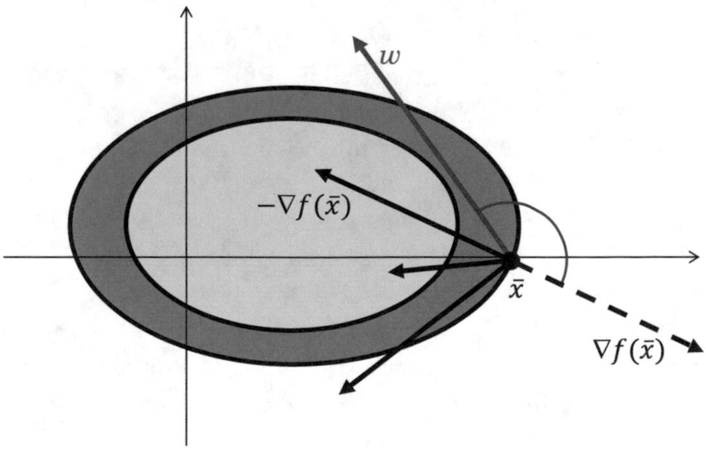


Fig. 1.18: Descent directions at $\bar{x}$ have an angle between 90 and 270 degrees with $\nabla f(\bar{x})$.

One can visualize the condition $\langle \nabla f(\bar{x}), w \rangle < 0$ as requiring the vector $w$ to form an angle with the vector $\nabla f(\bar{x})$ of between 90 and 270 degrees; see Figure 1.18. The latter vector points in a direction of "greatest steepness" of $f$ at $\bar{x}$ as seen as follows: For all $w \in \mathbb{R}^n$ with $\|w\|_2 = \|\nabla f(\bar{x})\|_2$, the Cauchy-Schwarz inequality[11] guarantees that

$$\vec{d}f(\bar{x}; w) = \langle \nabla f(\bar{x}), w \rangle \leq \|\nabla f(\bar{x})\|_2 \|w\|_2 = \|\nabla f(\bar{x})\|_2^2 = \vec{d}f(\bar{x}; \nabla f(\bar{x})).$$

So, $\nabla f(\bar{x})$ is a direction that maximizes the directional derivative $\vec{d}f(\bar{x}; w)$ over all directions $w$ with a normalized length. Similarly, $w = -\nabla f(\bar{x})$ minimizes the directional derivative and one refers to this $w$ as the *gradient descent direction*. Indeed, by 1.30, $w = -\nabla f(\bar{x})$ is a descent direction for $f$ at $\bar{x}$ when $\nabla f(\bar{x}) \neq 0$ as seen in Figure 1.18.

**Gradient Descent Method.**

Data. $x^0 \in \mathbb{R}^n$.
Step 0. Set $\nu = 0$.
Step 1. Stop if $\nabla f(x^\nu) = 0$, otherwise set

$$w^\nu = -\nabla f(x^\nu).$$

Step 2. Compute $\lambda^\nu \in \operatorname{argmin}_{\lambda \geq 0} f(x^\nu + \lambda w^\nu)$.
Step 3. Set $x^{\nu+1} = x^\nu + \lambda^\nu w^\nu$, replace $\nu$ by $\nu + 1$ and go to Step 1.

If the algorithm stops in Step 1, then we've obtained a point that satisfies the requirements of the Oresme and Fermat rules. The challenge is to determine what happens if it keeps on iterating and constructs a sequence $\{x^1, x^2, x^3, \dots\}$ or, more concisely, $\{x^\nu, \nu \in \mathbb{N}\}$, where

$$\mathbb{N} = \{1, 2, 3, \dots\}$$

is the set of *natural numbers*. This requires some terminology.

**Convergence and Limits.** The convergence of a sequence of points is intrinsically based on a proximity notion often expressed in terms of a norm. We might perceive proximity in terms of the Euclidean norm: $x$ is considered to be "near" $\bar{x}$ when $\|x - \bar{x}\|_2$ is small. However, any other norm identifies proximity equally well. A *neighborhood* of $\bar{x}$ is then all points that have a small distance to $\bar{x}$. The Euclidean ball $\mathbb{B}(\bar{x}, \varepsilon)$ is one example provided that $\varepsilon > 0$.

We refer to $\bar{x} \in \mathbb{R}^n$ as the *limit* or *limit point* of a sequence $\{x^\nu \in \mathbb{R}^n, \nu \in \mathbb{N}\}$ if $\|x^\nu - \bar{x}\| \to 0$ as $\nu \to \infty$, written $x^\nu \to \bar{x}$ or $\bar{x} = \lim x^\nu$, and we then say that the sequence of points *converges* to $\bar{x}$. Mostly, we let $\nu$ (Greek "nu") index sequences and one can assume that it's $\nu$ that tends to infinity in limiting expressions unless specified otherwise. It follows that if $x^\nu \to x$, then given any $\varepsilon > 0$, one can find a natural number $\bar{\nu}$ such that $x^\nu \in \mathbb{B}(\bar{x}, \varepsilon)$ for all $\nu \geq \bar{\nu}$.

[11] For vectors $x, y \in \mathbb{R}^n$, the *Cauchy-Schwarz inequality* asserts that $|\langle x, y \rangle| \leq \|x\|_2 \|y\|_2$.

**Cluster Points.** A sequence may not converge, but it could still have some inherent trends. For example, if $x^\nu = 1/\nu$ for $\nu$ odd and $x^\nu = 1 - 1/\nu$ for $\nu$ even, then $\{x^\nu, \nu \in \mathbb{N}\}$ has no limit. However, the subsequence $\{x^1, x^3, x^5, \dots\}$ converges to 0 and the subsequence $\{x^2, x^4, x^6, \dots\}$ converges to 1. The points 0 and 1 are examples of cluster points. Generally, a *cluster point* $\bar{x} \in \mathbb{R}^n$ of a sequence $\{x^\nu \in \mathbb{R}^n, \nu \in \mathbb{N}\}$ is the limit point of a subsequence $\{x^\nu, \nu \in N\}$, where $N$ specifies the indices defining the subsequence, i.e.,

$$N \in \mathcal{N}_\infty^\# = \text{ all infinite collections of increasing natural numbers.}$$

The convergence to $\bar{x}$ along the subsequence indexed by $N$ is written as $x^\nu \to_N \bar{x}$. In the odd-even example, $N = \{1, 3, 5, \dots\}$ and $N = \{2, 4, 6, \dots\}$ for the two subsequences, respectively. A limit point is always a cluster point, but a sequence can have *many* cluster points and, then, no limit as exemplified above.

**Bounded Sequences.** A sequence $\{x^\nu \in \mathbb{R}^n, \nu \in \mathbb{N}\}$ is *bounded* if there's $\rho \in [0, \infty)$ such that $\|x^\nu\| \leq \rho$ for all $\nu$. A bounded sequence has at least one cluster point, i.e., there exist $N \in \mathcal{N}_\infty^\#$ and $\bar{x}$ such that $x^\nu \to_N \bar{x}$. In fact, every sequence $\{x^\nu \in \mathbb{R}^n, \nu \in \mathbb{N}\}$ either has a cluster point or it *escapes to the horizon*, i.e., $\|x^\nu\| \to \infty$.

**Convergence 1.31** (gradient descent). *Suppose that $f : \mathbb{R}^n \to \mathbb{R}$ is smooth and the gradient descent method has generated $\{x^\nu, \nu \in \mathbb{N}\}$. If $\bar{x}$ is a cluster point of this sequence, then*

$$\nabla f(\bar{x}) = 0 \quad \text{and} \quad \vec{d}f(\bar{x}; w) \geq 0 \;\; \forall w \in \mathbb{R}^n,$$

*i.e., $\bar{x}$ satisfies the necessary conditions for a minimizer according to the Oresme and Fermat rules.*

**Proof.** Since $\lambda = 0$ is a possibility in Step 2, $f(x^{\nu+1}) \leq f(x^\nu)$ for all $\nu$. Suppose that $x^\nu \to_N \bar{x}$ for some $N \in \mathcal{N}_\infty^\#$ and, for the sake of contradiction, $\bar{x} \in \mathbb{R}^n$ has $\nabla f(\bar{x}) \neq 0$. Then, there's $\gamma > 0$ such that $f(x^{\nu+1}) - f(x^\nu) \leq -\gamma$ for all $\nu \in N$ sufficiently large, which implies that $f(x^\nu) \to -\infty$. This contradicts, however, the fact that $f(x^\nu) \to_N f(\bar{x})$, which must hold by the continuity of $f$. Thus, $\nabla f(\bar{x}) = 0$ and the conclusion follows; see (1.3).

We establish the claim about $\gamma$ as follows: Let $\varepsilon = \|\nabla f(\bar{x})\|_2$. By continuity, there's $\delta \in (0, \infty)$ with $\|\nabla f(x) - \nabla f(\bar{x})\|_2 \leq \varepsilon/3$ for all $x \in \mathbb{B}(\bar{x}, \delta)$. For such $x$, $\|\nabla f(x)\|_2 \leq 4\varepsilon/3$. Set $\bar{\lambda} = \delta/(2\varepsilon)$. There's $\bar{\nu}$ such that $x^\nu \in \mathbb{B}(\bar{x}, \delta/3)$ for all $\nu \in N$ and $\nu \geq \bar{\nu}$. Let's fix $\nu \in N$, $\nu \geq \bar{\nu}$. The mean value theorem states that for any smooth function $g : \mathbb{R}^n \to \mathbb{R}$ and $x, w \in \mathbb{R}^n$, there's $\sigma \in [0, 1]$ such that

$$g(x + w) - g(x) = \big\langle \nabla g(x + \sigma w), w \big\rangle. \tag{1.7}$$

Applying the theorem here, we obtain $\sigma \in [0, 1]$ such that

$$f(x^\nu + \bar\lambda w^\nu) - f(x^\nu) = -\bar\lambda\langle\nabla f(z), \nabla f(x^\nu)\rangle, \text{ where } z = x^\nu + \sigma\bar\lambda w^\nu. \tag{1.8}$$

We seek to bound this right-hand side from above and start by noting that

$$\|z - \bar x\|_2 \le \|x^\nu - \bar x\|_2 + \sigma\bar\lambda\|\nabla f(x^\nu)\|_2 \le \tfrac{1}{3}\delta + \tfrac{4}{3}\sigma\bar\lambda\varepsilon \le \tfrac{1}{3}\delta + \tfrac{4}{3}\bar\lambda\varepsilon = \delta$$

and therefore $\|\nabla f(z) - \nabla f(\bar x)\|_2 \le \varepsilon/3$. Trivially,

$$\begin{aligned}\langle\nabla f(z), \nabla f(x^\nu)\rangle &= \langle\nabla f(z) - \nabla f(\bar x) + \nabla f(\bar x), \nabla f(x^\nu) - \nabla f(\bar x) + \nabla f(\bar x)\rangle \\ &= \langle\nabla f(z) - \nabla f(\bar x), \nabla f(x^\nu) - \nabla f(\bar x)\rangle + \langle\nabla f(z) - \nabla f(\bar x), \nabla f(\bar x)\rangle \\ &\quad + \langle\nabla f(x^\nu) - \nabla f(\bar x), \nabla f(\bar x)\rangle + \|\nabla f(\bar x)\|_2^2.\end{aligned}$$

These four terms on the right-hand side can be bounded using the Cauchy-Schwarz inequality. For the first term,

$$\langle\nabla f(z) - \nabla f(\bar x), \nabla f(x^\nu) - \nabla f(\bar x)\rangle \ge -\|\nabla f(z) - \nabla f(\bar x)\|_2\|\nabla f(x^\nu) - \nabla f(\bar x)\|_2 \ge -\tfrac{1}{9}\varepsilon^2.$$

Similarly, the next two terms are both bounded by $-\varepsilon^2/3$ and the last term $\|\nabla f(\bar x)\|_2^2 = \varepsilon^2$. Putting this together we obtain that

$$\langle\nabla f(z), \nabla f(x^\nu)\rangle \ge -\tfrac{1}{9}\varepsilon^2 - \tfrac{1}{3}\varepsilon^2 - \tfrac{1}{3}\varepsilon^2 + \varepsilon^2 = \tfrac{2}{9}\varepsilon^2.$$

Returning to (1.8), this implies that

$$f(x^\nu + \bar\lambda w^\nu) - f(x^\nu) \le -\tfrac{2}{9}\bar\lambda\varepsilon^2 = -\tfrac{1}{9}\varepsilon\delta.$$

The minimization in Step 2 has $\bar\lambda$ as a possibility. Thus,

$$f(x^{\nu+1}) - f(x^\nu) \le f(x^\nu + \bar\lambda w^\nu) - f(x^\nu) \le -\tfrac{1}{9}\varepsilon\delta.$$

Since $\varepsilon$ and $\delta$ are independent of $\nu$, the same holds for any other $\nu \in N, \nu \ge \bar\nu$ and the claim holds with $\gamma = \varepsilon\delta/9$. □

There are two possible outcomes from the gradient descent method that we haven't discussed. First, the generated sequence $\{x^\nu, \nu \in \mathbb{N}\}$ may not have any cluster points, which takes place only if the sequence escapes to the horizon. Second, $\text{argmin}_{\lambda\ge 0} f(x^\nu + \lambda w^\nu)$ could be empty in an iteration making the algorithm break down in Step 2. For a smooth $f$ this can only take place if $f(x^\nu + \lambda w^\nu)$ keeps on decreasing as $\lambda$ grows as would be the case, for example, when $f(x) = e^x$. Either case is ruled out if $f(x) \to \infty$ when $\|x\|_2 \to \infty$.

Step 2 of the gradient descent method implicitly assumes that we've at our disposal a *line search routine* that calculates exactly a minimizer of $\lambda \mapsto f(x^\nu + \lambda w^\nu)$ for $\lambda \ge 0$. That's an unrealistic premise and even unnecessary. A rather coarse approximation is acceptable, which reduces the computational demand in Step 2 significantly. One common approach is to adopt the following modified Step 2, where $\alpha, \beta \in (0,1)$ are parameters kept constant for all $\nu$:

**A-Step 2.** Compute $\lambda^\nu = \beta^k$ ($= \beta$ raised to the power $k$), with $k$ being the smallest nonnegative integer satisfying

$$f(x^\nu + \beta^k w^\nu) - f(x^\nu) \le \alpha\beta^k \langle \nabla f(x^\nu), w^\nu \rangle.$$

We then refer to $\lambda^\nu$ as the *Armijo step size*. Although the Armijo step size may not minimize $\lambda \mapsto f(x^\nu + \lambda w^\nu)$, it guarantees that

$$f(x^{\nu+1}) - f(x^\nu) \le \alpha\lambda^\nu \langle \nabla f(x^\nu), w^\nu \rangle = -\alpha\lambda^\nu \|\nabla f(x^\nu)\|_2^2,$$

with the right-hand side being a negative number. Thus, the algorithm indeed produces points with "descending" function values. It's also clear that A-Step 2 is implementable. When $\nabla f(x^\nu) \ne 0$ and $\alpha, \beta \in (0,1)$, the expression for directional derivatives in (1.3) ensures that

$$\frac{f(x^\nu + \beta^k w^\nu) - f(x^\nu)}{\beta^k} \le \alpha \langle \nabla f(x^\nu), w^\nu \rangle$$

for sufficiently large $k$. Thus, as long as $f$ is smooth, A-Step 2 needs to check at most a finite number of integers $k$. The convergence statement for the gradient descent method remains unchanged under A-Step 2, with the proof undergoing only minor adjustments; see [83, Theorem 1.3.4].

An alternative and even simpler possibility is to assign every $\lambda^\nu$ some pre-determined, sufficiently small value $\bar\lambda > 0$ resulting in the same convergence properties under an additional assumption on the gradients; see [115, Theorem 5.1]. The computational cost in Step 2 then becomes negligible, but the overall practical performance can be poor unless $\bar\lambda$ is selected just right.

The main advantage of the gradient descent method, with Armijo step size or some other simple rule, is its low computational cost per iteration. The steps don't require expensive subroutines and can be carried out even for large $n$. Still, it's often replaced by more sophisticated algorithms that avoid the high number of iterations typically needed by the gradient descent method due to its meandering behavior.

**Exercise 1.32** (zigzagging in gradient descent). Implement the gradient descent method with the Armijo step size using $\alpha = 0.25$ and $\beta = 0.8$ and apply it to $f(x) = x_1^2 + \sigma(x_2 - 1)^2$, with $\sigma \in (0, \infty)$. Starting at $x^0 = (2, 2)$, determine how many iterations it takes to obtain $x^\nu$ with $\|\nabla f(x^\nu)\|_2 \le 10^{-7}$ for $\sigma = 2, 20, 200, 2000$ and $20000$.

**Guide.** Regardless of $\sigma$, $\operatorname{argmin} f = \{(0,1)\}$ and $\inf f = 0$. A plot of the function shows that the level-sets $\{f \le \tau\}$ are elongated ellipses for large $\sigma$ causing the gradient descent method to take many steps. Tracing the trajectory of $\{x^\nu, \nu \in \mathbb{N}\}$ reveals a zigzagging behavior. □

## 1.G Newton's Method

The iterations of the gradient descent method can be interpreted as follows: At a current point $x^\nu$, the function $f : \mathbb{R}^n \to \mathbb{R}$ to be minimized is approximated by

$$f^\nu(x) = f(x^\nu) + \langle \nabla f(x^\nu), x - x^\nu \rangle + \tfrac{1}{2\lambda^\nu}\|x - x^\nu\|_2^2, \tag{1.9}$$

where $\lambda^\nu \in (0, \infty)$ is a parameter to be determined. The next point $x^{\nu+1}$ is obtained by minimizing $f^\nu$, which is strictly convex even if $f$ is nonconvex; see Figure 1.19. Thus, by optimality condition 1.9,

$$x^{\nu+1} \in \operatorname{argmin} f^\nu \iff \nabla f^\nu(x^{\nu+1}) = 0 \iff x^{\nu+1} = x^\nu - \lambda^\nu \nabla f(x^\nu).$$

We now see that the parameter $\lambda^\nu$ is indeed the step size in the gradient descent method. If $f^\nu$ turns out to be a good approximation of $f$, then we expect $x^{\nu+1}$ to be near argmin $f$. However, it's obvious that this can only be the case for a limited class of functions regardless of how skillfully $\lambda^\nu$ is selected.

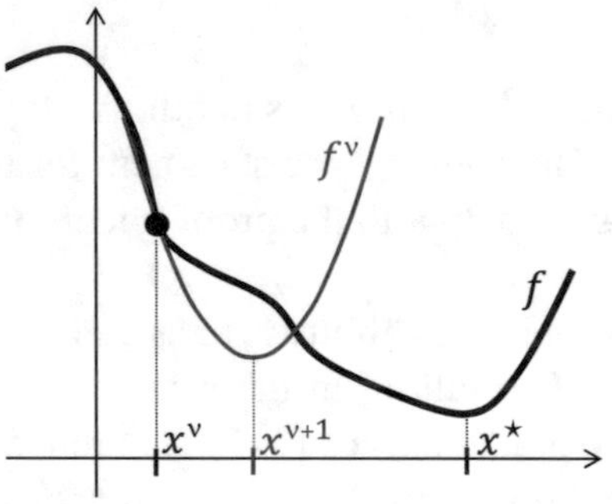


Fig. 1.19: A function $f$ is approximated by $f^\nu$ at a point $x^\nu$, with argmin $f^\nu$ producing a new point $x^{\nu+1}$.

Under the additional assumption that $f$ is twice smooth, a more trustworthy approximation of $f$ at $x^\nu$ is provided by

$$f^\nu(x) = f(x^\nu) + \langle \nabla f(x^\nu), x - x^\nu \rangle + \tfrac{1}{2\lambda^\nu}\langle x - x^\nu, \nabla^2 f(x^\nu)(x - x^\nu)\rangle, \tag{1.10}$$

again with $\lambda^\nu \in (0, \infty)$ being a parameter to be selected appropriately. When $\lambda^\nu = 1$, $f^\nu$ is the multi-variate Taylor expansion of $f$ at $x^\nu$ up to the second order. If $\nabla^2 f(x^\nu)$ is positive definite, then $f^\nu$ is strictly convex by 1.23. Again bringing in 1.9, we find that

$$x^{\nu+1} \in \operatorname{argmin} f^\nu \iff \lambda^\nu \nabla f(x^\nu) + \nabla^2 f(x^\nu)(x^{\nu+1} - x^\nu) = 0$$

and any such $x^{\nu+1}$ is unique; cf. 1.16. Since $\nabla^2 f(x^\nu)$ is positive definite, it's also invertible and the above equation has the unique solution

$$x^{\nu+1} = x^\nu - \lambda^\nu \big(\nabla^2 f(x^\nu)\big)^{-1} \nabla f(x^\nu),$$

which confirms the uniqueness of a minimizer of $f^\nu$. The suggested direction,

$$w^\nu = -\big(\nabla^2 f(x^\nu)\big)^{-1} \nabla f(x^\nu),$$

is known as *Newton's direction*. We can then follow the pattern of the gradient descent method, but utilizing Newton's direction instead of the gradient. This is also meaningful because Newton's direction is a descent direction by 1.30 as long as $\nabla f(x^\nu) \neq 0$ and $\nabla^2 f(x^\nu)$ is positive definite because then $(\nabla^2 f(x^\nu))^{-1}$ is positive definite and

$$\Big\langle \nabla f(x^\nu), \big(\nabla^2 f(x^\nu)\big)^{-1} \nabla f(x^\nu) \Big\rangle > 0.$$

In summary, this leads to an algorithm for minimizing a twice smooth function with positive definite Hessians.

**Newton's Method.**

Data. $x^0 \in \mathbb{R}^n$.
Step 0. Set $\nu = 0$.
Step 1. Stop if $\nabla f(x^\nu) = 0$, otherwise set

$$w^\nu = -\big(\nabla^2 f(x^\nu)\big)^{-1} \nabla f(x^\nu).$$

Step 2. Compute $\lambda^\nu \in \operatorname{argmin}_{\lambda \geq 0} f(x^\nu + \lambda w^\nu)$.
Step 3. Set $x^{\nu+1} = x^\nu + \lambda^\nu w^\nu$, replace $\nu$ by $\nu + 1$ and go to Step 1.

A convergence statement for Newton's method is essentially the same as that for the gradient descent method, but requires the assumptions that $f$ is twice smooth and the matrices $\nabla^2 f(x^\nu)$ are positive definite for all $\nu \in \mathbb{N}$; see [115, Theorem 5.12]. In implementations, one would again replace Step 2 by some step-size rule such as A-Step 2.

Newton's method has desirable *local* convergence properties. If the initial point $x^0$ is sufficiently close to a minimizer $x^\star$ with a positive definite $\nabla^2 f(x^\star)$, then $x^\nu$ converges "fast" to $x^\star$ even without having to compute a step size. That is, we may consider replacing Step 2 by the following fixed step-size rule, which is hinted to in the above Taylor expansion:

**F-Step 2.** Set $\lambda^\nu = 1$.

A precise statement about local convergence under this step-size rule makes use of the *Frobenius norm* of an $m \times n$-matrix $A = [A_{ij}]_{i,j=1}^{m,n}$ given by

$$\|A\|_F = \Big(\sum\nolimits_{i=1}^m \sum\nolimits_{j=1}^n A_{ij}^2\Big)^{1/2},$$

which has the property that $\|Ax\|_2 \leq \|A\|_F \|x\|_2$ for all $x \in \mathbb{R}^n$.

**Convergence 1.33** (Newton's method; local convergence). *Suppose that* $f : \mathbb{R}^n \to \mathbb{R}$ *is twice smooth,* $\nabla f(x^\star) = 0$, $\nabla^2 f(x^\star)$ *is positive definite and there exist* $\rho \in (0, \infty)$ *and* $\kappa \in [0, \infty)$ *such that*

$$\big\|\nabla^2 f(x) - \nabla^2 f(\bar{x})\big\|_F \le \kappa \|x - \bar{x}\|_2 \;\; \forall x, \bar{x} \in \mathbb{B}(x^\star, \rho).$$

*Let* $\{x^\nu, \nu \in \mathbb{N}\}$ *be constructed by Newton's method using F-Step 2. Then, there's* $\bar{\rho} > 0$ *such that when the initial point* $x^0 \in \mathbb{B}(x^\star, \bar{\rho})$*, one has*

$$x^\nu \to x^\star \quad \text{and} \quad \|x^{\nu+1} - x^\star\|_2 \le \beta\kappa \|x^\nu - x^\star\|_2^2 \;\; \forall \nu,$$

*where* $\beta = \|(\nabla^2 f(x^\star))^{-1}\|_F$.

**Proof.** Since the matrix $\nabla^2 f(x)$ is positive definite for $x = x^\star$ and varies continuously, $\nabla^2 f(x)$ is not only invertible at $x^\star$ but also in a neighborhood of this point. Consequently, there's $\hat{\rho} \in (0, \rho]$ such that

$$\Big\|\big(\nabla^2 f(x)\big)^{-1}\Big\|_F \le 2\beta \quad \forall x \in \mathbb{B}(x^\star, \hat{\rho}).$$

Suppose that $x^\nu \in \mathbb{B}(x^\star, \hat{\rho})$. Since $\nabla f(x^\star) = 0$,

$$\begin{aligned}
\|x^{\nu+1} - x^\star\|_2 &= \Big\|x^\nu - \big(\nabla^2 f(x^\nu)\big)^{-1}\nabla f(x^\nu) - x^\star + \big(\nabla^2 f(x^\nu)\big)^{-1}\nabla f(x^\star)\Big\|_2 \\
&= \Big\|\big(\nabla^2 f(x^\nu)\big)^{-1}\big(\nabla f(x^\star) - \nabla f(x^\nu) + \nabla^2 f(x^\nu)(x^\nu - x^\star)\big)\Big\|_2 \\
&\le \Big\|\big(\nabla^2 f(x^\nu)\big)^{-1}\Big\|_F \Big\|\nabla f(x^\star) - \nabla f(x^\nu) + \nabla^2 f(x^\nu)(x^\nu - x^\star)\Big\|_2.
\end{aligned}$$

The mean value theorem for vector-valued mappings states that for any smooth $F : \mathbb{R}^n \to \mathbb{R}^m$ and $x, w \in \mathbb{R}^n$,

$$F(x + w) - F(x) = \int_0^1 \nabla F(x + \sigma w) w \, d\sigma, \tag{1.11}$$

where $\nabla F(\bar{x})$ is the $m \times n$-Jacobian matrix of $F$ at $\bar{x}$ consisting of all its partial derivatives at $\bar{x}$. Applying this here, we find that

$$\begin{aligned}
&\nabla f(x^\star) - \nabla f(x^\nu) + \nabla^2 f(x^\nu)(x^\nu - x^\star) \\
&= -\int_0^1 \nabla^2 f\big(x^\star + \sigma(x^\nu - x^\star)\big)(x^\nu - x^\star)\, d\sigma + \nabla^2 f(x^\nu)(x^\nu - x^\star) \\
&= \int_0^1 \Big(\nabla^2 f(x^\nu) - \nabla^2 f\big(x^\star + \sigma(x^\nu - x^\star)\big)\Big)(x^\nu - x^\star)\, d\sigma.
\end{aligned}$$

When combined with the previous inequality, this yields

$$
\begin{aligned}
&\|x^{\nu+1} - x^\star\|_2 \\
&\le \left\|\left(\nabla^2 f(x^\nu)\right)^{-1}\right\|_F \int_0^1 \left\|\nabla^2 f(x^\nu) - \nabla^2 f\left(x^\star + \sigma(x^\nu - x^\star)\right)\right\|_F \|x^\nu - x^\star\|_2 \, d\sigma \\
&\le \left\|\left(\nabla^2 f(x^\nu)\right)^{-1}\right\|_F \int_0^1 \kappa(1-\sigma)\|x^\nu - x^\star\|_2^2 \, d\sigma \le 2\beta\tfrac{1}{2}\kappa\|x^\nu - x^\star\|_2^2
\end{aligned}
$$

and also $\|x^{\nu+1} - x^\star\|_2 \le \|x^\nu - x^\star\|_2$ provided that $\beta\kappa\|x^\nu - x^\star\|_2 \le 1$. Arguing by induction, we conclude that if

$$\|x^0 - x^\star\|_2 \le \bar{\rho} = \min\left\{\hat{\rho}, 1/(2\beta\kappa)\right\},$$

then $x^{\nu+1} \in \mathbb{B}(x^\star, \hat{\rho})$ for all $\nu$ and $x^\nu \to x^\star$. □

The given *rate of convergence* of Newton's method is *quadratic*, which indeed is "fast." The above recursion gives that

$$\|x^\nu - x^\star\|_2 \le \left(\beta\kappa\|x^0 - x^\star\|_2\right)^{2^\nu - 1}\|x^0 - x^\star\|_2.$$

For example, if $\beta\kappa\|x^0 - x^\star\|_2 = 0.5$, then the initial distance to $x^\star$ is reduced with a factor of $10^{-1}$, $10^{-2}$, $10^{-5}$ and $10^{-10}$ after 2, 3, 4 and 5 iterations, respectively. The number of correct digits in the approximating solution $x^\nu$ is roughly doubled in each iteration. In 1.32, which causes difficulties for the gradient descent method, Newton's method requires only a single iteration regardless of $x^0$. We realize this by recalling that in each iteration, Newton's method finds the minimizer of a strictly convex quadratic approximation. Since in 1.32 the function is already strictly convex and quadratic, the approximation turns out to be exact.

The improved performance of Newton's method relative to the gradient descent method could be expected as it leverages second-order information (Hessian matrices) about the function to be minimized and not only first-order information (gradients). However, the advantage diminishes and might vanish if the algorithms start relatively far away from a minimizer; the analysis in 1.33 is local and only relies on properties of $f$ near $x^\star$. Newton's method also breaks down when $\nabla^2 f(x^\nu)$ isn't positive definite and some alternative strategy needs to be devised at such points. In Figure 1.19, this would be the case if $x^\nu$ had been further to the left.

Newton's method might require few iterations, but each one of them is much more computationally costly than those of the gradient descent method. The most efficient way of computing $w^\nu$ in Step 1 isn't to determine the inverse $(\nabla^2 f(x^\nu))^{-1}$ but rather to solve the system of equations

$$\nabla^2 f(x^\nu) w^\nu = -\nabla f(x^\nu).$$

Typically, this is achieved by Cholesky decomposition, which still requires order $n^3$ operations unless the Hessian matrix is sparse (i.e., contains many zeros). For large-scale problems without much sparsity, Newton's method becomes less appealing.

Our strategy thus far has been to mimic the gradient descent method, while introducing, hopefully, a better direction $w^\nu$. As alluded to earlier, another possibility is to directly solve the system of equations $\nabla f(x) = 0$, which by 1.9 is equivalent to minimizing a convex $f$. This perspective also brings us to Newton's method.

**Solution of Nonlinear Equations.** Given the smooth functions $g_j : \mathbb{R}^n \to \mathbb{R}$, $j = 1, \dots, n$, the system of equations

$$g_1(x) = 0, \dots, g_n(x) = 0$$

can be written concisely as $G(x) = 0$, where $G : \mathbb{R}^n \to \mathbb{R}^n$ is a mapping with $G(x) = (g_1(x), \dots, g_n(x))$. The *Jacobian* of $G$ at $x$ is the $n\times n$-matrix $\nabla G(x)$ with gradients $\nabla g_j(x)$, $j = 1, \dots, n$ as rows. A first-order Taylor expansion of $G$ at $x^\nu$ gives the approximation

$$G^\nu(x) = G(x^\nu) + \nabla G(x^\nu)(x - x^\nu).$$

If $\nabla G(x^\nu)$ is invertible, then the solution of the approximating equation $G^\nu(x) = 0$ is

$$x^{\nu+1} = x^\nu - \big(\nabla G(x^\nu)\big)^{-1} G(x^\nu).$$

This provides the foundation for an iterative method for solving $G(x) = 0$.

**Newton-Raphson Algorithm.**

Data. $x^0 \in \mathbb{R}^n$.
Step 0. Set $\nu = 0$.
Step 1. Stop if $G(x^\nu) = 0$, otherwise set

$$w^\nu = -\big(\nabla G(x^\nu)\big)^{-1} G(x^\nu).$$

Step 2. Set $x^{\nu+1} = x^\nu + w^\nu$, replace $\nu$ by $\nu + 1$ and go to Step 1.

We realize that if $G(x) = \nabla f(x)$ for some function $f : \mathbb{R}^n \to \mathbb{R}$ to be minimized, then the Newton-Raphson algorithm coincides with Newton's method under F-Step 2. Consequently, after making the appropriate change of notation, one can follow step by step the proof of 1.33 to obtain a quadratic convergence rate for the Newton-Raphson algorithm as long as it's initialized sufficiently close to a solution $x^\star$ with $\nabla G(x^\star)$ invertible. The insight that Newton's method can be viewed as a procedure for computing a point satisfying the optimality condition $\nabla f(x) = 0$ reveals vividly the important connections between optimization problems and equation solving.

**Example 1.34** (divergence in Newton-Raphson). The need for initializing the Newton-Raphson algorithm sufficiently close to a solution is illustrated by

$$G(x) = \begin{cases} x^{1/3} & \text{if } x > 1 \\ -\frac{1}{3}x^3 + \frac{4}{3}x & \text{if } x \in [-1, 1] \\ -(-x)^{1/3} & \text{otherwise.} \end{cases}$$

If $x^0 \in [-1, 1]$, then the algorithm converges to the unique solution $x^\star = 0$ but that's not the case for other starting points. The Newton-Raphson algorithm is therefore often enhanced with a line search step parallel to Step 2 in Newton's method, which ensures that only steps with $\|G(x^{\nu+1})\|_2$ sufficiently below $\|G(x^\nu)\|_2$ are accepted.

**Detail.** We immediately find that

$$\nabla G(x) = \begin{cases} \frac{1}{3}x^{-2/3} & \text{if } x > 1 \\ -x^2 + \frac{4}{3} & \text{if } x \in [-1, 1] \\ \frac{1}{3}(-x)^{-2/3} & \text{otherwise} \end{cases}$$

and, for $x^\nu \notin [-1, 1]$,

$$x^{\nu+1} = x^\nu - \big(\nabla G(x^\nu)\big)^{-1} G(x^\nu) = -2x^\nu.$$

Consequently, the Newton-Raphson algorithm started from a point outside $[-1, 1]$ fails because the generated sequence becomes $x^0, -2x^0, 4x^0, -8x^0, \ldots$. □

**Exercise 1.35** (computation with Newton's). Implement Newton's method both with fixed step size ($\lambda^\nu = 1$) and with Armijo step size ($\alpha, \beta = 0.8$) and apply it to the Rosenbrock function $f(x) = 100(x_2 - x_1^2)^2 + (1 - x_1)^2$. Starting at $x^0 = (-1.2, 1)$, determine how many iterations it takes to obtain $x^\nu$ with $\|\nabla f(x^\nu)\|_2 \le 10^{-7}$. Also apply the gradient descent method with Armijo step size to the same function and compare the results.

**Guide.** The Rosenbrock function has a unique minimizer at (1, 1). □

## 1.H Acceleration and Regularization

The gradient descent method relies on inexpensive but potentially numerous iterations. One might wonder if it's possible to reduce the iteration count by clever adjustments without incurring the full cost of Newton's method. We'll examine two ideas in this direction and also introduce regularization as a means to construct algorithms.

**Conjugate Gradient Method.** A *basis* for $\mathbb{R}^n$ is a collection of $n$ linearly independent vectors. Any point in $\mathbb{R}^n$ can then be expressed as a linear combination of these vectors. The unit vector[12] $\{e^1, e^2, \ldots, e^n\}$ is a basis and any $x = (x_1, \ldots, x_n)$ can be expressed as $\sum_{j=1}^n x_j e^j$. However, this isn't the only possible basis. A choice tied to the geometry of the function to be minimized can simplify calculations and lead to fewer iterations.

For a symmetric positive definite $n \times n$-matrix $Q$ and $c \in \mathbb{R}^n$, suppose that $f : \mathbb{R}^n \to \mathbb{R}$ is of the form

$$f(x) = \tfrac{1}{2}\langle x, Qx\rangle + \langle c, x\rangle. \tag{1.12}$$

It can be beneficial to choose a basis that's related to $Q$. We say that a set of nonzero vectors $\{v^j \in \mathbb{R}^n, j = 1, \ldots, n\}$ is *Q-conjugate* if $\langle v^i, Qv^j\rangle = 0$ for all $i \neq j$. One can show

[12] $e^j \in \mathbb{R}^n$ has 1 in the $j$th position and 0 elsewhere.

that a $Q$-conjugate set of vectors is a basis for $\mathbb{R}^n$ and the problem of minimizing $f$ can therefore be written completely in terms of these vectors. Suppose that $\{w^0, w^1, \ldots, w^{n-1}\}$ is $Q$-conjugate. Then, given $x^0 \in \mathbb{R}^n$, any point

$$x = x^0 + \sum\nolimits_{\nu=0}^{n-1} \lambda^\nu w^\nu \ \text{ for some } \lambda^0, \ldots, \lambda^{n-1} \in \mathbb{R},$$

$$f(x) = f(x^0) + \sum\nolimits_{\nu=0}^{n-1} \tfrac{1}{2}\langle w^\nu, Qw^\nu\rangle(\lambda^\nu)^2 + \langle c + Qx^0, w^\nu\rangle\lambda^\nu$$

and the problem of minimizing $f$ can be replaced by minimizing this right-hand side with respect to $\lambda^0, \ldots, \lambda^{n-1}$. Interestingly, the latter problem decomposes into $n$ one-dimensional problems:

$$\Big\{ \underset{\lambda^\nu \in \mathbb{R}}{\text{minimize}} \ \tfrac{1}{2}\langle w^\nu, Qw^\nu\rangle(\lambda^\nu)^2 + \langle c + Qx^0, w^\nu\rangle\lambda^\nu, \quad \nu = 0, 1, \ldots, n-1 \Big\}.$$

Since $Q$ is positive definite, $\langle w^\nu, Qw^\nu\rangle > 0$ and the objective functions in these problems are even strictly convex. The Fermat rule gives the unique solution (cf. 1.9 and 1.16)

$$\bar{\lambda}^\nu = -\frac{\langle c + Qx^0, w^\nu\rangle}{\langle w^\nu, Qw^\nu\rangle}, \quad \nu = 0, 1, \ldots, n-1.$$

The unique minimizer of $f$ is therefore

$$x^\star = x^0 + \sum\nolimits_{\nu=0}^{n-1} \bar{\lambda}^\nu w^\nu,$$

which also can be computed by the recursion

$$x^{\nu+1} = x^\nu + \bar{\lambda}^\nu w^\nu, \quad \nu = 0, 1, \ldots, n-1, \quad \text{with } x^\star = x^n.$$

We now see that the computations are parallel to those of the gradient descent method and Newton's method: $w^\nu$ can be interpreted as an alternative direction and $\bar{\lambda}^\nu$ as a step size. Remarkably, one can show that $\bar{\lambda}^\nu \in \operatorname{argmin}_{\lambda \geq 0} f(x^\nu + \lambda w^\nu)$; the chosen step size is actually the one obtained by exact line search; cf. [83, Proposition 1.5.3]. The vectors $w^\nu$ can also be computed iteratively.

**Conjugate Gradient Method.**

Data. $x^0 \in \mathbb{R}^n$.
Step 0. Set $\nu = 0$.
Step 1. If $\nabla f(x^\nu) = 0$, then stop.
Step 2. If $\nu = 0$, then set $w^\nu = -\nabla f(x^\nu)$. Otherwise, set

$$w^\nu = -\nabla f(x^\nu) + \gamma^\nu w^{\nu-1}, \ \text{ where } \gamma^\nu = \frac{\langle Qw^{\nu-1}, \nabla f(x^\nu)\rangle}{\langle Qw^{\nu-1}, w^{\nu-1}\rangle}.$$

Step 3. Compute $\lambda^\nu \in \operatorname{argmin}_{\lambda \geq 0} f(x^\nu + \lambda w^\nu)$.
Step 4. Set $x^{\nu+1} = x^\nu + \lambda^\nu w^\nu$, replace $\nu$ by $\nu + 1$ and go to Step 1.

One can show that for $f$ in (1.12) the set $\{w^0, w^1, \ldots, w^{n-1}\}$ produced by the algorithm is indeed $Q$-conjugate and that at most $n$ iterations are needed to obtain the minimizer; see [83, Theorem 1.5.5]. Interestingly, Step 2 computes a direction $w^\nu$ that's a weighted average of the one in the gradient descent method and the one used in the previous iteration. This induces a "smoothing" of the trajectory followed by the algorithm and the typical zigzagging in the gradient descent method is avoided. For example, the thousands of iterations required in 1.32 for large $\sigma$ reduce to only two.

A comparison with Newton's method might at first appear less favorable since it minimizes (1.12) in one iteration. However, that iteration needs to solve

$$Qw = -Qx^0 - c,$$

which requires on the order of $n^3$ operations unless $Q$ is sparse. The $n$ iterations of the conjugate gradient method involves only matrix multiplication and the overall computing effort tends to be less than for Newton's method when $n$ is large. In fact, state-of-the-art iterative methods for large-scale linear systems $Ax = b$, with symmetric positive definite $A$, solve

$$\operatorname*{minimize}_{x \in \mathbb{R}^n} \tfrac{1}{2}\langle x, Ax\rangle - \langle b, x\rangle$$

using the conjugate gradient method. This optimization problem is indeed equivalent to solving the system of equations because the Fermat rule produces $Ax = b$ as an optimality condition for the problem. We refer to [83, Section 1.5] for extensions of the conjugate gradient method to general functions.

**Accelerated Gradient Descent.** The idea of "smoothing" the iterations of the gradient descent method isn't unique to the conjugate gradient method. Suppose that the function $f : \mathbb{R}^n \to \mathbb{R}$ to be minimized is convex and smooth, with gradients satisfying the *Lipschitz continuity property*: for some $\kappa \in (0, \infty)$,

$$\big\|\nabla f(x) - \nabla f(\bar{x})\big\|_2 \le \kappa\|x - \bar{x}\|_2 \quad \forall x, \bar{x} \in \mathbb{R}^n. \tag{1.13}$$

The scalar $\kappa$ is the *Lipschitz modulus* of $\nabla f : \mathbb{R}^n \to \mathbb{R}^n$. These assumptions enable us to construct relatively accurate approximations of $f$; 1.22 already furnishes a lower bound on $f$. For $x^\nu \in \mathbb{R}^n$, the upper bound

$$f(x) \le f(x^\nu) + \big\langle \nabla f(x^\nu), x - x^\nu \big\rangle + \tfrac{\kappa}{2}\|x - x^\nu\|_2^2 \quad \forall x \in \mathbb{R}^n$$

follows by an application of the mean value theorem (1.11). The right-hand side defines a strictly convex function with minimizer

$$x^{\nu+1} = x^\nu - \tfrac{1}{\kappa}\nabla f(x^\nu).$$

Consequently, minimization of this upper bound recursively leads to the gradient descent method, but with a fixed step size given by $1/\kappa$. Although simple, the resulting algorithm is usually slow since the step size tends to be tiny. In practice, this is further aggravated by the lack of knowledge about $\kappa$ and the use of conservative estimates. Analysis shows that the gradient descent method with fixed step size $\lambda^\nu = 1/\kappa$ for all $\nu$ constructs iterates satisfying the error bound

$$f(x^\nu) - f(x^\star) \leq \tfrac{1}{2}\kappa\nu^{-1}\|x^0 - x^\star\|_2^2 \quad \text{for any } x^\star \in \operatorname{argmin} f \text{ and } \nu \in \mathbb{N};$$

see [60, Theorem 3.3]. The rate of convergence $1/\nu$ is *sublinear* and much slower than that of Newton's method. An improved version, however, relies on a certain averaging process and performs better in theory and practice.

**Accelerated Gradient Method.**

Data. $x^0 \in \mathbb{R}^n$, $\alpha^\nu \in (0, 1]$, $\lambda^\nu \in (0, \infty)$ for $\nu = 0, 1, 2, \ldots$.
Step 0. Set $y^0 = x^0$ and $\nu = 0$.
Step 1. Stop if $\nabla f(y^\nu) = 0$, otherwise set

$$w^\nu = -\nabla f\big((1-\alpha^\nu)y^\nu + \alpha^\nu x^\nu\big).$$

Step 2. Compute $x^{\nu+1} = x^\nu + \lambda^\nu w^\nu$.
Step 3. Set $y^{\nu+1} = (1-\alpha^\nu)y^\nu + \alpha^\nu x^{\nu+1}$, replace $\nu$ by $\nu + 1$ and go to Step 1.

The algorithm deviates from the gradient descent method by producing the usual sequence $\{x^\nu, \nu \in \mathbb{N}\}$ as well as a second sequence $\{y^\nu, \nu \in \mathbb{N}\}$ and this helps smooth out the trajectory of the algorithm. The direction $w^\nu$ isn't computed at $x^\nu$ but rather at a weighted average of $x^\nu$ and $y^\nu$. Since $y^\nu$ depends on $y^{\nu-1}$, it tends to lag "behind" $x^\nu$ and induces a smoother trajectory for the algorithm. The effect is remarkable as it produces an algorithm with rate of convergence $1/\nu^2$.

**Convergence 1.36** (accelerated gradient method). *If $f : \mathbb{R}^n \to \mathbb{R}$ is convex, smooth and has Lipschitz continuous gradients with modulus $\kappa \in (0, \infty)$ as in* (1.13)*, then a sequence $\{y^\nu, \nu \in \mathbb{N}\}$ generated by the accelerated gradient method using $\alpha^\nu = 2/(\nu + 2)$ and $\lambda^\nu = (\nu + 1)/(2\kappa)$ satisfies*

$$f(y^\nu) - \inf f \leq \frac{2\kappa}{\nu(\nu+1)}\|x^0 - x^\star\|_2^2$$

*for any $x^\star \in \operatorname{argmin} f$ and $\nu \in \mathbb{N}$.*

**Proof.** The result is a special case of [60, Theorem 3.6]. □

We note the carefully selected choice of $\alpha^\nu$ and $\lambda^\nu$. The step size $\lambda^\nu$ grows and becomes much larger than $1/\kappa$, while $\alpha^\nu$ vanishes and the algorithm thereby places more emphasis on the $y^\nu$-sequence in Step 3. Still, in practice, one usually benefits from tuning these parameters on the models of interest to improve efficiency. In view of its simplicity, it's remarkable that the accelerated gradient method is optimal in the sense that no other algorithm can be constructed that converges faster for this class of functions using only gradient information [67, Theorem 2.1.7].

**Proximal Point Method.** The solution of many optimization problems can be accomplished by solving a sequence of simpler, approximating problems. The gradient descent method and Newton's method rely on smoothness to construct approximations, but this might not be viable as we move to more complicated problems. Another rather general approach to construct approximating problems is that of *regularization*. Suppose that we would like to minimize $f : \mathbb{R}^n \to \overline{\mathbb{R}}$. For some $\lambda \in (0, \infty)$ and a current point $x^\nu \in \mathbb{R}^n$, we might instead solve the approximating problem

$$\underset{x \in \mathbb{R}^n}{\text{minimize}}\ f(x) + \tfrac{1}{2\lambda}\|x - x^\nu\|_2^2.$$

If $f$ is convex and smooth, then the objective function in the approximating problem is strictly convex by 1.18(b) and the Fermat rule gives its unique minimizer as a point $x^{\nu+1}$ that satisfies

$$x^{\nu+1} = x^\nu - \lambda \nabla f(x^{\nu+1});$$

see 1.9 and 1.16. This reminds us of the gradient descent method, but with $\nabla f(x^\nu)$ replaced by $\nabla f(x^{\nu+1})$. However, the approach goes much beyond convex and smooth functions and leads to a variety of algorithms. The term $\frac{1}{2}\|x - x^\nu\|_2^2$ is referred to as a *regularizer*, with $\lambda$ being a parameter that determines the weight assigned to the term relative to $f(x)$. It promotes existence and uniqueness of solutions. For convex $f$, argmin $f$ can still be empty or have many minimizers. Neither can occur in the approximating problem. The approximation may also be convex when $f$ isn't. As a simple example, consider the nonconvex quadratic function given by $f(x) = -\alpha x^2$ for some positive $\alpha$. Then,

$$f(x) + \tfrac{1}{2\lambda}\|x - \bar{x}\|_2^2 = \big(\tfrac{1}{2\lambda} - \alpha\big)x^2 - \tfrac{1}{\lambda}x\bar{x} + \tfrac{1}{2\lambda}\bar{x}^2$$

defines a strictly convex function regardless of $\bar{x}$ when $\lambda \in (0, 1/(2\alpha))$.

Generally, for $\rho \in (0, \infty)$, we say that $f : \mathbb{R}^n \to \overline{\mathbb{R}}$ is *$\rho$-weakly convex* if the function

$$x \mapsto f(x) + \tfrac{1}{2}\rho\|x\|_2^2$$

is convex. For such $f$, the objective function in the approximating problem is convex as long as $\lambda \le 1/\rho$.

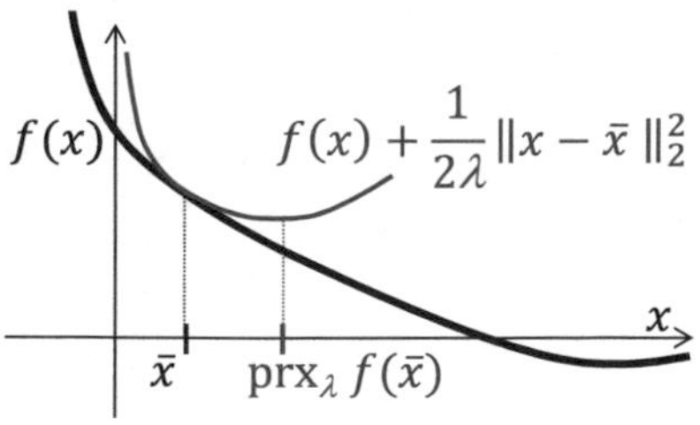


Fig. 1.20: A proximal point is a minimizer of $f$ augmented with a quadratic penalty term.

These factors tend to make the approximating problem simpler to solve than the actual one, but clearly much rely on the structure of $f$. The resulting method, which repeatedly solves the approximating problem, is therefore largely conceptual, but provides fertile ground for algorithmic developments.

For $\bar{x} \in \mathbb{R}^n$, $f : \mathbb{R}^n \to \overline{\mathbb{R}}$ and $\lambda \in (0, \infty)$, let's define the *set of proximal points*

$$\operatorname{prx}_\lambda f(\bar{x}) = \operatorname{argmin}_{x \in \mathbb{R}^n} f(x) + \tfrac{1}{2\lambda}\|x - \bar{x}\|_2^2,$$

which then gives the minimizers in the approximating problem; see Figure 1.20. In general, the set could be empty or contain multiple minimizers and certainly may lack an explicit characterization.

**Proximal Point Method.**

Data. $x^0 \in \mathbb{R}^n$ and $\lambda \in (0, \infty)$.
Step 0. Set $\nu = 0$.
Step 1. Compute $x^{\nu+1} \in \operatorname{prx}_\lambda f(x^\nu)$.
Step 2. Replace $\nu$ by $\nu + 1$ and go to Step 1.

Since $\operatorname{prx}_\lambda f(x^\nu)$ could be empty, additional assumptions are needed to ensure that the algorithm doesn't jam in Step 1. For example, $f$ being convex and not always $\infty$ or, alternatively, smooth and bounded from below suffice. Moreover, there's the possibility that the generated sequence $\{x^\nu, \nu \in \mathbb{N}\}$ isn't bounded. This is ruled out, however, when the level-set $\{f \le f(x^0)\}$ is bounded because then the descent property $f(x^{\nu+1}) \le f(x^\nu)$, established below, ensures that the whole sequence remains in that level-set.

**Convergence 1.37** (proximal point method). *Suppose that the proximal point method has constructed $\{x^\nu, \nu \in \mathbb{N}\}$ when applied to $f : \mathbb{R}^n \to \overline{\mathbb{R}}$. Then,*

$$f(x^{\nu+1}) \le f(x^\nu) \quad \forall \nu.$$

*Moreover, if $f$ is smooth and bounded from below, i.e., there's $\alpha \in \mathbb{R}$ such that $f(x) \ge \alpha$ for all $x \in \mathbb{R}^n$, and $\bar{x}$ is a cluster point of $\{x^\nu, \nu \in \mathbb{N}\}$, then $\nabla f(\bar{x}) = 0$.*

**Proof.** Since $x^{\nu+1}$ minimizes $x \mapsto f(x) + \frac{1}{2\lambda}\|x - x^\nu\|_2^2$, one has

$$f(x^{\nu+1}) \le f(x^{\nu+1}) + \tfrac{1}{2\lambda}\|x^{\nu+1} - x^\nu\|_2^2 \le f(x^\nu) + \tfrac{1}{2\lambda}\|x^\nu - x^\nu\|_2^2 = f(x^\nu),$$

which establishes the first claim. Next, for the sake of contradiction, suppose that there are $N \in \mathcal{N}_\infty^\#$ and $\delta \in (0, \infty)$ such that $\|x^{\nu+1} - x^\nu\|_2 \ge \delta$ for all $\nu \in N$. The above inequalities then imply that

$$f(x^{\nu+1}) - f(x^\nu) \le -\tfrac{1}{2\lambda}\delta^2 \quad \forall \nu \in N$$

so that $f(x^\nu) \to -\infty$. This contradicts the fact that $f$ is bounded from below. Thus, for all $\delta \in (0, \infty)$, there exists $\bar{\nu}$ such that $\|x^{\nu+1} - x^\nu\|_2 < \delta$ for all $\nu \ge \bar{\nu}$. Let $\delta \in (0, \infty)$ be arbitrary and $\bar{\nu}$ the corresponding index. When $f$ is smooth, the Fermat rule applied to the approximating problem guarantees that $\lambda \nabla f(x^{\nu+1}) = x^\nu - x^{\nu+1}$. Thus, for all $\nu \ge \bar{\nu}$,

$$\lambda\|\nabla f(x^{\nu+1})\|_2 = \|x^{\nu+1} - x^\nu\|_2 < \delta.$$

Hence, $\nabla f(x^\nu) \to 0$ and every cluster point $\bar{x}$ of $\{x^\nu, \nu \in \mathbb{N}\}$ satisfies $\nabla f(\bar{x}) = 0$. □

Several other convergence statements for the proximal point method are also possible, but the method is mostly a framework from which algorithms for more specific problems can be developed. We return to this in §2.D, §6.F and most prominently in Chap. 10; see also [76, 91] and [8, Chapter 27]. Presently, let's consider one example.

**Example 1.38** (iterative refinement). For a symmetric positive semidefinite $n \times n$-matrix $A$, the system of equations $Ax = b$ can't be solved by Cholesky decomposition or other efficient techniques if $A$ isn't invertible. One possibility is to consider the reformulation

$$A\bar{x} = b \iff \nabla f(\bar{x}) = 0 \iff \bar{x} \in \operatorname{argmin} f, \quad \text{where } f(x) = \tfrac{1}{2}\langle x, Ax\rangle - \langle b, x\rangle,$$

which holds by 1.9 and 1.24, and then apply the proximal point method to $f$.

**Detail.** For $x^\nu \in \mathbb{R}^n$, we can apply convexity and the Fermat rule to obtain

$$\begin{aligned}\operatorname{prx}_\lambda f(x^\nu) &= \operatorname{argmin}_{x\in\mathbb{R}^n} \tfrac{1}{2}\langle x, Ax\rangle - \langle b, x\rangle + \tfrac{1}{2\lambda}\|x - x^\nu\|_2^2\\ &= \big(A + \tfrac{1}{\lambda}I\big)^{-1}\big(b + \tfrac{1}{\lambda}x^\nu\big),\end{aligned}$$

where $I$ is the $n \times n$ identity matrix. For any $\lambda \in (0, \infty)$, $A + I/\lambda$ is positive definite and we can obtain $x^{\nu+1} \in \operatorname{prx}_\lambda f(x^\nu)$ by solving the system of equation

$$\big(A + \tfrac{1}{\lambda}I\big)x = b + \tfrac{1}{\lambda}x^\nu,$$

for example, by Cholesky decomposition. The approach is justified by 1.37, which applies as $f$ is smooth and also bounded from below by $f(x^\star)$, where $x^\star$ is a solution of $Ax = b$.

When $A$ is invertible but poorly conditioned, this approach might also be beneficial as it's typically better to solve a sequence of approximating well-conditioned problems than a single poorly conditioned one. □

**Exercise 1.39** (weak convexity). For a twice smooth function $f : \mathbb{R}^n \to \mathbb{R}$, show that $f$ is $\rho$-weakly convex if and only if $\nabla^2 f(x) + \rho I$ is positive semidefinite for all $x \in \mathbb{R}^n$, where $I$ is the $n \times n$ identity matrix.

## 1.I Quasi-Newton Methods

The fast convergence of Newton's method has to be treasured, but its need for storing $n \times n$-Hessian matrices and solving $n$ equations in $n$ variables make it less attractive for large-scale problems. A fundamental idea is to approximate the Hessian matrices in an economical manner that also facilitates the inevitable equation solving. In light of (1.9) and (1.10), the gradient descent method attempts this, but only with the coarsest possible approximation: a scaled identity matrix. Approximations based on gradients computed in previous iterations are much more accurate and lead to some of the most effective algorithms for unconstrained optimization.

**Quasi-Newton Method.**

Data. $x^0 \in \mathbb{R}^n$ and $n \times n$-matrix $B^0$.
Step 0. Set $\nu = 0$.
Step 1. Stop if $\nabla f(x^\nu) = 0$, otherwise obtain $w^\nu$ by solving

$$B^\nu w = -\nabla f(x^\nu).$$

Step 2. Compute $\lambda^\nu \in \operatorname{argmin}_{\lambda \geq 0} f(x^\nu + \lambda w^\nu)$.
Step 3. Set $x^{\nu+1} = x^\nu + \lambda^\nu w^\nu$, select $B^{\nu+1}$, replace $\nu$ by $\nu + 1$ and go to Step 1.

The matrix $B^\nu$ is the current approximation of $\nabla^2 f(x^\nu)$ and tries to capture the adjustment that needs to be made in the gradient descent direction on the basis of the local curvature of the function $f$. The method is also known as the *variable metric method* because we can think of $w^\nu$ as the gradient descent direction with respect to a different notion of distance (also called a metric) on $\mathbb{R}^n$ than the usual Euclidean norm. The actual behavior of the quasi-Newton method is determined by the choice of $B^\nu$.

Given a current approximation $B^\nu$, an idea is to select $B^{\nu+1}$ in a manner that reflects the curvature along the direction $w^\nu$ (from $x^\nu$ to $x^{\nu+1}$), i.e., by requiring

$$B^{\nu+1}(x^{\nu+1} - x^\nu) = \nabla f(x^{\nu+1}) - \nabla f(x^\nu).$$

In one dimension, this requirement corresponds to the most natural approximation of the second-order derivative of a function $g$ in terms of derivatives:

$$g''(x^{\nu+1}) \approx \frac{g'(x^{\nu+1}) - g'(x^\nu)}{x^{\nu+1} - x^\nu}.$$

Using an updating matrix $U^\nu$ such that $B^{\nu+1} = B^\nu + U^\nu$, the requirement is equivalent to having

$$U^\nu s^\nu = c^\nu - B^\nu s^\nu, \quad \text{with } s^\nu = x^{\nu+1} - x^\nu, \; c^\nu = \nabla f(x^{\nu+1}) - \nabla f(x^\nu).$$

This can be achieved by means of a matrix of rank 1 of the type $U^\nu = uv^\top$, where $u$ and $v$ are $n$-dimensional column vectors that need to be determined. Specifically, we seek column vectors $u$ and $v$ such that

$$uv^\top s^\nu = \langle v, s^\nu \rangle u = c^\nu - B^\nu s^\nu,$$

which implies that $u$ needs to be a multiple of $(c^\nu - B^\nu s^\nu)$. If $c^\nu = B^\nu s^\nu$, then $u = 0$ applies and we obtain $B^{\nu+1} = B^\nu$. Thus, let's consider the more challenging case with $c^\nu \neq B^\nu s^\nu$. Then, we can set

$$u = \frac{1}{\langle v, s^\nu \rangle}(c^\nu - B^\nu s^\nu) \quad \text{and} \quad U^\nu = \frac{1}{\langle v, s^\nu \rangle}(c^\nu - B^\nu s^\nu) v^\top$$

provided that $v$ is selected such that $\langle v, s^\nu \rangle \neq 0$. The flexibility regarding $v$ can be used to induce desirable properties. For example, we may wish to have a symmetric matrix $B^\nu$; recall that the Hessian $\nabla^2 f(x^\nu)$ is symmetric. The choice $v = c^\nu - B^\nu s^\nu$ yields

$$B^{\nu+1} = B^\nu + \frac{1}{\langle v, s^\nu \rangle}(c^\nu - B^\nu s^\nu)(c^\nu - B^\nu s^\nu)^\top,$$

which is symmetric when $B^\nu$ is symmetric. Moreover, the Sherman-Morrison-Woodbury formula provides a parallel updating for the inverses:

$$(B^{\nu+1})^{-1} = (B^\nu)^{-1} - \frac{1}{1 + \langle v, (B^\nu)^{-1} u \rangle}(B^\nu)^{-1} u v^\top (B^\nu)^{-1}$$

for these choices of $u$ and $v$. This is most convenient because it's the inverse that can be utilized in Step 1 of the algorithm to compute $w^\nu$. Thus, we obtain the direction without having to bring in Cholesky decomposition or some other subroutine for equation solving, the main bottleneck in Newton's method.

Two issues remain, however. It turns out that $B^{\nu+1}$ may not be positive definite even though $B^\nu$ has that property. This means that the quadratic approximation constructed isn't strictly convex. Another issue is the distinct possibility that

$$\langle v, s^\nu \rangle = \langle c^\nu - B^\nu s^\nu, s^\nu \rangle = 0,$$

which invalidates this choice of $(u, v)$ as the formulae break down. Although safeguards can be brought in (see, for example, [69, Section 6.2]), these concerns motivate more refined rank-two updating schemes.

One such alternative that's numerically reliable and well regarded is the *BFGS updating scheme*:[13]

$$B^{\nu+1} = B^\nu + \frac{1}{\langle c^\nu, s^\nu \rangle} c^\nu (c^\nu)^\top - \frac{1}{\langle B^\nu s^\nu, s^\nu \rangle} B^\nu s^\nu (B^\nu s^\nu)^\top.$$

Again, it's the inverse that's most useful in Step 1 of the quasi-Newton method:

$$(B^{\nu+1})^{-1} = \left(I - \frac{s^\nu (c^\nu)^\top}{\langle c^\nu, s^\nu \rangle}\right)(B^\nu)^{-1}\left(I - \frac{c^\nu (s^\nu)^\top}{\langle c^\nu, s^\nu \rangle}\right) + \frac{s^\nu (s^\nu)^\top}{\langle c^\nu, s^\nu \rangle}.$$

It can be shown both empirically and theoretically that the quasi-Newton method under this updating scheme, even with an Armijo-type step size, requires only moderately more iterations than Newton's method with a dramatic reduction in per-iteration computational cost; see [69, Section 6.1] for further details.

---

13 BFGS = Broyden-Fletcher-Goldfarb-Shanno who, independently, proposed and examined this formula.

## 1.J Coordinate Descent Algorithms

For very large-scale problems, even the iterations of the gradient descent method are costly. We might find it difficult to compute and store the full gradient and cheaper alternatives become necessary. A simple idea for minimizing $f : \mathbb{R}^n \to \overline{\mathbb{R}}$ in such situations is to optimize a group of variables at a time while keeping the other fixed. We can hope to reach argmin $f$ through the solution of a number of these low-dimensional minimization problems. Although often a viable heuristic[14], a naive algorithm based on this principle often falls short of the goal. In this section, we'll discuss conditions under which certain versions perform reasonably well.

A method based on selecting, at each step, the "steepest" coordinate direction does converge for the class of smooth convex functions with partial derivatives satisfying the *componentwise Lipschitz condition*: for every $j = 1, \dots, n$, there's a modulus $\kappa_j \in [0, \infty)$ such that[15]

$$\left| \frac{\partial f(x + \eta e^j)}{\partial x_j} - \frac{\partial f(x)}{\partial x_j} \right| \leq \kappa_j |\eta| \quad \forall \eta \in \mathbb{R}, x \in \mathbb{R}^n.$$

For example, $f(x) = \frac{1}{2}\langle x, Qx\rangle + \langle c, x\rangle$ satisfies the condition when $Q$ is a symmetric positive semidefinite matrix and $\kappa_j = Q_{jj}$, the $j$th diagonal element of $Q$.

**Steepest Coordinate Descent Algorithm.**

Data. $x^0 \in \mathbb{R}^n$ and $\kappa \in (0, \infty)$.
Step 0. Set $\nu = 0$.
Step 1. Let $j^\nu$ be the index $j = 1, \dots, n$ that maximizes $|\partial f(x^\nu)/\partial x_j|$.
Step 2. Set

$$x^{\nu+1} = x^\nu - \frac{1}{\kappa} \frac{\partial f(x^\nu)}{\partial x_{j^\nu}} e^{j^\nu}.$$

Step 3. Replace $\nu$ by $\nu + 1$ and go to Step 1.

**Convergence 1.40** (steepest coordinate descent). *Suppose that $f : \mathbb{R}^n \to \mathbb{R}$ is a smooth convex function with partial derivatives satisfying the componentwise Lipschitz condition with moduli $\{\kappa_j, j = 1, \dots, n\}$. If $\{x^\nu, \nu \in \mathbb{N}\}$ is generated by the steepest coordinate descent algorithm with $\kappa \geq \max_{j=1,\dots,n} \kappa_j$, then*

$$f(x^\nu) - \inf f \leq \frac{2\kappa n}{\nu + 4} \|x^0 - x^\star\|_2^2$$

*for any $x^\star \in$ argmin $f$ and $\nu \in \mathbb{N}$.*

**Proof.** By the mean value theorem (1.11) and the fact that $\kappa \geq \kappa_{j^\nu}$, we obtain

[14] An algorithm is a *heuristic* if it isn't associated with (substantive) mathematical guarantees about its performance. Heuristics can be useful for obtaining a preliminary solution quickly and to tackle exceptionally difficult problems.

[15] Again, $e^j \in \mathbb{R}^n$ is the vector with 1 in the $j$th position and 0 elsewhere.

$$
\begin{aligned}
f(x^{\nu+1}) - f(x^\nu) &\le \langle \nabla f(x^\nu), x^{\nu+1} - x^\nu \rangle + \tfrac{1}{2}\kappa_{j^\nu} \|x^{\nu+1} - x^\nu\|_2^2 \\
&= -\frac{\partial f(x^\nu)}{\partial x_{j^\nu}} \frac{1}{\kappa} \frac{\partial f(x^\nu)}{\partial x_{j^\nu}} + \frac{\kappa_{j^\nu}}{2} \left( \frac{-1}{\kappa} \frac{\partial f(x^\nu)}{\partial x_{j^\nu}} \right)^2 \\
&\le -\frac{1}{2\kappa} \left( \frac{\partial f(x^\nu)}{\partial x_{j^\nu}} \right)^2 \le -\frac{1}{2n\kappa} \|\nabla f(x^\nu)\|_2^2,
\end{aligned}
$$

where the last inequality follows because

$$
\left| \frac{\partial f(x^\nu)}{\partial x_{j^\nu}} \right| \ge \left| \frac{\partial f(x^\nu)}{\partial x_j} \right| \quad \text{for any } j = 1, \dots, n.
$$

The gradient inequality 1.22 and the Cauchy-Schwarz inequality establish that

$$
f(x^\nu) - f(x^\star) \le \langle \nabla f(x^\nu), x^\nu - x^\star \rangle \le \|\nabla f(x^\nu)\|_2 \|x^\nu - x^\star\|_2
$$

for $x^\star \in \operatorname{argmin} f$. Combining the results thus far, we obtain that

$$
f(x^\nu) - f(x^{\nu+1}) \ge \frac{\big(f(x^\nu) - f(x^\star)\big)^2}{2n\kappa \|x^\nu - x^\star\|_2^2},
$$

which provides a bound on the amount of progress that's made in each iteration. When this progress is summed up appropriately (see, for example, [68]), we reach the conclusion. □

Although one could tinker a bit with the procedure to make it applicable to a slightly larger class of functions, the algorithm's efficiency is hampered by the need to compute, at each iteration, the full gradient of $f$, which might be exactly what we want to avoid. If we settle for computing the full gradient, then we can just as well adopt the accelerated gradient method (§1.H), which anyhow is more effective in theory and practice. However, a variant of the algorithm turns out to be better than the competition in certain situations, in particular, when $n$ is huge as is often the case in data analytics.

In each iteration, let's restrict the attention to a few variables, selected randomly, and compute their partial derivatives. To carry out this scheme, partition the 1-to-$n$ indices in a number of subcollections, say $N_1, \dots, N_m$; conceivably, a subcollection could consist of a single index. The *partial gradient* at $x$ for $N_i$, denoted by $\nabla_i f(x)$, is the $n$-dimensional vector of partial derivatives with respect to the indices in $N_i$, with zeros filling in the rest, i.e., the $j$th component of $\nabla_i f(x)$ is

$$
\big(\nabla_i f(x)\big)_j = \frac{\partial f(x)}{\partial x_j} \quad \text{if} \quad j \in N_i, \qquad \text{and zero otherwise.}
$$

These partial gradients determine the directions in which the algorithm moves.

**Random Coordinate Descent Algorithm.**

Data. $\quad x^0 \in \mathbb{R}^n$, $\kappa_i \in (0, \infty)$, $N_i \subset \{1, \dots, n\}$, $p_i \in (0, 1)$, $i = 1, \dots, m$.

Step 0. Set $\nu = 0$.
Step 1. Randomly sample $i \in \{1,\dots,m\}$ according to $\{p_1,\dots,p_m\}$.
Step 2. Set

$$x^{\nu+1} = x^\nu - \tfrac{1}{\kappa_i}\nabla_i f(x^\nu).$$

Step 3. Replace $\nu$ by $\nu+1$ and go to Step 1.

The main advantage of the algorithm is its ease of implementation even for very large-scale problems; the choice of subcollections can be tailored to a particular application and account for data location and access speed. For a smooth convex function $f : \mathbb{R}^n \to \mathbb{R}$, one can show that, on average, the error after $\nu$ iterations is essentially the same as that in 1.40 for the steepest coordinate descent algorithm as long as the partial gradients satisfy the Lipschitz property:

$$\text{for } i = 1,\dots,m: \quad \big\|\nabla_i f(x+v) - \nabla_i f(x)\big\|_2 \le \kappa_i \|v\|_2 \quad \forall v \in V_i,$$

where

$$V_i = \Big\{\sum\nolimits_{j\in N_i} \beta_j e^j \;\Big|\; \beta_j \in \mathbb{R},\ j \in N_i\Big\}$$

is the subspace corresponding to $N_i$, and $\kappa_i$ in the algorithm coincides with these Lipschitz moduli. Moreover, the subcollections need to cover all the dimensions:

$$\bigcup\nolimits_{i=1}^m N_i = \{1,\dots,n\} \text{ and } N_i \cap N_j = \emptyset \;\; \forall i \ne j,$$

and the probabilities $\{p_1,\dots,p_m\}$ should satisfy

$$p_i = (\kappa_i)^\alpha / \kappa_\alpha, \quad \text{where } \kappa_\alpha = \sum\nolimits_{i=1}^m (\kappa_i)^\alpha$$

for some $\alpha \in \mathbb{R}$. In particular, $\alpha = 0$ is a possibility and this produces the same probability for each subcollection. We refer to [68] for precise statements and refinements; see also [88] for extensions to broader classes of functions. Analogous to acceleration of the gradient descent method, the random coordinate descent algorithm can also be "accelerated" as in [3, 68]. Chapter 10 returns to decomposition methods of this kind and then with technology that takes us beyond the convex and smooth cases.

**Example 1.41** (PageRank). In the early days, Google's search engine relied largely on its *PageRank* algorithm (named after Google's co-founder Larry Page) to rank the importance of various web pages. This allowed Google to present users with the more important (and typically most relevant and helpful) pages first. If the web of interest is composed of $n$ pages, labeled $j = 1,\dots,n$, we can model this web as a *directed graph*, where pages are the vertices and an edge points from vertex $j_1$ to vertex $j_2$ if the web page $j_1$ contains a link to $j_2$; see Figure 1.21. The relevance score of page $j$, denoted by $x_j$, can be determined by solving a minimization problem.

**Detail.** Let $n_j$ be the number of outgoing edges from vertex $j$ (i.e., the number of links to other pages on page $j$). We don't allow edges from a page to itself and dangling pages, with no outgoing edges. Hence, $n_j > 0$ for all $j$. The score $x_j$ represents the "voting power" of

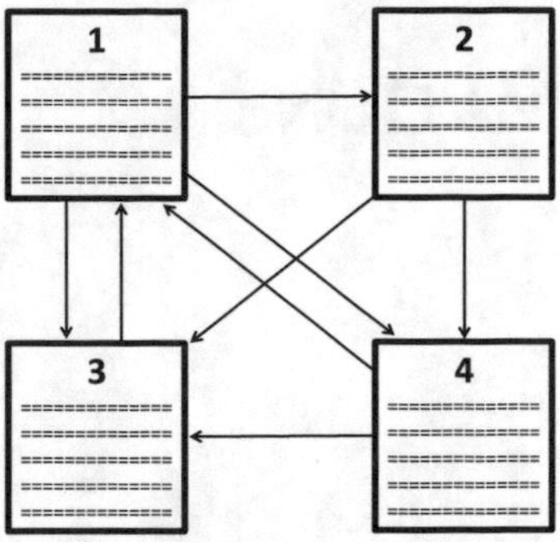


Fig. 1.21: Web pages and their links represented as a directed graph.

vertex $j$, which is to be subdivided among the $n_j$ outgoing edges; each outgoing edge thus carries $x_j/n_j$ vote-units. In Figure 1.21, $n_1 = 3, n_2 = 2, n_3 = 1$ and $n_4 = 2$. Let $B_k$ denote the indices of the pages that point to page $k$. Then, the score of page $k$ is obtained as

$$x_k = \sum_{j\in B_k} x_j/n_j, \quad k = 1,\ldots,n.$$

This is a system of equations more compactly expressed by

$$x = Ax, \quad \text{with } A = \begin{bmatrix} 0 & 0 & 1 & 1/2 \\ 1/3 & 0 & 0 & 0 \\ 1/3 & 1/2 & 0 & 1/2 \\ 1/3 & 1/2 & 0 & 0 \end{bmatrix}.$$

Computing the scores amounts to solving this system of equations. In the terminology of linear algebra, this is a problem of determining an eigenvector of $A$ corresponding to an eigenvalue of one. However, it can also be determined by solving the optimization problem

$$\underset{x\in\mathbb{R}^n}{\text{minimize}}\ f(x) = \tfrac{1}{2}\|Ax - x\|_2^2 + \tfrac{1}{2}\theta\Big(\sum_{j=1}^{n} x_j - 1\Big)^2,$$

where $\theta \in (0,\infty)$ is a parameter. The first term in the objective function is zero for any $x$ that satisfies the system of equations. If there's such a solution, then there's an infinite number of them as they can be scaled arbitrarily. The second term in the objective function provides a numerically useful "regularization" of the problem by giving preference to those solutions that sum to one. In the real world and certainly at Google, $n$ is huge. However, $A$ is usually sparse because the number of outgoing edges at each vertex is relatively small. In this case, the computation of a small number of partial derivatives of $f$ is cheap, but the cost of the full gradient could be prohibitive. Hence, one might expect that the random coordinate descent algorithm outperforms other procedures on large-scale problems and indeed computational experiments in [3, 68] point in this direction.

Carrying out the calculation for our example yields $x^\star = (12, 4, 9, 6)$, which then suggests that Page 1 is the most relevant. □

# Chapter 2
# CONVEX OPTIMIZATION

Optimization problems are often specified in terms of an objective function to be minimized and a set that defines admissible decisions. The newsvendor in §1.C may not be able to order more than $\alpha$ newspapers and would need to select an order quantity from the set $[0, \alpha]$. A maximum likelihood problem (§1.E) permits only certain types of estimates. This chapter initiates our treatment of such optimization problems with *constraints*. As we'll see, a problem of this general type can be formulated in terms of a modified function that incorporates the effect of both the set of admissible decisions and the original objective function. This perspective will allow us to achieve far-reaching extensions of the principles laid out in Chap. 1 and address nonsmooth problems broadly. Since it's most accessible, we'll concentrate on the convex case and leave a detailed treatment of more general settings for subsequent chapters.

## 2.A Formulations

In applications, we might be given an objective function $f_0 : \mathbb{R}^n \to \overline{\mathbb{R}}$ representing cost, loss, error or some other quantity that we would like to minimize as well as a set $C \subset \mathbb{R}^n$; any decision not in $C$ is deemed unacceptable. This leads to the minimization problem

$$\operatorname*{minimize}_{x \in \mathbb{R}^n} f_0(x) \ \text{ subject to } \ x \in C \quad \text{or, more concisely,} \quad \operatorname*{minimize}_{x \in C} f_0(x).$$

The *feasible set* of the problem is

$$C \cap \operatorname{dom} f_0$$

and contains the *feasible points (decisions)*. *Constraints* are conditions that make the feasible set a strict subset of $\mathbb{R}^n$. The problem is *unconstrained* when $C \cap \operatorname{dom} f_0 = \mathbb{R}^n$ and *constrained* otherwise. It's *feasible* when $C \cap \operatorname{dom} f_0 \neq \emptyset$.

**Example 2.1** (constrained data smoothing). As seen in §1.B, we can fit a polynomial $q$ to given data by minimizing

$$f_0(c) = \|Dc - y\|_2^2.$$



J. O. Royset and R. J-B Wets, *An Optimization Primer*, Springer Series in Operations Research and Financial Engineering, https://doi.org/10.1007/978-3-030-76275-9_2

Now, suppose that additional information dictates that $q(0) \in [\alpha, \beta]$ and $q(1) = \gamma$ for some given $\alpha \le \beta$ and $\gamma \in \mathbb{R}$. This generates constraints.

**Detail.** The requirements lead to the feasible set

$$C = \{c \in \mathbb{R}^{1+n} \mid c_0 \ge \alpha,\ c_0 \le \beta,\ c_n + \cdots + c_1 + c_0 = \gamma\}$$

and the optimization model

$$\begin{aligned} &\underset{c \in \mathbb{R}^{1+n}}{\text{minimize}}\ f_0(c) \\ &\text{subject to}\quad c_0 \ge \alpha,\ c_0 \le \beta,\ c_n + \cdots + c_1 + c_0 = \gamma. \end{aligned}$$

The need for data smoothing arises especially in time-series analysis, which examines how a phenomenon evolves over time. For example, we might be interested in the level of activity in the housing market, the evolution of consumer preferences, the spreading of an epidemic and so on. An autoregressive (AR) model describes, or predicts, by means of a (linear) transfer function, the random state $\boldsymbol{\xi}_t$ of a system at time $t$ in terms of the states at times $t-1, t-2, \ldots, t-n$:

$$\boldsymbol{\xi}_t = c_0 + c_1 \boldsymbol{\xi}_{t-1} + \cdots + c_n \boldsymbol{\xi}_{t-n} + \boldsymbol{\eta}_t,$$

where $\boldsymbol{\eta}_t$ is a random variable that accounts for disturbances not captured by the transfer function. Based on observations $\xi_{1-n}, \ldots, \xi_m$ of past states at times $1-n, \ldots, m$, respectively, we're looking for estimates of the coefficients $c = (c_0, c_1, \ldots, c_n)$ that minimize the role played by the disturbances. A least-squares approach leads to the model

$$\underset{c \in \mathbb{R}^{1+n}}{\text{minimize}}\ \frac{1}{m} \sum_{i=1}^{m} (c_0 + c_1 \xi_{i-1} + \cdots + c_n \xi_{i-n} - \xi_i)^2.$$

Suppose that on the basis of physical laws, economic considerations or simply common sense, we would like the coefficients to satisfy $c_1 \ge \cdots \ge c_n \ge 0$, i.e., the influence on a state by a prior state tends to be less if the latter is more in the past. Figure 2.1 illustrates the effect of including such constraints, where an AR model with $n = 20$ is fitted to $m = 137$ data points about annual average global temperatures (given in terms of deviation from the average during 1901–2000) up to 2016. The figure shows an upward trend for the last 57 years of record. The solid and dashed lines give the model's predictions with and without constraints, respectively. The constraints dampen undesirable oscillations in the predictions. □

Constraints that are listed in a model formulation are called *explicit constraint*; the example above has constraints of this type. However, an objective function $f_0$ might bring its own *induced constraints* that make dom $f_0$ a strict subset of $\mathbb{R}^n$. The example above has no induced constraints because its objective function is real-valued. However, §1.E describes an objective function involving terms of the form $-\ln \alpha$, which implies the induced constraint $\alpha > 0$. Expressions involving expected values, integrals and suprema might have $\infty$ as value and likewise cause induced constraints.

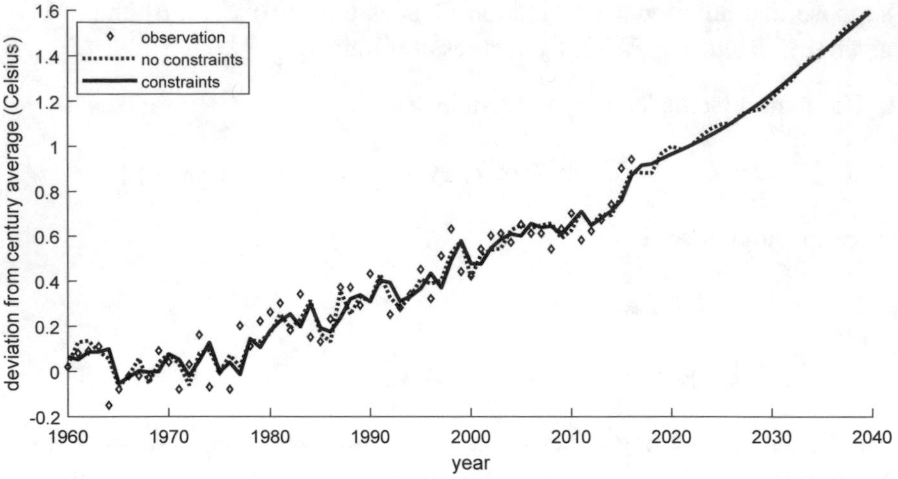


Fig. 2.1: Predictions of rise in global temperatures using an AR model.

A degenerate situation arises when $C \cap \operatorname{dom} f_0 = \emptyset$: no decision is acceptable and $\operatorname{argmin}_C f_0 = \emptyset$, in which case the corresponding minimization problem is *infeasible*. Infeasibility may occur due to modeling errors or because we insist on too many conditions. Although an important issue in some applications, infeasibility is usually identified and addressed by changing a model in such a way that feasibility is guaranteed; §2.E furnishes an example. Thus, much of our development can be focused on feasible problems.

Since we permit functions to have the value $\infty$, a minimization problem involving $f_0$ and $C$ can just as well be stated in terms of a function that incorporates the effect of both. In addition to the simplified notation resulting from this perspective, we'll be able to naturally extend the Oresme and Fermat rules from Chap. 1, develop an approximation theory and generally address multiple challenges in the subsequent chapters. Specifically, given $f_0 : \mathbb{R}^n \to \overline{\mathbb{R}}$ and a set $C \subset \mathbb{R}^n$, we can always define $f : \mathbb{R}^n \to \overline{\mathbb{R}}$ by setting

$$f(x) = \begin{cases} f_0(x) & \text{if } x \in C \\ \infty & \text{otherwise,} \end{cases}$$

which results in

$$\operatorname{argmin} f = \operatorname{argmin}_C f_0, \quad \inf f = \inf_C f_0, \quad \operatorname{dom} f = C \cap \operatorname{dom} f_0.$$

We adopt the terminology that a minimization problem has a certain property when its defining function has that property. For example, minimizing $f$ is a *convex problem* when $f$ is convex, minimizers of a problem are the minimizers of the defining function, etc.

The reformulation in terms of $f$ highlights the inherent lack of smoothness in the presence of constraints. As $x$ moves from being in dom $f$ to being outside, $f(x)$ may jump from some finite value (or $-\infty$) to infinity making $f$ neither continuous nor smooth.

**Example 2.2** (online retail). An online retailer seeks to optimize the prices of $n$ items under an *elastic demand* model so that its cost is minimized and the revenue is at least $\tau \in \mathbb{R}$. The problem can be formulated in terms of a single function.

**Detail.** Suppose that the demand for item $j$ under price vector $x = (x_1, \ldots, x_n) \in \mathbb{R}^n$ is

$$d_j(x) = \beta_j - \alpha_j(x_j - p_j), \quad j = 1, \ldots, n,$$

where $\beta_j$ is a baseline demand, $\alpha_j > 0$ reflects the fact that demand decreases as the price increases and $p_j$ is a reference price for item $j$. Let $d(x)$ be the vector with components $d_j(x)$. The revenue is obtained by summing up "price times demand," i.e., $\langle x, d(x)\rangle$, and the cost by $\langle c - x, d(x)\rangle$, where $c = (c_1, \ldots, c_n)$ is the vector of per-item expenses incurred by the retailer so that $x_j - c_j$ is the margin for item $j$. (A negative cost corresponds to making money.) Under the bounds $a_j \leq x_j \leq b_j$, this leads to the model[1]

$$\underset{x \in \mathbb{R}^n}{\text{minimize}} \ \big\langle c - x, d(x)\big\rangle \ \text{subject to} \ \big\langle x, d(x)\big\rangle \geq \tau, \quad a \leq x \leq b,$$

where $a = (a_1, \ldots, a_n)$ and $b = (b_1, \ldots, b_n)$. Equivalently, the problem is that of minimizing $f : \mathbb{R}^n \to \overline{\mathbb{R}}$ with

$$f(x) = \begin{cases} \big\langle c - x, d(x)\big\rangle & \text{if } \big\langle x, d(x)\big\rangle \geq \tau, \ a \leq x \leq b \\ \infty & \text{otherwise.} \end{cases}$$

Thus, dom $f = \{x \in \mathbb{R}^n \mid \langle x, d(x)\rangle \geq \tau, \ a \leq x \leq b\}$. □

**Equivalent Problems.** The above discussion illustrates that a particular challenge always has multiple formulations. We can use this to our advantage and seek those that are computationally and/or theoretically more tractable. Two problems are *equivalent* if we can obtain minimizers of one from those of the other with only minor effort and vice versa. What constitutes "minor effort" depends on the circumstances, but it usually means just some arithmetic operations or possibly the solution of linear equations. For example, a problem in the decision variable $x$ with the constraint $g(x) \leq 0$ can be stated equivalently as a problem in the decision variables $x$ and $y$ with the constraints $g(x) + y = 0$ and $y \geq 0$. The $x$-component of a minimizer of the latter problem furnishes a minimizer of the original problem. A minimizer $\bar{x}$ of the original problem specifies $(\bar{x}, -g(\bar{x}))$ as a minimizer of the other problem. Generally, a problem *reformulation* implies that we've obtained an equivalent problem, without introducing any approximations.

[1] Throughout, an inequality between two vectors is interpreted componentwise, i.e., for $x, y \in \mathbb{R}^n$: $x \leq y \iff x_j \leq y_j, j = 1, \ldots, n$.

The two examples above result in convex problems. We establish this by looking at a broader class involving the *indicator function* $\iota_C : \mathbb{R}^n \to \overline{\mathbb{R}}$ of a set $C \subset \mathbb{R}^n$ with values

$$\iota_C(x) = \begin{cases} 0 & \text{for } x \in C \\ \infty & \text{otherwise.} \end{cases}$$

Figure 2.2 illustrates its epigraph. Certainly, $\operatorname{dom} \iota_C = C$. Indicator functions allow us to reformulate a minimization problem given in terms of an objective function $f_0$ and a set $C$ into one with a single function through addition: $f = f_0 + \iota_C$. The resulting function has $f(x) = f_0(x) + \iota_C(x) = \infty$ whenever $x \notin C$, which then rules out such points as minimizers. This holds even if $f_0(x) = -\infty$ because of the convention $-\infty + \infty = \infty$.

**Proposition 2.3** (convexity of indicator function). *For $C \subset \mathbb{R}^n$, $\iota_C$ is convex if and only if $C$ is convex. Moreover, if $f_0 : \mathbb{R}^n \to \overline{\mathbb{R}}$ and $C$ are convex, then $f_0 + \iota_C$ is convex.*

**Proof.** If $\iota_C$ is convex, then $C$, as a level-set of a convex function, is also convex by 1.11. The converse holds by 1.12(b) because $\operatorname{epi} \iota_C = C \times [0, \infty)$; see Figure 2.2. A sum of convex functions is convex by 1.18(b). □

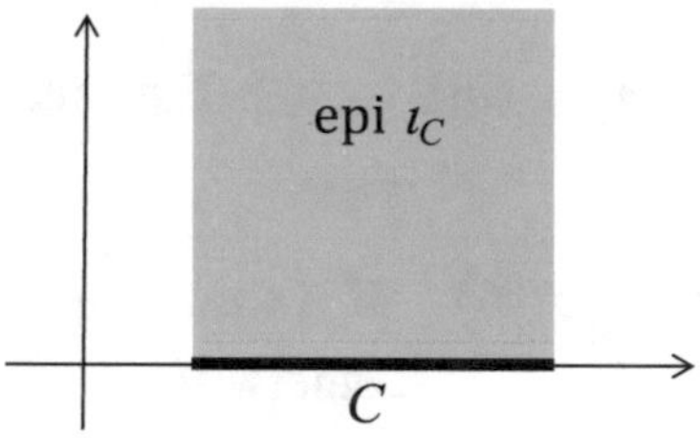


Fig. 2.2: The epigraph of an indicator function $\iota_C$.

In view of this proposition, we can establish the convexity of problems by examining the defining functions and sets separately. Let's consider some common constraints that include those of the above examples. Vectors $A_i \in \mathbb{R}^n$ and scalars $b_i$, $i = 1, \dots, m$, give rise to the collection of *linear equations*

$$\langle A_i, x \rangle = b_i, \quad i = 1, \dots, m, \qquad \text{or, equivalently,} \qquad Ax = b,$$

where $A$ is the $m \times n$-matrix with rows $A_1, \dots, A_m$ and $b = (b_1, \dots, b_m)$. The set of solutions $\{x \in \mathbb{R}^n \mid Ax = b\}$ is convex as already noted after 1.6. If $A_i$ isn't the zero vector, then $\{x \in \mathbb{R}^n \mid \langle A_i, x \rangle = b_i\}$ is called a *hyperplane* as it extends the notion of a plane in three-dimensional space.

Companions to linear equations are *linear inequalities* obtained by replacing "=" by "≤" above. If $A_i$ isn't the zero vector, then $\{x \in \mathbb{R}^n \mid \langle A_i, x \rangle \leq b_i\}$ is called a *halfspace* as it specifies the "half" of $\mathbb{R}^n$ on one side of the hyperplane $\{x \in \mathbb{R}^n \mid \langle A_i, x \rangle = b_i\}$; see Figure 2.3. Since $x \mapsto \langle A_i, x \rangle$ is linear and thus convex, $\{x \in \mathbb{R}^n \mid \langle A_i, x \rangle \leq b_i\}$ is convex by virtue of being a level-set of a convex function; cf. 1.11. Also,

$$\{x \in \mathbb{R}^n \mid Ax \le b\} = \bigcap\nolimits_{i=1}^{m} \{x \in \mathbb{R}^n \mid \langle A_i, x\rangle \le b_i\}$$

and is thus convex by the intersection rule 1.12(a).

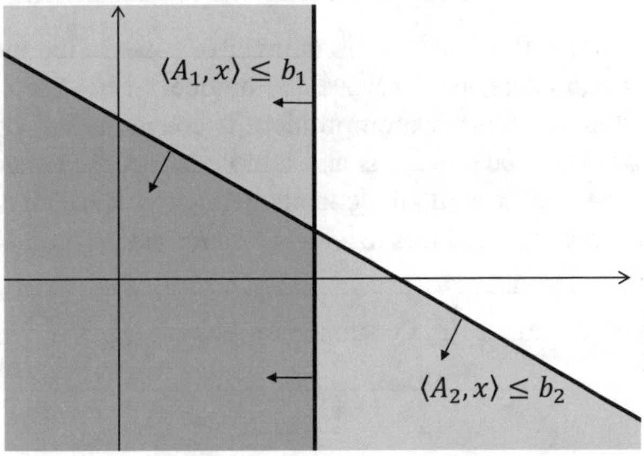


Fig. 2.3: Linear inequalities defining two halfspaces and their intersection.

A set $C \subset \mathbb{R}^n$ defined by those points satisfying a finite number of linear equalities and linear inequalities is called a *polyhedron* or a *polyhedral set*; see Figure 2.4. In particular, $C = \emptyset$ and $C = \mathbb{R}^n$ are polyhedral sets. Since a linear equality $\langle a, x\rangle = \alpha$ can be written as the two inequalities $\langle a, x\rangle \le \alpha$ and $\langle a, x\rangle \ge \alpha$, a polyhedron can always be expressed in terms of a finite number of inequalities. Although sometimes convenient theoretically, a reformulation of linear equalities in this manner is rarely beneficial computationally.

A polyhedral set is an intersection of convex sets and thus convex by 1.12(a); the feasible set in 2.1 furnishes an example. In 2.2, the feasible set can be expressed as the intersection of a polyhedron and the convex set $\{x \mid \langle x, d(x)\rangle \ge \tau\}$, and must then be convex; see 1.24 and 1.12(a).

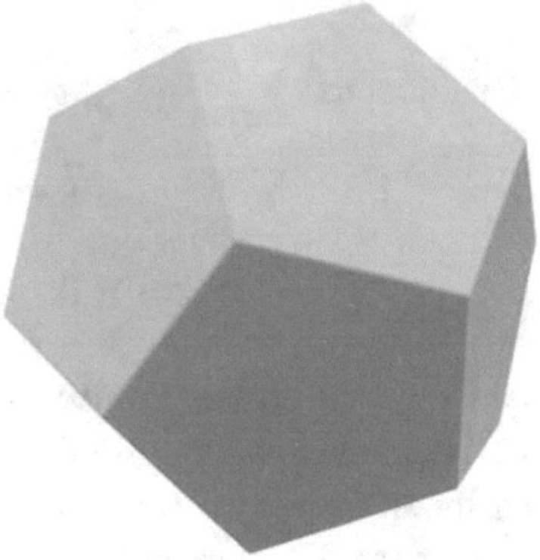

Fig. 2.4: Polyhedral set in $\mathbb{R}^3$.

A *quadratic optimization problem* takes the form

$$\underset{x\in\mathbb{R}^n}{\text{minimize}}\ \tfrac{1}{2}\langle x, Qx\rangle + \langle c, x\rangle \ \text{ subject to } Ax = b, Dx \le d$$

where $Q$ is a symmetric $n \times n$-matrix. The number of rows in the matrix $A$ determines the number of linear equalities and, likewise, the number of rows in $D$ gives the number of inequalities. A quadratic optimization problem is convex if and only if $Q$ is positive semidefinite; cf. 1.24. The model in 2.1 is an example of a convex quadratic optimization problem. If $Q = 0$ (the matrix with all elements being zero), which certainly is positive semidefinite, then the problem reduces to a *linear optimization problem*:

$$\underset{x\in\mathbb{R}^n}{\text{minimize}}\ \langle c, x\rangle \ \text{ subject to } Ax = b, Dx \le d$$

**Example 2.4** (transportation problem). An online retailer has one warehouse in Washington State (WA), with 355 units of a particular product in stock, and one in North Carolina (NC), with 590 units. Customers in San Francisco, Chicago and New York demand 320, 305 and 270 units of the product, respectively; see Figure 2.5. The per-unit costs of shipping from WA to San Francisco, Chicago and New York are \$1.20, \$1.70 and \$2.80, respectively. The corresponding numbers from NC are \$1.90, \$2.90 and \$1.20. We would like to decide the number of units to ship from each warehouse to each location such that the total shipping cost is minimized. This is a *transportation problem*.

**Detail.** Let $x_{ij}$ be the number of products to be shipped from warehouse $i$ to location $j$. Here, $i$ is WA or NC, but in general we may have $m$ warehouses, i.e., $i = 1, 2, \ldots, m$. Similarly, $j$ is San Francisco, Chicago or New York, but in general, $j = 1, 2, \ldots, n$. We denote by $c_{ij}$ the per-unit transportation cost from warehouse $i$ to location $j$, by $w_i$ the stock level in warehouse $i$ and by $b_j$ the demand at location $j$. With $x = (x_{ij}, i = 1, \ldots, m,\ j = 1, \ldots, n) \in \mathbb{R}^{mn}$, we obtain the model

$$\begin{aligned}\underset{x\in\mathbb{R}^{mn}}{\text{minimize}}\ \sum\nolimits_{i=1}^{m}\sum\nolimits_{j=1}^{n} c_{ij}x_{ij} \ \text{ subject to }\ & \sum\nolimits_{j=1}^{n} x_{ij} \le w_i,\quad i = 1, \ldots, m\\ & \sum\nolimits_{i=1}^{m} x_{ij} = b_j,\quad j = 1, \ldots, n\\ & x_{ij} \ge 0,\quad i = 1, \ldots, m;\ j = 1, \ldots, n.\end{aligned}$$

The first $m$ linear inequalities ensure that no warehouse exceeds its inventory and the $n$ linear equalities guarantee that all demand is being satisfied. The feasible set is polyhedral and can be described concisely by $Ax = b$ and $Dx \le d$, where $A$ is an $n \times mn$-matrix consisting of 0 and 1; $b$ is an $n$-vector with elements $b_j$; $D$ is an $(m + mn) \times mn$-matrix with 1, $-1$ and 0; and $d$ is an $(m + mn)$-vector with $w_i$ and 0 as elements. Thus, the model is a linear optimization problem. Table 2.1 specifies a minimizer, with minimum value \$1415.50, which can be obtained using the algorithms of §2.F and §2.G. The minimizer provides the insight that it's better to avoid the intuitive solution of supplying

San Francisco entirely from WA and New York from NC. Although WA and NC are the most cost efficient warehouses for San Francisco and New York, respectively, such an intuitive solution (also given in Table 2.1) results in expensive shipping to Chicago from NC with a total cost of \$1550.50, which is 9% higher than the minimum value. In industries where margins are small such as retail, reducing cost with a few percent is often highly significant and optimization is the technology with which this can be achieved. □

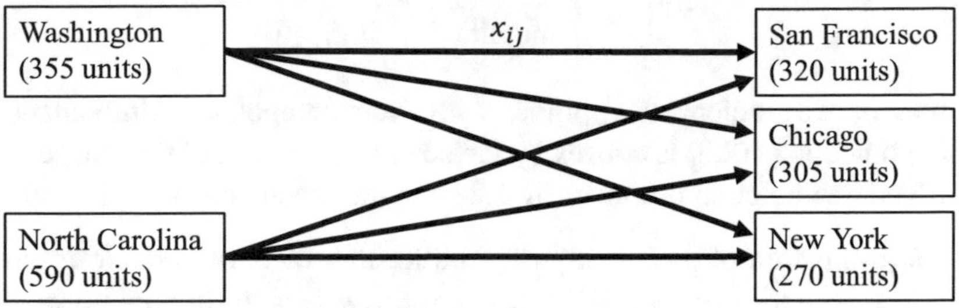


Fig. 2.5: Warehouses and demand locations in transportation problem.

Table 2.1: Shipping quantities for two different decisions: a minimizer (top) and an intuitive but expensive one (bottom).

| | Demand locations | | | Cost |
|---|---|---|---|---|
| Warehouses | San Francisco | Chicago | New York | |
| Washington State | 50 | 305 | 0 | \$1415.50 |
| North Carolina | 270 | 0 | 270 | |
| Washington State | 320 | 35 | 0 | \$1550.50 |
| North Carolina | 0 | 270 | 270 | |

Constraints introduce nonsmoothness by assigning certain decisions the value $\infty$. Smoothness also fails when the graph of a real-valued function has kinks, as would be the case if it represents a physical phenomenon with abrupt changes. Several commonly used functions are nonsmooth as well.

**Example 2.5** (lasso and $\ell^1$-regularization). In the data smoothing of §1.B, we may prefer a "simple" polynomial that has only a few nonzero coefficients. This is often desirable because it would highlight the important terms and avoid an unnecessarily intricate fit. One way to encourage sparsity of this kind is to augment the model in §1.B and consider

$$\underset{c\in\mathbb{R}^{1+n}}{\text{minimize}}\ \|Dc - y\|_2^2 + \theta\|c\|_1, \tag{2.3}$$

with $\theta \in (0, \infty)$. We can think of the added term as a penalty for having a nonzero $c$. Thus, the optimization problem aims to find a balance between a good fit with a low value of $\|Dc - y\|_2^2$ and a sparse fit with many zeros in $c$ so that $\theta\|c\|_1$ is also low. This general approach, known as *lasso regression*, is often used in data analytics and learning since it also tends to perform well from a statistical point of view. However, the objective function is now nonsmooth.

**Detail.** For $g(c) = \|c\|_1 = \sum_{j=0}^{n} |c_j|$, the partial derivative

$$\frac{\partial g(c)}{\partial c_j} = \begin{cases} 1 & \text{if } c_j > 0 \\ -1 & \text{if } c_j < 0 \\ \text{undefined} & \text{if } c_j = 0. \end{cases}$$

Thus, $g$ fails to have a gradient at any point $c$ with a zero component. Although nonsmooth, the objective function in (2.3) is convex by virtue of being a sum of two convex functions; cf. 1.18(b). The first function is convex by 1.24 and the second one by 1.15 and 1.18(d). □

**Exercise 2.6** (estimation of probability distribution). A random variable $\xi$ has possible values $\xi_1, \dots, \xi_n$, but the corresponding probabilities $p_1, \dots, p_n$ are unknown. Formulate the problem of finding $p_1, \dots, p_n$ such that the variance of $\xi$ is maximized, the expected value of $\xi$ is between $\alpha$ and $\beta$, the probabilities sum to one and no probability is less than $0.01/n$. Reformulate the resulting model as a minimization problem and check convexity.

**Guide.** The variance of $\xi$ is $\sum_{j=1}^{n} p_j \xi_j^2 - (\sum_{j=1}^{n} p_j \xi_j)^2$. □

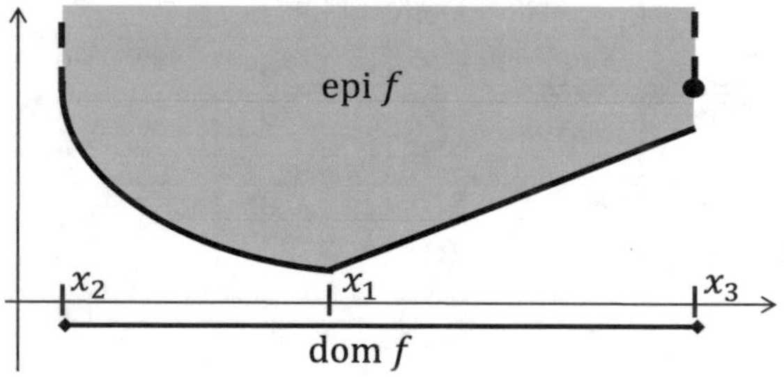


Fig. 2.6: The function $f$ isn't smooth at $x_1$, $x_2$ and $x_3$.

## 2.B Subderivatives and Subgradients

Since optimization models often involve nonsmooth functions, it becomes essential to extend the definitions and rules of differential calculus. We need to replace the notions of derivatives and gradients by appropriate substitutes, potentially a bit less exacting, that still have a potent *calculus*, can be the ingredients of algorithms and allow for generalizations of the Oresme and Fermat rules. This leads us to *subderivatives* and *subgradients*.

Even for convex functions, nonsmoothness materializes in several ways. At the point $x_1$ in Figure 2.6, $f$ has no derivative due to a kink in its graph. The point $x_2$ is on the boundary of dom $f$ and it becomes meaningless to talk about a derivative as the function isn't even continuous at $x_2$; its value jumps to $\infty$ to the left of $x_2$. At the point $x_3$, which also is on the boundary of dom $f$, the graph of $f$ "jumps up" as $x \nearrow x_3$. In fact, regardless of the side from which we approach $x_3$, function values don't tend to $f(x_3)$.

**Continuity and Smoothness.** For $f : \mathbb{R}^n \to \overline{\mathbb{R}}$, $\bar{x} \in \mathbb{R}^n$ and $\alpha \in \overline{\mathbb{R}}$, the expressions $f(x) \to \alpha$ as $x \to \bar{x}$ and $\lim_{x \to \bar{x}} f(x) = \alpha$ mean that $f(x)$ is close to $\alpha$ when $x$ is close to $\bar{x}$, i.e., for any $\varepsilon > 0$, there's $\delta > 0$ such that for all $x \in \mathbb{B}(\bar{x}, \delta)$:

$$\left|f(x) - \alpha\right| \leq \varepsilon \text{ if } \alpha \in \mathbb{R}, \quad f(x) \geq 1/\varepsilon \text{ if } \alpha = \infty, \quad f(x) \leq -1/\varepsilon \text{ if } \alpha = -\infty.$$

If the expressions only need to hold for $x \in \mathbb{B}(\bar{x}, \delta) \setminus \{\bar{x}\}$, then we specify $x \neq \bar{x}$.

A function $f : \mathbb{R}^n \to \overline{\mathbb{R}}$ is *continuous* at $\bar{x}$ if $f(x) \to f(\bar{x})$ as $x \to \bar{x}$. If $f$ is continuous at all $\bar{x} \in \mathbb{R}^n$, then we say it's continuous. The function in Figure 2.6 is continuous at every point except $x_2$ and $x_3$.

A mapping $x \mapsto F(x) = (f_1(x), \ldots, f_m(x))$, which has $n$ inputs and $m$ outputs, is *continuous* (at $\bar{x}$) if every component function $f_i : \mathbb{R}^n \to \mathbb{R}$ is continuous (at $\bar{x}$).

A function $f : \mathbb{R}^n \to \overline{\mathbb{R}}$ is *differentiable* at $\bar{x}$ if

$$f(x) - f(\bar{x}) = \langle v, x - \bar{x} \rangle + o\big(\|x - \bar{x}\|_2\big) \tag{2.4}$$

for some unique $v \in \mathbb{R}^n$, which then must be the gradient $\nabla f(\bar{x})$. Necessarily, this means that $f$ is finite and continuous at $\bar{x}$. The "little-o" notation $o(\alpha)$ represents a quantity that satisfies $o(\alpha)/\alpha \to 0$ as $\alpha \searrow 0$, which makes (2.4) just a short-hand notation for having

$$\lim_{\substack{x \to \bar{x} \\ x \neq \bar{x}}} \frac{f(x) - f(\bar{x}) - \langle v, x - \bar{x} \rangle}{\|x - \bar{x}\|_2} = 0.$$

A function $f : \mathbb{R}^n \to \overline{\mathbb{R}}$ is *smooth* at $\bar{x}$ if it's differentiable in a neighborhood of $\bar{x}$ and the mapping $x \mapsto \nabla f(x)$, usually just written $\nabla f$, is continuous at $\bar{x}$. Thus, $f$ can't be smooth at a point outside of dom $f$ or on its boundary. We say that $f$ is smooth, if it's smooth at all $\bar{x} \in \mathbb{R}^n$. Otherwise, it's *nonsmooth*.

Likewise, a mapping $F : \mathbb{R}^n \to \mathbb{R}^m$ is *differentiable* at $\bar{x}$ if

$$F(x) - F(\bar{x}) = V(x - \bar{x}) + o\big(\|x - \bar{x}\|_2\big) \tag{2.5}$$

for some unique $m \times n$-matrix $V$, which then must be the Jacobian $\nabla F(\bar{x})$. The mapping $F : \mathbb{R}^n \to \mathbb{R}^m$ is *smooth* at $\bar{x}$ if it's differentiable in a neighborhood of $\bar{x}$ and the mapping $x \mapsto \nabla F(x)$ is continuous at $\bar{x}$. We say that $F$ is smooth, if it's smooth at all $\bar{x} \in \mathbb{R}^n$.

All these properties are local so the definitions extend to functions and mappings that are defined only in a neighborhood of $\bar{x}$.

Although smoothness of a function may hold at "most" points, it often breaks down exactly at points of greatest interest: the minimizers of the function; see $x_1$ in Figure 2.6. So we can't simply ignore the issue of nonsmoothness. Our point of departure from differential calculus is the notion of directional derivatives in §1.A. Directional derivatives are more versatile than gradients as they can be defined at a point for which the function isn't differentiable. For example, $f(x) = |x|$ has directional derivatives at $x = 0$ but no gradient at that point. Nevertheless, directional derivatives are insufficient for a robust theory that can address both convex and nonconvex functions. A function may simply not have directional derivatives at a point because their defining limits don't exist.

**Example 2.7** (lack of directional derivative). The function on $\mathbb{R}$ defined by

$$f(x) = \begin{cases} 0 & \text{if } x = 0 \\ x \sin x^{-1} & \text{otherwise} \end{cases}$$

is continuous but fails to have directional derivatives at 0.

**Detail.** Since $\sin x^{-1} \in [-1, 1]$ for all $x \neq 0$, $f(x) \to 0$ as $x \to 0$ and $f$ is continuous at 0. For $\tau > 0$ and $\bar{w} \neq 0$,

$$\frac{f(0 + \tau\bar{w}) - f(0)}{\tau} = \bar{w} \sin(\tau\bar{w})^{-1},$$

which has no limit as $\tau \searrow 0$. Thus, $\vec{d}f(0; \bar{w})$ doesn't exist. □

The difficulty associated with directional derivatives can be addressed by replacing a limit of quotients by a limit of *infima* of quotients. This leads to a robust extension that applies to arbitrary functions.

**Definition 2.8** (subderivative). For $f : \mathbb{R}^n \to \overline{\mathbb{R}}$ and a point $\bar{x}$ with $f(\bar{x})$ finite, the *subderivative* of $f$ at $\bar{x}$ is the function

$$\bar{w} \mapsto df(\bar{x}; \bar{w}) = \lim_{\delta \searrow 0} \ \inf \left\{ \frac{f(\bar{x} + \tau w) - f(\bar{x})}{\tau} \ \middle| \ \tau \in (0, \delta],\ w \in \mathbb{B}(\bar{w}, \delta) \right\}.$$

In addition to considering the infimum of a quotient, we now also include the possibility of approaching $\bar{x}$ along a curve that eventually lines up with $\bar{w}$; see Figure 2.7. This enables us to handle functions that vary dramatically in a neighborhood of $\bar{x}$ as would be the case when $\bar{x}$ is on the boundary of dom $f$. For example, the oscillating function in 2.7 has $df(0; \bar{w}) = -|\bar{w}|$ because the minimum value of $w \sin(\tau w)^{-1}$ with respect to $\tau$ near zero is $-w$ for $w > 0$ and is $w$ for $w < 0$.

A subderivative is well defined because the infima in the expression are nondecreasing as $\delta$ decreases. Thus, the limit is taken over nondecreasing numbers, which either exists, is $\infty$ or is $-\infty$ because all the numbers are $-\infty$. In any case, $df(\bar{x}; \bar{w})$ is assigned a unique value in $\overline{\mathbb{R}}$ that represents the rate of change of $f(x)$ as $x$ moves away from $\bar{x}$ in the general direction of $\bar{w}$.

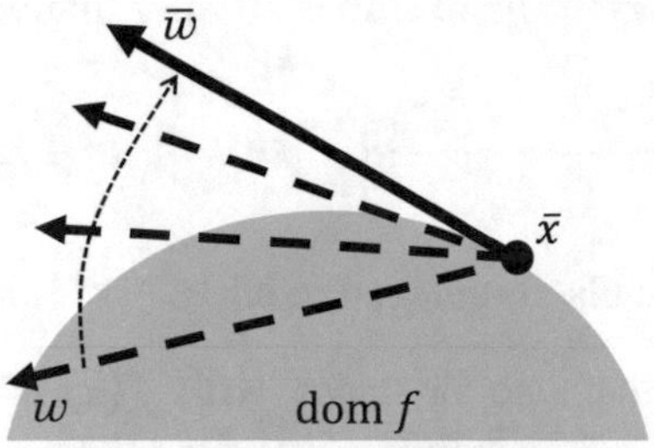


Fig. 2.7: The values of $f : \mathbb{R}^2 \to \overline{\mathbb{R}}$ are dramatically different when approaching $\bar{x}$ along $\bar{w}$ compared to along $w$: $(f(\bar{x} + \tau\bar{w}) - f(\bar{x}))/\tau = \infty$ for all $\tau > 0$. However, $f(\bar{x} + \tau w)$ is close to $f(\bar{x})$ for $\tau$ near zero and $w$ near $\bar{w}$, and $df(\bar{x}; \bar{w})$ is finite.

**Proposition 2.9** (subderivative of differentiable function). *If $f : \mathbb{R}^n \to \overline{\mathbb{R}}$ is differentiable at $\bar{x}$, then*

$$df(\bar{x}; \bar{w}) = \vec{d}f(\bar{x}; \bar{w}) = \langle \nabla f(\bar{x}), \bar{w} \rangle \ \ \forall \bar{w} \in \mathbb{R}^n.$$

**Proof.** Since $f$ is differentiable, (2.4) holds and

$$\frac{f(\bar{x} + \tau w) - f(\bar{x})}{\tau} = \frac{\langle \nabla f(\bar{x}), \tau w \rangle + o(\tau \|w\|_2)}{\tau} = \langle \nabla f(\bar{x}), w \rangle + \|w\|_2 \frac{o(\tau)}{\tau}.$$

The conclusion then follows because $w \to \bar{w}$ and $o(\tau)/\tau \to 0$. □

Immediately, we obtain a new, far-reaching Oresme rule: Regardless of convexity, smoothness, continuity or any other property, nonnegativity of the subderivative of $f$ at $x^\star$ is a necessary condition for $x^\star$ to be a local minimizer of $f$ when $f(x^\star)$ is finite.

**Theorem 2.10** (Oresme rule). *For $f : \mathbb{R}^n \to \overline{\mathbb{R}}$ and $x^\star$ with $f(x^\star) \in \mathbb{R}$, one has*

$$\boxed{x^\star \text{ local minimizer of } f \implies df(x^\star; w) \geq 0 \ \ \forall w \in \mathbb{R}^n}$$

**Proof.** Since $x^\star$ is a local minimizer, there exists $\varepsilon > 0$ such that $f(x) \geq f(x^\star)$ for all $x \in \mathbb{B}(x^\star, \varepsilon)$. Thus, for $\|\tau w\|_2 \leq \varepsilon$ and $\tau \in (0, \infty)$,

$$\frac{f(x^\star + \tau w) - f(x^\star)}{\tau} \geq 0.$$

Let $\bar{w} \in \mathbb{R}^n$ and $\delta > 0$. The infimum in the definition of subderivatives is over $\tau \in (0, \delta]$ and $w \in \mathbb{B}(\bar{w}, \delta)$. For such $\tau$ and $w$,

$$\|\tau w\|_2 \leq \tau(\|w - \bar{w}\|_2 + \|\bar{w}\|_2) \leq \delta^2 + \delta \|\bar{w}\|_2.$$

For sufficiently small $\delta$, this right-hand side is smaller than $\varepsilon$ and

$$\inf \left\{ \frac{f(x^\star + \tau w) - f(x^\star)}{\tau} \;\middle|\; \tau \in (0, \delta],\; w \in \mathbb{B}(\bar{w}, \delta) \right\} \geq 0.$$

The limit as $\delta \searrow 0$ is therefore also nonnegative and $df(x^\star; \bar{w}) \geq 0$. □

It's already clear from the smooth cases with $f(x) = -x^2$ and $f(x) = x^3$ that subderivatives could very well be nonnegative for other points than minimizers. Still, one can use the Oresme rule to confirm that a candidate solution isn't a local minimizer. Finding a point that satisfies an optimality condition, sometimes called a *stationary point*, can also be a computationally tractable substitute for minimization.

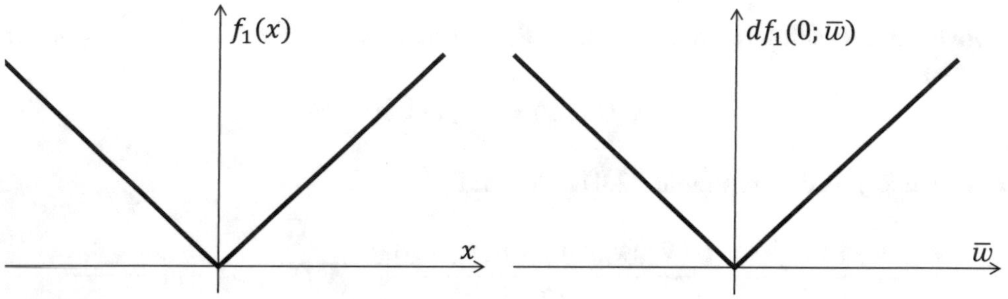


Fig. 2.8: Function $f_1$ of 2.11 and a subderivative.

**Example 2.11** (subderivatives). If $f_1(x) = |x|$, then for any $\bar{w}$ one has

$$df_1(\bar{x}; \bar{w}) = \begin{cases} \bar{w} & \text{if } \bar{x} > 0 \\ |\bar{w}| & \text{if } \bar{x} = 0 \\ -\bar{w} & \text{otherwise.} \end{cases}$$

**Detail.** For $\bar{x} \neq 0$, $f_1$ is smooth at $\bar{x}$ (see Figure 2.8(left)) and we can apply 2.9. For $\bar{x} = 0$, we appeal to the definition. Let $w \in \mathbb{R}$ and $\tau \in (0, \infty)$. Then,

$$\frac{f_1(\bar{x} + \tau w) - f_1(\bar{x})}{\tau} = \frac{|\tau w|}{\tau} = |w|$$

and we obtain $df_1(0; \bar{w}) = |\bar{w}|$; see Figure 2.8(right). At $\bar{x} = 0$, the subderivative is a nonlinear function, while elsewhere it's linear. As seen in 2.9, a subderivative of a function is always linear at a point of smoothness of the function. □

**Exercise 2.12** (subderivatives). Compute the subderivatives for the following two functions defined on $\mathbb{R}$: $f_2(x) = -x$ if $x \leq 0$ and $f_2(x) = x^2$ otherwise, and $f_3(x) = 2x$ if $x \geq 0$ and $f_3(x) = \infty$ otherwise. Draw the graphs of the functions as well as those of their subderivatives.

**Guide.** For $\bar{w} \in \mathbb{R}$, one has

$$df_2(\bar{x};\bar{w}) = \begin{cases} 2\bar{x}\bar{w} & \text{if } \bar{x} > 0 \\ \max\{0, -\bar{w}\} & \text{if } \bar{x} = 0 \\ -\bar{w} & \text{if } \bar{x} < 0 \end{cases} \qquad df_3(\bar{x};\bar{w}) = \begin{cases} 2\bar{w} & \text{if } \bar{x} > 0, \bar{w} \in \mathbb{R} \\ 2\bar{w} & \text{if } \bar{x} = 0, \bar{w} \geq 0 \\ \infty & \text{if } \bar{x} = 0, \bar{w} < 0. \end{cases}$$

Note that $f_3(\bar{x})$ isn't finite for $\bar{x} < 0$ so $df_3(\bar{x};\bar{w})$ is undefined for such $\bar{x}$. □

The functions in 2.11 and 2.12 have $\operatorname{argmin} f_1 = \operatorname{argmin} f_2 = \operatorname{argmin} f_3 = \{0\}$ and, indeed, 0 is correctly singled out by the Oresme rule 2.10. Here, the curly brackets around "0" remind us that $\operatorname{argmin} f_i$ is a *set* of points, but in this case it contains only zero.

Without being restricted to smooth functions, the Oresme rule 2.10 furnishes a basis for descent methods: From a current point $x^\nu$, a vector $w^\nu$ with $df(x^\nu; w^\nu) < 0$ indicates a direction nearby which we can find an improved point $x^{\nu+1}$ with $f(x^{\nu+1}) < f(x^\nu)$. If there's no such $w^\nu$, then $x^\nu$ satisfies the condition required by the Oresme rule and $x^\nu$ may suffice as a solution of the problem.

While this is a possible approach to algorithms, let's follow an alternative path motivated by the Fermat rule that's often easier to execute in practice. For convex functions, as in the smooth case, the two principal rules lead to equivalent optimality conditions. We return to the Oresme rule in Chap. 4, where it emerges as the stronger of the two in general.

An extension of the Fermat rule to nonsmooth functions requires a generalization of gradients. It turns out that the rates of change specified by a subderivative are exactly the quantities needed for a broad class of functions. The key insight is that the gradient $\nabla f(\bar{x})$ of a function $f$ at a point $\bar{x}$ of differentiability furnishes the slope coefficients of the subderivative at $\bar{x}$, which then is a linear function:

$$\bar{w} \mapsto df(\bar{x};\bar{w}) = \big\langle \nabla f(\bar{x}), \bar{w} \big\rangle.$$

This fact follows by 2.9 and is illustrated in Figure 2.9(left) for the function in 2.11. When the subderivative isn't linear, there might still be vectors $v$ such that

$$\bar{w} \mapsto df(\bar{x};\bar{w}) \text{ is approximated from below by } \bar{w} \mapsto \langle v, \bar{w} \rangle.$$

In Figure 2.9(right), any $v \in [-1, 1]$ produces a linear function $\bar{w} \mapsto \langle v, \bar{w} \rangle$ that sits below the subderivative. Vectors $v$ of this kind are subgradients and, despite lacking the "exactness" of gradients, provide useful information about the local behavior of $f$ near $\bar{x}$. We concentrate on the convex case, but return to general functions in Chap. 4.

**Definition 2.13** (subgradient). For a convex function $f : \mathbb{R}^n \to \overline{\mathbb{R}}$ and a point $\bar{x}$ with $f(\bar{x})$ finite, $v \in \mathbb{R}^n$ is a *subgradient* of $f$ at $\bar{x}$ when

$$df(\bar{x}; w) \geq \langle v, w \rangle \;\; \forall w \in \mathbb{R}^n.$$

The set of all subgradients at $\bar{x}$ is denoted by $\partial f(\bar{x})$.

Since $df(\bar{x}; w) = \langle \nabla f(\bar{x}), w \rangle$ when $f$ is smooth at $\bar{x}$, it's immediate from the definition that subgradients reduce to gradients in this case.

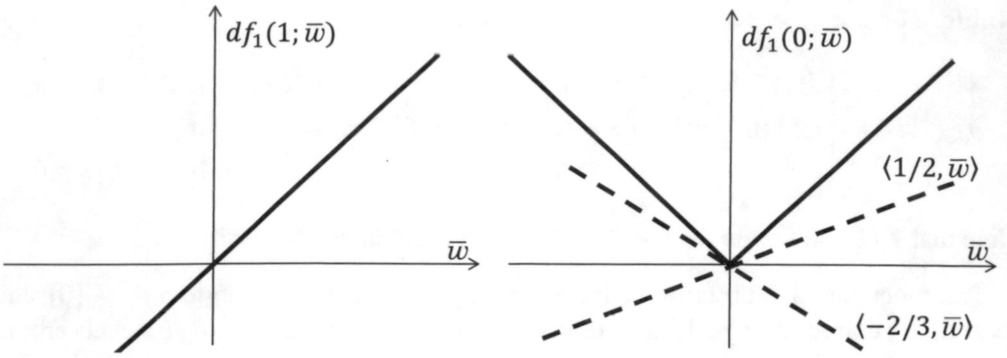


Fig. 2.9: Subderivatives $\bar{w} \mapsto df_1(1;\bar{w})$ (left) and $\bar{w} \mapsto df_1(0;\bar{w})$ with lower bounding approximations (right); $v = 1/2$ and $-2/3$ are subgradients of $f_1$ at zero.

**Proposition 2.14** (subgradients under smoothness). *For a convex function $f : \mathbb{R}^n \to \overline{\mathbb{R}}$ and a point $\bar{x}$ at which $f$ is smooth, one has*

$$\partial f(\bar{x}) = \{\nabla f(\bar{x})\}.$$

In general, for a convex function $f : \mathbb{R}^n \to \overline{\mathbb{R}}$ and a point $\bar{x}$ with $f(\bar{x})$ finite, $\partial f(\bar{x})$ is a convex set, which can be seen directly from the definition.

**Example 2.15** (subgradients). For the functions in 2.11 and 2.12, we've

$$\partial f_1(x) = \begin{cases} \{-1\} & \text{if } x < 0 \\ [-1,1] & \text{if } x = 0 \\ \{1\} & \text{otherwise} \end{cases} \qquad \partial f_2(x) = \begin{cases} \{-1\} & \text{if } x < 0 \\ [-1,0] & \text{if } x = 0 \\ \{2x\} & \text{otherwise.} \end{cases}$$

Function $f_3$ is finite only for $x \geq 0$ and has subgradients

$$\partial f_3(x) = \begin{cases} (-\infty, 2] & \text{if } x = 0 \\ \{2\} & \text{if } x > 0. \end{cases}$$

**Detail.** These subgradients, shown in Figure 2.10, are obtained by 2.14 or directly from the definition. For instance, $\partial f_2(0)$ consists of $v$ such that

$$df_2(0;w) = \max\{0, -w\} \geq \langle v, w\rangle \;\; \forall w \in \mathbb{R}.$$

The graph of $w \mapsto \max\{0, -w\}$ convinces us that we need $v \in [-1, 0]$. □

**Exercise 2.16** (subdifferentiability). For the convex function given by $f(x) = \max\{x_1 + x_2 - 1, 0\}$, determine its subderivatives and subgradients using their definitions.

Since a subgradient of a function at a point $\bar{x}$ reflects the rate of change of the function as captured by its subderivative at $\bar{x}$, it's not surprising that a subgradient can be used to

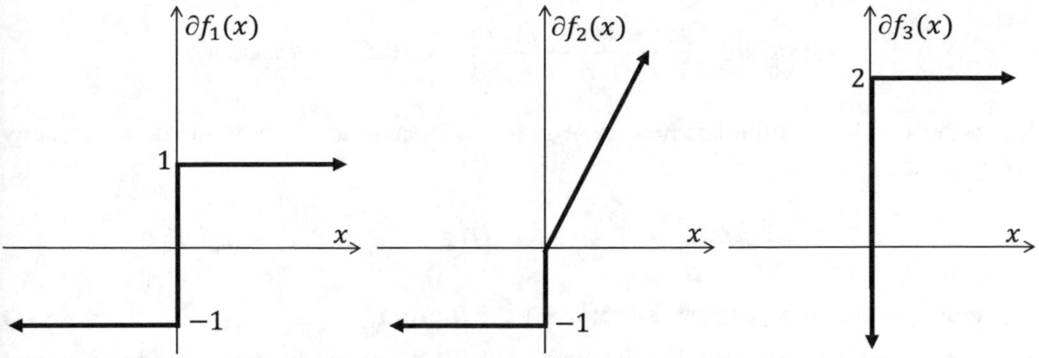


Fig. 2.10: Examples of subgradients from 2.15.

approximate a function locally near $\bar{x}$ in a manner similar to what's achieved by a gradient; see (2.4). In fact, we obtain a lower bounding approximation that generalizes 1.22.

**Theorem 2.17** (subgradient inequality). *For a convex function $f : \mathbb{R}^n \to \overline{\mathbb{R}}$ and a point $\bar{x}$ with $f(\bar{x})$ finite, one has*

$$v \in \partial f(\bar{x}) \iff f(x) \geq f(\bar{x}) + \langle v, x - \bar{x} \rangle \ \forall x \in \mathbb{R}^n.$$

**Proof.** Suppose that $\bar{w} \in \mathbb{R}^n$ and $f(x) \geq f(\bar{x}) + \langle v, x - \bar{x} \rangle$ for all $x \in \mathbb{R}^n$. Then, for $\tau, \delta > 0$ and $w \in \mathbb{B}(\bar{w}, \delta)$, one has

$$\frac{f(\bar{x} + \tau w) - f(\bar{x})}{\tau} \geq \frac{\langle v, \tau w \rangle}{\tau} = \langle v, \bar{w} \rangle + \langle v, w - \bar{w} \rangle \geq \langle v, \bar{w} \rangle - \delta \|v\|_2.$$

The infimum of the left-hand side over $\tau \in (0, \delta]$ and $w \in \mathbb{B}(\bar{w}, \delta)$ is then also bounded from below by $\langle v, \bar{w} \rangle - \delta \|v\|_2$. As $\delta \searrow 0$, we obtain that $df(\bar{x}; \bar{w}) \geq \langle v, \bar{w} \rangle$. Since $\bar{w}$ is arbitrary, $v \in \partial f(\bar{x})$.

For the converse, the convexity inequality 1.10 implies that with $0 < \sigma \leq \tau$ and $w \in \mathbb{R}^n$, one has

$$f(\bar{x} + \sigma w) \leq \frac{\sigma}{\tau} f(\bar{x} + \tau w) + \left(1 - \frac{\sigma}{\tau}\right) f(\bar{x}).$$

Hence,

$$\frac{f(\bar{x} + \sigma w) - f(\bar{x})}{\sigma} \leq \frac{f(\bar{x} + \tau w) - f(\bar{x})}{\tau}$$

and the quotient $(f(\bar{x} + \tau w) - f(\bar{x}))/\tau$ is nonincreasing as $\tau \searrow 0$. Suppose that $v \in \partial f(\bar{x})$ so that, for all $\bar{w} \in \mathbb{R}^n$, one has

$$df(\bar{x}; \bar{w}) = \lim_{\delta \searrow 0} \inf \left\{ \frac{f(\bar{x} + \tau w) - f(\bar{x})}{\tau} \;\middle|\; \tau \in (0, \delta],\ w \in \mathbb{B}(\bar{w}, \delta) \right\} \geq \langle v, \bar{w} \rangle.$$

The restriction to $w = \bar{w}$ preserves this inequality. Thus,

$$\lim_{\delta \searrow 0} \inf \left\{ \frac{f(\bar{x} + \tau\bar{w}) - f(\bar{x})}{\tau} \;\middle|\; \tau \in (0, \delta] \right\} \geq \langle v, \bar{w} \rangle.$$

Since the quotient on the left-hand side is nonincreasing as already noted, we actually have for all $\tau > 0$ that

$$\frac{f(\bar{x} + \tau\bar{w}) - f(\bar{x})}{\tau} \geq \langle v, \bar{w} \rangle, \text{ i.e., } f(\bar{x} + \tau\bar{w}) \geq f(\bar{x}) + \langle v, \tau\bar{w} \rangle.$$

The conclusion follows because any $x = \bar{x} + \tau\bar{w}$ for some $\bar{w}$. □

**Exercise 2.18** (linearization). For the model in 1.19, construct an approximating problem of the form minimizing $\max_{i=1,\dots,m}\langle a^i, x\rangle + b_i$ subject to $x \in C$ using the subgradient inequality 2.17. Write explicit expressions for $a^i \in \mathbb{R}^n$ and $b_i \in \mathbb{R}$. Show that the resulting minimum value is no greater than that in 1.19 and is nondecreasing as $m$ grows. Reformulate the problem as an equivalent linear optimization problem.

In addition to algorithmic possibilities based on affine approximations of convex functions, the subgradient inequality 2.17 leads to an extension of the Fermat rule for nonsmooth functions. The resulting optimality condition is equivalent to the condition obtained by the Oresme rule in the convex case.

**Theorem 2.19** (optimality for convex functions). *For a convex function $f : \mathbb{R}^n \to \overline{\mathbb{R}}$ and a point $x^\star$ at which $f$ is finite, one has*

$$x^\star \in \operatorname{argmin} f \iff df(x^\star; w) \geq 0 \;\forall w \in \mathbb{R}^n \iff 0 \in \partial f(x^\star).$$

**Proof.** Suppose that $0 \in \partial f(x^\star)$. By the subgradient inequality 2.17, $f(x) \geq f(x^\star) + \langle 0, x - x^\star \rangle$ for all $x \in \mathbb{R}^n$, which implies that $x^\star \in \operatorname{argmin} f$. The converse holds by reversing this argument. By the definition of subgradients, $df(x^\star; w) \geq 0$ for all $w \in \mathbb{R}^n$ implies that $0 \in \partial f(x^\star)$. If $x^\star \in \operatorname{argmin} f$, then $df(x^\star; w) \geq 0$ for all $w \in \mathbb{R}^n$ by the Oresme rule 2.10. □

The theorem extends 1.9 to the nonsmooth case by replacing directional derivatives by subderivatives and, more significantly, gradients by subgradients, with the earlier result now becoming a special case; see 2.9 and 2.14. While in the smooth case solving the equation $0 = \nabla f(x)$ is a viable approach to minimizing $f$, we're now presented with the possibility of solving the *generalized equation* $0 \in \partial f(x)$.

The functions in 2.11 and 2.12 are convex and satisfy the optimality conditions in 2.19 at $x = 0$, which then confirms that $\operatorname{argmin} f_1 = \operatorname{argmin} f_2 = \operatorname{argmin} f_3 = \{0\}$; see also 2.15.

## 2.C Subgradient Calculus

The most expedient way to obtain expressions for subgradients is to leverage calculus rules in the same manner as we compute gradients in differential calculus. This section provides some basic rules.

**Proposition 2.20** (subgradients for separable functions). *For convex* $f_i : \mathbb{R}^{n_i} \to \overline{\mathbb{R}}$, $i = 1, 2, \dots, m$, *the function* $f : \mathbb{R}^n \to \overline{\mathbb{R}}$ *defined by*

$$f(x) = \sum\nolimits_{i=1}^{m} f_i(x_i),$$

*where* $n = \sum_{i=1}^{m} n_i$, *has at a point* $\bar{x} = (\bar{x}_1, \dots, \bar{x}_m)$ *with finite* $f(\bar{x})$:

$$\partial f(\bar{x}) = \partial f_1(\bar{x}_1) \times \cdots \times \partial f_m(\bar{x}_m) = \big\{(v_1, \dots, v_m) \,\big|\, v_i \in \partial f_i(\bar{x}_i),\ i = 1, \dots, m\big\}.$$

**Proof.** Suppose that $v_i \in \partial f_i(\bar{x}_i), i = 1, \dots, m$. Then, $f_i(x_i) \geq f_i(\bar{x}_i) + \langle v_i, x_i - \bar{x}_i \rangle$ for all $x_i \in \mathbb{R}^{n_i}$ by the subgradient inequality 2.17 and

$$f(x) = \sum\nolimits_{i=1}^{m} f_i(x_i) \geq \sum\nolimits_{i=1}^{m} f_i(\bar{x}_i) + \sum\nolimits_{i=1}^{m} \langle v_i, x_i - \bar{x}_i \rangle = f(\bar{x}) + \langle v, x - \bar{x} \rangle$$

for $x = (x_1, \dots, x_m) \in \mathbb{R}^n$ and $v = (v_1, \dots, v_m) \in \mathbb{R}^n$. Since $x$ is arbitrary, $v \in \partial f(\bar{x})$ by 2.17 and

$$\partial f(\bar{x}) \supset \partial f_1(\bar{x}_1) \times \cdots \times \partial f_m(\bar{x}_m).$$

Next, suppose that $v = (v_1, \dots, v_m) \in \partial f(\bar{x})$. Then, by 2.17,

$$\sum\nolimits_{i=1}^{m} f_i(x_i) = f(x) \geq f(\bar{x}) + \langle v, x - \bar{x} \rangle = \sum\nolimits_{i=1}^{m} f_i(\bar{x}_i) + \sum\nolimits_{i=1}^{m} \langle v_i, x_i - \bar{x}_i \rangle$$

holds for all $x = (x_1, \dots, x_m) \in \mathbb{R}^n$ including $x = (x_1, \bar{x}_2, \dots, \bar{x}_m)$. Inserting this $x$ leads to $f_1(x_1) \geq f_1(\bar{x}_1) + \langle v_1, x_1 - \bar{x}_1 \rangle$. Thus, $v_1 \in \partial f_1(x_1)$ by 2.17. Instead of singling out the index $i = 1$, we can repeat the argument for any other index to conclude that $v_i \in \partial f_i(x_i)$ for all $i$ and, in fact,

$$\partial f(\bar{x}) \subset \partial f_1(\bar{x}_1) \times \cdots \times \partial f_m(\bar{x}_m).$$

This completes the proof. □

**Example 2.21** (subgradients of $\ell^1$-norm). The function given by $g(x) = \|x\|_1$ is convex and, alternatively, can be expressed as $g(x) = \sum_{j=1}^{n} g_j(x_j)$, with $g_j(x_j) = |x_j|$ for all $j$. Thus, 2.20 can be brought in to compute subgradients.

**Detail.** From 2.15, $\partial g_j(\bar{x}_j) = C_j$, where

$$C_j = \begin{cases} \{-1\} & \text{if } \bar{x}_j < 0 \\ [-1, 1] & \text{if } \bar{x}_j = 0 \\ \{1\} & \text{otherwise.} \end{cases}$$

Thus, by 2.20,

$$\partial g(\bar{x}) = C_1 \times \cdots \times C_n$$

for any $\bar{x} = (\bar{x}_1, \dots, \bar{x}_n)$. □

Since many minimization problems can be formulated in terms of a function of the form $f = f_0 + \iota_C$ (see §2.A), it's convenient to have expressions for subgradients of sums of functions. In particular, we see below that under mild assumptions $\partial f(x) = \partial f_0(x) + \partial\iota_C(x)$. This expression involves the *sum of two sets*, which we always interpret in the sense already encountered in 1.12(c) and visualized by Figure 1.12:

$$C_1 + C_2 = \{x_1 + x_2 \mid x_1 \in C_1, x_2 \in C_2\}$$

for any $C_1, C_2 \subset \mathbb{R}^n$. We start by examining indicator functions more closely.

**Example 2.22** (subgradients of indicator functions). For a convex set $C \subset \mathbb{R}^n$ and $\bar{x} \in C$, one has

$$\partial\iota_C(\bar{x}) = \big\{v \in \mathbb{R}^n \;\big|\; \langle v, x - \bar{x}\rangle \le 0 \;\; \forall x \in C\big\}.$$

**Detail.** The expression follows by the subgradient inequality 2.17 because $\iota_C(\bar{x}) = 0$ and $\iota_C(x) = \infty$ for $x \notin C$; see Figure 2.11. □

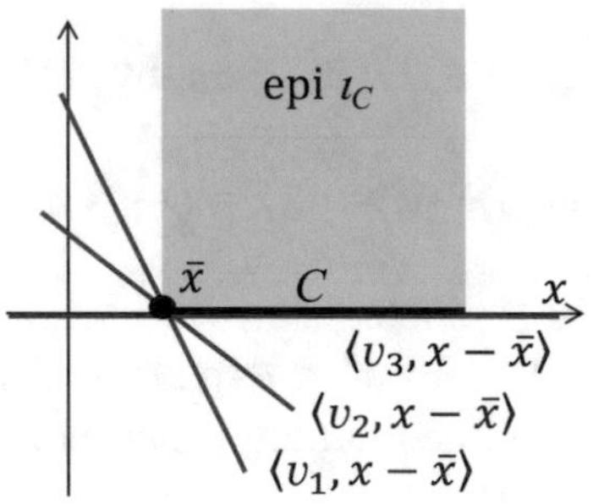


Fig. 2.11: At $\bar{x}$, $v_1$, $v_2$ and $v_3$ are examples of subgradients of $\iota_C$. In fact, $\partial\iota_C(\bar{x}) = (-\infty, 0]$.

The set of subgradients of an indicator function has a geometric interpretation, which motivates the following terminology.

**Definition 2.23** (normal cones of convex sets). For a nonempty convex set $C \subset \mathbb{R}^n$, $v \in \mathbb{R}^n$ is a *normal vector* to $C$ at $\bar{x} \in C$ when $v \in \partial\iota_C(\bar{x})$. The set of such $v$, denoted by $N_C(\bar{x})$, is the *normal cone* to $C$ at $\bar{x}$, i.e.,

$$N_C(\bar{x}) = \partial\iota_C(\bar{x}) = \big\{v \in \mathbb{R}^n \;\big|\; \langle v, x - \bar{x}\rangle \le 0 \;\; \forall x \in C\big\}.$$

If $C$ is a hyperplane, then $N_C(\bar{x})$ is simply the normal space from linear algebra consisting of vectors perpendicular to the hyperplane. Specifically, $v \in N_C(\bar{x})$ if and only if $\langle v, x - \bar{x}\rangle = 0$ for all $x \in C$. Figure 2.12 illustrates the situation for more general convex sets using the geometric interpretation that the vectors $x - \bar{x}$ need to be between 90 and 270 degrees away from a normal vector $v$ for any choice of $x \in C$. This means that if $C$ has a "smooth" boundary at $\bar{x}$ as in Figure 2.12(left), then all $v \in N_C(\bar{x})$ must point in the same direction. If $C$ has a "corner" at $\bar{x}$ as in Figure 2.12(right), then $v$ fans out in different directions. In general, as long as it contains not only the zero vector, $N_C(\bar{x})$ has vectors of any length: If $v \in N_C(\bar{x})$, then $\lambda v \in N_C(\bar{x})$ for any $\lambda > 0$.

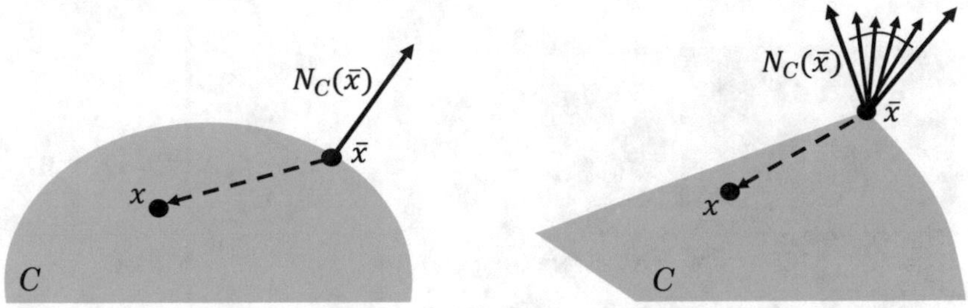


Fig. 2.12: Normal cones to convex sets illustrated as arrows emanating from $\bar{x}$.

**Proposition 2.24** (subgradients of sum of convex functions). *For convex $f, g : \mathbb{R}^n \to \overline{\mathbb{R}}$ and $\bar{x}$ with $f(\bar{x})$ and $g(\bar{x})$ finite, one has*[2]

$$\partial(f+g)(\bar{x}) \supset \partial f(\bar{x}) + \partial g(\bar{x}).$$

**Proof.** For $w \in \partial f(\bar{x}) + \partial g(\bar{x})$, there exist $u \in \partial f(\bar{x})$ and $v \in \partial g(\bar{x})$ such that $w = u + v$. Let $x \in \mathbb{R}^n$. By the subgradient inequality 2.17,

$$\langle u, x - \bar{x} \rangle \le f(x) - f(\bar{x}) \quad \text{and} \quad \langle v, x - \bar{x} \rangle \le g(x) - g(\bar{x})$$

so that

$$\langle u + v, x - \bar{x} \rangle \le (f+g)(x) - (f+g)(\bar{x}).$$

In view of the subgradient inequality 2.17, this implies that $u + v = w \in \partial(f+g)(\bar{x})$ and the inclusion holds. □

Equality in the proposition is common but certainly not universal. Let's consider the counterexample

$$f(x) = \begin{cases} -\sqrt{1-(x+1)^2} & \text{if } x \in [-2, 0] \\ \infty & \text{otherwise} \end{cases} \tag{2.6}$$

and $g = \iota_{[0,1]}$; see Figure 2.13. Then,

$$\partial f(0) = \emptyset, \quad \partial g(0) = (-\infty, 0], \qquad \text{but} \quad \partial(f+g)(0) = \mathbb{R}.$$

The difficulty may also occur when both sets of subgradients are nonempty. For example, let $f = \iota_C$ and $g = \iota_D$, with $C = \mathbb{B}((-1,0),1) \subset \mathbb{R}^2$ and $D = \mathbb{B}((1,0),1) \subset \mathbb{R}^2$. Then,

$$\partial f(0,0) = [0,\infty) \times \{0\}, \;\; \partial g(0,0) = (-\infty, 0] \times \{0\}, \;\; \partial f(0,0) + \partial g(0,0) = \mathbb{R} \times \{0\},$$

[2] For $f, g : \mathbb{R}^n \to \overline{\mathbb{R}}$, $f + g$ or $(f + g)$ is the function with values $(f+g)(x) = f(x) + g(x)$.

but $\partial(f+g)(0,0) = \mathbb{R}^2$ because $f+g = \iota_{\{(0,0)\}}$. These simple examples pinpoint the cause of the problem: dom $f$ and dom $g$ don't overlap significantly. Their intersection consists exclusively of boundary points of these sets.

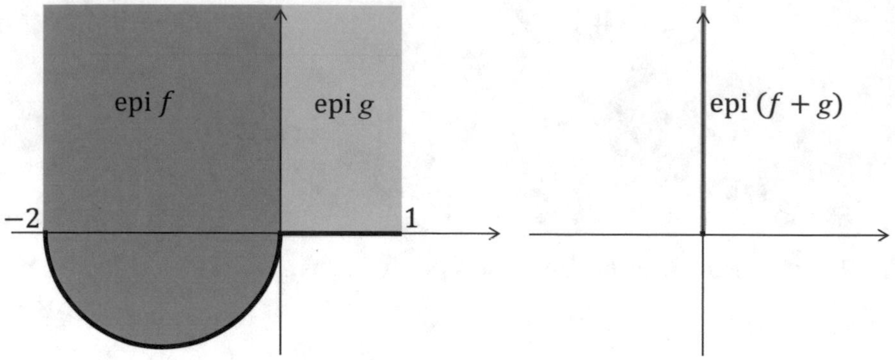


Fig. 2.13: A situation when the sum rule inclusion in 2.24 is strict.

**Closed and Open Sets.** It's often necessary to identify whether a set includes its "boundary." If it does, we say it's *closed*. More precisely, a set $C \subset \mathbb{R}^n$ is closed if every convergent sequence in $C$ has a limit point in $C$, i.e., $x^\nu \in C \to x$ implies $x \in C$. The set $(0,1)$ fails the condition because the sequence consisting of $x^\nu = 1/\nu$ has limit 0 outside the set. The ball $\mathbb{B}(\bar{x}, \rho)$ is a closed set and the same holds for the empty set and $\mathbb{R}^n$. A set $C$ is *open* if its *complement* $\mathbb{R}^n \setminus C$ is closed. Equivalently, $C$ is open if for all $x \in C$, there exists $\rho > 0$ such that $\mathbb{B}(x, \rho) \subset C$. For example, $C = (0,1)$ is open and the empty set and $\mathbb{R}^n$ are as well.

The *interior* of $C$, denoted by int $C$, consists of every point $x \in C$ for which there exists $\rho > 0$ such that $\mathbb{B}(x, \rho) \subset C$. Thus, the interior of $C$ is essentially $C$ without its "boundary." If $C = [0,1]$, then int $C = (0,1)$. Certainly, $C = \operatorname{int} C$ for an open set $C$.

Formally, we can then define the *boundary* of $C$, denoted by bdry $C$, as $C \setminus \operatorname{int} C$. For example, if $C = [0,1]$, then $\operatorname{bdry} C = \{0,1\}$. A set that isn't closed can be enlarged to form a closed set: The *closure* of $C$, denoted by cl $C$, is the union of $C$ and all limit points of sequences $\{x^\nu \in C, \nu \in \mathbb{N}\}$. For example, if $C = (0,1]$, then $\operatorname{cl} C = [0,1]$.

We now see from 2.22 that when $\bar{x} \in \operatorname{int} C$,

$$\partial \iota_C(\bar{x}) = N_C(\bar{x}) = \{0\}.$$

In the interior of the domain of a general convex function, we also have subgradients as formalized next.

**Proposition 2.25** (properties of subgradients). *For a convex function $f : \mathbb{R}^n \to \overline{\mathbb{R}}$ and a point $\bar{x}$ at which $f(\bar{x})$ is finite, $\partial f(\bar{x})$ is closed. If $\bar{x} \in \operatorname{int}(\operatorname{dom} f)$, then $f$ is continuous at $\bar{x}$ and $\partial f(\bar{x})$ is nonempty.*

**Proof.** Suppose that $v^\nu \in \partial f(\bar{x}) \to \bar{v}$. Let $x \in \mathbb{R}^n$. By the subgradient inequality 2.17, $f(x) - f(\bar{x}) \geq \langle v^\nu, x - \bar{x}\rangle$ for all $\nu$, which implies that $f(x) - f(\bar{x}) \geq \langle \bar{v}, x - \bar{x}\rangle$. The first assertion holds by 2.17. For the two other claims, we refer to [105, Theorem 2.35 and Corollary 8.10]. □

The potential difficulty at the boundary of the domain of a function is highlighted by (2.6), where $\partial f(0) = \emptyset$.

**Proper Functions.** A function $f : \mathbb{R}^n \to \overline{\mathbb{R}}$ is *proper* if $f(x) > -\infty$ for all $x \in \mathbb{R}^n$ and $f(x) < \infty$ for some $x \in \mathbb{R}^n$. Thus, a proper function has finite values on its domain, which must be nonempty. Although it's useful to permit $\infty$ for the purpose of expressing constraints, $-\infty$ is less critical because this value makes the minimum and minimizers somewhat trivial. In fact, a "cost" of $-\infty$ in an application is probably too good to be true and must be due to a modeling error. Consequently, we often restrict the attention to proper functions.

**Theorem 2.26** (Moreau-Rockafellar sum rule). *For proper convex functions $f, g : \mathbb{R}^n \to \overline{\mathbb{R}}$ and $\bar{x} \in \operatorname{dom} f \cap \operatorname{dom} g$,*

$$\partial(f + g)(\bar{x}) = \partial f(\bar{x}) + \partial g(\bar{x})$$

*provided that* $\operatorname{int}(\operatorname{dom} f) \cap \operatorname{dom} g \neq \emptyset$.

**Proof.** This fact can be deduced from [89, Theorem 23.8]; see also our development in §4.I. □

The assumption about "sufficient" overlap between the domains in the theorem rules out the pathological cases discussed above. It's immediate from the theorem that if $f$ is a real-valued convex function and $C$ is a nonempty convex set, then

$$\partial(f + \iota_C)(x) = \partial f(x) + N_C(x)$$

at every point $x \in C$ because in this case

$$\operatorname{int}(\operatorname{dom} f) \cap \operatorname{dom} \iota_C = \mathbb{R}^n \cap C = C$$

and both $f$ and $\iota_C$ are proper functions.

**Theorem 2.27** (optimality for constrained convex problems). *For a convex function $f : \mathbb{R}^n \to \mathbb{R}$ and a nonempty convex set $C \subset \mathbb{R}^n$, one has*

$$\begin{aligned} x^\star \in \operatorname{argmin}_C f \quad &\Longleftrightarrow \quad 0 \in \partial f(x^\star) + N_C(x^\star) \\ &\Longleftrightarrow \quad \textit{there's } v \in \partial f(x^\star) \textit{ such that } -v \in N_C(x^\star). \end{aligned}$$

**Proof.** By the Moreau-Rockafellar sum rule 2.26, $\partial(f + \iota_C)(x) = \partial f(x) + N_C(x)$ when $x \in C$ as discussed above. The conclusion then follows by the optimality condition 2.19. □

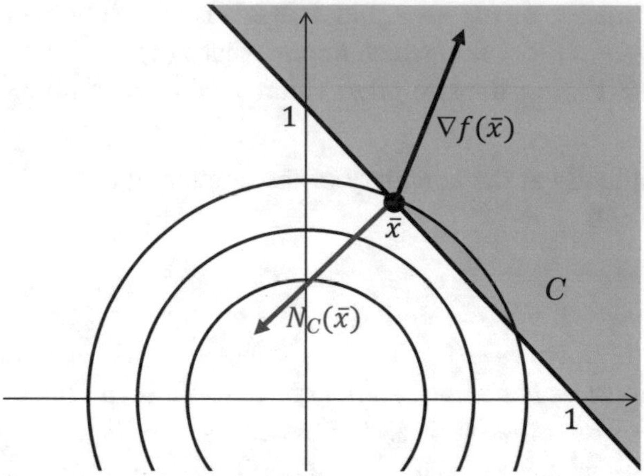


Fig. 2.14: Illustration of normal cone and gradient in 2.28.

**Exercise 2.28** (optimality for convex functions). For $f(x) = x_1^2 + x_2^2$ and $C = \{x \in \mathbb{R}^2 \mid x_1 + x_2 \geq 1\}$, (a) write a formula for the normal cones to $C$ and (b) use 2.27 to determine $\operatorname{argmin}_C f$.

**Guide.** Figure 2.14 indicates the nature of the normal cone to $C$ at a point on the boundary of $C$. □

**Example 2.29** (optimality in lasso regression). Solutions of the lasso regression problem in 2.5 can be fully characterized by $0 \in \partial f(c)$, where $f(c) = \|Dc - y\|_2^2 + \theta\|c\|_1$ and $\theta \in (0, \infty)$. An expression for the subgradients in this case follows by 2.21 and the Moreau-Rockafellar sum rule 2.26.

**Detail.** The smooth function $c \mapsto \|Dc - y\|_2^2$ has gradient $c \mapsto 2D^\top(Dc - y)$. Slightly modifying the argument in 2.21, we can show that the subgradients of $c \mapsto \theta\|c\|_1$ are

$$C_0 \times C_1 \times \cdots \times C_n, \text{ where } C_j = \begin{cases} \{-\theta\} & \text{if } c_j < 0 \\ [-\theta, \theta] & \text{if } c_j = 0 \\ \{\theta\} & \text{otherwise.} \end{cases}$$

Hence, by optimality condition 2.27, $c^\star$ is a minimizer of $f$ if and only if

$$\forall j = 0, 1, \ldots, n: \quad 2(D^\top y - D^\top Dc^\star)_j \in \begin{cases} \{-\theta\} & \text{if } c_j^\star < 0 \\ [-\theta, \theta] & \text{if } c_j^\star = 0 \\ \{\theta\} & \text{otherwise,} \end{cases}$$

where $(u)_j$ is the $j$th component of a vector $u$. □

## 2.D Proximal Gradient Methods

An objective function can often be expressed as the sum of two functions, which we should exploit algorithmically. Let's consider the problem

$$\operatorname*{minimize}_{x\in\mathbb{R}^n} f(x) + g(x),$$

where $f : \mathbb{R}^n \to \mathbb{R}$ is smooth and convex and $g : \mathbb{R}^n \to \overline{\mathbb{R}}$ is proper and convex. Most prominently, we might have $g = \iota_C$ for some nonempty convex set $C \subset \mathbb{R}^n$ and this enables us to account for constraints. Lasso regression and other problems also fit the mold. From a current point $x^\nu$, the idea is to compute $x^\nu - \lambda\nabla f(x^\nu)$ as in the gradient descent method (§1.F), but then apply an adjustment to account for $g$ as indicated by the proximal point method (§1.H). Specifically, we recall from the discussion around (1.9) that the gradient descent method applied to $f$ involves computing

$$x^{\nu+1} \in \operatorname{argmin}_{x\in\mathbb{R}^n} f(x^\nu) + \langle\nabla f(x^\nu), x - x^\nu\rangle + \tfrac{1}{2\lambda}\|x - x^\nu\|_2^2$$

for some step size $\lambda \in (0,\infty)$. When passing from $f$ to $f + g$, it's then natural to simply add $g$ in this iterative scheme, i.e.,

$$x^{\nu+1} \in \operatorname{argmin}_{x\in\mathbb{R}^n} f(x^\nu) + \langle\nabla f(x^\nu), x - x^\nu\rangle + \tfrac{1}{2\lambda}\|x - x^\nu\|_2^2 + g(x),$$

where the function being minimized can be viewed as an approximation of $f + g$. Equivalently, this can be written as

$$x^{\nu+1} \in \operatorname{argmin}_{x\in\mathbb{R}^n} g(x) + \tfrac{1}{2\lambda}\big\|x - \big(x^\nu - \lambda\nabla f(x^\nu)\big)\big\|_2^2 = \operatorname{prx}_\lambda g\big(x^\nu - \lambda\nabla f(x^\nu)\big)$$

because

$$\big\|x - \big(x^\nu - \lambda\nabla f(x^\nu)\big)\big\|_2^2 = \|x - x^\nu\|_2^2 + 2\lambda\langle\nabla f(x^\nu), x - x^\nu\rangle + \lambda^2\big\|\nabla f(x^\nu)\big\|_2^2$$

and a constant term can be dropped without affecting the set of minimizers. This leads to a versatile algorithm.

**Proximal Gradient Method.**

Data. $x^0 \in \mathbb{R}^n$ and $\lambda \in (0,\infty)$.
Step 0. Set $\nu = 0$.
Step 1. Compute

$$x^{\nu+1} \in \operatorname{prx}_\lambda g\big(x^\nu - \lambda\nabla f(x^\nu)\big).$$

Step 2. Replace $\nu$ by $\nu + 1$ and go to Step 1.

In view of 1.24 and 1.18, the functions

$$x \mapsto \tfrac{1}{2\lambda}\|x - \bar{x}\|_2^2 \quad\text{and}\quad x \mapsto g(x) + \tfrac{1}{2\lambda}\|x - \bar{x}\|_2^2$$

are strictly convex for any $\bar{x} \in \mathbb{R}^n$. Thus, the minimization carried out in Step 1 has at most one minimizer; cf. 1.16. There's such a minimizer when, in addition, epi $g$ is closed [105, Theorem 2.26(a)]. (We treat existence of minimizers systematically in §4.A.) Under these conditions, Step 1 is well defined, but there could still be challenges associated with computing $x^{\nu+1}$ in practice.

The proximal gradient method is sometimes called the *forward-backward algorithm* with the calculation of $x^\nu - \lambda\nabla f(x^\nu)$ representing the "forward step," aiming at reducing the value of $f$, and the computation of the proximal point $x^{\nu+1}$ as the "backward step."

Before examining the convergence properties, let's consider some specific instances. If $f(x) = 0$ for all $x$, then the algorithm reduces to the proximal point method. If $g(x) = 0$ for all $x$, then

$$\operatorname{prx}_\lambda g\big(x^\nu - \lambda\nabla f(x^\nu)\big) = \operatorname{argmin}_{x\in\mathbb{R}^n} \tfrac{1}{2\lambda}\big\|x - \big(x^\nu - \lambda\nabla f(x^\nu)\big)\big\|_2^2 = \big\{x^\nu - \lambda\nabla f(x^\nu)\big\}$$

by the optimality condition 2.19 and we return to the gradient descent method. When $g = \iota_C$ for some nonempty, closed and convex set $C \subset \mathbb{R}^n$, then Step 1 amounts to a projection of $x^\nu - \lambda\nabla f(x^\nu)$ on $C$.

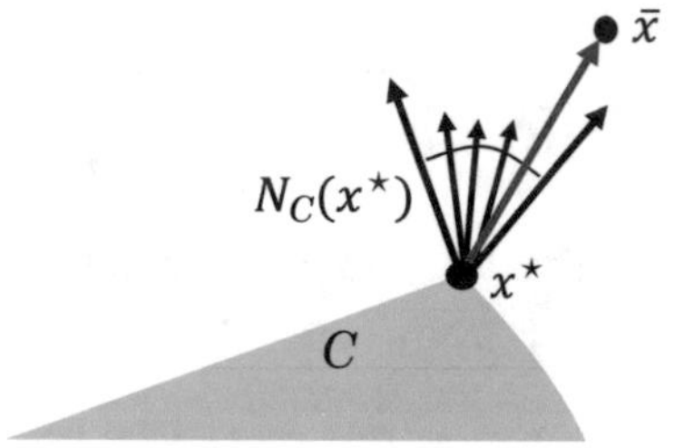


Fig. 2.15: For $x^\star$ to be a projection of $\bar{x}$ on $C$, the vector $\bar{x} - x^\star$ must be in the normal cone $N_C(x^\star)$.

**Example 2.30** (projection). Given a nonempty closed set $C \subset \mathbb{R}^n$ and a point $\bar{x} \in \mathbb{R}^n$, the closest points in $C$ from $\bar{x}$ is the *projection* of $\bar{x}$ on $C$ and is denoted by $\operatorname{prj}_C(\bar{x})$, i.e.,

$$\operatorname{prj}_C(\bar{x}) = \operatorname{argmin}_{x\in C} \tfrac{1}{2}\|x - \bar{x}\|_2^2.$$

For any $\lambda \in (0, \infty)$, $\operatorname{prj}_C(\bar{x}) = \operatorname{prx}_\lambda \iota_C(\bar{x})$, which is nonempty, and reduces to a single point when $C$ is convex.

**Detail.** The fact that we minimize $\frac{1}{2}\|x - \bar{x}\|_2^2$ instead of $\|x - \bar{x}\|_2$ is immaterial as the minimizers remain the same. The presence of $\lambda$ in the definition of $\operatorname{prx}_\lambda \iota_C(\bar{x})$ is likewise insignificant. Thus,

$$\operatorname{prj}_C(\bar{x}) = \operatorname{prx}_\lambda \iota_C(\bar{x}).$$

The objective function $x \mapsto \frac{1}{2}\|x - \bar{x}\|_2^2$ in the projection problem is quadratic and strictly convex; see 1.24. Since it tends to $\infty$ as $\|x\|_2 \to \infty$, $\operatorname{prj}_C(\bar{x})$ is nonempty; a formal argument for this fact leverages 4.9. When $C$ is convex, it follows from the discussion after the algorithm that $\text{prj}_C(\bar{x})$ is a unique point $x^\star$ because epi $\iota_C$ is closed. Moreover, by optimality condition 2.27, $x^\star$ then satisfies

$$\bar{x} - x^\star \in N_C(x^\star),$$

which is illustrated in Figure 2.15. In view of the definition of normal cones 2.23, this inclusion is equivalent to having

$$\langle \bar{x} - x^\star, x - x^\star \rangle \leq 0 \quad \forall x \in C,$$

i.e., the angle between $\bar{x} - x^\star$ and $x - x^\star$ is between 90 and 270 degrees. □

As seen in this example, the proximal gradient method applied to $g = \iota_C$ leads to a fundamental algorithm for constrained optimization that combines the idea of the gradient descent method with a projection step.

**Projected Gradient Method.**

Data. $x^0 \in \mathbb{R}^n$ and $\lambda \in (0, \infty)$.
Step 0. Set $\nu = 0$.
Step 1. Compute

$$x^{\nu+1} \in \text{prj}_C\big(x^\nu - \lambda \nabla f(x^\nu)\big).$$

Step 2. Replace $\nu$ by $\nu + 1$ and go to Step 1.

A projection can be relatively expensive to compute as it involves solving a constrained minimization problem. Consequently, the projected gradient method is typically only viable if this can be achieved expeditiously due to the special structure of $C$.

**Example 2.31** (projection on a box). For a set

$$C = \{x \in \mathbb{R}^n \mid \alpha_j \leq x_j \leq \beta_j,\ j = 1, \dots, n\},$$

with $-\infty \leq \alpha_j \leq \beta_j \leq \infty$, and a point $\bar{x} \in \mathbb{R}^n$, one has $x^\star \in \text{prj}_C(\bar{x})$ if and only if $\bar{x} - x^\star \in N_C(x^\star)$ by 2.30, which reduces to the componentwise checks

$$\bar{x}_j - x_j^\star \in N_{C_j}(x_j^\star), \quad j = 1, \dots, n,$$

where $C_j = \{x_j \in \mathbb{R} \mid \alpha_j \leq x_j \leq \beta_j\}$ and

$$N_{C_j}(x_j^\star) = \begin{cases} (-\infty, \infty) & \text{if } \alpha_j = x_j^\star = \beta_j \\ \{0\} & \text{if } \alpha_j < x_j^\star < \beta_j \\ [0, \infty) & \text{if } \alpha_j < x_j^\star = \beta_j \\ (-\infty, 0] & \text{if } \alpha_j = x_j^\star < \beta_j. \end{cases}$$

The decoupling into one-dimensional calculations keeps the computational expense low even for large $n$.

**Detail.** Since $C = C_1 \times C_2 \times \cdots \times C_n$ and

$$\iota_C(x) = \sum_{j=1}^{n} \iota_{C_j}(x_j),$$

we can apply 2.20 to obtain

$$N_C(x) = \partial\iota_C(x) = N_{C_1}(x_1) \times \cdots \times N_{C_n}(x_n)$$

at any point $x \in C$. The expressions for $N_{C_j}(x_j)$ follow directly from 2.23. □

**Convergence 2.32** (proximal gradient method). *Suppose that $f : \mathbb{R}^n \to \mathbb{R}$ is convex and smooth with gradients satisfying the Lipschitz condition: there's $\kappa \in (0, \infty)$ such that*

$$\left\|\nabla f(x) - \nabla f(\bar{x})\right\|_2 \le \kappa\|x - \bar{x}\|_2 \quad \forall x, \bar{x} \in \mathbb{R}^n.$$

*Moreover, suppose that $g : \mathbb{R}^n \to \overline{\mathbb{R}}$ is proper and convex, with* epi $g$ *closed, and* argmin $f + g \neq \emptyset$. *If the proximal gradient method has constructed $\{x^\nu, \nu \in \mathbb{N}\}$ using $\lambda \in (0, 2/\kappa)$, then $x^\nu$ converges to some minimizer of $f + g$ and, under the specific choice $\lambda = 1/\kappa$, one has*

$$f(x^\nu) - \inf f \le \frac{\kappa}{2\nu}\|x^0 - x^\star\|_2^2$$

*for any $x^\star \in$* argmin $f + g$.

**Proof.** The claim about convergence is found in [8, Corollary 27.9], with the rate recorded by [10, Theorem 10.21]. □

In practice, an implementation of the proximal gradient method requires either an adaptive rule for computing the step size $\lambda$ similar to the Armijo step size of §1.F, which then makes $\lambda$ vary from iteration to iteration, or some heuristic rule that starts with a large $\lambda$ and reduces the value whenever the algorithm stops making progress.

The rate of convergence in 2.32 matches that of the gradient descent method but falls short of the rate $\nu^{-2}$ of the accelerated gradient method (§1.H). However, the ideas of acceleration remains valid and in fact leads to *accelerated proximal gradient methods* for minimizing $f + g$ with rate $\nu^{-2}$ under similar assumptions; see [10, Section 10.7].

Further improvement in the rate is achieved under an additional assumption. For $\sigma > 0$, we say that a function $h : \mathbb{R}^n \to \overline{\mathbb{R}}$ is *$\sigma$-strongly convex* if

$$x \mapsto h(x) - \tfrac{1}{2}\sigma\|x\|_2^2$$

is convex. For example, the quadratic function with $h(x) = \frac{1}{2}\langle x, Qx\rangle$ and a symmetric positive definite matrix $Q$ is $\sigma_1$-strongly convex, where $\sigma_1$ is the smallest eigenvalue of $Q$.

**Convergence 2.33** (linear rate in proximal gradient method). *Suppose that $f : \mathbb{R}^n \to \mathbb{R}$ is $\sigma$-strongly convex and smooth with $\sigma > 0$ and gradients satisfying the Lipschitz condition: there's $\kappa \in (\sigma, \infty)$ such that*

$$\|\nabla f(x) - \nabla f(\bar{x})\|_2 \le \kappa \|x - \bar{x}\|_2 \quad \forall x, \bar{x} \in \mathbb{R}^n.$$

*Moreover, suppose that* $g : \mathbb{R}^n \to \overline{\mathbb{R}}$ *is proper and convex, with* epi $g$ *closed. If the proximal gradient method has constructed* $\{x^\nu, \nu \in \mathbb{N}\}$ *using* $\lambda = 1/\kappa$*, then*

$$f(x^\nu) - \inf f \le \tfrac{1}{2}\kappa\big(1 - \sigma/\kappa\big)^\nu \|x^0 - x^\star\|_2^2$$

*for any* $x^\star \in \operatorname{argmin} f + g$.

**Proof.** This result can be found in [10, Theorem 10.29]. □

The rate of convergence is *linear*, which is indeed much faster than the sublinear rate $1/\nu$ of 2.32. For example, if $1 - \sigma/\kappa = 0.8$, then $(1 - \sigma/\kappa)^\nu = 0.8^\nu = 2 \cdot 10^{-10}$ after $\nu = 100$ iterations. In contrast, $1/\nu = 1 \cdot 10^{-2}$.

The discussion above views the proximal gradient method as an extension of the gradient descent method; a move is made in the direction of $-\nabla f(x^\nu)$ with adjustments to account for $g$. However, another perspective related to the direct solution of the optimality condition 2.19 presents itself too.

**Proximal Gradient Method as Fixed-Point Iterations.** For $F : \mathbb{R}^n \to \mathbb{R}^n$, finding $x$ such that $x = F(x)$ is a *fixed-point problem* with a solution being referred to as a *fixed point*. The *fixed-point algorithm* simply iterates $x^{\nu+1} = F(x^\nu)$ and applies in a variety of contexts. We can interpret the proximal gradient method as an instance of the fixed-point algorithm applied to an optimality condition of the problem at hand. By 2.19 and the Moreau-Rockafellar sum rule 2.26,

$$x^\star \in \operatorname{argmin} f + g \iff 0 \in \nabla f(x^\star) + \partial g(x^\star)$$

for smooth convex $f : \mathbb{R}^n \to \mathbb{R}$ and proper convex $g : \mathbb{R}^n \to \overline{\mathbb{R}}$. Given $\lambda \in (0, \infty)$, the condition on the right-hand side can equivalently be stated as

$$0 \in \partial g(x^\star) + \tfrac{1}{\lambda}\big(x^\star - x^\star + \lambda \nabla f(x^\star)\big).$$

By 2.19, this is nothing but the optimality condition for the problem

$$\underset{x \in \mathbb{R}^n}{\text{minimize}}\; g(x) + \tfrac{1}{2\lambda}\big\|x - x^\star + \lambda \nabla f(x^\star)\big\|_2^2.$$

Hence, equivalently, we've

$$x^\star \in \operatorname{argmin}_{x \in \mathbb{R}^n} g(x) + \tfrac{1}{2\lambda}\big\|x - \big(x^\star - \lambda \nabla f(x^\star)\big)\big\|_2^2 = \operatorname{prx}_\lambda g\big(x^\star - \lambda \nabla f(x^\star)\big).$$

In summary, $x^\star$ minimizes $f + g$ if and only if $x^\star$ is a fixed point of the mapping

$$x \mapsto \operatorname{prx}_\lambda g\big(x - \lambda \nabla f(x)\big).$$

The proximal gradient method is then simply the fixed-point algorithm applied to this mapping.

The proximal gradient method needs an efficient way of computing $x^{\nu+1}$ in Step 1 to be viable. In addition to 2.31, we give an example from data analytics; the proximity-operator repository (`proximity-operator.net`) furnishes numerous other cases.

**Example 2.34** (proximal gradient method for lasso regression). When applied to the regression problem in 2.5 (see also 2.29), the proximal gradient method is implementable and scalable as Step 1 simplifies. We consider the slightly more general problem,

$$\underset{x\in\mathbb{R}^n}{\text{minimize}}\ \|Ax - b\|_2^2 + \theta\|x\|_1,$$

where $A$ is an $m\times n$-matrix, $b \in \mathbb{R}^m$ and $\theta \in (0,\infty)$. In this instance, Step 1 of the proximal gradient method amounts to finding $x^{\nu+1} = (x_1^{\nu+1},\dots,x_n^{\nu+1})$, where

$$\forall j = 1,\dots,n: \quad x_j^{\nu+1} = \begin{cases} \lambda\theta(w_j^\nu - 1) & \text{if } w_j^\nu > 1 \\ \lambda\theta(w_j^\nu + 1) & \text{if } w_j^\nu < -1 \\ 0 & \text{otherwise} \end{cases}$$

and

$$w^\nu = (w_1^\nu,\dots,w_n^\nu) = \frac{1}{\lambda\theta}(I - 2\lambda A^\top A)x^\nu + \frac{2}{\theta}A^\top b.$$

Moreover, the Lipschitz modulus $\kappa = \|2A^\top A\|_F$.

**Detail.** After dividing through by $\theta$, Step 1 of the proximal gradient method solves

$$\underset{x\in\mathbb{R}^n}{\text{minimize}}\ \|x\|_1 + \tfrac{1}{2\lambda\theta}\big\|x - x^\nu + 2\lambda(A^\top Ax^\nu - A^\top b)\big\|_2^2.$$

By 2.19 and the Moreau-Rockafellar sum rule 2.26, the optimality condition for this subproblem becomes

$$0 = v + \tfrac{1}{\lambda\theta}x - w^\nu,$$

where $v$ is a subgradient of the $\ell^1$-norm at $x$. This can be checked componentwise:

$$\forall j = 1,\dots,n: \quad 0 = v_j + \tfrac{1}{\lambda\theta}x_j - w_j^\nu.$$

Using the subgradient expression for the $\ell^1$-norm in 2.21, we obtain the stated procedure.

The smooth portion of the problem is $f(x) = \|Ax-b\|_2^2$, with $\nabla f(x) = 2(A^\top Ax - A^\top b)$ as already utilized. Thus,

$$\big\|\nabla f(x) - \nabla f(\bar{x})\big\|_2 = \big\|2A^\top A(x-\bar{x})\big\|_2 \le \|2A^\top A\|_F\|x-\bar{x}\|_2,$$

where $\|B\|_F$ denotes the Frobenius norm of a matrix $B$; cf. §1.G. □

**Exercise 2.35** (data smoothing). With data from 1.3, minimize $\|Dc - y\|_2^2 + \theta\|c\|_1$ to determine polynomials of degree $n = 5, 10, 20$ and 30 using the proximal gradient method as laid out in 2.34. In each case, show the effect of varying $\theta$ from zero and up to a large number. Graph the resulting polynomials.

**Guide.** A large $\theta$ penalizes heavily any nonzero coefficient, which results in sparse polynomials with few nonzero coefficients. □

## 2.E Linear Constraints

Polyhedral feasible sets allow us to compute projections, normal cones and other quantities of interest relatively easily. Since they're specified by linear equalities and inequalities, properties and procedures from linear algebra become central.

Suppose that a feasible set is described by the system of equations

$$Ax = b,$$

where $A$ is an $m \times n$-matrix. Is an optimization problem involving such a set feasible? A fundamental result of linear algebra is that there's a solution to $Ax = b$ if and only if $\operatorname{rank}[A\ b] = \operatorname{rank} A$, where $[A\ b]$ is the $m \times (n+1)$-matrix obtained by augmenting $A$ with the additional column $b$. Since the rank of a matrix can be computed by standard techniques, this check is often easily carried out. If a solution exists, then the set of *all* solutions is

$$\{x \in \mathbb{R}^n \mid x = \bar{x} + y,\ y \in \operatorname{null} A\},$$

where $\bar{x}$ is any solution satisfying $A\bar{x} = b$ and $\operatorname{null} A = \{x \in \mathbb{R}^n \mid Ax = 0\}$. Thus, a solution is unique if and only if the nullspace of $A$ is simply $\{0\}$.

**Example 2.36** (polynomial interpolation). The problem of *interpolating* a given set of points $\{(x_i, y_i) \in \mathbb{R}^2, i = 1, 2, \dots, m\}$ by a polynomial of degree $n$ amounts to finding a solution to a system of linear equations. This problem is generally different from that of finding a curve that's as close as possible to the points; cf. §1.B.

**Detail.** A polynomial of degree $n$ takes the form

$$q(x) = c_n x^n + \cdots + c_2 x^2 + c_1 x + c_0,$$

where $c = (c_0, c_1, \dots, c_n) \in \mathbb{R}^{1+n}$ are coefficients to be determined such that

$$q(x_i) = y_i, \quad i = 1, \dots, m.$$

Thus, we need to solve the equations

$$\begin{aligned}
c_0 + c_1x_1 + c_2x_1^2 + \cdots + c_nx_1^n &= y_1\\
c_0 + c_1x_2 + c_2x_2^2 + \cdots + c_nx_2^n &= y_2\\
\vdots \qquad\qquad \ddots \qquad\qquad & \vdots\\
c_0 + c_1x_m + c_2x_m^2 + \cdots + c_nx_m^n &= y_m,
\end{aligned}$$

or $Dc = y$ in the notation of §1.B. In particular, interpolation of the points $(0,1), (1,0), (2,1)$ by a polynomial of degree 2 requires us to solve

$$\begin{bmatrix}1 & 0 & 0\\ 1 & 1 & 1\\ 1 & 2 & 4\end{bmatrix}\begin{bmatrix}c_0\\ c_1\\ c_2\end{bmatrix} = \begin{bmatrix}1\\ 0\\ 1\end{bmatrix}.$$

The rank of the matrix is 3 in this case and the rank remains 3 after augmenting the matrix with the right-hand side vector. Thus, a solution exists and, in fact, it's unique because $\operatorname{null} D = \{0\}$. If the degree $n = 1$, then there would be no solution and no interpolating polynomial. □

**Example 2.37** (least-squares problems). When $A$ is an $m \times n$-matrix that isn't necessarily invertible, finding a solution to $Ax = b$ or verify that none exists can be achieved by solving the *least-squares problem*

$$\underset{x\in\mathbb{R}^n}{\text{minimize}}\ \|Ax - b\|_2^2.$$

The optimality condition 2.19 allows us to reduce this problem to one of linear algebra.

**Detail.** If the minimum value of the problem turns out to be zero, then any minimizer furnishes a solution to $Ax = b$. If the minimum value is positive, then there's no solution to the system of equations by the property of norms; cf. 1.15. In view of 1.24, the objective function is convex so the optimality condition 2.19, which reduces to $A^\top Ax^\star = A^\top b$, characterizes a minimizer. This new system of equations always has a solution. When $A$ has rank $n$, then $A^\top A$ is invertible and Cholesky decomposition or the conjugate gradient method (§1.H) can be applied. Regardless of the rank of $A$, $A^\top A$ is positive semidefinite so the proximal point method via iterative refinement can be brought in; see 1.38. □

**Exercise 2.38** (least-squares problems). Let $A$ be an $m \times n$-matrix and $b \in \mathbb{R}^m$. (a) An $\ell^2$-regularized least-squares problem, which encourages solutions that have small Euclidean norms, is of the form

$$\underset{x\in\mathbb{R}^n}{\text{minimize}}\ \|Ax - b\|_2^2 + \theta\|x\|_2^2,$$

for some $\theta \in (0, \infty)$. Show that the problem can be reformulated as a least-squares problem from 2.37. (b) A weighted least-squares problem is of the form

$$\underset{x\in\mathbb{R}^n}{\text{minimize}}\ \big\|W(Ax - b)\big\|_2^2,$$

where $W$ is a diagonal $m \times m$-matrix[3] that potentially weighs the components of the vector $Ax - b$ differently. Show that the problem can be reformulated as a least-squares problem; see 2.37. (c) On the data from 1.3, numerically demonstrate the effect of changing $\theta$ and $W$ in the $\ell^2$-regularized and weighted least-squares problems.

When equalities are supplemented by inequalities, linear algebra doesn't fully suffice to determine whether there's a solution. However, the issue can be settled by solving a linear optimization problem that, in fact, generates a solution if there's one. Specifically, consider the polyhedral set

$$C = \{x \in \mathbb{R}^n \mid Ax = b, Dx \le d\},$$

where $A$ and $D$ are $m \times n$ and $q \times n$ matrices, respectively. We would like to find $x \in C$ or verify that none exists. We achieve this by formulating the model

$$\begin{aligned}\underset{(x,u,v,w)\in\mathbb{R}^{n+2m+q}}{\text{minimize}} \quad & \sum_{i=1}^{m}(u_i + v_i) + \sum_{i=1}^{q} w_i \\ \text{subject to} \quad & Ax = b + u - v \\ & Dx \le d + w \\ & u, v, w \ge 0,\end{aligned}$$

which is a linear optimization problem. This problem has a nonempty feasible set; $x = 0$, $u_i = \max\{0, -b_i\}$, $v_i = \max\{0, b_i\}$ for $i = 1, \dots, m$ and $w_i = \max\{0, -d_i\}$ for $i = 1, \dots, q$ furnish a feasible point. If the minimum value is zero, then the $x$-component of a minimizer is a point in $C$. Otherwise, $C$ is empty. As we discuss in §2.F and §2.G, linear optimization problems are easily solved unless exceptionally large.

Application of the optimality condition 2.27 for a function $f = f_0 + \iota_C$ requires expressions for the normal cones to $C$. Sometimes these are immediately available from linear algebra.

**Range and Nullspace.** For an $m \times n$-matrix $A$,

$$\text{rge } A = \{Ax \mid x \in \mathbb{R}^n\} \text{ is the } \textit{range} \text{ of } A.$$

The *fundamental theorem of linear algebra* states that for $z \in \text{null } A$ and $x \in \text{rge } A^\top$, $\langle x, z\rangle = 0$, i.e., the vectors are *orthogonal*. This holds because $x \in \text{rge } A^\top$ implies that there's $y \in \mathbb{R}^m$ such that $x = A^\top y$. Thus, $\langle x, z\rangle = y^\top Az = 0$. In addition, the theorem states that the only vectors in $\mathbb{R}^n$ that are orthogonal to all $z \in \text{null } A$ are those in $\text{rge } A^\top$.

**Proposition 2.39** (normal cone to affine set). *For an $m \times n$-matrix $A$ and $b \in \mathbb{R}^m$, the set*

$$C = \{x \in \mathbb{R}^n \mid Ax = b\}$$

[3] An $n \times n$ matrix is *diagonal* if all its off-diagonal elements are zero.

*has its normal cone at $x \in C$ given by*

$$N_C(x) = \{A^\top y \mid y \in \mathbb{R}^m\}.$$

**Proof.** Let $\bar{x} \in C$. By definition, $v \in N_C(\bar{x})$ if and only if $\langle v, x - \bar{x}\rangle \le 0$ for all $x \in C$. Since $C = \bar{x} + \text{null}\, A$, this inequality is equivalent to having $\langle v, z\rangle \le 0$ for all $z \in \text{null}\, A$. If $z \in \text{null}\, A$, we also must have $-z \in \text{null}\, A$. Thus, $\langle v, z\rangle \le 0$ for all $z \in \text{null}\, A$ actually amounts to having $\langle v, z\rangle = 0$ for all $z \in \text{null}\, A$. As discussed in the paragraph above, the vectors $v$ that are orthogonal to null $A$ indeed make up rge $A^\top$. □

Figure 2.16 illustrates the situation in the proposition for the special case with a single linear equality, i.e., $C = \{x \mid \langle a, x\rangle = b\}$. Then, the normal cone comprises the vectors $ya$ for any scalar $y$.

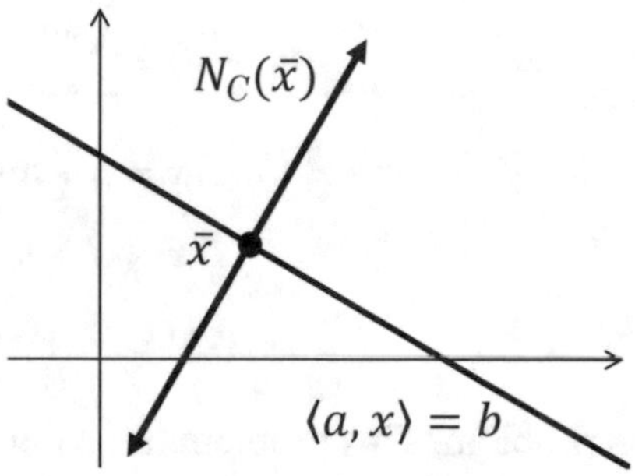


Fig. 2.16: Normal cone to an affine set.

**Example 2.40** (minimum norm problems). When there are multiple solutions to $Ax = b$, we may want to identify the "best" one in some sense. For some symmetric positive semidefinite $n \times n$-matrix $Q$, this might lead to the quadratic optimization problem

$$\underset{x \in \mathbb{R}^n}{\text{minimize}}\ \tfrac{1}{2}\langle x, Qx\rangle \ \text{ subject to } \ Ax = b.$$

If $Q = I$, the identity matrix, then the objective function takes the form $\frac{1}{2}\|x\|_2^2$ and the problem identifies the smallest solution. This resembles the $\ell^2$-regularized least-squares problem of 2.38, but balances the concerns about smallness and feasibility differently. The optimality condition 2.27 reduces in this case to a system of linear equations involving an auxiliary vector that needs to be determined together with $x$.

**Detail.** In view of 2.39, the optimality condition 2.27 ensures that $x$ is a minimizer if and only if $0 = Qx + A^\top y$ for some $y \in \mathbb{R}^m$. This equation together with the feasibility condition $Ax = b$ lead to the linear system of equations

$$\begin{bmatrix} Q & A^\top \\ A & 0 \end{bmatrix} \begin{bmatrix} x \\ y \end{bmatrix} = \begin{bmatrix} 0 \\ b \end{bmatrix}.$$

If $Q = I$ and $\operatorname{rank} A = m$, then we obtain the solution $y = -(AA^\top)^{-1}b$ and $x = A^\top(AA^\top)^{-1}b$. □

The intermediate vector $y$ plays a seemingly minor role in the example above, but can trace its history back to the eminent mathematician and astronomer J.-L. Lagrange (1736-1813). In fact, the components of such vectors are called *Lagrange multipliers* or simply multipliers. Lagrange used such vectors to develop optimality conditions for the more general problem of minimizing $f_0(x)$ subject to $f_1(x) = 0, \dots, f_m(x) = 0$, with all the functions being smooth. His fundamental insight was that optimization subject to equality constraints reduces (at least in the convex case) to solving a system of equations, linear or nonlinear, that's analogous to the Fermat rule of Chap. 1 for unconstrained optimization. However, the system of equations now involves one additional variable per constraint, the multipliers, that also need to be determined. It will quickly become clear that multipliers are central to the development of broad classes of optimality conditions. Still, we'll have to wait to Chap. 5 before we see their full significance. There, multipliers emerge as equal partners with the decision variables and, in fact, become solutions to related, dual optimization problems.

We might be hopeful that one can extend the successful treatment of linear equality constraints in 2.40 to also cover inequality constraints. However, this doesn't come without complications.

**Example 2.41** (optimization over a box). Consider the problem of minimizing a smooth convex function $f$ over a box

$$C = \{x \in \mathbb{R}^n \mid \alpha_j \le x_j \le \beta_j,\ j = 1, \dots, n\},$$

where $-\infty \le \alpha_j \le \beta_j \le \infty$. In view of the expression for normal cones to a box (see 2.31), the optimality condition 2.27 specializes to

$$\forall j = 1, \dots, n: \qquad -\frac{\partial f(x)}{\partial x_j} \in \begin{cases} (-\infty, \infty) & \text{if } \alpha_j = x_j = \beta_j \\ \{0\} & \text{if } \alpha_j < x_j < \beta_j \\ [0, \infty) & \text{if } \alpha_j < x_j = \beta_j \\ (-\infty, 0] & \text{if } \alpha_j = x_j < \beta_j. \end{cases}$$

In contrast to 2.40, these conditions aren't equations. If $f$ is quadratic, then the partial derivatives become linear but the issue lingers as the right-hand side remains conditional on $x$.

**Detail.** The difficulty stems from the fact that for each $j$ we usually don't know ahead of time which of the four cases in the normal cone formula apply; we need to know $x_j$ to make the right choice. Let's ignore the trivial case with $\alpha_j = \beta_j$, which we can identify from the problem data, and suppose that $\alpha_j < \beta_j$, with $\alpha_j, \beta_j \in \mathbb{R}$, for all $j = 1, \dots, n$. Then, we're down to three cases for each $j$, a total of $3^n$ possibilities. One such possibility amounts to an assumption about which case each component of a solution $x$ satisfies. Suppose that we *assume* that a solution $x$ satisfies $\alpha_j < x_j < \beta_j$ for $j \in J_2$, $x_j = \beta_j$ for

$j \in J_3$ and $x_j = \alpha_j$ for $J_4$, with $J_2 \cup J_3 \cup J_4 = \{1, 2, \ldots, n\}$. Then, we only need to set $x_j = \beta_j$ for $j \in J_3$ and $x_j = \alpha_j$ for $j \in J_4$ and solve the equations

$$-\frac{\partial f(x)}{\partial x_j} = 0 \quad \forall j \in J_2$$

to determine $\{x_j, j \in J_2\}$, which all together produce a vector $x \in \mathbb{R}^n$. Finally, we verify that $\alpha_j < x_j < \beta_j$ for $j \in J_2$ and check that

$$\frac{\partial f(x)}{\partial x_j} \leq 0 \quad \forall j \in J_3 \quad \text{and} \quad \frac{\partial f(x)}{\partial x_j} \geq 0 \quad \forall j \in J_4.$$

If this goes well, the obtained $x$ satisfies the optimality condition. The issue is that most often we reach a contradiction: the assumption isn't compatible with the checks. Thus, in the worst case, one needs to cycle through $3^n$ possible assumptions and this quickly becomes intractable. □

**Exercise 2.42** (optimization over a box). In the setting of 2.41, let $f(x) = 2(x_1 - 1)^2 + 3(x_2 + 1)^2 + 4x_3^2 - x_1 x_2/2$ and $C = [1, 2] \times [0, \infty) \times [-1, 1]$. Determine $\operatorname{argmin}_C f$ using the optimality condition in 2.41.

## 2.F Karush-Kuhn-Tucker Condition

If a feasible set is a polyhedron, then the optimality condition 2.27 takes an explicit form that leads to computational procedures. The key challenge is to determine the normal cones to the feasible set.

**Example 2.43** (normal cone to halfspace). For $a \in \mathbb{R}^n$ and $\alpha \in \mathbb{R}$, let

$$C = \big\{x \in \mathbb{R}^n \,\big|\, \langle a, x \rangle \leq \alpha \big\}.$$

Then, for any $\bar{x} \in C$,

$$N_C(\bar{x}) = \big\{az \,\big|\, z \in [0, \infty) \text{ if } \langle a, \bar{x} \rangle = \alpha, \;\; z = 0 \text{ otherwise}\big\}.$$

In contrast to the situation for affine sets, only nonnegative scalars $z$ need to be considered as negative ones would have produced vectors pointing "into" $C$; cf. Figure 2.17

**Detail.** If $\langle a, \bar{x} \rangle < \alpha$, then there's $\delta > 0$ such that $\langle a, x \rangle \leq \alpha$ for all $x \in \mathbb{B}(\bar{x}, \delta)$. Thus, $\langle v, x - \bar{x} \rangle \leq 0$ for all $x \in C$ only if $v = 0$. This means that $N_C(\bar{x}) = \{0\}$; cf. 2.23. When $\langle a, \bar{x} \rangle = \alpha$ and $a \neq 0$, then a normal vector $v$ needs to be perpendicular to the hyperplane specified by that equation, i.e., of the form $az$ for $z \in \mathbb{R}$. Moreover, to satisfy $\langle v, x - \bar{x} \rangle \leq 0$ for all $x \in C$, $z$ can't be negative. If $a = 0$, then $\alpha \geq 0$ because otherwise $C$ would have been empty. Thus, $C = \mathbb{R}^n$ and the zero vector is the only normal vector. □

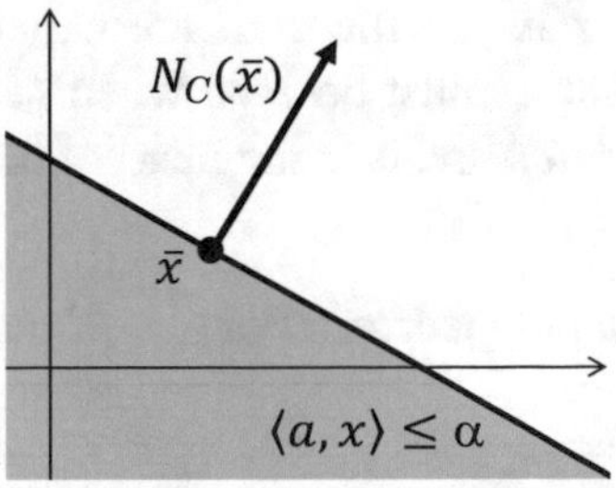


Fig. 2.17: Normal cone to a halfspace.

In the presence of multiple inequality constraints, possibly combined with equalities, we need to merge this example with 2.39 to obtain an expression for normal cones to polyhedral sets. It turns out that a normal vector to such an intersection of sets is simply the sum of normal vectors to the individual sets; we postpone the proof to §4.H.

**Proposition 2.44** (normal cone to polyhedral set). *Suppose that*

$$C = \{x \in \mathbb{R}^n \mid Ax = b,\ Dx \leq d\},$$

*where $A$ and $D$ are $m \times n$ and $q \times n$ matrices, respectively. For any $\bar{x} \in C$,*

$$N_C(\bar{x}) = \left\{A^\top y + D^\top z \,\middle|\, y \in \mathbb{R}^m;\ \ z_i \geq 0 \text{ if } i \in \mathbb{A}(\bar{x}),\ \ z_i = 0 \text{ otherwise}\right\}.$$

*Here, $\mathbb{A}(\bar{x}) = \{i \mid \langle D_i, \bar{x}\rangle = d_i\}$, $D_i$ is the ith row of $D$ and $d_i$ is the ith component of $d$.*

The set $\mathbb{A}(\bar{x})$ specifies the inequality constraints that are *active* at $\bar{x}$. Inequalities that fall outside this set are *inactive* and don't influence $N_C(\bar{x})$ because the corresponding $z_i = 0$. This is meaningful because inactive constraints at $\bar{x}$ have no impact on the local geometry of $C$ near $\bar{x}$.

We can now state the celebrated *Karush-Kuhn-Tucker* (KKT) condition for optimality of a smooth convex function under linear equality and inequality constraints; its validity is immediate from 2.44 and optimality condition 2.27.

**Theorem 2.45** (KKT condition for polyhedral sets). *Suppose that $f : \mathbb{R}^n \to \mathbb{R}$ is smooth and convex and*

$$C = \{x \in \mathbb{R}^n \mid Ax = b,\ Dx \leq d\},$$

*where $A$ and $D$ are $m \times n$ and $q \times n$ matrices, respectively. Then,*

$$\begin{aligned} x^\star \in \operatorname{argmin}_C f \iff\ & x^\star \in C \text{ and there are} \\ & y \in \mathbb{R}^m \text{ and } z_i \geq 0,\ i \in \mathbb{A}(x^\star),\ \ z_i = 0,\ i \notin \mathbb{A}(x^\star) \\ & \text{such that } -\nabla f(x^\star) = A^\top y + D^\top z. \end{aligned}$$

*Here, $\mathbb{A}(x^\star) = \{i \mid \langle D_i, x^\star\rangle = d_i\}$, $D_i$ is the ith row of $D$ and $d_i$ is the ith component of $d$.*

The components of $y$ and $z$ in the theorem are called *multipliers*; there's one per constraint. The requirement that $z_i$ must be zero when $\langle D_i, x^\star\rangle < d_i$ is referred to as a *complementary slackness condition* and is at the heart of the difficulty of solving the KKT condition.

**Example 2.46** (projection on a polyhedron). For $\bar{x} \in \mathbb{R}^n$ and a nonempty polyhedral set

$$C = \{x \in \mathbb{R}^n \mid Ax = b,\ Dx \le d\},$$

the projection $\mathrm{prj}_C(\bar{x})$ is a unique point (cf. 2.30) characterized by the KKT condition 2.45.

**Detail.** By 2.45, $x^\star \in \mathrm{prj}_C(\bar{x})$ if and only if $Ax^\star = b$, $Dx^\star \le d$ and

$$\bar{x} - x^\star = A^\top y + D^\top z, \ \text{ with } \ y \in \mathbb{R}^m \ \text{ and } \ z_i \ge 0,\ i \in \mathbb{A}(x^\star),\ \ z_i = 0,\ i \notin \mathbb{A}(x^\star).$$

In contrast to the situation in 2.40, this isn't a system of equations as it also involves inequalities and a complementary slackness condition. As in 2.41, we could *assume* the active constraints at a minimizer, say $\mathbb{A}_0 \subset \{1, \dots, q\}$, set $z_i = 0$ for $i \notin \mathbb{A}_0$, and solve

$$\begin{bmatrix} I & A^\top & \hat{D}^\top \\ A & 0 & 0 \\ \hat{D} & 0 & 0 \end{bmatrix} \begin{bmatrix} x \\ y \\ \hat{z} \end{bmatrix} = \begin{bmatrix} \bar{x} \\ b \\ \hat{d} \end{bmatrix},$$

where $\hat{D}$ consists of the rows $\{D_i, i \in \mathbb{A}_0\}$ of $D$ and $\hat{d}$ consists of the elements $\{d_i, i \in \mathbb{A}_0\}$ of $d$. If there's a solution $(x^\star, y^\star, \hat{z}^\star)$ of this system of equations with $\hat{z}^\star \ge 0$ and the set of active inequalities at $x^\star$ coincides with the assumed $\mathbb{A}_0$, then $x^\star$ satisfies the KKT condition and is a minimizer. However, if there's no such $(x^\star, y^\star, \hat{z}^\star)$, then the process needs to be repeated with another assumption about $\mathbb{A}_0$; there are $2^q$ possibilities as each inequality constraint can be either active or inactive. We find the (unique) minimizer in this manner, but at a substantial computational cost. This highlights the combinatorial aspect that arises in the case of inequality constraints.

Of course, if we simply want to assess whether a given $x^\star$ is the projection of $\bar{x}$ on $C$, then it's easy to determine the corresponding $\mathbb{A}(x^\star)$, the assumption about $\mathbb{A}_0$ can be made correctly and there's no need to cycle through a potentially exponentially large number of possibilities. □

The example points to the possibilities and challenges associated with using the KKT condition 2.45 to determine a minimizer of a convex quadratic optimization problem. The idea of cycling through the possible active constraints is the underlying principle behind the *simplex method* for solving linear optimization problems; see [69, Chapter 13]. Even though the method has to examine many possible sets of active constraints, with the number of possibilities being exponentially large in the number of constraints, it can be remarkably efficient in practice, in part due to convenient shortcuts available in the linear case. For general convex quadratic problems, the active-set approach can also lead to viable algorithms as long as the number of inequalities isn't too large; see [69, Section

16.4]. An approach that avoids enumerating through a large number of possible active sets, but rely on approximations instead, is laid out in the next section.

**Exercise 2.47** (KKT condition). For $C = \{x \in \mathbb{R}^2 \mid x_1+x_2 = 1, x_1 \geq 0, x_2 \geq 0\}$, determine and visualize the normal cone $N_C(x)$ at every point $x \in C$. Use the KKT condition to determine $\operatorname{argmin}_C f$ when $f(x) = (x_1 - 1)^2 + x_2$.

## 2.G Interior-Point Method

Let's consider the convex quadratic optimization problem

$$\underset{x \in \mathbb{R}^n}{\text{minimize}} \ \tfrac{1}{2}\langle x, Qx\rangle + \langle c, x\rangle \ \text{ subject to } \ Dx \leq d, \tag{2.7}$$

where $Q$ is a symmetric positive semidefinite $n \times n$-matrix, $D$ is a $q \times n$-matrix, with row vectors $D_i, i = 1, \ldots, q$, and $d = (d_1, \ldots, d_q)$. We omit equalities $Ax = b$, but this is just to simplify the exposition. They're easily incorporated in the following method, both theoretically and practically. Our strategy for obtaining a minimizer of the problem is to solve the corresponding KKT condition approximately using the Newton-Raphson algorithm of §1.G. Since $Q$ could very well be the zero matrix, the resulting method also solves linear optimization problems.

The KKT condition 2.45 for the problem becomes

$$Qx + D^\top z + c = 0, \quad d - Dx \geq 0, \quad z \geq 0, \quad \big(d_i - \langle D_i, x\rangle\big) z_i = 0, \ i = 1, \ldots, q,$$

where the complementary slackness condition $z_i = 0$ for inactive constraints is equivalently stated as $(d_i - \langle D_i, x\rangle)z_i = 0$. Since the Newton-Raphson algorithm treats equations and not inequalities, let's rewrite the feasibility condition as $-Dx - s + d = 0$ in terms of an additional vector $s = (s_1, \ldots, s_q)$ that necessarily needs to be nonnegative. With the exception of $z \geq 0$ and $s \geq 0$, these requirements can be written as

$$G(x, z, s) = \begin{bmatrix} Qx + D^\top z + c \\ -Dx - s + d \\ \operatorname{diag}(s)z - \sigma\mu e \end{bmatrix} = 0,$$

when the parameter $\sigma = 0$, where for any $a \in \mathbb{R}^q$, $\operatorname{diag}(a)$ is the $q\times q$-diagonal matrix with entries of $a$ on the diagonal. However, we permit $\sigma$ to be an adjustable parameter with value in $[0, 1]$, typically positive. This brings into play the term $\sigma\mu e$ with another parameter $\mu$, which is usually set to the current value of $\langle z, s\rangle/q$ and thereby measures how well the current point satisfies the complementary slackness condition. Here, $e = (1, \ldots, 1) \in \mathbb{R}^q$. We can then approach a minimizer of the convex quadratic optimization problem by approximately solving $G(x, z, s) = 0$ with successively smaller values of $\sigma\mu$.

An algorithm now emerges: From a current point $(x^\nu, z^\nu, s^\nu)$ that satisfies $z_i^\nu, s_i^\nu > 0$ for all $i$, carry out one iteration of the Newton-Raphson algorithm applied to $G(x, z, s) = 0$, but adjust the step size so that the next point $(x^{\nu+1}, z^{\nu+1}, s^{\nu+1})$ also has $z_i^{\nu+1}, s_i^{\nu+1} > 0$ for all $i$. The fact that these variables remain positive, and thus are always in the interior of the feasible set, motivates the name "interior-point method." The approach requires us to solve the system of equations

$$\nabla G(x^\nu, z^\nu, s^\nu)(\bar{x}, \bar{z}, \bar{s}) = -G(x^\nu, z^\nu, s^\nu)$$

in Step 1 of the Newton-Raphson algorithm, which amounts to

$$\begin{bmatrix} Q & D^\top & 0 \\ -D & 0 & -I \\ 0 & \operatorname{diag}(s^\nu) & \operatorname{diag}(z^\nu) \end{bmatrix} \begin{bmatrix} \bar{x} \\ \bar{z} \\ \bar{s} \end{bmatrix} = - \begin{bmatrix} Qx^\nu + D^\top z^\nu + c \\ -Dx^\nu - s^\nu + d \\ \operatorname{diag}(s^\nu)z^\nu - \sigma\mu e \end{bmatrix}. \tag{2.8}$$

The next point is then

$$(x^{\nu+1}, z^{\nu+1}, s^{\nu+1}) = (x^\nu, z^\nu, s^\nu) + \lambda^\nu(\bar{x}, \bar{z}, \bar{s}),$$

where $\lambda^\nu$ is some appropriately selected step size. The main computational cost of the algorithm comes from solving (2.8), which has $n + 2q$ unknowns, and some adjustments are needed to ensure efficiency such as eliminating the slack vector $s$. There's also the issue of adjusting $\sigma$; a value too close to zero might force small step sizes and slow convergence. A practical method that addresses these factors solves two systems of equations in each iteration. The idea is to improve the linear predictions inherent to the Newton-Raphson algorithm (§1.G) by carrying out a "corrector step."

**Interior-Point Method.**

Data. $(x^0, z^0, s^0) \in \mathbb{R}^{n+2q}$, with $z_i^0, s_i^0 > 0$ for $i = 1, \dots, q$.
Step 0. Set $\nu = 0$.
Step 1. Solve (2.8) with $\sigma = 0$ and obtain $(\bar{x}, \bar{z}, \bar{s})$.
Step 2. Compute the parameters

$$\begin{aligned} \mu &= \langle z^\nu, s^\nu \rangle / q \\ \hat{\lambda} &= \max\left\{\lambda \in (0, 1] \,\middle|\, (z^\nu, s^\nu) + \lambda(\bar{z}, \bar{s}) \geq 0\right\} \\ \hat{\mu} &= \left\langle z^\nu + \hat{\lambda}\bar{z}, s^\nu + \hat{\lambda}\bar{s}\right\rangle / q \\ \sigma &= (\hat{\mu}/\mu)^3. \end{aligned}$$

Step 3. Obtain $(\breve{x}, \breve{z}, \breve{s})$ by resolving (2.8), but now with the right-hand side replaced by

$$- \begin{bmatrix} Qx^\nu + D^\top z^\nu + c \\ -Dx^\nu - s^\nu + d \\ \operatorname{diag}(s^\nu)z^\nu + \operatorname{diag}(\bar{s})\bar{z} - \sigma\mu e \end{bmatrix}.$$

Step 4. Set $\tau^\nu \in (0, 1)$ and

$$\lambda^\nu = \min\Big\{\max\big\{\lambda \in (0,1] \,\big|\, z^\nu + \lambda\check{z} \ge (1-\tau^\nu)z^\nu\big\}$$
$$\max\big\{\lambda \in (0,1] \,\big|\, s^\nu + \lambda\check{s} \ge (1-\tau^\nu)s^\nu\big\}\Big\}.$$

Step 5. Set $(x^{\nu+1}, z^{\nu+1}, s^{\nu+1}) = (x^\nu, z^\nu, s^\nu) + \lambda^\nu(\check{x}, \check{z}, \check{s})$, replace $\nu$ by $\nu + 1$ and go to Step 1.

Step 2 computes $\hat{\mu}$ as a quantification of the violation of the requirement $z_i s_i = 0$ at a next tentative point as predicted by the linearization inherent to the Newton-Raphson algorithm. This furnishes through a heuristic rule (involving $\hat{\mu}/\mu$ raised to power 3) an adaptive value for the critical parameter $\sigma$. The second system of equations solved in Step 3 aims to correct for the fact that $z_i s_i = 0$ fails to hold at the tentative point through adjustments of the right-hand side. Since the two systems of equations only differ in their right-hand side, a factor method such as Cholesky decomposition is often productively employed. It's usually advantageous to let the parameter $\tau^\nu$ tend to 1 as $\nu$ grows. The interior-point method has strong practical performance even in the case of many inequality constraints. For proofs of convergence as well as other versions of the method; see §4.K and [69, Section 16.6].

If the quadratic optimization problem (2.7) also includes $m$ linear equality constraints, then the only substantive change is that the above systems of equations now include $m$ additional rows corresponding to these constraints and $m$ additional columns related to their multipliers.

Table 2.2: Data for 2.48, where $c = (c_1, c_2) \in \mathbb{R}^{14}$, $Q = \mathrm{diag}((q_1, q_2))$, $u = (u_1, u_2)$ and $D = [D^1\ D^2]$, which is a $5 \times 14$-matrix.

| | | | | | | | |
|---|---|---|---|---|---|---|---|
| $c_1$ | 1 | 2 | 1 | 4 | 1 | 3 | 0 |
| $c_2$ | 0 | 0 | 0 | 60 | 75 | 80 | 120 |
| $q_1$ | 1 | 2 | 1 | 0.25 | 1 | 0.5 | 120 |
| $q_2$ | 187.5 | 800/3 | 24000 | 0 | 0 | 0 | 0 |
| | 1 | 1.5 | 0.5 | 2 | 1 | 1 | 0 |
| | −0.29 | −0.4 | 0 | −0.11 | 0 | 0 | −1 |
| $D^1$ | −0.1 | −0.0975 | −0.315 | −0.51 | 0 | 0 | 0 |
| | 0 | 0 | 0 | −0.2 | −0.4875 | −0.1925 | 0 |
| | 0 | 0 | 0 | 0 | −0.3267 | −0.4833 | 0 |
| | 0 | 0 | 0 | 0 | 0 | 0 | 0 |
| | 0 | 0 | 0 | −1 | 0 | 0 | 0 |
| $D^2$ | −1 | 0 | 0 | 0 | −1 | 0 | 0 |
| | 0 | −1 | 0 | 0 | 0 | −1 | 0 |
| | 0 | 0 | 0 | 0 | 0 | 0 | −1 |
| $u_1$ | 4 | 20 | 4 | 10 | 3 | 2 | 0.5 |
| $u_2$ | 0.4 | 0.3 | 0.005 | $\infty$ | $\infty$ | $\infty$ | $\infty$ |

**Exercise 2.48** (computation with interior point). Implement the interior-point method. With $d = (24, 3.825, 0.9667, 3.1, 1.5)$ and the data of Table 2.2, reformulate the problem

$$\underset{x\in\mathbb{R}^{14}}{\text{minimize}}\ \tfrac{1}{2}\langle x, Qx\rangle + \langle c, x\rangle \ \text{ subject to } \ Dx \le d,\ 0 \le x \le u$$

such that it takes the form needed by the interior-point method. Solve this instance.

**Guide.** The reformulation requires an augmented constraint matrix of dimension $(5 + 14 + 14) \times 14$. A commercial implementation wouldn't carry out this reformulation as it increases the size of the systems of equations being solved. Instead, variable bounds are treated similarly to those on $z$ and $s$. □

## 2.H Support Vector Machines

The binary classification problem aims to predict whether an observation belongs to one category (labeled $-1$) or another (labeled 1) based on its characteristics and can be addressed using logistic regression; cf. 1.29. An alternative formulation of the problem leads to *support vector machines*.

Suppose that we've observations $\{x^i \in \mathbb{R}^n,\ i = 1,\dots,m\}$ and corresponding labels $\{y_i \in \{-1, 1\},\ i = 1,\dots,m\}$; see Figure 2.18 for a case with $n = 2$ and $m = 26$. A support vector machine aims to identify a function $f : \mathbb{R}^n \to \mathbb{R}$, a classifier, that "best" predicts the unknown label of a new observation $\bar{x} \in \mathbb{R}^n$. We adopt the following prediction rule:

$$\text{predict label 1 if } f(\bar{x}) > 0 \qquad\qquad \text{predict label } -1 \text{ if } f(\bar{x}) < 0$$

and leave it undetermined if $f(\bar{x}) = 0$. Let's consider affine functions with $f(x) = \langle a, x\rangle + \alpha$ and the problem then reduces to that of finding $(a, \alpha)$.

Figure 2.18 shows a line defined by $\langle a, x\rangle + \alpha = 0$ representing the boundary between the observations that we predict as 1 and those as $-1$. At an observation $\bar{x}$ to the left of the line, $\langle a, \bar{x}\rangle + \alpha < 0$ and we predict label $-1$. To the right, values are positive and the prediction is 1. The $a$ and $\alpha$ defining the line in Figure 2.18 seem reasonable; only three misclassifications are made across the dataset. How can we determine the "best" $a$ and $\alpha$?

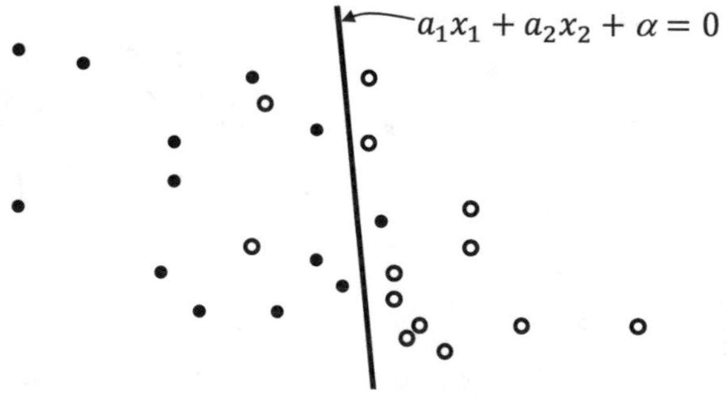


Fig. 2.18: Classification data with label $-1$ (dots) and label 1 (circles).

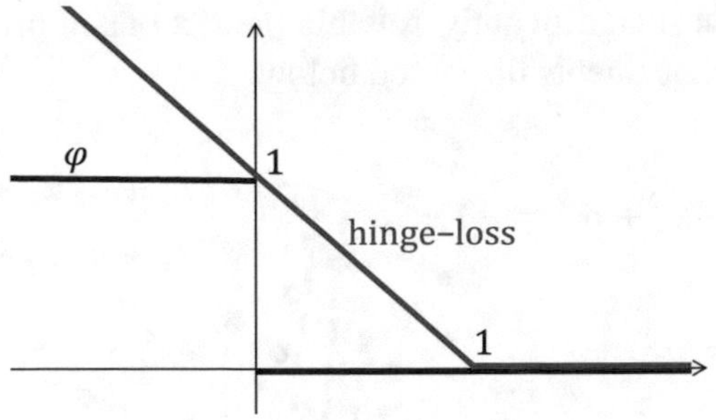


Fig. 2.19: The counting function $\varphi$ and its convex hinge-loss surrogate.

A goal might be to find $a$ and $\alpha$ that minimize the average number of misclassifications across the data as obtained by solving

$$\underset{(a,\alpha)\in\mathbb{R}^{n+1}}{\text{minimize}} \frac{1}{m}\sum_{i=1}^{m} \varphi\Big(y_i\big(\langle a, x^i\rangle + \alpha\big)\Big), \quad \text{where } \varphi(\gamma) = \begin{cases} 1 & \text{if } \gamma \le 0 \\ 0 & \text{otherwise.} \end{cases}$$

However, this is a nonconvex problem; the objective function isn't even continuous. The problem is therefore hard to solve at the scale needed in practice. As a surrogate, we consider the convex hinge-loss function $\gamma \mapsto \max\{0, 1-\gamma\}$ (see Figure 2.19) and this leads to the *hinge-loss problem*

$$\underset{(a,\alpha)\in\mathbb{R}^{n+1}}{\text{minimize}} \frac{1}{m}\sum_{i=1}^{m} \max\Big\{0, 1 - y_i\big(\langle a, x^i\rangle + \alpha\big)\Big\},$$

with a convex but nonsmooth objective function; cf. 1.13 and 1.18. Let

$$\mathbb{A}^+ = \{i \mid y_i = 1\} \quad \text{and} \quad \mathbb{A}^- = \{i \mid y_i = -1\}.$$

If the minimum value of the hinge-loss problem is zero with minimizer $(a^\star, \alpha^\star)$, then

$$\langle a^\star, x^i\rangle + \alpha^\star \ge 1 \;\; \forall i \in \mathbb{A}^+ \quad \text{and} \quad \langle a^\star, x^i\rangle + \alpha^\star \le -1 \;\; \forall i \in \mathbb{A}^-.$$

This means that

$$y_i\big(\langle a^\star, x^i\rangle + \alpha^\star\big) \ge 1, \quad i = 1, \dots, m$$

so the average number of misclassifications is zero. Thus, $(a^\star, \alpha^\star)$ is also a minimizer of the actual problem. This situation has a geometric interpretation: $(a^\star, \alpha^\star)$ defines a hyperplane that separates $\{x^i, i \in \mathbb{A}^+\}$ from $\{x^i, i \in \mathbb{A}^-\}$. There's even a buffer; no data point has $\langle a^\star, x^i\rangle + \alpha^\star \in (-1, 1)$. Figure 2.20 illustrates the situation, which is more clear-cut than in Figure 2.18.

In general, a minimizer of the hinge-loss problem may not minimize the actual one, but empirically we find that the hinge-loss problem often produces few misclassifications across a dataset and, most significantly, reliable predictions in practice, especially when combined with the enhancements discussed below.

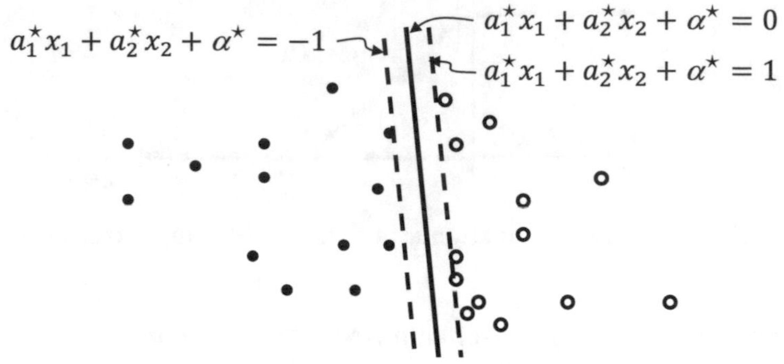


Fig. 2.20: Separation of data with label −1 (dots) and label 1 (circles) by a solid line, even leaving a buffer as indicated by the dashed lines.

**Example 2.49** (subgradients of hinge loss). For $x \in \mathbb{R}^n$, $y \in \mathbb{R}$ and $m \in \mathbb{N}$, let $h : \mathbb{R}^{n+1} \to \mathbb{R}$ be the function with

$$h(a, \alpha) = \tfrac{1}{m} \max\Big\{0, 1 - y\big(\langle a, x\rangle + \alpha\big)\Big\}.$$

Its subgradients at $(a, \alpha)$ are given by

$$\partial h(a, \alpha) = \begin{cases} \{0\} & \text{if } 1 - y\big(\langle a, x\rangle + \alpha\big) < 0 \\ \big\{(-yx/m, -y/m)\big\} & \text{if } 1 - y\big(\langle a, x\rangle + \alpha\big) > 0 \\ \big\{\beta(-yx/m, -y/m) \mid \beta \in [0, 1]\big\} & \text{otherwise.} \end{cases}$$

The objective function in the hinge-loss problem is the sum of such $h$ with $x = x^i$ and $y = y_i$. Thus, the subgradients of the objective function at $(a, \alpha)$ are simply the sum of subgradients from the corresponding $\partial h(a, \alpha)$ by the Moreau-Rockafellar sum rule 2.26.

**Detail.** If $1 - y(\langle \bar{a}, x\rangle + \bar{\alpha}) < 0$, then in a neighborhood of $(\bar{a}, \bar{\alpha})$ the function $h$ is zero. Thus, $h$ is smooth at $(\bar{a}, \bar{\alpha})$ and $\partial h(\bar{a}, \bar{\alpha}) = \{0\}$; cf. 2.14. If $1 - y(\langle \bar{a}, x\rangle + \bar{\alpha}) > 0$, then for $(a, \alpha)$ in a neighborhood of $(\bar{a}, \bar{\alpha})$ the function $h$ has values $(1 - y(\langle a, x\rangle + \alpha))/m$ so it again is smooth and the set of subgradients reduces to the gradient.

The function $h$ is nonsmooth, however, at a point $(\bar{a}, \bar{\alpha})$ with $1 - y(\langle \bar{a}, x\rangle + \bar{\alpha}) = 0$. By the subgradient inequality 2.17, $(v, \omega)$ is a subgradient at $(\bar{a}, \bar{\alpha})$ if and only if

$$\tfrac{1}{m} \max\Big\{0, 1 - y\big(\langle a, x\rangle + \alpha\big)\Big\} \geq \langle v, a - \bar{a}\rangle + \omega(\alpha - \bar{\alpha}) \;\; \forall (a, \alpha) \in \mathbb{R}^{n+1}.$$

Certainly, $(v, \omega) = (0, 0)$ satisfies this inequality. For $\beta \in [0, 1]$ and $(v, \omega) = \beta(-yx/m, -y/m)$, one has

$$\langle v, a - \bar{a} \rangle + \omega(\alpha - \bar{\alpha}) = \frac{\beta}{m}\Big(1 - y\big(\langle a, x \rangle + \alpha\big)\Big)$$

and the inequality is also satisfied. This establishes the last formula. □

As in regression and least-squares problems (cf. 2.5 and 2.38), the practical performance of support vector machines can be improved by adding a regularization term of the form $\theta\|a\|_1$ or $\theta\|a\|_2^2$, with $\theta \in (0, \infty)$. For the $\ell^1$-norm, this leads to the *$\ell^1$-regularized hinge-loss problem*

$$\underset{(a,\alpha)\in\mathbb{R}^{n+1}}{\text{minimize}}\ \frac{1}{m}\sum\nolimits_{i=1}^{m} \max\Big\{0, 1 - y_i\big(\langle a, x^i \rangle + \alpha\big)\Big\} + \theta\|a\|_1.$$

The prior example together with 2.29 and the Moreau-Rockafellar sum rule 2.26 yield expressions for subgradients of the new objective function.

Although nonsmooth, the $\ell^1$-regularized hinge-loss problem can be reformulated as the equivalent linear optimization problem

$$\begin{aligned} \underset{(a,\alpha)\in\mathbb{R}^{n+1}, u\in\mathbb{R}^n, v\in\mathbb{R}^m}{\text{minimize}}\ & \frac{1}{m}\sum\nolimits_{i=1}^{m} v_i + \theta\sum\nolimits_{j=1}^{n} u_j \\ \text{subject to}\ & 1 - y_i\big(\langle a, x^i \rangle + \alpha\big) \leq v_i,\ i = 1, \dots, m \\ & 0 \leq v_i,\ i = 1, \dots, m \\ & -a_j \leq u_j,\ a_j \leq u_j,\ j = 1, \dots, n. \end{aligned}$$

If $(a^\star, \alpha^\star, u^\star, v^\star)$ minimizes this linear optimization problem, then $(a^\star, \alpha^\star)$ minimizes the $\ell^1$-regularized hinge-loss problem. Conversely, if $(a^\star, \alpha^\star)$ minimizes the $\ell^1$-regularized hinge-loss problem and $u^\star$ and $v^\star$ have components

$$v_i^\star = \max\Big\{0, 1 - y_i\big(\langle a^\star, x^i \rangle + \alpha^\star\big)\Big\} \quad \text{and} \quad u_j^\star = |a_j^\star|,$$

then $(a^\star, \alpha^\star, u^\star, v^\star)$ minimizes the linear optimization problem. Thus, in principle, the $\ell^1$-regularized hinge-loss problem can be solved, via this reformulation, by the interior-point method (§2.G) as well as the simplex method (§2.F). However, the reformulation lifts the problem from having $n+1$ variables and no constraints to having $m+2n+1$ variables and $2m+2n$ constraints, which might be excessive in real-world applications. For large-scale instances, it's better to tackle the $\ell^1$-regularized hinge-loss problem directly by leveraging the expressions for subgradients in 2.49. One concrete possibility is laid out in the next section.

**Exercise 2.50** (alternative support vector machine). For given $(a^\star, \alpha^\star)$, the distance between the two sets $\{x \mid \langle a^\star, x \rangle + \alpha^\star \geq 1\}$ and $\{x \mid \langle a^\star, x \rangle + \alpha^\star \leq -1\}$ is $2/\|a^\star\|_2$. (In Figure 2.20, this is the distance between the dashed lines.) Since the first set corresponds to the points where one hopes to have all the positively labeled observations and similarly with the second set for the negatively labeled observations, we may seek to separate these sets as much as possible to reduce the chance of misclassification of future observations.

This leads to a formulation that penalizes large $\|a\|_2$:

$$\underset{(a,\alpha)\in\mathbb{R}^{n+1}}{\text{minimize}}\ \frac{1}{m}\sum_{i=1}^{m} \max\Big\{0, 1 - y_i\big(\langle a, x^i\rangle + \alpha\big)\Big\} + \theta\|a\|_2^2.$$

Derive an equivalent quadratic optimization problem.

**Guide.** The resulting problem involves $m$ additional variables and $2m$ constraints. □

## 2.I Subgradient Method

The gradient descent method (§1.F) minimizes a smooth function $f : \mathbb{R}^n \to \mathbb{R}$ by moving from a current point $x^\nu$ in the direction $-\nabla f(x^\nu)$. It works because $-\nabla f(x^\nu)$ is a descent direction unless $x^\nu$ is already satisfying the requirements of the Oresme and Fermat rules. Consequently, with an appropriately small step size $\lambda$, we've

$$f\big(x^\nu - \lambda\nabla f(x^\nu)\big) < f(x^\nu).$$

An idea for extending the method to nonsmooth functions would be to simply replace $\nabla f(x^\nu)$ by a subgradient of $f$ at $x^\nu$. At least when $f$ is convex, this indeed leads to viable algorithms but not without potential pitfalls.

Figure 2.21 illustrates the fundamental shift taking place when we pass from smooth to nonsmooth functions. On the left, $-\nabla f(\bar{x})$ provides a descent direction at $\bar{x}$ because $f$ is smooth. On the right, we've a subgradient $v$ for which

$$f(\bar{x} - \lambda v) > f(\bar{x})$$

regardless of $\lambda > 0$. Consequently, $-v$ isn't a descent direction for $f$ at $\bar{x}$.

Despite this disappointment, subgradients are still useful. For the function in Figure 2.21(right), the minimizer is $x^\star = (0, 0)$. Thus, movement along the direction $-v$ from $\bar{x}$ does get us *closer* to $x^\star$ as long as we don't move too far. This is an example of a general situation for convex functions.

**Lemma 2.51** *Suppose that $f : \mathbb{R}^n \to \overline{\mathbb{R}}$ is convex, $f(\bar{x})$ is finite and $v \in \partial f(\bar{x})$. Then, for any $x \in \mathbb{R}^n$ and $\lambda \in [0, \infty)$,*

$$\|\bar{x} - \lambda v - x\|_2^2 \le \|\bar{x} - x\|_2^2 - 2\lambda\big(f(\bar{x}) - f(x)\big) + \lambda^2\|v\|_2^2.$$

*Thus, when $f(x) < f(\bar{x})$,*

$$\|\bar{x} - \lambda v - x\|_2 < \|\bar{x} - x\|_2 \quad \forall \lambda \in \left(0, \frac{f(\bar{x}) - f(x)}{\frac{1}{2}\|v\|_2^2}\right).$$

**Proof.** By the subgradient inequality 2.17, $f(x) - f(\bar{x}) \ge \langle v, x - \bar{x}\rangle$ so that

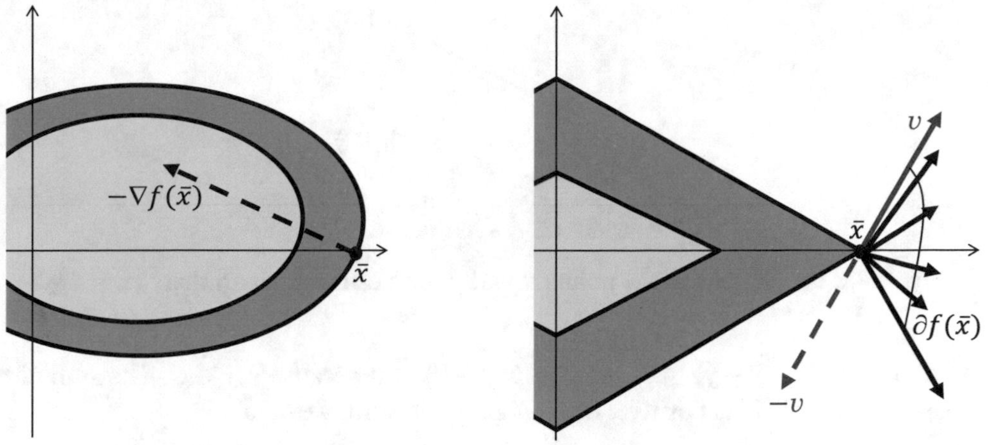


Fig. 2.21: Contours of a smooth function $f : \mathbb{R}^2 \to \mathbb{R}$ (left), for which $-\nabla f(\bar{x})$ provides a descent direction, and of $f(x) = \max\{x_1/2 + x_2, x_1/2 - x_2, -x_1/2 + x_2, -x_1/2 - x_2\} - 1$ (right), which at $\bar{x} = (2, 0)$ has $\partial f(\bar{x}) = \{(v_1, v_2) \mid v_1 = 1/2, v_2 \in [-1, 1]\}$. The subgradient $v = (1/2, 1)$ gives $f(\bar{x} - \lambda v) = 3\lambda/4 > f(\bar{x}) = 0$.

$$\begin{aligned}\|\bar{x} - \lambda v - x\|_2^2 &= \|\bar{x} - x\|_2^2 - 2\lambda\langle v, \bar{x} - x\rangle + \lambda^2\|v\|_2^2 \\ &\leq \|\bar{x} - x\|_2^2 - 2\lambda\big(f(\bar{x}) - f(x)\big) + \lambda^2\|v\|_2^2.\end{aligned}$$

We've $\|\bar{x} - \lambda v - x\|_2 < \|\bar{x} - x\|_2$ provided that $-2\lambda(f(\bar{x}) - f(x)) + \lambda^2\|v\|_2^2 < 0$. Since our interest is in $\lambda > 0$ and $f(x) < f(\bar{x})$, we can divide by $\lambda$ and obtain the stated requirement on $\lambda$. □

If we apply the lemma with $x \in \operatorname{argmin} f$, then we see that indeed $\bar{x} - \lambda v$ is closer to such a minimizer than $\bar{x}$ as long as the step size $\lambda$ is small enough. In view of 2.19, this means that $\bar{x} - \lambda v$ is closer to a solution of the optimality condition $0 \in \partial f(x)$ than $\bar{x}$ for such step sizes.

Unfortunately, the bound on the step size depends on $f(x) = \inf f$, which usually isn't known. Assuming that this challenge can be overcome, there's also the issue that $\bar{x} - \lambda v$ may not be in $\operatorname{dom} f$. We can correct for this by projecting $\bar{x} - \lambda v$ onto $\operatorname{dom} f$ as in the projected gradient method (§2.D). This works because a projection is nonexpansive as illustrated in Figure 2.22 and the next lemma.

**Lemma 2.52** *For a nonempty, closed and convex set $C \subset \mathbb{R}^n$ and $x, \bar{x} \in \mathbb{R}^n$, one has*

$$\|\operatorname{prj}_C(x) - \operatorname{prj}_C(\bar{x})\|_2 \leq \|x - \bar{x}\|_2.$$

**Proof.** We recall from 2.30 that both $\operatorname{prj}_C(x)$ and $\operatorname{prj}_C(\bar{x})$ contain a single point, say $y$ and $\bar{y}$, respectively, which then must satisfy the optimality conditions

$$\langle x - y, z - y\rangle \leq 0 \ \ \forall z \in C \quad \text{and} \quad \langle \bar{x} - \bar{y}, \bar{z} - \bar{y}\rangle \leq 0 \ \ \forall \bar{z} \in C.$$

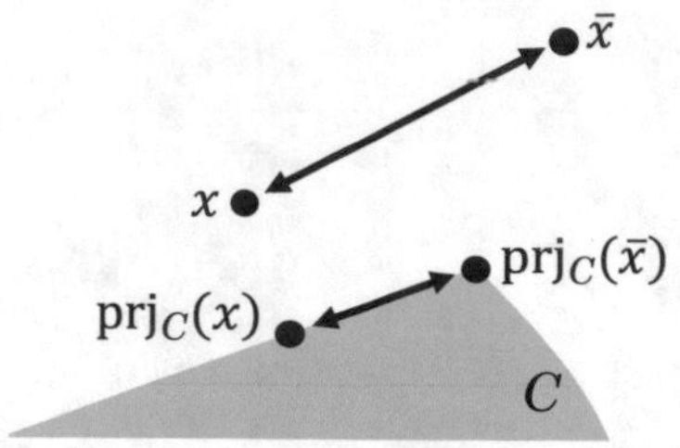


Fig. 2.22: Projections of the points $x$ and $\bar{x}$ are no further apart than $\|x - \bar{x}\|_2$.

In particular, $\langle x - y, \bar{y} - y\rangle \leq 0$ and $\langle \bar{x} - \bar{y}, y - \bar{y}\rangle \leq 0$ because $y, \bar{y} \in C$. The sum of the left-hand sides is also nonpositive. Rearranging that sum, we obtain

$$\|\bar{y} - y\|_2^2 = \langle \bar{y} - y, \bar{y} - y\rangle \leq \langle \bar{x} - x, \bar{y} - y\rangle \leq \|\bar{x} - x\|_2 \|\bar{y} - y\|_2,$$

where the last step is using the Cauchy-Schwarz inequality. □

In view of the lemma, we see that

$$\left\| \mathrm{prj}_{\mathrm{dom}\, f}(\bar{x} - \lambda v) - x^\star \right\|_2 \leq \|\bar{x} - \lambda v - x^\star\|_2$$

for any $x^\star \in \operatorname{argmin} f$ as long as $\mathrm{dom}\, f$ is closed. Leveraging these ideas, we obtain an algorithm for

$$\underset{x \in C}{\text{minimize}}\ f_0(x),$$

where $f_0 : \mathbb{R}^n \to \mathbb{R}$ is convex and $C \subset \mathbb{R}^n$ is nonempty, closed and convex. It's *not* a descent method (as motivated by the Oresme rule), but rather a procedure for solving the generalized equation $0 \in \partial f_0(x) + N_C(x)$ by moving steadily closer to its solutions. Thus, the algorithm is related to Newton's method with its focus on the Fermat rule. By the optimality condition 2.27, the generalized equation is of course equivalent to solving the minimization problem.

**Subgradient Method.**

Data. $x^0 \in C$ and $\lambda^\nu \in (0, \infty)$, $\nu = 0, 1, 2, \dots$.
Step 0. Set $\nu = 0$.
Step 1. Compute $v^\nu \in \partial f_0(x^\nu)$.
Step 2. Set

$$x^{\nu+1} \in \mathrm{prj}_C(x^\nu - \lambda^\nu v^\nu).$$

Step 3. Replace $\nu$ by $\nu + 1$ and go to Step 1.

**Convergence 2.53** (subgradient method). *For convex $f_0 : \mathbb{R}^n \to \mathbb{R}$ and nonempty, closed and convex $C \subset \mathbb{R}^n$, suppose that there's $\beta < \infty$ such that*

$$\sup\left\{\|v\|_2 \mid v \in \partial f_0(x),\ x \in C\right\} \leq \beta$$

*and the subgradient method has generated* $\{x^\nu, \nu = 0, 1, \dots \bar{\nu}\}$ *using step size* $\lambda^\nu = \lambda \in (0, \infty)$ *for all* $\nu$. *Then, for any* $x^\star \in \operatorname{argmin}_C f_0$ *and* $\varepsilon > 0$,

$$\min_{\nu=0,1,\dots,\bar{\nu}} f_0(x^\nu) - \inf_C f_0 \leq \frac{\lambda\beta^2 + \varepsilon}{2}, \quad \textit{when } \bar{\nu} \geq \left\lfloor \frac{\|x^0 - x^\star\|_2^2}{\varepsilon\lambda} \right\rfloor,$$

*where* $\lfloor \alpha \rfloor$ *is the largest integer no greater than* $\alpha$.

**Proof.** Applying 2.51 with $x = x^\star$ and 2.52, we obtain that

$$\|x^{\nu+1} - x^\star\|_2^2 \leq \|x^\nu - x^\star\|_2^2 - 2\lambda\big(f_0(x^\nu) - \inf_C f_0\big) + \lambda^2\beta^2.$$

Suppose for the sake of contradiction that

$$f_0(x^\nu) - \inf_C f_0 > \frac{\lambda\beta^2 + \varepsilon}{2}, \quad \nu = 0, 1, \dots, \bar{\nu}.$$

Then, for $\nu = 0, 1, \dots, \bar{\nu}$,

$$\|x^{\nu+1} - x^\star\|_2^2 < \|x^\nu - x^\star\|_2^2 - \lambda^2\beta^2 - \varepsilon\lambda + \lambda^2\beta^2 = \|x^\nu - x^\star\|_2^2 - \varepsilon\lambda.$$

Since this takes place over all $\nu = 0, 1, \dots, \bar{\nu}$, one has

$$\|x^{\nu+1} - x^\star\|_2^2 \leq \|x^0 - x^\star\|_2^2 - (\bar{\nu} + 1)\varepsilon\lambda.$$

The right-hand side is nonnegative and $\bar{\nu} + 1 \leq \|x^0 - x^\star\|_2^2/\varepsilon\lambda$, but this contradicts the definition of $\bar{\nu}$. □

The "last" iteration of the algorithm may not furnish the lowest function value because there's no guaranteed descent in each iteration. This motivates the check across all iterations in the convergence statement.

If $\varepsilon > 0$ is an acceptable tolerance, then the particular, and in some sense "best," step size $\lambda = \varepsilon/\beta^2$ gives

$$\min_{\nu=0,1,\dots,\bar{\nu}} f_0(x^\nu) - \inf_C f_0 \leq \varepsilon \quad \text{when } \bar{\nu} \geq \left\lfloor \frac{\beta^2\|x^0 - x^\star\|_2^2}{\varepsilon^2} \right\rfloor.$$

Alternatively, if the number of iterations $\bar{\nu}$ is pre-selected and

$$\lambda = \frac{\|x^0 - x^\star\|_2}{\beta\bar{\nu}^{1/2}},$$

then

$$\min_{\nu=0,1,\dots,\bar{\nu}} f_0(x^\nu) - \inf_C f_0 \leq \frac{\beta\|x^0 - x^\star\|_2}{\bar{\nu}^{1/2}}.$$

In contrast to the error after $\nu$ iterations of the gradient descent method and projected gradient method (proportional to $\nu^{-1}$) and the accelerated gradient method ($\nu^{-2}$), the subgradient method has a slower rate of convergence ($\nu^{-1/2}$). However, the former methods rely on smoothness. The main advantage of the subgradient method is its broad applicability to nonsmooth problems. It also requires little memory and has low computational cost per iteration provided that the projection step can be carried out easily; cf. 2.30, 2.31 and 2.46. The rate of convergence is also independent of $n$, the dimension of the problem, making the method attractive for large-scale problems. In fact, the rate is the best any algorithm can achieve for this broad class of convex functions [67, Section 3.2.3]. Still, the subgradient method tends to be sensitive to the choice of step size and some tuning is needed as the theoretically recommended value might not be known and anyhow is often on the small side in practice.

We end the discussion of the subgradient method by examining the requirement of having bounded subgradients in 2.53.

**Bounded and Compact Sets.** A set $C \subset \mathbb{R}^n$ is *bounded* if there's $\delta < \infty$ such that $\|x\|_2 \leq \delta$ for all $x \in C$. Otherwise, it's *unbounded*. If the set is both bounded and closed, then it's *compact*. Importantly, every sequence in a compact set $C$ has a cluster point in $C$.

**Proposition 2.54** (bounded subgradients). *For a proper convex function $f : \mathbb{R}^n \to \overline{\mathbb{R}}$ and a compact set $C \subset \text{int}(\text{dom}\, f)$, there's $\beta < \infty$ such that*

$$\sup \left\{ \|v\|_2 \;\middle|\; v \in \partial f(x),\; x \in C \right\} \leq \beta.$$

*Thus, in particular, for any $\bar{x} \in \text{int}(\text{dom}\, f)$, $\partial f(\bar{x})$ is compact.*

*If $f$ is also real-valued and satisfies the Lipschitz condition*

$$\left| f(x) - f(\bar{x}) \right| \leq \kappa \|x - \bar{x}\|_2 \quad \forall x, \bar{x} \in \mathbb{R}^n,$$

*with $\kappa \in [0, \infty)$, then*

$$\sup \left\{ \|v\|_2 \;\middle|\; v \in \partial f(x),\; x \in \mathbb{R}^n \right\} \leq \kappa.$$

**Proof.** For the sake of contradiction, suppose that there exist

$$\left\{ x^\nu \in C, v^\nu \in \partial f(x^\nu),\; \nu \in \mathbb{N} \right\}$$

such that $\|v^\nu\|_2 \to \infty$. Since $C$ is compact, there are $\bar{x} \in C$ and $N \in \mathcal{N}_\infty^\#$ such that $x^\nu \to_N \bar{x}$. Since $\bar{x} \in \text{int}(\text{dom}\, f)$, there's $\delta > 0$ such that $\mathbb{B}(\bar{x}, 2\delta) \subset \text{dom}\, f$. Let $\bar{\nu}$ be such that $v^\nu$ is nonzero for all $\nu \geq \bar{\nu}$ and define

$$\left\{ w^\nu = \delta v^\nu / \|v^\nu\|_2,\; \nu = \bar{\nu}, \bar{\nu} + 1, \ldots \right\},$$

which is a bounded sequence because $\|w^\nu\|_2 \leq \delta$ for all $\nu \geq \bar{\nu}$. Thus, there exist $\bar{w} \in \mathbb{B}(0, \delta)$ and a subsequence of $N$, which we also denote by $N$, such that $w^\nu \to_N \bar{w}$. By the subgradient inequality 2.17, this leads to

$$f(x^\nu + w^\nu) \geq f(x^\nu) + \langle v^\nu, w^\nu \rangle = f(x^\nu) + \delta\|v^\nu\|_2 \quad \forall \nu \geq \bar{\nu}.$$

Since both $\bar{x} + \bar{w}$ and $\bar{x}$ are contained in int(dom $f$), $f$ is continuous at these points by 2.25. The prior inequality then establishes that

$$f(\bar{x} + \bar{w}) - f(\bar{x}) \geq \tfrac{1}{2}\delta\|v^\nu\|_2$$

for sufficiently large $\nu \in N$. This contradicts the assertion of $\|v^\nu\|_2 \to \infty$. Since $\partial f(\bar{x})$ is closed by 2.25, the first conclusion follows.

Under the Lipschitz condition, the subgradient inequality yields

$$\kappa\|w\|_2 \geq f(x + w) - f(x) \geq \langle w, w \rangle = \|w\|_2^2$$

for $w \in \partial f(x)$ and the second conclusion follows. □

**Exercise 2.55** (subgradient method for support vector machines). Implement the subgradient method using a fixed step size $\lambda$ and apply it to the $\ell^1$-regularized hinge-loss problem in §2.H for some dataset. For a given number of iterations (such as $\bar{\nu} = 100000$), compare the smallest objective function value obtained during the iterations for different step sizes $\lambda$.

**Guide.** Leverage 2.49 to obtain expressions for the subgradients as well as the bound $\beta$ in 2.53, which can be used to guide the selection of $\lambda$. Ultimately, it's not only a low objective function value that matters. As an extra exercise, test the classifier produced by the solution $(a, \alpha)$ with the lowest objective function value on a separate but similar dataset and compute the number of correct classifications. Repeat the calculations for several values of $\theta$. □

## 2.J Conic Constraints

A set $C \subset \mathbb{R}^n$ is a *cone* if $0 \in C$ and $\lambda x \in C$ for all $\lambda > 0$ and $x \in C$. Cones are sets that like polyhedra exhibit convenient properties. They emerge as a principal way of describing the local geometry of a set and as constraints in applications.

We've already encountered one type of cones: the normal cone $N_C(x)$ to a convex set $C$ at a point $x \in C$; cf. 2.23. It follows directly from the definition that $N_C(x)$ is a closed convex cone. However, not all cones are convex as shown in Figure 2.23.

In applications, the *second-order cone*

$$C = \left\{(x, \alpha) \in \mathbb{R}^{n+1} \;\middle|\; \|x\|_2 \leq \alpha\right\}$$

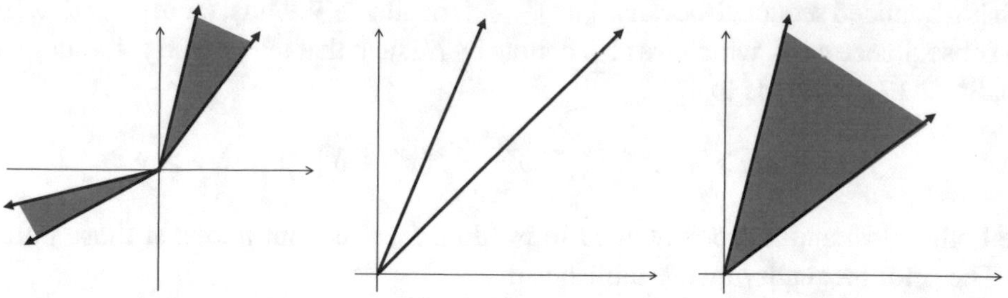

Fig. 2.23: Nonconvex cones (left, middle) and a convex cone (right) in $\mathbb{R}^2$.

arises in several contexts including models with geometric considerations. For $n = 2$, it has the shape of an ice cream cone; see Figure 2.24. Certainly, $C$ is a cone because $0 \in C$ and for $\lambda > 0$ and $(x, \alpha) \in C$, we've $\|\lambda x\|_2 = \lambda\|x\|_2 \leq \lambda\alpha$, which shows that $\lambda(x, \alpha) \in C$. As the epigraph of the $\ell^2$-norm, it's also convex by 1.15.

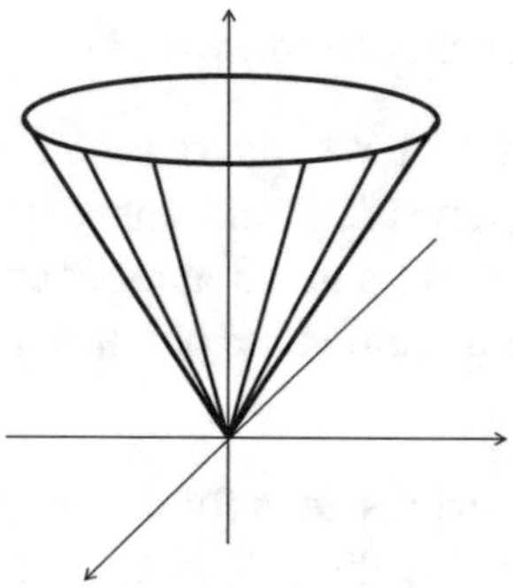

Fig. 2.24: Second-order cone.

**Example 2.56** (projection on second-order cone). For a point $(\bar{x}, \bar{\alpha}) \in \mathbb{R}^{n+1}$ and

$$C = \big\{(x, \alpha) \in \mathbb{R}^{n+1} \;\big|\; \|x\|_2 \leq \alpha\big\},$$

the projection

$$\operatorname{prj}_C\big((\bar{x}, \bar{\alpha})\big) = \big\{\tfrac{1}{2}(\bar{x}, \bar{\alpha}) + \tfrac{1}{2}(\bar{\alpha}\bar{x}/\sigma, \sigma)\big\},$$

where $0/0$ is interpreted as 0 and

$$\sigma = \sqrt{\tfrac{1}{2}\big\|(\bar{x}, \bar{\alpha})\big\|_2^2 + \tfrac{1}{2}\sqrt{\|(\bar{x}, \bar{\alpha})\|_2^4 - \|2\bar{\alpha}\bar{x}\|_2^2}}.$$

**Detail.** The second-order cone is a rare instance of a set with an explicit expression for projections; see [44] for a derivation. (Another instance is a box as shown in 2.31.) Thus, it's easily handled by the projected gradient method (§2.D), subgradient method (§2.I) and similar algorithms. □

We illustrate second-order cones with two applications, where they arise in combination with affine mappings and result in *second-order cone constraints* of the form

$$\|Ax - b\|_2 \le \langle c, x\rangle + \gamma,$$

where $A$ is an $m \times n$-matrix, $b \in \mathbb{R}^m$, $c \in \mathbb{R}^n$ and $\gamma \in \mathbb{R}$. Except in special cases such as in 2.56, there's no explicit formula for projection on a feasible set defined by one or more second-order cone constraints making projection-based algorithms unattractive. Extension of the interior-point method (§2.G), however, results in efficient algorithms [2]. It's also possible to approximate such feasible sets by polyhedra [12] and thereby allowing application of several algorithms from this chapter. For a general treatment of conic optimization, we refer to [11].

**Example 2.57** (chance constraints). Suppose that $x_j$ is the dollar amount you invest in stock $j$ and $\xi_j$ is the yield of stock $j$ between now and the time of your retirement, with, for example, $\xi_j = 1.2$ meaning that the stock goes up 20% during the time period. Then,

$$\sum\nolimits_{j=1}^{n} \xi_j x_j = \langle \xi, x\rangle$$

is the total value of your investment portfolio at the time of your retirement. Usually, the yields are unknown, but we can model them by a random vector $\boldsymbol{\xi} = (\boldsymbol{\xi}_1, \dots, \boldsymbol{\xi}_n)$, which we assume has a normal distribution. Similar to §1.C, we can then maximize the expected value $\mathbb{E}[\langle \boldsymbol{\xi}, x\rangle]$ or, equivalently, minimize $-\mathbb{E}[\langle \boldsymbol{\xi}, x\rangle]$. However, we want to impose a constraint that ensures a low probability of suffering a major loss. Let $\beta$ be a dollar amount that we consider a minimum requirement; a value of the portfolio below $\beta$ is considered a major loss. For some probability $\alpha \in (0.5, 1)$, typically close to one, this leads to the *chance constraint*

$$\text{prob}\big\{ \langle \boldsymbol{\xi}, x\rangle \ge \beta \big\} \ge \alpha,$$

where "prob" abbreviates "probability of." This constraint can be written as a second-order cone constraint.

**Detail.** Suppose that the normal random vector $\boldsymbol{\xi}$ has expected value and variance-covariance matrix

$$\bar{\xi} = \big(\mathbb{E}[\boldsymbol{\xi}_1], \dots, \mathbb{E}[\boldsymbol{\xi}_n]\big) \quad \text{and} \quad V = \mathbb{E}\big[(\boldsymbol{\xi} - \bar{\xi})(\boldsymbol{\xi} - \bar{\xi})^\top\big].$$

Then, one can show that $\langle \boldsymbol{\xi}, -x\rangle$ is a normal random variable with expected value and variance

$$\mu = \langle \bar{\xi}, -x\rangle \quad \text{and} \quad \sigma^2 = \text{var}\big(\langle \boldsymbol{\xi}, -x\rangle\big) = \langle x, Vx\rangle.$$

Consequently, when $\sigma > 0$,

$$\text{prob}\big\{ \langle \boldsymbol{\xi}, -x\rangle \le -\beta \big\} = \Phi\left(\frac{-\beta - \mu}{\sigma}\right),$$

where $\Phi$ is the distribution function of a standard normal random variable. The chance constraint then simplifies:

$$\text{prob}\{\langle \xi, x\rangle \geq \beta\} \geq \alpha \iff \frac{-\beta-\mu}{\sigma} \geq \Phi^{-1}(\alpha) \iff -\langle \bar{\xi}, x\rangle + \Phi^{-1}(\alpha)\sqrt{\langle x, Vx\rangle} \leq -\beta.$$

For the case $\sigma = 0$, $\langle \xi, x\rangle$ takes the value $\langle \bar{\xi}, x\rangle$ with probability one. Hence, $\text{prob}\{\langle \xi, x\rangle \geq \beta\} = 1$ if $\langle \bar{\xi}, x\rangle \geq \beta$ and the probability is zero otherwise. The above equivalence therefore holds even if $\sigma = 0$.

Since $V$ is a variance-covariance matrix, it's symmetric and positive semidefinite. A fact from linear algebra is that a symmetric matrix $A$ is positive semidefinite if and only if there exists a matrix $B$ such that $A = B^\top B$, where $B$ is the *square root* of $A$ and is computable by spectral decomposition. Let's denote by $V^{1/2}$ the square root of $V$. Then,

$$\sqrt{\langle x, Vx\rangle} = \sqrt{\langle V^{1/2}x, V^{1/2}x\rangle} = \|V^{1/2}x\|_2$$

and

$$\text{prob}\{\langle \xi, x\rangle \geq \beta\} \geq \alpha \iff -\langle \bar{\xi}, x\rangle + \Phi^{-1}(\alpha)\|V^{1/2}x\|_2 \leq -\beta,$$

which is a second-order cone constraint. □

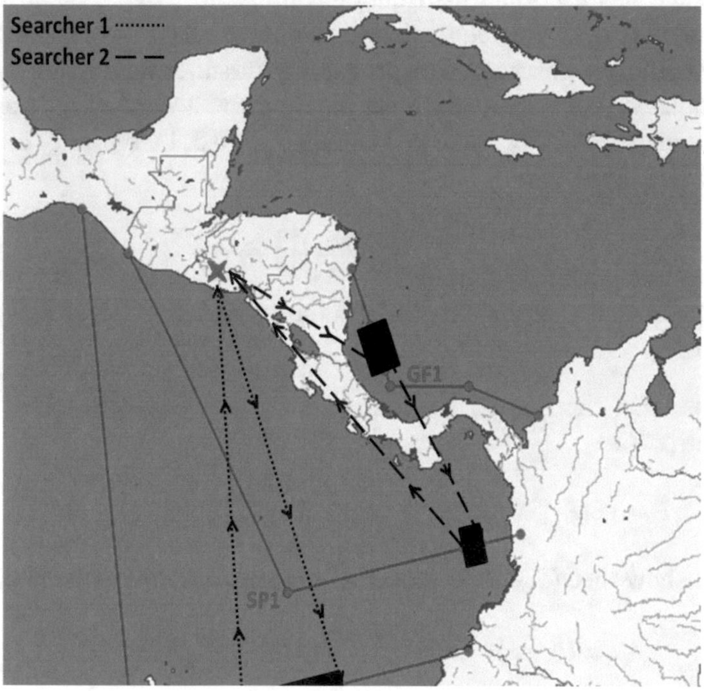


Fig. 2.25: Two search aircraft looking for three smugglers.

**Example 2.58** (smuggler interdiction). A multi-national unit is tasked with searching for smugglers at sea. Each morning a planner assigns a number of aircraft to search certain areas based on estimates of the smugglers' movement. Figure 2.25 illustrates a situation where three smugglers are believed to approximately follow the gray lines labeled GF1, SP1 and GF3. Two search aircraft are routed from an airbase (marked with a star) to search areas (black boxes) and then return to base. Searcher 1 is assigned to GF3 only, while Searcher 2 first looks for GF1 and then for SP1. A model involving second-order cone constraints helps the planner to determine optimal routes for the aircraft while accounting for their endurance.

**Detail.** For simplicity, let's assume that there are $n$ smugglers, but only one searcher (see [81, 82] for extensions). The $j$th smuggler moves with constant speed $\sigma_j$ along a line segment that starts at $x^j \in \mathbb{R}^2$ and ends at $y^j \in \mathbb{R}^2$, with departure at time $\tau_j$. (The extensions include piecewise linear smuggler paths as in Figure 2.25.) Suppose that the order in which the searcher looks for the smugglers is pre-determined and let that order be smuggler 1 first, then smuggler 2, etc. The planner needs to determine when the searcher should depart from its base, which is located at $x^0 \in \mathbb{R}^2$, where it should intercept the assumed path of the first smuggler and how much time it should spend searching for the first smuggler before moving on to the next one. Let $t_j$ be the travel time for the searcher to move from the $(j-1)$th smuggler to the $j$th smuggler, $a_j$ be the time when the searcher has reached the nominal location of smuggler $j$ and $d_j$ be the time spent searching for that smuggler. With a maximum allowed flight time $\bar{\tau}$, the decision variables $d_j$ and $t_j$ must obey the constraint

$$\sum\nolimits_{j=1}^{n} d_j + \sum\nolimits_{j=1}^{n+1} t_j \le \bar{\tau},$$

where $t_1$ is the time needed to move from the base to the first smuggler and $t_{n+1}$ represents the flight time from the $n$th smuggler back to the base.

The searcher can't start searching for the $j$th smuggler before it's finished with the previous smuggler and the necessary travel time has passed:

$$a_0 + t_1 \le a_1 \quad \text{and} \quad a_{j-1} + d_{j-1} + t_j \le a_j, \quad j = 2, 3, \dots, n,$$

where $a_0$ is the time of departure from the base, which could be restricted to a certain time window. The location of the $j$th smuggler is

$$x^j + (\tau - \tau_j)u^j \ \text{ at time } \tau \in \big[\tau_j,\ \tau_j + \|y^j - x^j\|_2/\sigma_j\big], \quad \text{where } u^j = \sigma_j \frac{y^j - x^j}{\|y^j - x^j\|_2}.$$

Thus, the location of the $j$th smuggler when the searcher arrives at time $a_j$ is $x^j + (a_j - \tau_j)u^j$. The location of the previous smuggler when the searcher departs at time $a_{j-1} + d_{j-1}$ is

$$x^{j-1} + (a_{j-1} + d_{j-1} - \tau_{j-1})u^{j-1}.$$

The distance between these two locations dictates the travel time $t_j$:

$$\big\|x^j + (a_j - \tau_j)u^j - x^{j-1} - (a_{j-1} + d_{j-1} - \tau_{j-1})u^{j-1}\big\|_2 \le \alpha t_j, \quad j = 2, \dots, n,$$

where $\alpha > \sigma_j$ is the searcher's speed. The travel time to the first smuggler is bounded by

$$\left\|x^1 + (a_1 - \tau_1)u^1 - x^0\right\|_2 \leq \alpha t_1$$

and the time back to base is restricted by

$$\left\|x^0 - x^n - (a_n + d_n - \tau_n)u^n\right\|_2 \leq \alpha t_{n+1}.$$

These $n + 1$ inequalities are second-order cone constraints.

The searcher can't search for the target before it departs, i.e., $a_j \geq \tau_j$, or after it arrives at its destination, i.e.,

$$a_j + d_j \leq \tau_j + \|y^j - x^j\|_2/\sigma_j.$$

The objective function is the expected value of contraband *not* detected, i.e.,

$$\sum_{j=1}^{n} q_j e^{-\beta_j d_j},$$

where $q_j > 0$ is the value of contraband onboard smuggler $j$ and $\beta_j > 0$ is a parameter reflecting the uncertainty associated with that smuggler. Combining all these elements, we reach a model with decision variables $\{a_j, j = 0, 1, \ldots, n\}$, $\{d_j, j = 1, \ldots, n\}$ and $\{t_j, j = 1, \ldots, n+1\}$, linear as well as second-order cone constraints and a convex objective function. □

## 2.K Polyhedral Analysis

Since polyhedral sets appear in numerous optimization models, it's worthwhile to take a deeper look and identify properties that can be exploited in algorithms.

The *positive hull* of a finite number of points $\{a^i \in \mathbb{R}^n,\ i = 1, \ldots, m\}$ is the set formed by nonnegatively weighted combinations of these points, i.e.,

$$\text{pos}\{a^i, \ldots, a^m\} = \left\{\sum_{i=1}^{m} \lambda_i a^i \,\middle|\, \lambda_i \geq 0,\ i = 1, \ldots, m\right\}.$$

For example, $\text{pos}\{(1,0), (1,1), (0,2)\}$ is the first quadrant of $\mathbb{R}^2$.

**Theorem 2.59** (Weyl's). *For* $\{a^i \in \mathbb{R}^n,\ i = 1, \ldots, m\}$, *there exists a collection* $\{d^i \in \mathbb{R}^n, i = 1, \ldots, q\}$ *such that*

$$\text{pos}\{a^1, \ldots, a^m\} = \left\{x \in \mathbb{R}^n \,\middle|\, \langle d^i, x\rangle \leq 0,\ i = 1, \ldots, q\right\}.$$

*Consequently,* $\text{pos}\{a^1, \ldots, a^m\}$ *is a polyhedral cone and thus closed and convex.*

**Proof.** Let's augment $\{a^i \in \mathbb{R}^n,\ i = 1, \ldots, m\}$ with $a^0 = 0$, which doesn't have any effect on its positive hull. For some finite index set $\mathbb{I}^{\nu-1}$ and $\{d^i \in \mathbb{R}^n, i \in \mathbb{I}^{\nu-1}\}$, suppose that

$$\text{pos}\{a^0, \ldots, a^{\nu-1}\} = \left\{x \in \mathbb{R}^n \,\middle|\, \langle d^i, x\rangle \leq 0,\ i \in \mathbb{I}^{\nu-1}\right\}. \tag{2.9}$$

This is certainly the case when $\nu = 1$ because

$$\mathrm{pos}\{a^0\} = \{0\} = \big\{x \in \mathbb{R}^n \,\big|\, \langle e^j, x\rangle \le 0,\ \langle -e^j, x\rangle \le 0,\ \ j = 1, \dots, n\big\},$$

where $e^j$ is the unit vector with 1 as its $j$th entry; these unit vectors constitute our initial collection $\{d^i,\ i \in \mathbb{I}^0\}$ with $\mathbb{I}^0 = \{1, \dots, 2n\}$ and $d^i = e^i$ and $d^{n+i} = -e^i$ for $i = 1, \dots, n$. Thus, it suffices to show that $\mathrm{pos}\{a^0, \dots, a^\nu\}$ admits a similar representation. For $i \in \mathbb{I}^{\nu-1}$, let $\alpha_i = \langle d^i, a^\nu\rangle$ and define

$$\mathbb{I}_0 = \big\{i \in \mathbb{I}^{\nu-1} \,\big|\, \alpha_i = 0\big\}, \quad \mathbb{I}_- = \big\{i \in \mathbb{I}^{\nu-1} \,\big|\, \alpha_i < 0\big\}, \quad \mathbb{I}_+ = \big\{i \in \mathbb{I}^{\nu-1} \,\big|\, \alpha_i > 0\big\}$$

and for all $i \in \mathbb{I}_+$, $j \in \mathbb{I}_-$, set

$$d^{ij} = d^i - \alpha_i d^j / \alpha_j.$$

It turns out that

$$\mathrm{pos}\{a^0, \dots, a^\nu\} = \big\{x \in \mathbb{R}^n \,\big|\, \langle d^i, x\rangle \le 0,\ i \in \mathbb{I}_0 \cup \mathbb{I}_-,\ \ \langle d^{ij}, x\rangle \le 0,\ i \in \mathbb{I}_+, j \in \mathbb{I}_-\big\}.$$

To show this claim, let $C^\nu$ be the set on the right-hand side. First, suppose that $\bar{x} \in \mathrm{pos}\{a^0, \dots, a^\nu\}$. Then, $\bar{x} = \bar{y} + \theta a^\nu$ for some $\bar{y} \in \mathrm{pos}\{a^0, \dots, a^{\nu-1}\}$ and $\theta \ge 0$. For $i \in \mathbb{I}_0 \cup \mathbb{I}_-$, one has

$$\langle d^i, \bar{x}\rangle = \langle d^i, \bar{y}\rangle + \theta\langle d^i, a^\nu\rangle \le 0$$

by also invoking (2.9). For $i \in \mathbb{I}_+$, $j \in \mathbb{I}_-$, one has

$$\begin{aligned}\langle d^{ij}, \bar{x}\rangle &= \langle d^i, \bar{y}\rangle - \frac{\alpha_i}{\alpha_j}\langle d^j, \bar{y}\rangle + \theta\langle d^i, a^\nu\rangle - \theta\frac{\alpha_i}{\alpha_j}\langle d^j, a^\nu\rangle \\ &= \langle d^i, \bar{y}\rangle - \frac{\alpha_i}{\alpha_j}\langle d^j, \bar{y}\rangle + \theta\alpha_i - \theta\alpha_i \le 0,\end{aligned}$$

again invoking (2.9) and recognizing that $\alpha_i > 0$ and $\alpha_j < 0$ in this case. Thus, $\bar{x} \in C^\nu$ and we've established that $\mathrm{pos}\{a^0, \dots, a^\nu\} \subset C^\nu$.

Second, let $\bar{x} \in C^\nu$. It suffices to show that there's $\theta \ge 0$ such that

$$\bar{x} - \theta a^\nu \in \mathrm{pos}\{a^0, \dots, a^{\nu-1}\}$$

because this means that $\bar{x} - \theta a^\nu = \sum_{i=1}^{\nu-1} \lambda_i a^i$ for some $\lambda_i \ge 0$ and then also

$$\bar{x} = \sum\nolimits_{i=1}^{\nu-1} \lambda_i a^i + \theta a^\nu \in \mathrm{pos}\{a^0, \dots, a^\nu\}.$$

In view of (2.9), our goal is therefore to construct $\theta \ge 0$ such that

$$\langle d^i, \bar{x} - \theta a^\nu\rangle = \langle d^i, \bar{x}\rangle - \theta\alpha_i \le 0 \quad \forall i \in \mathbb{I}^{\nu-1}. \tag{2.10}$$

If $\mathbb{I}_-$ is empty, then

$$\theta = \max\big\{0,\ \ \max_{i \in \mathbb{I}_+} \langle d^i, \bar{x}\rangle / \alpha_i\big\}$$

can be used, where the maximum over an empty set is interpreted as $-\infty$. To see this, note that for $i \in \mathbb{I}_0$, (2.10) certainly holds in view of the definition of $C^\nu$. For $i \in \mathbb{I}_+$,

$$\langle d^i, \bar{x}\rangle - \theta\alpha_i \le \langle d^i, \bar{x}\rangle - \frac{\alpha_i}{\alpha_i}\langle d^i, \bar{x}\rangle = 0$$

and (2.10) holds again. If $\mathbb{I}_- \neq \emptyset$, then

$$\theta = \min_{i\in\mathbb{I}_-} \langle d^i, \bar{x}\rangle/\alpha_i,$$

which is nonnegative, can be used because for $i \in \mathbb{I}_0$, (2.10) holds trivially. Given $i \in \mathbb{I}_-$,

$$\langle d^i, \bar{x}\rangle - \theta\alpha_i \le \langle d^i, \bar{x}\rangle - \frac{\alpha_i}{\alpha_i}\langle d^i, \bar{x}\rangle = 0$$

so (2.10) holds as well. For $i \in \mathbb{I}_+$ and any $j^\star \in \operatorname{argmin}_{j\in\mathbb{I}_-} \langle d^j, \bar{x}\rangle/\alpha_j$, one has

$$\langle d^i, \bar{x}\rangle - \theta\alpha_i = \langle d^i, \bar{x}\rangle - \frac{\alpha_i}{\alpha_{j^\star}}\langle d^{j^\star}, \bar{x}\rangle \le 0$$

by virtue of having $\bar{x} \in C^\nu$ so (2.10) holds again. Thus, we've established that $\bar{x} \in \operatorname{pos}\{a^0, \ldots, a^\nu\}$ and, in fact, $\operatorname{pos}\{a^0, \ldots, a^\nu\} \supset C^\nu$. □

The converse of Weyl's theorem relies on polarity. For a cone $C \subset \mathbb{R}^n$, the *polar* of $C$ is the cone

$$\operatorname{pol} C = \{v \in \mathbb{R}^n \mid \langle v, x\rangle \le 0 \ \forall x \in C\} \tag{2.11}$$

as illustrated in Figure 2.26(left). Since $\operatorname{pol} C = \cap_{x\in C}\{v \mid \langle v, x\rangle \le 0\}$, it's an intersection of closed convex sets and thus always closed and convex; see 1.12(a).

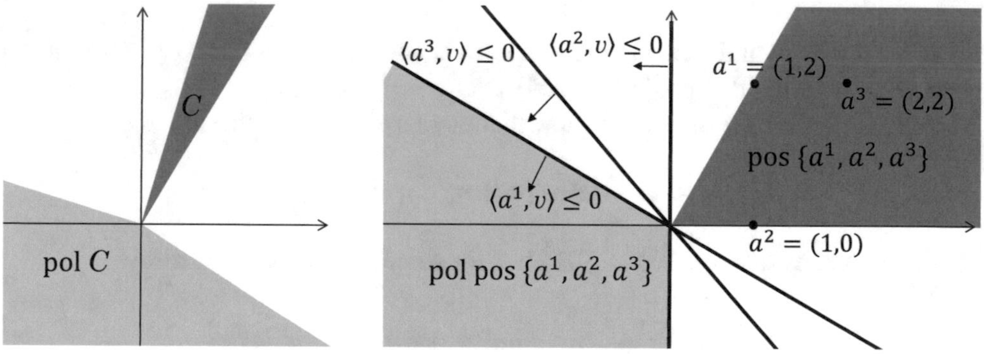


Fig. 2.26: Cones and their polars.

For a closed convex cone $C \subset \mathbb{R}^n$, the situation is perfectly symmetric:

$$\operatorname{pol}(\operatorname{pol} C) = C \tag{2.12}$$

by [105, Corollary 6.21]. This is utilized in the next proposition, which is equivalent to the celebrated Farkas' lemma and is explained in Figure 2.26(right).

**Proposition 2.60** (polar of polyhedral cone). *For $\{a^i \in \mathbb{R}^n, i = 1, \dots, m\}$, one has*

$$\operatorname{pol}\big(\operatorname{pos}\{a^1, \dots, a^m\}\big) = \big\{v \in \mathbb{R}^n \,\big|\, \langle a^i, v\rangle \le 0,\ \ i = 1, \dots, m\big\}$$
$$\operatorname{pos}\{a^1, \dots, a^m\} = \operatorname{pol}\big\{v \in \mathbb{R}^n \,\big|\, \langle a^i, v\rangle \le 0,\ \ i = 1, \dots, m\big\}.$$

**Proof.** Since $\{a^1, \dots, a^m\} \subset \operatorname{pos}\{a^1, \dots, a^m\}$, it follows directly from the definition of polars that

$$\operatorname{pol}\big(\operatorname{pos}\{a^1, \dots, a^m\}\big) \subset \big\{v \in \mathbb{R}^n \,\big|\, \langle a^i, v\rangle \le 0,\ \ i = 1, \dots, m\big\}.$$

Suppose that $\bar v$ satisfies $\langle \bar v, a^i\rangle \le 0$ for all $i$, which means that

$$\Big\langle \bar v, \sum_{i=1}^m \lambda_i a^i \Big\rangle = \sum_{i=1}^m \lambda_i \langle \bar v, a^i\rangle \le 0 \ \ \forall \lambda_i \ge 0,\ i = 1, \dots, m.$$

Thus, $\bar v \in \operatorname{pol}(\operatorname{pos}\{a^1, \dots, a^m\})$ and the first equation of the proposition holds. By (2.12),

$$\operatorname{pol}\Big(\operatorname{pol}\big(\operatorname{pos}\{a^1, \dots, a^m\}\big)\Big) = \operatorname{pos}\{a^1, \dots, a^m\}.$$

Thus, the second equation holds as well. □

**Theorem 2.61** (Minkowski's). *Every polyhedral cone is the positive hull of a finite collection of points.*

**Proof.** Let $C \subset \mathbb{R}^n$ be a polyhedral cone and thus expressible in terms of a finite number of linear equalities and inequalities. Since every equality can be written as two inequalities, we can assume without loss of generality that

$$C = \big\{x \in \mathbb{R}^n \,\big|\, \langle d^i, x\rangle \le \delta_i,\ i = 1, \dots, q\big\}$$

for some $\{d^i \in \mathbb{R}^n, \delta_i \in \mathbb{R}, i = 1, \dots, q\}$. Since $C$ is a cone, we can set all $\delta_i = 0$ because an inequality $\langle d^i, x\rangle \le \delta_i$ that is active at some point $\bar x \in C$ would need to have

$$\langle d^i, \bar x\rangle = \delta_i \quad \text{and} \quad \lambda\langle d^i, \bar x\rangle \le \delta_i$$

for all $\lambda \in [0, \infty)$ and that's not possible if $\delta_i \ne 0$.

Weyl's theorem 2.59 and 2.60 establish that

$$\operatorname{pol} C = \operatorname{pos}\{d^1, \dots, d^q\} = \big\{v \in \mathbb{R}^n \,\big|\, \langle a^i, v\rangle \le 0,\ i = 1, \dots, m\big\}$$
$$= \operatorname{pol}\big(\operatorname{pos}\{a^1, \dots, a^m\}\big)$$

for some $\{a^i \in \mathbb{R}^n, i = 1, \dots, m\}$. By (2.12), this means that $C = \operatorname{pos}\{a^1, \dots, a^m\}$ and the conclusion follows. □

A *convex combination* of $\{a^i \in \mathbb{R}^n,\ i = 1,\ldots,m\}$ is a point $\sum_{i=1}^m \lambda_i a^i$ with $\sum_{i=1}^m \lambda_i = 1$ and $\lambda_i \geq 0$ for all $i$. The *convex hull* of $C \subset \mathbb{R}^n$, denoted by con $C$, is the set of all convex combinations formed by points in $C$, i.e.,

$$\operatorname{con} C = \left\{\sum\nolimits_{i=1}^m \lambda_i a^i \;\middle|\; m \in \mathbb{N},\ \sum\nolimits_{i=1}^m \lambda_i = 1,\ \lambda_i \geq 0,\ a^i \in C,\ i = 1,\ldots,m\right\}.$$

In particular, for $\{a^i \in \mathbb{R}^n,\ i = 1,\ldots,m\}$,

$$\operatorname{con}\{a^1,\ldots,a^m\} = \left\{\sum\nolimits_{i=1}^m \lambda_i a^i \;\middle|\; \sum\nolimits_{i=1}^m \lambda_i = 1,\ \lambda_i \geq 0,\ i = 1,\ldots,m\right\}.$$

A convex hull is always a convex set; see Figure 2.27.

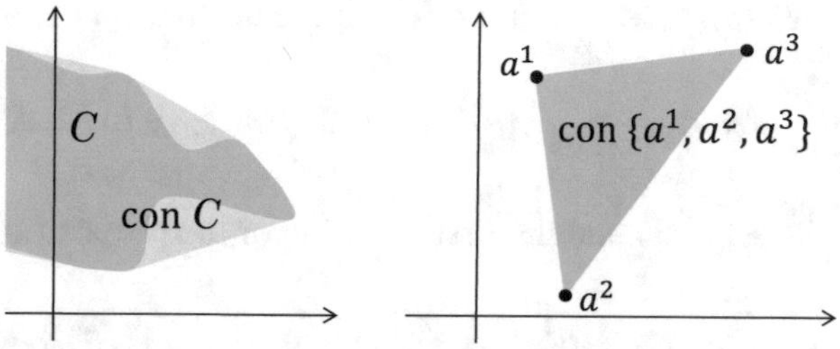


Fig. 2.27: Convex hulls.

A property of a polyhedral set is that it can be generated by two finite collections of points.

**Theorem 2.62** (finite generation). *For a nonempty set $C \subset \mathbb{R}^n$, there are $\{a^i \in \mathbb{R}^n, i = 1,\ldots,m\}$ and $\{c^i \in \mathbb{R}^n, i = 1,\ldots,q\}$ such that*

$$C = \operatorname{con}\{a^1,\ldots,a^m\} + \operatorname{pos}\{c^1,\ldots,c^q\}$$

*if and only if $C$ is polyhedral.*

**Proof.** Suppose that $C = \operatorname{con}\{a^1,\ldots,a^m\} + \operatorname{pos}\{c^1,\ldots,c^q\}$. Then,

$$\begin{aligned} C =& \left\{x \in \mathbb{R}^n \;\middle|\; \begin{bmatrix} x \\ 1 \end{bmatrix} = \sum\nolimits_{i=1}^m \lambda_i \begin{bmatrix} a^i \\ 1 \end{bmatrix} + \sum\nolimits_{i=1}^q \mu_i \begin{bmatrix} c^i \\ 0 \end{bmatrix},\ \lambda_i \geq 0, \mu_i \geq 0\ \forall i\right\} \\ =& \left\{x \in \mathbb{R}^n \;\middle|\; \begin{bmatrix} x \\ \theta \end{bmatrix} \in \operatorname{pos}\left\{\begin{bmatrix} a^1 \\ 1 \end{bmatrix},\ldots,\begin{bmatrix} a^m \\ 1 \end{bmatrix},\begin{bmatrix} c^1 \\ 0 \end{bmatrix},\ldots,\begin{bmatrix} c^q \\ 0 \end{bmatrix}\right\},\ \theta = 1\right\}. \end{aligned}$$

By Weyl's theorem 2.59, there's a collection $\{(d^i, \delta_i) \in \mathbb{R}^n \times \mathbb{R}, i = 1,\ldots,r\}$ such that

$$\begin{aligned} &\operatorname{pos}\left\{\begin{bmatrix} a^1 \\ 1 \end{bmatrix},\ldots,\begin{bmatrix} a^m \\ 1 \end{bmatrix},\begin{bmatrix} c^1 \\ 0 \end{bmatrix},\ldots,\begin{bmatrix} c^q \\ 0 \end{bmatrix}\right\} \\ &\qquad = \left\{(x,\theta) \in \mathbb{R}^n \times \mathbb{R} \;\middle|\; \langle d^i, x\rangle + \delta_i\theta \leq 0,\ i = 1,\ldots,r\right\}. \end{aligned}$$

Consequently, $C = \{x \in \mathbb{R}^n \mid \langle d^i, x\rangle \le -\delta_i,\ i = 1, \dots, r\}$, which is a polyhedral set.

For the converse, let

$$\begin{aligned} C &= \{x \in \mathbb{R}^n \mid \langle d^i, x\rangle \le \delta_i,\ i = 1, \dots, r\} \\ &= \{x \in \mathbb{R}^n \mid \langle d^i, x\rangle - \delta_i\theta \le 0,\ i = 1, \dots, r,\ -\theta \le 0,\ \theta = 1\} \\ &= \left\{x \in \mathbb{R}^n \,\middle|\, \begin{bmatrix} x \\ \theta \end{bmatrix} \in \text{pos}\left\{\begin{bmatrix} b^1 \\ \beta_1 \end{bmatrix}, \dots, \begin{bmatrix} b^m \\ \beta_m \end{bmatrix}\right\},\ \theta = 1\right\}, \end{aligned}$$

with the last equality following from Minkowski's theorem 2.61 for some $\{(b^i, \beta_i) \in \mathbb{R}^n \times \mathbb{R}, i = 1, \dots, m\}$. Since the theorem is applied to a set with the last coordinate being nonnegative ($\theta \ge 0$), one necessarily has $\beta_i \ge 0$ for all $i$. Without loss of generality, we can assume that for all $i$, $\beta_i = 1$ or $\beta_i = 0$; if not, simply multiply $(b^i, \beta_i)$ by $\beta_i^{-1}$ whenever $1 \ne \beta_i > 0$. This has no effect on the positive hull. Let $\mathbb{I}_1 = \{i \mid \beta_i = 1\}$ and $\mathbb{I}_0 = \{i \mid \beta_i = 0\}$. Then,

$$\begin{aligned} C &= \left\{x \in \mathbb{R}^n \,\middle|\, \begin{bmatrix} x \\ 1 \end{bmatrix} = \sum\nolimits_{i \in \mathbb{I}_0} \lambda_i \begin{bmatrix} b^i \\ 0 \end{bmatrix} + \sum\nolimits_{i \in \mathbb{I}_1} \mu_i \begin{bmatrix} b^i \\ 1 \end{bmatrix},\ \lambda_i \ge 0, \mu_i \ge 0\ \forall i\right\} \\ &= \left\{x \in \mathbb{R}^n \,\middle|\, x = \sum\nolimits_{i \in \mathbb{I}_0} \lambda_i b^i + \sum\nolimits_{i \in \mathbb{I}_1} \mu_i b^i,\ \sum\nolimits_{i \in \mathbb{I}_1} \mu_i = 1,\ \lambda_i \ge 0,\ \mu_i \ge 0\ \forall i\right\}. \end{aligned}$$

Thus, $C = \text{con}\{b^i, i \in \mathbb{I}_1\} + \text{pos}\{b^i, i \in \mathbb{I}_0\}$. □

The fact that a polyhedral set can either be represented by a finite number of linear inequalities or by a finite number of points can be exploited computationally.

We next turn to rules for checking whether a set is polyhedral.

**Proposition 2.63** (polyhedral sets under affine mapping). *For a polyhedron $C \subset \mathbb{R}^n$ and an affine mapping $F : \mathbb{R}^n \to \mathbb{R}^m$, the set*

$$F(C) = \{F(x) \mid x \in C\}$$

*is also polyhedral. In particular,*

$$D = \{x \in \mathbb{R}^n \mid Ux + Vy \le d \text{ for some } y \in Y\}$$

*is a polyhedral set when $Y \subset \mathbb{R}^q$ is polyhedral, $d \in \mathbb{R}^m$, $U$ is an $m \times n$-matrix and $V$ is an $m \times q$-matrix.*

**Proof.** If $C$ is empty, then $F(C)$ is also empty and thus polyhedral. If $C$ is nonempty, then by 2.62 there are $\{a^i \in \mathbb{R}^n, i = 1, \dots, m\}$ and $\{c^i \in \mathbb{R}^n, i = 1, \dots, q\}$ such that

$$C = \text{con}\{a^1, \dots, a^m\} + \text{pos}\{c^1, \dots, c^q\},$$

which implies that

$$F(C) = \text{con}\{F(a^1), \dots, F(a^m)\} + \text{pos}\{F(c^1), \dots, F(c^q)\}.$$

The second part holds because $D$ is obtained by mapping the polyhedral set

$$\{(x, y) \in \mathbb{R}^{n+m} \mid Ux + Vy \le d,\ y \in Y\}$$

using $F(x, y) = x$. □

**Proposition 2.64** (sum of polyhedra). *For polyhedra $\{C_i \subset \mathbb{R}^n,\ i = 1, \ldots, m\}$, the set $C = C_1 + \cdots + C_m$ is polyhedral.*

**Proof.** By 2.62, we can generated each nonempty $C_i$ using two finite collections of points. It's immediately clear that $C$ can be expressed similarly if nonempty. □

Polyhedral sets can be used to define a tractable class of functions that extends the reach of algorithms for linear optimization problems. A function $f : \mathbb{R}^n \to \overline{\mathbb{R}}$ is said to be *epi-polyhedral* if epi $f$ is a polyhedron.

From the definition, it follows immediately that an epi-polyhedral function $f : \mathbb{R}^n \to \overline{\mathbb{R}}$ is convex and dom $f$ is a polyhedral set by 2.63. Its epigraph must be of the form

$$\operatorname{epi} f = \{(x, \alpha) \in \mathbb{R}^{n+1} \mid \langle a^i, x\rangle + \eta_i\alpha \le \beta_i,\ i = 1, \ldots, q\}$$

and

$$f(x) = \inf_{\alpha \in \mathbb{R}} \{\alpha \mid \langle a^i, x\rangle + \eta_i\alpha \le \beta_i,\ i = 1, \ldots, q\}\ \ \forall x \in \mathbb{R}^n. \tag{2.13}$$

Consequently,

$$\begin{aligned} & x^\star \in \operatorname{argmin} f,\ \alpha^\star = \inf f \in \mathbb{R} \\ \Longleftrightarrow\ & (x^\star, \alpha^\star) \in \operatorname{argmin} \{\alpha \mid (x, \alpha) \in \operatorname{epi} f\} \\ \Longleftrightarrow\ & (x^\star, \alpha^\star) \in \operatorname{argmin} \{\alpha \mid \langle a^i, x\rangle + \eta_i\alpha \le \beta_i,\ \ i = 1, \ldots, q\} \end{aligned}$$

so every epi-polyhedral function can be minimized by solving a linear optimization problem. This fact already underpins the reformulation of the hinge-loss problem in §2.H.

**Example 2.65** (max of epi-polyhedral functions). If $f_i : \mathbb{R}^n \to \overline{\mathbb{R}}$, $i = 1, \ldots, m$, are epi-polyhedral functions, then

$$f(x) = \max\nolimits_{i=1,\ldots,m} f_i(x)$$

defines another epi-polyhedral function.

**Detail.** Since epi $f = \cap_{i=1}^m$ epi $f_i$ and an intersection of polyhedral sets is polyhedral, the conclusion follows. □

**Proposition 2.66** (epi-polyhedral inf-projection). *For epi-polyhedral $f : \mathbb{R}^m \times \mathbb{R}^n \to \overline{\mathbb{R}}$, the inf-projection $p : \mathbb{R}^m \to \overline{\mathbb{R}}$ given by*

$$p(u) = \inf \{f(u, x) \mid x \in \mathbb{R}^n\}$$

*is epi-polyhedral.*

**Proof.** Suppose that epi $p = F(\text{epi } f)$ for the affine mapping $F(u, x, \alpha) = (u, \alpha)$. Then, the conclusion is immediate by 2.63. The required relationship between the epigraphs is seen as follows. Since $(u, \alpha) \in F(\text{epi } f)$ means that $f(u, \bar{x}) \leq \alpha$ for some $\bar{x} \in \mathbb{R}^n$, we also have that

$$p(u) = \inf \{ f(u, x) \mid x \in \mathbb{R}^n \} \leq \alpha.$$

Consequently, $F(\text{epi } f) \subset \text{epi } p$. Next, let $(u, \alpha) \in \text{epi } p$. Then, $\inf\{f(u, x) \mid x \in \mathbb{R}^n\} \leq \alpha$. If this infimum is $-\infty$, then there exists $\bar{x}$ such that $f(u, \bar{x}) \leq \alpha$. Otherwise, when the infimum is a real number, a fact about linear optimization problems [105, Corollary 11.16] asserts that

$$\bar{x} \in \operatorname{argmin} \{ f(u, x) \mid x \in \mathbb{R}^n \}$$

and again $f(u, \bar{x}) \leq \alpha$. In either case, this means that $(u, \bar{x}, \alpha) \in \text{epi } f$ and then also $(u, \alpha) \in F(\text{epi } f)$. □

**Proposition 2.67** (sum of epi-polyhedral functions). *If* $\{f_i : \mathbb{R}^n \to \overline{\mathbb{R}}, i = 1, \dots, m\}$ *are epi-polyhedral functions, then*

$$f(x) = \sum\nolimits_{i=1}^m f_i(x)$$

*defines another epi-polyhedral function.*

**Proof.** Since each epi $f_i$ is a polyhedron, $f_i$ can be written in the form (2.13), i.e.,

$$f_i(x) = \inf_{\alpha_i \in \mathbb{R}} \{ \alpha_i \mid \langle a^{ki}, x \rangle + \eta_{ki}\alpha_i \leq \beta_{ki}, \ k = 1, \dots, q_i \}.$$

Thus,

$$f(x) = \inf_{\alpha \in \mathbb{R}^m} \Big\{ \sum\nolimits_{i=1}^m \alpha_i \ \Big| \ \langle a^{ki}, x \rangle + \eta_{ki}\alpha_i \leq \beta_{ki}, \ k = 1, \dots, q_i, \ i = 1, \dots, m \Big\}.$$

Since the function on $\mathbb{R}^m \times \mathbb{R}^n$ defined by

$$g(\alpha, x) = \begin{cases} \sum_{i=1}^m \alpha_i & \text{if } \langle a^{ki}, x \rangle + \eta_{ki}\alpha_i \leq \beta_{ki}, \ k = 1, \dots, q_i, \ i = 1, \dots, m \\ \infty & \text{otherwise} \end{cases}$$

is epi-polyhedral, it follows from 2.66 that $f$ is epi-polyhedral. □

**Example 2.68** (representation of epi-polyhedral functions). Suppose that $f : \mathbb{R} \to \mathbb{R}$ is defined for all $x \in \mathbb{R}$ by

$$f(x) = \begin{cases} \alpha_0 + \beta_0 x & \text{when } x < \tau_1 \\ \alpha_i + \beta_i x & \text{when } \tau_i \leq x < \tau_{i+1}, \ i = 1, \dots, q-1 \\ \alpha_q + \beta_q x & \text{when } x \geq \tau_q, \end{cases}$$

where $q \in \mathbb{N}$ and $\tau_1 < \tau_2 < \cdots < \tau_q$, $\alpha_0, \alpha_1, \dots, \alpha_q$ as well as $\beta_0 \leq \beta_1 \leq \cdots \leq \beta_q$ are all real numbers, with

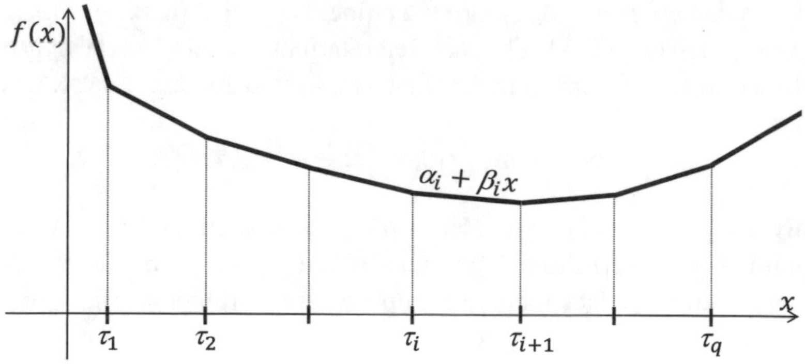


Fig. 2.28: Representation of epi-polyhedral function.

$$\alpha_i = \alpha_{i-1} + (\beta_{i-1} - \beta_i)\tau_i, \quad i = 1, \ldots, q.$$

Thus, $f$ is continuous. Since $\beta_i$ is nondecreasing as $i$ increases, one has

$$f(x) = \max_{i=0,1,\ldots,q} \alpha_i + \beta_i x$$

and we see that it's epi-polyhedral. Figure 2.28 illustrates one possibility. We can minimize $f$ by solving the linear optimization problem

$$\underset{x,\gamma\in\mathbb{R}}{\text{minimize}} \ \gamma \ \text{ subject to } \alpha_i + \beta_i x \leq \gamma, \ \ i = 0, 1, \ldots, q.$$

This reformulation could be beneficial if $f$ is part of a more complicated model.

**Detail.** As an alternative to the max-expression for $f$, we've

$$f(x) = \alpha_0 + \beta_0\tau_1 + \min_{z\in C}\Big\{ -\beta_0 z_0 + \sum\nolimits_{i=1}^{q} \beta_i z_i \,\Big|\, x - \tau_1 = -z_0 + \sum\nolimits_{i=1}^{q} z_i \Big\},$$

where $z = (z_0, z_1, \ldots, z_q)$ and

$$C = \big\{z \in \mathbb{R}^{1+q} \,\big|\, 0 \leq z_0, \ \ 0 \leq z_i \leq \tau_{i+1} - \tau_i, \ \ i = 1, \ldots, q-1, \ \ 0 \leq z_q\big\}.$$

This expression may be less intuitive but can sometimes be beneficial. It helps to think of $z_0$ as measuring how far to the left of $\tau_1$ the point $x$ is located and $z_1 + \cdots + z_q$ as quantifying how far to the right of $\tau_1$ the point might be. In particular, it can be shown that the minimization problem in the expression for $f$ has the following $z^\star$ as minimizer:

$$
\begin{aligned}
&\text{if } x < \tau_1: \quad z_0^\star = \tau_1 - x; \quad z_i^\star = 0, \ i = 1, \dots, q\\
&\text{if } \tau_i \le x < \tau_{i+1} \text{ and } i = 1, \dots, q-1: \quad z_0^\star = 0\\
&\qquad z_j^\star = \tau_{j+1} - \tau_j, \ j = 1, \dots, i-1\\
&\qquad z_i^\star = x - \tau_i\\
&\qquad z_j^\star = 0, \ j = i+1, \dots, q\\
&\text{if } x \ge \tau_q: \quad z_0^\star = 0; \quad z_q^\star = x - \tau_q; \quad z_i^\star = \tau_{i+1} - \tau_i, \ i = 1, \dots, q-1.
\end{aligned}
$$

Consequently, we can minimize $f$ by solving the equivalent problem

$$
\begin{aligned}
\underset{x\in\mathbb{R}, z\in\mathbb{R}^{1+q}}{\text{minimize}} \quad & \alpha_0 + \beta_0\tau_1 - \beta_0 z_0 + \sum_{i=1}^{q} \beta_i z_i\\
\text{subject to} \quad & x - \tau_1 = -z_0 + \sum_{i=1}^{q} z_i\\
& 0 \le z_0\\
& 0 \le z_i \le \tau_{i+1} - \tau_i, \ i = 1, \dots, q-1\\
& 0 \le z_q,
\end{aligned}
$$

which involves $q + 2$ variables, $2q$ inequalities and one equality constraint. Although of larger size, this second formulation might be superior to the first one due to its simpler constraints. □

# Chapter 3
# OPTIMIZATION UNDER UNCERTAINTY

We often encounter optimization problems with parameters that are unknown at the time when a solution needs to be obtained. A store doesn't know the future demand when its inventory is re-stocked. A classifier is fitted to data that may not represent the context in which it will be applied. Under such circumstances, we would like to formulate a model that accounts for uncertainty and produces good decisions even though future conditions are unknown. This leads to problems of *optimization under uncertainty*. We'll now examine several formulations of such problems and construct algorithms that address the often large-scale nature of the resulting models.

## 3.A Product Mix Optimization

A firm manufactures and sells four types of wooden surf boards. Each type requires $t_{1j}$ hours for carpentry and $t_{2j}$ hours for finishing, $j = 1, \ldots, 4$. In a time period, there are $d_1 = 6000$ hours available for carpentry and $d_2 = 4000$ for finishing. There's a profit per sale of \$12, \$25, \$21 and \$40 for the four types, respectively. Under the assumption that all units produced can be sold, the goal is to determine how much to manufacture of each type of board during a time period such that the total profit is maximized and the labor constraints are satisfied. This an example of the challenges in the area of *production planning*.

With $d = (d_1, d_2)$ and

$$T = \begin{bmatrix} t_{11} & t_{12} & t_{13} & t_{14} \\ t_{21} & t_{22} & t_{23} & t_{24} \end{bmatrix} = \begin{bmatrix} 4 & 9 & 7 & 10 \\ 1 & 1 & 3 & 40 \end{bmatrix},$$

this leads to the linear optimization problem

$$\underset{x \in \mathbb{R}^4}{\text{minimize}}\ \langle c, x \rangle \quad \text{subject to} \quad Tx \le d, \quad x \ge 0, \tag{3.1}$$



J. O. Royset and R. J-B Wets, *An Optimization Primer*, Springer Series in Operations Research and Financial Engineering, https://doi.org/10.1007/978-3-030-76275-9_3

where $c = (-12, -25, -21, -40)$ due to the switch from maximization to minimization. A minimizer[1]

$$x^\star = (1333.3,\ 0,\ 0,\ 66.7), \quad \text{with minimum value } -\$18667,$$

can be obtained by the interior-point method (§2.G). The main takeaway from this solution is that surf boards of type 2 and 3 shouldn't be produced at all. Maybe the firm should get rid of those two production lines?

Before we jump to such a drastic conclusion, let's get a bit more realistic and account for the fact that the number of hours needed to produce each board may not be known with certainty. Each entry in the matrix $T$ now becomes a random variable and we denote by $\boldsymbol{T}$ the resulting random matrix. For simplicity's sake, let's assume that each entry of $\boldsymbol{T}$ takes on four possible values with equal probability (1/4) and that these entries are statistically independent of one another; see Table 3.1.

Table 3.1: Possible values for each entry in the random matrix $\boldsymbol{T}$.

| entry | possible values | | | |
|---|---|---|---|---|
| $\boldsymbol{t}_{11}$ | 3.60 | 3.90 | 4.10 | 4.40 |
| $\boldsymbol{t}_{12}$ | 8.25 | 8.75 | 9.25 | 9.75 |
| $\boldsymbol{t}_{13}$ | 6.85 | 6.95 | 7.05 | 7.15 |
| $\boldsymbol{t}_{14}$ | 9.25 | 9.75 | 10.25 | 10.75 |
| $\boldsymbol{t}_{21}$ | 0.85 | 0.95 | 1.05 | 1.15 |
| $\boldsymbol{t}_{22}$ | 0.85 | 0.95 | 1.05 | 1.15 |
| $\boldsymbol{t}_{23}$ | 2.60 | 2.90 | 3.10 | 3.40 |
| $\boldsymbol{t}_{24}$ | 37.0 | 39.0 | 41.0 | 43.0 |

We've eight random variables, each taking four possible values. This yields a total of $4^8 = 65536$ possible $T$ matrices (outcomes) and each one of these has an equal probability of occurring. Let's denote by $p_i$ the probability of a particular outcome $T^i, i = 1, \ldots, 65536$, which then becomes $1/65536$.

Since the firm must decide on the production plan before the number of hours required for carpentry and finishing are known with certainty, there's the possibility that the plan actually exceeds the number of hours available. Therefore, we must factor in the possibility of having to pay for overtime. This recourse cost is incurred at $a_1 = 5$ dollars per extra carpentry hour and $a_2 = 10$ dollars per extra finishing hour and enters into play after the production plan $x$ has been selected and the time required, $T^i$, for each task, has been observed. At least potentially, the firm makes a different decision about overtime in each one of the 65536 possible outcomes for $\boldsymbol{T}$. Let $y_1^i$ and $y_2^i$ denote the number of overtime hours hired for carpentry and finishing, respectively, when the matrix $\boldsymbol{T}$ turns out to be $T^i$, and let

[1] There's no immediate concern that the minimizer contains non-integers; a model of this kind usually supports long-term planning where insight is more important than exact production numbers. It's reasonable to make some compromises to ensure convexity. Anyhow, the relatively large values in $x^\star$ make rounding to the nearest integer a satisfactory remedy.

$$y = \big(y_1^1, y_2^1, y_1^2, y_2^2, \ldots, y_1^{65536}, y_2^{65536}\big).$$

Then, we obtain the model

$$\begin{aligned}
\underset{x\in\mathbb{R}^4, y\in\mathbb{R}^{131072}}{\text{minimize}} \ & \sum\nolimits_{j=1}^{4} c_j x_j + \sum\nolimits_{i=1}^{65536} p_i (a_1 y_1^i + a_2 y_2^i) \\
\text{subject to} \ & \sum\nolimits_{j=1}^{4} t_{1j}^i x_j - y_1^i \leq d_1, \quad i = 1, \ldots, 65536 \\
& \sum\nolimits_{j=1}^{4} t_{2j}^i x_j - y_2^i \leq d_2, \quad i = 1, \ldots, 65536 \\
& x_j, y_1^i, y_2^i \geq 0, \quad j = 1, \ldots, 4; \ i = 1, \ldots, 65536,
\end{aligned}$$

which also is a linear optimization problem but of a larger size than the original one due to the introduction of $y$.

The objective function now being minimized is the sum of the immediate cost (actually, the negative profits) and the *expected recourse cost* because one must consider 65536 possible outcomes. The constraints involving random quantities are written out explicitly for all the outcomes:

$$\text{carpentry hours} \sum\nolimits_{j=1}^{4} t_{1j}^i x_j \ \text{shouldn't exceed regular plus overtime hours } d_1 + y_1^i,$$

and finishing is restricted similarly. Since there's the possibility of making a *recourse decision* $y^i = (y_1^i, y_2^i)$ that depends on the outcomes of the random variables, this is a *stochastic optimization problem with recourse*.

Oftentimes, there's more than one source of uncertainty in a problem. For example, due to employee absence, the available hours for carpentry and finishing may also have to be modeled as random variables, say, $\boldsymbol{d}_1$ and $\boldsymbol{d}_2$. Let's adopt the values listed in Table 3.2 as possibilities for these quantities.

Table 3.2: Possible values for $\boldsymbol{d}_1$, each occurring with probability 1/4, and for $\boldsymbol{d}_2$, each also occurring with probability 1/4.

| entry | possible values | | | |
|---|---|---|---|---|
| $\boldsymbol{d}_1$ | 5873 | 5967 | 6033 | 6127 |
| $\boldsymbol{d}_2$ | 3936 | 3984 | 4016 | 4064 |

We now need to replace $d$ by $d^i = (d_1^i, d_2^i)$ and account for $4^2 = 16$ possible $d^i$ vectors. This gives a total of $\nu = 1048576$ possible outcomes of $(\boldsymbol{T}, \boldsymbol{d})$. With $p_i = 1/\nu$ and $a = (a_1, a_2)$, the model becomes

$$\begin{array}{llllll}
\text{minimize} & \langle c, x\rangle & +p_1\langle a, y^1\rangle & +p_2\langle a, y^2\rangle & + \cdots + & p_\nu\langle a, y^\nu\rangle & \\
\text{subject to} & T^1 x & -y^1 & & & & \le d^1 \\
& T^2 x & & -y^2 & & & \le d^2 \\
& \vdots & & & \ddots & & \vdots \\
& T^\nu x & & & & -y^\nu & \le d^\nu \\
& x \ge 0, & y^1 \ge 0, & y^2 \ge 0, & \ldots & y^\nu \ge 0. &
\end{array}$$

Although still linear, the model is even larger than the previous one. The sparse, block-angular structure, however, is beneficial for the interior-point method (§2.G) in general and can also be leveraged more specifically; see §5.H. A minimizer is

$$x^{\star\star} = (257,\ 0,\ 665.2,\ 33.8), \text{ with minimum value } -\$18051.$$

This minimizer is *robust* because it's constructed by a process that examines all the 1048576 possibilities and accounts for the resulting recourse costs and the associated probabilities of having to pay these costs. In contrast, $x^\star$ doesn't fare well when assessed in the last model:

$$\langle c, x^\star\rangle + \sum_{i=1}^{\nu} p_i\langle a, y^i\rangle = -\$16942,$$

where $y^i = (y_1^i, y_2^i)$ is the overtime hours required under outcome $i$, i.e.,

$$y_1^i = \max\Big\{0, \sum_{j=1}^{4} t_{1j}^i x_j^\star - d_1^i\Big\} \quad \text{and} \quad y_2^i = \max\Big\{0, \sum_{j=1}^{4} t_{2j}^i x_j^\star - d_2^i\Big\}.$$

The expected profit of \$16942 is significantly worse than that from $x^{\star\star}$ (\$18051). We note that the profit initially stipulated for $x^\star$ (\$18667) is misleading because it doesn't account for potential overtime cost, which in fact turns out to be \$1725, on average, across the possible outcomes. Thus, the expected profit drops to \$16942.

Moreover, $x^\star$ provides the wrong strategic insight by suggesting that the production line for type 3 can be closed down. In reality, boards of type 3 play a central role by filling in the production schedule for some outcomes and therefore should be produced in large numbers. If the production line for such boards had been closed down, then the firm wouldn't have had this cost-saving option.

**Exercise 3.1** (stochastic resources). Consider the product mix problem when the only uncertainty is about the number of hours available for carpentry and finishing; $c$, $a$ and $T$ are as above. The random variable $\boldsymbol{d}_1$ takes on the values 4800, 5500, 6050 and 6150, each with probability 1/4, while the random variable $\boldsymbol{d}_2$ has values 3936, 3984, 4016 and 4064, also with probability 1/4. The random variables are statistically independent. Solve both the stochastic optimization problem and the deterministic version (3.1), with $d = (5625, 4000)$, the expected values of $\boldsymbol{d}_1$ and $\boldsymbol{d}_2$. Compare the minimizers and minimum values.

**Guide.** The number of possible outcomes $\nu = 16$ so in addition to the nonnegativity constraints, the stochastic optimization problem has 32 constraints and a minimum value −\$15900. The expected cost associated with the decision obtained by the deterministic version is −\$15687.5. □

## 3.B Expectation Functions

The modeling of unknown parameters as random variables leads to optimization problems expressed in terms of expected values. In this section, we'll provide background material on random variables and derive conditions under which *expectation functions* of the form

$$x \mapsto \mathbb{E}\big[f(\boldsymbol{\xi}, x)\big]$$

are well defined and convex. Here, $\boldsymbol{\xi}$ is a random vector with a *probability distribution* that quantifies the likelihood of the possible outcomes $\xi \in \Xi \subset \mathbb{R}^m$. The probability distribution is usually specified by a set of probabilities $\{p_\xi \in (0,1], \xi \in \Xi\}$, a density function $p : \mathbb{R}^m \to [0, \infty)$ and/or a distribution function $P : \mathbb{R}^m \to [0, 1]$.

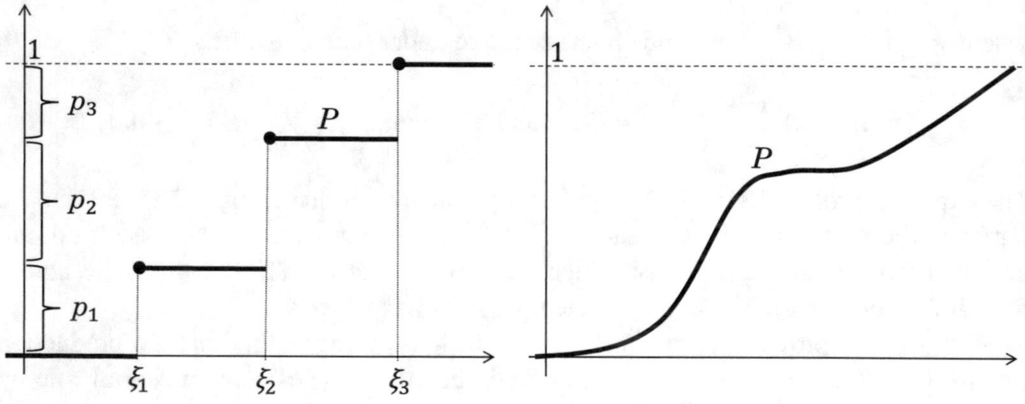


Fig. 3.1: Possible distribution functions of a random variable $\boldsymbol{\xi}$: finite (left) and continuous (right) distributions.

**Finite Distribution.** A random vector $\boldsymbol{\xi}$ is *finitely distributed* when the number of possible values (outcomes) for $\boldsymbol{\xi}$ is finite as in §3.A. Let $\Xi \subset \mathbb{R}^m$ be the set of these possible values, which then becomes the *support* of $\boldsymbol{\xi}$, and let $\{p_\xi > 0, \xi \in \Xi\}$ be the corresponding probabilities, naturally with $\sum_{\xi \in \Xi} p_\xi = 1$. In this case, the distribution function $P$ of $\boldsymbol{\xi}$ is a "staircase" as shown in Figure 3.1(left), with

$$P(\eta) = \sum\nolimits_{\xi \leq \eta} p_\xi \quad \forall \eta \in \mathbb{R}^m.$$

Moreover, $\mathbb{E}[\boldsymbol{\xi}] = \sum_{\xi \in \Xi} p_\xi \xi$ is the *expectation* (expected value, average, mean) of $\boldsymbol{\xi}$. We often construct new random variables from existing ones. For a function $h : \mathbb{R}^m \to \mathbb{R}$, the value of $h$ is uncertain when its argument is the unknown value of $\boldsymbol{\xi}$. We think of this uncertain quantity as a random variable, denoted by $h(\boldsymbol{\xi})$, with values $\{h(\xi), \xi \in \Xi\}$ and corresponding probabilities $\{p_\xi, \xi \in \Xi\}$. The expectation of $h(\boldsymbol{\xi})$ is then

$$\mathbb{E}\big[h(\boldsymbol{\xi})\big] = \sum\nolimits_{\xi \in \Xi} p_\xi h(\xi).$$

In stochastic optimization problems, we contend with an uncertain cost of some sort as well as how it's affected by different decisions. Thus, the cost is usually defined by a function that depends on both parameters and decision variables. For a given function $f : \mathbb{R}^m \times \mathbb{R}^n \to \mathbb{R}$ and a finitely distributed random vector $\boldsymbol{\xi}$ with support $\Xi \subset \mathbb{R}^m$, the *expectation function*

$$Ef : \mathbb{R}^n \to \mathbb{R}, \text{ with } Ef(x) = \mathbb{E}\big[f(\boldsymbol{\xi}, x)\big]$$

is well defined because for each $x \in \mathbb{R}^n$, $\mathbb{E}[f(\boldsymbol{\xi}, x)]$ is simply a finite sum of real numbers. The notation "$Ef$" indicates a function with values computed by taking expectations of $f$ with respect to the first argument. In view of 1.18(b), $Ef$ is convex when[2] $f(\xi, \cdot\,)$ is convex for all $\xi \in \Xi$. It's strictly convex if in addition $f(\xi, \cdot\,)$ is strictly convex for at least one $\xi \in \Xi$. Even though $\boldsymbol{\xi}$ is a random vector, $Ef$ is a well-defined (deterministic) function. Thus, the stochastic optimization problem of minimizing $Ef$ isn't actually "stochastic," but just a problem defined with a special kind of objective function that accounts for parameter uncertainty.

When we move beyond a finitely distributed random vector, we start dealing with sums of an infinite number of values, which might not be well defined. The next paragraphs review the necessary precautions that need to be taken.

**Discrete Distribution.** A random vector $\boldsymbol{\xi}$ is *discretely distributed* when the number of possible outcomes for $\boldsymbol{\xi}$ is countable so that the support $\Xi \subset \mathbb{R}^m$ contains a countable number of points, each associated with a corresponding probability $p_\xi > 0$ such that $\sum_{\xi \in \Xi} p_\xi = 1$. A Poisson distributed random variable is an example with $\Xi = \{0, 1, 2, \dots\}$ and

$$p_k = \lambda^k \exp(-\lambda)/k!, \quad k \in \Xi,$$

where $\lambda \in (0, \infty)$ is given and raised to power $k$ in the expression for $p_k$. All finitely distributed random vectors are certainly discretely distributed.

For any function $h : \mathbb{R}^m \to \mathbb{R}$, we can define a new discretely distributed random variable, denoted by $h(\boldsymbol{\xi})$, that takes values $\{h(\xi), \xi \in \Xi\}$, with corresponding probabilities $\{p_\xi, \xi \in \Xi\}$. A complication arises, however, when we would like to define expectations. Now, $\sum_{\xi \in \Xi} p_\xi \xi$ might not be finite. In fact, we run into the ambiguity associated with a series that consists of two subsequences, one with positive numbers tending to $\infty$ and one with negative numbers tending to $-\infty$. For example, suppose that

[2] A function in many variables defines a function in fewer variables if some of the variables are fixed. We indicate this by specifying the fixed variables and leaving a dot in place of the other variables. For fixed $\xi$, $f(\xi, \cdot\,)$ is here a function of the decision variables only.

$$\Xi = \{1, -1, 2, -2, \dots\} \quad \text{and} \quad p_\xi = \frac{3}{\pi^2}\frac{1}{\xi^2}.$$

Recalling the sum of p-series, we've that $\sum_{\xi\in\Xi} p_\xi = 1$, but the sum of $\xi p_\xi$ over the positive $\xi \in \Xi$ gives $\infty$ and over the negative ones gives $-\infty$. These "competing infinities" are left undefined in some texts, but in our context of minimization problems it's convenient to define $\infty - \infty = \infty$ as §1.D clarifies. We can then define the expectation for any discretely distributed random vector $\boldsymbol{\xi}$ and $h : \mathbb{R}^m \to \mathbb{R}$ as

$$\mathbb{E}\big[h(\boldsymbol{\xi})\big] = \mathbb{E}\big[\max\{0, h(\boldsymbol{\xi})\}\big] - \mathbb{E}\big[\max\{0, -h(\boldsymbol{\xi})\}\big].$$

Since both terms on the right-hand side sums up nonnegative numbers, they're well defined but possibly infinite. With the convention $\infty - \infty = \infty$, $\mathbb{E}[h(\boldsymbol{\xi})]$ becomes well defined.

**Continuous Distribution.** A random vector $\boldsymbol{\xi}$ is *continuously distributed* if its distribution function $P : \mathbb{R}^m \to [0, 1]$ is absolutely continuous[3] as seen in Figure 3.1(right), in which case it can be expressed by a *density function* $p : \mathbb{R}^m \to [0, \infty)$ and the $m$-dimensional integral

$$P(\eta) = \int_{\{\xi\le\eta\}} p(\xi)\, d\xi = \int_{-\infty}^{\eta_m} \cdots \int_{-\infty}^{\eta_2} \int_{-\infty}^{\eta_1} p(\xi)\, d\xi_1 d\xi_2 \cdots d\xi_m \quad \forall \eta \in \mathbb{R}^m.$$

The *support* of $\boldsymbol{\xi}$ is defined as $\operatorname{cl}\{\xi \in \mathbb{R}^m \mid p(\xi) > 0\}$, i.e., the closure of the points at which the density function is positive. A difficulty arises when we attempt to construct a new random variable $h(\boldsymbol{\xi})$ by means of a function $h : \mathbb{R}^m \to \mathbb{R}$. If $\boldsymbol{\xi}$ were discretely distributed with support $\Xi$, then it's clear what probability to assign to the various values $\{h(\xi), \xi \in \Xi\}$ so that $h(\boldsymbol{\xi})$ indeed is a well-defined random variable. When $\boldsymbol{\xi}$ is continuously distributed with density function $p$ and support $\Xi$, we need to be able to express the probability of the outcomes $\{\xi \in \Xi \mid h(\xi) \le \alpha\}$ in terms of $p$. This is well understood, but requires advanced mathematics better avoided at this stage. It suffices to realize that $h$ needs to be *measurable*[4], which simply means that it's sufficiently "nice" so that the necessary outcomes can be quantified by $p$. We note that nearly all functions arising in practice are indeed measurable. For example, if $h$ is continuous, or epi $h$ is closed, or hypo $h$ is closed, then $h$ is measurable and $h(\boldsymbol{\xi})$ becomes a well-defined random variable.

For a measurable function $h : \mathbb{R}^m \to \mathbb{R}$ and a continuously distributed random vector $\boldsymbol{\xi}$ with density function $p$, the expectation

$$\mathbb{E}\big[h(\boldsymbol{\xi})\big] = \int \max\{0, h(\xi)\}p(\xi)\, d\xi - \int \max\{0, -h(\xi)\}p(\xi)\, d\xi$$

is well defined, but possibly not finite. Again, we're taking the precaution to separate the positive values from the negative values and using the convention $\infty - \infty = \infty$. If $h$ is continuous, then introductory integration theory gives the exact meaning of the integrals

[3] We omit the nuance about absolutely continuous functions and simply recall that they're indeed continuous.

[4] For readers familiar with measure theory: We adopt the Borel sigma-algebra on $\mathbb{R}^m$.

and useful calculus rules. The same holds for more general $h$, but then measure theory enters and we omit the details.

**General Distribution.** Even if it doesn't fall in the special cases described above, every $m$-dimensional random vector $\boldsymbol{\xi}$ has a distribution function $P$ from which its expectation $\mathbb{E}[\boldsymbol{\xi}]$ is well defined but may not consist entirely of finite values. Its *support* is the smallest closed subset of $\mathbb{R}^m$ such that the probability of $\boldsymbol{\xi}$ taking values in the set is one. Moreover, a measurable function $h : \mathbb{R}^m \to \mathbb{R}$ gives a new random variable $h(\boldsymbol{\xi})$, with a well-defined *expectation* $\mathbb{E}[h(\boldsymbol{\xi})]$, which again might not be finite. Competing infinities are settled as above. Regardless of the type of probability distribution for $\boldsymbol{\xi}$, we say that $h(\boldsymbol{\xi})$ is *integrable* if $\mathbb{E}[h(\boldsymbol{\xi})]$ is finite, which necessarily means that $h$ is measurable. A vector of integrable random variables is an integrable random vector. Moreover, $f : \mathbb{R}^m \times \mathbb{R}^n \to \mathbb{R}$ defines an *expectation function*

$$Ef : \mathbb{R}^n \to \overline{\mathbb{R}}, \text{ with } Ef(x) = \mathbb{E}\big[f(\boldsymbol{\xi}, x)\big],$$

provided that $f(\cdot, x)$ is measurable for all $x \in \mathbb{R}^n$ because then we can view $f(\boldsymbol{\xi}, x)$ as a random variable for each $x \in \mathbb{R}^n$. If $f(\boldsymbol{\xi}, x)$ is integrable for all $x \in \mathbb{R}^n$, then $Ef$ is real-valued and $\operatorname{dom} Ef = \mathbb{R}^n$. In lieu of a full treatment of expectations for general distributions, we give two facts that extend the familiar properties of linearity and order preserving for sums and (Riemann) integrals; see §8.G and [14] for additional details.

**Proposition 3.2** (properties of expectations). *For a random vector $\boldsymbol{\xi}$ with support $\Xi \subset \mathbb{R}^m$ and $h, g : \mathbb{R}^m \to \mathbb{R}$ with $h(\boldsymbol{\xi})$ and $g(\boldsymbol{\xi})$ integrable, the following hold:*

(a) $h(\xi) \leq g(\xi)\ \forall \xi \in \Xi \implies \mathbb{E}[h(\boldsymbol{\xi})] \leq \mathbb{E}[g(\boldsymbol{\xi})]$.
(b) $\forall \alpha, \beta \in \mathbb{R} : \quad \mathbb{E}[\alpha h(\boldsymbol{\xi}) + \beta g(\boldsymbol{\xi})] = \alpha\mathbb{E}[h(\boldsymbol{\xi})] + \beta\mathbb{E}[g(\boldsymbol{\xi})]$.

**Proposition 3.3** (convexity of expectation functions). *For a random vector $\boldsymbol{\xi}$ with support $\Xi \subset \mathbb{R}^m$ and $f : \mathbb{R}^m \times \mathbb{R}^n \to \mathbb{R}$, suppose that*

(a) *$f(\boldsymbol{\xi}, x)$ is integrable for all $x \in \mathbb{R}^n$*
(b) *$f(\xi, \cdot)$ is convex for all $\xi \in \Xi$.*

*Then, $Ef : \mathbb{R}^n \to \mathbb{R}$ is well defined and convex.*

**Proof.** As discussed above, $Ef(x)$ is well defined and real-valued for all $x \in \mathbb{R}^n$ due to the integrability assumption. Suppose that $x^0, x^1 \in \mathbb{R}^n$ and $\lambda \in [0, 1]$. By (b) and 3.2,

$$\begin{aligned}\mathbb{E}\big[f(\boldsymbol{\xi}, (1-\lambda)x^0 + \lambda x^1)\big] &\leq \mathbb{E}\big[(1-\lambda)f(\boldsymbol{\xi}, x^0) + \lambda f(\boldsymbol{\xi}, x^1)\big]\\ &= (1-\lambda)\mathbb{E}\big[f(\boldsymbol{\xi}, x^0)\big] + \lambda\mathbb{E}\big[f(\boldsymbol{\xi}, x^1)\big]\end{aligned}$$

and $Ef$ is convex by the convexity inequality 1.10. □

**Exercise 3.4** (support vector machine). Let $\boldsymbol{\xi} = (\boldsymbol{y}, \boldsymbol{x})$ be a random vector with $\boldsymbol{y}$ taking the values $-1$ and $1$ and $\boldsymbol{x}$ taking values in $\mathbb{R}^n$. Suppose that $\boldsymbol{\xi}$ models future observations and corresponding labels in a binary classification setting; see §2.H. In contrast to the earlier section, the probability distribution of $\xi$ may not be finite. The hinge-loss problem then takes the form

$$\underset{(a,\alpha)\in\mathbb{R}^{n+1}}{\text{minimize}}\ \mathbb{E}\Big[\max\big\{0, 1 - y\big(\langle a, x\rangle + \alpha\big)\big\}\Big].$$

Show that the objective function in this problem is convex provided that $x$ is integrable.

**Exercise 3.5** (Jensen's inequality). For a convex function $f : \mathbb{R}^n \to \mathbb{R}$ and an $n$-dimensional integrable random vector $\xi$, show that $f(\mathbb{E}[\xi]) \leq \mathbb{E}[f(\xi)]$.

**Guide.** Utilize the subgradient inequality 2.17 with a subgradient of $f$ at $\mathbb{E}[\xi]$. □

## 3.C Risk Modeling

Expectation functions arise in formulations where the goal is to optimize the *average* performance across different outcomes, but we may also be concerned about "worst case" performance. For example, it would be undesirable to have minimum expected wait time as design criterion for a call center when customers with above-average wait times might simply hang-up, write angry reviews and take their business elsewhere. Similarly, an aeronautical engineer may not want to minimize the average displacement at the tip of an aircraft wing, but instead account for nearly "all possible" displacement levels. We'll develop an approach to optimization under uncertainty that includes expectation-minimization as a special case but also adapts to concerns about risk, safety, reliability and worst-case performance.

In an application, a random variable $\xi$ might model an unknown quantity such as a future cost or loss, which implies a particular concern about high values of $\xi$; a low value of $\xi$ would be "good news." With this perspective, an assessment of $\xi$ could be based on its 10% or 1% highest outcomes. These worst-case outcomes would then supplement the expected value in our optimization of decisions involving such future costs and losses. Quantiles and superquantiles formalize assessments of this kind.

**Definition 3.6** (quantiles and superquantiles). For a random variable $\xi$ with distribution function $P$ and $\alpha \in (0,1)$, we define

$$\alpha\text{-}quantile\text{:}\quad Q(\alpha) = \min\big\{\xi \in \mathbb{R} \,\big|\, P(\xi) \geq \alpha\big\}$$

$$\alpha\text{-}superquantile\text{:}\quad \bar{Q}(\alpha) = Q(\alpha) + \frac{1}{1-\alpha}\mathbb{E}\big[\max\{0, \xi - Q(\alpha)\}\big].$$

Moreover, we define $\bar{Q}(0) = \mathbb{E}[\xi]$ and $\bar{Q}(1) = \lim_{\alpha \nearrow 1} \bar{Q}(\alpha)$.

For $\alpha$ in Figure 3.2, the equation $P(\xi) = \alpha$ has a unique solution, which is the quantile $Q(\alpha)$. For $\beta$ in the figure, $P(\xi) = \beta$ has no solution, and one takes the lowest $\xi$ that satisfies the relation $P(\xi) \geq \beta$ as the $\beta$-quantile. For $\gamma$, there's a whole range of $\xi$ that solves $P(\xi) = \gamma$ and the lowest one is defined as the $\gamma$-quantile. In any case, an $\alpha$-quantile is the dividing line between the best $\alpha$100% outcomes and the worst $(1-\alpha)$100% outcomes of a random variable.

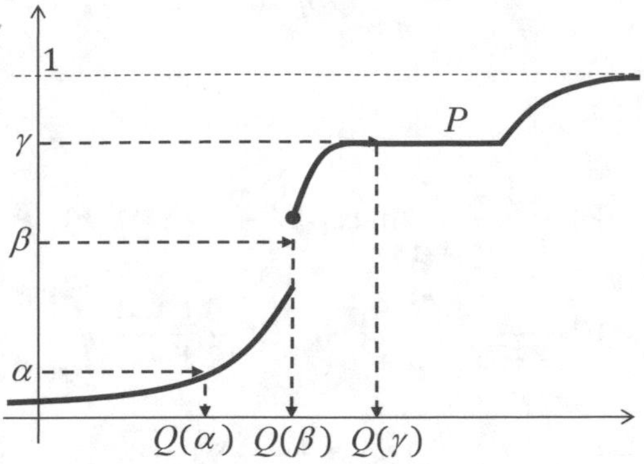


Fig. 3.2: Quantiles for distribution function $P$.

An $\alpha$-superquantile amounts to the corresponding $\alpha$-quantile *plus* a nonnegative term expressed by the expected value of the excess of $\boldsymbol{\xi}$ over the quantile. In effect,

$$\alpha\text{-superquantile of } \boldsymbol{\xi} \;=\; \text{average of the worst } (1-\alpha)100\% \text{ outcomes of } \boldsymbol{\xi}.$$

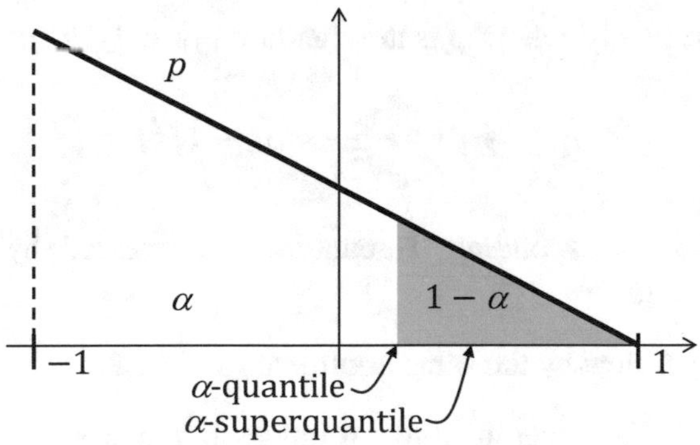


Fig. 3.3: Quantile and superquantile of density function $p$.

**Example 3.7** (triangular distribution). The $\alpha$-quantiles and $\alpha$-superquantiles of a random variable $\boldsymbol{\xi}$ with the triangular density function $p$ in Figure 3.3 are

$$Q(\alpha) = 1 - 2\sqrt{1-\alpha} \quad \text{and} \quad \bar{Q}(\alpha) = 1 - \tfrac{4}{3}\sqrt{1-\alpha}$$

for $\alpha \in (0, 1)$, respectively.

**Detail.** Let $P$ be the distribution function of $\boldsymbol{\xi}$. Since $p(\xi) = -\xi/2 + 1/2$ for $\xi \in [-1, 1]$, the solution of the equation

$$P(\xi) = \int_{-1}^{\xi} p(\eta)\, d\eta = \alpha$$

is $Q(\alpha) = 1 - 2\sqrt{1-\alpha}$. Thus,

$$\begin{aligned}\bar{Q}(\alpha) &= Q(\alpha) + \frac{1}{1-\alpha}\int_{-1}^{1} \max\{0, \xi - Q(\alpha)\} p(\xi)\, d\xi \\ &= Q(\alpha) + \frac{1}{1-\alpha}\int_{Q(\alpha)}^{1} \xi p(\xi)\, d\xi - \frac{Q(\alpha)}{1-\alpha}\int_{Q(\alpha)}^{1} p(\xi)\, d\xi \\ &= \frac{1}{1-\alpha}\int_{Q(\alpha)}^{1} \xi p(\xi)\, d\xi = 1 - \tfrac{4}{3}\sqrt{1-\alpha},\end{aligned}$$

after invoking the fact that $\int_{Q(\alpha)}^{1} p(\xi)\, d\xi = 1 - \alpha$ and the formula for $Q(\alpha)$. □

**Example 3.8** (normal distribution). The $\alpha$-quantiles and $\alpha$-superquantiles of a random variable $\boldsymbol{\xi}$ that's normally distributed with mean $\mu$ and variance $\sigma^2$ are

$$Q(\alpha) = \mu + \sigma\Phi^{-1}(\alpha) \quad \text{and} \quad \bar{Q}(\alpha) = \mu + \frac{\sigma\varphi\big(\Phi^{-1}(\alpha)\big)}{1-\alpha}$$

for $\alpha \in (0, 1)$, respectively, where $\varphi$ is the standard normal density function given by

$$\varphi(\xi) = \frac{1}{\sqrt{2\pi}} \exp\big(-\tfrac{1}{2}\xi^2\big)$$

and $\Phi^{-1}(\alpha)$ is the corresponding $\alpha$-quantile as expressed by the standard normal distribution function $\Phi$.

**Detail.** The results follow by the same arguments as in 3.7. □

We refer to [70] for superquantile formulae in the case of many other common probability distributions.

The boundary cases $\alpha = 0$ and $\alpha = 1$ represent extreme positions that we may take in assessing a random variable $\boldsymbol{\xi}$: $\bar{Q}(0)$ conveys information only about the average performance while $\bar{Q}(1)$ truly represents the worst outcome; $\bar{Q}(1) = 1$ in 3.7 and $\bar{Q}(1) = \infty$ in 3.8.

Superquantiles appear at first to be more cumbersome than quantiles as they're define in terms of the latter. However, they can be expressed independently of quantiles by a minimization formula that turns out to be highly compatible with optimization models.

**Proposition 3.9** (quantiles and superquantiles via minimization). *For $\alpha \in (0, 1)$ and an integrable random variable $\boldsymbol{\xi}$, with $\alpha$-quantile and $\alpha$-superquantile denoted by $Q(\alpha)$ and $\bar{Q}(\alpha)$, respectively, one has*

$$Q(\alpha) \in \operatorname{argmin}_{\gamma \in \mathbb{R}} \left\{ \gamma + \frac{1}{1-\alpha} \mathbb{E}\big[ \max\{0, \boldsymbol{\xi} - \gamma\} \big] \right\}$$

$$\bar{Q}(\alpha) = \min_{\gamma \in \mathbb{R}} \left\{ \gamma + \frac{1}{1-\alpha} \mathbb{E}\big[ \max\{0, \boldsymbol{\xi} - \gamma\} \big] \right\},$$

*with $\bar{Q}(\alpha)$ being finite. In fact, $Q(\alpha)$ is the lowest minimizer in case there are more than one.*

*If $\boldsymbol{\xi}$ is also continuously distributed with density function p, then*

$$\bar{Q}(\alpha) = \frac{1}{1-\alpha} \int_{Q(\alpha)}^{\infty} \xi p(\xi)\, d\xi.$$

**Proof.** Let $\Xi$ be the support of $\boldsymbol{\xi}$. The function $f : \mathbb{R} \times \mathbb{R} \to \mathbb{R}$ with

$$f(\xi, \gamma) = \gamma + \frac{1}{1-\alpha} \max\{0, \xi - \gamma\}$$

defines for any $\gamma \in \mathbb{R}$ an integrable random variable $f(\boldsymbol{\xi}, \gamma)$ because $\boldsymbol{\xi}$ is integrable. Since $f(\xi, \cdot\,)$ is convex for all $\xi \in \Xi$ by 1.18, we obtain from 3.3 that $Ef\colon \mathbb{R} \to \mathbb{R}$ is well defined and convex. The optimality condition $0 \in \partial Ef(\gamma)$ (cf. 2.19) then gives that $Q(\alpha)$ is the lowest minimizer of $Ef$; details are deferred to §3.F. The asserted minimization formulae for $Q(\alpha)$ and $\bar{Q}(\alpha)$ follow immediately because $Ef$ coincides with the objective function in those formulae. The claim about finiteness of $\bar{Q}(\alpha)$ is a consequence of the integrability of $\boldsymbol{\xi}$.

When $\boldsymbol{\xi}$ is continuously distributed, we obtain via 3.2 that

$$\begin{aligned}
\bar{Q}(\alpha) &= Q(\alpha) + \frac{1}{1-\alpha} \int_{-\infty}^{\infty} \max\big\{0, \xi - Q(\alpha)\big\} p(\xi)\, d\xi \\
&= Q(\alpha) + \frac{1}{1-\alpha} \int_{Q(\alpha)}^{\infty} \xi p(\xi)\, d\xi - \frac{Q(\alpha)}{1-\alpha} \int_{Q(\alpha)}^{\infty} p(\xi)\, d\xi,
\end{aligned}$$

which simplifies to the stated expression because $\int_{Q(\alpha)}^{\infty} p(\xi)\, d\xi = 1 - \alpha$. □

The focus on integrable random variables ensures that $\bar{Q}(\alpha)$ is finite for $\alpha \in (0, 1)$, while still addressing most applications. The multitude of minimizers alluded to in the proposition emerges when the distribution function $P$ of $\boldsymbol{\xi}$ has a "flat stretch" to the right of $Q(\alpha)$ and $P(Q(\alpha)) = \alpha$; see §3.F for details. The alternative formula in the case of a continuous distribution can be interpreted as expressing a conditional expectation.

In risk analysis, the $\alpha$-superquantile of $\boldsymbol{\xi}$ is referred to as the *superquantile risk* (s-risk) or *conditional value-at-risk* (cvar)[5] of $\boldsymbol{\xi}$ and we write

$$\text{s-rsk}_{\alpha}(\boldsymbol{\xi}) = \bar{Q}(\alpha)$$

to highlight the dependence on both $\boldsymbol{\xi}$ and $\alpha$.

[5] Average value-at-risk (avar) and expected shortfall are also used in the literature.

Let's consider optimization models involving superquantiles. Suppose that we're faced with a cost, loss or damage that depends on the value of an $m$-dimensional random vector $\boldsymbol{\xi}$ and a decision vector $x \in \mathbb{R}^n$, and is expressed by the function $f : \mathbb{R}^m \times \mathbb{R}^n \to \mathbb{R}$. Now, the random variables of interest are $\{f(\boldsymbol{\xi}, x), x \in \mathbb{R}^n\}$; they represent the uncertain cost for each decision $x$. Then, for a given $\alpha \in (0, 1)$, we may want to minimize the superquantile risk by considering the model

$$\underset{x \in C}{\text{minimize}} \ \ \text{s-rsk}_\alpha \big(f(\boldsymbol{\xi}, x)\big), \tag{3.2}$$

where $C \subset \mathbb{R}^n$ specifies constraints. That is, the goal is to determine a decision with the lowest possible cost on average across the worst $(1 - \alpha)100\%$ outcomes. The resulting decision would typically be rather different than those obtained by minimizing the expected cost; decisions with the possibility of high values tend to be avoided.

**Proposition 3.10** (convexity of superquantile functions). *For a random vector $\boldsymbol{\xi}$ with support $\Xi \subset \mathbb{R}^m$ and $f : \mathbb{R}^m \times \mathbb{R}^n \to \mathbb{R}$, suppose that*

(a) *$f(\boldsymbol{\xi}, x)$ is integrable for all $x \in \mathbb{R}^n$*
(b) *$f(\xi, \cdot)$ is convex for all $\xi \in \Xi$.*

*Then, for any $\alpha \in [0, 1)$, the function $x \mapsto \text{s-rsk}_\alpha(f(\boldsymbol{\xi}, x))$ is convex and real-valued.*

**Proof.** For the case $\alpha = 0$, the conclusion follows directly from 3.3 because then $\text{s-rsk}_\alpha(f(\boldsymbol{\xi}, x)) = \mathbb{E}[f(\boldsymbol{\xi}, x)]$. For $\alpha \in (0, 1)$, the function

$$(x, \gamma) \mapsto \gamma + \frac{1}{1 - \alpha} \max\big\{0, f(\xi, x) - \gamma\big\}$$

is convex for all $\xi \in \Xi$ by 1.18 and, for all $(x, \gamma)$, the random variable

$$\gamma + \frac{1}{1 - \alpha} \max\big\{0, f(\boldsymbol{\xi}, x) - \gamma\big\}$$

is integrable because $f(\boldsymbol{\xi}, x)$ is integrable. Thus, the function

$$(x, \gamma) \mapsto \gamma + \frac{1}{1 - \alpha} \mathbb{E}\big[\max\{0, f(\boldsymbol{\xi}, x) - \gamma\}\big]$$

is convex by 3.3. The conclusion then follows from the convexity of inf-projections by 1.21 and the minimization formula in 3.9. The real-valuedness is guaranteed by 3.9. □

If we ignore the uncertainty associated with $\boldsymbol{\xi}$ and simply consider minimizing $f(\hat{\xi}, x)$ subject to $x \in C$ for some nominal value $\hat{\xi}$, then we would face a convex problem as long as $f(\hat{\xi}, \cdot)$ and $C$ are convex; see 2.3. The proposition asserts that (3.2), which treats uncertainty much more comprehensively, is convex too when $f(\xi, \cdot)$ is convex for all $\xi$. Thus, optimization under uncertainty carried out in this manner doesn't muddle up convexity that might be present in $f$.

The minimization formula for superquantiles in 3.9 is useful computationally as it frees us from having to compute quantiles within an algorithm for superquantile-risk minimization. After inserting the formula, we realize that (3.2), with $\alpha \in (0,1)$, can be reformulated as

$$\underset{x\in C,\gamma\in\mathbb{R}}{\text{minimize}}\ \mathbb{E}\Big[\gamma + \frac{1}{1-\alpha}\max\{0, f(\xi,x)-\gamma\}\Big],$$

which brings us back to minimizing an expectation function. Thus, superquantiles are easily incorporated in optimization models. The problem is convex under the assumptions of 3.10 and the convexity of $C$, a fact we can deduce from the proof of that proposition. If $\xi$ is finitely distributed with support $\Xi$ and corresponding probabilities $\{p_\xi > 0, \xi \in \Xi\}$, then the problem simplifies to

$$\underset{x\in C,\gamma\in\mathbb{R}}{\text{minimize}}\ \gamma + \frac{1}{1-\alpha}\sum\nolimits_{\xi\in\Xi} p_\xi \max\{0, f(\xi,x)-\gamma\}, \tag{3.3}$$

which in turn can be reformulated as

$$\begin{aligned} &\underset{x\in C,\gamma\in\mathbb{R},z\in\mathbb{R}^\nu}{\text{minimize}}\ \gamma + \frac{1}{1-\alpha}\sum\nolimits_{\xi\in\Xi} p_\xi z_\xi \\ &\text{subject to}\ f(\xi,x)-\gamma \le z_\xi\ \text{and}\ 0 \le z_\xi\ \ \forall \xi\in\Xi, \end{aligned} \tag{3.4}$$

where $z = (z_\xi, \xi \in \Xi)$ are additional variables and $\nu$ is the cardinality of $\Xi$. The last problem is also convex as long as $C$ is convex and $f(\xi,\cdot)$ is convex for all $\xi \in \Xi$ because the feasible set

$$\big\{(x,\gamma,z) \in C\times\mathbb{R}\times\mathbb{R}^\nu \ \big|\ f(\xi,x)-\gamma \le z_\xi,\ 0 \le z_\xi\ \ \forall\xi\in\Xi\big\}$$

is convex by 1.12(a) and 1.11 and we can invoke 2.3. In fact, if $f(\xi,\cdot)$ is epi-polyhedral for all $\xi \in \Xi$ and $C$ is polyhedral, then the problem can be stated in terms of an epi-polyhedral function and, consequently, is equivalent to a linear optimization problem. In any case, we can either solve (3.3), which is nonsmooth due to the max-term, or (3.4) that might be better structured for available algorithms but involves $\nu$ additional decision variables and $2\nu$ additional constraints.

**Exercise 3.11** (investment planning). The amount of money you'll need in retirement is uncertain as it depends on your health, longevity and many other factors. Let's model it by a random variable $\boldsymbol{\xi}_0$ given in millions of current dollars. Presently, you've have saved 2 million dollars. (For simplicity, we assume you'll not be able to contribute more to your savings before your retirement.) You can place your savings in the bank, with a fixed yield, and/or invest in two possible funds with uncertain yield. We model these yields by the random variables $\boldsymbol{\xi}_1, \boldsymbol{\xi}_2, \boldsymbol{\xi}_3$. There are 12 possible values of the random vector $(\boldsymbol{\xi}_0, \boldsymbol{\xi}_1, \boldsymbol{\xi}_2, \boldsymbol{\xi}_3)$, each equally likely; see Table 3.3. Your goal is to determine an *investment plan* $x = (x_1, x_2, x_3)$ that minimizes the superquantile risk of the shortfall defined as $\boldsymbol{\xi}_0 - \boldsymbol{\xi}_1 x_1 - \boldsymbol{\xi}_2 x_2 - \boldsymbol{\xi}_3 x_3$, where $x_1$, $x_2$ and $x_3$ are the allocations to bank deposit, the first fund and the second fund, respectively.

**Guide.** Formulate and solve a model of the form (3.3) or (3.4). Vary $\alpha \in [0, 1)$ and see how the optimized investment plan changes. For each obtained plan, compute the outcomes of the shortfall and the probability of going broke, i.e., having a positive shortfall. □

Table 3.3: The 12 possible values for the random vector in 3.11.

| | possible values | | | | | | | | | | | |
|---|---|---|---|---|---|---|---|---|---|---|---|---|
| $\xi_0$ | 2.3 | 2.2 | 1.9 | 2.0 | 2.1 | 2.2 | 1.9 | 2.0 | 2.0 | 2.1 | 1.9 | 2.3 |
| $\xi_1$ | 1.0 | 1.0 | 1.0 | 1.0 | 1.0 | 1.0 | 1.0 | 1.0 | 1.0 | 1.0 | 1.0 | 1.0 |
| $\xi_2$ | 1.6 | 1.5 | 1.4 | 1.4 | 1.3 | 1.3 | 1.2 | 1.1 | 0.9 | 0.8 | 0.7 | 0.7 |
| $\xi_3$ | 0.9 | 0.9 | 1.0 | 0.9 | 1.0 | 1.1 | 1.1 | 1.2 | 1.1 | 1.2 | 1.1 | 1.2 |

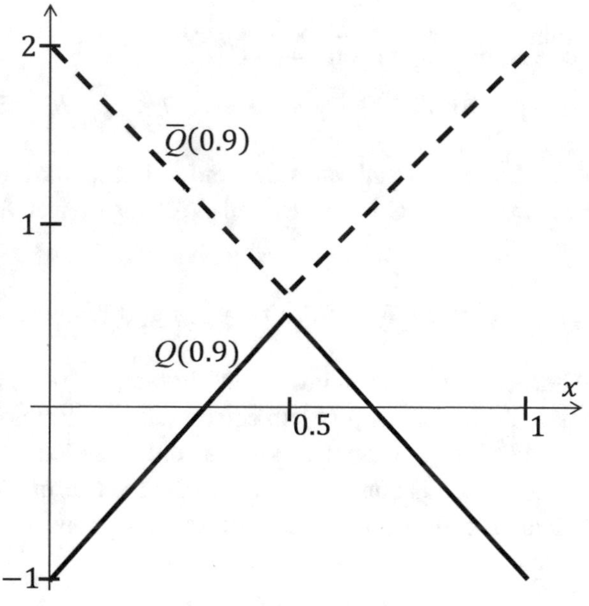


Fig. 3.4: Quantile $Q(0.9)$ (solid line) and superquantile $\bar{Q}(0.9)$ (dashed) of $f(\xi, x)$ in 3.12 as functions of $x$.

**Example 3.12** (quantile vs superquantile minimization). Suppose that we face a difficult decision regarding two financial instruments, both associated with uncertainty. The first instrument requires, with probability 0.1, a payment of 2 dollars per dollar committed and, with probability 0.9, yields one dollar per dollar committed. Let the random variable $\xi_1$ model these losses; it takes the value 2 with probability 0.1 and the value −1 with probability 0.9. The second instrument is modeled by the random variable $\xi_2$, which has the same probability distribution as that of $\xi_1$ but the two random variables are statistically independent. The instruments aren't lucrative: most likely we regain our commitment, but with a small probability we face a large loss. Suppose that we need to allocate 1 million dollars between these two instruments. Since the instruments appear equally unappealing, we might be led to believe that any allocation is fine. This would be a big mistake, but one that remains hidden for an analyst examining quantiles. Superquantiles reveal that the best strategy would be to allocate half a million to each instrument.

**Detail.** Let $x \in [0, 1]$ be the fraction of our million dollars allocated to the first instrument, the remainder is allocated to the second instrument. Then, the random variable describing our loss is

$$f(\boldsymbol{\xi}, x) = x\boldsymbol{\xi}_1 + (1 - x)\boldsymbol{\xi}_2.$$

Since there are only four possible outcomes of $\boldsymbol{\xi} = (\boldsymbol{\xi}_1, \boldsymbol{\xi}_2)$, we find that

$$f(\boldsymbol{\xi}, x) = \begin{cases} -1 & \text{with probability } 0.81 \\ 2 - 3x & \text{with probability } 0.09 \\ 3x - 1 & \text{with probability } 0.09 \\ 2 & \text{with probability } 0.01. \end{cases}$$

Let's now examine the $\alpha$-quantile and $\alpha$-superquantile of $f(\boldsymbol{\xi}, x)$ for $\alpha = 0.9$. The above probabilities establish that

$$Q(0.9) = \begin{cases} 3x - 1 & \text{if } x \in [0, 1/2] \\ 2 - 3x & \text{otherwise,} \end{cases}$$

which is visualized in Figure 3.4. With the quantiles known, we use

$$\bar{Q}(0.9) = Q(0.9) + 10\mathbb{E}\big[\max\{0, f(\boldsymbol{\xi}, x) - Q(0.9)\}\big]$$

to compute superquantiles. This results in

$$\bar{Q}(0.9) = \begin{cases} -2.7x + 2 & \text{if } x \in [0, 1/2] \\ 2.7x - 0.7 & \text{otherwise,} \end{cases}$$

which is also shown in Figure 3.4. The superquantiles define a convex function in $x$, while the quantiles don't. The minimization of superquantiles has $x^\star = 0.5$ as minimizer, but the minimization of quantiles results in $x = 0$ and $x = 1$. The decision $x^\star$ involves *hedging*, a well-known strategy in finance to reduce risk. The decision lowers the probability of a loss of 2 million dollars from 0.1 to 0.01 compared to the choice $x = 0$ or $x = 1$. This comes at the expense of reducing the probability of a loss of $-1$ million dollars from 0.9 to 0.81. The decision $x^\star$ also results in the possibility of a loss of 0.5 million dollars (with probability 0.18), but this might be much more palatable than a 2-million-dollar loss. The minimum value of the 0.9-superquantiles is 0.65 million dollars, which is a more reasonable and conservative assessment of the uncertain future loss than the wildly optimistic $-1$ million provided by the minimum value of the 0.9-quantiles.

Of course, there could be different opinions about these outcomes, but it's clear that, in general, superquantile minimization has two advantages over quantile minimization:

the former preserves convexity in the problem formulation and tends to bring forward nontrivial solutions such as those relying on hedging, while the latter may not do either. □

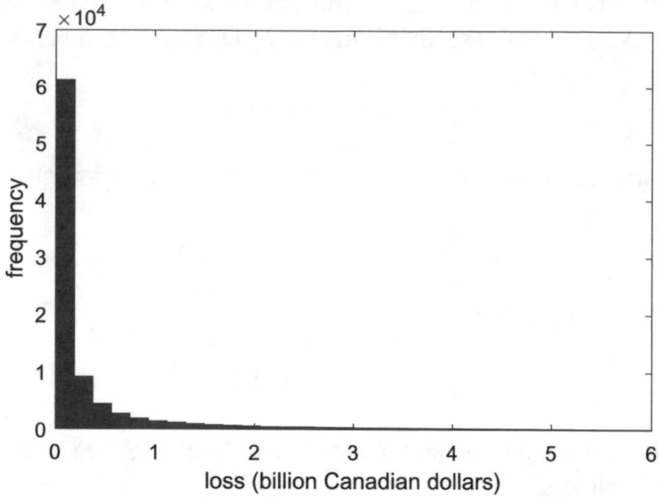


Fig. 3.5: Histogram of 100000 outcomes of the cumulative loss due to earthquake damage in the Vancouver region of Canada.

**Example 3.13** (resilience and superquantiles). Superquantiles have the ability to highlight the level of *resilience* in a system and identify how "bad" it can get. This is especially important in assessment of earthquake damage where losses are usually moderate, but they can become catastrophic. From a study of the cumulative loss due to earthquake damage during the next 50 years in the greater Vancouver region in Canada [63], we obtain 100000 loss values that one can view as outcomes for a random variable modeling the cumulative loss; see the histogram of Figure 3.5. The cumulative loss is most like less than one billion Canadian dollars. In fact, 90% of the 100000 outcomes are below 4.39 billions, i.e., the 0.9-quantile of the cumulative loss is 4.39 billions. Although sizable, it seems that the region is quite resilient to earthquakes. It turns out that this conclusion is flawed and a study of superquantiles paints a gloomier picture.

**Detail.** The largest outcome is actually 373 billions; Figure 3.5 should have been extended much to the right. Superquantiles quantify this long right-tail. Specifically, the 0.9-superquantile of the cumulative loss is 28.92 billions. This means that when the cumulative loss exceeds 4.39 billons, it does so substantially resulting in a loss of 28.92 billions on average. The region might not be resilient to earthquake losses after all. The picture remains the same if we consider the 0.95-quantile and the 0.95-superquantile, which are 14.00 and 49.88 billions, respectively.

A difference between quantiles and superquantiles emerges: A quantile is determined by the number of outcomes to its right, but the distances to the right can be arbitrary without changing the quantile. In contrast, a superquantile equals the corresponding quantile plus a term that averages the outcomes to the right of that quantile and thus is sensitive to values far to the right. □

**Exercise 3.14** (scale invariance). For an integrable random variable $\xi$ and $\alpha \in [0, 1]$, show that $\text{s-rsk}_\alpha(\lambda\xi) = \lambda\, \text{s-rsk}_\alpha(\xi)$ for any $\lambda \in [0, \infty)$. (Thus, converting losses from, say, dollar to yen doesn't fundamentally change the risk.)

## 3.D Models of Uncertainty

A central component of a stochastic optimization problem is the probability distribution of the random vector $\xi$ defining expectation functions and superquantiles. Since it determines the possible values of $\xi$ and the associated probabilities, the choice of probability distribution typically influences the resulting decision and becomes an important part of the modeling process. The choice of probability distribution is usually informed by data about the random phenomena that $\xi$ represents. However, when the data is biased, corrupted or simply unavailable in sufficient numbers, we need to introduce assumptions corresponding to our understanding of the situation. In this section, we'll construct models of uncertainty by estimating probability density functions and also discuss how they can be incorporated with optimization procedures.

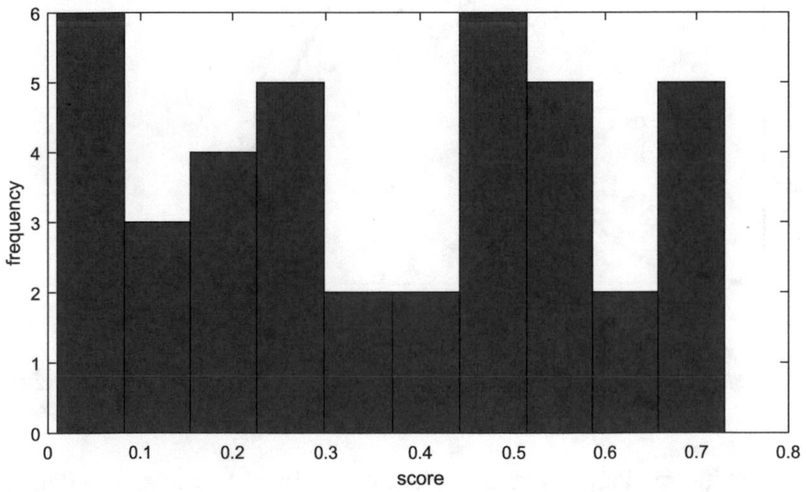


Fig. 3.6: Histogram of 40 scores about the performance of a technology.

**Example 3.15** (uncertainty modeling in the US Army). Periodically, the US Army decides in which promising new technologies it should invest. The technologies could be hardware such as improved night-vision goggles, means to extract water from air, body sensors and other futuristic gadgets. The decision needs to be made before the actual performance and usefulness of the technologies are fully known. The US Army, as many large organizations, relies on subject-matter experts to predict parameter values for which there's little or no data. Suppose that 40 experts are given a presentation about a technology and then asked to individually assign a score between 0.00 (poor) and 1.00 (excellent) that reflects the technology's future performance; see Figure 3.6 for a histogram summarizing the resulting 40 data points. Clearly, the opinions about the technology vary greatly, with the lowest score being 0.01 and the highest 0.73, but the data still provides a starting point from which we can develop a probability distribution for a random variable $\boldsymbol{\xi}$ that represents our (uncertain) assessment of the technology.

**Detail.** Given the data $\xi^1, \ldots, \xi^{40} \in [0, 1]$, we could have modeled $\boldsymbol{\xi}$ as having a finite distribution with $\Xi = \{\xi^1, \ldots, \xi^{40}\}$ and equal probabilities $p_i = 1/40$. This model of uncertainty assumes that one of the subject-matter experts is correct; a value not proposed by any expert can't occur. Generally, a model that *only* relies on historical data fails to account for the possibility that the future might not be an instance of the past! A possible approach to address this concern and to ensure that we plan for outcomes that appear reasonable, but may never have occurred, is to select a density function for $\boldsymbol{\xi}$ *informed* by the data.

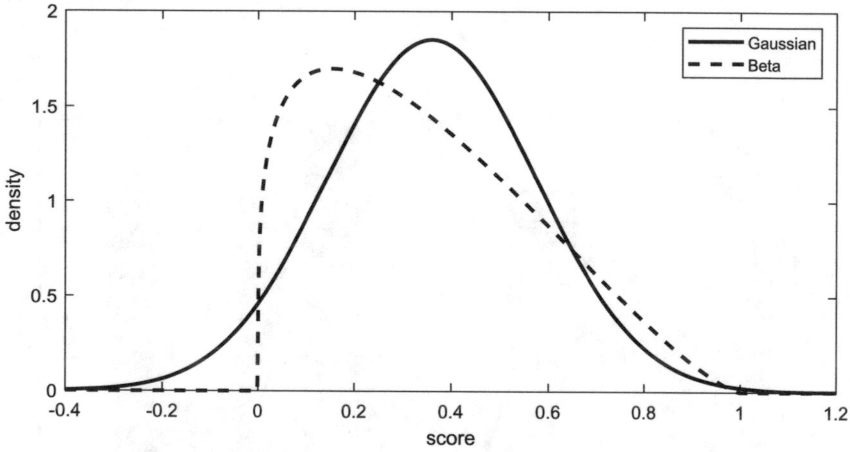


Fig. 3.7: Density functions modeling the uncertain performance of a technology.

Using the maximum likelihood criterion as laid out in §1.E, we can fit a density function to the 40 data points. The best-fitting Gaussian and beta density functions are shown by solid and dashed lines, respectively, in Figure 3.7. Although one can debate which of the two density functions is more appropriate, they both assign positive probabilities to outcomes that the finite-distribution model deems impossible to occur. A stochastic optimization model incorporating one of these density functions would account for a wider array of possible performance values for the technology, which might produce a better decision in practice; see [123].

Still, the obtained density functions exhibit some questionable characteristics. The Gaussian density function assigns positive probabilities to values below zero and above one, which contradicts the fact that scores are certainly between 0.00 and 1.00. Neither the Gaussian nor the beta density function reflects the presence of several peaks in Figure 3.6. □

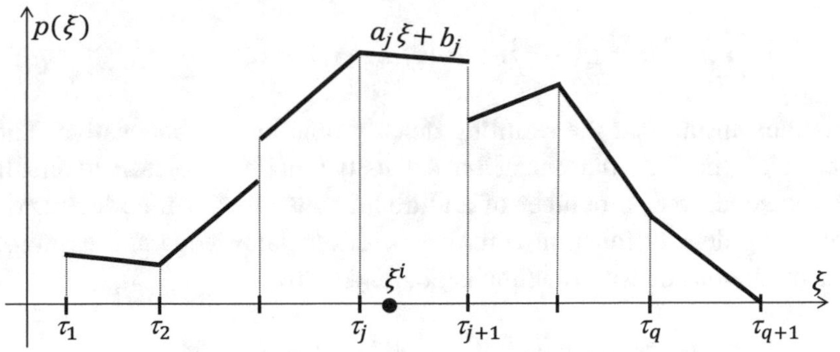


Fig. 3.8: Density function expressed as an epi-spline.

As a model of uncertainty about a parameter, we would like to choose a density function that's closely tailored to the available information and assumptions. The maximum likelihood approach provides a general framework for estimating density functions and permits a variety of side conditions. Given data $\xi^1, \ldots, \xi^\nu \in \mathbb{R}$, we recall from §1.E that a maximum likelihood estimate is a minimizer of the model

$$\underset{p \in \mathcal{P}}{\text{minimize}} \; -\frac{1}{\nu}\sum\nolimits_{i=1}^{\nu} \ln p(\xi^i), \tag{3.5}$$

where $\mathcal{P}$ is a class of candidate density functions. The usual classes of density functions such as Gaussian, beta, exponential and so forth may not satisfy our assumptions about the random phenomenon at hand. In 3.15, neither the Gaussian nor the beta density function meets our expectations. A more flexible class of functions are the first-order *epi-splines*, which are piecewise affine functions of the form

$$p(\xi) = a_j\xi + b_j \quad \text{when } \xi \in (\tau_j, \tau_{j+1}), \;\; j = 1, \ldots, q,$$

where $-\infty < \tau_1 < \tau_2 < \cdots < \tau_{q+1} < \infty$ are mesh points and $\{a_j, b_j, \; j = 1, \ldots, q\}$ are coefficients to be determined; see Figure 3.8. The value of $p$ exactly at a mesh point $\tau_j$ is usually pinned down by insisting on the density function being continuous or some other rule. The problem of minimizing over $p \in \mathcal{P}$ then becomes one of minimizing over the vectors of coefficients $a = (a_1, \ldots, a_q)$ and $b = (b_1, \ldots, b_q)$. Let $j_i$ be the index $j$ corresponding to the interval in which $\xi^i$ is located (cf. Figure 3.8). Then, $p(\xi^i) = a_{j_i}\xi^i + b_{j_i}$ and (3.5) specializes to

$$\begin{aligned}
&\underset{a,b\in\mathbb{R}^q}{\text{minimize}} -\frac{1}{\nu}\sum\nolimits_{i=1}^{\nu} \ln\left(a_{j_i}\xi^i + b_{j_i}\right)\\
&\text{subject to} \qquad a_j\tau_j + b_j \geq 0,\;\; j = 1,\dots,q\\
&\qquad\qquad\qquad a_j\tau_{j+1} + b_j \geq 0,\;\; j = 1,\dots,q\\
&\sum\nolimits_{j=1}^{q} \tfrac{1}{2}a_j\left(\tau_{j+1}^2 - \tau_j^2\right) + b_j\left(\tau_{j+1} - \tau_j\right) = 1.
\end{aligned}$$

The inequalities ensure that the resulting density function is nonnegative. The equality constraint implements the requirement that a density function integrates to one. In addition to these basic conditions, a number of additional constraints can be included to ensure that the resulting density function complies with our knowledge and assumptions. For example, a continuous density function is guaranteed by

$$a_{j-1}\tau_j + b_{j-1} = a_j\tau_j + b_j,\;\; j = 2,\dots,q.$$

The slope in one interval changes with at most $\delta \geq 0$ relative to the next one when

$$a_j - \delta \leq a_{j-1} \leq a_j + \delta,\;\; j = 2,\dots,q$$

and this introduces a "smooth" look when $q$ is large and $\delta$ is small. If already continuous, the density function becomes nonincreasing if $a_j \leq 0$ for all $j = 1,\dots,q$. Since all these constraints are linear equations and inequalities, they define a polyhedral feasible set. The objective function is convex because $\alpha \mapsto -\ln\alpha$ is convex (under the convention that $-\ln\alpha = \infty$ when $\alpha \leq 0$) and we can invoke 1.18. For further theoretical and computational considerations regarding epi-splines, we refer to [112, 113].

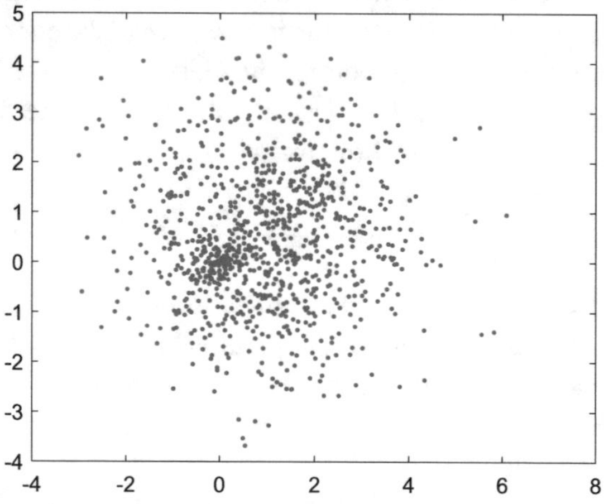


Fig. 3.9: Positions of 1000 robotic bugs. Axes are labeled in miles.

**Example 3.16** (robotic swarm). Consider a group (swarm) of 1000 aerial autonomous systems. These may be tiny "bugs" that are pollinating fruit trees or sensing air quality in an urban area. We would like to determine a probability distribution of the Euclidean distance from every bug to its nearest neighboring bug. Figure 3.9 shows the positions of the bugs at one time. With complete knowledge about everybody's position, it's easy to compute the nearest-neighbor distance for each bug; Figure 3.10 summarizes the distances. Nearly all bugs are closer than 0.2 miles of some other bug.

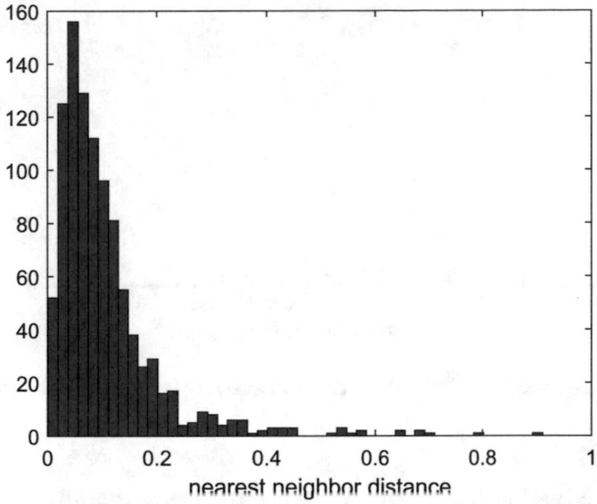


Fig. 3.10: Histogram of nearest-neighbor distance for robotic bugs. The horizontal axis has unit of miles and the vertical axis gives the number of bugs.

The challenge in practice is that the bugs are tiny and carry small batteries and no GPS device. They're programmed to only report nearest-neighbor distances occasionally. Our goal is to determine a density function that resembles Figure 3.10 using just a small number of such reports.

**Detail.** We follow the approach laid out before the example. Distances are nonnegative so we set $\tau_1 = 0$. Our prior experience indicates that distances are typically less than one mile and we set $\tau_{q+1} = 1$. Moreover, let $\tau_j = (j-1)/q$, with $q = 1000$. Using $\nu = 100$ reports, the maximum likelihood model, with continuity and change-in-slope (smoothing) constraints ($\delta = 0.5$), results in the density function in Figure 3.11, which is quite close to the "truth" in Figure 3.10.

With only 10 reports, however, the model results in the density function illustrated with a black line in Figure 3.12. There's a large bump around 0.65 miles, which we can safely say is "incorrect" based on our (perfect) knowledge from Figure 3.10. We seem to have *over-fitted* the data. An approach to combat over-fitting is to introduce additional constraints. Let's assume that a mode (peak) of the density function is located at $\xi^\star = 0.05$ miles. This requirement is enforced by the constraints

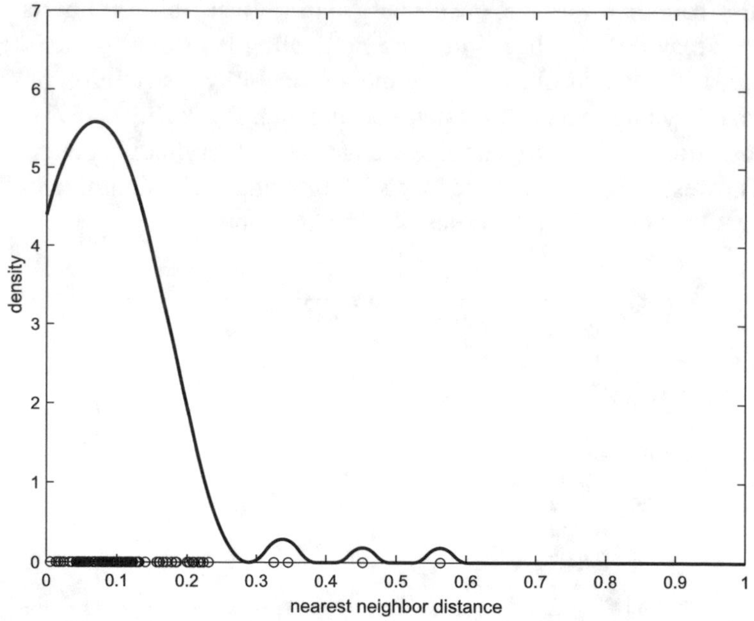


Fig. 3.11: Estimate of nearest-neighbor distance using 100 reports indicated by circles.

$$a_j\tau_j + b_j \le a_{j^\star}\xi^\star + b_{j^\star} \quad \text{and} \quad a_j\tau_{j+1} + b_j \le a_{j^\star}\xi^\star + b_{j^\star}, \quad j = 1, \ldots, q,$$

where $j^\star$ is the interval in which $\xi^\star$ is located. Moreover, since there seems to be more rapid changes in slope to the left of the mode than to the right, we modify the change-in-slope (smoothing) constraints to have $\delta = 5$ for intervals to the left of $\xi^\star$ and $\delta = 0.5$ to the right of $\xi^\star$. The resulting estimate is the red line in Figure 3.12. Although the bump at 0.65 miles persists, the density function drops to zero at zero miles. This seems reasonable as the bugs are programmed to avoid collision. With this insight, we impose $p(0) = 0$, i.e., $b_1 = 0$. We also enforce unimodality, i.e., the slope needs to be nonnegative to the left of $\xi^\star$ and nonpositive to the right:

$$a_j \ge 0 \text{ when } \tau_j < \xi^\star \quad \text{and} \quad a_j \le 0 \text{ when } \tau_j > \xi^\star.$$

The resulting estimate is given by the blue line in Figure 3.12. Indeed, the density function is unimodal, with a mode at $\xi^\star$, and we now have a reasonable tail, with no misleading bump. □

The example illustrates that it's possible to obtain quality estimates of a density function even with little data as long as additional constraints are brought in. Often, experienced analysts can identify reasonable assumptions that would guide the choice of constraints. Or, we can adopt a process of trial and error on different datasets to determine the more suitable set of constraints (which is called cross-validation in statistics).

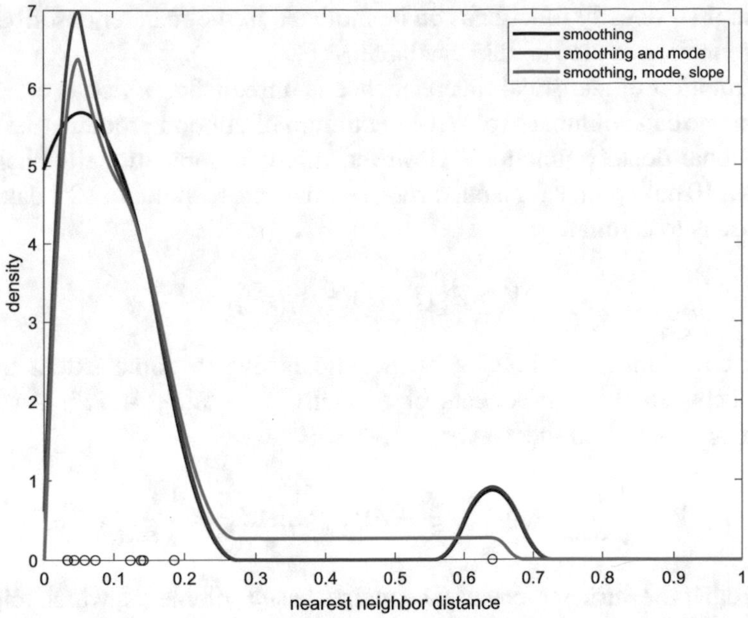


Fig. 3.12: Estimates of nearest-neighbor distance using 10 reports indicated by circles.

**Exercise 3.17** (distribution of age at failure). Let's carry out *failure analysis* for a type of engineering system, which has a lifetime of 10 years, but occasionally experiences repairable failures during that time. We would like to estimate the chances for a failure when the system is of a certain age using the following historical data of the age (in years) of a system at failure $\{0.1, 0.02, 9.3, 8.9, 3.2, 7.8, 9.5, 1.1, 9.7, 0.5\}$. The data as well as engineering insight tell us that failure is more likely for a young system (due to burn-in issues) and also for an aging system (due to wear and tear). For this reason, let's estimate a u-shaped density function as a model of uncertainty about the age at failure. Using epi-splines and the maximum likelihood procedure laid out above, write constraints that ensure a u-shaped density function. Solve the resulting estimation problem.

**Guide.** The 10-year lifetime of the system stipulates that it suffices to consider density functions on $[0, 10]$. A u-shaped function must be convex. This can be achieved by imposing continuity constraints as well as conditions on the slopes in each interval. □

In most applications, we're faced with multiple parameters and the need for modeling the probability distribution of a random vector $\boldsymbol{\xi} = (\boldsymbol{\xi}_1, \ldots, \boldsymbol{\xi}_m)$. The simplest multivariate model of uncertainty assumes that the components of the random vector are statistically independent in which case the marginal density functions $p_i : \mathbb{R} \to [0, \infty)$ for $\boldsymbol{\xi}_i$, $i = 1, \ldots, m$, fully describe the joint density function $p : \mathbb{R}^m \to [0, \infty)$ for $\boldsymbol{\xi}$ through

$$p(\xi) = \prod_{i=1}^{m} p_i(\xi_i).$$

Since the marginal density functions can be modeled in the manner described above, the joint density function is also readily available.

If the assumption of statistical independence is unrealistic, which is often the case in practice, then one can attempt to solve the maximum likelihood problem (3.5) over a class of $m$-dimensional density functions. However, this is computationally challenging and data intensive: 10 data points for a univariate estimate correspond to $10^m$ data points now. A compromise is to estimate

$$V = \mathbb{E}\big[\, (\boldsymbol{\xi} - \bar{\xi})(\boldsymbol{\xi} - \bar{\xi})^\top \,\big],$$

the variance-covariance matrix of $\boldsymbol{\xi}$, which quantifies to some extent the statistical dependence between the components of $\boldsymbol{\xi}$. Here, $\bar{\xi} = \mathbb{E}[\boldsymbol{\xi}] \in \mathbb{R}^m$. An estimate for this $m \times m$-matrix based on the data $\xi^1, \dots, \xi^\nu \in \mathbb{R}^m$ is

$$\hat{V} = \frac{1}{\nu} \sum\nolimits_{i=1}^{\nu} (\xi^i - \hat{\xi})(\xi^i - \hat{\xi})^\top, \quad \text{where } \hat{\xi} = \frac{1}{\nu} \sum\nolimits_{i=1}^{\nu} \xi^i.$$

A difficulty is that the variance-covariance matrix has $m^2$ elements, which might be larger than the sample size $\nu$, causing the estimate to be poor. In practice, one might compute the above estimates and then adjust the values subjectively, recalling that a variance-covariance matrix is always positive semidefinite. A model of the joint density function can then be constructed by assuming that it takes the form of a *Nataf density function*:

$$p(\xi) = \frac{\prod_{i=1}^m p_i(\xi_i)}{\prod_{i=1}^m \varphi(\eta_i)} \varphi_m(\eta, R) \quad \forall \xi = (\xi_1, \dots, \xi_m) \in \mathbb{R}^m,$$

where $p_i$ is the (estimated) marginal density function of $\boldsymbol{\xi}_i$, with corresponding distribution function $P_i$, $\eta = (\eta_1, \dots, \eta_m)$ with $\eta_i = \Phi^{-1}(P_i(\xi_i))$, $\varphi$ and $\Phi$ are the standard normal univariate density and distribution functions and

$$\varphi_m(\eta, R) = \frac{1}{\sqrt{(2\pi)^m \det R}} \exp\big( -\tfrac{1}{2} \langle \eta, R^{-1}\eta \rangle \big)$$

is the $m$-variate standard normal density function expressed in terms of the determinant (det) of the correlation matrix $R$, which has entries

$$R_{ij} = \frac{V_{ij}}{\sqrt{V_{ii} V_{jj}}}, \quad i, j = 1, \dots, m,$$

produced by the elements of $V$. The Nataf density function only requires knowledge of the marginal density functions, which can be modeled as discussed above, and the variance-covariance matrix $V$, usually substituted by its estimate $\hat{V}$. The matrix $R$ can be adjusted to improve the range of possible covariances that can be addressed; see [30, Chapter 17]. For other multi-variate models of uncertainty, we refer to [65].

**Approximation of Integrals.** Modeling of uncertainty in the manner sketched out above produces a density function $p : \mathbb{R}^m \to [0, \infty)$ that, hopefully, represents possible values of the random phenomena, but it results in a computational challenge: The expectation

$$\mathbb{E}\big[f(\boldsymbol{\xi}, x)\big] = \int f(\xi, x)p(\xi)\,d\xi$$

isn't easily computed for most $f : \mathbb{R}^m \times \mathbb{R}^n \to \mathbb{R}$. This $m$-dimensional integral over the support $\Xi$ is rarely solvable analytically. If $m \leq 5$ (or a bit higher if special sparse grid techniques are used), then application of *numerical integration* such as Gaussian quadrature rules approximates the expectation by

$$\sum\nolimits_{i=1}^{\nu} w_i f(\xi^i, x)p(\xi^i),$$

where $\xi^1, \dots, \xi^\nu \in \Xi$ and $w_1, \dots, w_\nu \in \mathbb{R}$ are points and weights, respectively, specified by the numerical integration procedure. Such approximations are usually quite accurate and any error vanishes as $\nu \to \infty$ under mild assumptions on $f$. If $m > 5$, then *Monte Carlo sampling* is typically a better approach to approximate an expectation.

**Monte Carlo Sampling.** For $f : \mathbb{R}^m \times \mathbb{R}^n \to \mathbb{R}$ and density function $p : \mathbb{R}^m \to [0, \infty)$, the *sample average*

$$\frac{1}{\nu}\sum\nolimits_{i=1}^{\nu} f(\xi^i, x) \approx \int f(\xi, x)p(\xi)\,d\xi,$$

where $\xi^1, \dots, \xi^\nu$ is a *sample* generated according to $p$. If $p$ is rather standard, then mathematical software packages generate a sample conveniently. Otherwise, the acceptance-rejection method and related techniques can be brought in. For a given $x$, the approximation error vanishes as $\nu \to \infty$ when $\mathbb{E}[f(\boldsymbol{\xi}, x)]$ is finite and the sample is generated independently. However, the accuracy can be poor even for relatively large $\nu$. *Importance sampling* may improve the accuracy by rewriting the expectation in terms of another density function $p_0 : \mathbb{R}^m \to [0, \infty)$ that for all $\xi$ has $p_0(\xi) > 0$ whenever $p(\xi) > 0$. Specifically,

$$\int f(\xi, x)p(\xi)\,d\xi = \int f(\xi, x)\frac{p(\xi)}{p_0(\xi)}p_0(\xi)\,d\xi.$$

The second expression is of the same form as the first one with $f(\xi, x)$ having been replaced by $f(\xi, x)p(\xi)/p_0(\xi)$. Monte Carlo sampling can be applied to this modified integral, but now with the sample being generated according to $p_0$. Thus, importance sampling also circumvents any difficulty associated with sampling according to $p$. A clever choice of $p_0$ improves the accuracy, but this requires insight about the problem at hand; see [4]. We return with a formal treatment of sample averages in §8.G and §8.H.

**Example 3.18** (exact evaluation of expectation function). Integration can sometimes be carried out explicitly. Let's consider a case that generalizes the one in §1.C. Suppose that $\gamma \leq \delta$,

$$f(\xi, x) = \max\{\gamma(\xi - x), \delta(\xi - x)\} \quad \forall \xi, x \in \mathbb{R}$$

and the random variable $\boldsymbol{\xi}$ has a piecewise affine distribution function $P$ defined by

$$P(\xi) = \begin{cases} 0 & \text{when } \xi < \tau_1 \\ \mu_i + \sigma_i(\xi - \tau_i) & \text{when } \tau_i \leq \xi < \tau_{i+1}, \quad i = 1, \dots, q-1 \\ 1 & \text{when } \tau_q \leq \xi, \end{cases}$$

where $-\infty < \tau_1 < \tau_2 < \cdots < \tau_q < \infty$,

$$\mu_{i+1} = \mu_i + \sigma_i(\tau_{i+1} - \tau_i),$$

$\sigma_i \geq 0$, $i = 1, \dots, q-1$, $\mu_1 = 0$ and $\mu_q = 1$ are given; see Figure 3.13. Thus, $P$ is continuous. The resulting expectation function $Ef$ arises in situations where a decision needs to "match" the random variable $\boldsymbol{\xi}$; per-unit shortfall $(\xi - x)$ is penalized by $\delta$ and per-unit excess $(x - \xi)$ is penalized by $-\gamma$. It has an explicit form.

**Detail.** The distribution function $P$ has a piecewise constant density function given by the slope coefficients $\sigma_1, \dots, \sigma_{q-1}$. Thus,

$$\mathbb{E}\big[f(\boldsymbol{\xi}, x)\big] = \sum_{i=1}^{q-1} \int_{\tau_i}^{\tau_{i+1}} \sigma_i \max\{\gamma(\xi - x), \delta(\xi - x)\}\, d\xi.$$

For $i = 1, \dots, q-1$, we consider three cases depending on the location of $x$ and utilize the fact that $\gamma \leq \delta$. If $x \leq \tau_i$, then

$$\int_{\tau_i}^{\tau_{i+1}} \max\{\gamma(\xi - x), \delta(\xi - x)\}\, d\xi = \int_{\tau_i}^{\tau_{i+1}} \delta(\xi - x)\, d\xi = \delta\varphi_i(x),$$

where $\varphi_i(x) = \frac{1}{2}\tau_{i+1}^2 - x\tau_{i+1} - \frac{1}{2}\tau_i^2 + x\tau_i$. If $x \in [\tau_i, \tau_{i+1}]$, then

$$\int_{\tau_i}^{\tau_{i+1}} \max\{\gamma(\xi - x), \delta(\xi - x)\}\, d\xi = \int_{\tau_i}^{x} \gamma(\xi - x)\, d\xi + \int_{x}^{\tau_{i+1}} \delta(\xi - x)\, d\xi$$
$$= \psi_i(x) = \tfrac{1}{2}(\delta - \gamma)x^2 + (\gamma\tau_i - \delta\tau_{i+1})x + \tfrac{1}{2}(\delta\tau_{i+1}^2 - \gamma\tau_i^2).$$

If $x \geq \tau_{i+1}$, then

$$\int_{\tau_i}^{\tau_{i+1}} \max\{\gamma(\xi - x), \delta(\xi - x)\}\, d\xi = \int_{\tau_i}^{\tau_{i+1}} \gamma(\xi - x)\, d\xi = \gamma\varphi_i(x).$$

We can now provide a piecewise description of $\mathbb{E}[f(\boldsymbol{\xi}, x)]$:

$$\mathbb{E}\big[f(\xi,x)\big] = \begin{cases} \delta\sum_{i=1}^{q-1}\sigma_i\varphi_i(x) & \text{if } x \le \tau_1 \\ \gamma\sum_{j=1}^{i-1}\sigma_j\varphi_j(x) + \sigma_i\psi_i(x) + \delta\sum_{j=i+1}^{q-1}\sigma_j\varphi_j(x) & \\ & \text{if } x \in (\tau_i,\tau_{i+1}],\ i = 1,\dots,q-1 \\ \gamma\sum_{i=1}^{q-1}\sigma_i\varphi_i(x) & \text{if } x > \tau_q. \end{cases}$$

By 3.3, $Ef$ is real-valued and convex. This can also be verified directly by examining the piecewise description: $Ef$ is affine for $x \le \tau_1$ and $x > \tau_q$, convex quadratic on each interval $(\tau_i, \tau_{i+1}]$ and the pieces are joined together in a continuous manner for both the function and the derivatives. □

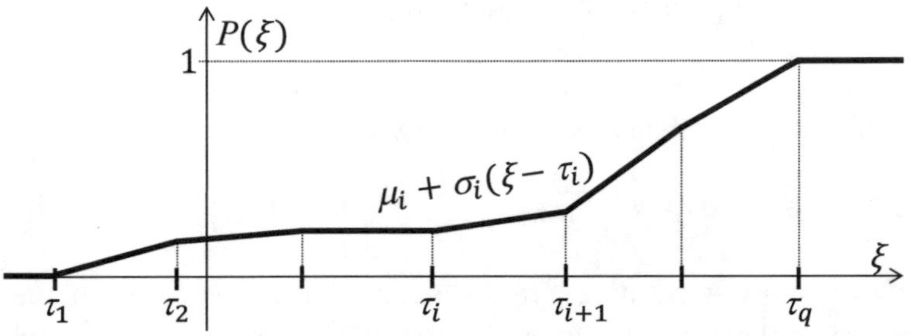


Fig. 3.13: Representation of piecewise affine distribution function.

## 3.E Risk-Adaptive Design

*Engineering design* is an area where risk considerations of the kind described in §3.C are critical. A typical goal is to make a system as safe as possible while not exceeding a monetary budget. As an illustration, consider the truss in Figure 3.14, which may be the load carrying part of a bridge. Let's optimize the cross-sectional areas of the seven members of the truss. The load and material properties are uncertain. Let $\boldsymbol{\xi}_j$ be the yield stress of member $j$. Members 1 and 2 have lognormally distributed yield stresses (in units of Newton (N) per square millimeter (mm)) with expectation 100 and variance 400. The other members have lognormally distributed yield stresses with expectation 200 and variance 1600. The yield stresses aren't statistically independent and their dependence is modeled using a Nataf density function; see §3.D. The truss is subject to a load (in kN) at its mid-span described by the random variable $\boldsymbol{\xi}_8$, statistically independent of the other ones, that's lognormally distributed with expectation 1000 and variance 160000. Let $\boldsymbol{\xi} = (\boldsymbol{\xi}_1, \dots, \boldsymbol{\xi}_8)$.

The decision vector $x = (x_1, \dots, x_7)$, where $x_j$ is the cross-sectional area (in 1000 mm$^2$) of member $j$. A concern is the force in each member, which is $\xi_8/\beta_j$ for member $j$ under load $\xi_8$, where $\beta_j$ is a factor given by the geometry of the truss: $\beta_j = \sqrt{3}/6$ for $j = 1, 2$ and $\beta_j = 1/\sqrt{3}$ for $j = 3, 4, \dots, 7$. The strength of member $j$ is $\xi_j x_j$. Thus, a

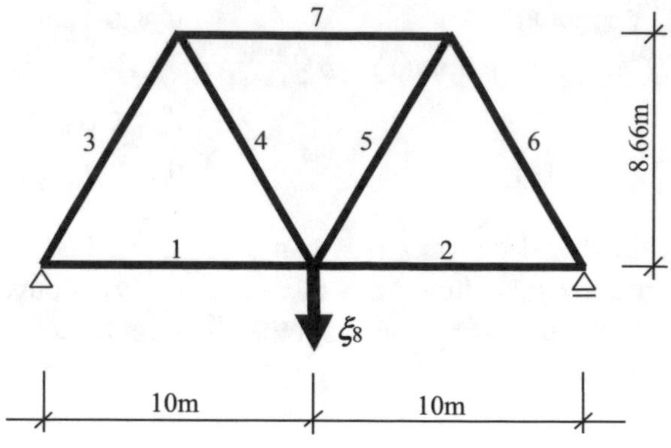


Fig. 3.14: Truss under uncertain load $\boldsymbol{\xi}_8$.

measure of performance of the truss is the load violation

$$f(\xi, x) = \max_{j=1,\dots,7} \{\xi_8/\beta_j - \xi_j x_j\},$$

which has kN as units. A positive $f(\xi, x)$ means that at least one of the members is experiencing a force that's greater than its strength. We seek a design $x$ with the lowest possible $\alpha$-superquantile risk, i.e., minimize s-rsk$_\alpha(f(\boldsymbol{\xi}, x))$. We impose the constraints $0.5 \le x_j \le 2, j = 1, 2, \dots, 7$, and thereby limit the cross-sectional area of each member to be between 500 mm$^2$ and 2000 mm$^2$. The material cost of the truss is proportional to $\sum_{j=1}^{7} x_j$, which is constrained by 8000 mm$^2$.

Since $\boldsymbol{\xi}$ has a continuous distribution, computation of the required expectations would involve eight-dimensional integrals and this is costly; see §3.D. We adopt an approximation based on Monte Carlo sampling and generate a sample $\xi^1, \dots, \xi^\nu$ according to the probability distribution of $\boldsymbol{\xi}$. In view of (3.4), the problem of minimizing s-rsk$_\alpha(f(\boldsymbol{\xi}, x))$ under the stated constraints becomes the linear optimization problem

$$\begin{aligned} \underset{x\in\mathbb{R}^7, \gamma\in\mathbb{R}, z\in\mathbb{R}^\nu}{\text{minimize}} \quad & \gamma + \frac{1}{(1-\alpha)\nu} \sum_{i=1}^{\nu} z_i && (3.6)\\ \text{subject to} \quad & \xi_8^i/\beta_j - \xi_j^i x_j - \gamma \le z_i, \quad i = 1, \dots, \nu, \ j = 1, \dots, 7 \\ & 0 \le z_i, \quad i = 1, \dots, \nu \\ & 0.5 \le x_j \le 2, \quad j = 1, \dots, 7 \\ & x_1 + \cdots + x_7 \le 8. \end{aligned}$$

Table 3.4 provides minimizers and minima for $\alpha = 0.9$. As $\nu$ grows, the Monte Carlo sampling approximation improves (cf. §3.D) and this is also reflected in the table. The run times for solving the linear optimization problem by the interior-point method (§2.G) are nearly zero for $\nu \le 10^4$, but grow to 18 and 271 seconds for the last two rows in the table, respectively. The run times can be reduced by curbing $\nu$ using Importance Sampling (cf.

§3.D), switching to the subgradient method of §2.I applied to the alternative formulation (3.3) and/or adopting the active-set strategy of 6.16.

For the most accurate approximation in the last row of the table, the resulting design has a minimum superquantile risk of −50.21, which means that on average in the worst 10% outcomes the load violation is about −50 kN. Thus, there's a small residual capacity. In contrast, a design obtained by minimizing the expected load violation, which corresponds to setting $\alpha = 0$, is distinctly different and the resulting 0.9-superquantile of the load violation turns out to be −46; the residual capacity is reduced with about 10%.

Table 3.4: Truss designs using various sample sizes $\nu$.

| $\nu$ | Design $x^\star$ (in mm$^2$) | | | | | | | s-rsk$_{0.9}$ $(f(\xi, x^\star))$ |
|---|---|---|---|---|---|---|---|---|
| $10^1$ | 1334 | 1469 | 941 | 1024 | 1079 | 1054 | 1098 | −80.23 |
| $10^2$ | 1423 | 1470 | 1037 | 1071 | 1001 | 978 | 1019 | −63.21 |
| $10^3$ | 1446 | 1437 | 1017 | 1047 | 1010 | 1037 | 1005 | −54.04 |
| $10^4$ | 1431 | 1433 | 1028 | 1033 | 1026 | 1026 | 1023 | −51.80 |
| $10^5$ | 1428 | 1426 | 1030 | 1028 | 1028 | 1030 | 1029 | −50.28 |
| $5\cdot 10^5$ | 1426 | 1425 | 1030 | 1028 | 1029 | 1030 | 1031 | −50.21 |

These results can also be interpreted in terms of probabilities. For $\alpha \in (0,1)$, $\tau \in \mathbb{R}$ and $x$, it's clear from the definition of the $\alpha$-quantile of $f(\xi, x)$, denoted by $Q(\alpha)$, that

$$Q(\alpha) \leq \tau \iff \text{prob}\big\{f(\xi, x) > \tau\big\} \leq 1 - \alpha. \tag{3.7}$$

Since $\bar{Q}(\alpha) \geq Q(\alpha)$, as seen directly from the definition of superquantiles in 3.6, we also have

$$\text{s-rsk}_\alpha\big(f(\xi, x)\big) \leq \tau \implies \text{prob}\big\{f(\xi, x) > \tau\big\} \leq 1 - \alpha. \tag{3.8}$$

Thus, for the last row in Table 3.4, this implies that the probability for $f(\xi, x^\star)$ to exceed $\tau = -50.21$ is at most $1 - \alpha = 0.1$.

In general, the equivalence in (3.7) between quantiles and probabilities of exceeding a threshold motivates a parallel relation between superquantiles and corresponding probabilities.

**Definition 3.19** (buffered probability). For an integrable random variable $\xi$ with superquantiles $\bar{Q}(\alpha)$, $\alpha \in [0,1]$, the *buffered probability* of exceeding $\tau \in \mathbb{R}$ is

$$\text{b-prob}\{\xi > \tau\} = \begin{cases} 0 & \text{if } \tau \geq \bar{Q}(1) \\ 1 - \bar{Q}^{-1}(\tau) & \text{if } \bar{Q}(0) < \tau < \bar{Q}(1) \\ 1 & \text{otherwise.} \end{cases}$$

Here, $\bar{Q}^{-1}(\tau)$ is the solution to $\tau = \bar{Q}(\alpha)$.

The definition is illustrated in Figure 3.15. For $\tau \in (\bar{Q}(0), \bar{Q}(1))$, $\bar{Q}^{-1}(\tau)$ is the solution to $\tau = \bar{Q}(\alpha)$, which indeed is unique because $\bar{Q}$ is a continuous function on $(0, 1)$ (as discussed in 5.9; see also [97, Theorem 2]) and $\bar{Q}$ is increasing on $(0, \text{prob}\{\boldsymbol{\xi} < \bar{Q}(1)\})$ by [62, Proposition A.1].

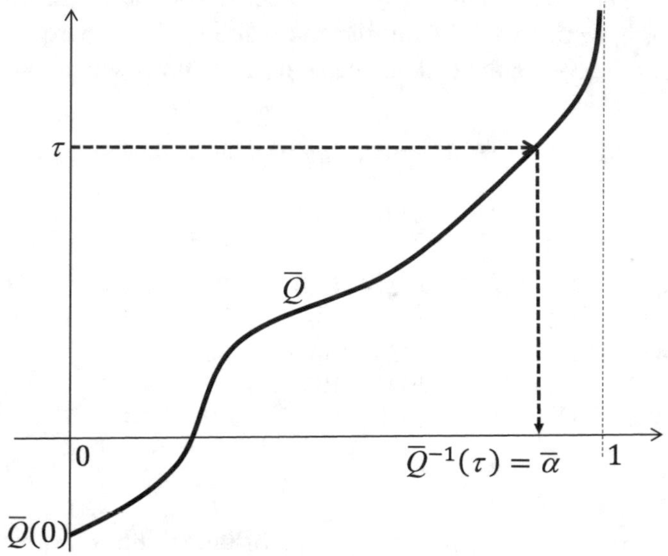


Fig. 3.15: Superquantile $\bar{Q}(\alpha)$ as a function of $\alpha \in [0, 1]$.

For an integrable random variable $\boldsymbol{\xi}$, $\alpha \in (0, 1]$ and $\tau \in \mathbb{R}$, we've that

$$\text{s-rsk}_\alpha(\boldsymbol{\xi}) \le \tau \iff \text{b-prob}\{\boldsymbol{\xi} > \tau\} \le 1 - \alpha. \tag{3.9}$$

Thus, for the last row in Table 3.4, this implies that the buffered probability of exceeding $\tau = -50.21$ is $1 - \alpha = 0.1$. In general, buffered probabilities become a useful way of interpreting results about superquantile risk and also offer alternative model formulations; see [71, 96]. We refer to [70] for buffered probability formulae in the case of many common probability distributions and to [109] for expressions of subgradients.

**Example 3.20** (chance constraints). Engineering design is often carried out with the goal of minimizing cost subject to a reliability constraint. For a random vector $\boldsymbol{\xi}$ with support $\Xi \subset \mathbb{R}^m$ and a function $f : \mathbb{R}^m \times \mathbb{R}^n \to \mathbb{R}$ such that $f(\xi, x)$ represents the performance of a system under outcome $\xi$ and design $x$, reliability could be expressed using the chance constraint

$$\text{prob}\big\{f(\boldsymbol{\xi}, x) > 0\big\} \le 1 - \alpha \tag{3.10}$$

for some $\alpha \in (0, 1)$, typically near one. Here, we assume that $f$ has been oriented and normalized such that a value above zero is deemed a system failure. While there are situations under which this constraint is theoretically and computationally attractive (cf. 2.57 and its normality assumption), the function of $x$ on the left-hand side in (3.10) is generally neither convex nor continuous even if $f$ has those properties.

It's usually better to quantify reliability in terms of buffered probabilities and adopt the constraint

$$\text{b-prob}\{f(\boldsymbol{\xi}, x) > 0\} \le 1 - \alpha, \tag{3.11}$$

where we assume that $f(\boldsymbol{\xi}, x)$ is integrable for all $x \in \mathbb{R}^n$. By (3.8) and (3.9), the buffered probability constraint is more conservative than the chance constraint: If $x$ satisfies (3.11), then it also satisfies (3.10), but the converse doesn't hold. Moreover, the buffered probability constraint can be implemented effectively using its connection to superquantiles.

**Detail.** In light of (3.9), the buffered probability constraint (3.11) is equally well expressed as

$$\text{s-rsk}_\alpha\big(f(\boldsymbol{\xi}, x)\big) \le 0.$$

Using the minimization formula in 3.9, we find that the problem of minimizing $f_0 : \mathbb{R}^n \to \overline{\mathbb{R}}$ subject to (3.11) is equivalent to

$$\underset{x\in\mathbb{R}^n, \gamma\in\mathbb{R}}{\text{minimize}}\ f_0(x)\ \text{ subject to }\ \gamma + \frac{1}{1-\alpha}\mathbb{E}\big[\max\{0, f(\boldsymbol{\xi}, x) - \gamma\}\big] \le 0.$$

Further properties and reformulations are then available following the pattern in §3.C. In particular, this inequality defines a convex set if $f(\xi, \cdot)$ is convex for all $\xi \in \Xi$. □

## 3.F Optimality in Stochastic Optimization

Algorithms and optimality conditions for stochastic optimization problems require subgradients of expectation functions. For finitely distributed random vectors, such functions involve finite sums and we can leverage the Moreau-Rockafellar sum rule 2.26 recursively. However, even in this case, the assumption of the sum rule indicates that we can't expect to always have the subgradients of an expectation function be equal to a sum of subgradients. For more general probability distributions, we also need to contend with the meaning of "summing up" an infinite number of subgradient sets. We'll now obtain an extension of the sum rule that furnishes subgradients in many applications.

For a random vector $\boldsymbol{\xi}$ with support $\Xi \subset \mathbb{R}^m$, a function $f : \mathbb{R}^m \times \mathbb{R}^n \to \mathbb{R}$ and a point $\bar{x} \in \mathbb{R}^n$, suppose that $f(\cdot, \bar{x})$ is measurable so that we can view $f(\boldsymbol{\xi}, \bar{x})$ as a random variable; cf. §3.B. Then, the expectation function given by

$$Ef(\bar{x}) = \mathbb{E}\big[f(\boldsymbol{\xi}, \bar{x})\big]$$

is well defined. Now, let's extend this thinking to subgradients with the goal of obtaining an expression for $\partial Ef(\bar{x})$. Suppose that $f(\xi, \cdot)$ is convex for all $\xi \in \Xi$. These convex functions have subgradients $\{\partial_x f(\xi, \bar{x}),\ \xi \in \Xi\}$, where $\partial_x f(\xi, \bar{x})$ is the set of subgradients of $f(\xi, \cdot)$ at $\bar{x}$. If the value of $\xi$ is uncertain, then $\partial_x f(\xi, \bar{x})$ is also uncertain, but $\partial_x f(\boldsymbol{\xi}, \bar{x})$ isn't a random vector because $\partial_x f(\xi, \bar{x})$ is generally a subset of $\mathbb{R}^n$ and not a single point. This requires new terminology.

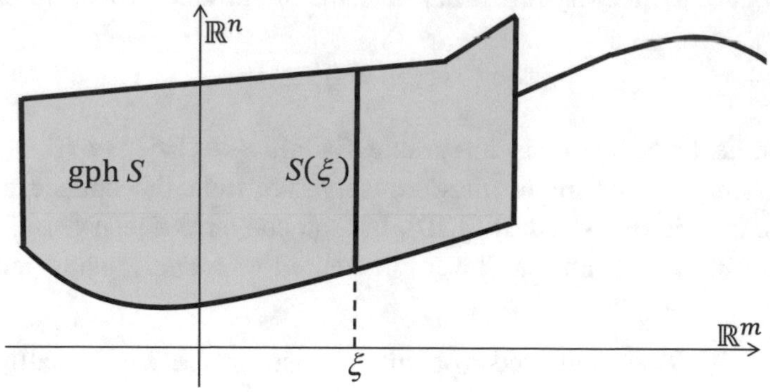


Fig. 3.16: Set-valued mapping $S : \mathbb{R}^m \rightrightarrows \mathbb{R}^n$ visualized by its graph, i.e., gph $S = \{(\xi, v) \in \mathbb{R}^m \times \mathbb{R}^n \mid v \in S(\xi)\}$.

**Set-Valued Mappings.** A *set-valued mapping* $S : \mathbb{R}^m \rightrightarrows \mathbb{R}^n$ takes as input a vector $\xi \in \mathbb{R}^m$ and returns a set $S(\xi) \subset \mathbb{R}^n$ as output; see Figure 3.16. The double arrow distinguishes set-valued mappings from a function $h : \mathbb{R}^m \to \overline{\mathbb{R}}$, which returns points in $\overline{\mathbb{R}}$, and from a *vector-valued mapping* $F : \mathbb{R}^m \to \mathbb{R}^n$, which defines points in $\mathbb{R}^n$. Still, we don't rule out that the set-valued mapping $S$ may have $S(\xi) = \{v\}$ for some $\xi$, i.e., $S(\xi)$ is a set consisting of a single point $v \in \mathbb{R}^n$ as is the case to the right in Figure 3.16. If this takes place for all $\xi$, then $S$ is *point-valued*. We may even have that $S(\xi) = \emptyset$, which takes place for $\xi$ to the left in Figure 3.16. The set-valued mapping is *convex-valued* (*closed-valued*) if $S(\xi)$ is a convex (closed) set for all $\xi \in \mathbb{R}^m$. For example, if $\bar{x} \in \mathbb{R}^n$ and $f : \mathbb{R}^m \times \mathbb{R}^n \to \mathbb{R}$ defines a convex function $f(\xi, \cdot)$ for all $\xi \in \mathbb{R}^m$, then the set-valued mapping $S : \mathbb{R}^m \rightrightarrows \mathbb{R}^n$ given by $S(\xi) = \partial_x f(\xi, \bar{x})$ is convex-valued (as mentioned after 2.14) and closed-valued by 2.25; see Figure 3.17 for a concrete example.

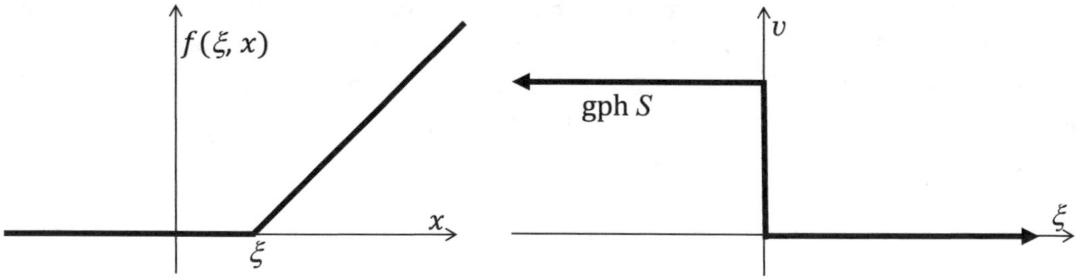


Fig. 3.17: Left: graph of function $f(\xi, \cdot)$ with $f(\xi, x) = \max\{0, x - \xi\}$. Right: gph $S = \{(\xi, v) \mid v \in S(\xi)\}$, where $S(\xi) = \{1\}$ if $\xi < 0$, $S(\xi) = \{0\}$ if $\xi > 0$ and $S(0) = [0, 1]$, i.e., $S(\xi) = \partial_x f(\xi, 0)$.

**Random Sets.** For a random vector $\boldsymbol{\xi}$ with support $\Xi \subset \mathbb{R}^m$ and a set-valued mapping $S : \mathbb{R}^m \rightrightarrows \mathbb{R}^n$, there's uncertainty about the subset of $\mathbb{R}^n$ produced by $S$ when its argument is uncertain. As in the case of random variables, we would like to quantify the probability of different outcomes using the probability distribution of $\boldsymbol{\xi}$. Thus, we require that $S$ is sufficiently "nice," or *measurable*. If $\boldsymbol{\xi}$ is discretely distributed, then every closed-valued $S : \mathbb{R}^m \rightrightarrows \mathbb{R}^n$ is measurable, but many other possibilities exist as well; see §8.G. If $S$ is a measurable set-valued mapping, then we say that $S(\boldsymbol{\xi})$ is a *random set*, with possible outcomes $\{S(\xi), \xi \in \Xi\}$ occurring according to the probability distribution of $\boldsymbol{\xi}$. Its *expectation* is defined as

$$\mathbb{E}\big[S(\boldsymbol{\xi})\big] = \Big\{\mathbb{E}\big[v(\boldsymbol{\xi})\big] \;\Big|\; v : \mathbb{R}^m \to \mathbb{R}^n,\ v(\xi) \in S(\xi)\ \forall \xi \in \Xi,\ v(\boldsymbol{\xi}) \text{ integrable}\Big\}.$$

In words, $\mathbb{E}[S(\boldsymbol{\xi})]$ is the set of all points that can be obtained by taking expectation of integrable random vectors whose defining mappings have values contained in those of $S$. For example, suppose that $f : \mathbb{R}^m \times \mathbb{R}^n \to \mathbb{R}$ and $f(\xi, \cdot)$ is smooth at $\bar{x}$ with gradient $\nabla_x f(\xi, \bar{x})$ for all $\xi \in \Xi$. If $\nabla_x f(\boldsymbol{\xi}, \bar{x})$ is an integrable random vector, then the set-valued mapping $S : \mathbb{R}^m \rightrightarrows \mathbb{R}^n$ with $S(\xi) = \{\nabla_x f(\xi, \bar{x})\}$ for all $\xi \in \Xi$ defines a random set $S(\boldsymbol{\xi})$ with expectation

$$\mathbb{E}\big[S(\boldsymbol{\xi})\big] = \Big\{\mathbb{E}\big[\nabla_x f(\boldsymbol{\xi}, \bar{x})\big]\Big\}.$$

In this case, the only vector-valued mapping that can be selected from $S$ is $\nabla_x f(\cdot, \bar{x})$ because $S(\xi)$ is a single point for all $\xi \in \Xi$. Thus, $\mathbb{E}[S(\boldsymbol{\xi})]$ is the single point $\mathbb{E}[\nabla_x f(\boldsymbol{\xi}, \bar{x})]$.

**Example 3.21** (subgradients of expectation function). For $f : \mathbb{R} \times \mathbb{R} \to \mathbb{R}$ with $f(\xi, x) = \max\{0, x - \xi\}$, the function $f(\xi, \cdot)$ is nonsmooth regardless of $\xi$; see Figure 3.18(left). Let $\boldsymbol{\xi}$ be a random variable with support $\Xi = \{-1, 0, 2\}$ and corresponding probabilities $p_{-1} = 0.2$, $p_0 = 0.5$ and $p_2 = 0.3$. Then, the expectation function given by $Ef(x) = \mathbb{E}[f(\boldsymbol{\xi}, x)]$ has well-defined subgradients

$$\partial Ef(x) = \mathbb{E}\big[\partial_x f(\boldsymbol{\xi}, x)\big]$$

at every point $x \in \mathbb{R}$; see Figure 3.18(right).

**Detail.** Since $Ef(x) = \sum_{\xi \in \{-1,0,2\}} p_\xi \max\{0, x - \xi\}$, the Moreau-Rockafellar sum rule 2.26 directly leads to $\partial Ef(x)$ as depicted in Figure 3.18(right). To see that this coincides with $\mathbb{E}[\partial_x f(\boldsymbol{\xi}, x)]$, let's concentrate on $x = 0$. We need $v(\xi) \in \partial_x f(\xi, 0)$:

$$v(-1) \in \partial_x f(-1, 0) = \{1\}, \quad v(0) \in \partial_x f(0, 0) = [0, 1], \quad v(2) \in \partial_x f(2, 0) = \{0\}.$$

Thus,

$$0.2v(-1) + 0.5v(0) + 0.3v(2) \in \mathbb{E}\big[\partial_x f(\boldsymbol{\xi}, 0)\big] = [0.2, 0.7].$$

Expressions for $\mathbb{E}[\partial_x f(\xi, x)]$ at other points are calculated similarly and they indeed coincide with $\partial Ef(x)$. □

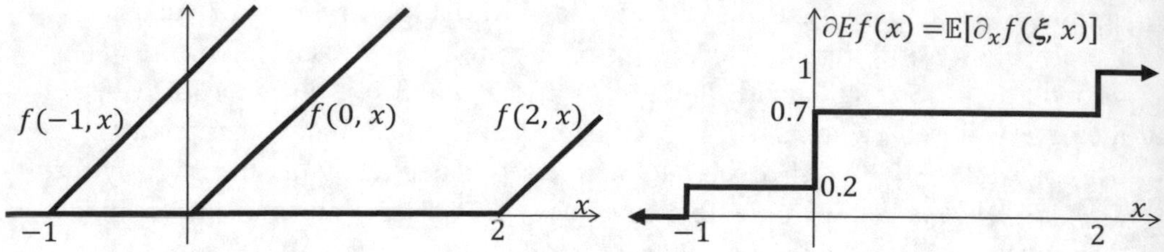


Fig. 3.18: Left: $f(\xi, x) = \max\{0, x - \xi\}$. Right: expectation of $\partial_x f(\xi, x)$ at different $x$.

**Proposition 3.22** (convexity of expectations of random sets). *Suppose that $\boldsymbol{\xi}$ is an $m$-dimensional random vector and $S : \mathbb{R}^m \rightrightarrows \mathbb{R}^n$ is a measurable convex-valued mapping. Then, $\mathbb{E}[S(\boldsymbol{\xi})]$ is a convex subset of $\mathbb{R}^n$.*

**Proof.** Let $\Xi$ be the support of $\boldsymbol{\xi}$, $\lambda \in [0, 1]$ and $s^0, s^1 \in \mathbb{E}[S(\boldsymbol{\xi})]$. Then, there are mappings $v^0, v^1 : \mathbb{R}^m \to \mathbb{R}^n$, with $v^0(\xi)$ and $v^1(\xi)$ in $S(\xi)$ for all $\xi \in \Xi$, that define integrable random vectors $v^0(\boldsymbol{\xi})$ and $v^1(\boldsymbol{\xi})$ so that

$$s^0 = \mathbb{E}\big[v^0(\boldsymbol{\xi})\big] \text{ and } s^1 = \mathbb{E}\big[v^1(\boldsymbol{\xi})\big].$$

Since $S$ is convex-valued,

$$v^\lambda(\xi) = (1 - \lambda)v^0(\xi) + \lambda v^1(\xi) \in S(\xi) \;\; \forall \xi \in \Xi$$

and the resulting function $v^\lambda : \mathbb{R}^m \to \mathbb{R}^n$ defines an integrable random vector $v^\lambda(\boldsymbol{\xi})$ with $\mathbb{E}[v^\lambda(\boldsymbol{\xi})] \in \mathbb{E}[S(\boldsymbol{\xi})]$. By 3.2(b),

$$\mathbb{E}\big[v^\lambda(\boldsymbol{\xi})\big] = (1 - \lambda)\mathbb{E}\big[v^0(\boldsymbol{\xi})\big] + \lambda\mathbb{E}\big[v^1(\boldsymbol{\xi})\big]$$

so then $(1 - \lambda)s^0 + \lambda s^1 \in \mathbb{E}[S(\boldsymbol{\xi})]$. □

The sets of subgradients of an expectation function turn out to be expectations of random sets generally as already experienced in 3.21. Informally, a subgradient of an expectation function is the expectation of subgradients.

**Proposition 3.23** (subgradients of expectation functions). *For a random vector $\boldsymbol{\xi}$ with support $\Xi \subset \mathbb{R}^m$ and $f : \mathbb{R}^m \times \mathbb{R}^n \to \mathbb{R}$, suppose that*

(a) *$f(\boldsymbol{\xi}, x)$ is integrable for all $x \in \mathbb{R}^n$*
(b) *$f(\xi, \cdot)$ is convex for all $\xi \in \Xi$.*

*Then, for all $x \in \mathbb{R}^n$, $\partial_x f(\boldsymbol{\xi}, x)$ is a random set and*

$$\partial Ef(x) = \mathbb{E}\big[\partial_x f(\boldsymbol{\xi}, x)\big]$$
$$= \Big\{\mathbb{E}\big[v(\boldsymbol{\xi})\big] \;\Big|\; v : \mathbb{R}^m \to \mathbb{R}^n,\; v(\xi) \in \partial_x f(\xi, x)\; \forall \xi \in \Xi,\; v(\boldsymbol{\xi}) \textit{ integrable}\Big\}.$$

*If in addition $f(\xi, \cdot)$ is smooth for all $\xi \in \Xi$, then*

$$\partial Ef(x) = \Big\{\mathbb{E}\big[\nabla_x f(\boldsymbol{\xi}, x)\big]\Big\}$$

*and $Ef$ is actually smooth with $\nabla Ef(x) = \mathbb{E}[\nabla_x f(\boldsymbol{\xi}, x)]$.*

**Proof.** Let $\bar{x} \in \mathbb{R}^n$. By [105, Exercise 14.29, Theorem 14.56], $\partial_x f(\cdot, \bar{x})$ is measurable so that $\partial_x f(\boldsymbol{\xi}, \bar{x})$ is a random set and the expectation $\mathbb{E}[\partial_x f(\boldsymbol{\xi}, \bar{x})]$ is well defined. We show that

$$\mathbb{E}\big[\partial_x f(\boldsymbol{\xi}, \bar{x})\big] \subset \partial Ef(\bar{x})$$

as follows. If $\bar{v} \in \mathbb{E}[\partial_x f(\boldsymbol{\xi}, \bar{x})]$, then there's $v : \mathbb{R}^m \to \mathbb{R}^n$ such that $v(\boldsymbol{\xi})$ is integrable with $\bar{v} = \mathbb{E}[v(\boldsymbol{\xi})]$ and, by the subgradient inequality 2.17, one has

$$\big\langle v(\xi), x - \bar{x}\big\rangle \le f(\xi, x) - f(\xi, \bar{x}), \quad \forall x \in \mathbb{R}^n,\; \xi \in \Xi.$$

After taking the expectation on both sides and invoking 3.2, one finds that

$$\langle \bar{v}, x - \bar{x}\rangle \le Ef(x) - Ef(\bar{x}), \quad \forall x \in \mathbb{R}^n.$$

Thus, $\bar{v} \in \partial Ef(\bar{x})$ by 2.17. It's more complicated to show the inclusion the other way, and we refer to [90, Theorem 23]. A direct proof for smooth functions is given by [118, Theorem 7.46]. □

The proposition exemplifies a situation where expectation and differentiation can be interchanged. It can be improved further by relaxing the assumption of smoothness when the probability distribution of $\boldsymbol{\xi}$ is continuous; see 3.33.

We can now bring in the optimality condition 2.19 to obtain a fundamental result in stochastic optimization.

**Theorem 3.24** (minimizers of convex expectation functions). *For a random vector $\boldsymbol{\xi}$ with support $\Xi \subset \mathbb{R}^m$ and $f : \mathbb{R}^m \times \mathbb{R}^n \to \mathbb{R}$, suppose that*

*(a) $f(\boldsymbol{\xi}, x)$ is integrable for all $x \in \mathbb{R}^n$*
*(b) $f(\xi, \cdot)$ is convex for all $\xi \in \Xi$.*

*Then, $x^\star \in \operatorname{argmin} Ef$ if and only if there exists*

$$v : \mathbb{R}^m \to \mathbb{R}^n \textit{ such that } \mathbb{E}\big[v(\boldsymbol{\xi})\big] = 0 \textit{ and } v(\xi) \in \partial_x f(\xi, x^\star)\; \forall \xi \in \Xi.$$

*The subgradient condition is equivalently expressed as*

$$\forall \xi \in \Xi : \quad x^\star \in \operatorname{argmin}_{x \in \mathbb{R}^n} f(\xi, x) - \big\langle v(\xi), x\big\rangle.$$

**Proof.** If a function $v$ exists with the stated properties, then it follows from 3.23 that $0 \in \partial Ef(x^\star)$, which is equivalent to having $x^\star \in \operatorname{argmin} Ef$ by the optimality condition 2.19. If $0 \in \partial Ef(x^\star)$, the existence of a function $v$ such that $\mathbb{E}[v(\boldsymbol{\xi})] = 0$ and $v(\xi) \in \partial_x f(\xi, x^\star)$ for all $\xi \in \Xi$ is guaranteed by 3.23.

The alternative expression follows because $0 \in \partial_x f(\xi, x) - v(\xi)$ is an optimality condition for the function $x \mapsto f(\xi, x) - \langle v(\xi), x\rangle$ by 2.19 and the Moreau-Rockafellar sum rule 2.26. □

In 3.21, $Ef(x) = \sum_{\xi \in \{-1,0,2\}} p_\xi \max\{0, x - \xi\}$ and $x^\star = -1$ is a minimizer because we can take $v(\xi) = 0$ for $\xi = -1, 0, 2$; see also Figure 3.18. A general implication of the theorem is that one can find a minimizer of an expectation function $Ef$ by minimizing $f(\xi, x)$, perturbed by the appropriate linear term $\langle v(\xi), x\rangle$, for some outcome $\xi$ with $v(\xi)$ conveying the effect of the other outcomes. Thus, if we just had known $v : \mathbb{R}^m \to \mathbb{R}^n$, stochastic optimization could be achieved without (costly) expectations. The idea of solving subproblems tailored to specific outcomes underpins decoupling algorithms; see §10.E.

The theorem can be used to construct optimality conditions for a variety of problems. We employ it to address superquantiles. The following arguments also provide the path to a solution of the newsvendor problem (§1.C).

**Proof of 3.9.** We recall that $f(\xi, \gamma) = \gamma + \max\{0, \xi - \gamma\}/(1 - \alpha)$ defines a convex function in $\gamma$ for all $\xi \in \Xi$ and the random variable $f(\boldsymbol{\xi}, \gamma)$ is integrable for all $\gamma$. Thus, the optimality condition 3.24 applies. For $\xi, \gamma \in \mathbb{R}$, the set of subgradients computed with respect to $\gamma$ is

$$\partial_\gamma f(\xi, \gamma) = \begin{cases} \{1\} & \text{if } \gamma > \xi \\ \big[1 - 1/(1-\alpha), 1\big] & \text{if } \gamma = \xi \\ \{1 - 1/(1-\alpha)\} & \text{otherwise.} \end{cases}$$

Let's fix $\gamma \in \mathbb{R}$. Every $v : \mathbb{R} \to \mathbb{R}$ with $v(\xi) \in \partial_\gamma f(\xi, \gamma)$ for all $\xi \in \Xi$ is then of the form

$$v(\xi) = \begin{cases} 1 & \text{if } \gamma > \xi \\ 1 - \lambda/(1-\alpha) & \text{if } \gamma = \xi \\ 1 - 1/(1-\alpha) & \text{otherwise} \end{cases}$$

for some $\lambda \in [0, 1]$. The random variable $v(\boldsymbol{\xi})$ has the value $1 - 1/(1 - \alpha)$ with probability $1 - P(\gamma)$, where $P$ is the distribution function of $\boldsymbol{\xi}$. It has the value $1 - \lambda/(1 - \alpha)$ with the same probability that $\boldsymbol{\xi}$ has the value $\gamma$. If $P$ is continuous, then that probability is zero. However, if $P$ has a discontinuity at $\gamma$, then the probability is the size of the "jump" of $P$ at $\gamma$. With $P(\gamma_-) = \lim_{\bar\gamma \nearrow \gamma} P(\bar\gamma)$, this probability is then $P(\gamma) - P(\gamma_-)$. Finally, $v(\boldsymbol{\xi})$ has the value 1 with probability $P(\gamma_-)$. Thus,

$$\mathbb{E}\big[v(\boldsymbol{\xi})\big] = P(\gamma_-) + \Big(1 - \frac{\lambda}{1-\alpha}\Big)\big(P(\gamma) - P(\gamma_-)\big) + \Big(1 - \frac{1}{1-\alpha}\Big)\big(1 - P(\gamma)\big)$$

and $v(\boldsymbol{\xi})$ is indeed integrable. Setting this expectation equal to zero amounts to

$$\lambda\big(P(\gamma) - P(\gamma_-)\big) = P(\gamma) - \alpha \text{ or, equivalently, } P(\gamma_-) \le \alpha \le P(\gamma).$$

In view of the optimality condition 3.24, $\gamma \in \operatorname{argmin} Ef$ if and only if $\gamma$ satisfies these inequalities. Certainly, $\gamma = Q(\alpha)$, the $\alpha$-quantile of $\boldsymbol{\xi}$, satisfies the inequalities. In fact, it's the lowest possible solution. If $P$ increases to the right of $Q(\alpha)$, then $Q(\alpha)$ is the unique solution. In general, $\operatorname{argmin} Ef = [Q(\alpha), Q^+(\alpha)]$, where $Q^+(\alpha) = \max\{\gamma \in \mathbb{R} \mid P(\gamma_-) \le \alpha\}$; the endpoint of any "flat stretch" of $P$ to the right $Q(\alpha)$ when $P(Q(\alpha)) = \alpha$. □

**Exercise 3.25** (newsvendor problem). For parameters $\delta > \gamma > 0$ and an integrable random variable $\boldsymbol{\xi}$, the newsvendor problem of 1.4 takes the form of minimizing $Ef(x) = \mathbb{E}[f(\boldsymbol{\xi}, x)]$, where

$$f(\xi, x) = (\gamma - \delta)x + \max\big\{0, \delta(x - \xi)\big\}.$$

(a) Derive an expression for $\partial Ef$ using 3.23. (b) Let $\gamma = 3$, $\delta = 7$ and $\boldsymbol{\xi}$ have a uniform density function with support $[10, 20]$. Draw the graph of $\partial Ef$. (c) Repeat (b) with the same $\gamma$ and $\delta$, but now letting $\boldsymbol{\xi}$ have a finite distribution with support $\{10, 11, \ldots, 20\}$ and equal probabilities. (d) For both probability distributions, determine the minimizers of $Ef$ using the optimality condition 3.24.

**Guide.** The exercise repeats 1.5 to some extent, but now with subgradients in the central role. □

## 3.G Stochastic Gradient Descent

Stochastic optimization problems quickly become large scale due to the need for considering many possible values of random vectors, for example, as produced by Monte Carlo sampling or a sizeable dataset. Although the algorithms of Chaps. 1 and 2 might apply, it's convenient and sometimes essential to have fast specialized algorithms that utilize the structure of these problems. The remainder of the chapter discusses a series of such algorithms.

For a function $f : \mathbb{R}^m \times \mathbb{R}^n \to \mathbb{R}$, a nonempty, closed and convex set $C \subset \mathbb{R}^n$ and a random vector $\boldsymbol{\xi}$ with support $\Xi \subset \mathbb{R}^m$, let's consider

$$\underset{x \in C}{\text{minimize}}\; Ef(x) = \mathbb{E}\big[f(\boldsymbol{\xi}, x)\big]. \tag{3.12}$$

Under the assumptions that $f(\boldsymbol{\xi}, x)$ is integrable for all $x \in \mathbb{R}^n$ and $f(\xi, \cdot)$ is convex for all $\xi \in \Xi$, $Ef$ is real-valued and convex by 3.3 and

$$\partial Ef(x) = \mathbb{E}\big[\partial_x f(\boldsymbol{\xi}, x)\big] \quad \forall x \in \mathbb{R}^n$$

by 3.23. Still, any algorithm for this problem needs to contend with the difficulty of computing integrals as discussed in §3.D. Even if $\boldsymbol{\xi}$ is finitely distributed, a large number of possible outcomes of this random vector slows down most algorithms; see §3.E for an example. This includes the subgradient method (§2.I), which in each iteration needs a subgradient $v \in \mathbb{E}[\partial_x f(\boldsymbol{\xi}, x^\nu)]$ at the current point $x^\nu$. We'll develop extensions of the subgradient method that implement a remarkable coarse approximation of this expectation.

We recall that the subgradient method applied to (3.12) computes

$$x^{\nu+1} \in \text{prj}_C(x^\nu - \lambda v), \quad \text{where } v \in \mathbb{E}\big[\partial_x f(\boldsymbol{\xi}, x^\nu)\big] \text{ and } \lambda \in (0, \infty).$$

As an approximation, we now simply set $v \in \partial_x f(\xi^\nu, x^\nu)$, where $\xi^\nu$ is a sample point generated according to the probability distribution of $\boldsymbol{\xi}$. Thus, we approximate an expectation by a single outcome. At first, this seems hopelessly inaccurate, but in fact the resulting method can work.

**Example 3.26** (expectation by stochastic gradient descent). Suppose that $\boldsymbol{\xi}$ is a random vector with support $\Xi \subset \mathbb{R}^n$,

$$f(\xi, x) = \tfrac{1}{2}\|\xi - x\|_2^2 \quad \forall \xi, x \in \mathbb{R}^n$$

and $f(\boldsymbol{\xi}, x)$ is integrable for all $x \in \mathbb{R}^n$. Let's attempt to minimize $Ef$ by iterating as suggested above:

$$x^{\nu+1} = x^\nu - \lambda^\nu v, \quad \text{where } v \in \partial_x f(\xi^\nu, x^\nu), \quad \nu = 0, 1, 2, \dots,$$

and $\{\xi^0, \xi^1, \dots\}$ is a sample generated independently according to the probability distribution of $\boldsymbol{\xi}$. Then, $x^\nu \to \mathbb{E}[\boldsymbol{\xi}]$ if $\lambda^\nu$ is selected appropriately, where $\mathbb{E}[\boldsymbol{\xi}]$ indeed minimizes $Ef$.

**Detail.** In this case, we obtain the subgradient

$$\partial_x f(\xi, x) = \{x - \xi\}$$

so that

$$x^{\nu+1} = x^\nu - \lambda^\nu(x^\nu - \xi^\nu) = \lambda^\nu \xi^\nu + (1 - \lambda^\nu)x^\nu.$$

Applying this expression recursively, we obtain that

$$\begin{aligned} x^{\nu+1} &= \lambda^\nu \xi^\nu \\ &\quad + (1-\lambda^\nu)\lambda^{\nu-1}\xi^{\nu-1} \\ &\quad + (1-\lambda^\nu)(1-\lambda^{\nu-1})\lambda^{\nu-2}\xi^{\nu-2} \\ &\quad \vdots \\ &\quad + (1-\lambda^\nu)(1-\lambda^{\nu-1})\cdot \ldots \cdot (1-\lambda^1)\lambda^0\xi^0 \\ &\quad + (1-\lambda^\nu)(1-\lambda^{\nu-1})\cdot \ldots \cdot (1-\lambda^1)(1-\lambda^0)x^0. \end{aligned}$$

Suppose that $\lambda^\nu = 1/(\nu + 1)$ and $x^0 = 0$. Then, for $i = 1, 2, \dots, \nu$,

$$(1-\lambda^\nu)(1-\lambda^{\nu-1})\cdot \ldots \cdot (1-\lambda^{\nu-i+1})\lambda^{\nu-i} = \frac{\nu}{\nu+1}\frac{\nu-1}{\nu}\cdot \ldots \cdot \frac{\nu-i+1}{\nu-i+2}\frac{1}{\nu-i+1}$$
$$= \frac{1}{\nu+1},$$

which together with the expression for $x^{\nu+1}$ give

$$x^{\nu+1} = \frac{1}{\nu+1}\sum\nolimits_{i=0}^{\nu} \xi^i. \tag{3.13}$$

Thus, the method simply computes the average of the sample points, which as discussed in §3.D, converges to $\mathbb{E}[\boldsymbol{\xi}]$. Interestingly, the method estimates $\mathbb{E}[\boldsymbol{\xi}]$ iteratively without the need for storing the sample. As soon as $x^{\nu+1}$ has been computed, $\xi^\nu$ (and $x^\nu$) can be discarded. This is a significant advantage over the formula (3.13) when $n$ and $\nu$ are large.

The fact that $x^\star = \mathbb{E}[\boldsymbol{\xi}]$ is a minimizer of $Ef$ follows by the optimality condition 3.24, where $v(\xi) = x - \xi$. Thus,

$$0 = \mathbb{E}\big[v(\boldsymbol{\xi})\big] = x - \mathbb{E}[\boldsymbol{\xi}]$$

and $x^\star$ is actually the only point satisfying this equation. □

These ideas extend much beyond the setting of the example. In particular, (3.12) can very well involve a nonsmooth function despite the fact that the deep-rooted name stochastic *gradient* descent method indicates otherwise.

**Stochastic Gradient Descent Method.**

Data. $x^0 \in C$ and $\lambda^\nu \in (0,\infty)$, $\nu = 0, 1, 2, \ldots$.
Step 0. Set $\nu = 0$.
Step 1. Generate a sample $\xi^\nu$ according to the distribution of $\boldsymbol{\xi}$.
Step 2. With $v \in \partial_x f(\xi^\nu, x^\nu)$, set

$$x^{\nu+1} \in \mathrm{prj}_C(x^\nu - \lambda^\nu v).$$

Step 3. Replace $\nu$ by $\nu + 1$ and go to Step 1.

There are several possible convergence results that can be derived for this algorithm. The classical treatment adopts a step size $\lambda^\nu$ proportional to $1/\nu$. As illustrated in 3.26, a diminishing step size of this kind suppresses the inherent noise in the subgradient estimates, but results in "small" step sizes quickly and poor practical performance. The theoretical performance deteriorates dramatically as well when the step size is scaled slightly or when the problem isn't strongly convex. With a fixed step size, we obtain "longer" step sizes but then need to suppress noise by considering the *average* of iterates. The main insight is that one should separate the exploration of the space by the algorithm from the final estimate of a solution. This strategy results in remarkable theoretical properties: Even with the coarse approximations of subgradients, the stochastic gradient descent method performs as the subgradient method.

We don't know the sample obtained in Step 1 of the stochastic gradient descent method a priori. Thus, we need to analyze the method from a stochastic point of view. Let's consider $\{\xi^0, \xi^1, \dots\}$ obtained in Step 1 as outcomes of the iid random vectors $\{\boldsymbol{\xi}^0, \boldsymbol{\xi}^1, \dots\}$, distributed as $\boldsymbol{\xi}$. Then, the iterates $\{x^\nu, \nu \in \mathbb{N}\}$ and their averages become random vectors as indicated by switching to bold face: $\{\boldsymbol{x}^\nu, \nu \in \mathbb{N}\}$. As seen next, the average of the first $\bar{\nu}$ iterates is a near-minimizer with a specific tolerance that's guaranteed in expectation.

**Convergence 3.27** (stochastic gradient descent). *For a random vector $\boldsymbol{\xi}$ with support $\Xi \subset \mathbb{R}^m$, a function $f : \mathbb{R}^m \times \mathbb{R}^n \to \mathbb{R}$ and a nonempty, closed and convex set $C \subset \mathbb{R}^n$, suppose that*

(a) *$f(\boldsymbol{\xi}, x)$ is integrable for all $x \in \mathbb{R}^n$*
(b) *$f(\xi, \cdot)$ is convex for all $\xi \in \Xi$*
(c) *for some $\beta \in \mathbb{R}$,*

$$\sup\left\{\mathbb{E}\big[\|v(\boldsymbol{\xi})\|_2^2\big] \;\middle|\; v(\boldsymbol{\xi}) \text{ integrable, } v(\xi) \in \partial_x f(\xi, x)\ \forall \xi \in \Xi,\ x \in C\right\} \le \beta^2.$$

*If the stochastic gradient descent method has generated $\{\boldsymbol{x}^\nu, \nu = 1, 2, \dots, \bar{\nu}\}$ using step size $\lambda^\nu = \lambda \in (0, \infty)$ for all $\nu$, then*

$$\mathbb{E}\big[Ef(\bar{\boldsymbol{x}})\big] - \inf_C Ef \le \frac{\|x^0 - x^\star\|_2^2 + (\bar{\nu} + 1)\lambda^2\beta^2}{2\lambda(\bar{\nu} + 1)},$$

*where $x^\star \in \mathop{\mathrm{argmin}}_C Ef$ and $\bar{\boldsymbol{x}} = \frac{1}{\bar{\nu}+1}\big(\sum_{\nu=1}^{\bar{\nu}} \boldsymbol{x}^\nu + x^0\big)$.*

**Proof.** For any $\bar{x}, \bar{v} \in \mathbb{R}^n$, we obtain via 2.52 that

$$\tfrac{1}{2}\big\|\operatorname{prj}_C(\bar{x} - \lambda\bar{v}) - x^\star\big\|_2^2 \le \tfrac{1}{2}\|\bar{x} - \lambda\bar{v} - x^\star\|_2^2 = \tfrac{1}{2}\|\bar{x} - x^\star\|_2^2 - \lambda\langle\bar{v}, \bar{x} - x^\star\rangle + \tfrac{1}{2}\lambda^2\|\bar{v}\|_2^2. \quad (3.14)$$

If $v(\boldsymbol{\xi}^\nu)$ is an integrable random vector with $v(\xi) \in \partial_x f(\xi, \bar{x})$ for all $\xi \in \Xi$, then

$$\mathbb{E}\big[v(\boldsymbol{\xi}^\nu)\big] \in \partial Ef(\bar{x})$$

by 3.23. (Informally, this means that subgradients of $f(\xi, \cdot)$ at $\bar{x}$, taken with $\xi$ being the outcome of $\boldsymbol{\xi}^\nu$, produce via an expectation a subgradient of $Ef(\bar{x})$.) Thus, there's a subgradient $\bar{v} \in \partial Ef(\bar{x})$ such that

$$\mathbb{E}\big[\big\langle v(\boldsymbol{\xi}^\nu), \bar{x} - x^\star\big\rangle\big] = \langle\bar{v}, \bar{x} - x^\star\rangle \quad (3.15)$$

because we can select $\bar{v} = \mathbb{E}[v(\boldsymbol{\xi}^\nu)]$. In the algorithm, the point of subdifferentiation isn't fixed at $\bar{x}$ but is rather the random vector $\boldsymbol{x}^\nu$. Let $v(\boldsymbol{\xi}^\nu, \boldsymbol{x}^\nu)$ be the random vector generated in Step 2, where the dependence on $\boldsymbol{x}^\nu$ is highlighted. (We write $\boldsymbol{x}^0$ instead of $x^0$ to simplify expressions.) The random vector $\boldsymbol{x}^\nu$ depends on $\boldsymbol{\xi}^0, \boldsymbol{\xi}^1, \dots, \boldsymbol{\xi}^{\nu-1}$, but is statistically independent of $\boldsymbol{\xi}^\nu$. Then, (3.15) generalizes to[6]

[6] This statement can be formalized using conditional expectations; see [66].

$$\mathbb{E}\big[\langle v(\boldsymbol{\xi}^\nu, \boldsymbol{x}^\nu), \boldsymbol{x}^\nu - x^\star\rangle\big] = \mathbb{E}\big[\langle \bar{v}(\boldsymbol{x}^\nu), \boldsymbol{x}^\nu - x^\star\rangle\big],$$

where $\bar{v}(x) \in \partial Ef(x)$ for every outcome $x$ of the random vector $\boldsymbol{x}^\nu$. In view of the subgradient inequality 2.17 and 3.2(a), this implies that

$$\mathbb{E}\big[\langle v(\boldsymbol{\xi}^\nu, \boldsymbol{x}^\nu), \boldsymbol{x}^\nu - x^\star\rangle\big] \geq \mathbb{E}\big[Ef(\boldsymbol{x}^\nu) - Ef(x^\star)\big].$$

We now leverage (3.14), take expectations on each side (cf. 3.2(a)) and bring in the previous inequality as well as the bound on subgradients to obtain

$$\mathbb{E}\Big[\tfrac{1}{2}\|\operatorname{prj}_C(\boldsymbol{x}^\nu - \lambda v(\boldsymbol{\xi}^\nu, \boldsymbol{x}^\nu)) - x^\star\|_2^2\Big]$$
$$\leq \mathbb{E}\big[\tfrac{1}{2}\|\boldsymbol{x}^\nu - x^\star\|_2^2\big] - \lambda\mathbb{E}\big[Ef(\boldsymbol{x}^\nu) - Ef(x^\star)\big] + \tfrac{1}{2}\lambda^2\beta^2$$

and, equivalently,

$$\lambda\mathbb{E}\big[Ef(\boldsymbol{x}^\nu) - Ef(x^\star)\big] \leq \mathbb{E}\big[\tfrac{1}{2}\|\boldsymbol{x}^\nu - x^\star\|_2^2\big] - \mathbb{E}\big[\tfrac{1}{2}\|\boldsymbol{x}^{\nu+1} - x^\star\|_2^2\big] + \tfrac{1}{2}\lambda^2\beta^2.$$

Summing up both sides from $\nu = 0$ to $\nu = \bar{\nu}$ gives

$$\lambda\sum\nolimits_{\nu=0}^{\bar{\nu}}\mathbb{E}\big[Ef(\boldsymbol{x}^\nu) - Ef(x^\star)\big] \leq \tfrac{1}{2}\|x^0 - x^\star\|_2^2 + \tfrac{1}{2}(\bar{\nu}+1)\lambda^2\beta^2$$

and dividing each side by $\lambda(\bar{\nu}+1)$ results in

$$\mathbb{E}\bigg[\frac{1}{\bar{\nu}+1}\sum\nolimits_{\nu=0}^{\bar{\nu}} Ef(\boldsymbol{x}^\nu)\bigg] - Ef(x^\star) \leq \frac{\tfrac{1}{2}\|x^0 - x^\star\|_2^2 + \tfrac{1}{2}(\bar{\nu}+1)\lambda^2\beta^2}{\lambda(\bar{\nu}+1)}.$$

Since $Ef$ is convex, Jensen's inequality 3.5 and 3.2(a) ensure that

$$\mathbb{E}\big[Ef(\bar{\boldsymbol{x}})\big] \leq \mathbb{E}\bigg[\frac{1}{\bar{\nu}+1}\sum\nolimits_{\nu=0}^{\bar{\nu}} Ef(\boldsymbol{x}^\nu)\bigg],$$

which combined with the previous inequality establish the conclusion. □

The convergence statement guarantees that

$$\mathbb{E}\big[Ef(\bar{\boldsymbol{x}})\big] - \inf_C Ef \leq \frac{\beta\|x^0 - x^\star\|_2}{\sqrt{\bar{\nu}+1}} \quad \text{when } \lambda = \frac{\|x^0 - x^\star\|_2}{\beta\sqrt{\bar{\nu}+1}},$$

which is the best possible choice of step size according to these derivations. Remarkably, this is the same rate of convergence as for the subgradient method. The rather coarse approximation of subgradients doesn't slow down the algorithm, at least not on average. (Of course, because of the inherent randomness, the output of the algorithm tends to vary for each run.) While the step size is fixed during the iterations, its value decays slowly as the pre-determined total number of iterations grows. We refer to [66] for further refinements; see also [60, Chapter 4].

The stochastic gradient descent method is a main algorithm for solving large-scale problems arising in least-squares regression, support vector machine, logistic regression and several other areas of data analytics and machine learning where expectation functions are often costly to evaluate and projection on a feasible set is relatively simple or even unnecessary.

**Example 3.28** (support vector machines). Classification (§2.H) can be carried out by means of the hinge-loss problem

$$\underset{(a,\alpha)\in\mathbb{R}^{n+1}}{\text{minimize}} \ \frac{1}{m}\sum_{i=1}^{m} f_i(a,\alpha), \quad \text{where } f_i(a,\alpha) = \max\Big\{0, 1 - y_i\big(\langle a, x^i\rangle + \alpha\big)\Big\}$$

and therefore involves an expectation function defined by a finite distribution. Application of the stochastic gradient descent method to this problem requires a subgradient $v \in \partial f_i(a,\alpha)$ for some randomly selected $i \in \{1, 2, \dots, m\}$. We deduce from 2.49 that

$$\partial f_i(a,\alpha) = \begin{cases} \{0\} & \text{if } 1 - y_i\big(\langle a, x^i\rangle + \alpha\big) < 0 \\ \big\{(-y_i x^i, -y_i)\big\} & \text{if } 1 - y_i\big(\langle a, x^i\rangle + \alpha\big) > 0 \\ \big\{\delta(-y_i x^i, -y_i) \,\big|\, \delta \in [0,1]\big\} & \text{otherwise.} \end{cases}$$

**Detail.** All the assumptions of 3.27 hold. In particular, $\beta^2$ can be set to $m^{-1}\sum_{i=1}^{m}\|x^i\|_2^2 + 1$. The stochastic gradient descent method may require many iterations but each one is remarkably cheap, both in computing time and memory requirement. For example, the various data points can even be stored on different servers without difficulty as only one data point is accessed at a time. In contrast, the subgradient method requires access to all the data in each iteration.

For the $\ell^1$-regularized hinge-loss problem, the objective function has the additional term $g(a,\alpha) = \theta\|a\|_1$. In view of 2.29, the set of subgradients of $g$ at $(a,\alpha) \in \mathbb{R}^{n+1}$ is

$$\partial g(a,\alpha) = C_1 \times \cdots \times C_n \times \{0\}, \quad \text{where } C_j = \begin{cases} \{-\theta\} & \text{if } a_j < 0 \\ [-\theta, \theta] & \text{if } a_j = 0 \\ \{\theta\} & \text{otherwise.} \end{cases}$$

Since $g$ doesn't depend on $i$, one would use $v = v^i + w$ in Step 2 of the stochastic gradient descent method, where $v^i \in \partial f_i(a^\nu, \alpha^\nu)$ and $w \in \partial g(a^\nu, \alpha^\nu)$. □

**Example 3.29** (text analytics). Classification problems appear in the analysis of large collections of documents. For instance, a social network may need to monitor posts for offensive language, hate speech and other expressions that violate their member policy. The text in a post can be associated with a vector $x \in \mathbb{R}^n$, where each component $x_j$ represents the frequency of a particular word or phrase. If a dataset for which we know the labels (violation, not-violation) is available, then a support vector machine (cf. §2.H and 3.28) can be constructed that categorizes any body of text.

**Detail.** As an illustration, consider the blog post:

I like optimization theory more than I thought. It is the most useful class I am taking this semester. I certainly like it more than statistics.

Let $j = 1, 2, 3, 4$ correspond to "like," "more," "I" and "class," respectively. These words appear 2, 2, 4 and 1 times. Let $x_j$ be the relative frequency of the $j$th word so that $x = (2/9, 2/9, 4/9, 1/9)$. This is called a "bag-of-words" model of the text. In this case, a human "judge" will probably label the body of text as not-violation. If a large collection of such documents has been labeled, then a support vector machine can use the resulting dataset to construct a function $x \mapsto \langle a, x\rangle + \alpha$ that predicts the label of any document, of course, possibly with some errors. In practice, the size of $x$ is tens of thousands and the dataset may contain millions of labeled documents. Thus, an algorithm that can handle large-scale problems becomes essential. The stochastic gradient descent method is attractive because its rate of convergence doesn't depend on the number of terms summed up in an expectation function and it accesses only one document in each iteration. □

Although the analysis of the stochastic gradient descent method gives guidance regarding the choice of step size, it still needs to be tuned in practice. A common approach is to hold the step size fixed for a number of iterations and then reduce it by a factor, and this process is repeated 3–10 times. We can also attempt to automatically adapt the algorithm to the inherent variability and scale of a problem instance. An example of such a strategy follows below, where we combine the ideas of the accelerated gradient method of §1.H, the sampling of the stochastic gradient descent method and an adaptive scaling, and this often works well in practice; see [57].

**Accelerated Stochastic Gradient Descent with ADAM.**

Data. $x^0 \in C$, $\lambda = 10^{-3}$, $\alpha = 0.9$, $\beta = 0.999$, $\varepsilon = 10^{-8}$.

Step 0. Set $\nu = 0$ and $\mu_j^0 = \sigma_j^0 = 0$, $j = 1, \dots, n$.

Step 1. Generate a sample $\xi^\nu$ according to the distribution of $\boldsymbol{\xi}$.

Step 2. Set $v = (v_1, \dots, v_n) \in \partial_x f(\xi^\nu, x^\nu)$.

Step 3. For $j = 1, \dots, n$, set

$$\mu_j^{\nu+1} = \frac{\alpha\mu_j^\nu + (1-\alpha)v_j}{1-\alpha^{\nu+1}} \quad \text{and} \quad \sigma_j^{\nu+1} = \frac{\beta\sigma_j^\nu + (1-\beta)(v_j)^2}{1-\beta^{\nu+1}}.$$

Step 4. Set $w = (w_1, \dots, w_n)$, with

$$w_j = \frac{\mu_j^{\nu+1}}{\sqrt{\sigma_j^{\nu+1}} + \varepsilon}, \quad j = 1, \dots, n.$$

Step 5. Set $x^{\nu+1} \in \operatorname{prj}_C(x^\nu - \lambda w)$, replace $\nu$ by $\nu + 1$ and go to Step 1.

Here, $\alpha^{\nu+1}$ is indeed $\alpha$ raised to power $\nu+1$ and $\beta^{\nu+1}$ is similarly understood; $(v_j)^2$ means $v_j$ squared. Steps 3 and 4 compute a weighted and scaled average of previous subgradients in an effort to modulate their often volatile values. The scaling can be viewed as an attempt to compute expectation and variance of each component of the subgradient. This motivates the acronym ADAM, which stands for Adaptive Moment Estimation.

**Exercise 3.30** (stochastic gradient descent for support vector machines). Implement the stochastic gradient descent method using a fixed step size $\lambda$ and apply it to the $\ell^1$-regularized hinge-loss problem in §2.H for some dataset. For a given number of iterations (such as $\bar{\nu} = 100000$), compute the objective function value (see 3.28)

$$\frac{1}{m}\sum\nolimits_{i=1}^{m} f_i(\bar{a}, \bar{\alpha}) + \theta\|\bar{a}\|_1, \quad \text{where } (\bar{a}, \bar{\alpha}) = \frac{1}{\bar{\nu}+1}\sum\nolimits_{\nu=0}^{\bar{\nu}}(a^\nu, \alpha^\nu)$$

is the average across all the iterations. Repeat the calculations 30 times and compute the average of the 30 objective function values. Replicate the process for different $\lambda$ and $\theta$. Test the classifiers on a separate but similar dataset and compute the number of correct classifications for each one.

**Guide.** Leverage 3.28 to obtain expressions for the subgradients as well as the bound $\beta$ in 3.27, which can be used to guide the selection of $\lambda$. □

## 3.H Simple Recourse Problems

Optimization problems frequently include constraints of a more complicated kind than bounds on the decision variables. This causes difficulty for the stochastic gradient descent method and related algorithms as projection on the feasible set quickly becomes costly. There's also hope that the special structure of a problem can be utilized to construct efficient algorithms that improve on the rather general approach of gradient descent. We'll now examine one such circumstance.

In applications, we often end up with problems whose decision processes fit the following pattern:

$$\text{decision: } x \rightsquigarrow \text{observation: } \xi \rightsquigarrow \text{recourse cost evaluation.}$$

The recourse cost evaluation may or may not be "simple" and it's useful to make this distinction in the selection of solution procedures. In some instances, the recourse cost evaluation requires solving another optimization problem, which we return to in §3.J. If it doesn't and the evaluation can be expressed in terms of an explicit formula, then the problem is said to be one with *simple recourse*. More precisely, a *stochastic optimization problem with simple recourse* takes the form

$$\underset{x \in C \subset \mathbb{R}^n}{\text{minimize}}\ f_0(x) + \mathbb{E}\big[f(\boldsymbol{\xi}, x)\big]$$

where $f : \mathbb{R}^m \times \mathbb{R}^n \to \mathbb{R}$ is given by an explicitly formula that's easy to evaluate. The newsvendor problem (§1.C) is of this type with

$$f_0(x) = (\gamma - \delta)x, \quad C = \mathbb{R}, \quad f(\xi, x) = \max\big\{0, \delta(x - \xi)\big\}.$$

The product mix problem (§3.A) has $f_0(x) = \langle c, x\rangle$, $C = [0, \infty)^4$ and

$$f(\xi, x) = a_1 \max\Big\{0, \sum_{j=1}^{4} t_{1j}x_j - d_1\Big\} + a_2 \max\Big\{0, \sum_{j=1}^{4} t_{2j}x_j - d_2\Big\},$$

with $\xi$ now being the vector with elements defined by the uncertain parameters $t_{1j}, t_{2j}, d_1, d_2$. (In §3.A, the problem is formulated in terms of recourse variables $y^1, \dots, y^\nu$, but these can be eliminated because it's trivial to determine the amount of overtime needed and we obtain a problem with simple recourse.) In general, the second part of the objective function in a stochastic optimization problem with simple recourse is, hopefully, a well-defined expectation function $Ef : \mathbb{R}^n \to \overline{\mathbb{R}}$ with values $Ef(x) = \mathbb{E}[f(\boldsymbol{\xi}, x)]$.

We focus on a computationally attractive class of problems that stems from the linear optimization problem

$$\underset{x\in\mathbb{R}^n}{\text{minimize}}\ \langle c, x\rangle \ \text{ subject to } \ Ax = b,\ Tx = \xi,\ x \geq 0,$$

where $A$ and $T$ are $m_1 \times n$ and $m_2 \times n$-matrices, respectively, and $\xi$ is an $m_2$-dimensional vector of parameters. The difficulty is, however, that the value of $\xi$ is unknown. To model this situation, we assign an additional cost to a decision $x$ should $Tx$ fail to match the actual value of $\xi$. The cost is described by a convex function $h : \mathbb{R}^{m_2} \to \mathbb{R}$ that's referred to as the *recourse cost function*. We assume that the vector of parameters is modeled by the random vector $\boldsymbol{\xi}$ with support $\Xi \subset \mathbb{R}^{m_2}$ and a given probability distribution. This leads to the simple recourse problem

$$\underset{x\in C}{\text{minimize}}\ \langle c, x\rangle + Ef(x), \tag{3.16}$$
$$\text{with } C = \{x \in \mathbb{R}^n \mid Ax = b, x \geq 0\} \ \text{ and } \ f(\xi, x) = h(\xi - Tx),$$

which is our focus in this section. We're again faced with the challenge of developing a formula for subgradients.

**Proposition 3.31** (basic composition rule for subgradients). *For a convex function $g : \mathbb{R}^m \to \mathbb{R}$, an $m \times n$-matrix $A$ and $b \in \mathbb{R}^m$, the function $f : \mathbb{R}^n \to \mathbb{R}$ given by*

$$f(x) = g(Ax + b)$$

*has subgradients*

$$\partial f(x) = A^\top \partial g(Ax + b) = \big\{A^\top v \mid v \in \partial g(Ax + b)\big\}.$$

**Proof.** This is a special case of the chain rule 4.64. □

**Proposition 3.32** (KKT condition for simple recourse problem). *Consider the problem (3.16) with convex $h : \mathbb{R}^{m_2} \to \mathbb{R}$ and integrable $h(\boldsymbol{\xi} - Tx)$ for all $x \in \mathbb{R}^n$. Then, $x^\star$ is a minimizer if and only if there are $y \in \mathbb{R}^{m_1}$, $z \in \mathbb{R}^n$ and $v : \mathbb{R}^{m_2} \to \mathbb{R}^{m_2}$, with $v(\boldsymbol{\xi})$ integrable, such that*

(a) $x^\star \geq 0,\ Ax^\star = b$
(b) $\forall j = 1, \dots, n$: $\quad$ *$z_j \geq 0$ if $x_j^\star = 0$ and $z_j = 0$ otherwise*

(c) $\forall \xi \in \Xi$: $\quad v(\xi) \in \partial h(\xi - Tx^\star)$
(d) $T^\top \mathbb{E}[v(\boldsymbol{\xi})] - c = A^\top y - z$.

**Proof.** Let $f_0 : \mathbb{R}^n \to \mathbb{R}$ be the objective function in (3.16), i.e.,

$$f_0(x) = \langle c, x \rangle + Ef(x),$$

which indeed is real-valued because $h(\boldsymbol{\xi} - Tx)$ is integrable for all $x \in \mathbb{R}^n$. By 1.18(c) and 3.3, $f_0$ is convex so the optimality condition 2.27 applies and it suffices to verify that the stated conditions are equivalent to $-u^\star \in N_C(x^\star)$ for some $u^\star \in \partial f_0(x^\star)$. In view of 3.23 and 3.31, $w \in \partial Ef(x^\star)$ if and only if there's $v : \mathbb{R}^{m_2} \to \mathbb{R}^{m_2}$, with $v(\boldsymbol{\xi})$ integrable, and $v(\xi) \in \partial h(\xi - Tx^\star)$ for all $\xi \in \Xi$ such that $w = -T^\top \mathbb{E}[v(\boldsymbol{\xi})]$. Then,

$$u^\star = c - T^\top \mathbb{E}\big[v(\boldsymbol{\xi})\big] \in \partial f_0(x^\star)$$

by the Moreau-Rockafellar sum rule 2.26. This fact together with the expression for $N_C(x)$ from 2.44 establish the claim. □

As in the KKT condition 2.45, the vectors $y$ and $z$ are multiplier vectors. The function $v : \mathbb{R}^{m_2} \to \mathbb{R}^{m_2}$ can be viewed as a multiplier *function* because it enters as an auxiliary quantity that helps us to characterize a minimizer. Except when $\boldsymbol{\xi}$ has a finite distribution, there's an infinite number of multipliers $\{v(\xi), \xi \in \Xi\}$ of this kind. Although (3.16) only involves a finite number of decision variables and constraints, algorithms that attempt to obtain a solution by levering this optimality condition need to contend with this complication. Even when $\boldsymbol{\xi}$ is finitely distributed, there could be a large number of multipliers, which again would cause computational difficulties. We now turn to specializations that alleviate these challenges.

We can reformulate (3.16) in a way that might be analytical and computational beneficial. If $\boldsymbol{\xi}$ represents demand for the output of certain economic activities, then $\chi = Tx$ is the vector being *tendered* by the decision-maker to match the uncertain demand $\boldsymbol{\xi}$. We refer to $\chi$ as a *tender*. Introducing the tender as a vector of decision variables in the formulation, we obtain

$$\underset{x\in\mathbb{R}^n,\chi\in\mathbb{R}^{m_2}}{\text{minimize}} \ \langle c, x \rangle + E\psi(\chi) \ \text{ subject to } \ Ax = b, \ Tx = \chi, \ x \geq 0, \tag{3.17}$$

which involves the function $\psi : \mathbb{R}^{m_2} \times \mathbb{R}^{m_2} \to \mathbb{R}$ defined by $\psi(\xi, \chi) = h(\xi - \chi)$. Obviously, $\psi(\xi, Tx) = f(\xi, x)$, which means that $E\psi$ and $Ef$ have essentially the same properties. So, why bother with this reformulation?

In some applications it turns out that $\psi$ and $E\psi$ are *separable* while $Ef$ isn't, i.e., $\psi$ is of the form

$$\psi(\xi, \chi) = \sum_{i=1}^{m_2} \psi_i(\xi_i, \chi_i) = \sum_{i=1}^{m_2} h_i(\xi_i - \chi_i) \quad \text{and} \quad E\psi(\chi) = \sum_{i=1}^{m_2} E\psi_i(\chi_i)$$

for univariate functions $\psi_i : \mathbb{R} \times \mathbb{R} \to \mathbb{R}$ and $h_i : \mathbb{R} \to \mathbb{R}$. Separability is quite common in practice because each component $\boldsymbol{\xi}_i$ of $\boldsymbol{\xi}$ might represent the demand for a specific product and the associated cost of not meeting this demand could be independent of those for other products.

The newsvendor problem of §1.C falls, trivially, in this category and so does the product mix problem of §3.A when the only uncertainty is about the number of hours of carpentry and finishing; cf. 3.1. Indeed, with $\xi = (\xi_1, \xi_2) = (d_1, d_2)$ and $\chi = (\chi_1, \chi_2)$, the product mix problem has

$$\psi(\xi, \chi) = \psi_1(\xi_1, \chi_1) + \psi_2(\xi_2, \chi_2),$$

where

$$\psi_1(\xi_1, \chi_1) = a_1 \max\{0, \chi_1 - \xi_1\} \quad \text{and} \quad \psi_2(\xi_2, \chi_2) = a_2 \max\{0, \chi_2 - \xi_2\}.$$

With deterministic $T_1 = (4, 9, 7, 10)$ and $T_2 = (1, 1, 3, 40)$, the product mix problem then takes the form:

$$\begin{aligned}&\underset{x\in\mathbb{R}^4, \chi\in\mathbb{R}^2}{\text{minimize}} \ \langle c, x\rangle + E\psi_1(\chi_1) + E\psi_2(\chi_2)\\ &\quad\text{subject to } \ \langle T_1, x\rangle = \chi_1, \ \langle T_2, x\rangle = \chi_2, \ x \geq 0.\end{aligned}$$

Separability renders all operations much easier because one is dealing with a juxtaposition of one-dimensional cases. Subgradients can be calculated with the help of 2.20 and expectations require only the (marginal) probability distribution of each random variable $\boldsymbol{\xi}_i$ rather than the joint distribution of the random vector $\boldsymbol{\xi}$. One-dimensional integrals might even be available explicitly; see 3.18 and the following example.

**Example 3.33** (expected recourse cost; continuous distribution). Suppose that $\boldsymbol{\xi}$ is an integrable random variable with density function $p : \mathbb{R} \to [0, \infty)$ and the recourse cost function $h : \mathbb{R} \to \mathbb{R}$ has $h(y) = \max\{\gamma y, \delta y\}$ with $\gamma \leq \delta$. For

$$\psi(\xi, \chi) = h(\xi - \chi),$$

the expectation function $E\psi$ can be expressed as

$$\begin{aligned}E\psi(\chi) &= \gamma \int_{-\infty}^{\chi} (\xi - \chi)p(\xi)\, d\xi + \delta \int_{\chi}^{\infty} (\xi - \chi)p(\xi)\, d\xi\\ &= \delta\big(\mathbb{E}[\boldsymbol{\xi}] - \chi\big) + (\delta - \gamma)\Big(\chi P(\chi) - \int_{-\infty}^{\chi} \xi p(\xi)\, d\xi\Big),\end{aligned}$$

where $P$ is the distribution function of $\boldsymbol{\xi}$. Moreover,

$$\partial E\psi(\chi) = \big\{(\delta - \gamma)P(\chi) - \delta\big\}$$

and $E\psi$ is actually smooth.

**Detail.** Recalling that $P(\chi) = \int_{-\infty}^{\chi} p(\xi)\, d\xi$, the expression for $E\psi(\chi)$ follows directly. Still, numerical integration may be needed to evaluate $E\psi(\chi)$, but this is easily done with high precision because the integral is one-dimensional; cf. §3.D. We obtain from 2.17 and 3.31 that

$$\partial h(y) = \begin{cases} \gamma & \text{if } y < 0 \\ [\gamma, \delta] & \text{if } y = 0 \\ \delta & \text{otherwise} \end{cases} \qquad \partial_\chi \psi(\xi, \chi) = \begin{cases} -\gamma & \text{if } \chi > \xi \\ [-\delta, -\gamma] & \text{if } \chi = \xi \\ -\delta & \text{otherwise.} \end{cases}$$

Since $\mathbb{E}[\boldsymbol{\xi}]$ is finite, $\psi(\boldsymbol{\xi}, \chi)$ is integrable for all $\chi$ and $\psi(\xi, \cdot)$ is convex by 1.13 for all $\xi \in \Xi$ so that 3.23 applies. Any $v : \mathbb{R} \to \mathbb{R}$ with $v(\xi) \in \partial_\chi \psi(\xi, \chi)$ for all $\xi \in \mathbb{R}$ has

$$\mathbb{E}\big[v(\boldsymbol{\xi})\big] = \int_{-\infty}^{\infty} v(\xi)p(\xi)\, d\xi = -\int_{-\infty}^{\chi} \gamma p(\xi)\, d\xi - \int_{\chi}^{\infty} \delta p(\xi)\, d\xi$$

because the value of $v$ at $\chi$ is immaterial for the integral. This gives the asserted formula for $\partial E\psi(\chi)$. Interestingly, even if $h$ is nonsmooth, $E\psi$ is smooth because $\partial E\psi(\chi)$ contains a single subgradient for each $\chi$ and the resulting gradient depends continuously on $\chi$. This shows that the smoothness assumption in 3.23 is by no means necessary for an expectation function to be smooth. □

**Example 3.34** (aircraft allocation[7]). Facing uncertain passenger demand, an airline seeks to allocate aircraft of various types among its routes in such a way as to minimize operating cost plus expected revenue loss from flying with less than full planes. Let $i = 1, \dots, m_1$ and $j = 1, \dots, m_2$ represent aircraft types and airline routes, respectively. The operating cost of a full aircraft of type $i$ on route $j$ is

$$c_{ij} = -t_{ij}\rho_j + o_{ij},$$

where $o_{ij}$ is the fixed expense of flying an aircraft of type $i$ on route $j$, $t_{ij}$ is the corresponding passenger capacity and $\rho_j$ is the revenue per passenger on route $j$. Let $b_i$ be the number of available aircraft of type $i$.

The passenger demand on the various routes is uncertain and let's model it by a random vector $\boldsymbol{\xi}$ assumed to be finitely distributed with support $\Xi$ and associated probabilities $p_\xi > 0, \xi \in \Xi$. Following the approach in §3.A, we obtain a *resource allocation problem* expressed in terms of decision variables $x_{ij}$, representing the number of aircraft of type $i$ on route $j$, and $y_j(\xi)$, the number of empty seats on route $j$ under demand $\xi = (\xi_1, \dots, \xi_{m_2})$:

$$\begin{aligned} \text{minimize } & \sum_{i=1}^{m_1}\sum_{j=1}^{m_2} c_{ij}x_{ij} + \sum_{\xi\in\Xi} p_\xi \sum_{j=1}^{m_2} \rho_j y_j(\xi) \\ \text{subject to } & \sum_{j=1}^{m_2} x_{ij} \le b_i, \;\; i = 1, \dots, m_1 \\ & \sum_{i=1}^{m_1} t_{ij}x_{ij} - y_j(\xi) \le \xi_j, \;\; j = 1, \dots, m_2, \; \xi \in \Xi \\ & x_{ij} \ge 0, \;\; i = 1, \dots, m_1, \; j = 1, \dots, m_2 \\ & y_j(\xi) \ge 0, \;\; j = 1, \dots, m_2, \; \xi \in \Xi. \end{aligned}$$

[7] This example summarizes the first published study of a nontrivial stochastic optimization problem with recourse [37]; the authors computed a solution for the data in 3.38 by hand!

Let $\nu$ be the number of possible outcomes in $\Xi$. Then, the model has $m_1 + m_2\nu$ linear inequalities and $m_1 m_2 + m_2\nu$ variables subject to nonnegativity constraints. Since $\nu$ is the number of possible outcomes for an $m_2$-dimensional random vector, it can easily be astronomical in which case a direct solution using the interior-point method of §2.G becomes impossible. Fortunately, we can obtain an alternative, separable formulation with number of variables and constraints independent of $\nu$.

**Detail.** With $\psi_j(\xi_j, \chi_j) = \max\{-\rho_j(\xi_j - \chi_j), 0\}$, the problem is equivalently stated as

$$\begin{aligned} \text{minimize } & \sum_{i=1}^{m_1}\sum_{j=1}^{m_2} c_{ij}x_{ij} + \sum_{j=1}^{m_2}\sum_{\xi\in\Xi} p_\xi \psi_j(\xi_j, \chi_j) \qquad (3.18)\\ \text{subject to } & \sum_{j=1}^{m_2} x_{ij} \le b_i, \; i = 1, \dots, m_1\\ & \sum_{i=1}^{m_1} t_{ij}x_{ij} - \chi_j = 0, \; j = 1, \dots, m_2\\ & x_{ij} \ge 0, \; i = 1, \dots, m_1, \; j = 1, \dots, m_2. \end{aligned}$$

The tenders $\chi_j$ represent the total number of seats made available on route $j$. There's no revenue loss if the demand $\xi_j$ on route $j$ exceeds $\chi_j$, but for each empty seat there's a cost $\rho_j$ representing lost revenue. The model has $m_1$ linear inequalities, $m_2$ linear equalities and $m_1 m_2 + m_2$ variables, some of which are nonnegative. The dependence on $\nu$ has been removed. The objective function is of the separable form discussed above, which has the added benefit that there's no need for the joint probability distribution of $\xi$; only the marginal ones for the demands on the individual routes enter. Thus,

$$\sum_{j=1}^{m_2}\sum_{\xi\in\Xi} p_\xi \psi_j(\xi_j, \chi_j) = \sum_{j=1}^{m_2}\sum_{k=1}^{\nu_j} p_j^k \psi_j(\xi_j^k, \chi_j),$$

where $\{\xi_j^k, k = 1, \dots, \nu_j\}$ are the possible demands for route $j$ and $\{p_j^k > 0, k = 1, \dots, \nu_j\}$ are the corresponding probabilities. When introduced in (3.18), this leads to many fewer terms in the objective function.

A disadvantage of the model is its nonlinearity, but this is compensated by the massive reduction in model size when $\nu$ is large. We might consider applying the subgradient method (§2.I) or stochastic gradient descent method (§3.G). However, they require time-consuming projections on a polyhedral set, which could make them less competitive. Another possibility is laid out below.

We've ignored the fact that a solution may include values that aren't integers, which of course implies that the resulting plan isn't immediately implementable. This may not be of major concern if the model is used for strategic planning and/or a solution can be rounded meaningfully to the nearest integer; see also 4.3. Still, we can handle the situation when an aircraft of type $i$ can't fly a specific route $j$, for example due to endurance, by fixing $x_{ij} = 0$. □

This example involves an expectation function defined by a finite distribution and then is expressed as a simple sum. Still, it's worthwhile computationally to consider an equivalent expression, parallel to those in 3.18 and 3.33, which also furnishes additional insight.

**Example 3.35** (expected recourse cost; finite distribution). Suppose that $\boldsymbol{\xi}$ is a random variable with values $\xi^1 < \cdots < \xi^\nu$ and corresponding probabilities $p_i > 0, i = 1, \dots, \nu$. If the recourse cost function $h : \mathbb{R} \to \mathbb{R}$ has $h(y) = \max\{\gamma y, \delta y\}$, with $\gamma \leq \delta$, and $\psi(\xi, \chi) = h(\xi - \chi)$, then

$$E\psi(\chi) = \begin{cases} \delta\big(\mathbb{E}[\boldsymbol{\xi}] - \chi\big) & \text{if } \chi < \xi^1 \\ \delta\mathbb{E}[\boldsymbol{\xi}] - (\delta - \gamma)\bar{\xi}^i + \big((\delta - \gamma)\bar{p}_i - \delta\big)\chi & \text{if } \xi^i \leq \chi < \xi^{i+1},\ i = 1, \dots, \nu - 1 \\ \gamma\big(\mathbb{E}[\boldsymbol{\xi}] - \chi\big) & \text{if } \xi^\nu \leq \chi, \end{cases}$$

where $\bar{p}_i = \sum_{j=1}^{i} p_j$ and $\bar{\xi}^i = \sum_{j=1}^{i} p_j \xi^j$. This shows that $E\psi$ is convex and piecewise affine.

**Detail.** The expression is a simplification of

$$\sum\nolimits_{i=1}^{\nu} p_i \max\big\{\gamma(\xi^i - \chi), \delta(\xi^i - \chi)\big\},$$

taking advantage of the fact that $\gamma \leq \delta$. For large $\nu$, the expression is faster to use than simply summing up a large number of terms because $\mathbb{E}[\boldsymbol{\xi}]$, $\bar{p}_i$ and $\bar{\xi}^i$ can be pre-computed. □

**Exercise 3.36** (subgradients for expected recourse cost functions). Write a formula for the subgradients of $E\psi$ in 3.35 using a similar piecewise formulation as for $E\psi$.

**Guide.** Since $E\psi$ is piecewise affine, with nonsmoothness at $\xi^i$, the subgradients coincide with gradients at all points except $\xi^1, \dots, \xi^\nu$. □

**Example 3.37** (aircraft allocation; cont.). Under the assumption that $\xi_j^1 < \cdots < \xi_j^{\nu_j}$ for all $j$, we can leverage 3.35 to reformulate (3.18) as

$$\begin{aligned} \text{minimize}\ & \sum\nolimits_{i=1}^{m_1} \sum\nolimits_{j=1}^{m_2} c_{ij} x_{ij} + \sum\nolimits_{j=1}^{m_2} z_j \\ \text{subject to}\ & \sum\nolimits_{j=1}^{m_2} x_{ij} \leq b_i,\ \ i = 1, \dots, m_1 \\ & -\rho_j \bar{\xi}_j^k + \rho_j \bar{p}_j^k \sum\nolimits_{i=1}^{m_1} t_{ij} x_{ij} \leq z_j,\ \ j = 1, \dots, m_2,\ k = 1, \dots, \nu_j - 1 \\ & -\rho_j \bar{\xi}_j + \rho_j \sum\nolimits_{i=1}^{m_1} t_{ij} x_{ij} \leq z_j,\ \ j = 1, \dots, m_2 \\ & x_{ij} \geq 0,\ z_j \geq 0,\ \ i = 1, \dots, m_1,\ j = 1, \dots, m_2, \end{aligned}$$

where $\bar{p}_j^k = \sum_{l=1}^{k} p_j^l$, $\bar{\xi}_j^k = \sum_{l=1}^{k} p_j^l \xi_j^l$ and $\bar{\xi}_j = \sum_{l=1}^{\nu_j} p_j^l \xi_j^l$ for all $j = 1, \dots, m_2$.

**Detail.** For each $j$, we use 3.35 with $\gamma = -\rho_j$ and $\delta = 0$ to obtain

$$E\psi_j(\chi_j) = \begin{cases} 0 & \text{if } \chi_j < \xi_j^1 \\ -\rho_j \bar{\xi}_j^k + \rho_j \bar{p}_j^k \chi_j & \text{if } \xi_j^k \leq \chi_j < \xi_j^{k+1}, \quad k = 1, \dots, \nu_j - 1 \\ -\rho_j\big(\bar{\xi}_j - \chi_j\big) & \text{if } \xi_j^{\nu_j} \leq \chi_j. \end{cases}$$

The reformulation of (3.18) emerges then as an application of the first idea in 2.68 to the present context and also using the fact that $\sum_{i=1}^{m_1} t_{ij} x_{ij} = \chi_j$.

The linear optimization problem has $m_1 + \sum_{j=1}^{m_2} \nu_j$ inequalities, $(m_1 + 1)m_2$ variables and some bounds. The situation is dramatically improved compared to the starting point in 3.34. Although it's larger than (3.18), the nonlinearity has been removed and we can employ the interior-point method (§2.G). The choice between a small nonlinear model and a larger linear one is usually determined by the specifics of the problem instance and the type of hardware and software available. □

**Exercise 3.38** (aircraft allocation; computations). Using the formulation in 3.37, solve the aircraft allocation problem with four aircraft types ($i = 1, \ldots, 4$), five routes ($j = 1, \ldots, 5$), $b = (10, 19, 25, 15)$, $\rho = (13, 13, 7, 7, 1)$ and the data given in Tables 3.5 and 3.6. Formulate and solve the corresponding problem when uncertainty is ignored and the demand is set to the expected demand. Compare the two solutions and their costs.

**Guide.** A minimizer for the first problem involves $x^\star_{11} = 10$, $x^\star_{22} = 12.85$, $x^\star_{23} = 0.82$, $x^\star_{24} = 5.33$, $x^\star_{32} = 4.31$, $x^\star_{35} = 20.69$, $x^\star_{41} = 7.34$ and $x^\star_{43} = 7.66$, with other $x^\star_{ij} = 0$. □

Table 3.5: Capacities and operating costs, with "–" indicating unavailable.

| aircraft | capacities: $t_{ij}$ | | | | |
|---|---|---|---|---|---|
| 1 | 16 | 15 | 28 | 23 | 81 |
| 2 | – | 10 | 14 | 15 | 57 |
| 3 | – | 5 | – | 7 | 29 |
| 4 | 9 | 11 | 22 | 17 | 55 |
| route: | 1 | 2 | 3 | 4 | 5 |

| aircraft | operating cost: $o_{ij}$ | | | | |
|---|---|---|---|---|---|
| 1 | 18 | 21 | 18 | 16 | 10 |
| 2 | – | 15 | 16 | 14 | 9 |
| 3 | – | 10 | – | 9 | 6 |
| 4 | 17 | 16 | 17 | 15 | 10 |
| route: | 1 | 2 | 3 | 4 | 5 |

Table 3.6: Demand $\xi_j^k$ and probability $p_j^k$, $k = 1, \ldots, \nu_j$, for all $j = 1, \ldots, 5$.

| route | demand : probability | | | | |
|---|---|---|---|---|---|
| 1 | 200 : 0.2 | 220 : 0.05 | 250 : 0.35 | 270 : 0.2 | 300 : 0.2 |
| 2 | 50 : 0.3 | 150 : 0.7 | | | |
| 3 | 140 : 0.1 | 160 : 0.2 | 180 : 0.4 | 200 : 0.2 | 220 : 0.1 |
| 4 | 10 : 0.2 | 50 : 0.2 | 80 : 0.3 | 100 : 0.2 | 340 : 0.1 |
| 5 | 580 : 0.1 | 600 : 0.8 | 620 : 0.1 | | |

## 3.I Control of Water Pollution

Eutrophication management deals with pollution control of lakes and rivers. Lake Stoöpt[8] has four sections: Section 1 in the west is surrounded by recreational areas, Sections 2 and 3 are subject to run-off from fertilized farm land and Section 4 in the east is bordered by an industrial zone; see Figure 3.19. To reduce pollution, we can install sewage treatment plants along the rivers feeding the lake and build reed basins, which absorb cl-phosphates.

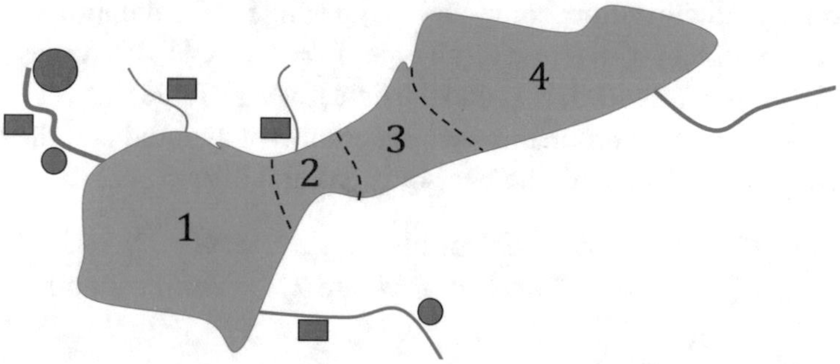


Fig. 3.19: Lake Stoöpt with treatment plants (rectangles) and reed basins (circles).

Pollution depends on the atmospheric conditions about which there's enough data to reliably estimate probability distributions. In the lake, transport of the pollution occurs mostly from west to east as the major inlets are in the western part of the lake and the only outlet is at the eastern end.

For $j = 1, \dots, n$, let $x_j$ be the size of the treatment plant or reed basin at location $j$; $s_j$ is an upper bound on $x_j$. Let $d_i \in \mathbb{R}$ be the amount of exogenous pollution entering Section $i$ of the lake and $\langle T_i, x \rangle$ be the net amount of pollution removed from Section $i$ by hydro-dynamical transport, with $T_i \in \mathbb{R}^n$. Thus, $d_i - \langle T_i, x \rangle$ is the amount of pollution in Section $i$, which we assume has been "normalized" so that negative values are considered excellent water quality levels. Since $d = (d_1, \dots, d_4)$ and $T_1, \dots, T_4$ depend on the weather, they're highly uncertain and we model them using the random vector

$$\boldsymbol{\xi} = (\boldsymbol{T}_1, \dots, \boldsymbol{T}_4, \boldsymbol{d}).$$

Similarly, an outcome is denoted by $\xi = (T_1, \dots, T_4, d)$. Other limitations on the choice of $x$ is modeled by the constraint $\langle a, x \rangle \le \alpha$.

In this engineering design setting and many other real-world problems, there's a desired state (excellent water quality), but decision-makers realize that it might neither be possible nor economical to insist on achieving that state under all circumstances (weather pattern). We would like to explore the trade-off between cost and system performance in the presence of uncertainty. One approach could be to formulate a model in which the 0.95-

[8] Actually, the lake in question is Lake Balaton in Hungary, but because it has been the motivation for a number of stochastic optimization problems it's sometimes referred to as "Lake Stochastic Optimization," here shortened to Lake Stoöpt.

superquantile of $\boldsymbol{d}_i - \langle \boldsymbol{T}_i, x\rangle$ is constrained to be below a threshold. This would result in decisions that guarantee certain levels of water quality in the worst 5% weather conditions; see §3.C. If strict regulatory requirements specify such thresholds, then this approach would be reasonable. In the following we develop a "softer" approach where poor water quality is permitted but penalized. This is meaningful if the responsible authority for the lake is simply fined, and not prosecuted, for low water quality.

For each $i = 1, \ldots, 4$ and $\beta_i, \tau_i \in (0, \infty)$, we adopt the recourse cost function

$$h_i(y) = \begin{cases} 0 & \text{if } y < 0 \quad \text{(excellent water quality)} \\ (2\beta_i)^{-1} y^2 & \text{if } y \in [0, \beta_i\tau_i] \quad \text{(acceptable water quality)} \\ \tau_i y - \frac{1}{2}\beta_i\tau_i^2 & \text{if } y > \beta_i\tau_i \quad \text{(unacceptable water quality).} \end{cases}$$

These recourse cost functions are easily interpreted (see Figure 3.20) and can be tied to fines levied in case of poor water quality. Let

$$\psi_i(\xi, x) = h_i\big(d_i - \langle T_i, x\rangle\big), \;\; i = 1, \ldots, 4,$$

which then is zero when the water quality is excellent in Section $i$ under weather condition $\xi$ but quantifies an increasingly large penalty (cost) as the water quality level drops. The total expected penalty (cost) across all sections is $\sum_{i=1}^{4} \mathbb{E}[\psi_i(\boldsymbol{\xi}, x)]$.

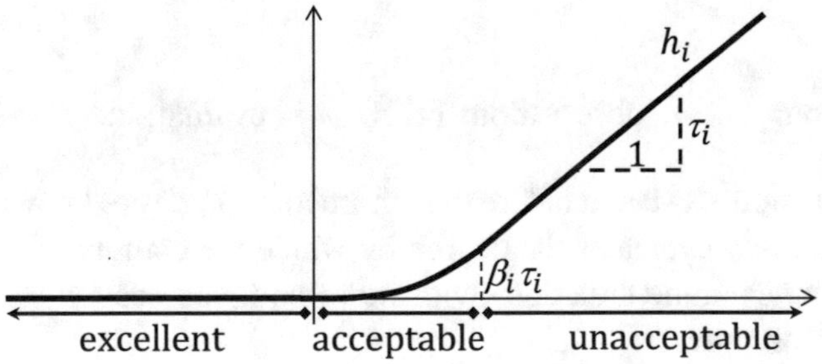


Fig. 3.20: Recourse cost function modeling penalties for low water quality.

The derivative of $h_i$ is

$$h_i'(y) = \begin{cases} 0 & \text{if } y < 0 \\ \beta_i^{-1} y & \text{if } y \in [0, \beta_i\tau_i] \\ \tau_i & \text{otherwise,} \end{cases}$$

which is continuous so $h_i$ is actually smooth and also convex. By 3.23 and 3.31,

$$\nabla E\psi_i(x) = -\mathbb{E}\Big[\boldsymbol{T}_i h_i'\big(\boldsymbol{d}_i - \langle \boldsymbol{T}_i, x\rangle\big)\Big]$$

as long as $\boldsymbol{d}_i$ and $\boldsymbol{T}_i$ are integrable because then $\psi_i(\boldsymbol{\xi}, x)$ is integrable for all $x$.

In addition to the expected penalty, there are direct costs associated with building the treatment plants and reed basins given by $f_0 : \mathbb{R}^n \to \mathbb{R}$. This leads to the formulation

$$\underset{x\in\mathbb{R}^n}{\text{minimize}}\ f_0(x) + \sum\nolimits_{i=1}^{4} \mathbb{E}\big[\psi_i(\boldsymbol{\xi}, x)\big]\ \text{ subject to }\ \langle a, x\rangle \le \alpha,\ 0 \le x \le s,$$

which is a stochastic optimization problem with simple recourse. The objective function is convex (smooth) whenever $f_0$ is convex (smooth). Since the vectors $\boldsymbol{T}_1, \dots, \boldsymbol{T}_4$ are random, the resulting *technology matrix* $\boldsymbol{T}$, with these as rows, is random and this prevents an effective reformulations in terms of tenders. The evaluation of the expected recourse cost now requires multi-dimensional integration; it's no longer possible to simplify with a series of one-dimensional integrals. Generally, the nature of the technology matrix, so named because it quantifies how an input $x$ is converted into an output $\boldsymbol{T}x$, therefore largely governs the difficulty of simple recourse problems.

## 3.J Linear Recourse Problems

Stochastic optimization problems with simple recourse may involve a random vector $\boldsymbol{d}$, representing uncertain demand, and a random technology matrix $\boldsymbol{T}$ that transforms the decision vector $x$ into a random vector $\boldsymbol{T}x$ to be compared with $\boldsymbol{d}$. This leads to a decision process in two stages:

$$\text{decision: } x \ \rightsquigarrow\ \text{observation: } (T, d)\ \rightsquigarrow\ \text{evaluation: } y = d - Tx.$$

The first-stage decision $x$ is taken before the parameters $(T, d)$ are known and the second-stage decision $y$ simply evaluates the degree by which we comply with the requirement $Tx = d$. Let's now examine situations with more complex second-stage decisions and consider the decision process:

$$\text{decision: } x\ \rightsquigarrow\ \text{observation: } (W, T, d)\ \rightsquigarrow\ \text{decision: } y \ge 0,\ Wy = d - Tx,$$

where the second-stage decision $y$ isn't simply about discrepancy between $Tx$ and $d$, but also depends on an uncertain *recourse matrix* $W$. In fact, there might be many such satisfactory $y$ among which we would like to take the one with the lowest cost, a quantity that also might be uncertain. This results in the *stochastic optimization problem with linear recourse*:

$$\underset{x\in C_1}{\text{minimize}}\ f_0(x) + \mathbb{E}\big[f(\boldsymbol{\xi}, x)\big], \tag{3.19}$$

where $C_1 \subset \mathbb{R}^{n_1}$, $f_0 : \mathbb{R}^{n_1} \to \mathbb{R}$ is the first-stage cost and $f : \mathbb{R}^q \times \mathbb{R}^{n_1} \to \overline{\mathbb{R}}$ has

$$f(\xi, x) = \inf\nolimits_y \big\{\langle a, y\rangle \,\big|\, Wy = d - Tx,\ \ y \ge 0\big\} \tag{3.20}$$

for $a \in \mathbb{R}^{n_2}$, $m_2 \times n_2$-matrix $W$, $m_2 \times n_1$-matrix $T$, $d \in \mathbb{R}^{m_2}$ and

$$\xi = (a, W_1, \ldots, W_{m_2}, T_1, \ldots, T_{m_2}, d),$$

with $W_i$ and $T_i$ being the $i$th rows of $W$ and $T$, respectively. We refer to $f$ also as the *recourse cost function*. The random vector $\boldsymbol{\xi}$ is then of dimension $q = n_2 + (1 + n_2)m_2 + n_1 m_2$.

The second-stage decision is now associated with its own cost and is determined by solving a potentially nontrivial linear optimization problem. The overall problem (3.19) incorporates both anticipation *and* adaptation: The first-stage decision is selected while anticipating future conditions and the second-stage decision adapts to the revealed value of $\boldsymbol{\xi}$. For example, there might be a trade-off between capital investments and the cost of day-to-day operations, between the design and layout of an overall system and the adjustments necessary to respond to particular situations.

A stochastic optimization problem with simple recourse is the special case when the linear optimization problem has a minimizer that can be obtain explicitly for every $(\xi, x)$ without having to invoke the interior-point method or similar algorithms.

**Exercise 3.39** (from simple to linear recourse). Show that a simple recourse problem with

$$f(\xi, x) = \sum\nolimits_{i=1}^{m_2} \max\Big\{\gamma_i\big(\xi_i - \langle T_i, x\rangle\big),\ \delta_i\big(\xi_i - \langle T_i, x\rangle\big)\Big\},$$

where $\xi = (\xi_1, \ldots, \xi_{m_2})$, $T_i \in \mathbb{R}^{n_1}$ and $\gamma_i \leq \delta_i$, can be reformulated as a problem with linear recourse.

**Guide.** Define the second-stage decision vector $y = (y_1^+, y_1^-, \ldots, y_{m_2}^+, y_{m_2}^-)$ and write

$$f(\xi, x) = \inf_y \Big\{\sum\nolimits_{i=1}^{m_2} \delta_i y_i^+ - \gamma_i y_i^- \;\Big|\; y_i^+ - y_i^- = \xi_i - \langle T_i, x\rangle, y_i^+ \geq 0, y_i^- \geq 0, i = 1, \ldots, m_2\Big\}$$

from which one can recognize $a$, $W$, $d$ and $T$ in (3.20). □

A main concern in stochastic optimization problems with linear recourse is that certain first-stage decisions (capital investments) could cause difficulties at the second stage; day-to-day operations may become infeasible, i.e., there are no feasible solutions in (3.20). Moreover, there could be a shortage of constraints, permitting the objective function value to be arbitrarily low. In either case, the recourse cost function ceases to be real-valued and $f(\xi, x) \in \overline{\mathbb{R}}$. This raises the issue of what's the meaning of the expectation of an extended real-valued random variable, but that's quickly settled as one can resolve the issue of competing infinities by following §3.B; §8.G gives a formal treatment. Let's focus on identifying when the expected recourse cost is finite and thereby avoiding unpleasant surprises in the second stage.

To understand the situation, we need to examine the expectation function $Ef : \mathbb{R}^{n_1} \to \overline{\mathbb{R}}$ given by

$$Ef(x) = \mathbb{E}\big[f(\boldsymbol{\xi}, x)\big]$$

and its domain. For fixed $\bar{x}$, if $f(\boldsymbol{\xi}, \bar{x})$ is integrable, then $\mathbb{E}[f(\boldsymbol{\xi}, \bar{x})]$ is finite and $\bar{x} \in \operatorname{dom} Ef$. Thus, the difficulty arises when $f(\boldsymbol{\xi}, \bar{x})$ isn't integrable. This would certainly be the case if $f(\xi, \bar{x}) = \infty$ for some outcome

$$\xi = (a, W_1, \ldots, W_{m_2}, T_1, \ldots, T_{m_2}, d)$$

with positive probability. For $f(\xi, \bar{x}) = \infty$ to happen, there must be no $y \geq 0$ satisfying $Wy = d - T\bar{x}$, i.e., (3.20) is infeasible. Then, $\mathbb{E}[f(\boldsymbol{\xi}, \bar{x})] = \infty$ and $\bar{x} \notin \operatorname{dom} Ef$.

The random variable $f(\boldsymbol{\xi}, \bar{x})$ could also fail to be integrable when $f(\xi, \bar{x})$ is finite for all $\xi \in \Xi$, the support of $\boldsymbol{\xi}$, as discussed in §3.B or when $f(\xi, \bar{x}) = -\infty$. While $\mathbb{E}[f(\boldsymbol{\xi}, \bar{x})] = -\infty$ is concerning from a modeling point of view as a cost of $-\infty$ seems too good to be true, we permit such $\bar{x}$ if it otherwise satisfies the constraints specified by $C_1$.

Generally, for a linear recourse problem, the set of *induced constraints* is given by

$$C_2 = \operatorname{dom} Ef = \left\{x \in \mathbb{R}^{n_1} \,\middle|\, \mathbb{E}\big[f(\boldsymbol{\xi}, x)\big] < \infty\right\}.$$

Together with $C_1$, the induced constraints furnish the feasible set $C_1 \cap C_2$ of the problem. In contrast to $C_1$, which we assume is explicitly specified, for example in terms of equality and inequality constraints, $C_2$ isn't always easily characterized. It depends on an intricate interplay between the various components of (3.20) as well as the probability distribution of $\boldsymbol{\xi}$.

A problem has *complete recourse* if $\operatorname{dom} Ef = \mathbb{R}^{n_1}$, i.e., there are no induced constraints. It has *relatively complete recourse* if $\operatorname{dom} Ef \supset C_1$, in which case there might be induced constraints, but they're superfluous due to the explicit constraints.

**Example 3.40** (induced constraints). Consider a recourse cost function given by

$$f(\xi, x) = \inf_y\{y \mid y = x - \xi,\; y \geq 0\}$$

with $\boldsymbol{\xi}$ uniformly distributed on $[1, 2]$. The set of induced constraints is then $C_2 = [2, \infty)$.

**Detail.** When $x \geq \xi$, $f(\xi, x) = x - \xi$, but $f(\xi, x) = \infty$ otherwise. Thus, $\mathbb{E}[f(\boldsymbol{\xi}, x)] = \infty$ unless $x \geq 2$, in which case $\mathbb{E}[f(\boldsymbol{\xi}, x)] = x - 3/2$, i.e.,

$$Ef(x) = \begin{cases} x - 3/2 & \text{if } x \in [2, \infty) \\ \infty & \text{otherwise.} \end{cases}$$

Then, the problem of minimizing $Ef(x)$ subject to $x \in C_1 = [0, \infty)$ has neither complete nor relatively complete recourse. However, minimize $Ef(x)$ subject to $x \in [2, \infty)$ has the same minimizer and minimum value as well as relatively complete recourse. In fact, every stochastic optimization problem with linear recourse can in principle be reformulated as one with relatively complete recourse by incorporating the induced constraints in $C_1$. However, there might be no viable procedure for carrying out this reformulation. □

In the absence of complete recourse, a main theoretical and computational challenge is to characterize, at least partially, the induced constraints. We illustrate the analysis for the special case with deterministic recourse matrix $W$ and recourse cost vector $a$ while the right-hand side $d$ and the technology matrix $T$ remain uncertain. We refer to this case as a problem with *fixed recourse*.

**Proposition 3.41** (sufficiency for feasibility). *Suppose that the stochastic optimization problem (3.19) has fixed recourse, i.e., a and W are deterministic, and* $\mathbb{E}[\boldsymbol{\xi}]$ *is finite. For* $\bar{x} \in \mathbb{R}^n$*, if* $f(\xi, \bar{x}) < \infty$ *for all* $\xi \in \Xi$*, then* $\bar{x} \in \operatorname{dom} Ef$*.*

**Proof.** Let $(d_\xi, T_\xi)$ be the outcome of the demand vector and technology matrix corresponding to $\xi \in \Xi$. The function

$$(y, \xi) \mapsto \langle a, y \rangle + \iota_C(y, \xi), \quad \text{with } C = \{(y, \xi) \mid Wy - d_\xi + T_\xi \bar{x} = 0, y \geq 0\} \neq \emptyset,$$

is epi-polyhedral because it's affine on its domain, which is a polyhedral set. The inf-projection of this function, i.e,

$$\xi \mapsto f(\xi, \bar{x}) = \inf_y \langle a, y \rangle + \iota_C(y, \xi),$$

is also epi-polyhedral by 2.66 with domain $\mathbb{R}^q$. Consequently, $\mathbb{R}^q$ can be partitioned into a finite number of subsets. On each, the function can be bounded from above by an affine function. Thus, $\mathbb{E}[f(\boldsymbol{\xi}, x)]$ is bounded from above by a sum of expectations involving these affine functions and the conclusion follows because $\mathbb{E}[\boldsymbol{\xi}]$ is finite. □

The proposition shows that in the case of fixed recourse, there's a clear path to avoid infinite expected recourse cost: Make sure that (3.20) is feasible for every $(\xi, \bar{x})$ and $\mathbb{E}[\boldsymbol{\xi}]$ is finite, which anyhow is typical in applications. When feasibility can only be ensured for $\bar{x} \in C_1$, we still have relatively complete recourse.

To further characterize $\operatorname{dom} Ef$, let's examine for a given $\xi \in \Xi$ the set

$$C_2(\xi) = \operatorname{dom} f(\xi, \cdot) = \{x \in \mathbb{R}^{n_1} \mid \text{there's } y \geq 0 \text{ such that } Wy = d_\xi - T_\xi x\},$$

where $d_\xi$ and $T_\xi$ are outcomes corresponding to $\xi \in \Xi$ for $\boldsymbol{d}$ and $\boldsymbol{T}$, respectively. If $\bar{x} \notin C_2(\xi)$, then $f(\xi, \bar{x}) = \infty$ and we may hope in view of 3.41 that the induced constraints can be described in terms of $C_2(\xi), \xi \in \Xi$. This is indeed the case, but the proof requires some additional terminology.

Let's extend the notion of a positive hull from §2.K; The *positive hull of a matrix* $A$, denoted by $\operatorname{pos} A$, is the positive hull of its column vectors. With $W^j$, $j = 1, \ldots, n_2$, being the column vectors of the recourse matrix $W$, one then has

$$v \in \operatorname{pos} W \iff \text{there are } y_1, \ldots, y_{n_2} \geq 0 \text{ such that } v = \sum_{j=1}^{n_2} y_j W^j = Wy.$$

In this notation,

$$C_2(\xi) = \{x \in \mathbb{R}^{n_1} \mid d_\xi - T_\xi x \in \operatorname{pos} W\}.$$

If $\operatorname{pos} W = \mathbb{R}^{m_2}$, then $C_2(\xi) = \mathbb{R}^{n_1}$ for all $\xi \in \Xi$ and the problem has complete recourse. To get further, we need to characterize $\operatorname{pos} W$.

A *frame* of a polyhedral cone $C \subset \mathbb{R}^n$ is a finite collection of vectors $\{a^i \in \mathbb{R}^n, i = 1, \ldots, m\}$ such that $\operatorname{pos}\{a^1, \ldots, a^m\} = C$ and no vector $a^i$ can be obtained as a positive linear combination of the others, i.e.,

$$a^i \notin \operatorname{pos}\left(\{a^1, \ldots, a^m\} \setminus a^i\right), \quad i = 1, \ldots, m.$$

Thus, a frame of $C$ is a minimal set of "generators" for $C$; every point in $C$ can be represented by the frame in an economical manner.

**Example 3.42** (frames of polyhedral cones). There's never a unique frame for a polyhedral cone $C$, except when $C = \{0\}$.

**Detail.** If $C$ contains a subspace, its frames may not have the same number of elements; $\{(1,0),(0,1),(-1,0),(0,-1)\}$ and $\{(1,-4),(2,1),(-1,1)\}$ are frames for $\mathbb{R}^2$. □

A *polar matrix* of an $m \times n$-matrix $A$ is a $q \times m$-matrix $A^*$ whose rows constitute a frame for pol(pos $A$). Every $m \times n$-matrix $A$ has a polar matrix, which is seen as follows. With $A^1, \dots, A^n$ being the columns of $A$, we recall from 2.60 that

$$\operatorname{pol}(\operatorname{pos} A) = \left\{y \in \mathbb{R}^m \;\middle|\; \langle A^j, y\rangle \le 0,\ j = 1, \dots, n\right\},$$

which is a polyhedral cone and thus can be written as $\operatorname{pos}\{A_1^*, \dots, A_q^*\}$ for some $A_1^*, \dots, A_q^* \in \mathbb{R}^m$ by Minkowski's theorem 2.61. If these vectors are selected to be "minimal" so that they form a frame of pol(pos $A$), then they satisfy the requirement in the definition.

The number of rows of $A^*$ isn't commensurate with the number of columns in $A$. For example, $A^*$ might have just one row of zeros (when $\operatorname{pos} A = \mathbb{R}^m$) or there could be a large number of rows.

**Example 3.43** (polar matrix of an invertible matrix). If $A$ is an invertible $n \times n$-matrix, then a polar matrix $A^* = -A^{-1}$.

**Detail.** Let $A_1^*, \dots, A_n^* \in \mathbb{R}^n$ be the rows of $A^*$; the suggested formula indicates that there are indeed $n$ rows. Since they constitute a frame for pol(pos $A$), we've

$$\operatorname{pos}\{A_1^*, \dots, A_n^*\} = \operatorname{pol}(\operatorname{pos} A). \tag{3.21}$$

The left-hand side is $\{vA^* \mid v \ge 0\}$, while the right-hand side is

$$\left\{y \in \mathbb{R}^n \;\middle|\; \langle A^j, y\rangle \le 0,\ j = 1, \dots, n\right\}$$

by 2.60, where $A^1, \dots, A^n$ are the columns of $A$. Equivalently, this is also expressed by $\{y \in \mathbb{R}^n \mid yA \le 0\}$. The suggested formula $A^* = -A^{-1}$ implies that if $y = vA^*$ for some $v \ge 0$, then $v = -yA$ and also $yA \le 0$. If $y$ satisfies $yA \le 0$, then $v = -yA \ge 0$ and $vA^* = -yAA^* = y$. Thus, the suggested formula satisfies (3.21). The rows of $-A^{-1}$ indeed form a frame of pol(pos $A$) because they're linearly independent and thus no row can be expressed as a linear combination of the others, let alone a positive combination. □

**Lemma 3.44** *For an $m \times n$-matrix $A$, if $A^*$ is a polar matrix of $A$, then*

$$\operatorname{pos} A = \left\{y \in \mathbb{R}^m \;\middle|\; A^* y \le 0\right\}.$$

**Proof.** From the definition of polar matrices, we've

$$\text{pol}\left(\text{pos}\{A^1, \dots, A^n\}\right) = \text{pos}\{A_1^*, \dots, A_q^*\},$$

where $A^j$, $j = 1, \dots, n$, are the columns of $A$ and $A_i^*, i = 1, \dots, q$, are the rows of $A^*$. Since $\text{pol}(\text{pol}\, C) = C$ for a closed convex cone $C$ by (2.12), it follows by also invoking 2.60 that

$$\begin{aligned}\text{pos}\{A^1, \dots, A^n\} &= \text{pol}\left(\text{pos}\{A_1^*, \dots, A_q^*\}\right)\\ &= \left\{y \in \mathbb{R}^m \,\middle|\, \langle A_i^*, y\rangle \le 0,\ i = 1, \dots, q\right\} = \{y \in \mathbb{R}^m \mid A^* y \le 0\}\end{aligned}$$

and the claim is established. □

Returning to the stochastic optimization problem with fixed recourse, it now follows directly from this lemma that

$$C_2(\xi) = \left\{x \in \mathbb{R}^{n_1} \,\middle|\, W^*(d_\xi - T_\xi x) \le 0\right\},$$

where $W^*$ is a polar matrix of the recourse matrix $W$. Since $C_2(\xi)$ is determined by a finite number of inequalities (the number of rows in $W^*$), it's a polyhedral set regardless of $\xi \in \Xi$.

**Theorem 3.45** (induced constraints). *If the stochastic optimization problem (3.19) has fixed recourse, i.e., a and W are deterministic, and $\mathbb{E}[\boldsymbol{\xi}]$ is finite, then*

$$C_2 = \bigcap\nolimits_{\xi \in \Xi} C_2(\xi),$$

*which must be closed and convex. Moreover, if $\Xi$ is finite and/or the technology matrix $T$ is deterministic, then $C_2$ is a polyhedral set.*

**Proof.** If $x \in \bigcap_{\xi \in \Xi} C_2(\xi)$, then $f(\xi, x) < \infty$ for all $\xi \in \Xi$ and $x \in \text{dom}\, Ef = C_2$ by 3.41. If $x \in C_2$, then $\widetilde{\Xi} = \{\xi \in \Xi \mid f(\xi, x) < \infty\}$ must occur with probability one because otherwise $Ef(x) = \infty$. Since $f(\cdot, x)$ is epi-polyhedral (see the argument in the proof of 3.41), its domain $\widetilde{\Xi}$ is closed. As the support of $\boldsymbol{\xi}$, $\Xi$ is the smallest closed set of points that occurs with probability one. Thus, $\widetilde{\Xi} = \Xi$ and $x \in \bigcap_{\xi \in \Xi} C_2(\xi)$. We've shown that $C_2 = \bigcap_{\xi \in \Xi} C_2(\xi)$.

Since $C_2(\xi)$, $\xi \in \Xi$, are polyhedral sets (and consequently closed and convex), it immediately follows that $C_2$ is a closed convex set; cf. 1.12(a). If $\Xi$ is a finite set, this also implies that $C_2$ is polyhedral. If $T$ doesn't depend on $\xi$ so that $\xi$ consists simply of $d$, then

$$C_2(\xi) = \left\{x \in \mathbb{R}^{n_1} \,\middle|\, W^* T x \ge W^* d_\xi\right\},$$

where $W^*$ is a polar matrix of $W$ and $d_\xi$ is the outcome corresponding to $\xi$. Let $(W^*T)_1, \dots, (W^*T)_q$ be the rows of $W^*T$ and $W_1^*, \dots, W_q^*$ be the rows of $W^*$. In this notation,

$$C_2(\xi) = \left\{x \in \mathbb{R}^{n_1} \,\middle|\, \left\langle (W^*T)_i, x\right\rangle \ge \langle W_i^*, d_\xi\rangle,\ i = 1, \dots, q\right\}.$$

With $\hat{d}_i = \sup_{\xi \in \Xi} \langle W_i^*, d_\xi\rangle$, this means that

$$C_2 = \{x \in \mathbb{R}^{n_1} \mid \langle (W^*T)_i, x\rangle \geq \hat{d}_i,\ i = 1, \ldots, q\}.$$

If $\hat{d}_i = \infty$ for some $i$, then $C_2$ is empty and trivially polyhedral. Otherwise $C_2$ is determined by a finite number of linear constraints. □

Although the theorem establishes that in the case of fixed recourse the set of induced constraints is convex (and possibly even polyhedral), it can be challenging to determine a computationally convenient description of $C_2$. If the polar matrix $W^*$ is known, then the theorem ensures that

$$C_2 = \{x \in \mathbb{R}^{n_1} \mid W^*(d_\xi - T_\xi x) \leq 0,\ \xi \in \Xi\}.$$

In principle, one can then add the constraints defining $C_2$ to $C_1$ and obtain a problem with relatively complete recourse. That problem, however, has an infinite number of constraints if $\Xi$ isn't finite and subsequent approximations are required. There's also the need for computing $W^*$, which generally requires a construction as in the proof of Weyl's theorem 2.59, possibly resulting in a matrix with many rows; see [124] for further details.

**Proposition 3.46** (convexity of expected recourse cost). *If the stochastic optimization problem* (3.19) *has fixed recourse, i.e., $a$ and $W$ are deterministic, and* $\mathbb{E}[\boldsymbol{\xi}]$ *is finite, then $Ef$ is convex.*

**Proof.** Let's fix $\xi \in \Xi$ and set $d_\xi$ and $T_\xi$ to be the corresponding outcomes. The function

$$(x, y) \mapsto \langle a, y\rangle + \iota_C(x, y), \text{ with } C = \{(x, y) \mid Wy - d_\xi + T_\xi x = 0, y \geq 0\},$$

is epi-polyhedral because it's affine on its domain, which is a polyhedral set. The inf-projection of this function,

$$x \mapsto f(\xi, x) = \inf_y \langle a, y\rangle + \iota_C(x, y),$$

is also epi-polyhedral by 2.66 and thus convex. If $\mathbb{E}[f(\boldsymbol{\xi}, x)] > -\infty$ for all $x \in \operatorname{dom} Ef$, then we can follow the arguments in the proof of 3.3 to establish that $Ef$ is convex because only points inside the convex set $\operatorname{dom} Ef$ need to be checked; cf. 3.45. If $\mathbb{E}[f(\boldsymbol{\xi}, x)] = -\infty$ for some $x \in \operatorname{dom} Ef$, then it turns out that $\mathbb{E}[f(\boldsymbol{\xi}, x)] = -\infty$ for all $x \in \operatorname{dom} Ef$ (cf. [124, Theorem 7.6]) and $Ef$ is again convex. □

We leave further details about algorithms for solving the stochastic optimization problem with linear recourse (3.19) to §5.H. However, when $\Xi$ is finite, the path followed in §3.A permits the reformulation

$$\underset{x \in C_1, y_\xi, \xi \in \Xi}{\text{minimize}}\ f_0(x) + \sum\nolimits_{\xi \in \Xi} p_\xi \langle a_\xi, y_\xi\rangle \tag{3.22}$$

$$\text{subject to } Wy_\xi = d_\xi - T_\xi x,\ y_\xi \geq 0\ \ \forall \xi \in \Xi.$$

If $f_0$ and $C_1$ are convex, then this problem is convex and computationally tractable as long as $\Xi$ isn't too large.

## 3.K Network Capacity Expansion

In a power transmission network, there's net demand $\xi_i$ at vertex $i \in V = \{1, \ldots, m\}$, which could be negative reflecting excess supply of power. Overall in the network, supply exceeds demand but the capacities of the transmission lines, represented by edges in the network, need to be expanded from the current levels. Let $a_e$ be the current capacity on edge $e$ and $x_e$ be the corresponding planned capacity expansion, which is limited by $v_e$ and associated with the cost $c_e x_e$. Our goal is to optimize the capacity expansion while accounting for a budget of $\beta$. The problem is historically motivated: In the 2001 California energy crisis, some of the blackouts were blamed on the lack of capacity of the transmission lines between Southern and Northern California.

The power flow in a transmission network is governed by laws of physics that are reasonably well approximated by a linear relation between the net demand vector $\xi = (\xi_1, \ldots, \xi_m)$ and a vector of voltage angles $y = (y_1, \ldots, y_m)$ as expressed via the vectors $b^i \in \mathbb{R}^m$, $i \in V$. Specifically, we would need the voltage angles to satisfy

$$\langle b^i, y \rangle \geq \xi_i \quad \forall i \in V. \tag{3.23}$$

A difference in voltage angle between two vertices connected by an edge requires a certain capacity that's captured with the constraints

$$|y_i - y_j| \leq a_e + x_e \quad \forall e = (i, j) \in E,$$

where $E$ is the set of $n$ edges in the network. The net demand is uncertain due to variability in consumption and renewable power generation and we model it as a random vector $\boldsymbol{\xi}$ with a finite distribution; outcomes are given by $\Xi$ and $\{p_\xi > 0, \xi \in \Xi\}$ are the corresponding probabilities. Since it wouldn't be economical to attempt to prevent violation of the constraints (3.23) for every outcome $\xi \in \Xi$, we adopt a nonnegative penalty of $\delta_i$ per unit violation of the constraint at vertex $i$. This leads to the formulation

$$\operatorname*{minimize}_{x \in \mathbb{R}^n} \sum\nolimits_{\xi \in \Xi} p_\xi f(\xi, x) \ \text{ subject to } \sum\nolimits_{e \in E} c_e x_e \leq \beta, \ 0 \leq x_e \leq v_e \ \forall e \in E,$$

where the recourse cost function

$$f(\xi, x) = \inf_{y \in \mathbb{R}^m} \Big\{ \sum\nolimits_{i \in V} \delta_i \max\{0, \xi_i - \langle b^i, y \rangle\} \ \Big| \ |y_i - y_j| \leq a_e + x_e \ \forall e = (i, j) \in E \Big\}.$$

We seek to determine a first-stage decision $x$ that minimizes expected penalties across all outcomes of $\boldsymbol{\xi}$. The second-stage decision about voltage angles is made after we observe the demand. Although not exactly in the form (3.20) of a problem with linear recourse, we can accomplish this through a reformulation.

Let's introduce the additional variables $z_i^+$ and $z_i^-$ to represent the positive and negative parts of $\xi_i - \langle b^i, y \rangle$ and these can be used to construct a linear objective function. We can express the variable $y_i$ in terms of its positive part $y_i^+$ and negative part $y_i^-$. Let $y^+$ and $y^-$ be the vectors of these positive and negative parts so that $y = y^+ - y^-$. After using $w_e^+$ and $w_e^-$ to convert the constraints on voltage angles into equalities, we obtain

$$f(\xi, x) = \inf\Big\{\sum_{i\in V} \delta_i z_i^+ \;\Big|\; \xi_i - \langle b^i, y^+ - y^-\rangle = z_i^+ - z_i^- \;\; \forall i \in V$$

$$y_i^+ - y_i^- - y_j^+ + y_j^- + w_e^+ = a_e + x_e \;\; \forall e = (i, j) \in E$$

$$y_i^+ - y_i^- - y_j^+ + y_j^- - w_e^- = -a_e - x_e \;\; \forall e = (i, j) \in E$$

$$y_i^+, y_i^-, z_i^+, z_i^-, w_e^+, w_e^- \ge 0 \;\; \forall i \in V,\; e \in E\Big\}.$$

The minimization is now over nonnegative variables subject to equality constraints and we recognize that it's of the form (3.20); the recourse cost vector $a$ in the notation of that equation consists of $\delta_i$ and zeros, the recourse matrix $W$ collects all the coefficients in front of the variables and is deterministic, the demand vector $d$ consists of $\xi_i$ and $a_e$ and the technology matrix $T$ is deterministic with one and zero elements. Thus, we've a problem with fixed recourse.

For any $\xi$ and $x \ge 0$, $f(\xi, x) < \infty$ because one can set

$$z_i^+ - z_i^- = \xi_i, \quad y_i^+ = y_i^- = 0 \;\; \forall i \in V; \qquad w_e^+ = w_e^- = a_e + x_e \;\; \forall e \in E$$

to produce a feasible point in the minimization defining $f$. Since $\boldsymbol{\xi}$ is finitely distributed, this means that $\mathbb{E}[f(\boldsymbol{\xi}, x)] < \infty$ as well; cf. 3.41. (In fact, the objective function is always nonnegative, which means that $\mathbb{E}[f(\boldsymbol{\xi}, x)]$ is finite.) Thus, dom $Ef$ contains the set defined by the explicit constraints and the problem has relatively complete recourse. The problem doesn't have complete recourse, however, because for $x_e < -a_e$ there's no feasible second-stage decision, which can already be seen by the constraint $|y_i - y_j| \le a_e + x_e$.

The problem can be reformulated as a linear optimization problem following the principle behind (3.22): each of the second-stage variables are replicated to allow them to depend on the outcome of $\boldsymbol{\xi}$. This results in the formulation

$$\text{minimize} \sum_{\xi\in\Xi} p_\xi \sum_{i\in V} \delta_i z_i^+(\xi) \;\; \text{subject to} \sum_{e\in E} c_e x_e \le \beta$$

$$x_e \le v_e \quad \forall e \in E$$

$$\xi_i - \big\langle b^i, y^+(\xi) - y^-(\xi)\big\rangle = z_i^+(\xi) - z_i^-(\xi) \quad \forall i \in V,\; \xi \in \Xi$$

$$y_i^+(\xi) - y_i^-(\xi) - y_j^+(\xi) + y_j^-(\xi) + w_e^+(\xi) = a_e + x_e \quad \forall e = (i, j) \in E,\; \xi \in \Xi$$

$$y_i^+(\xi) - y_i^-(\xi) - y_j^+(\xi) + y_j^-(\xi) - w_e^-(\xi) = -a_e - x_e \quad \forall e = (i, j) \in E,\; \xi \in \Xi$$

$$x_e, y_i^+(\xi), y_i^-(\xi), z_i^+(\xi), z_i^-(\xi), w_e^+(\xi), w_e^-(\xi) \ge 0 \quad \forall i \in V,\; e \in E,\; \xi \in \Xi.$$

Although we've ended up with a linear optimization problem, its size quickly becomes a concern as power transmission networks are large making the replication of variables and constraints across outcomes even more detrimental. Modern implementations of the interior-point method (§2.G) and other algorithms scan a formulation for redundancies and may reduce the size by eliminating some variables and constraints. This extends their reach somewhat, but it's often necessary to leverage the structure of the problem in more detail as discussed in the following chapters; see, in particular, §5.H.

# Chapter 4
# MINIMIZATION PROBLEMS

Applications may lead us beyond convex optimization problems as we seek to model physical phenomena, dynamical systems, logical conditions, complicated relations between data and numerous other real-world considerations. In this chapter, we'll start in earnest the development of theory and algorithms for minimizing nonconvex functions, but in the process also expand on tools for the convex case. The guiding principle will be to view functions through their epigraphs and this will enable us to maintain a geometrical perspective throughout. We'll see that epigraphs of functions and their approximations provide a main path to minimization algorithms. The normal vectors to an epigraph turn out to specify subgradients of the corresponding function and thus become central to the construction of optimality conditions and associated algorithms.

## 4.A Formulations

The difficulty of finding a minimizer of a function $f : \mathbb{R}^n \to \overline{\mathbb{R}}$ depends largely on the *problem structure* and our ability to exploit it. Thus far, we've utilized structural properties such as smoothness, linearity, sparsity and, most prominently, convexity to construct efficient algorithms. For example, the accelerated gradient method is guaranteed by 1.36 to obtain a near-minimizer of a smooth convex function on $\mathbb{R}^n$ within a number of iterations that's *independent* of $n$ and therefore can handle large-scale problems. In the context of linear and quadratic problems, sparsity (i.e., the presence of many zero elements in vectors and matrices) reduces computing times in algorithms such as the interior-point method (§2.G). As we move beyond convex problems, structural properties remain crucial. Models and algorithms are developed with the goal of identifying and leveraging advantageous problem structure.

The importance of structural properties is highlighted by examining the situation in their absence. Suppose that $f : \mathbb{R}^n \to \mathbb{R}$ is truly arbitrary in the sense that knowledge

 
J. O. Royset and R. J-B Wets, *An Optimization Primer*, Springer Series in Operations Research and Financial Engineering, https://doi.org/10.1007/978-3-030-76275-9_4

of $f(\bar{x})$ provides no information about $f(x)$, regardless of how close $x$ might be to $\bar{x}$; there's no smoothness or continuity anywhere. Then, there's no algorithm for minimizing $f$ that's consistently better than the most naive one: select $x \in \mathbb{R}^n$, randomly or according to some pattern, obtain $f(x)$, and repeat. This is a costly process. We need to evaluate the function $\varepsilon^{-n}$ times to cover a grid pattern of $[0,1]^n$ with spacing $\varepsilon > 0$. Even then, the minimum value of $f$ on $[0,1]^n$ could be any amount below the lowest value seen after all these evaluations.

Fortunately, many—if not most—problems arising in practice have at least some structure. The problem structure can be hidden, rather convoluted and not easily exploited. For example, if the function isn't specified by some explicit formula but rather implicitly through an algorithmic procedure, then we may not immediately be able to identify convexity, continuity and smoothness but these properties could still be present at least locally in parts of the space of decision variables. When successfully identified and exploited, problem structure can be tremendously beneficial.

**Example 4.1** (nonconvexity in neural networks). Image and speech recognition are highly challenging learning problems due to the vast variety of situations that an automated system may encounter. Even in a limited context such as understanding handwritten text, a system needs to tackle the diversity of personal "styles" our elementary school teachers failed to eradicate! Image and speech recognition systems based on *neural networks* solve highly structured nonconvex optimization problems.

**Detail.** The situation resembles the data fitting in §1.B and classification in 1.29, but now the data $\{x^i, y^i\}_{i=1}^{\nu}$ may represent a collection of black-and-white images so that $y^i$ is a vector predicting the label of the $i$th image (e.g., dog, cat, person, etc.) and $x^i$ is a vector specifying the gray-scale for each pixel. Again, the goal is to fit a function to the data, but now using a sophisticated class of functions, much beyond polynomials. These functions are defined in layers that successively transform an input vector $x$ to an output vector $h(x)$. For example, the functions may be of the form

$$h(x) = x_q, \quad \text{with } x_k = \varphi_k(A_k x_{k-1} - b_k), \quad k = 1, \dots, q, \quad \text{and} \quad x_0 = x,$$

where $A_k$ is a $d_k \times d_{k-1}$-matrix and $b_k \in \mathbb{R}^{d_k}$, both varying in size across the layers $k = 1, \dots, q$, are to be optimized in the same manner as the polynomial coefficients in §1.B; see Figure 4.1. The functions $\varphi_k : \mathbb{R}^{d_k} \to \mathbb{R}^{d_k}$ are given and typically relatively simple. They may transform each component of their input according to $\alpha \mapsto \max\{0, \alpha\}$ or $\alpha \mapsto 1/(1 + e^{-\alpha})$. If the fitting criteria is "average loss" expressed by some function $f : \mathbb{R}^{d_q} \times \mathbb{R}^{d_q} \to \overline{\mathbb{R}}$, then the resulting optimization problem becomes

$$\underset{A_1,\dots,A_q,b_1,\dots,b_q}{\text{minimize}} \ \frac{1}{\nu} \sum\nolimits_{i=1}^{\nu} f\big(h(x^i), y^i\big),$$

where the decision variables $A_1, \dots, A_q, b_1, \dots, b_q$ determine $h$. This objective function isn't convex except in rather specific instances such as when $q = 1$, $\varphi_1$ is affine and $f$ is convex in its first argument, the situation in §1.B. Still, problems of this general form are solved, at least *approximately*, on a routine basis across a range of application domains.

A main reason for this is that the problem possesses structure that allows the stochastic gradient descent method (§3.G) and related algorithms to be quite effective. □

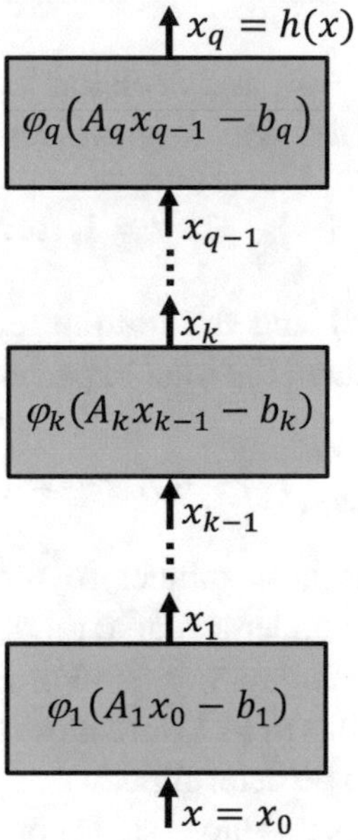


Fig. 4.1. Structure of neural network.

**Example 4.2** (decision rules in inventory management). Each morning an online retailer observes the demand for a product over the last 24 hours and then determines how much to order of that product to avoid running out. On day $t$, the retailer can order $x_t$ units at cost $\gamma$ per unit, but these arrive at the warehouse the next day, and also order $y_t$ units at an expedited unit cost $\alpha > \gamma$ and have them delivered to the warehouse immediately. At the end of the day, the retailer ships the product to satisfy the demand. If the inventory is too low to satisfy the demand, a backlog is registered at a per-unit cost $\beta$. If the inventory is too high, the retailer incurs a per-unit holding cost $\delta$. The retailer aims to minimize ordering, backlog and holding costs over a planning horizon of $\tau$ days by relying on a forecast of the demand for the next $\tau$ days. Since the forecast is uncertain, it's given in terms of $\nu$ scenarios:

$$\left\{\xi^i = (\xi_1^i, \dots, \xi_\tau^i) \in \mathbb{R}^\tau,\ i = 1, \dots, \nu\right\},$$

where $\xi_t^i$ is the demand forecast for day $t$ in scenario $i$. Scenario $i$ occurs with probability $p_i > 0$.

When making an order decision today, one should ideally account for all possible scenarios and the associated information that would be available when making ordering decisions in the future. An expansion of the number of variables as in §3.A and (3.22) is possible conceptually, but leads to extremely large-scale problems. In practice, an approach based on decision rules is more viable.

**Detail.** Let $s_t^i$ be the number of units available at the end of day $t$ in scenario $i$ and $(x_t^i, y_t^i)$ be the decision of day $t$ in scenario $i$. Then, each day $t$ unfolds as follows:

$$\text{inventory: } s_{t-1}^i \rightsquigarrow \text{ observation: } \xi_t^i \rightsquigarrow \text{ decision: } (x_t^i, y_t^i) \rightsquigarrow \text{ inventory: } s_t^i.$$

Starting from given inventory $s_0^i = \bar{s}_0$ and decision $x_0^i = \bar{x}_0$, which are the same for all scenarios because they're known, the inventory evolves according to the balance equation

$$s_t^i = s_{t-1}^i + x_{t-1}^i + y_t^i - \xi_t^i, \quad t = 1, \ldots, \tau, \quad i = 1, \ldots, \nu. \tag{4.1}$$

The backlog cost is $\max\{0, -\beta s_t^i\}$ and the holding cost is $\max\{0, \delta s_t^i\}$, which can be combined into $\max\{-\beta s_t^i, \delta s_t^i\}$. Then, the total expected cost becomes

$$\sum\nolimits_{i=1}^{\nu} p_i \sum\nolimits_{t=1}^{\tau} \gamma x_t^i + \alpha y_t^i + \max\{-\beta s_t^i, \delta s_t^i\}.$$

The problem of minimizing this cost subject to (4.1) as well as other constraints[1] quickly becomes intractable. Even under a coarse model of uncertainty stipulating "low," "medium" and "high" demand each day, $\nu = 3^\tau$. For example, $\tau = 10$ gives $\nu = 59049$ and then there are 590490 constraints in (4.1) and more than 1.5 million decision variables because $(x_t^i, y_t^i, s_t^i, \forall t, i)$ all need to be determined.

An alternative approach replaces this large-scale convex problem by a small nonconvex problem. Suppose that we're willing to only consider ordering decisions of the form

$$(x_t^i, y_t^i) = F(\xi_t^i, s_{t-1}^i + x_{t-1}^i),$$

where $F : \mathbb{R}^2 \to \mathbb{R}^2$ is a mapping to be determined. We refer to $F$ as a *decision rule* because it furnishes ordering quantities based on available information. While one can conceive situations where today's decisions are influenced by the whole history of demands up to the present, this decision rule only depends on today's demand ($\xi_t^i$) and the current inventory ($s_{t-1}^i + x_{t-1}^i$). This seems reasonable and may be easily accepted and understood by practitioners. One possibility is to let each component of $F$ be piecewise affine with

$$F(\eta, \sigma; z) = \begin{bmatrix} f_1(\eta, \sigma; z) \\ f_2(\eta, \sigma; z) \end{bmatrix} = \begin{bmatrix} \max\{0,\ z_1(\eta - \sigma) + z_2,\ z_3(\eta - \sigma) + z_4\} \\ \max\{0,\ z_5(\eta - \sigma) + z_6,\ z_7(\eta - \sigma) + z_8\} \end{bmatrix},$$

where $z = (z_1, \ldots, z_8)$ are coefficients to be determined. The expression is relatively simply and the coefficients can be related to how we react to various demand-inventory situations. Under the restriction to decision rules of this form, the problem reduces to that of finding $z \in \mathbb{R}^8$ such that

$$\varphi(z) = \sum\nolimits_{i=1}^{\nu} p_i \sum\nolimits_{t=1}^{\tau} \gamma f_1(\xi_t^i, s_{t-1}^i + x_{t-1}^i; z) + \alpha f_2(\xi_t^i, s_{t-1}^i + x_{t-1}^i; z) + \max\{-\beta s_t^i, \delta s_t^i\}$$

[1] One would need to make sure that $x_t^i, y_t^i$ depend only on information that's available on day $t$. We discuss this systematically in §10.E.

is minimized. This is a well-defined objective function: for each $z$, $\varphi(z)$ can be computed using (4.1) and the decision rule specified by $z$. While the problem has only 8 variables, it's nonconvex. □

**Example 4.3** (integer constraints). A constraint specifying that a variable can only take integer values is a particular source of nonconvexity. We recall the situation from 2.4: a minimizer dictates a low shipping volume of 50 units from Washington State to San Francisco. This might be undesirable for a variety of practical reasons not accounted for in the model and management may require that any shipment involves at least 100 units. We can incorporate this change using the constraints

$$w_i y_{ij} \geq x_{ij} \geq 100 y_{ij} \quad \text{and} \quad y_{ij} \in \{0, 1\}, \;\; i = 1, \dots, m, \;\; j = 1, \dots, n,$$

where $(y_{ij}, i = 1, \dots, m, j = 1, \dots, n)$ are additional decision variables. They've the interpretation that $y_{ij} = 1$ if units move between warehouse $i$ and demand location $j$. Then, the right-most inequality guarantees that $x_{ij} \geq 100$ as requested by management. The left-most inequality is of no importance because $x_{ij}$ is already limited by $w_i$. When $y_{ij} = 0$, no movement between warehouse $i$ and location $j$ is selected and the inequalities ensure that $x_{ij} = 0$. We refer to $y_{ij} \in \{0, 1\}$ and other restrictions to integer values as *integer constraints*.

**Detail.** The feasible set defined by the constraints in 2.4 and the additional constraints introduced here is nonconvex due to the integer constraints. However, the problem remains highly structured and this can be utilized by algorithms. A main idea is to replace $y_{ij} \in \{0, 1\}$ by $y_{ij} \in [0, 1]$ for some selected pairs $(i, j)$, by $y_{ij} = 0$ for some other pairs and by $y_{ij} = 1$ for the rest of the pairs, and solve the resulting problem, which then has no integer constraints. The process is repeated for other selections and leads to *branch-and-bound algorithms*. Since one may need to consider (nearly) all possible combinations of integer values for the variables subject to integer constraints, these algorithms can be computationally costly. However in many practical settings, problems with integer constraints solve surprisingly fast due to the underlying structure; see [125] for a systematic treatment. In any case, the solution of problems with integer constraints is largely contingent on efficient algorithms for solving the corresponding problems without such constraints.

Table 4.1 provides a minimizer for the specific problem at hand under the additional constraints. Its cost is slightly higher than that of Table 2.1 (without integer constraints), but now we avoid shipping low volumes. □

Table 4.1: Shipping quantities subject to volume constraint.

| | Demand locations | | | Cost |
|---|---|---|---|---|
| Warehouses | San Francisco | Chicago | New York | |
| Washington State | 0 | 305 | 0 | $1450.50 |
| North Carolina | 320 | 0 | 270 | |

**Exercise 4.4** (aircraft allocation with integer constraints). Solve the aircraft allocation problem in 3.38, but now with the additional constraints that every $x_{ij}$ must be an integer. Consider both uncertain demand and expected demand. Compare solutions as well as computing times with those from 3.38.

**Guide.** One would need to employ a branch-and-bound algorithm to systematically enumerate through the possible integer values for the variables. □

With the realization that structure is critical for successful solution of optimization problems, we're faced with the question of how to effectively formulate and examine a problem. We would like to highlight linear, convex and smooth "components," even though the resulting function may not possess any of these properties. Efficient computational tools can then be brought in to tackle these simpler components, while approximations and iterations may be needed for the remaining ones as well as the overall problem. We've already seen the benefits of this approach in 2.34 for the problem of minimizing $\|Ax - b\|_2^2 + \theta\|x\|_1$. Since the second term is nonsmooth, one might simply settle for a "general" algorithm for nonsmooth problems such as the subgradient method (§2.I). However, the proximal gradient method (§2.D) leverages the fact that the function is the sum of two terms, with only the second one being nonsmooth and then even of a special kind. The difference in rate of convergence of the two methods ($\nu^{-1/2}$ versus $\nu^{-1}$) highlights the benefits of exploiting structure.

It's common to write down optimization problems in the form minimizing $f(x)$ subject to $x \in C$, with $C$ often being expressed in terms of equality and inequality constraints. Historically, one required the objective and constraint functions to be smooth. The restriction to smooth functions is too limiting and reformulations that attempt to avoid the issue often increase the problem size substantially. For example, a reformulation to avoid $\|x\|_1$ in §2.H adds $n$ variables and $2n$ constraints.

A formulation in terms of *one* objective function and *many* constraint functions is fundamentally unbalanced. It would be useful to specify the objective function in more detail too; §2.D furnishes an example. From a modeling point of view, there's also something artificial about singling out one function as objective and the rest as constraints. For example, when designing a system, we might seek a low cost and a low risk of poor performance. To insist on having a single objective function would force us to prioritize cost over risk or vice versa. Moreover, the quantity, being it cost or risk, that has been downgraded to a constraint is subject to a strict requirement. In reality, violations might be acceptable if appropriately penalized. Chapter 3 promotes this perspective and uses it to address situations with uncertainty, where constraint violation might be unavoidable.

In view of these factors, it's often beneficial to express a minimization problem in terms of a *composite function* and a relatively simple set $X$:

$$\operatorname*{minimize}_{x \in X \subset \mathbb{R}^n} h\big(F(x)\big), \quad \text{with } h : \mathbb{R}^m \to \overline{\mathbb{R}} \text{ and } F : \mathbb{R}^n \to \mathbb{R}^m. \tag{4.2}$$

The various quantities of interest (cost, risk, etc.), expressed by $f_i : \mathbb{R}^n \to \mathbb{R}$, $i = 1, \ldots, m$, and collected as $F = (f_1, \ldots, f_m)$, can be treated equally or individually as the circumstances may dictate by adjusting $h$. The inequalities $f_i(x) \le 0$, $i = 1, \ldots, m$, are captured by

$$h(u) = \iota_{(-\infty,0]^m}(u): \qquad h\big(F(x)\big) = \begin{cases} 0 & \text{if } f_i(x) \le 0,\ i = 1,\dots,m \\ \infty & \text{otherwise.} \end{cases}$$

A per-unit penalty of $\theta \in (0, \infty)$ for $f_i$ exceeding zero is modeled by

$$h(u) = \sum_{i=1}^m \theta \max\{0, u_i\}: \qquad h\big(F(x)\big) = \sum_{i=1}^m \theta \max\big\{0, f_i(x)\big\}.$$

The constraints singled out in $X$ are typically bounds on the variables or at most linear equation and inequalities so that $X$ is a polyhedral set and, preferably, a simple one for which projections are carried out efficiently. A goal could be to have $h$ convex and $F$ smooth. Although not necessarily convex, a problem of this form is certainly highly structured and, as we'll see in the following sections and chapters, is quite tractable theoretically and computationally.

**Example 4.5** (nonconvex regularization in regression). The composite form (4.2) highlights the various components of an objective function as already leveraged in 2.34 to develop a specialized version of the proximal gradient method for lasso regression. There, $f(x) = \|Ax - b\|_2^2 + \theta\|x\|_1$ is viewed as the sum of two terms, but it can be broken down further. A detailed description allows us to examine the effect of nonconvex modifications.

**Detail.** With $A_i$ being the $i$th row of $m \times n$-matrix $A$, one has

$$f(x) = \|Ax - b\|_2^2 + \theta\|x\|_1 = h\big(F(x)\big),$$

where

$$\begin{aligned} h(u) &= \sum_{i=1}^m u_i^2 + \theta \sum_{j=1}^n |u_{m+j}|, \quad u = (u_1, \dots, u_{m+n}) \\ F(x) &= \big(f_1(x), \dots, f_{m+n}(x)\big) \\ f_i(x) &= \langle A_i, x\rangle - b_i, \qquad i = 1, \dots, m \\ f_{m+j}(x) &= x_j, \qquad j = 1, \dots, n. \end{aligned}$$

In this case, $h$ is convex and $F$ is affine. Empirical evidence indicates that the results improve from a statistical point of view when the linear $f_{m+j}$ is replaced by a nonlinear but smooth function. One possibility is to set

$$f_{m+j}(x) = \begin{cases} 2 - \exp(1 - x_j) & \text{if } x_j \in (1, \infty) \\ x_j & \text{if } x_j \in [-1, 1] \\ \exp(1 + x_j) - 2 & \text{otherwise.} \end{cases}$$

Thus, $f_{m+j}$ remains unchanged near the origin, but tapers off further out; see Figure 4.2. The problem of minimizing $f$ remains in the composite form (4.2) with convex $h$ and smooth $F$, but is now nonconvex. Still, it retains much structure that can be utilized algorithmically. □

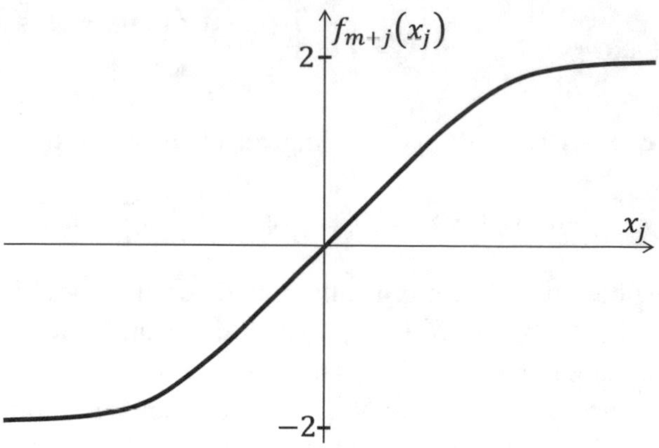


Fig. 4.2: Nonconvex regularization function.

**Example 4.6** (goal optimization). Consider a situation with several quantities of interest (cost, risk, damage, etc.). We would like to settle on a decision for which all the quantities are low. Suppose that $f_i : \mathbb{R}^n \to \mathbb{R}$ and $\tau_i \in \mathbb{R}$, $i = 1, \dots, m$, describe the quantities of interest and their associated goals (targets), respectively. We seek a decision $\bar{x}$ such that $f_i(\bar{x}) \leq \tau_i$ for all $i = 1, \dots, m$, but realize that our goals might be overly ambitious and impose a penalty $\theta_i \geq 0$ per unit shortfall in the $i$th quantity. This problem can be formulated in the composite form (4.2) resulting in a viable approach to *multi-objective optimization*.

**Detail.** Let $F(x) = (f_1(x), \dots, f_m(x))$ and

$$h(u) = \sum\nolimits_{i=1}^{m} \theta_i \max\{0, u_i - \tau_i\},$$

which defines a convex function by 1.13 and 1.18(b,d). Then, minimizing $f(x) = h(F(x))$ over $x \in \mathbb{R}^n$ leads to a decision that's as "close" as possible to satisfying all the goals. When $f_1, \dots, f_m$ are smooth, the composite function $f$ possesses significant structure albeit it may not be convex. □

**Exercise 4.7** (building layout). A rectangular warehouse, partially underground, is to be designed with three quantities of interest in mind: the total volume of the warehouse, the volume of the portion of the warehouse underground and the exposed area of the warehouse above ground, i.e., the area of the roof plus the area of the walls above ground. We would like the total volume to be high (ideally above 50000 cubic meters), the underground volume should be low (ideally below 38000 cubic meters), as this corresponds to the amount of excavation needed, and the exposed area should be low (preferably below 2250 square meters) because it's proportional to the heating and cooling cost. Formulate three different optimization models that determines $x = (x_1, x_2, x_3, x_4)$ by minimizing $h(F(x))$ subject to $x \geq 0$, where $x_1$ and $x_2$ are the length and width of the warehouse, respectively, $x_3$ and $x_4$ are the height below and above ground, respectively, and $F : \mathbb{R}^4 \to \mathbb{R}^3$ represents the three quantities of interest. The three models should have different $h$. Obtain solutions of the three models and compare the results.

Bringing forward structure is important theoretically and computationally, and allows us to make an assessment whether a formulation is suitable and attractive. One aspect that also enters in such an assessment is whether a solution exists. We can't expect an algorithm to find a minimizer if there's none. More precisely, for $f : \mathbb{R}^n \to \overline{\mathbb{R}}$, will the set of minimizers

$$\operatorname{argmin} f = \{x \in \operatorname{dom} f \mid f(x) = \inf f\} \neq \emptyset\ ?$$

When the problem is infeasible, i.e., $\operatorname{dom} f = \emptyset$, the answer is obviously no. If we're suspicious that a problem may be infeasible, then we would typically relax some constraints and/or replace hard requirements by penalties for violations as in 4.6. There's also no minimizer if a proper function is unbounded from below as for $f(x) = 1 - x$; see the left portion of Figure 4.3. In general, we say that the problem of minimizing $f$ is *unbounded* if $\inf f = -\infty$. Even if the function is bounded from below there might be no minimizer when it keeps on improving in a certain direction as illustrated in the middle portion of the figure with $f(x) = \exp(-x)$. The right portion of Figure 4.3 highlights a possibility where a discontinuity of $f$ causes the absence of minimizers: $f(x)$ steadily improves as $x$ moves from the right toward the discontinuity, but then jumps up. A rigorous treatment of this case requires us to bring in a new concept.

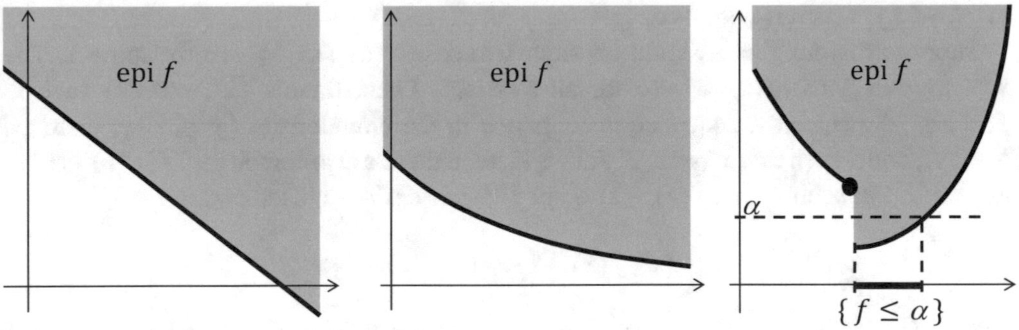


Fig. 4.3: Examples of functions for which argmin $f$ is empty. Left: $f$ unbounded. Middle: $f$ bounded from below. Right: $f$ discontinuous.

A function $f : \mathbb{R}^n \to \overline{\mathbb{R}}$ is *lower semicontinuous* (lsc) when epi $f$ is a closed set. This property is characterized by level-sets.

**Proposition 4.8** (lower semicontinuity). *For $f : \mathbb{R}^n \to \overline{\mathbb{R}}$, one has*

$$f \text{ is lsc} \iff \operatorname{epi} f \text{ is closed} \iff \forall \alpha \in \mathbb{R},\ \{f \leq \alpha\} \text{ is closed.}$$

**Proof.** The first equivalence is simply the definition of lsc. Next, suppose that epi $f$ is closed. Then, $\operatorname{epi} f \cap (\mathbb{R}^n \times \{\alpha\})$ is closed and this intersection specifies the same points in $\mathbb{R}^n$ as $\{f \leq \alpha\}$, which must also be closed. For the converse, let $\varepsilon > 0$ and $(x^\nu, \alpha^\nu) \in \operatorname{epi} f \to (x, \alpha)$. Then, there exists $\bar{\nu}$ such that for all $\nu \geq \bar{\nu}$, $\alpha^\nu \leq \alpha + \varepsilon$.

Consequently, $x^\nu \in \{f \le \alpha + \varepsilon\}$ for all such $\nu$. The level-sets of $f$ are closed, which implies that $f(x) \le \alpha + \varepsilon$. Since $\varepsilon$ is arbitrary, $f(x) \le \alpha$ and the conclusion follows. □

The source of difficulty in Figure 4.3(right) is that $f$ isn't lsc; epi $f$ isn't closed because some boundary points are "missing." Equivalently, $\{f \le \alpha\}$ in the figure is a halfopen interval and thus not closed. If the function value at the point of discontinuity (the black dot) had been on the right "branch," then the function would have been lsc and the point of discontinuity would have been the minimizer. From this we realize that $f$ being continuous implies that it's lsc, but the converse fails.

We say that a function $f : \mathbb{R}^n \to \overline{\mathbb{R}}$ is *level-bounded* if $\{f \le \alpha\}$ is bounded for all $\alpha \in \mathbb{R}$.

**Theorem 4.9** (existence of minimizer). *For a lsc and level-bounded function* $f : \mathbb{R}^n \to \overline{\mathbb{R}}$ *with nonempty epigraph,* argmin $f$ *is nonempty and compact.*

**Proof.** Suppose that $\inf f \in \mathbb{R}$. Then, there exists $\{x^\nu, \nu \in \mathbb{N}\}$ such that $f(x^\nu) \le \inf f + 1/\nu$. Since $\{x^\nu, \nu \in \mathbb{N}\}$ is contained in the bounded set $\{f \le \inf f + 1\}$, there exist $N \in \mathcal{N}_\infty^\#$ and $\bar{x}$ such that $x^\nu \to_N \bar{x}$. Thus,

$$(x^\nu, \inf f + 1/\nu) \in \operatorname{epi} f \to_N (\bar{x}, \inf f).$$

Since epi $f$ is closed (cf. 4.8), $(\bar{x}, \inf f) \in \operatorname{epi} f$ and $\bar{x} \in \operatorname{argmin} f$. Moreover, $\operatorname{argmin} f = \{f \le \inf f\}$, which is compact.

Suppose that $\inf f = -\infty$ and consider two cases: (a) $f(\bar{x}) = -\infty$ for some $\bar{x}$. Then, $\bar{x} \in \operatorname{argmin} f$. (b) $f(x) > -\infty$ for all $x \in \mathbb{R}^n$. Then, there's $\{x^\nu, \nu \in \mathbb{N}\}$ such that $f(x^\nu) \le -\nu$. Since the sequence is contained in the bounded set $\{f \le -1\}$, there exist $N \in \mathcal{N}_\infty^\#$ and $\bar{x}$ such that $x^\nu \to_N \bar{x}$. If $f(\bar{x}) \in \mathbb{R}$, then there's also $\bar{\nu}$ such that $f(x^\nu) \le f(\bar{x}) - 1$ for $\nu \ge \bar{\nu}$. For such $\nu$, $(x^\nu, f(\bar{x}) - 1) \in \operatorname{epi} f$. Since epi $f$ is closed and

$$\big(x^\nu, f(\bar{x}) - 1\big) \to_N \big(\bar{x}, f(\bar{x}) - 1\big),$$

one has $(\bar{x}, f(\bar{x}) - 1) \in \operatorname{epi} f$ as well, but this is a contradiction. If $f(\bar{x}) = \infty$, then we again reach a contradiction because $(x^\nu, 0) \in \operatorname{epi} f$ and epi $f$ is closed so that $(\bar{x}, 0) \in \operatorname{epi} f$. Thus, case (b) isn't possible. Consequently, $\operatorname{argmin} f = \{x \in \mathbb{R}^n \mid f(x) = -\infty\}$. Since this set is contained in $\{f \le 0\}$, it's bounded. It's also closed because

$$\big\{x \in \mathbb{R}^n \,\big|\, f(x) = -\infty\big\} = \bigcap\nolimits_{\alpha \in \mathbb{R}} \{f \le \alpha\}$$

and the fact that an intersection of closed sets is closed. □

Bounded level-sets are ensured if dom $f$ is bounded, which might be obvious due to bounds on the variables, or if $f(x) \to \infty$ whenever $\|x\|_2 \to \infty$ as in the beginning of 4.5.

**Exercise 4.10** (existence of minimizers). For $f : \mathbb{R} \to \overline{\mathbb{R}}$, (a) find an example of a lsc and level-bounded function $f$ with $\operatorname{argmin} f = \emptyset$; (b) find continuous $f$ with a local minimizer and $\operatorname{argmin} f = \emptyset$; and (c) find lsc $f$ with $\operatorname{argmin} f \ne \emptyset$ and unbounded $\{f \le \alpha\}$ for all $\alpha \in \mathbb{R}$.

## 4.B Network Design and Operation

A county needs to locate $n$ waste disposal sites that will service its $m$ towns. We can think of the towns as vertices in a network to which we would like to add $n$ vertices. The $m$ original vertices (towns) demand waste removal and the $n$ new vertices (sites) process the waste. We would like to locate the new vertices such that the cost of transporting the waste from the towns to the disposal sites is low, the weekly demand $d_i$ in town $i$ is (mostly) satisfied for all $i = 1, \ldots, m$, and the weekly processing capacity $c_j$ at site $j$ is preferably not exceeded for any $j = 1, \ldots, n$. Let

$$t^i = (t^i_1, t^i_2) \in \mathbb{R}^2 \quad \text{and} \quad x^j = (x^j_1, x^j_2) \in \mathbb{R}^2$$

be the coordinates for the location of town $i$ and site $j$, respectively. In addition to deciding $x = (x^1, \ldots, x^n)$, we also need to determine the amount of waste per week to be transported (flowed) from town $i$ to site $j$, which we denote by $w_{ij}$ and

$$w = (w_{ij}, i = 1, \ldots, m, j = 1, \ldots, n).$$

Since we're optimizing both the network itself and the flow in the network, this is a *network design and operation problem*.

As indicated, a decision needs to account for several quantities of interest. The first quantity is the cost of transportation, which we model as

$$f_0(x, w) = \sum\nolimits_{i=1}^{m} \sum\nolimits_{j=1}^{n} w_{ij} \|x^j - t^i\|_2.$$

Since $w_{ij}$ is the weekly amount of waste hauled from $i$ to $j$ and $\|x^j - t^i\|_2$ is the Euclidean distance between the two vertices, $f_0(x, w)$ is the total "ton-miles" hauled each week. Of course, this is just a surrogate for the actual cost, which involves considerations about half-full trucks and more accurate estimates of driving distances, but it may suffice in strategic planning for the county. A second group of quantities of interest is the amount of unfulfilled demand in the towns, i.e., demand minus waste hauled away:

$$f_i(w) = d_i - \sum\nolimits_{j=1}^{n} w_{ij}, \quad i = 1, \ldots, m.$$

The third group of quantities of interest is the amount of undercapacity at the sites, i.e., the received waste minus processing capacity:

$$f_{m+j}(w) = \sum\nolimits_{i=1}^{m} w_{ij} - c_j, \quad j = 1, \ldots, n.$$

With $X = \mathbb{R}^{2n} \times [0, \infty)^{mn}$ capturing the trivial fact that $w_{ij}$ is nonnegative, the problem can be formulated as

$$\underset{(x,w)\in X}{\text{minimize}} \; h_0\big(f_0(x, w)\big) + \sum\nolimits_{i=1}^{m} h_i\big(f_i(w)\big) + \sum\nolimits_{j=1}^{n} h_{m+j}\big(f_{m+j}(w)\big) \tag{4.3}$$

for some functions $h_0, \ldots, h_{m+n} : \mathbb{R} \to \overline{\mathbb{R}}$, which should reflect the relative importance of the quantities of interest. We may consider several possibilities. If $h_0(\alpha) = \alpha$ and $h_i(\alpha) = h_{m+j}(\alpha) = \max\{\gamma\alpha, \delta\alpha\}$, then we would minimize the total ton-miles plus penalties (rewards) for positive (negative) unfulfilled demand and undercapacity. The per-unit penalty $\delta \geq 0$ could be rather high to reflect the true cost of failing to satisfy demand and overloading the sites. The per-unit reward $\gamma \in [0, \delta)$ for being able to move more waste than needed and for leaving site capacity unused might very well be zero to reflect that there's no real benefit to such a situation. Alternatively, $h_i(\alpha) = h_{m+j}(\alpha) = \iota_{(-\infty,0]}(\alpha)$ results in the constraints

$$\sum\nolimits_{j=1}^{n} w_{ij} \geq d_i, \; i = 1, \ldots, m; \qquad \sum\nolimits_{i=1}^{m} w_{ij} \leq c_j, \; j = 1, \ldots, n.$$

In these and similar examples, $h_0, \ldots, h_{m+n}$ are convex. Since $f_1, \ldots, f_{m+n}$ are affine, the problem is highly structured with only $f_0$ causing some difficulty as it's neither convex nor smooth. However, $f_0(\cdot, w)$ is convex for any fixed $w \geq 0$ and $f_0(x, \cdot)$ is linear for any fixed $x$; cf. 1.15 and 1.18. How can we leverage these structural properties in computations?

One possibility is to adopt the ideas of the coordinate descent algorithm (§1.J) and alternate between fixing $w$ and minimizing with respect to $x$ and fixing $x$ and minimizing with respect to $w$. This requires the solution of only convex problems, but the assumptions for the coordinate descent algorithm don't hold and the approach would be a heuristic.

Another possibility is to remove the nonsmoothness in $f_0$ by reformulating the problem using more variables. Let the additional decision variable $z_{ij}$ represent $\|x^j - t^i\|_2$ and

$$z = (z_{ij}, i = 1, \ldots, m, j = 1, \ldots, n).$$

Since $w_{ij}$ is nonnegative, $f_0(x, w) \leq \alpha$ if and only if

$$\tilde{f}_0(w, z) = \sum\nolimits_{i=1}^{m} \sum\nolimits_{j=1}^{n} w_{ij} z_{ij} \leq \alpha \quad \text{and} \quad \|x^j - t^i\|_2 \leq z_{ij} \;\; \forall i, j.$$

When $h_0$ is nondecreasing, this leads to the reformulation

$$\begin{aligned} \underset{(x,w,z)\in X}{\text{minimize}} \; & h_0\big(\tilde{f}_0(w, z)\big) + \sum\nolimits_{i=1}^{m} \sum\nolimits_{j=1}^{n} \iota_{(-\infty,0]}\big(\|x^j - t^i\|_2 - z_{ij}\big) \\ & + \sum\nolimits_{i=1}^{m} h_i\big(f_i(w)\big) + \sum\nolimits_{j=1}^{n} h_{m+j}\big(f_{m+j}(w)\big), \end{aligned}$$

where now $X = \mathbb{R}^{2n} \times [0, \infty)^{mn} \times [0, \infty)^{mn}$. Since they simply implement the second-order cone constraints $\|x^j - t^i\|_2 \leq z_{ij}$ (cf. §2.J), the terms with indicator functions define a convex set. If $h_0$ is smooth and convex and $h_1, \ldots, h_{m+n}$ are convex, then the objective function is a composition of a convex function with a smooth function plus the sum of many convex functions. Although still nonconvex, the nonconvexity is only present in the first term, which anyhow is smooth and of a bilinear form. The only nonsmoothness is now due to indicator functions.

In general, there are often several equivalent models for a given problem, and even more possibilities if we permit approximations. The "best" choice in any particular situation depends on the available algorithms, the problem structure, the number of decision variables and many other factors. In the nonconvex case, it becomes especially important to experiment with various problem formulations to identify one that's more easily solved and provides better decisions. What constitute a "solution" is debatable and context dependent: Sometimes a local minimizer or a point satisfying an optimality condition is acceptable. Other times, a point with a relatively low objective function value is adopted as a solution.

Table 4.2: Data about towns in 4.11.

| town: | 1 | 2 | 3 | 4 | 5 |
|---|---|---|---|---|---|
| demand $d_i$ | 62 | 95 | 32 | 80 | 85 |
| coordinate $t_1^i$ | 0 | 4 | 30 | 20 | 16 |
| coordinate $t_2^i$ | 0 | 30 | 8 | 17 | 15 |

**Example 4.11** (random initiation of algorithms). The difficulty with the network design and operation problem and other nonconvex problems is that a (global) minimizer is much harder to obtain than in the convex case because the objective function has many "valleys." Random initiation of algorithms helps us to explore the space of decision variables more widely.

**Detail.** The gradient descent method (§1.F) and similar algorithms guarantee nonincreasing function values across their iterations, i.e., $f(x^{\nu+1}) \leq f(x^\nu)$ for all $\nu$, as they minimize some function $f$. When $f$ is nonconvex, this has the consequence that the points $\{x^\nu, \nu \in \mathbb{N}\}$ tend to remain near the initial point $x^0$. Visualizing the graph of $f$ as a landscape with many valleys, this means that $x^\nu$ is often in the same valley as $x^0$ and thus the algorithm fails to explore other valleys with potentially lower function values. The situation remains similar in the presence of constraints. An approach to explore the feasible set more broadly is to view the initial point $x^0$ in an algorithm as a random vector to which we assign a probability distribution with support that contains (parts of) the space of decision variables. One can then generate a sample according to the distribution and obtain a set of initial points. From each one, we can start the algorithm, let it run until a stopping criterion is satisfied and collect the solution. From the resulting set of candidate solutions we can select the most promising one. While a heuristic in general, the approach is simple to implement and works well if the "deep valleys" are also quite wide.

Suppose that $n = 2$, $m = 5$, $c_1 = 220$, $c_2 = 180$, $h_0(\alpha) = \alpha$ and $h_i(\alpha) = \iota_{(-\infty,0]}(\alpha)$, $i = 1, \dots, m+n$, in (4.3). Table 4.2 gives additional data; see also Figure 4.4. We randomly select a point in $X$ and use it as an initial point for an algorithm[2] applied to (4.3). Repeating this process ten times, we obtain ten candidate solutions that all satisfy the demand and capacity constraints but have total ton-miles varying between 2543 and 3760. Among these, the solution with the lowest total ton-miles places the larger site at coordinates

[2] Algorithms for nonconvex problems with constraints are covered later in this chapter and in Chaps. 5 and 6.

(16, 15) and the smaller at (4, 30). The worse solution among the ten specifies the larger site at (20, 17) and the smaller at (16, 15). Figure 4.4 illustrates these solutions with red and blue triangles, respectively. We could repeat this process for many more initial points and possibly obtain a solution with even lower total ton-miles. □

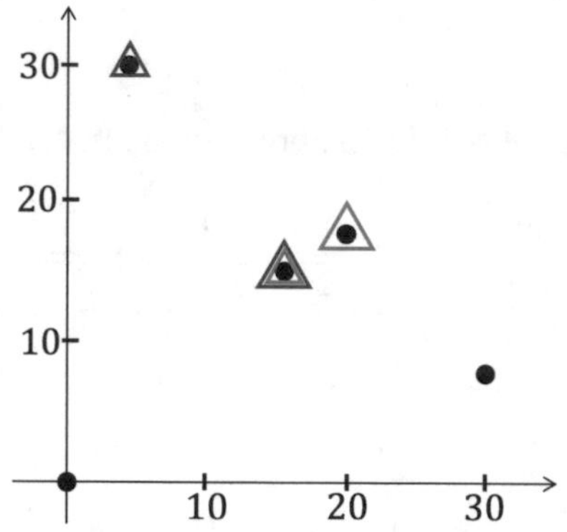


Fig. 4.4: Location of towns (black dots), best candidate solution (red triangles) and worst candidate solution (blue triangles).

## 4.C Epigraphical Approximation Algorithm

An effective approach for developing new optimization algorithms is to leverage existing ones. At a conceptual level, an algorithm for the problem at hand may consist of two steps: (a) construct an approximating problem of a type that's tractable for an existing algorithm and (b) solve the approximating problem using the existing algorithm. The fact that the new algorithm may only produce an approximating solution isn't a major concern as long as the level of accuracy can be controlled and be made arbitrarily high. In fact, almost everything in computational mathematics is carried out approximately, including representation of real numbers and evaluation of logarithms.

In a minimization setting, we think of a function $g$ as approximating another function $f$ if $\inf g$ is close to $\inf f$ and $\operatorname{argmin} g$ is somehow close to $\operatorname{argmin} f$. In what sense should $g$ be near $f$ to ensure these properties? We'll develop an approximation theory for minimization problems that answers this question and leads to new algorithms.

**Example 4.12** (penalization of constraints). For $f_0, f_1 : \mathbb{R}^n \to \mathbb{R}$, suppose that we're faced with the problem

$$\underset{x\in\mathbb{R}^n}{\text{minimize}}\; f(x) = f_0(x) + \iota_{\{0\}}\big(f_1(x)\big).$$

We saw in §4.A that penalties can be attractive alternatives to constraints for purely modeling reasons. Let's now utilize the same idea in an effort to construct an algorithm for this constrained problem. For $\theta^\nu \in (0,\infty)$, the alternative problems

$$\Big\{ \underset{x\in\mathbb{R}^n}{\text{minimize}}\, f^\nu(x) = f_0(x) + \theta^\nu\big(f_1(x)\big)^2, \quad \nu \in \mathbb{N} \Big\}$$

are unconstrained and also smooth when $f_0$ and $f_1$ are smooth. Thus, algorithms from Chap. 1 apply. We can be hopeful that the alternative functions $f^\nu$ approximate appropriately the actual function $f$ because

$$\theta^\nu\big(f_1(x)\big)^2 = \iota_{\{0\}}\big(f_1(x)\big) = 0 \quad \text{when } f_1(x) = 0$$

and otherwise $\theta^\nu(f_1(x))^2$ is large for large $\theta^\nu$ and thus "close" to $\iota_{\{0\}}(f_1(x)) = \infty$. Still, are we sure that the penalty term in $f^\nu$ outweighs any low value of $f_0$ at points violation $f_1(x) = 0$?

**Detail.** For a simple instance, let $f_0(x) = (x+1)^2$ and $f_1(x) = x$, $x \in \mathbb{R}$, so that $\operatorname{argmin} f = \{0\}$ and $\inf f = 1$. By the optimality condition 2.19,

$$\operatorname{argmin} f^\nu = \left\{\frac{-1}{1+\theta^\nu}\right\} \quad \text{and} \quad \inf f^\nu = \frac{\theta^\nu}{1+\theta^\nu}.$$

Although no finite $\theta^\nu$ is large enough to recover $\operatorname{argmin} f$ and $\inf f$ exactly, these quantities are approached as $\theta^\nu \to \infty$. The subsequent development achieves similar convergence in much more general settings. □

Let's examine in more detail what it takes for an alternative function to approximate an actual function in the sense that matters for minimization problems. One might think that the notions of pointwise and uniform convergence provide the answer.

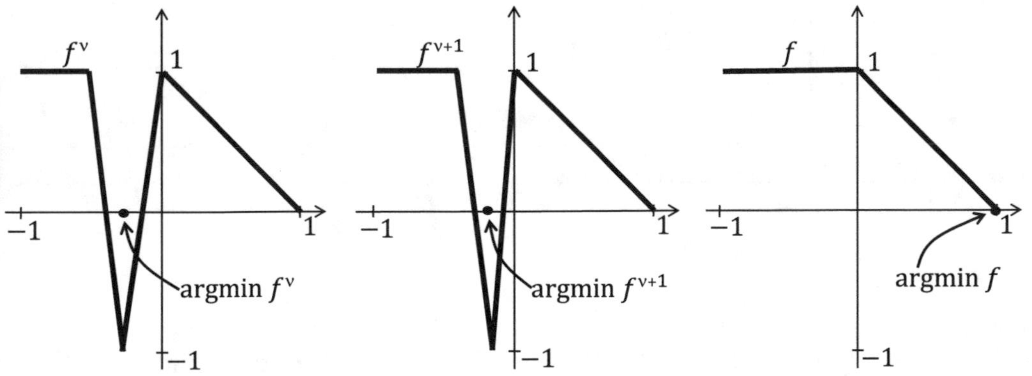


Fig. 4.5: Let $f^\nu(x) = \min\{1-x, 1, 2\nu|x+1/\nu|-1\}$ and $f(x) = \min\{1-x, 1\}$ if $x \in [-1,1]$, but otherwise they're $\infty$. While $f^\nu$ converges pointwise to $f$, $\operatorname{argmin} f^\nu = \{-1/\nu\}$ and $\inf f^\nu = -1$ fail to converge to $\operatorname{argmin} f = \{1\}$ and $\inf f = 0$, respectively.

**Pointwise and Uniform Convergence.** We say that $f^\nu : \mathbb{R}^n \to \overline{\mathbb{R}}$ *converges pointwise* to $f : \mathbb{R}^n \to \overline{\mathbb{R}}$ when

$$f^\nu(x) \to f(x) \quad \forall x \in \mathbb{R}^n.$$

This is indeed the case in 4.12. However, pointwise convergence is generally not enough to ensure convergence of minima and minimizers. Figure 4.5 furnishes a counterexample.

We say that $f^\nu : \mathbb{R}^n \to \overline{\mathbb{R}}$ *converges uniformly* to $f : \mathbb{R}^n \to \overline{\mathbb{R}}$ on $C \subset \mathbb{R}^n$ if

$$\sup_{x\in C} |f^\nu(x) - f(x)| \to 0.$$

If $C = \mathbb{R}^n$, then we say that $f^\nu$ converges uniformly to $f$. A difficulty here is the unboundedness of many functions. In 4.12, $|f^\nu(x) - f(x)| = \infty$ when $f_1(x) \neq 0$ and uniform convergence fails on any set $C$ that contains such points. Since minima and minimizers converge in 4.12, uniform convergence emerges as an unnecessarily strong condition that isn't fully needed. For the structured problem of minimizing a linear function subject to an inequality given by a smooth constraint function, uniform convergence of approximating objective and constraint functions to their actual counterparts isn't sufficient to guarantee convergence of minima and minimizers; see Figure 4.6.

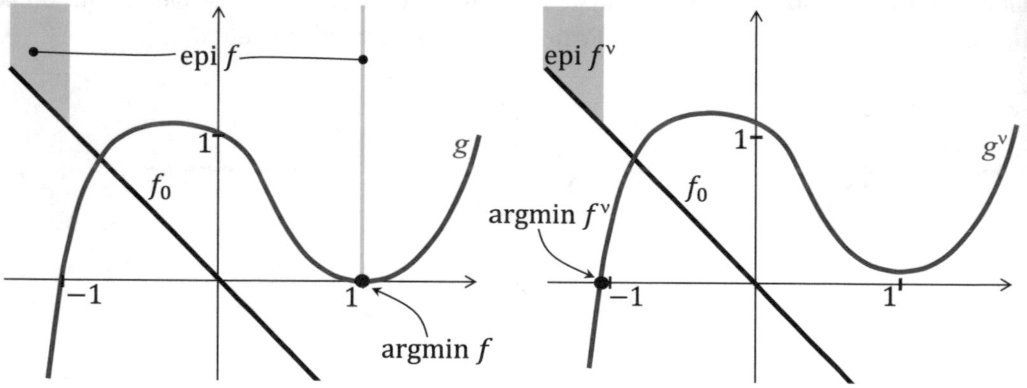


Fig. 4.6: Let $f_0(x) = -x$, $g(x) = x^3 - x^2 - x + 1$, $f(x) = f_0(x) + \iota_{(-\infty,0]}(g(x))$ and $f^\nu$ be defined similarly with $g$ replaced by $g^\nu = g + 1/\nu$. Then, $g^\nu$ converges uniformly to $g$, but minimum and minimizer of $f^\nu$ aren't close to those of $f$.

The prior discussion reveals that neither pointwise nor uniform convergence of functions furnishes a suitable condition that guarantees convergence of minima and minimizers. Convergence of epigraphs, however, provides a robust notion of approximation in tune with the needs of minimization problems. As will emerge gradually, if epi $f^\nu$ approaches epi $f$, then argmin $f^\nu$ and inf $f^\nu$ appropriately converge to argmin $f$ and inf $f$, respectively.

To make precise the meaning of one epigraph approaching another, we define the *point-to-set distance* between $\bar{x} \in \mathbb{R}^n$ and $C \subset \mathbb{R}^n$ as

$$\operatorname{dist}(\bar{x}, C) = \inf_{x \in C} \|x - \bar{x}\|_2 \quad \text{when} \quad C \neq \emptyset \quad \text{and} \quad \operatorname{dist}(\bar{x}, \emptyset) = \infty. \tag{4.4}$$

**Definition 4.13** (epi-convergence). For functions $f, f^\nu : \mathbb{R}^n \to \overline{\mathbb{R}}$, $f^\nu$ *epi-converges* to $f$, written $f^\nu \xrightarrow{e} f$, when

$$\operatorname{epi} f \text{ is closed} \quad \text{and} \quad \operatorname{dist}(z, \operatorname{epi} f^\nu) \to \operatorname{dist}(z, \operatorname{epi} f) \;\; \forall z \in \mathbb{R}^{n+1}.$$

Figure 4.7 visualizes the situation in 4.12: $f^\nu$ epi-converges to $f$, and minima and minimizers converge. In Figure 4.6, epi $f^\nu$ remains away from epi $f$, there's no epi-convergence and minima and minimizers fail to converge. In general, when approximating functions epi-converge to an actual function, an algorithm for minimizing the latter is immediately available, at least conceptually.

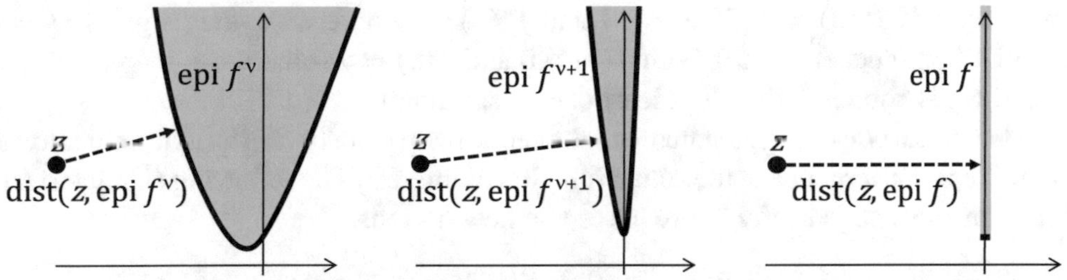


Fig. 4.7: The epigraph of $f^\nu$ approaches epi $f = \{0\} \times [1, \infty)$ in 4.12.

For $f : \mathbb{R}^n \to \overline{\mathbb{R}}$ and $\varepsilon \geq 0$, we denote the set of *near-minimizers* of $f$ by

$$\varepsilon\text{-argmin} f = \big\{ x \in \operatorname{dom} f \;\big|\; f(x) \leq \inf f + \varepsilon \big\}.$$

**Epigraphical Approximation Algorithm.**

Data. Functions $f^\nu : \mathbb{R}^n \to \overline{\mathbb{R}}$ and tolerances $\varepsilon^\nu \geq 0$, $\nu \in \mathbb{N}$.
Step 0. Set $\nu = 1$.
Step 1. Compute

$$x^\nu \in \varepsilon^\nu\text{-argmin} f^\nu.$$

Step 2. Replace $\nu$ by $\nu + 1$ and go to Step 1.

**Convergence 4.14** (epigraphical approximation algorithm). *Suppose that* $\{x^\nu, \nu \in \mathbb{N}\}$ *is generated by the epigraphical approximation algorithm,* $f^\nu$ *in the algorithm epi-converges to a proper function* $f : \mathbb{R}^n \to \overline{\mathbb{R}}$ *and* $\varepsilon^\nu \to 0$.

*Then, every cluster point of* $\{x^\nu, \nu \in \mathbb{N}\}$ *is a minimizer of* $f$ *and the corresponding* $f^\nu(x^\nu)$ *tends to* $\inf f$, *i.e., if there's* $N \in \mathcal{N}_\infty^\#$ *and* $\bar{x}$ *such that* $x^\nu \to_N \bar{x}$, *then*

$$\bar{x} \in \operatorname{argmin} f \quad \textit{and} \quad f^\nu(x^\nu) \to_N \inf f.$$

We omit the proof as a more general result emerges in 5.5. The assertion about function values shows that $f^\nu(x^\nu)$, presumably computed in Step 1 of the algorithm, furnishes an estimate of the minimum value of the actual problem of interest, at least for $\nu \in N$; from §1.F, recall that "$\to_N$" indicates the consideration of only indices $\nu$ in the subsequence $N$.

The epigraphical approximation algorithm provides a roadmap toward a wide variety of computational methods for minimizing $f$ by means of minimizing approximating functions $f^\nu$. However, numerous implementation details need to be settled such as how to solve the approximating problems and with what tolerance. One might also want to leverage the fact that the selected algorithm for the approximating problems solves a sequence of presumably similar problems so that Step 1 can be *warm-started* with quantities (candidate solution, step size, etc.) calculated in earlier iterations.

We note that $f^\nu \xrightarrow{e} f$ doesn't automatically imply that $\{x^\nu, \nu \in \mathbb{N}\}$ produced by the epigraphical approximation algorithm has a cluster point even when $f$ has a minimizer. For example, if $f^\nu(0) = 0$, $f^\nu(\nu) = -1$ and $f^\nu(x) = \infty$ otherwise, then $\operatorname{argmin} f^\nu = \{\nu\}$ but still $f^\nu$ epi-converges to $f$ with $f(0) = 0$ and $f(x) = \infty$ otherwise. In general, when $\{x^\nu, \nu \in \mathbb{N}\}$ is bounded, there's at least one cluster point.

In view of the convergence statement, the central step in justifying an approximation is to establish epi-convergence to the actual function of interest. The following characterization of epi-convergence is useful but requires two new notions.

**Lower and Upper Limits.** The lower limit of a sequence $\{\alpha^\nu \in \overline{\mathbb{R}}, \nu \in \mathbb{N}\}$ is given as

$$\operatorname{liminf} \alpha^\nu = \lim_{\nu \to \infty} \big( \inf_{k \ge \nu} \alpha^k \big).$$

For example, if $\alpha^\nu = 1/\nu$ when $\nu$ is odd and $\alpha^\nu = 1 - 1/\nu$ otherwise, then the sequence has no limit but $\operatorname{liminf} \alpha^\nu = 0$. Generally, if $\{\alpha^\nu \in \overline{\mathbb{R}}, \nu \in \mathbb{N}\}$ is bounded from below and has a nonempty set of cluster points $C \subset \mathbb{R}$, then $\operatorname{liminf} \alpha^\nu = \inf C$. Similarly, the upper limit of $\{\alpha^\nu \in \overline{\mathbb{R}}, \nu \in \mathbb{N}\}$ is

$$\operatorname{limsup} \alpha^\nu = \lim_{\nu \to \infty} \Big( \sup_{k \ge \nu} \alpha^k \Big).$$

In the odd-even example, $\operatorname{limsup} \alpha^\nu = 1$, which is a cluster point of the sequence. Generally, if $\{\alpha^\nu \in \overline{\mathbb{R}}, \nu \in \mathbb{N}\}$ is bounded from above and has a nonempty set of cluster points $C \subset \mathbb{R}$, then $\operatorname{limsup} \alpha^\nu = \sup C$. In contrast to the limit, the lower limit and upper limit of a sequence are always defined, but could be $\infty$ or $-\infty$. Certainly, we've that $\operatorname{liminf} \alpha^\nu \le \operatorname{limsup} \alpha^\nu$. The two are equal and finite if and only if $\{\alpha^\nu \in \overline{\mathbb{R}}, \nu \in \mathbb{N}\}$ has a limit, in which case $\operatorname{liminf} \alpha^\nu = \operatorname{limsup} \alpha^\nu = \lim \alpha^\nu$.

Occasionally, we abbreviate "*there exist(s)*" by the symbol ∃.

**Theorem 4.15** (characterization of epi-convergence). *For* $f, f^\nu : \mathbb{R}^n \to \overline{\mathbb{R}}$, $f^\nu \xrightarrow{e} f$ *if and only if*

(a) $\forall x^\nu \to x$, $\liminf f^\nu(x^\nu) \geq f(x)$
(b) $\forall x$, $\exists x^\nu \to x$ *such that* $\limsup f^\nu(x^\nu) \leq f(x)$.

The theorem, proven in §4.E, characterizes epi-convergence in two parts. For (a), $f^\nu$ must be "high enough" regardless of how we approach an arbitrary point $x$. For (b), there must be a "path" to $x$ along which $f^\nu$ is "low enough." Figure 4.8 illustrates this in detail with

$$f(x) = \begin{cases} -1 & \text{if } x \leq 0 \\ 1 & \text{otherwise} \end{cases} \qquad f^\nu(x) = \begin{cases} -1 & \text{if } x \leq -1/\nu \\ -\sqrt{-\nu x} & \text{if } x \in (-1/\nu, 0] \\ \sqrt{\nu x} & \text{if } x \in (0, 1/\nu] \\ 1 & \text{otherwise.} \end{cases}$$

To verify epi-convergence, let's first consider (a) in the theorem. For any $x^\nu \to x \leq 0$, $f^\nu(x^\nu) \geq -1 = f(x)$ for all $\nu$. For any $x^\nu \to x > 0$, $f^\nu(x^\nu) = 1 = f(x)$ for sufficiently large $\nu$. Thus: (a) holds. Second, let's consider (b). For $x \neq 0$, we can take $x^\nu = x$ because $f^\nu(x^\nu) = f^\nu(x) \to f(x)$. It only remains to show that for some $x^\nu \to 0$, $\limsup f^\nu(x^\nu) \leq f(0) = -1$. Here, we need to be careful: $x^\nu$ too close to 0 has $f^\nu(x^\nu)$ much above $-1$. However, $x^\nu = -1/\nu$ results in $f^\nu(x^\nu) = -1$ for all $\nu$ and we still have $x^\nu \to 0$ so (b) holds.

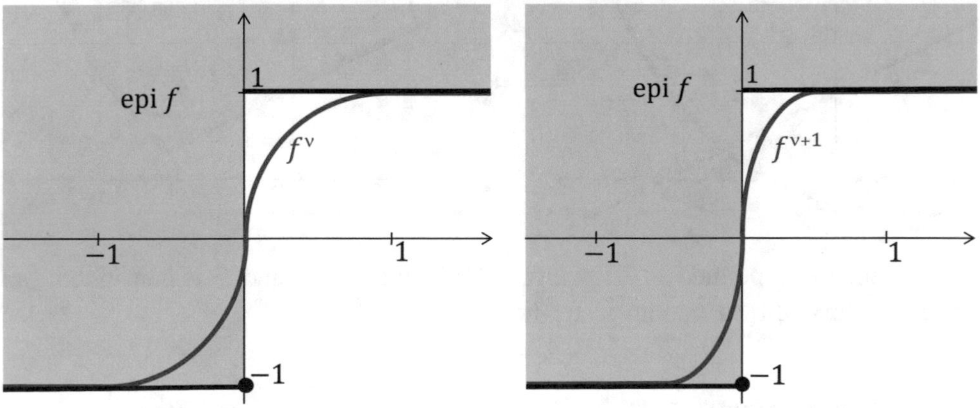


Fig. 4.8: Functions $f^\nu$ (red) epi-converge to $f$ (black).

**Example 4.16** (smoothing of max-function). For a smooth mapping $F : \mathbb{R}^n \to \mathbb{R}^m$, the composite function given by

$$f(x) = h\big(F(x)\big), \quad \text{where } h(u) = \max_{i=1,\dots,m} u_i$$

tends to be nonsmooth; see Figure 4.9. An approach for minimizing $f$ that leverages the structure is based on smooth approximations.

**Detail.** There are several possible approximating functions, but we concentrate on $f^\nu : \mathbb{R}^n \to \mathbb{R}$ defined by

$$f^\nu(x) = \frac{1}{\theta^\nu} \ln\Big(\sum_{i=1}^m \exp\big(\theta^\nu f_i(x)\big)\Big),$$

where $\theta^\nu \in (0, \infty)$ and $f_1, \dots, f_m$ are the components of $F$. It's apparent that $f^\nu$ is smooth and, in fact,

$$\nabla f^\nu(x) = \sum_{i=1}^m \mu_i^\nu(x) \nabla f_i(x), \quad \text{with } \mu_i^\nu(x) = \frac{\exp\big(\theta^\nu(f_i(x) - f(x))\big)}{\sum_{k=1}^m \exp\big(\theta^\nu(f_k(x) - f(x))\big)}.$$

If $f_i, i = 1, \dots, m$, are twice smooth, then $f^\nu$ is also twice smooth with

$$\nabla^2 f^\nu(x) = \sum_{i=1}^m \mu_i^\nu(x) \nabla^2 f_i(x) + \theta^\nu \sum_{i=1}^m \mu_i^\nu(x) \nabla f_i(x) \nabla f_i(x)^\top$$
$$- \theta^\nu \Big(\sum_{i=1}^m \mu_i^\nu(x) \nabla f_i(x)\Big)\Big(\sum_{i=1}^m \mu_i^\nu(x) \nabla f_i(x)\Big)^\top.$$

If $f_i, i = 1, \dots, m$, are convex, then $f$ is convex by 1.13 and this property carries over to $f^\nu$ as well. (One can show that $\nabla^2 f^\nu(x)$ is positive semidefinite when $\nabla^2 f_i(x)$ has that property for all $i = 1, \dots, m$; cf. 1.23 and [18, Example 3.14].)

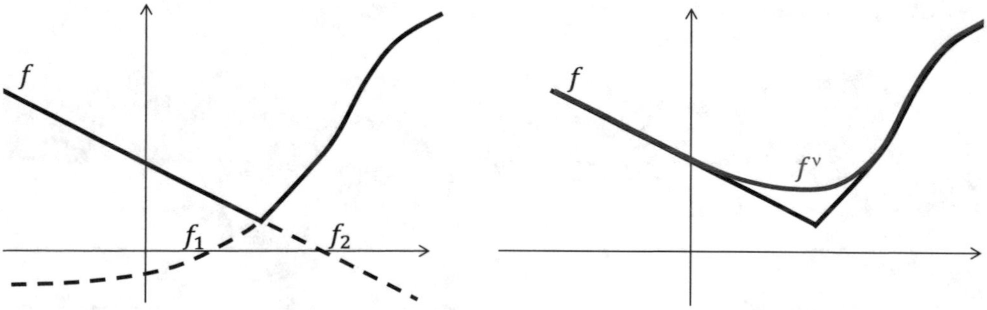


Fig. 4.9: Function $f$ defined as the pointwise maximum of $f_1$ and $f_2$ is nonsmooth (left), but is approximated by a smooth $f^\nu$ (right).

The alternative expression

$$f^\nu(x) = f(x) + \frac{1}{\theta^\nu} \ln\Big(\sum_{i=1}^m \exp\Big(\theta^\nu\big(f_i(x) - f(x)\big)\Big)\Big)$$

is useful in implementation (to avoid overflow when computing the exponential function) and also leads to the bounds

$$0 \le f^\nu(x) - f(x) \le \frac{\ln m}{\theta^\nu} \quad \forall x \in \mathbb{R}^n.$$

This implies epi $f^\nu \subset$ epi $f$; see Figure 4.9. Moreover, for any $x^\nu \to x$,

$$\liminf f^\nu(x^\nu) \ge \liminf f(x^\nu) = f(x)$$

because $f$ is continuous so 4.15(a) holds. Given $x \in \mathbb{R}^n$, set $x^\nu = x$. Then,

$$\limsup f^\nu(x^\nu) = \limsup f^\nu(x) \le \limsup \left( f(x) + \frac{\ln m}{\theta^\nu} \right) = f(x)$$

and 4.15(b) holds too when $\theta^\nu \to \infty$. We've established that $f^\nu \xrightarrow{e} f$ and the approximation is justified provided that $\theta^\nu \to \infty$. The implementation of the epigraphical approximation algorithm in this case typically benefits from a gradual increase in $\theta^\nu$ as ill-conditioning tends to increase with $\theta^\nu$ and causes slower rate of convergence in Newton's method or any other algorithm used to minimize $f^\nu$; see [77, 85]. □

**Exercise 4.17** (smoothing in neural networks). The functions $\varphi_k : \mathbb{R}^{d_k} \to \mathbb{R}^{d_k}$ in 4.1 are often of the ReLU kind:

$$\varphi_k(x) = \big( \max\{0, x_1\}, \dots, \max\{0, x_{d_k}\} \big) \quad \text{for} \quad x = (x_1, \dots, x_{d_k}).$$

Consider the component function $\alpha \mapsto \max\{0, \alpha\}$ and develop a smooth approximation for it using the principle laid out in 4.16. Give expressions for approximation error, derivatives and second-order derivatives.

For convex functions, epi-convergence reduces to pointwise convergence when the limiting function is real-valued.

**Proposition 4.18** (epi-convergence for convex functions). *For convex functions $f, f^\nu : \mathbb{R}^n \to \overline{\mathbb{R}}$ with $f$ real-valued, one has*

$$f^\nu \xrightarrow{e} f \iff f^\nu(x) \to f(x) \ \forall x \in \mathbb{R}^n.$$

*Under that circumstance, $-f^\nu \xrightarrow{e} -f$ and $\sup_{x \in B} |f^\nu(x) - f(x)| \to 0$ for any compact $B \subset \mathbb{R}^n$ as well.*

**Proof.** A good exercise or refer to [105, Theorem 7.17], which includes refinements. □

For composite functions, epi-convergence can often be established by checking the properties of the various components of the functions. This might offer an easier path than working directly with 4.15.

**Proposition 4.19** (epi-convergence of sums). *For $f, f^\nu, g, g^\nu : \mathbb{R}^n \to \overline{\mathbb{R}}$, suppose that $f^\nu \xrightarrow{e} f$, $g^\nu \xrightarrow{e} g$ and $f(x), g(x) > -\infty$ for all $x \in \mathbb{R}^n$. Then,*

$$f^\nu + g^\nu \xrightarrow{e} f + g$$

*under either one of the following two conditions:*

(a) $-g^\nu \xrightarrow{e} -g$.

(b) $g^\nu(x) \to g(x)$ *and* $f^\nu(x) \to f(x)$ *for all* $x \in \mathbb{R}^n$.

**Proof.** For sequences $\{\alpha^\nu, \beta^\nu \in \overline{\mathbb{R}}, \nu \in \mathbb{N}\}$, one has

$$\liminf(\alpha^\nu + \beta^\nu) \geq \liminf \alpha^\nu + \liminf \beta^\nu$$

provided that the right-hand side isn't $\infty - \infty$ or $-\infty + \infty$. Similarly,

$$\limsup(\alpha^\nu + \beta^\nu) \leq \limsup \alpha^\nu + \limsup \beta^\nu$$

and this holds without exceptions because $\infty - \infty = \infty$ or $-\infty + \infty = \infty$ by convention.

Let $x^\nu \to x$. Then,

$$\liminf f^\nu(x^\nu) \geq f(x) > -\infty \quad \text{and} \quad \liminf g^\nu(x^\nu) \geq g(x) > -\infty$$

by 4.15(a). This implies that

$$\liminf \big(f^\nu(x^\nu) + g^\nu(x^\nu)\big) \geq \liminf f^\nu(x^\nu) + \liminf g^\nu(x^\nu) \geq f(x) + g(x)$$

and 4.15(a) holds for the sum functions without either condition (a) or (b). Given $x \in \mathbb{R}^n$, there exists $x^\nu \to x$ such that $\limsup f^\nu(x^\nu) \leq f(x)$. Consider (a). Since $-g^\nu \xrightarrow{e} -g$ and $g^\nu \xrightarrow{e} g$, one has $g^\nu(x^\nu) \to g(x)$ and then also

$$\limsup \big(f^\nu(x^\nu) + g^\nu(x^\nu)\big) \leq \limsup f^\nu(x^\nu) + \limsup g^\nu(x^\nu) \leq f(x) + g(x).$$

Thus, 4.15(b) holds for the sum functions. Under condition (b), set $x^\nu = x$ and 4.15(b) holds again. □

We note that having $g^\nu \xrightarrow{e} g$ and $-g^\nu \xrightarrow{e} -g$ is equivalent to having $g^\nu(x^\nu) \to g(x)$ for all $x^\nu \to x$. Thus, condition (a) of the proposition holds, for example, when $g^\nu = g$ and $g$ is continuous. For further connections between pointwise convergence (as in condition (b)) and epi-convergence, we refer to [105, Section 7.B] and [31].

It's convenient to denote by $h \circ F$ a composite function $f$ constructed from $h : \mathbb{R}^m \to \overline{\mathbb{R}}$ and $F : \mathbb{R}^n \to \mathbb{R}^m$ by setting $f(x) = h(F(x))$, i.e.,

$$(h \circ F)(x) = h\big(F(x)\big).$$

We adopt this notation also when $h : \overline{\mathbb{R}} \to \overline{\mathbb{R}}$ and $F : \mathbb{R}^n \to \overline{\mathbb{R}}$.

**Proposition 4.20** (epi-convergence of composite functions). *For* $h, h^\nu : \mathbb{R} \to \overline{\mathbb{R}}$ *and* $f, f^\nu : \mathbb{R}^n \to \overline{\mathbb{R}}$, *the following hold:*

(a) *If* $h^\nu \xrightarrow{e} h$, $h^\nu(\alpha) \to h(\alpha)$ *for all* $\alpha \in \mathbb{R}$ *and* $f$ *is real-valued and continuous, then*

$$h^\nu \circ f \xrightarrow{e} h \circ f.$$

(b) *If* $f^\nu \xrightarrow{e} f$ *and* $h$ *is continuous, nondecreasing and extended to* $\overline{\mathbb{R}}$ *with the conventions* $h(-\infty) = \inf_{\alpha \in \mathbb{R}} h(\alpha)$ *and* $h(\infty) = \sup_{\alpha \in \mathbb{R}} h(\alpha)$, *then*

$$h \circ f^\nu \xrightarrow{e} h \circ f.$$

**Proof.** For (a), let $x^\nu \to x$, which ensures that $f(x^\nu) \to f(x)$. Since $\liminf h^\nu(\alpha^\nu) \geq h(\alpha)$ whenever $\alpha^\nu \to \alpha$ by 4.15(a), one has

$$\liminf h^\nu\big(f(x^\nu)\big) \geq h\big(f(x)\big).$$

Given $x \in \mathbb{R}^n$, set $x^\nu = x$. Then,

$$\limsup h^\nu\big(f(x^\nu)\big) = \limsup h^\nu\big(f(x)\big) = h\big(f(x)\big).$$

We've established both conditions of 4.15 and (a) has been proven.

For (b), let $x^\nu \to x$, which implies that $\liminf f^\nu(x^\nu) \geq f(x)$ by 4.15(a). Fix $\nu$ and let $\varepsilon > 0$. Suppose that

$$\alpha^\nu = \inf_{k \geq \nu} h\big(f^k(x^k)\big) \in \mathbb{R}.$$

Then, there exists $k^* \geq \nu$ such that

$$\alpha^\nu \geq h\big(f^{k^*}(x^{k^*})\big) - \varepsilon \geq h\big(\inf_{k \geq \nu} f^k(x^k)\big) - \varepsilon.$$

The last inequality holds because $h$ is nondecreasing. Since $\varepsilon$ is arbitrary,

$$\alpha^\nu \geq h\big(\inf_{k \geq \nu} f^k(x^k)\big).$$

A similar argument leads to the same inequality if $\alpha^\nu = -\infty$ and, trivially, also when $\alpha^\nu = \infty$. Since the inequality holds for all $\nu$, it follows by the continuity of $h$ that

$$\begin{aligned}\liminf h\big(f^\nu(x^\nu)\big) &= \lim_{\nu \to \infty} \Big(\inf_{k \geq \nu} h\big(f^k(x^k)\big)\Big) \geq \lim_{\nu \to \infty} h\big(\inf_{k \geq \nu} f^k(x^k)\big) \\ &= h\Big(\lim_{\nu \to \infty} \big(\inf_{k \geq \nu} f^k(x^k)\big)\Big) = h\big(\liminf f^\nu(x^\nu)\big) \geq h\big(f(x)\big).\end{aligned}$$

Given $x \in \mathbb{R}^n$, there exists $x^\nu \to x$ such that $f^\nu(x^\nu) \to f(x)$ by 4.15. Since $h$ is continuous, this implies $h(f^\nu(x^\nu)) \to h(f(x))$ and in view of 4.15 we've established (b). □

**Example 4.21** (penalty method for constrained optimization). We can now formalize and extend the ideas introduced in 4.12. Let's consider

$$\begin{aligned}\operatorname*{minimize}_{x \in X \subset \mathbb{R}^n} f_0(x) \quad \text{subject to} \quad & f_i(x) = 0, \; i = 1, \dots, m \\ & g_i(x) \leq 0, \; i = 1, \dots, q.\end{aligned}$$

The goal is to construct an approximation that can be solved by an existing algorithm. The next step emerges as we reformulate the problem in terms of a composite function $f : \mathbb{R}^n \to \overline{\mathbb{R}}$:

$$\operatorname*{minimize}_{x \in \mathbb{R}^n} f(x) = \iota_X(x) + f_0(x) + \sum\nolimits_{i=1}^m \iota_{\{0\}}\big(f_i(x)\big) + \sum\nolimits_{i=1}^q \iota_{(-\infty,0]}\big(g_i(x)\big).$$

The difficulty is caused by the indicator functions, but they can be approximated by continuous real-valued functions. As already suggested in 4.12, $\iota_{\{0\}}$ can be approximated by $\alpha \mapsto \theta^\nu \alpha^2$, with $\theta^\nu \in (0, \infty)$. Similarly, $\iota_{(-\infty,0]}$ is approximated by $\alpha \mapsto \theta^\nu(\max\{0, \alpha\})^2$ and this leads to the approximating problem

$$\underset{x\in\mathbb{R}^n}{\text{minimize}}\, f^\nu(x) = \iota_X(x) + f_0(x) + \sum_{i=1}^m \theta^\nu\big(f_i(x)\big)^2 + \sum_{i=1}^q \theta^\nu\big(\max\{0, g_i(x)\}\big)^2.$$

If $f_0, f_1, \dots, f_m$ and $g_1, \dots, g_q$ are smooth functions and $X = \mathbb{R}^n$ or $X$ is relatively simple so that projections can be carried efficiently, then methods from Chaps. 1 and 2 can be brought in productively. In fact, if $X = \mathbb{R}^n$, then $f^\nu$ is smooth. We can show that $f^\nu \xrightarrow{e} f$ as long as $\theta^\nu \to \infty$ so the epigraphical approximation algorithm is justified, which then is called the *penalty method*.

**Detail.** Let's establish that $f^\nu \xrightarrow{e} f$ under the relaxed assumption that $X$ is closed, $f_0$ is proper and lsc and $f_1, \dots, f_m$ as well as $g_1, \dots, g_q$ are real-valued and continuous. Let

$$h_1^\nu(\alpha) = \theta^\nu \alpha^2 \quad \text{and} \quad h_2^\nu(\alpha) = \theta^\nu\big(\max\{0, \alpha\}\big)^2.$$

Using 4.15, we can show that $h_1^\nu \xrightarrow{e} \iota_{\{0\}}$ and $h_2^\nu \xrightarrow{e} \iota_{(-\infty,0]}$. Moreover,

$$h_1^\nu(\alpha) \to \iota_{\{0\}}(\alpha) \quad \text{and} \quad h_2^\nu(\alpha) \to \iota_{(-\infty,0]}(\alpha) \quad \forall \alpha \in \mathbb{R}.$$

Then, for all $i$,

$$h_1^\nu \circ f_i \xrightarrow{e} \iota_{\{0\}} \circ f_i \quad \text{and} \quad h_2^\nu \circ g_i \xrightarrow{e} \iota_{(-\infty,0]} \circ g_i$$

by the composition rule 4.20(a). Since these functions also converge pointwise, we can invoke the sum rule 4.19(b) repeatedly because each term in the definitions of $f$ and $f^\nu$ is never $-\infty$; the lsc of $f_0$ ensures that $f_0 \xrightarrow{e} f_0$ and the closedness of $X$ guarantees that $\iota_X \xrightarrow{e} \iota_X$ by 4.13. Thus, $f^\nu \xrightarrow{e} f$.

The analysis indicates that $\theta^\nu$ needs to be high eventually, but this can cause numerical difficulties, ill-conditioning and a slow-down of the algorithm employed to solve the approximating problems. Thus, in practice, $\theta^\nu$ is increased only cautiously in the penalty method. □

**Exercise 4.22** (penalty method). Solve the approximating problem in 4.21 with $f_0(x) = x_1 + x_2$ and $f_1(x) = 100(x_2 - x_1^2)^2 + (1 - x_1)^2 - 0.1$ for $\theta^\nu = 1, 10, 100, 1000, \dots$ using an algorithm for unconstrained optimization. Discuss how the accuracy improves with larger $\theta^\nu$, but also how the computations eventually break down.

**Guide.** A minimizer of $f_0 + \iota_{\{0\}}(f_1(\cdot))$ is approximately (0.684, 0.467). □

**Example 4.23** (estimation using mixtures). Given data $\xi^1, \dots, \xi^n \in \mathbb{R}^m$, the Kiefer-Wolfowitz maximum likelihood estimate of a density function on $\mathbb{R}^m$, used in clustering and denoising, is a minimizer

$$\hat{p} \in \operatorname{argmin}_{p\in\mathcal{P}} -\frac{1}{n}\sum_{j=1}^n \ln p(\xi^j),$$

where $\mathcal{P}$ is the family of location mixtures of normal density functions; see §1.E and [117] for background. The standard normal density function on $\mathbb{R}^m$ is given by

$$\varphi(\xi) = (2\pi)^{-m/2} \exp\big( - \tfrac{1}{2}\|\xi\|_2^2\big).$$

The location mixture of two such density functions is

$$p(\xi) = w\varphi(\xi - z) + \bar{w}\varphi(\xi - \bar{z}) \quad \forall \xi \in \mathbb{R}^m,$$

where $z, \bar{z} \in \mathbb{R}^m$ determine the location of the two density functions and $w, \bar{w} \geq 0$, which sum to one, furnish the weights. One produces $\mathcal{P}$ by allowing for any weights and location points and, in fact, any number of terms.[3] The problem can be reformulated as

$$\mathop{\text{minimize}}_{x\in C} -\frac{1}{n}\sum\nolimits_{j=1}^{n} \ln x_j, \text{ where } C = \text{con}\big\{\big(\varphi(\xi^1 - z), \dots, \varphi(\xi^n - z)\big) \,\big|\, z \in \mathbb{R}^m\big\},$$

which is the convex hull of an infinite number points and needs to be approximated.

**Detail.** We can approximate the infinite number of points by the finite collection $\bar{z}^1, \dots, \bar{z}^\nu \in \mathbb{R}^m$ and construct

$$C^\nu = \text{con}\big\{\big(\varphi(\xi^1 - \bar{z}^k), \dots, \varphi(\xi^n - \bar{z}^k)\big) \,\big|\, k = 1, \dots, \nu\big\}.$$

The approximating problem with $C$ replaced by $C^\nu$ is tractable because it's equivalent to the convex problem

$$\mathop{\text{minimize}}_{y\in\mathbb{R}^\nu} -\frac{1}{n}\sum\nolimits_{j=1}^{n} \ln\Big(\sum\nolimits_{k=1}^{\nu} y_k\varphi(\xi^j - \bar{z}^k)\Big) \text{ subject to } \sum\nolimits_{k=1}^{\nu} y_k = 1,\ y_k \geq 0\ \forall k,$$

where we use the fact that elements of a convex hull of points can be expressed as a weighted average of those points.

To justify the approximation, let

$$f(x) = -\frac{1}{n}\sum\nolimits_{j=1}^{n} \ln x_j + \iota_C(x) \quad \text{and} \quad f^\nu(x) = -\frac{1}{n}\sum\nolimits_{j=1}^{n} \ln x_j + \iota_{C^\nu}(x),$$

which represent the actual and approximating problems, respectively. (Here, $-\ln\alpha = \infty$ for $\alpha \leq 0$.) Let's show that $f^\nu \xrightarrow{e} f$ when $\{\bar{z}^k, k \in \mathbb{N}\}$ has the property: for any $\bar{z} \in \mathbb{R}^m$, there exist points in $\{\bar{z}^k, k \in \mathbb{N}\}$ converging to $\bar{z}$.

First, consider 4.15(a) and let $x^\nu \to x$. Since $C^\nu \subset C$, $\liminf f^\nu(x^\nu) \geq \liminf f(x^\nu)$. This quantity is also bounded from below by $f(x)$, as we see by considering two cases: (a) If $x \in C$, then

$$\liminf f(x^\nu) \geq \liminf\Big( -\frac{1}{n}\sum\nolimits_{j=1}^{n} \ln x_j^\nu\Big) = -\frac{1}{n}\sum\nolimits_{j=1}^{n} \ln x_j = f(x)$$

[3] The statement can be formalized by introducing "infinite sums" given by integrals, but this isn't necessary at this stage.

because the ln-function is continuous. (b) If $x \notin C$, then $\liminf f(x^\nu) \to \infty$ when any $x_j \leq 0$. Thus, we can concentrate on the situation with $x_j > 0$ for all $j$. We trivially have $\liminf f(x^\nu) \to \infty$ if $x \notin \operatorname{cl} C$. If $x \in \operatorname{cl} C \setminus C$, then by *Caratheodory's theorem*[4] there are

$$\{z_j^\nu \in \mathbb{R}^m, \lambda_j^\nu \geq 0,\ j = 0, 1, \ldots, n,\ \nu \in \mathbb{N}\}$$

such that $\sum_{j=0}^n \lambda_j^\nu = 1$ for all $\nu$ and

$$\sum\nolimits_{j=0}^n \lambda_j^\nu \begin{bmatrix} \varphi(\xi^1 - z_j^\nu) \\ \vdots \\ \varphi(\xi^n - z_j^\nu) \end{bmatrix} \to x.$$

Since every $x_j > 0$, all $z_j^\nu$ can be taken from a bounded set. Thus, there are cluster points $\lambda_j^*, z_j^*$ of the sequences $\{\lambda_j^\nu, \nu \in \mathbb{N}\}$ and $\{z_j^\nu, \nu \in \mathbb{N}\}$, respectively, with $\sum_{j=0}^n \lambda_j^* = 1$ and $\lambda_j^* \geq 0$, and

$$x = \sum\nolimits_{j=0}^n \lambda_j^* \begin{bmatrix} \varphi(\xi^1 - z_j^*) \\ \vdots \\ \varphi(\xi^n - z_j^*) \end{bmatrix} \in C.$$

This contradicts the assumption that $x \in \operatorname{cl} C \setminus C$, which then isn't possible. We've established that $\liminf f(x^\nu) \geq f(x)$.

Second, consider 4.15(b) and let $x \in C$. We need to construct $x^\nu \in C^\nu \to x$. By Caratheodory's theorem, there are

$$\{z_j \in \mathbb{R}^m, \lambda_j \geq 0,\ j = 0, 1, \ldots, n\}$$

such that $\sum_{j=0}^n \lambda_j = 1$ and

$$x = \sum\nolimits_{j=0}^n \lambda_j \begin{bmatrix} \varphi(\xi^1 - z_j) \\ \vdots \\ \varphi(\xi^n - z_j) \end{bmatrix}.$$

Let's set

$$x^\nu = \sum\nolimits_{j=0}^n \lambda_j \begin{bmatrix} \varphi(\xi^1 - u_j^\nu) \\ \vdots \\ \varphi(\xi^n - u_j^\nu) \end{bmatrix}, \quad \text{with } u_j^\nu \in \operatorname{argmin}_{z \in \{\bar{z}^1, \ldots, \bar{z}^\nu\}} \|z_j - z\|_2.$$

Certainly, $x^\nu \in C^\nu$. Since $\{\bar{z}^k, k \in \mathbb{N}\}$ has points that are arbitrarily close to any $z_j$ by assumption, $u_j^\nu \to z_j$ as $\nu \to \infty$ for all $j$. By the continuity of $\varphi$, we've that $x^\nu \to x$ and also $f^\nu(x^\nu) \to f(x)$ by the continuity of the ln-function. Thus, the conditions of 4.15 hold and $f^\nu \xrightarrow{e} f$. □

[4] If $x \in \operatorname{con} X \subset \mathbb{R}^n$, then $x$ is the convex combination of at most $n + 1$ points in $X$, i.e., $x = \sum_{j=0}^n \lambda_j x^j$ for some $x^j \in X$ and $\lambda_j \geq 0$ with $\sum_{j=0}^n \lambda_j = 1$.

## 4.D Constraint Softening

The previous section brought forward the troubling fact that if a constraint function is replaced by a nearly identical function, the effect on the corresponding epigraphs—and thus also on minimizers—can be dramatic; see Figure 4.6. Generally, we prefer optimization models where minimizers and other quantities of interest change just a little under (small) perturbations of their functions and parameters. Such *stable* models are more trustworthy than others because in practice we almost always have lingering doubt about the "correct" function and parameter choices. In this section, we'll use insight from the epigraphical perspective to construct stable models. Thus, in contrast to the previous section and its focus on epigraphical approximations for computational reasons, our goal now is to assess whether a model is "good" in the sense of being stable under perturbations.

For concreteness, let's consider the problem

$$\underset{x\in\mathbb{R}^n}{\text{minimize}}\, f_0(x) \ \text{ subject to } \ g_i(x) \le 0,\ i = 1,\dots,q. \tag{4.5}$$

Underlying these functions there are modeling choices and parameter values not fully settled. If the minima and minimizers of the problem vary greatly with the assumptions, then the merit of a decision derived from the problem is questionable. In fact, a solution can be misleading and possibly an artifact of a particular assumption.

To formally analyze the situation, let $\{f_0^\nu, g_i^\nu, \dots, g_q^\nu,\ \nu \in \mathbb{N}\}$ be alternative functions that are close to the original ones in the sense that, for $\{\delta^\nu \ge 0, \nu \in \mathbb{N}\} \to 0$,

$$\sup_{x\in\mathbb{R}^n} \big|f_0^\nu(x) - f_0(x)\big| \le \delta^\nu \quad \text{and} \quad \sup_{x\in\mathbb{R}^n} \max_{i=1,\dots,q} \big|g_i^\nu(x) - g_i(x)\big| \le \delta^\nu.$$

We can think of $\delta^\nu$ as quantifying the changes in the objective and constraint functions across a range of modeling choices and parameter values. In practice, the exact value of $\delta^\nu$ might be unknown and the alternative functions could be largely conceptual. We are primarily interested in qualitative insight: Would a problem with the alternative functions produce solutions that are close to those of (4.5) when $\delta^\nu$ is small?

In general, the naive formulation

$$\underset{x\in\mathbb{R}^n}{\text{minimize}}\, f_0^\nu(x) \ \text{ subject to } \ g_i^\nu(x) \le 0,\ i = 1,\dots,q$$

may not have minimizers close to those of (4.5) even for arbitrarily small $\delta^\nu$ as illustrated in Figure 4.6. A better formulation emerges by *softening* the constraints and using a penalty $\theta^\nu \in (0,\infty)$:

$$\underset{x\in\mathbb{R}^n, y\in\mathbb{R}^q}{\text{minimize}}\ f_0^\nu(x) + \theta^\nu \sum\nolimits_{i=1}^q y_i \ \text{ subject to } \ g_i^\nu(x) \le y_i,\ 0 \le y_i,\ i = 1,\dots,q. \tag{4.6}$$

In this approximating problem constraint violations are permitted, but incur a per-unit penalty of $\theta^\nu$. Compared to (4.5), the formulation seems at first to be even more different than the naive one. However, this isn't the case. We can show that it's epigraphically close to (4.5) and thus its minimizers are also close by 4.14. Consequently, if the functions in

(4.5) are unsettled in some way or need to be approximated, then a formulation of the kind in (4.6) tends to be better than the naive one as it promotes stability.

To show the epigraphical relation between (4.5) and (4.6), let's write the approximating problem compactly as minimizing

$$f^\nu(x,y) = f_0^\nu(x) + h^\nu(x,y), \text{ with } h^\nu(x,y) = \sum\nolimits_{i=1}^q \varphi^\nu(y_i) + \sum\nolimits_{i=1}^q \iota_{(-\infty,0]}\big(g_i^\nu(x) - y_i\big)$$

and $\varphi^\nu(\alpha) = \theta^\nu \alpha$ if $\alpha \geq 0$ and $\varphi^\nu(\alpha) = \infty$ otherwise. The actual problem is expressed in terms of $x$ only, but to allow comparison with the approximation we artificially introduce $y$ and state it as minimizing

$$f(x,y) = f_0(x) + h(x,y), \text{ with } h(x,y) = \sum\nolimits_{i=1}^q \iota_{\{0\}}(y_i) + \sum\nolimits_{i=1}^q \iota_{(-\infty,0]}\big(g_i(x) - y_i\big).$$

Trivially, $y$ must be zero for $f$ to be finite so this is indeed an equivalent formulation of the actual problem. The similarities between $f^\nu$ and $f$ are clear. In particular, $\varphi^\nu$ is epigraphically similar to $\iota_{\{0\}}$.

Let's show that $f^\nu \xrightarrow{e} f$ when $f_0, f_0^\nu, g_i, g_i^\nu : \mathbb{R}^n \to \mathbb{R}$, $f_0$ is continuous, $g_i, i = 1, \dots, q$, are lsc and

$$\theta^\nu \to \infty, \qquad \delta^\nu \to 0, \qquad \theta^\nu \delta^\nu \to 0.$$

We start by establishing the conditions of 4.15 for $h^\nu$ and $h$. Suppose that $(x^\nu, y^\nu) \to (x,y)$. Let's first prove that $\liminf h^\nu(x^\nu, y^\nu) \geq h(x,y)$. Since $h^\nu$ is nonnegative, we only need to check when $h(x,y) = \infty$ and consider three cases:

First, suppose that there exists $i$ with $y_i > 0$. Then, there's $\bar\nu$ such that for all $\nu \geq \bar\nu$, one has $y_i^\nu \geq \frac{1}{2} y_i > 0$. Consequently, for such $\nu$,

$$h^\nu(x^\nu, y^\nu) \geq \theta^\nu y_i^\nu \geq \tfrac{1}{2}\theta^\nu y_i \to \infty.$$

Second, suppose that there exists $i$ with $y_i < 0$. Then, for sufficiently large $\nu$, $y_i^\nu < 0$ and $h^\nu(x^\nu, y^\nu) = \infty$.

Third, suppose that there exists $i$ with $g_i(x) - y_i > 0$ and $y_i = 0$. Then, $g_i(x) > 0$. Since $y_i^\nu < 0$ and/or $g_i^\nu(x^\nu) - y_i^\nu > 0$ imply $h^\nu(x^\nu, y^\nu) = \infty$, we can assume without loss of generality that $y_i^\nu \geq 0$ and $g_i^\nu(x^\nu) - y_i^\nu \leq 0$. Since $g_i$ is lsc, there exists $\bar\nu$ such that for all $\nu \geq \bar\nu$, one has $g_i(x^\nu) \geq g_i(x)/2$. Then, for such $\nu$,

$$h^\nu(x^\nu, y^\nu) \geq \theta^\nu y_i^\nu \geq \theta^\nu g_i^\nu(x^\nu) \geq \theta^\nu\big(g_i(x^\nu) - \delta^\nu\big) \geq \theta^\nu\big(\tfrac{1}{2}g_i(x) - \delta^\nu\big) \to \infty$$

because $\delta^\nu \to 0$. We've satisfied 4.15(a).

For 4.15(b), we only need to check $(x,y)$ with $y_i = 0$ and $g_i(x) \leq 0$ for all $i$. Set $x^\nu = x$ and

$$y_i^\nu = \max\big\{0, g_i^\nu(x)\big\} \leq \delta^\nu, \;\; i = 1, \dots, q.$$

Certainly, $(x^\nu, y^\nu) \to (x,y)$. Moreover, $g_i^\nu(x^\nu) - y_i^\nu \leq 0$ for all $i$ so that

$$h^\nu(x^\nu, y^\nu) = \sum\nolimits_{i=1}^q \varphi^\nu(y_i^\nu) = \sum\nolimits_{i=1}^q \theta^\nu y_i^\nu \leq \sum\nolimits_{i=1}^q \theta^\nu \delta^\nu \to 0 = h(x,y).$$

We've proven that $h^\nu \xrightarrow{e} h$.

The uniform convergence of $f_0^\nu$ to the continuous function $f_0$ establishes via 4.15 that $f_0^\nu \xrightarrow{e} f$ and $-f_0^\nu \xrightarrow{e} -f$, and these facts trivially extend when $f_0^\nu$ and $f_0$ are viewed as functions on $\mathbb{R}^n \times \mathbb{R}^q$ and not only $\mathbb{R}^n$. We can then bring in the sum rule 4.19(a) to finalize that

$$f^\nu = f_0^\nu + h^\nu \xrightarrow{e} f = f_0 + h.$$

The parameter $\theta^\nu$ can't grow arbitrarily fast for epi-convergence to take place. For example, one may take $\theta^\nu$ proportional to $(\delta^\nu)^{\sigma-1}$ for some $\sigma \in (0,1)$ so that $\theta^\nu \to \infty$, but $\theta^\nu\delta^\nu \to 0$. In fact, it isn't necessary to know the exact error bound $\delta^\nu$, but just the rate with which it vanishes because that's what dictates $\theta^\nu$.

Practitioners often soften constraints using penalties as in (4.6) and the above analysis supports that modeling choice. The resulting problem becomes a valid approximation of an actual problem, which may not be fully known, and thus promotes stability in the presence of unsettled assumptions and parameter values. In contrast, the naive formulation can be arbitrarily poor. This discussion highlights the importance of considering different problem formulations and examining not only their computational attractiveness but also their stability. We return to the subject in Chap. 5.

## 4.E Set Analysis

Epi-convergence emerges from the previous sections as the primary tool in analysis of approximations of minimization problems. The notion is defined as having the epigraphs of approximating functions approach the epigraph of an actual function. Convergence of sets in this sense arises more broadly and plays important roles in analysis of feasibility, perturbations, optimality, subdifferentiability and variational problems generally. In this section, we'll formalize what it means for sets to converge and thereby lay the foundation of an approximation theory for sets.

**Definition 4.24** (set-convergence). For $C, C^\nu \subset \mathbb{R}^n$, we say that $C^\nu$ *set-converges* to $C$, written $C^\nu \xrightarrow{s} C$ or $\operatorname{Lim} C^\nu = C$, when

$$C \text{ is closed} \quad \text{and} \quad \operatorname{dist}(x, C^\nu) \to \operatorname{dist}(x, C) \ \ \forall x \in \mathbb{R}^n.$$

If the sets in the definition are epigraphs, then we return to the definition of epi-convergence in 4.13, which immediately leads to the fact:

**Proposition 4.25** (epi-convergence as set-convergence). *For $f, f^\nu : \mathbb{R}^n \to \overline{\mathbb{R}}$, one has*

$$f^\nu \xrightarrow{e} f \iff \operatorname{epi} f^\nu \xrightarrow{s} \operatorname{epi} f.$$

As in the case of epi-convergence, it becomes important to have rules for verifying set-convergence that leverage knowledge of set-convergence of some underlying sets. In fact, such rules are building blocks toward new rules about epi-convergence. We start with a result about product sets, which is immediate from the definition.

**Proposition 4.26** (convergence of product sets). *For* $\{C_i, C_i^\nu \subset \mathbb{R}^{n_i}, \nu \in \mathbb{N}, i = 1, \dots, m\}$, *one has*

$$C_i^\nu \xrightarrow{s} C_i, \ i = 1, \dots, m \implies C_1^\nu \times \cdots \times C_m^\nu \xrightarrow{s} C_1 \times \cdots \times C_m.$$

A consequence of the proposition and 4.25 is that if $C^\nu \xrightarrow{s} C$, then the indicator functions $\iota_{C^\nu} \xrightarrow{e} \iota_C$ because

$$\operatorname{epi} \iota_{C^\nu} = C^\nu \times [0, \infty) \quad \text{and} \quad \operatorname{epi} \iota_C = C \times [0, \infty).$$

Hence, in view of the epigraphical sum rule 4.19(a), we've

$$f + \iota_{C^\nu} \xrightarrow{e} f + \iota_C$$

for any continuous $f : \mathbb{R}^n \to \mathbb{R}$. One way of constructing an approximation for a problem is therefore to approximate its feasible set.

Two concepts that together characterize set-convergence help us to identify what's actually needed.

The *inner limit* of $\{C^\nu \subset \mathbb{R}^n, \nu \in \mathbb{N}\}$, denoted by $\operatorname{LimInn} C^\nu$, is the collection of limit points to which sequences of points selected from the sets converge. Specifically,

$$\operatorname{LimInn} C^\nu = \{x \in \mathbb{R}^n \mid \exists x^\nu \in C^\nu \to x\}.$$

For example, the inner limit of the line segments $[1/\nu, 2/\nu]$ is $\{0\}$. A slight clarification regarding empty sets is in place. Suppose that $C^1 = \emptyset$, but $C^\nu = [1/\nu, 2/\nu]$ for $\nu \geq 2$. Still, $\operatorname{LimInn} C^\nu = \{0\}$. Since convergence of a sequence is always determined by its tail and not a finite number of its elements, we aren't concerned about not being able to select a point from $C^1$. It suffices in the definition of inner limits to have points in $C^\nu$ converging to $x$ for all $\nu$ sufficiently large.

Figure 4.10 gives two more examples. On the left, the third quadrant is the inner limit of sets that include points in the first quadrant. On the right, a collection of vertical line segments has a rectangle as inner limit.

Let's examine a less well-behaved case: If $C^\nu = \{0\}$ when $\nu$ is odd and $C^\nu = [0, 1]$ when $\nu$ is even, then $\operatorname{LimInn} C^\nu = \{0\}$. The inner limit isn't $[0, 1]$ because we can't construct $x^\nu \in C^\nu$ converging to any $x \in (0, 1]$. This brings us to the second concept.

The *outer limit* of $\{C^\nu \subset \mathbb{R}^n, \nu \in \mathbb{N}\}$, denoted by $\operatorname{LimOut} C^\nu$, is the collection of cluster points to which subsequences of points selected from the sets converge. Specifically,

$$\operatorname{LimOut} C^\nu = \left\{x \in \mathbb{R}^n \;\middle|\; \exists N \in \mathcal{N}_\infty^\# \text{ and } x^\nu \in C^\nu \xrightarrow[N]{} x\right\}.$$

We recall that $\mathcal{N}_\infty^\#$ is the set of all infinite collections of increasing indices. Hence, $x \in \operatorname{LimOut} C^\nu$ if we can select a subsequence of indices $N$, say the even numbers so that $N = \{2, 4, \dots\}$, and then select $\{x^\nu \in C^\nu, \nu \in N\}$ converging to $x$. In the example, $C^\nu = \{0\}$ when $\nu$ is odd and $C^\nu = [0, 1]$ otherwise, one has $\operatorname{LimOut} C^\nu = [0, 1]$ because for any $x \in [0, 1]$, we can rely on $N = \{2, 4, \dots\}$ and select $x^\nu = x$ for $\nu \in N$. This

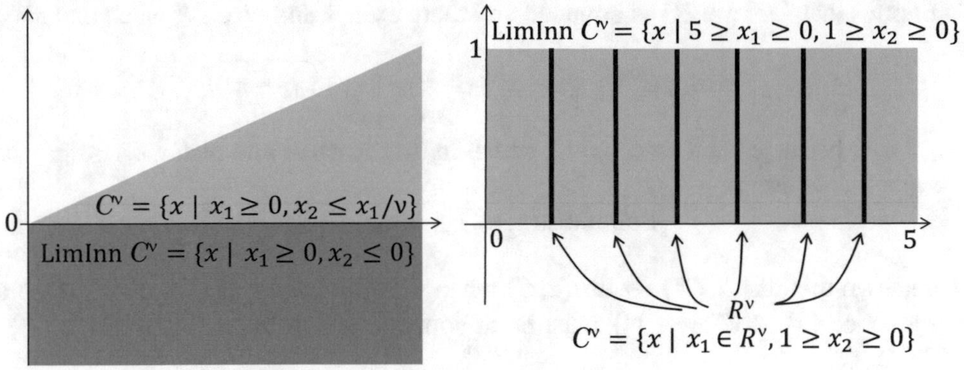


Fig. 4.10: Examples of inner limits; $R^\nu$ is the first $\nu$ numbers in an ordered list of the rational numbers in [0, 5].

example motivates the terminology "inner limit" and "outer limit." Generally,

$$\text{LimInn}\, C^\nu \subset \text{LimOut}\, C^\nu.$$

Although $\text{Lim}\, C^\nu$ may not exist (as is clear from the "odd-even" example), the inner limit and the outer limit always exist but could be empty. Moreover, they're closed sets.[5]

**Proposition 4.27** (inner, outer and set limits). *For $C, C^\nu \subset \mathbb{R}^n$, one has*

$$\text{Lim}\, C^\nu = C \iff \text{LimInn}\, C^\nu = \text{LimOut}\, C^\nu = C.$$

**Proof.** Suppose that $C$ is nonempty, $x \in \mathbb{R}^n$ and the right-hand side portion of the equivalence holds. Then, there's $\bar{x} \in C$ such that

$$\text{dist}(x, C) = \|x - \bar{x}\|_2$$

because $C$ is closed (see 2.30). There's also $x^\nu \in C^\nu \to \bar{x}$ by the definition of inner limits. The triangle inequality yields

$$\text{dist}(x, C^\nu) \leq \|x - x^\nu\|_2 \leq \|x - \bar{x}\|_2 + \|\bar{x} - x^\nu\|_2 \to \text{dist}(x, C).$$

That is, $\limsup \text{dist}(x, C^\nu) \leq \text{dist}(x, C)$. Moreover, for any $\varepsilon > 0$, there exists $x^\nu \in C^\nu$ such that

$$\text{dist}(x, C^\nu) \geq \|x^\nu - x\|_2 - \varepsilon.$$

Since $C$ is nonempty, one can select $\bar{y} \in C$ and $y^\nu \in C^\nu \to \bar{y}$ because $\text{LimInn}\, C^\nu = C$. Consequently, again invoking the triangle inequality, one has

$$\|x^\nu - x\|_2 - \varepsilon \leq \text{dist}(x, C^\nu) \leq \|x - y^\nu\|_2 \leq \|x - \bar{y}\|_2 + \|\bar{y} - y^\nu\|_2 \to \|x - \bar{y}\|_2.$$

[5] In other texts, the inner limit and the outer limit are denoted by liminf and limsup, respectively; see [105].

This implies that $\{x^\nu, \nu \in \mathbb{N}\}$ is bounded and there exist $\bar{x}$ and $N \in \mathcal{N}_\infty^\#$ such that $x^\nu \to_N \bar{x}$ and

$$\mathrm{dist}(x, C^\nu) \geq \|x - x^\nu\|_2 - \varepsilon \to_N \|x - \bar{x}\|_2 - \varepsilon.$$

Thus, $\bar{x} \in C$ because $\mathrm{LimOut}\, C^\nu = C$, which in turn implies that $\mathrm{dist}(x, C) \leq \|x - \bar{x}\|_2$. Since $\varepsilon > 0$ is arbitrary,

$$\liminf \mathrm{dist}(x, C^\nu) \geq \mathrm{dist}(x, C).$$

We've shown that $\mathrm{dist}(x, C^\nu) \to \mathrm{dist}(x, C)$ when $C$ is nonempty. If $C$ is empty, then any (sub)sequence $\{x^\nu \in C^\nu, \nu \in \mathbb{N}\}$ must be unbounded which implies that $\mathrm{dist}(x, C^\nu) \to \infty = \mathrm{dist}(x, C)$.

For the converse, suppose that $\mathrm{dist}(x, C^\nu) \to \mathrm{dist}(x, C)$ for all $x \in \mathbb{R}^n$ and $C$ is closed. Let $\bar{x} \in C$. Then, $\mathrm{dist}(\bar{x}, C^\nu) \to 0$ and there's $x^\nu \in C^\nu \to \bar{x}$, which means that $\bar{x} \in \mathrm{LimInn}\, C^\nu$. Consequently,

$$C \subset \mathrm{LimInn}\, C^\nu.$$

Next, suppose that $\bar{x} \in \mathrm{LimOut}\, C^\nu$. Then, there exist $N \in \mathcal{N}_\infty^\#$ and $\{x^\nu \in C^\nu, \nu \in N\} \to \bar{x}$. Since $\mathrm{dist}(\bar{x}, C^\nu) \leq \|\bar{x} - x^\nu\|_2$ for $\nu \in N$, one has

$$\mathrm{dist}(\bar{x}, C) = \mathrm{liminf}_{\nu \in N}\, \mathrm{dist}(\bar{x}, C^\nu) \leq \mathrm{liminf}_{\nu \in N}\, \|\bar{x} - x^\nu\|_2 = 0.$$

Thus, $\bar{x} \in C$ because $C$ is closed and

$$\mathrm{LimOut}\, C^\nu \subset C.$$

From the general fact that $\mathrm{LimInn}\, C^\nu \subset \mathrm{LimOut}\, C^\nu$, we conclude that both the inner limit and the outer limit coincide with $C$. □

**Example 4.28** (feasibility problem). Given a closed set $C \subset \mathbb{R}^n$, suppose that we're faced with the feasibility problem of finding $x \in C$. If $C$ is complicated, then we may instead attempt to find $x^\nu \in C^\nu$, where $C^\nu$ is an approximation of $C$. Set-convergence provides a sufficient condition for $x^\nu$ to be approximately feasible.

**Detail.** If $\mathrm{LimOut}\, C^\nu \subset C$, then every cluster point $\bar{x}$ of $\{x^\nu \in C^\nu, \nu \in \mathbb{N}\}$ is contained in $C$, i.e., $\bar{x}$ is a solution of the actual feasibility problem. By 4.27, a sufficient condition for $\mathrm{LimOut}\, C^\nu \subset C$ is that $C^\nu \xrightarrow{s} C$. □

We're now in a position to prove 4.15, which is nothing but a specialization of the definitions of the inner limit and outer limit to epigraphs.

**Proof of 4.15.** Suppose that LimOut (epi $f^\nu$) $\subset$ epi $f$ and $x^\nu \to x$. If liminf $f^\nu(x^\nu) \in \mathbb{R}$, then for some $N \in \mathcal{N}_\infty^\#$ we've $(x^\nu, f^\nu(x^\nu)) \in \operatorname{epi} f^\nu$ for $\nu \in N$ and

$$\big(x^\nu, f^\nu(x^\nu)\big) \underset{N}{\to} \big(x, \operatorname{liminf} f^\nu(x^\nu)\big) \subset \operatorname{epi} f.$$

Thus, liminf $f^\nu(x^\nu) \geq f(x)$. If liminf $f^\nu(x^\nu) = \infty$, the same inequality holds trivially. If liminf $f^\nu(x^\nu) = -\infty$, then for any $\beta \in \mathbb{R}$, one can construct $\alpha^\nu \to \beta$ with $(x^\nu, \alpha^\nu) \in \operatorname{epi} f^\nu$. The limit of this sequence is $(x, \beta)$, which then is in epi $f$. Thus, $f(x) \leq \beta$. Since $\beta$ is arbitrary, $f(x) = -\infty$ and again liminf $f^\nu(x^\nu) \geq f(x)$. We've established 4.15(a).

Suppose that LimInn (epi $f^\nu$) $\supset$ epi $f$ and $x \in \mathbb{R}^n$. If $f(x) \in \mathbb{R}$, then

$$\big(x, f(x)\big) \in \operatorname{LimInn}(\operatorname{epi} f^\nu).$$

Thus, there exist $x^\nu \to x$ and $\alpha^\nu \to f(x)$, with $f^\nu(x^\nu) \leq \alpha^\nu$. This implies that limsup $f^\nu(x^\nu) \leq f(x)$. If $f(x) = -\infty$, then for $\alpha \in \mathbb{R}$ we've

$$(x, \alpha) \in \operatorname{LimInn}(\operatorname{epi} f^\nu).$$

Thus, there exist $x^\nu \to x$ and $\alpha^\nu \to \alpha$, with $f^\nu(x^\nu) \leq \alpha^\nu$. This implies that limsup $f^\nu(x^\nu) \leq \alpha$. Since $\alpha$ is arbitrary, we can conclude that limsup $f^\nu(x^\nu) = -\infty$. Consequently, 4.15(b) holds because the last case $f(x) = \infty$ is trivial.

For the converse, suppose that 4.15(a) holds and $(x, \alpha) \in$ LimOut (epi $f^\nu$). Then, there exist $N \in \mathcal{N}_\infty^\#$ and $(x^\nu, \alpha^\nu) \in \operatorname{epi} f^\nu \underset{N}{\to} (x, \alpha)$. Thus,

$$\alpha = \operatorname{liminf}_{\nu \in N} \alpha^\nu \geq \operatorname{liminf}_{\nu \in N} f^\nu(x^\nu) \geq f(x)$$

and $(x, \alpha) \in$ epi $f$. This implies that LimOut (epi $f^\nu$) $\subset$ epi $f$.

Next, suppose that 4.15(b) holds and $(x, \alpha) \in$ epi $f$. There exists $x^\nu \to x$ with limsup $f^\nu(x^\nu) \leq f(x)$. Take $\alpha^\nu = \max\{\alpha, f^\nu(x^\nu)\}$, which converges to $\alpha$. Hence,

$$(x^\nu, \alpha^\nu) \in \operatorname{epi} f^\nu \to (x, \alpha).$$

This implies that $(x, \alpha) \in$ LimInn (epi $f^\nu$). Thus, LimInn (epi $f^\nu$) $\supset$ epi $f$. The conclusion then follows by the general fact that inner limits are contained in outer limits. □

A feasible set is often specified as the intersection and/or union of a collection of other sets, each representing some requirement, which may need approximation. The question then becomes: how do approximations of the "component" sets affect the feasible set?

**Proposition 4.29** (set-convergence of intersection and union). *For $C^\nu, D^\nu \subset \mathbb{R}^n$, one has*

$$\operatorname{LimOut}\big(C^\nu \cap D^\nu\big) \subset \big(\operatorname{LimOut} C^\nu\big) \cap \big(\operatorname{LimOut} D^\nu\big).$$

*Moreover, suppose that $C^\nu \xrightarrow{s} C$ and $D^\nu \xrightarrow{s} D$. Then, the following hold:*

(a) *If all $C^\nu, D^\nu$ are convex and* $(\operatorname{int} C) \cap D \neq \emptyset$, *then* $\operatorname{Lim}\big(C^\nu \cap D^\nu\big) = C \cap D$.
(b) $\operatorname{Lim}\big(C^\nu \cup D^\nu\big) = C \cup D$.

**Proof.** The first and third statements follow readily from the definitions of inner limits and outer limits. For the convex case, consult [105, Theorem 4.32]. □

The situation is less than ideal for intersections: LimOut $(C^\nu \cap D^\nu)$ can be strictly contained in LimOut $C^\nu \cap$ LimOut $D^\nu$ even in the convex case. For example, $C^\nu = [-1, -1/\nu]$ and $D^\nu = [1/\nu, 1]$ have $\mathrm{Lim}(C^\nu \cap D^\nu) = \emptyset$, while $\mathrm{Lim}\, C^\nu \cap \mathrm{Lim}\, D^\nu = \{0\}$. This highlights the challenge in constrained optimization already discussed in §4.C and §4.D: A tiny change of the components of a problem may have large effects on the set of minimizers. Set-convergence helps us to identify the trouble. If a feasible set $C \cap D$ isn't stable under perturbations of $C$ and $D$ in the sense that approximations $C^\nu \cap D^\nu$ fail to set-converge to $C \cap D$ despite $C^\nu \xrightarrow{s} C$ and $D^\nu \xrightarrow{s} D$, then optimization over $C \cap D$ is somehow ill-posed. In the convex case, we achieve such stability if $C$ and $D$ overlap "sufficiently" as stated in the proposition. A similar overlap condition is also needed when computing subgradients of sums of two functions; cf. the Moreau-Rockafellar sum rule 2.26.

In contrast to intersections, unions of sets produce no difficulty and this fact translates into a result about certain inf-projections.

**Example 4.30** (epi-convergence of min-functions). For $f_i, f_i^\nu : \mathbb{R}^n \to \overline{\mathbb{R}}$, suppose that $f$ and $f^\nu$ are *min-functions* given by

$$f(x) = \min_{i=1,\dots,m} f_i(x) \quad \text{and} \quad f^\nu(x) = \min_{i=1,\dots,m} f_i^\nu(x).$$

Then, $f^\nu \xrightarrow{e} f$ whenever $f_i^\nu \xrightarrow{e} f_i$ for all $i$.

**Detail.** Since $\mathrm{epi}\, f = \cup_{i=1}^m \mathrm{epi}\, f_i$ as seen in Figure 4.11, 4.29 immediately gives the conclusion. □

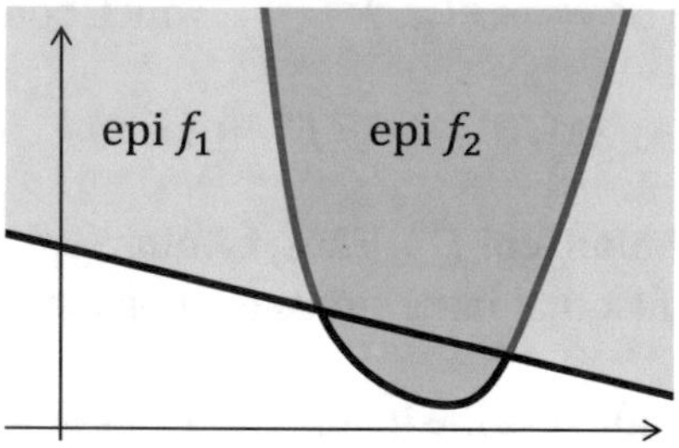


Fig. 4.11: The epigraph of the min-function given by $f(x) = \min\{f_1(x), f_2(x)\}$ is the union of epi $f_1$ and epi $f_2$.

Convexity is preserved under set-convergence, a fact that can be established by direct appeal to the definitions.

**Proposition 4.31** (convergence of convex sets). *If* $\{C^\nu \subset \mathbb{R}^n, \nu \in \mathbb{N}\}$ *is a sequence of convex sets, then* LimInn $C^\nu$ *is convex and so is* Lim $C^\nu$ *if it exists.*

We note that LimOut $C^\nu$ may not be convex even if $C^\nu$ is convex. For example, if $C^\nu = [-1,1] \times \{0\} \subset \mathbb{R}^2$ when $\nu$ is odd and $C^\nu = \{0\} \times [-1,1] \subset \mathbb{R}^2$ when $\nu$ is even, then $C^\nu$ and LimInn $C^\nu = \{0\}$ are convex. However,

$$\text{LimOut}\, C^\nu = \big([-1,1] \times \{0\}\big) \cup \big(\{0\} \times [-1,1]\big)$$

is a cross and thus nonconvex.

**Exercise 4.32** (set-convergence for outer approximations). Show that for $C^\nu, D^\nu \subset \mathbb{R}^n$ set-converging to $C$ and $D$, respectively, one has $\text{Lim}\,(C^\nu \cap D^\nu) = C \cap D$ whenever $C^\nu \supset C$ and $D^\nu \supset D$ for all $\nu$.

**Guide.** A starting point is provided by 4.29. □

## 4.F Robotic Path Planning

We would like to design a robot that maneuvers a corridor of width $\alpha > 0$ without hitting its walls. The corridor is described by the set

$$X = \big\{x \in \mathbb{R}^2 \;\big|\; 0 \le x_1 \le \alpha,\ 0 \le x_2 \le \alpha + \beta \quad \text{or} \quad \alpha < x_1,\ \beta \le x_2 \le \alpha + \beta\big\},$$

where $\beta > 0$ is the distance to the corner of the corridor; see Figure 4.12. Starting from the initial position $(\alpha/2, 0) \in \mathbb{R}^2$, the robot moves with constant speed $\sigma = 1$. Its position and heading at time $t$ are denoted by $(x_1(t), x_2(t)) \in \mathbb{R}^2$ and $x_3(t) \in \mathbb{R}$, respectively, with the latter being measured in radians counterclockwise relative to the horizontal axis. The initial heading is $\pi/3$; see Figure 4.12.

At any time $t$, we steer the robot by controlling the rate of change $u(t)$ of the heading. The relationship between this control input and the position and heading of the robot is modeled by the differential equation

$$x_1'(t) = \sigma \cos x_3(t), \qquad x_2'(t) = \sigma \sin x_3(t), \qquad x_3'(t) = u(t),$$

with initial conditions $x_1(0) = \alpha/2$, $x_2(0) = 0$ and $x_3(0) = \pi/3$. A robot following this differential equation is called a *Dubin's vehicle*. For simplicity, we assume that

$$u(t) = \begin{cases} u_1 & \text{for } t \in \big[0, \tfrac{1}{3}\tau\big] \\ u_2 & \text{for } t \in \big(\tfrac{1}{3}\tau, \tfrac{2}{3}\tau\big] \\ 0 & \text{for } t \in \big(\tfrac{2}{3}\tau, \tau\big], \end{cases}$$

where $\tau$ is the mission duration. Then, we only need to optimize $u_1, u_2$. (In practice, one would select a much more general class of functions $u : [0,\tau] \to \mathbb{R}$ over which to optimize.) We limit the possible control input to $u_1, u_2 \in [-1,1]$ as the robot can't achieve a larger rate of change of the heading. The theory of differential equations tells us that for every such control input $u = (u_1, u_2)$ the differential equation has a unique solution, which we denote by

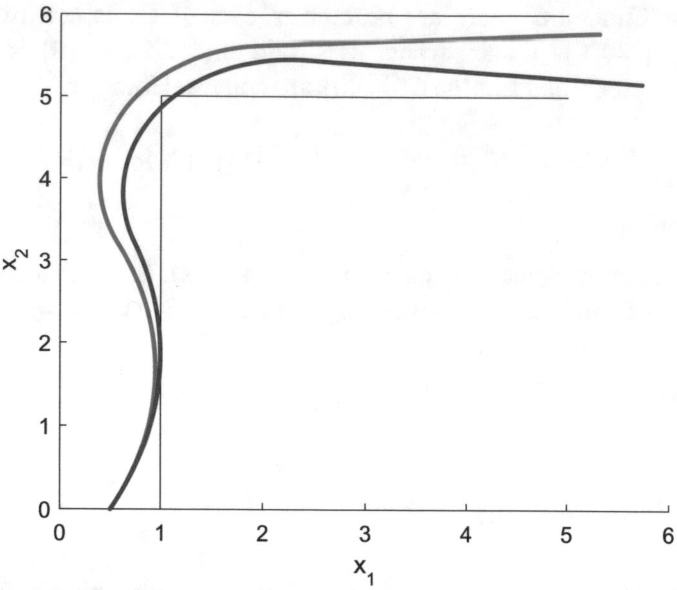


Fig. 4.12: Feasible (blue) and infeasible (red) trajectories of robot in corridor $X$ starting at $(\alpha/2, 0)$ with $\alpha = 1$ and $\beta = 5$.

$$\{x_1(t;u),\ x_2(t;u),\ x_3(t;u),\quad 0 \le t \le \tau\}$$

and compute with high precision using a Runga-Kutta method. With $\tau = 10$, Figure 4.12 shows in blue the first two components of the solution for $u' = (0.3, -0.6)$ and in red for $\bar{u} = (0.27, -0.61)$, i.e.,

$$\Big\{\big(x_1(t;u'),\ x_2(t;u')\big),\quad 0 \le t \le \tau\Big\} \text{ and } \Big\{\big(x_1(t;\bar{u}),\ x_2(t;\bar{u})\big),\quad 0 \le t \le \tau\Big\}.$$

We would like the robot to apply as little control input as possible while remaining in the corridor. This leads to the model

$$\underset{u\in U}{\text{minimize}}\ \|u\|_2^2, \quad \text{with } U = \Big\{u \in [-1,1]^2 \ \Big|\ \big(x_1(t;u), x_2(t;u)\big) \in X \ \ \forall t \in [0,\tau]\Big\}.$$

This feasible set, illustrated in black by Figure 4.13, is described by an infinite number of constraints, one for each $t \in [0, \tau]$. Thus, the model is an example of a *semi-infinite optimization problem*, with "semi" indicating that there are still only a finite number of decision variables. Such problems pose significant computational challenges and in fact the illustration in Figure 4.13 is simply a high-precision approximation of the actual feasible set.

An approximation of the feasible set can be achieved by checking the position of the robot at a finite number of points in time. Specifically, let

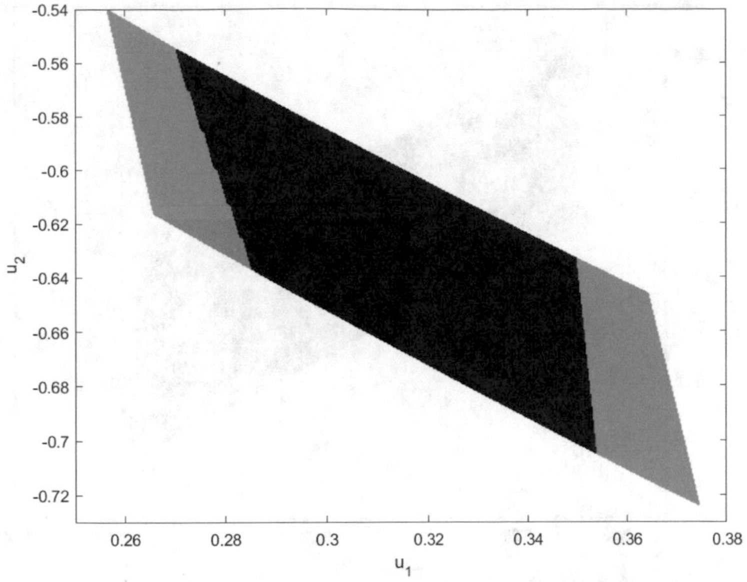


Fig. 4.13: Feasible set $U$ (black) and approximating feasible set $U^2$ (gray and black) when position is only checked at time $t = 5$ and $t = 10$.

$$U^\nu = \left\{u \in [-1,1]^2 \;\middle|\; \big(x_1(t;u), x_2(t;u)\big) \in X \;\; \forall t \in T^\nu\right\},$$

where

$$T^\nu = \{\tau/\nu, 2\tau/\nu, \dots, (\nu-1)\tau/\nu, \tau\} \quad \forall \nu \in \mathbb{N}.$$

Since $T^\nu \subset [0,\tau]$, $U^\nu \supset U$ for all $\nu$. Figure 4.13 shows $U^2$, which is significantly larger than $U$. (The illustration is actually of $U^{1000}$ and not the inaccessible $U$.) Accuracy improves with $U^{10}$ in Figure 4.14, but the red trajectory in Figure 4.12 with control input $\bar{u} = (0.27, -0.61) \in U^{10}$ is still infeasible relative to $U$.

The accuracy of the approximating feasible set improves further as $\nu$ increases and, in fact, $U^\nu \xrightarrow{s} U$ as can be seen by the following argument. Let

$$\varphi(u,t) = \operatorname{dist}\Big(\big(x_1(t;u), x_2(t,u)\big), X\Big),$$

which defines a continuous function; see for example [83, Section 5.6]. If the robot is in the corridor at time $t$, then $\varphi(u,t) = 0$ and otherwise $\varphi(u,t) > 0$. Since $U^\nu \supset U$, we only need to establish that $\operatorname{LimOut} U^\nu \subset U$. Suppose that $u \in \operatorname{LimOut} U^\nu$. Then, there exist $N \in \mathcal{N}_\infty^\#$ and $u^\nu \in U^\nu \xrightarrow[N]{} u$. Let $t^\star \in [0,\tau]$ be such that

$$\varphi(u,t^\star) = \sup_{t\in[0,\tau]} \varphi(u,t),$$

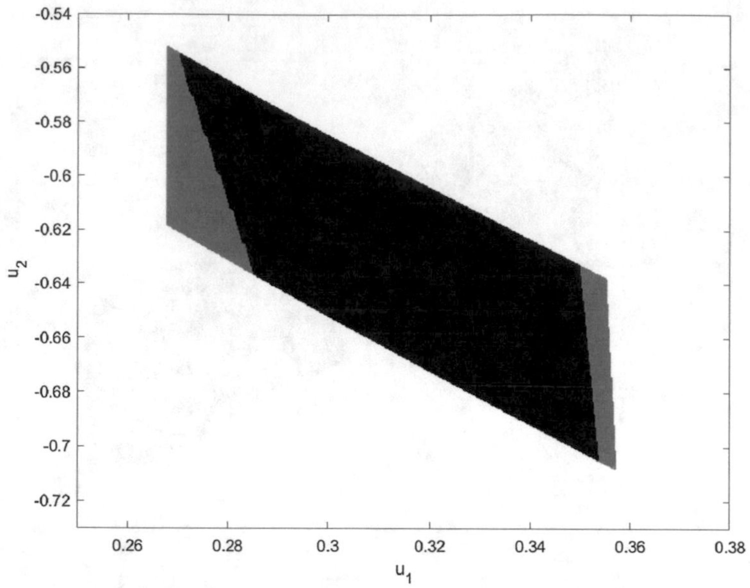


Fig. 4.14: Feasible set $U$ (black) and approximating feasible set $U^{10}$ (gray and black) when position is checked 10 times.

which exists by 4.9. For all $\nu \in N$, there's $t^\nu \in T^\nu$ such that $|t^\nu - t^\star| \leq 1/\nu$. Since $\varphi(u^\nu, t^\nu) = 0$ for all $\nu \in N$ and $\varphi$ is continuous at $(u, \tau^\star)$, one has

$$\sup_{t \in [0,\tau]} \varphi(u, t) = \varphi(u, t^\star) = 0.$$

Hence, $u \in U$ and we've shown that $U^\nu \xrightarrow{s} U$.

The objective function of the problem is continuous, which allows us to establish that

$$\| \cdot \|_2^2 + \iota_{U^\nu} \xrightarrow{e} \| \cdot \|_2^2 + \iota_U;$$

see the discussion after 4.26. We conclude from this development that the time-discretization approach results in a valid approximation and the epigraphical approximation algorithm is justified when applied in this context.

## 4.G Tangent and Normal Cones I

The path toward optimality conditions for nonconvex problems mirrors that taken in Chap. 2. The conditions stemming from the Fermat rule will again be in terms of subgradients and normal cones. However, normal cones take on an even more prominent role as they can in fact be used to define subgradients. We'll also define tangent cones, which furnish a complementary view of the local geometry of a set and lead to additional optimality conditions via the Oresme rule.

Normal and tangent cones extend the classical notions of normal and tangent spaces. Informally, we recall that for a smooth mapping $F : \mathbb{R}^n \to \mathbb{R}^m$ the tangent space to

$$C = \{x \in \mathbb{R}^n \mid F(x) = 0\}$$

at a point $\bar{x} \in C$ is the collection of vectors that are "tangential" to $C$ at $\bar{x}$ and the normal space to $C$ at $\bar{x}$ is the collection of vectors that are perpendicular to all vectors in the tangent space; see Figure 4.15. In optimization and variational problems, nonsmooth functions and inequality systems are prevalent, which require us to go beyond these notions.

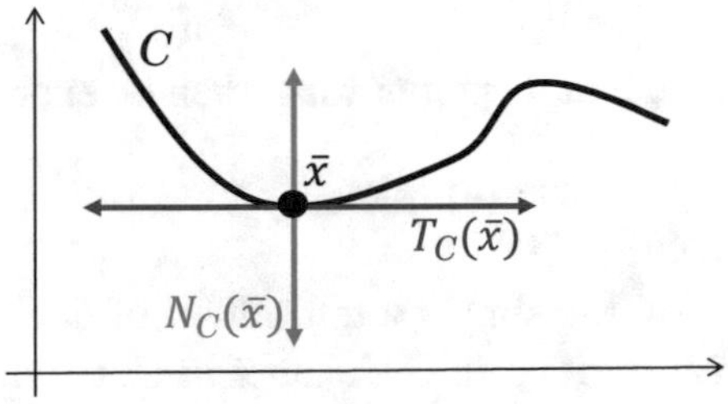


Fig. 4.15: Tangent space (red) and normal space (blue) to a set $\{x \in \mathbb{R}^2 \mid F(x) = 0\}$ at a point $\bar{x}$ as defined by a smooth mapping $F : \mathbb{R}^2 \to \mathbb{R}$.

**Definition 4.33** (tangent cone). A vector $w \in \mathbb{R}^n$ is *tangent* to a set $C \subset \mathbb{R}^n$ at $\bar{x} \in C$ if

$$(x^\nu - \bar{x})/\tau^\nu \to w \text{ for some } x^\nu \in C \to \bar{x} \text{ and } \tau^\nu \searrow 0.$$

The set of all such tangent vectors is $T_C(\bar{x})$, the *tangent cone* to $C$ at $\bar{x}$.

Figure 4.16 illustrates typical situations. In the left portion, $C$ has a "smooth" boundary at $\bar{x}$ and $T_C(\bar{x})$ is simply a halfspace formed by the usual tangent space to the boundary of $C$ at $\bar{x}$. When $\bar{x}$ is at a "corner," as seen in the middle portion of the figure, the tangent cone narrows. At the "inward kink" of the heart-shaped set to the right, the tangent vectors fan out in all directions except some pointing upward. In this figure and elsewhere, tangent vectors are drawn after being translated by $\bar{x}$, i.e., the beginning of an arrow visualizing a tangent vector $w$ is moved from the origin to $\bar{x}$. This is consistent with our understanding of a vector as uniquely defined by a direction and a magnitude.

It's clear from the definition of tangent cones that $T_C(\bar{x}) = \mathbb{R}^n$ when $\bar{x} \in \operatorname{int} C$. In the special case when $C = \{\bar{x}\}$, a single point, the zero vector is the only tangent vector. And, indeed, $T_C(\bar{x})$ is always a cone; see the definition in §2.J.

Tangent cones can be understood through set-convergence. We can view a tangent cone to $C$ at one of its points $\bar{x}$ as a "magnified" look at $C$ near $\bar{x}$. In general, magnification (or shrinking) of a set is obtained by *scalar multiplication*: For $C \subset \mathbb{R}^n$ and $\lambda \in \mathbb{R}$, the set

$$\lambda C = \{\lambda x \mid x \in C\}. \tag{4.7}$$

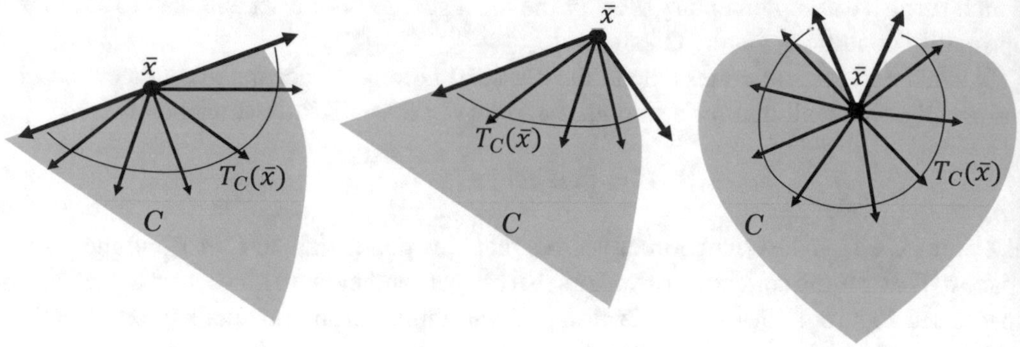


Fig. 4.16: Tangent vectors and tangent cones at $\bar{x}$, all translated to $\bar{x}$.

The tangent cone to $C$ at one of its points $\bar{x}$ can then be expressed equivalently as

$$T_C(\bar{x}) = \text{LimOut}\ \nu(C - \bar{x}).$$

Since $C - \bar{x} = \{x - \bar{x} \mid x \in C\}$ is simply a translation of $C$ toward the origin such that 0 occupies the same relative position in $C - \bar{x}$ as $\bar{x}$ in $C$, the alternative formula for the tangent cone states that $T_C(\bar{x})$ is the outer limit of this translation, magnified by $\nu$; see Figure 4.17. From this perspective, it's immediately clear that $T_C(\bar{x})$ is closed by virtue of being an outer limit.

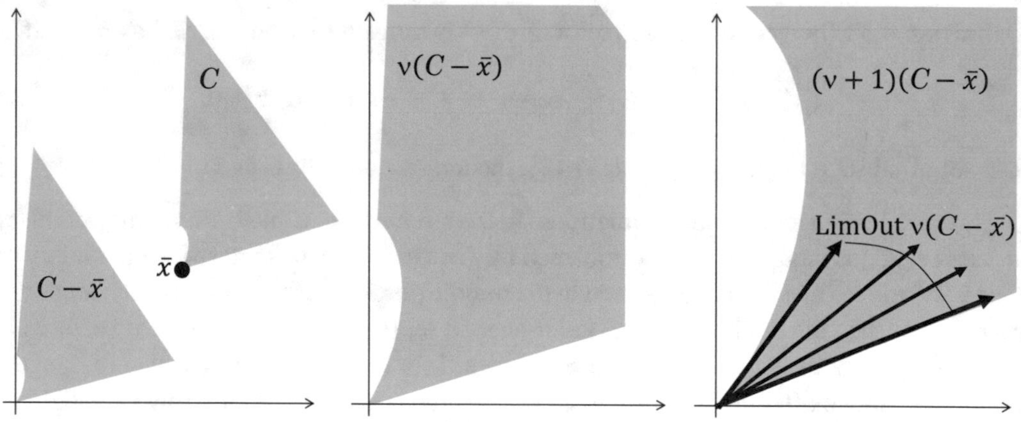


Fig. 4.17: Translation and magnification lead to the tangent cone.

**Proposition 4.34** (tangent cone to convex sets). *For a convex set $C \subset \mathbb{R}^n$ and one of its points $\bar{x}$,*

$$T_C(\bar{x}) = \text{cl}\left\{w \in \mathbb{R}^n \,\middle|\, \exists \lambda > 0 \text{ such that } \bar{x} + \lambda w \in C\right\}.$$

**Proof.** Let $K = \{w \mid \exists\lambda > 0 \text{ such that } \bar{x} + \lambda w \in C\}$. Since $C$ contains any line segment between a point in $C$ and $\bar{x}$, one has

$$K = \{\lambda(x - \bar{x}) \mid \lambda > 0, x \in C\}.$$

This implies that $K \subset T_C(\bar{x})$. Since $T_C(\bar{x})$ is closed, $\operatorname{cl} K \subset T_C(\bar{x})$. Moreover, $T_C(\bar{x}) \subset \operatorname{cl} K$ because for $w \in T_C(\bar{x})$ the definition of tangent cones guarantees the existence of $\tau^\nu \searrow 0$ and $x^\nu \in C \to \bar{x}$ such that $w^\nu = (x^\nu \to \bar{x})/\tau^\nu \to w$ and $w^\nu \in K$. □

Figure 4.18 illustrates the importance of taking the closure in the proposition. If "cl" had been left out, then the tangent vector $w$ in the figure would have been omitted from the formula as it doesn't point inside $C$ and no $\lambda > 0$ exists.

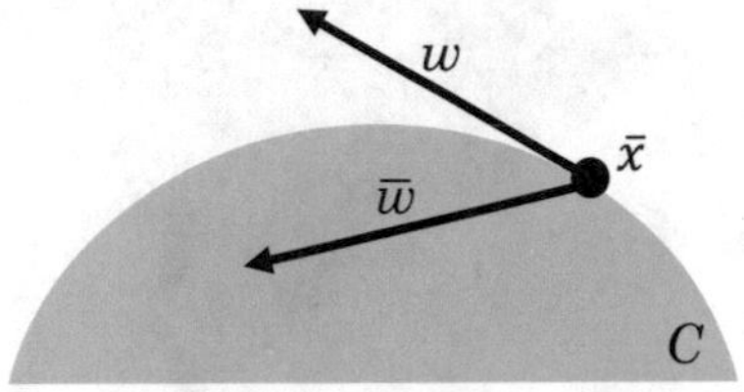


Fig. 4.18: Two tangent vectors to a convex set $C$ at $\bar{x}$.

The expression in 4.34 doesn't apply for general sets. For example, let's consider

$$C = \{x \in \mathbb{R}^2 \mid x_2 = 0\} \cup \{x \in \mathbb{R}^2 \mid x_1 = 0\}.$$

At $\bar{x} = (1, 0)$, $T_C(\bar{x}) = \{w \mid w_2 = 0\}$ while 4.34 stipulates the set

$$\{w \in \mathbb{R}^2 \mid w_2 = 0\} \cup \{w \in \mathbb{R}^2 \mid w_1 \le 0\}.$$

Chapter 2 defines the normal cone to a convex set $C \subset \mathbb{R}^n$ at a point as the set of subgradients of the indicator function $\iota_C$ at the point. But, normal cones are more fundamental, extend beyond the convex case and, in fact, furnish the basis for subgradients of general functions.

**Definition 4.35** (normal cone). For $\bar{x} \in C \subset \mathbb{R}^n$, a vector $v \in \mathbb{R}^n$ is *regular normal* to $C$ at $\bar{x}$ if

$$\langle v, w \rangle \le 0 \quad \forall w \in T_C(\bar{x}).$$

The set of all such regular normal vectors is denoted by $\widehat{N}_C(\bar{x})$.

A vector $v \in \mathbb{R}^n$ is *normal* to $C$ at $\bar{x}$ if

$$v^\nu \to v \text{ for some } v^\nu \in \widehat{N}_C(x^\nu) \text{ and } x^\nu \in C \to \bar{x}.$$

The set of all such normal vectors is $N_C(\bar{x})$, the *normal cone* to $C$ at $\bar{x}$.

We recall that $\langle v, w \rangle = \|v\|_2 \|w\|_2 \cos\alpha$, where $\alpha$ is the angle between the vectors $v$ and $w$. Thus, a regular normal vector forms an angle of between 90 and 270 degrees with every tangent vector and thereby extends the "perpendicular relationship" of normal and tangent spaces from differential geometry. Figure 4.19 illustrates regular normal vectors corresponding to the tangent cones in Figure 4.16. On the left, where the boundary is smooth at $\bar{x}$, the regular normal vectors all point in the same direction. At a kink (middle portion of the figure), the regular normal vectors fan out as the tangent cone is smaller.

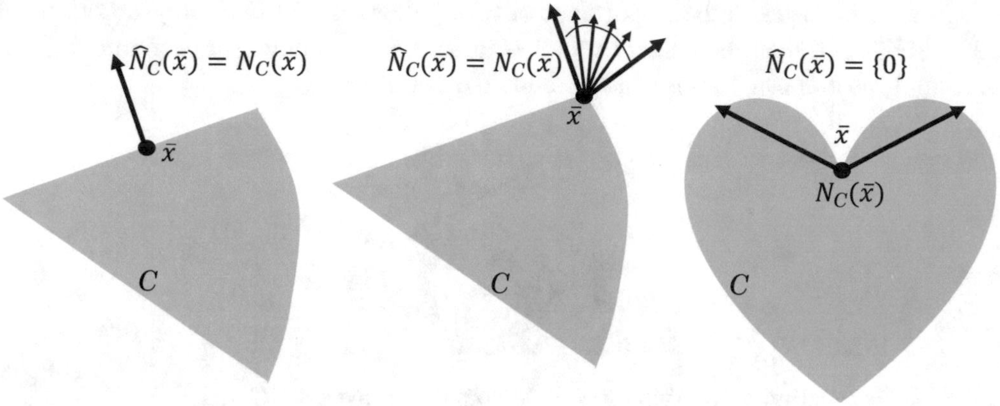


Fig. 4.19: Normal vectors and normal cones at $\bar{x}$, all translated to $\bar{x}$.

On their own, regular normal vectors fail to provide a solid basis for the treatment of "complicated" sets with inward kinks and other irregularities as seen in the right portion of Figure 4.19. At the inward kink in this heart-shaped set, $\widehat{N}_C(\bar{x}) = \{0\}$ since $T_C(\bar{x})$ fans out in nearly all directions and it becomes impossible to find a nonzero vector that forms an angle of at least 90 degrees with all tangent vectors. We obtain a robust notion of "normality" by considering the situation at points *near* $\bar{x}$ and this leads to the enrichment of vectors in $N_C(\bar{x})$ that aren't regular normals. Figure 4.20 shows regular normal vectors $v^\nu$ at points $x^\nu \in C$ approaching $\bar{x}$ from the left. In the limit, these vectors give rise to the normal vector $\bar{v}$ in the right-most portion of the figure. Likewise, points $x^\nu \in C$ approaching $\bar{x}$ from the right result in the normal vectors at $\bar{x}$ pointing northwest. It might be counterintuitive that the normal vectors at $\bar{x}$ are directed *into* the set, but this is indeed representative of the local geometry at this point and allows us to treat such cases in a meaningful way.

It follows directly from the definition that $\widehat{N}_C(\bar{x})$ is generally a convex cone (cf. §2.J) because it's the intersection of linear inequalities (cf. 1.12(a)), but as seen in Figure 4.19(right) convexity may fail for a normal cone. Still, $N_C(\bar{x})$ is indeed a cone and always contains the zero vector with this being its only element if $\bar{x} \in \operatorname{int} C$ because then $T_C(\bar{x}) = \mathbb{R}^n$. If $C = \{\bar{x}\}$, then $T_C(\bar{x}) = \{0\}$ and $N_C(\bar{x}) = \mathbb{R}^n$. We note that both $N_C(\bar{x})$ and $T_C(\bar{x})$ depend only on the properties of $C$ near $\bar{x}$. We can alter $C$ arbitrarily outside a ball $\mathbb{B}(\bar{x}, \varepsilon)$ with $\varepsilon > 0$ and retain $N_C(\bar{x})$ as well as $T_C(\bar{x})$ unchanged.

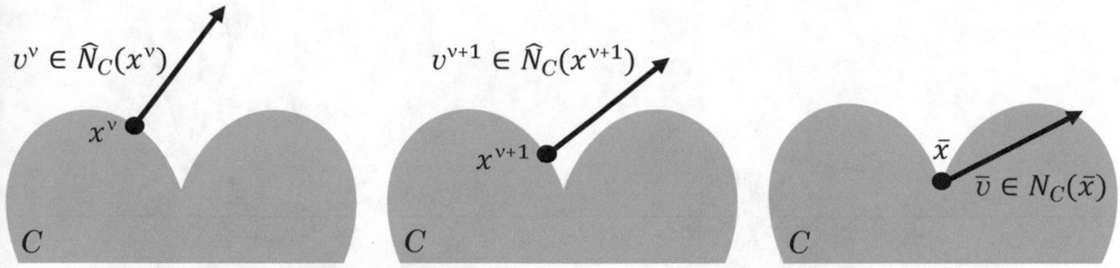


Fig. 4.20: A normal vector $\bar{v}$ as the limit of regular normal vectors.

**Exercise 4.36** (tangent and normal cones). For $C = \{x \in \mathbb{R}^2 \mid x_1 + x_2 \geq 0, -x_1^2 + x_2 \geq 0\}$, use the definitions of tangent vectors, regular normal vectors and normal vectors to derive expressions for $T_C(\bar{x})$, $\widehat{N}_C(\bar{x})$ and $N_C(\bar{x})$ at $\bar{x} = (0, 0)$. Draw $C$ as well as the vectors.

**Guide.** Directly from the definition and verified by 4.34, one can show that $T_C(\bar{x}) = \{w \mid w_1 + w_2 \geq 0, w_2 \geq 0\}$. □

We immediately obtain optimality conditions that extend 2.27 beyond the convex case.

**Proposition 4.37** (basic optimality conditions). *For $f : \mathbb{R}^n \to \overline{\mathbb{R}}$ and $C \subset \mathbb{R}^n$, suppose that $x^\star$ is a local minimizer of $f$ over $C$ and $f$ is differentiable at $x^\star$. Then,*

$$\langle \nabla f(x^\star), w \rangle \geq 0 \ \forall w \in T_C(x^\star) \quad \text{and} \quad -\nabla f(x^\star) \in \widehat{N}_C(x^\star),$$

*with the two conditions being equivalent and implying $-\nabla f(x^\star) \in N_C(x^\star)$.*

**Proof.** Since $x^\star$ is a local minimizer, $f(x) - f(x^\star) \geq 0$ for every $x \in C$ in a neighborhood of $x^\star$. By (2.4),

$$f(x) - f(x^\star) = \langle \nabla f(x^\star), x - x^\star \rangle + o(\|x - x^\star\|_2).$$

Thus,

$$\langle -\nabla f(x^\star), x - x^\star \rangle \leq o(\|x - x^\star\|_2) \ \text{ for } x \in C.$$

Let $w \in T_C(x^\star)$. By the definition of tangent vectors, there are $\tau^\nu \searrow 0$ and $x^\nu \in C \to x^\star$ such that $(x^\nu - x^\star)/\tau^\nu \to w$. Consequently,

$$\langle -\nabla f(x^\star), (x^\nu - x^\star)/\tau^\nu \rangle \leq \frac{o(\|x^\nu - x^\star\|_2)}{\|x^\nu - x^\star\|_2} \frac{\|x^\nu - x^\star\|_2}{\tau^\nu} \to 0$$

because $\|x^\nu - x^\star\|_2/\tau^\nu \to \|w\|_2$. Thus, $\langle -\nabla f(x^\star), w \rangle \leq 0$. Since $w$ is arbitrary, we've shown the asserted tangent cone expression and then also $-\nabla f(x^\star) \in \widehat{N}_C(x^\star)$ due to the definition of regular normal vectors. This also establishes the equivalence between the two conditions. Since $\widehat{N}_C(x^\star) \subset N_C(x^\star)$, the further implication holds trivially. □

The first optimality condition supplements the Oresme rule 2.10 by explicitly accounting for constraints. Instead of having to check all directions $w$, we now only

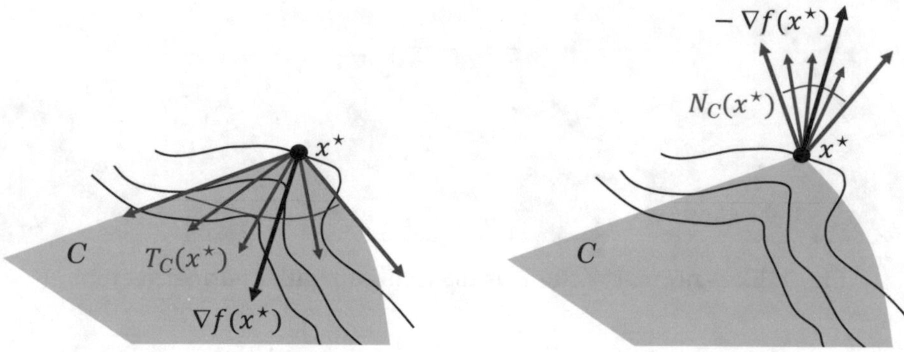


Fig. 4.21: The optimality conditions in 4.37 check the relation between $\nabla f(x^\star)$ and a tangent cone (left) and a normal cone (right).

need to consider those in the tangent cone to the feasible set at the current point as all others surely lead to infeasible points, at least for a small step size; see Figure 4.21(left). The second optimality condition is in the spirit of the Fermat rule (cf. §1.A), but accounts for constraints by replacing zero on the right-hand side by the normal cone to the feasible set at the current point; see Figure 4.21(right). If $C = \mathbb{R}^n$, then $N_C(x) = \{0\}$ for all $x \in \mathbb{R}^n$ and the condition reduces to the Fermat rule. Both optimality conditions are only *necessary* for a local minimizers, in contrast to those in 2.27, which are both necessary and sufficient. In the absence of convexity, there are no general sufficient conditions for (global) minimizers.

**Exercise 4.38** (optimality conditions). For $C$ in 4.36 and $f(x_1, x_2) = (x_2 - 1)^3$, determine all points that satisfy the optimality conditions in 4.37.

**Guide.** The minimizer is the origin, but the optimality conditions identify numerous other points as well. Thus, the conditions are only necessary for a local minimizer. □

The definition of a normal vector to $C$ at $\bar{x}$, as the limit of regular normal vectors at neighboring points, can be expressed equivalently by

$$N_C(\bar{x}) = \bigcup_{x^\nu \in C \to \bar{x}} \operatorname{LimOut} \widehat{N}_C(x^\nu),$$

which confirms that $N_C(\bar{x})$ is closed. Moreover, if $x^\nu \in C \to \bar{x}$ and $v^\nu \in \widehat{N}_C(x^\nu) \to \bar{v}$, then $\bar{v} \in N_C(\bar{x})$; see also Figure 4.20. In fact, directly from the definitions, we've obtain the following result.

**Proposition 4.39** (limits of normal vectors). *For $C \subset \mathbb{R}^n$, one has*

$$x^\nu \in C \to \bar{x} \in C \ \textit{and}\ v^\nu \in N_C(x^\nu) \to \bar{v} \implies \bar{v} \in N_C(\bar{x}).$$

**Example 4.40** (approximate solution of optimality condition). For smooth $f : \mathbb{R}^n \to \mathbb{R}$ and closed $C \subset \mathbb{R}^n$, consider the problem of minimizing $f$ over $C$. Suppose that a numerical procedure generates the sequence $\{x^\nu \in C, \nu \in \mathbb{N}\}$ with

$$\operatorname{dist}\big( - \nabla f(x^\nu), N_C(x^\nu)\big) \le \varepsilon^\nu$$

for some tolerance $\varepsilon^\nu \searrow 0$. That is, at each iteration the optimality condition $-\nabla f(x) \in N_C(x)$ of 4.37 is solved approximately to produce $x^\nu$. Then, 4.39 ensures that any cluster point $\bar{x}$ of the sequence satisfies $-\nabla f(\bar{x}) \in N_C(\bar{x})$.

**Detail.** Suppose that $x^\nu \xrightarrow[N]{} \bar{x}$ for some $N \in \mathcal{N}_\infty^\#$. By continuity, $\nabla f(x^\nu) \xrightarrow[N]{} \nabla f(\bar{x})$. Let

$$v^\nu \in \operatorname{argmin}\Big\{ \|v - \nabla f(x^\nu)\|_2 \;\Big|\; -v \in N_C(x^\nu) \Big\}.$$

Then, $\|v^\nu - \nabla f(x^\nu)\|_2 \le \varepsilon^\nu$, which implies that $v^\nu \xrightarrow[N]{} \nabla f(\bar{x})$. Since

$$-v^\nu \in N_C(x^\nu) \xrightarrow[N]{} -\nabla f(\bar{x}),$$

we must have $-\nabla f(\bar{x}) \in N_C(\bar{x})$ by 4.39. □

The inward kink in Figure 4.20 is an example of "irregularity" that when absent leads to significant simplifications. The key question is whether looking at nearby points adds any normal vectors as is the case at the inward kink of the heart-shaped set.

**Definition 4.41** (Clarke regular). A set $C \subset \mathbb{R}^n$ is *Clarke regular* at one of its points $\bar{x}$ if

$$N_C(\bar{x}) = \widehat{N}_C(\bar{x})$$

and $C \cap \mathbb{B}(\bar{x}, \varepsilon)$ is closed for some $\varepsilon > 0$.

At any point $x \in C$ but $\bar{x}$, $N_C(x) = \widehat{N}_C(x)$ in Figure 4.20; the normal cone is characterized fully by considering $x$ only and $C$, assumed to be closed, is Clarke regular at such points. However, $\widehat{N}_C(\bar{x}) = \{0\}$ and $C$ fails to be Clarke regular at $\bar{x}$. The normal cone at $\bar{x}$ is indeed influenced by the geometry at nearby points.

For convex sets, our definition of normal cones encapsulates 2.23 and the distinction between normal vectors and regular normal vectors vanishes.

**Proposition 4.42** (normal cone to convex sets). *For a convex set $C \subset \mathbb{R}^n$ and one of its points $\bar{x}$,*

$$N_C(\bar{x}) = \widehat{N}_C(\bar{x}) = \big\{ v \in \mathbb{R}^n \;\big|\; \langle v, x - \bar{x}\rangle \le 0 \;\; \forall x \in C \big\}.$$

*If $C$ is closed, then it's Clarke regular at $\bar{x}$.*

**Proof.** In the notation of the proof of 4.34,

$$\widehat{N}_C(\bar{x}) = \big\{ v \in \mathbb{R}^n \;\big|\; \langle v, w\rangle \le 0 \;\; \forall w \in \operatorname{cl} K \big\} = \big\{ v \in \mathbb{R}^n \;\big|\; \langle v, w\rangle \le 0 \;\; \forall w \in K \big\}$$

because the closure operation doesn't imply any additional restrictions on $v$. This establishes the right-most equality. Generally, $N_C(\bar{x}) \supset \widehat{N}_C(\bar{x})$ and we only need to establish the inclusion the other way for the left-most equality to hold. Suppose that $v \in N_C(\bar{x})$. Let $x \in C$ be arbitrary. Then, there exist $x^\nu \in C \to \bar{x}$ and $v^\nu \in \widehat{N}_C(x^\nu) \to v$. By the expression for regular normal vectors just established, $\langle v^\nu, x - x^\nu\rangle \le 0$. This implies that $\langle v, x - \bar{x}\rangle \le 0$ as well. Since $x$ is arbitrary, we conclude that $v \in \widehat{N}_C(\bar{x})$. □

Under Clarke regularity, there's perfect symmetry between normal and tangent cones and they're polar to each other; cf. (2.11). Consequently, an expression for one immediately gives an expression for the other as illustrated in Figure 4.22.

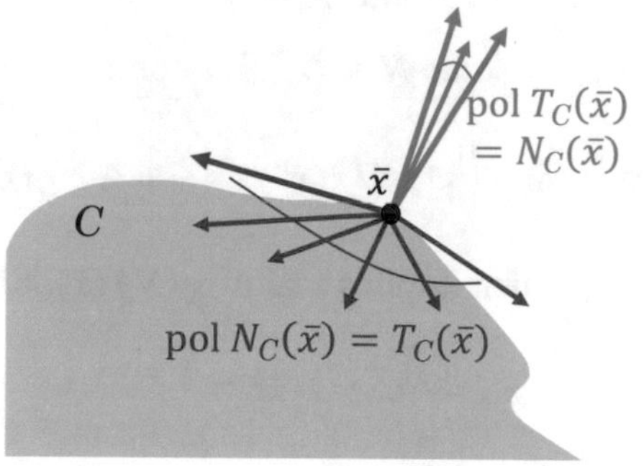


Fig. 4.22: The polarity relation between normal and tangent cones in 4.43.

**Proposition 4.43** (polarity between tangent and normal cones). *For $C \subset \mathbb{R}^n$ and one of its points $\bar{x}$ at which it's Clarke regular,*

$$N_C(\bar{x}) = \{v \in \mathbb{R}^n \mid \langle v, w\rangle \leq 0 \;\; \forall w \in T_C(\bar{x})\} = \operatorname{pol} T_C(\bar{x})$$
$$T_C(\bar{x}) = \{w \in \mathbb{R}^n \mid \langle v, w\rangle \leq 0 \;\; \forall v \in N_C(\bar{x})\} = \operatorname{pol} N_C(\bar{x}),$$

*with both cones being nonempty, closed and convex.*

**Proof.** The above discussion covers the claims about the cones being nonempty and closed. The first assertion simply reiterates the definition of normal cones under Clarke regularity. This in turn immediately establishes that $\subset$ holds in the second assertion and we only need to prove the inclusion the other way. Suppose that $\bar{w} \notin T_C(\bar{x})$. Our goal is to establish that there's a vector $\bar{v} \in N_C(\bar{x})$ such that $\langle \bar{v}, \bar{w}\rangle > 0$ because then the inclusion follows.

Since normal and tangent vectors at $\bar{x}$ are determined by the properties of $C$ in a neighborhood of $\bar{x}$, we can assume without loss of generality that $C$ is compact. A tangent cone is closed so there's $\rho > 0$ such that $\mathbb{B}(\bar{w}, \rho) \cap T_C(\bar{x}) = \emptyset$. Moreover, there's $\varepsilon > 0$ such that the compact convex set

$$B = \{\bar{x} + \tau w \mid w \in \mathbb{B}(\bar{w}, \rho), \; \tau \in [0, \varepsilon]\}$$

meets $C$ only at $\bar{x}$. Consider a sequence $\varepsilon^\nu \searrow 0$ with $\varepsilon^\nu < \varepsilon$ and

$$B^\nu = \{\bar{x} + \tau w \mid w \in \mathbb{B}(\bar{w}, \rho), \; \tau \in [\varepsilon^\nu, \varepsilon]\},$$

which is compact and convex and doesn't intersect with $C$. The function given by

$$h(x, u) = \tfrac{1}{2}\|x - u\|_2^2$$

attains its minimum over $C \times B^\nu$ at some point $(x^\nu, u^\nu)$ by 4.9. Thus, $x^\nu$ minimizes $h(\cdot, u^\nu)$ over $C$. By 4.37, this implies[6] $-\nabla_x h(x^\nu, u^\nu) \in N_C(x^\nu)$. Likewise, $-\nabla_u h(x^\nu, u^\nu) \in N_{B^\nu}(u^\nu)$ and we've confirmed the two conditions

$$u^\nu - x^\nu \in N_C(x^\nu) \quad \text{and} \quad x^\nu - u^\nu \in N_{B^\nu}(u^\nu).$$

Necessarily, $x^\nu \neq u^\nu$ because $C \cap B^\nu = \emptyset$, but $x^\nu \to \bar{x}$ and $u^\nu \to \bar{x}$ because the sets $B^\nu$ increase to $B$ (the closure of their union), and $C \cap B = \{\bar{x}\}$.

Let's define

$$v^\nu = \frac{u^\nu - x^\nu}{\|u^\nu - x^\nu\|_2},$$

so $\|v^\nu\|_2 = 1$, $v^\nu \in N_C(x^\nu)$ and $-v^\nu \in N_{B^\nu}(u^\nu)$. The sequence of vectors $\{v^\nu, \nu \in \mathbb{N}\}$ is bounded and therefore has a cluster point $\bar{v}$ with unit length. Possibly after passing to a subsequence, we can assume without loss of generality that $v^\nu \to \bar{v}$. In view of having $v^\nu \in N_C(x^\nu)$ and $x^\nu \to \bar{x}$, this implies that $\bar{v} \in N_C(\bar{x})$ by 4.39. We also have $\langle -v^\nu, u - u^\nu \rangle \leq 0$ for all $u \in B^\nu$ because $-v^\nu \in N_{B^\nu}(u^\nu)$ and $B^\nu$ is convex; see 4.42. Since $B^\nu$ increases to $B$ while $u^\nu \to \bar{x}$, we obtain in the limit that

$$\langle -\bar{v}, u - \bar{x} \rangle \leq 0 \quad \forall u \in B. \tag{4.8}$$

Among the vectors $u \in B$ are all those of the form $\bar{x} + \varepsilon w$, with $w = \bar{w} + \rho z$ and $\|z\|_2 \leq 1$. Plugging these expressions into (4.8), we get

$$\big\langle -\bar{v}, \varepsilon(\bar{w} + \rho z)\big\rangle \leq 0$$

for all such $z$. We can divide by $\varepsilon$ and take $z = -\bar{v}$ to obtain

$$\langle -\bar{v}, \bar{w} \rangle + \rho\|\bar{v}\|_2^2 \leq 0.$$

Since $\bar{v}$ has unit length, we've confirmed that $\langle \bar{v}, \bar{w} \rangle \geq \rho > 0$ as needed.

By the established formulae and 1.12(a), the cones are convex. □

Next we turn to practical formulae for normal and tangent cones expressed in terms of quantities typically used to specify a set such as constraint functions. As indicated by 4.37, this leads to numerous optimality conditions.

**Proposition 4.44** (product sets). *For closed sets $C_i \subset \mathbb{R}^{n_i}$, $i = 1, \dots, m$, and*

$$C = C_1 \times \cdots \times C_m \subset \mathbb{R}^n,$$

*where $\sum_{i=1}^m n_i = n$, consider $\bar{x} = (\bar{x}_1, \dots, \bar{x}_m)$, with $\bar{x}_i \in C_i$. Then,*

$$N_C(\bar{x}) = N_{C_1}(\bar{x}_1) \times \cdots \times N_{C_m}(\bar{x}_m)$$
$$T_C(\bar{x}) \subset T_{C_1}(\bar{x}_1) \times \cdots \times T_{C_m}(\bar{x}_m).$$

[6] For $h$ taking two arguments $x$ and $u$, we indicate a gradient with respect $x$ by $\nabla_x h$ and with respect to $u$ by $\nabla_u h$.

*Moreover, $C$ is Clarke regular at $\bar{x}$ if and only if each $C_i$ is Clarke regular at $\bar{x}_i$. In this case, the inclusion is replaced by an equality.*

**Proof.** If $(w_1, \ldots, w_m) \in T_C(\bar{x})$, with $w_i \in \mathbb{R}^{n_i}$, then, by the definition of tangent vectors in 4.33, there are

$$(x_1^\nu, \ldots, x_m^\nu) \in C \to (\bar{x}_1, \ldots, \bar{x}_m)$$

and $\tau^\nu \searrow 0$ such that $(x_i^\nu - \bar{x}_i)/\tau^\nu \to w_i$ for all $i = 1, \ldots, m$. Since $x_i^\nu \in C_i \to \bar{x}_i$ as well, $w_i \in T_{C_i}(\bar{x}_i)$. We've established the inclusion for tangent cones. For the remaining parts, see [105, Proposition 6.41]. □

**Example 4.45** (normal and tangent cones to a box). For the box

$$C = \prod\nolimits_{j=1}^{n} C_j, \quad \text{where } C_j = \{\gamma \in \mathbb{R} \mid \alpha_j \le \gamma \le \beta_j\}$$

and $-\infty \le \alpha_j \le \beta_j \le \infty$, suppose that $\bar{x} = (\bar{x}_1, \ldots, \bar{x}_n) \in C$. Then,

$$N_C(\bar{x}) = N_{C_1}(\bar{x}_1) \times \cdots \times N_{C_n}(\bar{x}_n) \qquad T_C(\bar{x}) = T_{C_1}(\bar{x}_1) \times \cdots \times T_{C_n}(\bar{x}_n)$$

$$N_{C_j}(\bar{x}_j) = \begin{cases} (-\infty, \infty) & \text{if } \alpha_j = \bar{x}_j = \beta_j \\ \{0\} & \text{if } \alpha_j < \bar{x}_j < \beta_j \\ [0, \infty) & \text{if } \alpha_j < \bar{x}_j = \beta_j \\ (-\infty, 0] & \text{if } \alpha_j = \bar{x}_j < \beta_j \end{cases} \qquad T_{C_j}(\bar{x}_j) = \begin{cases} \{0\} & \text{if } \alpha_j = \bar{x}_j = \beta_j \\ (-\infty, \infty) & \text{if } \alpha_j < \bar{x}_j < \beta_j \\ (-\infty, 0] & \text{if } \alpha_j < \bar{x}_j = \beta_j \\ [0, \infty) & \text{if } \alpha_j = \bar{x}_j < \beta_j. \end{cases}$$

**Detail.** The expressions for the normal cones are furnished by 2.31. The tangent cones follow by 4.44 and 4.43. □

The next result provides far-reaching expressions for normal and tangent cones and is the foundation from which we derive many specialized formulae. We put off the proof to the next section.

**Theorem 4.46** (tangent and normal cones for constraint systems). *For $F : \mathbb{R}^n \to \mathbb{R}^m$ and closed sets $X \subset \mathbb{R}^n$ and $D \subset \mathbb{R}^m$, let*

$$C = \{x \in X \mid F(x) \in D\}.$$

*Suppose that $F$ is smooth in a neighborhood of $\bar{x} \in C$ and the following qualification holds:*

$$y \in N_D\big(F(\bar{x})\big) \quad \textit{and} \quad -\nabla F(\bar{x})^\top y \in N_X(\bar{x}) \implies y = 0. \tag{4.9}$$

*Then,*

$$N_C(\bar{x}) \subset \big\{\nabla F(\bar{x})^\top y + z \mid y \in N_D\big(F(\bar{x})\big),\ z \in N_X(\bar{x})\big\}.$$

*If in addition $X$ and $D$ are Clarke regular at $\bar{x}$ and $F(\bar{x})$, respectively, then $C$ is Clarke regular at $\bar{x}$, the inclusion for the normal cone holds with equality and*

$$T_C(\bar{x}) = \big\{w \in T_X(\bar{x}) \mid \nabla F(\bar{x})w \in T_D\big(F(\bar{x})\big)\big\}.$$

**Example 4.47** (equality constraint). At $\bar{x} = (1, 1)$, the set $C = \{x \mid F(x) = x_1^2 - x_2 = 0\}$ has

$$N_C(\bar{x}) = \big\{(2, -1)y \,\big|\, y \in \mathbb{R}\big\} \quad \text{and} \quad T_C(\bar{x}) = \big\{(w_1, w_2) \,\big|\, 2w_1 = w_2\big\}.$$

**Detail.** In 4.46, we can set $X = \mathbb{R}^2$ and $D = \{0\} \subset \mathbb{R}$, which both are closed convex sets and thus are Clarke regular at all their points by 4.42. Consequently, $N_X(\bar{x}) = \{(0, 0)\}$ and $N_D(0) = \mathbb{R}$. The Jacobian $\nabla F(x) = (2x_1, -1)$. This leads to the asserted expression for $N_C(\bar{x})$. Since $T_X(\bar{x}) = \mathbb{R}^2$ and $T_D(0) = \{0\}$, we obtain the expression for $T_C(\bar{x})$. The set $C$ is a smooth curve. Thus, $N_C(\bar{x})$ and $T_C(\bar{x})$ coincide with the classical normal space and tangent space to $C$ at $\bar{x}$, respectively.

The obtained formulae are contingent on the qualification (4.9). In this case, it amounts to checking whether $y \in \mathbb{R}$ and $(-2, 1)y = (0, 0)$ imply $y = 0$. Since $-2y = 0$ and $1y = 0$, this is certainly the case. □

**Exercise 4.48** (inequality constraints). For the data in 4.36, apply 4.46 and obtain expressions for $N_C(\bar{x})$ and $T_C(\bar{x})$. Verify the qualification (4.9).

**Guide.** Set $F(x) = (x_1 + x_2, -x_1^2 + x_2)$, $D = [0, \infty)^2$ and $X = \mathbb{R}^2$. □

**Example 4.49** (equalities and inequalities). For $f_i : \mathbb{R}^n \to \mathbb{R}$, $i = 1, \dots, m$, and $g_i : \mathbb{R}^n \to \mathbb{R}$, $i = 1, \dots, q$, all smooth functions, let's consider

$$C = \big\{x \in \mathbb{R}^n \,\big|\, f_i(x) = 0,\ i = 1, \dots, m;\ \ g_i(x) \le 0,\ i = 1, \dots, q\big\}.$$

If the qualification (4.9) holds at $\bar{x} \in C$, then

$$\begin{aligned} N_C(\bar{x}) = \Big\{\textstyle\sum_{i=1}^m y_i \nabla f_i(\bar{x}) + \sum_{i=1}^q z_i \nabla g_i(\bar{x}) \,\Big|\, & \forall i = 1, \dots, m:\ \ y_i \in \mathbb{R} \\ & \forall i = 1, \dots, q:\ \ z_i = 0 \text{ if } g_i(\bar{x}) < 0;\ \ z_i \ge 0 \text{ otherwise}\Big\} \\ T_C(\bar{x}) = \Big\{w \in \mathbb{R}^n \,\Big|\, & \forall i = 1, \dots, m:\ \ \big\langle \nabla f_i(\bar{x}), w\big\rangle = 0 \\ & \forall i = 1, \dots, q:\ \ \big\langle \nabla g_i(\bar{x}), w\big\rangle \le 0 \text{ if } g_i(\bar{x}) = 0\Big\}. \end{aligned}$$

**Detail.** In 4.46, we set $X = \mathbb{R}^n$,

$$F(x) = \big(f_1(x), \dots, f_m(x), g_1(x), \dots, g_q(x)\big)$$

and $D = D_1 \times \cdots \times D_{m+q}$, where $D_i = \{0\}$ for $i = 1, \dots, m$ and $D_{m+i} = (-\infty, 0]$ for $i = 1, \dots, q$. Then, $N_X(\bar{x}) = \{0\}$ and $T_X(\bar{x}) = \mathbb{R}^n$. Since both $X$ and $D$ are closed and convex, they're Clarke regular by 4.42. In view of 4.45,

$$N_D\big(F(\bar{x})\big) = N_{D_1}\big(f_1(\bar{x})\big) \times \cdots \times N_{D_m}\big(f_m(\bar{x})\big) \times N_{D_{m+1}}\big(g_1(\bar{x})\big) \times \cdots \times N_{D_{m+q}}\big(g_q(\bar{x})\big),$$

where for $i = 1, \dots, m$, $N_{D_i}(f_i(\bar{x})) = \mathbb{R}$ and for $i = 1, \dots, q$,

$$N_{D_{m+i}}\big(g_i(\bar{x})\big) = \begin{cases} \{0\} & \text{if } g_i(\bar{x}) < 0 \\ [0, \infty) & \text{otherwise.} \end{cases}$$

Likewise, the tangent cones become

$$T_D\big(F(\bar{x})\big) = T_{D_1}\big(f_1(\bar{x})\big) \times \cdots \times T_{D_m}\big(f_m(\bar{x})\big) \times T_{D_{m+1}}\big(g_1(\bar{x})\big) \times \cdots \times T_{D_{m+q}}\big(g_q(\bar{x})\big),$$

where for $i = 1, \dots, m$, $T_{D_i}(f_i(\bar{x})) = \{0\}$ and for $i = 1, \dots, q$,

$$T_{D_{m+i}}\big(g_i(\bar{x})\big) = \begin{cases} \mathbb{R} & \text{if } g_i(\bar{x}) < 0 \\ (-\infty, 0] & \text{otherwise.} \end{cases}$$

With these specializations, $z$ in 4.46 is zero while $y \in \mathbb{R}^{m+q}$ consists of two parts. In a change of notation, let this two-part vector be denoted by $(y, z)$, with $y \in \mathbb{R}^m$ and $z \in \mathbb{R}^q$ corresponding to the equality and inequality constraints, respectively. This leads to the asserted expressions. For example, $(y, z) \in N_D(F(\bar{x}))$ imposes no requirement on $y$ but restricts the sign of $z$.

For this constraint systems, the qualification (4.9) amounts to

$$\begin{aligned} &\sum\nolimits_{i=1}^{m} y_i \nabla f_i(\bar{x}) + \sum\nolimits_{i \in \mathbb{A}(\bar{x})} z_i \nabla g_i(\bar{x}) = 0; \;\; z_i \geq 0, \; i \in \mathbb{A}(\bar{x}) \\ &\qquad \Longrightarrow \; y_i = 0, \; i = 1, \dots, m; \;\; z_i = 0, \; i \in \mathbb{A}(\bar{x}), \end{aligned} \tag{4.10}$$

where $\mathbb{A}(\bar{x}) = \{i \mid g_i(\bar{x}) = 0\}$, and is called the *Mangasarian-Fromovitz constraint qualification*. A sufficient condition for it to hold is that $\{\nabla f_i(\bar{x}), i = 1, \dots, m; \nabla g_i(\bar{x}), i \in \mathbb{A}(\bar{x})\}$ are linearly independent vectors. □

**Exercise 4.50** (canonical inequality constraints). For the data in 4.36, reformulate $C$ such that it fits the format of 4.49 and use the formulae provided there to obtain expressions for $N_C(\bar{x})$ and $T_C(\bar{x})$. Verify the Mangasarian-Fromovitz constraint qualification.

**Guide.** Set $g_1(x) = -x_1 - x_2$ and $g_2(x) = x_1^2 - x_2$. □

**Example 4.51** (composite constraints). For lsc convex functions $h_i : \mathbb{R}^m \to \overline{\mathbb{R}}, i = 1, \dots, q$, and a smooth mapping $F : \mathbb{R}^n \to \mathbb{R}^m$, let

$$C = \big\{x \in \mathbb{R}^n \;\big|\; h_i\big(F(x)\big) \leq 0, \; i = 1, \dots, q\big\}.$$

At first, it may seem that $C$ falls outside 4.46 because $h_i$ could be nonsmooth. However, we can equivalently write

$$C = \big\{x \in \mathbb{R}^n \;\big|\; F(x) \in D\big\}, \quad \text{where } D = \big\{u \in \mathbb{R}^m \;\big|\; h_i(u) \leq 0, \; i = 1, \dots, q\big\}.$$

Since $D$ is closed and convex by 4.8 and 1.11, it's Clarke regular at all its points (cf. 4.42) and we can apply 4.46.

**Detail.** Verifying whether a vector $v \in N_C(\bar{x})$ for some $\bar{x} \in C$ would then amount to finding $y \in N_D(F(\bar{x}))$ with $v = \nabla F(\bar{x})^\top y$. An expression for $N_D(F(\bar{x}))$ may be readily available from Chap. 2; see also 6.20. □

It's sometimes convenient to build up formulae for normal and tangent cones for constraint systems using the cones for individual constraints.

**Proposition 4.52** (cones for intersection of sets). *For $C = C_1 \cap \cdots \cap C_m$ and $\bar{x} \in C$, suppose that each $C_i \subset \mathbb{R}^n$ is closed and*

$$y_i \in N_{C_i}(\bar{x}),\ i = 1, \ldots, m, \quad \text{and} \quad \sum\nolimits_{i=1}^m y_i = 0 \quad \Longrightarrow \quad y_i = 0,\ i = 1, \ldots, m.$$

*Then,*

$$N_C(\bar{x}) \subset N_{C_1}(\bar{x}) + \cdots + N_{C_m}(\bar{x}).$$

*If in addition every $C_i$ is Clarke regular at $\bar{x}$, then $C$ is Clarke regular at $\bar{x}$, the above inclusion holds with equality and*

$$T_C(\bar{x}) = T_{C_1}(\bar{x}) \cap \cdots \cap T_{C_m}(\bar{x}).$$

**Proof.** Let $X = \mathbb{R}^n$, $D = C_1 \times \cdots \times C_m \subset (\mathbb{R}^n)^m$ and $F : \mathbb{R}^n \to (\mathbb{R}^n)^m$ be defined by $F(x) = (x, \ldots, x)$. Then, $\nabla F(\bar{x})^\top y = y_1 + \cdots + y_m$, with $y = (y_1, \ldots, y_m) \in (\mathbb{R}^n)^m$. By 4.44, one has

$$N_D\big(F(\bar{x})\big) = N_{C_1}(\bar{x}) \times \cdots \times N_{C_m}(\bar{x}) \quad \text{and} \quad T_D\big(F(\bar{x})\big) = T_{C_1}(\bar{x}) \times \cdots \times T_{C_m}(\bar{x}),$$

the latter requiring the additional assumption of Clarke regularity. By 4.46,

$$N_C(\bar{x}) \subset \big\{y_1 + \cdots + y_m \,\big|\, y_i \in N_{C_i}(\bar{x})\big\}$$

with a strengthening to an equality under Clarke regularity. Similarly, $\nabla F(\bar{x})w \in T_D(F(\bar{x}))$ simplifies to $w \in T_{C_i}(\bar{x})$ for all $i$.

The qualification (4.9) translates directly to the condition indicated. □

## 4.H Tangent and Normal Cones II

We'll now prove the asserted cone expressions for polyhedral sets in 2.44 and for more general constraint systems in 4.46. The first proof is rather straightforward and also leads to a formula for tangent cones to polyhedral sets.

**Proof of 2.44.** Let $\lambda > 0$ and $w \in \mathbb{R}^n$. For $i \in \mathbb{A}(\bar{x})$, a point $\bar{x} + \lambda w$ satisfies the inequality $\langle D_i, x\rangle \le d_i$ if and only if $\langle D_i, w\rangle \le 0$. If $i \notin \mathbb{A}(\bar{x})$, then the inequality is satisfied automatically if $\lambda$ is sufficiently small. Moreover, $\bar{x} + \lambda w$ satisfies the equality $Ax = b$ if and only if $Aw = 0$. By the expression for tangent cones to convex sets in 4.34, we've that

$$T_C(\bar{x}) = \{w \in \mathbb{R}^n \mid Aw = 0, \ \langle D_i, w\rangle \le 0 \ \forall i \in \mathbb{A}(\bar{x})\}. \tag{4.11}$$

Since $C$ is polyhedral and thus Clarke regular at $\bar{x}$ (cf. 4.42), 4.43 applies and $N_C(\bar{x}) = \operatorname{pol} T_C(\bar{x})$. Let's write $Aw = 0$ as $\langle A_i, w\rangle \le 0$ and $\langle -A_i, w\rangle \le 0$, $i = 1, \dots, m$, where $A_i$ is the $i$th row of $A$. Thus, $T_C(\bar{x})$ can be expressed in terms of inequalities only and we apply 2.60 to obtain

$$\begin{aligned} N_C(\bar{x}) &= \operatorname{pos}\{A_1, \dots, A_m, -A_1, \dots, -A_m, D_i, i \in \mathbb{A}(\bar{x})\} \\ &= \Big\{\sum\nolimits_{i=1}^m (y_i - y_{m+i})A_i + \sum\nolimits_{i\in\mathbb{A}(\bar{x})} z_i D_i \ \Big|\ y_i \ge 0,\ i = 1, \dots, 2m;\ \ z_i \ge 0,\ i \in \mathbb{A}(\bar{x})\Big\}. \end{aligned}$$

The conclusion then follows by introducing $z_i = 0$ for $i \notin \mathbb{A}(\bar{x})$ and replacing $y_i - y_{m+i}$ by $y_i$ without sign restriction. □

The following proof of 4.46 is lengthy, but it doesn't use any advanced mathematical tools; just facts about sequences and continuity. It's also possible to prove 2.44 along similar lines, which in particular bypasses 2.60 used above. One can establish that the qualification (4.9) holds automatically for polyhedral sets so that expressions for normal and tangent cones of equality and inequality systems in 4.49 apply and these then reduce to 2.44. The insight from such developments establishes that in the presence of a *mix* of linear constraints and smooth constraint functions, we can relax the qualification (4.9) to largely only involve the smooth functions. Moreover, if the smooth functions are convex, then an alternative Slater constraint qualification can be substituted in; see 5.47 and [92, Theorem 4.4] for such refinements.

Let's start with an intermediate concept. For a closed set $C \subset \mathbb{R}^n$ and $\bar{x} \in C$, a vector $v \in \mathbb{R}^n$ is a *proximal normal vector* to $C$ at $\bar{x}$ when $\bar{x} \in \operatorname{prj}_C(\bar{x} + \tau v)$ for some $\tau \in (0, \infty)$. When this holds, $\operatorname{prj}_C(\bar{x} + \tau' v) = \{\bar{x}\}$ for every $\tau' \in (0, \tau)$, which follows by the triangle inequality.

Following the reasoning in 2.30 and invoking the optimality condition 4.37,

$$\begin{aligned} \bar{x} \in \operatorname{prj}_C(\bar{x} + \tau v) &= \operatorname{argmin}_{x\in C} \tfrac{1}{2}\|x - \bar{x} - \tau v\|_2^2 = \operatorname{argmin}_{x\in C} \tfrac{1}{2\tau}\|x - \bar{x} - \tau v\|_2^2 \\ &\Longrightarrow\ v \in \widehat{N}_C(\bar{x}). \end{aligned}$$

Thus, the proximal normal vectors to $C$ at $\bar{x}$ form a subset of $\widehat{N}_C(\bar{x})$. The subset can be strict as seen in Figure 4.23 for the example $C = \operatorname{epi} f$, where

$$f(\alpha) = \begin{cases} \alpha^{3/5} & \text{if } \alpha \ge 0 \\ -(-\alpha)^{3/5} & \text{otherwise.} \end{cases}$$

**Lemma 4.53** *For a closed set $C \subset \mathbb{R}^n$ and $\bar{x} \in C$, suppose that $v \in N_C(\bar{x})$. Then, there exist $x^\nu \in C \to \bar{x}$ and $v^\nu \to v$, with $v^\nu$ being a proximal normal vector to $C$ at $x^\nu$.*

**Proof.** Let $\tau^\nu \searrow 0$ and

$$\varphi^\nu(x) = \tfrac{1}{2\tau^\nu}\|x - \bar{x} - \tau^\nu v\|_2^2.$$

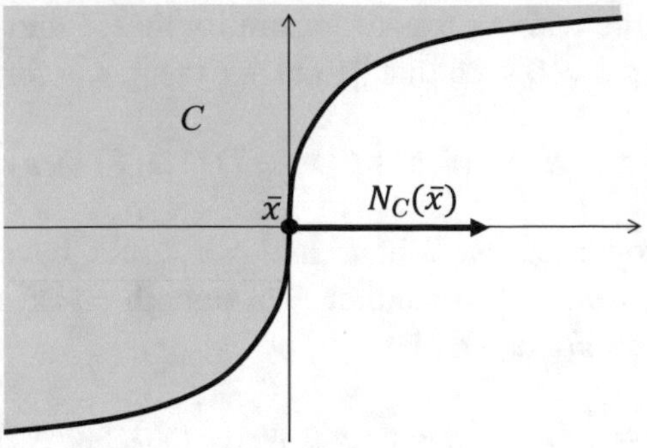


Fig. 4.23: At $\bar{x} = (0, 0)$, $N_C(\bar{x}) = \widehat{N}_C(\bar{x}) = \{\lambda(1, 0) \mid \lambda \geq 0\}$ but the only proximal normal vector at that point is $(0, 0)$.

For every $\nu$, $\varphi^\nu(x) \to \infty$ as $\|x\|_2 \to \infty$ and it follows by 4.9 that there's $x^\nu \in \operatorname{argmin}_C \varphi^\nu$. Since

$$x - \bar{x} - \tau^\nu v = x - x^\nu - \tau^\nu(v - w^\nu), \text{ with } w^\nu = \tfrac{1}{\tau^\nu}(x^\nu - \bar{x}),$$

we also have

$$x^\nu \in \operatorname{prj}_C \left(x^\nu + \tau^\nu(v - w^\nu)\right)$$

so, by definition, $v - w^\nu$ is a proximal normal vector to $C$ at $x^\nu$. In view of the optimality of $x^\nu$, one has

$$\tfrac{1}{2}\tau^\nu \|v\|_2^2 = \varphi^\nu(\bar{x}) \geq \varphi^\nu(x^\nu) = \tfrac{1}{2}\tau^\nu \|v - w^\nu\|_2^2$$

and also

$$\|v\|_2^2 \geq \|v\|_2^2 - 2\langle v, w^\nu\rangle + \|w^\nu\|_2^2.$$

This implies further that

$$\|w^\nu\|_2^2 \leq 2\langle v, w^\nu\rangle \leq 2\|v\|_2\|w^\nu\|_2 \quad \text{and} \quad \|w^\nu\|_2 \leq 2\|v\|_2.$$

Since $\{w^\nu, \nu \in \mathbb{N}\}$ is bounded, it has a cluster point $w$ and $x^\nu \to \bar{x}$ because $\tau^\nu \searrow 0$. Since $w^\nu$ is a quotient of the kind required in the definition of tangent vectors (see 4.33), $w \in T_C(\bar{x})$ and then $\langle v, w\rangle \leq 0$ when $v$ is a regular normal vector. Recalling that $\|v\|_2^2 \geq \|v - w^\nu\|_2^2$, we must also have $\|v\|_2^2 \geq \|v - w\|_2^2$ and then

$$\|w\|_2^2 \leq 2\langle v, w\rangle \leq 0$$

when $v$ is a regular normal vector. Hence, $w = 0$ and the only possible cluster point of $\{w^\nu, \nu \in \mathbb{N}\}$ is zero; the assertion holds with $v^\nu = v - w^\nu$ if $v$ is a regular normal vector. Since a general normal vector is by definition the limit of regular normal vectors, each of which can be approximated by proximal normal vectors as just established, it follows immediately that the result holds for any normal vector. □

**Proof of 4.46.** We assume without loss of generality that $X$ and $D$ are compact because for any $\varepsilon \in (0, \infty)$ there's $\delta > 0$ such that $\|F(x) - F(\bar{x})\|_2 \leq \varepsilon$ for all $x \in \mathbb{B}(\bar{x}, \delta)$ and then

$$\{x \in X \cap \mathbb{B}(\bar{x}, \delta) \mid F(x) \in D \cap \mathbb{B}(F(\bar{x}), \varepsilon)\}$$

has the same local properties as $C$ at $\bar{x}$ and we could have proceeded with these intersections. Likewise, we can assume that $F$ is smooth on $\mathbb{R}^n$ as a similar localization can always be carried out. For $x \in C$, let

$$S(x) = \{\nabla F(x)^\top y + z \mid y \in N_D(F(x)), z \in N_X(x)\}.$$

The proof consists of two parts: The first one leading to $N_C(\bar{x}) \subset S(\bar{x})$ and the second one to the remaining assertions under the additional assumption of Clarke regularity.

**Part 1.** This part consists of three steps. First, we show that the qualification (4.9) holding at $\bar{x}$ implies that it must also hold at all points $x \in C$ in a neighborhood of $\bar{x}$. Second, we establish that $v \in S(\bar{x})$ holds for proximal normal vectors $v$. Third, we bring in 4.53 to confirm that the inclusion holds for all normal vectors.

**Part 1a.** For the sake of contradiction, suppose that there are points $x^\nu \in C \to \bar{x}$ such that the qualification fails at every $x^\nu$, i.e., there exist $\{y^\nu \neq 0, \nu \in \mathbb{N}\}$ such that

$$y^\nu \in N_D(F(x^\nu)) \quad \text{and} \quad -\nabla F(x^\nu)^\top y^\nu \in N_X(x^\nu).$$

Since $N_D(F(x^\nu))$ and $N_X(x^\nu)$ are cones, the scaled vectors $\bar{y}^\nu = y^\nu / \|y^\nu\|_2$ can be substituted for $y^\nu$. Moreover, $\{\bar{y}^\nu, \nu \in \mathbb{N}\}$ is bounded and therefore has a cluster point $\bar{y}$. By 4.39, this implies that

$$\bar{y} \in N_D(F(\bar{x})) \quad \text{and} \quad -\nabla F(\bar{x})^\top \bar{y} \in N_X(\bar{x}),$$

which contradict the qualification (4.9) because $\bar{y}$ is nonzero. Consequently, the qualification holds also for all $x \in C$ in a neighborhood of $\bar{x}$.

**Part 1b.** Suppose that $v$ is a proximal normal vector to $C$ at $\bar{x}$. Then, there's $\tau \in (0, \infty)$ such that $\{\bar{x}\} = \mathrm{prj}_C(\bar{x} + \tau v)$. Let's define

$$h(x) = \tfrac{1}{2\tau}\|x - \bar{x} - \tau v\|_2^2.$$

In view of the discussion prior to 4.53, $\{\bar{x}\} = \mathrm{argmin}_C\, h$ and $\nabla h(\bar{x}) = -v$. Let $\tau^\nu \searrow 0$ and define the smooth functions $\varphi^\nu$ on $\mathbb{R}^n \times \mathbb{R}^m$ by

$$\varphi^\nu(x, u) = h(x) + \tfrac{1}{2\tau^\nu}\|F(x) - u\|_2^2.$$

Let's analyze the problem of minimizing $\varphi^\nu$ over the compact set $X \times D$. By 4.9, there's

$$(x^\nu, u^\nu) \in \mathrm{argmin}_{X \times D}\, \varphi^\nu$$

for each $\nu$. Minimizing $\varphi^\nu$ over $X \times D$ can be viewed as arising from the penalty method for the problem of minimizing $h(x)$ subject to $F(x) = u$, $x \in X$, $u \in D$. Consequently, every cluster point of $\{(x^\nu, u^\nu), \nu \in \mathbb{N}\}$ is a minimizer of the latter problem by 4.21 and the convergence statement 4.14 for the epigraphical approximation algorithm. That problem has a unique solution $(\bar{x}, F(\bar{x}))$ since it's equivalent to minimizing $h$ over $C$. Consequently, $(x^\nu, u^\nu) \to (\bar{x}, F(\bar{x}))$ because the sequence is contained in a compact set.

The minimum of $\varphi^\nu(x^\nu, \cdot\,)$ over $D$ is attained at $u^\nu$, whereas the minimum of $\varphi^\nu(\cdot\,, u^\nu)$ over $X$ is attained at $x^\nu$. In view of the optimality condition 4.37, this implies that

$$-\nabla_u \varphi^\nu(x^\nu, u^\nu) \in N_D(u^\nu) \quad \text{and} \quad -\nabla_x \varphi^\nu(x^\nu, u^\nu) \in N_X(x^\nu).$$

Let's define

$$y^\nu = -\nabla_u \varphi^\nu(x^\nu, u^\nu) = \big(F(x^\nu) - u^\nu\big)/\tau^\nu$$
$$z^\nu = -\nabla_x \varphi^\nu(x^\nu, u^\nu) = -\nabla h(x^\nu) - \nabla F(x^\nu)^\top y^\nu.$$

Moreover, $\nabla F(x^\nu) \to \nabla F(\bar{x})$ and

$$\nabla h(x^\nu) = \tfrac{1}{\tau}(x^\nu - \bar{x} - \tau v) \to -v = \nabla h(\bar{x}).$$

There are two cases to distinguish: either (i) $\{y^\nu, \nu \in \mathbb{N}\}$ has a cluster point $y$ or (ii) it escapes to the horizon.

For (i), 4.39 guarantees that $y \in N_D(F(\bar{x}))$ and then $z^\nu \to z = v - \nabla F(\bar{x})^\top y$ after passing to a subsequence. Again by 4.39, this means that $z \in N_X(\bar{x})$ and $v \in S(\bar{x})$.

For (ii), the vectors $\bar{y}^\nu = y^\nu/\|y^\nu\|_2$, which like $y^\nu$ belong to $N_D(F(x^\nu))$, are bounded because $\|\bar{y}^\nu\|_2 = 1$ so then must have a cluster point $\bar{y}$ with $\|\bar{y}\|_2 = 1$. Let $\bar{z}^\nu = z^\nu/\|y^\nu\|_2$, which like $z^\nu$ belongs to $N_X(x^\nu)$. Then,

$$\bar{z}^\nu = -\nabla h(x^\nu)\frac{1}{\|y^\nu\|_2} - \nabla F(x^\nu)^\top \bar{y}^\nu \to 0 - \nabla F(\bar{x})^\top \bar{y}$$

after passing to a subsequence because $\|y^\nu\|_2 \to \infty$. By 4.39,

$$\bar{y} \in N_D\big(F(\bar{x})\big) \quad \text{and} \quad -\nabla F(\bar{x})^\top \bar{y} \in N_X(\bar{x}).$$

But $\bar{y} \neq 0$, so this is impossible under the qualification (4.9). Thus, only case (i) is viable and we've shown that $v \in S(\bar{x})$.

**Part 1c.** The arguments in Part 1b rely on the qualification (4.9) holding at $\bar{x}$, but as established in Part 1a the qualification also holds at $x \in C$ in a neighborhood of $\bar{x}$. Consequently, we can repeat the arguments in Part 1b for such $x$ and conclude that if $v$ is a proximal normal vector to $C$ at $x$, then $v \in S(x)$.

Now, let $v \in N_C(\bar{x})$. By 4.53, there exist $x^\nu \in C \to \bar{x}$ and $v^\nu \to v$ with $v^\nu$ being a proximal normal vector to $C$ at $x^\nu$, so then, as just argued, $v^\nu \in S(x^\nu)$, i.e.,

$$v^\nu = \nabla F(x^\nu)^\top y^\nu + z^\nu \;\text{ for some }\; y^\nu \in N_D\big(F(x^\nu)\big) \;\text{ and }\; z^\nu \in N_X(x^\nu).$$

Again we consider two cases: either (i) $\{y^\nu, \nu \in \mathbb{N}\}$ has a cluster point $y$ or (ii) it escapes to the horizon.

For (i), 4.39 guarantees that $y \in N_D(F(\bar{x}))$ and, after passing to subsequences,

$$z^\nu = v^\nu - \nabla F(x^\nu)^\top y^\nu \to v - \nabla F(\bar{x})^\top y$$

with that limit, denoted by $z$, then satisfying $z \in N_X(\bar{x})$ by 4.39. Consequently, $v \in S(\bar{x})$.

For (ii), the vectors $\bar{y}^\nu = y^\nu / \|y^\nu\|_2$, which like $y^\nu$ belong to $N_D(F(x^\nu))$, are bounded because $\|\bar{y}^\nu\|_2 = 1$ so then must have a cluster point $\bar{y}$ with $\|\bar{y}\|_2 = 1$. Let $\bar{z}^\nu = z^\nu / \|y^\nu\|_2$, which like $z^\nu$ belongs to $N_X(x^\nu)$. Then, after passing to subsequences,

$$\bar{z}^\nu = \frac{v^\nu}{\|y^\nu\|_2} - \nabla F(x^\nu)^\top \bar{y}^\nu \to 0 - \nabla F(\bar{x})^\top \bar{y}.$$

By 4.39,

$$\bar{y} \in N_D\big(F(\bar{x})\big) \quad \text{and} \quad -\nabla F(\bar{x})^\top \bar{y} \in N_X(\bar{x}).$$

But $\bar{y} \neq 0$, so this is impossible under the qualification (4.9). Thus, only case (i) is viable.

**Part 2.** We now bring in the regularity assumption and proceed in three steps. First, we show that the expression for the tangent cone holds with $\subset$. Second, we establish that $N_C(\bar{x}) \supset S(\bar{x})$, which together with Part 1 implies equality. From this we're also able to conclude that $C$ is regular. Third, we prove that $\supset$ holds in the tangent cone expression.

**Part 2a.** Let $w \in T_C(\bar{x})$. Then, there exist points $x^\nu \in C \to \bar{x}$ and scalars $\tau^\nu \searrow 0$ with

$$w^\nu = (x^\nu - \bar{x})/\tau^\nu \to w,$$

which can be seen from the definition of tangent vectors in 4.33. Since $\bar{x}, x^\nu \in X$, we've $w \in T_X(\bar{x})$ again by the definition. To show that $\nabla F(\bar{x})w \in T_D(F(\bar{x}))$, we need to construct $u^\nu \in D \to F(\bar{x})$ and $\lambda^\nu \searrow 0$ such that

$$\frac{u^\nu - F(\bar{x})}{\lambda^\nu} \to \nabla F(\bar{x})w.$$

Let $u^\nu = F(x^\nu)$ and $\lambda^\nu = \tau^\nu$. Then, $u^\nu \in D$. Since $F$ is smooth, one has

$$F(x^\nu) \to F(\bar{x}), \quad \text{with} \quad F(x^\nu) = F(\bar{x}) + \nabla F(\bar{x})(x^\nu - \bar{x}) + o\big(\|x^\nu - \bar{x}\|_2\big)$$

by (2.5). Hence,

$$\frac{u^\nu - F(\bar{x})}{\lambda^\nu} = \frac{F(x^\nu) - F(\bar{x})}{\tau^\nu} = \nabla F(\bar{x})w^\nu + \frac{o\big(\|x^\nu - \bar{x}\|_2\big)}{\|x^\nu - \bar{x}\|_2}\|w^\nu\|_2 \to \nabla F(\bar{x})w$$

and $w$ satisfies the conditions of the right-hand side of the tangent cone expression. This holds without invoking the qualification (4.9) or regularity.

**Part 2b.** Suppose that $v \in S(\bar{x})$. Then,

$$v = \nabla F(\bar{x})^\top y + z \;\text{ for some }\; y \in N_D\big(F(\bar{x})\big) \;\text{ and }\; z \in N_X(\bar{x}).$$

We plan to show that $v \in \widehat{N}_C(\bar{x})$. Since $\widehat{N}_C(\bar{x}) \subset N_C(\bar{x})$ generally, we then have $v \in N_C(\bar{x})$ but also, when combined with Part 1, $\widehat{N}_C(\bar{x}) = N_C(\bar{x})$. Consequently, we can conclude that $C$ is Clarke regular at $\bar{x}$. With this in mind, we need to verify that $\langle v, w\rangle \le 0$ for every $w \in T_C(\bar{x})$; cf. 4.35. Let $w \in T_C(\bar{x})$. Then, by Part 2a,

$$w \in T_X(\bar{x}) \quad\text{and}\quad \nabla F(\bar{x})w \in T_D\big(F(\bar{x})\big).$$

Since $D$ and $X$ are Clarke regular at $F(\bar{x})$ and $\bar{x}$, respectively, $y \in \widehat{N}_D(F(\bar{x}))$ and $z \in \widehat{N}_X(\bar{x})$. Again recalling the definition of these regular normal vectors in 4.35, we've in particular that $\langle y, \nabla F(\bar{x})w\rangle \le 0$ and $\langle z, w\rangle \le 0$. Combining these facts, one obtains

$$\langle v, w\rangle = \big\langle \nabla F(\bar{x})^\top y + z, w\big\rangle = \big\langle y, \nabla F(\bar{x})w\big\rangle + \langle z, w\rangle \le 0.$$

Since $w \in T_C(\bar{x})$ is arbitrary, we've established that $v \in \widehat{N}_C(\bar{x})$.

**Part 2c.** Suppose that $w \in T_X(\bar{x})$ satisfies $\nabla F(\bar{x})w \in T_D(F(\bar{x}))$. Since $C$ is Clarke regular at $\bar{x}$ by Part 2b, the polarity relation 4.43 applies and it suffices to show that $\langle v, w\rangle \le 0$ for every $v \in N_C(\bar{x})$. Let $v \in N_C(\bar{x})$. From Part 1, $v - \nabla F(\bar{x})^\top y + z$ for some $y \in N_D(F(\bar{x}))$ and $z \in N_X(\bar{x})$. We can then conclude that

$$\langle v, w\rangle = \big\langle y, \nabla F(\bar{x})w\big\rangle + \langle z, w\rangle \le 0$$

because $y \in \widehat{N}_D(F(\bar{x}))$ and $z \in \widehat{N}_X(\bar{x})$ by Clarke regularity of $D$ at $F(\bar{x})$ and of $X$ at $\bar{x}$ so $\langle y, \nabla F(\bar{x})w\rangle \le 0$ and $\langle z, w\rangle \le 0$; cf. 4.35. Since $v$ is arbitrary, we've $w \in T_C(\bar{x})$. □

## 4.I Subdifferentiability

Subderivatives and subgradients of a function are intimately tied to tangent and normal cones to the epigraph of the function and this geometrical perspective enables us to cut the umbilical cord to convexity completely. It also leads fruitfully to expressions for subderivatives and subgradients from those for tangent and normal cones of the preceding sections. We start with subderivatives, which were already brought forth in 2.8 and led to the far-reaching Oresme rule 2.10. While subgradients will remain our main tool for expressing optimality conditions, subderivatives support these developments through their tight connection with subgradients akin to the polar relation between normal and tangent cones (cf. 4.43). In the end, however, subderivatives will emerge as the more potent concept for difficult nonconvex functions, producing "stronger" optimality conditions than those obtained using subgradients.

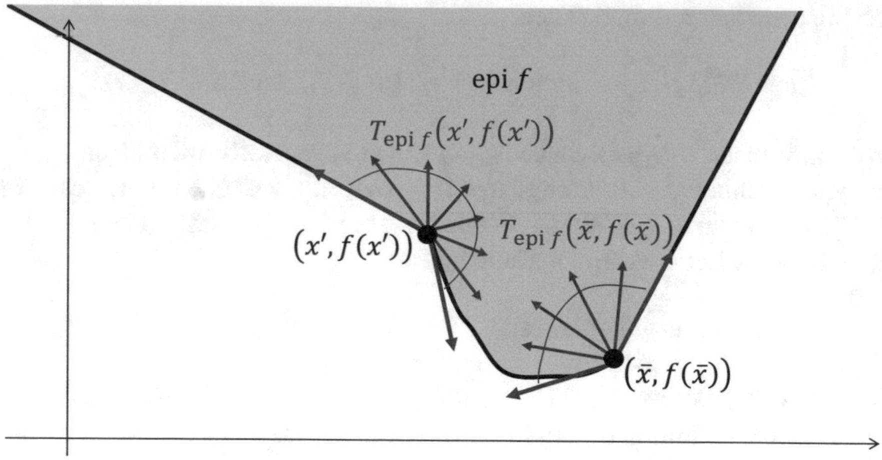


Fig. 4.24: Tangent cones to the epigraph of a function.

**Proposition 4.54** (subderivatives from tangent cones). *For $f : \mathbb{R}^n \to \overline{\mathbb{R}}$ and $\bar{x}$ with $f(\bar{x})$ finite, one has*

$$\operatorname{epi} df(\bar{x};\cdot) = T_{\operatorname{epi} f}\big(\bar{x}, f(\bar{x})\big).$$

**Proof.** For $\tau > 0$, let $\Delta_\tau f(\bar{x};\cdot) : \mathbb{R}^n \to \overline{\mathbb{R}}$ be given by

$$\Delta_\tau f(\bar{x}; w) = \frac{f(\bar{x} + \tau w) - f(\bar{x})}{\tau}, \quad \text{with } \operatorname{epi} \Delta_\tau f(\bar{x};\cdot) = \frac{1}{\tau}\Big(\operatorname{epi} f - \big(\bar{x}, f(\bar{x})\big)\Big).$$

In this notation, it's actually immediate from the definition of $df(\bar{x};\cdot)$ (cf. 2.8) that

$$\operatorname{epi} df(\bar{x};\cdot) = \bigcup_{\tau^\nu \searrow 0} \operatorname{LimOut}\big(\operatorname{epi} \Delta_{\tau^\nu} f(\bar{x};\cdot)\big).$$

In view of the discussion prior to 4.34, the conclusion follows. □

Figure 4.24 shows the epigraph of a function $f$ and its tangent cones at $(\bar{x}, f(\bar{x}))$ and $(x', f(x'))$, which in view of the proposition are the epigraphs of $df(\bar{x};\cdot)$ and $df(x';\cdot)$, respectively. We can view $df(\bar{x};\bar{w})$ as a local approximation of $f$ constructed at $\bar{x}$ such that

$$f(\bar{x} + \bar{w}) \approx f(\bar{x}) + df(\bar{x};\bar{w}).$$

While the tangent cones to an epigraph specify subderivatives, the normal cones define subgradients of an arbitrary function.

**Definition 4.55** (subgradients). For $f : \mathbb{R}^n \to \overline{\mathbb{R}}$ and $\bar{x}$ with $f(\bar{x})$ finite, a vector $v$ is a *subgradient* of $f$ at $\bar{x}$ if

$$(v, -1) \in N_{\operatorname{epi} f}\big(\bar{x}, f(\bar{x})\big).$$

The set of all subgradients of $f$ at $\bar{x}$ is denoted by $\partial f(\bar{x})$.

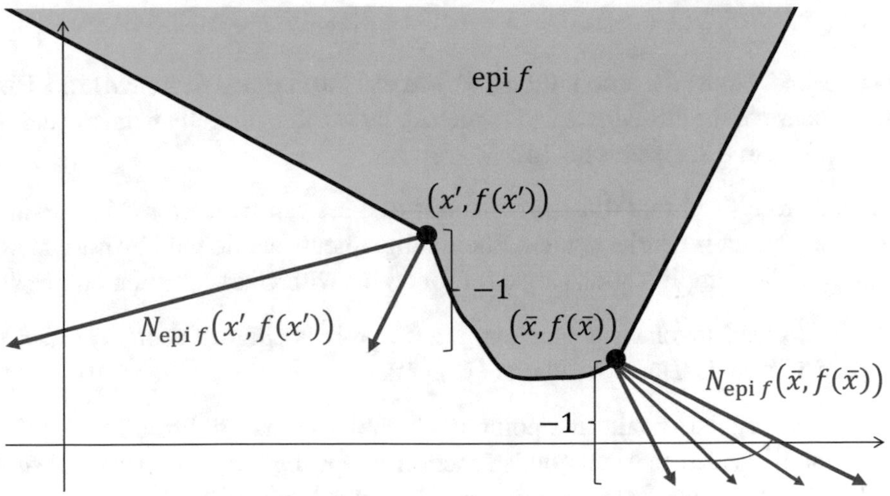


Fig. 4.25: Normal cones to the epigraph of a function.

Figure 4.25 illustrates the normal cones corresponding to the tangent cones in Figure 4.24. Graphically, it seems like $N_{\text{epi}\,f}(\bar{x}, f(\bar{x}))$ contains vectors of the form $\lambda(v, -1)$ for $\lambda \geq 0$ and $v \in [1/2, 2]$. This means that such $v$ are subgradients of $f$ at $\bar{x}$. Note that 2 is the slope of $f$ to the right of $\bar{x}$ and $1/2$ is the slope to the left. At $(x', f(x'))$, normal vectors are of the form $\lambda(-4, -1)$ and $\lambda(-1/2, -1)$ for $\lambda \geq 0$. Then, $v = -4$ and $v = -1/2$ are subgradients and in fact the only subgradients because the normal cone at $(x', f(x'))$ consists of only two rays. Again, the subgradients represent the slope of $f$ as we approach $x'$ from the left and from the right.

As an introductory numerical example, let's consider $f(x) = x^2$. Then, $N_{\text{epi}\,f}(1, f(1)) = \{(2, -1)z \mid z \geq 0\}$ by 4.49 because

$$\text{epi}\, f = \{(x, \alpha) \in \mathbb{R}^2 \mid g(x, \alpha) = x^2 - \alpha \leq 0\}$$

and $\nabla g(x, \alpha) = (2, -1)$. Thus, $(v, -1) \in N_{\text{epi}\,f}(1, f(1))$ if and only if $v = 2$. This means that $\partial f(1) = \{2\}$ as expected.

**Example 4.56** (indicator function). For $C \subset \mathbb{R}^n$, $\bar{x} \in C$ and $K = T_C(\bar{x})$, one has

$$\partial \iota_C(\bar{x}) = N_C(\bar{x}) \quad \text{and} \quad d\iota_C(\bar{x}; w) = \iota_K(w) \quad \forall w \in \mathbb{R}^n.$$

Definition 2.23 of normal cones in the convex case is therefore compatible with the present development.

**Detail.** From the definition of subgradients, $v \in \partial \iota_C(\bar{x})$ if and only if $(v, -1) \in N_{\text{epi}\,\iota_C}(\bar{x}, 0)$. Since $\text{epi}\,\iota_C = C \times [0, \infty)$, it follows by 4.44 that

$$N_{\text{epi}\,\iota_C}(\bar{x}, 0) = N_C(\bar{x}) \times N_{[0,\infty)}(0) = N_C(\bar{x}) \times (-\infty, 0]$$

as long as $C$ is closed. If $C$ is also Clarke regular at $\bar{x}$, then

$$\operatorname{epi} d\iota_C(\bar{x};\cdot) = T_{\operatorname{epi}\iota_C}(\bar{x},0) = T_C(\bar{x}) \times T_{[0,\infty)}(0) = T_C(\bar{x}) \times [0,\infty)$$

by 4.54 and 4.44. Under these conditions, we see that the claimed formulae hold. However, one can dispense of the closedness and regularity by working directly from the definitions; see [105, Theorem 8.2, Exercise 8.14]. □

We recall from §4.G that the normal cones to a set can be expressed in a simplified manner when the set is Clarke regular. Since subgradients are defined by normal cones to epigraphs, parallel simplifications arise for functions with Clarke regular epigraphs.

**Definition 4.57** (epi-regular). A function $f : \mathbb{R}^n \to \overline{\mathbb{R}}$ is *epigraphically regular*, or *epi-regular*, at $\bar{x}$ if epi $f$ is Clarke regular at $(\bar{x}, f(\bar{x}))$.

Since Clarke regularity fails at a point at which the set has an inward kink (cf. §4.G), we immediately realize that a smooth function is epi-regular at every point and a lsc convex functions is epi-regular at every point at which it's finite. In fact, a function $f$ can't be epi-regular at a point $\bar{x}$ with $f(\bar{x}) \notin \mathbb{R}$ because then $(\bar{x}, f(\bar{x}))$ isn't a point in epi $f \subset \mathbb{R}^n \times \mathbb{R}$. In Figure 4.25, $f$ is epi-regular at $\bar{x}$ but not at $x'$.

**Proposition 4.58** (subgradients and gradients). *For $f, g : \mathbb{R}^n \to \overline{\mathbb{R}}$ and a point $\bar{x}$, the following hold:*

(a) *If $f$ is differentiable at $\bar{x}$, then $\nabla f(\bar{x}) \in \partial f(\bar{x})$.*
(b) *If $f$ is differentiable and epi-regular at $\bar{x}$, then $\partial f(\bar{x}) = \{\nabla f(\bar{x})\}$.*
(c) *If $f$ is smooth in a neighborhood of $\bar{x}$, then*

$$\partial f(\bar{x}) = \{\nabla f(\bar{x})\} \quad \text{and} \quad \partial(f+g)(\bar{x}) = \nabla f(\bar{x}) + \partial g(\bar{x})$$

*provided that $g(\bar{x})$ is finite.*

**Proof.** A simple proof is sketched out in [105, Exercise 8.8]. □

Part (c) of the proposition furnishes a sum rule that partially extends the Moreau-Rockafellar sum rule 2.26. In the important special case of $g = \iota_C$ with $C \subset \mathbb{R}^n$, we obtain for $\bar{x} \in C$ that

$$\partial(f + \iota_C)(\bar{x}) = \nabla f(\bar{x}) + N_C(\bar{x})$$

via 4.56, regardless of convexity.

Let's recall the subtle difference between differentiable and smooth functions; see the discussion around (2.4). For example, suppose that

$$f(x) = \begin{cases} x^2(1 + \sin x^{-1}) & \text{if } x \neq 0 \\ 0 & \text{otherwise.} \end{cases}$$

This function is continuous, but its values oscillate increasingly as $x$ approaches zero; see Figure 4.26. Since

$$\frac{f(x) - f(0) - vx}{|x|} = |x|\big(1 + \sin x^{-1}\big) - v\frac{x}{|x|} \to 0 \quad \text{as } x \to 0 \text{ with } x \neq 0$$

only when $v = 0$, $f$ is differentiable at zero with derivative $f'(0) = 0$. However, the derivative

$$f'(x) = 2x\big(1 + \sin x^{-1}\big) - \cos x^{-1} \;\; \forall x \neq 0,$$

which doesn't have a limit as $x$ tends to zero. Thus, $f$ isn't smooth at zero because its derivative isn't continuous at that point. The effect on subgradients is seen through the normal vectors to epi $f$. For $\bar{x} \neq 0$, $f$ is smooth and $\partial f(\bar{x}) = \{f'(\bar{x})\}$ by 4.58(c). Alternatively, we can use 4.49 and the description $\operatorname{epi} f = \{(x, \alpha) \mid f(x) - \alpha \leq 0\}$ to obtain

$$N_{\operatorname{epi} f}\big(\bar{x}, f(\bar{x})\big) = \big\{z\big(f'(\bar{x}), -1\big) \,\big|\, z \geq 0\big\}$$
$$T_{\operatorname{epi} f}\big(\bar{x}, f(\bar{x})\big) = \big\{(w_1, w_2) \,\big|\, f'(\bar{x})w_1 - w_2 \leq 0\big\};$$

see Figure 4.26(right). Since $(v, -1) \in N_{\operatorname{epi} f}(\bar{x}, f(\bar{x}))$ if and only if $v = f'(\bar{x})$, we've confirmed that $\partial f(\bar{x}) = \{f'(\bar{x})\}$.

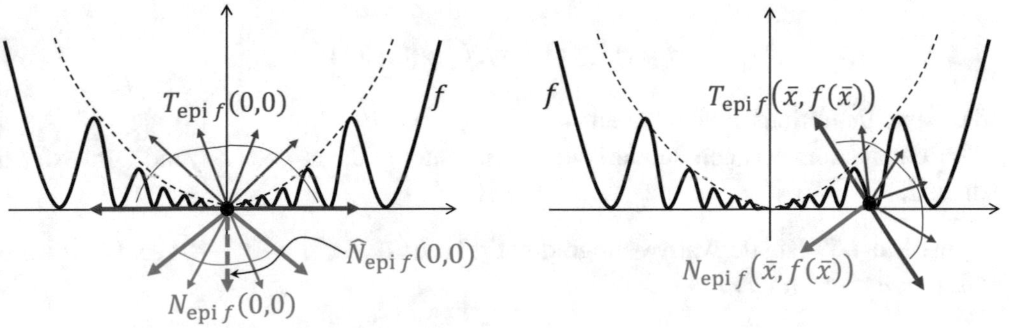


Fig. 4.26: Normal and tangent vectors to the epigraph of an oscillating function.

The situation at zero is different, however. The tangent cone

$$T_{\operatorname{epi} f}(0,0) = \big\{(w_1, w_2) \,\big|\, w_2 \geq 0\big\}$$

as can be seen directly from its definition and then

$$\widehat{N}_{\operatorname{epi} f}(0,0) = \big\{(v_1, v_2) \,\big|\, v_1 = 0,\; v_2 \leq 0\big\};$$

see Figure 4.26(left). (We can't use 4.49 because $f$ isn't smooth at zero.) These regular normal vectors are then augmented by the limits of vectors in $N_{\operatorname{epi} f}(x^\nu, f(x^\nu))$ as $x^\nu \to 0$ to produce

$$N_{\operatorname{epi} f}(0,0) = \big\{(v_1, v_2) \,\big|\, v_2 \leq -|v_1|\big\}$$

because $f'(x^\nu)$ can be forced to converge to any number in $[-1, 1]$ by a clever choice of $x^\nu \to 0$. Thus, $\partial f(0) = [-1, 1]$, which shows that the inclusion in 4.58(a) can be strict. The source of difficulty is that $f$ isn't epi-regular at zero. This example is somewhat pathological but serves as a warning that we can't take epi-regularity for granted.

We need to reconcile the definition of subgradients with our initial treatment in Chap. 2. If $f : \mathbb{R}^n \to \overline{\mathbb{R}}$ is a convex function that's finite at $\bar{x}$, then a vector $v \in \mathbb{R}^n$ is a subgradient of $f$ at $\bar{x}$ according to 2.13 when

$$\langle v, w \rangle \le df(\bar{x}; w) \;\; \forall w \in \mathbb{R}^n. \tag{4.12}$$

This inequality is equivalent to having $\langle v, w \rangle \le \beta$ for all $w \in \mathbb{R}^n$ and $\beta \in \mathbb{R}$ satisfying $df(\bar{x}; w) \le \beta$. Thus, (4.12) holds if and only if

$$\big\langle (v, -1), (w, \beta) \big\rangle \le 0 \quad \forall (w, \beta) \in \operatorname{epi} df(\bar{x}; \cdot\,) = T_{\operatorname{epi} f}\big(\bar{x}, f(\bar{x})\big),$$

where the tangent cone expression follows by 4.54. In view of the definition of regular normal vectors from 4.35, this implies that (4.12) holds if and only if

$$(v, -1) \in \widehat{N}_{\operatorname{epi} f}\big(\bar{x}, f(\bar{x})\big).$$

Since epi $f$ is convex, we see from 4.42 that

$$\widehat{N}_{\operatorname{epi} f}\big(\bar{x}, f(\bar{x})\big) = N_{\operatorname{epi} f}\big(\bar{x}, f(\bar{x})\big)$$

and the two definitions are equivalent.

The connection between subderivatives and subgradients goes beyond convexity as seen next.

**Proposition 4.59** (subderivative-subgradient relations). *For $f : \mathbb{R}^n \to \overline{\mathbb{R}}$ and $\bar{x}$ at which the function is epi-regular, one has*

$$\partial f(\bar{x}) \ne \emptyset \iff df(\bar{x}; 0) \ne -\infty.$$

*Moreover, under these conditions,*

$$\partial f(\bar{x}) = \big\{ v \in \mathbb{R}^n \;\big|\; \langle v, w \rangle \le df(\bar{x}; w) \;\; \forall w \in \mathbb{R}^n \big\}$$
$$df(\bar{x}; w) = \sup \big\{ \langle v, w \rangle \;\big|\; v \in \partial f(\bar{x}) \big\} \qquad \forall w \in \mathbb{R}^n.$$

**Proof.** Since epi $f$ is Clarke regular at $(\bar{x}, f(\bar{x}))$, the polar relationship in 4.43 for $C = \operatorname{epi} f$ holds at this point. Hence, $N_{\operatorname{epi} f}(\bar{x}, f(\bar{x}))$ contains only horizontal vectors of the form $(v, 0)$ if and only if $T_{\operatorname{epi} f}(\bar{x}, f(\bar{x}))$ includes the entire vertical line $\{(0, \beta) \mid \beta \in \mathbb{R}\}$. The first condition corresponds to having $\partial f(\bar{x}) = \emptyset$ by definition 4.55 and the second one to $df(\bar{x}; 0) = -\infty$ in view of 4.54. This establishes the asserted equivalence.

The expression for $\partial f(\bar{x})$ reiterates the prior discussion, which also confirms that

$$df(\bar{x}; w) \ge \sup\nolimits_{v \in \partial f(\bar{x})} \langle v, w \rangle \;\; \forall w \in \mathbb{R}^n.$$

The perpendicular relationship between $(w, df(\bar{x}; w))$, a boundary point in $T_{\operatorname{epi} f}(\bar{x}, f(\bar{x}))$, and some normal vector in $N_{\operatorname{epi} f}(\bar{x}, f(\bar{x}))$ is guaranteed by the polarity in 4.43. This ensures that the inequality holds with equality. □

A special situation occurs when a function becomes "infinitely steep" at a point as takes place for $f(x) = \sqrt{|x|}$ at zero; see Figure 4.27(left). There,

$$N_{\mathrm{epi}\, f}(0,0) = \{(v_1, v_2) \mid v_2 \leq 0\}$$

and $(v_1, -1) \in N_{\mathrm{epi}\, f}(0,0)$ for any $v_1 \in \mathbb{R}$ so that $\partial f(0) = \mathbb{R}$. For functions that aren't real-valued, it also occurs on the boundary of their domain as at $(x', f(x'))$ in Figure 4.27(right). If a function $f$ gets arbitrarily steep in the manner taking place in Figure 4.27(right) at $\bar{x}$, then $N_{\mathrm{epi}\, f}(\bar{x}, f(\bar{x}))$ contains only "horizontal" vectors, i.e., all normal vectors are of the form $(v, 0)$ for some $v \in \mathbb{R}^n$, and $\partial f(\bar{x}) = \emptyset$. We need to watch for these circumstances and, in fact, think of $v \in \mathbb{R}^n$ satisfying $(v, 0) \in N_{\mathrm{epi}\, f}(\bar{x}, f(\bar{x}))$ as a supplementary kind of subgradient.

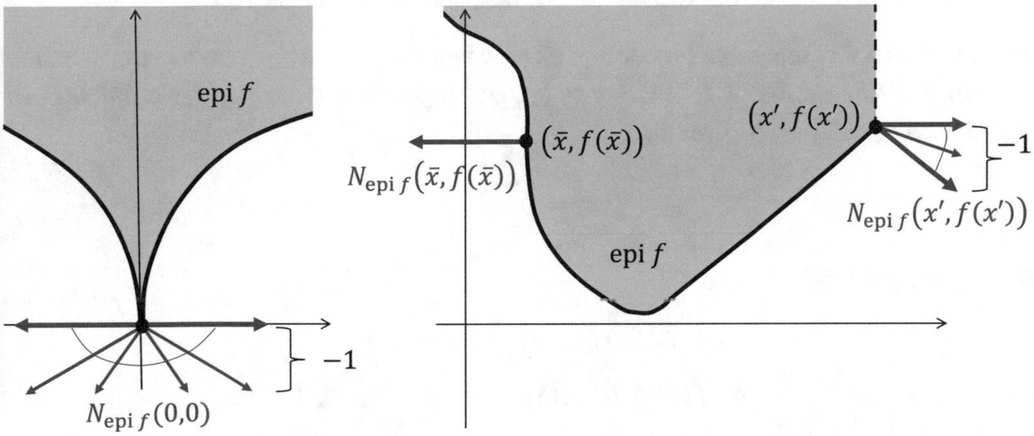


Fig. 4.27: Normal vectors $(v, 0)$ to epigraphs produce horizon subgradients $v$.

**Definition 4.60** (horizon subgradients). For a lsc function $f : \mathbb{R}^n \to \overline{\mathbb{R}}$ and a point $\bar{x}$ with $f(\bar{x})$ finite, a vector $v$ is a *horizon subgradient* of $f$ at $\bar{x}$ if

$$(v, 0) \in N_{\mathrm{epi}\, f}\big(\bar{x}, f(\bar{x})\big).$$

The set of all horizon subgradients of $f$ at $\bar{x}$ is denoted by $\partial^\infty f(\bar{x})$.

In Figure 4.27, $\partial^\infty f(0) = \mathbb{R}$ on the left and $\partial^\infty f(\bar{x}) = (-\infty, 0]$ and $\partial^\infty f(x') = [0, \infty)$ on the right. Horizon subgradients can be defined for functions that aren't lsc (see [105, Section 8.B]), but we omit the details.

**Proposition 4.61** (horizon subgradients for smooth functions). *If $f : \mathbb{R}^n \to \overline{\mathbb{R}}$ is smooth in a neighborhood of $\bar{x}$, then $\partial^\infty f(\bar{x}) = \{0\}$.*

**Proof.** This is immediately clear from the definition because $f$ can't be arbitrarily steep near $\bar{x}$. □

**Exercise 4.62** (subgradients and horizon subgradients). For $f : \mathbb{R} \to \overline{\mathbb{R}}$ given by $f(x) = x^2$ when $x \leq 1$ and $f(x) = \infty$ otherwise, compute $\partial f(1)$ and $\partial^\infty f(1)$.

**Guide.** Use the expression $\text{epi}\, f = \{(x, \alpha) \mid x^2 - \alpha \le 0, x - 1 \le 0\}$ together with 4.49 to derive a formula for $N_{\text{epi}\, f}(1, 1)$. □

We next turn to calculus rules for subderivatives and subgradients. For $f : \mathbb{R}^n \to \overline{\mathbb{R}}$ and $\lambda \in (0, \infty)$, let $\lambda f$ or $(\lambda f)$ be the function with value $\lambda f(x)$ at $x$. Then, for a point $\bar{x}$ where $f$ is finite, one has

$$d(\lambda f)(\bar{x}; w) = \lambda df(\bar{x}; w) \;\; \forall w \in \mathbb{R}^n$$

as is immediate from the definition of subderivatives in 2.8. Moreover,

$$\partial(\lambda f)(\bar{x}) = \lambda \partial f(\bar{x}) \quad \text{and} \quad \partial^\infty(\lambda f)(\bar{x}) = \partial^\infty f(\bar{x}), \tag{4.13}$$

which can also be traced back to the definitions; see [105, Section 10.D].

**Proposition 4.63** (separable functions). *For lsc functions $f_i : \mathbb{R}^{n_i} \to \overline{\mathbb{R}}$, $i = 1, \dots, m$, and a point $\bar{x} = (\bar{x}_1, \dots, \bar{x}_m) \in \mathbb{R}^n$, with $n = \sum_{i=1}^m n_i$, suppose that each $f_i(\bar{x}_i)$ is finite. Then, the function $f : \mathbb{R}^n \to \overline{\mathbb{R}}$ given by*

$$f(x) = \sum\nolimits_{i=1}^m f_i(x_i)$$

*has the expressions*

$$\partial f(\bar{x}) = \partial f_1(\bar{x}_1) \times \cdots \times \partial f_m(\bar{x}_m)$$
$$\partial^\infty f(\bar{x}) \subset \partial^\infty f_1(\bar{x}_1) \times \cdots \times \partial^\infty f_m(\bar{x}_m).$$

*Suppose that $df_i(\bar{x}_i; 0) \neq -\infty$ for all $i = 1, \dots, m$. Then,*

$$df(\bar{x}; w) \ge df_1(\bar{x}_1; w_1) + \cdots + df_m(\bar{x}_m; w_m) \;\; \forall w = (w_1, \dots, w_m).$$

*If in addition each $f_i$ is epi-regular at $\bar{x}_i$, then $f$ is epi-regular at $\bar{x}$ and the inclusion and the inequality hold with equality.*

**Proof.** We refer to [105, Proposition 10.5] for a proof, which seemly requires the assumption $df_i(\bar{x}_i; 0) \neq -\infty$ for the two initial assertions. However, a closer examination of the proof shows that the assumption is superfluous. We also note that $df_i(\bar{x}_i; 0) \neq -\infty$ is equivalent to having $df_i(\bar{x}_i; 0) = 0$ because the epigraph of $df_i(\bar{x}_i; \cdot)$ is a cone by 4.54 and thus includes the origin as well as every ray extending from there to a point in that epigraph. □

The condition $df_i(\bar{x}_i; 0) \neq -\infty$ weeds out pathological cases of the kind seen in Figure 4.27(right). There $df(\bar{x}; 0) = -\infty$ because one can approach $(\bar{x}, f(\bar{x}))$ along an "infinitely steep uphill." At $x'$, there's no issue and $df(x'; 0) \neq -\infty$ and likewise at 0 in the left portion of the figure; recall that generally $\text{epi}\, df(x; \cdot) = T_{\text{epi}\, f}(x, f(x))$ by 4.54. In situations with epi-regularity, one can leverage 4.59 to reach such conclusions.

**Theorem 4.64** (chain rule). *For a proper lsc function* $h : \mathbb{R}^m \to \overline{\mathbb{R}}$ *and a smooth mapping* $F : \mathbb{R}^n \to \mathbb{R}^m$, *let* $f = h \circ F$ *and* $\bar{x} \in \operatorname{dom} f$. *Then,*

$$df(\bar{x}; w) \geq dh\big(F(\bar{x}); \nabla F(\bar{x})w\big) \quad \forall w \in \mathbb{R}^n.$$

*Suppose that the following qualification holds:*

$$y \in \partial^\infty h\big(F(\bar{x})\big) \;\; and \;\; \nabla F(\bar{x})^\top y = 0 \quad \Longrightarrow \quad y = 0. \tag{4.14}$$

*Then,*

$$\partial f(\bar{x}) \subset \nabla F(\bar{x})^\top \partial h\big(F(\bar{x})\big) \quad and \quad \partial^\infty f(\bar{x}) \subset \nabla F(\bar{x})^\top \partial^\infty h\big(F(\bar{x})\big).$$

*If in addition h is epi-regular at* $F(\bar{x})$, *then f is epi-regular at* $\bar{x}$ *and the inequality and the inclusions hold with equality.*

**Proof.** Let $\bar{F} : \mathbb{R}^n \times \mathbb{R} \to \mathbb{R}^m \times \mathbb{R}$ be defined by $\bar{F}(x, \alpha) = (F(x), \alpha)$. Then,

$$\operatorname{epi} f = \big\{(x, \alpha) \in \mathbb{R}^n \times \mathbb{R} \;\big|\; \bar{F}(x, \alpha) \in \operatorname{epi} h\big\}$$

and we can invoke 4.46 to compute the normal cones and tangent cones to epi $f$. As seen from Step 2a in the proof of 4.46, neither regularity nor a qualification is needed to obtain the tangent cone inclusion:

$$T_{\operatorname{epi} f}\big(\bar{x}, f(\bar{x})\big) \subset \Big\{\tilde{w} \in \mathbb{R}^n \times \mathbb{R} \;\Big|\; \nabla \bar{F}\big(\bar{x}, f(\bar{x})\big)\tilde{w} \in T_{\operatorname{epi} h}\Big(\bar{F}\big(\bar{x}, f(\bar{x})\big)\Big)\Big\}.$$

By 4.54, $(w, \beta) \in T_{\operatorname{epi} f}(\bar{x}, f(\bar{x}))$ if and only if $(w, \beta) \in \operatorname{epi} df(\bar{x}; \cdot)$, which via the tangent cone inclusion implies

$$\big(\nabla F(\bar{x})w, \beta\big) \in T_{\operatorname{epi} h}\big(F(\bar{x}), f(\bar{x})\big) \;\; \text{because} \;\; \nabla \bar{F}\big(\bar{x}, f(\bar{x})\big) = \begin{bmatrix} \nabla F(\bar{x}) & 0 \\ 0 & 1 \end{bmatrix}.$$

Again invoking 4.54, this means that $(\nabla F(\bar{x})w, \beta) \in \operatorname{epi} dh(F(\bar{x}); \cdot)$. We've established the asserted subderivative inequality.

Turning the attention to subgradients, we need to verify the qualification (4.9), which now amounts to

$$\tilde{y} \in N_{\operatorname{epi} h}\Big(\bar{F}\big(\bar{x}, f(\bar{x})\big)\Big) \quad \text{and} \quad \nabla \bar{F}\big(\bar{x}, f(\bar{x})\big)^\top \tilde{y} = 0 \quad \Longrightarrow \quad \tilde{y} = 0.$$

With $\tilde{y} = (y, \beta) \in \mathbb{R}^m \times \mathbb{R}$, this specializes further to

$$(y, \beta) \in N_{\operatorname{epi} h}\big(F(\bar{x}), f(\bar{x})\big), \;\; \nabla F(\bar{x})^\top y = 0, \;\; \beta = 0 \quad \Longrightarrow \quad y = 0, \; \beta = 0.$$

Since $\beta = 0$ is the only value to be considered and $(y, 0) \in N_{\operatorname{epi} h}(F(\bar{x}), f(\bar{x}))$ is the same as $y \in \partial^\infty h(F(\bar{x}))$ by definition 4.60, qualification (4.9) reduces to (4.14). We can then invoke the normal cone inclusion from 4.46:

$$N_{\mathrm{epi}\,f}\big(\bar{x}, f(\bar{x})\big) \subset \Big\{\nabla\bar{F}\big(\bar{x}, f(\bar{x})\big)^\top\bar{y} \;\Big|\; \bar{y} \in N_{\mathrm{epi}\,h}\Big(\bar{F}\big(\bar{x}, f(\bar{x})\big)\Big)\Big\}.$$

Thus, for $(v, -1) \in N_{\mathrm{epi}\,f}(\bar{x}, f(\bar{x}))$, there's $y \in \mathbb{R}^m$ such that

$$v = \nabla F(\bar{x})^\top y \quad \text{and} \quad (y, -1) \in N_{\mathrm{epi}\,h}\big(F(\bar{x}), f(\bar{x})\big).$$

The latter condition is equivalent to having $y \in \partial h(F(\bar{x}))$ by definition 4.55 and this establishes the inclusion for subgradients. The same argument with $-1$ replaced by 0 leads to the expression for horizon subgradients. If $h$ is epi-regular at $F(\bar{x})$, then epi $h$ is Clarke regular at $(F(\bar{x}), f(\bar{x}))$. Thus, we've equality in the tangent and normal cone expressions in 4.46. This leads to the remaining conclusions. □

We note that the multiplication of a vector with a set in the theorem has a meaning parallel to the definition of scalar multiplication with a set (see (4.7)):

$$\nabla F(x)^\top\partial h\big(F(\bar{x})\big) = \Big\{\nabla F(x)^\top y \;\Big|\; y \in \partial h\big(F(\bar{x})\big)\Big\}.$$

Thus, if $h$ is smooth, then we recover the well-known chain rule:

$$\nabla f(\bar{x}) = \nabla F(x)^\top \nabla h\big(F(\bar{x})\big)$$

because

$$\partial h\big(F(\bar{x})\big) = \Big\{\nabla h\big(F(\bar{x})\big)\Big\}$$

in that case by 4.58.

The qualification (4.14) is satisfied when $\nabla F(\bar{x})$ has linearly independent rows because then $\nabla F(\bar{x})^\top y = 0$ would ensure that $y = 0$. However, (4.14) offers another possibility as well: the only horizon subgradient of $h$ at $F(\bar{x})$ is 0. This would require epi $h$ to have no "horizontal" normal vectors at $(F(\bar{x}), f(\bar{x}))$ of the kind illustrated in Figure 4.27. In the theorem, the requirements of "no horizontal normal vectors" and linear independence work in tandem so that essentially both must fail for the qualification to be violated.

**Proposition 4.65** (expressions for horizon subgradients). *If $C \subset \mathbb{R}^n$ and $\bar{x} \in C$, then*

$$\partial^\infty \iota_C(\bar{x}) = N_C(\bar{x}).$$

*If $h : \mathbb{R}^m \to \overline{\mathbb{R}}$ is proper, lsc and convex and $\bar{u} \in \mathrm{dom}\, h$, then*

$$\partial^\infty h(\bar{u}) = N_{\mathrm{dom}\,h}(\bar{u}).$$

**Proof.** The first assertion follows from the definitions; see [105, Exercise 8.14]. For the second one, the definition of normal cones in the convex case (cf. 2.23) implies that

$$N_{\mathrm{epi}\,h}\big(\bar{u}, h(\bar{u})\big) = \Big\{(y, \beta) \;\Big|\; \big\langle (y, \beta), (u, \alpha) - \big(\bar{u}, h(\bar{u})\big)\big\rangle \le 0 \;\; \forall (u, \alpha) \in \mathrm{epi}\, h\Big\}.$$

Thus, $(y, 0) \in N_{\operatorname{epi} h}(\bar u, h(\bar u))$ if and only if $\langle y, u - \bar u\rangle \le 0$ whenever $h(u) \le \alpha$, which establishes the assertion in view of 2.23. □

The proposition helps us to verify the qualification (4.14) by simply checking the domain of $h$. In particular, if $h$ is also real-valued in addition to being convex, then $\operatorname{dom} h = \mathbb{R}^m$, $N_{\operatorname{dom} h}(\bar u) = \{0\}$ for all $\bar u \in \mathbb{R}^m$ and (4.14) holds trivially. This also leads to a proof of 3.31. As seen in Figure 4.28, the second part of the proposition doesn't hold in the absence of convexity.

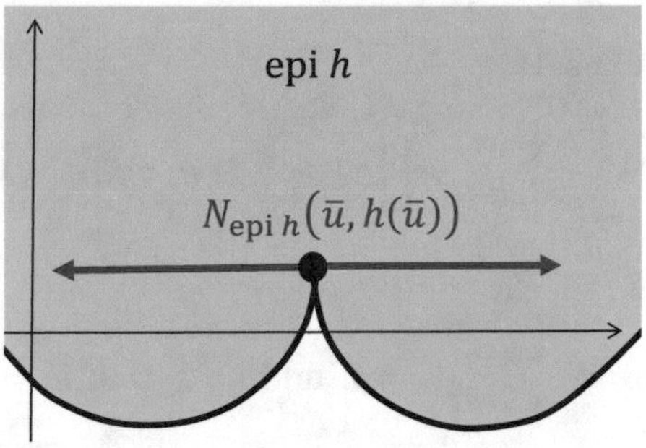


Fig. 4.28: A nonconvex function $h$ with $\partial^\infty h(\bar u) = \mathbb{R} \neq N_{\operatorname{dom} h}(\bar u) = \{0\}$.

We end the section with applications of the chain rule 4.64.

**Proposition 4.66** (subdifferentiability of max-function). *For smooth $f_i : \mathbb{R}^n \to \mathbb{R}$, $i = 1, \dots, m$, the function $f : \mathbb{R}^n \to \mathbb{R}$ defined by*

$$f(x) = \max_{i=1,\dots,m} f_i(x)$$

*is continuous and epi-regular at any $\bar x \in \mathbb{R}^n$ and*

$$\partial f(\bar x) = \operatorname{con}\big\{\nabla f_i(\bar x), i \in \mathbb{A}(\bar x)\big\}$$
$$df(\bar x; w) = \max_{i\in\mathbb{A}(\bar x)} \big\langle \nabla f_i(\bar x), w\big\rangle \;\; \forall w \in \mathbb{R}^n,$$

*where $\mathbb{A}(\bar x) = \{i \mid f_i(\bar x) = f(\bar x)\}$.*

**Proof.** Let $x^\nu \to \bar x$. For $i \in \mathbb{A}(\bar x)$, one has

$$f(x^\nu) \ge f_i(x^\nu) \to f_i(\bar x) = f(\bar x).$$

Thus, $\liminf f(x^\nu) \ge f(\bar x)$. Let $\varepsilon > 0$. There's $\delta > 0$ such that

$$f_i(x) \le f_i(\bar x) + \varepsilon \le f(\bar x) + \varepsilon \;\; \forall x \in \mathbb{B}(\bar x, \delta), \;\; i = 1, \dots, m.$$

Thus, $f(x) \leq f(\bar{x}) + \varepsilon$ for $x \in \mathbb{B}(\bar{x}, \delta)$. Since $\varepsilon$ is arbitrary, $\limsup f(x^\nu) \leq f(\bar{x})$. We've established that $f(x^\nu) \to f(\bar{x})$ so $f$ is continuous at $\bar{x}$.

The formulae are obtained via the chain rule 4.64. Let

$$h(u) = \max\{u_1, \dots, u_m\} \quad \text{and} \quad F(x) = \big(f_1(x), \dots, f_m(x)\big).$$

Since $h$ is real-valued and convex, it's continuous and epi-regular at every point, with

$$\operatorname{epi} h = \big\{(u, \alpha) \in \mathbb{R}^m \times \mathbb{R} \bigm| u_i - \alpha \leq 0,\ i = 1, \dots, m\big\}.$$

By the normal cone formula 2.44,

$$N_{\operatorname{epi} h}(\bar{u}, \bar{\alpha}) = \Big\{\Big(y, \ -\sum\nolimits_{i=1}^{m} y_i\Big) \Bigm| y_i \geq 0 \ \text{ if } \bar{u}_i = \bar{\alpha}, \ \ y_i = 0 \ \text{ otherwise}\Big\}.$$

Thus,

$$(y, -1) \in N_{\operatorname{epi} h}\big(\bar{u}, h(\bar{u})\big) \iff \sum\nolimits_{i=1}^{m} y_i = 1 \ \text{ and } \ y_i \geq 0 \ \text{ if } \bar{u}_i = h(\bar{u}), \ \ y_i = 0 \ \text{ otherwise},$$

which specify the subgradients of $h$ at $\bar{u}$. We can then invoke 4.59 to obtain that

$$dh(\bar{u}; w) = \sup\big\{\langle y, w\rangle \bigm| y \in \partial h(\bar{u})\big\} = \max\nolimits_{i=1,\dots,m}\big\{w_i \bigm| \bar{u}_i = h(\bar{u})\big\}.$$

As discussed prior to the proposition, the qualification (4.14) holds because $h$ is convex and real-valued, and the chain rule 4.64 applies. This yields

$$\partial f(\bar{x}) = \Big\{\nabla F(\bar{x})^\top y \Bigm| \sum\nolimits_{i=1}^{m} y_i = 1; \ \ y_i \geq 0 \ \text{ if } i \in \mathbb{A}(\bar{x}), \ \ y_i = 0 \ \text{ otherwise}\Big\}$$
$$df(\bar{x}; w) = \max\nolimits_{i=1,\dots,m}\big\{\langle \nabla f_i(\bar{x}), w\rangle \bigm| i \in \mathbb{A}(\bar{x})\big\},$$

which simplify to the asserted formulae. □

**Proposition 4.67** (subgradients of sums). *For proper lsc functions* $f_i : \mathbb{R}^n \to \overline{\mathbb{R}}$, $i = 1, \dots, m$, *and a point* $\bar{x}$ *where each of them is finite, consider the function with*

$$f(x) = \sum\nolimits_{i=1}^{m} f_i(x).$$

*Suppose that the following qualification holds:*

$$v_i \in \partial^\infty f_i(\bar{x}),\ i = 1, \dots, m \quad \textit{and} \quad \sum\nolimits_{i=1}^{m} v_i = 0 \quad \Longrightarrow \quad v_i = 0,\ i = 1, \dots, m. \tag{4.15}$$

*Then,*

$$\partial f(\bar{x}) \subset \sum\nolimits_{i=1}^{m} \partial f_i(\bar{x}) \quad \textit{and} \quad \partial^\infty f(\bar{x}) \subset \sum\nolimits_{i=1}^{m} \partial^\infty f_i(\bar{x}).$$

*If in addition each* $f_i$ *is epi-regular at* $\bar{x}$ *and* $df_i(\bar{x}; 0) \neq -\infty$, *then* $f$ *is epi-regular at* $\bar{x}$ *and the inclusions hold with equality.*

**Proof.** We apply the chain rule 4.64 with $F : \mathbb{R}^n \to (\mathbb{R}^n)^m$ and $h : (\mathbb{R}^n)^m \to \overline{\mathbb{R}}$ defined by $F(x) = (x, \ldots, x)$ and

$$h(x_1, \ldots, x_m) = \sum\nolimits_{i=1}^{m} f_i(x_i).$$

Certainly, $f = h \circ F$. Using 4.63, we obtain

$$\partial h\big(F(\bar{x})\big) = \partial f_1(\bar{x}) \times \cdots \times \partial f_m(\bar{x})$$
$$\partial^\infty h\big(F(\bar{x})\big) \subset \partial^\infty f_1(\bar{x}) \times \cdots \times \partial^\infty f_m(\bar{x}).$$

The qualification (4.14) then translates into the present assumption and the inclusions hold. Under the additional assumptions, $h$ is epi-regular at $(\bar{x}, \ldots, \bar{x})$ by 4.63 and we can bring in the stronger conclusions from 4.64. □

A property that's helpful in verifying the qualification (4.15) is a local version of Lipschitz continuity. For $f : \mathbb{R}^n \to \overline{\mathbb{R}}$ and $\bar{x}$, $f$ is *locally Lipschitz continuous* at $\bar{x}$ when there are $\delta \in (0, \infty)$ and $\kappa \in [0, \infty)$ such that

$$\big|f(x) - f(x')\big| \leq \kappa \|x - x'\|_2 \quad \forall x, x' \in \mathbb{B}(\bar{x}, \delta).$$

If $f$ is locally Lipschitz continuous at every $\bar{x} \in \mathbb{R}^n$, then $f$ is locally Lipschitz continuous.

Informally, if $f$ is locally Lipschitz continuous at $\bar{x}$, then the rate of change of $f(x)$ for $x$ near $\bar{x}$ isn't arbitrarily large and this, in particular, means that $\bar{x} \subset \text{int}(\text{dom} f)$. For example, the functions in Figure 4.27 aren't locally Lipschitz continuous at the points 0 (left figure) and $\bar{x}, x'$ (right figure).

**Proposition 4.68** (local Lipschitz continuity). *For $f : \mathbb{R}^n \to \overline{\mathbb{R}}$, each condition below suffices for $f$ to be locally Lipschitz continuous at $\bar{x} \in \mathbb{R}^n$:*

(a) $\bar{x} \in \text{int}(\text{dom} f)$ *and $f$ is proper and convex.*
(b) *$f$ is smooth in a neighborhood of $\bar{x}$.*
(c) $f(x) = \max_{i=1,\ldots,m} f_i(x)$ *for all $x$ in a neighborhood of $\bar{x}$ and $f_i : \mathbb{R}^n \to \overline{\mathbb{R}}$, $i = 1, \ldots, m$, are locally Lipschitz continuous at $\bar{x}$.*
(d) $f(x) = \min_{i=1,\ldots,m} f_i(x)$ *for all $x$ in a neighborhood of $\bar{x}$, with $f_i$ as in (c).*

**Proof.** These results can be deduced from Propositions 9.7 and 9.10 and Example 9.14 in [105]. □

A significant simplification associated with local Lipschitz continuity of $f$ at $\bar{x}$ is the fact that there are no "horizontal" normal vectors to epi $f$ at $(\bar{x}, f(\bar{x}))$. In fact, this characterizes the property fully.

**Proposition 4.69** (subgradients and local Lipschitz continuity). *Suppose that $f : \mathbb{R}^n \to \overline{\mathbb{R}}$ is lsc and $\bar{x}$ is a point at which $f$ is finite. Then,*

$$f \text{ is locally Lipschitz continuous at } \bar{x} \iff \partial^\infty f(\bar{x}) = \{0\}.$$

*Under these circumstances, $\partial f(\bar{x})$ is nonempty and compact.*

**Proof.** We refer to [105, Theorem 9.13] for a proof. □

**Example 4.70** (sum of locally Lipschitz functions). For a closed set $C \subset \mathbb{R}^n$, proper lsc functions $f_i : \mathbb{R}^n \to \overline{\mathbb{R}}$, $i = 1, \ldots, m$, and a point $\bar{x} \in C$ with all $f_i(\bar{x})$ finite, suppose that every $f_i$ is locally Lipschitz continuous at $\bar{x}$. Then, the function given by

$$f(x) = \iota_C(x) + \sum\nolimits_{i=1}^m f_i(x)$$

has subgradients and horizon subgradients satisfying

$$\partial f(\bar{x}) \subset N_C(\bar{x}) + \sum\nolimits_{i=1}^m \partial f_i(\bar{x})$$
$$\partial^\infty f(\bar{x}) \subset N_C(\bar{x})$$

If in addition each $f_i$ is epi-regular at $\bar{x}$ and $C$ is Clarke regular at $\bar{x}$, then $f$ is epi-regular at $\bar{x}$ and the inclusions hold with equality.

**Detail.** We can leverage the sum rule 4.67 for the functions $\iota_C, f_1, \ldots, f_m$. The qualification (4.15) holds because $v^i \in \partial^\infty f_i(\bar{x}) = \{0\}$ by 4.69 so $v^0 \in \partial^\infty \iota_C(\bar{x})$ and $v^0 + \sum_{i=1}^m v^i = 0$ automatically imply that $v^0 = 0$ as well. The asserted inclusions then follow from 4.67; the passing to normal cones are justified by 4.56 and 4.65. To obtain equalities, we need to verify that neither $d\iota_C(\bar{x}; 0)$ nor $df_i(\bar{x}; 0)$ is $-\infty$. Working directly from the definition of subderivatives, we obtain that these quantities are in fact zero. □

**Exercise 4.71** (nonconvex regularization). For an $m \times n$-matrix $A$ and $b \in \mathbb{R}^m$, consider the function with

$$f(x) = \|Ax - b\|_2^2 + \sum\nolimits_{j=1}^n h\big(g(x_j)\big),$$

where $h(\alpha) = \theta|\alpha|$, $\theta \in [0, \infty)$ and

$$g(\beta) = \begin{cases} 2 - \exp(1 - \beta) & \text{if } \beta \in (1, \infty) \\ \beta & \text{if } \beta \in [-1, 1] \\ \exp(1 + \beta) - 2 & \text{otherwise.} \end{cases}$$

As discussed in 4.5, $f$ appears in data analytics. Derive expressions for the subgradients as well as horizon subgradients of $f$. Confirm that $f$ is epi-regular at all points.

**Guide.** Leverage 4.58, 4.63 and 4.64. □

**Exercise 4.72** (superquantile-risk minimization). For $\alpha \in (0, 1)$, $f : \mathbb{R}^m \times \mathbb{R}^n \to \mathbb{R}$ and a random vector $\boldsymbol{\xi}$ with finite distribution, the problem of minimizing the superquantile risk of $f(\boldsymbol{\xi}, x)$ over $x \in C$, a closed subset of $\mathbb{R}^n$, can be formulated as

$$\underset{x \in \mathbb{R}^n, \gamma \in \mathbb{R}}{\text{minimize}}\ \varphi(x, \gamma) = \iota_{C \times \mathbb{R}}(x, \gamma) + \gamma + \frac{1}{1 - \alpha} \sum\nolimits_{\xi \in \Xi} p_\xi \max\big\{0, f(\xi, x) - \gamma\big\},$$

where $\Xi$ is the support of $\boldsymbol{\xi}$ and $\{p_\xi > 0, \xi \in \Xi\}$ are the corresponding probabilities. Under the assumption that $f(\xi, \cdot)$ is smooth for all $\xi \in \Xi$ and $(\bar{x}, \bar{\gamma}) \in C \times \mathbb{R}$, obtain a set that contains $\partial\varphi(\bar{x}, \bar{y})$. Discuss what assumption can be imposed on $C$ to ensure that the obtained set coincides with $\partial\varphi(\bar{x}, \bar{y})$.

**Guide.** Leverage the sum rule 4.67 among other facts. □

## 4.J Optimality Conditions

We're now in position to obtain a far-reaching extension of the Fermat rule at a level of generality parallel to the Oresme rule in 2.10. In this section, we'll also show how the structure of a problem can be brought forward in specific optimality conditions.

**Theorem 4.73** (Fermat rule). *For $f : \mathbb{R}^n \to \overline{\mathbb{R}}$ and $x^\star$ with $f(x^\star) \in \mathbb{R}$, one has*

$$\boxed{x^\star \text{ local minimizer of } f \implies 0 \in \partial f(x^\star)}$$

*In general, the Oresme rule implies the Fermat rule, i.e.,*

$$df(x^\star; w) \geq 0 \;\; \forall w \in \mathbb{R}^n \quad \implies \quad 0 \in \partial f(x^\star).$$

*If $f$ is epi-regular at $x^\star$, then the rules are equivalent.*

**Proof.** Let $(w, \beta) \in T_{\text{epi} f}(x^\star, f(x^\star))$. Then, $df(x^\star; w) \leq \beta$ by 4.54. Since the Oresme rule 2.10 guarantees that $df(x^\star; w) \geq 0$, one has $\beta \geq 0$ so then also

$$\big\langle (0, -1), (w, \beta) \big\rangle \leq 0.$$

In view of the definition of regular normal vectors from 4.35, we've

$$(0, -1) \in \widehat{N}_{\text{epi} f}\big(x^\star, f(x^\star)\big) \subset N_{\text{epi} f}\big(x^\star, f(x^\star)\big)$$

and the first conclusion follows as well as the implication from the Oresme rule to the Fermat rule.

If $f$ is epi-regular at $x^\star$, then epi $f$ is Clarke regular at $(x^\star, f(x^\star))$ and the polarity relation 4.43 holds. Thus, when $0 \in \partial f(x^\star)$ or, equivalently, $(0, -1) \in N_{\text{epi} f}(x^\star, f(x^\star))$, we've

$$\big\langle (0, -1), (w, \beta) \big\rangle \leq 0 \;\; \forall (w, \beta) \in T_{\text{epi} f}\big(x^\star, f(x^\star)\big).$$

Bringing in 4.54, this means that for every $(w, \beta)$ satisfying $df(x^\star; w) \leq \beta$, we must have $\beta \geq 0$ but that holds only if $df(x^\star; w) \geq 0$ for all $w \in \mathbb{R}^n$. □

We can now pause for a moment and recognize the leap made since §1.A. There, the classical Oresme and Fermat rules, defined in terms of directional derivatives and gradients, provide necessary conditions for minimizers of smooth functions and some others in the case of Oresme. In contrast, the Oresme rule 2.10 and the Fermat rule 4.73 specify necessary conditions for arbitrary functions using subderivatives and subgradients.

These concepts are defined for all functions and supported by an extensive calculus that enables practitioners to deal with a large number of cases. Although the Oresme and Fermat rules only specify necessary conditions for a minimizer, they're central to many algorithms. In particular, they justify substituting the actual minimization problem by the task of finding a point that satisfies such a condition.

The example $f(x) = \min\{x, x^2\}$ for $x \in \mathbb{R}$ illustrates the role of epi-regularity. At the point $(0, 0)$, epi $f$ has an inward kink and thus fails to be Clarke regular; Figure 4.29 gives the normal and tangent cones at that point and then also the subgradients and subderivatives by 4.55 and 4.54. The optimality condition specified by the Fermat rule holds at zero because $\partial f(0) = \{0, 1\}$. However, $df(0; -1) < 0$ and we conclude that 0 isn't a minimizer by the Oresme rule. Thus, the Fermat rule in this case identifies a spurious point.

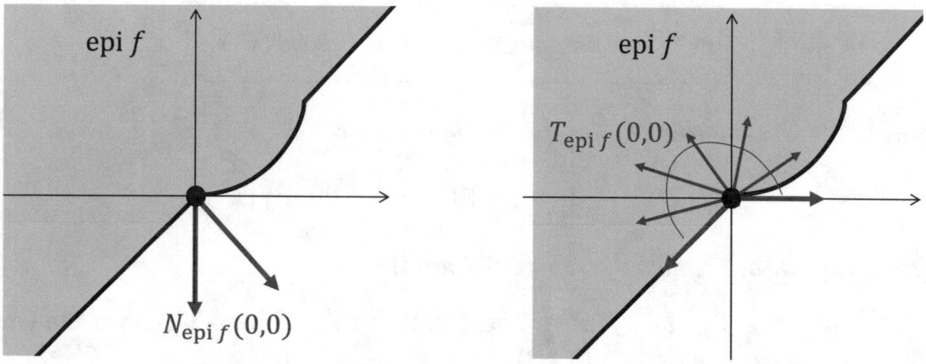


Fig. 4.29: For $f(x) = \min\{x, x^2\}$, the optimality condition from the Fermat rule is satisfied at zero but not the one from the Oresme rule.

The implication in the Fermat rule 4.73 can be extended further by considering any outer approximation of $\partial f(x^\star)$. In particular, for any $S(\bar{x}) \supset \partial f(\bar{x})$, the condition $0 \in S(\bar{x})$ is necessary for $\bar{x}$ to be a local minimizer. This thinking can be combined with the outer approximations of $\partial f(\bar{x})$ furnished by the chain rule 4.64 to construct optimality conditions under less demanding assumptions.

Of course, as we move further along in such a string of implications, we tend to end up with *weaker* optimality conditions, i.e., conditions that are satisfied by more and more points that aren't local minimizers. In Figure 4.29, the Fermat rule is indeed weaker than the Oresme rule. In the convex case, there's no daylight between the Oresme and Fermat rules, with both furnishing a complete characterization of minimizers; cf. 2.19. At a point of epi-regularity, there's still no difference between the two rules by 4.73. Moving beyond these cases, however, there are many possible optimality conditions due to discrepancy between the rules as well as different ways of bounding subderivatives and subgradients.

Generally, we would like to adopt an optimality condition that's reasonably strong but also computationally attractive. Conditions based on the Fermat rule are usually easier to develop and solve due to an extensive collection of outer approximations of subgradients; see §4.I. The Oresme rule would require upper bounds on the subderivatives, which can be more involved. Qualifications in the form of normal cones (as in 4.46) are also more easily satisfied than counterparts in terms of tangent cones; see the discussion around Exercise 6.39 in [105]. Still, the Oresme rule can furnish a viable approach. For example, the basic optimality condition in terms of $T_C(x^\star)$ (see 4.37) can be computationally attractive and significantly stronger than the corresponding normal cone condition.

From a numerical point of view, however, the weakening associated with passing from the Oresme rule to the Fermat rule isn't as significant as it may seem. If $0 \in \partial f(\bar{x})$, then there exist $x^\nu \to \bar{x}$ and $\varepsilon^\nu \searrow 0$, with $f(x^\nu) \to f(\bar{x})$ and

$$df(x^\nu; w) \geq -\varepsilon^\nu \|w\|_2 \quad \forall w \in \mathbb{R}^n.$$

(We can see this by recalling the definition of subgradients, passing to approximating regular normal vectors and invoking 4.54.) Thus, a point that satisfies the Fermat rule is close to points that nearly satisfy the Oresme rule. Since actual computations always involve tolerances, the practical difference between the two rules is less significant.

An algorithm may generate $x^\nu \to \bar{x}$ with $\partial f(x^\nu)$ getting closer to satisfy the Fermat rule 4.73 as $\nu \to \infty$, i.e.,

$$\operatorname{dist}\big(0, \partial f(x^\nu)\big) \to 0.$$

Then, $0 \in \partial f(\bar{x})$ as one would hope provided that $f(x^\nu) \to f(\bar{x})$. This is confirmed by the next proposition with further details emerging in §7.G.

**Proposition 4.74** (limits of subgradients). *For $f : \mathbb{R}^n \to \overline{\mathbb{R}}$ and $\bar{x}$ with $f(\bar{x}) \in \mathbb{R}$, one has*

$$x^\nu \to \bar{x}, \;\; f(x^\nu) \to f(\bar{x}), \;\; v^\nu \in \partial f(x^\nu) \to \bar{v} \quad \Longrightarrow \quad \bar{v} \in \partial f(\bar{x}).$$

**Proof.** This follows by the corresponding property for normal cones in 4.39. □

Let's now see how the Fermat rule can be used to construct optimality conditions for specific problems.

**Theorem 4.75** (optimality for composite function). *For smooth $f_0 : \mathbb{R}^n \to \mathbb{R}$, smooth $F : \mathbb{R}^n \to \mathbb{R}^m$, closed $X \subset \mathbb{R}^n$ and proper, lsc and convex $h : \mathbb{R}^m \to \overline{\mathbb{R}}$, suppose that the following qualification holds at $x^\star$:*

$$y \in N_{\operatorname{dom} h}\big(F(x^\star)\big) \;\; \text{and} \;\; -\nabla F(x^\star)^\top y \in N_X(x^\star) \quad \Longrightarrow \quad y = 0. \tag{4.16}$$

*If $x^\star$ is a local minimizer of the problem*

$$\operatorname*{minimize}_{x \in X} f_0(x) + h\big(F(x)\big),$$

*then*

$$\exists y \in \partial h\big(F(x^\star)\big) \;\; \text{such that} \;\; -\nabla f_0(x^\star) - \nabla F(x^\star)^\top y \in N_X(x^\star).$$

**Proof.** Let's define $G : \mathbb{R}^n \to \mathbb{R}^{n+m}$ by $G(x) = (x, F(x))$ and $g : \mathbb{R}^{n+m} \to \overline{\mathbb{R}}$ by

$$g(u) = \iota_X(u_1, \dots, u_n) + h(u_{n+1}, \dots, u_{n+m}).$$

Then,

$$f(x) = \iota_X(x) + h\big(F(x)\big) = g\big(G(x)\big),$$

with $g$ being proper and lsc, and $G$ being smooth. Consequently,

$$\partial f(x^\star) \subset \nabla G(x^\star)^\top \partial g\big(G(x^\star)\big)$$

by the chain rule 4.64, assuming the qualification (4.14) is met. For a point $u$ at which $g$ is finite,

$$\partial g(u) = N_X(u_1, \dots, u_n) \times \partial h(u_{n+1}, \dots, u_{n+m})$$

by 4.63 and 4.56. Moreover,

$$\nabla G(x^\star)^\top \begin{bmatrix} z \\ y \end{bmatrix} = z + \nabla F(x^\star)^\top y.$$

By the Fermat rule 4.73 and 4.58(c),

$$0 \in \partial(f_0 + f)(x^\star) = \nabla f_0(x^\star) + \partial f(x^\star) \subset \nabla f_0(x^\star) + \nabla G(x^\star)^\top \partial g\big(G(x^\star)\big),$$

which then implies that

$$0 = \nabla f_0(x^\star) + z + \nabla F(x^\star)^\top y \quad \text{for some } z \in N_X(x^\star) \text{ and } y \in \partial h\big(F(x^\star)\big).$$

This simplifies to the asserted expression. It only remains to verify the qualification (4.14). Since the first part of (4.14) becomes

$$(z, y) \in \partial^\infty g\big(G(x^\star)\big) \subset \partial^\infty \iota_X(x^\star) \times \partial^\infty h\big(F(x^\star)\big) = N_X(x^\star) \times N_{\mathrm{dom}\, h}\big(F(x^\star)\big)$$

by 4.63 and 4.65 and the second part reduces to $0 = z + \nabla F(x^\star)^\top y$ as already identified, the assumed (4.16) suffices. □

A main insight from this optimality condition is that one can pass beneficially from a problem in just $x$ to one of finding a pair $(x, y)$, with $y$ being a *multiplier vector*. As seen in earlier chapters, this broader view reveals many computational possibilities including the interior-point method (§2.G). For the problem in 4.75, we may go further and introduce another vector $z \in \mathbb{R}^m$ that can be tied to "feasibility" so that the optimality condition amounts to finding

$$(x, y, z) \quad \text{such that} \quad z = F(x), \quad y \in \partial h(z), \quad -\nabla f_0(x) - \nabla F(x)^\top y \in N_X(x).$$

The first requirement is a nonlinear equation approachable by the Newton-Raphson algorithm (§1.G). The last requirement is similar to the basic optimality condition 4.37 and is especially approachable when $X$ is polyhedral; see 2.44. The middle requirement is equivalent to having

$$z \in \operatorname{argmin}\big\{h - \langle y, \cdot \rangle\big\}$$

by the optimality condition 2.19 and thus involves solving a convex optimization problem. One can imagine iterative procedures that aim to solve for $(x, y, z)$ by utilizing existing algorithms for these requirements. Although not necessarily a minimizer of the composite function of interest, a resulting $x$ obtained in this manner might be deemed acceptable as a solution of the minimization problem. Of course, there are many details to be filled in and much depends on the nature of the convex function $h$. We'll see several concrete developments in the next sections and chapters.

**Example 4.76** (smooth objective and equality constraint functions). For smooth functions $f_i : \mathbb{R}^n \to \mathbb{R}$, $i = 0, 1, \ldots, m$, consider the problem

$$\underset{x\in\mathbb{R}^n}{\text{minimize}}\ f_0(x) \ \text{ subject to } \ f_i(x) = 0,\ i = 1, \ldots, m.$$

If $x^\star$ is a local minimizer with $\{\nabla f_i(x^\star), i = 1, \ldots, m\}$ linearly independent, then[7]

$$\nabla f_0(x^\star) + \sum\nolimits_{i=1}^m y_i \nabla f_i(x^\star) = 0, \quad f_i(x^\star) = 0, \quad y_i \in \mathbb{R}, \ i = 1, \ldots, m.$$

This amounts to $n + m$ nonlinear equations and the Newton-Raphson algorithm (§1.G) applies as long as the functions are twice smooth.

**Detail.** We can obtain the optimality condition by considering $f = f_0 + \iota_C$, where

$$C = \{x \in \mathbb{R}^n \mid f_i(x) = 0, \ i = 1, \ldots, m\},$$

and then use the basic optimality condition 4.37; $N_C(x^\star)$ is known via 4.49. Alternatively, one can leverage the sum rule 4.67 as utilized in 4.70 and then also the formula for $N_C(x^\star)$.

We work through 4.75 and set

$$h(u_1, \ldots, u_m) = \sum\nolimits_{i=1}^m \iota_{\{0\}}(u_i),$$

which is proper, lsc and convex, $F(x) = (f_1(x), \ldots, f_m(x))$ and $X = \mathbb{R}^n$. By 4.63, $\partial h(0) = \mathbb{R}^m$. Since $N_X(x) = \{0\}$, we obtain the stated condition assuming that the qualification (4.16) holds. In this case, the qualification amounts to checking whether $y \in \mathbb{R}^m$ and $\nabla F(x^\star)^\top y = 0$ imply $y = 0$. But, this is just the linear independence assumption.

The need for a qualification is highlighted by the instance

$$f_1(x) = (x_1 + 1)^2 + x_2^2 - 1 \quad \text{and} \quad f_2(x) = (x_1 - 1)^2 + x_2^2 - 1;$$

see Figure 4.30. Since $x^\star = (0, 0)$ is the only feasible point, it's also a minimizer regardless of $f_0$. However, $\nabla f_1(x^\star) = (2, 0)$ and $\nabla f_2(x^\star) = (-2, 0)$, which aren't linearly independent. Thus, the qualification fails at $x^\star$ and the derived optimality condition doesn't apply. Still, for $f_0(x) = x_1 + x_2$, it would have amounted to

[7] J.-L. Lagrange derived this optimality condition, which led to the name "Lagrange multipliers" for the components of $y$; see §2.E.

$$\begin{bmatrix}1\\1\end{bmatrix} + y_1 \begin{bmatrix}2(x_1+1)\\2x_2\end{bmatrix} + y_2 \begin{bmatrix}2(x_1-1)\\2x_2\end{bmatrix} = \begin{bmatrix}0\\0\end{bmatrix},$$

which has no solution at the feasible point. This might have led us to conclude that the problem has no local minimizer, but that's erroneous because the optimality condition is invoked without verifying the qualification. In general, if $\bar{x}$ is a point at which (4.16) holds and $\bar{x}$ doesn't satisfy the optimality condition 4.75, then $\bar{x}$ isn't a local minimizer of $\iota_X + f_0 + h \circ F$. □

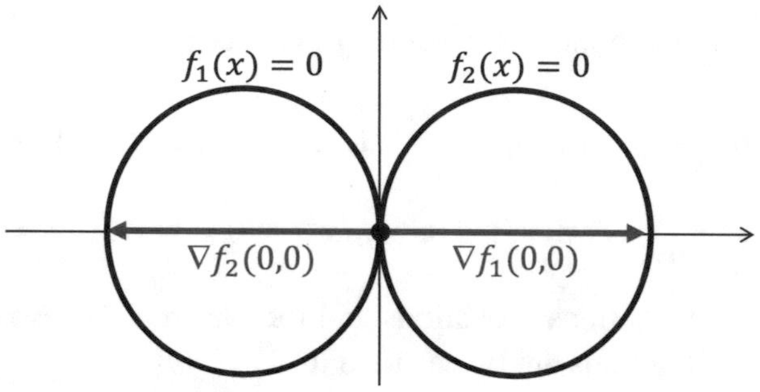


Fig. 4.30: Failure of qualification at (0, 0).

We next extend the KKT condition 2.45 from a polyhedral feasible set to one with equalities and inequalities specified by smooth functions. While historically significant, its inability to highlight other structural properties than smoothness reduces the resulting KKT condition to a supporting role with 4.75 being more versatile for modern applications.

**Theorem 4.77** (KKT condition for optimality). *For $f_i : \mathbb{R}^n \to \mathbb{R}$, $i = 0, 1, \ldots, m$, and $g_i : \mathbb{R}^n \to \mathbb{R}$, $i = 1, \ldots, q$, all smooth, suppose that the Mangasarian-Fromovitz constraint qualification (4.10) holds at $x^\star$. If $x^\star$ is a local minimizer of the problem*

$$\text{minimize } f_0(x) \text{ subject to } f_i(x) = 0,\ i = 1, \ldots, m; \quad g_i(x) \le 0,\ i = 1, \ldots, q,$$

*then*

$$\exists y \in \mathbb{R}^m \text{ and } z \in \mathbb{R}^q \text{ such that } z_i \ge 0,\ i \in \mathbb{A}(x^\star), \quad z_i = 0,\ i \notin \mathbb{A}(x^\star)$$
$$-\nabla f_0(x^\star) = \sum_{i=1}^m y_i \nabla f_i(x^\star) + \sum_{i=1}^q z_i \nabla g_i(x^\star),$$

*where $\mathbb{A}(x^\star) = \{i \mid g_i(x^\star) = 0\}$.*

**Proof.** This follows directly from the basic optimality condition 4.37 and the normal cone formula in 4.49. □

The requirement that a multiplier needs to be zero for inactive constraints is referred to as a *complementary slackness condition*. It complicates the task of solving the KKT condition as compared to the case with equality constraints only in 4.76.

**Exercise 4.78** (KKT condition). For $f_0(x) = x_1 + x_2$ and $g_1(x) = (x_1 - 1)^2 + x_2^2 - 1 \leq 0$, determine all points that satisfy the KKT condition and check whether the Mangasarian-Fromovitz constraint qualification holds.

**Guide.** Try $\bar{x} = (1 - 1/\sqrt{2},\ -1/\sqrt{2})$ and $z_1 = 1/\sqrt{2}$. □

**Example 4.79** (equality as two inequalities). For smooth $f_0, f_1 : \mathbb{R}^n \to \mathbb{R}$, the problem

$$\underset{x \in \mathbb{R}^n}{\text{minimize}}\ f_0(x) \quad \text{subject to} \quad f_1(x) \leq 0,\ -f_1(x) \leq 0$$

fails the Mangasarian-Fromovitz constraint qualification (4.10) at every feasible point. An equivalent and better formulation is to simply minimize $f_0(x)$ subject to $f_1(x) = 0$.

**Detail.** At a feasible point $\bar{x}$, the Mangasarian-Fromovitz constraint qualification amounts to

$$z_1 \nabla f_1(\bar{x}) - z_2 \nabla f_1(\bar{x}) = 0 \quad \text{and} \quad z_1, z_2 \geq 0 \quad \Longrightarrow \quad z_1 = z_2 = 0,$$

which doesn't hold. In the alternative formulation with an equality constraint, the Mangasarian-Fromovitz constraint qualification is simply that $y_1 \nabla f_1(\bar{x}) = 0$ should imply $y_1 = 0$. This holds as long as $\nabla f_1(\bar{x}) \neq 0$. □

**Example 4.80** (integer constraints). For $X \subset \mathbb{R}^n$ and $c \in \mathbb{R}^n$, the problem

$$\underset{x \in X}{\text{minimize}}\ \langle c, x \rangle \quad \text{subject to} \quad x_j \in \{0, 1\},\ j = 1, \dots, n$$

can equivalently be formulated as

$$\underset{x \in X}{\text{minimize}}\ \langle c, x \rangle \quad \text{subject to} \quad f_j(x) = x_j(1 - x_j) = 0,\ j = 1, \dots, n.$$

However, the latter formulation hides the structure of the problem and can make every feasible point satisfy the KKT condition 4.77.

**Detail.** Suppose that every feasible $\bar{x}$ is in the interior of $X$ so we can ignore the set for the sake of optimality conditions and concentrate on the KKT condition with equality constraints only. Then, the Mangasarian-Fromovitz constraint qualification (4.10) holds at every feasible point because

$$\sum_{j=1}^{n} y_j \nabla f_j(\bar{x}) = 0 \iff y_j(1 - 2\bar{x}_j) = 0,\ j = 1, \dots, n.$$

The KKT condition 4.77 becomes $c_j + y_j(1 - 2\bar{x}_j) = 0$ for $j = 1, \dots n$, which holds with $y_j = -c_j$ when $\bar{x}_j = 0$ and $y_j = c_j$ when $\bar{x}_j = 1$. Consequently, the KKT condition is much too weak to be helpful for the reformulated problem. As discussed in 4.3, problems with integer constraints are better addressed directly by taking advantage of their structure. □

**Example 4.81** (minimizers of max-function). For smooth functions $f_i : \mathbb{R}^n \to \mathbb{R}$, $i = 1, \dots, m$, consider

$$\underset{x \in \mathbb{R}^n}{\text{minimize}}\ f(x) = \max_{i=1,\dots,m} f_i(x).$$

If $x^\star$ is a local minimizer of the problem, then

$$\exists y \in \mathbb{R}^m \text{ such that } y_i \geq 0,\ i \in \mathbb{A}(x^\star),\ \ y_i = 0,\ i \notin \mathbb{A}(x^\star),$$
$$\sum_{i=1}^m y_i = 1 \quad \text{and} \quad \sum_{i=1}^m y_i \nabla f_i(x^\star) = 0,$$

where $\mathbb{A}(x^\star) = \{i \mid f_i(x^\star) = f(x^\star)\}$.

**Detail.** The Fermat rule 4.73 and the expression for subgradients in 4.66 give the necessary condition

$$0 \in \operatorname{con}\{\nabla f_i(x^\star), i \in \mathbb{A}(x^\star)\},$$

which is equivalent to having

$$0 = \sum_{i \in \mathbb{A}(x^\star)} y_i \nabla f_i(x^\star)$$

for some $y_i \geq 0, i \in \mathbb{A}(x^\star)$, with $\sum_{i \in \mathbb{A}(x^\star)} y_i = 1$. □

The example produces an alternative to the KKT condition 4.77 for inequality constraints through the observation: If $x^\star$ is a local minimizer of

$$\operatorname*{minimize}_{x \in \mathbb{R}^n} f_0(x) \ \text{ subject to } \ g_i(x) \leq 0,\ i = 1, \dots, q, \tag{4.17}$$

then it's also a local minimizer of

$$\operatorname*{minimize}_{x \in \mathbb{R}^n} f(x) = \max\{f_0(x) - f_0(x^\star), g_1(x), \dots, g_q(x)\}. \tag{4.18}$$

To see this, note that when $x^\star$ is a local minimizer of (4.17), there's $\varepsilon > 0$ such that $f_0(x) \geq f_0(x^\star)$ for all $x \in \mathbb{B}(x^\star, \varepsilon)$ with $g_i(x) \leq 0$ for all $i$. Certainly, $f(x) \geq 0$ for all $x \in \mathbb{B}(x^\star, \varepsilon)$ and $f(x^\star) = 0$, which implies that $x^\star$ is a local minimizer of (4.18). Consequently, any necessary optimality condition for (4.18) is also a necessary optimality condition for (4.17). Let's develop this in detail.

**Proposition 4.82** (Fritz-John condition for optimality). *For smooth $f_0, g_i : \mathbb{R}^n \to \mathbb{R}$, $i = 1, \dots, q$, suppose that $x^\star$ is a local minimizer of (4.17). Then,*

$$\exists (y_0, y_1, \dots, y_q) \in \mathbb{R}^{1+q} \text{ such that } y_0 \geq 0,\ y_i \geq 0,\ i \in \mathbb{A}(x^\star),\ y_i = 0,\ i \notin \mathbb{A}(x^\star),$$
$$\sum_{i=0}^q y_i = 1 \quad \text{and} \quad y_0 \nabla f_0(x^\star) + \sum_{i=1}^q y_i \nabla g_i(x^\star) = 0,$$

*where $\mathbb{A}(x^\star) = \{i \mid g_i(x^\star) = 0\}$.*

**Proof.** In view of the prior discussion, we obtain an optimality condition by applying 4.81 to the function given by

$$f(x) = \max\{f_0(x) - f_0(x^\star), g_1(x), \dots, g_q(x)\}.$$

Thus, there exists $(y_0, \ldots, y_q) \in \mathbb{R}^{1+q}$ with $y_0 \geq 0$ when $f_0(x^\star) - f_0(x^\star) = f(x^\star)$, which always is the case because $f(x^\star) = 0$. Moreover, $y_i \geq 0$ for $i = 1, \ldots, q$ with $g_i(x^\star) = 0$ and $y_i = 0$ otherwise. In addition to summing to one, the multipliers satisfy the stated gradient condition. □

In contrast to the KKT condition 4.77, the Fritz-John condition places an extra multiplier in front of the gradient of the objective function. Significantly, it avoids *any* qualification, which could be beneficial in cases such as the one at the end of 4.76. Let's pass from equalities to inequality constraints in that example, which doesn't change the conclusion: the minimizer violates the KKT condition. Still, one can show that it satisfies the Fritz-John condition with $y = (0, 1/2, 1/2)$.

**Exercise 4.83** (weakness of Fritz-John condition). Show that the Fritz-John condition is weaker than the KKT condition. Specifically, (a) prove that if a point $\bar{x}$ satisfies the KKT condition 4.77 for an inequality constrained problem with multipliers $z_1, \ldots, z_q$, then $\bar{x}$ also satisfies the Fritz-John condition 4.82 with multipliers

$$y_0 = \frac{1}{1 + \sum_{j=1}^q z_j} \quad \text{and} \quad y_i = \frac{z_i}{1 + \sum_{j=1}^q z_j}.$$

(b) Prove that the converse holds if $y_0 > 0$. (c) For the problem of minimizing $f_0(x) = x^2$ subject to $g_1(x) = \sin x - 1 \leq 0$, determine all local minimizers, all points satisfying the Fritz-John condition, all points satisfying the KKT condition and all points satisfying the Mangasarian-Fromovitz constraint qualification (4.10).

## 4.K SQP and Interior-Point Methods

The various optimality conditions of the previous section serve as the foundation for many algorithms. They immediately provide means to check whether a particular candidate solution can be ruled out as a minimizer and thus emerge as key components of stopping criteria for algorithms. An optimality condition also facilitates the shift from the possibly difficult task of finding a minimizer to that of solving a set of equations, inequalities and/or inclusions. While the shift typically entails a relaxation of the requirements of a minimizer because the optimality condition is only *necessary*, it's usually a meaningful and computationally attractive compromise. In practice, a solution to such an optimality condition is quite often a local minimizer or a point with a low objective function value as we'll see in §6.A. In this section, we'll sketch out two algorithms for solving the KKT condition 4.77.

We recall from §1.G that Newton's method for minimizing a twice smooth function $f : \mathbb{R}^n \to \mathbb{R}$ can be interpreted as an algorithm for finding a solution to the optimality condition $\nabla f(x) = 0$. The interior-point method of §2.G extends this idea by solving the KKT condition for a convex quadratic function subject to linear constraints. Let's develop this further and address the problem

$$\underset{x\in\mathbb{R}^n}{\text{minimize}}\ f_0(x) \ \text{ subject to } \ f_i(x)=0,\ i=1,\dots,m;\ \ g_i(x)\le 0,\ i=1,\dots,q, \tag{4.19}$$

where $f_i : \mathbb{R}^n \to \mathbb{R}$, $i = 0, 1, \dots, m$, and $g_i : \mathbb{R}^n \to \mathbb{R}$, $i = 1, \dots, q$ are twice smooth. Our goal is to construct an algorithm that computes $(x, y, z)$ satisfying the KKT condition 4.77.

Let's initially consider the simpler case without inequalities and introduce the *Lagrangian* $l : \mathbb{R}^n \times \mathbb{R}^m \to \mathbb{R}$ defined by

$$l(x, y) = f_0(x) + \langle F(x), y \rangle,$$

where $F(x) = (f_1(x), \dots, f_m(x))$. This function will gradually grow in importance, but for now it simply serves as a notational shortcut. In light of 4.76, the KKT condition, now written in terms of the Lagrangian, reduces to $\nabla l(x, y) = 0$ because

$$\nabla_x l(x, y) = \nabla f_0(x) + \nabla F(x)^\top y \quad \text{and} \quad \nabla_y l(x, y) = F(x).$$

This is a system of $n + m$ nonlinear equations with unknowns $x \in \mathbb{R}^n$ and $y \in \mathbb{R}^m$, and the Newton-Raphson algorithm of §1.G can be brought in. At a current point $(x^\nu, y^\nu)$, that algorithm replaces $\nabla l(x, y) = 0$ by the affine approximation

$$\nabla l(x^\nu, y^\nu) + \nabla^2 l(x^\nu, y^\nu)(v, w) = 0,$$

which amounts to

$$\begin{bmatrix} \nabla^2_{xx} l(x^\nu, y^\nu) & \nabla F(x^\nu)^\top \\ \nabla F(x^\nu) & 0 \end{bmatrix} \begin{bmatrix} v \\ w \end{bmatrix} = \begin{bmatrix} -\nabla f_0(x^\nu) - \nabla F(x^\nu)^\top y^\nu \\ -F(x^\nu) \end{bmatrix}, \tag{4.20}$$

where $\nabla^2_{xx} l(x^\nu, y^\nu)$ is the Hessian of $l(\cdot, y^\nu)$ evaluated at $x^\nu$. As long as the matrix on the left-hand side is invertible, this linear system of equations has a unique solution $(v^\nu, w^\nu)$, which then defines a new point

$$(x^{\nu+1}, y^{\nu+1}) = (x^\nu + v^\nu, y^\nu + w^\nu)$$

and the Newton-Raphson algorithm can proceed. The matrix is indeed invertible if

(a) $\nabla f_1(x^\nu), \dots, \nabla f_m(x^\nu)$ are linearly independent
(b) $\langle v, \nabla^2_{xx} l(x^\nu, y^\nu) v \rangle > 0$ for all $v \neq 0$ with $\nabla F(x^\nu) v = 0$.

The first condition is the qualification in 4.76. The second condition states that $\nabla^2_{xx} l(x^\nu, y^\nu)$ must be positive definite on the tangent cone to the feasible set at $x^\nu$. Under slightly stronger assumptions, one can show that when started sufficiently close to a solution of $\nabla l(x, y) = 0$, the Newton-Raphson algorithm converges, in fact quadratically, to that solution; see [69, Sections 18.1, 18.7] as well as §1.G.

As in the case of Newton's method, the introduction of a line search procedure allows us to start arbitrarily far from a solution; cf. 1.34. It's also possible to devise strategies for modifying a matrix that fails to be invertible so that the Newton-Raphson algorithm can proceed in such cases as well. There are challenges, however, associated with the presence of *inequality* constraints because then the KKT condition includes equations as well as inequalities and the multipliers need to satisfy complementary slackness. We discuss two main ideas.

**Sequential Quadratic Programming.** The algorithm just described for equality-constrained problems can be viewed from another angle that leads to a natural extension for inequality constraints. For given $(x^\nu, y^\nu)$, consider the quadratic optimization problem

$$\begin{aligned}&\underset{v\in\mathbb{R}^n}{\text{minimize}}\ l(x^\nu, y^\nu) + \langle \nabla_x l(x^\nu, y^\nu), v\rangle + \tfrac{1}{2}\langle v, \nabla^2_{xx} l(x^\nu, y^\nu)v\rangle\\ &\text{subject to } F(x^\nu) + \nabla F(x^\nu)v = 0.\end{aligned}$$

The objective function here is a quadratic approximation of $v \mapsto l(x^\nu + v, y^\nu)$ and the constraint involves a linear approximation of $v \mapsto F(x^\nu + v)$. Thus, the quadratic problem approximates that of minimizing $l(x, y)$ over $x$ subject to $F(x) = 0$, which is nothing but the equality-constrained problem regardless of $y$. Under assumptions (a,b), the quadratic problem has a unique minimizer $v^\nu$, which together with a unique multiplier vector $w^\nu$, is given as the solution of its KKT condition 2.45:

$$\begin{aligned}\nabla_x l(x^\nu, y^\nu) + \nabla^2_{xx} l(x^\nu, y^\nu)v + \nabla F(x^\nu)^\top w = 0\\ F(x^\nu) + \nabla F(x^\nu)v = 0.\end{aligned}$$

After writing this in matrix form, we realize that it coincides with (4.20). Consequently, under assumptions (a,b), $v^\nu$ minimizes the above quadratic problem with $w^\nu$ as the corresponding multiplier vector if and only if $(v^\nu, w^\nu)$ solves (4.20). We've gained the insight that the Newton-Raphson algorithm applied to the KKT condition of an equality-constrained problem is equivalent to solving a sequence of approximating problems involving a quadratic approximation of the Lagrangian and a linear approximation of the constraints. This perspective motivates the name sequential quadratic programming (SQP). More significantly, it points to a path for handling inequality constraints.

Let $G(x) = (g_1(x), \ldots, g_q(x))$. Again, a Lagrangian $l : \mathbb{R}^n \times \mathbb{R}^m \times \mathbb{R}^q \to \mathbb{R}$, now expanded to

$$l(x, y, z) = f_0(x) + \langle F(x), y\rangle + \langle G(x), z\rangle,$$

helps us to express the ideas compactly. To address inequalities, we modify the quadratic optimization problem to take the form

$$\begin{aligned}&\underset{v\in\mathbb{R}^n}{\text{minimize}}\ l(x^\nu, y^\nu, z^\nu) + \langle \nabla_x l(x^\nu, y^\nu, z^\nu), v\rangle + \tfrac{1}{2}\langle v, \nabla^2_{xx} l(x^\nu, y^\nu, z^\nu)v\rangle\\ &\text{subject to } \ F(x^\nu) + \nabla F(x^\nu)v = 0 \quad G(x^\nu) + \nabla G(x^\nu)v \le 0,\end{aligned}$$

where $(x^\nu, y^\nu, z^\nu)$ are the current vectors. Under an extension of assumptions (a,b), the quadratic problem is strictly convex. It involves linear inequalities, however, and this adds significantly to the computational burden per iteration. (In addition to the interior-point method of §2.G, we refer to [69] for specialized algorithms for solving this quadratic problem.) Still, when paired with a line search procedure and regularization of the quadratic problem when (strict) convexity fails, the resulting algorithm is highly competitive especially when the number of active constraints at a solution is near the number of variables in the original problem; see [69, Chap. 18] for details.

**Interior-Point Method.** The interior-point method of §2.G can be extended to the present setting. The KKT condition 4.77 for (4.19) amounts to

$$\nabla_x l(x,y,z) = 0, \quad F(x) = 0, \quad G(x) \leq 0, \quad z \geq 0, \quad g_i(x)z_i = 0, \; i = 1, \dots q,$$

where the complementary slackness condition $z_i = 0$ when $g_i(x) < 0$ is written in equation form. A vector of nonnegative slack variables $s \in \mathbb{R}^q$ can be brought in to state these conditions equivalently as

$$\begin{bmatrix} \nabla_x l(x,y,z) \\ F(x) \\ G(x) + s \\ \operatorname{diag}(s)z - \mu e \end{bmatrix} = 0 \quad \text{and} \quad s \geq 0, \; z \geq 0,$$

when $\mu = 0$. Here, $\operatorname{diag}(s)$ is the $q \times q$-matrix with $s$ along its diagonal and zeros elsewhere and $e = (1, \dots, 1) \in \mathbb{R}^q$. Similar to §2.G, we can obtain a solution to these conditions with $\mu = 0$ by initially considering a positive $\mu$ and therefore positive $s$ and $z$ components. From a starting point $(x, y, z, s)$ with positive components in $z$ and $s$, one can apply the Newton-Raphson algorithm to the above equality system for a number of iterations while ensuring that $s$ and $z$ remain positive. Then, $\mu$ is reduced to a positive number closer to zero and the Newton-Raphson iterations are repeated. Under certain assumptions, this approach converges to a solution of the KKT condition for (4.19) if initialized sufficiently near a solution. An arbitrary starting point is permitted when the algorithm includes a line search. There's also a need for modifications when the matrix encountered in a Newton-Raphson iteration isn't invertible; details can be found in [69, Chap. 19]. In problems with many variables but relatively few constraints active at a solution, interior-point methods of this kind are especially competitive.

# Chapter 5
# PERTURBATION AND DUALITY

An optimization problem that turns out to be infeasible or unbounded prompts us to go back and examine the underlying model assumptions. Even when a solution is obtained, we typically would like to assess the validity of the model. Is the solution insensitive to changes in model parameters? This is important because the exact values of parameters may not be known and one would like to avoid being misled by a solution obtained using incorrect values. Thus, it's rarely enough to address an application by formulating a model, solving the resulting optimization problem and presenting the solution as *the* answer. One would need to confirm that the model is suitable and this can, at least in part, be achieved by considering a *family of optimization problems* constructed by perturbing parameters of concern. The resulting sensitivity analysis uncovers troubling situations with unstable solutions and indicates better formulations.

Embedding an actual problem of interest within a family of problems is also a primary path to optimality conditions as well as computationally attractive, alternative problems, which under ideal circumstances, and when properly tuned, may even furnish the minimum value of the actual problem. The tuning of these alternative problems turns out to be intimately tied to finding multipliers in optimality conditions and thus emerges as a main component of several optimization algorithms. In fact, the tuning amounts to solving certain *dual* optimization problems. We'll now turn to the opportunities afforded by this broad perspective.

## 5.A Rockafellians

Suppose that we've formulated an optimization model to address a particular application. The model involves parameters with values that are somewhat unsettled. The parameters might represent the assumed budget of resources available, the probabilities of various outcomes, weights in an additive objective function or targets in goal optimization. A preliminary study would involve the solution of the resulting problem with nominal values for the parameters. However, a more comprehensive approach considers different values of the parameters and their effect on minima and minimizers in an effort to validate

 
J. O. Royset and R. J-B Wets, *An Optimization Primer*, Springer Series in Operations Research and Financial Engineering, https://doi.org/10.1007/978-3-030-76275-9_5

the model. To formalize this thinking, let's introduce a function that fully represents the actual objective function as well as parametric perturbations.

**Definition 5.1** (Rockafellian). For the problem of minimizing $f_0 : \mathbb{R}^n \to \overline{\mathbb{R}}$, we say that $f : \mathbb{R}^m \times \mathbb{R}^n \to \overline{\mathbb{R}}$ is a *Rockafellian* with *anchor* at $\bar{u} \in \mathbb{R}^m$ if

$$f(\bar{u}, x) = f_0(x) \quad \forall x \in \mathbb{R}^n.$$

Suppose that $f_0 : \mathbb{R}^n \to \overline{\mathbb{R}}$ represents the actual problem of interest. Its minimization might bring forth a "good" decision, but when examined in isolation $f_0$ fails to identify the effect of changing parameter values. An associated Rockafellian $f : \mathbb{R}^m \times \mathbb{R}^n \to \overline{\mathbb{R}}$ explicitly specifies the dependence on $m$ parameters and defines the *family of problems*

$$\Big\{ \operatorname*{minimize}_{x \in \mathbb{R}^n} f(u, x), \quad u \in \mathbb{R}^m \Big\}.$$

The minimum values $p(u) = \inf f(u, \cdot)$, the sets of minimizers $P(u) = \operatorname{argmin} f(u, \cdot)$ and other quantities can then be examined as they vary with the *perturbation vector* $u \in \mathbb{R}^m$. For example, in goal optimization as laid out by 4.6, we may face the problem

$$\operatorname*{minimize}_{x \in \mathbb{R}^n} f_0(x) = \sum\nolimits_{i=1}^m \theta_i \max\big\{0, f_i(x) - \bar{u}_i\big\},$$

where $f_i : \mathbb{R}^n \to \mathbb{R}$ models a quantity of interest and $\theta_i$ and $\bar{u}_i$ specify the corresponding weight and target. The exact targets might be under discussion and $\bar{u} = (\bar{u}_1, \dots, \bar{u}_m)$ is just a tentative choice. It would be prudent to consider a range of possible target values. Thus, we may adopt the Rockafellian with

$$f(u, x) = \sum\nolimits_{i=1}^m \theta_i \max\big\{0, f_i(x) - u_i\big\},$$

which then has anchor at $\bar{u}$.

In budgeting, we may seek to minimize $f_0(x)$ subject to the constraint that $g(x)$ shouldn't exceed a budget $\bar{u}$. With a focus on the budget level, we could consider a Rockafellian with

$$f(u, x) = f_0(x) + \iota_{(-\infty, 0]}\big(g(x) + u\big)$$

and thus capture sensitivity of the solution to budgetary perturbations.

We typically assume that all perturbation vectors $u \in \mathbb{R}^m$ are of interest and not only those in some set $U \subset \mathbb{R}^m$ such as the nonnegative vectors. This simplifies the development, but refinements follow by nearly identical arguments to those detailed below.

By considering a Rockafellian, we can carry out *sensitivity analysis* and answer the fundamental question: Would we make a significantly different decision (or assessment or prediction) if the parameters were perturbed? If the answer is yes, then it would be prudent to reevaluate model assumptions and parameter values. At the minimum, any recommendation derived from the analysis should be qualified as being sensitive to assumptions.

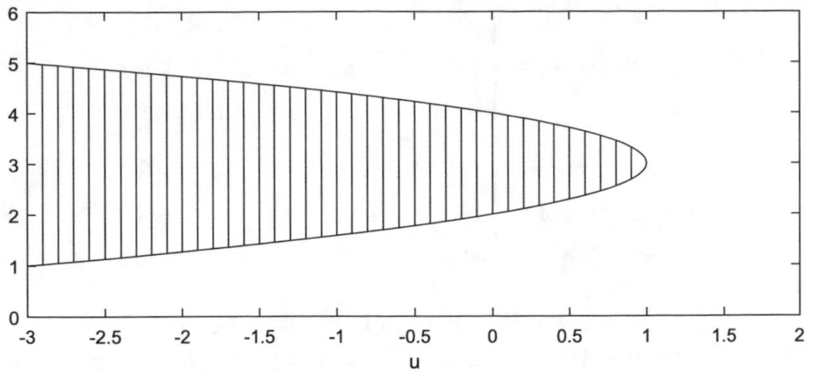


Fig. 5.1: The feasible set $C(u)$ for varying $u$ in 5.2.

There are many different Rockafellians that can be associated with a minimization problem as they represent different ways the problem can be changed. This offers much flexibility in modeling and computations. The choice of a particular Rockafellian reflects concerns beyond getting a solution and in some way completes the formulation of the problem. These concerns could be about the implications of shifting, restructuring and removing constraints. But by no means are the choice of Rockafellians limited to those affecting only the constraints. One might consider perturbations that alter the rewards associated with some individual or a particular combination of decisions; the goal optimization example above illustrates a possibility. When dealing with stochastic optimization problems, a Rockafellian might be selected to reflect dependence of the solutions on the distribution of the random components, for example.

**Example 5.2** (perturbation). The problem of minimizing $x^2+1$ subject to $(x-2)(x-4)+1 \leq 0$ can be associated with a Rockafellian defined by

$$f(u,x) = x^2 + 1 + \iota_{(-\infty,0]}\big(g(u,x)\big) \quad \text{and} \quad g(u,x) = (x-2)(x-4) + u,$$

with anchor at $\bar{u} = 1$. The minimum value $p(u) = \inf f(u, \cdot\,)$ doesn't vary continuously in $u$ at 1. Among the numerous Rockafellians that might be associated with the problem, this particular one highlights the sensitivity to changes on the right-hand side of the constraint.

**Detail.** Since the feasible set $C(u) = \operatorname{dom} f(u, \cdot\,)$ is given by the constraint $(x-2)(x-4) + u \leq 0$, we obtain as seen in Figure 5.1:

$$C(u) = \begin{cases} \big[\, 3 - \sqrt{1-u},\ 3 + \sqrt{1-u}\,\big] & \text{if } u \leq 1 \\ \emptyset & \text{otherwise.} \end{cases}$$

As functions of $u$, the minimizers and minimum values become

$$\operatorname{argmin} f(u,\cdot) = \begin{cases} \{0\} & \text{if } u < -8 \\ \{3 - \sqrt{1-u}\} & \text{if } -8 \le u \le 1 \\ \emptyset & \text{otherwise} \end{cases}$$

$$p(u) = \inf f(u,\cdot) = \begin{cases} 1 & \text{if } u < -8 \\ 11 - u - 6\sqrt{1-u} & \text{if } -8 \le u \le 1 \\ \infty & \text{otherwise.} \end{cases}$$

At every $u < 1$, $p$ is continuous. However, $p(1) = 10$ and $p(u) = \infty$ for $u > 1$, which makes $p$ discontinuous at 1. By looking at the level-sets $\{p \le \alpha\}$, we realize that they're closed for all $\alpha \in \mathbb{R}$. Thus, $p$ is lsc by 4.8. □

A Rockafellian $f : \mathbb{R}^m \times \mathbb{R}^n \to \overline{\mathbb{R}}$ together with the epigraphical approximation theory of §4.C provide the basis for sensitivity analysis. In particular, they offer a means to examine continuity properties of the *inf-projection*

$$u \mapsto p(u) = \inf f(u,\cdot),$$

which we refer to as the *min-value function*, and related properties for the set-valued mapping $u \mapsto P(u) = \operatorname{argmin} f(u,\cdot)$ representing minimizers. However, a cautionary lesson is furnished by the example $f(u,x) = \max\{-1, ux\}$, which defines a convex function in $x$ for any $u \in \mathbb{R}$. As indicated by Figure 5.2, $f(1/\nu,\cdot) \xrightarrow{e} f(0,\cdot)$ but $p(1/\nu) = -1$ for all $\nu \in \mathbb{N}$ and certainly fails to converge to $p(0) = 0$. The trouble in this example is that minimizers from $\operatorname{argmin} f(1/\nu,\cdot)$ don't have a cluster point as assumed in 4.14. This pathological situation is eliminated under tightness.

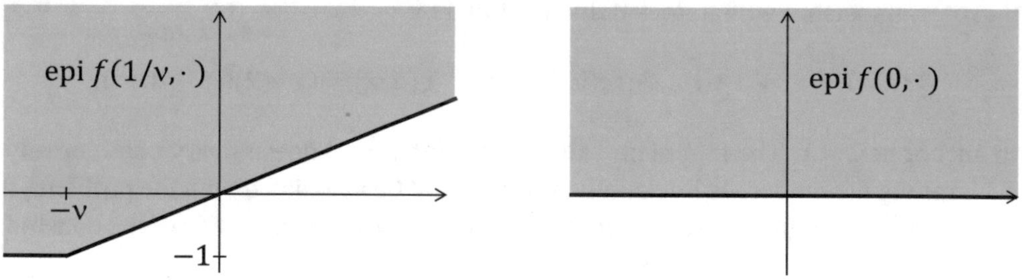


Fig. 5.2: The minimum value of $f(1/\nu,\cdot)$ doesn't converge to that of $f(0,\cdot)$ even though $f(1/\nu,\cdot)$ epi-converges to $f(0,\cdot)$.

**Definition 5.3** (tightness). The functions $\{f^\nu : \mathbb{R}^n \to \overline{\mathbb{R}}, \nu \in \mathbb{N}\}$ are *tight* if for all $\varepsilon > 0$, there are compact $B_\varepsilon \subset \mathbb{R}^n$ and $\nu_\varepsilon \in \mathbb{N}$ such that

$$\inf_{B_\varepsilon} f^\nu \le \inf f^\nu + \varepsilon \quad \forall \nu \ge \nu_\varepsilon.$$

The functions *epi-converge tightly* if in addition to being tight they also epi-converge to some function.

The restriction to some compact set $B_\varepsilon$ in the definition produces the same or a higher minimum value. Tightness is about limiting the increase to an arbitrarily small $\varepsilon$ by choosing $B_\varepsilon$ large enough. For a single function, this is always possible if its infimum isn't $-\infty$. For a collection of functions to be tight, the set $B_\varepsilon$ needs to work for all but a finite number of the functions. In the discussion prior to the definition, tightness fails because for any compact set $B$, $\inf_B f(1/\nu, \cdot) \geq -1/2$ for sufficiently large $\nu$ and $\inf f(1/\nu, \cdot) = -1$.

A collection $\{f^\nu : \mathbb{R}^n \to \overline{\mathbb{R}}, \nu \in \mathbb{N}\}$ is tight if $\cup_{\nu\in\mathbb{N}} \operatorname{dom} f^\nu$ is contained in a bounded set. However, many other possibilities exist. For example, if there's a compact set $B \subset \mathbb{R}^n$ such that $B \cap \operatorname{argmin} f^\nu$ is nonempty for all $\nu$, then the condition of the definition also holds.

In the context of a function $f : \mathbb{R}^m \times \mathbb{R}^n \to \overline{\mathbb{R}}$ with values $f(u, x)$, tightness of a collection $\{f(u^\nu, \cdot), \nu \in \mathbb{N}\}$ is closely related to whether the function is *level-bounded in x locally uniformly in u* by which we mean the following property:

$$\forall \bar{u} \in \mathbb{R}^m \text{ and } \alpha \in \mathbb{R} \quad \exists \varepsilon > 0 \text{ and a bounded set } B \subset \mathbb{R}^n$$
$$\text{such that } \{f(u, \cdot) \leq \alpha\} \subset B \ \ \forall u \in \mathbb{B}(\bar{u}, \varepsilon).$$

Informally, the property amounts to having, for each $\bar{u}$ and $\alpha$, a bounded level-set $\{f(\bar{u}, \cdot) \leq \alpha\}$ with the bound remaining valid under perturbation around $\bar{u}$.

**Proposition 5.4** (tightness from uniform level-boundedness). *For a function $f : \mathbb{R}^m \times \mathbb{R}^n \to \overline{\mathbb{R}}$ with values $f(u, x)$, suppose that $f$ is level-bounded in $x$ locally uniformly in $u$. If $\{u^\nu \in \mathbb{R}^m, \nu \in \mathbb{N}\}$ has a limit and $\{\inf f(u^\nu, \cdot), \nu \in \mathbb{N}\}$ is bounded from above, then $\{f(u^\nu, \cdot), \nu \in \mathbb{N}\}$ is tight.*

**Proof.** Let $\varepsilon \in (0, \infty)$, $\bar{u}$ be the limit of $\{u^\nu, \nu \in \mathbb{N}\}$ and $\alpha \in \mathbb{R}$ be an upper bound on $\{\inf f(u^\nu, \cdot), \nu \in \mathbb{N}\}$. Since $f$ is level-bounded in $x$ locally uniformly in $u$, there's $\delta > 0$ and a bounded set $B \subset \mathbb{R}^n$ such that

$$\{f(u, \cdot) \leq \alpha + \varepsilon\} \subset B \quad \forall u \in \mathbb{B}(\bar{u}, \delta).$$

Let $\nu_\varepsilon$ be such that $u^\nu \in \mathbb{B}(\bar{u}, \delta)$ for all $\nu \geq \nu_\varepsilon$. Set $B_\varepsilon = \operatorname{cl} B$, which is compact. Then, for $\nu \geq \nu_\varepsilon$,

$$\{f(u^\nu, \cdot) \leq \alpha + \varepsilon\} \subset B_\varepsilon.$$

Fix $\nu \geq \nu_\varepsilon$. First, suppose that $\inf f(u^\nu, \cdot) \in \mathbb{R}$. Then, there's $x^\nu$ such that

$$f(u^\nu, x^\nu) \leq \inf f(u^\nu, \cdot) + \varepsilon \leq \alpha + \varepsilon.$$

Consequently,

$$\inf_{B_\varepsilon} f(u^\nu, \cdot) \leq f(u^\nu, x^\nu) \leq \inf f(u^\nu, \cdot) + \varepsilon.$$

Second, suppose that $\inf f(u^\nu, \cdot) = -\infty$. Let $\mu \leq \alpha + \varepsilon$. Then, there's $x^\nu$ such that $f(u^\nu, x^\nu) \leq \mu \leq \alpha + \varepsilon$. Thus,

$$\inf_{B_\varepsilon} f(u^\nu, \cdot) \leq f(u^\nu, x^\nu) \leq \mu.$$

Since $\mu$ is arbitrary, we've established that $\inf_{B_\varepsilon} f(u^\nu, \cdot) = -\infty$. This means that

$$\inf_{B_\varepsilon} f(u^\nu, \cdot) \leq \inf f(u^\nu, \cdot) + \varepsilon$$

holds in both cases. □

The upper bound on $\{\inf f(u^\nu, \cdot), \nu \in \mathbb{N}\}$ in the proposition rules out $f : \mathbb{R} \times \mathbb{R} \to \overline{\mathbb{R}}$ with $f(u, x) = 1/|u|$ if $x = 1/|u|$ and $f(u, x) = \infty$ otherwise, which is level-bounded in $x$ locally uniformly in $u$ but $\{f(1/\nu, \cdot), \nu \in \mathbb{N}\}$ isn't tight; see Figure 5.3.

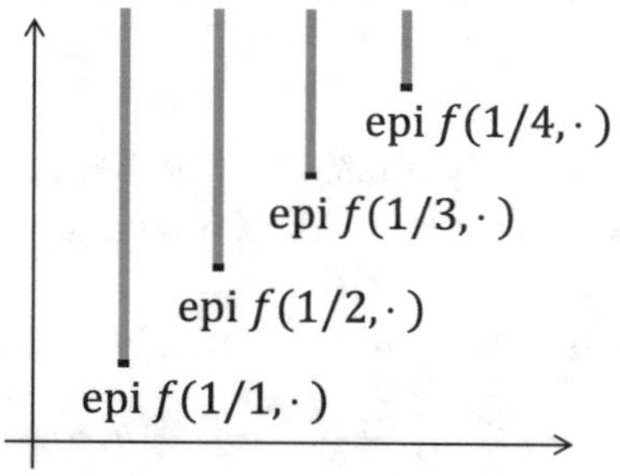


Fig. 5.3: Example of a function $f$ that's level-bounded in $x$ locally uniformly in $u$ but fails to produce a tight collection of functions.

With the refinement of tightness, we obtain the following consequences of epi-convergence, which also furnish the proof of 4.14.

**Theorem 5.5** (consequences of epi-convergence). *Suppose that $f^\nu : \mathbb{R}^n \to \overline{\mathbb{R}}$ epi-converges to a proper function $f : \mathbb{R}^n \to \overline{\mathbb{R}}$. Then, the following hold for any $\varepsilon \in [0, \infty)$:*

(a) $\limsup\,(\inf f^\nu) \leq \inf f$.
(b) *If $\varepsilon^\nu \in [0, \infty) \to \varepsilon$, then*

$$\text{LimOut}\left(\varepsilon^\nu\text{-argmin}\, f^\nu\right) \subset \varepsilon\text{-argmin}\, f.$$

(c) *If $\{x^\nu \in \varepsilon^\nu\text{-argmin}\, f^\nu, \nu \in N\}$ converges for some $N \in \mathcal{N}_\infty^\#$ and $\varepsilon^\nu \to 0$, then*

$$\lim_{\nu \in N}(\inf f^\nu) = \inf f.$$

(d) $\inf f^\nu \to \inf f > -\infty \iff \{f^\nu, \nu \in \mathbb{N}\}$ *is tight.*
(e) $\inf f^\nu \to \inf f \implies \exists\, \varepsilon^\nu \in [0, \infty) \to \varepsilon$ *such that*

$$\text{LimInn}\left(\varepsilon^\nu\text{-argmin}\, f^\nu\right) \supset \varepsilon\text{-argmin}\, f.$$

**Proof.** We leverage the characterization 4.15 of epi-convergence. For (a), suppose first that $\inf f$ is finite and let $\gamma \in (0, \infty)$. There are $x \in \mathbb{R}^n$ such that $f(x) \leq \inf f + \gamma$ and also, by 4.15(b), $x^\nu \to x$ such that $\limsup f^\nu(x^\nu) \leq f(x)$. Thus,

$$\limsup\,(\inf f^\nu) \leq \limsup f^\nu(x^\nu) \leq f(x) \leq \inf f + \gamma.$$

Second, suppose that $\inf f = -\infty$. Then, there are $x \in \mathbb{R}^n$ such that $f(x) \le -\gamma$ and also, by 4.15(b), $x^\nu \to x$ such that $\limsup f^\nu(x^\nu) \le f(x)$. Thus,

$$\limsup\,(\inf f^\nu) \le \limsup f^\nu(x^\nu) \le f(x) \le -\gamma.$$

Since $\gamma$ is arbitrary, $\limsup\,(\inf f^\nu) \le \inf f$ holds in both cases.

For (b), suppose that $\bar{x} \in \mathrm{LimOut}(\varepsilon^\nu\text{-}\mathrm{argmin}\, f^\nu)$. Then, by the definition of outer limits, there are $N \in \mathcal{N}_\infty^\#$ and $x^\nu \in \varepsilon^\nu\text{-}\mathrm{argmin}\, f^\nu \to_N \bar{x}$. Thus,

$$\limsup_{\nu\in N} f^\nu(x^\nu) \le \limsup_{\nu\in N}(\inf f^\nu + \varepsilon^\nu) \le \inf f + \varepsilon,$$

where the last inequality follows by (a). In view of 4.15(a), this implies that

$$f(\bar{x}) \le \liminf_{\nu\in N} f^\nu(x^\nu) \le \limsup_{\nu\in N} f^\nu(x^\nu) \le \inf f + \varepsilon.$$

Since $f$ is proper, $\inf f < \infty$ and then $\bar{x} \in \mathrm{dom}\, f$. Thus, $\bar{x} \in \varepsilon\text{-}\mathrm{argmin}\, f$.

For (c), let $\bar{x}$ be the limit of $\{x^\nu, \nu \in N\}$. By (b), $\bar{x} \in \mathrm{argmin}\, f$. Then, 4.15(a) implies that

$$\liminf_{\nu\in N}(\inf f^\nu + \varepsilon^\nu) \ge \liminf_{\nu\in N} f^\nu(x^\nu) \ge f(\bar{x}) = \inf f.$$

Since $\limsup(\inf f^\nu) \le \inf f$ holds from (a), the conclusion follows.

For (d), suppose that $\inf f^\nu \to \inf f > -\infty$ and let $\gamma > 0$. Then, there are $\nu_1 \in \mathbb{N}$ such that $\inf f \le \inf f^\nu + \gamma/3$ for all $\nu \ge \nu_1$ and also $\bar{x} \in \mathbb{R}^n$ such that $f(\bar{x}) \le \inf f + \gamma/3$. By 4.15(b), there are $x^\nu \to \bar{x}$ and $\nu_2 \ge \nu_1$ such that $f^\nu(x^\nu) \le f(\bar{x}) + \gamma/3$ for all $\nu \ge \nu_2$. Let $B$ be a compact set containing $\{x^\nu, \nu \in \mathbb{N}\}$. Thus, for $\nu \ge \nu_2$,

$$\inf_B f^\nu \le f^\nu(x^\nu) \le f(\bar{x}) + \tfrac{1}{3}\gamma \le \inf f + \tfrac{2}{3}\gamma \le \inf f^\nu + \gamma.$$

We've shown that $\nu_2$ and $B$ furnish the required index and set in the definition of tightness; see 5.3.

For the converse, we first rule out the possibility $\inf f = -\infty$. If this were the case, then $\inf f^\nu \to -\infty$ by (a) and, for some compact set $B \subset \mathbb{R}^n$, $\inf_B f^\nu \to -\infty$ because of tightness. We must then also have $\{x^\nu \in B, \nu \in \mathbb{N}\}$ such that $f^\nu(x^\nu) \to -\infty$. Since $B$ is compact, there are $N \in \mathcal{N}_\infty^\#$ and $\bar{x}$ such that $x^\nu \to_N \bar{x}$. By 4.15(a),

$$\liminf_{\nu\in N} f^\nu(x^\nu) \ge f(\bar{x}) > -\infty$$

because $f$ is proper. This contradicts $f^\nu(x^\nu) \to -\infty$ and thus $\inf f > -\infty$.

Second, we show that for any compact set $B \subset \mathbb{R}^n$, there's $\bar{\nu}$ such that $\{\inf_B f^\nu, \nu \ge \bar{\nu}\}$ is bounded from below. For the sake of contradiction, suppose that there's $N \in \mathcal{N}_\infty^\#$ such that $\inf_B f^\nu \to_N -\infty$. Since $B$ is compact, this implies the existence of $\{x^\nu \in B, \nu \in N\}$ with $f^\nu(x^\nu) \to_N -\infty$, another subsequence $N' \subset N$ and a limit $\bar{x}$ of $\{x^\nu, \nu \in N'\}$. By 4.15(a),

$$\liminf_{\nu\in N'} f^\nu(x^\nu) \ge f(\bar{x}) > -\infty$$

because $f$ is proper. This contradicts $f^\nu(x^\nu) \to_N -\infty$.

Third, for a compact set $B$, we establish that $\text{liminf}(\inf_B f^\nu) \geq \inf_B f$. We can assume without loss of generality that $B$ is nonempty because otherwise the statement holds trivially. Let $\alpha = \text{liminf}(\inf_B f^\nu)$. If $\alpha = \infty$, the claim holds trivially. The case $\alpha = -\infty$ is ruled out by the prior paragraph. For $\alpha \in \mathbb{R}$, the prior paragraph ensures that $\inf_B f^\nu > -\infty$ for sufficiently large $\nu$. For such $\nu$, there's $x^\nu \in B$ such that $f^\nu(x^\nu) \leq \inf_B f^\nu + \nu^{-1}$. Consequently, there exist also $\bar{x} \in B$ and $N \in \mathcal{N}_\infty^\#$ such that $x^\nu \underset{N}{\to} \bar{x}$. By 4.15(a), $\text{liminf}_{\nu\in N} f^\nu(x^\nu) \geq f(\bar{x})$ and then

$$\text{liminf}_{\nu\in N} \big(\inf_B f^\nu\big) \geq \text{liminf}_{\nu\in N} \big(f^\nu(x^\nu) - \nu^{-1}\big) \geq f(\bar{x}) \geq \inf_B f.$$

To remove the restriction to $N$, suppose for the sake of contradiction that $\alpha < \inf_B f$. Since $\alpha \in \mathbb{R}$, there's $N' \in \mathcal{N}_\infty^\#$ such that $\{\inf_B f^\nu, \nu \in N'\}$ has $\alpha$ as limit. We can then repeat the arguments above for this subsequence and conclude that

$$\text{liminf}_{\nu\in N''} \big(\inf_B f^\nu\big) \geq \inf_B f \;\; \text{for some subsequence } N'' \subset N'.$$

But, $\alpha$ is also the limit of $\{\inf_B f^\nu, \nu \in N''\}$ and this contradicts the assumption $\alpha < \inf_B f$.

Fourth, let $\varepsilon > 0$ and $B_\varepsilon$ be the corresponding compact set according to 5.3. Then, by tightness and the previous paragraph,

$$\text{liminf} \big(\inf f^\nu\big) + \varepsilon \geq \text{liminf} \big(\inf_{B_\varepsilon} f^\nu\big) \geq \inf_{B_\varepsilon} f \geq \inf f.$$

Since $\varepsilon$ is arbitrary, we've established that $\text{liminf}(\inf f^\nu) \geq \inf f$ and then also $\inf f^\nu \to \inf f$ by (a).

For (e), let $\bar{x} \in \varepsilon\text{-argmin} f$. Then, $f(\bar{x}) < \infty$. By 4.15(b), there's $x^\nu \to \bar{x}$ such that $\text{limsup} f^\nu(x^\nu) \leq f(\bar{x})$. First, suppose that $\inf f \in \mathbb{R}$. Then,

$$\begin{aligned} f^\nu(x^\nu) - \inf f^\nu &= f^\nu(x^\nu) - f(\bar{x}) + f(\bar{x}) - \inf f + \inf f - \inf f^\nu \\ &\leq f^\nu(x^\nu) - f(\bar{x}) + \varepsilon + \inf f - \inf f^\nu. \end{aligned}$$

We've shown that $x^\nu \in \varepsilon^\nu\text{-argmin} f^\nu$, where

$$\varepsilon^\nu = \varepsilon + \max\big\{0, f^\nu(x^\nu) - f(\bar{x}) + \inf f - \inf f^\nu\big\}.$$

Thus, $\bar{x} \in \text{LimInn}(\varepsilon^\nu\text{-argmin} f^\nu)$. Since $\varepsilon^\nu \to \varepsilon$, the assertion holds. Second, suppose that $\inf f = -\infty$. Then, $f(\bar{x}) = -\infty$, but that contradicts the fact that $f$ is proper. □

While $\text{LimOut}(\text{argmin} f^\nu) \subset \text{argmin} f$ holds by item (b) of the theorem, the inclusion can be strict as illustrated in Figure 5.4, where $f(x) = 0$ and $f^\nu(x) = x^2/\nu$ if $x \in [-1, 1]$ and $f(x) = f^\nu(x) = \infty$ otherwise. Then, $\text{argmin} f^\nu = \{0\}$ for all $\nu$ but $\text{argmin} f = [-1, 1]$. Still, item (e) shows that if the tolerances $\varepsilon^\nu$ in the approximating problems vanish sufficiently slowly, then

$$\varepsilon^\nu\text{-argmin} f^\nu \xrightarrow{s} \text{argmin} f.$$

In Figure 5.4, one can select $\varepsilon^\nu = 1/\nu$ because then $\varepsilon^\nu$-argmin $f^\nu = [-1, 1]$, which actually coincides with argmin $f$ regardless of $\nu$.

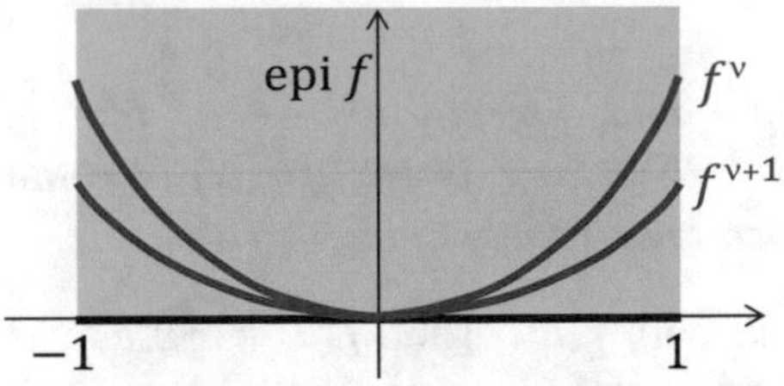


Fig. 5.4: Set-convergence of near-minimizers of $f^\nu$ to argmin $f$.

Before applying the theorem in the context of Rockafellians, we refine our terminology regarding semicontinuity. We say that $f : \mathbb{R}^n \to \overline{\mathbb{R}}$ is *lsc at* $\bar{x}$ if

$$x^\nu \to \bar{x} \quad \Longrightarrow \quad \operatorname{liminf} f(x^\nu) \geq f(\bar{x}). \tag{5.1}$$

This holds for all $\bar{x} \in \mathbb{R}^n$ if and only if $f$ is lsc by [105, Lemma 1.7]. Thus, this sequential condition supplements the other characterizations in 4.8. Likewise, $f$ is *upper semicontinuous (usc) at* $\bar{x}$ if $-f$ is lsc at $\bar{x}$ or, equivalently,

$$x^\nu \to \bar{x} \quad \Longrightarrow \quad \operatorname{limsup} f(x^\nu) \leq f(\bar{x}).$$

It's *usc* if this holds for all $\bar{x} \in \mathbb{R}^n$. Consequently, $f$ is continuous at $\bar{x}$ if and only if it's both lsc and usc at $\bar{x}$.

We can now state the main result about stability of minima and minimizers under perturbations as defined by a Rockafellian. Typically, we would like the min-value function to be continuous. Then small changes in the perturbation vector imply small changes in the minimum value. Sensitivity analysis often revolves around checking whether this indeed is the case. In the absence of continuity, our stipulation of the minimum "cost" might be highly sensitive to small changes in modeling assumptions. For example, if the min-value function is only lsc at $\bar{u}$, then the minimum cost could become much higher under a minor perturbation away from $\bar{u}$, possibly resulting in an unwelcome surprise. A min-value function that's usc at $\bar{u}$ may jump down under perturbation away from $\bar{u}$, which is more desirable as it represents unexpected improvements in the minimum value.

**Theorem 5.6** (stability). *For a proper function* $f : \mathbb{R}^m \times \mathbb{R}^n \to \overline{\mathbb{R}}$, *let*

$$p(u) = \inf f(u, \cdot), \qquad P(u) = \operatorname{argmin} f(u, \cdot), \qquad P_\varepsilon(u) = \varepsilon\text{-argmin} f(u, \cdot)$$

*with* $u \in \mathbb{R}^m$ *and* $\varepsilon \in (0, \infty)$. *Given* $\bar{u} \in \mathbb{R}^m$, *the following hold:*

(a) *$p$ is lsc at $\bar{u}$ when $f$ is lsc and, for any $u^\nu \to \bar{u}$, $\{f(u^\nu, \cdot), \nu \in \mathbb{N}\}$ is tight.*
(b) *$p$ is usc at $\bar{u}$ when, for any $u^\nu \to \bar{u}$ and $\bar{x} \in \mathbb{R}^n$, there's $x^\nu \to \bar{x}$ such that*

$$\operatorname{limsup} f(u^\nu, x^\nu) \leq f(\bar{u}, \bar{x}),$$

*which in particular holds when $f$ is usc.*

(c) *$p$ is continuous at $\bar{u}$ and, for $\varepsilon \in (0, \infty)$,*

$$\bigcup_{u^\nu \to \bar{u}} \operatorname{LimOut} P(u^\nu) \subset P(\bar{u}) \quad \textit{and} \quad \bigcap_{u^\nu \to \bar{u}} \operatorname{LimInn} P_\varepsilon(u^\nu) \supset P(\bar{u})$$

*when $f(\bar{u}, \cdot)$ is proper and, for any $u^\nu \to \bar{u}$, $f(u^\nu, \cdot) \xrightarrow{e} f(\bar{u}, \cdot)$ tightly. These requirements hold if $P(\bar{u}) \neq \emptyset$, $f$ is lsc, $f(\cdot, x)$ is continuous for all $x$ and $f$ is level-bounded in $x$ locally uniformly in $u$.*

**Proof.** For (a), let $u^\nu \to \bar{u}$. With $f(u^\nu, \cdot)$ and $f(\bar{u}, \cdot)$ in the roles of $f^\nu$ and $f$, respectively, we can repeat the proof of 5.5, part (d), steps 2–4, and conclude that $\liminf p(u^\nu) \geq p(\bar{u})$. A closer examination shows that the requirement of $f(\bar{u}, \cdot)$ being proper can be relaxed to $f(\bar{u}, x) > -\infty$ for all $x$, which holds because $f$ is proper. Moreover, $f(u^\nu, \cdot) \xrightarrow{e} f(\bar{u}, \cdot)$ can be relaxed to $\liminf f(u^\nu, x^\nu) \geq f(\bar{u}, x)$ for $x^\nu \to x$ and this holds by lsc of $f$.

For (b), one can follow the arguments in the proof of 5.5(a). Part (c) is a direct consequence of 5.5, with the sufficient condition seen as follows: If there's $\bar{x} \in \operatorname{argmin} f(\bar{u}, \cdot)$, then $f(\bar{u}, \bar{x}) < \infty$ and $f(\bar{u}, \cdot)$ is proper. The characterization of epi-convergence in 4.15 and the continuity assumptions establish that $f(u^\nu, \cdot) \xrightarrow{e} f(\bar{u}, \cdot)$. Then, 5.5(a) ensures that $\limsup p(u^\nu) \leq p(\bar{u}) = f(\bar{u}, \bar{x}) < \infty$, which in turn establishes tightness via 5.4. □

In 5.2, $f$ is proper and lsc. For $u^\nu \to \bar{u}$, $\{C(u^\nu), \nu \in \mathbb{N}\}$ is bounded so that $\{f(u^\nu, \cdot), \nu \in \mathbb{N}\}$ is tight. Thus, 5.6(a) establishes that $p$ is lsc in that example.

**Example 5.7** (regularization as perturbation). In data analytics (cf. 2.5, 2.38) and approaches related to the proximal point method (§1.H), a continuous objective function $f_0 : \mathbb{R}^n \to \mathbb{R}$ is often augmented with a regularization term of the form $\theta\|x\|_1$ or $\frac{1}{2\lambda}\|x - \bar{x}\|_2^2$. The weight ($\theta$ or $1/\lambda$) associated with the term is an important modeling choice and we examine its perturbation. Specifically, let's consider the problem

$$\operatorname*{minimize}_{x \in C} f_0(x) + |\bar{u}| r(x),$$

where $r : \mathbb{R}^n \to [0, \infty)$ is a continuous function representing regularization and $C \subset \mathbb{R}^n$ is nonempty and closed. We use the absolute value of the perturbation parameter to ensure nonnegativity. A Rockafellian for the problem is given by

$$f(u, x) = \iota_C(x) + f_0(x) + |u| r(x).$$

Under the assumption that $\iota_C(x) + f_0(x) \to \infty$ when $\|x\|_2 \to \infty$, the stability theorem 5.6(c) applies with $\bar{u} = 0$ and we conclude that the minimum value and minimizers of $f(u, \cdot)$ change "continuously" as $u$ shifts from a small number to 0 as specified in that theorem.

**Detail.** Since $\iota_C + f_0$ is lsc, level-bounded and proper, 4.9 ensures that it has a minimizer. This level-boundedness implies that $f$ is level-bounded in $x$ locally uniformly in $u$. Since $f$ is lsc and $f(\cdot, x)$ is continuous for all $x$, the sufficient condition in 5.6(c) holds. □

**Example 5.8** (perturbation of inequalities). For lsc $f_0, g_i : \mathbb{R}^n \to \mathbb{R}$, $i = 1, \dots, q$, let's consider the problem

$$\underset{x \in \mathbb{R}^n}{\text{minimize}}\, f_0(x) + \iota_{(-\infty,0]^q}\big(G(x)\big),$$

where $G(x) = (g_1(x), \dots, g_q(x))$ and the Rockafellian given by

$$f(u, x) = f_0(x) + \iota_{(-\infty,0]^q}\big(G(x) + u\big).$$

The min-value function given by $p(u) = \inf f(u, \cdot\,)$ is lsc under a tightness assumption, but isn't necessarily continuous at 0. Thus, the minimum value obtained from solving the actual problem should be presented with the qualification that it could easily be much higher if the constraints are changed slightly.

**Detail.** To show that $f$ is lsc, let $u^\nu \to \bar{u}$, $x^\nu \to \bar{x}$ and $\alpha \in \mathbb{R}$ such that $f(u^\nu, x^\nu) \leq \alpha$ for all $\nu$. Then, $f_0(x^\nu) \leq \alpha$ and $g_i(x^\nu) + u^\nu \leq 0$ for all $i$, which imply that $f_0(\bar{x}) \leq \alpha$ and $g_i(\bar{x}) + \bar{u} \leq 0$ as well by the lsc property. Thus, $f(\bar{u}, \bar{x}) \leq \alpha$ and $f$ is lsc by 4.8. We can then bring in the stability theorem 5.6(a) to conclude that $p$ is lsc as long as the required tightness assumption is satisfied. For instance, we achieve this when the feasible sets $\{x \mid G(x) + u^\nu \leq 0\}$ are uniformly bounded as $\{u^\nu, \nu \in \mathbb{N}\}$ converges.

Continuity of the min-value function $p$ is a much more delicate issue as already brought forth by 5.2, which is of the present form with $f_0(x) = x^2 + 1$ and $g_1(x) = (x-2)(x-4)+1$. Since $p$ is lsc in that case as argued below the stability theorem 5.6, let's examine usc at 0 and the assumption in 5.6(b). We realize that, for $u^\nu \searrow 0$ and $\bar{x} = 3$, there's no $x^\nu \to \bar{x}$ such that $\limsup f(u^\nu, x^\nu) \leq f(0, \bar{x})$; we always have $f(u^\nu, x^\nu) = \infty$ and the assumption in 5.6(b) fails. However, the difficulty is unique to $u = 0$. The min-value function is continuous at $u \neq 0$. □

**Exercise 5.9** (change in risk averseness). For $f : \mathbb{R}^m \times \mathbb{R}^n \to \mathbb{R}$ and a nonempty compact set $C \subset \mathbb{R}^n$, consider the superquantile-risk minimization problem (3.3) and its minimum value as a function of the parameter $\alpha$. (A high $\alpha$ implies a cautious approach that attempts to avoid decisions with high "cost." This is referred to as having high risk averseness. In practice, it's often difficult to determine the right level of risk averseness.) Identify a Rockafellian that represents changes in $\alpha$. Under the assumption that $f(\xi, \cdot\,)$ is continuous for all $\xi \in \Xi$, show that the min-value function defined by the Rockafellian is continuous at any $\bar{\alpha} \in (0, 1)$.

**Guide.** Express $\alpha = e^u/(1 + e^u)$ and define a Rockafellian of the form

$$\varphi\big(u, (x, \gamma)\big) = \iota_C(x) + \gamma + (1 + e^u)\sum\nolimits_{\xi \in \Xi} p_\xi \max\big\{0, f(\xi, x) - \gamma\big\}.$$

Confirm that it's level-bounded in $(x, \gamma)$ locally uniformly in $u$ and leverage the stability theorem 5.6. □

## 5.B Quantitative Stability

Although continuity of a min-value function conveys a sense of stability for the family of problems defined by a Rockafellian, it doesn't quantify *how much* the minimum value changes under a particular perturbation. We'll now make a step in that direction and compute subgradients of min-value functions, which represent first-order estimates of this change. En route to formulae for such subgradients, we'll discover that a Rockafellian not only defines a family of alternative problems but also an optimality condition, with an associated multiplier vector, for any one of the individual problems. Thus, embedding a problem within a family via a Rockafellian has benefits beyond sensitivity analysis and emerges as a main approach to constructing optimality conditions and computational procedures.

Let's start by looking at optimality conditions for the problem of minimizing $f_0 : \mathbb{R}^n \to \overline{\mathbb{R}}$. The condition $0 \in \partial f_0(x)$ is necessary for a local minimizer of $f_0$ by the Fermat rule 4.73. A Rockafellian $f : \mathbb{R}^m \times \mathbb{R}^n \to \overline{\mathbb{R}}$ for the problem, with anchor at $\bar{u}$, has $f(\bar{u}, x) = f_0(x)$ for all $x \in \mathbb{R}^n$ by definition, but also gives rise to the alternative optimality condition

$$(y, 0) \in \partial f(\bar{u}, x)$$

for the actual problem under a qualification as stated in the next theorem. The auxiliary vector $y \in \mathbb{R}^m$ is associated with the perturbation vector $u$ and emerges as a key quantity in sensitivity analysis. While this introduces additional unknowns compared to $0 \in \partial f_0(x)$, it's often easier to solve for $(x, y)$ in the alternative condition than finding just $x$ in the original one. We refer to $y$ as a *multiplier vector*.

**Theorem 5.10** (Rockafellar condition for optimality). *For the problem of minimizing $f_0 : \mathbb{R}^n \to \overline{\mathbb{R}}$, suppose that $\bar{x} \in \mathbb{R}^n$ is a local minimizer, $f : \mathbb{R}^m \times \mathbb{R}^n \to \overline{\mathbb{R}}$ is a proper lsc Rockafellian with anchor at $\bar{u} \in \mathbb{R}^m$ and the following qualification holds:*

$$(y, 0) \in \partial^\infty f(\bar{u}, \bar{x}) \implies y = 0. \tag{5.2}$$

*Then,*

$$\exists \bar{y} \in \mathbb{R}^m \text{ such that } (\bar{y}, 0) \in \partial f(\bar{u}, \bar{x}).$$

*This condition is sufficient for $\bar{x}$ to be a (global) minimizer of $f_0$ when $f$ is epi-regular at $(\bar{u}, \bar{x})$ and $f_0$ is convex.*

**Proof.** Since $f_0(x) = f(\bar{u}, x) = f(F(x))$ with $F(x) = (\bar{u}, x)$, we can apply the chain rule 4.64 to obtain

$$\partial f_0(\bar{x}) \subset \nabla F(\bar{x})^\top \partial f(\bar{u}, \bar{x}) \tag{5.3}$$

as long as the qualification (4.14) holds. Since $\nabla F(\bar{x})$ is simply the $(m + n) \times n$-matrix consisting of the $m \times n$ zero matrix stacked on top of the $n \times n$ identity matrix, the right-hand side in this inclusion amounts to those $v \in \mathbb{R}^n$ such that $(y, v) \in \partial f(\bar{u}, \bar{x})$ for some $y \in \mathbb{R}^m$. The Fermat rule 4.73 ensures that $0 \in \partial f_0(\bar{x})$ and, when combined with (5.3), also the first conclusion. Still, it remains to verify (4.14), which now reads:

$$w \in \partial^\infty f(\bar{u}, \bar{x}) \text{ and } \nabla F(\bar{x})^\top w = 0 \implies w = 0.$$

Let's say that $w = (y, v)$ with $y \in \mathbb{R}^m$ and $v \in \mathbb{R}^n$. Then, $\nabla F(\bar{x})^\top w = v$ so the requirement that $v = 0$ holds trivially. The additional requirement that $y = 0$ is satisfied due to (5.2).

When $f$ is epi-regular at $(\bar{u}, \bar{x})$, the chain rule 4.64 yields equality in (5.3). By the optimality condition 2.19 for convex functions, the second conclusion follows. □

A vast number of possibilities emerge from the theorem, with the Rockafellian now possibly being chosen based on computational concerns more than those of sensitivity analysis. The perspective brings a broader view of multipliers beyond a specific constraint structure and the associated variational geometry discussed in §4.J. In fact, every Rockafellian defines an optimality condition for the actual problem expressed by a multiplier vector, at least as long as the qualification (5.2) holds. However, two examples show that the multipliers emerging from the Rockafellar condition 5.10 are very much in tune with those in earlier developments.

**Example 5.11** (connection with equality constraints). For the problem of minimizing $f_0 : \mathbb{R}^n \to \overline{\mathbb{R}}$, suppose that $f : \mathbb{R}^m \times \mathbb{R}^n \to \overline{\mathbb{R}}$ is a proper lsc Rockafellian with anchor at $\bar{u} \in \mathbb{R}^m$. Then, the problem can be reformulated as

$$\underset{u \in \mathbb{R}^m, x \in \mathbb{R}^n}{\text{minimize}} \ f(u, x) \ \text{ subject to } \ f_i(u, x) = \bar{u}_i - u_i = 0, \ \ i = 1, \dots, m.$$

The multipliers associated with these equality constraints in the sense of 4.76 coincide with $y$ emerging from the Rockafellar condition 5.10.

**Detail.** When $f$ is smooth, the reformulation has the optimality condition

$$\nabla f(u, x) + \sum\nolimits_{i=1}^{m} y_i \nabla f_i(u, x) = 0$$

by 4.76. This simplifies to $y = \nabla_u f(u, x)$ and $\nabla_x f(u, x) = 0$, which together with the feasibility condition $\bar{u} = u$, produce $(y, 0) = \nabla f(\bar{u}, x)$. Thus, the multiplier vector $y$ from 4.76 coincides with the multiplier vector in the Rockafellar condition 5.10.

Without smoothness, a similar connection holds. The equality constraints form a set $C \subset \mathbb{R}^m \times \mathbb{R}^n$ with normal cone (cf. 4.49)

$$N_C(\bar{u}, x) = \left\{ \sum\nolimits_{i=1}^{m} y_i \nabla f_i(\bar{u}, x) \,\middle|\, y \in \mathbb{R}^m \right\} = \mathbb{R}^m \times \{0\}^n.$$

Assuming that the sum rule 4.67 applies at $(\bar{u}, x)$, the Fermat rule 4.73 then produces the optimality condition

$$0 \in \partial f(\bar{u}, x) + N_C(\bar{u}, x) = \partial f(\bar{u}, x) + \left(\mathbb{R}^m \times \{0\}^n\right)$$

for the reformulation. This is equivalent to having $(y, 0) \in \partial f(\bar{u}, x)$ for some $y \in \mathbb{R}^m$ and we've recovered the Rockafellar condition 5.10. □

**Example 5.12** (multipliers for max-functions). For smooth $f_i : \mathbb{R}^n \to \mathbb{R}$, $i = 1, \dots, m$, the problem

$$\underset{x \in \mathbb{R}^n}{\text{minimize}} \ \max\nolimits_{i=1,\dots,m} f_i(x)$$

can be associated with the Rockafellian $f : \mathbb{R}^m \times \mathbb{R}^n \to \mathbb{R}$ given by

$$f(u,x) = h\big(F(x) + u\big),$$

where $h(z) = \max\{z_1, \ldots, z_m\}$ and $F(x) = (f_1(x), \ldots, f_m(x))$. The actual problem is recovered by setting $u = 0$, which then is the anchor of this Rockafellian. For example, the perturbation vector $u$ may represent changes to the targets in goal optimization; cf. 4.6. The multipliers arising from the Rockafellar condition 5.10 coincide with those of 4.81.

**Detail.** We compute $\partial f(u,x)$ by the chain rule 4.64 and let $\hat{F}(u,x) = F(x) + u$ so that $f(u,x) = h(\hat{F}(x,u))$. Thus, $\nabla \hat{F}(u,x) = (I, \nabla F(x))$ is an $m \times (m+n)$-matrix, where $I$ is the $m \times m$ identity matrix. Since $h$ is epi-regular at every point by virtue of being convex and real-valued,

$$\partial f(u,x) = \nabla \hat{F}(u,x)^\top \partial h\big(\hat{F}(u,x)\big),$$

where the subgradients of $h$ at a point $\bar{z} = (\bar{z}_1, \ldots, \bar{z}_m)$ are given in 4.66 as $y \in \mathbb{R}^m$ satisfying $\sum_{i=1}^m y_i = 1$, $y_i \geq 0$ if $\bar{z}_i = h(\bar{z})$ and $y_i = 0$ otherwise. The qualification (4.14) holds because the only horizon subgradient of $h$ at any point is 0 by 4.65. Hence,

$$(\tilde{y}, 0) \in \partial f(u,x) \iff \begin{bmatrix} \tilde{y} \\ 0 \end{bmatrix} = \begin{bmatrix} I \\ \nabla F(x)^\top \end{bmatrix} y \text{ for some } y \in \mathbb{R}^m \text{ with } \sum\nolimits_{i=1}^m y_i = 1;$$

$$y_i \geq 0 \text{ if } f_i(x) + u_i = h\big(F(x) + u\big), \quad y_i = 0 \text{ otherwise.}$$

Certainly, $\tilde{y} = y$ and we recover the optimality condition 4.81 if $u = 0$. □

Let's now return to sensitivity analysis and show that the multiplier vectors emerging from the Rockafellar condition 5.10 are essentially the subgradients of the corresponding min-value function. Consequently, passing from a minimization problem in $x$ via a Rockafellian to a family of problems and an optimality condition involving a pair $(x, y)$ isn't only computationally advantageous but also furnishes important insight.

**Theorem 5.13** (subgradients of min-value function). *For a proper lsc function $f : \mathbb{R}^m \times \mathbb{R}^n \to \overline{\mathbb{R}}$, with $f(u,x)$ level-bounded in $x$ locally uniformly in $u$, let*

$$p(u) = \inf f(u, \cdot) \quad \textit{and} \quad P(u) = \operatorname{argmin} f(u, \cdot) \;\; \forall u \in \mathbb{R}^m.$$

*Then, at any $\bar{u} \in \operatorname{dom} p$,*

$$\partial p(\bar{u}) \subset \bigcup_{\bar{x} \in P(\bar{u})} \big\{ y \in \mathbb{R}^m \;\big|\; (y, 0) \in \partial f(\bar{u}, \bar{x}) \big\}.$$

*If $f$ is convex, then $p$ is convex, the inclusion holds with equality and the sets in the union coincide.*

**Proof.** For $F : \mathbb{R}^m \times \mathbb{R}^n \times \mathbb{R} \to \mathbb{R}^m \times \mathbb{R}$ with $F(u,x,\alpha) = (u,\alpha)$, we've that[1] $\operatorname{epi} p = F(\operatorname{epi} f)$. To see this, note that $(u,\alpha) \in \operatorname{epi} p$ implies $\inf f(u,\cdot) \leq \alpha < \infty$. Since $f(u,\cdot)$ is lsc with bounded level-sets and $p(u) < \infty$, there's $x \in \operatorname{argmin} f(u,\cdot)$ by 4.9 and $f(u,x) \leq \alpha$. If $(u,\alpha) \in F(\operatorname{epi} f)$, then $p(u) \leq f(u,x) \leq \alpha < \infty$ for some $x$.

[1] For $C \subset \mathbb{R}^n$ and $F : \mathbb{R}^n \to \mathbb{R}^m$, we recall that $F(C) = \{F(x) \mid x \in C\}$.

For $\bar{u} \in \operatorname{dom} p$, $p(\bar{u}) \in \mathbb{R}$ because $\operatorname{argmin} f(\bar{u}, \cdot)$ is nonempty by 4.9 and $f$ is proper. By definition, $v \in \partial p(\bar{u})$ if and only if $(v, -1) \in N_{\operatorname{epi} p}(\bar{u}, p(\bar{u}))$ and our goal becomes to determine $N_{F(\operatorname{epi} f)}(\bar{u}, p(\bar{u}))$. In this regard, we can leverage [105, Theorem 6.43], which states that for closed $C \subset \mathbb{R}^n$, smooth $G : \mathbb{R}^n \to \mathbb{R}^m$ and $\bar{u} \in G(C)$, one has

$$N_{G(C)}(\bar{u}) \subset \bigcup_{\bar{x} \in G^{-1}(\bar{u}) \cap C} \left\{ y \in \mathbb{R}^m \mid \nabla G(\bar{x})^\top y \in N_C(\bar{x}) \right\}$$

provided that $G^{-1}(\mathbb{B}(\bar{u}, \varepsilon)) \cap C$ is bounded for some $\varepsilon > 0$. Applying this fact with $G = F$ and $C = \operatorname{epi} f$, we obtain for $(\bar{u}, \bar{\alpha}) \in F(\operatorname{epi} f)$ that

$$N_{F(\operatorname{epi} f)}(\bar{u}, \bar{\alpha}) \subset \bigcup_{(u,x,\beta)} \left\{ (y, \gamma) \in \mathbb{R}^m \times \mathbb{R} \mid (y, 0, \gamma) \in N_{\operatorname{epi} f}(u, x, \beta) \right\}$$

because $\nabla F(u, x, \alpha)$ is the $(m + 1) \times (m + n + 1)$-matrix with an $m \times m$ identity matrix in the upper left corner, 1 in the bottom right corner and 0 elsewhere. The union is computed over $(u, x, \beta) \in F^{-1}(\bar{u}, \bar{\alpha}) \cap \operatorname{epi} f$, but this reduces to having $x$ with $(\bar{u}, x, \bar{\alpha}) \in \operatorname{epi} f$. When $\bar{\alpha} = p(\bar{u})$, this reduces further to $x \in P(\bar{u})$. Consequently,

$$N_{\operatorname{epi} p}\big(\bar{u}, p(\bar{u})\big) \subset \bigcup_{\bar{x} \in P(\bar{u})} \Big\{ (y, \gamma) \in \mathbb{R}^m \times \mathbb{R} \,\Big|\, (y, 0, \gamma) \in N_{\operatorname{epi} f}\big(\bar{u}, \bar{x}, p(\bar{u})\big) \Big\}.$$

The boundedness assumption required in [105, Theorem 6.43] holds because $f$ is level-bounded in $x$ locally uniformly in $u$. We can now conclude that

$$\begin{aligned} y \in \partial p(\bar{u}) &\iff (y, -1) \in N_{\operatorname{epi} p}\big(\bar{u}, p(\bar{u})\big) \implies (y, 0, -1) \in N_{\operatorname{epi} f}\big(\bar{u}, \bar{x}, p(\bar{u})\big) \\ &\iff (y, 0) \in \partial f(\bar{u}, \bar{x}) \;\text{ for some }\; \bar{x} \in P(\bar{u}). \end{aligned}$$

This proves the first assertion. Under convexity, $p$ is convex by 1.21 and [105, Theorem 6.43] furnishes a refinement that ensures equality in the above inclusion and that the union can be dropped. □

We can view $f : \mathbb{R}^m \times \mathbb{R}^n \to \overline{\mathbb{R}}$ in the theorem as a Rockafellian with anchor at $\bar{u}$ for the problem of minimizing $f_0 : \mathbb{R}^n \to \overline{\mathbb{R}}$, where $f_0(x) = f(\bar{u}, x)$ for all $x \in \mathbb{R}^n$. Then, $p(\bar{u})$ is the minimum value of the problem and $\partial p(\bar{u})$ estimates the effect of perturbation on the minimum value. The theorem goes beyond the continuity properties in the stability theorem 5.6 by quantifying the rate of change. Under the assumption that the algorithm employed to solve the problem furnishes both a minimizer of $f_0$ and associated multipliers in the sense of the Rockafellar condition, we're now able to quickly estimate the effect of a particular perturbation by invoking the theorem. More significantly, however, we can identify a poorly formulated problem. If the multiplier vector $y$ is relatively large in magnitude, then the minimum value tends to change substantially even under small changes to $u$ away from $\bar{u}$ and this calls into question the suitability of the formulation. One can also examine the different components of $y$ and see what aspect of a perturbation is more influential.

**Example 5.14** (perturbation of inequalities). Let's return to 5.2 and the Rockafellian given by

$$f(u,x) = x^2 + 1 + \iota_{(-\infty,0]}(g(u,x)), \text{ with } g(u,x) = (x-2)(x-4) + u.$$

We can utilize 5.13 to estimate how the min-value function $p$ changes near $u = 0$.

**Detail.** Since $f$ is proper and lsc, we can apply the Rockafellar condition 5.10 to the actual problem of minimizing $f(0,\cdot)$. By 4.58 and the chain rule 4.64,

$$\partial f(u,x) = (0,\ 2x) + \nabla g(u,x) N_{(-\infty,0]}\big(g(u,x)\big) = (0,\ 2x) + (1,\ 2x-6)Y(u,x),$$

for $(u,x) \in \operatorname{dom} f$, where $Y(u,x) = [0,\infty)$ if $g(u,x) = 0$ and $Y(u,x) = \{0\}$ otherwise. We note that the qualification (4.14) holds because

$$\nabla g(u,x)y = (1,\ 2x-6)y = (0,0) \implies y = 0.$$

Thus, $(y,0) \in \partial f(0,x)$ if and only if $0 = 2x + y(2x-6)$ for some $y \in Y(0,x)$. We see immediately that $(x^\star, y^\star) = (2,2)$ is the unique solution of the condition. Since $f$ is lsc and convex, it's epi-regular at all points of finiteness. The qualification (5.2) also hold, which can be established by utilizing 4.65. We then conclude that $x^\star$ is the unique minimizer of the actual problem by the Rockafellar condition 5.10. (This could equally well have been obtained from the KKT condition 4.77.) Moveover, $\partial p(0) = \{y^\star\}$ by 5.13. Thus, $p$ is actually differentiable at 0 with gradient 2. A small perturbation $u$ is then expected to change the minimum value with approximately $2u$. Since $p$ is convex by 1.21, it follows via the subgradient inequality 2.17 that

$$p(u) \geq p(0) + 2(u-0) = 5 + 2u \ \ \forall u \in \mathbb{R}.$$

The right-hand side furnishes a locally accurate approximation of $p$ near 0; see 5.2 as well as Figure 5.5.

While $u = 1$ is still in $\operatorname{dom} p$, $p$ has a vertical slope at that point and $\partial p(1) = \emptyset$. This can be discovered through 5.13 as well because there's no $y$ with $(y,0) \in \partial f(1,\bar{x})$ when $\bar{x} = 3$, the minimizer of $f(1,\cdot)$; see also 5.8. □

**Exercise 5.15** (perturbation in projections). For a given $\bar{x} \in \mathbb{R}^n$, consider the projection problem of minimizing $\frac{1}{2}\|x-\bar{x}\|_2^2$ subject to $Ax = b$, where $A$ is an $m \times n$-matrix with rank $m$. Use 5.13 to estimate the effect of changing $\bar{x}$ on the minimum value.

**Guide.** Adopt the Rockafellian $f(u,x) = \iota_X(x) + \frac{1}{2}\|x - (\bar{x}+u)\|_2^2$, with $X = \{x \mid Ax = b\}$, and compute its subgradients and then use the Rockafellar condition 5.10 to compute minimizers. This resembles the calculations in 2.40. Then, invoke 5.13. □

The subgradient expressions in 5.13 are far-reaching but don't explicitly bring forward any structure that might be present in a model. As an example, let's consider a composite function that covers many practical situations. In the process, we'll confirm that the multipliers arising in Chap. 4 from the variational geometry of normal cones indeed correspond to those of the Rockafellar condition 5.10 under the choice of a particular Rockafellian. This generalizes the connections in 5.11 and 5.12.

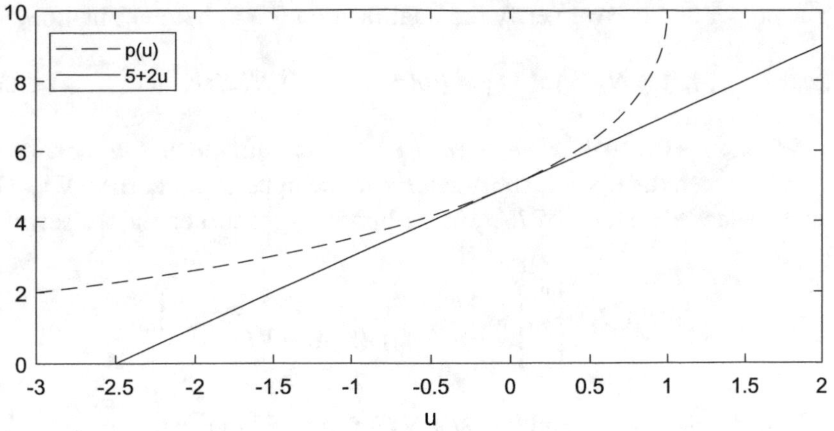


Fig. 5.5: Min-value function $p$ and its estimate in 5.14.

**Proposition 5.16** (multiplier vectors as subgradients). *For smooth* $f_0 : \mathbb{R}^n \to \mathbb{R}$, *smooth* $F : \mathbb{R}^n \to \mathbb{R}^m$, *proper, lsc and convex* $h : \mathbb{R}^m \to \overline{\mathbb{R}}$ *and nonempty closed* $X \subset \mathbb{R}^n$, *consider the problem*

$$\underset{x \in X}{\text{minimize}}\ f_0(x) + h\big(F(x)\big)$$

*and the associated Rockafellian* $f : \mathbb{R}^m \times \mathbb{R}^n \to \overline{\mathbb{R}}$ *given by*

$$f(u,x) = \iota_X(x) + f_0(x) + h\big(F(x) + u\big).$$

*Suppose that* $f(u,x)$ *is level-bounded in* $x$ *locally uniformly in* $u$. *Let* $p(u) = \inf f(u, \cdot)$, $P(u) = \operatorname{argmin} f(u, \cdot)$ *and, for* $x \in P(u)$,

$$Y(u,x) = \big\{ y \in \partial h\big(F(x) + u\big) \;\big|\; -\nabla f_0(x) - \nabla F(x)^\top y \in N_X(x) \big\}.$$

*Then, for* $\bar{u} \in \operatorname{dom} p$, *one has*

$$\partial p(\bar{u}) \subset \bigcup_{\bar{x} \in P(\bar{u})} Y(\bar{u}, \bar{x}).$$

**Proof.** Suppose that $\bar{y} \in \partial p(\bar{u})$. Then, by 5.13, there's $\bar{x} \in P(\bar{u})$ such that $(\bar{y}, 0) \in \partial f(\bar{u}, \bar{x})$. We derive an outer approximation of this set of subgradients. Let

$$g(x, z_0, z_1, \dots, z_m) = \iota_X(x) + z_0 + h(z_1, \dots, z_m)$$
$$G(u,x) = \big(x, f_0(x), f_1(x) + u_1, \dots, f_m(x) + u_m\big),$$

where $f_1(x), \dots, f_m(x)$ are the components of $F(x)$. Then, $f(u,x) = g(G(u,x))$ and the chain rule 4.64 leads to

$$\partial f(\bar{u}, \bar{x}) \subset \nabla G(\bar{u}, \bar{x})^\top \partial g\big(G(\bar{u}, \bar{x})\big)$$

because $g$ is proper and lsc; we verify the qualification (4.14) below. Via 4.63,

$$\partial g(x, z_0, z_1, \ldots, z_m) = N_X(x) \times \{1\} \times \partial h(z_1, \ldots, z_m) \;\; \forall (x, z_0, z_1, \ldots, z_m) \in \operatorname{dom} g.$$

Moreover, $\nabla G(u, x)$ is the $(n + 1 + m) \times (m + n)$-matrix with the $m \times m$ identity matrix in the bottom left corner, the $n \times n$ identity matrix in the upper right corner, $\nabla f_0(x)$, viewed as a row vector, stacked on top of $\nabla F(x)$ in the bottom right corner and 0 elsewhere. Thus,

$$\nabla G(\bar{u}, \bar{x})^\top \begin{bmatrix} w \\ y_0 \\ y \end{bmatrix} = \begin{bmatrix} y \\ w + \nabla f_0(\bar{x}) y_0 + \nabla F(\bar{x})^\top y \end{bmatrix}. \tag{5.4}$$

Then, $(\bar{y}, 0)$ must satisfy $\bar{y} = y$ and $0 = w + \nabla f_0(\bar{x}) y_0 + \nabla F(\bar{x})^\top y$ for some

$$w \in N_X(\bar{x}), \;\; y_0 = 1, \;\; y \in \partial h\big(F(\bar{x}) + \bar{u}\big).$$

These expressions simplify to

$$-\nabla f_0(\bar{x}) - \nabla F(\bar{x})^\top \bar{y} \in N_X(\bar{x}) \quad \text{and} \quad \bar{y} \in \partial h\big(F(\bar{x}) + \bar{u}\big).$$

Consequently, $\bar{y} \in Y(\bar{u}, \bar{x})$. It remains to verify the qualification (4.14). By 4.63 and 4.65,

$$\partial^\infty g\big(G(\bar{u}, \bar{x})\big) \subset N_X(\bar{x}) \times \{0\} \times N_{\operatorname{dom} h}\big(F(\bar{x}) + \bar{u}\big).$$

If $(w, y_0, y) \in \partial^\infty g(G(\bar{u}, \bar{x}))$ and this vector makes (5.4) vanish, then $(w, y_0, y) = (0, 0, 0)$ and the qualification holds. □

In the broad context of minimization of a composite function, the proposition shows that under mild assumptions the min-value function associated with a typical Rockafellian has all its subgradients being multiplier vectors associated with some minimizer of the actual problem. This fact holds even when the actual problem is nonconvex and nonsmooth.

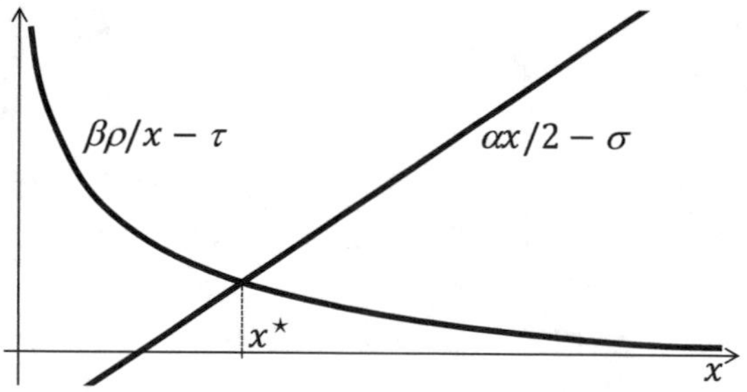


Fig. 5.6: Objective function to determine an economic order quantity.

**Example 5.17** (economic order quantity in inventory management). A store manager needs to determine how many units to order of a given product each time the product runs out. The number of units sold per year is $\rho > 0$, which means that an order quantity of $x$ units results in $\rho/x$ orders per year. Suppose that each one of these orders has a fixed cost of $\beta > 0$. (The cost of the units doesn't factor in here because the total number of units ordered across the whole year is always $\rho$.) The manager also faces an inventory cost of $\alpha > 0$ per unit and year. With an order quantity $x$, the average inventory is $x/2$ so the annual inventory cost becomes $\alpha x/2$. The manager would like to determine an order quantity $x$ such that both the ordering cost $\beta\rho/x$ and the inventory cost $\alpha x/2$ are low. Let's adopt the goal optimization model (cf. 4.6)

$$\underset{x \in X \subset \mathbb{R}}{\text{minimize}} \ \max\left\{\beta\rho/x - \tau,\ \tfrac{1}{2}\alpha x - \sigma\right\},$$

where $\tau, \sigma$ are the goals for the ordering cost and the inventory cost, respectively; see Figure 5.6. What's the effect of changing the goals?

**Detail.** For perturbation $u \in \mathbb{R}^2$ of the goals, let's adopt the Rockafellian given by

$$f(u, x) = \iota_X(x) + h\big(F(x) + u\big),$$

where $h(z) = \max\{z_1, z_2\}$ and

$$F(x) = \big(f_1(x), f_2(x)\big) = \big(\beta\rho/x - \tau, \tfrac{1}{2}\alpha x - \sigma\big).$$

Then, $f$ has anchor at $\bar{u} = 0$ and the actual model corresponds to minimizing $f(0, \cdot)$. Assuming that $X$ is a nonempty closed subset of the positive numbers, then the assumptions of 5.16 are satisfied as long as $f_1$ is extended in a smooth manner from being defined on the positive numbers to all of $\mathbb{R}$. (This is only a technical issue as we don't consider order quantities below one anyways.) Since the subgradients of $h$ are known from 4.66, we obtain in the notation of 5.16 that

$$Y(u, x) = \Big\{y \in \mathbb{R}^2 \;\Big|\; \beta\rho y_1 x^{-2} - \tfrac{1}{2}\alpha y_2 \in N_X(x)$$
$$y_1 + y_2 = 1;\ y_i \geq 0 \text{ if } f_i(x) + u_i = f(u, x),\ y_i = 0 \text{ otherwise; } i = 1, 2\Big\}$$

for $x \in P(u)$. In general, one would employ an algorithm to compute a minimizer $x^\star$ of the model at hand under $\bar{u}$, which hopefully also computes the corresponding multipliers in $Y(\bar{u}, x^\star)$. If we assume that $\sigma = \tau = 0$ and $X$ isn't active at a minimizer, then we obtain the unique minimizer analytically by solving $f_1(x) = f_2(x)$, which produces

$$x^\star = \sqrt{2\beta\rho/\alpha}, \ \text{ with } Y(0, x^\star) = \left\{\left(\tfrac{1}{2}, \tfrac{1}{2}\right)\right\}.$$

Then, by 5.16, the minimum value function $p(u) = \inf f(u, \cdot)$ is differentiable at $u = 0$ with $\nabla p(0) = (1/2, 1/2)$. This provides the insight that if the current goal of zero ordering cost is changed to a small positive number $\tau$, which corresponds to setting $u = (-\tau, 0)$, then the change in minimum value is approximately $\langle \nabla p(0), u\rangle = -\tau/2$. The negative value is reasonable as raising the goal value reduces the shortfall. □

## 5.C Lagrangians and Dual Problems

In addition to their role in optimality conditions and sensitivity analysis, Rockafellians for a minimization problem furnish a principal path to constructing *problem relaxations*. These are alternative problems with minimum values no higher than that of the actual one. Relaxations are computationally useful and, when properly tuned, may even reproduce the minimum value of the actual problem. The process of tuning the relaxations is in itself an optimization problem, which we refer to as a *dual problem*. We'll eventually see that dual problems furnish yet another way of computing multipliers.

For $f_0 : \mathbb{R}^n \to \overline{\mathbb{R}}$, let's consider the problem

$$\underset{x\in\mathbb{R}^n}{\text{minimize}}\ f_0(x)$$

and an associated Rockafellian $f : \mathbb{R}^m \times \mathbb{R}^n \to \overline{\mathbb{R}}$ with anchor at 0. (The focus on 0 instead of a more general $\bar{u}$ promotes symmetry below, without much loss of generality because one can always shift the perturbation vector by redefining $f$.) The Rockafellian defines a family of problems, among which

$$\underset{x\in\mathbb{R}^n}{\text{minimize}}\ f(0,x)$$

is the actual problem of interest because $f(0,x) = f_0(x)$ for all $x$. The choice of Rockafellian could be dictated by a concern about certain model parameters as part of a sensitivity analysis, but just as well by the need for constructing tractable relaxations of the actual problem and this is the goal now.

We can't do worse if permitted to *optimize* the perturbation vector together with $x$. Thus, the problem of minimizing $f(u,x)$ over $(u,x) \in \mathbb{R}^m \times \mathbb{R}^n$ is a relaxation of the actual problem. Regardless of $y \in \mathbb{R}^m$, this is also the case for minimizing $f(u,x) - \langle y, u\rangle$ over $(u,x) \in \mathbb{R}^m \times \mathbb{R}^n$ as $u = 0$ remains a possibility. Consequently, for any $y \in \mathbb{R}^m$,

$$\begin{aligned} \inf\nolimits_{x\in\mathbb{R}^n} f_0(x) &= \inf\nolimits_{x\in\mathbb{R}^n} f(0,x) \\ &\geq \inf\nolimits_{(u,x)\in\mathbb{R}^m\times\mathbb{R}^n} \big\{f(u,x) - \langle y,u\rangle\big\} = \inf\nolimits_{x\in\mathbb{R}^n} l(x,y), \end{aligned} \tag{5.5}$$

where we use the short-hand notation

$$l(x,y) = \inf\nolimits_{u\in\mathbb{R}^m} \big\{f(u,x) - \langle y,u\rangle\big\}. \tag{5.6}$$

The function $l : \mathbb{R}^n \times \mathbb{R}^m \to \overline{\mathbb{R}}$ given by this formula is the *Lagrangian* of $f$.

A benefit from bringing in a vector $y$ is that we now can tune the relaxation by selecting $y$ appropriately. We interpret $y_i$ as the "price" associated with setting $u_i$ nonzero so that $-y_i u_i$ is the additional cost for such $u_i$ to be accrued on top of $f(u,x)$. Then, $\inf_x l(x,y)$ is the lowest possible cost that can be achieved when one is permitted to "buy" perturbations at prices stipulated by $y$. The notation $y$ for this price vector isn't arbitrary; deep connections with multipliers emerge below. However, we start by examining Lagrangians arising from typical Rockafellians.

**Example 5.18** (Lagrangian for equalities and inequalities). For $f_i : \mathbb{R}^n \to \mathbb{R}$, $i = 0, 1, \dots, m$ and $g_i : \mathbb{R}^n \to \mathbb{R}$, $i = 1, \dots, q$, let's consider the problem

$$\operatorname*{minimize}_{x \in \mathbb{R}^n} f_0(x) \text{ subject to } f_i(x) = 0,\ i = 1, \dots, m;\ g_i(x) \le 0,\ i = 1, \dots, q.$$

Among many possibilities, a Rockafellian is obtained by perturbing the right-hand side of the constraints. While such perturbations can be viewed from the angle of sensitivity analysis as changing "budgetary thresholds," they can also be introduced purely for the purpose of developing optimality conditions and/or algorithms. Specifically, let $D = \{0\}^m \times (-\infty, 0]^q$ and

$$F(x) = \big(f_1(x), \dots, f_m(x), g_1(x), \dots, g_q(x)\big).$$

A Rockafellian $f : \mathbb{R}^{m+q} \times \mathbb{R}^n \to \overline{\mathbb{R}}$ for the problem is defined by

$$f(u, x) = f_0(x) + \iota_D\big(F(x) + u\big).$$

The actual problem is then to minimizing $f(0, \cdot)$ and the Lagrangian has

$$l(x, y) = \begin{cases} f_0(x) + \langle F(x), y \rangle & \text{if } y_{m+1}, \dots, y_{m+q} \ge 0 \\ -\infty & \text{otherwise.} \end{cases}$$

Moreover, the actual problem is equivalently stated as

$$\operatorname*{minimize}_{x \in \mathbb{R}^n} \ \sup_{y \in \mathbb{R}^{m+q}} l(x, y).$$

**Detail.** With $x \in \mathbb{R}^n$ and $y \in \mathbb{R}^{m+q}$, the Rockafellian $f$ produces the Lagrangian given by

$$l(x, y) = \inf_{u \in \mathbb{R}^{m+q}} \big\{ f_0(x) + \iota_D\big(F(x) + u\big) - \langle y, u \rangle \big\}.$$

If there's $y_i < 0$ for some $i \in \{m+1, \dots, m+q\}$, then we can select $u_j = -f_j(x)$ for all $j \in \{1, \dots, m\}$ and $u_j = -g_j(x)$ for all $j \in \{m+1, \dots, m+q\} \setminus \{i\}$ so that $\iota_D(F(x) + u)$ remains 0 as $u_i \to -\infty$. But, then

$$f_0(x) + \iota_D\big(F(x) + u\big) - \langle y, u \rangle \to -\infty$$

and $l(x, y) = -\infty$.

If $y_i \ge 0$ for all $i \in \{m+1, \dots, m+q\}$, then $\bar{u}$, with components $\bar{u}_j = -f_j(x)$ for all $j \in \{1, \dots, m\}$ and $\bar{u}_j = -g_j(x)$ for all $j \in \{m+1, \dots, m+q\}$, solves

$$\operatorname*{minimize}_{u \in \mathbb{R}^{m+q}} f_0(x) + \iota_D\big(F(x) + u\big) - \langle y, u \rangle$$

and this results in

$$l(x, y) = f_0(x) - \langle y, \bar{u} \rangle = f_0(x) + \big\langle y, F(x) \big\rangle.$$

With the expression for $l$ established, we also see that

$$\sup_{y\in\mathbb{R}^{m+q}} l(x,y) = \sup\big\{f_0(x) + \langle F(x),y\rangle \bigm| y_{m+1},\dots,y_{m+q} \geq 0\big\} = f(0,x)$$

because if $f_i(x) = 0$ for all $i = 1,\dots,m$ and $g_i(x) \leq 0$ for all $i = 1,\dots,q$, then $y = 0$ is a maximizer. If $g_i(x) > 0$ for some $i$, then $y_i \to \infty$ brings the supremum to infinity. If $f_i(x) \neq 0$ for some $i$, then $y_i \to \infty$ or $-\infty$ brings the supremum to infinity again. □

**Exercise 5.19** (Lagrangian). In the setting of 5.18 with $f_0(x) = \exp(x)$ and $g_1(x) = -x$, plot the Lagrangian as a function of $x$ for many different $y \in \mathbb{R}$ and compare it with $f_0 + \iota_{[0,\infty)}$. Confirm that (5.5) holds.

**Example 5.20** (Lagrangian for max-function). For $f_i : \mathbb{R}^n \to \mathbb{R}$, $i = 1,\dots,m$, let's consider the problem

$$\underset{x\in\mathbb{R}^n}{\text{minimize}}\ \max_{i=1,\dots,m} f_i(x)$$

and the associated Rockafellian $f : \mathbb{R}^m \times \mathbb{R}^n \to \overline{\mathbb{R}}$ expressed by

$$f(u,x) = h\big(F(x) + u\big),$$

where $F(x) = (f_1(x),\dots,f_m(x))$ and $h(z) = \max\{z_1,\dots,z_m\}$. We recover the actual problem by minimizing $f(0,\cdot)$. The corresponding Lagrangian has

$$l(x,y) = \begin{cases} \big\langle F(x),y\big\rangle & \text{if } y \geq 0,\ \sum_{i=1}^m y_i = 1 \\ -\infty & \text{otherwise.} \end{cases}$$

Moreover, the actual problem is equivalently stated as

$$\underset{x\in\mathbb{R}^n}{\text{minimize}}\ \sup_{y\in\mathbb{R}^m} l(x,y).$$

**Detail.** With $x \in \mathbb{R}^n$ and $y \in \mathbb{R}^m$, the Rockafellian $f$ leads to the Lagrangian given by

$$l(x,y) = \inf_{u\in\mathbb{R}^m} \big\{h\big(F(x)+u\big) - \langle y,u\rangle\big\}.$$

If $y_i < 0$ for some $i$, then we can select $u_j = 0$ for all $j \in \{1,\dots,m\}\setminus\{i\}$ and let $u_i \to -\infty$, which implies that $h(F(x)+u) - \langle y,u\rangle \to -\infty$ and $l(x,y) = -\infty$.

If $y_i \geq 0$ for all $i$ and $\sum_{i=1}^m y_i = \beta \neq 1$, then $u_i = \alpha$ for all $i$ gives that

$$h\big(F(x)+u\big) - \langle y,u\rangle = \max_{i=1,\dots,m}\big\{f_i(x) + \alpha\big\} - \alpha\beta = \max_{i=1,\dots,m} f_i(x) + \alpha(1-\beta).$$

If $\beta > 1$, then $\alpha \to \infty$ makes the previous term approach $-\infty$ and likewise for $\beta < 1$ with $\alpha \to -\infty$. In either case, $l(x,y) = -\infty$.

If $y_i \geq 0$ for all $i$ and $\sum_{i=1}^m y_i = 1$, then $\bar{u} = (-f_1(x),\dots,-f_m(x))$ solves

$$\underset{u\in\mathbb{R}^m}{\text{minimize}}\, h\big(F(x)+u\big) - \langle y,u\rangle$$

because increasing $u_i$ above $\bar{u}_i$ isn't beneficial due to $y_i \leq 1$. Likewise, reducing $u_i$ below $\bar{u}_i$ only benefits $h(F(x) + u)$ if all $u_1, \dots, u_m$ are reduced, but then $-\langle y, u\rangle$ grows a corresponding amount because $\sum_{i=1}^m y_i = 1$. The minimizer $\bar{u}$ then produces the formula for $l(x, y)$.

We see that

$$\sup_{y\in\mathbb{R}^m} l(x, y) = \max_{i=1,\dots,m} f_i(x) = f(0, x) \;\; \forall x \in \mathbb{R}^n$$

because, generally, $\max\{\langle z, y\rangle \mid y \geq 0, \sum_{i=1}^m y_i = 1\} = \max_{i=1,\dots,m} z_i$. □

The Rockafellians in these examples lead to explicit expressions for the corresponding Lagrangians. This is especially useful if they were to be employed within an algorithm for the purpose of computing lower bounds on the minimum values of the actual problems via (5.5). Still, one shouldn't feel confined to these choices; opportunities for innovative constructions abound.

The examples show that a problem, after being enriched with perturbations as defined by a Rockafellian, can be restated in terms of a Lagrangian: The actual problem can be viewed as minimizing the worst-case Lagrangian. This holds much beyond these examples. Before treating more general problems, however, we introduce a central concept from convex analysis that's hidden in the definition of Lagrangians.

**Definition 5.21** (conjugate function). For $h : \mathbb{R}^m \to \overline{\mathbb{R}}$, the function $h^* : \mathbb{R}^m \to \overline{\mathbb{R}}$ defined by

$$h^*(v) = \sup_{u\in\mathbb{R}^m} \big\{\langle v, u\rangle - h(u)\big\}$$

is the *conjugate* of $h$. The mapping of a function $h$ into its conjugate is referred to as the *Legendre-Fenchel transform.*

For a Rockafellian $f : \mathbb{R}^m \times \mathbb{R}^n \to \overline{\mathbb{R}}$ and its Lagrangian $l$ defined by (5.6), we see that

$$-l(\bar{x}, \cdot\,) \text{ is the conjugate of } f(\cdot\,, \bar{x})$$

regardless of $\bar{x} \in \mathbb{R}^n$. From this angle, a conjugate function is an economical way of expressing a Lagrangian associated with a particular Rockafellian. Geometrically, the conjugate of a function $h : \mathbb{R}^m \to \overline{\mathbb{R}}$ defines the affine functions with epigraphs containing the epigraph of $h$. From Figure 5.7, we see specifically that

$$(v, \beta) \in \operatorname{epi} h^* \iff \sup_{u\in\mathbb{R}^m} \big\{\langle v, u\rangle - h(u)\big\} \leq \beta \tag{5.7}$$
$$\iff \langle v, u\rangle - h(u) \leq \beta \;\; \forall u \in \mathbb{R}^m \iff \langle v, u\rangle - \beta \leq h(u) \;\; \forall u \in \mathbb{R}^m.$$

Thus, $\operatorname{epi} h \subset \operatorname{epi}(\langle v, \cdot\,\rangle - \beta)$ as illustrated in the figure, where in fact

$$h(u) = \begin{cases} \frac{1}{2}u^2 & \text{if } u \leq 0 \\ 0 & \text{if } u \in (0, 1] \\ \infty & \text{otherwise} \end{cases} \qquad h^*(v) = \begin{cases} \frac{1}{2}v^2 & \text{if } v \leq 0 \\ v & \text{otherwise.} \end{cases}$$

The expression for $h^*$ follows directly from the definition.

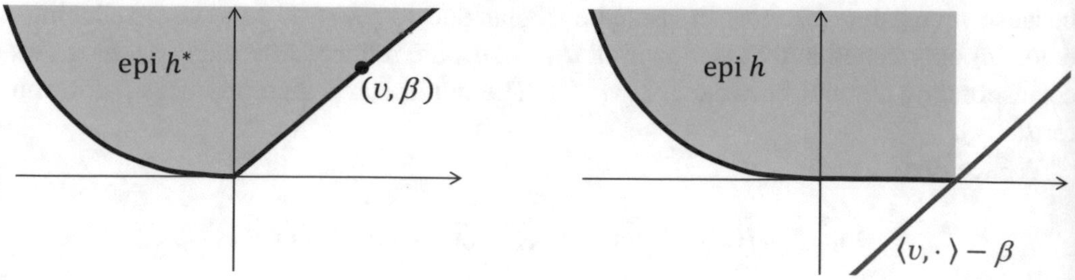


Fig. 5.7: A convex function and its conjugate.

**Exercise 5.22** (conjugate of quadratic function). Determine the conjugate of the quadratic function given by

$$h(u) = \langle c, u \rangle + \tfrac{1}{2}\langle u, Qu \rangle,$$

where $Q$ is a symmetric positive definite $m \times m$-matrix. Specialize the resulting formula for the conjugate in the case of $c = 0$ and $Q = I$, the identity matrix.

**Guide.** Write $h^*(v) = -\inf_{u \in \mathbb{R}^m}\{h(u) - \langle u, v \rangle\}$ and determine a minimizer using the optimality condition 2.19. □

One might hope that applying the Legendre-Fenchel transform twice returns the original function. As stated next, this is indeed the case under convexity with minor exceptions.

**Theorem 5.23** (Fenchel-Moreau). *For $h : \mathbb{R}^m \to \overline{\mathbb{R}}$, one has*

$$(h^*)^*(u) \le h(u) \;\; \forall u \in \mathbb{R}^m,$$

*with equality holding when $h$ is proper, lsc and convex and then $h^*$ is also proper, lsc and convex.*

**Proof.** Let $\bar{u} \in \mathbb{R}^m$. We first establish that $(h^*)^*(\bar{u}) \le h(\bar{u})$ or, equivalently, that

$$\langle v, \bar{u} \rangle - h^*(v) \le h(\bar{u}) \;\; \forall v \in \mathbb{R}^m.$$

Let $\bar{v} \in \mathbb{R}^m$ be arbitrary. Thus, it suffices to show that

$$\langle \bar{v}, \bar{u} \rangle - h^*(\bar{v}) \le h(\bar{u}). \tag{5.8}$$

If $h^*(\bar{v}) = \infty$, then the relation holds trivially. If $h^*(\bar{v}) = -\infty$, then for any $\beta \in \mathbb{R}$, $(\bar{v}, \beta) \in \operatorname{epi} h^*$ so that $\langle \bar{v}, \bar{u} \rangle - \beta \le h(\bar{u})$ by (5.7). Thus, $h(\bar{u}) = \infty$ and the relation holds again. If $h^*(\bar{v}) \in \mathbb{R}$, then $(\bar{v}, h^*(\bar{v})) \in \operatorname{epi} h^*$ so that (5.8) holds by (5.7).

Second, suppose that $\bar{u} \in \operatorname{int}(\operatorname{dom} h)$. Since $h$ is proper and convex, $h(\bar{u})$ is finite and by 2.25 there's a subgradient $\bar{v} \in \partial h(\bar{u})$. By the optimality condition 2.19, $\bar{u} \in \operatorname{argmin}\{h - \langle \bar{v}, \cdot \rangle\} = \operatorname{argmax}\{\langle \bar{v}, \cdot \rangle - h\}$. Thus,

$$h^*(\bar{v}) = \sup_{u \in \mathbb{R}^m} \big\{\langle \bar{v}, u \rangle - h(u)\big\} = \langle \bar{v}, \bar{u} \rangle - h(\bar{u}).$$

Using this fact, we obtain that

$$(h^*)^*(\bar{u}) = \sup_{v \in \mathbb{R}^m} \{\langle \bar{u}, v\rangle - h^*(v)\} \geq \langle \bar{u}, \bar{v}\rangle - h^*(\bar{v}) = \langle \bar{u}, \bar{v}\rangle - \langle \bar{u}, \bar{v}\rangle + h(\bar{u}).$$

Thus, $(h^*)^*(\bar{u}) = h(\bar{u})$ when also invoking the first assertion. A similar argument takes care of $\bar{u} \notin \text{int}(\text{dom}\, h)$; see [105, Theorems 8.13, 11.1] for further details. □

The theorem allows us to establish that the actual objective function of interest is recovered by maximizing the Lagrangian as experienced in the above examples.

**Proposition 5.24** (reformulation in terms of a Lagrangian). *For the problem of minimizing* $f_0 : \mathbb{R}^n \to \overline{\mathbb{R}}$ *and a proper Rockafellian* $f : \mathbb{R}^m \times \mathbb{R}^n \to \overline{\mathbb{R}}$ *with anchor at* 0, *suppose that* $f(\cdot, x)$ *is lsc and convex for all* $x \in \mathbb{R}^n$. *If* $l : \mathbb{R}^n \times \mathbb{R}^m \to \overline{\mathbb{R}}$ *is the corresponding Lagrangian defined by (5.6), then*

$$f_0(x) = f(0, x) = \sup_{y \in \mathbb{R}^m} l(x, y) \;\; \forall x \in \mathbb{R}^n.$$

**Proof.** Let $\bar{x} \in \mathbb{R}^n$ be fixed. By definition,

$$-l(\bar{x}, y) = \sup_{u \in \mathbb{R}^m} \{\langle y, u\rangle - f(u, \bar{x})\} \;\; \forall y \in \mathbb{R}^m$$

so $-l(\bar{x}, \cdot\,)$ is the conjugate of $f(\cdot, \bar{x})$. If $f(\cdot, \bar{x})$ is proper, then the Fenchel-Moreau theorem 5.23 applies and

$$f(u, \bar{x}) = \sup_{y \in \mathbb{R}^m} \{\langle u, y\rangle + l(\bar{x}, y)\}.$$

In particular, $f(0, \bar{x}) = \sup_{y \in \mathbb{R}^m} l(\bar{x}, y)$. If $f(u, \bar{x}) = \infty$ for all $u$, then trivially $l(\bar{x}, y) = \infty$ for all $y$ so $\sup_{y \in \mathbb{R}^m} l(\bar{x}, y) = \infty$ and the claim holds again. □

Thus far, a Lagrangian has served in two capacities: Under mild assumptions on the underlying Rockafellian as laid out in 5.24, its maximization with respect to $y$ recovers the actual objective function and, without any assumptions, its minimization with respect to $x$ furnishes a lower bound on the minimum value of the actual problem via (5.5). This maximization and minimization define

$$\forall x \in \mathbb{R}^n : \quad \varphi(x) = \sup_{y \in \mathbb{R}^m} l(x, y)$$
$$\forall y \in \mathbb{R}^m : \quad \psi(y) = \inf_{x \in \mathbb{R}^n} l(x, y).$$

The function $\varphi$ is simply an equivalent expression for the actual objective function under the assumptions of 5.24 so its minimization coincides in that case with the actual problem. The function $\psi$ quantifies the lower bound in (5.5) so its maximization produces the *best possible* lower bound. We think of this maximization as an effort to tune the relaxations produced by the chosen Rockafellian. A Lagrangian therefore sets up a pair of optimization problems with profound connections.

Specifically, a Rockafellian $f : \mathbb{R}^m \times \mathbb{R}^n \to \overline{\mathbb{R}}$ associated with the problem of minimizing $f_0 : \mathbb{R}^n \to \overline{\mathbb{R}}$ and an anchor at 0 defines via its Lagrangian $l : \mathbb{R}^n \times \mathbb{R}^m \to \overline{\mathbb{R}}$, given by (5.6), a *pair* of *primal* and *dual problems*:

| | | | |
|---|---|---|---|
| (primal) | $\underset{x\in\mathbb{R}^n}{\text{minimize}}\ \varphi(x),$ | where | $\varphi(x) = \sup_{y\in\mathbb{R}^m} l(x,y)$ |
| (dual) | $\underset{y\in\mathbb{R}^m}{\text{maximize}}\ \psi(y),$ | where | $\psi(y) = \inf_{x\in\mathbb{R}^n} l(x,y)$ |

The name "primal problem" is motivated by its equivalence with the actual problem of interest under the assumptions of 5.24 because then

$$f_0(x) = f(0,x) = \varphi(x) \quad \forall x \in \mathbb{R}^n.$$

In parallel, we refer to $x$ and $y$ as *primal* and *dual variables*, respectively, and $\varphi$ and $\psi$ as *primal* and *dual objective functions*, respectively.

**Theorem 5.25** (weak duality). *For a Lagrangian $l : \mathbb{R}^n \times \mathbb{R}^m \to \overline{\mathbb{R}}$, regardless of its expression, and its primal and dual problems, one has*

$$\inf_{x\in\mathbb{R}^n} \varphi(x) \geq \sup_{y\in\mathbb{R}^m} \psi(y) \geq \psi(\bar{y}) \quad \forall \bar{y} \in \mathbb{R}^m.$$

**Proof.** For $\bar{x} \in \mathbb{R}^n$ and $\bar{y} \in \mathbb{R}^m$,

$$\varphi(\bar{x}) = \sup_{y\in\mathbb{R}^m} l(\bar{x},y) \geq l(\bar{x},\bar{y}) \geq \inf_{x\in\mathbb{R}^n} l(x,\bar{y}) = \psi(\bar{y}).$$

Thus, $\inf_{x\in\mathbb{R}^n} \varphi(x) \geq \psi(\bar{y})$ and the conclusion follows since $\bar{y}$ is arbitrary. □

The theorem establishes that a dual problem is always a relaxation of its primal problem, which reinforces the fact from (5.5) that the minimum value of a Lagrangian bounds from below the minimum value of the actual problem. This can be exploited algorithmically to address difficult optimization problems as we'll see in the next section. Before we enter into details, however, let's consider some common examples of primal and dual problems.

**Example 5.26** (dual linear problems). For $c \in \mathbb{R}^n$, $b \in \mathbb{R}^m$ and an $m \times n$-matrix $A$, the linear optimization problem

$$\underset{x\in\mathbb{R}^n}{\text{minimize}}\ \langle c,x\rangle \ \text{subject to}\ Ax = b,\ x \geq 0$$

coincides with the primal problem produced by the Rockafellian expressed as

$$f(u,x) = \langle c,x\rangle + \iota_{[0,\infty)^n}(x) + \iota_{\{0\}^m}(b - Ax + u),$$

which then defines the Lagrangian given by

$$l(x,y) = \begin{cases} \langle c,x\rangle - \langle y, Ax - b\rangle & \text{if } x \geq 0 \\ \infty & \text{otherwise.} \end{cases}$$

The corresponding dual problem is

$$\underset{y\in\mathbb{R}^m}{\text{maximize}}\ \langle b, y\rangle \ \text{subject to}\ A^\top y \le c.$$

**Detail.** In an argument similar to those in 5.18, we see that

$$l(x,y) = \inf_{u\in\mathbb{R}^m}\big\{f(u,x) - \langle y,u\rangle\big\} = \begin{cases}\langle b,y\rangle + \langle c - A^\top y, x\rangle & \text{if } x \ge 0\\ \infty & \text{otherwise,}\end{cases}$$

which is equivalent to the asserted expression. From 5.24, $f(0,x) = \sup_{y\in\mathbb{R}^m} l(x,y)$, but this can be seen directly as well. Thus, the actual problem coincides with the primal problem produced by the Rockafellian via this Lagrangian. The dual objective function has

$$\psi(y) = \inf_{x\in\mathbb{R}^n} l(x,y) = \begin{cases}\langle b,y\rangle & \text{if } c - A^\top y \ge 0\\ -\infty & \text{otherwise}\end{cases}$$

and this yields the asserted dual problem.

The process is symmetric. If we start with the problem of minimizing $\langle -b, y\rangle$ subject to $A^\top y \le c$ and adopting the Rockafellian

$$\tilde f(u,y) = \langle -b, y\rangle + \iota_{(-\infty,0]^n}(A^\top y - c + u),$$

then we obtain the Lagrangian

$$\tilde l(y,x) = \begin{cases}\langle -c, x\rangle + \big\langle Ax - b, y\big\rangle & \text{if } x \ge 0\\ -\infty & \text{otherwise}\end{cases}$$

and

$$\tilde\psi(x) = \inf_{y\in\mathbb{R}^m} \tilde l(y,x) = \begin{cases}\langle -c, x\rangle & \text{if } Ax = b,\ x \ge 0\\ -\infty & \text{otherwise.}\end{cases}$$

The dual problem to that of minimizing $\langle -b, y\rangle$ subject to $A^\top y \le c$ is therefore maximizing $\langle -c, x\rangle$ subject to $Ax = b$ and $x \ge 0$.

The above choice of Rockafellian has the advantage that it leads to an explicit expression for the dual problem. However, there are many alternatives. For example, one might only be concerned about changes to one of the equality constraints. This leads to a different Lagrangian and, in turn, another dual problem. Specifically, let's consider $Ax = b$ *and* $\langle a, x\rangle = \alpha$, the latter being of concern. This leads to the problem

$$\underset{x\in\mathbb{R}^n}{\text{minimize}}\ \langle c, x\rangle \ \text{subject to}\ Ax = b,\ \langle a, x\rangle = \alpha,\ x \ge 0,$$

which coincides with the primal problem produced by an alternative Rockafellian expressed as

$$f(u,x) = \langle c, x\rangle + \iota_{[0,\infty)^n}(x) + \iota_{\{0\}^m}(b - Ax) + \iota_{\{0\}}\big(\alpha - \langle a, x\rangle + u\big).$$

Now, only one of the equality constraints is perturbed. This Rockafellian produces an alternative Lagrangian given by

$$l(x,y) = \inf_{u\in\mathbb{R}} \big\{ f(u,x) - yu \big\} = \begin{cases} \alpha y + \langle c - ay, x\rangle & \text{if } Ax = b,\ x \geq 0 \\ \infty & \text{otherwise,} \end{cases}$$

which defines a dual problem of maximizing $\psi(y) = \inf_x l(x,y)$ over $y \in \mathbb{R}$. While this dual problem has only a single variable, its objective function isn't generally available in an explicit form; for each $y$, one needs to solve a linear optimization problem to obtain $\psi(y)$. Still, insight as well as computational benefits may emerge from this approach. □

**Example 5.27** (dual quadratic problems). For $c \in \mathbb{R}^n$, $b \in \mathbb{R}^m$, an $m \times n$-matrix $A$ and a symmetric positive definite $n \times n$-matrix $Q$, the quadratic optimization problem

$$\operatorname*{minimize}_{x\in\mathbb{R}^n} \ \langle c, x\rangle + \tfrac{1}{2}\langle x, Qx\rangle \ \text{ subject to } \ Ax = b,\ x \geq 0$$

coincides with the primal problem produced by the Rockafellian given as

$$f(u,v,x) = \langle c, x\rangle + \tfrac{1}{2}\langle x, Qx\rangle + \iota_{\{0\}^m}\big(b - Ax + u\big) + \iota_{(-\infty,0]^n}(-x + v),$$

which then defines the Lagrangian expressed as

$$l(x,y,z) = \begin{cases} \langle c, x\rangle + \tfrac{1}{2}\langle x, Qx\rangle + \langle b - Ax, y\rangle + \langle -x, z\rangle & \text{if } z \geq 0 \\ -\infty & \text{otherwise.} \end{cases}$$

The corresponding dual problem is

$$\operatorname*{maximize}_{y\in\mathbb{R}^m, z\in\mathbb{R}^n} \ \alpha + \langle \tilde{b}, y\rangle + \langle Q^{-1}c, z\rangle - \tfrac{1}{2}\big\langle Q^{-1}(A^\top y + z), A^\top y + z\big\rangle \ \text{ subject to } \ z \geq 0,$$

where $\alpha = -\frac{1}{2}\langle c, Q^{-1}c\rangle$ and $\tilde{b} = b + AQ^{-1}c$, which is a quadratic problem.

**Detail.** The quadratic problem fits the framework of 5.18. From the discussion there, the actual problem coincides with the primal problem produced by this Lagrangian. Suppose that $z \geq 0$. Since $Q$ is positive definite and symmetric, it follows that $l(\cdot, y, z)$ is strictly convex by 1.24. The optimality condition 2.19 establishes that

$$\bar{x} = Q^{-1}(A^\top y + z - c)$$

is the minimizer of $l(\cdot, y, z)$ because the inverse $Q^{-1}$ exists and, in fact, is also positive definite and symmetric. The dual objective function has $\psi(y,z) = l(\bar{x}, y, z)$, which simplifies to the stated expression after some algebra. Since $Q^{-1}$ is positive definite, it has a square root $R$ such that $Q^{-1} = R^\top R$. The dual objective function can then be expressed as

$$\alpha + \langle \tilde{b}, y\rangle + \big\langle Q^{-1}c, z\big\rangle - \tfrac{1}{2}\big\|R(A^\top y + z)\big\|_2^2,$$

which defines a concave function in $(y, z)$; cf. 1.18(c). □

These examples give dual problems with concave objective functions. This is always the case: For a Rockafellian $f : \mathbb{R}^m \times \mathbb{R}^n \to \overline{\mathbb{R}}$, the dual objective function $\psi$ has

$$-\psi(y) = -\inf_{x,u} \{f(u,x) - \langle y,u \rangle\} = \sup_{x,u} \{\langle y,u \rangle - f(u,x)\}.$$

Thus, $-\psi$ is convex by 1.18(a). Interestingly, regardless of convexity in a primal problem and in any of the underlying functions, the dual objective function $\psi$ is concave and thus its maximization can be achieved by minimizing the convex function $-\psi$. Although the dual problem may only provide a lower bound on the minimum value of the actual problem at hand, it could be more tractable due to this property.

A versatile proposition for problems with a composite structure, which includes the earlier examples, is central to the later developments. The resulting Lagrangian and corresponding dual problem are the foundations for several algorithms.

**Proposition 5.28** (Lagrangian for composite function). *For $f_0 : \mathbb{R}^n \to \mathbb{R}$, $F : \mathbb{R}^n \to \mathbb{R}^m$ and proper, lsc and convex $h : \mathbb{R}^m \to \overline{\mathbb{R}}$, consider the problem*

$$\underset{x \in X \subset \mathbb{R}^n}{\text{minimize}}\ f_0(x) + h\big(F(x)\big).$$

*The Rockafellian given by*

$$f(u,x) = \iota_X(x) + f_0(x) + h\big(F(x) + u\big)$$

*recovers the actual problem as minimizing $f(0,\cdot)$ and produces a Lagrangian with*

$$l(x,y) = \iota_X(x) + f_0(x) + \big\langle F(x), y \big\rangle - h^*(y).$$

*Moreover, the actual problem is equivalently stated as*

$$\underset{x \in \mathbb{R}^n}{\text{minimize}}\ \sup_{y \in \mathbb{R}^m} l(x,y).$$

**Proof.** By (5.6), the Lagrangian takes the form

$$\begin{aligned} l(x,y) &= \inf_{u \in \mathbb{R}^m} \big\{\iota_X(x) + f_0(x) + h\big(F(x) + u\big) - \langle y,u \rangle\big\} \\ &= \iota_X(x) + f_0(x) - \sup_{u \in \mathbb{R}^m} \big\{\langle y,u \rangle - h\big(F(x) + u\big)\big\}. \end{aligned}$$

For $a \in \mathbb{R}^m$, we deduce from its definition 5.21 that the conjugate of $u \mapsto h(u+a)$ is the function $v \mapsto h^*(v) - \langle a,v \rangle$. Thus, the above sup-expression equals $h^*(y) - \langle F(x), y \rangle$ and we obtain the formula for $l$. If $f$ is proper, then 5.24 applies because $f(\cdot,x)$ is lsc and convex for all $x$; see in part 1.18(c). In that case, we obtain

$$f(0,x) = \sup_{y \in \mathbb{R}^m} l(x,y).$$

If $f$ isn't proper, then it must have $f(u,x) = \infty$ for all $u,x$. Thus, $l(x,y) = \infty$ for all $x,y$ and the same expression holds for $f(0,x)$ in this case as well. □

We recall the convention $\infty - \infty = \infty$. So, in the proposition, $l(x, y) = \infty$ for $x \notin X$ even if $h^*(y) = \infty$. The expression for the Lagrangian is made concrete by explicit formulae for conjugate functions. Let's record some common cases.

**Example 5.29** (conjugate functions). For $U \subset \mathbb{R}^m$, one has

$$h(u) = \iota_U(u) \implies h^*(v) = \sup_{u \in U} \langle u, v \rangle,$$

which, if $U$ is a cone, specializes to

$$h(u) = \iota_U(u) \implies h^*(v) = \iota_{\operatorname{pol} U}(v),$$

where $\operatorname{pol} U = \{v \mid \langle u, v \rangle \le 0 \ \forall u \in U\}$ is the polar to $U$; see (2.11) and Figure 2.26.

For any $V \subset \mathbb{R}^m$, one has

$$h(u) = \sup_{v \in V} \langle u, v \rangle \implies h^*(v) = \iota_{\operatorname{cl} \operatorname{con} V}(v)$$

and then $h^*(v) = \iota_V(v)$ when $V$ is closed and convex.

Moreover,

$$h(u) = \max\{u_1, \dots, u_m\} \implies h^*(v) = \begin{cases} 0 & \text{if } v \ge 0, \ \sum_{i=1}^m v_i = 1 \\ \infty & \text{otherwise.} \end{cases}$$

**Detail.** The definition of conjugate functions immediately leads to the formulae in the case of $h = \iota_U$. For $h = \sup_{v \in V} \langle \cdot, v \rangle$, we refer to [105, Theorem 8.24]. The last claim is implicitly derived in 5.20. Additional formulae are given in [105, Section 11.F]. □

**Exercise 5.30** (logarithm). (a) For $f(x) = -\ln x$ if $x > 0$ and $f(x) = \infty$ otherwise, show that $f^*(y) = -\ln(-y) - 1$ if $y < 0$ and $f^*(y) = \infty$ otherwise. (b) For $f(x) = x \ln x$ if $x > 0$, $f(0) = 0$ and $f(x) = \infty$ otherwise, show that $f^*(y) = \exp(y - 1)$.

**Exercise 5.31** (exponential). For $f(x) = \exp(x)$, show that $f^*(y) = y \ln y - y$ if $y > 0$, $f^*(0) = 0$ and $f^*(y) = \infty$ otherwise.

**Exercise 5.32** (inverse). For $f(x) = 1/x$ if $x > 0$ and $f(x) = \infty$ otherwise, show that $f^*(y) = -2\sqrt{-y}$ if $y \le 0$ and $f^*(y) = \infty$ otherwise.

In addition to being helpful in expressing Lagrangians as in the context of 5.28, conjugate functions actually underpin dual problems even more fundamentally. For any Rockafellian $f : \mathbb{R}^m \times \mathbb{R}^n \to \overline{\mathbb{R}}$ and the corresponding Lagrangian $l$, the dual objective function has

$$\begin{aligned} \psi(y) &= \inf_x l(x, y) = \inf_{u,x} \{f(u, x) - \langle u, y \rangle\} \\ &= -\sup_{u,x} \{\langle u, y \rangle + \langle x, 0 \rangle - f(u, x)\} = -f^*(y, 0). \end{aligned}$$

Consequently, the dual objective function is immediately available from any explicit formula for the conjugate of $f$ that might be available and this holds regardless of the actual problem and its perturbation as expressed by $f$.

**Exercise 5.33** (linear constraints). For $f_0 : \mathbb{R}^n \to \mathbb{R}$, $b \in \mathbb{R}^m$, $d \in \mathbb{R}^q$ as well as $m \times n$ and $q \times n$ matrices $A$ and $D$, consider the problem of minimizing $f_0(x)$ subject to $Ax = b$ and $Dx \le d$ and the Rockafellian $f : \mathbb{R}^{m+q} \times \mathbb{R}^n \to \overline{\mathbb{R}}$ given by

$$f(u, v, x) = f_0(x) + \iota_{\{0\}^m}(Ax - b + u) + \iota_{(-\infty,0]^q}(Dx - d + v).$$

Show that the corresponding dual objective function has

$$\psi(y, z) = \begin{cases} -\langle b, y\rangle - \langle d, z\rangle - f_0^*\big( - A^\top y - D^\top z\big) & \text{if } z \ge 0 \\ -\infty & \text{otherwise.} \end{cases}$$

**Guide.** The Rockafellian fits the framework of 5.18, which then furnishes the expression for the corresponding Lagrangian. □

## 5.D Lagrangian Relaxation

It's important to be able to assess the quality of a candidate solution to a problem and thereby determine if further calculations are needed. Optimality conditions serve this purpose, but dual problems also offer possibilities that are often viable in practice. For the problem of minimizing $f_0 : \mathbb{R}^n \to \overline{\mathbb{R}}$ and a candidate solution $\bar{x}$, the *optimality gap* of $\bar{x}$ is given by

$$f_0(\bar{x}) - \inf f_0.$$

A relatively small optimality gap indicates that $\bar{x}$ is rather good in the sense that the associated "cost" isn't much higher than the minimum value. This could convince a decision-maker to adopt $\bar{x}$, making further efforts to find an even better solution superfluous. An optimality gap is measured in the units of the objective function and is thus well understood by a decision-maker in most cases. The meaning of an optimality gap of \$1 million is clear: We're potentially leaving this amount on the table, which might be small or large depending on the circumstances.

Since there are often application-dependent heuristic algorithms for finding a candidate solution $\bar{x}$ of a difficult problem, the challenging part in calculating the optimality gap is to estimate $\inf f_0$. Typically, we prefer a lower bound on $\inf f_0$ as it produces a conservative estimate of the optimality gap and this is where a dual problem can be brought in.

The steps are as follows: First, express the actual objective function $f_0$ via a Rockafellian $f : \mathbb{R}^m \times \mathbb{R}^n \to \overline{\mathbb{R}}$ with an anchor at 0 so that $f(0, x) = f_0(x)$ for all $x \in \mathbb{R}^n$. Second, form the corresponding Lagrangian $l : \mathbb{R}^n \times \mathbb{R}^m \to \overline{\mathbb{R}}$ by (5.6). Third, using any $\bar{y} \in \mathbb{R}^m$, obtain a lower bound

$$\psi(\bar{y}) = \inf_{x\in\mathbb{R}^n} l(x, \bar{y}) \le \inf f(0, \cdot\,) = \inf f_0$$

by minimizing the Lagrangian; see (5.5). These steps are referred to as *Lagrangian relaxation* and are especially productive when the actual problem is difficult, but $l(\cdot, y)$

is relatively simple to minimize. The best possible lower bound is achieved by tuning $y$, i.e., solving the corresponding dual problem to obtain

$$\sup_{y\in\mathbb{R}^m} \psi(y) \leq \inf f_0.$$

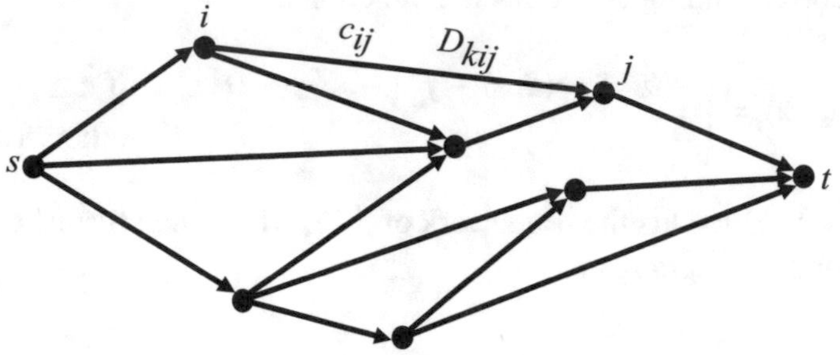


Fig. 5.8: Directed graph with vertices and edges.

**Example 5.34** (constrained shortest path problem). Let $(V, E)$ be a directed graph with vertex set $V$ and edge set $E$. Each edge $(i, j) \in E$ connects distinct vertices $i, j \in V$, and it possesses length $c_{ij} \in [0, \infty)$ and weights $D_{kij} \in [0, \infty)$ for $k = 1, \dots, q$; see Figure 5.8. A directed $s$-$t$ path is an ordered set of edges of the form $\{(s, i_1), (i_1, i_2), \dots, (i_{\nu-1}, t)\}$ for some $\nu \in \mathbb{N}$. Given two distinct vertices $s, t \in V$, the *shortest-path problem* seeks to determine a directed $s$-$t$ path such that the sum of the edge lengths along the path is minimized. This is a well-studied problem that can be solved efficiently using specialized algorithms; see [1, Chapters 4 and 5].

For nonnegative $d_k$, $k = 1, \dots, q$, the task becomes significantly harder if the sum of the weights $D_{kij}$ along the path can't exceed $d_k$ for each $k$. This is the *constrained shortest-path problem*, which can be addressed by Lagrangian relaxation. In routing of a drone through a discretized three-dimensional airspace, the weights $D_{1ij}$ might represent fuel consumption along edge $(i, j)$, which can't exceed a capacity $d_1$. Figure 5.9 illustrates a route satisfying such a fuel constraint while minimizing exposure to enemy radars expressed by $c_{ij}$; cf. [110].

**Detail.** The constrained shortest-path problem is formulated as follows. Suppose that $m$ is the number of vertices in $V$ and $n$ is the number of edges in $E$. Let $A$ denote the $m \times n$-vertex-edge incidence matrix such that if edge $e = (i, j)$, then $A_{ie} = 1$, $A_{je} = -1$ and $A_{i'e} = 0$ for any $i' \neq i, j$. Also, let $b_s = 1$, $b_t = -1$ and $b_i = 0$ for $i \in V \setminus \{s, t\}$ and collect them in the vector $b$. For each $k = 1, \dots, q$, we place the edge weights $\{D_{kij}, (i, j) \in E\}$ in the vector $D_k$. The $q \times n$-matrix $D$ has $D_k$ as its $k$th row. Let $d = (d_1, \dots, d_q)$. With $c$ being the vector of $\{c_{ij}, (i, j) \in E\}$, the constrained shortest-path problem may then be formulated as (cf. [1, p. 599])

$$\text{minimize } \langle c, x\rangle \quad \text{subject to } Ax = b,\ Dx \leq d,\ x \in \{0, 1\}^n.$$

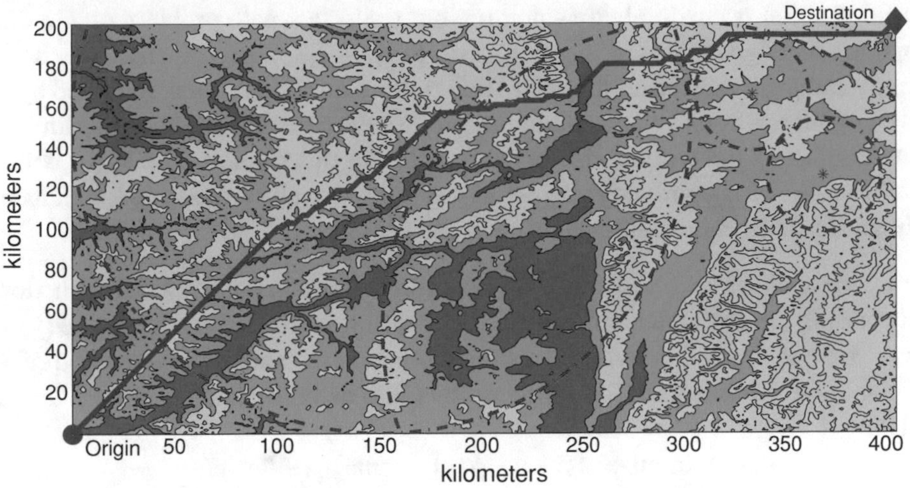


Fig. 5.9: Route for a drone through a three-dimensional airspace to a destination (blue line) that minimizes exposure to enemy radars (red circles) while satisfying a fuel constraint. Altitude changes to leverage terrain masking aren't shown.

A point $\bar{x} \in \{0,1\}^n$ satisfying $Ax = b$ corresponds to an $s$-$t$ path with $\bar{x}_{ij} = 1$ if edge $(i,j)$ is on the path and $\bar{x}_{ij} = 0$ otherwise. We assume there's at least one such path. Then, $\langle c, \bar{x}\rangle$ gives the length of the path and $\langle D_k, \bar{x}\rangle$ the $k$th weight of the path.

In the absence of the weight-constraint $Dx \leq d$, the model reduces to a shortest path problem and this opens up an opportunity for applying Lagrangian relaxation via 5.28. Let

$$X = \{x \in \{0,1\}^n \mid Ax = b\},$$

which is nonempty by assumption, and consider the Rockafellian given by

$$f(u,x) = \iota_X(x) + \langle c, x\rangle + \iota_{(-\infty,0]^q}(Dx - d + u).$$

The actual problem corresponds to minimizing $f(0,\cdot)$ over $\mathbb{R}^n$. These definitions fit the setting of 5.28 and

$$l(x,y) = \iota_X(x) + \langle c, x\rangle + \langle Dx - d, y\rangle - \iota_Y(y), \quad \text{where } Y = \{y \in \mathbb{R}^q \mid y \geq 0\}$$

by 5.29. Thus, the chosen Rockafellian results in a Lagrangian without the constraint $Dx \leq d$. The corresponding dual problem is

$$\underset{y\in\mathbb{R}^q}{\text{maximize}}\ \psi(y) = \inf_{x\in\mathbb{R}^n} l(x,y).$$

For $y \geq 0$, which are the only values of concern because $\psi(y) = -\infty$ otherwise, we obtain explicitly that

$$\psi(y) = -\langle d, y\rangle + \inf_{x\in X}\langle c + D^\top y, x\rangle.$$

The minimization occurring here is nothing but a shortest-path problem on the directed graph, but with edge lengths changed from $c_{ij}$ to $c_{ij} + \sum_{k=1}^{q} D_{kij} y_k$, which can be solved efficiently using specialized algorithms. The resulting minimum value, modified by $\langle d, y\rangle$, yields $\psi(y)$, a lower bound on the minimum value of the constrained shortest-path problem as seen from (5.5) and the introductory discussion in this section. The lower bound can be used to assess the optimality gap for any candidate path, for example, obtained by a greedy search or an enumeration algorithm; see [23]. □

For a Lagrangian $l : \mathbb{R}^n \times \mathbb{R}^m \to \overline{\mathbb{R}}$ and the corresponding dual objective function $\psi$, it would certainly be desirable to obtain the best possible lower bound $\sup_{y\in\mathbb{R}^m} \psi(y)$. Since $\psi$ is concave and typically possesses a simple domain, the subgradient method of §2.I is often viable for this purpose. Returning to the familiar minimization setting, the task at hand is to

$$\underset{y\in\mathbb{R}^m}{\text{minimize}}\ \tilde{\psi}(y) = -\psi(y) = \sup\nolimits_{x\in\mathbb{R}^n} -l(x, y).$$

The subgradient method would need a description of $\operatorname{dom}\tilde{\psi}$ and a way to compute subgradients of the convex function $\tilde{\psi}$. We develop general formulae for this in §6.E, but some cases are approachable with the existing machinery.

**Example 5.35** (constrained shortest path problem; cont.). Solving the dual problem in 5.34 reduces to

$$\underset{y\in Y}{\text{minimize}}\ \tilde{\psi}(y) = \sup\nolimits_{x\in X} \big\{ -\langle c, x\rangle - \langle Dx - d, y\rangle \big\}.$$

Since $X$ contains only a finite number of points, subgradients of $\tilde{\psi}$ can be obtained from 4.66 for any $\bar{y} \geq 0$:

$$v \in \partial\tilde{\psi}(\bar{y}) \iff v = \sum\nolimits_{x\in X^\star(\bar{y})} z_x(d - Dx) \text{ for some } z_x \geq 0 \text{ with } \sum\nolimits_{x\in X^\star(\bar{y})} z_x = 1,$$

where

$$X^\star(\bar{y}) = \operatorname{argmin}_{x\in X} \big\{\langle c, x\rangle + \langle Dx - d, \bar{y}\rangle\big\}.$$

Consequently, we obtain a subgradient $v$ of $\tilde{\psi}$ at $\bar{y}$ by computing a minimizer $\bar{x} \in X^\star(\bar{y})$, which can be achieved by a shortest-path algorithm (see [1, Chapters 4 and 5]), and then setting $v = d - D\bar{x}$. Moreover, $\operatorname{dom}\tilde{\psi} = Y = [0, \infty)^q$ is easily projected onto and this makes the subgradient method viable.

## 5.E Saddle Points

In the setting of equality and inequality constraints, we saw already in §4.K that a Lagrangian provides an efficient way of expressing optimality conditions and we'll now extend this idea to many more problems. The resulting insight connects multipliers with dual variables and also leads to an interpretation of a Lagrangian as the pay-off function in a game between two players.

**Proposition 5.36** (optimality condition in Lagrangian form). *For smooth* $f_0 : \mathbb{R}^n \to \mathbb{R}$, *smooth* $F : \mathbb{R}^n \to \mathbb{R}^m$, *nonempty closed* $X \subset \mathbb{R}^n$ *and proper, lsc and convex* $h : \mathbb{R}^m \to \overline{\mathbb{R}}$, *consider the problem*

$$\underset{x \in X}{\text{minimize}}\ f_0(x) + h\big(F(x)\big),$$

*the Rockafellian defined as*

$$f(u, x) = \iota_X(x) + f_0(x) + h\big(F(x) + u\big)$$

*and the associated Lagrangian given by*

$$l(x, y) = \iota_X(x) + f_0(x) + \big\langle F(x), y\big\rangle - h^*(y).$$

*Then, the optimality condition 4.75 for the problem, i.e.,*

$$\bar{y} \in \partial h\big(F(\bar{x})\big) \qquad -\nabla f_0(\bar{x}) - \nabla F(\bar{x})^\top \bar{y} \in N_X(\bar{x}), \tag{5.10}$$

*is equivalently expressed as*

$$0 \in \partial_x l(\bar{x}, \bar{y}) \qquad 0 \in \partial_y(-l)(\bar{x}, \bar{y}). \tag{5.11}$$

*If* $l(\cdot, y)$ *is convex for all* $y \in \mathbb{R}^m$, *then these conditions can also be stated as*

$$\bar{x} \in \operatorname{argmin} l(\cdot, \bar{y}) \qquad \bar{y} \in \operatorname{argmax} l(\bar{x}, \cdot).$$

Before proving the result, we observe that (5.11) appears in §4.K for the special case $X = \mathbb{R}^n$ and $h = \iota_{\{0\}^m}$. Then, $0 \in \partial_x l(\bar{x}, \bar{y})$ becomes $0 = \nabla f_0(\bar{x}) + \nabla F(\bar{x})^\top \bar{y}$ and $0 \in \partial_y(-l)(\bar{x}, \bar{y})$ reduces to $0 = F(\bar{x})$ because $h^*$ is identically equal to 0. Thus, (5.11) simply states that $\nabla l(\bar{x}, \bar{y}) = 0$.

The proposition confirms the connection between multipliers and dual variables: They're two names for the same object! The multiplier interpretation of $\bar{y}$ in the proposition is expressed in (5.10) and stems from the normal cones to epi $h$ as they define the subgradients of $h$. The dual variable perspective of (5.11) comes from the Lagrangian and its role in defining a dual problem.

The proof of the proposition is supported by the following general fact.

**Proposition 5.37** (inversion rule for subgradients). *For a proper, lsc and convex function* $h : \mathbb{R}^m \to \overline{\mathbb{R}}$ *and its conjugate* $h^* : \mathbb{R}^m \to \overline{\mathbb{R}}$, *one has*

$$v \in \partial h(u) \iff u \in \partial h^*(v) \iff h(u) + h^*(v) = \langle u, v\rangle.$$

**Proof.** Suppose that $\bar{u} \in \partial h^*(\bar{v})$. By the convexity of $h^*$, the optimality condition 2.19 and 4.58(c), this means that

$$\bar{v} \in \operatorname{argmin}\big\{h^* - \langle \bar{u}, \cdot\rangle\big\} = \operatorname{argmax}\big\{\langle \bar{u}, \cdot\rangle - h^*\big\}.$$

Thus, in view of the Fenchel-Moreau theorem 5.23, which applies because $h$ is proper, lsc and convex, one has

$$h(\bar{u}) = (h^*)^*(\bar{u}) = \sup_{v\in\mathbb{R}^m} \{\langle \bar{u}, v\rangle - h^*(v)\} = \langle \bar{u}, \bar{v}\rangle - h^*(\bar{v}).$$

Thus, we've proven "$\Longrightarrow$" in the second equivalence. Since

$$\sup_{u\in\mathbb{R}^m} \{\langle \bar{v}, u\rangle - h(u)\} = h^*(\bar{v}) = \langle \bar{v}, \bar{u}\rangle - h(\bar{u})$$

by the just established equality, we've that $\bar{u}$ achieves this maximum, i.e.,

$$\bar{u} \in \operatorname{argmax}\{\langle \bar{v}, \cdot\rangle - h\} = \operatorname{argmin}\{h - \langle \bar{v}, \cdot\rangle\}.$$

This in turn implies $\bar{v} \in \partial h(\bar{u})$ by the optimality condition 2.19 together with 4.58(c). We've established that "$\Longleftarrow$" holds in the first equivalence.

Suppose that $\bar{v} \in \partial h(\bar{u})$. Since $h^*$ is proper, lsc and convex by the Fenchel-Moreau theorem 5.23, the above arguments can be repeated with the roles of $h$ and $h^*$ reversed. This establishes "$\Longrightarrow$" in the first equivalence. Suppose that $h(\bar{u}) + h^*(\bar{v}) = \langle \bar{u}, \bar{v}\rangle$ so that $h^*(\bar{v}) = \langle \bar{u}, \bar{v}\rangle - h(\bar{u})$. Then, $\bar{u}$ must attain the maximum in the definition of $h^*(\bar{v})$. Again, this implies that $\bar{v} \in \partial h(\bar{u})$ and the conclusion follows. □

**Proof of 5.36.** The Lagrangian follows by 5.28. For $\bar{x} \in X$ and $\bar{y} \in \operatorname{dom} h^*$, 4.56 and 4.58 yield

$$\partial_x l(\bar{x}, \bar{y}) = N_X(\bar{x}) + \nabla f_0(\bar{x}) + \nabla F(\bar{x})^\top \bar{y}$$

so that $0 \in \partial_x l(\bar{x}, \bar{y})$ corresponds to $-\nabla f_0(\bar{x}) - \nabla F(\bar{x})^\top \bar{y} \in N_X(\bar{x})$. Similarly,

$$\partial_y(-l)(\bar{x}, \bar{y}) = -F(\bar{x}) + \partial h^*(\bar{y}).$$

Then, $0 \in \partial_y(-l)(\bar{x}, \bar{y})$ amounts to having $F(\bar{x}) \in \partial h^*(\bar{y})$ and, equivalently by 5.37, $\bar{y} \in \partial h(F(\bar{x}))$. The requirement $\bar{x} \in X$ is a necessity for $N_X(\bar{x})$ to contain a vector and likewise for $\partial_x l(\bar{x}, \bar{y})$; recall $l(\bar{x}, \bar{y})$ needs to be finite. Moreover, $\bar{y} \in \operatorname{dom} h^*$ is a necessity for $-l(\bar{x}, \bar{y})$ to be finite. Under the convexity assumption, the optimality condition 2.19 produces the final claim because $-l(\bar{x}, \cdot)$ is convex; see the discussion after 5.27. □

A function with two arguments that's convex in the first one and concave in the second one, which we refer to as a *convex-concave function*, can be visualized as having a graph that's shaped like a saddle. In view of 5.36, a Lagrangian with this property identifies a solution $\bar{x}$ and a corresponding multiplier vector $\bar{y}$ through minimization in its first argument and maximization in its second one. The corresponding $(\bar{x}, \bar{y})$ is an example of a saddle point.

For a function $g : \mathbb{R}^n \times \mathbb{R}^m \to \overline{\mathbb{R}}$ (which could be a Lagrangian), we say that $(\bar{x}, \bar{y}) \in \mathbb{R}^n \times \mathbb{R}^m$ is a *saddle point* of $g$ when

$$\boxed{\bar{x} \in \operatorname{argmin} g(\cdot, \bar{y}) \qquad \bar{y} \in \operatorname{argmax} g(\bar{x}, \cdot)}$$

A Lagrangian that's convex in its first argument is a convex-concave function because it's always concave in the second argument; see the discussion after 5.27. Consequently, for such Lagrangians, $(\bar{x}, \bar{y})$ satisfies the optimality condition (5.10) if and only if it's a saddle point by 5.36.

The saddle point condition can equivalently be expressed as

$$g(x, \bar{y}) \geq g(\bar{x}, \bar{y}) \geq g(\bar{x}, y) \quad \forall x \in \mathbb{R}^n, y \in \mathbb{R}^m,$$

which highlights the connection with game theory. Suppose that $g$ is the pay-off in a game between two non-cooperating players that proceeds as follows: Player 1 selects $x \in \mathbb{R}^n$ and Player 2 selects $y \in \mathbb{R}^m$, the choices are revealed simultaneously, and Player 1 pays $g(x, y)$ to Player 2; a negative amount indicates that Player 1 gets money. A *solution to this game* is a saddle point $(\bar{x}, \bar{y})$ of $g$. This is meaningful because then neither player has any incentive to deviate from their respective strategies $\bar{x}$ and $\bar{y}$. In selecting $\bar{x}$, Player 1 guarantees that the amount paid to Player 2 won't exceed $g(\bar{x}, \bar{y})$, even if Player 2 knew in advance that $\bar{x}$ would be chosen. This results from the right-most inequality above. At the same time, in selecting $\bar{y}$, Player 2 guarantees that the amount received from Player 1 won't fall short of $g(\bar{x}, \bar{y})$, regardless of whether Player 1 acts with knowledge of this choice or not, as seen from the left-most inequality above. The task of finding a solution to the game is an example from the broader class of *variational problems*.

**Example 5.38** (two-person zero-sum game). Two players, Minnie and Maximilian, choose strategies $x \in \mathbb{R}^n$ and $y \in \mathbb{R}^m$, respectively, to play repeatedly a game with pay-off:

$$-\langle y, Ax\rangle \text{ for Minnie} \qquad\qquad \langle y, Ax\rangle \text{ for Maximilian},$$

where $A$ is an $m \times n$-matrix to which one refers as the *pay-off matrix*.

**Detail.** In a repeated game, it becomes important not to be predictable. A way of accounting for this is to assume that the strategies $x$ and $y$ assign probabilities to the choice of a specific column (Minnie) and row (Maximilian) of $A$; i.e., Minnie chooses to play column $j$ with probability $x_j$, whereas Maximilian plays row $i$ with probability $y_i$. Thus, in these terms, Minnie has to find a strategy

$$x \in X = \left\{x \in \mathbb{R}^n \;\middle|\; \sum_{j=1}^{n} x_j = 1,\ x \geq 0\right\}$$

that on average maximizes her returns and Maximilian has to select a strategy

$$y \in Y = \left\{y \in \mathbb{R}^m \;\middle|\; \sum_{i=1}^{m} y_i = 1,\ y \geq 0\right\}$$

that maximizes his average returns. Since the payment to Maximilian, on average, is exactly $\langle y, Ax\rangle$, we've the pay-off as indicated initially. Moreover, a solution of the game is a saddle point of the convex-concave function $g : \mathbb{R}^n \times \mathbb{R}^m \to \overline{\mathbb{R}}$ given by

$$g(x, y) = \langle y, Ax\rangle + \iota_X(x) - \iota_Y(y).$$

Although, $g$ didn't arise in the context of a minimization problem, we can make the connection with 5.36 as follows: With $X$, $Y$ and $A$ as defined by the game, set $f_0$ to the zero function, $F(x) = Ax$ and $h(z) = \max\{z_1, \ldots, z_m\}$ in that proposition, invoke 5.29 and we obtain a Lagrangian of the form $\langle y, Ax\rangle + \iota_X(x) - \iota_Y(y)$, which coincides with the convex-concave function $g$ in the game. This perspective offers the possibility of solving the game by finding a solution to the optimality condition (5.10). Tracing further back, this optimality condition stems from the problem

$$\operatorname*{minimize}_{x\in X} h\big(F(x)\big) = \max\big\{\langle A_1, x\rangle, \ldots, \langle A_m, x\rangle\big\},$$

where $A_i$ is the $i$th row of $A$. But, this can be achieved by solving the linear optimization problem

$$\operatorname*{minimize}_{x\in X, \alpha\in\mathbb{R}} \alpha \ \text{ subject to } \ \langle A_i, x\rangle \leq \alpha, \ \ i = 1, \ldots, m.$$

An application of the interior-point method yields a minimizer $x^\star$ as well as the corresponding multiplier vector $y^\star$, which then satisfies (5.10); see §2.G. Regardless of how $x^\star$ is obtained, it's always possible to utilize (5.10) to recover $y^\star$. Another algorithmic strategy is laid in §5.J.

We note that there's symmetry in this game and our focus on Minnie over Maximilian is arbitrary. One could equally well have taken Maximillian's perspective, with his decision then becoming "primal" and that of Minnie being a multiplier vector. □

**Exercise 5.39** (penny-nickel-dime game). Suppose that in a game of two players each has a penny, a nickel and a dime. The game consists of each player selecting one of these coins and displaying it. If the sum of the cents on the two displayed coins is an odd number, then Player 1 wins Player 2's coin, but if the sum is even, Player 2 wins Player 1's coin. Set up a pay-off matrix $A$ for this game and solve it as indicated in 5.38 by determining strategies $x \in X$ and $y \in Y$ for Players 1 and 2, respectively.

**Guide.** The solution of the game is $x^\star = (1/2,\ 0,\ 1/2)$ and $y^\star = (10/11, 0, 1/11)$. Interestingly, neither player should ever play nickel. Also, $\langle y^\star, Ax^\star\rangle = 0$, which means the game is fair: neither player gets ahead in the long run if they follow these strategies. Since $(x^\star, y^\star)$ is a saddle point, any other strategy can't be better. For example, if Player 2 changes to $\bar{y} = (1/3, 1/3, 1/3)$, then $\langle \bar{y}, Ax^\star\rangle = -2/3$ and Player 1 makes approximately 66 cents per 100 games. □

We next turn the attention to the connection between saddle points of a Lagrangian and the primal and dual problems defined by the Lagrangian.

**Theorem 5.40** (saddle points). *For a proper function $f : \mathbb{R}^m \times \mathbb{R}^n \to \overline{\mathbb{R}}$, with $f(\cdot, x)$ lsc and convex for all $x \in \mathbb{R}^n$, and the corresponding Lagrangian $l$ given by (5.6), consider the primal and dual objective functions defined by*

$$\varphi(x) = \sup_{y\in\mathbb{R}^m} l(x, y) \quad \textit{and} \quad \psi(y) = \inf_{x\in\mathbb{R}^n} l(x, y).$$

*Then, we've the following relations:*

$$\left.\begin{aligned} &\bar{x} \in \operatorname{argmin} \varphi \\ &\bar{y} \in \operatorname{argmax} \psi \\ &\inf \varphi = \sup \psi \end{aligned}\right\} \iff (\bar{x}, \bar{y}) \textit{ is a saddle point of } l \iff \varphi(\bar{x}) = \psi(\bar{y}) = l(\bar{x}, \bar{y})$$

$$\implies 0 \in \partial_x l(\bar{x}, \bar{y}) \;\; \textit{and} \;\; 0 \in \partial_y(-l)(\bar{x}, \bar{y}).$$

*If $l(\cdot, y)$ is convex for all $y \in \mathbb{R}^m$, then the converse of the last implication also holds, with all these conditions being equivalent to*

$$(\bar{y}, 0) \in \partial f(0, \bar{x}).$$

**Proof.** The two equivalences hold trivially by the definition of a saddle point. A saddle point $(\bar{x}, \bar{y})$ of $l$ satisfies by definition $\bar{x} \in \operatorname{argmin} l(\cdot, \bar{y})$ and $\bar{y} \in \operatorname{argmax} l(\bar{x}, \cdot)$, which in turn implies that $0 \in \partial_x l(\bar{x}, \bar{y})$ and $0 \in \partial_y(-l)(\bar{x}, \bar{y})$ by the Fermat rule 4.73. If $l(\cdot, \bar{y})$ is convex, then the converse holds by optimality condition 2.19; recall that $-l(\bar{x}, \cdot)$ is convex.

For the final assertion, suppose that $(\bar{x}, \bar{y})$ satisfies $0 \in \partial_x l(\bar{x}, \bar{y})$ and $0 \in \partial_y(-l)(\bar{x}, \bar{y})$. The first inclusion implies that $l(x, \bar{y}) \geq l(\bar{x}, \bar{y})$ for all $x \in \mathbb{R}^n$. Moreover, the existence of a subgradient for $l(\cdot, \bar{y})$ at $\bar{x}$ ensures that $l(\bar{x}, \bar{y})$ is finite. Thus, $\varphi(\bar{x})$ as well as $f(0, \bar{x})$ are finite; see 5.24. We then have that $f(\cdot, \bar{x})$ is proper, lsc and convex. By the Fenchel-Moreau theorem 5.23, $-l(\bar{x}, \cdot)$ is proper, lsc and convex as well and 5.37 applies. In particular, $0 \in \partial_y(-l)(\bar{x}, \bar{y})$ implies that $f(0, \bar{x}) = l(\bar{x}, \bar{y})$. Collecting these facts, one obtains

$$l(x, \bar{y}) = \inf_u \big\{ f(u, x) - \langle \bar{y}, u \rangle \big\} \geq l(\bar{x}, \bar{y}) = f(0, \bar{x}) \;\; \forall x \in \mathbb{R}^n$$

and then also

$$f(u, x) \geq f(0, \bar{x}) + \langle \bar{y}, u - 0 \rangle + \langle 0, x - \bar{x} \rangle \;\; \forall u \in \mathbb{R}^m, x \in \mathbb{R}^n. \tag{5.12}$$

This means that $(\bar{y}, 0) \in \partial f(0, \bar{x})$ by the subgradient inequality 2.17.

For the converse, suppose that $(\bar{y}, 0) \in \partial f(0, \bar{x})$. Again by 2.17, (5.12) holds. But, this means that

$$f(u, \bar{x}) \geq f(0, \bar{x}) + \langle \bar{y}, u - 0 \rangle \;\; \forall u \in \mathbb{R}^m$$

so that $\bar{y} \in \partial_u f(0, \bar{x})$. Since $f(0, \bar{x})$ is finite, $f(\cdot, \bar{x})$ is proper, lsc and convex. Its conjugate is $-l(\bar{x}, \cdot)$. Thus, by 5.37, $\bar{y} \in \partial_u f(0, \bar{x})$ ensures that $0 \in \partial_y(-l)(\bar{x}, \bar{y})$ and $f(0, \bar{x}) = l(\bar{x}, \bar{y})$. From (5.12),

$$\inf_u \big\{ f(u, x) - \langle \bar{y}, u \rangle \big\} \geq f(0, \bar{x}) = l(\bar{x}, \bar{y}) \;\; \forall x \in \mathbb{R}^n.$$

On the left-hand side, we've $l(x, \bar{y})$ and the right-hand side can be written as $l(\bar{x}, \bar{y}) + \langle 0, x - \bar{x} \rangle$. By the subgradient inequality 2.17, this means that $0 \in \partial_x l(\bar{x}, \bar{y})$. □

The implications of this theorem are manifold, especially in view of the equivalence between the primal problem and an actual problem of interest; see 5.24. From a numerical point of view, it suggests the possibility of minimizing a function $f_0 : \mathbb{R}^n \to \overline{\mathbb{R}}$ by constructing a Rockafellian with anchor at 0, which then plays the role of $f$ in the theorem, and then find a saddle point of the associated Lagrangian. A minimization problem can therefore essentially always be reformulated as a game between a "primal player" and a "dual player," and this fact doesn't rely on convexity in the actual problem. Alternatively, one could solve the corresponding dual problem to obtain $\bar{y}$ and then somehow recover a primal solution, which might be immediately available as seen next.

**Example 5.41** (primal-dual relation for linear problems). From 5.26, we recall that

$$\begin{array}{ll} \underset{x\in\mathbb{R}^n}{\text{minimize}}\ \langle c, x\rangle & \underset{y\in\mathbb{R}^m}{\text{maximize}}\ \langle b, y\rangle \\ \text{subject to } Ax = b & \text{subject to } A^\top y \leq c \\ \qquad x \geq 0 & \end{array}$$

are primal-dual pairs under the particular Rockafellian considered there. The primal problem has a minimizer if and only if the dual problem has a maximizer, in which case their optimal values coincide. Moreover[2], one has

$$\begin{aligned} \text{primal problem unbounded} &\implies \text{dual problem infeasible} \\ \text{primal problem infeasible} &\implies \text{dual problem infeasible or unbounded} \\ \text{dual problem unbounded} &\implies \text{primal problem infeasible} \\ \text{dual problem infeasible} &\implies \text{primal problem infeasible or unbounded.} \end{aligned}$$

**Detail.** By 2.45, a minimizer of the primal problem is characterized by

$$\begin{aligned} &\forall j = 1, \dots, n: \quad z_j \geq 0 \text{ if } x_j = 0; \ \ z_j = 0 \text{ otherwise} \\ &-c = A^\top y - z, \ \ Ax = b, \ \ x \geq 0. \end{aligned}$$

After eliminating $z$ and flipping the sign of $y$, we obtain that

$$\begin{aligned} &\forall j = 1, \dots, n: \quad \text{if } x_j > 0, \text{ then } \langle A^j, y\rangle = c_j \\ &A^\top y \leq c, \ \ Ax = b, \ \ x \geq 0, \end{aligned} \tag{5.13}$$

where $A^j$ is the $j$th column of $A$. Again by 2.45, a maximizer of the dual problem or, equivalently, a minimizer of $\langle -b, y\rangle$ subject to $A^\top y \leq c$, is characterized by

$$\begin{aligned} &\forall j = 1, \dots, n: \quad x_j \geq 0 \text{ if } \langle A^j, y\rangle = c_j; \ \ x_j = 0 \text{ otherwise} \\ &b = Ax, \ \ A^\top y \leq c, \end{aligned}$$

where we pretentiously label the multiplier vector by $x$. We see that $(\bar{x}, \bar{y})$ satisfies these conditions if and only if the pair also satisfies (5.13). In view of 2.45, this means that $\bar{x}$ is

[2] The problem of minimizing a function $f$ is unbounded if $\inf f = -\infty$ (cf. §4.A) and the problem of maximizing $f$ is unbounded if $\sup f = \infty$.

a minimizer of the primal problem with corresponding multiplier vector $\bar{y}$ if and only if $\bar{y}$ is a maximizer of the dual problem with corresponding multiplier vector $\bar{x}$. In such a case,

$$\langle c, \bar{x}\rangle = \sum\nolimits_{j=1}^{n} c_j \bar{x}_j = \sum\nolimits_{j=1}^{n} \langle A^j, \bar{y}\rangle \bar{x}_j = \sum\nolimits_{i=1}^{m} \langle A_i, \bar{x}\rangle \bar{y}_i = \langle b, \bar{y}\rangle,$$

where $A_i$ is the $i$th row of $A$. Thus, the assertion about equal objective function values holds. The symmetry between the primal and dual problems implies that one can apply the interior-point method (§2.G) to the dual problem and then recover a primal solution by simply recording the corresponding multipliers.

If the primal problem is unbounded, then the dual problem is infeasible by weak duality 5.25. Likewise, if the dual problem is unbounded, then the primal problem must be infeasible. If one of the problems is infeasible, then the other one is either infeasible or unbounded; it can't have a solution as that would have implied one for the first problem as well. A linear optimization problem either has a minimizer, is infeasible or is unbounded. The case in Figure 4.3(middle) can't occur; see [105, Corollary 11.16]. Consequently, we've exhausted the possibilities.

An example of when both the primal and the dual problems are infeasible is furnished by

$$\underset{x \in \mathbb{R}^4}{\text{minimize}} \ -x_1 - x_2 \ \text{ subject to } \ x_1 - x_2 - x_3 = 1, \ -x_1 + x_2 - x_4 = 1, \ x \geq 0$$

$$\underset{y \in \mathbb{R}^2}{\text{maximize}} \ y_1 + y_2 \ \text{ subject to } \ y_1 - y_2 \leq -1, \ -y_1 + y_2 \leq -1, \ -y_1 \leq 0, \ -y_2 \leq 0.$$

Thus, the present choice of Rockafellian doesn't generally satisfy the desired property that the resulting dual problem has a maximum value reasonably near the minimum value of the primal problem. In this case, the difference between the two values is infinity. The next section discusses "ideal" Rockafellians that produce a difference of 0.

We can examine the various equivalences in the saddle point theorem 5.40. For example, if $\bar{x}$ minimizes the primal problem, $\bar{y}$ maximizes the dual problem and we also have the same optimal values, then the Lagrangian $l$ from 5.26 has

$$l(\bar{x}, \bar{y}) = \langle c, \bar{x}\rangle - \langle \bar{y}, A\bar{x} - b\rangle = \langle c, \bar{x}\rangle$$

because $\bar{x}$ is feasible in the primal problem. This confirms the right-most equivalence in 5.40. The other facts can be illustrated similarly. □

In a solution strategy for minimizing a function $f_0 : \mathbb{R}^n \to \overline{\mathbb{R}}$ based on constructing a Rockafellian, determining the Lagrangian $l$, maximizing the resulting dual objective function $\psi$ to obtain $\bar{y}$ and, finally, minimizing $l(\cdot, \bar{y})$ to gain $\bar{x}$, we always have that

$$\varphi(\bar{x}) \geq \psi(\bar{y}) = \inf l(\cdot, \bar{y}) = l(\bar{x}, \bar{y})$$

by weak duality 5.25, where $\varphi$ is the primal objective function. Thus, if $\varphi(\bar{x})$ turns out to be equal to $l(\bar{x}, \bar{y})$, then for $x \in \mathbb{R}^n$ one has

$$\varphi(x) = \sup l(x, \cdot) \geq l(x, \bar{y}) \geq l(\bar{x}, \bar{y}) = \varphi(\bar{x})$$

and $\bar{x}$ must be a minimizer of the primal problem; see also the saddle point theorem 5.40. In turn, this means that $\bar{x}$ is a minimizer of $f_0$ as well by 5.24, which we assume applies. Regardless of whether this works out perfectly, we obtain an upper bound on the optimality gap for $\bar{x}$ by the expression $f_0(\bar{x}) - l(\bar{x}, \bar{y})$; see (5.5). Since the dual problem and the Lagrangian problem may both be simpler than the actual one, this could be a viable strategy.

In the more specific case of 5.28, one has

$$\bar{x} \in \operatorname{argmin} l(\cdot, \bar{y}) \text{ and } \bar{y} \in \partial h\big(F(\bar{x})\big) \implies \bar{x} \in \operatorname{argmin} \iota_X + f_0 + h \circ F.$$

This follows by an application of 5.37:

$$\begin{aligned}\iota_X(\bar{x}) + f_0(\bar{x}) + h\big(F(\bar{x})\big) &= \iota_X(\bar{x}) + f_0(\bar{x}) + \big\langle F(\bar{x}), \bar{y}\big\rangle - h^*(\bar{y}) \\ &= \inf l(\cdot, \bar{y}) \leq \inf_x \sup_y l(x, y) = \inf \iota_X + f_0 + h \circ F.\end{aligned}$$

Thus, after minimizing $l(\cdot, \bar{y})$ and obtaining $\bar{x}$, if it turns out that $\bar{y} \in \partial h(F(\bar{x}))$, then $\bar{x}$ solves the actual problem.

**Example 5.42** (two-person zero-sum game; cont.). Let's return to the game in 5.38 and view it from the vantage point of the saddle point theorem 5.40. With $A_i, i = 1, \dots, m$, being the rows and $A^j, j = 1, \dots, n$, being the columns of the pay-off matrix, a solution of the game is obtained by having

$$\begin{aligned}&\text{Minnie solve} \quad \underset{x \in X}{\text{minimize}} \ \max_{i=1,\dots,m} \langle A_i, x\rangle \\ &\text{Maximilian solve} \quad \underset{y \in Y}{\text{maximize}} \ \min_{j=1,\dots,n} \langle A^j, y\rangle.\end{aligned}$$

Since both $X$ and $Y$ are compact, it follows by 4.9 that there are minimizers for these problems and there's a solution of the game regardless of the pay-off matrix.

**Detail.** Let $\varphi(x) = \sup_{y \in \mathbb{R}^m} l(x, y)$ and $\psi(y) = \inf_{x \in \mathbb{R}^n} l(x, y)$, with

$$l(x, y) = \langle y, Ax\rangle + \iota_X(x) - \iota_Y(y).$$

As argued in 5.38, Minnie's problem corresponds to the primal problem of minimizing $\varphi$. A parallel argument to the one carried out there establishes that Maximilian's problem is equivalent to maximizing $\psi$. Thus, Minnie and Maximilian solve the primal and dual problems in the saddle point theorem 5.40. The optimality conditions $0 \in \partial_x l(\bar{x}, \bar{y})$ and $0 \in \partial_y(-l)(\bar{x}, \bar{y})$, which specialize to $-A^\top \bar{y} \in N_X(\bar{x})$ and $A\bar{x} \in N_Y(\bar{y})$, respectively, then characterize a solution to the game. This can be used to verify the solution of the penny-nickel-dime game in 5.39. □

## 5.F Strong Duality

Fundamental to a dual problem is its lower bounding property that we refer to as weak duality; cf. 5.25. However, the practical usefulness of the property diminishes if the resulting bound is much below the minimum value of the primal problem. For linear optimization problems, with a particular Rockafellian, the bound is tight by 5.41: The maximum value of the dual problem reaches the whole way up to the minimum value of the primal problem when the latter isn't infinite.

Given a Lagrangian $l : \mathbb{R}^n \times \mathbb{R}^m \to \overline{\mathbb{R}}$, the primal-dual pair

$$\Big\{ \underset{x\in\mathbb{R}^n}{\text{minimize}}\ \varphi(x), \quad \underset{y\in\mathbb{R}^m}{\text{maximize}}\ \psi(y) \Big\},$$

with $\varphi(x) = \sup_{y\in\mathbb{R}^m} l(x, y)$ and $\psi(y) = \inf_{x\in\mathbb{R}^n} l(x, y)$, possesses *strong duality* when

$$\inf \varphi = \sup \psi.$$

In addition to the linear optimization setting, this best-case scenario takes place when the Lagrangian has a saddle point; see the saddle point theorem 5.40. However, it isn't automatic and we may have a *duality gap* given by

$$\inf \varphi - \sup \psi.$$

**Example 5.43** (failure of strong duality). For the problem of minimizing $x^3$ subject to $x \ge 0$ and the Rockafellian given by

$$f(u, x) = x^3 + \iota_{(-\infty,0]}(-x + u),$$

we obtain a Lagrangian of the form

$$l(x, y) = \begin{cases} x^3 - xy & \text{if } y \ge 0 \\ -\infty & \text{otherwise;} \end{cases}$$

see 5.18. The dual objective function $\psi(y) = -\infty$ for all $y \in \mathbb{R}$, while the minimum value of the corresponding primal problem is 0 so the duality gap is $\infty$.

**Detail.** In this case, the Lagrangian isn't convex in its first argument. However, strong duality may fail even under convexity. Consider the problem

$$\underset{x\in\mathbb{R}^2}{\text{minimize}}\ e^{-x_1} \text{ subject to } g(x) \le 0, \text{ where } g(x) = \begin{cases} x_1^2/x_2 & \text{if } x_2 > 0 \\ \infty & \text{otherwise,} \end{cases}$$

and a Rockafellian of the form

$$f(u, x) = \begin{cases} e^{-x_1} & \text{if } g(x) + u \le 0 \\ \infty & \text{otherwise.} \end{cases}$$

Similar to 5.18, this produces a Lagrangian with

$$l(x,y) = \begin{cases} e^{-x_1} + yg(x) & \text{if } x \in \operatorname{dom} g, y \geq 0 \\ \infty & \text{if } x \notin \operatorname{dom} g \\ -\infty & \text{otherwise.} \end{cases}$$

Consequently, the dual objective function $\psi(y) = 0$ if $y \geq 0$, but $\psi(y) = -\infty$ otherwise. The maximum value of the dual problem is therefore 0. The actual problem coincides with the primal problem of minimizing $\varphi(x) = \sup_y l(x,y)$, which has minimum value of 1. Thus, the duality gap equals 1 even though $l(\cdot, y)$ is convex regardless of $y \in \mathbb{R}$. □

Despite these discouraging examples, there's a large class of problems beyond linear optimization problems for which strong duality holds as we see next.

For the problem of minimizing $f_0 : \mathbb{R}^n \to \overline{\mathbb{R}}$ and a proper Rockafellian $f : \mathbb{R}^m \times \mathbb{R}^n \to \overline{\mathbb{R}}$ with anchor at 0, suppose that $f(\cdot, x)$ is lsc and convex for all $x \in \mathbb{R}^n$. We recall from 5.24 that in terms of the corresponding Lagrangian $l : \mathbb{R}^n \times \mathbb{R}^m \to \overline{\mathbb{R}}$, given by (5.6), the primal objective function has

$$\varphi(x) = \sup_{y \in \mathbb{R}^m} l(x,y) = f(0,x) = f_0(x).$$

Let $p(u) = \inf f(u, \cdot)$ so then $p(0) = \inf f_0 = \inf \varphi$. In contrast, by the definition of conjugates in 5.21,

$$\begin{aligned}
(p^*)^*(u) &= \sup_y \{\langle u,y\rangle - p^*(y)\} = \sup_y \Big\{\langle u,y\rangle - \sup_{\bar{u}} \{\langle \bar{u},y\rangle - p(\bar{u})\}\Big\} \\
&= \sup_y \Big\{\langle u,y\rangle + \inf_{\bar{u}} \{p(\bar{u}) - \langle \bar{u},y\rangle\}\Big\} \qquad (5.14) \\
&= \sup_y \Big\{\langle u,y\rangle + \inf_{x,\bar{u}} \{f(\bar{u},x) - \langle \bar{u},y\rangle\}\Big\} \\
&= \sup_y \{\langle u,y\rangle + \inf_x l(x,y)\} = \sup_y \{\langle u,y\rangle + \psi(y)\}
\end{aligned}$$

so that $(p^*)^*(0) = \sup \psi$. From this perspective, the question of strong duality boils down to whether $p(0) = (p^*)^*(0)$. We know from the Fenchel-Moreau theorem 5.23 that $p(0) \geq (p^*)^*(0)$ so we immediately obtain an alternative proof of weak duality 5.25. But, 5.23 also specifies a sufficient condition for strong duality: $p$ is proper, lsc and convex, which then implies that $p(u) = (p^*)^*(u)$ for all $u$ including 0.

Let's see how these requirements translate when the Rockafellian $f$ is proper and lsc. The stability theorem 5.6(a) ensures that $p$ is lsc if a tightness condition holds. In particular, $p$ is lsc if $f(u,x)$ is level-bounded in $x$ locally uniformly in $u$; see 5.4 for details. The same level-bounded condition ensures that $p$ is proper as well. By 1.21, $p$ is convex when $f$ is convex.

In the composite setting with smooth convex $f_0 : \mathbb{R}^n \to \mathbb{R}$, affine $F : \mathbb{R}^n \to \mathbb{R}^m$, nonempty, closed and convex $X \subset \mathbb{R}^n$ and proper, lsc and convex $h : \mathbb{R}^m \to \overline{\mathbb{R}}$, the Rockafellian given by

$$f(u,x) = \iota_X(x) + f_0(x) + h\big(F(x) + u\big)$$

is proper, lsc and convex; see 1.18. The assumption on $F$ can be relaxed to having each component function being convex when $h = \iota_{(-\infty,0]^m}$. In either case, $p$ is then convex by 1.21. One can build on this via some tightness assumption to confirm that $p$ is also proper and lsc. The following result summarizes key insights, with proof and more details available in [105, Theorem 11.39].

**Theorem 5.44** (strong duality). *For a proper, lsc and convex function $f : \mathbb{R}^m \times \mathbb{R}^n \to \overline{\mathbb{R}}$ with corresponding Lagrangian $l$ given by (5.6), the primal and dual problems*

$$\underset{x\in\mathbb{R}^n}{\text{minimize}}\ \varphi(x) = \sup_{y\in\mathbb{R}^m} l(x,y) \qquad \underset{y\in\mathbb{R}^m}{\text{maximize}}\ \psi(y) = \inf_{x\in\mathbb{R}^n} l(x,y)$$

*satisfy strong duality, i.e.,*

$$\inf \varphi = \sup \psi,$$

*provided that $0 \in \operatorname{int}(\operatorname{dom} p)$, where $p$ is the min-value function given by $p(u) = \inf f(u,\cdot)$.*
*If in addition $p(0) > -\infty$, then*

$$\partial p(0) = \operatorname{argmax} \psi,$$

*which must be a nonempty and bounded set.*

In 5.14, $\operatorname{dom} p = (-\infty, 1]$ so we certainly have $0 \in \operatorname{int}(\operatorname{dom} p)$ and strong duality holds; see Figure 5.5. Moreover, $p(0) = 5$ and $\partial p(0) = \{2\}$, which imply that the dual problem has 2 as its unique maximizer, with 5 as maximum value. We can determine all of this based on 5.44 without having a detailed formula for the dual problem. The formula for $\partial p(0)$ supplements 5.13, but most significantly it highlights the profound role played by a dual problem. Under the conditions of the theorem, solving the dual problem furnishes both the minimum value of the actual problem as well as its sensitivity to perturbations as defined by a Rockafellian.

**Exercise 5.45** (primal-dual pairs). For $f_0(x) = 2x_1^2 + x_2^2 + 3x_3^2$ and $g(x) = -2x_1 - 3x_2 - x_3 + 1$, consider the problem of minimizing $f_0(x)$ subject to $g(x) \leq 0$ and the Rockafellian defined by $f(u,x) = f_0(x) + \iota_{(-\infty,0]}(g(x) + u)$. Determine the corresponding Lagrangian as well as the primal and dual problems. Solve the problems and verify strong duality. Check the assumptions in the strong duality theorem 5.44.

**Guide.** The actual problem fits the setting of 5.28, with the conjugate following from 5.29, and this produces a Lagrangian $l$. Obtain a minimizer of $l(\cdot, y)$ for each $y$ analytically using the optimality condition 2.19 and derive a formula for the dual objective function. To apply 5.44, one can show that the feasible set remains nonempty under small changes to the right-hand side. □

The requirement $0 \in \operatorname{int}(\operatorname{dom} p)$ in the strong duality theorem 5.44 is assured when $f(u,\cdot)$ approximates $f(0,\cdot)$ in a certain sense for $u$ near 0. This is formalized in the next statement, which highlights the role of suitable approximations to ensure strong duality.

**Proposition 5.46** (strong duality from epigraphical inner limits). *For a proper, lsc and convex function $f : \mathbb{R}^m \times \mathbb{R}^n \to \overline{\mathbb{R}}$, suppose that*

$$\operatorname{LimInn}\big(\operatorname{epi} f(u^\nu,\cdot)\big) \supset \operatorname{epi} f(0,\cdot) \ \textit{ whenever } \ u^\nu \to 0.$$

*Let the primal and dual objective functions $\varphi$ and $\psi$ be constructed from $f$ as in 5.44 and $p$ be the min-value function given by $p(u) = \inf f(u, \cdot)$.*

*If $0 \in \operatorname{dom} p$, then strong duality holds, i.e., $\inf \varphi = \sup \psi$.*

**Proof.** Our goal is to show that the assumption implies $0 \in \operatorname{int}(\operatorname{dom} p)$ because then the conclusion follows from 5.44. For the sake of contradiction, suppose that $0 \notin \operatorname{int}(\operatorname{dom} p)$. Then, there's $\bar{u}^\nu \to 0$ such that $p(\bar{u}^\nu) = \infty$. Let $u^\nu \to 0$. We see from the proof of the characterization 4.15 of epi-convergence that

$$\operatorname{LimInn}\big(\operatorname{epi} f(u^\nu, \cdot)\big) \supset \operatorname{epi} f(0, \cdot)$$

is equivalent to 4.15(b), i.e., for every $x$, there's $x^\nu \to x$ such that $\limsup f(u^\nu, x^\nu) \leq f(0, x)$. Thus, the assumptions of 5.6(b) hold and $p$ is usc at 0. Since $p(\bar{u}^\nu) = \infty$, this implies that $p(0) = \infty$, which contradicts $0 \in \operatorname{dom} p$. □

The conclusion of the proposition isn't surprising. The strong duality theorem 5.44 requires that the min-value function doesn't jump to infinity as the perturbation vector $u$ departs from 0 and this can be avoided when the family of functions $\{f(u, \cdot), u \in \mathbb{R}^m\}$ defining the perturbations are epigraphically well behaved. After all, epi-convergence is the key ingredient for the minima of functions to converge to the minimum value of a limiting function. This insight provides guidance for selecting Rockafellians in the first place: the epigraphical behavior together with convexity are keys to achieving strong duality.

In a more specific setting, strong duality is guaranteed by ensuring that the constraint functions leave some "slack," which often is easily verified.

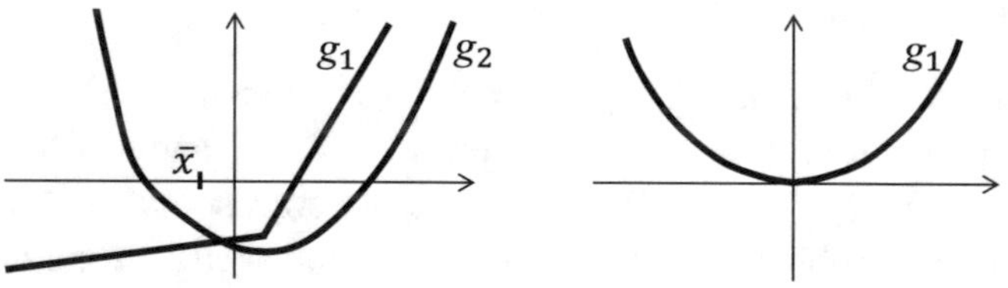


Fig. 5.10: Examples of when the Slater condition holds (left) and fails (right).

**Example 5.47** (Slater constraint qualification). For smooth convex functions $f_0, g_i : \mathbb{R}^n \to \mathbb{R}$, $i = 1, \dots, q$ and the problem

$$\underset{x \in \mathbb{R}^n}{\text{minimize}}\ f_0(x) \ \text{ subject to } \ g_i(x) \leq 0,\ i = 1, \dots, q,$$

let's consider the Rockafellian given by

$$f(u, x) = f_0(x) + \iota_{(-\infty,0]^q}\big(G(x) + u\big), \ \text{ with } G(x) = \big(g_1(x), \dots, g_q(x)\big).$$

The resulting primal and dual problems (as defined via the Lagrangian given by 5.18) satisfy strong duality provided that the following *Slater constraint qualification* holds:

$$\exists \bar{x} \text{ such that } g_i(\bar{x}) < 0, \;\; i = 1, \ldots, q.$$

Figure 5.10(left) illustrates a situation, with two constraints, that satisfies the qualification. If a constraint function doesn't reach below 0 as would be the case with $g_1(x) = x^2$, then the qualification fails; see Figure 5.10(right).

**Detail.** Let $p(u) = \inf f(u, \cdot)$. Under the Slater constraint qualification, there exist $\bar{x} \in \mathbb{R}^n$ and $\delta > 0$ such that $g_i(\bar{x}) + u_i \leq 0$ when $|u_i| \leq \delta$ for all $i$. Consequently, $p(u) \leq f_0(\bar{x}) \in \mathbb{R}$ when $\|u\|_\infty \leq \delta$, which means that $0 \in \operatorname{int}(\operatorname{dom} p)$ and strong duality holds by 5.44.

The Slater constraint qualification implies the Mangasarian-Fromovitz constraint qualification in (4.10) at every feasible point for this class of problems. To see this, let $\hat{x}$ be feasible, i.e., $g_i(\hat{x}) \leq 0$ for $i = 1, \ldots, q$, and let the multipliers $\{z_i, i \in \mathbb{A}(\hat{x})\}$ satisfy

$$\sum\nolimits_{i \in \mathbb{A}(\hat{x})} z_i \nabla g_i(\hat{x}) = 0 \quad \text{and} \quad z_i \geq 0 \;\; \forall i \in \mathbb{A}(\hat{x}),$$

where $\mathbb{A}(\hat{x}) = \{i \mid g_i(\hat{x}) = 0\}$. Our goal is to show that $z_i = 0$ for $i \in \mathbb{A}(\hat{x})$. Let $\bar{x}$ be a point that verifies the Slater constraint qualification. By the gradient inequality 1.22, one has

$$0 > g_i(\bar{x}) - g_i(\hat{x}) \geq \langle \nabla g_i(\hat{x}), \bar{x} - \hat{x} \rangle$$

for $i \in \mathbb{A}(\hat{x})$. Then,

$$0 = \Big\langle \sum\nolimits_{i \in \mathbb{A}(\hat{x})} z_i \nabla g_i(\hat{x}), \; \bar{x} - \hat{x} \Big\rangle = \sum\nolimits_{i \in \mathbb{A}(\hat{x})} z_i \langle \nabla g_i(\hat{x}), \bar{x} - \hat{x} \rangle.$$

Since the inner products on the right-hand side evaluate to negative numbers and $z_i \geq 0$, we must have $z_i = 0$ for all $i \in \mathbb{A}(\hat{x})$. □

**Example 5.48** (strong duality without Slater constraint qualification). In the setting of 5.47, the Slater constraint qualification is sufficient for strong duality but by no means necessary. For example, consider the problem of minimizing $x$ subject to $x^2 \leq 0$. The Slater constraint qualification fails, but strong duality holds.

**Detail.** The setting involves the Rockafellian given by

$$f(u, x) = x + \iota_{(-\infty, 0]}(x^2 + u).$$

We obtain directly that

$$p(u) = \inf f(u, \cdot) = \begin{cases} -\sqrt{-u} & \text{for } u \leq 0 \\ \infty & \text{otherwise} \end{cases}$$

so the requirement $0 \in \operatorname{int}(\operatorname{dom} p)$ of the strong duality theorem 5.44 doesn't hold. The corresponding Lagrangian has

$$l(x, y) = \begin{cases} x + yx^2 & \text{for } y \geq 0 \\ -\infty & \text{otherwise} \end{cases}$$

by 5.18 and the dual objective function has

$$\psi(y) = \begin{cases} -1/(4y) & \text{for } y > 0 \\ -\infty & \text{otherwise.} \end{cases}$$

Consequently, $p(0) = \sup \psi = 0$. □

An approach to settle border cases of the kind described in this example is to again return to an epigraphical analysis of thc underlying Rockafellian.

**Theorem 5.49** (strong duality from epi-convergence). *For the problem of minimizing $f_0 : \mathbb{R}^n \to \overline{\mathbb{R}}$ and a proper Rockafellian $f : \mathbb{R}^m \times \mathbb{R}^n \to \overline{\mathbb{R}}$ with anchor at 0, suppose that there are $u^\nu \to 0$ and $y^\nu \in \mathbb{R}^m$ such that*

(a) $f(u^\nu, \cdot) \xrightarrow{e} f(0, \cdot)$ *tightly*
(b) $\operatorname{liminf} \langle y^\nu, u^\nu \rangle \le 0$
(c) $\inf f(u^\nu, \cdot) = \sup \psi^\nu = \psi^\nu(y^\nu)$,

*where*

$$\psi^\nu(y) = \inf_x l^\nu(x, y) \;\; \text{and} \;\; l^\nu(x, y) = \inf_u \big\{ f(u^\nu + u, x) - \langle y, u \rangle \big\}.$$

*Let $\psi$ be the dual objective function produced by $f$ via (5.6). Then,*

$$\inf f_0 = \sup \psi > -\infty.$$

**Proof.** Let $p$ be the min-value function defined by $p(u) = \inf f(u, \cdot)$. Since $f$ is proper, $f(0, x) > -\infty$ for all $x \in \mathbb{R}^n$. Thus, $f(0, \cdot)$ is proper when $0 \in \operatorname{dom} p$ and we can apply 5.5(d) to establish that $p(u^\nu) \to p(0) > -\infty$. When $0 \notin \operatorname{dom} p$, we can modify the arguments in the proof of 5.5 as follows.

For any compact set $B \subset \mathbb{R}^n$, $\inf_B f(u^\nu, \cdot) \to \infty$. To see this, let's assume for the sake of contradiction, that there are $N \in \mathcal{N}_\infty^\#$ and $\alpha \in \mathbb{R}$ such that $\inf_B f(u^\nu, \cdot) \le \alpha$ for $\nu \in N$. Since $B$ is compact, this implies the existence of $\{x^\nu \in B, \nu \in N\}$ with $f(u^\nu, x^\nu) \le \alpha + 1$, a further subsequence $N' \subset N$ and a limit $\bar{x}$ of $\{x^\nu, \nu \in N'\}$. By the (partial) characterization of epi-convergence in 4.15(a),

$$\operatorname{liminf}_{\nu \in N'} f(u^\nu, x^\nu) \ge f(0, \bar{x}) = \infty$$

because $\inf f(0, \cdot) = \infty$. This contradicts that $f(u^\nu, x^\nu) \le \alpha + 1$ for $\nu \in N$.

Next, let $\varepsilon > 0$. By the tightness assumption, there's a compact set $B_\varepsilon$ such that

$$\inf f(u^\nu, \cdot) + \varepsilon \ge \inf_{B_\varepsilon} f(u^\nu, \cdot)$$

for sufficiently large $\nu$; see 5.3. Since the right-hand side tends to infinity as just established, $p(u^\nu) \to \infty$. Thus, $p(u^\nu) \to p(0)$ even when $0 \notin \operatorname{dom} p$.

Let $l$ be the Lagrangian of $f$ defined by (5.6). Note that

$$l^\nu(x, y) = \inf_w \big\{ f(w, x) - \langle y, w - u^\nu \rangle \big\} = l(x, y) + \langle y, u^\nu \rangle.$$

Thus, $\psi^\nu(y) = \psi(y) + \langle y, u^\nu \rangle$. This fact together with assumption (c) result in

$$p(u^\nu) = \sup \psi^\nu = \psi^\nu(y^\nu) = \psi(y^\nu) + \langle y^\nu, u^\nu \rangle \le \sup \psi + \langle y^\nu, u^\nu \rangle.$$

Consequently,

$$p(0) = \liminf p(u^\nu) \le \liminf \big( \sup \psi + \langle y^\nu, u^\nu \rangle \big) \le \sup \psi.$$

By also invoking the lower bound (5.5) and the fact that $f_0 = f(0, \cdot)$, the conclusion follows. □

The theorem reveals the following insight: If we can construct a Rockafellian that leads to a sequence of perturbed functions epi-converging tightly to the actual objective function and these perturbed functions on their own are associated with a strong duality property, then the resulting dual problem indeed reproduces the minimum value of the actual problem provided that assumption (b) also holds.

Since $u^\nu \to 0$, assumption (b) certainly holds when $\{y^\nu, \nu \in \mathbb{N}\}$ is bounded. One can view $\psi^\nu$ as a dual objective function produced by the Rockafellian of the form $f^\nu(u, x) = f(u^\nu + u, x)$. The vector $y^\nu$ then solves the dual problem associated with $f^\nu$. Thus, it's plausible that $\{y^\nu, \nu \in \mathbb{N}\}$ could be bounded.

In some cases, assumption (b) is automatic even when $\{y^\nu, \nu \in \mathbb{N}\}$ is unbounded. For example, if

$$f(u, x) = \hat{f}_0(x) + \iota_{(-\infty,0]^m}\big(F(x) + u\big)$$

for $\hat{f}_0 : \mathbb{R}^n \to \mathbb{R}$ and $F : \mathbb{R}^n \to \mathbb{R}^m$, both smooth, then

$$\psi^\nu(y) = \inf_{x \in \mathbb{R}^n} \hat{f}_0(x) + \big\langle F(x) + u^\nu, y \big\rangle - \iota_{[0,\infty)^m}(y)$$

by 5.28 and 5.29. Thus, a maximizer $y^\nu$ of $\psi^\nu$ is necessarily nonnegative. We can then choose $u^\nu \le 0$ so that $\langle y^\nu, u^\nu \rangle \le 0$ and $f(u^\nu, \cdot) \xrightarrow{e} f(0, \cdot)$, which can be seen by working directly from the characterization 4.15. (Tightness must be checked separately, but holds for instance when $\hat{f}_0(x) \to \infty$ as $\|x\|_2 \to \infty$.)

The theorem isn't restricted to any particular type of Rockafellian and may even go beyond the setting of 5.24. Still, in the convex case, several aspects simplify.

**Corollary 5.50** *For the problem of minimizing* $f_0 : \mathbb{R}^n \to \overline{\mathbb{R}}$ *and a proper, lsc and convex Rockafellian* $f : \mathbb{R}^m \times \mathbb{R}^n \to \overline{\mathbb{R}}$ *with anchor at* 0, *suppose that there's* $u^\nu \in \operatorname{int}(\operatorname{dom} p) \to 0$ *such that* $f(u^\nu, \cdot) \xrightarrow{e} f(0, \cdot)$ *tightly, where p is the min-value function given by* $p(u) = \inf f(u, \cdot)$. *Let* $\psi$ *be the dual objective function produced by f via (5.6).*

*If* $\inf f_0 < \infty$, *then*

$$\inf f_0 = \sup \psi$$

*and this value is finite.*

**Proof.** Since $0 \in \operatorname{dom} p$ because $\inf f_0 < \infty$, the initial argument in the proof of 5.49 establishes that $p(u^\nu) \to p(0) \in \mathbb{R}$ and then $p(u^\nu) > -\infty$ for sufficiently large $\nu$. Thus, the strong duality theorem 5.44 applies to the Rockafellian $f^\nu : \mathbb{R}^m \times \mathbb{R}^n \to \overline{\mathbb{R}}$ defined by

$$f^\nu(u,x) = f(u^\nu + u, x),$$

with a corresponding min-value function in the form

$$p^\nu(u) = \inf f^\nu(u,\cdot\,) = p(u^\nu + u).$$

In particular, we note that $0 \in \operatorname{int}(\operatorname{dom} p^\nu)$ because $u^\nu \in \operatorname{int}(\operatorname{dom} p)$ and, for sufficiently large $\nu$, $p^\nu(0) > -\infty$ because $p(u^\nu) > -\infty$. From these facts as well as 5.24, we realize that $p^\nu(0) = \sup \psi^\nu$ and that there's

$$y^\nu \in \operatorname{argmax} \psi^\nu = \partial p^\nu(0) = \partial p(u^\nu);$$

see 5.49 for the definition of $\psi^\nu$. Consequently, $p(u^\nu) = \sup \psi^\nu = \psi^\nu(y^\nu)$. The min-value function $p$ is convex by 1.21 so the subgradient inequality 2.17 establishes that

$$p(2u^\nu) - p(u^\nu) \geq \langle y^\nu, u^\nu \rangle.$$

Both terms on the left-hand side tend to the same real number. Thus, $\limsup \langle y^\nu, u^\nu \rangle \leq 0$. All the assumptions of 5.49 then hold and the conclusion follows. □

**Example 5.51** (strong duality without Slater constraint qualification; cont.). The corollary confirms the strong duality in 5.48 even though the Slater constraint qualification fails.

**Detail.** In this case, the Rockafellian, given by $f(u,x) = x + \iota_{(-\infty,0]}(x^2 + u)$, is proper, lsc and convex. Moreover, $p(u) = -\sqrt{-u}$ for $u \leq 0$ and $p(u) = \infty$ otherwise. Thus, one can take $u^\nu = -1/\nu$ in the corollary and then $f(u^\nu,\cdot\,) \xrightarrow{e} f(0,\cdot\,)$; the epi-convergence is actually tight since $\operatorname{dom} f(u^\nu,\cdot\,) \subset \operatorname{dom} f(u^1,\cdot\,)$. □

## 5.G Reformulations

Strong duality presents numerous possibilities for reformulations. Even for difficult problems without easily obtainable strong duality, there might be parts of the formulation that can be reworked using duality. We'll illustrate the possibilities with two examples involving an infinite number of constraints.

**Example 5.52** (linear constraints under uncertainty). For $f_0 : \mathbb{R}^n \to \mathbb{R}$, $X \subset \mathbb{R}^n$ and $\alpha \in \mathbb{R}$, consider the problem

$$\operatorname*{minimize}_{x \in X} f_0(x) \ \text{ subject to } \ \langle a, x \rangle \leq \alpha$$

and suppose that there's uncertainty about the vector $a$. As in Chap. 3, we could have viewed $a$ as a random vector with a probability distribution. However, when we lack data from which to build a distribution or when we seek guaranteed feasibility for a range of values of $a$, a reasonable alternative is to adopt the following simple model of uncertainty. Let $\bar{a} \in \mathbb{R}^n$ be a nominal value of $a$, $s \in \mathbb{R}^n$ be a nonnegative scaling vector and

$$\{a \in \mathbb{R}^n \mid a_j = \bar{a}_j + s_j\xi_j \ \forall \xi \in \Xi, \ j = 1, \dots, n\}$$

be the set of considered values of $a$, where

$$\Xi = \Big\{\xi \in \mathbb{R}^n \ \Big|\ \sum_{j=1}^n |\xi_j| \le \beta, \ |\xi_j| \le 1, \ j = 1, \dots, n\Big\}$$

for some $\beta \ge 0$. Thus, any value of $a_j$ between $\bar{a}_j - s_j$ and $\bar{a}_j + s_j$ might be considered, but this is further restricted by the size of $\beta$. In particular, if $\beta < n$, then not all components of $a$ can be at their extreme values.

We guarantee a solution satisfying $\langle a, x\rangle \le \alpha$ for every considered value of $a$ by imposing the constraints

$$\langle a(\xi), x\rangle \le \alpha \ \ \forall \xi \in \Xi, \quad \text{where } a(\xi) = (\bar{a}_1 + s_1\xi_1, \dots, \bar{a}_n + s_n\xi_n).$$

The apparent implementation challenge associated with the now infinite number of constraints can be overcome using strong duality.

**Detail.** The infinite collection of constraints is equivalent to the single constraint

$$\sup_{\xi\in\Xi} \langle a(\xi), x\rangle \le \alpha,$$

where the left-hand side can be expressed by

$$\sup_{\xi\in\Xi} \langle a(\xi), x\rangle = \sum_{j=1}^n \bar{a}_j x_j + \sup_{\xi\in\Xi} \sum_{j=1}^n s_j x_j \xi_j$$
$$= \langle \bar{a}, x\rangle + \sup\Big\{\sum_{j=1}^n s_j|x_j|\xi_j \ \Big|\ \sum_{j=1}^n \xi_j \le \beta, \ 0 \le \xi_j \le 1, \ j = 1, \dots, n\Big\}.$$

The change to only nonnegative $\xi_j$ is permitted due to the fact that $s_j \ge 0$. Let's fix $x \in \mathbb{R}^n$. The last supremum is the maximum value of a linear problem; a maximizer exists by 4.9. Moreover, by strong duality from 5.41 this can be computed just as well by solving the corresponding minimization problem, which is referred to as the primal problem in 5.41. Specifically, the supremum is achieved by

$$\underset{\xi\in\mathbb{R}^n}{\text{maximize}} \ \langle b, \xi\rangle \ \text{ subject to } \ A^\top\xi \le c, \tag{5.15}$$

where

$$b = \big(s_1|x_1|, \dots, s_n|x_n|\big), \qquad c = (\beta, 1, \dots, 1, 0, \dots, 0) \in \mathbb{R}^{1+2n}$$

and $A$ is an $n \times (1 + 2n)$-matrix with ones in the first column, the $n \times n$ identity matrix occupies the next $n$ columns and the negative of the $n \times n$ identity matrix fills out the rest. By 5.41, the primal problem is then

$$\underset{\eta\in\mathbb{R}^{1+2n}}{\text{minimize}} \ \langle c, \eta\rangle \ \text{ subject to } \ A\eta = b, \ \eta \ge 0.$$

In more detail, this becomes

$$\underset{\eta \geq 0}{\text{minimize}}\ \beta\eta_0 + \sum_{j=1}^{n} \eta_j \ \text{ subject to } \ \eta_0 + \eta_j - \eta_{n+j} = s_j|x_j|, \ \ j = 1, \dots, n.$$

We can eliminate $\eta_{n+1}, \dots, \eta_{2n}$ and obtain the equivalent formulation

$$\underset{\eta_0,\dots,\eta_n \geq 0}{\text{minimize}}\ \beta\eta_0 + \sum_{j=1}^{n} \eta_j \ \text{ subject to } \ \eta_0 + \eta_j \geq s_j|x_j|, \ \ j = 1, \dots, n.$$

Since the minimum value here matches the corresponding maximum value in (5.15) by 5.41, we obtain that

$$\begin{aligned}&\sup_{\xi\in\Xi} \langle a(\xi), x\rangle \\ &= \langle \bar{a}, x\rangle + \inf\Big\{\beta\eta_0 + \sum_{j=1}^{n} \eta_j \ \Big|\ \eta_0 \geq 0,\ \eta_0 + \eta_j \geq s_j|x_j|,\ \eta_j \geq 0,\ j = 1, \dots, n\Big\}.\end{aligned}$$

Thus, finding $x$ such that $\sup_{\xi\in\Xi}\langle a(\xi), x\rangle \leq \alpha$ is equivalent to finding $x$ and $\eta_0, \dots, \eta_n$ such that

$$\langle \bar{a}, x\rangle + \beta\eta_0 + \sum_{j=1}^{n} \eta_j \leq \alpha, \ \ \eta_0 \geq 0, \ \ \eta_0 + \eta_j \geq s_j|x_j|, \ \ \eta_j \geq 0, \ \ j = 1, \dots, n.$$

The only remaining nonlinear parts are the $|x_j|$-terms, but these can be reformulated by doubling the number of constraints because $s_j \geq 0$. The problem of minimizing $f_0(x)$ subject to $\langle a(\xi), x\rangle \leq \alpha$ for all $\xi \in \Xi$ becomes then

$$\begin{aligned}\underset{x\in\mathbb{R}^n,\eta\in\mathbb{R}^{1+n}}{\text{minimize}}\ f_0(x) \ \text{ subject to } \ &\langle \bar{a}, x\rangle + \beta\eta_0 + \sum_{j=1}^{n} \eta_j \leq \alpha \\ &\eta_0 + \eta_j - s_j x_j \geq 0, \ \ j = 1, \dots, n \\ &\eta_0 + \eta_j + s_j x_j \geq 0, \ \ j = 1, \dots, n \\ &x \in X, \ \ \eta_0, \eta_j \geq 0, \ \ j = 1, \dots, n.\end{aligned}$$

The infinite collection of constraints in $n$ variables is now expressed by $2 + 3n$ constraints using $1 + 2n$ variables. The reformulation remains valid for arbitrary $f_0$ and $X$ as they didn't enter the derivation. □

**Example 5.53** (maximum-flow interdiction). In a directed graph $(V, E)$ with vertex set $V$ and edge set $E$, there are two special vertices $s \neq t$ identified: $s$ is the source vertex and $t$ is the sink vertex. A network operator would like to maximize the flow of a commodity from $s$ to $t$ across the edges which have finite capacities. In the *maximum-flow network-interdiction problem*, an interdictor wishes to minimize that maximum flow by interdicting (destroying) edges. Using strong duality, the problem can be formulated as a reasonably tractable minimization problem.

**Detail.** Each edge $e \in E$ has a capacity $u_e > 0$ and interdiction cost $r_e > 0$; see Figure 5.11. The interdiction cost is the amount of some resource necessary to destroy edge $e$, i.e., reduce that edge's capacity to 0. We also define the outgoing and incoming edges from vertex $i$ as

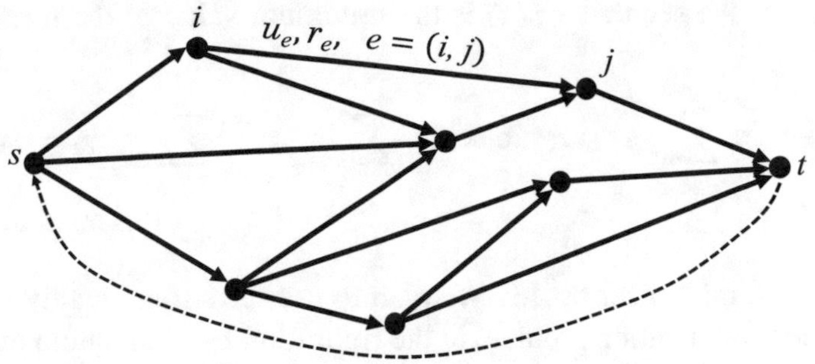


Fig. 5.11: Directed graph for the maximum-flow network-interdiction problem.

$$E_i^+ = \{(i, j) \in E \mid j \in V\} \quad \text{and} \quad E_i^- = \{(j, i) \in E \mid j \in V\},$$

respectively. Let $x_e = 1$ if edge $e$ is interdicted, and let $x_e = 0$ otherwise. The variable $y_e$ denotes the flow on edge $e \in E$ and $y_a$ the flow on an artificial "return edge" $a = (t, s) \notin E$; see Figure 5.11. Let $n$ be the number of edges in $E$ and $y = (y_e, e \in E)$. Then, for a given interdiction plan $x$, the maximum flow through the network is

$$f(x) = \sup_{y_a, y} \Big\{ y_a \,\Big|\, \sum_{e \in E_i^+} y_e - \sum_{e \in E_i^-} y_e = \delta_i y_a \ \forall i \in V, \\ 0 \le y_e \le u_e(1 - x_e) \ \forall e \in E \Big\},$$

with $\delta_s = 1$, $\delta_t = -1$ and $\delta_i = 0$ for all $i \in V \setminus \{s, t\}$; cf. [1, Chapters 6-7]. The equality constraints ensure flow balance at each vertex. The last set of inequalities forces the effective capacity of an edge to 0 if $x_e = 1$ and otherwise leaves it at the nominal $u_e$. Equivalently,

$$f(x) = \sup_{y_a, y} \Big\{ y_a - \sum_{e \in E} x_e y_e \,\Big|\, \sum_{e \in E_i^+} y_e - \sum_{e \in E_i^-} y_e = \delta_i y_a \ \forall i \in V, \\ 0 \le y_e \le u_e \ \forall e \in E \Big\}.$$

The *maximum-flow network-interdiction problem* then takes the form

$$\underset{x \in \mathbb{R}^n}{\text{minimize}}\ f(x) \ \text{ subject to } \sum_{e \in E} r_e x_e \le \alpha, \ \ x \in \{0, 1\}^n,$$

where $\alpha$ is the amount of interdiction resource available. In contrast to earlier examples with an objective function given as the maximum over a *finite* number of functions, we here face an uncountable number. Since minimizing $f$ is equivalent to minimizing a scalar $\tau$ subject to the constraint $f(x) \le \tau$, the maximum-flow network-interdiction problem can be viewed as resulting in an infinite number of constraints. However, using strong duality we can reformulate the problem in terms of a finite number of linear equalities and inequalities.

Let's fix $x \geq 0$. We see that $-f(x)$ is the minimum value of the linear optimization problem

$$\underset{y_a, y_1, \ldots, y_n}{\text{minimize}} \; -y_a + \sum\nolimits_{e \in E} x_e y_e \;\text{ subject to }\; \sum\nolimits_{e \in E_i^+} y_e - \sum\nolimits_{e \in E_i^-} y_e = \delta_i y_a \;\; \forall i \in V$$
$$0 \leq y_e \leq u_e \;\; \forall e \in E,$$

which always has a minimizer by 4.9. We plan to use the strong duality in 5.41 and just need to place the minimization problem in the right form by eliminating upper bounds on the variables. This leads to the reformulation

$$\underset{y_a, y_1, \ldots, y_{2n}}{\text{minimize}} \; -y_a + \sum\nolimits_{e \in E} x_e y_e \;\text{ subject to }\; \sum\nolimits_{e \in E_i^+} y_e - \sum\nolimits_{e \in E_i^-} y_e = \delta_i y_a \;\; \forall i \in V$$
$$y_a \geq 0, \;\; y_e + y_{n+e} = u_e, \;\; y_e \geq 0, \;\; y_{n+e} \geq 0 \;\; \forall e \in E,$$

where the addition of $y_a \geq 0$ doesn't change the minimum value. Let's redefine $y$ by setting $y = (y_a, y_1, \ldots, y_{2n})$. The reformulation is of the form minimizing $\langle c, y \rangle$ subject to $Ay = b$ and $y \geq 0$, where $c = (-1, x_1, \ldots, x_n, 0, \ldots, 0)$, $b = (0, \ldots, 0, u_1, \ldots, u_n) \in \mathbb{R}^{m+n}$, with $m$ being the number of vertices, and

$$A = \begin{bmatrix} -\delta_1 & a_{11} & \cdots & a_{1n} & 0 \cdots 0 \\ -\delta_2 & a_{21} & \cdots & a_{2n} & 0 \cdots 0 \\ \vdots & \vdots & \ddots & \vdots & \vdots \ddots \vdots \\ -\delta_m & a_{m1} & \cdots & a_{mn} & 0 \cdots 0 \\ 0 & 1 & \cdots & 0 & 1 \cdots 0 \\ \vdots & \vdots & \ddots & \vdots & \vdots \ddots \vdots \\ 0 & 0 & \cdots & 1 & 0 \cdots 1 \end{bmatrix}$$

is an $(m + n) \times (1 + 2n)$-matrix with $a_{ie} = 1$ if $e \in E_i^+$, $a_{ie} = -1$ if $e \in E_i^-$ and 0 otherwise. By 5.41, the corresponding dual problem is maximizing $\langle b, (v, w) \rangle$ subject to $A^\top(v, w) \leq c$, where $v \in \mathbb{R}^m$ and $w \in \mathbb{R}^n$. Explicitly, this amounts to

$$\underset{v, w}{\text{maximize}} \sum\nolimits_{e \in E} u_e w_e \;\text{ subject to }\; v_i - v_j + w_e \leq x_e \;\; \forall (i, j) = e \in E$$
$$v_t - v_s \leq -1, \;\; w_e \leq 0 \;\; \forall e \in E.$$

Since the maximum value here is $-f(x)$ by 5.41, $f(x)$ coincides with the minimum value of $\sum_{e \in E} -u_e w_e$ under the same constraints. The overall problem can then be formulated as that of minimizing $\sum_{e \in E} -u_e w_e$ over $x, v, w$ subject to these constraints as well as the original constraints on $x$. We can also replace "$\leq -1$" by "$= -1$" because of the structure of the problem and this eliminates redundance. Putting all together, the maximum-flow network-interdiction problem can equivalently be stated as

$$\underset{x,v,w}{\text{minimize}} \sum_{e\in E} -u_e w_e \text{ subject to } v_i - v_j + w_e \le x_e \;\; \forall (i,j) = e \in E$$
$$v_t - v_s = -1$$
$$\sum_{e\in E} r_e x_e \le \alpha$$
$$w_e \le 0, \;\; x_e \in \{0,1\} \;\; \forall e \in E.$$

Thus, an objective function involving a supremum over an uncountable number of possibilities can be reformulated into a linear optimization problem with integer constraints. As mentioned in 4.3, there are well-developed algorithms for such problems that rely on branch-and-bound techniques; see also [114] for refinements and [20] for related defense applications.

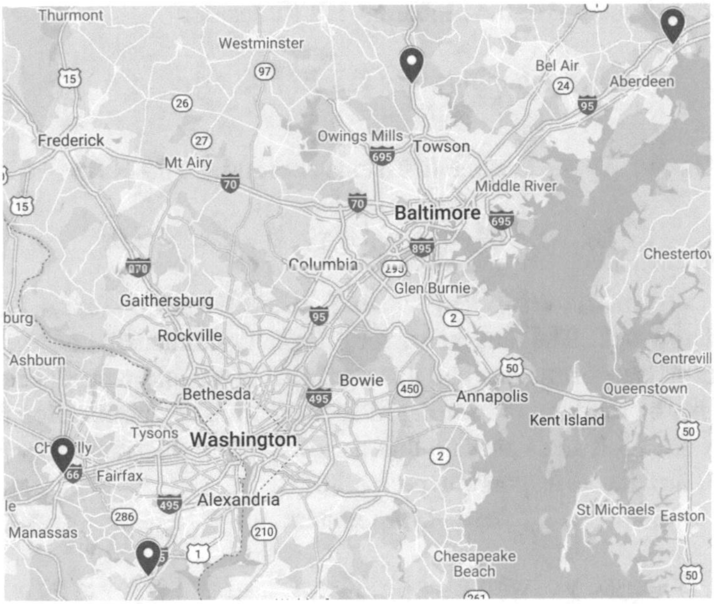


Fig. 5.12: Optimal points (marked) of interdiction in a maximum-flow network-interdiction problem involving 3672 vertices and 10031 edges.

As a concrete instance, let's consider the evacuation of Washington, D.C., after a terrorist attack and the amount of traffic that can "flow" out of Washington. A robust plan needs to consider the effect of destroyed roadways and bridges. The maximum-flow network-interdiction problem informs decision-makers about evacuation routes as well as points of vulnerability. A dataset involving all roads with speed limits of 30 miles per hour or higher in Maryland, Virginia and Washington, D.C., gives a network of vertices and edges with 152 vertices in the Baltimore-Washington region, to be aggregated into a source vertex $s$, representing points of departure. About the periphery 35 vertices representing evacuation centers are aggregated into a sink vertex $t$. The resulting network has 3672 vertices and 10031 edges. If $\alpha = 4$ so that only four edges can be interdicted, we obtain a minimizer of the maximum-flow network-interdiction problem visualized in Figure 5.12; four marked locations have been interdicted and the flow out of Washington to the evacuation centers is reduced from a nominal capacity of 194 to 146 units. The network is moderately resilient: the reduction from 194 to 146 is significant but not catastrophic. Moreover, the points of interdiction could be proposed as locations where hardening of the infrastructure might be especially beneficial; cf. [114] for further information. □

## 5.H L-Shaped Method

Stochastic optimization problems with linear recourse have large-scale formulations that may exceed our computational capacity; see §3.J and especially (3.22). However, duality theory can be brought in for the purpose of constructing tractable approximations that leverage the linear structure of the problems. Let's consider

$$\operatorname*{minimize}_{x \in C_1} \ f_0(x) + \mathbb{E}\big[f(\boldsymbol{\xi}, x)\big],$$

where $C_1 \subset \mathbb{R}^{n_1}$, $f_0 : \mathbb{R}^{n_1} \to \mathbb{R}$ is the first-stage cost and $f : \mathbb{R}^q \times \mathbb{R}^{n_1} \to \overline{\mathbb{R}}$ has

$$f(\xi, x) = \inf_y \big\{\langle a, y\rangle \,\big|\, Wy = d - Tx, \ \ y \geq 0\big\} \tag{5.16}$$

for $a \in \mathbb{R}^{n_2}$, $m_2 \times n_2$-matrix $W$, $m_2 \times n_1$-matrix $T$, $d \in \mathbb{R}^{m_2}$ and $\xi = (a, T_1, \dots, T_{m_2}, d)$, with $T_i$ being the $i$th row of $T$. The random vector $\boldsymbol{\xi}$ is then of dimension $q = n_2 + (1 + n_1)m_2$. We assume that $W$ is deterministic and this is critical in the following development. Since $a$ may or may not be random, our setting includes stochastic optimization problems with fixed recourse; see §3.J. Let's assume that the probability distribution of $\boldsymbol{\xi}$ is finite and $\{p_\xi > 0, \xi \in \Xi\}$ are the corresponding probabilities.

For fixed $\xi = (a, T_1, \dots, T_{m_2}, d) \in \mathbb{R}^q$, the function

$$(x, y) \mapsto \langle a, y\rangle + \iota_C(x, y), \ \ \text{with } C = \big\{(x, y) \in \mathbb{R}^{n_1} \times \mathbb{R}^{n_2} \,\big|\, Wy - d + Tx = 0, y \geq 0\big\}$$

is epi-polyhedral because it's affine on its domain, which is a polyhedral set. Then,

$$x \mapsto f(\xi, x) = \inf_y \langle a, y\rangle + \iota_C(x, y)$$

is also epi-polyhedral by 2.66 and thus convex. As seen from 2.67, this means that the expectation function $Ef$ is epi-polyhedral and convex.

The structure offers an opportunity for approximation: A subgradient of $f(\xi, \cdot)$ at a point $x^k$ where the function is finite leads to a lower bound on $f(\xi, \cdot)$ through the subgradient inequality 2.17. Specifically, let $w_\xi^k \in \partial_x f(\xi, x^k)$ and then

$$f(\xi, x) \geq f(\xi, x^k) + \langle w_\xi^k, x - x^k\rangle \ \ \forall x \in \mathbb{R}^{n_1}.$$

After repeating this for each $\xi \in \Xi$ and summing both sides, we obtain the lower bound

$$Ef(x) \geq \sum_{\xi\in\Xi} p_\xi f(\xi, x^k) + \left\langle \sum_{\xi\in\Xi} p_\xi w_\xi^k,\ x - x^k \right\rangle\ \forall x \in \mathbb{R}^{n_1}.$$

If the process is carried out at the points $x^1, \ldots, x^\nu$ and the lower bound is taken as the maximum value across the bounds, then we obtain an approximating problem

$$\underset{x\in C_1, \alpha\in\mathbb{R}}{\text{minimize}}\ f_0(x) + \alpha$$

$$\text{subject to } \sum_{\xi\in\Xi} p_\xi f(\xi, x^k) + \left\langle \sum_{\xi\in\Xi} p_\xi w_\xi^k, x - x^k \right\rangle \leq \alpha, \quad k = 1, \ldots, \nu.$$

In contrast to alternative solution approaches such as those based on (3.22), the size of this problem doesn't depend on $\Xi$. The cardinality of $\Xi$ enters only in the preprocessing of the data.

An algorithm emerges from this development: For an initial guess $x^1$, compute subgradients

$$\{w_\xi^1 \in \partial_x f(\xi, x^1),\ \xi \in \Xi\},$$

solve the approximating problem with $\nu = 1$ and hopefully obtain a minimizer $x^2$ of that problem. At $x^2$, compute subgradients

$$\{w_\xi^2 \in \partial_x f(\xi, x^2),\ \xi \in \Xi\},$$

solve the approximating problem with $\nu = 2$, obtain $x^3$ and so forth. This is the idea behind the *L-shaped method*. When $C_1$ is polyhedral and $f_0$ is linear, the approach requires solving only a sequence of linear optimization problems. Beyond such cases, the approximating problem may remain tractable (especially compared to the actual problem) even though it grows in size by one constraint per iteration.

**Proposition 5.54** (properties of recourse function). *For $\xi = (a, T_1, \ldots, T_{m_2}, d) \in \mathbb{R}^q$ and $m_2 \times n_2$-matrix $W$, let $f(\xi, \cdot)$ be given by (5.16). Then, $f(\xi, \cdot)$ is convex and*

$$f(\xi, x) \geq \sup_{v\in V} \langle d - Tx, v\rangle\ \ \forall x \in \mathbb{R}^{n_1}, \ \text{where } V = \{v \in \mathbb{R}^{m_2} \mid W^\top v \leq a\}.$$

*At a point $\bar{x}$ where $f(\xi, \cdot)$ is finite, one has*

$$f(\xi, \bar{x}) = \sup_{v\in V} \langle d - T\bar{x}, v\rangle$$
$$-T^\top \bar{v} \in \partial_x f(\xi, \bar{x}) \quad \forall \bar{v} \in \operatorname{argmax}_{v\in V} \langle d - T\bar{x}, v\rangle,$$

*with this set of maximizers being nonempty.*

**Proof.** The claim about convexity summarizes the discussion earlier in the section. The minimization problem defining $f(\xi, x)$ has maximizing $\langle d - Tx, v\rangle$ over $v \in V$ as a dual problem by 5.41, which also establishes the claimed inequality.

Since $f(\xi, \bar{x})$ is finite, one has

$$f(\xi, \bar{x}) = \sup_{v\in V} \langle d - T\bar{x}, v\rangle \quad \text{and} \quad \operatorname{argmax}_{v\in V} \langle d - T\bar{x}, v\rangle \neq \emptyset$$

by 5.41. This means that $V$ is nonempty and the function $h : \mathbb{R}^{m_2} \to \overline{\mathbb{R}}$ given by $h(u) = \sup_{v \in V} \langle u, v \rangle$ is proper. It's convex by 1.18(a) and lsc because epi $h$ is an intersection of closed sets. Then, by 5.37 and 2.19,

$$\bar{v} \in \partial h(\bar{u}) \iff \bar{u} \in \partial h^*(\bar{v}) \iff \bar{v} \in \operatorname{argmin}\{h^* - \langle \bar{u}, \cdot \rangle\}.$$

From 5.29, we see that $h^*(v) = \iota_V(v)$. Thus, for $\bar{u} \in \operatorname{dom} h$,

$$\partial h(\bar{u}) = \operatorname{argmin}\{h^* - \langle \bar{u}, \cdot \rangle\} = \operatorname{argmax}_{v \in V} \langle \bar{u}, v \rangle.$$

Let $\bar{w} \in -T^\top \bar{v}$, with $\bar{v} \in \operatorname{argmax}_{v \in V} \langle d - T\bar{x}, v \rangle$. Then, $\bar{v} \in \partial h(d - T\bar{x})$. Consequently, by the subgradient inequality 2.17 as applied to $h$, one has

$$\begin{aligned} f(\xi, x) \geq h(d - Tx) &\geq h(d - T\bar{x}) + \big\langle \bar{v}, d - Tx - (d - T\bar{x}) \big\rangle \\ &= f(\xi, \bar{x}) + \langle -T^\top \bar{v}, x - \bar{x} \rangle \;\; \forall x \in \mathbb{R}^{n_1}. \end{aligned}$$

This amounts to a subgradient inequality for $f(\xi, \cdot)$ and the claim holds by 2.17. □

While the proposition furnishes subgradients at points where $f(\xi, \cdot)$ is finite, the approximating problem may lead to other points too or even fail to produce a solution at all. An implementable algorithm needs to address these concerns. We make an assumption that helps us weed out some pathological cases:

$$f(\xi, x) > -\infty \quad \forall \xi \in \Xi, \; x \in C_1. \tag{5.17}$$

This implies that the second-stage problem isn't unbounded, which is reasonable because otherwise $\mathbb{E}[f(\boldsymbol{\xi}, \bar{x})] = -\infty$ for some $\bar{x} \in C_1$ and then the actual problem is unbounded. Still, we permit $f(\xi, \bar{x}) = \infty$, which corresponds to induced constraints; see §3.J. We may discover after the uncertainty is resolved that $\bar{x} \in C_1$ is infeasible due to the lack of an admissible second-stage decision $y$. Sometimes the second-stage model is formulated such that this obviously can't occur. However, in general, we need to check for such infeasibility and, in fact, the L-shaped method gradually learns the induced constraints and this is useful in many ways.

At a current point $x^\nu \in C_1$, the key quantity to examine is the feasible set

$$V(a) = \{v \in \mathbb{R}^{m_2} \mid W^\top v \leq a\}$$

in the dual problem

$$\underset{v \in V(a)}{\text{maximize}} \; \langle d - Tx^\nu, v \rangle \tag{5.18}$$

corresponding to the minimization problem behind $f(\xi, x^\nu)$, with $\xi = (a, T_1, \dots, T_{m_2}, d)$; see 5.54 and also 5.41.

**Infeasible dual problem.** If $V(a) = \emptyset$ for some $\xi = (a, T_1, \dots, T_{m_2}, d) \in \Xi$, then (5.18) is infeasible and $f(\xi, x) = \infty$ or $f(\xi, x) = -\infty$ by 5.41 for any $x \in \mathbb{R}^{n_1}$ because $V(a)$ is independent of $x$. Since the possibility $f(\xi, x) = -\infty$ for $x \in C_1$ violates (5.17), we must then have $f(\xi, x) = \infty$ for all $x \in C_1$ and the actual problem is infeasible due to induced constraints.

**Feasible dual problem.** If $V(a) \neq \emptyset$ for $\xi = (a, T_1, \ldots, T_{m_2}, d) \in \Xi$, then $f(\xi, x^\nu) = \infty$ when (5.18) is unbounded (cf. 5.41) and otherwise $f(\xi, x^\nu) \in \mathbb{R}$. The dual problem is unbounded for $\xi$ if there's a direction $\bar{z} \in \mathbb{R}^{m_2}$ and $\bar{v} \in V(a)$ such that

$$W^\top(\bar{v} + \lambda\bar{z}) \leq a \;\; \forall \lambda \in [0, \infty) \quad \text{and} \quad \langle d - Tx^\nu, \bar{z}\rangle > 0$$

because then the objective function value of the dual problem keeps on increasing as we move along the line from $\bar{v}$ in the direction of $\bar{z}$. Since the length of a direction doesn't matter and the feasibility condition reduces to $W^\top \bar{z} \leq 0$, we can check for unboundedness by solving for each $\xi = (a, T_1, \ldots, T_{m_2}, d) \in \Xi$:

$$\underset{z \in \mathbb{R}^{m_2}}{\text{maximize}} \; \langle d - Tx^\nu, z\rangle \text{ subject to } W^\top z \leq 0, \;\; z \in [-1, 1]^{m_2}. \tag{5.19}$$

If the maximum value here is 0 for all $\xi \in \Xi$, then the dual problems (5.18) across all $\xi \in \Xi$ aren't unbounded and $f(\xi, x^\nu) \in \mathbb{R}$ for all $\xi \in \Xi$. If there's $\xi \in \Xi$ producing a positive maximum value, then the corresponding (5.18) is unbounded, $f(\xi, x^\nu) = \infty$ and $Ef(x^\nu) = \infty$; the current point $x^\nu$ is infeasible due to an induced constraint. Let $z_\xi^\nu$ be a direction that attains such a positive maximum value. We can eliminate $x^\nu$ from further consideration as well as other $x$ that allows unbounded growth in (5.18) along $z_\xi^\nu$ by imposing the constraint

$$\langle d - Tx, z_\xi^\nu\rangle \leq 0 \quad \text{or, equivalently,} \quad \langle T^\top z_\xi^\nu, x\rangle \geq \langle d, z_\xi^\nu\rangle.$$

These refinements ensure a well-defined algorithm.

**L-Shaped Method.**

Data. Tolerance $\varepsilon \geq 0$.
Step 0. Set $\nu = \tau = \sigma = 0$.
Step 1. Replace $\nu$ by $\nu + 1$.
If $\sigma = 0$, then solve the *master problem*

$$\underset{x \in C_1}{\text{minimize}} \; f_0(x) \text{ subject to } \langle B_k, x\rangle \geq \beta_k, \quad k = 1, \ldots, \tau.$$

If this master problem doesn't have a minimizer, then stop.
Compute a minimizer $x^\nu$ of this master problem and set $\alpha^\nu = -\infty$.
Else, solve the *master problem*

$$\underset{x \in C_1, \alpha \in \mathbb{R}}{\text{minimize}} \; f_0(x) + \alpha \text{ subject to } \langle B_k, x\rangle \geq \beta_k, \quad k = 1, \ldots, \tau$$
$$\langle D_k, x\rangle + \alpha \geq \delta_k, \quad k = 1, \ldots, \sigma.$$

If this master problem doesn't have a minimizer, then stop.
Compute a minimizer $(x^\nu, \alpha^\nu)$ of this master problem.

Step 2. Solve (5.19) for $\xi = (a, T_1, \ldots, T_{m_2}, d) \in \Xi$ and obtain maximizers $\{z_\xi^\nu, \xi \in \Xi\}$.
If the corresponding maximum values are all 0, then go to Step 3.
For some $\xi = (a, T_1, \ldots, T_{m_2}, d)$ with a positive maximum value, set

$$B_{\tau+1} = T^\top z_\xi^\nu, \qquad \beta_{\tau+1} = \langle d, z_\xi^\nu \rangle,$$

replace $\tau$ by $\tau + 1$ and go to Step 1.

Step 3. Solve (5.18) for $\xi = (a, T_1, \ldots, T_{m_2}, d) \in \Xi$.
If no maximizers exist, then stop.
Let $\{v_\xi^\nu, \xi \in \Xi\}$ be maximizers and set

$$D_{\sigma+1} = \sum\nolimits_{\xi \in \Xi} p_\xi T_\xi^\top v_\xi^\nu, \qquad \delta_{\sigma+1} = \sum\nolimits_{\xi \in \Xi} p_\xi \langle d_\xi, v_\xi^\nu \rangle,$$

where $T_\xi$ and $d_\xi$ are the $T$-matrix and $d$-vector corresponding to $\xi$.
If

$$\delta_{\sigma+1} - \langle D_{\sigma+1}, x^\nu \rangle - \alpha^\nu \le \varepsilon,$$

then stop.
Replace $\sigma$ by $\sigma + 1$ and go to Step 1.

The master problems correspond to the approximating problem discussed above, with $\sigma$ constraints of the kind described there and also $\tau$ constraints deriving from induced constraints. If the algorithm stops in Step 1 because a master problem is infeasible, then the actual problem is infeasible because the master problems are relaxations of the actual problem. The algorithm may also stop in Step 1 because a master problem is unbounded or fails to have a minimizer for some other reason (cf. for example 4.9), but these situations are usually avoided by imposing assumptions on $C_1$ and $f_0$.

Step 2 checks for unboundedness in the dual problems as described around (5.19). If we discover unboundedness for some $\xi$, then we return to Step 1 with an additional constraint. Otherwise, we proceed to Step 3 with the confidence that (5.18) isn't unbounded for any $\xi \in \Xi$.

If we encounter $\xi \in \Xi$ without a maximizer in Step 3, then (5.18) is infeasible for this $\xi$ because unboundedness has been ruled out in the previous step. Since the feasible set in (5.18) is independent of $x^\nu$, this is then the case for any $x \in C_1$ and the actual problem is infeasible as discussed above. If every $\xi \in \Xi$ returns a maximizer, then we leverage 5.54 to compute subgradients of $f(\xi, \cdot)$ at $x^\nu$ and construct a constraint of the form in the approximating problem, which is then added to the master problems.

A master problem furnishes a lower bound $f_0(x^\nu) + \alpha^\nu$ on the minimum value of the actual problem because it relies on a lower approximation of $f(\xi, \cdot)$. Step 3 provides an upper bound on that minimum value because $x^\nu$ is feasible in the actual problem whenever we get past the initial stopping condition. This upper bound is

$$f_0(x^\nu) + \sum\nolimits_{\xi \in \Xi} p_\xi f(\xi, x^\nu) = f_0(x^\nu) + \delta_{\sigma+1} - \langle D_{\sigma+1}, x^\nu \rangle.$$

Thus, the optimality gap is bounded by $\delta_{\sigma+1} - \langle D_{\sigma+1}, x^\nu \rangle - \alpha^\nu$ and the algorithm terminates if this is sufficiently small.

Compared to the expanded formulation (3.22), the L-shaped method considers many smaller problems repeatedly: linear optimization problems in Steps 2 and 3 and then the slowly growing master problems. An important advantage is that the upper and lower bounds provide an estimate of the optimality gap that may allow us to terminate after just a few iterations. The need for obtaining (global) minimizers of the master problems is a bottleneck but can be relaxed to near-minimizers with only minor adjustments to the lower bound calculation. The linear optimization problems are usually quick to solve, but a large number of $\xi \in \Xi$ may require further refinements. For example, in Step 2, the only variation between the linear problems is in the coefficients of the objective function; the feasible sets remain the same and this can be utilized. We refer to [119] for such developments and for further theory in the case of linear $f_0$ and polyhedral $C_1$.

**Example 5.55** (L-shaped method for simple recourse). Let's consider the special case when $f(\xi, x)$ in (5.16) is defined by

$$f(\xi, x) = \inf_y \big\{ \langle a^+, y^+ \rangle + \langle a^-, y^- \rangle \,\big|\, Iy^+ - Iy^- = d - Tx,\ y^+ \geq 0,\ y^- \geq 0 \big\},$$

where $\xi = (T_1, \dots, T_{m_2}, d) \in \mathbb{R}^{m_2(n_1+1)}$ and $T_i$ is the $i$th row of $T$. The second-stage decision vector $y = (y^+, y^-) \in \mathbb{R}^{n_2}$ is broken into two equally long vectors. Thus, the identity matrix $I$ is of dimension $m_2 \times n_2/2$ so we necessarily have $m_2 = n_2/2$. In the mold of the earlier development, $a = (a^+, a^-)$ is deterministic and $W = (I, -I)$. This results in simple recourse by 3.39. However, let's reach the same conclusion using duality theory and also examine the implications for the L-shaped method.

**Detail.** For any $\xi = (T_1, \dots, T_{m_2}, d)$ and $x$, it's always possible to find a feasible $y$ so $f(\xi, x) < \infty$, i.e., there are no induces constraints. Moreover, (5.18) specializes to

$$\underset{v}{\text{maximize}}\ \langle d - Tx^\nu, v \rangle \ \text{subject to} \quad -a^- \leq v \leq a^+.$$

Via 5.41, $f(\xi, x^\nu) < \infty$ implies that the maximization problem can't be unbounded and Step 2 in the L-shaped method can be skipped altogether. The maximization problem could still be infeasible, but this is avoided if $-a^- \leq a^+$, which is easily checked and corresponds to the condition $\gamma_i \leq \delta_i$ imposed in 3.39. Under this assumption, $f(\xi, x)$ is always finite. In Step 3, a maximizer of the dual problem is analytically available. □

**Exercise 5.56** (implementation of L-shaped method). A grocer sells four types of fruits—nectarines, bananas, oranges and apples—that are obtained from two different suppliers. The shipments from Supplier 1 come in lots of 4 kg of nectarines, 9 kg of bananas, 3 kg of oranges and 10 kg of apples, while those from Supplier 2 consist of 1 kg of nectarines, 1 kg of bananas, 1.85 kg of oranges and 2.7 kg of apples. Supplier 1 can deliver up to 40 lots at \$5 per lot if the order is placed early enough and, if additional supplies are required, Supplier 1 can guarantee same-day delivery of another 40 lots at \$8 per lot. Supplier 2 can deliver up to 800 lots at \$7 per lot if the order is placed early enough, and can also guarantee same-day delivery of another 200 lots at \$15 per lot. The objective of the grocer is to satisfy the demand for these four types of fruits at a minimum expected cost. The uncertain (daily) demand is given by Table 5.1, with each of the $3^4 = 81$ possible values of the demand $\xi \in \mathbb{R}^4$ having equal probability of occurring. Formulate and solve the grocer's problem using the L-shaped method.

Table 5.1: Demand levels in the grocer's problem.

| | demand (kg) | | |
|---|---|---|---|
| nectarines | 1050 | 1150 | 1200 |
| bananas | 1000 | 1250 | 1500 |
| oranges | 1250 | 1750 | 2000 |
| apples | 2500 | 3000 | 3250 |

**Guide.** The first-stage variables are $x_1$ and $x_2$, with $x_i$ representing the number of lots ordered "early" from Supplier $i$. The second-stage variables $y_1$ and $y_2$ specify the number of same-day lots ordered from the two suppliers. Let's ignore any need for ordering only integer number of lots and allow these variables to be a nonnegative number. (Integer restrictions on the first-stage variables are easily incorporated into $C_1$ and only make the master problems a bit harder to solve.) □

## 5.I Monitoring Functions

A problem needs to be sufficiently structured to be tractable theoretically and computationally. A promising kind in this regard is

$$\underset{x \in X}{\text{minimize}}\ f_0(x) + h\big(F(x)\big) \tag{5.20}$$

for some closed $X \subset \mathbb{R}^n$, proper, lsc and convex $h : \mathbb{R}^m \to \overline{\mathbb{R}}$, smooth $f_0 : \mathbb{R}^n \to \mathbb{R}$ and smooth $F : \mathbb{R}^n \to \mathbb{R}^m$. However, a bottleneck might be aspects related to the function $h$. For example, the Rockafellian in 5.28 produces a Lagrangian with

$$l(x, y) = \iota_X(x) + f_0(x) + \big\langle F(x), y\big\rangle - h^*(y)$$

and we then need an expression for the conjugate of $h$. Let's examine choices of $h$ that result in explicit expressions for the corresponding conjugate functions and dual problems.

For a nonempty polyhedral set $Y \subset \mathbb{R}^m$ and a symmetric positive semidefinite $m \times m$-matrix $B$, we say that $h_{Y,B} : \mathbb{R}^m \to \overline{\mathbb{R}}$ is a *monitoring function* when

$$h_{Y,B}(u) = \sup_{y \in Y} \big\{\langle u, y\rangle - \tfrac{1}{2}\langle y, By\rangle\big\}.$$

Monitoring functions track quantities of interest and appropriately assign penalties to undesirable values. Setting $h = h_{Y,B}$ in (5.20) is often a viable modeling choice and turns out to be computationally tractable. Common situations are covered by specific choices of $Y$ and $B$:

$$h_{Y,B}(u) = \max_{i=1,\dots,m} u_i \quad \text{for } Y = \left\{ y \in \mathbb{R}^m \,\middle|\, \sum_{i=1}^m y_i = 1,\ y \geq 0 \right\} \text{ and } B = 0$$

$$h_{Y,B}(u) = \iota_D(u) \quad \text{for } D = \text{pol}\, Y,\ \text{polyhedral cone } Y \text{ and } B = 0.$$

The first choice can be used to penalize various values of $F(x)$, while the second one enforces the constraint $F(x) \in D$.

A monitoring function is convex by 1.18(a), lsc by virtue of having an epigraph that's the intersection of closed sets and also proper because $Y$ is nonempty and then $h_{Y,B}(u) > -\infty$ for all $u$ and

$$h_{Y,B}(0) = \sup_{y \in Y} -\tfrac{1}{2}\langle y, By \rangle \leq 0$$

by the positive semidefiniteness of $B$. Moreover,

$$h^*_{Y,B}(y) = \tfrac{1}{2}\langle y, By \rangle + \iota_Y(y)$$

as can be seen from the definition of a conjugate function and the Fenchel-Moreau theorem 5.23. Thus, under the choice of $h = h_{Y,B}$ in (5.20), the Lagrangian from 5.28 has

$$l(x, y) = \iota_X(x) + f_0(x) + \big\langle F(x), y \big\rangle - \tfrac{1}{2}\langle y, By \rangle - \iota_Y(y),$$

which is rather explicit since $Y$ is presumably a known polyhedral set and $X$ is usually also simple. Via the inverse rule 5.37 and the optimality condition 2.19, we obtain that

$$\partial h_{Y,B}(u) = \text{argmin}_{y \in Y}\, \tfrac{1}{2}\langle y, By \rangle - \langle u, y \rangle \ \ \forall u \in \text{dom}\, h_{Y,B}.$$

The horizon subgradients, important in the chain rule 4.64 and elsewhere, are given by (see [105, Example 11.18])

$$\partial^\infty h_{Y,B}(u) = \left\{ y \in Y^\infty \cap \text{null}\, B \,\middle|\, \langle y, u \rangle = 0 \right\} \ \ \forall u \in \text{dom}\, h_{Y,B},$$

where for an arbitrarily selected $\bar{y} \in Y$,

$$Y^\infty = \left\{ y \in \mathbb{R}^m \,\middle|\, \bar{y} + \lambda y \in Y \ \ \forall \lambda \in [0, \infty) \right\}.$$

For example, if $Y$ is bounded, then $Y^\infty = \{0\}$ and $\partial^\infty h_{Y,B}(u) = \{0\}$ for all $u \in \text{dom}\, h_{Y,B}$. Alternatively, if $B$ has full rank, then null $B = \{0\}$ and we reach the same conclusion.

**Example 5.57** (duality for quadratic problem with monitoring). For a nonempty polyhedral set $X \subset \mathbb{R}^n$, a symmetric positive semidefinite $n \times n$-matrix $Q$, an $m \times n$ matrix $A$, $c \in \mathbb{R}^n$ and $b \in \mathbb{R}^m$, consider the problem

$$\underset{x \in X}{\text{minimize}}\ \langle c, x \rangle + \tfrac{1}{2}\langle x, Qx \rangle + h_{Y,B}(b - Ax),$$

where $h_{Y,B}$ is a monitoring function, which necessarily means that $Y$ is nonempty and polyhedral and $B$ is symmetric and positive semidefinite. For a Rockafellian given by

$$f(u,x) = \iota_X(x) + \langle c, x\rangle + \tfrac{1}{2}\langle x, Qx\rangle + h_{Y,B}(b - Ax + u),$$

which recovers the actual problem as minimizing $f(0,\cdot)$, the dual problem is

$$\operatorname*{maximize}_{y\in Y} \ \langle b, y\rangle - \tfrac{1}{2}\langle y, By\rangle - h_{X,Q}(A^\top y - c).$$

If either the actual problem or the dual problem has a finite optimal value, then the other one has the same value and strong duality holds. Under such circumstances, there's a minimizer for the actual problem and a maximizer for the dual problem.

**Detail.** By 5.28, the Lagrangian in this case has

$$l(x,y) = \iota_X(x) + \langle c, x\rangle + \tfrac{1}{2}\langle x, Qx\rangle + \langle b - Ax, y\rangle - \tfrac{1}{2}\langle y, By\rangle - \iota_Y(y).$$

The minimization of the Lagrangian produces a dual objective function with

$$\begin{aligned}\psi(y) &= -\iota_Y(y) + \langle b, y\rangle - \tfrac{1}{2}\langle y, By\rangle - \sup_x \left\{\langle A^\top y - c, x\rangle - \iota_X(x) - \tfrac{1}{2}\langle x, Qx\rangle\right\}\\ &= -\iota_Y(y) + \langle b, y\rangle - \tfrac{1}{2}\langle y, By\rangle - h_{X,Q}(A^\top y - c).\end{aligned}$$

The claim about strong duality is supported by [105, Theorem 11.42, Example 11.43]. □

If $h$ in (5.20) is a monitoring function that penalizes certain values of $F$ and these penalties are assigned componentwise, then it suffices to consider a separable monitoring function, i.e.,

$$h_{Y,B}(u) = \sum_{i=1}^m h_{Y_i,\beta_i}(u_i) = \sum_{i=1}^m \sup_{y_i\in Y_i}\left\{u_i y_i - \tfrac{1}{2}\beta_i y_i^2\right\}$$

for $B$, a diagonal matrix with nonnegative elements $\beta_1,\dots,\beta_m$, and $Y = Y_1\times\cdots\times Y_m$, with $Y_i$ being a nonempty closed interval.

**Example 5.58** (one-dimensional monitoring function). If $\beta > 0$ and $Y = [\sigma,\tau]$ for $\sigma,\tau\in\overline{\mathbb{R}}$ with $\sigma\le\tau$, then we obtain a smooth monitoring function given by

$$h_{Y,\beta}(u) = \begin{cases}\sigma u - \tfrac{1}{2}\beta\sigma^2 & \text{if } u < \sigma\beta\\ \tfrac{1}{2\beta}u^2 & \text{if } \sigma\beta\le u\le\tau\beta\\ \tau u - \tfrac{1}{2}\beta\tau^2 & \text{if } u > \tau\beta.\end{cases}$$

Figure 5.13(left) illustrates the possibilities.

**Detail.** In particular, $h_{[0,\infty),\beta}(u) = 0$ if $u\le 0$ and $(2\beta)^{-1}u^2$ otherwise and thus penalizes positive values. For $\tau\in[0,\infty)$, a nonsmooth monitoring function is given by

$$h_{[0,\tau],0}(u) = \begin{cases}0 & \text{if } u\le 0\\ \tau u & \text{if } u > 0;\end{cases}$$

see Figure 5.13(right). An actual constraint is produced by $h_{[0,\infty),0}(u) = \iota_{(-\infty,0]}(u)$. □

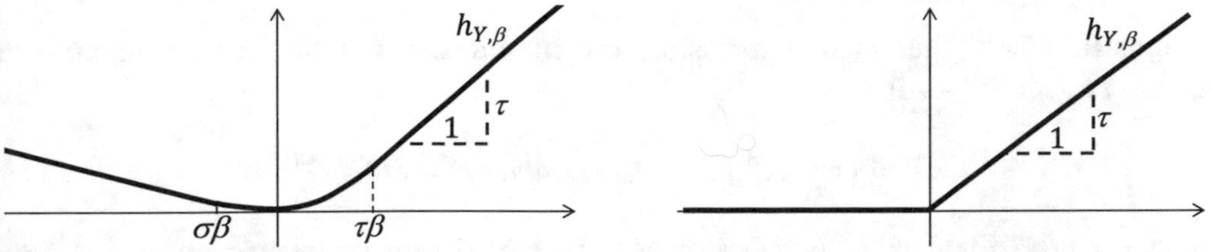


Fig. 5.13: Monitoring functions $h_{Y,\beta}$ in 5.58 with $\beta > 0$ and $Y = [\sigma, \tau] \subset \mathbb{R}$ (left); $\beta = 0$ and $Y = [0, \tau] \subset \mathbb{R}$ (right).

**Exercise 5.59** (duality with separable monitoring). For an $m_1 \times n$-matrix $A$, with rows $A_1, \dots, A_{m_1}$, an $m_2 \times n$-matrix $T$, with rows $T_1, \dots, T_{m_2}$, nonnegative scalars $q_1, \dots, q_n$, positive scalars $\beta_1, \dots, \beta_{m_2}$, $r_1, \dots, r_{m_2}$ as well as $s, c \in \mathbb{R}^n$, $b \in \mathbb{R}^{m_1}$ and $d \in \mathbb{R}^{m_2}$, consider the problem

$$\begin{aligned} &\underset{x \in \mathbb{R}^n}{\text{minimize}}\ \langle c, x\rangle + \tfrac{1}{2}\sum\nolimits_{j=1}^{n} q_j x_j^2 + \sum\nolimits_{i=1}^{m_2} h_{[0,r_i],\beta_i}\big(d_i - \langle T_i, x\rangle\big) \\ &\text{subject to}\ \ Ax \ge b, \quad 0 \le x \le s. \end{aligned}$$

Show that the dual problem corresponding to a Rockafellian of the kind supporting 5.57 can be written as

$$\begin{aligned} &\underset{v \in \mathbb{R}^{m_1}, w \in \mathbb{R}^{m_2}}{\text{maximize}}\ \langle b, v\rangle + \langle d, w\rangle - \tfrac{1}{2}\sum\nolimits_{i=1}^{m_2} \beta_i w_i^2 - \sum\nolimits_{j=1}^{n} h_{[0,s_j],q_j}\big(\langle A^j, v\rangle + \langle T^j, w\rangle - c_j\big) \\ &\text{subject to}\ \ v \ge 0, \ \ 0 \le w \le r, \end{aligned}$$

where $A^1, \dots, A^n$ and $T^1, \dots, T^n$ are the columns of $A$ and $T$, respectively.

**Guide.** Following 5.57, set $X = \{x \in \mathbb{R}^n \mid 0 \le x \le s\}$ and $Q$ to be the diagonal matrix with elements $q_1, \dots, q_n$. The effect of the monitoring functions and the constraint $Ax \ge b$ can be incorporated by setting

$$h_{Y,B}\big(\tilde{b} - \tilde{A}x\big), \ \text{ with } Y = [0,\infty)^{m_1} \times [0, r_1] \times \cdots \times [0, r_{m_2}],$$

$B$ being the diagonal $(m_1+m_2)\times(m_1+m_2)$-matrix with elements $0, \dots, 0, \beta_1, \dots, \beta_{m_2}$, $\tilde{b} = (b, d) \in \mathbb{R}^{m_1}\times\mathbb{R}^{m_2}$ and $\tilde{A}$ being the $(m_1+m_2)\times n$-matrix with rows $A_1, \dots, A_{m_1}, T_1, \dots, T_{m_2}$. By 5.57, the dual problem is

$$\underset{y \in Y}{\text{maximize}}\ \langle \tilde{b}, y\rangle - \tfrac{1}{2}\langle y, By\rangle - h_{X,Q}(\tilde{A}^\top y - c),$$

which can be simplified as indicated using $y = (v, w)$. □

There are many solution approaches for (5.20) when $h$ is a monitoring function, even if the dual problem may not be quite as explicit as in 5.57 and 5.59. One possibility that leverages existing algorithms for equality constraints is supported by the following fact.

**Proposition 5.60** (alternative expression for monitoring function). *If a monitoring function* $h_{Y,B} : \mathbb{R}^m \to \overline{\mathbb{R}}$ *has*

$$Y = \{y \in \mathbb{R}^m \mid A^\top y \leq b\} \quad \textit{and} \quad B = DJ^{-1}D^\top$$

*for some* $m \times q$*-matrix* $A$, $b \in \mathbb{R}^q$, $m \times m$*-matrix* $D$ *and symmetric positive definite* $m \times m$*-matrix* $J$, *then*

$$h_{Y,B}(u) = \inf_{v,w} \left\{ \langle b, v\rangle + \tfrac{1}{2}\langle w, Jw\rangle \;\middle|\; Av + Dw = u,\ v \geq 0 \right\} \ \forall u \in \mathbb{R}^m.$$

**Proof.** Fix $u \in \mathbb{R}^m$. Let's view the minimization problem in the asserted formula for $h_{Y,B}$ as an instance of the problem examined in 5.28 with $u - Av - Dw$ and $[0,\infty)^q \times \mathbb{R}^m$ playing the roles of $F(x)$ and $X$, respectively. Under the Rockafellian adopted there, we obtain a Lagrangian of the form

$$l(v, w, y) = \iota_{[0,\infty)^q \times \mathbb{R}^m}(v, w) + \langle b, v\rangle + \tfrac{1}{2}\langle w, Jw\rangle + \langle u - Av - Dw, y\rangle.$$

We note that the term appearing at the end of the Lagrangian in 5.28 vanishes in this case because it represents the conjugate of $\iota_{\{0\}^m}$; see 5.29. The corresponding dual objective function then has

$$\psi(y) = \inf_{v,w} l(v, w, y) = \langle u, y\rangle + \inf_{v \geq 0, w} \left\{ \langle v, b - A^\top y\rangle + \tfrac{1}{2}\langle w, Jw\rangle - \langle w, D^\top y\rangle \right\}.$$

For $y \in Y$, the minimization in $v$ brings the term $\langle v, b - A^\top y\rangle$ to 0, while minimization in $w$ is achieved at $J^{-1}D^\top y$ using the optimality condition 2.19. For $y \notin Y$, $\psi(y) = -\infty$. This results in

$$\psi(y) = -\iota_Y(y) + \langle u, y\rangle - \tfrac{1}{2}\langle D^\top y, J^{-1}D^\top y\rangle.$$

Thus, the optimal dual value

$$\sup_{y \in Y} \left\{ \langle u, y\rangle - \tfrac{1}{2}\langle y, DJ^{-1}D^\top y\rangle \right\} = h_{Y,B}(u).$$

Since the minimization problem in the asserted formula for $h_{Y,B}$ and the maximization of $\psi$ fit the setting of 5.57, which in fact we could have used to obtain the expression for the dual problem, strong duality holds when the dual optimal value is finite. Since it defines a monitoring function, $Y$ is nonempty. Thus, the dual problem is always feasible and its maximum value is either finite or the problem is unbounded. In the latter case, by weak duality 5.25, the minimization problem is infeasible and the asserted formula holds even in that case. □

The required representation of $B$ isn't a significant limitation. If $B = 0$, we can select $D = 0$ and $J = I$, the identity matrix. If $B$ is positive definite, then spectral decomposition gives $D$ and $J$. Specifically, there are an $m \times m$-matrix $D$ (consisting of orthonormal eigenvectors) and an $m \times m$-matrix $\Lambda$, with the eigenvalues of $B$ along its diagonal and 0 elsewhere, such that $B = D\Lambda D^\top$; see [22, Theorem 4.2]. Since $B$ is symmetric and positive definite, its eigenvalues are positive and we can take $J$ to be $\Lambda^{-1}$.

**Example 5.61** (reformulation of problems with monitoring). For $f_0 : \mathbb{R}^n \to \mathbb{R}$, $F : \mathbb{R}^n \to \mathbb{R}^m$ and $X \subset \mathbb{R}^n$, consider the problem

$$\underset{x \in X}{\text{minimize}}\ f_0(x) + h_{Y,B}\big(F(x)\big),$$

with $Y = \{y \in \mathbb{R}^m \mid A^\top y \le b\}$ and $B = DJ^{-1}D^\top$ for some $m \times q$-matrix $A$, $b \in \mathbb{R}^q$, $m \times m$-matrix $D$ and symmetric positive definite $m \times m$-matrix $J$. Then, the problem is equivalently stated as

$$\underset{x \in X, v \in \mathbb{R}^q, w \in \mathbb{R}^m}{\text{minimize}}\ f_0(x) + \langle b, v\rangle + \tfrac{1}{2}\langle w, Jw\rangle \ \text{ subject to } \ Av + Dw = F(x),\ v \ge 0.$$

**Detail.** The reformulation follows from 5.60. If $f_0$ and $F$ are smooth and $X$ is described by smooth constraint functions, then we can apply SQP and interior-point methods (§4.K) to the reformulation. Although such algorithms generally only obtain solutions satisfying KKT conditions, they're available in many software packages. □

## 5.J Lagrangian Finite-Generation Method

Large-scale problems aren't easily solved due to their excessive computing times and memory requirements. For example, the reformulation in 5.61 may simply involve too many constraints for SQP and interior-point methods. We then need to examine the structural properties of the problem and attempt to identify tractable approximations. Separability is one such property that dramatically improves tractability. However, the property may not be immediately present: a Lagrangian can be separable with neither the primal nor the dual problem possessing this property. We'll now examine such a situation, which leads to an algorithm for a class of large-scale problems.

For a nonempty polyhedral set $X \subset \mathbb{R}^n$, a symmetric positive semidefinite $n \times n$-matrix $Q$, an $m \times n$ matrix $A$, $c \in \mathbb{R}^n$ and $b \in \mathbb{R}^m$, let's consider the problem

$$\underset{x \in \mathbb{R}^n}{\text{minimize}}\ \varphi(x) = \iota_X(x) + \langle c, x\rangle + \tfrac{1}{2}\langle x, Qx\rangle + h_{Y,B}(b - Ax),$$

where $h_{Y,B}$ is a monitoring function, $B$ is a diagonal matrix with nonnegative diagonal elements $\beta_1, \dots, \beta_m$ and $Y = Y_1 \times \cdots \times Y_m$ is a box, with each $Y_i$ being a nonempty closed interval, potentially unbounded. Suppose that $m$ is large, making a direct solution impractical or even impossible. In the setting of 5.57, the actual problem can be viewed as a primal problem paired with the dual problem

$$\underset{y \in \mathbb{R}^m}{\text{maximize}}\ \psi(y) = -\iota_Y(y) + \langle b, y\rangle - \tfrac{1}{2}\sum\nolimits_{i=1}^{m} \beta_i y_i^2 - h_{X,Q}(A^\top y - c).$$

We might attempt to solve the dual problem in lieu of the primal one, but there's no clear advantage to this strategy. In particular, neither problem is separable because of $X$ and $Q$. As a way forward, let's consider an approximation.

Suppose that $Y$ is approximated by

$$Y^\nu = \operatorname{con}\{\bar{y}^1, \dots, \bar{y}^\nu\},$$

the convex hull of a finite collection of points $\bar{y}^1, \dots, \bar{y}^\nu \in Y$, where $\nu$ is typically much less than $m$. This leads to the approximating dual problem

$$\underset{y\in\mathbb{R}^m}{\text{maximize}}\ \psi^\nu(y) = -\iota_{Y^\nu}(y) + \langle b, y\rangle - \tfrac{1}{2}\sum\nolimits_{i=1}^{m} \beta_i y_i^2 - h_{X,Q}(A^\top y - c). \tag{5.21}$$

As seen from 5.57, this maximization problem can be viewed, in turn, as a dual problem of the approximating primal problem

$$\underset{x\in\mathbb{R}^n}{\text{minimize}}\ \varphi^\nu(x) = \iota_X(x) + \langle c, x\rangle + \tfrac{1}{2}\langle x, Qx\rangle + h_{Y^\nu,B}(b - Ax).$$

Thus, we've replaced the actual primal-dual pair with an approximating primal-dual pair. The idea of the *Lagrangian finite-generation method* is to obtain a saddle point for the underlying Lagrangian of the approximating pair and use this as an approximation of a saddle point for the Lagrangian of the actual pair, which in turn furnishes a minimizer of the actual problem; see the saddle point theorem 5.40.

By 5.57, the Lagrangians of the actual and approximating problems are

$$l(x, y) = \iota_X(x) + \langle c, x\rangle + \tfrac{1}{2}\langle x, Qx\rangle + \langle b - Ax, y\rangle - \tfrac{1}{2}\langle y, By\rangle - \iota_Y(y)$$
$$l^\nu(x, y) = \iota_X(x) + \langle c, x\rangle + \tfrac{1}{2}\langle x, Qx\rangle + \langle b - Ax, y\rangle - \tfrac{1}{2}\langle y, By\rangle - \iota_{Y^\nu}(y).$$

Suppose that $(x^\nu, y^\nu)$ is a saddle point of $l^\nu$. Is it one for $l$ as well? Since $Y^\nu \subset Y$, $l^\nu(x, y) = l(x, y)$ for all $x \in \mathbb{R}^n$ and $y \in Y^\nu$. Moreover,

$$x^\nu \in \operatorname{argmin} l^\nu(\cdot, y^\nu) = \operatorname{argmin} l(\cdot, y^\nu).$$

Thus, we've satisfied one "half" of the saddle point condition for $l$. The other "half," $y^\nu \in \operatorname{argmax} l(x^\nu, \cdot\,)$, is typically *not* satisfied, however.

Let's quantify by how much $(x^\nu, y^\nu)$ falls short under the assumption that strong duality holds for the actual primal-dual pair, i.e., $\inf \varphi = \sup \psi$. Since $(x^\nu, y^\nu)$ is a saddle point of $l^\nu$, the saddle point theorem 5.40 ensures that

$$x^\nu \in \operatorname{argmin} \varphi^\nu, \quad y^\nu \in \operatorname{argmax} \psi^\nu, \quad \varphi^\nu(x^\nu) = \psi^\nu(y^\nu) = l^\nu(x^\nu, y^\nu).$$

The latter quantities equal $l(x^\nu, y^\nu)$ as well. Since $\psi(y) \geq \psi^\nu(y)$ for all $y$,

$$\inf \varphi = \sup \psi \geq \psi^\nu(y^\nu) = l(x^\nu, y^\nu).$$

An upper bound on the minimum value of the actual problem follows by

$$\inf \varphi \le \varphi(x^\nu) = \sup l(x^\nu, \cdot) = l(x^\nu, \bar{y}^{\nu+1}) \ \text{ for } \ \bar{y}^{\nu+1} \in \operatorname{argmax} l(x^\nu, \cdot),$$

where the first equality is a consequence of 5.24. Hence, the optimality gaps of $x^\nu$ and $y^\nu$ can be quantified as

$$\varphi(x^\nu) - \inf \varphi \le \varepsilon^\nu, \quad \sup \psi - \psi(y^\nu) \le \varepsilon^\nu, \quad \text{where } \varepsilon^\nu = l(x^\nu, \bar{y}^{\nu+1}) - l(x^\nu, y^\nu).$$

These calculations require maximization over the high-dimensional $Y$ to obtain $\bar{y}^{\nu+1}$, which seems to bring us back to the original concern about a large $m$. However, now we maximize the Lagrangian $l(x^\nu, \cdot)$ and not $\psi$. This makes a big difference: The former is separable, but the latter isn't. Specifically,

$$(\bar{y}_1^{\nu+1}, \dots, \bar{y}_m^{\nu+1}) \in \operatorname{argmax} l(x^\nu, \cdot) = \operatorname{argmax}_{y \in Y} \left\{ \langle b - Ax^\nu, y \rangle - \tfrac{1}{2}\langle y, By \rangle \right\}$$

$$\Longleftrightarrow \bar{y}_i^{\nu+1} \in \operatorname{argmax}_{y \in Y_i} \left\{ \big(b_i - \langle A_i, x^\nu \rangle\big) y_i - \tfrac{1}{2}\beta_i y_i^2 \right\}, \ i = 1, \dots, m, \tag{5.22}$$

where $A_i$ is the $i$th row of $A$. Thus, the formidable optimization of $y$ is reduced to many trivial problems.

If $\varepsilon^\nu$ isn't sufficiently small, we've in the process identified a point $\bar{y}^{\nu+1} \notin Y^\nu$ that at least for $x^\nu$ is most sorely missed in the approximating $Y^\nu$. Naturally, we amend the approximation by setting $Y^{\nu+1} = \operatorname{con}\{\bar{y}^1, \dots, \bar{y}^\nu, \bar{y}^{\nu+1}\}$ and the process is repeated.

**Lagrangian Finite-Generation Method.**

Data. $\bar{y}^1 \in Y$, $\varepsilon \in [0, \infty)$.
Step 0. Set $Y^1 = \{\bar{y}^1\}$ and $\nu = 1$.
Step 1. Compute a saddle point $(x^\nu, y^\nu)$ of $l^\nu$.
Step 2. Compute $\bar{y}^{\nu+1} = (\bar{y}_1^{\nu+1}, \dots, \bar{y}_m^{\nu+1})$ using (5.22).
Step 3. If $l(x^\nu, \bar{y}^{\nu+1}) - l(x^\nu, y^\nu) \le \varepsilon$, then stop.
Step 4. Set

$$Y^{\nu+1} = \operatorname{con}\{\bar{y}^1, \dots, \bar{y}^{\nu+1}\}.$$

Step 5. Replace $\nu$ by $\nu + 1$ and go to Step 1.

Under the assumptions that there always exist a saddle point $(x^\nu, y^\nu)$ in Step 1 and a maximizer $\bar{y}^{\nu+1}$ in Step 2 and that strong duality holds for the actual primal-dual pair, the algorithm is well defined and terminates after a finite number of iterations with a solution satisfying any positive tolerance $\varepsilon$; see [103] for a proof and further details. It's clear what it takes to have a maximizer in Step 2: Each $i$ must have either $\beta_i > 0$ or $Y_i$ bounded. Thus, let's focus the discussion on Step 1.

Finding a saddle point of $l^\nu$ involves the high-dimensional vector $y$, which is problematic. However, a reformulation leveraging the structure of $Y^\nu$ overcomes this difficulty. Every $y \in Y^\nu$ can be expressed as

$$y = \sum\nolimits_{k=1}^{\nu} \lambda_k \bar{y}^k \ \text{ for some } (\lambda_1, \dots, \lambda_\nu) \in \mathbb{R}^\nu \ \text{ with } \sum\nolimits_{k=1}^{\nu} \lambda_k = 1, \ \lambda_k \ge 0 \ \forall k;$$

see the discussion above 2.62. Concisely, $y = \bar{Y}_\nu \lambda$, where $\lambda = (\lambda_1, \dots, \lambda_\nu)$ and $\bar{Y}_\nu$ is the $m \times \nu$-matrix with $\{\bar{y}^k, k = 1, \dots, \nu\}$ as columns. Thus,

$$\langle b, y\rangle = \langle b^\nu, \lambda\rangle, \qquad \text{where } b^\nu = \bar{Y}_\nu^\top b$$

$$\langle y, By\rangle = \langle \lambda, B_\nu\lambda\rangle, \qquad \text{where } B_\nu = \bar{Y}_\nu^\top B\bar{Y}_\nu$$

$$A^\top y = A_\nu^\top \lambda, \qquad \text{where } A_\nu^\top = A^\top \bar{Y}_\nu.$$

Implementing this change of variables, we obtain a reformulation of (5.21):

$$\operatorname*{maximize}_{\lambda\in\Lambda} \ \langle b^\nu, \lambda\rangle - \tfrac{1}{2}\langle \lambda, B_\nu\lambda\rangle - h_{X,Q}(A_\nu^\top\lambda - c), \tag{5.23}$$

where

$$\Lambda = \Big\{\lambda \in \mathbb{R}^\nu \ \Big|\ \sum_{k=1}^{\nu} \lambda_k = 1,\ \lambda \geq 0\Big\}.$$

By 5.57, the reformulation can be viewed as a dual problem of

$$\operatorname*{minimize}_{x\in X} \ \langle c, x\rangle + \tfrac{1}{2}\langle x, Qx\rangle + h_{\Lambda,B_\nu}(b^\nu - A_\nu x) \tag{5.24}$$

and this primal-dual pair corresponds to the Lagrangian

$$\hat{l}^\nu(x,\lambda) = \iota_X(x) + \langle c, x\rangle + \tfrac{1}{2}\langle x, Qx\rangle + \langle b^\nu - A_\nu x, \lambda\rangle - \tfrac{1}{2}\langle \lambda, B_\nu\lambda\rangle - \iota_\Lambda(\lambda).$$

Thus, we can carry out Step 1 by obtaining a saddle point $(x^\nu, \lambda^\nu)$ of $\hat{l}^\nu$ and then setting $y^\nu = \bar{Y}_\nu\lambda^\nu$. Since $\nu$ is much smaller than $m$, it's typically easier to find a saddle point of $\hat{l}^\nu$ than of $l^\nu$.

The problem (5.24) is feasible because $X$ is nonempty and $h_{\Lambda,B_\nu}$ is real-valued. If $X$ is bounded, then (5.24) isn't unbounded either, which implies via 5.57 that both (5.23) and (5.24) have solutions and the pair satisfies strong duality. Consequently, there's a saddle point of $\hat{l}^\nu$, and then also for $l^\nu$, furnished by the pair of solutions; cf. the saddle point theorem 5.40. Step 1 therefore amounts to solving (5.23) and (5.24), for example leveraging 5.61. Typically, the solution of one of the problems also furnishes a multiplier vector that's a solution of the other problem, but the ease by which the multipliers can be accessed depends on the algorithm (software) used. In any case, (5.23) and (5.24) share by 5.36 the optimality condition

$$-c - Qx + A_\nu^\top\lambda \in N_X(x) \qquad b^\nu - A_\nu x - B_\nu\lambda \in N_\Lambda(\lambda).$$

Thus, when $x^\nu$ solves (5.24), then these relations with $x = x^\nu$ furnish $\lambda^\nu$, which is computable by solving a quadratic optimization problem. Similarly, the relations recover $x^\nu$ from a solution $\lambda^\nu$ of (5.23).

**Exercise 5.62** (Lagrangian finite-generation method). Implement the algorithm and solve the instance with

$$X = \{x \in \mathbb{R}^6 \mid \langle a, x\rangle \le 24,\ 0 \le x \le s\}$$
$$a = (1, 1.5, 0.5, 2, 1, 1),\ \ s = (4, 20, 4, 10, 3, 2),\ \ c = (1, 2, 1, 4, 1, 3)$$
$$B = \mathrm{diag}(0.5/60, 0.4/75, 0.3/80, 0.005/120),\ \ Q = \mathrm{diag}(1, 2, 1, 0.25, 1, 0.5)$$
$$Y = [0, 60] \times [0, 75] \times [0, 80] \times [0, 120]$$

$$A = \begin{bmatrix} 0.29 & 0.4 & 0 & 0.11 & 0 & 0 \\ 0.1 & 0.0975 & 0.315 & 0.51 & 0 & 0 \\ 0 & 0 & 0 & 0.2 & 0.4875 & 0.1925 \\ 0 & 0 & 0 & 0 & 0.3267 & 0.4833 \end{bmatrix}, \qquad b = \begin{bmatrix} 3.825 \\ 0.9667 \\ 3.1 \\ 1.5 \end{bmatrix}.$$

Here, $\mathrm{diag}(v)$ is the diagonal matrix with the vector $v$ as its diagonal elements.

**Guide.** The minimizer is $x^\star = (4, 4.25, 0, 4.31, 3, 2)$, with $y^\star = (58.87, 0, 80, 0)$. □

**Example 5.63** (control of water pollution). In the model described in §3.I for controlling water pollution, let's assume that the direct cost associated with decision $x$ is

$$f_0(x) = \langle c, x\rangle + \tfrac{1}{2}\sum\nolimits_{j=1}^{n} q_j x_j^2$$

and this leads to

$$\underset{x \in X}{\text{minimize}}\ \langle c, x\rangle + \tfrac{1}{2}\sum\nolimits_{j=1}^{n} q_j x_j^2 + \mathbb{E}\Big[\sum\nolimits_{i=1}^{m_2} h_{[0,\tau_i],\beta_i}\big(\boldsymbol{d}_i - \langle \boldsymbol{T}_{i,x}\rangle\big)\Big],$$

where $X = \{x \in \mathbb{R}^n \mid \langle a, x\rangle \le \alpha,\ 0 \le x \le s\}$ and the rest of the data is explained in §3.I. In contrast to that section, we here use the notation for monitoring functions (see 5.58 specifically) and also generalize slightly by having $m_2$ terms in the second sum; $m_2 = 4$ in §3.I. If the probability distribution of $(\boldsymbol{d}, \boldsymbol{T}_1, \dots, \boldsymbol{T}_{m_2})$ is finite, then the Lagrangian finite-generation method applies.

**Detail.** For each $i = 1, \dots, m_2$, suppose that $p_i^k > 0$ is the probability that $(\boldsymbol{d}_i, \boldsymbol{T}_i)$ takes the value $(d_i^k, T_{ik})$, $k = 1, \dots, \nu_i$. Then, $\sum_{k=1}^{\nu_i} p_i^k = 1$ and

$$\mathbb{E}\Big[\sum\nolimits_{i=1}^{m_2} h_{[0,\tau_i],\beta_i}\big(\boldsymbol{d}_i - \langle \boldsymbol{T}_{i,x}\rangle\big)\Big] = \sum\nolimits_{i=1}^{m_2}\sum\nolimits_{k=1}^{\nu_i} p_i^k h_{[0,\tau_i],\beta_i}\big(d_i^k - \langle T_i^k, x\rangle\big)$$
$$= \sum\nolimits_{i=1}^{m_2}\sum\nolimits_{k=1}^{\nu_i} h_{[0,\tau_i^k],\beta_i^k}\big(d_i^k - \langle T_i^k, x\rangle\big),$$

where $\tau_i^k = p_i^k \tau_i$ and $\beta_i^k = \beta_i / p_i^k$. Thus, the term is indeed of the form $h_{Y,B}$, where $Y$ is a box and $B$ is a diagonal matrix. The dimension of $Y$ is $\sum_{i=1}^{m_2} \nu_i$, with $\nu_i$ being the number of possible outcomes of $(\boldsymbol{d}_i, \boldsymbol{T}_i)$, and thus tends to be large. This is exactly a situation in which the Lagrangian finite-generation method can be efficient. □

**Exercise 5.64** (pollution control). Using the Lagrangian finite-generation method, solve the problem in 5.63 with the data $X$, $c$, $q_1, \dots, q_6$ (diagonal elements of $Q$) and $\beta_1, \dots, \beta_4$ (diagonal elements of $B$) as in 5.62. Moreover $(\tau_1, \dots, \tau_4) = (60, 75, 80, 120)$ and the probability distribution of $(\boldsymbol{d}_i, \boldsymbol{T}_i)$ is listed below.

| $d_1^k$ | $T_1^k$ | $p_1^k$ |
|---|---|---|
| 3.5 | (0.26 0.35 0 0.08 0 0) | 1/12 |
| 3.8 | (0.26 0.35 0 0.08 0 0) | 1/12 |
| 4 | (0.26 0.35 0 0.08 0 0) | 1/6 |
| 3.5 | (0.28 0.40 0 0.10 0 0) | 1/24 |
| 3.8 | (0.28 0.40 0 0.10 0 0) | 1/24 |
| 4 | (0.28 0.40 0 0.10 0 0) | 1/12 |
| 3.5 | (0.30 0.40 0 0.10 0 0) | 1/24 |
| 3.8 | (0.30 0.40 0 0.10 0 0) | 1/24 |
| 4 | (0.30 0.40 0 0.10 0 0) | 1/12 |
| 3.5 | (0.32 0.45 0 0.15 0 0) | 1/12 |
| 3.8 | (0.32 0.45 0 0.15 0 0) | 1/12 |
| 4 | (0.32 0.45 0 0.15 0 0) | 1/6 |

| $d_2^k$ | $T_2^k$ | $p_2^k$ |
|---|---|---|
| 0.8 | (0.10 0.05 0.27 0.48 0 0) | 1/8 |
| 1 | (0.10 0.05 0.27 0.48 0 0) | 1/12 |
| 1.4 | (0.10 0.05 0.27 0.48 0 0) | 1/24 |
| 0.8 | (0.10 0.10 0.30 0.52 0 0) | 1/8 |
| 1 | (0.10 0.10 0.30 0.52 0 0) | 1/12 |
| 1.4 | (0.10 0.10 0.30 0.52 0 0) | 1/24 |
| 0.8 | (0.10 0.12 0.33 0.52 0 0) | 1/8 |
| 1 | (0.10 0.12 0.33 0.52 0 0) | 1/12 |
| 1.4 | (0.10 0.12 0.33 0.52 0 0) | 1/24 |
| 0.8 | (0.10 0.12 0.36 0.52 0 0) | 1/8 |
| 1 | (0.10 0.12 0.36 0.52 0 0) | 1/12 |
| 1.4 | (0.10 0.12 0.36 0.52 0 0) | 1/24 |

| $d_3^k$ | $T_3^k$ | $p_3^k$ |
|---|---|---|
| 3 | (0 0 0 0.10 0.40 0.15) | 1/8 |
| 3.2 | (0 0 0 0.10 0.40 0.15) | 1/8 |
| 3 | (0 0 0 0.20 0.50 0.20) | 1/4 |
| 3.2 | (0 0 0 0.20 0.50 0.20) | 1/4 |
| 3 | (0 0 0 0.30 0.55 0.22) | 1/8 |
| 3.2 | (0 0 0 0.30 0.55 0.22) | 1/8 |

| $d_4^k$ | $T_4^k$ | $p_4^k$ |
|---|---|---|
| 0.5 | (0 0 0 0 0.28 0.40) | 1/12 |
| 1.5 | (0 0 0 0 0.28 0.40) | 1/6 |
| 2.5 | (0 0 0 0 0.28 0.40) | 1/12 |
| 0.5 | (0 0 0 0 0.30 0.50) | 1/12 |
| 1.5 | (0 0 0 0 0.30 0.50) | 1/6 |
| 2.5 | (0 0 0 0 0.30 0.50) | 1/12 |
| 0.5 | (0 0 0 0 0.40 0.55) | 1/12 |
| 1.5 | (0 0 0 0 0.40 0.55) | 1/6 |
| 2.5 | (0 0 0 0 0.40 0.55) | 1/12 |

**Guide.** The minimizer is $x^\star = (4, 4.10, 0, 4.43, 3, 2)$. □

# Chapter 6
# WITHOUT CONVEXITY OR SMOOTHNESS

While we're well along in our treatment of nonconvex and nonsmooth problems, some important subjects remain. In this chapter, we'll examine conditions that are *sufficient* for a point to be a local minimizer, even without convexity, and thereby supplementing the necessary conditions furnished by the Oresme and Fermat rules. We'll extend the duality theory of Chap. 5 and close any duality gap that may be present in the absence of convexity. This will lead to a successful class of algorithms for nonconvex problems. Optimality conditions without leveraging smoothness, functions lacking epi-regularity and error analysis will also be covered.

## 6.A Second-Order Analysis

In the absence of convexity, there's no easy check that can be used to certify that a point is a (global) minimizer. However, verification of a *local* minimizer is often possible, at least when the function is well structured and second-order information is available.

**Twice Smooth.** A function $f : \mathbb{R}^n \to \overline{\mathbb{R}}$ is *twice differentiable* at $\bar{x}$ if $f$ is differentiable in a neighborhood of $\bar{x}$ and its gradient mapping $x \mapsto \nabla f(x)$ is differentiable at $\bar{x}$. Then,

$$f(x) - f(\bar{x}) = \langle \nabla f(\bar{x}), x - \bar{x} \rangle + \tfrac{1}{2}\langle x - \bar{x}, \nabla^2 f(\bar{x})(x - \bar{x}) \rangle + o(\|x - \bar{x}\|_2^2). \quad (6.1)$$

The function is *twice smooth* at $\bar{x}$ if it's twice differentiable in a neighborhood of $\bar{x}$ and the mapping $x \mapsto \nabla^2 f(x)$ is continuous at $\bar{x}$. We say that $f$ is twice smooth if it's twice smooth at all $\bar{x} \in \mathbb{R}^n$. Similarly, a mapping $F : \mathbb{R}^n \to \mathbb{R}^m$ given by $F(x) = (f_1(x), \dots, f_m(x))$ for $f_i : \mathbb{R}^n \to \mathbb{R}$ is *twice smooth* if each function $f_i$ is twice smooth. These definitions extend to functions and mappings that are defined only in a neighborhood of $\bar{x}$.



J. O. Royset and R. J-B Wets, *An Optimization Primer*, Springer Series in Operations Research and Financial Engineering, https://doi.org/10.1007/978-3-030-76275-9_6

**Proposition 6.1** (local optimality for twice differentiable function). *For $f : \mathbb{R}^n \to \overline{\mathbb{R}}$, consider the problem of minimizing $f$ on $\mathbb{R}^n$. If $f$ is twice differentiable at a point $\bar{x} \in \mathbb{R}^n$, then the following hold:*

(a) *If $\bar{x}$ is a local minimizer, then $\nabla f(\bar{x}) = 0$ and $\nabla^2 f(\bar{x})$ is positive semidefinite.*
(b) *If $\nabla f(\bar{x}) = 0$ and $\nabla^2 f(\bar{x})$ is positive definite, then $\bar{x}$ is a local minimizer and, in fact, there are $\delta, \gamma > 0$ such that*

$$f(x) \geq f(\bar{x}) + \gamma \|x - \bar{x}\|_2^2 \ \ \forall x \in \mathbb{B}(\bar{x}, \delta).$$

**Proof.** The gradient condition in (a) is a special case of 4.37. For the Hessian condition, note that $f(x) - f(\bar{x}) \geq 0$ for all $x \in \mathbb{R}^n$ in a neighborhood of $\bar{x}$. Fix a nonzero $w \in \mathbb{R}^n$. By (6.1),

$$f(\bar{x} + \lambda w) - f(\bar{x}) = \tfrac{1}{2}\big\langle \lambda w, \nabla^2 f(\bar{x})\lambda w\big\rangle + o\big(\|\lambda w\|_2^2\big).$$

Thus,

$$-\lambda^2\big\langle w, \nabla^2 f(\bar{x}) w\big\rangle \leq o\big(\lambda^2 \|w\|_2^2\big)$$

and

$$-\big\langle w, \nabla^2 f(\bar{x}) w\big\rangle \leq \frac{o\big(\lambda^2\|w\|_2^2\big)}{\lambda^2\|w\|_2^2}\|w\|_2^2 \to 0 \ \text{ as } \ \lambda \searrow 0.$$

We've established that $\langle w, \nabla^2 f(\bar{x}) w\rangle \geq 0$. Since $w$ is arbitrary, this is equivalent to having $\nabla^2 f(\bar{x})$ positive semidefinite.

For (b), the problem

$$\underset{w \in \mathbb{R}^n}{\text{minimize}} \ \tfrac{1}{2}\big\langle w, \nabla^2 f(\bar{x}) w\big\rangle \ \ \text{subject to} \ \ \|w\|_2 = 1$$

has a minimizer by 4.9. Let the minimum value be $\alpha$, which is positive because $\langle w, \nabla^2 f(\bar{x}) w\rangle > 0$ for all nonzero $w$ by the positive definiteness of $\nabla^2 f(\bar{x})$. Since

$$\tfrac{1}{2}\big\langle w, \nabla^2 f(\bar{x}) w\big\rangle \geq \alpha \ \ \forall w \ \text{ with } \ \|w\|_2 = 1,$$

they also satisfy

$$\tfrac{1}{2}\big\langle \lambda w, \nabla^2 f(\bar{x})\lambda w\big\rangle \geq \alpha\lambda^2 \ \ \forall \lambda \in \mathbb{R}. \tag{6.2}$$

Let $\varepsilon \in (0, \alpha)$. Since $\nabla f(\bar{x}) = 0$, (6.1) implies that there's $\delta > 0$ such that

$$\frac{\big|f(x) - f(\bar{x}) - \frac{1}{2}\langle x - \bar{x}, \nabla^2 f(\bar{x})(x - \bar{x})\rangle\big|}{\|x - \bar{x}\|_2^2} \leq \varepsilon \ \ \forall x \in \mathbb{B}(\bar{x}, \delta), \ x \neq \bar{x}.$$

Then, for $x \in \mathbb{B}(\bar{x}, \delta)$,

$$f(x) - f(\bar{x}) \geq \tfrac{1}{2}\big\langle x - \bar{x}, \nabla^2 f(\bar{x})(x - \bar{x})\big\rangle - \varepsilon\|x - \bar{x}\|_2^2.$$

This inequality and (6.2), with $w = (x - \bar{x})/\|x - \bar{x}\|_2$ and $\lambda = \|x - \bar{x}\|_2$, give

$$f(x) - f(\bar{x}) \geq \alpha\|x - \bar{x}\|_2^2 - \varepsilon\|x - \bar{x}\|_2^2 = (\alpha - \varepsilon)\|x - \bar{x}\|_2^2.$$

Thus, $\bar{x}$ is a local minimizer and $f$ grows quadratically near $\bar{x}$. □

The proposition enables us to rule out 0 as a local minimizer of $f(x) = -x^2$ even though $f'(0) = 0$. The slight "gap" in the proposition is unavoidable: The converse of (a) fails because $f(x) = x^3$ has both $f'(0) = 0$ and $f''(0) = 0$, but no minimizer at 0. The converse of (b) fails because $f(x) = x^4$ has a local minimizer at 0, but $f''(0) = 0$ isn't positive definite.

While settling for a local minimizer might at first be disappointing, there are many applications where such a solution is still useful. For example, the goal could be to simply improve on a current decision. A local minimizer $x^\star$ might then suffice; it comes with the property that no additional small change generates further improvement. The focus on small changes is especially meaningful for organizations and systems that prefer evolution in their decisions as compared to a dramatic shift.

There's also empirical evidence that certain high-dimensional problems have "few" local minimizers that aren't global minimizers. To make this plausible, suppose that a twice smooth function has a Hessian matrix $H$ at $\bar{x} \in \mathbb{R}^n$ that's of the form $H = (A+A^\top)/2$, where $A$ is an $n \times n$-matrix with coefficients randomly generated according to a standard normal probability distribution. (The averaging of $A$ and $A^\top$ makes $H$ symmetric.) What is the probability that $H$ is positive semidefinite? By 6.1, this is a requirement for $\bar{x}$ to be a minimizer. It turns out that the probability is 0.5 if $n = 1$, 0.15 if $n = 2$ and $10^{-4}$ if $n = 5$; the probability vanishes quickly as $n$ grows. In many dimensions, it's unlikely that there's not a single direction that "leads" downhill. Thus, local minimizers are rare in problems with such a structure. A particular setting where all or most local minimizers are global minimizers is in the training of certain neural networks; see for example [53, 54].

We recall that $f : \mathbb{R}^n \to \overline{\mathbb{R}}$ is quasi-convex when $\{f \leq \alpha\}$ is convex for all $\alpha \in \mathbb{R}$. This is a class of functions for which 6.1(b) is sufficient for a (global) minimizer.

**Proposition 6.2** (quasi-convex functions). *For a quasi-convex function $f : \mathbb{R}^n \to \overline{\mathbb{R}}$ and a point $\bar{x}$ at which $f$ is twice differentiable, one has*

$$\nabla f(\bar{x}) = 0 \quad \textit{and} \quad \nabla^2 f(\bar{x}) \ \textit{positive definite} \quad \Longrightarrow \quad \bar{x} \in \operatorname{argmin} f.$$

**Proof.** By 6.1(b), there's $\delta > 0$ such that $f(x) > f(\bar{x})$ for all $x \in \mathbb{B}(\bar{x}, \delta) \setminus \{\bar{x}\}$. For the sake of contradiction, suppose that there exists $\hat{x} \in \mathbb{R}^n$ with $f(\hat{x}) < f(\bar{x})$. Then, there's $\lambda \in (0, 1)$ such that

$$x^\lambda = (1 - \lambda)\hat{x} + \lambda\bar{x} \in \mathbb{B}(\bar{x}, \delta).$$

Since $\hat{x}, \bar{x} \in \{f \leq f(\bar{x})\}$ and this level-set is convex, every point on the line segment between $\hat{x}$ and $\bar{x}$ is also contained in the level-set. In particular, $x^\lambda \in \{f \leq f(\bar{x})\}$, but this contradicts that $f(x^\lambda) > f(\bar{x})$. □

**Example 6.3** (estimating a normal distribution; cont.). Let's return to 1.28 and estimate both the mean and variance of a normal random variable. By (1.6), the problem can be formulated as

$$\underset{\mu\in\mathbb{R},\sigma>0}{\text{minimize}} -\frac{1}{m}\sum\nolimits_{i=1}^{m} \ln\left(\frac{1}{\sqrt{2\pi}\sigma}\exp\left(-\frac{(\xi_i-\mu)^2}{2\sigma^2}\right)\right).$$

While the objective function isn't convex, it's quasi-convex and 6.2 applies.

**Detail.** Let $f:\mathbb{R}^2\to\overline{\mathbb{R}}$ be the objective function, i.e., for $\mu\in\mathbb{R}$ and $\sigma>0$,

$$f(\mu,\sigma)=\frac{1}{m}\sum\nolimits_{i=1}^{m}\varphi_i(\mu,\sigma),\quad\text{where } \varphi_i(\mu,\sigma)=\frac{(\mu-\xi_i)^2}{2\sigma^2}+\ln\sigma+\ln\sqrt{2\pi},$$

and $f(\mu,\sigma)=\infty$ otherwise. When $\sigma>0$, differentiation gives us

$$\nabla\varphi_i(\mu,\sigma)=\begin{bmatrix}(\mu-\xi_i)/\sigma^2\\ 1/\sigma-(\mu-\xi_i)^2/\sigma^3\end{bmatrix}$$

$$\nabla^2\varphi_i(\mu,\sigma)=\begin{bmatrix}1/\sigma^2 & -2(\mu-\xi_i)/\sigma^3\\ -2(\mu-\xi_i)/\sigma^3 & (3(\mu-\xi_i)^2/\sigma^2-1)/\sigma^2\end{bmatrix}.$$

Averaging these matrices, we see that the matrix $\nabla^2 f(\mu,\sigma)$ isn't positive semidefinite[1] for all $(\mu,\sigma)\in\mathbb{R}\times(0,\infty)$. This is a necessity for the convexity of $f$; cf. 1.23, which extends to functions that are twice smooth on an open convex set and has the value $\infty$ elsewhere [105, Theorem 2.14]. We can show that $f$ is quasi-convex, however. Consider

$$\psi_i(\mu,\sigma)=\tfrac{1}{2}(\mu-\xi_i)^2/\sigma+\sigma\ln\sigma,$$

with

$$\nabla^2\psi_i(\mu,\sigma)=\begin{bmatrix}1/\sigma & -(\mu-\xi_i)/\sigma^2\\ -(\mu-\xi_i)/\sigma^2 & (\mu-\xi_i)^2/\sigma^3+1/\sigma\end{bmatrix},$$

which is positive semidefinite for any $(\mu,\sigma)$ in the convex set $\mathbb{R}\times(0,\infty)$. Again invoking the extension of 1.23, we conclude that $\psi_i$ is convex when set to infinity outside $\mathbb{R}\times(0,\infty)$. Let $\alpha\in\mathbb{R}$. Then,

$$\{f\le\alpha\}=\left\{(\mu,\sigma)\in\mathbb{R}\times(0,\infty)\;\middle|\;\sum\nolimits_{i=1}^{m}\psi_i(\mu,\sigma)+m\sigma\ln\sqrt{2\pi}-m\sigma\alpha\le 0\right\}.$$

In view of the convexity of $\psi_i$ and 1.18(b), the right-hand side in this equation is a level-set of a convex function and must then be a convex set by 1.11. Thus, $f$ is quasi-convex.

Next, the optimality condition $\nabla f(\mu,\sigma)=0$ has the unique solution

$$\bar\mu=\frac{1}{m}\sum\nolimits_{i=1}^{m}\xi_i\quad\text{and}\quad\bar\sigma^2=\frac{1}{m}\sum\nolimits_{i=1}^{m}(\xi_i-\bar\mu)^2.$$

Provided that $\bar\sigma>0$, i.e., the data points aren't all identical, $\nabla^2 f(\bar\mu,\bar\sigma)$ is the diagonal matrix with elements $1/\bar\sigma^2$ and $2/\bar\sigma^2$, which is positive definite. Thus, $(\bar\mu,\bar\sigma)$ is a minimizer by 6.2 and, in fact, the only minimizer because $\nabla f(\mu,\sigma)=0$ has a unique solution. □

---

[1] A symmetric $2\times 2$-matrix is positive semidefinite if and only if its determinant as well as its diagonal elements are nonnegative.

**Theorem 6.4** (second-order conditions for local minimizers). *For $f_0 : \mathbb{R}^n \to \mathbb{R}$ and $F : \mathbb{R}^n \to \mathbb{R}^m$, both twice smooth, and nonempty polyhedral sets $X \subset \mathbb{R}^n$ and $D \subset \mathbb{R}^m$, suppose that the following qualification holds at $\bar{x}$:*

$$y \in N_D\big(F(\bar{x})\big) \;\; \text{and} \;\; -\nabla F(\bar{x})^\top y \in N_X(\bar{x}) \implies y = 0. \tag{6.3}$$

*Consider the problem*

$$\operatorname*{minimize}_{x \in X} f_0(x) + \iota_D\big(F(x)\big)$$

*and adopt the notation*

$$Y(\bar{x}) = \big\{y \in N_D\big(F(\bar{x})\big) \;\big|\; -\nabla f_0(\bar{x}) - \nabla F(\bar{x})^\top y \in N_X(\bar{x})\big\}$$
$$K(\bar{x}) = \big\{w \in T_X(\bar{x}) \;\big|\; \nabla F(\bar{x})w \in T_D\big(F(\bar{x})\big),\; \big\langle \nabla f_0(\bar{x}), w\big\rangle = 0\big\}.$$

*Then, with $F(x) = (f_1(x), \dots, f_m(x))$, the following hold:*

(a) *If $\bar{x}$ is a local minimizer, then $Y(\bar{x}) \neq \emptyset$ and*

$$\max_{y \in Y(\bar{x})} \Big\langle w, \Big(\nabla^2 f_0(\bar{x}) + \sum\nolimits_{i=1}^m y_i \nabla^2 f_i(\bar{x})\Big) w\Big\rangle \geq 0 \;\; \forall w \in K(\bar{x}). \tag{6.4}$$

(b) *If $Y(\bar{x}) \neq \emptyset$, (6.4) holds and the inequality is strict for $w \neq 0$, then $\bar{x}$ is a local minimizer and there are $\delta, \gamma > 0$ such that*

$$f_0(x) \geq f_0(\bar{x}) + \gamma \|x - \bar{x}\|_2^2 \;\; \forall x \in \mathbb{B}(\bar{x}, \delta) \cap X \;\; \text{with} \;\; F(x) \in D.$$

**Proof.** We only prove the case with $D = \mathbb{R}^m$ and refer to [105, Example 13.25] for the full result. In that case, $F$ is immaterial and (a) simplifies to having $-\nabla f_0(\bar{x}) \in N_X(\bar{x})$ and

$$\big\langle w, \nabla^2 f_0(\bar{x}) w\big\rangle \geq 0 \;\; \forall w \in T_X(\bar{x}) \;\text{with}\; \big\langle \nabla f_0(\bar{x}), w\big\rangle = 0.$$

The first part about $Y(\bar{x}) \neq \emptyset$ is immediate from the basic optimality condition 4.37 and we concentrate on the second part. Suppose that $w \in T_X(\bar{x})$ with $\langle \nabla f_0(\bar{x}), w\rangle = 0$. If $w = 0$, then the inequality involving the Hessian holds trivially. If $w \neq 0$, then we can assume without loss of generality that $\|w\|_2 = 1$ because when the inequality holds for such a vector it also holds for $\lambda w$ as long as $\lambda \geq 0$. Since $\bar{x}$ is a local minimizer, there's $\delta > 0$ such that

$$f_0(\bar{x} + \lambda w) - f_0(\bar{x}) \geq 0 \;\text{ whenever } \lambda \in [0, \delta] \text{ and } \bar{x} + \lambda w \in X.$$

Moreover, there's $\bar{\delta} \in (0, \delta]$ such that $\bar{x} + \lambda w \in X$ for all $\lambda \in [0, \bar{\delta}]$. To prove this claim, recall that

$$X = \big\{x \in \mathbb{R}^n \;\big|\; \langle D_i, x\rangle \leq d_i,\; i = 1, \dots, q\big\}$$

for some $\{D_i \in \mathbb{R}^n, d_i \in \mathbb{R}, i = 1, \dots, q\}$. Thus, $\bar{x} + \lambda w \in X$ requires that

$$\langle D_i, \bar{x}\rangle + \lambda\langle D_i, w\rangle \le d_i \tag{6.5}$$

for all $i$. If the $i$th inequality constraint in the expression for $X$ is active at $\bar{x}$, then $\langle D_i, \bar{x}\rangle = d_i$ and we only need $\langle D_i, w\rangle \le 0$. This holds because $w \in T_X(\bar{x})$; see (4.11). If the $i$th inequality constraint isn't active at $\bar{x}$, then $\langle D_i, w\rangle < d_i$. Thus, for sufficiently small $\lambda$, (6.5) is satisfied again and the claim holds. We've established that $f_0(\bar{x}+\lambda w) - f_0(\bar{x}) \ge 0$ for all $\lambda \in [0, \bar{\delta}]$.

Let $\varepsilon > 0$. Since $f_0$ is twice differentiable at $\bar{x}$, (6.1) holds. Consequently, there's $\bar{\lambda} \in (0, \bar{\delta}]$ such that

$$\left| f_0(\bar{x} + \lambda w) - f_0(\bar{x}) - \lambda\big\langle \nabla f_0(\bar{x}), w\big\rangle - \tfrac{1}{2}\lambda^2\big\langle w, \nabla^2 f_0(\bar{x})w\big\rangle \right| \le \varepsilon\lambda^2 \quad \forall \lambda \in [0, \bar{\lambda}].$$

Since $\big\langle \nabla f_0(\bar{x}), w\big\rangle = 0$, we've by the earlier developments that

$$0 \le f_0(\bar{x} + \lambda w) - f_0(\bar{x}) \le \tfrac{1}{2}\lambda^2\big\langle w, \nabla^2 f_0(\bar{x})w\big\rangle + \varepsilon\lambda^2 \quad \forall \lambda \in [0, \bar{\lambda}].$$

In particular, the inequalities hold for a positive $\lambda$ and then imply that

$$\big\langle w, \nabla^2 f_0(\bar{x})w\big\rangle + 2\varepsilon \ge 0.$$

Since $\varepsilon > 0$ is arbitrary, $\langle w, \nabla^2 f_0(\bar{x})w\rangle \ge 0$ as required.

For (b), the assumption guarantees that $-\nabla f_0(\bar{x}) \in N_X(\bar{x})$ and

$$\big\langle w, \nabla^2 f_0(\bar{x})w\big\rangle > 0 \quad \forall w \in T_X(\bar{x}) \text{ with } \big\langle \nabla f_0(\bar{x}), w\big\rangle = 0 \text{ and } w \ne 0.$$

One can now proceed as in the proof of 6.1(b) by minimizing $\langle w, \nabla^2 f_0(\bar{x})w\rangle$ subject to the constraints $w \in T_X(\bar{x})$, $\langle \nabla f_0(\bar{x}), w\rangle = 0$ and $\|w\|_2 = 1$ and then use the fact, directly from the definition of a tangent cone in 4.33, that $x \in X$ can be expressed as $\bar{x} + \lambda w$ for some $\lambda \ge 0$ and $w \in T_X(\bar{x})$ with $\|w\|_2 = 1$. □

**Example 6.5** (second-order condition for equality constraints). In the setting of 6.4, suppose that $D = \{0\}^m$ and $X = \mathbb{R}^n$ so the problem under consideration becomes

$$\underset{x \in \mathbb{R}^n}{\text{minimize}}\ f_0(x) \ \text{ subject to } \ f_i(x) = 0,\ i = 1, \dots, m.$$

Suppose that at the point $\bar{x} \in \mathbb{R}^n$, $\{\nabla f_i(\bar{x}), i = 1, \dots, m\}$ are linearly independent. Let

$$l(x, y) = f_0(x) + \sum\nolimits_{i=1}^m y_i f_i(x),$$

which by 5.18 defines the Lagrangian under a particular Rockafellian. However, here we simply use it as a short-hand notation.

If $\bar{x}$ is a local minimizer, then there's $y \in \mathbb{R}^m$ such that

$$f_i(\bar{x}) = 0, \ i = 1, \dots, m; \quad \nabla_x l(\bar{x}, y) = 0$$
$$\big\langle w, \nabla^2_{xx} l(\bar{x}, y) w \big\rangle \ge 0 \ \ \forall w \in \mathbb{R}^n \ \text{ with } \big\langle \nabla f_i(\bar{x}), w \big\rangle = 0, \ i = 1, \dots, m.$$

If there's $y \in \mathbb{R}^m$ such that

$$f_i(\bar{x}) = 0, \ i = 1, \dots, m; \quad \nabla_x l(\bar{x}, y) = 0 \tag{6.6}$$
$$\big\langle w, \nabla^2_{xx} l(\bar{x}, y) w \big\rangle > 0 \ \ \forall w \neq 0 \ \text{ with } \big\langle \nabla f_i(\bar{x}), w \big\rangle = 0, \ i = 1, \dots, m,$$

then $\bar{x}$ is a local minimizer.

**Detail.** Suppose that $\bar{x}$ is feasible. We invoke 6.4, but now with $N_D(0) = \mathbb{R}^m$, $T_D(0) = \{0\}^m$, $N_X(\bar{x}) = \{0\}^n$ and $T_X(\bar{x}) = \mathbb{R}^n$. Moreover, $y \in Y(\bar{x})$ means that $f_i(\bar{x}) = 0$ for all $i = 1, \dots, m$ and $\nabla_x l(\bar{x}, y) = 0$. We also have the simplification

$$K(\bar{x}) = \big\{ w \in \mathbb{R}^n \,\big|\, \big\langle \nabla f_i(\bar{x}), w \big\rangle = 0, \ i = 0, 1, \dots, m \big\}.$$

The qualification (6.3) holds because $\{\nabla f_i(\bar{x}), \ i = 1, \dots, m\}$ are linearly independent. This gives the asserted expressions except that we seem to have left out $\langle \nabla f_0(\bar{x}), w \rangle = 0$ above. However, in view of the gradient condition $\nabla_x l(\bar{x}, y) = 0$, we've for any $w \in \mathbb{R}^n$ that

$$\big\langle \nabla f_0(\bar{x}), w \big\rangle + \sum\nolimits_{i=1}^m y_i \big\langle \nabla f_i(\bar{x}), w \big\rangle = 0.$$

Since we already require that $\langle \nabla f_i(\bar{x}), w \rangle = 0$ for $i = 1, \dots, m$, $\langle \nabla f_0(\bar{x}), w \rangle = 0$ automatically.

The linear independence assumption makes $\nabla F(\bar{x}) \nabla F(\bar{x})^\top$ invertible. The condition $\nabla_x l(\bar{x}, y) = 0$ then implies that

$$y = -\big(\nabla F(\bar{x}) \nabla F(\bar{x})^\top\big)^{-1} \nabla F(\bar{x}) \nabla f_0(\bar{x}).$$

Thus, in this case, the set $Y(\bar{x})$ consists of a single point and the maximization over $Y(\bar{x})$ in (6.4) is trivial.

The linear independence assumption is actually not needed for the sufficiency condition and $\bar{x}$ is a local minimizer even without it; see for example [13, Proposition 3.2.1]. □

**Exercise 6.6** (second-order conditions). With $f_0(x) = x_1 + x_2$, $F(x) = x_1^2 + x_2^2$, $X = [0, \infty) \times \mathbb{R}$ and $D = (-\infty, 1]$, use 6.4 to check whether $\bar{x} = (0, -1)$ is a local minimizer.

## 6.B Augmented Lagrangians

The closest relation between primal and dual problems materialized in Chap. 5 under convexity. We'll now see how one can broaden the reach of duality theory by adopting more creative choices of Rockafellians. This results in new classes of Lagrangians and efficient algorithms for many nonconvex problems.

Let's recall the setting in §5.F: We are faced with the problem of minimizing $f_0 : \mathbb{R}^n \to \overline{\mathbb{R}}$ and define a proper Rockafellian $f : \mathbb{R}^m \times \mathbb{R}^n \to \overline{\mathbb{R}}$, with anchor at 0 and lsc convex $f(\cdot, x)$ for all $x \in \mathbb{R}^n$. This leads to a Lagrangian $l$ given by (5.6) and a primal objective function $\varphi$ with

$$\varphi(x) = \sup_{y \in \mathbb{R}^m} l(x, y) = f(0, x) = f_0(x)$$

as seen from 5.24. In terms of $p(u) = \inf f(u, \cdot)$, one has $\inf \varphi = p(0) = \inf f_0$. By (5.14),

$$(p^*)^*(u) = \sup_{y \in \mathbb{R}^m} \big\{\langle u, y\rangle + \psi(y)\big\},$$

where $\psi$ is the objective function of the dual problem, i.e.,

$$\psi(y) = \inf_{x \in \mathbb{R}^n} l(x, y).$$

Thus, $(p^*)^*(0) = \sup \psi$. As discussed in §5.F, when $p$ isn't proper, lsc and convex, strong duality may fail, i.e., $p(0) > (p^*)^*(0)$, leaving the dual problem less informative than one might have hoped. Figure 6.1(left) illustrates the difficulty by plotting epi $p$ together with the graphs of two of the affine functions defining $(p^*)^*$. By the Fenchel-Moreau theorem 5.23, $p(u) \geq (p^*)^*(u)$ for all $u \in \mathbb{R}^m$. Hence, regardless of $y \in \mathbb{R}^m$, the affine function

$$u \mapsto \langle u, y\rangle + \psi(y)$$

lies below $p$. As indicated in the figure, the highest level these affine functions may reach at $u = 0$ is simply the highest intercept $\psi(y)$, which is furnished by a maximizer $\bar{y}$ of $\psi$. However, if epi $p$ bulges in at $(0, p(0))$ as in the figure, then $\psi(\bar{y}) = (p^*)^*(0)$ remains below $p(0)$ and there's a positive duality gap

$$p(0) - (p^*)^*(0).$$

Still, under rather general circumstances, we can close the duality gap if we employ more flexible functions than affine ones as indicated in Figure 6.1(right), where a quadratic function is able to reach up to $p(0)$ while remaining below $p$.

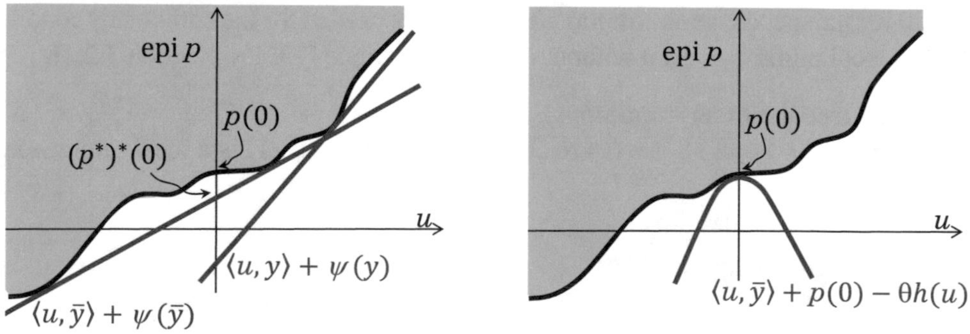


Fig. 6.1: Example of duality gap $p(0) - (p^*)^*(0)$ (left) and how it can vanish when using a quadratic augmenting function $h$ and parameter $\theta$ (right).

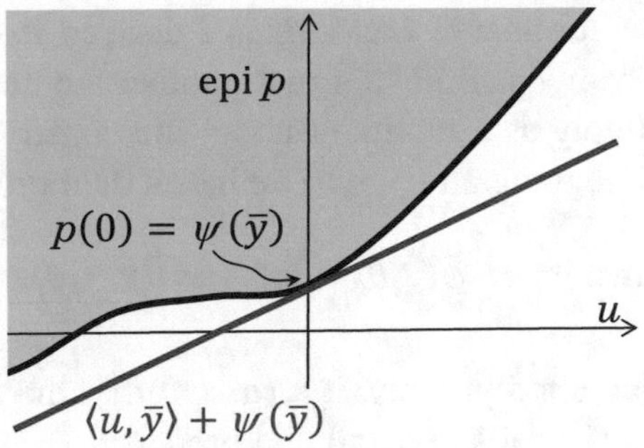


Fig. 6.2: Example of no duality gap with dual maximizer $\bar{y}$ furnishing a function $u \mapsto \langle \bar{y}, u\rangle + \psi(\bar{y})$ that bounds $p$ from below.

In the ideal situation with strong duality and $\bar{y}$ attaining the dual maximum value,

$$p(u) \geq (p^*)^*(u) \geq \langle u, \bar{y}\rangle + \psi(\bar{y}) = \langle u, \bar{y}\rangle + p(0) \ \ \forall u \in \mathbb{R}^m$$

and $p(0) > -\infty$ because having a maximizer $\bar{y}$ rules out that $\sup \psi = -\infty$; see Figure 6.2. Conversely, if

$$p(u) \geq \langle u, \bar{y}\rangle + p(0) \ \ \forall u \in \mathbb{R}^m \quad \text{and} \quad p(0) > -\infty,$$

then

$$\psi(\bar{y}) = \inf_{x,u} \big\{f(u,x) - \langle \bar{y}, u\rangle\big\} = \inf_u \big\{p(u) - \langle \bar{y}, u\rangle\big\} \geq p(0) > -\infty.$$

Since $\psi(y) \leq p(0)$ for all $y \in \mathbb{R}^m$ by weak duality 5.25, this implies $\psi(\bar{y}) = p(0)$ so that $\bar{y} \in \operatorname{argmax} \psi$. In summary, we've shown

$$\left.\begin{array}{l} \inf \varphi = \sup \psi \\ \bar{y} \in \operatorname{argmax} \psi \end{array}\right\} \iff \left\{\begin{array}{l} p(u) \geq \langle u, \bar{y}\rangle + p(0) \ \ \forall u \in \mathbb{R}^m \\ p(0) > -\infty. \end{array}\right. \tag{6.7}$$

This insight helps us achieve the situation in Figure 6.1(right). Let's formalize the ideas. A proper, lsc and convex function $h : \mathbb{R}^m \to \overline{\mathbb{R}}$ with $\inf h = 0$ and $\operatorname{argmin} h = \{0\}$ is said to be an *augmenting function*. An augmenting function $h$ together with a penalty parameter $\theta \in (0, \infty)$ define a new Rockafellian given by

$$\bar{f}(u,x) = f(u,x) + \theta h(u),$$

also with anchor at 0. From the properties of $f$ and $h$, $\bar{f}(\cdot, x)$ is convex and lsc for all $x \in \mathbb{R}^n$. If $f(0, \cdot)$ is finite at one point or $h$ is real-valued, then $\bar{f}$ is proper as well. As is the case for any Rockafellian, $\bar{f}$ produces a Lagrangian via (5.6) with values

$$\bar{l}(x, y, \theta) = \inf_{u \in \mathbb{R}^m} \big\{\bar{f}(u,x) - \langle y, u\rangle\big\} = \inf_{u \in \mathbb{R}^m} \big\{f(u,x) + \theta h(u) - \langle y, u\rangle\big\}.$$

To distinguish it from the "ordinary" Lagrangian $l$ derived from $f$, $\bar{l}(\cdot,\cdot,\theta)$ is referred to as an *augmented Lagrangian*. Still, it's just another Lagrangian, with the qualifier "augmented" included simply as a reminder about the construction via an augmenting function. The augmented Lagrangian $\bar{l}(\cdot,\cdot,\theta)$ defines a dual problem:

$$\underset{y\in\mathbb{R}^m}{\text{maximize}}\ \bar{\psi}(y,\theta) = \inf_{x\in\mathbb{R}^n} \bar{l}(x,y,\theta).$$

When $\theta$ and $h$ are selected appropriately, it turns out that the new dual problem might satisfy strong duality even if $f$ fails to entail that property.

Under the new Rockafellian $\bar{f}$, the min-value function becomes

$$\bar{p}(u) = \inf_{x\in\mathbb{R}^n} \bar{f}(u,x) = \inf_{x\in\mathbb{R}^n} \big\{f(u,x) + \theta h(u)\big\} = p(u) + \theta h(u),$$

but still $\bar{p}(0) = p(0)$ because $h(0) = 0$. The condition for strong duality in (6.7) then translates to

$$\left.\begin{array}{l} \inf\varphi = \sup\bar{\psi}(\cdot,\theta) \\ \bar{y}\in\operatorname{argmax}\bar{\psi}(\cdot,\theta) \end{array}\right\} \iff \left\{\begin{array}{l} p(u) \geq \langle u,\bar{y}\rangle + p(0) - \theta h(u)\ \ \forall u\in\mathbb{R}^m \\ p(0) > -\infty. \end{array}\right.$$

Thus, the new dual objective function $\bar{\psi}(\cdot,\theta)$ achieves strong duality if we can select a Rockafellian $f$, a modification in terms of an augmenting function $h$ and a penalty parameter $\theta$ such that $\langle\cdot,\bar{y}\rangle + p(0) - \theta h$ remains below $p$ for some $\bar{y}$. This is the situation visualized in Figure 6.1(right).

Let's summarize our findings. Although the duality theory in Chap. 5 permits a wide variety of Rockafellians, we focused on some specific types that resulted in quite explicit expressions for the corresponding dual problems. Those Rockafellians often lead to duality gaps for nonconvex problems. A more refined choice using augmenting functions can address that shortcoming.

The penalty parameter $\theta$ can also be tuned so we may as well view $\bar{l}$ as a function from $\mathbb{R}^n\times\mathbb{R}^m\times(0,\infty)$ to $\overline{\mathbb{R}}$ and include $\theta$ in the maximization. This leads to the *augmented dual problem*

$$\boxed{\underset{y\in\mathbb{R}^m,\theta\in(0,\infty)}{\text{maximize}}\ \bar{\psi}(y,\theta) = \inf_{x\in\mathbb{R}^n} \bar{l}(x,y,\theta)}$$

where $\bar{\psi}$ is the *augmented dual objective function*. Since

$$\inf f_0 = \inf f(0,\cdot) \geq \bar{\psi}(y,\theta)\quad \forall y\in\mathbb{R}^m,\ \theta\in(0,\infty),$$

the maximum value in the augmented dual problem is a lower bound on the minimum value of the actual problem. More significantly, it coincides with that minimum value under remarkably broad circumstances. We make this clear after an example.

**Example 6.7** (augmented Lagrangians for composite functions). For $f_0:\mathbb{R}^n\to\mathbb{R}$, $F:\mathbb{R}^n\to\mathbb{R}^m$ and a proper, lsc and convex function $h:\mathbb{R}^m\to\overline{\mathbb{R}}$, let's consider the problem

$$\underset{x \in X \subset \mathbb{R}^n}{\text{minimize}}\ f_0(x) + h\big(F(x)\big)$$

and the Rockafellian expressed by

$$f(u, x) = \iota_X(x) + f_0(x) + h\big(F(x) + u\big).$$

We choose the augmenting function $\frac{1}{2}\|\cdot\|_2^2$. This produces a new Rockafellian given by

$$\bar{f}(u, x) = \iota_X(x) + f_0(x) + h\big(F(x) + u\big) + \tfrac{1}{2}\theta\|u\|_2^2$$

and a corresponding augmented Lagrangian with

$$\bar{l}(x, y, \theta) = \iota_X(x) + f_0(x) + \big\langle F(x), y\big\rangle + \tfrac{1}{2}\theta\big\|F(x)\big\|_2^2 - h_\theta^*\big(y + \theta F(x)\big),$$

where for $\theta \in (0, \infty)$, $h_\theta^*$ is the conjugate of the function defined by

$$h_\theta(u) = h(u) + \tfrac{1}{2}\theta\|u\|_2^2,$$

which thus has the form

$$h_\theta^*(v) = \min_{w \in \mathbb{R}^m} h^*(w) + \tfrac{1}{2\theta}\|w - v\|_2^2.$$

The actual problem is equivalently stated as

$$\underset{x \in \mathbb{R}^n}{\text{minimize}}\ \sup_{y \in \mathbb{R}^m, \theta \in (0,\infty)} \bar{l}(x, y, \theta).$$

In the special case with equality constraints only, i.e., $X = \mathbb{R}^n$ and $h = \iota_{\{0\}^m}$, the augmented Lagrangian simplifies to

$$\bar{l}(x, y, \theta) = f_0(x) + \big\langle F(x), y\big\rangle + \tfrac{1}{2}\theta\big\|F(x)\big\|_2^2. \tag{6.9}$$

**Detail.** Let's fix $x \in \mathbb{R}^n$, $y \in \mathbb{R}^m$ and $\theta \in (0, \infty)$ and define a proper, lsc and convex function as

$$\tilde{h}(u) = h\big(F(x) + u\big) + \tfrac{1}{2}\theta\|u\|_2^2.$$

Then, we obtain

$$\begin{aligned}\bar{l}(x, y, \theta) &= \inf_{u \in \mathbb{R}^m} \Big\{\iota_X(x) + f_0(x) + h\big(F(x) + u\big) + \tfrac{1}{2}\theta\|u\|_2^2 - \langle y, u\rangle\Big\}\\ &= \iota_X(x) + f_0(x) - \sup_{u \in \mathbb{R}^m} \big\{\langle y, u\rangle - \tilde{h}(u)\big\}\\ &= \iota_X(x) + f_0(x) - \tilde{h}^*(y)\end{aligned}$$

and it only remains to determine the conjugate of $\tilde{h}$. Following the strategy in 5.22, the conjugate of $\frac{1}{2}\theta\|\cdot\|_2^2$ is $\frac{1}{2\theta}\|\cdot\|_2^2$. The conjugate of $u \mapsto h(F(x)+u)$ is $y \mapsto h^*(y) - \langle F(x), y\rangle$ as can be seen from the definition of conjugate functions in 5.21. Thus, the conjugates of the two parts of $\tilde{h}$ are known. They combine to form the conjugate of $\tilde{h}$ by the rule in [105, Theorem 11.23]. Specifically,

$$\tilde{h}^*(y) = \min_{w \in \mathbb{R}^m} \Big\{ h^*(w) - \big\langle F(x), w \big\rangle + \tfrac{1}{2\theta}\|y - w\|_2^2 \Big\}.$$

To make the connection with the Lagrangian found in §5.I, which involves the term $\langle F(x), y \rangle$, we note that

$$\begin{aligned}
\tfrac{1}{2\theta}\|y - w\|_2^2 &= \tfrac{1}{2\theta}\big\|w - y - \theta F(x) + \theta F(x)\big\|_2^2 \\
&= \big\langle F(x), w \big\rangle + \tfrac{1}{2\theta}\big\|w - y - \theta F(x)\big\|_2^2 - \big\langle F(x), y \big\rangle - \tfrac{1}{2}\theta\big\|F(x)\big\|_2^2.
\end{aligned}$$

Inserting this expression into the formula for $\tilde{h}^*(y)$, we obtain the asserted form of $\bar{l}(x, y, \theta)$.

The expression for $\bar{l}(x, y, \theta)$ is remarkably similar to that for $l(x, y)$ in §5.I, but there $h^*$ instead of $h_\theta^*$ furnishes the term to be subtracted at the end. Since the former could easily be discontinuous while the latter is smooth by [105, Theorem 2.26], the passing to an augmented Lagrangian could make computations easier.

For the special case with equality constraints only, we obtain the simplification $h_\theta = \iota_{\{0\}^m}$ so that $h_\theta^*(v) = 0$ for all $v$ and (6.9) holds. For other specialization, we refer to [105, Example 11.57].

Next, we consider the alternative formulation of the actual problem. Let $x \in \mathbb{R}^n$ and $\theta \in (0, \infty)$ be fixed. By definition,

$$-\bar{l}(x, y, \theta) = \sup_{u \in \mathbb{R}^m} \big\{ \langle y, u \rangle - \bar{f}(u, x) \big\}$$

so $-\bar{l}(x, \cdot, \theta)$ is the conjugate of $\bar{f}(\cdot, x)$. Certainly, $\bar{f}(\cdot, x)$ is lsc and convex as these properties transfer from $h$. If $\bar{f}(\cdot, x)$ is also proper, then the Fenchel-Moreau theorem 5.23 applies and

$$\bar{f}(u, x) = \sup_{y \in \mathbb{R}^m} \big\{ \langle u, y \rangle + \bar{l}(x, y, \theta) \big\}.$$

In particular, $\bar{f}(0, x) = \sup_{y \in \mathbb{R}^m} \bar{l}(x, y, \theta)$. If $\bar{f}(\cdot, x)$ isn't proper, then we must have $\bar{f}(u, x) = \infty$ for all $u$ and, trivially, $\bar{l}(x, y, \theta) = \infty$ for all $y$ so $\sup_{y \in \mathbb{R}^m} \bar{l}(x, y, \theta) = \infty$. Thus, in either case, we recover the objective function of the actual problem:

$$\iota_X(x) + f_0(x) + h\big(F(x)\big) = \sup_{y \in \mathbb{R}^m} \bar{l}(x, y, \theta).$$

Since this holds for any $\theta \in (0, \infty)$, we can also include maximization over positive $\theta$ without changing the result. □

**Theorem 6.8** (duality without convexity). *For a proper lsc function $f : \mathbb{R}^m \times \mathbb{R}^n \to \overline{\mathbb{R}}$, with $f(\cdot, x)$ convex for all $x \in \mathbb{R}^n$, an augmenting function $h$ and the corresponding augmented Lagrangian $\bar{l}$ and augmented dual objective function $\bar{\psi}$ given by*

$$\bar{l}(x, y, \theta) = \inf_{u \in \mathbb{R}^m} \big\{ f(u, x) + \theta h(u) - \langle y, u \rangle \big\} \quad \textit{and} \quad \bar{\psi}(y, \theta) = \inf_{x \in \mathbb{R}^n} \bar{l}(x, y, \theta),$$

*suppose that* $f(u,x)$ *is level-bounded in* $x$ *locally uniformly in* $u$ *and* $\inf_x \bar{l}(x,y,\theta) > -\infty$ *for at least one* $(y,\theta) \in \mathbb{R}^m \times (0,\infty)$. *Let*

$$\bar{\varphi}(x) = \sup_{y\in\mathbb{R}^m,\theta\in(0,\infty)} \bar{l}(x,y,\theta).$$

*Then, one has*

$$f(0,x) = \bar{\varphi}(x) = \sup_{y\in\mathbb{R}^m} \bar{l}(x,y,\theta) \;\; \forall x \in \mathbb{R}^n,\; \theta \in (0,\infty)$$
$$\inf \bar{\varphi} = \sup \bar{\psi}$$

*so that* $\inf f(0,\cdot) = \sup \bar{\psi}$. *Moreover,*

$$\left.\begin{array}{l}\bar{x} \in \operatorname{argmin} \bar{\varphi}\\ (\bar{y},\bar{\theta}) \in \operatorname{argmax} \bar{\psi}\end{array}\right\} \iff \left\{\begin{array}{l}\bar{x} \in \operatorname{argmin} \bar{l}(\cdot,\bar{y},\bar{\theta})\\ (\bar{y},\bar{\theta}) \in \operatorname{argmax} \bar{l}(\bar{x},\cdot,\cdot),\end{array}\right.$$

*with* $(\bar{y},\bar{\theta}) \in \operatorname{argmax} \bar{\psi}$ *being characterized as those pairs* $(\bar{y},\bar{\theta})$ *such that*

$$p(u) \;\geq\; \langle u,\bar{y}\rangle + p(0) - \bar{\theta}h(u) \quad \forall u \in \mathbb{R}^m,$$

*where* $p(u) = \inf f(u,\cdot)$.

**Proof.** Since an augmenting function is nonnegative, $\bar{l}(x,y,\theta)$ is nondecreasing in $\theta$. This fact, the prior discussion and the saddle point theorem 5.40 establish most of these assertions. It only remains to show that the equality $\inf \bar{\varphi} = \sup \bar{\psi}$ holds and not only the automatic inequality.

The assumption ensures that there's $(\tilde{y},\tilde{\theta})$ such that $\bar{\psi}(\tilde{y},\tilde{\theta})$ is finite. Since $p(0) = \inf \bar{\varphi}$, we only need to show that $\bar{\psi}(\tilde{y},\theta) \to p(0)$ as $\theta \to \infty$. Let

$$\tilde{p}(u) = p(u) + \tilde{\theta}h(u) - \langle \tilde{y},u\rangle,$$

which implies that $\tilde{p}(0) = p(0)$. The function $\tilde{p}$ is bounded from below by $\bar{\psi}(\tilde{y},\tilde{\theta})$ because the latter is its infimum. Since

$$\bar{\psi}(\tilde{y},\tilde{\theta}+\nu) = \inf_u \big\{\tilde{p}(u) + \nu h(u)\big\} \;\; \forall \nu \in \mathbb{N},$$

it suffices to prove that $\bar{\psi}(\tilde{y},\tilde{\theta}+\nu) \to \tilde{p}(0)$ as $\nu \to \infty$. Via the stability theorem 5.6(a) and 5.4, $p$ is lsc and then $\tilde{p}$ is also lsc. Using the characterization of epi-convergence in 4.15, we can show that

$$\tilde{p} + \nu h \xrightarrow{e} \tilde{p}(0) + \iota_{\{0\}}$$

as $\nu \to \infty$. Since $h$ is convex with $\operatorname{argmin} h = \{0\}$, it's level-bounded and this is also the case for $\tilde{p} + \nu h$ because $\tilde{p}$ is bounded from below. The epi-convergence must then be tight and we conclude from 5.5(d) that

$$\inf_u \big\{\tilde{p}(u) + \nu h(u)\big\} \to \inf_u \big\{\tilde{p}(0) + \iota_{\{0\}}(u)\big\}.$$

Hence, $\bar{\psi}(\tilde{y}, \tilde{\theta} + \nu) \to \tilde{p}(0)$ as $\nu \to \infty$. □

The theorem shows that even though the underlying Rockafellian $f$ might not be convex in its second argument ($x$), the resulting augmented dual problem achieves a result parallel to strong duality: its maximum value coincides with the minimum value of the actual problem. Moreover, a saddle point of the augmented Lagrangian furnishes a minimizer of the actual problem as well as a maximizer of the augmented dual problem.

An augmented dual objective function $\bar{\psi}$ is concave[2] because $-\bar{\psi}$ is convex by 1.18(a), which matches the situation for "ordinary" dual objective functions. However, in contrast to some of the common dual problems in Chap. 5, we can't expect to obtain an explicit expression for $\bar{\psi}$ without assuming that the underlying Rockafellian $f$ is rather specific and then we probably wouldn't have needed the augmentation in the first place. This means that solving the augmented dual problem is challenging in general. Still, the insight from 6.8 offers opportunities for reformulations.

**Example 6.9** (reformulation of sup-projection). For $g : \mathbb{R}^q \times \mathbb{R}^n \to \mathbb{R}$, $F : \mathbb{R}^q \times \mathbb{R}^n \to \mathbb{R}^m$ and $\Xi \subset \mathbb{R}^q$, let's consider the problem

$$\operatorname*{minimize}_{x \in X \subset \mathbb{R}^n} f_0(x) = \sup_{\xi \in \Xi} \big\{ g(\xi, x) \bigm| F(\xi, x) = 0 \big\},$$

which may arise in modeling of a strict performance requirement for all values of a parameter vector $\xi$; see §5.G. The function $f_0$ is referred to as a *sup-projection*. In contrast to the earlier encounters, we now face complications caused by the constraint $F(\xi, x) = 0$: The set of $\xi$ over which the maximization takes place depends on $x$. For example, the convexity of $g(\xi, \cdot)$ for all $\xi \in \Xi$ doesn't guarantee the convexity of $f_0$ as would have been the case if $F$ didn't depend on $x$; cf. 1.18(a). This represents a significant theoretical and computation challenge, but it can be overcome in part using an augmented Lagrangian.

**Detail.** For a fixed $x \in \mathbb{R}^n$, let's write

$$f_0(x) = -\hat{f}_0(x), \quad \text{with } \hat{f}_0(x) = \inf_{\xi \in \Xi} \big\{ -g(\xi, x) \bigm| F(\xi, x) = 0 \big\}.$$

This minimization in $\xi$ is of the form addressed in 6.7 and we obtain an augmented Lagrangian given by

$$\bar{l}(\xi, y, \theta) = \iota_\Xi(\xi) - g(\xi, x) + \big\langle F(\xi, x), y \big\rangle + \tfrac{1}{2}\theta \big\| F(\xi, x) \big\|_2^2,$$

where we suppress the dependence on $x$ in $\bar{l}$. Now, minimizing $-g(\xi, x)$ over $\xi \in \Xi$ with the constraint $F(\xi, x) = 0$ is equivalent to

$$\operatorname*{minimize}_{\xi \in \mathbb{R}^q} \bar{\varphi}(\xi) = \sup_{y, \theta} \bar{l}(\xi, y, \theta)$$

and the augmented dual problem amounts to

[2] We only define $\bar{\psi}$ on $\mathbb{R}^m \times (0, \infty)$ and thereby deviate from our standard approach of defining functions on the whole space. This requires minor adjustments to earlier definitions and results, but we omit the details.

$$\underset{y\in\mathbb{R}^m,\theta\in(0,\infty)}{\text{maximize}}\ \bar\psi(y,\theta) = \inf_\xi \bar l(\xi, y, \theta).$$

As long as the required assumptions hold, 6.8 guarantees that $\inf\bar\varphi = \sup\bar\psi$. Since $\inf\bar\varphi = \hat f_0(x)$, one has

$$\hat f_0(x) = \sup_{y,\theta}\inf_\xi \Big\{\iota_\Xi(\xi) - g(\xi,x) + \big\langle F(\xi,x), y\big\rangle + \tfrac{1}{2}\theta\big\|F(\xi,x)\big\|_2^2\Big\}.$$

We can now return to the actual problem. Since $f_0(x) = -\hat f_0(x)$ for all $x$, we obtain the reformulation

$$\underset{x\in X}{\text{minimize}} - \hat f_0(x) = \inf_{y,\theta}\sup_{\xi\in\Xi}\Big\{g(\xi,x) - \big\langle F(\xi,x), y\big\rangle - \tfrac{1}{2}\theta\big\|F(\xi,x)\big\|_2^2\Big\}$$

of the actual problem, which is more compactly written as

$$\underset{x\in X, y\in\mathbb{R}^m,\theta\in(0,\infty)}{\text{minimize}}\ \sup_{\xi\in\Xi}\Big\{g(\xi,x) - \big\langle F(\xi,x), y\big\rangle - \tfrac{1}{2}\theta\big\|F(\xi,x)\big\|_2^2\Big\}.$$

Although we don't have an explicit formula for the augmented dual objective function, the structure of the actual problem allows us to incorporate the dual elements $y, \theta$ with the "outer" decision vector $x$. At an expense of a slightly larger number of decision variables, we now minimize a sup-projection with an index set that's independent of $x$ (and $y$ and $\theta$). If $\Xi$ is finite or approximated by a finite set, then the reformulation can even be cast in terms of a finite number of constraints making SQP and interior-point methods (§4.K) available when $g(\xi,\cdot)$ and $F(\xi,\cdot)$ are smooth. We refer to [84] for developments of this approach in the context of inequality constraints; see also 9.23. □

The augmented Lagrangian for equality constraints in 6.7 resembles the approximating objective function in the penalty method; see 4.21. There the penalty parameter had to be driven to infinity to ensure epi-convergence, which may cause numerical ill-conditioning. It turns out that an approach based on an augmented Lagrangian can bypass this issue.

**Example 6.10** (augmented Lagrangian for equality constraints). In the setting of 6.7 with only equality constraints as described in (6.9) and under the assumption that all the functions are twice smooth, suppose that $(\bar x, \bar y)$ satisfies the second-order sufficient condition (6.6) for the constrained problem

$$\underset{x\in\mathbb{R}^n}{\text{minimize}}\ f_0(x)\ \text{ subject to }\ F(x) = 0. \tag{6.10}$$

Then, $\bar x$ also satisfies the second-order sufficient condition 6.1(b) for the unconstrained problem

$$\underset{x\in\mathbb{R}^n}{\text{minimize}}\ \bar l(x,\bar y,\theta) = f_0(x) + \big\langle F(x), \bar y\big\rangle + \tfrac{1}{2}\theta\big\|F(x)\big\|_2^2 \tag{6.11}$$

provided that $\theta$ is sufficiently large.

This furnishes a strong motivation for solving the unconstrained problem in lieu of the constrained one. If we obtain a point that satisfies 6.1(b) for the unconstrained problem, then at least we've satisfied a rather stringent necessary condition: Every local minimizer for the constrained problem of the "strong kind," i.e., the second-order sufficient condition (6.6) holds, must share the property 6.1(b) with the obtained point.

**Detail.** For the function $\bar{l}(\cdot, \bar{y}, \theta)$, 6.1(b) at $\bar{x}$ amounts to

$$\nabla_x \bar{l}(\bar{x}, \bar{y}, \theta) = 0 \quad \text{and} \quad \big\langle w, \nabla^2_{xx} \bar{l}(\bar{x}, \bar{y}, \theta) w \big\rangle > 0 \ \forall w \neq 0.$$

Let's show that this holds for sufficiently large $\theta$. The gradient

$$\nabla_x \bar{l}(\bar{x}, \bar{y}, \theta) = \nabla f_0(\bar{x}) + \sum\nolimits_{i=1}^m \bar{y}_i \nabla f_i(\bar{x}) + \theta \sum\nolimits_{i=1}^m \nabla f_i(\bar{x}) f_i(\bar{x}). \tag{6.12}$$

Since $f_i(\bar{x}) = 0$ for all $i = 1, \dots, m$, the last term is 0. The first two terms sum to 0 because $(\bar{x}, \bar{y})$ satisfies the second-order sufficient condition (6.6) for the constrained problem (6.10). Thus, the gradient $\nabla_x \bar{l}(\bar{x}, \bar{y}, \theta)$ vanishes for any $\theta \in (0, \infty)$. The Hessian

$$\begin{aligned}\nabla^2_{xx} \bar{l}(\bar{x}, \bar{y}, \theta) &= \nabla^2 f_0(\bar{x}) + \sum\nolimits_{i=1}^m \bar{y}_i \nabla^2 f_i(\bar{x}) + \theta \sum\nolimits_{i=1}^m \big(\nabla^2 f_i(\bar{x}) f_i(\bar{x}) + \nabla f_i(\bar{x}) \nabla f_i(\bar{x})^\top\big) \\ &= \nabla^2 f_0(\bar{x}) + \sum\nolimits_{i=1}^m \bar{y}_i \nabla^2 f_i(\bar{x}) + \theta \sum\nolimits_{i=1}^m \nabla f_i(\bar{x}) \nabla f_i(\bar{x})^\top\end{aligned}$$

again because $f_i(\bar{x}) = 0$. Let's adopt the short-hand notation

$$\nabla^2_{xx} l(\bar{x}, \bar{y}) = \nabla^2 f_0(\bar{x}) + \sum\nolimits_{i=1}^m \bar{y}_i \nabla^2 f_i(\bar{x}).$$

Then,

$$\big\langle w, \nabla^2_{xx} \bar{l}(\bar{x}, \bar{y}, \theta) w \big\rangle = \big\langle w, \nabla^2_{xx} l(\bar{x}, \bar{y}) w \big\rangle + \sum\nolimits_{i=1}^m \theta \Big( \big\langle w, \nabla f_i(\bar{x}) \big\rangle \Big)^2.$$

It suffices to consider $w$ with $\|w\|_2 = 1$ because positivity of this expression is unaffected by rescaling. For the sake of contradiction, suppose that for all $\nu \in \mathbb{N}$ there's $w^\nu$ with $\|w^\nu\|_2 = 1$ such that

$$\big\langle w^\nu, \nabla^2_{xx} l(\bar{x}, \bar{y}) w^\nu \big\rangle + \sum\nolimits_{i=1}^m \nu \Big( \big\langle w^\nu, \nabla f_i(\bar{x}) \big\rangle \Big)^2 \leq 0.$$

Since $\{w^\nu, \nu \in \mathbb{N}\}$ is bounded, there exist $N \in \mathcal{N}^\#_\infty$ and $\bar{w}$ such that $w^\nu \to_N \bar{w}$. Then, $\|\bar{w}\|_2 = 1$ and

$$\big\langle \bar{w}, \nabla^2_{xx} l(\bar{x}, \bar{y}) \bar{w} \big\rangle + \sum\nolimits_{i=1}^m \operatorname{limsup}_{\nu \in N} \nu \Big( \big\langle w^\nu, \nabla f_i(\bar{x}) \big\rangle \Big)^2 \leq 0. \tag{6.13}$$

Consequently, $\langle w^\nu, \nabla f_i(\bar{x}) \rangle \to_N 0$, which implies that $\langle \bar{w}, \nabla f_i(\bar{x}) \rangle = 0$. Since $(\bar{x}, \bar{y})$ satisfies the second-order sufficient condition (6.6) for the constrained problem (6.10), one has

$$\big\langle \bar{w}, \nabla^2_{xx} l(\bar{x}, \bar{y}) \bar{w} \big\rangle > 0.$$

This contradicts (6.13). We've shown that when $\theta$ is large enough,

$$\langle w, \nabla^2_{xx} \bar{l}(\bar{x}, \bar{y}, \theta) w \rangle > 0$$

for all nonzero $w$. □

**Exercise 6.11** (exactness of augmented Lagrangian). For the problem of minimizing $f_0(x) = x_1^2 + x_2^2$ subject to $F(x) = x_2 - 1 = 0$, consider the Rockafellian and augmenting function in 6.7 and 6.10, write the resulting augmented Lagrangian $\bar{l}$ and determine its minimizers with respect to $x$ for different $y$ and $\theta$. Show the effect of setting $y$ equal to a multiplier that together with some $\bar{x}$ satisfy the second-order sufficient condition (6.6) for the actual problem.

The insight from the example and the exercise leads to a simple but yet powerful algorithm for the equality constrained problem (6.10): Minimize the augmented Lagrangian $\bar{l}(\cdot, \bar{y}, \theta)$ for a sufficiently large $\theta$; see (6.11). In fact, we can solve this unconstrained problem multiple times using gradually higher values of $\theta$ with the expectation that it doesn't have to be driven arbitrarily high to achieve a certain equivalence with the constrained problem. A troubling aspect, however, is the unknown multiplier vector $\bar{y}$. While we don't know its value ahead of time, it can be estimated as follows. Given $y$, the first-order condition $\nabla_x \bar{l}(x, y, \theta) = 0$ for the unconstrained problem amounts to having

$$\nabla f_0(x) + \nabla F(x)^\top \big(y + \theta F(x)\big) = 0$$

as seen in (6.12), which resembles the first-order condition for the constrained problem with $y_i + \theta f_i(x)$ furnishing the value of the $i$th multiplier. Consequently, if we obtain a solution $x^\nu$ for the unconstrained problem (6.11) that satisfies

$$\nabla_x \bar{l}(x^\nu, y^\nu, \theta^\nu) = 0$$

for particular $y^\nu$ and $\theta^\nu$, then $y^{\nu+1} = y^\nu + \theta^\nu F(x^\nu)$ immediately provides an updated estimate of the multiplier vector. This approach, which is also called the *method of multipliers*, leads to the following algorithm for the equality constrained problem (6.10) under the assumption that $f_0$ and $F$ are smooth.

**Augmented Lagrangian Method for Equality Constraints.**

Data. $y^1 \in \mathbb{R}^m$, $\theta^\nu \in (0, \infty)$, $\varepsilon^\nu \geq 0$, $\nu \in \mathbb{N}$.
Step 0. Set $\nu = 1$.
Step 1. Determine $x^\nu$ such that

$$\big\|\nabla_x \bar{l}(x^\nu, y^\nu, \theta^\nu)\big\|_2 \leq \varepsilon^\nu.$$

Step 2. Compute $y^{\nu+1} = y^\nu + \theta^\nu F(x^\nu)$.
Step 3. Replace $\nu$ by $\nu + 1$ and go to Step 1.

The algorithm is attractive because Step 1 involves only a gradient condition for the augmented Lagrangian in (6.11), which can be addressed using methods from Chap. 1

with the previous solution $x^{\nu-1}$ usually furnishing an excellent starting point for the calculations. Moreover, its convergence properties don't rely on convexity. Suppose that $\theta^\nu \to \infty$ monotonically, $\varepsilon^\nu \to 0$ and $\{y^\nu, \nu \in \mathbb{N}\}$ is bounded in the algorithm. If $x^\nu \to_N \bar{x}$ for some $N \in \mathcal{N}_\infty^\#$, with $\nabla f_1(\bar{x}), \ldots, \nabla f_m(\bar{x})$ linearly independent, then there's $\bar{y}$ such that

$$y^\nu + \theta^\nu F(x^\nu) \to_N \bar{y}, \quad F(\bar{x}) = 0, \quad \nabla f_0(\bar{x}) + \nabla F(\bar{x})^\top \bar{y} = 0;$$

see [13, Proposition 4.2.2] for details. Consequently, the algorithm produces a solution as well as a multiplier vector that together satisfy the optimality condition 4.76 for the constrained problem (6.10). There are even circumstances under which $\theta^\nu$ doesn't have to be driven arbitrarily high; cf. 6.10.

In the presence of inequality constraints

$$g_i(x) \leq 0, \quad i = 1, \ldots, q$$

as well as equalities, we can specialize the expression in 6.7 and obtain an explicit formula for an augmented Lagrangian. Alternatively, we can restate the inequalities as the equality constraints

$$g_i(x) + z_i = 0, \quad i = 1, \ldots, q,$$

with nonnegative slack variables $z_1, \ldots, z_q$. After minor adjustments, the above development carries through but then Step 1 in the algorithm requires us to replace the unconstrained (6.10) by a problem with nonnegativity constraints. However, the more challenging subproblem can be addressed, for example, via projections. The augmented Lagrangian method for equality and inequality constraints leads to implementations that are comparable in practical performance to those of the SQP and interior-point methods of §4.K.

**Exercise 6.12** (augmented Lagrangian). Implement the augmented Lagrangian method, test it on the problem instance in 4.22 and compare the need for increasing the penalty parameter with that for the penalty method.

The "exactness" of the augmented Lagrangian in 6.10 can be extended much beyond the case of equality constraints. As already indicated in Figure 6.1(right), an augmented Lagrangian $\bar{l}(\cdot, \bar{y}, \theta)$ brings the duality gap to zero if $\bar{y}$, $\theta$ and the augmenting function $h$ produce a graph for $\langle \cdot, \bar{y} \rangle + p(0) - \theta h$ that "reaches" the point $(0, p(0))$. Figure 6.3 illustrates what might go wrong. In the left portion, $h$ is quadratic, say $\frac{1}{2}\|\cdot\|_2^2$ and $\bar{y}$ is appropriately selected. Still, $\theta$ is too small because the resulting graph intersects epi $p$. For the larger $\bar{\theta}$, however, the duality gap vanishes. In the right portion of the figure, epi $p$ has an inward kink at $(0, p(0))$ and an augmenting function of the quadratic kind can't achieve the desired result for any finite $\theta$. However, $h(u) = \|u\|_1$ achieves a zero duality gap for a finite $\bar{\theta}$ as illustrated. This points to the vast number of possible augmented Lagrangians, with "exactness" being achieved under general circumstances.

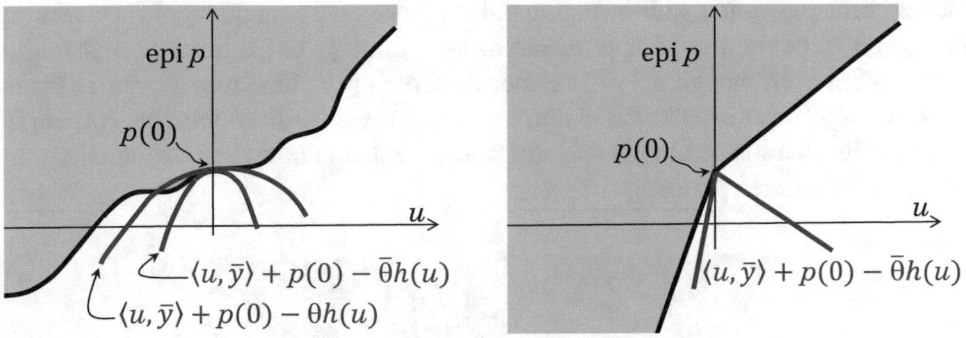


Fig. 6.3: The penalty parameter $\theta$ needs to be sufficient large (left) and the augmenting function $h$ might have to be nonsmooth (right).

**Proposition 6.13** (exactness of augmented Lagrangian). *For a proper lsc function* $f : \mathbb{R}^m \times \mathbb{R}^n \to \overline{\mathbb{R}}$, *with* $f(\cdot, x)$ *convex for all* $x \in \mathbb{R}^n$, *an augmenting function* $h$ *and the corresponding augmented Lagrangian* $\bar{l}$ *and augmented dual objective function* $\bar{\psi}$ *given by*

$$\bar{l}(x, y, \theta) = \inf_{u \in \mathbb{R}^m} \{f(u, x) + \theta h(u) - \langle y, u\rangle\} \quad \textit{and} \quad \bar{\psi}(y, \theta) = \inf_{x \in \mathbb{R}^n} \bar{l}(x, y, \theta),$$

*suppose that* $f(u, x)$ *is level-bounded in* $x$ *locally uniformly in* $u$ *and* $\inf_x \bar{l}(x, y, \theta) > -\infty$ *for at least one* $(y, \theta) \in \mathbb{R}^m \times (0, \infty)$. *Let* $p(u) = \inf f(u, \cdot)$.

*Then, for* $\bar{y} \in \mathbb{R}^m$,

$$\inf f(0, \cdot) = \inf \bar{l}(\cdot, \bar{y}, \theta) \quad \textit{and} \quad \operatorname{argmin} f(0, \cdot) = \operatorname{argmin} \bar{l}(\cdot, \bar{y}, \theta)$$

*for all* $\theta$ *sufficiently large if and only if*

$$\exists \varepsilon > 0, \hat{\theta} > 0 \;\; \textit{such that} \;\; p(u) \geq \langle u, \bar{y}\rangle + p(0) - \hat{\theta}h(u) \;\; \forall u \in \mathbb{B}(0, \varepsilon)$$

$$\Longleftrightarrow \;\; \exists \bar{\theta} > 0 \;\; \textit{such that} \;\; (\bar{y}, \bar{\theta}) \in \operatorname{argmax} \bar{\psi}.$$

The proposition confirms that our informal discussion about selecting $\theta$, $\bar{y}$ and $h$ such that $\langle u, \bar{y}\rangle + p(0) - \theta h(u)$ remains below $p(u)$ is valid quite generally. A proof is provided by [105, Theorem 11.61]. Most significantly, the proposition points to the potential benefits from bringing in a Rockafellian and an augmenting function as compared to the penalty method in 4.21: both a dual vector $y$ and a penalty parameter $\theta$ emerge and the latter doesn't have to be made arbitrarily high to address constraints. Moreover, the approach extends much beyond equality and inequality constraints, with much flexibility afforded through the choice of Rockafellian.

## 6.C Epigraphical Nesting

The epigraphical approximation algorithm of §4.C demands that the approximating functions epi-converge to the actual objective function. This amounts to a *global* approximation, but one might hope that a *local* approximation near a (local) minimizer of the actual function would suffice. Figure 6.4 illustrates the situation: epi $f^\nu$ is near epi $f$ with argmin $f^\nu$ being a good approximation of argmin $f$, but argmin $g^\nu$ is also a good approximation even though epi $g^\nu$ is rather far from epi $f$. One may therefore focus on approximating $f$ near argmin $f$ and don't waste effort away from this set. Although the idea is simple, its implementation is nontrivial as we don't know the location of argmin $f$, which must be learned gradually.

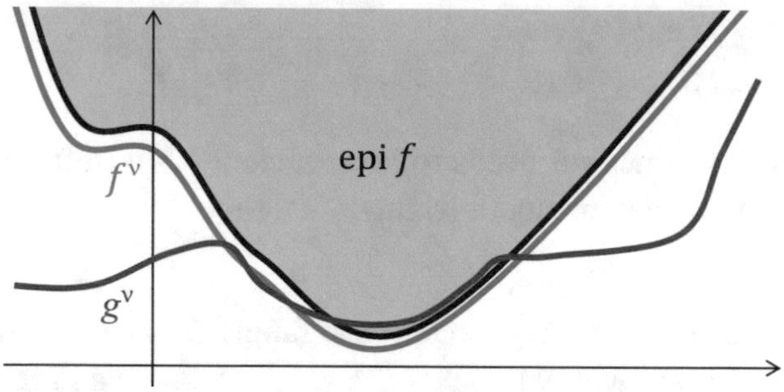


Fig. 6.4: Approximation of epi $f$ globally by epi $f^\nu$ (blue line) and locally near argmin $f$ by epi $g^\nu$ (red line).

**Proposition 6.14** (epigraphical nesting). *For $f^\nu, f : \mathbb{R}^n \to \overline{\mathbb{R}}$, suppose that*

$$\text{LimInn}\,(\text{epi}\, f^\nu) \supset \text{epi}\, f.$$

*Then,* $\limsup(\inf f^\nu) \le \inf f$. *If $\varepsilon^\nu \searrow 0$ and, for some $N \in \mathcal{N}_\infty^\#$,*

$$x^\nu \in \varepsilon^\nu\text{-argmin}\, f^\nu \underset{N}{\to} \bar{x} \quad \textit{and} \quad f^\nu(x^\nu) \underset{N}{\to} f(\bar{x}),$$

*then one also has* $\bar{x} \in \text{argmin}\, f$.

**Proof.** The first claim is essentially 5.5(a). An examination of the proof shows that it holds under the present assumptions. Moreover,

$$f(\bar{x}) = \limsup_{\nu \in N} f^\nu(x^\nu) \le \limsup_{\nu \in N}(\inf f^\nu + \varepsilon^\nu) \le \inf f$$

confirms the second claim. □

The proposition shows that "lower bounding" approximations $\{f^\nu, \nu \in \mathbb{N}\}$ can still be useful. They may not converge epigraphically, but do have epigraphs that in the limit contain epi $f$. We immediately achieve a lower bound on the actual minimum value, which is often algorithmically and practically important as it can be used to assess the optimality gap for a candidate solution. Moreover, the approximations may even produce points that converge to a minimizer of $f$ provided that the approximations become accurate at these points. They don't have to be accurate everywhere.

For $X \subset \mathbb{R}^n$, functions $f_0, g_\alpha : \mathbb{R}^n \to \mathbb{R}$ and an arbitrary nonempty set $\mathbb{A}$, we make these ideas concrete by considering the problem

$$\underset{x \in \mathbb{R}^n}{\text{minimize}}\, f(x) = f_0(x) + \iota_C(x), \text{ where } C = \{x \in X \mid g_\alpha(x) \le 0 \ \forall \alpha \in \mathbb{A}\}.$$

In particular, we're interested in cases with $\mathbb{A}$ containing a large or even infinite number of indices. An approximation of the problem is obtained by replacing $\mathbb{A}$ by a subset $\mathbb{A}^\nu$, which typically only contains a small number of indices, and this leads to the approximating problem

$$\underset{x \in \mathbb{R}^n}{\text{minimize}}\, f^\nu(x) = f_0(x) + \iota_{C^\nu}(x), \text{ where } C^\nu = \{x \in X \mid g_\alpha(x) \le 0 \ \forall \alpha \in \mathbb{A}^\nu\}.$$

When $\mathbb{A}^\nu \subset \mathbb{A}$, we obtain an outer approximation because epi $f^\nu \supset$ epi $f$ and the first requirement in 6.14 is satisfied. The outer approximation algorithm minimizes $f$ by adaptively constructing $\mathbb{A}^\nu$ and $f^\nu$ such that the second requirement in 6.14 also holds.

For any $\varphi : \mathbb{R}^n \to \overline{\mathbb{R}}$ and $\varepsilon \in [0, \infty)$, we denote the set of *near-maximizers* of $g$ by

$$\varepsilon\text{-argmax}\, \varphi = \{x \in \mathbb{R}^n \mid \varphi(x) \ge \sup \varphi - \varepsilon > -\infty\}.$$

Thus, $\varepsilon$-argmax $\varphi = \varepsilon$-argmin$(-\varphi)$.

**Outer Approximation Algorithm.**

Data. $x^0 \in X$, $\mathbb{A}^0 = \emptyset$, $\{\varepsilon^\nu, \delta^\nu \in [0, \infty), \nu \in \mathbb{N}\}$.
Step 0. Set $\nu = 1$.
Step 1. Set $\mathbb{A}^\nu = \mathbb{A}^{\nu-1} \cup \{\alpha^\nu\}$, where

$$\alpha^\nu \in \delta^\nu\text{-argmax}_{\alpha \in \mathbb{A}}\, g_\alpha(x^{\nu-1}).$$

Step 2. Compute $x^\nu \in \varepsilon^\nu$-argmin $f^\nu$.
Step 3. Replace $\nu$ by $\nu + 1$ and go to Step 1.

The algorithm improves the approximating function $f^\nu$ in each iteration by augmenting $\mathbb{A}^\nu$ with the index corresponding to the (nearly) most "critical" constraint at the current point. It turns out that this satisfies the hypotheses of 6.14.

Let's starts with some useful concepts. A family of functions $\{f_\alpha : \mathbb{R}^n \to \overline{\mathbb{R}}, \alpha \in \mathbb{A}\}$ is *locally bounded* at $\bar{x}$ if there are $\varepsilon, \rho \in (0, \infty)$ such that $|f_\alpha(x)| \le \rho$ for all $\alpha \in \mathbb{A}$ and $x \in \mathbb{B}(\bar{x}, \varepsilon)$. For a family $\{f_\alpha : \mathbb{R}^n \to \overline{\mathbb{R}}, \alpha \in \mathbb{A}\}$ that's locally bounded at $\bar{x}$, we say it's also *equicontinuous relative to* $X$ at $\bar{x}$ if for every $\varepsilon > 0$ there exists $\delta > 0$ such that

$$|f_\alpha(x) - f_\alpha(\bar{x})| \le \varepsilon \ \forall \alpha \in \mathbb{A} \text{ and } x \in X \cap \mathbb{B}(\bar{x}, \delta).$$

In particular, this implies that $f_\alpha$ is *continuous relative to* $X$ at $\bar{x}$ in the sense that

$$f_\alpha(x^\nu) \to f_\alpha(\bar{x}) \text{ when } x^\nu \in X \to \bar{x} \in X.$$

When these conditions hold for all $\bar{x} \in X$, we leave out "at $\bar{x}$" in the above statements. Moreover, if $X = \mathbb{R}^n$, then we also drop "relative to $X$."

**Convergence 6.15** (outer approximation algorithm). *For closed $X \subset \mathbb{R}^n$, continuous $f_0 : \mathbb{R}^n \to \mathbb{R}$ and nonempty $\mathbb{A}$, suppose that $\{g_\alpha : \mathbb{R}^n \to \mathbb{R}, \alpha \in \mathbb{A}\}$ is locally bounded and equicontinuous relative to $X$ and $\operatorname{argmax}_{\alpha \in \mathbb{A}} g_\alpha(x)$ is nonempty for all $x \in X$.*

*If $\{x^\nu, \nu \in \mathbb{N}\}$ is generated by the outer approximation algorithm, $x^\nu \underset{N}{\to} \bar{x}$ for some $N \in \mathcal{N}_\infty^\#$ and both $\varepsilon^\nu$ and $\delta^\nu$ vanish, then*

$$\bar{x} \in \operatorname{argmin} f \quad \text{and} \quad f^\nu(x^\nu) \underset{N}{\to} \inf f.$$

**Proof.** The function $g$ given by

$$g(x) = \sup_{\alpha \in \mathbb{A}} g_\alpha(x)$$

is real-valued on $X$ because $\operatorname{argmax}_{\alpha \in \mathbb{A}} g_\alpha(x)$ is nonempty for all $x \in X$. It's also continuous relative to $X$ as we prove in 6.27. Let

$$g^\nu(x) = \max_{\alpha \in \mathbb{A}^\nu} g_\alpha(x).$$

Since $X$ is closed, $\bar{x} \in X$. Our goal is to show that $g^\nu(x^\nu) \underset{N}{\to} g(\bar{x})$. Let $N = \{\nu_1, \nu_2, \dots\}$ and $\mu_k = \nu_{k-1} + 1$. In iteration $\nu_k$, the algorithm constructs $\mathbb{A}^{\nu_k} = \{\alpha^1, \alpha^2, \dots, \alpha^{\nu_k}\}$. Since $\mu_k \leq \nu_k$, it follows that $\alpha^{\mu_k} \in \mathbb{A}^{\nu_k}$. Hence,

$$g(x^{\nu_k}) \geq g^{\nu_k}(x^{\nu_k}) \geq g_{\alpha^{\mu_k}}(x^{\nu_k}).$$

Let $\varepsilon > 0$. The equicontinuity of $\{g_\alpha, \alpha \in \mathbb{A}\}$, the continuity of $g$ and the vanishing $\delta^\nu$ imply that there's $k_0 \in \mathbb{N}$ such that for $k \geq k_0$,

$$g_{\alpha^{\mu_k}}(x^{\nu_k}) \geq g_{\alpha^{\mu_k}}(x^{\nu_{k-1}}) - \tfrac{1}{3}\varepsilon, \quad g(x^{\nu_k}) \leq g(\bar{x}) + \varepsilon, \quad g(x^{\nu_{k-1}}) \geq g(\bar{x}) - \tfrac{1}{3}\varepsilon$$

and $\delta^{\mu_k} \leq \varepsilon/3$. The construction of $\alpha^{\mu_k}$ in Step 1 ensures that $g_{\alpha^{\mu_k}}(x^{\nu_{k-1}}) \geq g(x^{\nu_{k-1}}) - \delta^{\mu_k}$. These relations then imply that for $k \geq k_0$,

$$\begin{aligned} g(\bar{x}) + \varepsilon \geq g(x^{\nu_k}) \geq g^{\nu_k}(x^{\nu_k}) \geq g_{\alpha^{\mu_k}}(x^{\nu_k}) &\geq g_{\alpha^{\mu_k}}(x^{\nu_{k-1}}) - \tfrac{1}{3}\varepsilon \\ &\geq g(x^{\nu_{k-1}}) - \delta^{\mu_k} - \tfrac{1}{3}\varepsilon \geq g(\bar{x}) - \varepsilon. \end{aligned}$$

Since $\varepsilon > 0$ is arbitrary, we've established that $g^{\nu_k}(x^{\nu_k}) \to g(\bar{x})$ as $k \to \infty$. Then, $g(\bar{x}) \leq 0$ because $g^\nu(x^\nu) \leq 0$ for all $\nu$. By the continuity of $f_0$, we also obtain that

$$f^{\nu_k}(x^{\nu_k}) = f_0(x^{\nu_k}) \to f_0(\bar{x}) = f(\bar{x}) \quad \text{as } k \to \infty.$$

Since epi $f^\nu \supset$ epi $f$ trivially, 6.14 applies and the conclusion follows. □

The outer approximation algorithm provides a means to solve problems involving an infinite number of constraints. These include dual and augmented dual problems. For example, a dual problem of maximizing $\psi(y) = \inf_x l(x, y)$ from §5.C can be reformulated as

$$\underset{y \in \mathbb{R}^m, \beta \in \mathbb{R}}{\text{minimize}}\ \beta \ \text{ subject to } \ -l(x, y) \leq \beta \ \ \forall x \in \mathbb{R}^n,$$

which is of the required form. Since $l(x,\cdot)$ is often epi-polyhedral, Step 2 may reduce to solving a linear optimization problem. Step 1 requires us to minimize $l(\cdot,y)$, which presumably is much simpler than solving the actual problem. In 6.9, we ended up with a problem that can be formulated as

$$\underset{x,y,\theta,\beta}{\text{minimize}}\ \beta\ \ \text{subject to}\ \ g(\xi,x) - \langle F(\xi,x),y\rangle - \tfrac{1}{2}\theta\|F(\xi,x)\|_2^2 \le \beta\ \ \forall \xi \in \Xi.$$

Again, the outer approximation algorithm applies, but its implementation may involve further approximations due to the difficulty of finding near-maximizers and near-minimizers in Steps 1 and 2.

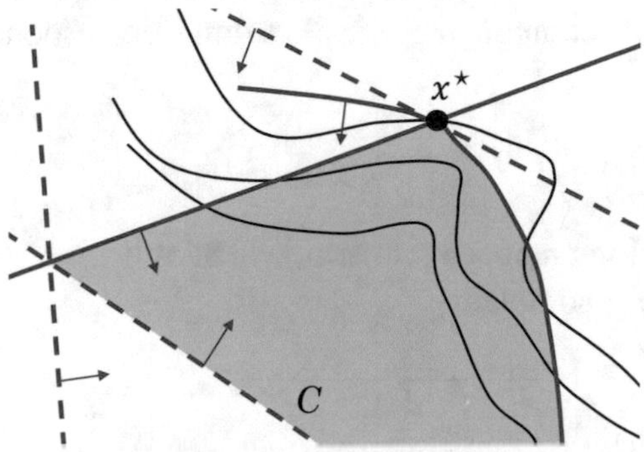


Fig. 6.5: A feasible set $C$ is defined by $q = 5$ inequalities in $n = 2$ dimensions, but only two inequalities (solid red lines) are actually needed to define a local minimizer $x^\star$.

**Example 6.16** (active-set strategy). In an optimization problem with $n$ variables and $q$ inequality constraints, with $q$ much larger than $n$, we can expect that the number of constraints that are active at a local minimizer is much less than $q$. That is, a large number of constraints could have been removed from the problem and the point would have remained a local minimizer; see Figure 6.5. The difficulty is that we rarely know which constraints can be removed a priori. The outer approximation algorithm addresses this situation by adaptively adding in constraints that appear to matter.

Let's examine the case of superquantile-risk minimization, which can be formulated as (cf. §3.C)

$$\underset{x\in X,\gamma\in\mathbb{R},z\ge 0}{\text{minimize}}\ f_0(\gamma,z)\ \text{subject to}\ f(\xi,x) - \gamma \le z_\xi\ \ \forall \xi \in \Xi,$$

where $X \subset \mathbb{R}^n$ and $\Xi$ is a finite set of $\bar\nu$ possible values for the parameter vector $\xi \in \mathbb{R}^m$ so that $z = (z_\xi, \xi \in \Xi) \in \mathbb{R}^{\bar\nu}$. Moreover, $f : \mathbb{R}^m \times \mathbb{R}^n \to \mathbb{R}$ and

$$f_0(\gamma,z) = \gamma + \frac{1}{1-\alpha}\sum\nolimits_{\xi\in\Xi} p_\xi z_\xi,$$

with $p_\xi > 0$ and $\alpha \in (0, 1)$. Since a refined model of uncertainty may require a large $\bar{\nu}$, the problem often becomes large scale. The outer approximation algorithm solves the problem via a series of smaller subproblems.

**Detail.** The problem at hand is of the form required by the algorithm, with $\Xi$ playing the role of $\mathbb{A}$. Let $\Xi^\nu$ be the subset of $\Xi$ corresponding to $\mathbb{A}^\nu$ in the algorithm. Then, Step 2 amounts to solving

$$\underset{x \in X, \gamma \in \mathbb{R}, z \geq 0}{\text{minimize}} \; f_0(\gamma, z) \; \text{subject to} \; f(\xi, x) - \gamma \leq z_\xi \;\; \forall \xi \in \Xi^\nu. \tag{6.14}$$

Since $\Xi^\nu$ contains only $\nu$ elements, (6.14) has significantly fewer constraints than the actual problem. Still, $z$ remains high-dimensional, but this can be addressed easily: In Step 2, simply ignore all components of $z$ that don't correspond to $\Xi^\nu$, i.e., $f_0(\gamma, z)$ is replaced by

$$\gamma + \frac{1}{1-\alpha} \sum\nolimits_{\xi \in \Xi^\nu} p_\xi z_\xi$$

in (6.14). Solve the resulting reduced problem, which only involves $n + 1 + \nu$ variables, to the required tolerance $\varepsilon^\nu$ and obtain

$$\big(x^\nu, \gamma^\nu, z^\nu_\xi, \xi \in \Xi^\nu\big).$$

Set $z^\nu_\xi = 0$ for all $\xi \in \Xi \setminus \Xi^\nu$. Since $p_\xi > 0$, there's no incentive in (6.14) to make $z_\xi > 0$ for $\xi \in \Xi \setminus \Xi^\nu$ and the constructed

$$(x^\nu, \gamma^\nu, z^\nu), \quad \text{with } z^\nu = (z^\nu_\xi, \xi \in \Xi),$$

solves (6.14) to the required tolerance. If warm-started with the point $(x^{\nu-1}, \gamma^{\nu-1}, z^{\nu-1})$ from the previous iteration, the optimization in Step 2 is often rather quick. Step 1 of the algorithm specializes to finding

$$\xi^\nu \in \delta^\nu\text{-argmax}_{\xi \in \Xi} \big\{ f(\xi, x^{\nu-1}) - \gamma^{\nu-1} - z^{\nu-1}_\xi \big\} \quad \text{and} \quad \Xi^\nu = \Xi^{\nu-1} \cup \{\xi^\nu\}.$$

The vector $\xi^\nu$ specifies the (nearly) worst possible values of the parameters for the current point $(x^{\nu-1}, \gamma^{\nu-1}, z^{\nu-1})$. The maximization is over a finite collection of numbers, which is easily carried out unless $\bar{\nu}$ is extremely large.

An examination of the convergence proof for the outer approximation algorithm in 6.15 shows that it's perfectly acceptable to add more than one item from $\Xi$ to $\Xi^\nu$ in every iteration. This makes the subproblem in Step 2 larger, but may potentially reduce the number of overall iterations substantially. For example, problem-specific insight may identify a moderate number of "likely critical" $\xi \in \Xi$, say 50, which form the initial $\Xi^0$. Subsequently, one might add the $\xi$-vectors corresponding to the five highest values in Step 1 for each iteration. Strategies for *dropping* constraints in the outer approximation algorithm are discussed in [47]. □

**Exercise 6.17** (risk-adaptive design). Consider the truss design problem in §3.E with the slight simplification that the random variables $\xi_1, \ldots, \xi_8$ are statistical independent.

Generate a sample $\{\xi^1, \ldots, \xi^{50000})$ from the probability distribution of $\xi = (\xi_1, \ldots, \xi_8)$ and solve (3.6) with $\nu = 50000$ and $\alpha = 0.9$. Implement the outer approximation algorithm as described in 6.16 and solve the same problem instance. Compare run times.

**Guide.** The solution of (3.6) may take quite a while because it involves 400008 constraints. For the outer approximation algorithm, think of clever ways of selecting the initial $\Xi^0$ and explore the possibility of adding more than one point to $\Xi^\nu$ in Step 1. □

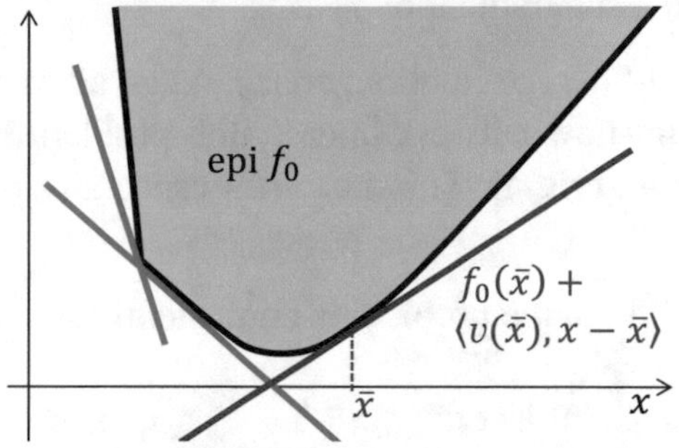


Fig. 6.6: The cutting plane method utilizes a sequence of polyhedral approximations constructed using the subgradient inequality.

**Example 6.18** (cutting plane method). A version of the outer approximation algorithm can be useful in the presence of convexity. For a convex function $f_0 : \mathbb{R}^n \to \mathbb{R}$ and a closed set $X \subset \mathbb{R}^n$, which could be nonconvex, we seek to

$$\underset{x \in X}{\text{minimize}}\ f_0(x).$$

The cutting plane method achieves this by considering polyhedral approximations of epi $f_0$.

**Detail.** Let $x, \bar{x} \in X$ and $v(\bar{x}) \in \partial f_0(\bar{x})$, which exists by 2.25. The subgradient inequality 2.17 leads to

$$f_0(x) \geq f_0(\bar{x}) + \big\langle v(\bar{x}), x - \bar{x}\big\rangle$$

as seen in Figure 6.6. Since this holds for every $\bar{x} \in X$,

$$f_0(x) \geq \sup_{\bar{x} \in X} \big\{ f_0(\bar{x}) + \big\langle v(\bar{x}), x - \bar{x}\big\rangle \big\}.$$

The choice $\bar{x} = x$ is permissible in the maximization so the right-hand side equals $f_0(x)$. Consequently, the actual problem is equivalent to

$$\underset{x \in X, \beta \in \mathbb{R}}{\text{minimize}}\ \beta \ \text{ subject to } \ f_0(\bar{x}) + \big\langle v(\bar{x}), x - \bar{x}\big\rangle \leq \beta \quad \forall \bar{x} \in X,$$

which is of the form addressed by the outer approximation algorithm with $X$ playing the role of $\mathbb{A}$. The *cutting plane method* is simply the outer approximation algorithm applied to this reformulation. Step 2 of the algorithm now involves the solution of a problem with a linear objective function, linear inequality constraints indexed by a finite set $X^\nu \subset X$ and the constraint set $X$, which presumably is well structured so this can be accomplished. Given $(x^{\nu-1}, \beta^{\nu-1})$ from the previous iteration, Step 1 involves computing

$$\bar{x}^\nu \in \operatorname{argmax}_{\bar{x}\in X} \left\{ f_0(\bar{x}) + \langle v(\bar{x}), x^{\nu-1} - \bar{x}\rangle - \beta^{\nu-1} \right\}.$$

This is trivial; we can select $\bar{x}^\nu = x^{\nu-1}$ because it attains $f_0(x^{\nu-1})$. Step 1 therefore reduces to obtaining $f_0(x^{\nu-1})$ and a subgradient of $f_0$ at $x^{\nu-1}$. □

**Example 6.19** (maximum-flow interdiction; cont.). As an alternative to the duality-based approach to the maximum-flow network-interdiction problem in 5.53, we can leverage the outer approximation algorithm. This becomes especially promising for large-scale instances.

**Detail.** In the notation of 5.53, the problem is equivalently stated as

$$\operatorname*{minimize}_{x\in X, \beta\in\mathbb{R}} \beta \text{ subject to } y_a - \sum\nolimits_{e\in E} x_e y_e \leq \beta \quad \forall y \in Y,$$

where

$$X = \left\{ x \in \{0,1\}^n \;\middle|\; \sum\nolimits_{e\in E} r_e x_e \leq \alpha \right\}$$

$$Y = \left\{ (y_a, y_e,\ e \in E) \;\middle|\; \sum\nolimits_{e\in E_i^+} y_e - \sum\nolimits_{e\in E_i^-} y_e = \delta_i y_a \ \ \forall i \in V,\ \ 0 \leq y_e \leq u_e \ \ \forall e \in E \right\}$$

and the outer approximation algorithm applies. Given $(x^{\nu-1}, \beta^{\nu-1})$, with $x^{\nu-1} = (x_e^{\nu-1}, e \in E)$, from the previous iteration, Step 1 amounts to determining

$$y^\nu \in \operatorname{argmax}_{y\in Y} \left\{ y_a - \sum\nolimits_{e\in E} x_e^{\nu-1} y_e - \beta^{\nu-1} \right\}.$$

Thus, we need to determine a maximum flow through the network in response to interdiction $x^{\nu-1}$. This can be computed quickly using specialized algorithms; see for example [1, Chaps. 6–7]. In Step 2, we solve the problem

$$\operatorname*{minimize}_{x\in X, \beta\in\mathbb{R}} \beta \text{ subject to } y_a - \sum\nolimits_{e\in E} x_e y_e \leq \beta \quad \forall y \in Y^\nu,$$

where $Y^\nu \subset Y$ is the collection of "best responses" from Step 1 accumulated thus far. This is a linear optimization problem with integer constraints and well-developed algorithms are available; see the discussion in 4.3. The problem can be viewed as that of finding the best interdiction decision $x$ given all the "known" network flows in $Y^\nu$. The network operator may have other flows to its disposal, but at least the interdictor considers all those that the network operator has considered "best" in the past. □

For further applications of epigraphical nesting, we refer to [50] and especially [49], which furnishes a refinement of the L-shaped method (§5.H) for solving stochastic optimization problems by introducing sampling.

## 6.D Optimality Conditions

We've already developed a strong foundation in subdifferential calculus and can obtain numerous optimality conditions based on the Fermat rule. In particular, the chain rule 4.64 leverages smoothness hidden "inside" a composite function, which may itself not be smooth. We'll now generalize such results to cases where smoothness isn't as readily available or is absent all together.

We start with a fundamental building block: the normal cone to a nearly arbitrary level-set. This requires a new concept visualized in Figure 6.7. For an arbitrary set $C \subset \mathbb{R}^n$, the *conic hull* of $C$ is the set

$$\operatorname{cnc} C = \{0\} \cup \{\lambda x \mid x \in C, \ \lambda > 0\}.$$

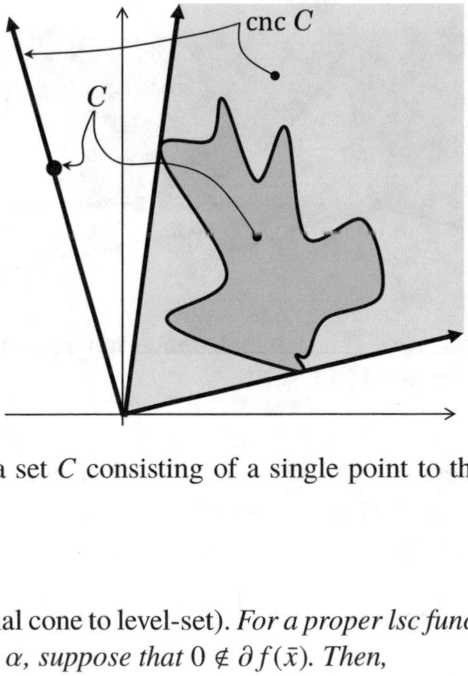


Fig. 6.7: Conic hull of a set $C$ consisting of a single point to the left and a leaf-shaped part to the right.

**Proposition 6.20** (normal cone to level-set). *For a proper lsc function $f : \mathbb{R}^n \to \overline{\mathbb{R}}$, $\alpha \in \mathbb{R}$ and $\bar{x} \in \mathbb{R}^n$ with $f(\bar{x}) = \alpha$, suppose that $0 \notin \partial f(\bar{x})$. Then,*

$$N_{\{f \leq \alpha\}}(\bar{x}) \subset \operatorname{cnc} \partial f(\bar{x}) \cup \partial^\infty f(\bar{x}).$$

*If in addition $f$ is epi-regular at $\bar{x}$, then $\{f \leq \alpha\}$ is Clarke regular at $\bar{x}$ and the above inclusion can be replaced by an equality.*

**Proof.** Since $\{f \leq \alpha\} = \{x \mid F(x) \in D\}$ with $F(x) = (x, \alpha)$ and $D = \operatorname{epi} f$, we can apply 4.46; $\nabla F(\bar{x})$ is the $(n + 1) \times n$-matrix consisting of the $n \times n$ identity matrix and a row of zeros at the bottom. Thus,

$$N_{\{f\leq\alpha\}}(\bar{x}) \subset \left\{\nabla F(\bar{x})^\top y \;\middle|\; y \in N_D\big(F(x)\big)\right\}$$
$$= \left\{v \in \mathbb{R}^n \;\middle|\; (v,\beta) \in N_{\operatorname{epi} f}\big(\bar{x}, f(\bar{x})\big) \text{ for some } \beta \in \mathbb{R}\right\},$$

with the inclusion being replaced by an equality if epi $f$ is Clarke regular at $(\bar{x}, f(\bar{x}))$. In view of the definitions of subgradients in 4.55 and of horizon subgradients in 4.60, this translates into the stated expression. It only remains to verify the qualification (4.9), which reduces to

$$(v,\beta) \in N_{\operatorname{epi} f}\big(\bar{x}, f(\bar{x})\big) \text{ and } v = 0 \implies (v,\beta) = 0.$$

More concisely, $(0,\beta) \in N_{\operatorname{epi} f}(\bar{x}, f(\bar{x}))$ must imply that $\beta = 0$. This is indeed the case when $0 \notin \partial f(\bar{x})$ as seen by the definition of subgradients. □

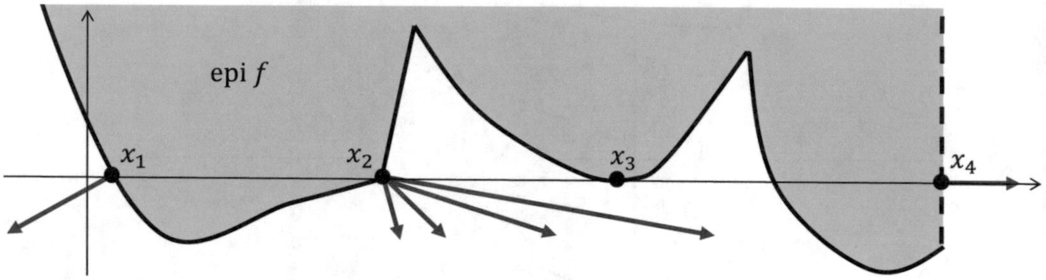


Fig. 6.8: Normal cones $N_{\operatorname{epi} f}(x_i, f(x_i))$, which define subgradients and horizon subgradients and then also normal cones to $\{f \leq 0\}$.

Figure 6.8 illustrates some possibilities. At $x_1$, $f$ is smooth so that the normal vectors to epi $f$ at $(x_1, f(x_1))$ all point in the same southwest direction and $\partial f(x_1)$ becomes some negative number, say $-1.5$. By the proposition,

$$N_{\{f\leq 0\}}(x_1) = \operatorname{cnc}\{-1.5\} = (-\infty, 0].$$

At $x_2$, $f$ is nonsmooth with $\partial f(x_2) = [0.2, 5]$, say, so then

$$N_{\{f\leq 0\}}(x_2) = \operatorname{cnc}[0.2, 5] = [0, \infty).$$

At $x_3$, $f$ is smooth but $\partial f(x_3) = \{0\}$ and the proposition doesn't apply. The condition $0 \notin \partial f(\bar{x})$ is a qualification that relates to the one for smooth constraint functions in 4.46: For the special case of $C = \{x \mid g_1(x) \leq 0\}$, with $g_1$ smooth, the requirement is $\nabla g_1(\bar{x}) \neq 0$; see also the Mangasarian-Fromovitz constraint qualification (4.10). At $x_4$, there are no subgradients of $f$ as all normal vectors are horizontal and cnc $\partial f(x_4) = \{0\}$. Still, $\partial^\infty f(x_4) = [0, \infty)$, which then implies that

$$N_{\{f\leq 0\}}(x_4) = [0, \infty).$$

**Example 6.21** (minimization with $\ell^1$-constraint). For a smooth function $f_0 : \mathbb{R}^n \to \mathbb{R}$ and $\tau > 0$, consider the problem

$$\operatorname*{minimize}_{x \in \mathbb{R}^n} f_0(x) \text{ subject to } \|x\|_1 \le \tau,$$

which may arise as an alternative to the lasso regression in 2.5. A necessary condition for a feasible $\bar{x}$ to be a minimizer of the problem is that

$$\exists \lambda \ge 0 \text{ and } v = (v_1, \dots, v_n), \text{ where } v_j \in \begin{cases} \{-1\} & \text{if } \bar{x}_j < 0 \\ [-1, 1] & \text{if } \bar{x}_j = 0 \\ \{1\} & \text{otherwise,} \end{cases}$$

such that $-\nabla f_0(\bar{x}) = \lambda v$ with $\lambda = 0$ when $\|\bar{x}\|_1 < \tau$.

**Detail.** Let $C = \{f \le \tau\}$ with $f(x) = \|x\|_1$. By 4.58,

$$\partial(f_0 + \iota_C)(\bar{x}) = \nabla f_0(\bar{x}) + N_C(\bar{x}).$$

The Fermat rule 4.73 ensures that $-\nabla f_0(\bar{x}) \in N_C(\bar{x})$ is a necessary condition. If $f(\bar{x}) < \tau$, then $\bar{x} \in \operatorname{int} C$ and $N_C(\bar{x}) = \{0\}$ and the assertion holds. If $f(\bar{x}) = \tau$, which is a positive number, then 6.20 applies because $0 \notin \partial f(\bar{x})$; an expression for $\partial f(\bar{x})$ is given by 2.21. By 4.65, $\partial^\infty f(\bar{x}) = \{0\}$. The stated condition then follows from the formula for $\partial f(\bar{x})$. □

A chain rule without smoothness altogether is immediately available. Later on in the section, we'll give a more refined version.

**Proposition 6.22** (chain rule without smoothness). *For proper lsc functions $f : \mathbb{R}^n \to \overline{\mathbb{R}}$ and $h : \mathbb{R} \to \overline{\mathbb{R}}$, suppose that $h$ is extended to $(-\infty, \infty]$ by setting $h(\infty) = \infty$, $\bar{x}$ is a point at which $h \circ f$ is finite and $h$ is nondecreasing with $h(\alpha) > h(f(\bar{x}))$ for all $\alpha > f(\bar{x})$ and $\sup h = \infty$. If*

$$\text{either } 0 \notin \partial f(\bar{x}) \text{ or } \partial^\infty h\big(f(\bar{x})\big) = \{0\},$$

*then*

$$\partial(h \circ f)(\bar{x}) \subset \big\{\lambda v \;\big|\; \lambda \in \partial h\big(f(\bar{x})\big),\ v \in \partial f(\bar{x})\big\} \cup \partial^\infty f(\bar{x}).$$

**Proof.** Let $g(x, \alpha) = h(\alpha) + \iota_{\operatorname{epi} f}(x, \alpha)$. Then,

$$h\big(f(x)\big) = \inf_{\alpha \in \mathbb{R}} g(x, \alpha)$$

and, for $x = \bar{x}$, the minimum is attained uniquely by $\bar{\alpha} = f(\bar{x})$. We can then bring in 5.13 and obtain that

$$\partial(h \circ f)(\bar{x}) \subset \big\{v \in \mathbb{R}^n \;\big|\; (v, 0) \in \partial g(\bar{x}, \bar{\alpha})\big\}.$$

The subgradients of $g$ can be computed by 4.63 and 4.67 as

$$\partial g(\bar{x}, \bar{\alpha}) \subset \big(\{0\} \times \partial h(\bar{\alpha})\big) + N_{\operatorname{epi} f}(\bar{x}, \bar{\alpha}),$$

which then leads to the stated formula. For verification of the assumptions in these supporting results, we refer to [105, Proposition 10.19]. □

To obtain a more general chain rule without having to rely on monotonicity as in 6.22, we assume local Lipschitz continuity. This results in a far-reaching extension of the chain rule 4.64. For a mapping $F : \mathbb{R}^n \to \mathbb{R}^m$, with $F(x) = (f_1(x), \dots, f_m(x))$, we say that $F$ is *locally Lipschitz continuous* if each $f_i : \mathbb{R}^n \to \mathbb{R}$ is locally Lipschitz continuous; see the discussion after 4.67.

**Theorem 6.23** (general chain rule). *For a proper lsc function $h : \mathbb{R}^m \to \overline{\mathbb{R}}$ and a locally Lipschitz continuous mapping $F : \mathbb{R}^n \to \mathbb{R}^m$, let $f : \mathbb{R}^n \times \mathbb{R}^m \to \mathbb{R}$ be given by*

$$f(x, y) = \langle F(x), y \rangle.$$

*Suppose that $\bar{x} \in \operatorname{dom}(h \circ F)$ and the following qualification holds:*

$$y \in \partial^\infty h\big(F(\bar{x})\big) \;\; \text{and} \;\; 0 \in \partial_x f(\bar{x}, y) \implies y = 0. \tag{6.15}$$

*Then,*

$$\partial(h \circ F)(\bar{x}) \subset \bigcup_{y \in \partial h(F(\bar{x}))} \partial_x f(\bar{x}, y).$$

*If in addition $h$ is epi-regular at $F(\bar{x})$ and $f(\cdot, y)$ is epi-regular at $\bar{x}$ for each $y \in \partial h(F(\bar{x}))$, then $h \circ F$ is epi-regular at $\bar{x}$ and the inclusion can be replaced with an equality.*

**Proof.** We refer to [105, Theorem 10.49] for a proof as it relies on concepts of differentiation of set-valued mappings. □

The connection with the chain rule 4.64 is clear: When $F$ is smooth,

$$\partial_x f(\bar{x}, y) = \nabla F(\bar{x})^\top y.$$

Thus, the condition $0 \in \partial_x f(\bar{x}, y)$ becomes simply $\nabla F(\bar{x})^\top y = 0$ and

$$\bigcup_{y \in \partial h(F(\bar{x}))} \partial_x f(\bar{x}, y) = \bigcup_{y \in \partial h(F(\bar{x}))} \big\{\nabla F(\bar{x})^\top y\big\} = \nabla F(\bar{x})^\top \partial h\big(F(\bar{x})\big).$$

If $F$ isn't smooth, then the challenge is to compute the subgradients of $f(\cdot, y)$. We recall that

$$f(x, y) = \sum\nolimits_{i=1}^m y_i f_i(x),$$

where $y = (y_1, \dots, y_m)$ and $F(x) = (f_1(x), \dots, f_m(x))$. Typically, one then proceeds in two steps. First, determine the subgradients of $y_i f_i$ using (4.13) and, second, apply the sum rule 4.67. When $y_i < 0$, (4.13) isn't directly applicable to $y_i f_i$ but one can consider $(-y_i)(-f_i)$. Thus, the focus switches to subgradients of the function $-f_i$ because

$$\partial(y_i f_i)(\bar{x}) = -y_i \partial(-f_i)(\bar{x})$$

by (4.13) in that case.

**Example 6.24** (normal cone for general constraint system). For a locally Lipschitz continuous mapping $F : \mathbb{R}^n \to \mathbb{R}^m$ and a closed set $D \subset \mathbb{R}^m$, let

$$C = \{x \in \mathbb{R}^n \mid F(x) \in D\},$$

$\bar{x} \in C$ and define $f : \mathbb{R}^n \times \mathbb{R}^m \to \mathbb{R}$ by

$$f(x, y) = \langle F(x), y \rangle.$$

Suppose that the following qualification holds:

$$y \in N_D\big(F(\bar{x})\big) \text{ and } 0 \in \partial_x f(\bar{x}, y) \implies y = 0.$$

Then,

$$N_C(\bar{x}) \subset \bigcup_{y \in N_D(F(\bar{x}))} \partial_x f(\bar{x}, y).$$

If in addition $D$ is Clarke regular at $F(\bar{x})$ and $f(\cdot, y)$ is epi-regular at $\bar{x}$ for each $y \in N_D(F(\bar{x}))$, then $C$ is Clarke regular at $\bar{x}$ and the inclusion can be replaced with an equality.

**Detail.** This follows directly from the general chain rule 6.23 with $h = \iota_D$ after recalling that in this case

$$\partial h\big(F(\bar{x})\big) = \partial^\infty h\big(F(\bar{x})\big) = N_D\big(F(\bar{x})\big)$$

by 4.56 and 4.65. The result is largely an extension of 4.46, but omits a generic set $X$. However, this is easily brought back by using the intersection rule in 4.52. The example highlights the fruitful interplay between normal cones to sets and subgradients of functions: The former can be used to derive the latter and vice versa. □

We've now reached a pinnacle and can state a broad optimality condition without a trace of smoothness. It extends 4.75 and can subsequently be specialized in numerous directions; see for example 6.42.

**Theorem 6.25** (optimality for general composite function). *For $f_0 : \mathbb{R}^n \to \mathbb{R}$ and $F : \mathbb{R}^n \to \mathbb{R}^m$, both locally Lipschitz continuous, as well as closed $X \subset \mathbb{R}^n$ and proper, lsc and convex $h : \mathbb{R}^m \to \overline{\mathbb{R}}$, let $f : \mathbb{R}^n \times \mathbb{R}^m \to \mathbb{R}$ and $l : \mathbb{R}^n \times \mathbb{R}^m \to \mathbb{R}$ be defined by*

$$f(x, y) = \langle F(x), y \rangle \quad \text{and} \quad l(x, y) = f_0(x) + \langle F(x), y \rangle.$$

*Suppose that the following qualification holds at $x^\star$:*

$$y \in N_{\mathrm{dom}\, h}\big(F(x^\star)\big) \text{ and } 0 \in \partial_x f(x^\star, y) + N_X(x^\star) \implies y = 0. \tag{6.16}$$

*If $x^\star$ is a local minimizer of the problem*

$$\operatorname*{minimize}_{x \in X} f_0(x) + h\big(F(x)\big),$$

*then*

$$\exists y \in \partial h\big(F(x^\star)\big) \;\; \text{such that} \;\; 0 \in \partial_x l(x^\star, y) + N_X(x^\star).$$

**Proof.** Let's define $G : \mathbb{R}^n \to \mathbb{R}^{n+1+m}$ by $G(x) = (x, f_0(x), F(x))$ and $g : \mathbb{R}^{n+1+m} \to \overline{\mathbb{R}}$ by

$$g(u) = \iota_X(u_1, \dots, u_n) + u_{n+1} + h(u_{n+2}, \dots, u_{n+1+m}).$$

Then,

$$\iota_X(x) + f_0(x) + h\big(F(x)\big) = g\big(G(x)\big)$$

with $g$ being proper and lsc, and $G$ being locally Lipschitz continuous.

We apply the general chain rule 6.23 to $g \circ G$ and first check the qualification (6.15). For $u \in \operatorname{dom} g$, one has

$$\partial^\infty g(u) \subset N_X(u_1, \dots, u_n) \times \{0\} \times N_{\operatorname{dom} h}(u_{n+2}, \dots, u_{n+1+m})$$

by 4.63, 4.56, 4.69 and 4.65. Let $\varphi_0 : \mathbb{R}^n \times \mathbb{R} \times \mathbb{R}^m \to \mathbb{R}$ and $\varphi : \mathbb{R}^n \times \mathbb{R}^n \times \mathbb{R} \times \mathbb{R}^m \to \mathbb{R}$ be defined by

$$\varphi_0(x, y_0, y) = f_0(x)y_0 + f(x, y) \quad \text{and} \quad \varphi(x, z, y_0, y) = \langle x, z\rangle + \varphi_0(x, y_0, y).$$

Using 4.58(c), we establish that

$$\partial_x \varphi(x^\star, z, y_0, y) = z + \partial_x \varphi_0(x^\star, y_0, y).$$

The scalar multiple of a locally Lipschitz continuous function as well as the sum of locally Lipschitz continuous functions are locally Lipschitz continuous. Thus, both $x \mapsto f_0(x)y_0$ and $f(\cdot, y)$ have this property. We can then apply 4.70 to obtain that

$$\partial_x \varphi_0(x^\star, y_0, y) \subset \partial(y_0 f_0)(x^\star) + \partial_x f(x^\star, y).$$

In the case of $g \circ G$, the qualification (6.15) becomes

$$(z, y_0, y) \in \partial^\infty g\big(G(x^\star)\big) \;\; \text{and} \;\; 0 \in \partial_x \varphi(x^\star, z, y_0, y) \implies z = 0, \; y_0 = 0, \; y = 0.$$

In view of the previous development, the conditions on the left-hand side mean that

$$z \in N_X(x^\star), \;\; y_0 = 0, \;\; y \in N_{\operatorname{dom} h}\big(F(x^\star)\big), \;\; 0 \in z + \partial_x f(x^\star, y).$$

Thus, $y = 0$ by (6.16). Since $f(\cdot, 0)$ is the zero function, $\partial_x f(x^\star, 0) = \{0\}$ and $0 \in z + \partial_x f(x^\star, 0)$ guarantees that $z = 0$ as well. Thus, the required qualification holds and we obtain from 6.23 that

$$\partial(g \circ G)(x^\star) \subset \bigcup_{(z, y_0, y) \in \partial g(G(x^\star))} \partial_x \varphi(x^\star, z, y_0, y).$$

For $u \in \operatorname{dom} g$, one has

$$\partial g(u) = N_X(u_1, \dots, u_n) \times \{1\} \times \partial h(u_{n+2}, \dots, u_{n+1+m})$$

by 4.63, 4.56 and 4.58. The Fermat rule 4.73 ensures that $0 \in \partial(g \circ G)(x^\star)$ and from this the conclusion follows. □

**Exercise 6.26** (max-function as constraint). For smooth $f_i : \mathbb{R}^n \to \mathbb{R}$, $i = 0, 1, \dots, m$, consider the problem

$$\underset{x \in \mathbb{R}^n}{\text{minimize}}\ f_0(x) + \iota_{(-\infty,0]}\big(\max_{i=1,\dots,m} f_i(x)\big).$$

Obtain an optimality condition for the problem by setting $h = \iota_{(-\infty,0]}$ and $F(x) = \max_{i=1,\dots,m} f_i(x)$ in 6.25. Compare the resulting condition with the KKT condition 4.77 for the equivalent problem of minimizing $f_0(x)$ subject to $f_i(x) \leq 0$, $i = 1, \dots, m$.

**Guide.** Use (4.13) and 4.66 to obtain subgradients of $x \mapsto y \max_{i=1,\dots,m} f_i(x)$ for positive values of $y$. □

## 6.E Sup-Projections

We've encountered on several occasions the need for minimizing functions of the form

$$f(x) = \sup_{\alpha \in \mathbb{A}} f_\alpha(x), \tag{6.17}$$

which are referred to as *sup-projections*. In the context of duality, the negative of a Lagrangian would furnish $f_\alpha(x)$. When a decision needs to be robust against a worst-case situation, the index set $\mathbb{A}$ could be a set of parameter values representing the relevant situations. Regardless of the context, it becomes necessary to establish continuity properties of sup-projections as well as formulae for subgradients; convexity of $f$ can be checked using 1.18(a). For the special case with a finite $\mathbb{A}$, we refer to 4.66.

**Proposition 6.27** (semicontinuity of sup-projection). *For a nonempty set $\mathbb{A}$ and a family of functions $\{f_\alpha : \mathbb{R}^n \to \mathbb{R}, \alpha \in \mathbb{A}\}$, the sup-projection $f : \mathbb{R}^n \to \overline{\mathbb{R}}$ given by (6.17) is lsc at $\bar{x}$ provided that $f_\alpha$ is lsc at $\bar{x}$ for all $\alpha \in \mathbb{A}$.*

*If the family is locally bounded and equicontinuous relative to $X \subset \mathbb{R}^n$ and* $\operatorname{argmax}_{\alpha \in \mathbb{A}} f_\alpha(x)$ *is nonempty for all $x \in X$, then $f$ is also continuous relative to $X$, i.e.,*

$$f(x^\nu) \to f(\bar{x}) \quad \text{when} \quad x^\nu \in X \to \bar{x} \in X.$$

**Proof.** Suppose that $x^\nu \to \bar{x}$. In view of (5.1), we need to show that $\liminf f(x^\nu) \geq f(\bar{x})$. Suppose that $f(\bar{x}) \in \mathbb{R}$ and let $\varepsilon > 0$. Then, there exists $\bar{\alpha} \in \mathbb{A}$ such that $f(\bar{x}) \leq f_{\bar{\alpha}}(\bar{x}) + \varepsilon$. There's also $\bar{\nu}$ such that $f_{\bar{\alpha}}(\bar{x}) \leq f_{\bar{\alpha}}(x^\nu) + \varepsilon$ for all $\nu \geq \bar{\nu}$ because $f_{\bar{\alpha}}$ is lsc at $\bar{x}$; see (5.1). Thus, for $\nu \geq \bar{\nu}$,

$$f(\bar{x}) \leq f_{\bar{\alpha}}(\bar{x}) + \varepsilon \leq f_{\bar{\alpha}}(x^\nu) + 2\varepsilon \leq f(x^\nu) + 2\varepsilon.$$

Since $\varepsilon > 0$ is arbitrary, $\liminf f(x^\nu) \geq f(\bar{x})$. If $f(\bar{x}) = \infty$, then a similar argument establishes that $f(x^\nu) \to \infty$. This confirms the first assertion.

Next, suppose that $x^\nu \in X \to \bar{x} \in X$. Since the equicontinuity assumption implies continuity of $f_\alpha$ relative to $X$, we've from the first result that $\liminf f(x^\nu) \geq f(\bar{x})$. Let

$$\bar{\alpha}^\nu \in \operatorname{argmax}_{\alpha \in \mathbb{A}} f_\alpha(x^\nu).$$

The equicontinuity assumption ensures that $f_{\bar{\alpha}^\nu}(x^\nu) - f_{\bar{\alpha}^\nu}(\bar{x})$ tends to 0. Thus,

$$\begin{aligned}\limsup f(x^\nu) &= \limsup \big(f_{\bar{\alpha}^\nu}(x^\nu) - f_{\bar{\alpha}^\nu}(\bar{x}) + f_{\bar{\alpha}^\nu}(\bar{x})\big)\\ &\leq \limsup \big(f_{\bar{\alpha}^\nu}(x^\nu) - f_{\bar{\alpha}^\nu}(\bar{x}) + f(\bar{x})\big) = f(\bar{x})\end{aligned}$$

because $f_{\bar{\alpha}^\nu}(\bar{x}) \leq f(\bar{x})$ and the conclusion follows. □

**Example 6.28** (absence of equicontinuity). Trouble arises when a family of functions $\{f_\alpha, \alpha \in \mathbb{A}\}$ isn't locally bounded and equicontinuous at a point. Let's consider $f_\alpha(x) = \alpha x$ for $\alpha \in \mathbb{A} = [0, \infty)$. Then,

$$f(x) = \sup_{\alpha \in \mathbb{A}} f_\alpha(x) = \begin{cases} 0 & \text{if } x \leq 0 \\ \infty & \text{otherwise.} \end{cases}$$

While $f$ is lsc by 6.27, which is also seen directly, it isn't continuous at 0.

**Detail.** The issue is that for $\bar{x} = 0$,

$$\big|f_\alpha(x) - f_\alpha(\bar{x})\big| = |\alpha x|$$

isn't uniformly small for all $\alpha \in [0, \infty)$ and $x$ near 0. Thus, $\{f_\alpha, \alpha \in \mathbb{A}\}$ isn't locally bounded and equicontinuous at 0. This takes place even though $(\alpha, x) \mapsto f_\alpha(x)$ is continuous.

The issue doesn't occur if $\mathbb{A}$ is replaced by the compact set $[0, 1]$. Then, $f(x) = 0$ if $x \leq 0$ and $f(x) = x$ otherwise. The assumptions of 6.27 hold and $f$ is continuous. □

When the index set $\mathbb{A}$ is a subset of $\mathbb{R}^m$, the (semi-)continuity of sup-projections can be established by drawing on earlier results for inf-projections through the relation

$$f(x) = \sup_{y \in Y} g(x, y) = -\inf_{y \in Y} -g(x, y).$$

**Proposition 6.29** (parametric sup-projections). *For a continuous function $g : \mathbb{R}^n \times \mathbb{R}^m \to \mathbb{R}$ and a nonempty compact set $Y \subset \mathbb{R}^m$, the family of functions $\{g(\cdot, y), y \in Y\}$ is locally bounded and equicontinuous. Moreover, the function given by*

$$f(x) = \sup_{y \in Y} g(x, y)$$

*is continuous.*

**Proof.** Continuity follows from 5.6(c) and the remark given before the proposition. To establish local boundedness, let $\bar{x} \in \mathbb{R}^n$. Since $g$ is continuous, $(x, y) \mapsto |g(x, y)|$ is also continuous and it follows by 4.9 that

$$\sup_{x \in \mathbb{B}(\bar{x},1), y \in Y} |g(x, y)|$$

is finite. Thus, the family is locally bounded at $\bar{x}$. For equicontinuity, let $\varepsilon > 0$ and

$$\varphi(x, y) = |g(x, y) - g(\bar{x}, y)|,$$

which defines a continuous function. The sup-projection given by

$$\psi(x) = \sup_{y \in Y} \varphi(x, y)$$

is continuous by application of the continuity-portion of the proposition. Consequently, there's $\delta > 0$ such that

$$\varepsilon \geq |\psi(x) - \psi(\bar{x})| = \sup_{y \in Y} \varphi(x, y) \quad \forall x \in \mathbb{B}(\bar{x}, \delta),$$

which is exactly what's needed. □

Typically, sup-projections aren't smooth. Even the simple example $f(x) = \sup_{y \in [-1,1]} xy$, which actually has $f(x) = |x|$ for all $x$, is nonsmooth. They still possess several useful properties and, in particular, approachable expressions for subgradients and subderivatives.

**Proposition 6.30** (subdifferentiablity of sup-projections). *For a continuous function $g : \mathbb{R}^n \times \mathbb{R}^m \to \mathbb{R}$ and a nonempty compact set $Y \subset \mathbb{R}^m$, suppose that $g(\cdot, y)$ is smooth for all $y \in Y$ and $(x, y) \mapsto \nabla_x g(x, y)$ is continuous relative to $\mathbb{R}^n \times Y$. Then, the sup-projection $f : \mathbb{R}^n \to \mathbb{R}$ given by*

$$f(x) = \sup_{y \in Y} g(x, y)$$

*is locally Lipschitz continuous and epi-regular at any $\bar{x} \in \mathbb{R}^n$ and, with $Y^\star(\bar{x}) = \operatorname{argmax}_{y \in Y} g(\bar{x}, y)$, one has*

$$\partial f(\bar{x}) = \operatorname{con}\{\nabla_x g(\bar{x}, y), \ y \in Y^\star(\bar{x})\}$$
$$df(\bar{x}; w) = \max_{y \in Y^\star(\bar{x})} \langle \nabla_x g(\bar{x}, y), w \rangle \quad \forall w \in \mathbb{R}^n.$$

**Proof.** Let $\bar{x} \in \mathbb{R}^n$. First, consider the Lipschitz property. Since the function $(x, y) \mapsto \|\nabla_x g(x, y)\|_2$ is continuous relative to $\mathbb{R}^n \times Y$ and $Y$ is compact, it follows by 4.9 that

$$\kappa = \sup \{\|\nabla_x g(x, y)\|_2 \mid y \in Y, \ x \in \mathbb{B}(\bar{x}, 1)\}$$

is finite. Let's fix $x, x' \in \mathbb{B}(\bar{x}, 1)$ and $\bar{y} \in Y$. By the mean value theorem (1.7), there's $\sigma \in [0, 1]$ such that

$$\begin{aligned}\big|g(x,\bar y)-g(x',\bar y)\big| &= \big|\langle\nabla_x g\big(x'+\sigma(x-x'),\bar y\big),x-x'\rangle\big|\\ &\le \big\|\nabla_x g\big(x'+\sigma(x-x'),\bar y\big)\big\|_2\|x-x'\|_2\le\kappa\|x-x'\|_2.\end{aligned}$$

Thus, $|g(x,y)-g(x',y)|\le\kappa\|x-x'\|_2$ for all $y\in Y$ and

$$f(x)=\sup_{y\in Y} g(x,y)\le\sup_{y\in Y} g(x',y)+\kappa\|x-x'\|_2=f(x')+\kappa\|x-x'\|_2.$$

Reversing the roles of $x$ and $x'$, we obtain that $|f(x)-f(x')|\le\kappa\|x-x'\|_2$. Since $x,x'\in\mathbb{B}(\bar x,1)$ are arbitrary, $f$ is locally Lipschitz continuous at $\bar x$.

Second, we turn to the formula for subderivatives. By 4.9, $Y^\star(x)=\operatorname{argmax}_{y\in Y} g(x,y)$ is compact and nonempty for all $x\in\mathbb{R}^n$. Let's fix $\bar w\in\mathbb{R}^n$. For $\tau\in(0,\infty)$ and $\hat y\in Y^\star(\bar x+\tau\bar w)$, one has $f(\bar x+\tau\bar w)=g(\bar x+\tau\bar w,\hat y)$ and $f(\bar x)\ge g(\bar x,\hat y)$ so that

$$\begin{aligned}\frac{f(\bar x+\tau\bar w)-f(\bar x)}{\tau} &\le \frac{g(\bar x+\tau\bar w,\hat y)-g(\bar x,\hat y)}{\tau}\\ &\le \max_{\sigma\in[0,1]}\langle\nabla_x g(\bar x+\sigma\tau\bar w,\hat y),\bar w\rangle\\ &\le \max_{\sigma\in[0,1]}\max_{y\in Y^\star(\bar x+\tau\bar w)}\langle\nabla_x g(\bar x+\sigma\tau\bar w,y),\bar w\rangle,\end{aligned}$$

where the second inequality follows by the mean value theorem (1.7). Let $\tau^\nu\searrow 0$. By definition 2.8 for subderivatives,

$$\begin{aligned} df(\bar x;\bar w) &= \lim_{\delta\searrow 0}\inf\left\{\frac{f(\bar x+\tau w)-f(\bar x)}{\tau}\;\middle|\;\tau\in(0,\delta],\ w\in\mathbb{B}(\bar w,\delta)\right\}\\ &\le \limsup\frac{f(\bar x+\tau^\nu\bar w)-f(\bar x)}{\tau^\nu}\\ &\le \limsup\max_{\sigma\in[0,1]}\max_{y\in Y^\star(\bar x+\tau\bar w)}\langle\nabla_x g(\bar x+\sigma\tau^\nu\bar w,y),\bar w\rangle. \end{aligned}\tag{6.18}$$

Let's consider the function $\varphi:(0,\infty)\to\mathbb{R}$ given by

$$\varphi(\tau)=\max_{\sigma\in[0,1]}\max_{y\in Y^\star(\bar x+\tau\bar w)}\langle\nabla_x g(\bar x+\sigma\tau\bar w,y),\bar w\rangle.$$

As an intermediate result, we establish that

$$\operatorname{LimOut} Y^\star(\bar x+\tau^\nu\bar w)\subset Y^\star(\bar x).\tag{6.19}$$

To see this, note that if $\bar y\in\operatorname{LimOut} Y^\star(\bar x+\tau^\nu\bar w)$, then there are $N\in\mathcal{N}_\infty^\#$ and $y^\nu\in Y^\star(\bar x+\tau^\nu\bar w)\to_N\bar y$. Since $Y$ is closed, $\bar y\in Y$. Let $\varepsilon>0$. By 6.29, $f$ is continuous, which together with the continuity of $g$, mean that there's $\bar\nu$ such that for all $\nu\ge\bar\nu$ and $\nu\in N$,

$$f(\bar x)\le f(\bar x+\tau^\nu\bar w)+\varepsilon\quad\text{and}\quad g(\bar x+\tau^\nu\bar w,y^\nu)\le g(\bar x,\bar y)+\varepsilon.$$

Consequently, for $\nu\ge\bar\nu$ and $\nu\in N$,

$$f(\bar{x}) \le f(\bar{x} + \tau^\nu \bar{w}) + \varepsilon = g(\bar{x} + \tau^\nu \bar{w}, y^\nu) + \varepsilon \le g(\bar{x}, \bar{y}) + 2\varepsilon.$$

Since $\varepsilon$ is arbitrary, this means that $f(\bar{x}) \le g(\bar{x}, \bar{y})$ and $\bar{y} \in Y^\star(\bar{x})$. We've established (6.19).

Next, let

$$(\sigma^\nu, y^\nu) \in \operatorname{argmax} \Big\{ \big\langle \nabla_x g(\bar{x} + \sigma\tau^\nu \bar{w}, y), \bar{w} \big\rangle \;\Big|\; \sigma \in [0, 1],\; y \in Y^\star(\bar{x} + \tau^\nu \bar{w}) \Big\},$$

which exists by 4.9. Since $\{\sigma^\nu, y^\nu, \nu \in \mathbb{N}\}$ are contained in the compact set $[0, 1] \times Y$, there are $\bar{\sigma} \in [0, 1]$, $\bar{y} \in Y$ and $N \in \mathcal{N}_\infty^\#$ such that $\sigma^\nu \to_N \bar{\sigma}$ and $y^\nu \to_N \bar{y}$. Consequently,

$$\varphi(\tau^\nu) = \big\langle \nabla_x g(\bar{x} + \sigma^\nu \tau^\nu \bar{w}, y^\nu), \bar{w} \big\rangle \to_N \big\langle \nabla_x g(\bar{x}, \bar{y}), \bar{w} \big\rangle.$$

Since (6.19) holds, $\bar{y} \in Y^\star(\bar{x})$ so that

$$\big\langle \nabla_x g(\bar{x}, \bar{y}), \bar{w} \big\rangle \le \max_{y \in Y^\star(\bar{x})} \big\langle \nabla_x g(\bar{x}, y), \bar{w} \big\rangle.$$

We've shown that

$$\operatorname{limsup}_{\nu \in N} \varphi(\tau^\nu) \le \max_{y \in Y^\star(\bar{x})} \big\langle \nabla_x g(\bar{x}, y), \bar{w} \big\rangle.$$

One can then argue by contradiction to establish that $\nu \in N$ can here be replaced by $\nu \in \mathbb{N}$. Using this fact and returning to (6.18), we obtain that

$$df(\bar{x}; \bar{w}) \le \max_{y \in Y^\star(\bar{x})} \big\langle \nabla_x g(\bar{x}, y), \bar{w} \big\rangle.$$

The corresponding lower bound is more easily achieved. Let $\bar{y} \in Y^\star(\bar{x})$. For $\tau \in (0, \infty)$ and $w \in \mathbb{R}^n$, one has $f(\bar{x} + \tau w) \ge g(\bar{x} + \tau w, \bar{y})$ and

$$\frac{f(\bar{x} + \tau w) - f(\bar{x})}{\tau} \ge \frac{g(\bar{x} + \tau w, \bar{y}) - g(\bar{x}, \bar{y})}{\tau}.$$

This fact together with the definition of subderivatives in 2.8 establish that

$$\begin{aligned} df(\bar{x}; \bar{w}) &\ge \lim_{\delta \searrow 0} \inf \left\{ \frac{g(\bar{x} + \tau w, \bar{y}) - g(\bar{x}, \bar{y})}{\tau} \;\middle|\; \tau \in (0, \delta],\; w \in \mathbb{B}(\bar{w}, \delta) \right\} \\ &= dg(\bar{x}, \bar{y}; \bar{w}) = \big\langle \nabla_x g(\bar{x}, \bar{y}), \bar{w} \big\rangle, \end{aligned}$$

where the last equality follows by 2.9. Since $\bar{y} \in Y^\star(\bar{x})$ is arbitrary, we also have

$$df(\bar{x}; \bar{w}) \ge \max_{y \in Y^\star(\bar{x})} \big\langle \nabla_x g(\bar{x}, y), \bar{w} \big\rangle$$

and the formula for subderivatives is established.

Third, we establish epi-regularity. Since $f$ is continuous, epi $f$ is closed. It suffices to show that

$$\widehat{N}_{\operatorname{epi} f}\big(\bar{x}, f(\bar{x})\big) \supset N_{\operatorname{epi} f}\big(\bar{x}, f(\bar{x})\big).$$

Let $(\bar{v}, \bar{\alpha}) \in N_{\mathrm{epi}\, f}(\bar{x}, f(\bar{x}))$. Then, by definition 4.35, there are $x^\nu \to \bar{x}$, $v^\nu \to \bar{v}$ and $\alpha^\nu \to \bar{\alpha}$ such that

$$\big\langle (v^\nu, \alpha^\nu),\ (w, \beta)\big\rangle \leq 0 \quad \forall (w, \beta) \in T_{\mathrm{epi}\, f}\big(x^\nu, f(x^\nu)\big) = \mathrm{epi}\, df(x^\nu; \cdot\,), \tag{6.20}$$

where the connection with subderivatives is furnished by 4.54. Next, we show that

$$\mathrm{LimInn}\big(\, \mathrm{epi}\, df(x^\nu; \cdot\,)\big) \supset \mathrm{epi}\, df(\bar{x}; \cdot\,). \tag{6.21}$$

Let $(\bar{w}, \bar{\beta}) \in \mathrm{epi}\, df(\bar{x}; \cdot\,)$ and construct $w^\nu = \bar{w}$ and $\beta^\nu = \max\{\bar{\beta}, df(x^\nu, \bar{w})\}$ for all $\nu$. Then,

$$(w^\nu, \beta^\nu) \in \mathrm{epi}\, df(x^\nu; \cdot\,).$$

By 4.9 and the expression for subderivatives just proven, there is a sequence $\{y^\nu \in Y^\star(x^\nu), \nu \in \mathbb{N}\}$ such that

$$df(x^\nu; \bar{w}) = \big\langle \nabla_x g(x^\nu, y^\nu), \bar{w} \big\rangle.$$

Since $Y$ is compact, there are $\bar{y} \in Y$ and $N \in \mathcal{N}_\infty^\#$ such that $y^\nu \to_N \bar{y}$. Then,

$$df(x^\nu; \bar{w}) = \big\langle \nabla_x g(x^\nu, y^\nu), \bar{w} \big\rangle \to_N \big\langle \nabla_x g(\bar{x}, \bar{y}), \bar{w} \big\rangle.$$

We can deduce from the earlier arguments that $\mathrm{LimOut}\, Y^\star(x^\nu) \subset Y^\star(\bar{x})$. Thus, $\bar{y} \in Y^\star(\bar{x})$ and

$$\mathrm{limsup}_{\nu \in N}\, df(x^\nu; \bar{w}) = \big\langle \nabla_x g(\bar{x}, \bar{y}), \bar{w} \big\rangle \leq \max_{y \in Y^\star(\bar{x})} \big\langle \nabla_x g(\bar{x}, y), \bar{w} \big\rangle = df(\bar{x}; \bar{w}).$$

Arguing by contradiction, we can establish that

$$\mathrm{limsup}\, df(x^\nu; \bar{w}) \leq df(\bar{x}; \bar{w})$$

as well. This means that $\beta^\nu \to \bar{\beta}$ because $df(\bar{x}; \bar{w}) \leq \bar{\beta}$. We've proven that

$$(\bar{w}, \bar{\beta}) \in \mathrm{LimInn}\,\big(\, \mathrm{epi}\, df(x^\nu; \cdot\,)\big)$$

and (6.21) holds.

Next, let

$$(\bar{w}, \bar{\beta}) \in T_{\mathrm{epi}\, f}\big(\bar{x}, f(\bar{x})\big) = \mathrm{epi}\, df(\bar{x}; \cdot\,).$$

From (6.21), there are $(\bar{w}^\nu, \bar{\beta}^\nu) \in \mathrm{epi}\, df(x^\nu; \cdot\,) \to (\bar{w}, \bar{\beta})$. It then follows from (6.20) that

$$\big\langle (v^\nu, \alpha^\nu),\ (\bar{w}^\nu, \bar{\beta}^\nu)\big\rangle \leq 0 \quad \forall \nu \in \mathbb{N}$$

and thus

$$\big\langle (\bar{v}, \bar{\alpha}),\ (\bar{w}, \bar{\beta})\big\rangle \leq 0.$$

Since $(\bar{w}, \bar{\beta}) \in T_{\mathrm{epi}\, f}(\bar{x}, f(\bar{x}))$ is arbitrary, we've shown that $(\bar{v}, \bar{\alpha}) \in \widehat{N}_{\mathrm{epi}\, f}(\bar{x}, f(\bar{x}))$.

Fourth, we consider subgradients. By 5.29, the conjugate of $\iota_{\partial f(\bar{x})}$ is the function

$$w \mapsto \sup_{v \in \partial f(\bar{x})} \langle v, w \rangle,$$

which we recognize from 4.59 as the subderivative $df(\bar{x}; \cdot)$ because $f$ is epi-regular at $\bar{x}$ as just established. The obtained formula for this subderivative can equivalently be stated as

$$w \mapsto \max_{v \in C} \langle v, w \rangle, \quad \text{where } C = \big\{\nabla_x g(\bar{x}, y) \,\big|\, y \in Y^\star(\bar{x})\big\}.$$

By 5.29, the conjugate of $w \mapsto \max_{v \in C} \langle v, w \rangle$ is $\iota_{\mathrm{cl\,con}\, C}$. From 4.69, we see that $\partial f(\bar{x})$ is nonempty and compact because $f$ is locally Lipschitz continuous. It's also convex by virtue of being the intersection of convex sets; see 4.59. Thus, $\iota_{\partial f(\bar{x})}$ is proper, lsc and convex and we can invoke the Fenchel-Moreau theorem 5.23 to confirm that the conjugate of the conjugate of $\iota_{\partial f(\bar{x})}$ is simply $\iota_{\partial f(\bar{x})}$. We've shown that

$$\iota_{\partial f(\bar{x})} = \iota_{\mathrm{cl\,con}\, C}.$$

Since $C$ is compact due to the compactness of $Y^\star(\bar{x})$, $\mathrm{con}\, C$ is closed (see [105, Theorem 2.30]) and we conclude that $\partial f(\bar{x}) = \mathrm{con}\, C$. □

Duality is a prime source of sup-projections. We recall from Chap. 5 that for a Lagrangian $l : \mathbb{R}^n \times \mathbb{R}^m \to \overline{\mathbb{R}}$, a dual problem takes the form of maximizing $\psi(y) = \inf_{x \in \mathbb{R}^n} l(x, y)$ over $y \in \mathbb{R}^m$. With a reorientation towards minimization, the goal becomes to

$$\underset{y \in \mathbb{R}^m}{\text{minimize}}\ \tilde{\psi}(y) = \sup_{x \in \mathbb{R}^n} -l(x, y), \tag{6.22}$$

where $\tilde{\psi}$ is a sup-projection. Since $\tilde{\psi}$ is convex (see the discussion after 5.27), the subgradient method of §2.I is a natural candidate for minimizing $\tilde{\psi}$ and then subgradients become essential. They can be computed via 6.30 and involve differentiation of $l$ with respect to $y$ as well as

$$\bar{x} \in \operatorname{argmax}_{x \in \mathbb{R}^n} -l(x, \bar{y}).$$

Thus, the subgradient method in this case requires us to minimize the Lagrangian for a candidate dual solution repeatedly. Let's make this concrete in the context of monitoring functions.

For nonempty compact $X \subset \mathbb{R}^n$, continuous $f_0 : \mathbb{R}^n \to \mathbb{R}$, continuous $F : \mathbb{R}^n \to \mathbb{R}^m$, nonempty polyhedral $Y \subset \mathbb{R}^m$ and a symmetric positive semidefinite $m \times m$-matrix $B$, suppose that the problem of interest is

$$\underset{x \in X}{\text{minimize}}\ f_0(x) + h_{Y,B}\big(F(x)\big), \tag{6.23}$$

which under a particular Rockafellian leads to the Lagrangian

$$l(x, y) = \iota_X(x) + f_0(x) + \big\langle F(x), y \big\rangle - \tfrac{1}{2}\langle y, By \rangle - \iota_Y(y)$$

by the arguments in §5.I. In this case, (6.22) becomes

$$\underset{y \in Y}{\text{minimize}} \ \sup_{x \in X} \left\{ -f_0(x) - \langle F(x), y \rangle + \tfrac{1}{2} \langle y, By \rangle \right\}$$

and the subgradient method applies. This results in an algorithm for (6.22) and thus also for the dual problem of (6.23).

**Dual Algorithm.**

Data. $y^0 \in Y$ and $\lambda^\nu > 0$, $\nu = 0, 1, 2, \ldots$.
Step 0. Set $\nu = 0$.
Step 1. Compute

$$x^\nu \in \operatorname{argmin}_{x \in X} f_0(x) + \langle F(x), y^\nu \rangle.$$

Step 2. Set $v = -F(x^\nu) + By^\nu$.
Step 3. Set

$$y^{\nu+1} \in \operatorname{prj}_Y(y^\nu - \lambda^\nu v).$$

Step 4. Replace $\nu$ by $\nu + 1$ and go to Step 1.

Certainly, the function

$$(x, y) \mapsto -f_0(x) - \langle F(x), y \rangle + \tfrac{1}{2} \langle y, By \rangle$$

is continuous because $f_0$ and $F$ are continuous. Likewise, as a function of $y$ only, it's also smooth and its gradient $-F(x) + By$ at $y$ is continuous, jointly in $(x, y)$, and the assumptions of 6.30 are satisfied for the sup-projection

$$y \mapsto \sup_{x \in X} \left\{ -f_0(x) - \langle F(x), y \rangle + \tfrac{1}{2} \langle y, By \rangle \right\}.$$

Step 1 of the algorithm computes a maximizer in the expression for the current $y = y^\nu$; the term $\frac{1}{2}\langle y, By \rangle$ is dropped since it doesn't influence the maximizer. Step 2 simply records a subgradient; see 6.30. Since $Y$ is polyhedral, the projection in Step 3 can be carried out by solving a quadratic optimization problem. Often $Y$ is simple and this can be achieved with little effort.

Typically, the main bottleneck is Step 1 and the details of a specific application determines whether the dual algorithm is competitive. For example, if $m$ is large, possibly representing many constraints, then the problem in Step 1 might be much simpler than the actual problem (6.23) because the high dimensional $F(x)$ is being collapsed into a single number $\langle F(x), y^\nu \rangle$. Regardless of ease of computation, every $(x^\nu, y^\nu)$ in the dual algorithm furnishes a lower bound

$$f_0(x^\nu) + \langle F(x^\nu), y^\nu \rangle - \tfrac{1}{2} \langle y^\nu, By^\nu \rangle$$

on the minimum value of the actual problem (6.23) by weak duality 5.25. An upper bound is given by $f_0(x^\nu) + h_{Y,B}(F(x^\nu))$. Thus, the optimality gap for $x^\nu$ is

$$\sup_{y \in Y} \left\{ \langle F(x^\nu), y \rangle - \tfrac{1}{2} \langle y, By \rangle \right\} - \langle F(x^\nu), y^\nu \rangle + \tfrac{1}{2} \langle y^\nu, By^\nu \rangle.$$

Moreover, as discussed prior to 5.42 and seen directly here, $x^\nu$ is a minimizer of (6.23) if $y^\nu$ maximizes $y \mapsto \langle F(x^\nu), y\rangle - \frac{1}{2}\langle y, By\rangle$ over $Y$.

**Exercise 6.31** (dual algorithm). For $f_0(x) = \frac{1}{2}x_1^2 + \frac{1}{2}x_2^2$, $F(x) = (x_1-2)^2 + x_2^2 - 1$, $Y = [0, \infty)$ and $B = 0$, solve (6.23) using the dual algorithm. Start with $y^0 = 0$ and explore the effect of different step sizes $\lambda^\nu$.

**Guide.** The unique minimizer is $(1, 0)$ with corresponding multiplier $1/2$. □

**Example 6.32** (saving the condor). From a study of the condor population in California, we've data about the bands of DNA fragments associated with each individual in a group of $m$ condors. Specifically, $\xi_{ij} = 1$ if condor $i$ has DNA band $j$, and $\xi_{ij} = 0$ otherwise. Let $\xi = (\xi_{ij}, i = 1, \dots, m, j = 1, \dots, n)$. A maximum likelihood estimate of the corresponding probability distribution can be obtained by minimizing the sup-projection given by

$$f(x) = \sup_{y \in Y} \frac{1}{nm}\big\{g(x, y) - h(y) - g(x, \xi)\big\},$$

where

$$x = (x_i, i = 1, \dots, m+n;\ x_{ik}, i = 1, \dots, m-1, k = i+1, \dots, m)$$

and, for $y = (y_{ij}, i = 1, \dots, m, j = 1, \dots, n) \in Y = [0, 1]^{nm}$,

$$\begin{aligned} g(x, y) &= \sum_{i=1}^{m}\sum_{j=1}^{n} x_i y_{ij} + \sum_{i=1}^{m}\sum_{j=1}^{n} x_{m+j} y_{ij} \\ &\quad + \frac{1}{m}\sum_{i=1}^{m-1}\sum_{k=i+1}^{m}\sum_{j=1}^{n} x_{ik} y_{ij} y_{kj} \\ h(y) &= \sum_{i=1}^{m}\sum_{j=1}^{n} y_{ij} \ln y_{ij} + (1 - y_{ij}) \ln(1 - y_{ij}). \end{aligned}$$

Here, $0 \ln 0 = 0$. Background information about the statistical model can be found in [24].

**Detail.** The function $g(\cdot, y)$ is linear for all $y \in Y$ so $f$ is convex by 1.18(a). The function given by

$$\tilde{g}(x, y) = g(x, y) - h(y) - g(x, \xi)$$

is continuous relative to $\mathbb{R}^d \times Y$, where $d$ is the dimension of $x$. If we set $\tilde{g}(x, y)$ to $g(x, y) - g(x, \xi)$ for $y \notin Y$, then $\tilde{g}$ is continuous. The function $\tilde{g}(\cdot, y)$ is smooth for all $y \in Y$ and $(x, y) \mapsto \nabla_x \tilde{g}(x, y)$ is continuous relative to $\mathbb{R}^d \times Y$. These facts together with the compactness of $Y$ ensure that all the assumptions in 6.30 hold. According to the resulting formula, a subgradient of $f$ at $x^\nu$ depends on a maximizer of $\tilde{g}(x^\nu, \cdot)$ over $Y$ and this can be challenging to compute because $g(x^\nu, \cdot)$ is quadratic but possibly not concave. □

**Example 6.33** (approximation of sup-projection). For continuous $g : \mathbb{R}^n \times \mathbb{R}^m \to \mathbb{R}$, nonempty compact $Y \subset \mathbb{R}^m$ and finite $Y^\nu \subset Y$, consider the sup-projections given by

$$f^\nu(x) = \sup_{y \in Y^\nu} g(x, y) \quad \text{and} \quad f(x) = \sup_{y \in Y} g(x, y).$$

This sets up a simple algorithm for minimizing $f$ via the epigraphical approximation algorithm. If for each $\bar{y} \in Y$ there's $y^\nu \in Y^\nu \to \bar{y}$, then $f^\nu \xrightarrow{e} f$.

**Detail.** Epi-convergence is established by the characterization 4.15. Suppose that $x^\nu \to x$. By 4.9, there's $\bar{y} \in \operatorname{argmax}_{y \in Y} g(x, y)$. The assumption on $Y^\nu$ ensures that there's $y^\nu \in Y^\nu \to \bar{y}$. Thus,

$$\liminf f^\nu(x^\nu) \geq \liminf g(x^\nu, y^\nu) = g(x, \bar{y}) = f(x)$$

because $g$ is continuous. We've established 4.15(a). For 4.15(b), we only need to recognize that $f^\nu(x) \leq f(x)$ for all $x \in \mathbb{R}^n$ because $Y^\nu \subset Y$. Consequently, $f^\nu \xrightarrow{e} f$.

There's much flexibility in how $Y^\nu$ is constructed. For example, if $Y = [0, 1]^m$, then $Y^\nu = \{1/\nu, 2/\nu, \ldots \nu/\nu\}^m$ satisfies the requirement that any point in $Y$ can be approached by a sequence of points in $\{Y^\nu, \nu \in \mathbb{N}\}$. Although $Y^\nu$ is finite, its cardinality is easily large when $m > 2$, which could make minimization of $f^\nu$ less straightforward and one might have to combine this approach with an active-set strategy; see 6.16. □

## 6.F Proximal Composite Method

The proximal point method serves as the basis for solving several convex problems, but can also be extended to the nonconvex setting. For a function $h : \mathbb{R}^m \to \overline{\mathbb{R}}$, a nonempty closed set $X \subset \mathbb{R}^n$ and a twice smooth mapping $F : \mathbb{R}^n \to \mathbb{R}^m$, let's consider the problem

$$\operatorname*{minimize}_{x \in X} h\big(F(x)\big). \tag{6.24}$$

Given $\lambda > 0$, the proximal point method for this problem (see §1.H) would involve computing

$$x^{\nu+1} \in \operatorname{argmin}_{x \in X} h\big(F(x)\big) + \tfrac{1}{2\lambda}\|x - x^\nu\|_2^2.$$

Typically, this step isn't implementable, even if $h$ and $X$ are convex, because the subproblem could very well be nonconvex. An important exception is furnished next.

**Proposition 6.34** (weak convexity of composite function). *If convex $h : \mathbb{R}^m \to \mathbb{R}$ and smooth $F : \mathbb{R}^n \to \mathbb{R}^m$ satisfy the Lipschitz conditions*

$$\big|h(u) - h(\bar{u})\big| \leq \kappa\|u - \bar{u}\|_2 \quad \forall u, \bar{u} \in \mathbb{R}^m$$
$$\big\|\nabla F(x) - \nabla F(\bar{x})\big\|_F \leq \mu\|x - \bar{x}\|_2 \quad \forall x, \bar{x} \in \mathbb{R}^n$$

*for $\kappa, \mu \in [0, \infty)$, then $f = h \circ F$ is $\kappa\mu$-weakly convex.*

**Proof.** By definition, $f$ is $\kappa\mu$-weakly convex if $\varphi = f + \frac{1}{2}\kappa\mu\|\cdot\|_2^2$ is convex. Since $\varphi$ is real-valued, we deduce from [105, Theorem 8.13], which is a nonsmooth version of 1.22, that it suffices to show

$$\varphi(x) \geq \varphi(\bar{x}) + \langle \bar{v}, x - \bar{x}\rangle \quad \forall x, \bar{x} \in \mathbb{R}^n \text{ and } \bar{v} \in \partial\varphi(\bar{x}).$$

Let's fix $x$, $\bar{x}$ and $\bar{v} \in \partial\varphi(\bar{x})$. By 4.58(c) and 4.64,

$$\bar{v} = \nabla F(\bar{x})^\top \bar{w} + \kappa\mu\bar{x}$$

with $\bar{w} \in \partial h(F(\bar{x}))$; the qualification (4.14) holds by 4.65. The Lipschitz condition on the Jacobian of $F$ and the mean value theorem (1.11) produce

$$\big\|F(x) - F(\bar{x}) - \nabla F(\bar{x})(x - \bar{x})\big\|_2 \leq \tfrac{1}{2}\mu\|x - \bar{x}\|_2^2.$$

Then, leveraging this inequality as well as the subgradient inequality 2.17 applied to $h$ yield

$$\begin{aligned}
f(x) &\geq f(\bar{x}) + \big\langle \bar{w}, F(x) - F(\bar{x})\big\rangle \\
&= f(\bar{x}) + \big\langle \bar{w}, F(x) - F(\bar{x}) - \nabla F(\bar{x})(x - \bar{x})\big\rangle + \big\langle \bar{w}, \nabla F(\bar{x})(x - \bar{x})\big\rangle \\
&\geq f(\bar{x}) + \big\langle \bar{w}, \nabla F(\bar{x})(x - \bar{x})\big\rangle - \tfrac{1}{2}\mu\|\bar{w}\|_2\|x - \bar{x}\|_2^2 \\
&\geq f(\bar{x}) + \big\langle \nabla F(\bar{x})^\top \bar{w}, x - \bar{x}\big\rangle - \tfrac{1}{2}\kappa\mu\|x - \bar{x}\|_2^2,
\end{aligned}$$

where the last inequality is a consequence of 2.54. After adding $\frac{1}{2}\kappa\mu\|x\|_2^2$ to each side and rearranging some terms, we obtain that

$$f(x) + \tfrac{1}{2}\kappa\mu\|x\|_2^2 \geq f(\bar{x}) + \tfrac{1}{2}\kappa\mu\|\bar{x}\|_2^2 + \big\langle \nabla F(\bar{x})^\top \bar{w} + \kappa\mu\bar{x},\, x - \bar{x}\big\rangle,$$

which simply means that $\varphi(x) \geq \varphi(\bar{x}) + \langle \bar{v}, x - \bar{x}\rangle$. □

Under the conditions of the proposition, the subproblem in the proximal point method is convex when $\lambda \leq 1/(\kappa\mu)$. Since the arguments in 1.37 essentially carries over to the present context, we would also have that any cluster point $\bar{x}$ of $\{x^\nu, \nu \in \mathbb{N}\}$ generated in this manner satisfies an optimality condition for (6.24) expressed by subgradients and normal cones. Although this approach possesses such a convergence property, it suffers from the common difficulty of estimating Lipschitz moduli. Often, we end up with wildly conservative estimates and this produces a tiny $\lambda$, which in turn results in little progress in each iteration. This is especially troubling since the effort required to solve each subproblem might be substantial.

An alternative and usually more efficient approach is to approximate $F$ with an affine mapping constructed at the current point $x^\nu$. The subproblem then simplifies to

$$x^{\nu+1} \in \operatorname{argmin}_{x \in X}\, h\big(F(x^\nu) + \nabla F(x^\nu)(x - x^\nu)\big) + \tfrac{1}{2\lambda}\|x - x^\nu\|_2^2,$$

which usually is tractable if $X$ is relatively simple, for example, polyhedral, and $h$ is convex. If $h$ is a monitoring function and $X$ is a polyhedral set, then the subproblem can also be solved via a dual problem; see 5.57. As long as $\lambda$ is adjusted appropriately, this basic scheme leads to a surprisingly versatile algorithm for (6.24).

Let's assume that $X$ is also convex and $h$ is convex and real-valued, in which case 4.75 furnishes the following optimality condition for $x^\star$ to be a local minimizer of (6.24):

$$\exists y \in \partial h\big(F(x^\star)\big) \text{ such that } -\nabla F(x^\star)^\top y \in N_X(x^\star). \tag{6.25}$$

The condition holds without any qualification because dom $h = \mathbb{R}^n$. Our goal is to compute a point that satisfies this condition.

**Proximal Composite Method.**

Data. $x^0 \in X$, $\tau \in (1,\infty)$, $\sigma \in (0,1)$, $\bar\lambda \in (0,\infty)$, $\lambda^0 \in (0,\bar\lambda]$.
Step 0. Set $\nu = 0$.
Step 1. Compute

$$z^\nu \in \operatorname{argmin}_{x\in X} h\big(F(x^\nu) + \nabla F(x^\nu)(x - x^\nu)\big) + \tfrac{1}{2\lambda^\nu}\|x - x^\nu\|_2^2.$$

If $z^\nu = x^\nu$, then stop.
Step 2. If

$$h\big(F(x^\nu)\big) - h\big(F(z^\nu)\big) \geq \sigma\Big(h\big(F(x^\nu)\big) - h\big(F(x^\nu) + \nabla F(x^\nu)(z^\nu - x^\nu)\big)\Big),$$

then set $\lambda^{\nu+1} = \min\{\tau\lambda^\nu, \bar\lambda\}$ and go to Step 3.
Else, replace $\lambda^\nu$ by $\lambda^\nu/\tau$ and go to Step 1.
Step 3. Set $x^{\nu+1} = z^\nu$, replace $\nu$ by $\nu + 1$ and go to Step 1.

Let's examine Step 1. The optimality condition 2.19 combined with 4.67 and 4.64 ensure that

$$0 \in \nabla F(x^\nu)^\top \partial h\big(F(x^\nu) + \nabla F(x^\nu)(z^\nu - x^\nu)\big) + \tfrac{1}{\lambda^\nu}(z^\nu - x^\nu) + N_X(z^\nu).$$

If $x^\nu$ satisfies (6.25), then $z^\nu = x^\nu$ solves the subproblem in Step 1. Moreover, since the subproblem has a unique minimizer by strict convexity (cf. 1.16), Step 1 is guaranteed to produce $x^\nu$ and the algorithm stops as it should.

Next, suppose that $x^\nu$ doesn't satisfy (6.25). Then, $z^\nu \neq x^\nu$ and

$$h\big(F(x^\nu) + \nabla F(x^\nu)(z^\nu - x^\nu)\big) + \tfrac{1}{2\lambda^\nu}\|z^\nu - x^\nu\|_2^2 < h\big(F(x^\nu)\big). \tag{6.26}$$

The first term on the left-hand side is an approximation of $h(F(z^\nu))$. If accurate enough, the inequality ensures that $h(F(z^\nu)) < h(F(x^\nu))$ because the second term on the left is positive. This would be ideal as it means that we've obtained a reduction in objective function value. The test in Step 2 serves as a check on the accuracy of the approximation. The left-hand side of the test specifies the actual decrease in value associated with a move from $x^\nu$ to $z^\nu$. This needs to be at least a fraction of the decrease stipulated by the approximation, which is positive by (6.26). If this is the case, the move to $z^\nu$ is accepted and $z^\nu$ becomes the next point $x^{\nu+1}$. After this successful reduction in objective function value, we increase the proximal parameter $\lambda^\nu$, which encourages a longer step next time. This is reasonable because the approximation is quite accurate and might remain so in an even larger neighborhood.

If the test in Step 2 fails, we deem the approximation too inaccurate and recompute $z^\nu$ with a smaller $\lambda^\nu$. This produces a new $z^\nu$ closer to $x^\nu$, where the accuracy of the approximation is usually better. The algorithm can't cycle indefinitely with such failed steps at a fixed $\nu$, which we confirm as follows.

Suppose for the sake of contradiction that $\{\lambda_k^\nu, k \in \mathbb{N}\}$ is a vanishing sequence of proximal parameters and $\{z_k^\nu, k \in \mathbb{N}\}$ are the corresponding minimizers in Step 1 produced during such an indefinite cycle of failed steps. Then, the functions in the subproblem epi-converge to $h(F(x^\nu)) + \iota_{\{x^\nu\}}$, which can be established by the characterization 4.15. This implies that $z_k^\nu \to x^\nu$ by 5.5(b) and the fact that the level-sets of the functions being minimized in Step 1 are uniformly bounded across $k$. (Alternatively, one can apply 9.40 to reach this conclusion.) Since $F$ is twice smooth, a consequence of (6.1) is that, for any $\varepsilon > 0$, there exists $\gamma \in [0, \infty)$ such that for all $z \in \mathbb{B}(x^\nu, \varepsilon)$,

$$\big\|F(x^\nu) + \nabla F(x^\nu)(z - x^\nu) - F(z)\big\|_2 \le \gamma \|z - x^\nu\|_2^2. \tag{6.27}$$

The function $h$ is locally Lipschitz continuous at $F(x^\nu)$ by virtue of being convex and real-valued; see 4.68. Consequently, there are also $\kappa \in [0, \infty)$ and $\delta > 0$ such that

$$\Big|h\big(F(x^\nu) + \nabla F(x^\nu)(z - x^\nu)\big) - h\big(F(z)\big)\Big| \le \kappa \|z - x^\nu\|_2^2 \quad \forall z \in \mathbb{B}(x^\nu, \delta).$$

In particular, for sufficiently large $k$,

$$h\big(F(z_k^\nu)\big) \le h\big(F(x^\nu) + \nabla F(x^\nu)(z_k^\nu - x^\nu)\big) + \kappa \|z_k^\nu - x^\nu\|_2^2.$$

After rearranging terms in (6.26), which holds with $\lambda^\nu = \lambda_k^\nu$ and $z^\nu = z_k^\nu$, we also have that

$$\|z_k^\nu - x^\nu\|_2^2 < 2\lambda_k^\nu \Big(h\big(F(x^\nu)\big) - h\big(F(x^\nu) + \nabla F(x^\nu)(z_k^\nu - x^\nu)\big)\Big).$$

Combining these two inequalities, we obtain for sufficiently large $k$ that

$$\begin{aligned} h\big(F(x^\nu)\big) - h\big(F(z_k^\nu)\big) &\ge h\big(F(x^\nu)\big) - h\big(F(x^\nu) + \nabla F(x^\nu)(z_k^\nu - x^\nu)\big) - \kappa \|z_k^\nu - x^\nu\|_2^2 \\ &\ge \big(1 - 2\lambda_k^\nu \kappa\big)\Big(h\big(F(x^\nu)\big) - h\big(F(x^\nu) + \nabla F(x^\nu)(z_k^\nu - x^\nu)\big)\Big). \end{aligned}$$

Since $\lambda_k^\nu \to 0$, eventually $(1 - 2\lambda_k^\nu \kappa) \ge \sigma$ and the test in Step 2 is satisfied. We've reached a contradiction. Consequently, the algorithm can't cycle indefinitely at any point $x^\nu$ and either stops in Step 1 with a point satisfying (6.25) or generates a sequence of points $\{x^\nu, \nu \in \mathbb{N}\}$.

**Convergence 6.35** (proximal composite method). *For closed convex $X \subset \mathbb{R}^n$, twice smooth $F : \mathbb{R}^n \to \mathbb{R}^m$ and convex $h : \mathbb{R}^m \to \mathbb{R}$, suppose that the proximal composite method has generated a sequence $\{x^\nu, \nu \in \mathbb{N}\}$ with a cluster point $x^\star$. Then, $x^\star$ satisfies the optimality condition (6.25).*

Let's look at some examples before we give a proof.

**Example 6.36** (superquantile risk minimization). For $X \subset \mathbb{R}^n$, $f : \mathbb{R}^m \times \mathbb{R}^n \to \mathbb{R}$ and $\alpha \in (0,1)$, we recall from §3.C that superquantile-risk minimization results in the problem

$$\underset{x\in X,\gamma\in\mathbb{R}}{\text{minimize}}\ \gamma + \frac{1}{1-\alpha}\sum\nolimits_{\xi\in\Xi} p_\xi \max\{0, f(\xi,x)-\gamma\},$$

where $\Xi$ is a finite set and $p_\xi > 0$. In many engineering applications, $f(\xi,x)$ and $\nabla_x f(\xi,x)$ are costly to compute because one would need to run a simulation or solve a large-scale system of equations. The proximal composite method may then be ideal as it allows us to work through a series of approximating subproblems without recomputing function and gradient values.

**Detail.** Let $\Xi = \{\xi^1,\dots,\xi^q\}$, $p_\xi = p_i$ for $\xi = \xi^i$, $F(\gamma,x) = (\gamma, f(\xi^1,x),\dots,f(\xi^q,x))$ and $h : \mathbb{R}^{1+q} \to \mathbb{R}$ be given by

$$h(u) = u_0 + \frac{1}{1-\alpha}\sum\nolimits_{i=1}^q p_i \max\{0, u_i - u_0\}, \quad \text{with } u = (u_0,u_1,\dots,u_q).$$

Then, the problem of interest becomes that of minimizing $h(F(\gamma,x))$ over $(\gamma,x) \in \mathbb{R}\times X$. Since $h$ is real-valued and convex by 1.18, the proximal composite method applies as long as $f(\xi^i,\cdot)$, $i=1,\dots,q$, are twice smooth and $X$ is closed and convex. After $f(\xi^i,x^\nu)$ and $\nabla_x f(\xi^i,x^\nu)$ have been evaluated for all $i=1,\dots,q$, the solution of the subproblem in Step 1, which may be repeated several times for successively smaller $\lambda^\nu$, requires no additional evaluation of $f$ and its gradient. In many engineering applications $n$ is small, which makes the dimension of the subproblems low. The feasible set $X$ is also often simple. Thus, we're left with solving many relatively easy convex problems. In contrast, the direct approach suggested in §3.C requires us to solve a single problem with $n+q+1$ variables and $q$ constraints involving potentially nonconvex functions. □

**Example 6.37** (exact penalty method). For $f_0 : \mathbb{R}^n \to \mathbb{R}$ and $F : \mathbb{R}^n \to \mathbb{R}^m$, both twice smooth, and closed convex $X \subset \mathbb{R}^n$, let's consider the problem

$$\underset{x\in X}{\text{minimize}}\ f_0(x) \ \text{ subject to } \ F(x) = 0.$$

The *exact penalty method* solves the approximation

$$\underset{x\in X}{\text{minimize}}\ f_0(x) + \theta\|F(x)\|_1,$$

where $\theta \in (0,\infty)$. The approach resembles the penalty method of 4.21, but there $\|F(x)\|_2^2$ takes the place of $\|F(x)\|_1$. The $\ell^2$-norm is sensible as it results in a smooth objective function in the approximating problem. With the proximal composite method, there's less concern about nonsmoothness and the $\ell^1$-norm has a distinct advantage: Under weak assumptions, $\theta$ doesn't have to be driven arbitrarily high, which avoids the numerical difficulties associated with the $\ell^2$-penalty.

**Detail.** The exact penalty method is closely related to augmented Lagrangians in §6.B: Set the multiplier vector to 0 and adopt the $\ell^1$-norm as the augmenting function. We deduce from the exactness proposition 6.13 for augmented Lagrangians that, quite generally,

$\theta$ only needs to exceed a certain level for the actual and approximating problems to have identical minimizers. Regardless of $\theta$, the approximating problem is of the form minimizing $g(G(x))$ subject to $x \in X$, where

$$G(x) = \big(f_0(x), F(x)\big) \quad \text{and} \quad g(u) = u_0 + \theta \sum_{i=1}^{m} |u_i|, \quad \text{with } u = (u_0, u_1, \dots, u_m).$$

The proximal composite method then applies. □

**Example 6.38** (phase retrieval). A measurement system is often limited to quantifying the magnitude squared of the Fourier transform of a signal. The problem of reconstructing a signal from recorded Fourier magnitudes is known as phase retrieval. Given squared Fourier magnitudes $b_1, \dots, b_m$, the goal becomes to determine $x$ such that $b_i = |\langle a^i, x\rangle|^2$, where $a^1, \dots, a^m$ are known. These vectors as well as $x$ are generally complex-valued, but let's ignored that here for simplicity. A formulation of this problem is then

$$\underset{x \in \mathbb{R}^n}{\text{minimize}} \ \frac{1}{m} \sum_{i=1}^{m} \Big| \langle a^i, x\rangle^2 - b_i \Big|,$$

which is of the form required by the proximal composite method.

**Detail.** Let $h(u) = \frac{1}{m}\sum_{i=1}^{m} |u_i|$ for $u = (u_1, \dots, u_m)$ and

$$F(x) = \big(\langle a^1, x\rangle^2 - b_1, \dots, \langle a^m, x\rangle^2 - b_m\big).$$

Then, $F$ is certainly twice smooth and $\nabla F(x)$ is the $m \times n$ matrix with $2a^i\langle a^i, x\rangle$ as the $i$th row. In fact, $h \circ F$ is $\mu$-weakly convex by 6.34, where

$$\mu = 2\Big(\sum_{i=1}^{m} \|a^i\|_2^4\Big)^{1/2}.$$

To see this, note that the Lipschitz modulus of $h$ is one and

$$\big\|\nabla F(x) - \nabla F(\bar{x})\big\|_F^2 = \sum_{i=1}^{m} \big\|2a^i\langle a^i, x - \bar{x}\rangle\big\|_2^2 \le 4\sum_{i=1}^{m} \|a^i\|_2^4 \|x - \bar{x}\|_2^2,$$

where we use the Cauchy-Schwarz inequality at the end. □

Our treatment of the proximal composite method is restricted to a convex and real-valued function $h$. However, the algorithm can be extended via penalties as described in 7.46; see also [61] for other generalizations. We end the section by proving 6.35.

**Lemma 6.39** *If $\tau \in (1, \infty)$, $\delta \in (0, \infty)$ and $n \in \mathbb{N}$, then*[3]

$$\inf\Big\{\sum_{j=1}^{n} x_j^2 \tau^j \ \Big|\ \sum_{j=1}^{n} x_j \ge \delta,\ x_j \ge 0,\ j = 1, \dots, n\Big\} > \delta^2(\tau - 1).$$

---

[3] Here, superscripts indeed indicate powers.

**Proof.** Since scaling is possible, we can assume without loss of generality that $\delta = 1$. The problem is convex and it follows by the KKT condition 2.45 that a minimizer $\bar{x} = (\bar{x}_1, \dots, \bar{x}_n)$ has

$$\bar{x}_j = \frac{\tau^{1-j} - \tau^{-j}}{1 - \tau^{-n}}, \quad j = 1, \dots, n.$$

Since $\bar{x}_j > 0$ for all $j$ and sum to one, this is most easily verified by ignoring the nonnegativity requirements and considering the equality constraint $\sum_{j=1}^{n} x_j = 1$. We obtain the minimum value by plugging these expressions into the objective function. □

**Proof of 6.35.** Suppose that $N \in \mathcal{N}_\infty^\#$ is such that $x^\nu \to_N x^\star$. Then,

$$h\big(F(x^\nu)\big) \to_N h\big(F(x^\star)\big)$$

because $h$ and $F$ are continuous. By (6.26), the right-hand side of the test in Step 2 is positive so $h(F(x^\nu)) - h(F(x^{\nu+1}))$ is also positive. This monotonicity establishes that the whole sequence

$$h\big(F(x^\nu)\big) \to h\big(F(x^\star)\big).$$

Again using (6.26) and the test,

$$\begin{aligned} h\big(F(x^{\nu+1})\big) &\le h\big(F(x^\nu)\big) - \sigma\Big(h\big(F(x^\nu)\big) - h\big(F(x^\nu) + \nabla F(x^\nu)(x^{\nu+1} - x^\nu)\big)\Big) \\ &\le h\big(F(x^\nu)\big) - \sigma \tfrac{1}{2\lambda^\nu} \|x^{\nu+1} - x^\nu\|_2^2. \end{aligned}$$

Consequently, one has

$$\begin{aligned} h\big(F(x^0)\big) - h\big(F(x^\star)\big) &\ge \sum\nolimits_{\nu=0}^{\infty} h\big(F(x^\nu)\big) - h\big(F(x^{\nu+1})\big) \\ &\ge \frac{\sigma}{2} \sum\nolimits_{\nu=1}^{\infty} \frac{1}{\lambda^\nu} \|x^{\nu+1} - x^\nu\|_2^2 \ge \frac{\sigma}{2\bar{\lambda}} \sum\nolimits_{\nu=1}^{\infty} \|x^{\nu+1} - x^\nu\|_2^2, \end{aligned}$$

which implies that $x^{\nu+1} - x^\nu \to 0$.

There are two possibilities. First, suppose that

$$\tfrac{1}{\lambda^\nu}(x^{\nu+1} - x^\nu) \to_N 0.$$

The optimality condition 4.75 for the subproblem in Step 1 gives that

$$0 \in \nabla F(x^\nu)^\top \partial h\big(F(x^\nu) + \nabla F(x^\nu)(x^{\nu+1} - x^\nu)\big) + \tfrac{1}{\lambda^\nu}(x^{\nu+1} - x^\nu) + N_X(x^{\nu+1}).$$

In view of 4.74 and 4.39, we find that this inclusion, which holds for all $\nu \in N$, implies (6.25). Thus, we can concentrate on the second possibility when the assumption of the first one doesn't hold. Since the same argument holds for a subsequence of $N$, we can proceed under the hypothesis that

$$\operatorname{liminf}_{\nu \in N} \tfrac{1}{\lambda^\nu} \|x^{\nu+1} - x^\nu\|_2 > 0.$$

We'll show that this possibility can't occur.

Now, $\lambda^\nu \to_N 0$ and we can assume that $\{\lambda^\nu, \nu \in N\}$ is decreasing because otherwise $N$ can be redefined while still retaining a subsequence. We consider two cases.

**Case A.** Suppose that there are infinitely many $\nu \in N$ such that the test in Step 2 failed at least once in iteration $\nu$. Without loss of generality, we can assume that such failure takes place for all $\nu \in N$ because otherwise we can redefine $N$ and still retain a subsequence. Thus, the proximal parameter $\tau\lambda^\nu$ would have produced a solution $\hat{z}^\nu$ of the subproblem in Step 1 that subsequently caused failure in Step 2.

If $\hat{z}^\nu - x^\nu \to_N 0$, then the facts that $F$ is twice smooth and $h$ is locally Lipschitz continuous by 4.68 ensure that there's $\kappa \in [0, \infty)$ such that for sufficiently large $\nu \in N$,

$$h\big(F(\hat{z}^\nu)\big) \le h\big(F(x^\nu) + \nabla F(x^\nu)(\hat{z}^\nu - x^\nu)\big) + \kappa \|\hat{z}^\nu - x^\nu\|_2^2;$$

see the arguments leading to (6.27). Parallel to (6.26), we also have that

$$\|\hat{z}^\nu - x^\nu\|_2^2 < 2\tau\lambda^\nu \Big(h\big(F(x^\nu)\big) - h\big(F(x^\nu) + \nabla F(x^\nu)(\hat{z}^\nu - x^\nu)\big)\Big). \tag{6.28}$$

Combining these two inequalities, we obtain for sufficiently large $\nu \in N$ that

$$\begin{aligned} h\big(F(x^\nu)\big) - h\big(F(\hat{z}^\nu)\big) &\ge h\big(F(x^\nu)\big) - h\big(F(x^\nu) + \nabla F(x^\nu)(\hat{z}^\nu - x^\nu)\big) - \kappa\|\hat{z}^\nu - x^\nu\|_2^2 \\ &\ge (1 - 2\tau\lambda^\nu\kappa)\Big(h\big(F(x^\nu)\big) - h\big(F(x^\nu) + \nabla F(x^\nu)(\hat{z}^\nu - x^\nu)\big)\Big). \end{aligned}$$

Since $\lambda^\nu \to_N 0$, eventually $(1 - 2\tau\lambda^\nu\kappa) \ge \sigma$ and the test in Step 2 is satisfied, which is a contradiction.

Thus, it suffices to examine when $\{\hat{z}^\nu - x^\nu, \nu \in N\}$ is bounded away from 0. The subgradient inequality 2.17 applied to $h$ ensures that for any $v^\nu \in \partial h(F(x^\nu))$,

$$h\big(F(x^\nu) + \nabla F(x^\nu)(\hat{z}^\nu - x^\nu)\big) \ge h\big(F(x^\nu)\big) - \|v^\nu\|_2 \big\|\nabla F(x^\nu)(\hat{z}^\nu - x^\nu)\big\|_2.$$

Since $\{\|v^\nu\|_2, \nu \in N\}$ is bounded as seen from 2.54 and $\{\hat{z}^\nu - x^\nu, \nu \in N\}$ is bounded away from 0, one has

$$\|v^\nu\|_2 \big\|\nabla F(x^\nu)(\hat{z}^\nu - x^\nu)\big\|_2 < \tfrac{1}{2\tau\lambda^\nu}\|\hat{z}^\nu - x^\nu\|_2^2$$

for sufficiently large $\nu \in N$. We can combine this with the previous inequality to obtain that

$$\begin{aligned} h\big(F(x^\nu) + \nabla F(x^\nu)(\hat{z}^\nu - x^\nu)\big) &> h\big(F(x^\nu)\big) - \tfrac{1}{2\tau\lambda^\nu}\|\hat{z}^\nu - x^\nu\|_2^2 \\ &> h\big(F(x^\nu) + \nabla F(x^\nu)(\hat{z}^\nu - x^\nu)\big) \end{aligned}$$

for such $\nu$, where the last inequality follows from the optimality of $\hat{z}^\nu$ in Step 1; cf. (6.28). We've constructed a strict inequality between a quantity and itself, which is a contradiction. So the possibility of $\{\hat{z}^\nu - x^\nu, \nu \in N\}$ being bounded away from 0 is also ruled out. We conclude that the hypothesis of Case A can't hold.

**Case B.** Suppose that there's a finite collection of $\nu \in N$ such that the test in Step 2 failed at least once in iteration $\nu$. We can remove these from $N$ and proceed under the assumption that for all $\nu \in N$, there's no failure in Step 2. Let $\nu^-$ be the index in $N$ prior to $\nu$.

For $\nu \in N$, let $\mu(\nu)$ be the last iteration prior to $\nu$ with failure in Step 2 so that $\mu(\nu) \notin N$. (Recall that $\lambda^\nu \to_N 0$ so there must be iterations with failure outside $N$.) In fact, we can assume without loss of generality that $\nu^- < \mu(\nu) < \nu$. Since there's no decrease in the proximal parameter during iterations $\mu(\nu) + 1, \ldots, \nu$, one has

$$\lambda^\nu = \tau^{\nu-\mu(\nu)} \lambda^{\mu(\nu)} \quad \forall \nu \in N, \tag{6.29}$$

where $\nu - \mu(\nu)$ is actually the power of $\tau$. Since $\lambda^\nu \to_N 0$, we can here assume without loss of generality that the upper bound $\bar{\lambda}$ doesn't kick in.

Let's now show that

$$x^{\mu(\nu)} - x^\nu \to_N 0.$$

Suppose for the sake of contradiction that this isn't the case. Then, there's $\delta > 0$ such that $\|x^{\mu(\nu)} - x^\nu\|_2 \geq \delta$ for infinitely many $\nu \in N$. Without loss of generality, let's assume we've this for all $\nu \in N$, i.e.,

$$\delta \leq \|x^{\mu(\nu)} - x^\nu\|_2 \leq \sum_{k=\mu(\nu)}^{\nu-1} \|x^{k+1} - x^k\|_2$$

for $\nu \in N$. Let's look at function values during these iterations. For $\nu \in N$,

$$h\big(F(x^{\mu(\nu)})\big) - h\big(F(x^\nu)\big) = \sum_{k=\mu(\nu)}^{\nu-1} h\big(F(x^k)\big) - h\big(F(x^{k+1})\big)$$

$$\geq \frac{\sigma}{2} \sum_{k=\mu(\nu)}^{\nu-1} \frac{1}{\lambda^k} \|x^{k+1} - x^k\|_2^2 = \frac{\sigma}{2\lambda^\nu} \sum_{k=\mu(\nu)}^{\nu-1} \tau^{\nu-k} \|x^{k+1} - x^k\|_2^2,$$

where the inequality is a consequence of the discussion around (6.26) and the fact that the test in Step 2 passes during these iterations. The expression also leverages a formula for $\lambda^k$ along the lines of (6.29). The sum can be bounded from below by using 6.39, which leads to

$$h\big(F(x^{\mu(\nu)})\big) - h\big(F(x^\nu)\big) \geq \frac{\sigma}{2\lambda^\nu} \delta^2(\tau - 1) \geq \frac{\sigma}{2\bar{\lambda}} \delta^2(\tau - 1).$$

Since the right-hand side is positive, this contradicts the fact that $h(F(x^\nu))$ tends to $h(F(x^\star))$. Thus, $x^{\mu(\nu)} - x^\nu \to_N 0$ and then also

$$\lim_{\nu \in N} x^{\mu(\nu)} = \lim_{\nu \in N} x^\nu + \lim_{\nu \in N} \big(x^{\mu(\nu)} - x^\nu\big) = x^\star.$$

We now have another subsequence of indices $\{\mu(\nu), \nu \in N\}$ for which the corresponding points converge to $x^\star$. We can then repeat the above arguments (see Case A) with this subsequence in the place of $N$ and conclude that the parameter value $\lambda^{\mu(\nu)}\tau$ attempted at iteration $\mu(\nu)$ would be accepted for all $\nu$ sufficiently large. This contradicts the definition of $\mu(\nu)$. Thus, Case B can't hold either. □

## 6.G Design of Multi-Component Systems

Engineering design of a multi-component system usually involves analyzing the overall performance as expressed in terms of capabilities of components. Even though the performance of the individual components might be described by convex and/or smooth functions, the system performance is only occasionally convex and rarely smooth.

Let's consider a vector $x \in \mathbb{R}^n$ of decision variables that describes the design of a system with $m$ components, where $g_i : \mathbb{R}^n \to \overline{\mathbb{R}}$ quantifies the performance of the $i$th component. Specifically, $g_i(x)$ is the performance of component $i$ under design $x$; §3.E furnishes an example. A system-level analysis, however, needs to consider how the different components contribute toward the overall performance of the system. Do all components need to perform well or just one of them? A *series system* is governed by its worst component. An example would be a chain attaching an anchor to a boat. The chain consists of many links (components), but if any one breaks the boat is loose and the system failed. Consistent with our orientation toward minimization, we assume that lower values of $g_i$ mean better performance. Then, in a series system, the performance of design $x$ is quantified by

$$\max_{i=1,\dots,m} g_i(x) \qquad \text{(series system).}$$

The truss in §3.E is such a system.

The performance of a *parallel system* is governed by the best component. The two pilots in a passenger jet can be thought of as two components of the aircraft control system. The system remains operational even if one pilot becomes incapacitated. The performance of design $x$ for a parallel system is therefore quantified by

$$\min_{i=1,\dots,m} g_i(x) \qquad \text{(parallel system).}$$

For a general multi-component system, the situation is more complicated. One subset of components might need to have all its components perform well, while another subset may only need one well-performing component. We represent this by identifying subsets $\mathbb{I}_k \subset \{1, 2, \dots, m\}$, $k = 1, \dots, q$, each containing a collection of components that somehow "work together" with the best one governing the whole collection's performance. The performance of design $x$ for such a *general system* is then quantified by

$$g(x) = \max_{k=1,\dots,q} \min_{i \in \mathbb{I}_k} g_i(x) \qquad \text{(general system).}$$

If $q = m$ and $\mathbb{I}_k = \{k\}$ for all $k = 1, \dots, q$, then we return to a series system. When $q = 1$ and $\mathbb{I}_1 = \{1, \dots, m\}$, we've the special case of a parallel system. If every $\mathbb{I}_k$ contains a single index and $g_i$ is convex, then $g$ is convex by 1.18(a). Smoothness of $g$ can't be expected at a point $\bar{x}$ with more than one $i$ satisfying $g(\bar{x}) = g_i(\bar{x})$ even if all $g_i$ are smooth at $\bar{x}$. For example, with $q = 1$, $\mathbb{I}_1 = \{1, 2\}$, $g_1(x) = 0$ and $g_2(x) = x$, $g$ is nonsmooth at 0.

This example also shows that epi-regularity is easily lost; the kink at 0 is of the inward kind as we discuss in more detail below. Despite these challenges, $g$ is well structured under the assumption that each $g_i$ is convex and/or smooth and the machinery developed over the last several sections enables us to address problems of the form

$$\underset{x \in X}{\text{minimize}}\, f_0(x) \ \text{ subject to } \ g(x) \le 0, \tag{6.30}$$

where $f_0 : \mathbb{R}^n \to \mathbb{R}$ represents the cost of the various designs, $X \subset \mathbb{R}^n$ is relatively simple, for example, specifying bounds on the variables, and $g$ is of the form described above.

**Example 6.40** (problem enumeration). The constraint $g(x) \le 0$ in (6.30) is equivalently stated as $\min_{i \in \mathbb{I}_k} g_i(x) \le 0$, $k = 1, \dots, q$. However, this doesn't help us significantly as the challenging functions $x \mapsto \min_{i \in \mathbb{I}_k} g_i(x)$ remain. An alternative is to select $i_1 \in \mathbb{I}_1, i_2 \in \mathbb{I}_2, \dots, i_q \in \mathbb{I}_q$ and consider the problem

$$P(i_1, \dots, i_q): \quad \underset{x \in X}{\text{minimize}}\, f_0(x) \ \text{ subject to } \ g_{i_k}(x) \le 0, \ \ k = 1, \dots, q.$$

The collection of all possible choices for $i_1, \dots, i_q$ produces the family of problems

$$\big\{P(i_1, \dots, i_q), \ \ i_k \in \mathbb{I}_k, \ k = 1, \dots, q\big\},$$

which indeed leads to a solution of the actual problem.

**Detail.** Each problem $P(i_1, \dots, i_q)$ has a pre-selected index $i_k$ for every $k = 1, \dots, q$, which can be thought of as a guess of a minimizing index in $\min_{i \in \mathbb{I}_k} g_i(\bar{x})$ at a solution $\bar{x}$. Presumably, $P(i_1, \dots, i_q)$ is computationally tractable. In fact, it's convex if $X$ is convex and $f_0$ as well as all $g_i$ are convex. If we consider all possible guesses of such minimizing indices and obtain a minimizer for each of the resulting problems, then the minimizer with the lowest value of $f_0$ is a minimizer of (6.30). Thus, we can solve the actual problem by solving every problem in the family $\{P(i_1, \dots, i_q), \ i_k \in \mathbb{I}_k, k = 1, \dots, q\}$.

To see this, let $\bar{x}$ be a minimizer of $P(i_1, \dots, i_q)$ and $x^\star$ be a minimizer of (6.30). Then, $f_0(\bar{x}) \ge f_0(x^\star)$ because the restriction to a specific choice $i_k$ can't result in an improvement over having a choice of $i \in \mathbb{I}_k$. There's for each $k$ an index $i_k^\star \in \mathbb{I}_k$ such that

$$g_{i_k^\star}(x^\star) = \min_{i \in \mathbb{I}_k} g_i(x^\star).$$

Since $x^\star$ is feasible relative to the constraints $x \in X$ and $g_{i_k^\star}(x) \le 0$, for all $k = 1, \dots, q$, we've that

$$f_0(x^\star) \ge \inf_{x \in X} \big\{ f_0(x) \,\big|\, g_{i_k^\star}(x) \le 0, \ k = 1, \dots, q \big\}.$$

The choice $(i_1, i_2, \dots, i_q)$ is arbitrary and can as well be $(i_1^\star, i_2^\star, \dots, i_q^\star)$. This implies that the previous right-hand side is also bounded from below by $f_0(x^\star)$. Thus, the infimum in that inequality is attained with $x^\star$. The problem $P(i_1^\star, \dots, i_q^\star)$ furnishes a minimizer of the actual problem in (6.30).

The trouble, of course, is that we don't know the right set of indices a priori and we need to solve one problem for each *possible* set; a total of $\prod_{k=1}^{q} m_k$ problems, where $m_k$ is the cardinality of $\mathbb{I}_k$. When $m_k = 5$ and $q = 5$, this results in 3125 problems, a manageable number. However, for large-scale models, this approach based on enumerating possibilities becomes untenable but illustrates how one sometimes can circumvent nonconvexity and/or nonsmoothness by enumeration. □

For series systems, we can fall back on properties of max-functions; see 1.13 and 4.66. In the case of general systems, we need to contend with the presence of "min" in the definition of $g$. Let's examine the resulting min-functions. Their local Lipschitz continuity is addressed by 4.68(d).

**Proposition 6.41** (min-functions). *For $g_i : \mathbb{R}^n \to \mathbb{R}$, $i = 1, \dots, m$ and $\bar{x} \in \mathbb{R}^n$, the function $f : \mathbb{R}^n \to \mathbb{R}$ given by*

$$f(x) = \min_{i=1,\dots,m} g_i(x)$$

*is continuous at $\bar{x}$ when each $g_i$ is continuous at $\bar{x}$. If each $g_i$ is smooth in a neighborhood of $\bar{x}$, then*

$$\partial f(\bar{x}) \subset \{\nabla g_i(\bar{x}),\ i \in \mathbb{A}(\bar{x})\},$$

*where $\mathbb{A}(\bar{x}) = \{i \mid g_i(\bar{x}) = f(\bar{x})\}$.*

**Proof.** Continuity follows essentially by the same arguments as for max-functions; see 4.66. Let

$$h(u) = \min_{i=1,\dots,m} u_i \quad \text{and} \quad G(x) = \big(g_1(x), \dots, g_m(x)\big).$$

Then, $f = h \circ G$ and we can apply the general chain rule 6.23 because $G$ is locally Lipschitz continuous by 4.68 and $y \in \partial^\infty h(G(x))$ implies $y = 0$ by 4.69. Let $\mathbb{A}_0(u) = \{i \mid u_i = h(u)\}$. Note that $h$ is differentiable at $u$ if and only if $\mathbb{A}_0(u)$ contains a single index, say $i$, and then $\nabla h(u) = e^i$, the unit vector with the $i$th component being 1 and the other components being 0. Suppose that $\bar{u} \in \mathbb{R}^m$ and $v \in \partial h(\bar{u})$. Since $h$ is continuous, there's $\delta > 0$ such that $\mathbb{A}_0(u) \subset \mathbb{A}_0(\bar{u})$ for all $u \in \mathbb{B}(\bar{u}, \delta)$. By the definition of subgradients in 4.55, there exist $u^\nu \to \bar{u}$ and regular normal vectors

$$(v^\nu, -1) \in \widehat{N}_{\operatorname{epi} h}\big(u^\nu, h(u^\nu)\big)$$

converging to $(v, -1)$. Since $-h$ is epi-regular and locally Lipschitz continuous by 4.66 and 4.68(c), existence of such normal vectors actually means that $h$ is differentiable at $u^\nu$ and $\nabla h(u^\nu) = v^\nu$ by [105, Corollary 9.21]. Thus, $v^\nu = e^{i(\nu)}$ for a unique $i(\nu) \in \mathbb{A}_0(u^\nu)$. Since $i(\nu)$ is taken from a finite set and $u^\nu$ is eventually in $\mathbb{B}(\bar{u}, \delta)$, there exist $N \in \mathcal{N}_\infty^\#$ and $i^\star \in \mathbb{A}_0(\bar{u})$ such that $i(\nu) = i^\star$ for all $\nu \in N$. Thus, $v^\nu = e^{i^\star}$ for all $\nu \in N$. We conclude that $v = e^{i^\star}$. We've shown that $\partial h(\bar{u}) \subset \{e^i, i \in \mathbb{A}_0(\bar{u})\}$.

The general chain rule 6.23 then asserts that

$$\partial f(\bar{x}) \subset \bigcup_{y \in \partial h(G(\bar{x}))} \{\nabla G(\bar{x})^\top y\} \subset \bigcup_{y \in \{e^i, i \in \mathbb{A}(\bar{x})\}} \{\nabla G(\bar{x})^\top y\},$$

which leads to the conclusion. □

The example $f(x) = \min\{0, x, -2x\}$ reveals that the inclusion in the proposition can't be replaced by an equality in general. A direct examination of the epigraph of $f$ and its normal cones confirms that $\partial f(0) = \{1, -2\}$ as seen in Figure 6.9. However, the gradients of the active functions at 0 amount to $\{0, 1, -2\}$ in this case.

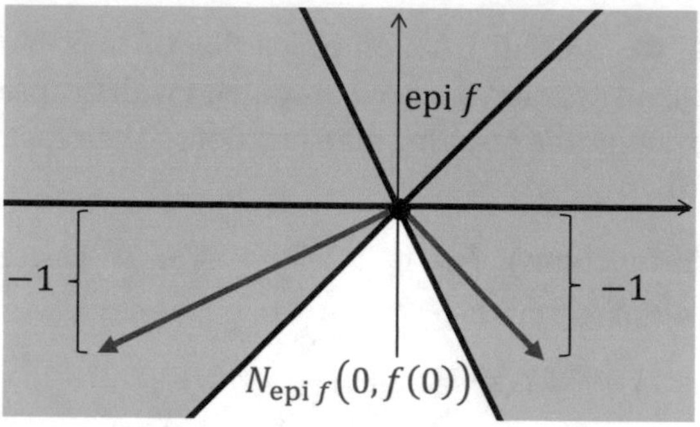


Fig. 6.9: Illustration of 6.41 with $f(x) = \min\{0, x, -2x\}$.

**Example 6.42** (optimality in multi-component design). We can now state optimality conditions for (6.30). Suppose that $X$ is closed and $f_0$ as well as all $g_i$ are smooth. Let

$$f_k(x) = \min_{i \in \mathbb{I}_k} g_i(x), \quad \mathbb{A}_k(x) = \big\{i \in \mathbb{I}_k \,\big|\, f_k(x) = g_i(x)\big\}, \ k = 1, \dots, q,$$

and $F(x) = (f_1(x), \dots, f_q(x))$. Suppose that the following qualification holds at $x^\star$:

$$\forall k = 1, \dots, q: \quad y_k = 0 \text{ if } f_k(x^\star) < 0; \quad y_k \geq 0 \text{ otherwise, and}$$
$$-\sum\nolimits_{k=1}^{q} y_k \nabla g_{i_k}(x^\star) \in N_X(x^\star), \text{ with } i_k \in \mathbb{A}_k(x^\star) \implies (y_1, \dots, y_q) = 0.$$

If $x^\star$ is a local minimizer of (6.30), then

$$\exists y \in \mathbb{R}^q, \text{ with } y_k = 0 \text{ if } f_k(x^\star) < 0; \quad y_k \geq 0 \text{ otherwise,}$$
$$\text{such that } -\nabla f_0(x^\star) \in N_X(x^\star) + \sum\nolimits_{k=1}^{q} \big\{y_k \nabla g_i(x^\star), \ i \in \mathbb{A}_k(x^\star)\big\}.$$

**Detail.** We can reformulate (6.30) as minimizing $f_0(x) + h(F(x))$ subject to $x \in X$, where $h(u) = \iota_{(-\infty,0]^q}(u)$. Since $h$ is proper, lsc and convex, we just need to verify the qualification (6.16) to invoke the optimality condition 6.25. With $f(x, y) = \langle F(x), y\rangle$, the qualification involves $\partial_x f(x^\star, y)$ for $y = (y_1, \dots, y_q) \geq 0$. Since $f_k$, $k = 1, \dots, q$, are locally Lipschitz continuous by 4.68(d), we realize from 4.70 and 6.41 that

$$\partial_x f(x^\star, y) \subset \sum\nolimits_{k=1}^{q} \partial(y_k f_k)(x^\star) \subset \sum\nolimits_{k=1}^{q} \big\{y_k \nabla g_i(x^\star), \ i \in \mathbb{A}_k(x^\star)\big\}. \tag{6.31}$$

Now, suppose that

$$y \in N_{(-\infty,0]^q}\big(F(x^\star)\big) \quad \text{and} \quad 0 \in \partial_x f(x^\star, y) + N_X(x^\star).$$

Then, there's $z \in N_X(x^\star)$ such that

$$-z \in \sum_{k=1}^{q} \{y_k \nabla g_i(x^\star),\ i \in \mathbb{A}_k(x^\star)\}.$$

This implies that there are indices $\{i_k \in \mathbb{A}_k(x^\star), k = 1, \dots, q\}$ such that

$$z = -\sum_{k=1}^{q} y_k \nabla g_{i_k}(x^\star).$$

The imposed qualification ensures that $y = 0$ and (6.16) holds. It only remains to specialize the optimality condition 6.25.

Let $l(x, y) = f_0(x) + \langle F(x), y\rangle$. By 4.58 and (6.31),

$$\partial_x l(x^\star, y) \subset \nabla f_0(x^\star) + \sum_{k=1}^{q} \{y_k \nabla g_i(x^\star),\ i \in \mathbb{A}_k(x^\star)\}$$

for $y \geq 0$ and the asserted condition follows. When $X = \mathbb{R}^n$, the qualification holds if $\{\nabla g_{i_k}(x^\star), k = 1, \dots, q\}$ are linearly independent regardless of how $i_k \in \mathbb{A}_k(x^\star)$ is selected. □

**Example 6.43** (epigraphical approximations). An algorithm for solving (6.30) can rely on the smoothing technique in 4.16 and the solution of the approximating problem

$$\underset{x \in \mathbb{R}^n}{\text{minimize}}\ f^\nu(x) = \iota_X(x) + f_0(x) + \iota_{(-\infty,0]^q}\big(F^\nu(x)\big)$$

where $F^\nu(x) = (f_1^\nu(x), \dots, f_q^\nu(x))$ and

$$f_k^\nu(x) = -\frac{1}{\theta^\nu} \ln\Big(\sum_{i \in \mathbb{I}_k} \exp\big(-\theta^\nu g_i(x)\big)\Big), \quad \text{with } \theta^\nu \in (0, \infty).$$

Suppose that $f_0, g_1, \dots, g_m$ are smooth and $X$ is closed. Then, each $f_k^\nu$ is smooth by 4.16 and SQP, interior-point and augmented Lagrangian methods of §4.K and §6.B are available for the approximating problem under the additional assumption that $X$ is relatively simple. Moreover, the approximation is justified by the fact that $f^\nu$ epi-converges to the function representing (6.30) when $\theta^\nu \to \infty$.

**Detail.** Let $f_k(x) = \min_{i \in \mathbb{I}_k} g_i(x)$, $k = 1, \dots, q$. We deduce from 4.16 that $f_k$ can be approximated by $f_k^\nu$ and

$$0 \leq f_k(x) - f_k^\nu(x) \leq \frac{\ln m_k}{\theta^\nu} \quad \forall x \in \mathbb{R}^n, \tag{6.32}$$

where $m_k$ is the cardinality of $\mathbb{I}_k$. Let

$$f(x) = \iota_X(x) + f_0(x) + \iota_{(-\infty,0]^q}\big(F(x)\big)$$

and $F(x) = (f_1(x), \dots, f_q(x))$. Certainly, (6.30) amounts to minimizing $f$.

Let's first leverage the characterization 4.15 to show that

$$\iota_X + \iota_{(-\infty,0]^q}\big(F^\nu(\cdot)\big) \xrightarrow{e} \iota_X + \iota_{(-\infty,0]^q}\big(F(\cdot)\big). \tag{6.33}$$

Suppose that $x^\nu \to x$. If $\iota_X(x) = \infty$, then $\iota_X(x^\nu) = \infty$ for sufficiently large $\nu$ because $X$ is closed. If $\iota_{(-\infty,0]^q}(F(x)) = \infty$, then there are $k$ and $\varepsilon > 0$ such that $f_k(x) = 3\varepsilon$. By (6.32) and the continuity of $f_k$, cf. 6.41, there's $\bar\nu$ such that $f_k^\nu(x^\nu) \geq f_k(x^\nu) - \varepsilon$ and $f_k(x^\nu) \geq f_k(x) - \varepsilon$ for all $\nu \geq \bar\nu$. Combining these inequalities, we find that

$$f_k^\nu(x^\nu) \geq f_k(x^\nu) - \varepsilon \geq f_k(x) - 2\varepsilon = \varepsilon \quad \forall \nu \geq \bar\nu.$$

Thus, $\iota_{(-\infty,0]^q}(F^\nu(x^\nu)) = \infty$ for such $\nu$. In either situation,

$$\operatorname{liminf} \Big(\iota_X(x^\nu) + \iota_{(-\infty,0]^q}\big(F^\nu(x^\nu)\big)\Big) = \infty.$$

If $\iota_X(x) = \iota_{(-\infty,0]^q}(F(x)) = 0$, then

$$\operatorname{liminf} \Big(\iota_X(x^\nu) + \iota_{(-\infty,0]^q}\big(F^\nu(x^\nu)\big)\Big) \geq 0$$

trivially. We've confirmed 4.15(a).

For 4.15(b), note that

$$\iota_X(x) + \iota_{(-\infty,0]^q}\big(F^\nu(x)\big) \leq \iota_X(x) + \iota_{(-\infty,0]^q}\big(F(x)\big) \quad \forall x \in \mathbb{R}^n$$

because of (6.32). We've established (6.33).

Since $f_0 \xrightarrow{e} f_0$ and $-f_0 \xrightarrow{e} -f_0$ by the continuity of $f_0$, the epigraphical sum rule 4.19(a) applies and this establishes that $f^\nu \xrightarrow{e} f$. □

**Exercise 6.44** (smoothing in superquantile-risk minimization). In practice, the performance of the various components of a system is often uncertain. Let $\Xi \subset \mathbb{R}^m$ be a finite set of possible values of a parameter vector $\xi$ and $g_i : \mathbb{R}^m \times \mathbb{R}^n \to \mathbb{R}$, $i = 1, \dots, r$, be twice smooth. Then,

$$g(\xi, x) = \max\nolimits_{k=1,\dots,q} \min\nolimits_{i \in \mathbb{I}_k} g_i(\xi, x)$$

is the performance of the system under design $x$ and parameter vector $\xi$. As above, $\mathbb{I}_k \subset \{1, \dots, r\}$ is a group of components. Let $p_\xi > 0$ be the probability of $\xi \in \Xi$. (a) Formulate the problem of minimizing the superquantile-risk of the system performance subject to the constraint that the design $x \in X \subset \mathbb{R}^n$, where $X$ is closed and convex. (b) Use a smooth approximation of $x \mapsto \min_{i \in \mathbb{I}_k} g_i(\xi, x)$ to construct an approximating problem of the form: minimize $h \circ F^\nu$ over a convex feasible set, where $h$ is real-valued and convex and $F^\nu$ is twice smooth. (c) Justify the approximation using epigraphical analysis. (d) Discuss how the proximal composite method would be implemented for the approximating problem.

**Guide.** One can largely proceed as in 6.36 and 6.43. □

## 6.H Difference-of-Convex Functions

While a composite function may not be convex, it's usually worthwhile to identify components that may have this property. A *difference-of-convex (dc) function* is of the form

$$f(x) = g(x) - h(x), \quad \text{where } g, h : \mathbb{R} \to \overline{\mathbb{R}} \text{ are convex,}$$

and thus features convexity. However, as $f(x) = 0 - \max\{0, x\}$ illustrates, a dc function is typically neither convex nor smooth nor epi-regular. The min-function $f(x) = \min_{i=1,\dots,m} g_i(x)$ from §6.G is a dc function when each $g_i$ is concave because $f(x) = 0 - \max_{i=1,\dots,m}\{-g_i(x)\}$. The class of dc functions is surprisingly large. It includes all twice smooth functions and can approximate to an arbitrary level of accuracy any lsc function; see [107]. In general, the dc-structure isn't always apparent and may need some creative reformulation.

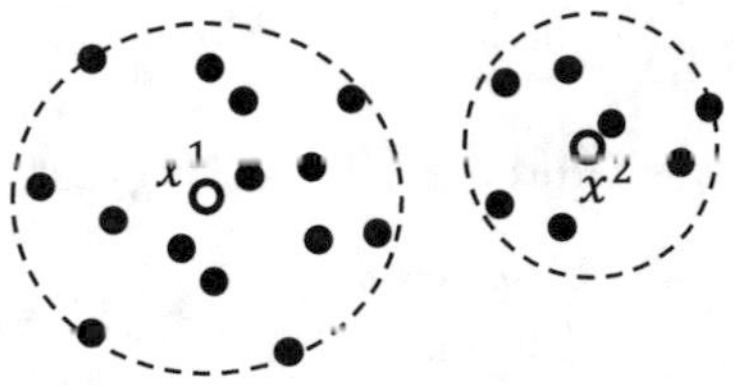


Fig. 6.10: Clustering of data points $\{\xi^i \in \mathbb{R}^n, i = 1, \dots, m\}$ (dots) into two groups with centers $x^1$ and $x^2$ (circles).

**Example 6.45** (clustering). A fundamental problem in data analytics is that of clustering: Partition the data points $\{\xi^i \in \mathbb{R}^n, i = 1, \dots, m\}$ into $q$ groups (clusters) such that the distance between points in a group is as small as possible. This problem is often formulated as

$$\underset{x^j \in \mathbb{R}^n, j=1,\dots,q}{\text{minimize}} \sum_{i=1}^{m} \min_{j=1,\dots,q} \|\xi^i - x^j\|_2^2,$$

where $x^j$ is the *center* of cluster $j$. Those data points that are closer to $x^j$ than to the other centers are considered part of the $j$th cluster; see Figure 6.10. The objective function can be written as a dc function.

**Detail.** The sum of a finite number of dc functions is a dc function so we can concentrate on the min-expression. For $\{\alpha_j, j = 1, \dots, q\}$, we've the elementary identity

$$\min_{j=1,\dots,q} \alpha_j = \sum_{j=1}^{q} \alpha_j - \max_{k=1,\dots,q} \sum_{j=1, j\neq k}^{q} \alpha_j.$$

Thus, the objective function can be written as

$$\sum_{i=1}^{m} \sum_{j=1}^{q} \|\xi^i - x^j\|_2^2 - \sum_{i=1}^{m} \max_{k=1,\dots,q} \sum_{j=1, j\neq k}^{q} \|\xi^i - x^j\|_2^2,$$

which indeed is the difference between two convex functions. □

**Example 6.46** (network design and operation). We recall from §4.B that a problem of network design and operation may lead to minimization of

$$f(x, w) = \sum_{i=1}^{m} \sum_{j=1}^{n} w_{ij} \|x^j - t^i\|_2,$$

where $t^i \in \mathbb{R}^2, i = 1, \dots, m$, are given. The function $f$ can be written as a dc function.

**Detail.** For $\alpha, \beta \in \mathbb{R}$, we've that

$$\alpha\beta = \tfrac{1}{2}(\alpha + \beta)^2 - \tfrac{1}{2}(\alpha^2 + \beta^2).$$

This can be used to write the product of two convex functions as the difference of two convex functions under some conditions. In the present case, we've a product of $w_{ij}$ and $\|x^j - t^i\|_2$. The above identity then implies that

$$w_{ij}\|x^j - t^i\|_2 = \tfrac{1}{2}\big(w_{ij} + \|x^j - t^i\|_2\big)^2 - \tfrac{1}{2}\big(w_{ij}^2 + \|x^j - t^i\|_2^2\big).$$

Incorporating the double sum, we find that $f(x, w) = g(x, w) - h(x, w)$, with

$$g(x, w) = \tfrac{1}{2} \sum_{i=1}^{m} \sum_{j=1}^{n} \big(w_{ij} + \|x^j - t^i\|_2\big)^2$$

$$h(x, w) = \tfrac{1}{2} \sum_{i=1}^{m} \sum_{j=1}^{n} w_{ij}^2 + \|x^j - t^i\|_2^2.$$

While $h$ is convex, the situation for $g$ is a bit unclear. Since $w_{ij}$ (the amount transported from $i$ to $j$) is nonnegative, we can just as well set

$$g(x, w) = \tfrac{1}{2} \sum_{i=1}^{m} \sum_{j=1}^{n} \Big( \max\big\{0,\ w_{ij} + \|x^j - t^i\|_2\big\}\Big)^2,$$

which is convex by 1.18(d) because $\alpha \mapsto (\max\{0, \alpha\})^2$ is nondecreasing and convex. □

**Example 6.47** (nonconvex regularization). Regression problems often benefit from having a regularizer added to a least-squares objective function; see 2.5 and 4.5. One possibility is to adopt the *minimax concave penalty regularizer*, which takes the form

$$r(\alpha) = \begin{cases} \lambda|\alpha| - \tfrac{1}{2}\alpha^2/\gamma & \text{when } |\alpha| \leq \gamma\lambda \\ \tfrac{1}{2}\gamma\lambda^2 & \text{otherwise,} \end{cases}$$

with $\gamma > 1$ and $\lambda \geq 0$ being parameters. This results in the regression problem

$$\operatorname*{minimize}_{x \in \mathbb{R}^n} f(x) = \|Ax - b\|_2^2 + \sum_{j=1}^{n} r(x_j).$$

Near zero, $r$ resembles the absolute value function, but its rate of increase gradually decreases as $\alpha$ moves away from zero with the graph of $r$ becoming flat for $|\alpha| > \gamma\lambda$. We can write $f$ as a dc function.

**Detail.** We find that

$$\sum_{j=1}^{n} r(x_j) = \tilde{g}(x) - h(x),$$

where

$$\tilde{g}(x) = \sum_{j=1}^{n} r(x_j) + \tfrac{1}{2\gamma}x_j^2 \quad \text{and} \quad h(x) = \sum_{j=1}^{n} \tfrac{1}{2\gamma}x_j^2,$$

which both are convex. Setting $g(x) = \|Ax - b\|_2^2 + \tilde{g}(x)$, we see that $f = g - h$ is a dc function.

This manner of constructing a dc function hints to the general fact that every $\rho$-weakly convex function $f$ is a dc function; $f + \frac{1}{2}\rho\|\cdot\|_2^2$ and $\frac{1}{2}\rho\|\cdot\|_2^2$ furnish the two convex functions. □

Let's consider optimality conditions for the problem of minimizing $f = g - h$ under the assumption that $g : \mathbb{R}^n \to \overline{\mathbb{R}}$ is proper, lsc and convex and $h : \mathbb{R}^n \to \mathbb{R}$ is convex. Since $f = g + (-h)$, we can apply the sum rule 4.67 to $g$ and $(-h)$ and obtain that at any point $\bar{x} \in \operatorname{dom} g$,

$$\partial f(\bar{x}) \subset \partial g(\bar{x}) + \partial(-h)(\bar{x}).$$

The qualification (4.15) in the sum rule holds because $h$, and then also $-h$, is locally Lipschitz continuous at $\bar{x}$ by 4.68(a) and $\partial^\infty(-h)(\bar{x}) = \{0\}$ by 4.69. The Fermat rule 4.73 now produces

$$0 \in \partial g(\bar{x}) + \partial(-h)(\bar{x})$$

as a necessary condition for $\bar{x} \in \operatorname{dom} g$ to be a local minimizer of $f$.

It would be natural to express an optimality condition in terms of $\partial h(\bar{x})$ and not $\partial(-h)(\bar{x})$. However, $\partial(-h)(\bar{x})$ could differ from $-\partial h(\bar{x})$ when $h$ is nonsmooth. For example, $h(x) = \max\{-x, 2x\}$ results in $\partial h(0) = [-1, 2]$, while $(-h)(x) = \min\{x, -2x\}$ gives $\partial(-h)(0) = \{-2, 1\}$. Thus, $\partial(-h)(0) \subset -\partial h(0)$. This relation holds in much more general cases.

**Proposition 6.48** (subgradients of negative-functions). *If $f : \mathbb{R}^n \to \overline{\mathbb{R}}$ is locally Lipschitz continuous and epi-regular at every point in a neighborhood of $\bar{x} \in \mathbb{R}^n$, then*

$$\partial(-f)(\bar{x}) \subset -\partial f(\bar{x}).$$

**Proof.** *Rademacher's theorem* [105, Theorem 9.60] asserts that locally Lipschitz continuous functions are differentiable at "most" points and this can be used to show that $-\partial(-f)(\bar{x})$ consists of the limits of gradients $\nabla f(x^\nu)$ at points of differentiability of $f$ as $x^\nu \to \bar{x}$. Thus, $-\partial(-f)(\bar{x}) \subset \partial f(\bar{x})$; see [105, Corollary 9.21] for details. □

Before we leverage this proposition in an optimality condition, let's consider a basic sum rule for subderivatives that brings into play the Oresme rule; refinements are found in [105, Corollary 10.9].

**Proposition 6.49** (sum rule for subderivatives). *For a proper lsc function* $f_1 : \mathbb{R}^n \to \overline{\mathbb{R}}$, *a convex function* $f_2 : \mathbb{R}^n \to \mathbb{R}$ *and a point* $\bar{x} \in \operatorname{dom} f_1$, *suppose that* $df_1(\bar{x}; 0) \neq -\infty$. *Then,*

$$d(f_1 + f_2)(\bar{x}; w) \geq df_1(\bar{x}; w) + df_2(\bar{x}; w) \;\; \forall w \in \mathbb{R}^n.$$

**Proof.** Let $F : \mathbb{R}^n \to \mathbb{R}^{2n}$ and $h : \mathbb{R}^{2n} \to \overline{\mathbb{R}}$ be given by $F(x) = (x, x)$ and $h(u_1, u_2) = f_1(u_1) + f_2(u_2)$. Since $h$ is proper and lsc and $F$ is smooth, the chain rule 4.64 produces

$$d(f_1 + f_2)(\bar{x}; w) \geq dh\big(F(\bar{x}); \nabla F(\bar{x})w\big) \geq df_1(\bar{x}; w) + df_2(\bar{x}; w),$$

where the last inequality is a consequence of 4.63 because $h$ is a separable function with $df_1(\bar{x}; 0) \neq -\infty$ and $df_2(\bar{x}; 0) \neq -\infty$; cf. 2.25 and 4.59. □

**Proposition 6.50** (optimality for dc functions). *For a proper, lsc and convex function* $g : \mathbb{R}^n \to \overline{\mathbb{R}}$, *a convex function* $h : \mathbb{R}^n \to \mathbb{R}$ *and* $x^\star \in \operatorname{dom} g$, *consider the problem*

$$\operatorname*{minimize}_{x \in \mathbb{R}^n} g(x) - h(x).$$

*If* $x^\star$ *is a local minimizer, then*

$$dg(x^\star; w) \geq dh(x^\star; w) \;\; \forall w \in \mathbb{R}^n \quad \text{or, equivalently,} \quad \partial h(x^\star) \subset \partial g(x^\star).$$

*These conditions in turn imply*

$$\partial g(x^\star) \cap \partial h(x^\star) \neq \emptyset.$$

**Proof.** By 4.68(a), $h$ is locally Lipschitz continuous at $x^\star$ and also epi-regular; see the discussion before 4.58. We can then combine 6.48 with the prior derivations to reach the fact that $x^\star$ must satisfy

$$0 \in \partial g(x^\star) - \partial h(x^\star),$$

which is equivalent to the last condition.

Next, let $f = g - h$, which is a proper lsc function. Then, $g = f + h$ because $h$ is real-valued. By the sum rule 6.49 for subderivatives applied to $g = f + h$,

$$dg(x^\star; w) \geq df(x^\star; w) + dh(x^\star; w) \;\; \forall w \in \mathbb{R}^n$$

provided that $df(x^\star; 0) \neq -\infty$. This is indeed the case because by the Oresme rule 2.10 one has $df(x^\star; w) \geq 0$ for all $w \in \mathbb{R}^n$ including 0. But, this also establishes

$$dg(x^\star; w) \geq dh(x^\star; w) \;\; \forall w \in \mathbb{R}^n$$

as asserted. The equivalence with $\partial h(x^\star) \subset \partial g(x^\star)$ is a consequence of 4.59. Note that $dg(x^\star; 0) \neq -\infty$ and $dh(x^\star; 0) \neq -\infty$ by 2.25 and 4.59.

Since $\partial h(x^\star)$ is nonempty, $\partial h(x^\star) \subset \partial g(x^\star)$ implies that $\partial h(x^\star) \cap \partial g(x^\star) \neq \emptyset$. Thus, the first paragraph of the proof is superfluous. □

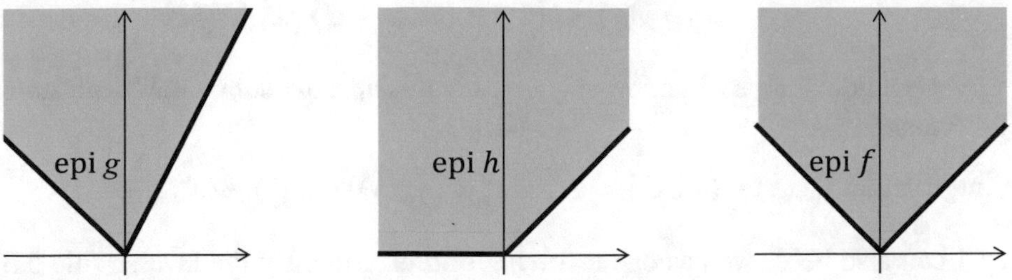


Fig. 6.11: The dc function $f = g - h$ given in 6.51 and its components.

**Example 6.51** (optimality for dc functions). For $g(x) = \max\{-x, 2x\}$ and $h(x) = \max\{0, x\}$, all the optimality conditions in 6.50 identify $\bar{x} = 0$, which in fact is the only minimizer of $g - h$; see Figure 6.11.

**Detail.** We find that

$$\partial g(x) = \begin{cases} \{-1\} & \text{for } x < 0 \\ [-1, 2] & \text{for } x = 0 \\ \{2\} & \text{otherwise} \end{cases} \qquad \partial h(x) = \begin{cases} \{0\} & \text{for } x < 0 \\ [0, 1] & \text{for } x = 0 \\ \{1\} & \text{otherwise.} \end{cases}$$

Thus, $\partial h(\bar{x}) \subset \partial g(\bar{x})$ and $\partial h(\bar{x}) \cap \partial g(\bar{x}) \neq \emptyset$ only for $\bar{x} = 0$. □

**Exercise 6.52** (optimality for dc functions). Use the optimality conditions in 6.50 to determine points that may be local minimizers for $g - h$, with (a) $g(x) = 0$ and $h(x) = \max\{0, x\}$ and (b) $g(x) = 0$ and $h(x) = \theta \max\{-x, x\}$, $\theta \in (0, \infty)$.

A broadly applicable algorithm for minimizing $g - h$ is now available when $g : \mathbb{R}^n \to \overline{\mathbb{R}}$ is proper, lsc and convex and $h : \mathbb{R}^n \to \mathbb{R}$ is convex. It involves solving a sequence of convex approximations of $g - h$ obtained adaptively by linearizing $h$ at the current point.

**DC Algorithm.**

Data. $x^1 \in \mathbb{R}^n$.
Step 0. Set $\nu = 1$.
Step 1. Determine $v^\nu \in \partial h(x^\nu)$.
Step 2. Compute

$$x^{\nu+1} \in \operatorname{argmin}_{x \in \mathbb{R}^n} g(x) - \langle v^\nu, x \rangle.$$

If $x^{\nu+1} = x^\nu$, then stop. Else, go to Step 3.
Step 3. Replace $\nu$ by $\nu + 1$ and go to Step 1.

The subproblem in Step 2 has a geometric interpretation. The subgradient inquality 2.17 gives that

$$h(x) \geq h(x^\nu) + \langle v^\nu, x - x^\nu \rangle \ \ \forall x \in \mathbb{R}^n.$$

Thus, the function

$$x \mapsto g(x) - h(x^\nu) - \langle v^\nu, x - x^\nu \rangle$$

is an upper bounding approximation of $g - h$. It's this approximation that's minimized in Step 2 because

$$\mathrm{argmin}_{x \in \mathbb{R}^n} \{g(x) - \langle v^\nu, x \rangle\} = \mathrm{argmin}_{x \in \mathbb{R}^n} \{g(x) - h(x^\nu) - \langle v^\nu, x - x^\nu \rangle\}.$$

Step 1 can also be viewed as one involving optimization. By the inverse rule 5.37 for subgradients, $v^\nu \in \partial h(x^\nu)$ if and only if $x^\nu \in \partial h^*(v^\nu)$. This in turn is equivalent to

$$v^\nu \in \mathrm{argmin}_{v \in \mathbb{R}^n} \{h^*(v) - \langle v, x^\nu \rangle\}$$

by the optimality condition 2.19 applied to the minimization of $h^* - \langle \cdot, x^\nu \rangle$. Since $h$ is real-valued, it has a subgradient at any point by 2.25 and the algorithm can't jam in this step.

The convex functions $g$ and $h$ that furnish the description of an actual function of interest $f$ through $f = g - h$ are never unique; one can always add the same real-valued convex function to both $g$ and $h$ and obtain another description. For example, it may be desirable to have a unique minimizer in Step 2 and this can be achieved by adding $\theta \| \cdot \|_2^2$ to both $g$ and $h$ for $\theta \in (0, \infty)$. This also helps with ensuring that the subproblem in Step 2 actually has a minimizer; cf. 4.9.

The optimality condition 2.19 applied to the subproblem in Step 2 establishes that $v^\nu \in \partial g(x^{\nu+1})$. Consequently, if the algorithm terminates in Step 2, then $v^\nu \in \partial g(x^\nu)$. Since we already have $v^\nu \in \partial h(x^\nu)$ from Step 1, $x^\nu$ satisfies the weaker condition in 6.50.

**Convergence 6.53** (dc algorithm). *For a proper, lsc and convex function $g : \mathbb{R}^n \to \overline{\mathbb{R}}$ and a convex function $h : \mathbb{R}^n \to \mathbb{R}$, suppose that $\{x^\nu, \nu \in \mathbb{N}\}$ is generated by the dc algorithm. Then, the following hold:*

(a) $g(x^{\nu+1}) - h(x^{\nu+1}) \leq g(x^\nu) - h(x^\nu)$ *for all* $\nu \in \mathbb{N}$.
(b) *If $\bar{x}$ is a cluster point of $\{x^\nu, \nu \in \mathbb{N}\}$, then $\partial g(\bar{x}) \cap \partial h(\bar{x}) \neq \emptyset$.*

**Proof.** For (a), we note that

$$g(x^{\nu+1}) - \langle v^\nu, x^{\nu+1} \rangle \leq g(x^\nu) - \langle v^\nu, x^\nu \rangle$$

because $x^{\nu+1}$ is a minimizer in Step 2. This fact and the convexity inequality 2.17 yield

$$\begin{aligned} g(x^{\nu+1}) - h(x^{\nu+1}) &\leq g(x^\nu) - h(x^{\nu+1}) + \langle v^\nu, x^{\nu+1} - x^\nu \rangle \\ &\leq g(x^\nu) - h(x^\nu) - \langle v^\nu, x^{\nu+1} - x^\nu \rangle + \langle v^\nu, x^{\nu+1} - x^\nu \rangle \\ &= g(x^\nu) - h(x^\nu). \end{aligned}$$

For (b), let $N \in \mathcal{N}_\infty^\#$ and $\bar{x}$ be such that $x^\nu \to_N \bar{x}$. Since $h$ is real-valued and convex, $\{\partial h(x^\nu), \nu \in N\}$ is contained in a compact set by 2.54 so there's a subsequence of $N$ such that $v^\nu$ converges along this subsequence to some $\bar{v}$. We see from 4.74 that $\bar{v} \in \partial h(\bar{x})$. Let's also denote this subsequence by $N$. The functions $g - \langle v^\nu, \cdot \rangle$ epi-converge to $g - \langle \bar{v}, \cdot \rangle$ along $\nu \in N$ by the epigraphical sum rule 4.19(a) or 4.19(b). Hence,

$$\bar{x} \in \operatorname{argmin}\{g - \langle \bar{v}, \cdot \rangle\}$$

as seen from 5.5(b). The optimality condition 2.19 then ensures that $\bar{v} \in \partial g(\bar{x})$ and the conclusion holds. □

The dc algorithm is remarkably simple and involves no hard-to-tune parameters such as step sizes. It's a descent algorithm in the sense that a computed point can't be worst than the previous one, which provides significant practical robustness to the algorithm. The main challenge is usually associated with constructing suitable convex functions $g$ and $h$ as they're never uniquely defined.

## 6.I DC in Regression and Classification

Regression and classification problems often aim to fit an affine function $\langle a, \cdot \rangle + \alpha$ to a dataset. However, in some cases one may obtain a better fit by using more flexible functions such as dc functions of the form

$$f(\cdot, w) = \max_{k=1,\dots,q} \{\langle a^k, \cdot \rangle + \alpha_k\} - \max_{k=1,\dots,r} \{\langle b^k, \cdot \rangle + \beta_k\}, \tag{6.34}$$

where $w = (a^1, \dots, a^q, \alpha_1, \dots, \alpha_q, b^1, \dots, b^r, \beta_1, \dots, \beta_r)$ is a parameter vector to be optimized. As we'll see, these functions can be fitted to a dataset using the dc algorithm.

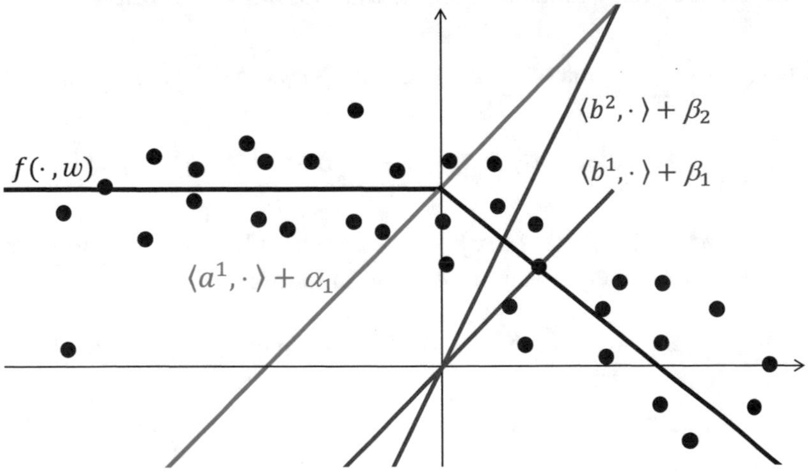


Fig. 6.12: A dc function $f(\cdot, w) = \langle a^1, \cdot \rangle + \alpha_1 - \max_{k=1,2}\{\langle b^k, \cdot \rangle + \beta_k\}$ is fitted to a cloud of data points using $a^1 = \alpha_1 = b^1 = 1$, $b^2 = 2$ and $\beta_1 = \beta_2 = 0$.

**Example 6.54** (median regression). Given a dataset $\{y_i \in \mathbb{R}, x^i \in \mathbb{R}^n,\ i = 1,\dots,m\}$, we would like to determine a function $f(\cdot, w) : \mathbb{R}^n \to \mathbb{R}$ such that the error $y_i - f(x^i, w)$ is small on average. Figure 6.12 illustrates a dataset, which can't be fitted well using an affine function. However, a dc function of the form (6.34) has the flexibility to capture the changing trend.

In 2.5, we measure discrepancy in terms of squared errors but absolute errors can also be used. The latter choice is less sensitive to outliers such as the data point in the lower left portion of Figure 6.12 and results in estimates of median values instead of average values. We recall that the median of a dataset is extremely robust in the sense that points above the median can be moved up arbitrarily and points below the median can be shifted down without affecting the median value. This robustness carries over to the regression setting and can be especially beneficial in situations with corrupted data.

The resulting *median regression* problem is formulated as

$$\underset{w\in\mathbb{R}^{(q+r)(n+1)}}{\text{minimize}}\ \frac{1}{m}\sum\nolimits_{i=1}^{m} \big|y_i - f(x^i, w)\big| + \|Dw\|_1,$$

where the dc function $f(\cdot, w)$ is given by (6.34) and $D$ is a diagonal matrix, which when nonzero encourages a sparse solution as in 2.5. The reason for the more refined choice $\|Dw\|_1$ over $\theta\|w\|_1$ is that we now may want to penalize nonzero values of the different components of $w$ with varying severity. For example, $\alpha_1$ may receive no penalty, while $\alpha_2, \dots, \alpha_q, \beta_1, \dots, \beta_r$ are associated with a positive penalty to reduce unnecessary redundancy in the model.

The problem can be reformulated as one of minimizing $g - h$, where

$$g(w) = \|Dw\|_1 + \frac{2}{m}\sum\nolimits_{i=1}^{m} \max\Big\{\max_{k=1,\dots,q} \big\{\langle a^k, x^i\rangle + \alpha_k - y_i\big\},\ \max_{k=1,\dots,r} \big\{\langle b^k, x^i\rangle + \beta_k\big\}\Big\}$$

$$h(w) = \frac{1}{m}\sum\nolimits_{i=1}^{m}\Big(\max_{k=1,\dots,q} \big\{\langle a^k, x^i\rangle + \alpha_k - y_i\big\} + \max_{k=1,\dots,r} \big\{\langle b^k, x^i\rangle + \beta_k\big\}\Big).$$

**Detail.** Trivially, $|\sigma - \tau| = 2\max\{\sigma, \tau\} - (\sigma + \tau)$. This identity allows us to reformulate the absolute value of a dc function as another dc function:

$$\begin{aligned}
\big|y - f(x, w)\big| &= \Big|\max_{k=1,\dots,q} \big\{\langle a^k, x\rangle + \alpha_k - y\big\} - \max_{k=1,\dots,r} \big\{\langle b^k, x\rangle + \beta_k\big\}\Big| \\
&= 2\max\Big\{\max_{k=1,\dots,q} \big\{\langle a^k, x\rangle + \alpha_k - y\big\},\ \max_{k=1,\dots,r} \big\{\langle b^k, x\rangle + \beta_k\big\}\Big\} \\
&\quad - \Big(\max_{k=1,\dots,q} \big\{\langle a^k, x\rangle + \alpha_k - y\big\} + \max_{k=1,\dots,r} \big\{\langle b^k, x\rangle + \beta_k\big\}\Big).
\end{aligned}$$

Since the objective function involves a sum of such differences, we can sum up each term in this difference to obtain a dc function; the regularizer $\|Dw\|_1$ is simply added to the first part.

We can minimize $g - h$ using the dc algorithm. This requires a subgradient of $h$ at a current point

$$\bar{w} = (\bar{a}^1, \ldots, \bar{a}^q, \bar{\alpha}_1, \ldots, \bar{\alpha}_q, \bar{b}^1, \ldots, \bar{b}^r, \bar{\beta}_1, \ldots, \bar{\beta}_r),$$

which can be computed as follows: For each $i = 1, \ldots, m$, let

$$k_i^\star \in \operatorname{argmax}_{k=1,\ldots,q} \{\langle \bar{a}^k, x^i \rangle + \bar{\alpha}_k\} \qquad k_i^{\star\star} \in \operatorname{argmax}_{k=1,\ldots,r} \{\langle \bar{b}^k, x^i \rangle + \bar{\beta}_k\}.$$

The sum rule 4.67, which applies because the summation is of real-valued convex functions (cf. 4.68(a) and 4.69), as well as 4.66 and (4.13) give that

$$\frac{1}{m} \sum_{i=1}^m v^i + u^i \in \partial h(\bar{w}),$$

where $v^i \in \mathbb{R}^{(q+r)(n+1)}$ is the vector with $x^i$ in the block that starts at element $n(k_i^\star - 1) + 1$ and ends at element $nk_i^\star$, with 1 at element $nq + k_i^\star$ and with 0 elsewhere. Moreover, $u^i \in \mathbb{R}^{(q+r)(n+1)}$ has $x^i$ in the block that starts at element $(n+1)q + n(k_i^{\star\star} - 1) + 1$ and ends at element $(n+1)q + nk_i^{\star\star}$, has 1 at element $(n+1)q + nr + k_i^{\star\star}$ and 0 elsewhere.

A similar approach in the context of least-squares regression is pursued in [28], which also discusses refinements of the dc algorithm. □

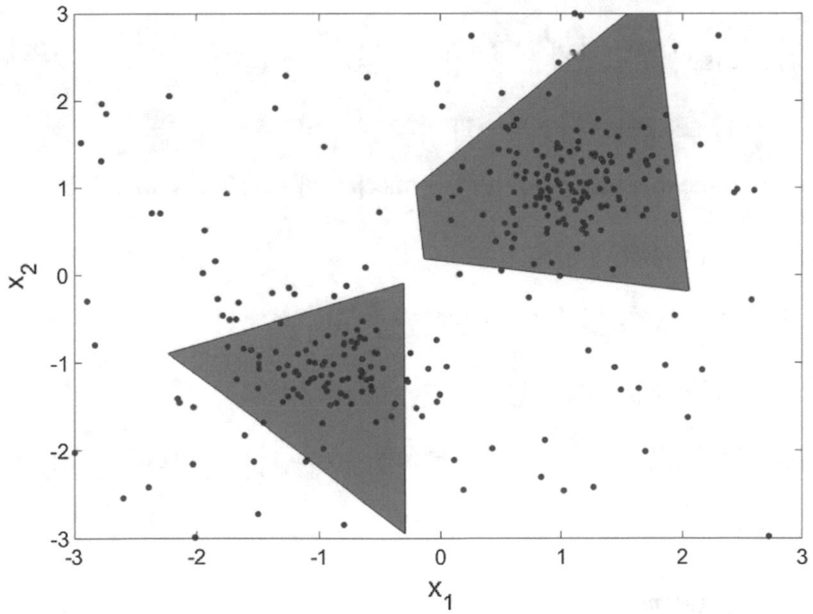


Fig. 6.13: Classification data with positive labels (marked with red dots) concentrated in two clusters and scattered negative labels (marked with blue dots). Still, the level-set of a dc function identifies the regions with positive labels quite well.

**Example 6.55** (support vector machine). Given data $\{y_i \in \{-1,1\}, x^i \in \mathbb{R}^n, i = 1,\dots,m\}$, we would like to construct a classifier that predicts the label $y \in \{-1,1\}$ of a future observation $x$. In §2.H, we leverage affine functions for this purpose. Figure 6.13 illustrates a situation with $n = 2$, where affine functions perform poorly; a straight line can't separate the positive labels (marked with red dots) from the negative ones (blue dots) even approximately. As an alternative, let's adopt a function of the form (6.34) and this leads to the following hinge-loss problem:

$$\underset{w\in\mathbb{R}^{(q+r)(n+1)}}{\text{minimize}}\ \frac{1}{m}\sum_{i=1}^{m} \max\{0, 1 - y_i f(x^i, w)\} + \|Dw\|_1.$$

We again permit penalties as expressed by a diagonal matrix $D$. This problem can be formulated as minimizing $g - h$, where

$$\begin{aligned} g(w) = {} & \frac{1}{m}\sum_{i\in\mathbb{I}^+} \max\Big\{\max_{k=1,\dots,q}\{\langle a^k, x^i\rangle + \alpha_k\},\ \max_{k=1,\dots,r}\{\langle b^k, x^i\rangle + \beta_k + 1\}\Big\} \\ & + \frac{1}{m}\sum_{i\in\mathbb{I}^-} \max\Big\{\max_{k=1,\dots,q}\{\langle a^k, x^i\rangle + \alpha_k + 1\},\ \max_{k=1,\dots,r}\{\langle b^k, x^i\rangle + \beta_k\}\Big\} \\ & + \|Dw\|_1 \\ h(w) = {} & \frac{1}{m}\sum_{i\in\mathbb{I}^+} \max_{k=1,\dots,q}\{\langle a^k, x^i\rangle + \alpha_k\} + \frac{1}{m}\sum_{i\in\mathbb{I}^-} \max_{k=1,\dots,r}\{\langle b^k, x^i\rangle + \beta_k\} \end{aligned}$$

and $\mathbb{I}^+ = \{i \mid y_i = 1\}$ and $\mathbb{I}^- = \{i \mid y_i = -1\}$.

**Detail.** It's easy to see that $\max\{0, \sigma - \tau\} = \max\{\sigma, \tau\} - \tau$. Thus, for $i \in \mathbb{I}^+$,

$$\begin{aligned} & \max\{0, 1 - y_i f(x^i, w)\} \\ & = \max\Big\{0, 1 - \max_k\{\langle a^k, x^i\rangle + \alpha_k\} + \max_k\{\langle b^k, x^i\rangle + \beta_k\}\Big\} \\ & = \max\Big\{0, \max_k\{\langle b^k, x^i\rangle + \beta_k + 1\} - \max_k\{\langle a^k, x^i\rangle + \alpha_k\}\Big\} \\ & = \max\Big\{\max_k\{\langle a^k, x^i\rangle + \alpha_k\},\ \max_k\{\langle b^k, x^i\rangle + \beta_k + 1\}\Big\} - \max_k\{\langle a^k, x^i\rangle + \alpha_k\}. \end{aligned}$$

Similarly, for $i \in \mathbb{I}^-$,

$$\begin{aligned} & \max\{0, 1 - y_i f(x^i, w)\} \\ & = \max\Big\{0, \max_k\{\langle a^k, x^i\rangle + \alpha_k + 1\} - \max_k\{\langle b^k, x^i\rangle + \beta_k\}\Big\} \\ & = \max\Big\{\max_k\{\langle a^k, x^i\rangle + \alpha_k + 1\},\ \max_k\{\langle b^k, x^i\rangle + \beta_k\}\Big\} - \max_k\{\langle b^k, x^i\rangle + \beta_k\}. \end{aligned}$$

The asserted formulae for $g$ and $h$ are obtained by summing up these expressions. Figure 6.13 illustrates $\{f(\cdot, \bar{w}) \geq 0\}$ for an optimized parameter vector $\bar{w}$ obtained using the dc algorithm. The level-set takes the form of the union of two polyhedra (colored orange) and predicts quite well the regions in which red dots are more likely than blue dots. □

## 6.J Approximation Errors

Minimization problems are approximated for computational reasons and to examine the effect of changing parameters. We've seen that epigraphical analysis validates an approximation scheme by establishing convergence of solutions of approximating problems to those of an actual problem. Still, the magnitude of the solution error for a given approximation has remained unsettled. Chapter 5 provides a step in this direction by estimating the effect of perturbations in terms of subgradients of min-value functions. This offers insight into accurate approximations of a problem but is less informative when the approximations are coarse. We'll now see that a notion of distance between two epigraphs bounds the discrepancy between the corresponding minima and near-minimizers, even if the two epigraphs are far apart.

Let's start by clarifying the meaning of a distance between two sets $C, D \subset \mathbb{R}^n$ and between a point and a set. We've consistently measured the distance between a point and $C$ using the Euclidean norm, but to allow for more flexibility we now permit the use of *any* norm $\|\cdot\|$ on $\mathbb{R}^n$. The *point-to-set distance under* $\|\cdot\|$ between $\bar{x} \in \mathbb{R}^n$ and $C \subset \mathbb{R}^n$ is

$$\operatorname{dist}_*(\bar{x}, C) = \inf_{x \in C} \|x - \bar{x}\| \text{ when } C \neq \emptyset \text{ and } \operatorname{dist}_*(\bar{x}, \emptyset) = \infty. \tag{6.35}$$

The asterisk subscript indicates that the distance is computed using some other norm than the Euclidean one, which underpins $\operatorname{dist}(\bar{x}, C)$; see (4.4). The *excess* of $C$ over $D$ is defined as

$$\operatorname{exs}(C; D) = \begin{cases} \sup_{x \in C} \operatorname{dist}_*(x, D) & \text{if } C \neq \emptyset, D \neq \emptyset \\ \infty & \text{if } C \neq \emptyset, D = \emptyset \\ 0 & \text{otherwise,} \end{cases}$$

which then depends on the chosen norm as well. We assume the Euclidean norm if no other choice is specified. Figure 6.14 shows $\operatorname{exs}(C; D)$ and $\operatorname{exs}(D; C)$ under the Euclidean norm and they're different. This asymmetry makes it natural to look at the larger of the two, which is the classical Hausdorff distance between $C$ and $D$. In our setting with unbounded sets commonly occurring, the Hausdorff distance could easily be infinity. Thus, we adopt a slightly refined measure of distance between sets that rely on a truncation using *balls*, which now are defined by

$$\mathbb{B}_*(\bar{x}, \rho) = \{x \in \mathbb{R}^n \mid \|x - \bar{x}\| \leq \rho\},$$

where again the norm is arbitrary as indicated by the asterisk subscript. For any $\rho \geq 0$, the *truncated Hausdorff distance* between $C$ and $D$ is

$$\hat{d\!l}_\rho(C, D) = \max\Big\{\operatorname{exs}\big(C \cap \mathbb{B}_*(0, \rho);\ D\big),\ \ \operatorname{exs}\big(D \cap \mathbb{B}_*(0, \rho);\ C\big)\Big\},$$

which depends on the norm underpinning the excess and the ball. (The norm is always the same for both.) Again, the Euclidean norm is the default. Figure 6.14 has $\operatorname{exs}(C \cap \mathbb{B}(0, \rho); D) = \operatorname{exs}(C; D)$ because the portion of $C$ furthest away from $D$ is contained in

$\mathbb{B}(0,\rho)$, while $\text{exs}(D \cap \mathbb{B}(0,\rho); C)$ is a bit smaller than $\text{exs}(D; C)$. Nevertheless, $\hat{dl}_\rho(C, D) = \text{exs}(C; D)$ in this case. Generally, $\hat{dl}_\rho(C, D)$ is finite for any nonempty $C, D$ and $\rho \in [0, \infty)$.

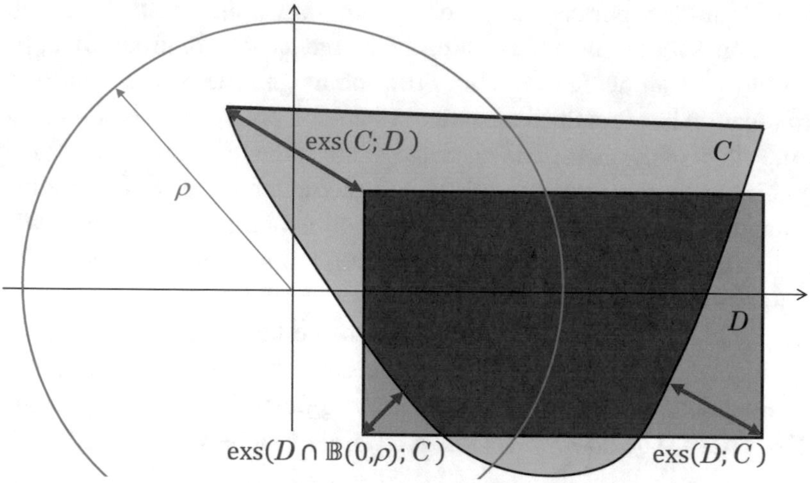


Fig. 6.14: The excess of $C$ over $D$ and vice versa as well as the effect of truncation.

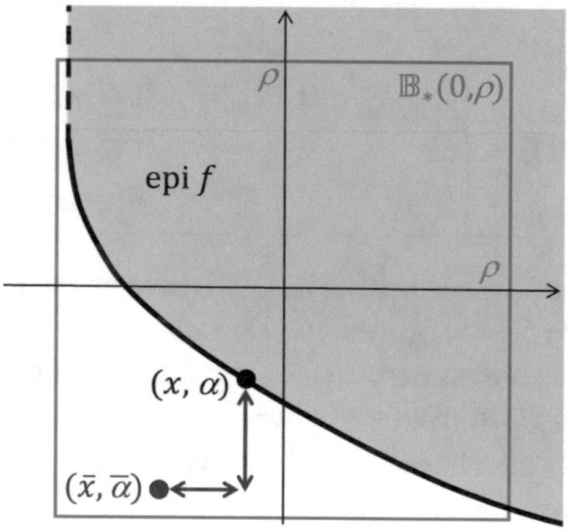


Fig. 6.15: The distance between $(\bar{x}, \bar{\alpha})$ and $(x, \alpha) \in \text{epi}\, f$ under the norm (6.36) is defined as the larger of the "horizontal" and "vertical" distances.

We're interested in distances between epigraphs as these relate to distances between minima and near-minimizers. For $f, g : \mathbb{R}^n \to \overline{\mathbb{R}}$, epi $f$ and epi $g$ are subsets of $\mathbb{R}^{n+1}$ and hence require a norm on $\mathbb{R}^{n+1}$. We consistently choose

$$\max\left\{\|x - \bar{x}\|_2, |\alpha - \bar{\alpha}|\right\}, \quad \text{with } (x, \alpha), (\bar{x}, \bar{\alpha}) \in \mathbb{R}^n \times \mathbb{R}, \tag{6.36}$$

which typically produces the simplest results. Figure 6.15 illustrates the point-to-set distance $\text{dist}_*((\bar{x}, \bar{\alpha}), \text{epi}\, f)$ under this norm; it's the minimum value of the larger of the "horizontal" and "vertical" distances to a point in epi $f$. In this case, the corresponding ball

$$\mathbb{B}_*(0,\rho) = \{(x,\alpha) \in \mathbb{R}^n \times \mathbb{R} \mid \|x\|_2 \le \rho, |\alpha| \le \rho\} = \mathbb{B}(0,\rho) \times [-\rho,\rho]. \tag{6.37}$$

It has the shape of a square in Figure 6.15 ($n = 1$), but is a cylinder for $n = 2$ and a hyper-cylinder more generally.

**Theorem 6.56** (minimization error). *For $f, g : \mathbb{R}^n \to \overline{\mathbb{R}}$, $\rho \in [0,\infty)$ and $\varepsilon \in [0, 2\rho]$, suppose that*

$$\inf f, \inf g \in [-\rho, \rho - \varepsilon], \qquad \operatorname{argmin} f \cap \mathbb{B}(0,\rho) \neq \emptyset, \qquad \operatorname{argmin} g \cap \mathbb{B}(0,\rho) \neq \emptyset.$$

*Adopt the norm (6.36) for $\mathbb{R}^{n+1}$ and let $\delta > \varepsilon + 2\hat{d\!l}_\rho(\operatorname{epi} f, \operatorname{epi} g)$. Then, one has*

$$|\inf f - \inf g| \le \hat{d\!l}_\rho(\operatorname{epi} f, \operatorname{epi} g)$$

$$\operatorname{exs}\big(\varepsilon\text{-}\operatorname{argmin} g \cap \mathbb{B}(0,\rho);\ \delta\text{-}\operatorname{argmin} f\big) \le \hat{d\!l}_\rho(\operatorname{epi} f, \operatorname{epi} g).$$

**Proof.** Let $\eta = \hat{d\!l}_\rho(\operatorname{epi} f, \operatorname{epi} g)$, $\gamma > 0$ and $\mathbb{B}_*(0,\rho)$ be a ball under the assumed norm on $\mathbb{R}^{n+1}$ as given by (6.37). Since $\operatorname{argmin} g \cap \mathbb{B}(0,\rho) \neq \emptyset$, there's $\bar{x} \in \mathbb{B}(0,\rho)$ such that $g(\bar{x}) = \inf g \in [-\rho, \rho - \varepsilon]$. Thus,

$$\big(\bar{x}, g(\bar{x})\big) \in \operatorname{epi} g \cap \mathbb{B}_*(0,\rho).$$

One can also find a point $(x,\alpha) \in \operatorname{epi} f$ such that

$$\max\big\{\|x - \bar{x}\|_2, |\alpha - g(\bar{x})|\big\} \le \operatorname{dist}_*\Big(\big(\bar{x}, g(\bar{x})\big), \operatorname{epi} f\Big) + \gamma, \tag{6.38}$$

where $\operatorname{dist}_*$ specifies the point-to-set distances under norm (6.36). Then,

$$\eta \ge \operatorname{exs}\big(\operatorname{epi} g \cap \mathbb{B}_*(0,\rho);\ \operatorname{epi} f\big) \ge \operatorname{dist}_*\Big(\big(\bar{x}, g(\bar{x})\big),\ \operatorname{epi} f\Big) \ge \big|\alpha - g(\bar{x})\big| - \gamma.$$

Collecting the above results, one finds that

$$\inf f \le f(x) \le \alpha \le g(\bar{x}) + \eta + \gamma = \inf g + \eta + \gamma.$$

Since $\gamma$ is arbitrary, we've established that

$$\inf f \le \inf g + \eta.$$

The same argument with the roles of $f$ and $g$ reversed leads to the first conclusion.

Next, let $\bar{x} \in \varepsilon$-argmin $g \cap \mathbb{B}(0, \rho)$. Then,

$$g(\bar{x}) \le \inf g + \varepsilon \le \rho, \qquad g(\bar{x}) \ge \inf g \ge -\rho, \qquad (\bar{x}, g(\bar{x})) \in \operatorname{epi} g \cap \mathbb{B}_*(0, \rho).$$

There's $(x, \alpha) \in \operatorname{epi} f$ such that (6.38) again holds. Similar to the previous paragraph, this means that

$$\eta \ge \|x - \bar{x}\|_2 - \gamma \quad \text{and} \quad \eta \ge \big|\alpha - g(\bar{x})\big| - \gamma.$$

These facts together with the first result establish that

$$f(x) \le \alpha \le g(\bar{x}) + \eta + \gamma \le \inf g + \varepsilon + \eta + \gamma \le \inf f + \varepsilon + 2\eta + \gamma.$$

Thus, $x \in (\varepsilon + 2\eta + \gamma)$-argmin $f$. Since $\|x - \bar{x}\|_2 \le \eta + \gamma$, this implies that

$$\operatorname{exs}\big(\varepsilon\text{-argmin}\, g \cap \mathbb{B}(0, \rho);\ (\varepsilon + 2\eta + \gamma)\text{-argmin}\, f\big) \le \eta + \gamma.$$

For $\gamma \in (0, \delta - \varepsilon - 2\eta]$, we've $\varepsilon + 2\eta + \gamma \le \delta$ and also obtain

$$\operatorname{exs}\big(\varepsilon\text{-argmin}\, g \cap \mathbb{B}(0, \rho);\ \delta\text{-argmin}\, f\big) \le \eta + \gamma.$$

Since $\gamma$ can be made arbitrarily near 0, the second conclusion follows. □

The theorem is valid for functions that are very different; there's no need for $g$ to be an accurate approximation of $f$. Moreover, it applies broadly as there are no assumptions on $f$ and $g$ except (loose) bounds on minima and minimizers, which translate into the size of the truncation radius $\rho$. The truncated Hausdorff distance is closely tied to epi-convergence: $f^\nu \xrightarrow{e} f$ if and only if $\hat{d\!l}_\rho(\operatorname{epi} f^\nu, \operatorname{epi} f) \to 0$ for all $\rho \ge 0$ by [105, Theorem 7.58]. Further connections with the Attouch-Wets distance (also called integrated set-distance) emerge as well; see [108] and references therein.

Figure 6.16(left) shows that the bound on minima is sharp: $g = \iota_{[0,2]}$ and $f = 1 + \iota_{[1,3]}$ have $\operatorname{exs}(\operatorname{epi} g \cap \mathbb{B}_*(0, \rho); \operatorname{epi} f) = 1$ for $\rho \in [0, \infty)$, where $\mathbb{B}_*(0, \rho)$ is a ball under the assumed norm (6.36) on $\mathbb{R}^{n+1}$ as given by (6.37). Moreover,

$$\operatorname{exs}\big(\operatorname{epi} f \cap \mathbb{B}_*(0, \rho);\ \operatorname{epi} g\big) = \begin{cases} 0 & \text{if } \rho \in [0, 2] \\ \rho - 2 & \text{if } \rho \in (2, 3] \\ 1 & \text{if } \rho \in (3, \infty). \end{cases}$$

Hence, $\hat{d\!l}_\rho(\operatorname{epi} f, \operatorname{epi} g) = 1$ for all $\rho \in [0, \infty)$, which coincides with $|\inf f - \inf g|$. In this case, the theorem applies for $\rho \ge 1$ and then can't be improved without additional assumptions because the given $f$ and $g$ result in the upper bound being attained.

The bound on near-minimizers is also sharp. Figure 6.16(middle) illustrates the same $g$, but now $f(1) = 2$, $f(x) = x$ for $x \in (1, 2)$, $f(2) = -1$ and $f(x) = \infty$ otherwise. Again, $\operatorname{exs}(\operatorname{epi} g \cap \mathbb{B}_*(0, \rho); \operatorname{epi} f) = 1$ for $\rho \in [0, \infty)$, but

$$\operatorname{exs}\big(\operatorname{epi} f \cap \mathbb{B}_*(0, \rho);\ \operatorname{epi} g\big) = \begin{cases} 0 & \text{if } \rho \in [0, 2) \\ 1 & \text{if } \rho \in [2, \infty) \end{cases}$$

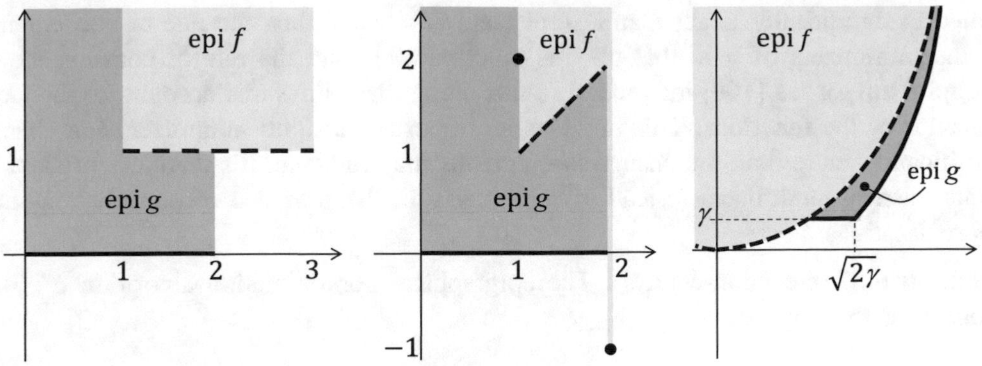


Fig. 6.16: The bounds on minima (left) and near-minimizers (middle) are sharp in the theorem of minimization error 6.56. The error for minimizers can be large (right).

so that $\hat{dl}_\rho(\operatorname{epi} f, \operatorname{epi} g) = 1$ for all $\rho \in [0, \infty)$. The theorem applies for $\rho \geq 2$, the bottleneck being that $\mathbb{B}(0, \rho)$ needs to intersect with $\operatorname{argmin} f = \{2\}$. For such $\rho$, the assertion is that the excess of $\varepsilon$-$\operatorname{argmin} g \cap \mathbb{B}(0, \rho) = [0, 2]$ over $\delta$-$\operatorname{argmin} f$ should be at most 1 for $\delta > \varepsilon + 2$. It turns out that

$$\delta\text{-}\operatorname{argmin} f = \begin{cases} \{2\} & \text{if } \delta \in [0, 2] \\ (1, \delta - 1] \cup \{2\} & \text{if } \delta \in (2, 3) \\ [1, 2] & \text{if } \delta \in [3, \infty). \end{cases}$$

Thus, for $\varepsilon = 0$, which is the most difficult case to pass,

$$\operatorname{exs}\big(\operatorname{argmin} g \cap \mathbb{B}(0, \rho);\ \delta\text{-}\operatorname{argmin} f\big) = \begin{cases} 2 & \text{if } \delta \in [0, 2] \\ 1 & \text{if } \delta \in (2, \infty). \end{cases}$$

The situation improves only marginally if the functions are lsc. For example, if $f$ is modified to have $f(1) = 1$, then

$$\delta = \varepsilon + 2\hat{dl}_\rho(\operatorname{epi} f, \operatorname{epi} g)$$

works as well and this, in fact, holds generally; see [106, Theorem 4.3].

One might hope that $\delta$ could be set to $\varepsilon$ in the theorem after some modification of the right-hand side bound. This is the case, but not without complications. Figure 6.16(right) illustrates the case when $f(x) = x^2$ so that $\operatorname{argmin} f = \{0\}$ and

$$g(x) = \begin{cases} \gamma & \text{if } x \in \big[-\sqrt{2\gamma}, \sqrt{2\gamma}\,\big] \\ x^2 - \gamma & \text{otherwise,} \end{cases} \qquad \text{with} \quad \operatorname{argmin} g = \big[-\sqrt{2\gamma}, \sqrt{2\gamma}\,\big],$$

where $\gamma > 0$. For any $\rho \geq 0$, $\hat{dl}_\rho(\operatorname{epi} f, \operatorname{epi} g) = \gamma$. Let $\rho \geq \sqrt{2}$ and $\gamma \in (0, 1]$. Then,

$$\operatorname{exs}\big(\operatorname{argmin} g \cap \mathbb{B}(0, \rho);\ \operatorname{argmin} f\big) = \sqrt{2\gamma},$$

which is significantly larger than $\hat{dl}_\rho(\text{epi } f, \text{epi } g) = \gamma$. In fact, the rate of convergence of the minimizers of $g$ to that of $f$ is much slower than the rate of convergence of $\hat{dl}_\rho(\text{epi } f, \text{epi } g)$; see [106] for general results along these lines that account for the local behavior of the functions. This provides an important insight: minimizers can change significantly more than the change in epigraphs may indicate. It's therefore prudent to focus on near-minimizers instead of minimizers in the presence of perturbations and approximations.

**Example 6.57** (precision-control). The epigraphical approximation algorithm of §4.C requires us to compute

$$x^\nu \in \varepsilon^\nu\text{-argmin } f^\nu,$$

where $f^\nu : \mathbb{R}^n \to \overline{\mathbb{R}}$ is an approximation of the actual function $f : \mathbb{R}^n \to \overline{\mathbb{R}}$. The knowledge of $\hat{dl}_\rho(\text{epi } f^\nu, \text{epi } f)$ provides guidance about the choice of $\varepsilon^\nu$ as well as rules for stopping the algorithm.

**Detail.** Let's assume that the theorem of minimization error 6.56 applies. Then, because

$$\inf f^\nu \leq f^\nu(x^\nu) \leq \inf f^\nu + \varepsilon^\nu,$$

we obtain bounds on the minimum value of the actual problem:

$$f^\nu(x^\nu) - \varepsilon^\nu - \hat{dl}_\rho(\text{epi } f^\nu, \text{epi } f) \leq \inf f \leq f^\nu(x^\nu) + \hat{dl}_\rho(\text{epi } f^\nu, \text{epi } f).$$

Although a smaller $\varepsilon^\nu$ allows us to make a slightly stronger conclusion from these inequalities, it's clear that $\varepsilon^\nu$ preferably should be proportional to $\hat{dl}_\rho(\text{epi } f^\nu, \text{epi } f)$. If $\varepsilon^\nu$ is set much lower, then we're probably wasting computational efforts in minimizing $f^\nu$ because our ability to pin down $\inf f$ is anyhow limited by the size of $\hat{dl}_\rho(\text{epi } f^\nu, \text{epi } f)$. Likewise, if $\varepsilon^\nu$ is set higher, then it becomes the dominant error term and dilutes the estimate of $\inf f$ even for accurate $f^\nu$. The situation is similar for near-minimizers. As long as $x^\nu \in \mathbb{B}(0, \rho)$, it follows by the theorem of minimization error 6.56 that

$$\text{dist}\big(x^\nu, \delta\text{-argmin } f\big) \leq \hat{dl}_\rho(\text{epi } f^\nu, \text{epi } f) \quad \text{when } \delta > \varepsilon^\nu + 2\hat{dl}_\rho(\text{epi } f^\nu, \text{epi } f)$$

and there's little benefit from setting $\varepsilon^\nu$ much below $\hat{dl}_\rho(\text{epi } f^\nu, \text{epi } f)$.

We refer to [111] for a more detailed analysis of how to adjust precision in various approximations. □

We next turn to computations of the truncated Hausdorff distance.

**Proposition 6.58** (Kenmochi condition). *For $f, g : \mathbb{R}^n \to \overline{\mathbb{R}}$, both with nonempty epigraphs, $\rho \in [0, \infty)$ and the norm (6.36) on $\mathbb{R}^{n+1}$, one has*

$$\hat{dl}_\rho(\text{epi } f, \text{epi } g) = \inf \Big\{ \eta \geq 0 \;\Big|\; \begin{array}{l} \inf_{\mathbb{B}(x,\eta)} g \leq \max\{f(x), -\rho\} + \eta \;\; \forall x \in \{f \leq \rho\} \cap \mathbb{B}(0,\rho) \\ \inf_{\mathbb{B}(x,\eta)} f \leq \max\{g(x), -\rho\} + \eta \;\; \forall x \in \{g \leq \rho\} \cap \mathbb{B}(0,\rho) \end{array} \Big\}.$$

**Proof.** Let $\bar\eta = \hat{d\!l}_\rho(\operatorname{epi} f, \operatorname{epi} g)$ and $\operatorname{dist}_*$ specify the point-to-set-distance on $\mathbb{R}^{n+1}$ under the assumed norm, with $\mathbb{B}_*(0,\rho)$ being a corresponding ball as given by (6.37). First, let's establish that the relation holds with $\geq$. Set $\varepsilon > 0$. Suppose that

$$x \in \{f \leq \rho\} \cap \mathbb{B}(0,\rho) \quad \text{and} \quad f(x) \geq -\rho.$$

Then, $(x, f(x)) \in \operatorname{epi} f \cap \mathbb{B}_*(0,\rho)$ and there exists $(\bar x, \bar\alpha) \in \operatorname{epi} g$ such that

$$\bar\eta \geq \operatorname{exs}\big(\operatorname{epi} f \cap \mathbb{B}_*(0,\rho);\ \operatorname{epi} g\big) \geq \operatorname{dist}_*\Big(\big(x, f(x)\big),\ \operatorname{epi} g\Big) \geq \big|f(x) - \bar\alpha\big| - \varepsilon$$

and, similarly, $\bar\eta \geq \|x - \bar x\|_2 - \varepsilon$. Thus,

$$\inf\nolimits_{\mathbb{B}(x,\bar\eta+\varepsilon)} g \leq g(\bar x) \leq \bar\alpha \leq f(x) + \bar\eta + \varepsilon \leq \max\big\{f(x), -\rho\big\} + \bar\eta + \varepsilon.$$

Next, suppose that

$$x \in \{f \leq \rho\} \cap \mathbb{B}(0,\rho) \quad \text{and} \quad f(x) < -\rho.$$

Then, $(x, -\rho) \in \operatorname{epi} f \cap \mathbb{B}_*(0,\rho)$ and there exists $(\bar x, \bar\alpha) \in \operatorname{epi} g$ such that

$$\bar\eta \geq \operatorname{exs}\big(\operatorname{epi} f \cap \mathbb{B}_*(0,\rho);\ \operatorname{epi} g\big) \geq \operatorname{dist}_*\big((x, -\rho),\ \operatorname{epi} g\big) \geq \big|\rho + \bar\alpha\big| - \varepsilon$$

and, similarly, $\bar\eta \geq \|x - \bar x\|_2 - \varepsilon$. Consequently,

$$\inf\nolimits_{\mathbb{B}(x,\bar\eta+\varepsilon)} g \leq g(\bar x) \leq \bar\alpha \leq -\rho + \bar\eta + \varepsilon \leq \max\big\{f(x), -\rho\big\} + \bar\eta + \varepsilon.$$

We've shown that the first set of constraints on the right-hand side in the asserted equality is satisfied with $\eta = \bar\eta + \varepsilon$. Repeating the arguments with the roles of $f$ and $g$ reversed, the same holds for the second set of constraints. Thus, the right-hand side does not exceed $\bar\eta + \varepsilon$. Since $\varepsilon$ is arbitrary, the relation holds with $\geq$.

Second, let's prove that the relation holds with $\leq$. Let

$$f_\rho(x) = \begin{cases} \max\big\{f(x), -\rho\big\} & \text{if } x \in \{f \leq \rho\} \cap \mathbb{B}(0,\rho) \\ \infty & \text{otherwise.} \end{cases}$$

Suppose that $\eta \in [0, \infty)$ satisfies

$$\inf\nolimits_{\mathbb{B}(x,\eta)} g \leq \max\big\{f(x), -\rho\big\} + \eta \quad \forall x \in \{f \leq \rho\} \cap \mathbb{B}(0,\rho),$$

which is the same as having

$$\inf\nolimits_{\mathbb{B}(x,\eta)} g \leq f_\rho(x) + \eta \quad \forall x \in \operatorname{dom} f_\rho.$$

Let $x \in \operatorname{dom} f_\rho$ and $\varepsilon > 0$. The previous inequality ensures that there's $\bar x \in \mathbb{R}^n$ such that

$$\|x - \bar x\|_2 \leq \eta \quad \text{and} \quad g(\bar x) \leq f_\rho(x) + \eta + \varepsilon.$$

Set $\bar{\alpha} = \max\{g(\bar{x}), f_\rho(x) - \eta - \varepsilon\}$. Thus, $g(\bar{x}) \le \bar{\alpha} < \infty$ and

$$-\eta - \varepsilon \le f_\rho(x) - \bar{\alpha} \le f_\rho(x) - \big(f_\rho(x) - \eta - \varepsilon\big) = \eta + \varepsilon.$$

We've constructed a point $(\bar{x}, \bar{\alpha}) \in \operatorname{epi} g$ with

$$\max\big\{\|x - \bar{x}\|_2, |f_\rho(x) - \bar{\alpha}|\big\} \le \eta + \varepsilon.$$

Since $\varepsilon > 0$ is arbitrary, this implies that

$$\operatorname{dist}_*\Big(\big(x, f_\rho(x)\big), \operatorname{epi} g\Big) \le \eta,$$

which then holds for all $x \in \operatorname{dom} f_\rho$. Let $(\hat{x}, \hat{\alpha}) \in \operatorname{epi} f \cap \mathbb{B}_*(0, \rho)$. Then, $\hat{x} \in \operatorname{dom} f_\rho$ because $f(\hat{x}) \le \hat{\alpha} \le \rho$ and $\|\hat{x}\|_2 \le \rho$. Consequently,

$$\operatorname{dist}_*\Big(\big(\hat{x}, f_\rho(\hat{x})\big), \operatorname{epi} g\Big) \le \eta.$$

Since $f_\rho(\hat{x}) \le \hat{\alpha}$ and the distance is computed to an epigraph, we also have

$$\operatorname{dist}_*\big((\hat{x}, \hat{\alpha}), \operatorname{epi} g\big) \le \eta.$$

In view of the arbitrary choice of $(\hat{x}, \hat{\alpha})$, this means that

$$\operatorname{exs}\big(\operatorname{epi} f \cap \mathbb{B}_*(0, \rho);\ \operatorname{epi} g\big) \le \eta.$$

We can repeat this argument with the roles of $f$ and $g$ reversed and conclude that $\bar{\eta} \le \eta$. Thus, the infimum over such $\eta$ can't be below $\bar{\eta}$. □

The Kenmochi condition shows that while the pointwise difference $|f(x) - g(x)|$ may enter in the calculation of $\hat{dl}_\rho(\operatorname{epi} f, \operatorname{epi} g)$, it isn't critical. In fact, the pointwise distance may be large and $\hat{dl}_\rho(\operatorname{epi} f, \operatorname{epi} g)$ could still be small. This is the case for $f = \iota_{(-\infty,0]}$ and $g = \iota_{(-\infty,\varepsilon]}$ with $\varepsilon > 0$. To see this, note that we don't need to compare $f(\varepsilon) = \infty$ and $g(\varepsilon) = 0$ in the Kenmochi condition but rather $\inf_{\mathbb{B}(\varepsilon,\eta)} f$ and $g(\varepsilon)$. For $\eta = \varepsilon$, the former matches the latter.

The insight from the Kenmochi condition leads to bounds on the truncated Hausdorff distance between epigraphs of various kinds. We adopt the following terminology: A function $f : \mathbb{R}^n \to \mathbb{R}$ is *Lipschitz continuous with modulus function* $\kappa$ if $\kappa : [0, \infty) \to [0, \infty)$ and

$$\big|f(x) - f(\bar{x})\big| \le \kappa(\rho)\|x - \bar{x}\|_2 \quad \forall x, \bar{x} \in \mathbb{B}(0, \rho), \quad \rho \in [0, \infty).$$

**Proposition 6.59** (estimates from sup-norm). *For functions $f, g : \mathbb{R}^n \to \overline{\mathbb{R}}$ with nonempty epigraphs and $\rho \in [0, \infty)$, adopt the norm (6.36) on $\mathbb{R}^{n+1}$ and let*

$$C_\rho = \big(\{f \le \rho\} \cup \{g \le \rho\}\big) \cap \mathbb{B}(0, \rho).$$

*If* $C_\rho \neq \emptyset$, *then*

$$\hat{dl}_\rho(\text{epi } f, \text{epi } g) \leq \sup_{x \in C_\rho} \big|f(x) - g(x)\big|.$$

*If* $C_\rho = \emptyset$, *then* $\hat{dl}_\rho(\text{epi } f, \text{epi } g) = 0$.

*Moreover, if* $f$ *and* $g$ *are real-valued and Lipschitz continuous, with common modulus function* $\kappa$, *and* $X \subset \mathbb{R}^n$ *is nonempty, then*

$$\hat{dl}_\rho(\text{epi } f, \text{epi } g) \leq \max\big\{1, \kappa(\hat{\rho})\big\} \text{exs}(C_\rho; X) + \sup_{x \in X} \big|f(x) - g(x)\big|$$

*provided that* $\hat{\rho} > \rho + \text{exs}(C_\rho; X)$.

**Proof.** The first assertion holds immediately by the Kenmochi condition 6.58 because $\inf_{\mathbb{B}(x,\eta)} g \leq g(x)$ and likewise for $f$. The case with $C_\rho = \emptyset$ is trivial as there are no constraints to check in the Kenmochi condition.

For the second assertion, set $\eta = \text{exs}(C_\rho; X)$ and let $\varepsilon \in (0, \hat{\rho} - \rho - \eta]$. Suppose that $x \in \{f \leq \rho\} \cap \mathbb{B}(0, \rho)$. Then, there's $\bar{x} \in X$ with $\|x - \bar{x}\|_2 \leq \eta + \varepsilon$ and

$$\inf_{\mathbb{B}(x,\eta+\varepsilon)} g \leq g(\bar{x}) \leq f(\bar{x}) + \sup_X |f - g| \leq \max\{f(x), -\rho\} + \kappa(\hat{\rho})(\eta + \varepsilon) + \sup_X |f - g|.$$

A similar result holds with the roles of $f$ and $g$ reversed. Thus, by the Kenmochi condition 6.58, one has

$$\hat{dl}_\rho(\text{epi } f, \text{epi } g) \leq \max\big\{\eta + \varepsilon,\ \kappa(\hat{\rho})(\eta + \varepsilon) + \sup_X |f - g|\big\}.$$

Since $\varepsilon$ is arbitrary, the second conclusion follows. □

**Example 6.60** (approximation of sup-projection; cont.). Let's return to the sup-projections

$$f^\nu(x) = \sup_{y \in Y^\nu} g(x, y) \quad \text{and} \quad f(x) = \sup_{y \in Y} g(x, y)$$

from 6.33 and quantify $\hat{dl}_\rho(\text{epi } f^\nu, \text{epi } f)$ under the norm (6.36) for $\mathbb{R}^{n+1}$.

For $\rho \in [0, \infty)$, suppose that there's $\kappa(\rho) \in [0, \infty)$ such that

$$\big|g(x, y) - g(x, \bar{y})\big| \leq \kappa(\rho)\|y - \bar{y}\|_2 \ \ \forall x \in \mathbb{B}(0, \rho), \quad y, \bar{y} \in Y.$$

Then,

$$\hat{dl}_\rho(\text{epi } f, \text{epi } f^\nu) \leq \kappa(\rho)\,\text{exs}(Y; Y^\nu).$$

**Detail.** Let $x \in \mathbb{B}(0, \rho)$. Since $Y^\nu \subset Y$, $f^\nu(x) \leq f(x)$ and we only need to concentrate on any discrepancy the other way around. Since $Y$ is compact and $g(x, \cdot)$ is continuous as laid out in 6.33, there are $\bar{y} \in \text{argmax}_{y \in Y} g(x, y)$ by 4.9 and also $y \in Y^\nu$ such that $\|y - \bar{y}\|_2 \leq \text{exs}(Y; Y^\nu)$. Thus,

$$f(x) = g(x, \bar{y}) \leq g(x, y) + \kappa(\rho)\|y - \bar{y}\|_2 \leq f^\nu(x) + \kappa(\rho)\,\text{exs}(Y; Y^\nu).$$

The assertion then follows from the first part of 6.59. If $Y = [0, 1]^m$ and

$$Y^\nu = \{1/\nu, 2/\nu, \ldots, \nu/\nu\}^m,$$

then $\text{exs}(Y; Y^\nu) = \sqrt{m}/\nu$. □

**Exercise 6.61** (approximation by inf-projection). For $f : \mathbb{R}^n \to \overline{\mathbb{R}}$ and $\varepsilon \in (0, \infty)$, consider the problem of minimizing $f$ and the alternative problem

$$\mathop{\text{minimize}}_{x \in \mathbb{R}^n} g(x) = \inf_{y \in \mathbb{B}(0,\varepsilon)} f(x + y).$$

The latter is equivalently stated as minimizing $f(x + y)$ over $(x, y) \in \mathbb{R}^n \times \mathbb{B}(0, \varepsilon)$, which involves more variables and can be easier to solve in situations with difficult-to-satisfy constraints. Determine an upper bound on $\hat{d\!l}_\rho(\text{epi}\, f, \text{epi}\, g)$.

**Guide.** Leverage the Kenmochi condition 6.58. □

**Proposition 6.62** (indicator functions). *For $C, D \subset \mathbb{R}^n$, $\rho \in [0, \infty)$ and the norm (6.36) on $\mathbb{R}^{n+1}$, one has*

$$\hat{d\!l}_\rho(\text{epi}\, \iota_C, \text{epi}\, \iota_D) = \hat{d\!l}_\rho(C, D).$$

**Proof.** This follows readily from the definition of the truncated Hausdorff distance and the fact that for $x \in \mathbb{R}^n$, $\alpha \in [0, \infty)$ and $X \subset \mathbb{R}^n$, one has

$$\text{dist}(x, X) = \text{dist}_* \big((x, \alpha), \text{epi}\, \iota_X\big),$$

where $\text{dist}_*$ gives the point-to-set distance under the norm (6.36). □

If $C, D \subset \mathbb{R}^n$ are simple, such as boxes, then a direct calculation of $\hat{d\!l}_\rho(C, D)$ is straightforward. However, if $C = \{f \le 0\}$ and $D = \{g \le 0\}$ for functions $f, g : \mathbb{R}^n \to \overline{\mathbb{R}}$, then the situation is more delicate. Even when $f$ is uniformly close to $g$, $\hat{d\!l}_\rho(C, D)$ can be large. For example, if $f(x) = x^2$ and $g(x) = x^2 + \varepsilon$, then $\hat{d\!l}_\rho(C, D) = \infty$ regardless of $\varepsilon > 0$ because $D = \emptyset$. This relates to the discussion around Figure 4.6. Still, in the convex case one can obtain a versatile bound under the assumption that the minimum values of $f$ and $g$ are strictly below 0, which eliminates the pathological case above.

**Proposition 6.63** (level-sets of convex functions). *For $\rho \in [0, \infty)$ and proper, lsc and convex $f, g : \mathbb{R}^n \to \overline{\mathbb{R}}$, suppose that*

$$\inf f < 0, \qquad \inf g < 0, \qquad \text{argmin}\, f \ne \emptyset, \qquad \text{argmin}\, g \ne \emptyset.$$

*If the norm on $\mathbb{R}^{n+1}$ is (6.36), then*

$$\hat{d\!l}_\rho\big(\{f \le 0\}, \{g \le 0\}\big) \le \left(1 - \frac{\bar{\rho} + \hat{\rho}}{\max\{\inf g, \inf f\}}\right) \hat{d\!l}_\rho(\text{epi}\, f, \text{epi}\, g)$$

*provided that*

$$\bar{\rho} \ge \max\big\{\text{dist}(0, \text{argmin}\, f),\ \text{dist}(0, \text{argmin}\, g)\big\}$$

$$\hat{\rho} \ge \max\big\{\bar{\rho},\ \rho + \hat{d\!l}_\rho(\text{epi}\, f, \text{epi}\, g)\big\}.$$

**Proof.** We obtain this result by specializing [108, Proposition 3.8]. □

The constant in front of $\hat{d\!l}_\rho(\operatorname{epi} f, \operatorname{epi} g)$ worsens as $\inf f$ or $\inf g$ gets close to 0 and we approach pathological cases of the kind described prior to the proposition. In an application, one may want to first minimize a constraint function to check whether it has enough "depth" for a certain threshold level to be imposed as a constraint. This would furnish at least a bound on $\max\{\inf g, \inf f\}$.

**Proposition 6.64** (composite functions). *For proper $f_0, g_0 : \mathbb{R}^n \to \overline{\mathbb{R}}$, consider the functions*

$$f = f_0 + h \circ F \quad \text{and} \quad g = g_0 + h \circ G,$$

*where $F(x) = (f_1(x), \ldots, f_m(x))$, $G(x) = (g_1(x), \ldots, g_m(x))$ and*

(a) *$h : \mathbb{R}^m \to \mathbb{R}$ is Lipschitz continuous with modulus function $\kappa$*
(b) *$f_i, g_i : \mathbb{R}^n \to \mathbb{R}$, $i = 1, \ldots, m$, are Lipschitz continuous with common modulus function $\lambda$.*

*If the norm on $\mathbb{R}^{n+1}$ is (6.36), then for $\rho \in [0, \infty)$ one has*

$$\begin{aligned}\hat{d\!l}_\rho(\operatorname{epi} f, \operatorname{epi} g) \leq & \big(1 + \sqrt{m}\kappa(\check{\rho})\lambda(\hat{\rho})\big)\hat{d\!l}_{\bar{\rho}}(\operatorname{epi} f_0, \operatorname{epi} g_0) \\ & + \kappa(\check{\rho}) \sup_{x \in \mathbb{B}(0,\rho)} \|F(x) - G(x)\|_2\end{aligned}$$

*provided that $\bar{\rho}$, $\hat{\rho}$ and $\check{\rho}$ are sufficiently large. Specifically,*

$$\begin{aligned}&\bar{\rho} \geq \rho + \max\Big\{\sup_{x \in \mathbb{B}(0,\rho)} \big|h\big(F(x)\big)\big|,\ \sup_{x \in \mathbb{B}(0,\rho)} \big|h\big(G(x)\big)\big|\Big\} \\ &\hat{\rho} > \rho + \hat{d\!l}_{\bar{\rho}}(\operatorname{epi} f_0, \operatorname{epi} g_0) \\ &\check{\rho} \geq \max\Big\{\sup_{x \in \mathbb{B}(0,\hat{\rho})} \|F(x)\|_2,\ \sup_{x \in \mathbb{B}(0,\hat{\rho})} \|G(x)\|_2\Big\}.\end{aligned}$$

**Proof.** We leverage the Kenmochi condition 6.58. Let $\mathbb{B}_*(0, \bar{\rho})$ be a ball under the assumed norm on $\mathbb{R}^{n+1}$; see (6.37). Set $\eta = \hat{d\!l}_{\bar{\rho}}(\operatorname{epi} f_0, \operatorname{epi} g_0)$, $\varepsilon \in (0, \hat{\rho} - \rho - \eta]$ and $\bar{x} \in \{f \leq \rho\} \cap \mathbb{B}(0, \rho)$. Then, $f_0(\bar{x}) \leq \rho - h(F(\bar{x})) \leq \bar{\rho}$.

First, suppose that $f_0(\bar{x}) \geq -\bar{\rho}$ so that

$$\big(\bar{x}, f_0(\bar{x})\big) \in \operatorname{epi} f_0 \cap \mathbb{B}_*(0, \bar{\rho}).$$

Consequently, there exists $(\hat{x}, \hat{\alpha}) \in \operatorname{epi} g_0$ with $\|\bar{x} - \hat{x}\|_2 \leq \eta + \varepsilon$ and $|\hat{\alpha} - f_0(\bar{x})| \leq \eta + \varepsilon$. Thus,

$$g_0(\hat{x}) \leq \hat{\alpha} \leq f_0(\bar{x}) + \eta + \varepsilon.$$

Combining the various facts, we obtain that

$$\begin{aligned}
&\inf_{x\in\mathbb{B}(\bar{x},\eta+\varepsilon)} g(x) \le g_0(\hat{x}) + h\big(G(\hat{x})\big)\\
&= g_0(\hat{x}) + h\big(F(\bar{x})\big) + h\big(G(\hat{x})\big) - h\big(G(\bar{x})\big) + h\big(G(\bar{x})\big) - h\big(F(\bar{x})\big)\\
&\le f_0(\bar{x}) + \eta + \varepsilon + h\big(F(\bar{x})\big) + \kappa(\breve{\rho})\big\|G(\hat{x}) - G(\bar{x})\big\|_2 + \kappa(\breve{\rho})\big\|F(\bar{x}) - G(\bar{x})\big\|_2\\
&\le f_0(\bar{x}) + h\big(F(\bar{x})\big) + \eta + \varepsilon + \kappa(\breve{\rho})\sqrt{m}\max_{i=1,\dots,m}\big|g_i(\hat{x}) - g_i(\bar{x})\big|\\
&\quad + \kappa(\breve{\rho})\sup_{x\in\mathbb{B}(0,\rho)}\big\|F(x) - G(x)\big\|_2\\
&\le f(\bar{x}) + \eta + \varepsilon + \kappa(\breve{\rho})\sqrt{m}\lambda(\hat{\rho})(\eta+\varepsilon) + \kappa(\breve{\rho})\sup_{x\in\mathbb{B}(0,\rho)}\big\|F(x) - G(x)\big\|_2\\
&\le \max\big\{f(\bar{x}), -\rho\big\} + \big(1 + \kappa(\breve{\rho})\sqrt{m}\lambda(\hat{\rho})\big)(\eta+\varepsilon) + \kappa(\breve{\rho})\sup_{x\in\mathbb{B}(0,\rho)}\big\|F(x) - G(x)\big\|_2.
\end{aligned}$$

Second, suppose that $f_0(\bar{x}) < -\bar{\rho}$. Then,

$$(\bar{x}, -\bar{\rho}) \in \operatorname{epi} f_0 \cap \mathbb{B}_*(0, \bar{\rho})$$

and there exists $(\hat{x}, \hat{\alpha}) \in \operatorname{epi} g_0$ with $\|\bar{x} - \hat{x}\|_2 \le \eta + \varepsilon$ and $|\hat{\alpha} + \bar{\rho}| \le \eta + \varepsilon$. Thus,

$$g_0(\hat{x}) \le \hat{\alpha} \le -\bar{\rho} + \eta + \varepsilon.$$

Similar to above, this leads to

$$\begin{aligned}
&\inf_{x\in\mathbb{B}(\bar{x},\eta+\varepsilon)} g(x)\\
&\le -\bar{\rho} + h\big(F(\bar{x})\big) + \eta + \varepsilon + \kappa(\breve{\rho})\sqrt{m}\lambda(\hat{\rho})(\eta+\varepsilon) + \kappa(\breve{\rho})\sup_{x\in\mathbb{B}(0,\rho)}\big\|F(x) - G(x)\big\|_2\\
&\le \max\big\{f(\bar{x}), -\rho\big\} + \big(1 + \kappa(\breve{\rho})\sqrt{m}\lambda(\hat{\rho})\big)(\eta+\varepsilon) + \kappa(\breve{\rho})\sup_{x\in\mathbb{B}(0,\rho)}\big\|F(x) - G(x)\big\|_2.
\end{aligned}$$

The second inequality follows because

$$-\bar{\rho} + h\big(F(\bar{x})\big) \le -\bar{\rho} + \sup_{x\in\mathbb{B}(0,\rho)}\big|h\big(F(x)\big)\big| \le -\rho.$$

Thus, in both cases, we obtain the same upper bound on $\inf_{x\in\mathbb{B}(\bar{x},\eta+\varepsilon)} g(x)$. Repeating these arguments with the roles of $f$ and $g$ switched, we obtain the result by the Kenmochi condition 6.58 after letting $\varepsilon$ tend to 0. □

**Example 6.65** (approximation of sup-projection; cont.). With the sup-projections $f$ and $f^\nu$ in 6.60, suppose that we seek to approximate

$$\operatorname*{minimize}_{x\in X} h\big(f(x)\big) \quad \text{by} \quad \operatorname*{minimize}_{x\in X^\nu} h\big(f^\nu(x)\big),$$

where $X^\nu, X \subset \mathbb{R}^n$ are nonempty and $h : \mathbb{R} \to \mathbb{R}$ is Lipschitz continuous with modulus function $\kappa$. In addition to $g$ being continuous, we assume that $g(\cdot, y)$ is smooth for all $y \in Y$ and $(x, y) \mapsto \nabla_x g(x, y)$ is continuous relative to $\mathbb{R}^n \times Y$ as well as $g(x, \cdot)$ is smooth for all $x \in \mathbb{R}^n$ and $(x, y) \mapsto \nabla_y g(x, y)$ is continuous relative to $\mathbb{R}^n \times \operatorname{con} Y$. Then, for $\rho \in [0, \infty)$,

$$\hat{d\!l}_\rho\big(\operatorname{epi}(\iota_X + h\circ f), \operatorname{epi}(\iota_{X^\nu} + h\circ f^\nu)\big) \le \big(1 + \kappa(\breve{\rho})\lambda(\hat{\rho})\big)\hat{d\!l}_{\bar{\rho}}(X, X^\nu) + \kappa(\breve{\rho})\mu(\rho)\operatorname{exs}(Y; Y^\nu)$$

provided that

$$\bar{\rho} \geq \rho + \max\Big\{ \sup_{x \in \mathbb{B}(0,\rho)} \big|h\big(f(x)\big)\big|,\ \sup_{x \in \mathbb{B}(0,\rho)} \big|h\big(f^\nu(x)\big)\big| \Big\}$$

$$\hat{\rho} > \rho + \hat{dl}_{\bar{\rho}}(X, X^\nu)$$

$$\breve{\rho} \geq \max\Big\{ \sup_{x \in \mathbb{B}(0,\hat{\rho})} \big|f(x)\big|,\ \sup_{x \in \mathbb{B}(0,\hat{\rho})} \big|f^\nu(x)\big| \Big\}.$$

The quantities $\lambda(\hat{\rho})$ and $\mu(\rho)$ are computed below.

**Detail.** We obtain the estimate through 6.64 with $f_0 = \iota_X$ and $g_0 = \iota_{X^\nu}$ so that $\hat{dl}_{\bar{\rho}}(\operatorname{epi} f_0, \operatorname{epi} g_0) = \hat{dl}_{\bar{\rho}}(X, X^\nu)$ by 6.62. Let's calculate the modulus function $\lambda$ for $f$. Since $(x, y) \mapsto \|\nabla_x g(x, y)\|_2$ is continuous relative to $\mathbb{R}^n \times Y$ and $Y$ is compact, it follows by 4.9 that

$$\lambda(\rho) = \sup\big\{ \|\nabla_x g(x, y)\|_2 \;\big|\; y \in Y,\ x \in \mathbb{B}(0, \rho) \big\}$$

is finite for $\rho \in [0, \infty)$. Let's fix $x, x' \in \mathbb{B}(0, \rho)$ and $\bar{y} \in Y$. By the mean value theorem (1.7), there's $\sigma \in [0, 1]$ such that

$$\begin{aligned} \big|g(x, \bar{y}) - g(x', \bar{y})\big| &= \big|\big\langle \nabla_x g\big(x' + \sigma(x - x'), \bar{y}\big), x - x' \big\rangle\big| \\ &\leq \big\|\nabla_x g\big(x' + \sigma(x - x'), \bar{y}\big)\big\|_2 \|x - x'\|_2 \leq \lambda(\rho)\|x - x'\|_2. \end{aligned}$$

Thus, $|g(x, y) - g(x', y)| \leq \lambda(\rho)\|x - x'\|_2$ for all $y \in Y$ and

$$f(x) = \sup_{y \in Y} g(x, y) \leq \sup_{y \in Y} g(x', y) + \lambda(\rho)\|x - x'\|_2 = f(x') + \lambda(\rho)\|x - x'\|_2.$$

Reversing the roles of $x$ and $x'$, we obtain

$$\big|f(x) - f(x')\big| \leq \lambda(\rho)\|x - x'\|_2.$$

The same argument with $Y^\nu$ replacing $Y$ shows that $\lambda(\rho)$ is also a modulus function for $f^\nu$.

Next, we bound $\sup_{x \in \mathbb{B}(0,\rho)} |f(x) - f^\nu(x)|$. Following the reasoning in 6.60, we need to obtain a modulus function $\mu$ for $g(x, \cdot)$. By [105, Corollary 2.30], $\operatorname{con} Y$ is compact because $Y$ is compact. Since $(x, y) \mapsto \|\nabla_y g(x, y)\|_2$ is continuous relative to $\mathbb{R}^n \times \operatorname{con} Y$ and $\operatorname{con} Y$ is compact, it follows by 4.9 that

$$\mu(\rho) = \sup\big\{ \|\nabla_y g(x, y)\|_2 \;\big|\; y \in \operatorname{con} Y,\ x \in \mathbb{B}(0, \rho) \big\}$$

is finite for $\rho \in [0, \infty)$. Let's fix $y, y' \in Y$ and $\bar{x} \in \mathbb{B}(0, \rho)$. By the mean value theorem (1.7), there's $\sigma \in [0, 1]$ such that

$$\begin{aligned} \big|g(\bar{x}, y) - g(\bar{x}, y')\big| &= \big|\big\langle \nabla_y g\big(\bar{x}, y' + \sigma(y - y')\big), y - y' \big\rangle\big| \\ &\leq \big\|\nabla_y g\big(\bar{x}, y' + \sigma(y - y')\big)\big\|_2 \|y - y'\|_2 \leq \mu(\rho)\|y - y'\|_2. \end{aligned}$$

Now, returning to the arguments in 6.60, we obtain

$$\sup_{x \in \mathbb{B}(0,\rho)} \left| f(x) - f^\nu(x) \right| \le \mu(\rho)\, \mathrm{exs}(Y; Y^\nu).$$

The asserted bound then follows by collecting these results and applying 6.64. □

The pointwise error bounds between two functions needed in 6.59 and 6.64 are satisfied, for example, by the smooth approximations in 4.16 and also by approximations obtained through linearizations.

**Example 6.66** (error in linearization). In the proximal composite method (§6.F) and elsewhere, we approximate a twice smooth mapping $F : \mathbb{R}^n \to \mathbb{R}^m$ by the linearization

$$\bar{F}(x) = F(\bar{x}) + \nabla F(\bar{x})(x - \bar{x})$$

obtained at a specific point $\bar{x} \in \mathbb{R}^n$. For any $\rho \in [0, \infty)$ with $\bar{x} \in \mathbb{B}(0, \rho)$, there's $\gamma \in [0, \infty)$ such that

$$\sup_{x \in \mathbb{B}(0,\rho)} \left\| F(x) - \bar{F}(x) \right\|_2 \le \gamma \rho^2.$$

**Detail.** This fact is immediate from (6.27). A slightly more refined analysis quantifies $\gamma$ as well: A second-order mean value theorem asserts that for a twice smooth $f : \mathbb{R}^n \to \mathbb{R}$ and $x, x' \in \mathbb{R}^n$, there's $\sigma \in [0, 1]$ such that

$$f(x) = f(x') + \left\langle \nabla f(x'), x - x' \right\rangle + \tfrac{1}{2}\left\langle x - x', \nabla^2 f\big(x' + \sigma(x - x')\big)(x - x') \right\rangle.$$

We apply this to each component $f_i$ of $F$. Let $\rho \in [0, \infty)$ and $x, \bar{x} \in \mathbb{B}(0, \rho)$. Moreover, set

$$\tau = \max_{i=1,\dots,m} \sup_{x' \in \mathbb{B}(0,\rho)} \left\| \nabla^2 f_i(x') \right\|_F,$$

which is finite because $\nabla^2 f_i$ is continuous and $\mathbb{B}(0, \rho)$ is compact; cf. 4.9. Then, for each $i = 1, \dots, m$, there's $\sigma_i \in [0, 1]$ such that

$$\begin{aligned} \left| f_i(\bar{x}) + \left\langle \nabla f_i(\bar{x}), x - \bar{x} \right\rangle - f_i(x) \right| &= \tfrac{1}{2} \left| \left\langle x - \bar{x}, \nabla^2 f_i\big(\bar{x} + \sigma_i(x - \bar{x})\big)(x - \bar{x}) \right\rangle \right| \\ &\le \tfrac{1}{2} \left\| \nabla^2 f_i\big(\bar{x} + \sigma_i(x - \bar{x})\big) \right\|_F \|x - \bar{x}\|_2^2 \le 2\tau\rho^2. \end{aligned}$$

Consequently,

$$\left\| F(x) - \bar{F}(x) \right\|_2 \le 2\tau\sqrt{m}\rho^2$$

and the asserted bound holds with $\gamma = 2\tau\sqrt{m}$, which also depends on $\rho$ through $\tau$. □

We've repeatedly assumed that a function is Lipschitz continuous with a certain modulus function. In the study of rates of convergence, it's often sufficient to know that there exists such a modulus function; its exact value might be secondary. For example, in 6.65,

$$\hat{dl}_\rho\big(\mathrm{epi}(\iota_X + h \circ f), \mathrm{epi}(\iota_{X^\nu} + h \circ f^\nu)\big)$$

vanishes at the sublinear rate $1/\nu$ under the assumption that $Y = [0, 1]^m$, $Y^\nu = \{1/\nu, 2/\nu, \dots \nu/\nu\}^m$ and $X = X^\nu$. To see this, just realize that $\kappa(\check{\rho})$ and $\mu(\rho)$ are constant, independent of $\nu$. Their exact values don't change the rate of convergence. Thus, it suffices to know that there's a modulus function $\kappa$ for $h$ and similarly regarding $\mu$. In such situations, the following fact is useful.

**Proposition 6.67** (modulus function from local Lipschitz continuity). *If $f : \mathbb{R}^n \to \mathbb{R}$ is locally Lipschitz continuous, then there exists a function $\kappa : [0,\infty) \to [0,\infty)$ such that*

$$\big|f(x) - f(\bar{x})\big| \leq \kappa(\rho)\|x - \bar{x}\|_2 \quad \forall x, \bar{x} \in \mathbb{B}(0,\rho), \quad \rho \in [0,\infty),$$

*i.e., $f$ is Lipschitz continuous with modulus function $\kappa$.*

**Proof.** Let $\rho \in [0,\infty)$. Suppose for the sake of contradiction that

$$\sup\left\{\frac{\big|f(x') - f(x'')\big|}{\|x' - x''\|_2} \;\middle|\; x', x'' \in \mathbb{B}(0,\rho),\ x' \neq x''\right\} = \infty. \tag{6.39}$$

Then, there are $x^\nu, \bar{x}^\nu \in \mathbb{B}(0,\rho)$ with $x^\nu \neq \bar{x}^\nu$ such that

$$\frac{\big|f(x^\nu) - f(\bar{x}^\nu)\big|}{\|x^\nu - \bar{x}^\nu\|_2} \to \infty.$$

Since $\mathbb{B}(0,\rho)$ is compact, there are $N \in \mathcal{N}_\infty^\#$ and $x, \bar{x} \in \mathbb{B}(0,\rho)$ such that $x^\nu \to_N x$ and $\bar{x}^\nu \to_N \bar{x}$. Let's consider two cases: Suppose that $x \neq \bar{x}$. Then, the above denominator is bounded away from 0 and we must have $|f(x^\nu) - f(\bar{x}^\nu)| \to \infty$. However, this isn't possible because $f$ is continuous and $x^\nu, \bar{x}^\nu$ are contained in the compact set $\mathbb{B}(0,\rho)$; see 4.9. We've reached a contradiction. Suppose next that $x = \bar{x}$. Since $f$ is locally Lipschitz continuous at $\bar{x}$, there are $\delta > 0$ and $\mu \in [0,\infty)$ such that

$$\big|f(x') - f(x'')\big| \leq \mu\|x' - x''\|_2 \quad \forall x', x'' \in \mathbb{B}(\bar{x},\delta).$$

Since $x^\nu \to_N \bar{x}$ and $\bar{x}^\nu \to_N \bar{x}$, there's $\bar{\nu}$ such that for all $\nu \in N$ with $\nu \geq \bar{\nu}$, one has $x^\nu, \bar{x}^\nu \in \mathbb{B}(\bar{x},\delta)$. For these $\nu$,

$$\frac{\big|f(x^\nu) - f(\bar{x}^\nu)\big|}{\|x^\nu - \bar{x}^\nu\|_2} \leq \mu,$$

which is a contradiction. Thus, the left-hand side of (6.39) is finite and the conclusion follows. □

**Exercise 6.68** (smoothing error). For smooth $g_i : \mathbb{R}^n \to \mathbb{R}$, $i = 1,\dots,m$, $X = [0,\infty)^n$ and $\mathbb{I}_k \subset \{1,\dots,m\}$, $k = 1,\dots,q$, consider the problem

$$\underset{x \in X}{\text{minimize}}\ f(x) = \max\nolimits_{k=1,\dots,q} \min\nolimits_{i \in \mathbb{I}_k} g_i(x)$$

and its approximation

$$\underset{x \in X^\nu}{\text{minimize}}\ f^\nu(x) = \max\nolimits_{k=1,\dots,q} f_k^\nu(x),$$

where $X^\nu = [0, \alpha^\nu]^n$, with $\alpha^\nu \in [0, \infty)$, and

$$f_k^\nu(x) = -\frac{1}{\theta^\nu} \ln\Big(\sum\nolimits_{i \in \mathbb{I}_k} \exp\big(-\theta^\nu g_i(x)\big)\Big), \text{ with } \theta^\nu \in (0, \infty).$$

Derive a bound on $\hat{d\!l}_\rho(\text{epi}(\iota_X + f), \text{epi}(\iota_{X^\nu} + f^\nu))$.

**Guide.** Apply 6.64 with $f_0 = \iota_X$, $g_0 = \iota_{X^\nu}$ and $h : \mathbb{R}^q \to \mathbb{R}$ having $h(u) = \max\{u_1, \ldots, u_q\}$. Leverage 6.43 to determine the difference between $\min_{i \in \mathbb{I}_k} g_i(x)$ and $f_k^\nu(x)$. The existence of Lipschitz modulus functions for $x \mapsto \min_{i \in \mathbb{I}_k} g_i(x)$ and $f_k^\nu$ can be settled by 6.67. For extra credit, determine specific formulae for the modulus functions. □

# Chapter 7
# GENERALIZED EQUATIONS

When modeling the interaction between several agents, each trying to achieve some objective, we often need to express their dependence using a mixture of equations, inequalities and more general inclusions. Even for minimization problems, we've seen that solutions can be characterized, at least in part, by inclusions derived from the Fermat rule and the Rockafellar condition. The resulting requirements in these diverse situations give rise to generalized equations, which emerge as a central tool in formulation and solution of optimization as well as variational problems. We'll now explore this broader landscape.

## 7.A Formulations

While we express a system of equations $F(x) = 0$ in terms of a mapping $F : \mathbb{R}^n \to \mathbb{R}^m$, a generalized equation is given by a *set-valued mapping* $S : \mathbb{R}^n \rightrightarrows \mathbb{R}^m$. We encountered set-valued mappings in §3.F when examining subgradients of expectation functions, but they arise much more broadly. The distinguishing feature relative to a (point-valued) mapping $F : \mathbb{R}^n \to \mathbb{R}^m$ is that we now allow for an input $x$ to produce an output $S(x)$ that's a subset of $\mathbb{R}^m$ and not necessarily a single point. In fact, we even allow the subset to be empty so that the *domain*

$$\operatorname{dom} S = \left\{x \in \mathbb{R}^n \;\middle|\; S(x) \neq \emptyset\right\}$$

of the set-valued mapping $S$ could be a strict subset of $\mathbb{R}^n$ as in Figure 7.1. Of course, $F$ can be viewed as the particular set-valued mapping with $S(x) = \{F(x)\}$ for all $x \in \mathbb{R}^n$ and then $\operatorname{dom} S = \mathbb{R}^n$.

The *graph* of a set-valued mapping $S : \mathbb{R}^n \rightrightarrows \mathbb{R}^m$ is the subset

$$\operatorname{gph} S = \left\{(x, y) \in \mathbb{R}^n \times \mathbb{R}^m \;\middle|\; y \in S(x)\right\},$$



J. O. Royset and R. J-B Wets, *An Optimization Primer*, Springer Series in Operations Research and Financial Engineering, https://doi.org/10.1007/978-3-030-76275-9_7

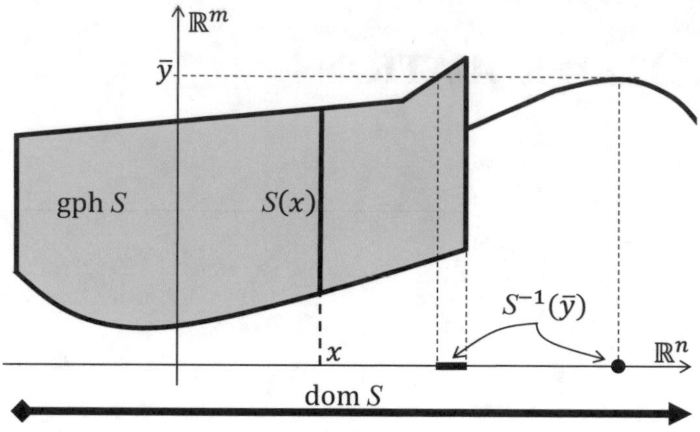


Fig. 7.1: Set-valued mapping $S : \mathbb{R}^n \rightrightarrows \mathbb{R}^m$, its domain $\operatorname{dom} S = \{x \mid S(x) \neq \emptyset\}$ and the solution set $S^{-1}(\bar{y})$ to the generalized equation $\bar{y} \in S(x)$.

which describes $S$ fully. Properties of a set-valued mapping can often be examined through its graph just as a function is studied via its epigraph. Given $\bar{y} \in \mathbb{R}^m$ and $S : \mathbb{R}^n \rightrightarrows \mathbb{R}^m$, a *generalized equation* is then

$$\bar{y} \in S(x), \quad \text{with set of solutions} \quad S^{-1}(\bar{y}) = \big\{x \in \mathbb{R}^n \bigm| \bar{y} \in S(x)\big\}$$

Figure 7.1 illustrates the graph of a set-valued mapping as well as the solution set for a resulting generalized equation.

**Example 7.1** (linear equations and inequalities). For an $m \times n$-matrix $A$ and $b \in \mathbb{R}^m$, the equation $Ax = b$ can be viewed as a generalized equation. Let $S : \mathbb{R}^n \rightrightarrows \mathbb{R}^m$ be the set-valued mapping given by

$$S(x) = \{Ax\}, \quad \text{with} \quad \operatorname{dom} S = \mathbb{R}^n.$$

Even though $S(x)$ is a single point for all $x$, the solution set

$$S^{-1}(b) = \{x \in \mathbb{R}^n \mid Ax = b\}$$

could be empty, a single point or contain an infinite number of solutions.

The inequalities $\langle a^1, x\rangle \leq b_1, \ldots, \langle a^m, x\rangle \leq b_m$ can be viewed as the generalized equation

$$(b_1, \ldots, b_m) \in \bar{S}(x), \text{ where } \bar{S}(x) = \big[\langle a^1, x\rangle, \infty\big) \times \cdots \times \big[\langle a^m, x\rangle, \infty\big).$$

**Detail.** The solution set of the equation $Ax = b$ as $b$ varies defines a set-valued mapping $T : \mathbb{R}^m \rightrightarrows \mathbb{R}^n$ with

$$T(b) = S^{-1}(b) \;\; \forall b \in \mathbb{R}^m.$$

If $n = m$ and $A$ is invertible, then $T(b) = \{A^{-1}b\}$, a single point, for every $b \in \mathbb{R}^m$ and $\operatorname{dom} T = \mathbb{R}^m$. However, we could equally well have $T(b) = \emptyset$, especially when $m > n$. Then, $\operatorname{dom} T$ certainly becomes a strict subset of $\mathbb{R}^m$. If $A\bar{x} = b$, then $T(b) = \bar{x} + \operatorname{null} A$ so $T$ could also return more than a single point.

Since set-valued mappings appear even in the simple case of linear systems of equations, point-valuedness emerges more like an exception that should be cherished but not insisted on. The set-valued perspective is a flexible approach to handle numerous situations that avoids getting us bogged down with concerns about existence and uniqueness. □

For $C \subset \mathbb{R}^n$, the normal cone $N_C(x)$ has thus far been defined only for $x \in C$. Let's adopt the convention:

$$N_C(x) = \emptyset \ \text{ if } x \notin C.$$

Then, we can view $N_C$ as a set-valued mapping from $\mathbb{R}^n$ to $\mathbb{R}^n$, which we refer to as the *normal cone mapping*, and $\operatorname{dom} N_C = C$. Likewise, for $f : \mathbb{R}^n \to \overline{\mathbb{R}}$, let's define

$$\partial f(x) = \emptyset \ \text{ if } f(x) \notin \mathbb{R},$$

which then allows us to view $\partial f$ as a set-valued mapping from $\mathbb{R}^n$ to $\mathbb{R}^n$ called the *subgradient mapping*. Since $\partial f(x)$ can be empty even if $f(x)$ is finite and it's certainly empty by the above convention if $f(x) = -\infty$, we realize that $\operatorname{dom} \partial f$ could be strictly contained in $\operatorname{dom} f$.

**Example 7.2** (optimality conditions). For a smooth function $f : \mathbb{R}^n \to \mathbb{R}$, the Fermat rule $\nabla f(x) = 0$ is formulated as an equation but this gives way to the inclusion $0 \in \partial f(x)$ for more general functions. These conditions and their numerous extensions can be viewed as generalized equations. For the problem of minimizing $f : \mathbb{R}^n \to \overline{\mathbb{R}}$ over $C \in \mathbb{R}^n$ and the optimality condition

$$0 \in \partial f(x) + N_C(x),$$

which is necessary for a minimizer under a qualification (cf. 4.67), is equivalently stated as $0 \in S(x)$, where $S : \mathbb{R}^n \rightrightarrows \mathbb{R}^n$ is the set-valued mapping with $S(x) = \partial f(x) + N_C(x)$.

**Detail.** For closed $X \subset \mathbb{R}^n$, proper, lsc and convex $h : \mathbb{R}^m \to \overline{\mathbb{R}}$ and smooth $F : \mathbb{R}^n \to \mathbb{R}^m$, 4.75 specifies the optimality condition

$$y \in \partial h\big(F(x)\big) \ \text{ and } \ -\nabla F(x)^\top y \in N_X(x).$$

This can be stated equivalently as $0 \in S(x, y)$ using the set-valued mapping $S : \mathbb{R}^{n+m} \rightrightarrows \mathbb{R}^{n+m}$ given by

$$S(x, y) = \big(\nabla F(x)^\top y + N_X(x)\big) \times \Big(-y + \partial h\big(F(x)\big)\Big).$$

If $h = h_{Y,B}$ as in §5.I, then

$$h^*(y) = \tfrac{1}{2}\langle y, By\rangle + \iota_Y(y) \quad \text{and} \quad \partial h^*(y) = By + N_Y(y).$$

Using the inversion rule 5.37, we obtain that

$$S(x, y) = \big(\nabla F(x)^\top y + N_X(x)\big) \times \big(By - F(x) + N_Y(y)\big),$$

with $\operatorname{dom} S = X \times Y$. □

Other set-valued mappings that we've encountered thus far include those defined by minimizers, projections, proximal points and level-sets:

$$S(u) = \operatorname{argmin} f(u, \cdot), \quad S(x) = \operatorname{prj}_C(x), \quad S(x) = \operatorname{prx}_\lambda f(x), \quad S(\alpha) = \{f \le \alpha\}.$$

Generalized equations also occur directly in applications and then often in the context of an equilibrium condition.

**Example 7.3** (spatial price equilibrium). Consider a product that's made by $m$ geographically dispersed manufacturer and bought by consumers in $n$ different regions. We would like to understand how many units $s_i$ that manufacturer $i$ will supply, the number of units $d_j$ that region $j$ will demand and the quantity that will be transported from $i$ to $j$. With $s = (s_1, \dots, s_m)$ and $d = (d_1, \dots, d_n)$, let's assume that the price by which manufacturer $i$ offers the product follows a known function of the supply from all the manufacturers, i.e., $p_i(s)$ is the price by manufacturer $i$ given supply vector $s$, with $p_i : \mathbb{R}^m \to \mathbb{R}$. Moreover, let $q_j : \mathbb{R}^n \to \mathbb{R}$ be a function that gives the price for the product in region $j$ so that $q_j(d)$ is the price in region $j$ under demand vector $d$. Since the transportation cost can be significant, the price of the product could be different among the regions and the manufacturers may offer the product at different price points.

A basic economic equilibrium principle is that if units are shipped from manufacturer $i$ to region $j$ at all, then the price at which manufacturer $i$ offers the product plus the transportation cost from manufacturer $i$ to region $j$ must equal the price in region $j$. Alternatively, if there are no shipments from $i$ to $j$, then the manufacturer's price plus the transportation cost must be greater than or equal to the price in region $j$. With $w_{ij}$ being the number of units shipped between manufacturer $i$ and region $j$ and $w = (w_{ij}, i = 1, \dots, m, j = 1, \dots, n)$, suppose that $c_{ij} : \mathbb{R}^{mn} \to \mathbb{R}$ is a given function so that $c_{ij}(w)$ is the cost of shipping from $i$ to $j$ under shipment vector $w$.

The *equilibrium condition* is that supply vector $s$, demand vector $d$ and shipment vector $w$ satisfy

$$\forall i, j: \quad p_i(s) + c_{ij}(w) \in \begin{cases} \{q_j(d)\} & \text{if } w_{ij} > 0 \\ \big[q_j(d), \infty\big) & \text{if } w_{ij} = 0. \end{cases}$$

We can write this condition as a generalized equation involving normal cones.

**Detail.** The set of admissible supply, demand and shipment vectors is

$$C = \Big\{(s, d, w) \in \mathbb{R}^m \times \mathbb{R}^n \times \mathbb{R}^{mn} \;\Big|\; w_{ij} \ge 0, \; \sum\nolimits_{j=1}^n w_{ij} = s_i, \; \sum\nolimits_{i=1}^m w_{ij} = d_j \;\; \forall i, j\Big\},$$

which simply ensures nonnegative shipment quantities and "supply equals demand." It turns out that $(s^\star, d^\star, w^\star) \in C$ satisfies the equilibrium condition if and only if

$$0 \in F(s^\star, d^\star, w^\star) + N_C(s^\star, d^\star, w^\star) \tag{7.1}$$

where $F : \mathbb{R}^m \times \mathbb{R}^n \times \mathbb{R}^{mn} \to \mathbb{R}^m \times \mathbb{R}^n \times \mathbb{R}^{mn}$ is given by

$$F(s, d, w) = \big(p_1(s), \dots, p_m(s), -q_1(d), \dots, -q_n(d), c_{11}(w), \dots, c_{mn}(w)\big).$$

To establish this fact, first suppose that $(s^\star, d^\star, w^\star) \in C$ satisfies the equilibrium condition. For any $i, j$ and nonnegative $w_{ij}$,

$$\big(p_i(s^\star) + c_{ij}(w^\star) - q_j(d^\star)\big)(w_{ij} - w_{ij}^\star) \geq 0$$

because either the first term in the product is 0 (when $w_{ij}^\star > 0$) or both terms are nonnegative (when $w_{ij}^\star = 0$). Summing over all $i$ and $j$ leads to

$$\sum_{i=1}^m \sum_{j=1}^n \big(p_i(s^\star) + c_{ij}(w^\star) - q_j(d^\star)\big)(w_{ij} - w_{ij}^\star) \geq 0. \tag{7.2}$$

This fact combined with the requirement $(s^\star, d^\star, w^\star) \in C$ as well as the notation

$$s_i = \sum_{j=1}^n w_{ij}, \quad s_i^\star = \sum_{j=1}^n w_{ij}^\star, \quad d_j = \sum_{i=1}^m w_{ij}, \quad d_j^\star = \sum_{i=1}^m w_{ij}^\star$$

result in

$$\sum_{i=1}^m p_i(s^\star)(s_i - s_i^\star) - \sum_{j=1}^n q_j(d^\star)(d_j - d_j^\star) + \sum_{i=1}^m \sum_{j=1}^n c_{ij}(w^\star)(w_{ij} - w_{ij}^\star) \geq 0.$$

Since $w_{ij} \geq 0$ is arbitrary, we've established that the previous inequality holds for all $(s, d, w) \in C$, where $s = (s_1, \dots, s_m)$ and $d = (d_1, \dots, d_n)$. By the definition of normal cones in the convex case (cf. 2.23), this in turn is equivalent to (7.1).

Second, suppose that (7.1) holds for $(s^\star, d^\star, w^\star) \in C$. By the insight just obtained, this means that (7.2) holds for all nonnegative $w_{ij}$. In particular, for fixed $i', j'$, (7.2) holds with $w_{ij} = w_{ij}^\star$ for all $(i, j) \neq (i', j')$, which implies

$$\big(p_{i'}(s^\star) + c_{i'j'}(w^\star) - q_{j'}(d^\star)\big)(w_{i'j'} - w_{i'j'}^\star) \geq 0.$$

Since $i', j'$ are arbitrary, we obtain the equilibrium condition for $(s^\star, d^\star, w^\star)$. □

In the example, an economic equilibrium principle results in a generalized equation

$$0 \in S(x) = F(x) + N_C(x)$$

for a mapping $F : \mathbb{R}^n \to \mathbb{R}^n$ and a set $C \subset \mathbb{R}^n$. This is indeed a common form in applications and coincides with the condition arising from the Fermat rule when minimizing a smooth function $f : \mathbb{R}^n \to \overline{\mathbb{R}}$ over $C$; see the basic optimality condition 4.37. In contrast to such minimization, where $F(x) = \nabla f(x)$, the example involves no apparent optimization problem and the generalized equation simply expresses an equilibrium condition.

Regardless of the origin of $F$, when $C$ is convex the generalized equation can be reformulated using the normal cone formula 2.23. This results in a *variational inequality*

$$\boxed{x \in C \quad \text{and} \quad \langle F(x), \bar{x} - x \rangle \geq 0 \ \forall \bar{x} \in C}$$

If $C = [0, \infty)^n$, then the variational inequality simplifies and gives rise to a *complementarity problem*: find $x \in \mathbb{R}^n$ such that

$$\boxed{x \geq 0, \quad F(x) \geq 0, \quad \langle F(x), x \rangle = 0}$$

We realize the equivalence by recalling from 2.31 that in this case

$$N_C(x) = N_{[0,\infty)}(x_1) \times \cdots \times N_{[0,\infty)}(x_n), \ \text{ with } N_{[0,\infty)}(\alpha) = \begin{cases} \{0\} & \text{if } \alpha > 0 \\ (-\infty, 0] & \text{if } \alpha = 0. \end{cases}$$

Thus, with $F(x) = (f_1(x), \ldots, f_n(x))$, the condition $0 \in F(x) + N_C(x)$ amounts to having $f_i(x) = 0$ when $x_i > 0$ and $f_i(x) \geq 0$ when $x_i = 0$, and also $x \geq 0$ because otherwise $N_C(x)$ is empty. This leads to the complementarity problem, where $\langle F(x), x \rangle = 0$ is simply a convenient way of expressing the required relation between $f_i(x)$ and $x_i$.

Complementarity problems as well as solving generalized equations and variational inequalities are examples of *variational problems*. As in the case of optimization problems, there can be merit to reformulating variational problems and putting them in a form that's compatible with existing algorithms. The next sections show that complementarity problems are especially tractable and we often seek to end up with that form; see [27, 35] for a more detailed treatment.

**Example 7.4** (reformulation as complementarity problem). For $F : \mathbb{R}^n \to \mathbb{R}^n$ and a smooth mapping $G : \mathbb{R}^n \to \mathbb{R}^m$, suppose that $C = \{x \in [0, \infty)^n \mid G(x) \leq 0\}$. The generalized equation

$$0 \in F(x) + N_C(x)$$

may not be written as a variational inequality because $C$ could be nonconvex. Likewise, $C$ isn't necessarily the nonnegative orthant and a reformulation as a complementary problem seems impossible. With the introduction of a multiplier vector, however, we can lift the problem to higher dimensions where indeed it can be stated as a complementarity problem under a qualification.

**Detail.** We obtain from 4.46 that for $x \in C$,

$$N_C(x) = \left\{ \nabla G(x)^\top y + z \,\middle|\, y \in N_{(-\infty,0]^m}\big(G(x)\big), \ z \in N_{[0,\infty)^n}(x) \right\}$$

as long as the qualification (4.9) holds. The condition $z \in N_{[0,\infty)^n}(x)$ is equivalent to having $x \geq 0$, $z \leq 0$ and $z_j x_j = 0$ for all $j = 1, \ldots, n$, with the latter condition also captured by $\langle z, x \rangle = 0$ because $x$ and $z$ are sign-restricted. Similarly, $y \in N_{(-\infty,0]^m}(G(x))$

means that $G(x) \leq 0$, $y \geq 0$ and $\langle y, G(x)\rangle = 0$. Under the assumption that the qualification (4.9) holds, the generalized equation then amounts to finding $(x, y, z)$ such that

$$\begin{aligned}
&x \geq 0, \quad y \geq 0, \quad z \leq 0, \quad G(x) \leq 0\\
&0 = F(x) + \nabla G(x)^\top y + z\\
&\big\langle y, G(x)\big\rangle = 0, \quad \langle z, x\rangle = 0.
\end{aligned}$$

By eliminating $z$, we obtain the simplification

$$\begin{aligned}
&x \geq 0, \quad y \geq 0\\
&F(x) + \nabla G(x)^\top y \geq 0, \qquad -G(x) \geq 0\\
&\big\langle x,\ F(x) + \nabla G(x)^\top y\big\rangle = 0, \quad \big\langle y, -G(x)\big\rangle = 0.
\end{aligned}$$

This is indeed a complementarity problem in the variables $(x, y)$ with a mapping $H : \mathbb{R}^{n+m} \to \mathbb{R}^{n+m}$ given by

$$H(x, y) = \begin{bmatrix} F(x) + \nabla G(x)^\top y \\ -G(x) \end{bmatrix}.$$

The original generalized equation has $n$ unknowns, but they aren't easily determined directly. Even though the obtained complementarity problem has $n + m$ unknowns, the structure of $C$ is brought out and this makes algorithms immediately available as seen in the next sections. □

**Example 7.5** (Walras barter model). Suppose that there are $m$ agents trading $n$ goods. Agent $i$ has an initial endowment $w_i \in [0, \infty)^n$ of these goods and seeks to minimize its "displeasure" by selling and buying goods. Presumably, the agent sells less desirable goods from its initial endowment and buys those that it prefers. Let $f_i : \mathbb{R}^n \to \mathbb{R}$ be a known smooth convex function that represents the agent's displeasure with various holdings of the goods, which typically is given by a utility function. Thus, agent $i$ seeks to determine a holding $x_i \in \mathbb{R}^n$ of the goods such that $f_i(x_i)$ is as low as possible. The prices of the goods aren't fixed, but are determined by a market place with all the agents. This means that neither the value of the initial endowment $w_i$ nor the cost of purchasing a particular holding $x_i$ is known a priori. We can determine the prices as well as the agents' holdings by solving a generalized equation.

**Detail.** Let the prices of the goods be denoted by the vector $p = (p_1, \ldots, p_n) \in \mathbb{R}^n$. Certainly, agent $i$ is constrained by $\langle p, x_i\rangle \leq \langle p, w_i\rangle$; the value of its holding can't exceed the value of the initial endowment. Let's also assume that the holdings need to be nonnegative. Thus, agent $i$ would like to determine

$$\bar{x}_i \in \operatorname{argmin}_{x_i \in C_i(p)} f_i(x_i), \quad \text{with } C_i(p) = \big\{x_i \in [0, \infty)^n \;\big|\; \langle p, x_i\rangle \leq \langle p, w_i\rangle\big\},$$

or, equivalently by 2.27, find $\bar{x}_i$ that solves

$$0 \in \nabla f_i(x_i) + N_{C_i(p)}(x_i). \tag{7.3}$$

The price $p_j$ is assumed to be one that ensures that the total demand for good $j$ doesn't exceed the total supply, which means that the market is in *equilibrium*. Specifically, with $w_i = (w_{i1}, \ldots, w_{in})$ and $x_i = (x_{i1}, \ldots, x_{in})$ so that $w_{ij}$ and $x_{ij}$ are initial endowment and holding of good $j$ for agent $i$, respectively, the *equilibrium condition* becomes

$$\forall j = 1, \ldots, n: \quad \sum_{i=1}^{m} (w_{ij} - x_{ij}) \in \begin{cases} \{0\} & \text{if } p_j > 0 \\ [0, \infty) & \text{if } p_j = 0. \end{cases}$$

That is, if the price of good $j$ is positive, then the total supply from everybody's endowments must equal the total holdings for good $j$, i.e., "supply equals demand." If the price is 0, then the total supply still bounds the total demand from above. This equilibrium condition is equivalently stated as

$$p \geq 0, \quad \sum_{i=1}^{m} (w_i - x_i) \geq 0, \quad \Big\langle \sum_{i=1}^{m} (w_i - x_i), p \Big\rangle = 0. \tag{7.4}$$

Given nonnegative decision vectors $x_1, \ldots, x_m$, finding a price vector $p$ that satisfies this equilibrium condition is a complementarity problem, which can also be written as the generalized equation

$$0 \in \sum_{i=1}^{m} (w_i - x_i) + N_{[0,\infty)^n}(p).$$

As seen, the decisions $x_1, \ldots, x_m$ must satisfy the $m$ generalized equations from (7.3). Consequently, $m + 1$ coupled generalized equations specify an *equilibrium price vector* and the agents' decisions. This can be written as the (large) generalized equation $0 \in S(x_1, \ldots, x_m, p)$, where

$$S(x_1, \ldots, x_m, p) = \Big( \sum_{i=1}^{m} (w_i - x_i) + N_{[0,\infty)^n}(p) \Big) \times \prod_{i=1}^{m} \big( \nabla f_i(x_i) + N_{C_i(p)}(x_i) \big).$$

Thus, $S(x_1, \ldots, x_m, p) \subset \mathbb{R}^{n(m+1)}$. □

**Exercise 7.6** (computing Walras equilibria). Consider the model in 7.5 with $w_1 = (1, 0)$, $w_2 = (1, 1)$, $f_1(x_1) = (x_{11} - 1)^2 + (x_{12} - 1)^2$ and $f_2(x_2) = (x_{21} - 2)^2 + (x_{22} - 1)^2$. For an array of different price vectors $p \in [0, 2]^2$, solve the agent problems

$$\underset{x_1 \in C_1(p)}{\text{minimize}}\ f_1(x_1) \qquad\qquad \underset{x_2 \in C_2(p)}{\text{minimize}}\ f_2(x_2)$$

and check whether the resulting decision vectors together with $p$ satisfy the equilibrium condition (7.4). Is there more than one $p$ that leads to an equilibrium?

**Guide.** For $p = (1, 1)$, one finds that $\bar{x}_1 = (0.5, 0.5)$ and $\bar{x}_2 = (1.5, 0.5)$ solve the agent problems and the equilibrium condition is also satisfied. □

## 7.B Equilibrium in Energy Markets

Electricity in a region or within a microgrid is usually produced by multiple providers. Their production needs to be adjusted to the demand for electricity and, ideally, be achieved at minimum cost. The situation is complicated by the increasing reliance on renewable energy sources, especially solar and wind, which make electricity production uncertain and easily out of tune with demand. Consumers are also gradually becoming more willing to modulate their usage in response to changes in price. We can design and analyze energy markets that account for these trends by formulating and solving generalized equations.

**Example 7.7** (microgrid). Let's consider a microgrid with four agents: a battery pack, a gasoline-fueled generator, a consumer and an array of solar panels. Planning for tomorrow, let $x_1$ be the amount of energy we put in the battery today and this costs $x_1^2/2$. Let $x_2$ be the energy produced by the generator tomorrow at cost $x_2^2$. The solar panels supply $\xi$ units tomorrow at no cost. (Later in the section, we'll account for uncertainty in solar production.) We denote by $x_3$ tomorrow's consumption, from which the consumer derives a utility of $16x_3 - x_3^2$. Suppose that a central planner decides storage, production and consumption by minimizing production cost minus utility under the constraint that consumption $x_3$ can't exceed production $x_1 + x_2 + \xi$. This is meaningful from a centralized perspective as long as the consumer's utility is calibrated appropriately. We've reached the problem

$$\underset{x_1,x_2,x_3 \geq 0}{\text{minimize}} \ \tfrac{1}{2}x_1^2 + x_2^2 - 16x_3 + x_3^2 \ \text{ subject to } \ -x_1 - x_2 + x_3 - \xi \leq 0. \tag{7.5}$$

An optimality condition for the problem gives rise to a multiplier that can be interpreted as a price of energy in a market where the agents make their own decisions without interference by a centralized authority. This decentralized approach can be modeled using a generalized equation.

**Detail.** An optimality condition for (7.5) is furnished by 4.75. Let $X = [0, \infty)^3$,

$$f_0(x) = \tfrac{1}{2}x_1^2 + x_2^2 - 16x_3 + x_3^2$$

and $g(x) = -x_1 - x_2 + x_3 - \xi$. Then, (7.5) takes the form

$$\underset{x \in \mathbb{R}^3}{\text{minimize}} \ \iota_X(x) + f_0(x) + \iota_{(-\infty,0]}\big(g(x)\big)$$

and 4.75 asserts the optimality condition:

$$y \in N_{(-\infty,0]}\big(g(x)\big) \quad \text{and} \quad -\nabla f_0(x) - y\nabla g(x) \in N_X(x). \tag{7.6}$$

The associated qualification holds and in fact isn't needed as the feasible set is polyhedral; see the discussion in §4.H. In view of the formula for normal cones to boxes (cf. 4.45), the second part amounts to the three conditions

$$-x_1 + y \in N_{[0,\infty)}(x_1), \quad -2x_2 + y \in N_{[0,\infty)}(x_2), \quad 16 - 2x_3 - y \in N_{[0,\infty)}(x_3).$$

In the case of $\xi = 1$, we obtain the solution $x^\star = (3.5,\ 1.75,\ 6.25)$ with $y^\star = 3.5$. Since the objective function is strictly convex and the feasible set is polyhedral, we conclude that $x^\star$ is the unique minimizer of (7.5); see 2.27 and 1.16. In the setting of a central planner, this might be all what's needed. But, the analysis also explains an energy market in which the agents operate in their own interests.

By 2.27, the normal cone conditions can be viewed as stemming from the problems

$$\underset{x_1 \geq 0}{\text{minimize}}\ \tfrac{1}{2}x_1^2 - yx_1, \qquad \underset{x_2 \geq 0}{\text{minimize}}\ x_2^2 - yx_2, \qquad \underset{x_3 \geq 0}{\text{minimize}}\ yx_3 - 16x_3 + x_3^2.$$

The first problem pertains to the battery, with $x_1^2/2$ being the cost of charging and storing $x_1$ units of energy. The term $yx_1$ is subtracted from the cost and this leads us to interpreting $y$ as the *price of energy*. The battery provides $x_1$ units of energy for which it's paid $yx_1$. For a fixed price $y$, the first problem is then to minimize cost minus income for the battery. Similarly, the second problem deals with the generator and also aspires to minimize cost minus income. In the third problem, the consumer faces a cost of $yx_3$ for its consumption of $x_3$ units of energy. The goal for the consumer becomes to minimize cost minus utility. With $y = 3.5$, we know already that 3.5, 1.75 and 6.25 are minimizers in the three problems, respectively, and this results in costs and incomes summarized in Table 7.1.

Table 7.1: Costs and incomes for agents in 7.7.

| agent | cost | income | profit |
|---|---|---|---|
| battery | 6.1250 | 12.2500 | 6.1250 |
| generator | 3.0625 | 6.1250 | 3.0625 |
| solar | 0 | 3.5000 | 3.5000 |
| consumer | 21.8750 | – | – |

Interestingly, the supply of energy $3.5 + 1.75 + 1$ by the three producers matches the consumption 6.25 even though the four agents operate according to their own self-interests; there's no central authority that insists on demand being met exactly. The reason, of course, is that the price is "right." How is the price determined? From (7.6), we see that $y \in N_{(-\infty,0]}(g(x))$, which amounts to having $y \geq 0$, $g(x) \leq 0$ and $y = 0$ if $g(x) < 0$. The last condition corresponds to overproduction, in which case the price drops to 0.

In summary, (7.6) can be viewed from two angles: First, it's the optimality condition for a centralized planning problem, in which case $y$ is simply a multiplier vector with the usual interpretation; cf. Chap. 5. Second, it specifies four generalized equations modeling selfish agents that are linked by a *market clearing condition* $y \in N_{(-\infty,0]}(g(x))$ from which the price $y$ of energy is determined.

Regardless of the perspective, (7.6) specifies a complementarity problem: find $x \in \mathbb{R}^3$ and $y \in \mathbb{R}$ such that

$$x, y \geq 0, \qquad F(x,y) \geq 0, \qquad \big\langle F(x,y), (x,y) \big\rangle = 0,$$

where $x = (x_1, x_2, x_3)$ and $F(x,y) = (\nabla f_0(x) + y\nabla g(x),\ -g(x))$. □

The example leads to a fundamental question: When will agents acting in their own interest, but linked through a price mechanism, settle on the same decisions as a central planner?

**Proposition 7.8** (social welfare). *For smooth $f_i, g_i : \mathbb{R}^{n_i} \to \mathbb{R}$ and closed $X_i \subset \mathbb{R}^{n_i}$, $i = 1, \dots, m$, consider the centralized planning problem*

$$\operatorname*{minimize}_{x \in X} \sum_{i=1}^m f_i(x_i) \text{ subject to } \sum_{i=1}^m g_i(x_i) \le 0, \tag{7.7}$$

*where $X = X_1 \times \cdots \times X_m$ and $x = (x_1, \dots, x_m)$, the market clearing condition*

$$y \ge 0, \quad \sum_{i=1}^m g_i(x_i) \le 0, \quad y = 0 \text{ if } \sum_{i=1}^m g_i(x_i) < 0 \tag{7.8}$$

*and the agent problems*

$$\Big\{ \operatorname*{minimize}_{x_i \in X_i} f_i(x_i) + y g_i(x_i), \quad i = 1, \dots, m \Big\}. \tag{7.9}$$

*If $x_1^\star, \dots, x_m^\star$ are local minimizers of the agent problems (7.9) under price $y = y^\star$ and these local minimizers together with the price satisfy the market clearing condition (7.8), then $x^\star = (x_1^\star, \dots, x_m^\star)$ is a local minimizer of the centralized planning problem (7.7).*

*If $x^\star = (x_1^\star, \dots, x_m^\star)$ is a local minimizer of the centralized planning problem (7.7) and $-\nabla g_i(x_i^\star) \notin N_{X_i}(x_i^\star)$ for at least one agent $i$, then there exists a price $y^\star$ that together with $x^\star$ satisfies the market clearing condition (7.8) and*

$$\forall i = 1, \dots, m: \quad -\nabla f_i(x_i^\star) - y^\star \nabla g_i(x_i^\star) \in N_{X_i}(x_i).$$

*When $f_i, g_i$ and $X_i$ are convex, this implies that $x_i^\star$ minimizes the ith agent problem in (7.9) under price $y = y^\star$.*

**Proof.** If $x_1^\star, \dots, x_m^\star$ are local minimizers in (7.9) under price $y^\star \ge 0$, then there's $\varepsilon > 0$ such that for all $i = 1, \dots, m$,

$$x_i^\star \in \operatorname{argmin} \big\{ f_i(x_i) + y^\star g_i(x_i) \,\big|\, x_i \in X_i \cap \mathbb{B}(x_i^\star, \varepsilon) \big\}$$

and we also have $x^\star = (x_1^\star, \dots, x_m^\star) \in \operatorname{argmin}_{\tilde{X}} f_0 + y^\star g$, where

$$f_0(x) = \sum_{i=1}^m f_i(x_i), \quad g(x) = \sum_{i=1}^m g_i(x_i), \quad \tilde{X} = X_1 \cap \mathbb{B}(x_1^\star, \varepsilon) \times \cdots \times X_m \cap \mathbb{B}(x_m^\star, \varepsilon).$$

This is equivalently stated as $x^\star \in \operatorname{argmin} l(\cdot, y^\star)$, with

$$l(x, y) = \iota_{\tilde{X}}(x) + f_0(x) + y g(x) - \iota_{[0,\infty)}(y).$$

Under the assumption that the market clearing condition (7.8) holds for $(x^\star, y^\star)$, one has $y^\star \in N_{(-\infty,0]}(g(x^\star))$. Consequently, by 2.23, 5.29 and 5.37, this leads to

$$\iota_{(-\infty,0]}\big(g(x^\star)\big) + \iota_{[0,\infty)}(y^\star) = y^\star g(x^\star).$$

Combining these facts, we obtain

$$\iota_{\tilde{X}}(x^\star) + f_0(x^\star) + \iota_{(-\infty,0]}\big(g(x^\star)\big) = \iota_{\tilde{X}}(x^\star) + f_0(x^\star) + y^\star g(x^\star) - \iota_{[0,\infty)}(y^\star)$$
$$= \inf l(\cdot, y^\star) \le \inf_x \sup_y l(x,y) = \inf \iota_{\tilde{X}} + f_0 + \iota_{(-\infty,0]} \circ g,$$

where the last equality follows by 5.28. We've shown that $x^\star$ is a local minimizer of (7.7).

If $x^\star = (x_1^\star, \ldots, x_m^\star)$ is a local minimizer in (7.7), then 4.75 implies that

$$\exists y^\star \in \partial\iota_{(-\infty,0]}\big(g(x^\star)\big) \;\; \text{such that} \;\; -\nabla f_0(x^\star) - y^\star \nabla g(x^\star) \in N_X(x^\star),$$

where the first part translates into (7.8) and the second part decomposes into the asserted conditions; cf. 4.44. This conclusion relies on the qualification (4.16), which presently specializes to

$$y \in N_{(-\infty,0]}\big(g(x^\star)\big) \;\; \text{and} \;\; -y\nabla g(x^\star) \in N_X(x^\star) \implies y = 0.$$

This automatically holds when $g(x^\star) < 0$ so let's concentrate on $g(x^\star) = 0$. We need $y \ge 0$ and $-y\nabla g_i(x_i^\star) \in N_{X_i}(x_i^\star)$ for all $i$ to imply that $y = 0$. Since by assumption there's $i^\star$ such that

$$-\nabla g_{i^\star}(x_{i^\star}^\star) \notin N_{X_{i^\star}}(x_{i^\star}^\star),$$

we must also have

$$-\lambda\nabla g_{i^\star}(x_{i^\star}^\star) \notin N_{X_{i^\star}}(x_{i^\star}^\star)$$

for positive $\lambda$. Consequently, $-y\nabla g_{i^\star}(x_{i^\star}^\star) \in N_{X_{i^\star}}(x_{i^\star}^\star)$ implies that $y = 0$.

The final conclusion for the convex case follows immediately from the optimality condition 2.27. □

The proposition provides a far-reaching extension of 7.7: Agent $i$ buys $g_i(x_i)$ units of a commodity, for which it pays $yg_i(x_i)$, and also faces a (production) cost of $f_i(x_i)$. A negative $g_i(x_i)$ corresponds to selling the commodity. As long as the market clearing condition is met, the agents' decisions are actually optimal from a centralized point of view. As we see from 7.7, this is a delicate matter. In that example, if the price $y = 10$, then the minimizers of the agent problems become $x_1^\star = 10$, $x_2^\star = 5$ and $x_3^\star = 3$. This results in too much energy production. In contrast, if the price is $y = 1$, then the production is too low. The second part of the proposition asserts that under a mild qualification, there's a "right" price.

The insight from the proposition allows us to design a market that makes the agents behave in a manner that optimizes the overall "social welfare." This may involve an auctioneer to which agents submit their purchasing (selling) amounts and proposed prices. The auctioneer checks supply versus demand and reports to each agent whether its proposed price is too high or too low. After some iterations, the auctioneer may be able to satisfy the market clearing condition (7.8) and the corresponding decisions are not only good for the agents individually but also for the overall system.

The proposition can be extended to situations with uncertainty (see, for example, [39, 80]), but we omit a comprehensive treatment and instead concentrate on a particular example.

**Example 7.9** (uncertain solar production). Let's return to 7.7, but now adopt the more realistic assumption that the amount of energy produced by the solar panels is random. The charging of the battery takes place before we know the solar production and this leads to a decision process in two stages:

$$\text{battery decision: } x_1 \rightsquigarrow \text{ observation of solar: } \xi \rightsquigarrow \text{ other decisions: } x_2, x_3.$$

The central planner now solves the problem

$$\underset{x_1 \geq 0}{\text{minimize}} \ \tfrac{1}{2}x_1^2 + \mathbb{E}\big[f(\boldsymbol{\xi}, x_1)\big], \tag{7.10}$$

where $\boldsymbol{\xi}$ is a random variable that models the uncertain solar production and $f : \mathbb{R}\times\mathbb{R} \to \mathbb{R}$ has

$$f(\xi, x_1) = \inf_{x_2, x_3 \geq 0} \big\{x_2^2 - 16x_3 + x_3^2 \ \big| \ -x_1 - x_2 + x_3 - \xi \leq 0\big\}.$$

The objective function consists of a first-stage cost due to the battery and an expected cost-minus-utility associated with the generator and the consumption. The problem is a *stochastic optimization problem with nonlinear recourse*. It has complete recourse in the sense that every $x_1$ permits a feasible second-stage decision. (This may change if $x_2$ is subject to a maximum production capacity and $x_3$ needs to exceed a certain consumption level.)

As in the deterministic case, the multiplier vector in an optimality condition can be interpreted as the price of energy.

**Detail.** Suppose that $\boldsymbol{\xi}$ has a finite distribution with support $\Xi$ and probabilities $p_\xi > 0$ for $\xi \in \Xi$. Let

$$f_1(x_1) = \tfrac{1}{2}x_1^2, \qquad f_2(x_2) = x_2^2, \qquad f_3(x_3) = -16x_3 + x_3^2$$

be the battery cost, generator cost and the negative utility for the consumer, respectively, as in 7.7. We can then write the problem explicitly as

$$\begin{aligned} &\text{minimize } f_1(x_1) + \textstyle\sum_{\xi\in\Xi} p_\xi\big(f_2(x_{2\xi}) + f_3(x_{3\xi})\big) \\ &\text{subject to } -x_1 - x_{2\xi} + x_{3\xi} - \xi \leq 0 \ \ \forall \xi \in \Xi \\ &\qquad\qquad x_1, x_{2\xi}, x_{3\xi} \geq 0 \ \ \forall \xi \in \Xi. \end{aligned}$$

Since the production by the generator and the consumption are decided after the solar production is revealed, those decisions are tailored to the particular outcomes. This results in a vector of decision variables

$$x = \big(x_1, \ x_{2\xi}, \ x_{3\xi}, \ \xi \in \Xi\big) \in \mathbb{R}^{1+2m},$$

where $m$ is the number of values of $\xi$. Moreover, let $X = [0,\infty)^{1+2m}$,

$$g(\xi, x) = -x_1 - x_{2\xi} + x_{3\xi} - \xi \quad \text{and} \quad G(x) = \big(g(\xi, x),\ \xi \in \Xi\big) \in \mathbb{R}^m.$$

In this notation, the problem becomes

$$\mathop{\text{minimize}}_{x\in\mathbb{R}^{1+2m}}\ \iota_X(x) + \varphi(x) + \iota_{(-\infty,0]^m}\big(G(x)\big),$$

where

$$\varphi(x) = f_1(x_1) + \sum\nolimits_{\xi\in\Xi} p_\xi\big(f_2(x_{2\xi}) + f_3(x_{3\xi})\big),$$

and we can leverage 4.75 to obtain the optimality condition:

$$\begin{aligned} y &\in N_{(-\infty,0]^m}\big(G(x)\big) \\ -\nabla\varphi(x) - \nabla G(x)^\top y &\in N_X(x). \end{aligned}$$

While this condition supports the solution of the central planner's problem, it also has an interpretation in terms of an energy market. Let $y = (y_\xi, \xi \in \Xi)$. Since $\nabla G(x)$ is an $m \times (1+2m)$-matrix with $-1$ in the first column, the negative $m \times m$ identity matrix in the next $m$ columns and the $m \times m$ identity matrix in the remaining columns, the second line of the optimality condition separates into

$$\begin{aligned} -f_1'(x_1) + \sum\nolimits_{\xi\in\Xi} y_\xi &\in N_{[0,\infty)}(x_1) \\ -p_\xi f_2'(x_{2\xi}) + y_\xi &\in N_{[0,\infty)}(x_{2\xi}) \quad \forall \xi \in \Xi \\ -p_\xi f_3'(x_{3\xi}) - y_\xi &\in N_{[0,\infty)}(x_{3\xi}) \quad \forall \xi \in \Xi. \end{aligned}$$

This is most easily interpreted if we scale $y_\xi$. Since $p_\xi > 0$, one has

$$-p_\xi f_2'(x_{2\xi}) + y_\xi \in N_{[0,\infty)}(x_{2\xi}) \iff -f_2'(x_{2\xi}) + \tilde{y}_\xi \in N_{[0,\infty)}(x_{2\xi}), \text{ with } \tilde{y}_\xi = y_\xi/p_\xi.$$

In the scaled multipliers, $\sum_{\xi\in\Xi} y_\xi$ becomes $\sum_{\xi\in\Xi} p_\xi \tilde{y}_\xi$. We proceed with these scaled multipliers, but drop the cumbersome "tilde" notation. This results in

$$\begin{aligned} -f_1'(x_1) + \sum\nolimits_{\xi\in\Xi} p_\xi y_\xi &\in N_{[0,\infty)}(x_1) \\ -f_2'(x_{2\xi}) + y_\xi &\in N_{[0,\infty)}(x_{2\xi}) \quad \forall \xi \in \Xi \\ -f_3'(x_{3\xi}) - y_\xi &\in N_{[0,\infty)}(x_{3\xi}) \quad \forall \xi \in \Xi. \end{aligned}$$

The requirement $y \in N_{(-\infty,0]^m}(G(x))$, after passing to the scaled multipliers, is the *market clearing condition*

$$y_\xi \geq 0, \quad g(\xi, x) \leq 0, \quad y_\xi g(\xi, x) = 0 \quad \forall \xi \in \Xi.$$

We now see that $y$ can be interpreted as a vector of prices, one for each outcome of the solar production. Moreover, the other requirements are optimality conditions for the agent problems

$$\underset{x_1 \geq 0}{\text{minimize}}\ \tfrac{1}{2}x_1^2 - \sum\nolimits_{\xi \in \Xi} p_\xi y_\xi x_1$$

$$\Big\{ \underset{x_{2\xi} \geq 0}{\text{minimize}}\ x_{2\xi}^2 - y_\xi x_{2\xi}, \qquad \xi \in \Xi \Big\}$$

$$\Big\{ \underset{x_{3\xi} \geq 0}{\text{minimize}}\ y_\xi x_{3\xi} - 16x_{3\xi} + x_{3\xi}^2, \quad \xi \in \Xi \Big\}.$$

The first problem corresponds to the battery, where $\sum_{\xi \in \Xi} p_\xi y_\xi x_1$ is the expected payment to the battery. The second set of problems pertains to the generator with the generator-agent solving one problem for each outcome of solar production. The third set of problems is for the consumer, which also tailors the decision to the solar production. In summary, the centralized planning problem (7.10) is equivalent to these agent problems in conjunction with the market clearing condition. Regardless of the perspective, the decisions and the prices solve a generalized equation and in fact also a complementary problem just as in the deterministic case; see 7.7.

In the specific case of $\Xi = \{1, 3\}$ and $p_1 = p_3 = 1/2$, we obtain the solution $x_1^\star = 3$ for the battery, $x_{21}^\star = 2$ and $x_{23}^\star = 1$ for the generator and $x_{31}^\star = 6$ and $x_{33}^\star = 7$ for the consumer, with prices being $y_1^\star = 4$ and $y_3^\star = 2$. Naturally, in the case of more solar energy, the price decreases and the consumption increases. Table 7.2 summarizes the situation. The market is revenue neutral in both solar scenarios: The payment of the consumer (24) matches the amount the producers should receive $(12 + 8 + 4)$ for $\xi = 1$ and likewise for $\xi = 3$ when $14 = 6 + 2 + 6$. □

Table 7.2: Costs and incomes in 7.9 for solar production $\xi = 1$ and $\xi = 3$.

| | $\xi = 1$ | | | $\xi = 3$ | | |
|---|---|---|---|---|---|---|
| agent | cost | income | profit | cost | income | profit |
| battery | 4.5 | 12.0 | 7.5 | 4.5 | 6.0 | 1.5 |
| generator | 4.0 | 8.0 | 4.0 | 1.0 | 2.0 | 1.0 |
| solar | 0 | 4.0 | 4.0 | 0 | 6.0 | 6.0 |
| consumer | 24.0 | – | – | 14.0 | – | – |

## 7.C Traffic Equilibrium

Let's consider the flow of traffic in a network modeled by a directed graph with set of vertices $V$ and set of edges $E$. Agents would like to move through the network as quickly as possible, but there's no central decision-maker that guides the flow. We can think of the mayhem unfolding on the freeways during rush hour with each vehicle being one agent. Despite the freedom each agent has in choosing a route, the amount of traffic on the various edges (road segments) often settles at surprisingly predictable levels. To model the situation, let $g_{ij} : \mathbb{R} \to (0, \infty)$ be a continuous function that specifies the time it takes to traverse edge $(i, j) \in E$. It's rarely constant and depends on the amount of flow (number of vehicles) $x_{ij}$ on edge $(i, j)$. For example,

$$g_{ij}(x_{ij}) = \tau_{ij}\big(1 + 0.1(x_{ij}/\gamma_{ij})^4\big),$$

where $\tau_{ij}$ is the nominal travel time and $\gamma_{ij}$ is the nominal capacity. Then, the travel time grows dramatically as the flow exceeds the nominal capacity.

One could imagine a central decision-maker determining the flow $x_{ij}$ on every edge with the goal of minimizing the total travel time across all the agents, but that isn't how many networks operate and certainly not road networks. Each agent determines its route as best as it can. Under some assumptions, we can start to understand the traffic flows in such networks by solving a complementarity problem.

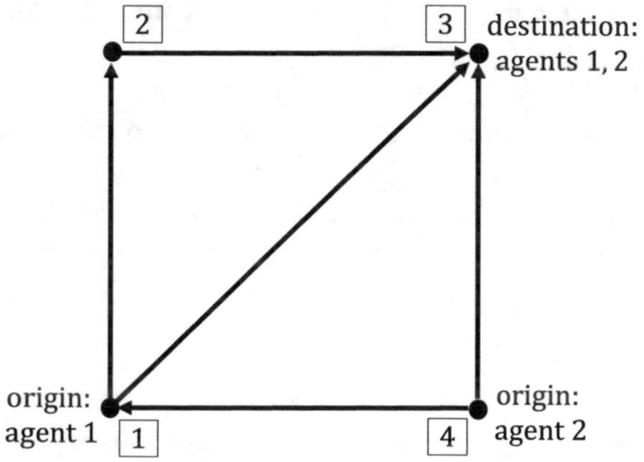


Fig. 7.2: Agents move from origin to destination vertices in a directed graph.

Suppose that we know the number of agents and for each one, its origin vertex and destination vertex; see Figure 7.2. Between an origin vertex $s \in V$ and a destination vertex $t \in V$, there are typically several possible directed paths, i.e., ordered sets of edges of the form $\{(s, i_1), (i_1, i_2), \ldots, (i_{\nu-1}, t)\}$ for some $\nu \in \mathbb{N}$. An agent that would like to travel from $s$ to $t$ selects one such path to follow. We ignore the possibility of rerouting during the transit from $s$ to $t$. Let $P_{st}$ be the set of all paths between $s$ and $t$, which can be quite a few but are certainly finite as we assume that there's only a finite number of edges. For path $p \in P_{st}$, let $w_p$ be the flow along path $p$, i.e., the number of agents that select $p$ as its path from $s$ to $t$. The travel time along $p$ is determined by $w_p$ and also the flow on other paths that overlap with $p$. To clarify this, let

$$a_{ijp} = \begin{cases} 1 & \text{if path } p \text{ includes edge } (i,j) \in E \\ 0 & \text{otherwise.} \end{cases}$$

Let $O$ be the set of all pairs of origin-destination vertices across all the agents; Figure 7.2 has two such pairs. Then, the flow on edge $(i, j) \in E$ becomes

$$x_{ij} = \sum\nolimits_{(s,t)\in O} \sum\nolimits_{p\in P_{st}} a_{ijp} w_p.$$

With $w = (w_p, p \in P_{st}, (s,t) \in O)$, the travel time along a particular path $p$ is

$$\varphi_p(w) = \sum\nolimits_{(i,j)\in E} a_{ijp}\, g_{ij}(x_{ij}) = \sum\nolimits_{(i,j)\in E} a_{ijp}\, g_{ij}\Big(\sum\nolimits_{(s,t)\in O} \sum\nolimits_{\bar{p}\in P_{st}} a_{ij\bar{p}} w_{\bar{p}}\Big).$$

The *Wardrop equilibrium principle* asserts that each agent chooses a path between its origin-destination vertices such that across all paths with positive flow between these vertices the travel time is equal; paths with higher travel times have no flow and there are no paths with lower travel times.

We can express this equilibrium principle by

$$w_p \geq 0, \quad \varphi_p(w) - v_{st} \geq 0, \quad w_p\big(\varphi_p(w) - v_{st}\big) = 0 \quad \forall p \in P_{st},\ (s,t) \in O, \tag{7.11}$$

where $v_{st}$ is the minimum travel time between $(s, t)$, which we need to determine together with $w$. The first inequality ensures that path flows are nonnegative and the second one that no path has travel time below the minimum value. The equality formulates the requirement that the flow $w_p = 0$ when the travel time is higher than the minimum, i.e., $\varphi_p(w) - v_{st} > 0$. Although the Wardrop equilibrium principle ignores dynamical effects, rerouting and other re-world considerations, it models the aggregated behavior of a traffic network fairly well.

Let $d_{st}$ be the given demand of flow between origin-destination pair $(s, t)$, i.e., the number of agents that need to flow between $s$ and $t$. Then, we also have the constraints

$$\sum\nolimits_{p\in P_{st}} w_p - d_{st} \geq 0 \ \ \forall (s,t) \in O.$$

It would have been more natural to state these demand constraints as equalities, but to mirror (7.11) we combined them with the minimum travel time $v_{st}$ in the conditions

$$v_{st} \geq 0, \quad \sum\nolimits_{p\in P_{st}} w_p - d_{st} \geq 0, \quad v_{st}\Big(\sum\nolimits_{p\in P_{st}} w_p - d_{st}\Big) = 0 \ \ \forall (s,t) \in O. \tag{7.12}$$

The minimum time $v_{st}$ is naturally nonnegative. If it's positive (which would be the typical case), then the demand inequalities must hold as equalities.

In summary, the Wardrop equilibrium principle states that the flow along the various paths $w$ and the corresponding minimum travel times $(v_{st}, (s,t) \in O)$ must satisfy (7.11) and (7.12). We can determine these quantities by solving a complementarity problem. To see this, let

$$u = \big(w_p, v_{st},\ p \in P_{st},\ (s,t) \in O\big)$$

be the vector with all the unknown variables and

$$\begin{aligned} f_{stp}(u) &= \varphi_p(w) - v_{st} \\ f_{st0}(u) &= \sum\nolimits_{p \in P_{st}} w_p - d_{st} \\ F(u) &= \big(f_{stp}(u), f_{st0}(u),\ p \in P_{st},\ (s,t) \in O\big). \end{aligned}$$

Then, assuming that the elements of $u$ and $F(u)$ have been organized in corresponding order, (7.11) and (7.12) can be stated as

$$u \geq 0, \quad F(u) \geq 0, \quad \langle F(u), u \rangle = 0.$$

This complementarity problem can equally well be stated as the generalized equation

$$0 \in F(u) + N_{[0,\infty)^{r+q}}(u),$$

where $r = \sum_{(s,t)\in O} m_{st}$, $m_{st}$ is the number of paths between $s$ and $t$, and $q$ is the number of origin-destination pairs in $O$. A solution strategy for a generalized equation of this kind is to attempt to find a function $f : \mathbb{R}^{r+q} \to \mathbb{R}$ with gradient $\nabla f(u) = F(u)$ and then minimize $f$ over $[0,\infty)^{r+q}$. If one can find such a function, then the basic optimality condition 4.37 furnishes the justification for this approach. However, the approach isn't immediately possible in the present case, but this changes if we bring forward $x_{ij}$.

Let $x \in \mathbb{R}^n$ be the vector with components $\{x_{ij}, (i,j) \in E\}$, where $n$ is the number of edges. Now, consider the problem

$$\begin{aligned} \underset{(x,w)\in\mathbb{R}^n\times\mathbb{R}^r}{\text{minimize}}\ f(x,w) &= \sum\nolimits_{(i,j)\in E} \int_0^{x_{ij}} g_{ij}(\alpha)\, d\alpha \qquad (7.13) \\ \text{subject to } x_{ij} - \sum\nolimits_{(s,t)\in O} \sum\nolimits_{p\in P_{st}} a_{ijp} w_p &= 0 \quad \forall (i,j) \in E \\ \sum\nolimits_{p\in P_{st}} w_p &= d_{st} \quad \forall (s,t) \in O \\ w_p &\geq 0 \quad \forall p \in P_{st},\ (s,t) \in O. \end{aligned}$$

An optimality condition for this problem is equivalent to (7.11) and (7.12) and consequently determines the behavior of all the agents. In contrast to the situation in §7.B, (7.13) doesn't arise naturally as the problem for a central planner. Nevertheless, (7.13) is valuable in a computational approach for solving the actual traffic problem as optimization algorithms from earlier chapters apply.

To show the asserted equivalence, let

$$C = \big\{(x,w) \in \mathbb{R}^n \times \mathbb{R}^r \ \big|\ A(x,w) = b, w \geq 0\big\}$$

be the feasible set in (7.13), where $A$ is an $(n+q)\times(n+r)$-matrix consisting of four parts. In the upper left corner, it has the $n \times n$ identity matrix. In the upper right corner, it has an $n \times r$ matrix with $-a_{ijp}$ in row $(i,j)$ and column $p$. The bottom left corner has only

0 elements. The bottom right corner comprises a $q \times r$-matrix with $-1$ in row $(s,t) \in O$ and column $p$ if $(s,t)$ is the origin-destination pair for path $p$, but is 0 otherwise. The $(n+q)$-dimensional vector $b$ starts with $n$ zeros and then contains $\{-d_{st}, (s,t) \in O\}$.

By the Leibniz rule for differentiation under the integral, $\nabla f(x,w)$ is the $(n+r)$-dimensional vector with $\{g_{ij}(x_{ij}), (i,j) \in E\}$ as the first $n$ elements and then 0 elsewhere. Thus, $f$ is smooth and 4.37 furnishes the optimality condition $0 \in \nabla f(x,w) + N_C(x,w)$, where 2.44 gives

$$N_C(x,w) = \big\{A^\top(y,v) - (0,z) \;\big|\; y \in \mathbb{R}^n,\ v \in \mathbb{R}^q,\ z \in \mathbb{R}^r;\quad z_p \geq 0 \text{ if } w_p = 0 \text{ and } z_p = 0 \text{ if } w_p > 0\big\}.$$

Writing out the optimality condition explicitly, we obtain

$$0 = g_{ij}(x_{ij}) + y_{ij} \quad \forall (i,j) \in E$$

$$0 = -\sum\nolimits_{(i,j)\in E} a_{ijp}\, y_{ij} - v_{st} - z_p \quad \forall p \in P_{st},\ (s,t) \in O$$

$$z_p \geq 0 \text{ if } w_p = 0 \text{ and } z_p = 0 \text{ if } w_p > 0.$$

This consolidates to

$$\forall p \in P_{st},\ (s,t) \in O: \sum\nolimits_{(i,j)\in E} a_{ijp}\, g_{ij}(x_{ij}) - v_{st} \in \begin{cases} [0,\infty) & \text{if } w_p = 0 \\ \{0\} & \text{if } w_p > 0. \end{cases}$$

It's now apparent that this condition, together with feasibility $(x,w) \in C$, is equivalent to (7.11) and (7.12).

Our formulations involve a variable $w_p$ for each path $p$ and this can easily result in a huge number of variables. We refer to [38] for refinements that overcome this bottleneck.

**Exercise 7.10** (Wardrop equilibrium). Determine a Wardrop equilibrium flow in Figure 7.2 by solving (7.13) with origin-destination pairs $O = \{(1,3), (4,3)\}$, $d_{13} = 40$, $d_{43} = 60$ and $g_{ij}(x_{ij}) = \tau_{ij}(1 + 0.1(x_{ij}/\gamma_{ij})^4)$. Here, $\tau_{12} = \tau_{23} = 3$, $\tau_{13} = \tau_{41} = 1$, $\tau_{43} = 2$ and $\gamma_{12} = \gamma_{23} = 10$, $\gamma_{13} = \gamma_{41} = 30$, $\gamma_{43} = 20$. Verify that the flow satisfies (7.11) and (7.12). Repeat the calculations with $d_{13} = 60$ and $d_{43} = 90$.

**Guide.** In the first case, the travel times for origin-destination pairs (1, 3) and (4, 3) are 3.067 and 4.108, respectively. In the second case, the times are 8.077 and 9.414. □

## 7.D Reformulation as Minimization Problems

The large variety of generalized equations rules out the possibility of having one universal algorithm for their solution. As in the case of optimization problems, much depends on the structure of a particular generalized equation. It's clear from the previous sections that sometimes a generalized equation can be converted into a minimization problem. This is the case for

$$0 \in F(x) + N_C(x)$$

when there's a differentiable function $f : \mathbb{R}^n \to \mathbb{R}$ with

$$\nabla f(x) = F(x) \quad \forall x \in \mathbb{R}^n.$$

Then, the generalized equation is nothing but an optimality condition for the problem of minimizing $f$ over $C$ (cf. the basic optimality condition 4.37), which optimization algorithms from the earlier chapters are designed to address. However, $\nabla F(x)$ may not be symmetric which rules out the existence of such $f$ because the Hessian $\nabla^2 f(x)$ is always symmetric. In fact, when $F$ is differentiable, $\nabla F(x)$ being symmetric for all $x \in \mathbb{R}^n$ is necessary and sufficient for such $f$ to exist [43].

The connection with optimization also raises the question: Is there a structural property for generalized equations analogous to convexity for minimization problems? This is indeed the case and the role is played by monotonicity. For $C \subset \mathbb{R}^n$, a mapping $F : \mathbb{R}^n \to \mathbb{R}^n$ is *monotone relative to C* if

$$\langle F(x) - F(\bar{x}), x - \bar{x} \rangle \geq 0 \quad \forall x, \bar{x} \in C. \tag{7.14}$$

It's *monotone* when this holds with $C = \mathbb{R}^n$.

For example, if $F : \mathbb{R}^n \to \mathbb{R}^n$ is smooth and $\nabla F(x)$ is positive semidefinite for all $x \in \mathbb{R}^n$, then $F$ is monotone; see [105, Proposition 12.3]. More concretely, $F(x, y) = (\nabla f_0(x) + y \nabla g(x), -g(x))$ from 7.7 produces the Jacobian

$$\nabla F(x, y) = \begin{bmatrix} 1 & 0 & 0 & -1 \\ 0 & 2 & 0 & -1 \\ 0 & 0 & 2 & 1 \\ 1 & 1 & -1 & 0 \end{bmatrix},$$

which indeed is positive semidefinite.

Monotonicity in a variational inequality implies certain properties for the corresponding set of solutions.

**Theorem 7.11** (monotone variational inequalities). *For a nonempty, closed and convex set $C \subset \mathbb{R}^n$ and a continuous mapping $F : \mathbb{R}^n \to \mathbb{R}^n$, which is also monotone relative to C, the set of solutions to the variational inequality*

$$x \in C \quad \text{and} \quad \langle F(x), \bar{x} - x \rangle \geq 0 \ \forall \bar{x} \in C$$

*is closed and convex but possibly empty.*

*The solution set is nonempty if there are $\bar{x} \in C$ and $\rho \geq 0$ such that*

$$\langle F(x), \bar{x} - x \rangle \leq 0 \ \forall x \in C \text{ with } \|x - \bar{x}\|_2 > \rho.$$

*In particular, the solution set is nonempty when C is nonempty and compact.*

**Proof.** Let $\bar{x} \in C$. For solutions $x^\nu$ of the variational inequality converging to $x$, one has $x^\nu \in C$ and $\langle F(x^\nu), \bar{x} - x^\nu \rangle \geq 0$. Since $F$ is continuous and $C$ is closed, we must have $x \in C$ and $\langle F(x), \bar{x} - x \rangle \geq 0$. We've shown that the solution set is closed because $\bar{x}$ is arbitrary. For the convexity and existence claims, we refer to [105, Example 12.48] and [105, Exercise 12.52], respectively; see also 9.11. □

Let's consider versatile approaches for converting a generalized equation into a minimization problem. They apply regardless of monotonicity, but simplifications accrue when that property holds. The basic idea extends that for linear equations in 2.37: With $F : \mathbb{R}^n \to \mathbb{R}^m$, the system of equations $F(x) = 0$ is often efficiently solved by minimizing $\|F(x)\|_2^2$, especially when $F(x) = Ax - b$.

**Example 7.12** (gap function for complementarity problems). For $F : \mathbb{R}^n \to \mathbb{R}^n$, consider the complementarity problem

$$x \geq 0, \quad F(x) \geq 0, \quad \big\langle F(x), x \big\rangle = 0.$$

With $F(x) = (f_1(x), \dots, f_n(x))$, $\bar{x}$ is a solution of this problem if and only if it's a minimizer of

$$\operatorname*{minimize}_{x \in \mathbb{R}^n} \tfrac{1}{2} \sum\nolimits_{i=1}^n \Big(\varphi\big(f_i(x), x_i\big)\Big)^2, \quad \text{where } \varphi(\alpha, \beta) = \sqrt{\alpha^2 + \beta^2} - \alpha - \beta.$$

The objective function here is referred to as a *gap function* for the complementarity problem as it measures the "amount" by which a point fails to solve the problem.

**Detail.** Let's examine the function $\varphi$. If $\varphi(\alpha, \beta) = 0$, then

$$\sqrt{\alpha^2 + \beta^2} = \alpha + \beta. \tag{7.15}$$

After squaring each side, we find that $\alpha^2 + \beta^2 = \alpha^2 + \beta^2 + 2\alpha\beta$. Thus, $\alpha\beta = 0$. Now, if one of the variables is negative, then the other must be 0. But, then $\alpha + \beta < 0$. This contradicts (7.15) and we realize that both variables are nonnegative. We've shown that $\varphi(\alpha, \beta) = 0$ implies $\alpha \geq 0, \beta \geq 0, \alpha\beta = 0$. For the converse, $\alpha\beta = 0$ ensures that

$$\alpha^2 + \beta^2 = \alpha^2 + \beta^2 + 2\alpha\beta = (\alpha + \beta)^2.$$

Since $\alpha, \beta \geq 0$, we then have that (7.15) holds and also $\varphi(\alpha, \beta) = 0$. In view of this equivalence, the complementarity problem can be reformulated as the equations $\varphi(f_i(x), x_i) = 0$, $i = 1, \dots, n$, and then also as the asserted minimization problem.

The objective function is smooth when $F$ is smooth, which we prove as follows. First, let's consider the function given by

$$h(\alpha, \beta) = h_0\big(\varphi(\alpha, \beta)\big), \quad \text{where } h_0(u) = \tfrac{1}{2}u^2.$$

If $(\alpha, \beta) \neq (0, 0)$, then $h$ is smooth at $(\alpha, \beta)$ with

$$\nabla h(\alpha,\beta) = \varphi(\alpha,\beta)\nabla\varphi(\alpha,\beta), \text{ where } \nabla\varphi(\alpha,\beta) = \begin{bmatrix} \frac{\alpha}{\|(\alpha,\beta)\|_2} - 1 \\ \frac{\beta}{\|(\alpha,\beta)\|_2} - 1 \end{bmatrix}.$$

Next, we compute $\partial h(0,0)$ using the general chain rule 6.23. Since

$$\varphi(\alpha,\beta) = \big\|(\alpha,\beta)\big\|_2 - \alpha - \beta,$$

we realize in light of 1.15 that $\varphi$ is convex. It's also real-valued and $\varphi$ is locally Lipschitz continuous by 4.68(a). The qualification (6.15) holds because $\partial^\infty h_0(u) = \{0\}$ by 4.61 for all $u \in \mathbb{R}$. Let $f(\alpha,\beta,y) = y\varphi(\alpha,\beta)$. Now, 6.23 establishes that

$$\partial h(0,0) \subset \bigcup_{y\in\partial h_0(\varphi(0,0))} \partial_{\alpha\beta} f(0,0,y) = \big\{(0,0)\big\}$$

because $\partial h_0(\varphi(0,0)) = \{\varphi(0,0)\} = \{0\}$. Since the additional epi-regularity requirements in 6.23 are also satisfied, this inclusion holds with equality. When $(\alpha,\beta) \to (0,0)$, $\nabla h(\alpha,\beta) \to (0,0)$, which establishes that $h$ is smooth.

For $g_i(x) = h(G(x))$, where $G(x) = (f_i(x), x_i)$, we obtain that

$$\nabla g_i(\bar{x}) = \nabla G(\bar{x})^\top \nabla h\big(G(\bar{x})\big).$$

In summary, if $\bar{x}$ is such that $(f_i(\bar{x}), \bar{x}_i) \neq (0,0)$, then

$$\nabla g_i(\bar{x}) = \varphi\big(f_i(\bar{x}), \bar{x}_i\big)\left(\left(\frac{f_i(\bar{x})}{\|(f_i(\bar{x}), \bar{x}_i)\|_2} - 1\right)\nabla f_i(\bar{x}) + \left(\frac{\bar{x}_i}{\|(f_i(\bar{x}), \bar{x}_i)\|_2} - 1\right)e^i\right),$$

where $e^i \in \mathbb{R}^n$ has 1 as its $i$th element and 0 elsewhere. If $\bar{x}$ is such that $(f_i(\bar{x}), \bar{x}_i) = (0,0)$, then $\nabla g_i(\bar{x}) = 0$.

Although we've converted the complementarity problem into a minimization problem, there's still a challenge. We need to obtain a minimizer of a potentially nonconvex objective function. When $F$ is smooth and monotone, however, this task simplifies greatly because then a point that makes the gradient of this objective function vanish is also a minimizer; see [36, Theorem 4.1]. In fact, this holds as well under the weaker condition:

$$\forall x \neq \bar{x} \in \mathbb{R}^n: \quad \exists i \text{ such that } x_i \neq \bar{x}_i \text{ and } \big(f_i(x) - f_i(\bar{x})\big)(x_i - \bar{x}_i) \geq 0.$$

Consequently, at least under these conditions, algorithms for minimizing a smooth function (see Chap. 1) apply and can be expected to compute a point that solves the complementarity problem. □

**Example 7.13** (Auslender gap function for variational inequalities). For $F : \mathbb{R}^n \to \mathbb{R}^n$ and a nonempty, closed and convex set $C \subset \mathbb{R}^n$, consider the variational inequality

$$x \in C \quad \text{and} \quad \big\langle F(x), \bar{x} - x\big\rangle \geq 0 \;\; \forall \bar{x} \in C.$$

The *Auslender gap function* takes the form

$$f(x) = \sup_{y\in C} \langle F(x), x - y\rangle.$$

Then, a point $x^\star$ solves the variational inequality if and only if $x^\star \in \operatorname{argmin}_{x\in C} f(x)$ and $f(x^\star) = 0$.

**Detail.** For every $x \in C$, $f(x) \geq 0$ because $y = x$ is always a possibility in the maximization. If $x^\star$ solves the variational inequality, then

$$\langle F(x^\star), x^\star - y\rangle \leq 0 \quad \forall y \in C.$$

Thus, $f(x^\star) \leq 0$ and $x^\star \in \operatorname{argmin}_{x\in C} f(x)$. For the converse, suppose that $f(x^\star) = 0$ and $x^\star \in C$. Then,

$$\langle F(x^\star), x^\star - x\rangle \leq 0 \quad \forall x \in C$$

so also $\langle F(x^\star), x - x^\star\rangle \geq 0$ for all $x \in C$ and $x^\star$ satisfies the variational inequality.

The Auslender gap function is nonsmooth and we need to fall back on facts about sup-projections in analysis and computations; see §6.E. □

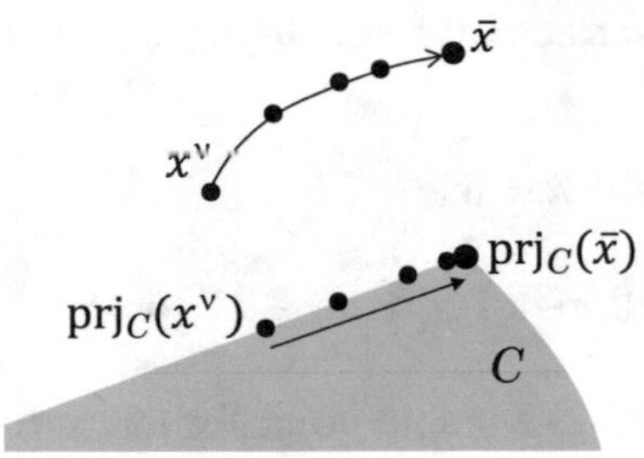


Fig. 7.3: Continuity of projection on a closed convex set.

The next gap function utilizes the following continuity property of projections, which is also of independent interest; see Figure 7.3.

**Proposition 7.14** (continuity of projection). *For a nonempty, closed and convex set $C \subset \mathbb{R}^n$, the mapping from $\mathbb{R}^n$ to $\mathbb{R}^n$ given by*

$$x \mapsto \mathrm{prj}_C(x)$$

*is continuous.*

**Proof.** For $x^\nu \to \bar{x}$, let

$$\varphi^\nu(y) = \iota_C(y) + \tfrac{1}{2}\|y - x^\nu\|_2^2 \quad \text{and} \quad \varphi(y) = \iota_C(y) + \tfrac{1}{2}\|y - \bar{x}\|_2^2.$$

There's a unique minimizer of $\varphi^\nu$ and this is the single point $\mathrm{prj}_C(x^\nu)$. Likewise, the unique minimizer of $\varphi$ is the single point $\mathrm{prj}_C(\bar{x})$; see 2.30. We can use the epigraphical sum rule 4.19(a) to establish that $\varphi^\nu \xrightarrow{e} \varphi$ because $\iota_C \xrightarrow{e} \iota_C$ and

$$\tfrac{1}{2}\|y^\nu - x^\nu\|_2^2 \to \tfrac{1}{2}\|\bar{y} - \bar{x}\|_2^2 \quad \text{whenever } y^\nu \to \bar{y}.$$

The sequence $\{\bar{y}^\nu \in \operatorname{argmin} \varphi^\nu, \nu \in \mathbb{N}\}$ exists and is bounded because

$$\tfrac{1}{2}\|\bar{y}^\nu - x^\nu\|_2^2 \le \tfrac{1}{2}\|\bar{y} - x^\nu\|_2^2$$

and $\{x^\nu, \nu \in \mathbb{N}\}$ is bounded, where $\bar{y} \in \operatorname{argmin} \varphi$. Thus, $\{\bar{y}^\nu, \nu \in \mathbb{N}\}$ has at least one cluster point and this point must minimize $\varphi$ by 5.5(b). Since argmin $\varphi$ consists of a single point, $\bar{y}^\nu \to \bar{y}$, i.e., $\operatorname{prj}_C(x^\nu) \to \operatorname{prj}_C(\bar{x})$. □

**Example 7.15** (Fukushima gap function for variational inequalities). For $F : \mathbb{R}^n \to \mathbb{R}^n$ and a nonempty, closed and convex set $C \subset \mathbb{R}^n$, consider the variational inequality

$$x \in C \quad \text{and} \quad \langle F(x), \bar{x} - x \rangle \ge 0 \;\; \forall \bar{x} \in C.$$

The *Fukushima gap function* takes the form

$$f(x) = -\big\langle F(x),\ \operatorname{prj}_C\big(x - F(x)\big) - x \big\rangle - \tfrac{1}{2}\big\|\operatorname{prj}_C\big(x - F(x)\big) - x\big\|_2^2.$$

Then, a point $x^\star$ solves the variational inequality if and only if $x^\star \in \operatorname{argmin}_{x \in C} f(x)$ and $f(x^\star) = 0$.

**Detail.** First, let's establish the fact that

$$x \in C \;\text{ and }\; \langle F(x), \bar{x} - x \rangle \ge 0 \;\; \forall \bar{x} \in C \iff x = \operatorname{prj}_C\big(x - F(x)\big).$$

To see this, recall that $\operatorname{prj}_C(z)$ is a single point for any $z \in \mathbb{R}^n$ because $C$ is nonempty, closed and convex; see 2.30. The point $x = \operatorname{prj}_C(z)$ is characterized by

$$\langle x - z, \bar{x} - x \rangle \ge 0 \;\; \forall \bar{x} \in C$$

as derived in 2.30. Setting $z = x - F(x)$ establishes the claim.

Second, by rearranging some terms, we can write

$$f(x) = \tfrac{1}{2}\big\|F(x)\big\|_2^2 - \tfrac{1}{2}\Big\|\operatorname{prj}_C\big(x - F(x)\big) - \big(x - F(x)\big)\Big\|_2^2.$$

Let $x \in C$. Since $\|F(x)\|_2$ is the distance between $x - F(x)$ and $x$ and

$$\Big\|\operatorname{prj}_C\big(x - F(x)\big) - \big(x - F(x)\big)\Big\|_2$$

is the distance between $x - F(x)$ and its projection on $C$, we must have $f(x) \ge 0$. Moreover, these distances must be equal if and only if

$$x = \operatorname{prj}_C\big(x - F(x)\big)$$

in which case $f(x) = 0$. In view of the first fact, we can summarize that $f(x) \geq 0$ for all $x \in C$ and also $f(x^\star) = 0$ if and only if $x^\star$ solves the variational inequality. We note that $\inf_C f > 0$ corresponds to a case with no solution to the variational inequality.

If $F$ is smooth, then $f$ is smooth with gradient

$$\nabla f(x) = F(x) - \big(\nabla F(x) - I\big)\Big(\operatorname{prj}_C\big(x - F(x)\big) - x\Big),$$

where $I$ is the $n \times n$ identity matrix. To establish this, let's fix $x \in \mathbb{R}^n$ and define

$$g(x, y) = \big\langle F(x), x - y\big\rangle - \tfrac{1}{2}\|y - x\|_2^2.$$

Then,

$$\begin{aligned} &\bar{y} \in \operatorname{argmax}_{y \in C}\, g(x, y) \iff \bar{y} \in \operatorname{argmin}_{y \in C} -g(x, y) \\ &\iff -F(x) - \bar{y} + x \in N_C(\bar{y}) \iff \bar{y} \in \operatorname{prj}_C\big(x - F(x)\big), \end{aligned}$$

where the second equivalence stems from the optimality condition 2.27 and the third one from the analysis of projections in 2.30. Consequently,

$$\begin{aligned} \max_{y \in C}\, g(x, y) &= g(x, \bar{y}) \\ &= \big\langle F(x), x - \operatorname{prj}_C\big(x - F(x)\big)\big\rangle - \tfrac{1}{2}\big\|\operatorname{prj}_C\big(x - F(x)\big) - x\big\|_2^2 = f(x). \end{aligned}$$

The difference relative to the Auslender gap function in 7.13 is now clear: that gap function doesn't include the regularization term $\frac{1}{2}\|y - x\|_2^2$ in the maximization. When we include the term, the maximization is always attained at a unique point because $C$ is nonempty, closed and convex. Next, we bring in 6.30 to determine the subgradients of $f$ using its expression as a sup-projection. In 6.30, the set over which the maximization is taking place needs to be compact while $C$ may not be. To overcome this difficulty, we use that $x \mapsto \operatorname{prj}_C(x - F(x))$ is continuous by 7.14. This continuity property then implies that at $\hat{x} \in \mathbb{R}^n$, there's $\delta > 0$ such that

$$\big\|\operatorname{prj}_C\big(x - F(x)\big) - \operatorname{prj}_C\big(\hat{x} - F(\hat{x})\big)\big\|_2 \leq \tfrac{1}{2} \quad \forall x \in \mathbb{B}(\hat{x}, \delta).$$

Consequently, for any $x \in \mathbb{B}(\hat{x}, \delta)$, we can repeat the above development with $C$ replaced by $C \cap \mathbb{B}(\operatorname{prj}_C(\hat{x} - F(\hat{x})), 1)$ and reach the same conclusions. We can then apply 6.30 and obtain

$$\partial f(x) = \Big\{\nabla_x g\Big(x, \operatorname{prj}_C\big(x - F(x)\big)\Big)\Big\} \quad \text{when} \quad x \in \mathbb{B}(\hat{x}, \delta).$$

The uniqueness of $\operatorname{prj}_C(x - F(x))$ ensures that there's only one subgradient at $x$. The asserted expression for the gradient follows and it varies continuously. Thus, $f$ is indeed smooth at every point $\mathbb{B}(\hat{x}, \delta)$. We can repeat this argument for any $\hat{x} \in \mathbb{R}^n$, which establishes that $f$ is smooth with the given gradient.

While $f$ is smooth when $F$ is smooth, there's still the concern that an algorithm assigned to minimize $f$ over $C$ may not achieve a minimizer as convexity of $f$ isn't guaranteed.

At least when $\nabla F(x)$ is positive definite for all $x \in C$, which means that $F$ is necessarily monotone relative to $C$ by [105, Proposition 12.3], this isn't an issue because then every point satisfying the optimality condition $-\nabla f(x) \in N_C(x)$ is actually a minimizer of $f$ over $C$; see [43, Theorem 3.3]. Algorithms such as SQP, interior-point and augmented Lagrangian methods (cf. §4.K and §6.B) are designed to reach points satisfying this optimality condition. □

**Example 7.16** (gap function for generalized equations). Given $\bar{v} \in \mathbb{R}^m$ and $S : \mathbb{R}^n \rightrightarrows \mathbb{R}^m$, closed-valued and convex-valued, consider the gap function $f : \mathbb{R}^n \to \overline{\mathbb{R}}$ with

$$f(x) = \sup_{y \in \mathbb{R}^n} \inf_{z \in S(x)} \langle \bar{v} - z, y \rangle.$$

Then, $x^\star$ solves the generalized equation $\bar{v} \in S(x)$ if and only if $x^\star \in \operatorname{argmin} f$ and $f(x^\star) = 0$.

**Detail.** Since $y = 0$ is a possibility in the maximization, $f(x) \geq 0$ for all $x \in \mathbb{R}^n$. If $x^\star$ solves the generalized equation, then $z = \bar{v}$ is a possibility in the inner minimization and

$$f(x^\star) \leq \sup_{y \in \mathbb{R}^n} \langle \bar{v} - \bar{v}, y \rangle = 0.$$

Thus, $x^\star \in \operatorname{argmin} f$. This holds even if $S$ isn't closed-valued and convex-valued.

For the converse, let's fix $x \in \mathbb{R}^n$ and utilize the alternative formula

$$f(x) = \sup_{y \in \mathbb{R}^n} \big\{ \langle \bar{v}, y \rangle - \sigma_{S(x)}(y) \big\}, \quad \text{where } \sigma_{S(x)}(y) = \sup_{z \in S(x)} \langle z, y \rangle$$

is the *support function* of the set $S(x)$. Thus, the conjugate of $\sigma_{S(x)}$ evaluated at $\bar{v}$ coincides with $f(x)$. By 5.29, the conjugate of $\sigma_D$ is $\iota_{\operatorname{cl} \operatorname{con} D}$ for any $D \subset \mathbb{R}^m$. This implies that $f(x) = \iota_{S(x)}(\bar{v})$ because $S(x)$ is closed and convex. Now, if $f(x^\star) = 0$, then $\iota_{S(x^\star)}(\bar{v}) = 0$, which confirms that $\bar{v} \in S(x^\star)$. We've established the asserted equivalence.

Even without the assumption about closed-valued and convex-valued $S$, if $f(x) > 0$, then there's $y \in \mathbb{R}^n$ such that

$$\inf_{z \in S(x)} \langle \bar{v} - z, y \rangle > 0.$$

Consequently, for all $z \in S(x)$, $\langle \bar{v} - z, y \rangle > 0$, which means that $\bar{v} \notin S(x)$.

The gap function $f$ isn't easily minimized and typically requires approximations as part of a computational procedure. □

**Exercise 7.17** (solution of energy problem by optimization). Use the gap function in 7.12 to solve the complementarity problem at the end of 7.7. Then, use the Fukushima gap function in 7.15 to solve the corresponding variational inequality. Compare the results and the computational effort required.

**Guide.** Since both gap functions are smooth and the associated constraints in the case of Fukushima are simple, their minimization can be carried out by optimization algorithms from earlier chapters. □

## 7.E Projection Methods

Projections play a central role in the Fukushima gap function of 7.15 and the developments there hint to the possibility of solving the variational inequality

$$x \in C \quad \text{and} \quad \langle F(x), \bar{x} - x \rangle \geq 0 \;\; \forall \bar{x} \in C \tag{7.16}$$

directly without passing through a gap function. As before, $F : \mathbb{R}^n \to \mathbb{R}^n$ and $C$ is nonempty, closed and convex. We know that $x$ is a solution of (7.16) if and only if $x = \text{prj}_C(x - \lambda F(x))$ for $\lambda = 1$; see 7.15. The arguments underpinning that result can be modified slightly to cover $\lambda > 0$ as well. It's therefore natural to propose the iterative scheme

$$x^{\nu+1} = \text{prj}_C\big(x^\nu - \lambda F(x^\nu)\big),$$

which is the fixed-point algorithm (§2.D) applied to the mapping $x \mapsto \text{prj}_C(x - \lambda F(x))$. It also mirrors the projected gradient method (§2.D) with $F$ playing the role of a gradient and then $\lambda$ emerges as a step size. One can show that this simple scheme indeed converges to a solution of (7.16) (see, for example, [74]) if $\lambda$ is selected appropriately and $F$ is *strongly monotone*, i.e., there's $\sigma \in [0, \infty)$ such that $F - \sigma I$ is monotone, where $I : \mathbb{R}^n \to \mathbb{R}^n$ is the identity mapping, i.e., $I(x) = x$. In practice, however, it isn't easy to select a suitable $\lambda$ as even its theoretical requirements depend on constants such as $\sigma$ that might not be known. A more robust approach is to split every iteration into two parts and to adjust $\lambda$ adaptively.

Specifically, for $\lambda^\nu > 0$, let's consider the iterations generated by

$$\begin{aligned} \bar{x}^\nu &\in \text{prj}_C\big(x^\nu - \lambda^\nu F(x^\nu)\big) \\ x^{\nu+1} &\in \text{prj}_C\big(x^\nu - \lambda^\nu F(\bar{x}^\nu)\big). \end{aligned} \tag{7.17}$$

Figure 7.4 illustrates the scheme together with a solution $x^\star$. The double projection enables us to adaptively select $\lambda^\nu$ and avoid the assumption about strong monotonicity.

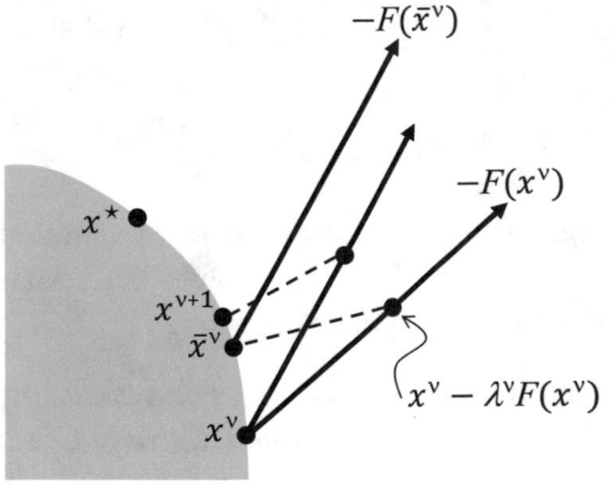


Fig. 7.4: Steps in the double-projection algorithm.

**Proposition 7.18** (double-projection iterations). *For a nonempty, closed and convex set $C \subset \mathbb{R}^n$ and a mapping $F : \mathbb{R}^n \to \mathbb{R}^n$, monotone relative to $C$, suppose that $\{x^\nu, \bar{x}^\nu, \nu \in \mathbb{N}\}$ is generated by (7.17) starting from $x^0 \in C$. Then, for any solution $x^\star$ of (7.16) and $\nu \in \mathbb{N}$, one has*

$$\|x^{\nu+1} - x^\star\|_2^2 \leq \|x^\nu - x^\star\|_2^2 - \|x^\nu - \bar{x}^\nu\|_2^2 \left(1 - (\lambda^\nu)^2 \frac{\|F(x^\nu) - F(\bar{x}^\nu)\|_2^2}{\|x^\nu - \bar{x}^\nu\|_2^2}\right).$$

**Proof.** We refer to [55] for a proof; see also [64]. □

The proposition establishes that if $\lambda^\nu$ is sufficiently small, then the expression in the parenthesis is positive and $\|x^\nu - x^\star\|_2$ decreases for all $\nu$. This implies that $\|x^\nu - \bar{x}^\nu\|_2 \to 0$. If $\lambda^\nu$ is also bounded away from 0, then $x^\nu \to x^\star$. These insights result in the following algorithm, which implements an adaptive procedure for $\lambda^\nu$.

**Double-Projection Algorithm.**

Data. $x^0 \in C$, $\tau \in (0,1)$ and $\lambda^0 \in (0, \infty)$.

Step 0. Set $\nu = 0$.

Step 1. Compute $\bar{x}^\nu \in \mathrm{prj}_C\big(x^\nu - \lambda^\nu F(x^\nu)\big)$.

Step 2. If $\bar{x}^\nu = x^\nu$, then stop.

Step 3. If

$$\frac{\lambda^\nu \|F(x^\nu) - F(\bar{x}^\nu)\|_2}{\|x^\nu - \bar{x}^\nu\|_2} \leq \tau,$$

then set $\lambda^{\nu+1} = \lambda^\nu$, compute

$$x^{\nu+1} \in \mathrm{prj}_C\big(x^\nu - \lambda^\nu F(\bar{x}^\nu)\big)$$

and go to Step 5.

Else, set $\lambda = \lambda^\nu$ and go to Step 4.

Step 4. Set

$$\lambda^\nu = \min\left\{\tfrac{1}{2}\lambda, \frac{\|x^\nu - \bar{x}^\nu\|_2}{\sqrt{2}\,\|F(x^\nu) - F(\bar{x}^\nu)\|_2}\right\}$$

and go to Step 1.

Step 5. Replace $\nu$ by $\nu + 1$ and go to Step 1.

If the algorithm terminates after a finite number of iterations, then the last iterate $x^\nu$ satisfies $x^\nu \in \mathrm{prj}_C(x^\nu - \lambda^\nu F(x^\nu))$ and thus solves the variational inequality (7.16). Otherwise, we've the following convergence property.

**Convergence 7.19** (double-projection algorithm). *For a nonempty, closed and convex set $C \subset \mathbb{R}^n$ and a mapping $F : \mathbb{R}^n \to \mathbb{R}^n$, monotone relative to $C$, suppose that $\{x^\nu, \nu \in \mathbb{N}\}$ is generated by the double-projection algorithm and $F$ satisfies the Lipschitz condition: for some $\kappa \in [0, \infty)$,*

$$\|F(x) - F(\bar{x})\|_2 \leq \kappa\|x - \bar{x}\|_2 \quad \forall x, \bar{x} \in C.$$

*Then, $x^\nu$ converges to a solution of (7.16) if one exists.*

**Proof.** The Lipschitz condition ensures that $\lambda^\nu$ is bounded away from 0. The conclusion then follows from 7.18 and the supporting discussion. □

An important aspect of the algorithm is that we don't need to know the Lipschitz modulus $\kappa$ in an implementation; the step size $\lambda^\nu$ is computed adaptively. The square root appearing in Step 4 emerges from an attempt to maximize the second term on the right-hand side in 7.18 but can be adjusted; see [64] for details.

The numerous projections on $C$ can become computationally costly unless $C$ has additional structural properties. For improvements of the above scheme with the goal of reducing the number of projections overall, we refer to [120] which also relaxes the monotonicity assumption.

**Exercise 7.20** (solution of energy problem using projection). Implement the double-projection algorithm and solve the variational inequality corresponding to the complementarity problem at the end of 7.7. Report the number of projections needed.

## 7.F Nonsmooth Newton-Raphson Algorithm

One might hope that generalized equations can be solved by a modified version of the Newton-Raphson algorithm (§1.G), which is a workhorse for usual equation solving. This turns out to be the case for at least some variational inequalities. Given $F : \mathbb{R}^n \to \mathbb{R}^n$ and a nonempty, closed and convex set $C \subset \mathbb{R}^n$, let's consider the variational inequality

$$x \in C \quad \text{and} \quad \langle F(x), \bar{x} - x \rangle \geq 0 \;\; \forall \bar{x} \in C.$$

We recall from 7.15 that $x$ solves the variational inequality if and only if $x = \text{prj}_C(x - F(x))$. With an auxiliary vector $z \in \mathbb{R}^n$, the variational inequality is therefore equivalent to the *equations*

$$x - z = F(x) \qquad x = \text{prj}_C(z).$$

We can eliminate $x$ by substituting $\text{prj}_C(z)$ in its place. This leads to

$$\text{prj}_C(z) - z = F\big(\text{prj}_C(z)\big).$$

Our goal is to solve for $z$ in this equation because then a solution of the variational inequality is readily available through a single projection. Thus, the variational inequality is effectively reduced to the equation

$$\tilde{F}(z) = F\big(\text{prj}_C(z)\big) + z - \text{prj}_C(z) = 0, \tag{7.18}$$

where $\tilde{F} : \mathbb{R}^n \to \mathbb{R}^n$ is the *normal map* of the variational inequality. The main challenge here is that $z \mapsto \text{prj}_C(z)$ typically isn't smooth so the Newton-Raphson algorithm may fail. In the simple case with $C = [0, \infty)$, one has $\mathrm{prj}_C(z) = 0$ for $z \leq 0$ and $\mathrm{prj}_C(z) = z$ otherwise, which results in a kink at $z = 0$ and loss of smoothness. Let's develop an extension of the Newton-Raphson algorithm that works for some nonsmooth mappings including the normal map $\tilde{F}$ under certain assumptions.

For $G : \mathbb{R}^n \to \mathbb{R}^n$, we seek to solve

$$G(x) = 0.$$

Let's recall the Newton-Raphson algorithm under the assumption that $G$ is smooth, but we'll soon dispense with that assumption. Given a current point $x^\nu \in \mathbb{R}^n$, the algorithm first constructs an affine approximation

$$G^\nu(x) = G(x^\nu) + \nabla G(x^\nu)(x - x^\nu).$$

Second, it computes a solution $\bar{x}^{\nu+1}$ to $G^\nu(x) = 0$. Third, the method either sets $x^{\nu+1} = \bar{x}^{\nu+1}$ or carries out a search along the line segment between $x^\nu$ and $\bar{x}^{\nu+1}$ to determine a better point according to some gap function, for example, given as $\|G(x)\|_2$. The purpose of the line search is to prevent breakdown of the method if initiated too far away from a solution; see 1.34.

Let's examine more closely why the Newton-Raphson algorithm is successful and assume that at each $x^\nu$ generated by the algorithm the Jacobian $\nabla G(x^\nu)$ is invertible and

$$\sup_{\nu \in \mathbb{N}} \left\| \left(\nabla G(x^\nu)\right)^{-1} \right\|_F < \infty.$$

Two aspects are key: First, $G^\nu$ is an accurate approximation of $G$ near $x^\nu$. Specifically, by (2.5), one has

$$G(x) = G^\nu(x) + o\big(\|x - x^\nu\|_2\big). \tag{7.19}$$

Second, $G^\nu(x)$ vanishes quickly as $x$ approaches $\bar{x}^{\nu+1}$ along the line segment from $x^\nu$. Let

$$p^\nu(\lambda) = x^\nu + \lambda(\bar{x}^{\nu+1} - x^\nu).$$

Then, $\{p^\nu(\lambda), \lambda \in [0, 1]\}$ is the line segment between $x^\nu$ and $\bar{x}^{\nu+1}$. We can think of the function $p^\nu : [0, 1] \to \mathbb{R}^n$ as specifying the straight-line path between these two points. The assertion about vanishing $G^\nu$ along this path can now be quantified by

$$G^\nu\big(p^\nu(\lambda)\big) = G(x^\nu) + \lambda \nabla G(x^\nu)(\bar{x}^{\nu+1} - x^\nu) = (1 - \lambda)G(x^\nu),$$

which holds because $\bar{x}^{\nu+1}$ solves $G^\nu(x) = 0$, i.e.,

$$\nabla G(x^\nu)(\bar{x}^{\nu+1} - x^\nu) = -G(x^\nu).$$

We can combine the expression for $G^\nu(p^\nu(\lambda))$ with (7.19) to obtain

$$G\big(p^\nu(\lambda)\big) = (1 - \lambda)G(x^\nu) + o\big(\lambda\|\bar{x}^{\nu+1} - x^\nu\|_2\big).$$

Let $\sigma \in (0, 1)$. When $G(x^\nu) \neq 0$, there must exist $\bar{\lambda} \in (0, 1]$ such that

$$\|G(p^\nu(\lambda))\|_2 \leq (1 - \sigma\lambda)\|G(x^\nu)\|_2 \quad \forall \lambda \in (0, \bar{\lambda}].$$

Thus, there are step sizes $\{\lambda^\nu, \nu \in \mathbb{N}\}$, bounded away from 0, such that $x^{\nu+1} = p^\nu(\lambda^\nu)$ produces

$$\|G(x^{\nu+1})\|_2 \leq (1 - \sigma\lambda^\nu)\|G(x^\nu)\|_2 \quad \forall \nu \in \mathbb{N}.$$

We've established that each iteration improves the gap function with a guaranteed amount. Eventually, the algorithm gets close enough to a solution, the step size $\lambda^\nu = 1$ becomes acceptable and the quadratic rate of convergence indicated in §1.G takes place.

We would like to mimic the three components of the Newton-Raphson algorithm (approximation, root-finding and line search) for a nonsmooth $G$, but immediately run into a difficulty: The affine approximation relies on the Jacobian of $G$, which now may not exist. Alternatives depend on the nature of $G$. If faced with $\tilde{F}(z) = 0$ from a variational inequality, then the common choice is

$$B^\nu \big( \text{prj}_C(z) - \text{prj}_C(z^\nu) \big) + c^\nu + z - \text{prj}_C(z) = 0, \tag{7.20}$$

where $z^\nu$ is the current point and

$$B^\nu = \nabla F\big(\text{prj}_C(z^\nu)\big)$$
$$c^\nu = F\big(\text{prj}_C(z^\nu)\big).$$

Thus, the approximation of $\tilde{F}$ is obtained by linearizing $F$ and keeping the other parts intact. Regardless, let $G^\nu$ be some approximation of $G$.

The second component is root-finding: $G^\nu(x) = 0$. In the smooth setting, this would involve solving a system of linear equations, which has a unique solution when $\nabla G(x^\nu)$ is invertible. Now, this may require additional effort, but the details depend on the nature of the approximation $G^\nu$. We describe one possibility below. Regardless of how it's computed, let $\bar{x}^{\nu+1}$ be a solution of $G^\nu(x) = 0$.

The third component also needs modification. A line search along the segment between $x^\nu$ and $\bar{x}^{\nu+1}$ doesn't suffice since $G^\nu$ may not be affine. (In fact, $G^\nu$ is rarely affine since it needs to approximate a nonsmooth function to a sufficiently high accuracy.) We could very well have

$$\big\|G^\nu\big(x^\nu + \lambda(\bar{x}^{\nu+1} - x^\nu)\big)\big\|_2 \quad \text{and} \quad \big\|G\big(x^\nu + \lambda(\bar{x}^{\nu+1} - x^\nu)\big)\big\|_2$$

both increase as $\lambda$ grows from 0. It becomes necessary to construct a *nonlinear* path $p^\nu : [0, 1] \to \mathbb{R}^n$ between $x^\nu$ and $\bar{x}^{\nu+1}$ along which $G^\nu$ vanishes quickly. The search for the next point $x^{\nu+1}$ takes place along this path.

Let's now put together these ideas in an algorithm that also incorporates the possibility that a path with the desired properties can't be extended the whole way to $\bar{x}^{\nu+1}$. Thus, we permit the path $p^\nu$ to only be defined on $[0, \bar{\lambda}^\nu]$ for some $\bar{\lambda}^\nu \in (0, 1]$, which can be computational much easier to achieve.

**Nonsmooth Newton-Raphson Algorithm.**

Data. $x^0 \in \mathbb{R}^n$ and $\sigma, \tau \in (0,1)$.
Step 0. Set $\nu = 0$.
Step 1. If $G(x^\nu) = 0$, then stop.
Step 2. Construct an approximation $G^\nu : \mathbb{R}^n \to \mathbb{R}^n$ of $G$ with the property that

$$G(x) = G^\nu(x) + o\big(\|x - x^\nu\|_2\big).$$

Step 3. Construct a continuous $p^\nu : [0, \bar\lambda^\nu] \to \mathbb{R}^n$ with $\bar\lambda^\nu \in [0,1]$ such that

$$p^\nu(0) = x^\nu, \qquad G^\nu\big(p^\nu(\lambda)\big) = (1-\lambda)G(x^\nu) \ \ \forall \lambda \in [0, \bar\lambda^\nu].$$

If $\bar\lambda^\nu < 1$, then $p^\nu$ must satisfy the additional condition:
Either $G^\nu$ isn't continuously invertible[1] near $p^\nu(\bar\lambda^\nu)$ or

$$\big\|G\big(p^\nu(\lambda)\big)\big\|_2 < (1 - \sigma\lambda)\big\|G(x^\nu)\big\|_2 \tag{7.21}$$

fails at $\lambda = \bar\lambda^\nu$.
Step 4. If (7.21) holds with $\lambda = \bar\lambda^\nu$, then set $\lambda^\nu = \bar\lambda^\nu$.
Else, select $\lambda^\nu \in [0, \bar\lambda^\nu)$ such that (7.21) holds with $\lambda = \lambda^\nu$ and

$$\lambda^\nu \geq \tau \sup\big\{\lambda_0 \in [0, \bar\lambda^\nu] \,\big|\, (7.21) \text{ holds } \ \forall \lambda \in [0, \lambda_0]\big\}.$$

Step 5. Set $x^{\nu+1} = p^\nu(\lambda^\nu)$, replace $\nu$ by $\nu + 1$ and go to Step 1.

We omit a full justification of the algorithm, but summarize the key facts; see [87] for details. If $G$ is continuous, the approximations in Step 2 have certain uniformity properties similar to those in the smooth case and some other technical conditions are satisfied, then either the algorithm terminates in iteration $\nu$ and $G(x^\nu) = 0$ or the generated sequence $x^\nu \to x^\star$ with $G(x^\star) = 0$. Moreover, there exists $\delta \in (0,1)$ such that

$$\big\|G(x^\nu)\big\|_2 \leq \delta^\nu \big\|G(x^0)\big\|_2 \ \ \forall \nu \in \mathbb{N},$$

where indeed $\delta^\nu$ means $\delta$ to power $\nu$. When $x^\nu$ is close enough to $x^\star$, the rate improves and for some $\gamma > 0$ and sufficiently large $\nu$, we obtain

$$\|x^{\nu+1} - x^\star\|_2 \leq \gamma \|x^\nu - x^\star\|_2^2,$$

which is the same fast rate established in §1.G for Newton's method. We refer to [122] for related approaches.

**Example 7.21** (mixed complementarity problems). It's clear that the nonsmooth Newton-Raphson algorithm is faced with several implementation challenges when applied to

[1] A mapping $H : \mathbb{R}^n \to \mathbb{R}^n$ is continuously invertible near $\bar x$ if there are neighborhoods $C$ and $D$ of $\bar x$ and $H(\bar x)$, respectively, such that the restriction of $H$ to $C$ produces a bijective mapping $H : C \to D$ and its inverse mapping $H^{-1} : D \to C$ is continuous.

general problems. However, these can be overcome for the important special case of finding zeros of the normal map of variational inequalities with smooth $F : \mathbb{R}^n \to \mathbb{R}^n$ and

$$C = \{x \in \mathbb{R}^n \mid \alpha_j \le x_j \le \beta_j,\ j = 1, \dots, n\}$$

for $-\infty \le \alpha_j \le \beta_j \le \infty$. Solving a variational inequality with such $C$ is a *mixed complementarity problem.*

**Detail.** A complementarity problem corresponds to the special case with $\alpha_j = 0$ and $\beta_j = \infty$. The equation $F(x) = 0$ corresponds to $\alpha_j = -\infty$ and $\beta_j = \infty$. With $F(x) = (f_1(x), \dots, f_n(x))$, a mixed complementarity problem amounts to solving the generalized equation

$$0 \in F(x) + N_C(x)$$

or, equivalently, finding $x$ such that

$$f_j(x) \in \begin{cases} (-\infty, \infty) & \text{if } \alpha_j = x_j = \beta_j \\ \{0\} & \text{if } \alpha_j < x_j < \beta_j \\ (-\infty, 0] & \text{if } \alpha_j < x_j = \beta_j \\ [0, \infty) & \text{if } \alpha_j = x_j < \beta_j, \end{cases}$$

where we leverage the normal cone formula in 2.31.

Solving the KKT condition 4.77 amounts to a mixed complementarity problem. Specifically, for smooth $f_i : \mathbb{R}^n \to \mathbb{R}$, $i = 0, 1, \dots, m$, and smooth $g_i : \mathbb{R}^n \to \mathbb{R}$, $i = 1, \dots, q$, the KKT condition can be written economically as

$$\begin{aligned} \nabla f_0(x) + \nabla F(x)^\top y + \nabla G(x)^\top z &\in N_{\mathbb{R}^n}(x) \\ F(x) &\in N_{\mathbb{R}^m}(y) \\ G(x) &\in N_{[0,\infty)^q}(z), \end{aligned}$$

where $F(x) = (f_1(x), \dots, f_m(x))$ and $G(x) = (g_1(x), \dots, g_q(x))$. Thus, solving the KKT condition corresponds to the mixed complementarity problem with mapping

$$(x, y, z) \mapsto -\big(\nabla f_0(x) + \nabla F(x)^\top y + \nabla G(x)^\top z,\ F(x),\ G(x)\big)$$

and $C = \mathbb{R}^n \times \mathbb{R}^m \times [0, \infty)^q$.

Since a mixed complementarity problem corresponds to a special variational inequality, it can be solved by finding a zero of a normal map; see (7.18). Moreover, a normal map has a viable approximation given by (7.20), which can be used in Step 2 of the nonsmooth Newton-Raphson algorithm. Step 3 in the algorithm is also implementable and in fact it suffices to consider piecewise affine paths because the approximation $G^\nu$ in this case is piecewise affine. The choice of step size (Step 4) is now relatively straightforward, but benefits can accrue from a modification allowing occasional *increases* in the gap function. We refer to [29] for implementation details used in the solver PATH. □

## 7.G Continuity of Set-Valued Mappings

The continuity of a mapping allows us to pass to the limit and obtain the "expected" result. For example, a numerical method for solving $F(x) = 0$ may produce a sequence $x^\nu \to \bar{x}$ with $\|F(x^\nu)\|_2 \to 0$. One would then hope that $F(\bar{x}) = 0$, but this might not be the case if $F$ isn't continuous; see Figure 7.5(left) where $F(x) = x$ for $x > 0$ and $F(x) = 1$ otherwise. The difficulty extends to the generalized equation $\bar{v} \in S(x)$. One might have

$$x^\nu \to \bar{x} \quad \text{with} \quad \operatorname{dist}\big(\bar{v}, S(x^\nu)\big) \to 0, \quad \text{but} \quad \bar{v} \notin S(\bar{x})$$

as seen in Figure 7.5(right). We'll now define what it means for a set-valued mapping to be (semi)continuous, which in turn leads to safeguards against the unfortunate cases in the figure. We'll also study the stability of feasible sets under perturbations.

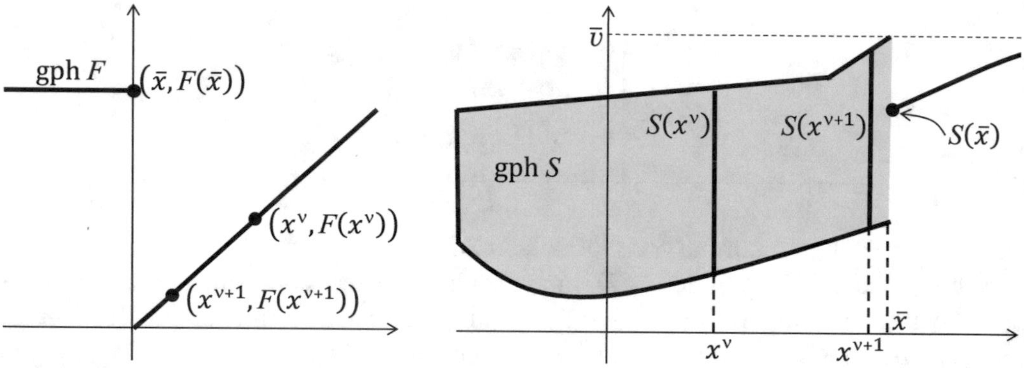


Fig. 7.5: A mapping $F : \mathbb{R} \to \mathbb{R}$ that isn't continuous at $\bar{x}$ (left); a set-valued mapping $S$ with $\bar{v} \notin S(\bar{x})$ despite $x^\nu \to \bar{x}$ and $\operatorname{dist}(\bar{v}, S(x^\nu)) \to 0$ (right).

For a set-valued mapping $S : \mathbb{R}^n \rightrightarrows \mathbb{R}^m$ and $C \subset \mathbb{R}^n$, $S$ is *outer semicontinuous* (osc) at $\bar{x} \in C$ relative to $C$ if

$$\bigcup_{x^\nu \in C \to \bar{x}} \operatorname{LimOut} S(x^\nu) \subset S(\bar{x}).$$

The mapping is osc relative to $C$ if this holds for all $\bar{x} \in C$. We omit "relative to $C$" when $C = \mathbb{R}^n$. In Figure 7.5(right), $S$ is osc at all points but $\bar{x}$. At $\bar{x}$, $S$ is osc relative to $[\bar{x}, \infty)$ because when $x$ approaches $\bar{x}$ from the right, $S(x)$ changes gradually. In contrast, when $x$ approaches $\bar{x}$ from the left, $S(x)$ changes abruptly from being a sizeable line segment to a single point $S(\bar{x})$.

**Proposition 7.22** (osc of normal cone and subgradient mappings). *For $C \subset \mathbb{R}^n$, the normal cone mapping $N_C : \mathbb{R}^n \rightrightarrows \mathbb{R}^n$ is osc relative to $C$ and, in fact, osc if $C$ is closed.*

*For $f : \mathbb{R}^n \to \overline{\mathbb{R}}$ and a point $\bar{x}$ where $f$ is finite, the subgradient mapping $\partial f : \mathbb{R}^n \rightrightarrows \mathbb{R}^n$ is osc at $\bar{x}$ relative to any set $C$ containing $\bar{x}$ and satisfying the property: $x^\nu \in C \to \bar{x}$ implies $f(x^\nu) \to f(\bar{x})$.*

**Proof.** These facts translate 4.39 and 4.74 into the terminology of osc. □

If $f$ is continuous at $\bar{x}$, then $C$ can be $\mathbb{R}^n$ in the statement about subgradient mappings. However, the condition is needed in general.

**Example 7.23** (lack of osc for subgradient mappings). Consider the function given by

$$f(x) = \begin{cases} x+1 & \text{if } x < 0 \\ \frac{1}{2}x & \text{otherwise} \end{cases}$$

as seen in Figure 7.6(left). Its subgradient mapping isn't osc at 0.

**Detail.** In this case, the subgradient mapping of $f$ takes the values

$$\partial f(x) = \begin{cases} \{1\} & \text{if } x < 0 \\ (-\infty, 1/2] & \text{if } x = 0 \\ \{1/2\} & \text{otherwise,} \end{cases}$$

which are illustrated in Figure 7.6(right). If $x^\nu \searrow 0$, then

$$\operatorname{LimOut} \partial f(x^\nu) = \{1/2\} \subset \partial f(0) = (-\infty, 1/2],$$

which means that $\partial f$ is osc relative to $[0, \infty)$. However, if $x^\nu \nearrow 0$, then $\operatorname{LimOut} \partial f(x^\nu) = \{1\}$ isn't contained in $\partial f(0)$ and $\partial f$ fails to be osc. This latter possibility is ruled out in 7.22 because $f(x^\nu)$ doesn't converge to $f(0)$. □

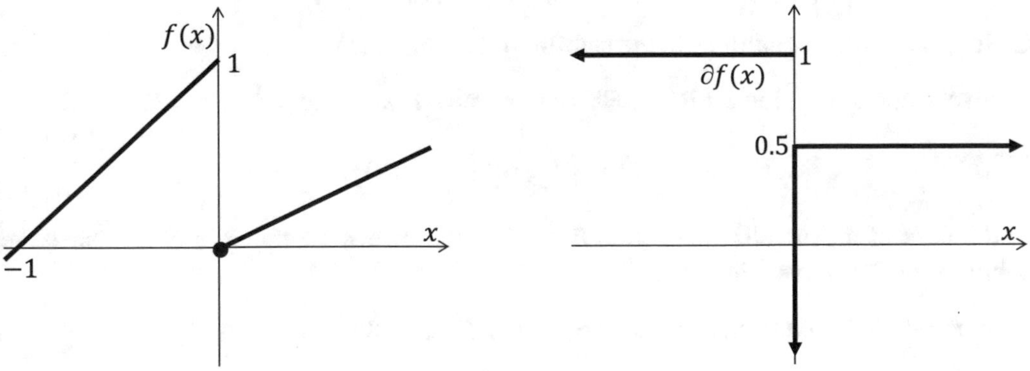


Fig. 7.6: A subgradient mapping that isn't osc at $x = 0$.

Our initial algorithmic concern is put to rest when $S : \mathbb{R}^n \rightrightarrows \mathbb{R}^m$ is osc. Approximating solutions $x^\nu \to \bar{x}$ with $\operatorname{dist}(\bar{v}, S(x^\nu)) \to 0$ must be accompanied by $v^\nu \in S(x^\nu) \to \bar{v}$. Thus, $\bar{v} \in \operatorname{LimOut} S(x^\nu)$ by the definition of outer limits. Since $S$ is osc, $\operatorname{LimOut} S(x^\nu)$ is contained in $S(\bar{x})$ and this in turn implies that $\bar{v} \in S(\bar{x})$.

It's quite common in applications for a set-valued mapping to be osc. For example, the stability theorem 5.6(c) provides conditions under which the set-valued mapping $P : \mathbb{R}^m \rightrightarrows \mathbb{R}^n$ given by $P(u) = \operatorname{argmin} f(u, \cdot)$ is osc. Let's examine some more illustrations.

**Example 7.24** (maximizers furnishing sup-projection value). For a continuous function $g : \mathbb{R}^n \times \mathbb{R}^m \to \mathbb{R}$, a nonempty compact set $Y \subset \mathbb{R}^m$ and the sup-projection given by

$$f(x) = \sup_{y \in Y} g(x, y),$$

one has that the set-valued mapping $Y^\star : \mathbb{R}^n \rightrightarrows \mathbb{R}^m$, with

$$Y^\star(x) = \operatorname{argmax}_{y \in Y} g(x, y),$$

is osc and $\operatorname{dom} Y^\star = \mathbb{R}^n$.

**Detail.** Our goal is to leverage the stability theorem 5.6(c). Let $\tilde{g} : \mathbb{R}^n \times \mathbb{R}^m \to \overline{\mathbb{R}}$ be given by

$$\tilde{g}(x, y) = -g(x, y) + \iota_Y(y).$$

Then, $\operatorname{argmin} \tilde{g}(x, \cdot) = Y^\star(x)$. Let $\bar{x} \in \mathbb{R}^n$. Since $g$ is real-valued and continuous and $Y$ is nonempty and compact, $\tilde{g}$ is proper and lsc. Moreover, $\operatorname{argmin} \tilde{g}(\bar{x}, \cdot)$ is nonempty by 4.9 because $Y$ is compact and $g(\bar{x}, \cdot)$ is continuous. The compactness of $Y$ also ensures that $\tilde{g}$ is level-bounded in $y$ locally uniformly in $x$. Since $\tilde{g}(\cdot, y)$ is continuous for every $y$, all the assumptions of the stability theorem 5.6(c) hold and one can conclude that $Y^\star$ is osc at $\bar{x}$. In fact, $Y^\star$ is osc and $\operatorname{dom} Y^\star = \mathbb{R}^n$ because $\bar{x}$ is arbitrary. □

**Exercise 7.25** (osc of projection). For a nonempty closed set $C \subset \mathbb{R}^n$, show that the set-valued mapping $S : \mathbb{R}^n \rightrightarrows \mathbb{R}^n$ given by $S(x) = \operatorname{prj}_C(x)$ is osc.

**Guide.** Verify the conditions in the stability theorem 5.6(c). □

A set-valued mapping $S : \mathbb{R}^n \rightrightarrows \mathbb{R}^m$ has an *inverse mapping* $S^{-1} : \mathbb{R}^m \rightrightarrows \mathbb{R}^n$ with

$$S^{-1}(v) = \{x \in \mathbb{R}^n \mid v \in S(x)\}.$$

In the case of a generalized equation $\bar{v} \in S(x)$, the inverse mapping of $S$ furnishes the solution set $S^{-1}(\bar{v})$; see Figure 7.1.

**Proposition 7.26** (characterization of osc). *For $S : \mathbb{R}^n \rightrightarrows \mathbb{R}^m$, one has*

$$\operatorname{gph} S \text{ is closed} \iff S \text{ is osc} \iff S^{-1} \text{ is osc.}$$

**Proof.** Since $\operatorname{gph} S$ is a subset of $\mathbb{R}^n \times \mathbb{R}^m$, its closedness means that

$$(x^\nu, v^\nu) \in \operatorname{gph} S \to (\bar{x}, \bar{v}) \implies (\bar{x}, \bar{v}) \in \operatorname{gph} S.$$

But, this is equivalent to

$$x^\nu \to \bar{x}, \ v^\nu \to \bar{v}, \ v^\nu \in S(x^\nu) \implies \bar{v} \in S(\bar{x}),$$

which is guaranteed by osc. The converse holds similarly. The statement about the inverse mapping $S^{-1}$ follows because $v \in S(x)$ if and only if $x \in S^{-1}(v)$. □

For an osc set-valued mapping $S : \mathbb{R}^n \rightrightarrows \mathbb{R}^m$, the proposition has the consequence that every solution of the generalized equation $v^\nu \in S(x)$ is near a solution to $\bar{v} \in S(x)$ when $v^\nu$ is near $\bar{v}$. Formally,

$$\text{LimOut}\, S^{-1}(v^\nu) \subset S^{-1}(\bar{v})$$

when $v^\nu \to \bar{v}$. This provides a certain stability to the solution sets of generalized equations expressed by $S$. After all, the exact value of the parameter vector $\bar{v}$ might not be known in practice and this stability property provides an assurance that the generalized equation is suitable, at least in this sense, to address an application; see the general discussion at the beginning of Chap. 5.

**Example 7.27** (osc of feasible-set mapping). For closed $X \subset \mathbb{R}^n$, continuous $f_i : \mathbb{R}^d \times \mathbb{R}^n \to \mathbb{R}$, $i = 1, \dots, m$, and continuous $g_i : \mathbb{R}^d \times \mathbb{R}^n \to \mathbb{R}$, $i = 1, \dots, q$, consider the set-valued mapping $S : \mathbb{R}^d \rightrightarrows \mathbb{R}^n$ given by

$$S(u) = \big\{x \in X \,\big|\, f_i(u, x) = 0,\ i = 1, \dots, m;\ \ g_i(u, x) \le 0,\ i = 1, \dots, q\big\},$$

which is of interest when we study stability properties of feasible sets. It turns out that $S$ is osc.

**Detail.** In view of 7.26, it suffices to show the closedness of the graph

$$\begin{aligned} \text{gph}\, S &= \big\{(u, x) \in \mathbb{R}^d \times \mathbb{R}^n \,\big|\, x \in S(u)\big\} \\ &= \bigcap_{i=1}^{m} \big\{(u, x) \,\big|\, f_i(u, x) = 0\big\} \bigcap_{i=1}^{q} \big\{(u, x) \,\big|\, g_i(u, x) \le 0\big\} \bigcap \big(\mathbb{R}^d \times X\big). \end{aligned}$$

Since $X$ is closed and the functions involved are continuous, this is an intersection of closed sets and is therefore closed. □

For a set-valued mapping $S : \mathbb{R}^n \rightrightarrows \mathbb{R}^m$ and $C \subset \mathbb{R}^n$, we say that $S$ is *inner semicontinuous* (isc) at $\bar{x} \in C$ relative to $C$ if

$$\bigcap_{x^\nu \in C \to \bar{x}} \text{LimInn}\, S(x^\nu) \supset S(\bar{x}).$$

The mapping is isc relative to $C$ if this holds for all $\bar{x} \in C$. We omit "relative to $C$" when $C = \mathbb{R}^n$. Moreover, $S$ is *continuous* (at $\bar{x}$, relative to $C$) if it's both osc and isc (at $\bar{x}$, relative to $C$). This means that

$$S(x^\nu) \xrightarrow{s} S(\bar{x}) \quad \text{when } S \text{ is continuous at } \bar{x} \text{ and } x^\nu \to \bar{x}.$$

Figure 7.5(right) illustrates a situation where $S$ is isc at $\bar{x}$. In applications, isc occurs less frequently than osc. For example, the set-valued mapping with $S(u) = \{x \in \mathbb{R} \mid x^3 - x^2 - x + 1 + u \le 0\}$ has $S(0) = (-\infty, -1] \cup \{1\}$, but the point $\{1\}$ "disappears" from $S(u)$ for $u > 0$; see Figure 4.6. Thus, $S$ is osc at 0 (as guaranteed by 7.27) but not isc.

The situation improves in the presence of convexity. If $S : \mathbb{R}^n \rightrightarrows \mathbb{R}^m$ has a convex graph, then $S$ is isc at any point $x \in \operatorname{int}(\operatorname{dom} S)$; see [105, Theorem 5.9b]. For nonempty, closed and convex $C \subset \mathbb{R}^n$, we also see from 7.14 that $\operatorname{prj}_C$ is continuous and then also isc.

**Example 7.28** (convex feasible-set mapping). For continuous $g_i : \mathbb{R}^d \times \mathbb{R}^n \to \mathbb{R}$, $i = 1, \dots, q$, with $g_i(u, \cdot)$ convex for all $u \in \mathbb{R}^d$, let's consider the set-valued mapping $S : \mathbb{R}^d \rightrightarrows \mathbb{R}^n$, where

$$S(u) = \{x \in \mathbb{R}^n \mid g_i(u, x) \le 0,\ i = 1, \dots, q\}.$$

At a point $\bar{u}$, suppose that there's $\bar{x}$ such that $g_i(\bar{u}, \bar{x}) < 0$ for $i = 1, \dots, q$. Then, $S$ is continuous at $\bar{u}$.

**Detail.** A proof of this fact is sketched out in [105, Exercise 5.10]. We here only note the importance of having a point $\bar{x}$ such that $g_i(\bar{u}, \bar{x}) < 0$; see also the closely related Slater condition in 5.47.

For $q = 1$ and $g_1(u, x) = x^2 + u$, we've

$$S(u) = \begin{cases} \emptyset & \text{if } u > 0 \\ \left[-\sqrt{-u}, \sqrt{-u}\,\right] & \text{otherwise} \end{cases}$$

and $S$ isn't isc at 0 because $\operatorname{LimInn} S(u^\nu) = \emptyset$ doesn't contain $S(0) = \{0\}$ for $u^\nu \searrow 0$. Certainly, with $\bar{u} = 0$, $g_1(\bar{u}, \bar{x}) < 0$ doesn't hold for any $\bar{x}$. □

Since we often face sums of set-valued mappings, such as $x \mapsto F(x) + N_C(x)$, it's useful to have a rule about preservation of (semi)continuity.

**Proposition 7.29** (continuity of sums of set-valued mappings). *For $S, T : \mathbb{R}^n \rightrightarrows \mathbb{R}^m$ and $C \subset \mathbb{R}^n$, suppose that $S$ is continuous at $\bar{x} \in C$ relative to $C$ and there's $\varepsilon > 0$ such that $\{S(x),\ x \in C \cap \mathbb{B}(\bar{x}, \varepsilon)\}$ are contained in a bounded set. If $T$ is osc, isc or continuous at $\bar{x}$ relative to $C$, the same property holds for $(S + T) : \mathbb{R}^n \rightrightarrows \mathbb{R}^m$.*

**Proof.** Suppose that $T$ is osc at $\bar{x}$ relative to $C$, $x^\nu \in C \to \bar{x}$ and

$$\bar{v} \in \operatorname{LimOut}\big(S(x^\nu) + T(x^\nu)\big).$$

For $S + T$ to be osc at $\bar{x}$ relative to $C$, it suffices to show that $\bar{v} \in S(\bar{x}) + T(\bar{x})$. By the definition of outer limits, there are $N \in \mathcal{N}_\infty^\#$ and

$$v^\nu \in S(x^\nu) + T(x^\nu)$$

such that $v^\nu \to_N \bar{v}$. We must then also have $u^\nu \in S(x^\nu)$ and $w^\nu \in T(x^\nu)$ with $v^\nu = u^\nu + w^\nu$. Since eventually $x^\nu \in \mathbb{B}(\bar{x}, \varepsilon)$,

$$\bigcup_{\nu \in N} S(x^\nu)$$

is contained in a bounded set. Thus, $\{u^\nu, \nu \in N\}$ has a cluster point $\bar{u}$ so that, after passing to a subsequence which we also denote by $N$, $u^\nu \to_N \bar{u}$. We already have $v^\nu \to_N \bar{v}$ so then

$w^\nu \xrightarrow[N]{} \bar{v} - \bar{u}$. Since both $S$ and $T$ are osc at $\bar{x}$ relative to $C$, $\bar{u} \in S(\bar{x})$ and $\bar{v} - \bar{u} \in T(\bar{x})$. This means that

$$\bar{v} = \bar{u} + (\bar{v} - \bar{u}) \in S(\bar{x}) + T(\bar{x}).$$

The other assertions follow by similar arguments. □

For $C \subset \mathbb{R}^n$, the set-valued mapping given by $F + N_C$ is osc relative to $C$ if $F : \mathbb{R}^n \to \mathbb{R}^n$ is continuous as seen by the proposition and 7.22. Consequently, a numerical procedure that generates $\{x^\nu \in C, \nu \in \mathbb{N}\}$, with

$$\operatorname{dist}\big(0,\ F(x^\nu) + N_C(x^\nu)\big) \to 0,$$

would have $0 \in F(\bar{x}) + N_C(\bar{x})$ for any cluster point $\bar{x} \in C$, which might be exactly what we seek to establish.

## 7.H Graphical Approximation Algorithm

Applications often give rise to generalized equations that can't be solved by existing algorithms, at least not in reasonable time, and it becomes necessary to consider approximations. Approximations may also be introduced artificially to study the effect of perturbations in a problem as part of a sensitivity analysis. Regardless of the circumstances, we would like to examine whether replacing a set-valued mapping by an alternative one results in significantly different solutions of the corresponding generalized equations.

**Example 7.30** (homotopy method). As a simple illustration, let's consider the problem of finding a solution to the equation

$$F(x) = x + \sin x + 1 = 0,$$

which may cause difficulties for the Newton-Raphson algorithm (§1.G) because the Jacobian $\nabla F(x) = 1 + \cos x$ isn't invertible when $x = \pi + 2\pi\nu$, $\nu \in \{\dots, -2, -1, 0, 1, 2, \dots\}$ and the algorithm breaks down at these points. However, solving the alternative equation

$$F_\lambda(x) = (1 - \lambda)(x + \sin x + 1) + \lambda x = 0$$

coincides with the actual problem when $\lambda = 0$ and for $\lambda \in (0, 1]$ appears to be a reasonable approximation that offers some advantages.

**Detail.** For $\lambda \in (0, 1]$, the Jacobian

$$\nabla F_\lambda(x) = 1 + (1 - \lambda)\cos x \geq \lambda > 0$$

and thus is invertible for all $x$. The Newton-Raphson algorithm tends to perform more reliably on the alternative problem than on the actual one. Moreover, for $\lambda = 1$, we see immediately that 0 solves $F_\lambda(x) = 0$. We may then adopt the following procedure: Given $\{\lambda^\nu \in (0, 1), \varepsilon^\nu > 0, \nu \in \mathbb{N}\}$ with $\lambda^\nu > \lambda^{\nu+1}$ and $\varepsilon^\nu > \varepsilon^{\nu+1}$, set $x^0 = 0$ (the solution of $F_1(x) = 0$) and, for $\nu \in \mathbb{N}$, apply the Newton-Raphson algorithm to

$$F_{\lambda^\nu}(x) = 0,$$

starting from $x^{\nu-1}$, until a point $x^\nu$ with $|F_{\lambda^\nu}(x^\nu)| \le \varepsilon^\nu$ is found. This procedure generates $\{x^\nu, \nu \in \mathbb{N}\}$. One can expect to compute $x^\nu$ quickly as the Newton-Raphson algorithm benefits from the starting point $x^{\nu-1}$, which we hope is close to $x^\nu$ when $\lambda^{\nu-1}$ is close to $\lambda^\nu$. Such "warm-starts" can be highly beneficial; we know from §1.G that the Newton-Raphson algorithm is fast if started close to a solution. A procedure that solves a sequence of problems in this manner is a *homotopy method*.

Under the assumption that $\lambda^\nu, \varepsilon^\nu \to 0$, does this procedure work in the sense that the generated sequence $\{x^\nu, \nu \in \mathbb{N}\}$ converges to a solution of the actual problem? After all, we're approximately solving a sequence of approximating problems. We'll be able to answer this in the affirmative even for quite general problems in this section. □

For $\bar{v} \in \mathbb{R}^m$ and $S : \mathbb{R}^n \rightrightarrows \mathbb{R}^m$, we seek to solve the generalized equation

$$\bar{v} \in S(x)$$

by considering alternative set-valued mappings $S^\nu : \mathbb{R}^n \rightrightarrows \mathbb{R}^m$ that presumably are more tractable or offer some other advantage. This basic idea leads to a broadly applicable algorithmic framework.

**Graphical Approximation Algorithm.**

Data. $S^\nu : \mathbb{R}^n \rightrightarrows \mathbb{R}^m$ and $\varepsilon^\nu \ge 0, \nu \in \mathbb{N}$.
Step 0. Set $\nu = 1$.
Step 1. Compute $x^\nu \in \mathbb{R}^n$ such that

$$\mathrm{dist}\big(\bar{v}, S^\nu(x^\nu)\big) \le \varepsilon^\nu.$$

Step 2. Replace $\nu$ by $\nu + 1$ and go to Step 1.

**Convergence 7.31** (graphical approximation algorithm). *For $S : \mathbb{R}^n \rightrightarrows \mathbb{R}^m$ and $\bar{v} \in \mathbb{R}^m$, suppose that $\{x^\nu, \nu \in \mathbb{N}\}$ is generated by the graphical approximation algorithm using $S^\nu : \mathbb{R}^n \rightrightarrows \mathbb{R}^m$ with*

$$\mathrm{LimOut}(\mathrm{gph}\, S^\nu) \subset \mathrm{gph}\, S.$$

*If $\varepsilon^\nu \to 0$, then every cluster point of $\{x^\nu, \nu \in \mathbb{N}\}$ solves $\bar{v} \in S(x)$.*

**Proof.** It follows from Step 1 that there's $v^\nu \in S^\nu(x^\nu)$ with $\|v^\nu - \bar{v}\|_2 \le \varepsilon^\nu + \nu^{-1}$. Since $\varepsilon^\nu \to 0$, this implies that $v^\nu \to \bar{v}$. Suppose that $\bar{x}$ is a cluster point of $\{x^\nu, \nu \in \mathbb{N}\}$. It follows by the definition of outer limits that

$$(\bar{x}, \bar{v}) \in \mathrm{LimOut}(\mathrm{gph}\, S^\nu)$$

and then also that $(\bar{x}, \bar{v}) \in \mathrm{gph}\, S$ as asserted. □

Figure 7.7 shows the graphs of the set-valued mappings $S_\lambda : \mathbb{R} \rightrightarrows \mathbb{R}$, $\lambda = 0, 0.1, 0.5, 1$, produced in 7.30 by setting

$$S_\lambda(x) = \{F_\lambda(x)\},$$

where in fact the mappings always return a single point. Visually, it appears that $\operatorname{gph} S_\lambda$ approaches $\operatorname{gph} S_0$ as $\lambda$ decreases and solutions of $0 \in S_\lambda(x)$ tend to that of $0 \in S_0(x)$. We can confirm this for any $\lambda^\nu \searrow 0$ using 7.31: If $(\bar{x}, \bar{v}) \in \operatorname{LimOut}(\operatorname{gph} S_{\lambda^\nu})$, then there are $N \in \mathcal{N}_\infty^\#$ and $x^\nu \to_N \bar{x}$ such that $F_{\lambda^\nu}(x^\nu) \to_N \bar{v}$. We would like to show that $(\bar{x}, \bar{v}) \in \operatorname{gph} S_0$ or, equivalently, $F_0(\bar{x}) = \bar{v}$, because then

$$\operatorname{LimOut}(\operatorname{gph} S_{\lambda^\nu}) \subset \operatorname{gph} S_0$$

and 7.31 applies. For this purpose,

$$|F_0(\bar{x}) - \bar{v}| \le |F_0(\bar{x}) - F_{\lambda^\nu}(x^\nu)| + |F_{\lambda^\nu}(x^\nu) - \bar{v}|.$$

The second term on the right-hand side vanishes and the first term equals

$$\left|\bar{x} + \sin\bar{x} + 1 - (1-\lambda^\nu)(x^\nu + \sin x^\nu + 1) - \lambda^\nu x^\nu\right| \to_N 0$$

and indeed $F_0(\bar{x}) = \bar{v}$. The alternative problems solved in 7.30 are justified in the sense that every cluster point of the generated sequence solves the actual problem.

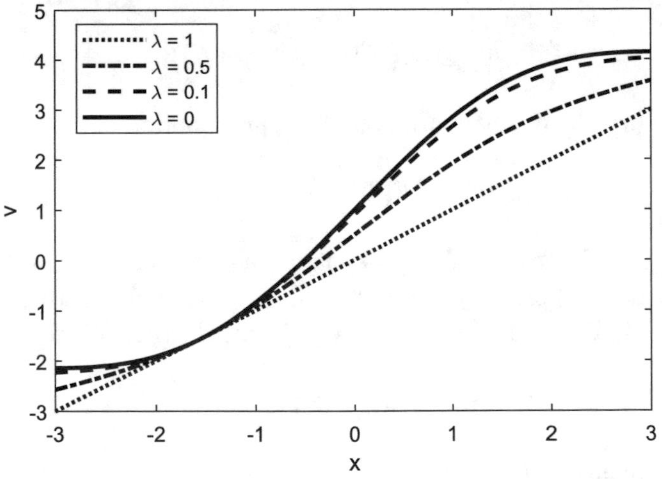


Fig. 7.7: Graphs of set-valued mappings given by $S_\lambda(x) = \{F_\lambda(x)\}$ in 7.30.

**Example 7.32** (smoothing algorithm for complementarity problems). For a smooth mapping $F : \mathbb{R}^n \to \mathbb{R}^n$, the complementarity problem

$$x \ge 0, \quad F(x) \ge 0, \quad \langle F(x), x\rangle = 0$$

corresponds to the generalized equation $0 \in F(x) + N_C(x)$, with $C = [0, \infty)^n$. Consequently, in view of the discussion surrounding the normal map in (7.18), the problem can in turn be expressed as

$$0 \in S(z) = \{F(\operatorname{prj}_C(z)) + z - \operatorname{prj}_C(z)\}.$$

Since the projection mapping $\operatorname{prj}_C$ is nonsmooth, this equation can't be solved by the usual Newton-Raphson algorithm (§1.G). However, a smooth approximation can be brought in and justified using 7.31; see also [86] for related approaches.

**Detail.** In this case, the projection

$$\operatorname{prj}_C(z) = (\max\{0, z_1\}, \ldots, \max\{0, z_n\})$$

has components that can be approximated in a smooth manner; see 4.16. Specifically, let

$$\varphi^\nu(\alpha) = \frac{1}{\theta^\nu} \ln\left(1 + e^{\alpha\theta^\nu}\right),$$

where $\theta^\nu \in (0, \infty)$ and

$$0 \le \varphi^\nu(\alpha) - \max\{0, \alpha\} \le \frac{\ln 2}{\theta^\nu} \quad \forall \alpha \in \mathbb{R}. \tag{7.22}$$

The projection mapping can be approximated by the smooth mapping $\Phi^\nu : \mathbb{R}^n \to \mathbb{R}^n$ with

$$\Phi^\nu(z) = (\varphi^\nu(z_1), \ldots, \varphi^\nu(z_n)).$$

A substitution of $\operatorname{prj}_C$ by $\Phi^\nu$ then results in the approximation

$$S^\nu(z) = \{F(\Phi^\nu(z)) + z - \Phi^\nu(z)\},$$

which we employ in the graphical approximation algorithm. There, we solve $0 \in S^\nu(z)$ approximately using successively larger $\theta^\nu$. This requires only root-finding of a smooth mapping in Step 1. By 7.31, this is indeed a justified approach for solving $0 \in S(z)$ provided that

$$\operatorname{LimOut}(\operatorname{gph} S^\nu) \subset \operatorname{gph} S.$$

To establish this inclusion, let

$$(\bar{z}, \bar{v}) \in \operatorname{LimOut}(\operatorname{gph} S^\nu)$$

so there are $N \in \mathcal{N}_\infty^\#$, $z^\nu \to_N \bar{z}$ and $v^\nu \to_N \bar{v}$ such that $v^\nu \in S^\nu(z^\nu)$ for $\nu \in N$. Then,

$$\begin{aligned}&\big\|F\big(\operatorname{prj}_C(\bar z)\big) + \bar z - \operatorname{prj}_C(\bar z) - \bar v\big\|_2\\ &\le \big\|F\big(\operatorname{prj}_C(\bar z)\big) + \bar z - \operatorname{prj}_C(\bar z) - F\big(\Phi^\nu(z^\nu)\big) - z^\nu + \Phi^\nu(z^\nu)\big\|_2\\ &\qquad + \big\|F\big(\Phi^\nu(z^\nu)\big) + z^\nu - \Phi^\nu(z^\nu) - \bar v\big\|_2.\end{aligned}$$

The second term on the right-hand side vanishes because

$$F\big(\Phi^\nu(z^\nu)\big) + z^\nu - \Phi^\nu(z^\nu) = v^\nu \underset{N}{\to} \bar v.$$

The first term is bounded from above by

$$\big\|F\big(\operatorname{prj}_C(\bar z)\big) - F\big(\Phi^\nu(z^\nu)\big)\big\|_2 + \|\bar z - z^\nu\|_2 + \big\|\operatorname{prj}_C(\bar z) - \Phi^\nu(z^\nu)\big\|_2,$$

which vanishes because $\Phi^\nu(z^\nu) \underset{N}{\to} \operatorname{prj}_C(\bar z)$ as seen from (7.22). Consequently,

$$F\big(\operatorname{prj}_C(\bar z)\big) + \bar z - \operatorname{prj}_C(\bar z) = \bar v$$

and $(\bar z, \bar v) \in \operatorname{gph} S$. This establishes the desired inclusion. □

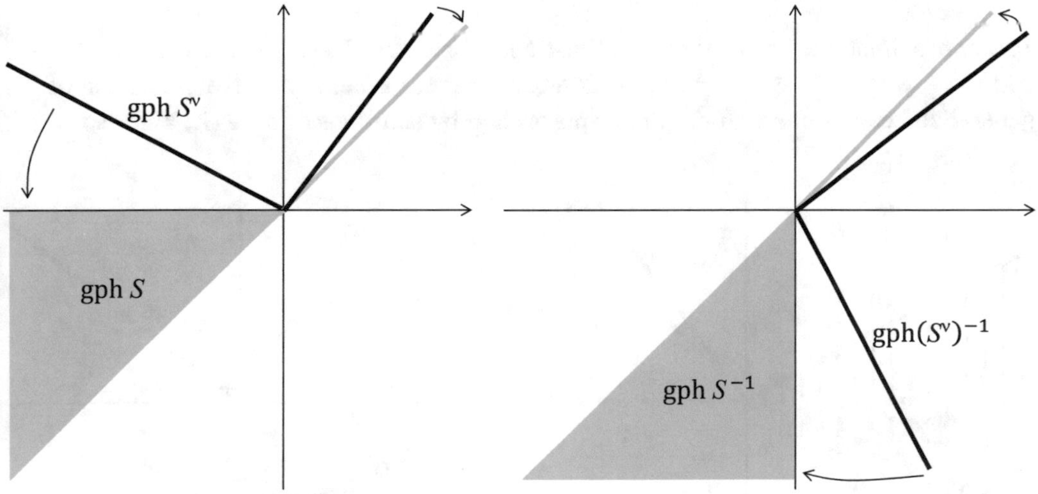


Fig. 7.8: Approximations satisfying LimOut(gph $S^\nu$) ⊂ gph $S$ (left) and their inverse mappings (right).

While there are favorable consequences from merely having the outer limit of gph $S^\nu$ being contained in gph $S$ as seen in 7.31, the approximations are somewhat fragile if

$$\operatorname{LimInn}(\operatorname{gph} S^\nu) \supset \operatorname{gph} S \ \text{ doesn't hold.}$$

Figure 7.8(left) shows gph $S^\nu$, shaped like a "V," with LimOut(gph $S^\nu$) ⊂ gph $S$. However, LimInn(gph $S^\nu$) ⊃ gph $S$ fails because gph $S^\nu$ never gets close to large portions of gph $S$ in the third quadrant. The right portion of the figure shows the graphs of the corresponding inverse mappings. We see that

$$(S^\nu)^{-1}(0) = \{0\}$$

coincides with *one* point in

$$S^{-1}(0) = (-\infty, 0],$$

but isn't close to the other ones. If the actual generalized equation is modified slightly from

$$0 \in S(x) \quad \text{to} \quad -\delta \in S(x)$$

for some small $\delta > 0$, then the situation changes for the worse because $S^{-1}(-\delta)$ is very different from $(S^\nu)^{-1}(-\delta) = \emptyset$.

In contrast, Figure 7.9 shows a situation with $\text{LimOut}(\text{gph}\, S^\nu) \subset \text{gph}\, S$ being supplemented by

$$\text{LimInn}(\text{gph}\, S^\nu) \supset \text{gph}\, S$$

so that $\text{gph}\, S^\nu \xrightarrow{s} \text{gph}\, S$. Then, in fact we've

$$\text{gph}(S^\nu)^{-1} \xrightarrow{s} \text{gph}\, S^{-1}.$$

The approximations are now more robust and, for every $\bar{x} \in S^{-1}(\bar{v})$, there are $x^\nu \to \bar{x}$ and $v^\nu \to \bar{v}$ with $x^\nu \in (S^\nu)^{-1}(v^\nu)$. This means that regardless of $\bar{v}$, every solution of the generalized equation $\bar{v} \in S(x)$ can be approached by solutions of $v^\nu \in S^\nu(x)$ for some $v^\nu$.

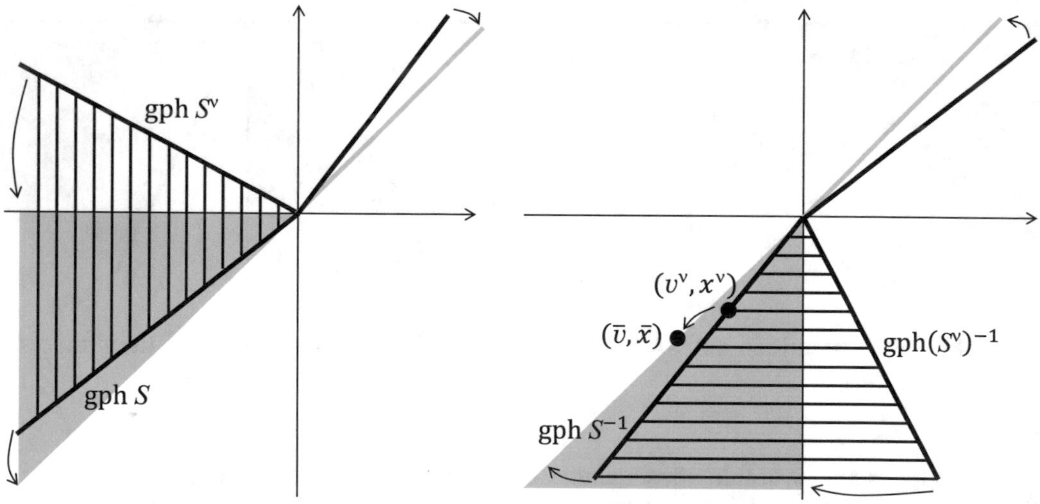


Fig. 7.9: Approximations $\text{gph}\, S^\nu$ set-converging to $\text{gph}\, S$ (left) and corresponding inverse mappings also set-converge (right).

**Definition 7.33** (graphical convergence). For set-valued mappings $S^\nu, S : \mathbb{R}^n \rightrightarrows \mathbb{R}^m$, we say that $S^\nu$ *converges graphically* to $S$, written $S^\nu \xrightarrow{g} S$, when

$$\operatorname{gph} S^\nu \xrightarrow{s} \operatorname{gph} S.$$

Since $v \in S(x)$ if and only if $x \in S^{-1}(v)$, the property alluded to before the definition holds generally:

$$S^\nu \xrightarrow{g} S \quad \Longleftrightarrow \quad (S^\nu)^{-1} \xrightarrow{g} S^{-1}.$$

To put this concept to use in the context of generalized equation, let's adopt the following terminology: For $S : \mathbb{R}^n \rightrightarrows \mathbb{R}^m$, $\bar{v} \in \mathbb{R}^m$ and $\varepsilon \geq 0$, the *set of $\varepsilon$-solutions* to the generalized equation $\bar{v} \in S(x)$ is defined as

$$S^{-1}\big(\mathbb{B}(\bar{v}, \varepsilon)\big) = \bigcup_{v \in \mathbb{B}(\bar{v}, \varepsilon)} S^{-1}(v). \tag{7.23}$$

Thus, an $\varepsilon$-solution $\hat{x} \in S^{-1}(\mathbb{B}(\bar{v}, \varepsilon))$ has $\operatorname{dist}(\bar{v}, S(\hat{x})) \leq \varepsilon$.

**Theorem 7.34** (consequences of graphical convergence). *For set-valued mappings $S, S^\nu : \mathbb{R}^n \rightrightarrows \mathbb{R}^m$, suppose that $S^\nu \xrightarrow{g} S$ and $\bar{v} \in \mathbb{R}^m$. Then, the following hold:*

(a) $\forall \{\varepsilon^\nu \geq 0, \nu \in \mathbb{N}\} \to 0$,

$$\operatorname{LimOut}(S^\nu)^{-1}\big(\mathbb{B}(\bar{v}, \varepsilon^\nu)\big) \subset S^{-1}(\bar{v}).$$

(b) $\exists \{\varepsilon^\nu \geq 0, \nu \in \mathbb{N}\} \to 0$ *such that*

$$\operatorname{LimInn}(S^\nu)^{-1}\big(\mathbb{B}(\bar{v}, \varepsilon^\nu)\big) \supset S^{-1}(\bar{v}).$$

**Proof.** For (a), we only utilize

$$\operatorname{LimOut}\big(\operatorname{gph}(S^\nu)^{-1}\big) \subset \operatorname{gph} S^{-1}.$$

Suppose that

$$\bar{x} \in \operatorname{LimOut}(S^\nu)^{-1}\big(\mathbb{B}(\bar{v}, \varepsilon^\nu)\big).$$

Then, there exist $N \in \mathcal{N}_\infty^\#$ and $x^\nu \xrightarrow{N} \bar{x}$ such that

$$x^\nu \in (S^\nu)^{-1}\big(\mathbb{B}(\bar{v}, \varepsilon^\nu)\big)$$

and, for some $v^\nu \in \mathbb{B}(\bar{v}, \varepsilon^\nu)$,

$$(v^\nu, x^\nu) \in \operatorname{gph}(S^\nu)^{-1}.$$

Since cluster points of such sequences are contained in $\operatorname{gph} S^{-1}$, one has $\bar{x} \in S^{-1}(\bar{v})$.

For (b), we use

$$\operatorname{LimInn}\big(\operatorname{gph}(S^\nu)^{-1}\big) \supset \operatorname{gph} S^{-1}.$$

Let $\bar{x} \in S^{-1}(\bar{v})$. Then,

$$(\bar{v}, \bar{x}) \in \operatorname{LimInn}\big(\operatorname{gph}(S^\nu)^{-1}\big)$$

and there exist $v^\nu \to \bar{v}$ and $x^\nu \to \bar{x}$ such that $x^\nu \in (S^\nu)^{-1}(v^\nu)$. We seek to establish that

$$\bar{x} \in \operatorname{LimInn}(S^\nu)^{-1}\big(\mathbb{B}(\bar{v}, \varepsilon^\nu)\big)$$

for some vanishing $\varepsilon^\nu$ and need to construct $\bar{x}^\nu \to \bar{x}$, $\bar{v}^\nu \to \bar{v}$ and $\varepsilon^\nu \to 0$ such that

$$\bar{x}^\nu \in (S^\nu)^{-1}(\bar{v}^\nu) \quad \text{and} \quad \bar{v}^\nu \in \mathbb{B}(\bar{v}, \varepsilon^\nu).$$

We see that $\bar{x}^\nu = x^\nu$, $\bar{v}^\nu = v^\nu$ and $\varepsilon^\nu = \|v^\nu - \bar{v}\|_2$ suffice. □

Since (a) is essentially equivalent to 7.31, the main novelty of the theorem is in (b). Every solution of the actual generalized equation $\bar{v} \in S(x)$ can be approached by certain $\varepsilon$-solutions of the approximating generalized equations. This formalizes the discussion around Figure 7.9.

**Example 7.35** (projection mappings). For nonempty closed sets $C, C^\nu \subset \mathbb{R}^n$, one has

$$C^\nu \xrightarrow{s} C \implies \operatorname{prj}_{C^\nu} \xrightarrow{g} \operatorname{prj}_C .$$

Thus, 7.34 applies and we can conclude that the set of points projecting to $y^\nu \in C^\nu$ has outer limit contained in the set of points projecting to $y \in C$ when $y^\nu \to y$.

**Detail.** For $x^\nu \to \bar{x}$, let

$$f^\nu(x) = \|x - x^\nu\|_2 + \iota_{C^\nu}(x) \quad \text{and} \quad f(x) = \|x - \bar{x}\|_2 + \iota_C(x).$$

Since $\operatorname{epi} \iota_{C^\nu} = C^\nu \times [0, \infty)$ and $\operatorname{epi} \iota_C = C \times [0, \infty)$, it follows from 4.26 that $\iota_{C^\nu} \xrightarrow{e} \iota_C$. We also have that $\|y^\nu - x^\nu\|_2 \to \|\bar{y} - \bar{x}\|_2$ whenever $y^\nu \to \bar{y}$. By the epigraphical sum rule 4.19(a), $f^\nu \xrightarrow{e} f$ and then

$$\operatorname{LimOut}\big(\operatorname{prj}_{C^\nu}(x^\nu)\big) = \operatorname{LimOut}\big(\operatorname{argmin} f^\nu\big) \subset \operatorname{argmin} f = \operatorname{prj}_C(\bar{x})$$

as seen from 5.5(b). First, we show that

$$\operatorname{LimOut}\big(\operatorname{gph} \operatorname{prj}_{C^\nu}\big) \subset \operatorname{gph} \operatorname{prj}_C .$$

Let $(\hat{x}, \hat{v}) \in \operatorname{LimOut}(\operatorname{gph} \operatorname{prj}_{C^\nu})$. Then, there are $N \in \mathcal{N}_\infty^\#$, $\hat{x}^\nu \to_N \hat{x}$ and $\hat{v}^\nu \to \hat{v}$ such that $\hat{v}^\nu \in \operatorname{prj}_{C^\nu}(\hat{x}^\nu)$. From the fact just developed, $\hat{v} \in \operatorname{prj}_C(\hat{x})$. Thus, $(\hat{x}, \hat{v}) \in \operatorname{gph} \operatorname{prj}_C$.

Second, we show that

$$\operatorname{LimInn}\big(\operatorname{gph} \operatorname{prj}_{C^\nu}\big) \supset \operatorname{gph} \operatorname{prj}_C .$$

Let $(x^0, x^1) \in \operatorname{gph} \operatorname{prj}_C$, $\lambda \in (0, 1)$ and $x^\lambda = (1 - \lambda)x^0 + \lambda x^1$. Since $x^1 \in \operatorname{prj}_C(x^0)$, $\{x^1\} = \operatorname{prj}_C(x^\lambda)$. In view of the above epi-convergence and 5.5(b),

$$\operatorname{LimOut}\big(\operatorname{prj}_{C^\nu}(x^\lambda)\big) \subset \operatorname{prj}_C(x^\lambda).$$

Since $\mathrm{prj}_C(x^\lambda)$ consists of a single point, the left-hand side in this inclusion must either be that point or be empty. The latter is ruled out because $C^\nu \xrightarrow{s} C$ implies the existence of $\bar\nu$ such that $C^\nu \cap \mathbb{B}(x^1, 1) \neq \emptyset$ for all $\nu \geq \bar\nu$ and then

$$\bigcup\nolimits_{\nu\in\mathbb{N}} \mathrm{prj}_{C^\nu}(x^\lambda) \subset \mathbb{B}(0,\rho)$$

for some $\rho \in [0,\infty)$. To see this, let $\nu \geq \bar\nu$. If $\bar x^\nu \in \mathrm{prj}_{C^\nu}(x^\lambda)$, then

$$\|\bar x^\nu - x^\lambda\|_2 \leq \|x - x^\lambda\|_2 \quad \forall x \in C^\nu.$$

In particular, there's $\hat x^\nu \in C^\nu \cap \mathbb{B}(x^1, 1)$ such that

$$\|\bar x^\nu - x^\lambda\|_2 \leq \|\hat x^\nu - x^\lambda\|_2.$$

This implies that

$$\begin{aligned}\|\bar x^\nu\|_2 &\leq \|\bar x^\nu - x^\lambda\|_2 + \|x^\lambda\|_2 \leq \|\hat x^\nu - x^\lambda\|_2 + \|x^\lambda\|_2\\ &\leq \|\hat x^\nu - x^1\|_2 + \|x^1 - x^\lambda\|_2 + \|x^\lambda\|_2 \leq 1 + \|x^1 - x^\lambda\|_2 + \|x^\lambda\|_2\end{aligned}$$

and $\rho$ can be taken as the right-hand side.

We've established that

$$\mathrm{LimOut}\big(\mathrm{prj}_{C^\nu}(x^\lambda)\big) = \mathrm{prj}_C(x^\lambda) = \{x^1\}.$$

This fact, the boundedness of $\{\mathrm{prj}_{C^\nu}(x^\lambda), \nu \in \mathbb{N}\}$ and the general relation

$$\mathrm{LimInn}\big(\mathrm{prj}_{C^\nu}(x^\lambda)\big) \subset \mathrm{LimOut}\big(\mathrm{prj}_{C^\nu}(x^\lambda)\big)$$

imply that $\mathrm{prj}_{C^\nu}(x^\lambda) \xrightarrow{s} \{x^1\}$. Thus, there's $v^\nu \in \mathrm{prj}_{C^\nu}(x^\lambda) \to x^1$. This means that

$$(x^\lambda, x^1) \in \mathrm{LimInn}\big(\mathrm{gph}\, \mathrm{prj}_{C^\nu}\big).$$

Since $\mathrm{LimInn}(\mathrm{gph}\,\mathrm{prj}_{C^\nu})$ is closed and $\lambda \in (0,1)$ is arbitrary, we also have

$$(x^0, x^1) \in \mathrm{LimInn}\big(\mathrm{gph}\, \mathrm{prj}_{C^\nu}\big),$$

which completes the argument.

It turns out that the converse holds as well:

$$\mathrm{prj}_{C^\nu} \xrightarrow{g} \mathrm{prj}_C \implies C^\nu \xrightarrow{s} C,$$

but we refer to [105, Example 5.35] for a proof. □

**Exercise 7.36** (relations to other modes of convergence). For set-valued mappings $S, S^\nu : \mathbb{R}^n \rightrightarrows \mathbb{R}^m$, (a) show that $S^\nu(x) \xrightarrow{s} S(x)$ for all $x \in \mathbb{R}^n$ doesn't guarantee $S^\nu \xrightarrow{g} S$ by providing a counterexample; (b) show that $S^\nu \xrightarrow{g} S$ doesn't guarantee $S^\nu(x) \xrightarrow{s} S(x)$ for all

$x \in \mathbb{R}^n$ by providing a counterexample; and (c) show that $S^\nu(x^\nu) \xrightarrow{s} S(x)$ for all $x^\nu \to x$ implies $S^\nu \xrightarrow{g} S$.

Let's now consider some operations that preserve graphical convergence.

**Proposition 7.37** (graphical convergence under scalar multiplication). *If $S : \mathbb{R}^n \rightrightarrows \mathbb{R}^m$ is osc and $\lambda^\nu \to \lambda \neq 0$, then $(\lambda^\nu S) \xrightarrow{g} (\lambda S)$.*

**Proof.** Suppose that $(\bar{x}, \bar{v}) \in \text{LimOut}(\text{gph}\, \lambda^\nu S)$. Then, there are $N \in \mathcal{N}_\infty^\#$, $x^\nu \to_N \bar{x}$ and $w^\nu \in S(x^\nu)$ such that $\lambda^\nu w^\nu \to_N \bar{v}$. Since $\lambda^\nu \to \lambda \neq 0$, $\{w^\nu, \nu \in N\}$ is bounded. After passing to another subsequence, which we also denote by $N$, we've $w^\nu \to_N \bar{w}$ for some $\bar{w} \in \mathbb{R}^m$. Then, $\bar{w} \in S(\bar{x})$ because $S$ is osc and, in turn, $\lambda \bar{w} = \bar{v}$. This means that $\bar{v} \in \lambda S(\bar{x})$ and also $(\bar{x}, \bar{v}) \in \text{gph}\, \lambda S$. We've shown that

$$\text{LimOut}(\text{gph}\, \lambda^\nu S) \subset \text{gph}\, \lambda S.$$

Next, suppose that $(\bar{x}, \bar{v}) \in \text{gph}\, \lambda S$ so that $\bar{v} = \lambda \bar{w}$ for some $\bar{w} \in S(\bar{x})$. Set $x^\nu = \bar{x}$ and $v^\nu = \lambda^\nu \bar{w}$. Then, $x^\nu \to \bar{x}$, $v^\nu \to \bar{v}$ and $v^\nu \in \lambda^\nu S(x^\nu)$. Thus,

$$(\bar{x}, \bar{v}) \in \text{LimInn}(\text{gph}\, \lambda^\nu S).$$

This confirms $\text{LimInn}(\text{gph}\, \lambda^\nu S) \supset \text{gph}\, \lambda S$. □

**Proposition 7.38** (graphical convergence for sums). *For set-valued mappings $S, S^\nu, T, T^\nu : \mathbb{R}^n \rightrightarrows \mathbb{R}^m$, suppose that $S^\nu(x^\nu) \xrightarrow{s} S(x)$ when $x^\nu \to x$ and for each $\bar{x} \in \mathbb{R}^n$, there's $\varepsilon > 0$ and $\rho \in [0, \infty)$ such that*

$$\bigcup_{x \in \mathbb{B}(\bar{x}, \varepsilon), \nu \in \mathbb{N}} S^\nu(x) \subset \mathbb{B}(0, \rho).$$

*If $T^\nu \xrightarrow{g} T$, then $(S^\nu + T^\nu) \xrightarrow{g} (S + T)$.*

**Proof.** Suppose that

$$(\bar{x}, \bar{w}) \in \text{LimOut}\big(\text{gph}(S^\nu + T^\nu)\big).$$

Then, there are $N \in \mathcal{N}_\infty^\#$, $x^\nu \to_N \bar{x}$, $u^\nu \in S^\nu(x^\nu)$ and $v^\nu \in T^\nu(x^\nu)$ such that $u^\nu + v^\nu \to_N \bar{w}$. The boundedness assumption ensures that $\{u^\nu, \nu \in N\}$ is contained in a bounded set. Thus, after passing to another subsequence, which we also denote by $N$, we've $u^\nu \to_N \bar{u}$ for some $\bar{u} \in \mathbb{R}^m$. Since

$$\text{LimOut}_{\nu \in N}\, S^\nu(x^\nu) \subset S(\bar{x}),$$

we must have $\bar{u} \in S(\bar{x})$. Moreover,

$$(x^\nu, v^\nu) \in \text{gph}\, T^\nu \to_N (\bar{x}, \bar{w} - \bar{u}).$$

The graphical convergence $T^\nu \xrightarrow{g} T$ then implies that $(\bar{x}, \bar{w} - \bar{u}) \in \text{gph}\, T$. Thus, $\bar{w} \in S(\bar{x}) + T(\bar{x})$ and then also $(\bar{x}, \bar{w}) \in \text{gph}(T + S)$. We've shown that

$$\text{LimOut}\big(\text{gph}(S^\nu + T^\nu)\big) \subset \text{gph}(S + T).$$

Next, suppose that $(\bar{x}, \bar{w}) \in \operatorname{gph}(S+T)$. Then, there are $\bar{u} \in S(\bar{x})$ and $\bar{v} \in T(\bar{x})$ such that $\bar{w} = \bar{u} + \bar{v}$. Since $\operatorname{gph} T^\nu \xrightarrow{s} \operatorname{gph} T$, there are $x^\nu \to \bar{x}$ and $v^\nu \to \bar{v}$ such that $v^\nu \in T^\nu(x^\nu)$. We also have $S^\nu(x^\nu) \xrightarrow{s} S(\bar{x})$, which guarantees the existence of $u^\nu \in S^\nu(x^\nu) \to \bar{u}$. Consequently,

$$(x^\nu, u^\nu + v^\nu) \in \operatorname{gph}(S^\nu + T^\nu) \to (\bar{x}, \bar{u} + \bar{v}).$$

This means that

$$(\bar{x}, \bar{w}) \in \operatorname{LimInn}\big(\operatorname{gph}(S^\nu + T^\nu)\big).$$

We've established that $\operatorname{LimInn}(\operatorname{gph}(S^\nu + T^\nu)) \supset \operatorname{gph}(S+T)$. □

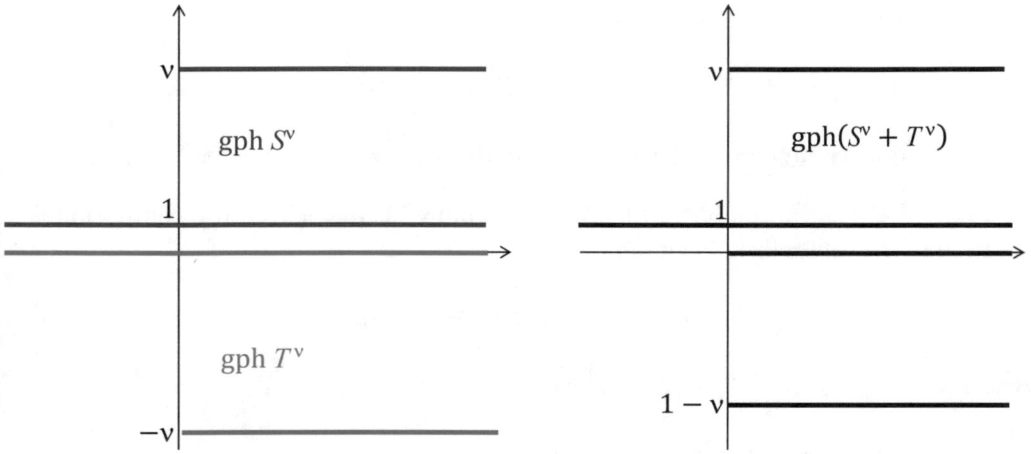


Fig. 7.10: Example of $\operatorname{Lim}(\operatorname{gph}(S^\nu + T^\nu)) \neq \operatorname{gph}(S+T)$.

The need for some assumption about boundedness in the proposition is highlighted by the example

$$S^\nu(x) = \begin{cases} \{1\} & \text{if } x < 0 \\ \{1, \nu\} & \text{otherwise} \end{cases} \qquad T^\nu(x) = \begin{cases} \{0\} & \text{if } x < 0 \\ \{0, -\nu\} & \text{otherwise;} \end{cases}$$

see Figure 7.10(left). Let $S(x) = \{1\}$ and $T(x) = \{0\}$ for all $x \in \mathbb{R}$. Then, $S^\nu(x^\nu) \xrightarrow{s} S(x)$ when $x^\nu \to x$ and $T^\nu \xrightarrow{g} T$. Figure 7.10(right) shows $\operatorname{gph}(S^\nu + T^\nu)$, where

$$S^\nu(x) + T^\nu(x) = \begin{cases} \{1\} & \text{if } x < 0 \\ \{0, 1, \nu, 1-\nu\} & \text{otherwise.} \end{cases}$$

Since $\operatorname{gph}(S+T)$ consists only of the horizontal line $\mathbb{R} \times \{1\}$, it's different than

$$\operatorname{Lim}\big(\operatorname{gph}(S^\nu + T^\nu)\big) = \big(\mathbb{R} \times \{1\}\big) \cup \big([0, \infty) \times \{0\}\big).$$

This takes place even though $S$ is continuous and $S(x)$ is bounded for each $x$. The boundedness assumption in 7.38 rules out cases of this kind.

**Example 7.39** (homotopy methods for generalized equations). Let's now extend 7.30. For an osc set-valued mapping $S : \mathbb{R}^n \rightrightarrows \mathbb{R}^n$ and $\bar{v} \in \mathbb{R}^n$, consider the generalized equation $\bar{v} \in S(x)$. A homotopy method defines a sequence of alternative problems in terms of $S^\nu : \mathbb{R}^n \rightrightarrows \mathbb{R}^n$, where

$$S^\nu(x) = (1 - \lambda^\nu)S(x) + \lambda^\nu x \quad \text{and} \quad \lambda^\nu \in (0, 1].$$

We can show that $S^\nu \xrightarrow{g} S$ when $\lambda^\nu \searrow 0$, which justifies the method via 7.34.

**Detail.** By 7.37, $(1-\lambda^\nu)S \xrightarrow{g} S$. Let $T^\nu(x) = \{\lambda^\nu x\}$ and $T(x) = \{0\}$. Then, $T^\nu(x^\nu) \xrightarrow{s} T(x)$ when $x^\nu \to x$. For any $\bar{x} \in \mathbb{R}^n$, the set

$$\{T^\nu(x) \mid x \in \mathbb{B}(\bar{x}, 1),\ \nu \in \mathbb{N}\}$$

is contained in a bounded set. Hence, 7.38 applies and $(1 - \lambda^\nu)S + T^\nu = S^\nu \xrightarrow{g} S$. □

**Example 7.40** (traffic equilibrium). We recall from §7.C that a Wardrop equilibrium is a solution of the generalized equation

$$0 \in S(u) = F(u) + N_{[0,\infty)^{r+q}}(u),$$

where $u = (w_p, v_{st},\ p \in P_{st}, (s,t) \in O)$ and

$$F(u) = \big(f_{stp}(u), f_{st0}(u),\ p \in P_{st},\ (s,t) \in O\big)$$

$$f_{stp}(u) = \varphi_p(w) - v_{st}, \quad \varphi_p(w) = \sum\nolimits_{(i,j)\in E} a_{ijp}\, g_{ij}\Big(\sum\nolimits_{(s,t)\in O} \sum\nolimits_{\bar{p}\in P_{st}} a_{ij\bar{p}} w_{\bar{p}}\Big);$$

see §7.C for additional definitions. Let's consider the effect of replacing the travel-time functions $g_{ij} : \mathbb{R} \to (0,\infty)$ by alternative functions $g_{ij}^\nu : \mathbb{R} \to (0,\infty)$. These alternative functions define

$$F^\nu(u) = \big(f_{stp}^\nu(u), f_{st0}(u),\ p \in P_{st},\ (s,t) \in O\big)$$

$$f_{stp}^\nu(u) = \varphi_p^\nu(w) - v_{st}, \quad \varphi_p^\nu(w) = \sum\nolimits_{(i,j)\in E} a_{ijp}\, g_{ij}^\nu\Big(\sum\nolimits_{(s,t)\in O} \sum\nolimits_{\bar{p}\in P_{st}} a_{ij\bar{p}} w_{\bar{p}}\Big)$$

as well as the alternative generalized equation

$$0 \in S^\nu(u) = F^\nu(u) + N_{[0,\infty)^{r+q}}(u).$$

It turns out that if $g_{ij}^\nu(\alpha^\nu) \to g_{ij}(\alpha)$ whenever $\alpha^\nu \to \alpha$, then $S^\nu \xrightarrow{g} S$. Consequently, by 7.34, the alternative generalized equations produce approximations of a Wardrop equilibrium corresponding to the actual problem, to an arbitrary level of accuracy, as $\nu \to \infty$.

**Detail.** With $T(u) = N_{[0,\infty)^{r+q}}(u)$, we certainly have $T \xrightarrow{g} T$ because $\operatorname{gph} T$ is closed. Let

$$u^\nu = \big(w_p^\nu, v_{st}^\nu,\ p \in P_{st},\ (s,t) \in O\big) \to u = \big(w_p, v_{st},\ p \in P_{st},\ (s,t) \in O\big).$$

Then, $F^\nu(u^\nu) \to F(u)$ because $\varphi_p^\nu(w^\nu) \to \varphi_p(w)$ and $f_{st0}$ is continuous. We intend to apply 7.38 and need to check its boundedness assumption. Since $g_{ij}^\nu(\alpha^\nu) \to g_{ij}(\alpha)$ whenever $\alpha^\nu \to \alpha$, we must have that for $\bar\alpha \in \mathbb{R}$ there's $\varepsilon > 0$ such that

$$B = \big\{g_{ij}^\nu(\alpha) \,\big|\, \alpha \in \mathbb{B}(\bar\alpha, \varepsilon),\ \nu \in \mathbb{N}\big\}$$

is contained in a bounded set. To see this, note that if it didn't hold, then there would exist $\alpha^\nu \to \bar\alpha$ such that $g_{ij}^\nu(\alpha^\nu) \to \infty$. However, this contracts the convergence of $g_{ij}^\nu(\alpha^\nu)$ to the real number $g_{ij}(\bar\alpha)$. The fact that $B$ is bounded then transitions up to $F^\nu$ and we conclude that 7.38 applies. □

A Wardrop equilibrium is characterized not only by a generalized equation, which we leverage in the previous example, but also by the minimization problem (7.13). As an alternative to the approach in the example, one could show epi-convergence of the functions corresponding to this problem and then invoke 5.5(b,e) to reach the same conclusion. Such a connection between graphical convergence of set-valued mappings representing optimality conditions and epi-convergence of the corresponding functions is generally present in the convex case.

**Theorem 7.41** (Attouch's). *For proper, lsc and convex functions $f^\nu, f : \mathbb{R}^n \to \overline{\mathbb{R}}$, consider their subgradient mappings $\partial f^\nu, \partial f : \mathbb{R}^n \rightrightarrows \mathbb{R}^n$. Then,*

$$f^\nu \xrightarrow{e} f \iff \partial f^\nu \xrightarrow{g} \partial f \text{ and } f^\nu(x^\nu) \to f(\bar x) \text{ for some } v^\nu \to \bar v,\ x^\nu \to \bar x$$
$$\text{with } v^\nu \in \partial f^\nu(x^\nu) \text{ and } \bar v \in \partial f(\bar x).$$

**Proof.** We refer to [105, Theorem 12.35] for details and only remark that the result can't hold in the nonconvex case. For example, consider the "saw-tooth" functions given by

$$f^\nu(x) = \begin{cases} x - (i-1)/\nu & \text{if } x \in \big[(i-1)/\nu,\ i/\nu\big),\ \ i = 1, \dots, \nu \\ 1/\nu & \text{if } x = 1 \\ \infty & \text{otherwise.} \end{cases}$$

Since $\sup_{x \in [0,1]} |f^\nu(x)| \leq 1/\nu$, one has $f^\nu \xrightarrow{e} \iota_{[0,1]}$. However, $\partial f^\nu(x) = \{1\}$ for every $x \in [0,1]$ except if $x = i/\nu$, $i = 0, 1, \dots, \nu$, and even then it doesn't match $\partial\iota_{[0,1]}(x)$. □

## 7.I Consistent Approximations

The epigraphical approximation algorithm of §4.C provides a path to address a difficult optimization problem: obtain near-minimizers of increasingly accurate approximating problems. However, in the nonconvex setting, a near-minimizer may be beyond our reach even for an approximating problem and we need to consider the corresponding optimality conditions. Graphical convergence of set-valued mappings representing optimality conditions then joins epi-convergence of objective functions as the key properties in a comprehensive approximation theory for minimization problems. In this section, we'll develop fully implementable algorithms that leverage our knowledge of these two notions of convergence.

For functions $f, f^\nu : \mathbb{R}^n \to \overline{\mathbb{R}}$, let's consider

$$\operatorname*{minimize}_{x\in\mathbb{R}^n} f(x) \quad \text{and} \quad \left\{ \operatorname*{minimize}_{x\in\mathbb{R}^n} f^\nu(x), \quad \nu \in \mathbb{N} \right\},$$

which are the actual problem of interest and a family of alternative problems, respectively. Our intent is to address the actual problem by solving the sequence of alternative problems. However, since the problems could be nonconvex, "solving" needs to be understood in the sense of achieving points that satisfy the associated generalized equations

$$0 \in S(x, y) \quad \text{and} \quad \left\{ 0 \in S^\nu(x, y), \quad \nu \in \mathbb{N} \right\},$$

where $S, S^\nu : \mathbb{R}^{n+m} \rightrightarrows \mathbb{R}^q$. Typically, the generalized equations express necessary conditions for minimizers of the actual and alternative problems, at least under suitable qualifications. The vector $y$ plays the role of multipliers, but could also be absent as in the case of the optimality condition $0 \in S(x) = \partial f(x)$.

An algorithm that requires a near-minimizer of a nonconvex $f^\nu$ in one of its steps is typically *conceptual* as an iteration can't be carried out in finite time and an implementation of the algorithm would require additional approximations. In contrast, solving $0 \in S^\nu(x, y)$ approximately is often possible in finite time and thus leads to an *implementable algorithm* for the actual problem. We therefore shift the focus away from the epigraphical approximation algorithm to the graphical approximation algorithm of §7.H as applied to set-valued mappings representing optimality conditions. Still, epi-convergence remains in play by furnishing one of the parts in the following comprehensive notion of approximation.

The pairs $\{(f^\nu, S^\nu), \nu \in \mathbb{N}\}$ are *weakly consistent approximations* of $(f, S)$ when

$$f^\nu \xrightarrow{e} f \quad \text{and} \quad \operatorname{LimOut}(\operatorname{gph} S^\nu) \subset \operatorname{gph} S.$$

If in addition $S^\nu \xrightarrow{g} S$, then $\{(f^\nu, S^\nu), \nu \in \mathbb{N}\}$ are *consistent approximations* of $(f, S)$.

In the presence of (weakly) consistent approximations, we're on solid ground for carrying out the planned scheme. We refer to [21, 33, 51, 79, 83] for related concepts of consistency and applications to a variety of problems including optimal control.

**Convergence 7.42** (consistent approximations). *In the notation of this section, suppose that $\{(f^\nu, S^\nu), \nu \in \mathbb{N}\}$ are weakly consistent approximations of $(f, S)$ and $\{(x^\nu, y^\nu), \nu \in \mathbb{N}\}$ is generated by the graphical approximation algorithm of §7.H when applied to $S^\nu$ with $\varepsilon^\nu \to 0$ and $\bar{v} = 0$, i.e.,*

$$\operatorname{dist}\big(0, S^\nu(x^\nu, y^\nu)\big) \le \varepsilon^\nu \quad \forall \nu \in \mathbb{N}.$$

*Then, the following hold:*

(a) *If there are* $N \in \mathcal{N}^{\#}_{\infty}$ *and* $(\bar{x}, \bar{y})$ *such that* $(x^\nu, y^\nu) \underset{N}{\to} (\bar{x}, \bar{y})$, *then*

$$0 \in S(\bar{x}, \bar{y}) \quad and \quad f(\bar{x}) \le \operatorname{liminf}_{\nu \in N} f^\nu(x^\nu).$$

(b) *For any* $\alpha < \beta$,

$$\operatorname{LimOut}\{f^\nu \le \alpha\} \subset \{f \le \alpha\} \subset \operatorname{LimInn}\{f^\nu \le \beta\}.$$

**Proof.** The assertion about $0 \in S(\bar{x}, \bar{y})$ holds by 7.31. Since $f^\nu \xrightarrow{e} f$, it follows by the characterization 4.15(a) that

$$f(\bar{x}) \le \operatorname{liminf}_{\nu \in N} f^\nu(x^\nu).$$

For (b), let

$$\hat{x} \in \operatorname{LimOut}\{f^\nu \le \alpha\}.$$

Then, there are $N \in \mathcal{N}^{\#}_{\infty}$ and $\hat{x}^\nu \underset{N}{\to} \hat{x}$ with $f^\nu(\hat{x}^\nu) \le \alpha$. It then follows from 4.15(a) that

$$f(\hat{x}) \le \operatorname{liminf}_{\nu \in N} f^\nu(\hat{x}^\nu) \le \alpha.$$

Thus, $\hat{x} \in \{f \le \alpha\}$ and the first inclusion holds. For the second one, let $\hat{x} \in \{f \le \alpha\}$. Since $f^\nu \xrightarrow{e} f$, there's $\hat{x}^\nu \to \hat{x}$ such that

$$\operatorname{limsup} f^\nu(\hat{x}^\nu) \le f(\hat{x}) \le \alpha$$

by 4.15(b) and then $f^\nu(\hat{x}^\nu) \le \beta$ for sufficiently large $\nu$. We've constructed a sequence $\hat{x}^\nu \to \hat{x}$ with $\hat{x}^\nu \in \{f^\nu \le \beta\}$ for sufficiently large $\nu$, which means that $\hat{x} \in \operatorname{LimInn}\{f^\nu \le \beta\}$. □

Part (a) guarantees that any cluster point produced by the algorithm satisfies the optimality condition we've associated with the actual problem. Moreover, the liminf-expression implies that $\bar{x}$ is at least as good (in actual objective function value) as the observed values $\{f^\nu(x^\nu), \nu \in N\}$ indicate. This holds even though $f$ is never evaluated as part of the algorithm. Part (b) provides assurance that $\{f^\nu \le \alpha\}$ approximates $\{f \le \alpha\}$, which is useful when examining the existence of minimizers and exploring the space for "good" points.

**Example 7.43** (smoothing as consistent approximations). For smooth $f_i : \mathbb{R}^n \to \mathbb{R}$, $i = 1, \dots, m$, let's consider

$$\underset{x \in \mathbb{R}^n}{\text{minimize}}\ f(x) = \max_{i=1,\dots,m} f_i(x), \quad \text{with optimality condition } 0 \in S(x) = \partial f(x),$$

and the alternatives

$$\underset{x \in \mathbb{R}^n}{\text{minimize}}\ f^\nu(x), \quad \text{with optimality condition } 0 \in S^\nu(x) = \big\{\nabla f^\nu(x)\big\},$$

where $f^\nu$ is the smooth approximation from 4.16, i.e.,

$$f^\nu(x) = \frac{1}{\theta^\nu} \ln\Big(\sum\nolimits_{i=1}^m \exp\big(\theta^\nu f_i(x)\big)\Big), \quad \text{with } \theta^\nu \in (0, \infty).$$

We can show that $\{(f^\nu, S^\nu), \nu \in \mathbb{N}\}$ are weakly consistent approximations of $(f, S)$ when $\theta^\nu \to \infty$. More general smoothing is discussed in [25].

Since $f^\nu$ is smooth, algorithms from Chap. 1 apply and, under mild assumptions, are guaranteed to produce $x^\nu$ with

$$\|\nabla f^\nu(x^\nu)\|_2 \leq \varepsilon^\nu$$

in finite computing time when $\varepsilon^\nu > 0$; see, for example, 1.31. Step 1 in the graphical approximation algorithm is therefore implementable.

**Detail.** We know from 4.16 that $f^\nu \xrightarrow{e} f$ and it only remains to show that $\mathrm{LimOut}(\operatorname{gph} S^\nu) \subset \operatorname{gph} S$. Suppose that

$$(\bar{x}, \bar{v}) \in \mathrm{LimOut}(\operatorname{gph} S^\nu).$$

Then, there are $N \in \mathcal{N}_\infty^\#$, $x^\nu \to_N \bar{x}$ and $v^\nu \to_N \bar{v}$ such that $(x^\nu, v^\nu) \in \operatorname{gph} S^\nu$ but this simply means that

$$\nabla f^\nu(x^\nu) \to_N \bar{v}.$$

Let's show that $\bar{v} \in \partial f(\bar{x})$. By 4.66,

$$\partial f(\bar{x}) = \operatorname{con}\big\{\nabla f_i(\bar{x}), i \in \mathbb{A}(\bar{x})\big\},$$

where $\mathbb{A}(\bar{x}) = \{i \mid f_i(\bar{x}) = f(\bar{x})\}$. From 4.16, we see that

$$\nabla f^\nu(x) = \sum\nolimits_{i=1}^m \mu_i^\nu(x) \nabla f_i(x), \quad \text{with } \mu_i^\nu(x) = \frac{\exp\big(\theta^\nu(f_i(x) - f(x))\big)}{\sum_{k=1}^m \exp\big(\theta^\nu(f_k(x) - f(x))\big)}.$$

If $i \notin \mathbb{A}(\bar{x})$, then there's $\delta \in (0, \infty)$ such that $f_i(\bar{x}) + \delta = f(\bar{x})$. Since $f$ is continuous by 4.66 and $f_i$ is smooth, there's $\bar{\nu} \in \mathbb{N}$ such that

$$f_i(x^\nu) \leq f_i(\bar{x}) + \tfrac{1}{3}\delta \quad \text{and} \quad f(x^\nu) \geq f(\bar{x}) - \tfrac{1}{3}\delta \;\; \forall \nu \geq \bar{\nu}, \nu \in N.$$

This means that

$$f_i(x^\nu) - f(x^\nu) \leq f_i(\bar{x}) - f(\bar{x}) + \tfrac{2}{3}\delta = -\tfrac{1}{3}\delta$$

for such $\nu$ and

$$\exp\Big(\theta^\nu\big(f_i(x^\nu) - f(x^\nu)\big)\Big) \to_N 0.$$

The denominator in the defining expression for $\mu_i^\nu(x^\nu)$ is greater than 1 because $f_k(x^\nu) - f(x^\nu) = 0$ for $k \in \mathbb{A}(x^\nu)$. Thus,

$$\mu_i^\nu(x^\nu) \to_N 0.$$

For any $i$, $\mu_i^\nu(x^\nu) \in (0,1)$. Consequently, after passing to another subsequence which we also denote by $N$, there are $\mu_i^\infty \in [0,1]$, $i = 1,\dots,m$, such that

$$\mu_i^\nu(x^\nu) \to_N \mu_i^\infty.$$

Since $\sum_{i=1}^m \mu_i^\nu(x^\nu) = 1$ for all $\nu$, we must also have $\sum_{i=1}^m \mu_i^\infty = 1$, with $\mu_i^\infty = 0$ if $i \notin \mathbb{A}(\bar{x})$ as already seen. We conclude that

$$\nabla f^\nu(x^\nu) = \sum\nolimits_{i=1}^m \mu_i^\nu(x^\nu)\nabla f_i(x^\nu) \to_N \bar{v} = \sum\nolimits_{i\in\mathbb{A}(\bar{x})} \mu_i^\infty \nabla f_i(\bar{x}) \in \partial f(\bar{x}).$$

We've established that $(\bar{x},\bar{v}) \in \operatorname{gph} S$ and this implies $\operatorname{LimOut}(\operatorname{gph} S^\nu) \subset \operatorname{gph} S$. □

In general, there's no clear relationship between the accuracy by which $\bar{x}$ satisfies an optimality condition for a function $f$ and the optimality gap $f(\bar{x}) - \inf f$. After all, one can imagine arbitrarily weak optimality conditions that have little in common with minimizers. In the convex case, however, the following fact is surprisingly general.

**Proposition 7.44** (optimality gap for convex functions). *For a proper, lsc and convex function* $f : \mathbb{R}^n \to \overline{\mathbb{R}}$ *with* $\operatorname{argmin} f \neq \emptyset$ *and a point* $\bar{x}$ *at which* $f$ *is finite, one has*

$$f(\bar{x}) - \inf f \le \operatorname{dist}(\bar{x}, \operatorname{argmin} f)\operatorname{dist}\big(0, \partial f(\bar{x})\big).$$

**Proof.** Let $\rho = \operatorname{dist}(0, \partial f(\bar{x}))$. If $\partial f(\bar{x}) = \emptyset$, then $\rho = \infty$ and the claim holds trivially. We can concentrate on the case with $\partial f(\bar{x}) \neq \emptyset$. Since $\partial f(\bar{x})$ is closed by 2.25, there's $\bar{v} \in \partial f(\bar{x})$ such that $\rho = \|\bar{v}\|_2$. This means that $0 \in \partial f(\bar{x}) + \mathbb{B}(0,\rho)$. Let

$$g(x) = f(x) + \rho\|x - \bar{x}\|_2.$$

We can show that

$$\partial g(\bar{x}) = \partial f(\bar{x}) + \mathbb{B}(0,\rho).$$

For this purpose, let $h(x) = \|x\|_2$ and then $h(x) = \sup_{y\in\mathbb{B}(0,1)}\langle x, y\rangle$ as well. It follows by 5.29 that the conjugate $h^* = \iota_{\mathbb{B}(0,1)}$. Since

$$y \in \partial h(x) \iff x \in \partial h^*(y) = N_{\mathbb{B}(0,1)}(y)$$

by the inverse rule 5.37, we see that $y \in \partial h(0)$ if and only if $0 \in N_{\mathbb{B}(0,1)}(y)$. This holds for all $y \in \mathbb{B}(0,1)$ and we conclude that $\partial h(0) = \mathbb{B}(0,1)$. Thus, $\mathbb{B}(0,\rho)$ is the set of subgradients of $x \mapsto \rho\|x - \bar{x}\|_2$ at $\bar{x}$. We can now invoke the sum rule 4.67 to obtain the claimed formula; the qualification (4.15) holds because $x \mapsto \rho\|x - \bar{x}\|_2$ is real-valued and convex so its horizon subgradients are always 0 by 4.65. Since $0 \in \partial g(\bar{x})$, it follows from 2.19 that $\bar{x} \in \operatorname{argmin} g$. We then also have

$$f(\bar{x}) = g(\bar{x}) \le g(x) = f(x) + \rho\|x - \bar{x}\|_2 \quad \forall x \in \mathbb{R}^n.$$

In particular, this holds for $x \in \operatorname{argmin} f$. Thus,

$$f(\bar{x}) \leq \inf f + \rho \|x - \bar{x}\|_2$$

and the conclusion follows. □

Since there's often a coarse bound available for $\mathrm{dist}(\bar{x}, \mathrm{argmin} f)$, the proposition enables us to conclude that the optimality gap is bounded by a quantity that's proportional to $\mathrm{dist}(0, \partial f(\bar{x}))$.

We next turn to the far-reaching class of composite problems and the associated optimality conditions stated in 4.75.

**Theorem 7.45** (consistent approximations in composite minimization). *For nonempty closed $X, X^\nu \subset \mathbb{R}^n$, smooth $f_0, f_0^\nu : \mathbb{R}^n \to \mathbb{R}$, smooth $F, F^\nu : \mathbb{R}^n \to \mathbb{R}^m$ and proper, lsc and convex $h, h^\nu : \mathbb{R}^m \to \overline{\mathbb{R}}$, consider the problems*

$$\underset{x \in \mathbb{R}^n}{\text{minimize}}\, f(x) = \iota_X(x) + f_0(x) + h\big(F(x)\big)$$

$$\underset{x \in \mathbb{R}^n}{\text{minimize}}\, f^\nu(x) = \iota_{X^\nu}(x) + f_0^\nu(x) + h^\nu\big(F^\nu(x)\big).$$

*Associated with these problems are the set-valued mappings $S, S^\nu : \mathbb{R}^{n+2m} \rightrightarrows \mathbb{R}^{2m+n}$ given by*

$$S(x, y, z) = \big\{F(x) - z\big\} \times \big\{\partial h(z) - y\big\} \times \big\{\nabla f_0(x) + \nabla F(x)^\top y + N_X(x)\big\}$$
$$S^\nu(x, y, z) = \big\{F^\nu(x) - z\big\} \times \big\{\partial h^\nu(z) - y\big\} \times \big\{\nabla f_0^\nu(x) + \nabla F^\nu(x)^\top y + N_{X^\nu}(x)\big\}.$$

*Suppose that $X^\nu \xrightarrow{s} X$, $h^\nu \xrightarrow{e} h$, $f_0^\nu(x^\nu) \to f_0(x)$, $\nabla f_0^\nu(x^\nu) \to \nabla f_0(x)$, $F^\nu(x^\nu) \to F(x)$ and $\nabla F^\nu(x^\nu) \to \nabla F(x)$ when $x^\nu \to x$. Then, the following hold:*

(a) *If $h$ is real-valued, then $f^\nu \xrightarrow{e} f$.*
(b) *If $h^\nu(u) \to h(u)$ for all $u \in \mathrm{dom}\, h$ and, for all $\nu$, $X^\nu = X$ and $F^\nu = F$, then $f^\nu \xrightarrow{e} f$.*
(c) *If $X^\nu$ is convex for all $\nu$, then $S^\nu \xrightarrow{g} S$.*
(d) *If $X^\nu = X$ for all $\nu$, then $S^\nu \xrightarrow{g} S$.*

*If either (a) or (b) holds and either (c) or (d) holds, then $\{(f^\nu, S^\nu), \nu \in \mathbb{N}\}$ are consistent approximations of $(f, S)$.*

**Proof.** For (a), we leverage the epigraphical sum rule 4.19(a). Since $X^\nu \xrightarrow{s} X$, $\iota_{X^\nu} \xrightarrow{e} \iota_X$; see the discussion after 4.26. Directly from the characterization 4.15, we also have $f_0^\nu \xrightarrow{e} f_0$ and $-f_0^\nu \xrightarrow{e} -f_0$. It then follows from 4.19(a) that

$$\iota_{X^\nu} + f_0^\nu \xrightarrow{e} \iota_X + f_0. \tag{7.24}$$

If $x^\nu \to x$, then $F^\nu(x^\nu) \to F(x)$ and also $h^\nu(F^\nu(x^\nu)) \to h(F(x))$ by 4.18 because $h$ is real-valued. This fact, together with 4.15, establishes that

$$h^\nu \circ F^\nu \xrightarrow{e} h \circ F \quad \text{and} \quad -h^\nu \circ F^\nu \xrightarrow{e} -h \circ F. \tag{7.25}$$

We can now combine (7.24) and (7.25) with the epigraphical sum rule 4.19(a) to reach the conclusion.

For (b), we've $\iota_X \xrightarrow{e} \iota_X$ and $\iota_X(x) \to \iota_X(x)$ for all $x \in \mathbb{R}^n$. In view of 4.19(b), we can conclude that

$$\iota_X + h^\nu \circ F \xrightarrow{e} \iota_X + h \circ F$$

if $h^\nu(F(x)) \to h(F(x))$ for all $x$ and $h^\nu \circ F \xrightarrow{e} h \circ F$. The first of the two requirements holds when $F(x) \in \operatorname{dom} h$ by the assumption on $h$. When $F(x) \notin \operatorname{dom} h$, 4.15(a) ensures that $h^\nu(F(x)) \to h(F(x))$ still holds. For the second requirement, we utilize 4.15. Let $x^\nu \to x$. Then, $F(x^\nu) \to F(x)$ and

$$\liminf h^\nu\big(F(x^\nu)\big) \geq h\big(F(x)\big)$$

because $h^\nu \xrightarrow{e} h$; cf. 4.15(a). Since we've established $h^\nu(F(x)) \to h(F(x))$ already, both parts of 4.15 hold and $h^\nu \circ F \xrightarrow{e} h \circ F$. We also need to bring in $f_0^\nu, f_0$. Since $f_0^\nu \xrightarrow{e} f_0$ and $-f_0^\nu \xrightarrow{e} -f_0$, this is accomplished with the epigraphical sum rule 4.19(a).

For (c,d), we start by showing $\operatorname{LimOut}(\operatorname{gph} S^\nu) \subset \operatorname{gph} S$ and let

$$(\bar{x}, \bar{y}, \bar{z}, \bar{u}, \bar{v}, \bar{w}) \in \operatorname{LimOut}(\operatorname{gph} S^\nu).$$

Then, there are $N \in \mathcal{N}_\infty^\#$, $x^\nu \to_N \bar{x}$, $y^\nu \to_N \bar{y}$, $z^\nu \to_N \bar{z}$, $u^\nu \to_N \bar{u}$, $v^\nu \to_N \bar{v}$ and $w^\nu \to_N \bar{w}$ with

$$(x^\nu, y^\nu, z^\nu, u^\nu, v^\nu, w^\nu) \in \operatorname{gph} S^\nu.$$

Consequently,

$$u^\nu = F^\nu(x^\nu) - z^\nu, \quad v^\nu \in \partial h^\nu(z^\nu) - y^\nu, \quad w^\nu \in \nabla f_0^\nu(x^\nu) + \nabla F^\nu(x^\nu)^\top y^\nu + N_{X^\nu}(x^\nu).$$

Since $F^\nu(x^\nu) \to_N F(\bar{x})$, $\bar{u} = F(\bar{x}) - \bar{z}$. Via Attouch's theorem 7.41, $h^\nu \xrightarrow{e} h$ implies that $\bar{v} \in \partial h(\bar{z}) - \bar{y}$.

Under assumption (c), $N_{X^\nu} \xrightarrow{g} N_X$ by Attouch's theorem 7.41 because $\iota_{X^\nu} \xrightarrow{e} \iota_X$ and these functions are convex. Note that $\iota_X$ is convex by 4.31. Consequently, the limit of vectors in $\operatorname{gph} N_{X^\nu}$ is contained in $\operatorname{gph} N_X$ and

$$\bar{w} - \nabla f_0(\bar{x}) - \nabla F(\bar{x})^\top \bar{y} \in N_X(\bar{x}). \tag{7.26}$$

Under assumption (d),

$$w^\nu - \nabla f_0^\nu(x^\nu) - \nabla F^\nu(x^\nu)^\top y^\nu \in N_X(x^\nu) \quad \forall \nu \in \mathbb{N}$$

and (7.26) holds as well by 4.39. In summary, we've shown that

$$(\bar{x}, \bar{y}, \bar{z}, \bar{u}, \bar{v}, \bar{w}) \in \operatorname{gph} S,$$

which then implies that $\operatorname{LimOut}(\operatorname{gph} S^\nu) \subset \operatorname{gph} S$.

Next, let's establish $\operatorname{LimInn}(\operatorname{gph} S^\nu) \supset \operatorname{gph} S$ and assume that $(\bar{x}, \bar{y}, \bar{z}, \bar{u}, \bar{v}, \bar{w}) \in \operatorname{gph} S$. This means that

$$\bar{u} = F(\bar{x}) - \bar{z}, \quad \bar{v} \in \partial h(\bar{z}) - \bar{y}, \quad \bar{w} \in \nabla f_0(\bar{x}) + \nabla F(\bar{x})^\top \bar{y} + N_X(\bar{x}).$$

Since $h^\nu \xrightarrow{e} h$, it follows by Attouch's theorem 7.41 that $\partial h^\nu \xrightarrow{g} \partial h$ and there are $v^\nu \to \bar{v}$ and $z^\nu \to \bar{z}$ such that $v^\nu + \bar{y} \in \partial h^\nu(z^\nu)$.

Under (c), $N_{X^\nu} \xrightarrow{g} N_X$ as argued above. Consequently, there's $(x^\nu, t^\nu) \in \operatorname{gph} N_{X^\nu}$ with $x^\nu \to \bar{x}$ and

$$t^\nu \to \bar{w} - \nabla f_0(\bar{x}) - \nabla F(\bar{x})^\top \bar{y}.$$

Let's construct

$$w^\nu = t^\nu + \nabla f_0^\nu(x^\nu) + \nabla F^\nu(x^\nu)^\top \bar{y}.$$

We then have $w^\nu \to \bar{w}$ and

$$w^\nu - \nabla f_0^\nu(x^\nu) - \nabla F^\nu(x^\nu)^\top \bar{y} \in N_{X^\nu}(x^\nu).$$

Under (d), construct $x^\nu = \bar{x}$ as well as

$$w^\nu = \bar{w} - \nabla f_0(\bar{x}) - \nabla F(\bar{x})^\top \bar{y} + \nabla f_0^\nu(x^\nu) + \nabla F^\nu(x^\nu)^\top \bar{y}.$$

Then, $w^\nu \to \bar{w}$ and

$$w^\nu \in \nabla f_0^\nu(x^\nu) + \nabla F^\nu(x^\nu)^\top \bar{y} + N_X(x^\nu).$$

Regardless of the two cases, let $u^\nu = F^\nu(x^\nu) - z^\nu$, which converges to $\bar{u}$. In summary, we've constructed

$$(x^\nu, \bar{y}, z^\nu, u^\nu, v^\nu, w^\nu) \in \operatorname{gph} S^\nu \to (\bar{x}, \bar{y}, \bar{z}, \bar{u}, \bar{v}, \bar{w}).$$

This means that $(\bar{x}, \bar{y}, \bar{z}, \bar{u}, \bar{v}, \bar{w}) \in \operatorname{LimInn}(\operatorname{gph} S^\nu)$ and, thus, $\operatorname{LimInn}(\operatorname{gph} S^\nu) \supset \operatorname{gph} S$ holds. □

The assumptions in (b) accommodate penalty methods. For example, if $h = \iota_{\{0\}^m}$ and $h^\nu = \theta^\nu \|\cdot\|_2^2$, then $h^\nu \xrightarrow{e} h$ and $h^\nu(u) \to h(u)$ for all $u \in \operatorname{dom} h$ provided that $\theta^\nu \to \infty$; see 4.21.

Interestingly, the assumptions in (c,d) might be perceived as milder than the ones in (a,b). This is caused by multiplier vectors in the optimality conditions; their presence furnishes more flexibility in absorbing inaccuracies.

**Example 7.46** (extension of proximal composite method). The proximal composite method in §6.F is limited to real-valued $h$. We can now overcome this by considering approximations. For nonempty, closed and convex $X \subset \mathbb{R}^n$, twice smooth $F : \mathbb{R}^n \to \mathbb{R}^m$, convex $h^\nu : \mathbb{R}^n \to \mathbb{R}$ and proper, lsc and convex $h : \mathbb{R}^m \to \overline{\mathbb{R}}$, consider the problems

$$\operatorname*{minimize}_{x \in \mathbb{R}^n} f(x) = \iota_X(x) + h\big(F(x)\big) \quad \text{and} \quad \operatorname*{minimize}_{x \in \mathbb{R}^n} f^\nu(x) = \iota_X(x) + h^\nu\big(F(x)\big).$$

Associated with these problems are the set-valued mappings $S, S^\nu : \mathbb{R}^{n+2m} \rightrightarrows \mathbb{R}^{2m+n}$ given by

$$S(x, y, z) = \{F(x) - z\} \times \{\partial h(z) - y\} \times \{\nabla F(x)^\top y + N_X(x)\}$$
$$S^\nu(x, y, z) = \{F(x) - z\} \times \{\partial h^\nu(z) - y\} \times \{\nabla F(x)^\top y + N_X(x)\}.$$

If $h^\nu \xrightarrow{e} h$ and $h^\nu(u) \to h(u)$ for all $u \in \operatorname{dom} h$, then $\{(f^\nu, S^\nu), \nu \in \mathbb{N}\}$ are consistent approximations of $(f, S)$.

Moreover, one can apply the graphical approximation algorithm to solve the generalized equation $0 \in S(x, y, z)$ via the approximate solution of $0 \in S^\nu(x, y, z)$. The latter can be carried out by the proximal composite method.

**Detail.** Consistent approximation is a consequence of 7.45. In the present context, Step 1 of the graphical approximation algorithm requires us to obtain $(x^\nu, y^\nu, z^\nu)$ such that

$$\operatorname{dist}\big(0, S^\nu(x^\nu, y^\nu, z^\nu)\big) \le \varepsilon^\nu.$$

We achieve this by applying the proximal composite method to the problem of minimizing $f^\nu$. Since the assumptions of the method hold, one of three outcomes is guaranteed:

(a) The proximal composite method stops after a finite number of iterations with $\hat{x}$ satisfying

$$\hat{y} \in \partial h^\nu\big(F(\hat{x})\big) \quad \text{and} \quad -\nabla F(\hat{x})^\top \hat{y} \in N_X(\hat{x}) \tag{7.27}$$

for some $\hat{y}$; see the discussion in §6.F. We can then set $x^\nu = \hat{x}$, $y^\nu = \hat{y}$ and $z^\nu = F(\hat{x})$ with the result that

$$\operatorname{dist}\big(0, S^\nu(x^\nu, y^\nu, z^\nu)\big) = 0.$$

(b) The proximal composite method generates a sequence $\{\hat{x}^k, k \in \mathbb{N}\}$ with a cluster point $\hat{x}$. For any positive $\varepsilon^\nu$, it turns out that for some $k^\star$ along this sequence we can set $x^\nu = \hat{x}^{k^\star+1}$ and construct $y^\nu$ and $z^\nu$ such that

$$\operatorname{dist}\big(0, S^\nu(x^\nu, y^\nu, z^\nu)\big) \le \varepsilon^\nu.$$

This means that the proximal composite method can be terminated after a finite number of iterations and Step 1 of the graphical approximation algorithm is again accomplished in finite computing time.

The stopping criterion for determining $k^\star$ and constructing $y^\nu$ and $z^\nu$ is as follows. From the proof of 6.35, we deduce that in the $k$th iteration of the proximal composite method it computes $\hat{x}^{k+1}$ such that

$$-\nabla F(\hat{x}^k)^\top \hat{y}^{k+1} - \tfrac{1}{\lambda^k}(\hat{x}^{k+1} - \hat{x}^k) \in N_X(\hat{x}^{k+1}) \tag{7.28}$$
$$\text{for some } \hat{y}^{k+1} \in \partial h^\nu\big(F(\hat{x}^k) + \nabla F(\hat{x}^k)(\hat{x}^{k+1} - \hat{x}^k)\big).$$

Moreover, $\hat{x}^{k+1} - \hat{x}^k \to 0$ as $k \to \infty$ and $(\hat{x}^{k+1} - \hat{x}^k)/\lambda^k \to 0$ as $k \to \infty$ along the subsequence corresponding to $\hat{x}$. Set

$$\begin{aligned}
\hat{z}^{k+1} &= F(\hat{x}^k) + \nabla F(\hat{x}^k)(\hat{x}^{k+1} - \hat{x}^k) \\
u^{k+1} &= F(\hat{x}^{k+1}) - \hat{z}^{k+1} \\
v^{k+1} &= 0 \\
w^{k+1} &= -\nabla F(\hat{x}^k)^\top \hat{y}^{k+1} + \nabla F(\hat{x}^{k+1})^\top \hat{y}^{k+1} - \tfrac{1}{\lambda^k}(\hat{x}^{k+1} - \hat{x}^k).
\end{aligned}$$

We see that

$$(u^{k+1}, v^{k+1}, w^{k+1}) \in S^\nu(\hat{x}^{k+1}, \hat{y}^{k+1}, \hat{z}^{k+1})$$

by direct substitution. Then, for $k$ along the subsequence corresponding to $\hat{x}$, $\hat{z}^{k+1} \to F(\hat{x})$, which means that the associated sets of subgradients $\partial h^\nu(\hat{z}^{k+1})$ are contained in a compact set by 2.54 and then the same holds for $\hat{y}^{k+1}$. This fact as well as the recognition that

$$F(\hat{x}^{k+1}) \to F(\hat{x}) \quad \text{and} \quad \nabla F(\hat{x}^{k+1}) - \nabla F(\hat{x}^k) \to 0$$

imply that $u^{k+1} \to 0$ and $w^{k+1} \to 0$ as $k \to \infty$ along the subsequence. Consequently, there's $k^\star$ such that

$$\big\|(u^{k^\star+1}, v^{k^\star+1}, w^{k^\star+1})\big\|_2 \leq \varepsilon^\nu.$$

At that iteration, we set $x^\nu = \hat{x}^{k^\star+1}$, $y^\nu = \hat{y}^{k^\star+1}$ and $z^\nu = \hat{z}^{k^\star+1}$ and then certainly have

$$\operatorname{dist}\big(0, S^\nu(x^\nu, y^\nu, z^\nu)\big) \leq \varepsilon^\nu.$$

The only nontrivial calculation required for this stopping criterion is to determine $\hat{y}^{k+1}$ in (7.28), which isn't explicitly produced by the proximal composite method. However, $\hat{x}^k$ and $\hat{x}^{k+1}$ are known and then

$$\partial h^\nu\big(F(\hat{x}^k) + \nabla F(\hat{x}^k)(\hat{x}^{k+1} - \hat{x}^k)\big) \quad \text{and} \quad N_X(\hat{x}^{k+1})$$

become fixed convex sets. We can therefore determine $\hat{y}^{k+1}$ by solving a convex optimization problem, which simplifies further, for example, when $h^\nu$ is a monitoring function; see §5.I.

(c) The last outcome is that the proximal composite method generates a sequence $\{\hat{x}^k, k \in \mathbb{N}\}$ with no cluster points. However, since the proximal composite method has the property that

$$\iota_X(\hat{x}^{k+1}) + h^\nu\big(F(\hat{x}^{k+1})\big) \leq \iota_X(\hat{x}^k) + h^\nu\big(F(\hat{x}^k)\big) \quad \forall k \in \mathbb{N},$$

this outcome can't occur if the level-set

$$\big\{\iota_X + h^\nu \circ F \leq \iota_X(\hat{x}^0) + h^\nu\big(F(\hat{x}^0)\big)\big\}$$

is bounded, where $\hat{x}^0$ is the starting point for the proximal composite method. This mild additional assumption can be related to the level-sets of $\iota_X + h \circ F$ via 7.42. □

**Exercise 7.47** (approximation of monitoring function). For nonempty polyhedral sets $Y, Y^\nu \subset \mathbb{R}^m$ and symmetric positive semidefinite $m \times m$-matrices $B, B^\nu$, consider the monitoring functions given by

$$h(u) = \sup_{y \in Y} \{\langle u, y\rangle - \tfrac{1}{2}\langle y, By\rangle\} \quad \text{and} \quad h^\nu(u) = \sup_{y \in Y^\nu} \{\langle u, y\rangle - \tfrac{1}{2}\langle y, B^\nu y\rangle\}.$$

These are typical candidates for the functions in 7.45 and 7.46, with significant computational advantages. (a) Show that if $B^\nu \to B$ and $Y^\nu \xrightarrow{s} Y$, then $h^\nu \xrightarrow{e} h$. (b) Show that if, in addition to the assumptions in (a), $B$ is positive definite or $Y$ is bounded, then $h$ is real-valued and $h^\nu(u) \to h(u)$ for $u \in \mathbb{R}^m$.

**Guide.** Use §5.I to determine the conjugate functions $h^*$ and $(h^\nu)^*$. A fact for proper, lsc and convex functions is that they epi-converge if and only if their conjugates epi-converge; see [105, Theorem 11.34]. Use this fact together with 4.26 and 4.19(a) to establish that $h^\nu \xrightarrow{e} h$. The final conclusion can then be reached via 4.18. □

## 7.J Approximation Errors

In addition to guarantees about vanishing errors in the limit, it's useful to quantify the error associated with any given approximating problem. We next turn to such quantification in the context of generalized equations. The development mirrors that in §6.J for minimization problems. Again, the truncated Hausdorff distance is the yardstick, but it's now applied to the graphs of set-valued mappings. We'll show that if the graphs of two set-valued mappings are close in the sense of this distance, the near-solutions of the corresponding generalized equations are also close.

For $S, T : \mathbb{R}^n \to \mathbb{R}^m$, gph $S$ and gph $T$ are subsets of $\mathbb{R}^{n+m}$ and hence the truncated Hausdorff distance between them requires a norm on $\mathbb{R}^{n+m}$; see §6.J for definitions. Typically, we choose

$$\max\{\|x - \bar{x}\|_2, \|v - \bar{v}\|_2\}, \quad \text{with } (x, v), (\bar{x}, \bar{v}) \in \mathbb{R}^n \times \mathbb{R}^m. \tag{7.29}$$

**Theorem 7.48** (solution error in generalized equations). *For $S, T : \mathbb{R}^n \rightrightarrows \mathbb{R}^m$ with nonempty graphs, suppose that $0 \le \varepsilon \le \rho < \infty$ and $\|\bar{v}\|_2 \le \rho - \varepsilon$. Then, under the norm (7.29) for $\mathbb{R}^{n+m}$, one has*

$$\text{exs}\Big(S^{-1}\big(\mathbb{B}(\bar{v}, \varepsilon)\big) \cap \mathbb{B}(0, \rho);\ T^{-1}\big(\mathbb{B}(\bar{v}, \delta)\big)\Big) \le \hat{d\!l}_\rho(\text{gph}\, S, \text{gph}\, T)$$

*provided that $\delta > \varepsilon + \hat{d\!l}_\rho(\text{gph}\, S, \text{gph}\, T)$.*

**Proof.** Let $\mathbb{B}_*(0, \rho)$ be the ball in $\mathbb{R}^{n+m}$, centered at the origin, with radius $\rho$ measured in the norm (7.29). Suppose that

$$\gamma \in \big(0, \delta - \varepsilon - \hat{d\!l}_\rho(\text{gph}\, S, \text{gph}\, T)\big]$$

and

$$x \in S^{-1}\big(\mathbb{B}(\bar{v}, \varepsilon)\big) \cap \mathbb{B}(0, \rho).$$

Then, there's $v \in S(x)$ with $\|v - \bar{v}\|_2 \le \varepsilon$ so that $(x, v) \in \mathbb{B}_*(0, \rho)$. Consequently, for some $(\hat{x}, \hat{v}) \in \operatorname{gph} T$,

$$\begin{aligned}\max\big\{\|x - \hat{x}\|_2, \|v - \hat{v}\|_2\big\} &\le \operatorname{dist}\big((x, v), \operatorname{gph} T\big) + \gamma \\ &\le \operatorname{exs}\big(\operatorname{gph} S \cap \mathbb{B}_*(0, \rho);\ \operatorname{gph} T\big) + \gamma \\ &\le \hat{d\!l}_\rho(\operatorname{gph} S, \operatorname{gph} T) + \gamma.\end{aligned}$$

Moreover, one has

$$\|\hat{v} - \bar{v}\|_2 \le \|\hat{v} - v\|_2 + \|v - \bar{v}\|_2 \le \hat{d\!l}_\rho(\operatorname{gph} S, \operatorname{gph} T) + \gamma + \varepsilon \le \delta,$$

which implies that $\hat{x} \in T^{-1}(\mathbb{B}(\bar{v}, \delta))$. We've established that

$$\operatorname{exs}\Big(S^{-1}\big(\mathbb{B}(\bar{v}, \varepsilon)\big) \cap \mathbb{B}(0, \rho);\ T^{-1}\big(\mathbb{B}(\bar{v}, \delta)\big)\Big) \le \hat{d\!l}_\rho(\operatorname{gph} S, \operatorname{gph} T) + \gamma.$$

Since $\gamma$ is arbitrary, the conclusion follows. □

We recall from (7.23) that $S^{-1}(\mathbb{B}(\bar{v}, \varepsilon))$ is the set of $\varepsilon$-solutions to the generalized equation $\bar{v} \in S(x)$ and, similarly, $T^{-1}(\mathbb{B}(\bar{v}, \delta))$ is the set of $\delta$-solutions of $\bar{v} \in T(x)$. Consequently, the theorem asserts that each one of those $\varepsilon$-solutions in $\mathbb{B}(0, \rho)$ is within a Euclidean distance equal to $\hat{d\!l}_\rho(\operatorname{gph} S, \operatorname{gph} T)$ from one of those $\delta$-solutions. In addition to minor restrictions on $\rho$, the result is contingent on $\delta$ being larger than $\varepsilon + \hat{d\!l}_\rho(\operatorname{gph} S, \operatorname{gph} T)$. In the case $\varepsilon = 0$, we can apply the theorem to establish that for any solution $\bar{x} \in \mathbb{B}(0, \rho)$ of the generalized equation $\bar{v} \in S(x)$, one has

$$\operatorname{dist}\Big(\bar{x},\ T^{-1}\big(\mathbb{B}(\bar{v}, \delta)\big)\Big) \le \hat{d\!l}_\rho(\operatorname{gph} S, \operatorname{gph} T)$$

provided that $\delta > \hat{d\!l}_\rho(\operatorname{gph} S, \operatorname{gph} T)$. Thus, $\bar{x}$ is "near a near-solution" to $\bar{v} \in T(x)$, with proximity being quantified as $\hat{d\!l}_\rho(\operatorname{gph} S, \operatorname{gph} T)$.

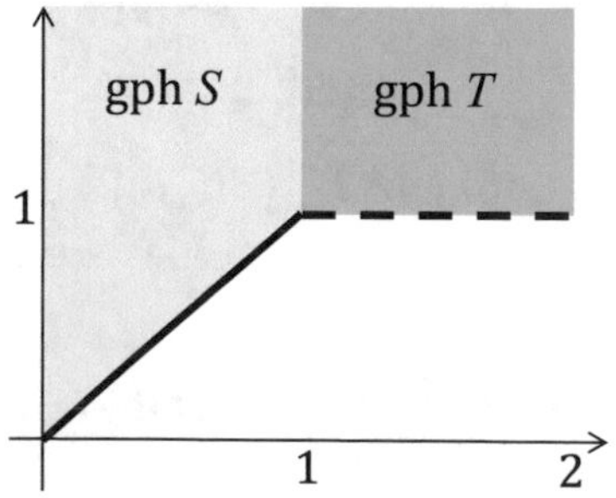


Fig. 7.11: The bound is sharp in the theorem of solution error 7.48.

The theorem doesn't permit the same tolerance for the two generalized equations (i.e., $\varepsilon = \delta$) and this restriction is unavoidable without introducing additional assumptions. In fact, the theorem is sharp as seen in Figure 7.11, where

$$S(x) = \begin{cases} [x, \infty) & \text{if } x \in [0,1] \\ \emptyset & \text{otherwise} \end{cases} \qquad T(x) = \begin{cases} (1, \infty) & \text{if } x \in [1,2] \\ \emptyset & \text{otherwise.} \end{cases}$$

Then for $\rho \geq 0$, $\hat{dl}_\rho(\operatorname{gph} S, \operatorname{gph} T) = 1$, $S^{-1}(0) = \{0\}$, $T^{-1}(\delta) = [1,2]$ and

$$\operatorname{exs}\Big(S^{-1}(0) \cap \mathbb{B}(0,\rho);\ T^{-1}\big(\mathbb{B}(0,\delta)\big)\Big) = 1$$

when $\delta > 1$. If $\delta \leq 1$, then the excess becomes infinity because $T^{-1}(\delta) = \emptyset$.

The truncated Hausdorff distance "almost" satisfies the triangle inequality and this presents opportunities for bounding a distance using two more accessible distances.

**Proposition 7.49** (triangle inequality, extended sense). *For $C_1, C_2, C_3 \subset \mathbb{R}^n$, $\rho \in [0, \infty)$, any norm on $\mathbb{R}^n$ and $\bar\rho > 2\rho + \max_{i=1,2,3} \operatorname{dist}_*(0, C_i)$, one has*

$$\hat{dl}_\rho(C_1, C_3) \leq \hat{dl}_{\bar\rho}(C_1, C_2) + \hat{dl}_{\bar\rho}(C_2, C_3).$$

**Proof.** Let $\mathbb{B}_*(\bar x, \delta) = \{x \in \mathbb{R}^n \mid \|x - \bar x\| \leq \delta\}$ be a ball in $\mathbb{R}^n$ under the norm $\|\cdot\|$ adopted in the proposition. For any $C, D \subset \mathbb{R}^n$, let's define

$$dl_\rho(C, D) = \sup_{\|x\| \leq \rho} \big|\operatorname{dist}_*(x, C) - \operatorname{dist}_*(x, D)\big|,$$

where $\operatorname{dist}_*$ is also defined in terms of $\|\cdot\|$. It turns out that $dl_\rho$ is closely related to $\hat{dl}_\rho$ and satisfies the triangle inequality

$$dl_\rho(C_1, C_3) \leq dl_\rho(C_1, C_2) + dl_\rho(C_2, C_3).$$

To establish this inequality, note that for any $x \in \mathbb{B}_*(0, \rho)$,

$$\big|\operatorname{dist}_*(x, C_1) - \operatorname{dist}_*(x, C_3)\big| \leq \big|\operatorname{dist}_*(x, C_1) - \operatorname{dist}_*(x, C_2)\big| + \big|\operatorname{dist}_*(x, C_2) - \operatorname{dist}_*(x, C_3)\big|.$$

A maximization of the terms on the right-hand side establishes that

$$\big|\operatorname{dist}_*(x, C_1) - \operatorname{dist}_*(x, C_3)\big| \leq dl_\rho(C_1, C_2) + dl_\rho(C_2, C_3).$$

Since $x$ is arbitrary, we've proven the triangle inequality for $dl_\rho$.

Next, we relate $\hat{dl}_\rho$ to $dl_\rho$. Suppose that $C, D \subset \mathbb{R}^n$ are nonempty and $\varepsilon \in (0, \infty)$. Then,

$$\begin{aligned} &\operatorname{dist}_*(x, D) \leq \operatorname{dist}_*(x, C) + \varepsilon \quad \forall x \in \mathbb{B}_*(0, \rho) \\ \Longrightarrow \quad & C \cap \mathbb{B}_*(0, \rho) \subset D_\varepsilon^+ = \big\{\bar x \in \mathbb{R}^n \;\big|\; \operatorname{dist}_*(\bar x, D) \leq \varepsilon\big\}. \end{aligned}$$

This holds because $\bar{x} \in C \cap \mathbb{B}_*(0,\rho)$ has $\operatorname{dist}_*(\bar{x}, C) = 0$. Let's apply this fact with $\varepsilon = d\!l_\rho(C, D)$. Then, $\bar{x} \in C \cap \mathbb{B}_*(0,\rho)$ implies that $\operatorname{dist}_*(\bar{x}, D) \leq d\!l_\rho(C, D)$ and also

$$\operatorname{exs}\big(C \cap \mathbb{B}_*(0,\rho); D\big) \leq d\!l_\rho(C, D).$$

Due to symmetry between $C$ and $D$, we conclude that

$$\hat{d\!l}_\rho(C, D) \leq d\!l_\rho(C, D).$$

For a relation the other way around, we start by proving that

$$C \cap \mathbb{B}_*(0,\bar{\rho}) \subset D_\varepsilon^+ \implies \operatorname{dist}_*(x, D) \leq \operatorname{dist}_*(x, C) + \varepsilon \quad \forall x \in \mathbb{B}_*(0,\rho). \tag{7.30}$$

For $x \in C \cap \mathbb{B}_*(0,\bar{\rho}) \subset D_\varepsilon^+$ and $\bar{x} \in \mathbb{R}^n$, one has

$$\operatorname{dist}_*(\bar{x}, D) \leq \|\bar{x} - x\| + \operatorname{dist}_*(x, D) \leq \|\bar{x} - x\| + \varepsilon.$$

The minimization over $x \in C \cap \mathbb{B}_*(0,\bar{\rho})$ gives that

$$\operatorname{dist}_*(\bar{x}, D) \leq \operatorname{dist}_*\big(\bar{x}, C \cap \mathbb{B}_*(0,\bar{\rho})\big) + \varepsilon. \tag{7.31}$$

This holds trivially if $C \cap \mathbb{B}_*(0,\bar{\rho}) = \emptyset$. Suppose that $\bar{x} \in \mathbb{B}_*(0,\rho)$. For every $\nu$, there's $\bar{y}^\nu \in C$ such that $\|\bar{x} - \bar{y}^\nu\| \leq \operatorname{dist}_*(\bar{x}, C) + \nu^{-1}$. Moreover,

$$\|\bar{y}^\nu\| \leq \|\bar{x}\| + \|\bar{x} - \bar{y}^\nu\| \leq \|\bar{x}\| + \operatorname{dist}_*(\bar{x}, C) + \nu^{-1} \leq 2\rho + \operatorname{dist}_*(0, C) + \nu^{-1}.$$

Let's assume that $\bar{\rho} > 2\rho + \operatorname{dist}_*(0, C)$. Then, there's $\bar{\nu}$ such that $\bar{y}^\nu \in C \cap \mathbb{B}_*(0,\bar{\rho})$ for all $\nu \geq \bar{\nu}$. For such $\nu$,

$$\operatorname{dist}_*\big(\bar{x}, C \cap \mathbb{B}_*(0,\bar{\rho})\big) \leq \|\bar{x} - \bar{y}^\nu\| \leq \operatorname{dist}_*(\bar{x}, C) + \nu^{-1}.$$

Letting $\nu \to \infty$ in this expression and observing that $\operatorname{dist}_*(\bar{x}, C \cap \mathbb{B}_*(0,\bar{\rho})) \geq \operatorname{dist}_*(\bar{x}, C)$ generally, we obtain that

$$\operatorname{dist}_*\big(\bar{x}, C \cap \mathbb{B}_*(0,\bar{\rho})\big) = \operatorname{dist}_*(\bar{x}, C),$$

which together with (7.31) establishes (7.30). The implication in (7.30) then ensures that

$$\hat{d\!l}_{\bar{\rho}}(C, D) \geq d\!l_\rho(C, D).$$

The sets $C_1, C_2, C_3$ are nonempty due to the requirement on $\bar{\rho}$ and they can play the roles of $C$ and $D$ in the above expressions. This leads to

$$\hat{d\!l}_\rho(C_1, C_3) \leq d\!l_\rho(C_1, C_3) \leq d\!l_\rho(C_1, C_2) + d\!l_\rho(C_2, C_3) \leq \hat{d\!l}_{\bar{\rho}}(C_1, C_2) + \hat{d\!l}_{\bar{\rho}}(C_2, C_3)$$

as claimed. □

The proposition is a key component in the proof of the following sum rule, where we say that $S : \mathbb{R}^n \rightrightarrows \mathbb{R}^m$ is *Lipschitz continuous with modulus function* $\kappa$ *at level* $\check{\rho} \in [0, \infty]$ if $\kappa : [0, \infty) \to [0, \infty)$ and

$$\hat{dl}_{\check{\rho}}\big(S(x), S(\bar{x})\big) \leq \kappa(\rho)\|x - \bar{x}\|_2 \quad \forall x, \bar{x} \in \mathbb{B}(0, \rho) \text{ and } \rho \in [0, \infty).$$

Here, the truncated Hausdorff distance utilizes the Euclidean norm for $\mathbb{R}^m$.

**Proposition 7.50** (sum rule for set-valued mappings). *For* $S_i, T_i : \mathbb{R}^n \rightrightarrows \mathbb{R}^m$, $i = 1, 2$, *suppose that* $\operatorname{dom} S_1 = \operatorname{dom} T_1 = \mathbb{R}^n$, $\operatorname{dom} S_2$ *and* $\operatorname{dom} T_2$ *are nonempty,* $\rho \in [0, \infty)$ *and* $S_1$ *as well as* $T_1$ *are Lipschitz continuous with common modulus function* $\kappa$ *at level* $\check{\rho} \in [0, \infty]$. *Then, under the norm (7.29) for* $\mathbb{R}^{n+m}$, *one has*

$$\hat{dl}_{\rho}\big(\operatorname{gph}(S_1 + S_2), \operatorname{gph}(T_1 + T_2)\big) \leq \sup_{\|x\|_2 \leq \rho} \hat{dl}_{\check{\rho}}\big(S_1(x), T_1(x)\big) + \big(1 + \kappa(\hat{\rho})\big)\hat{dl}_{\bar{\rho}}(\operatorname{gph} S_2, \operatorname{gph} T_2)$$

*provided that*

$$\bar{\rho} \geq \rho + \rho', \quad \text{with } \rho' \text{ such that } S_1(x), T_1(x) \subset \mathbb{B}(0, \rho') \ \forall x \in \mathbb{B}(0, \rho)$$
$$\hat{\rho} > \rho + \hat{dl}_{\bar{\rho}}(\operatorname{gph} S_2, \operatorname{gph} T_2)$$
$$\check{\rho} > 3\rho' + \kappa(\hat{\rho})(\hat{\rho} - \rho).$$

**Proof.** Let $\mathbb{B}_*(\bar{z}, \delta) = \{z \in \mathbb{R}^{n+m} \mid \|z - \bar{z}\| \leq \delta\}$, where $\|\cdot\|$ is the norm (7.29). Suppose that

$$(x, y) \in \operatorname{gph}(T_1 + T_2) \cap \mathbb{B}_*(0, \rho).$$

Thus, for some $y_1 \in T_1(x)$ and $y_2 \in T_2(x)$, we've $y = y_1 + y_2$ and

$$\|y_2\|_2 \leq \|y\|_2 + \|y_1\|_2 \leq \rho + \rho' \leq \bar{\rho}.$$

Let

$$\varepsilon \in \big(0, \hat{\rho} - \rho - \hat{dl}_{\bar{\rho}}(\operatorname{gph} S_2, \operatorname{gph} T_2)\big].$$

Now, $(x, y_2) \in \operatorname{gph} T_2 \cap \mathbb{B}_*(0, \bar{\rho})$ so there's $(\bar{x}, \bar{y}_2) \in \operatorname{gph} S_2$ with

$$\max\big\{\|x - \bar{x}\|_2, \|y_2 - \bar{y}_2\|_2\big\} \leq \hat{dl}_{\bar{\rho}}(\operatorname{gph} S_2, \operatorname{gph} T_2) + \varepsilon \leq \hat{\rho} - \rho,$$

which ensures that

$$\|\bar{x}\|_2 \leq \|x - \bar{x}\|_2 + \|x\|_2 \leq \hat{\rho} - \rho + \rho \leq \hat{\rho}.$$

Since $\operatorname{dom} S_1 = \mathbb{R}^n$, there's $\bar{y}_1 \in S_1(\bar{x})$ such that

$$\operatorname{dist}\big(y_1, S_1(\bar{x})\big) \geq \|y_1 - \bar{y}_1\|_2 - \varepsilon.$$

Then, $(\bar{x}, \bar{y}_1 + \bar{y}_2) \in \operatorname{gph}(S_1 + S_2)$. Since $\|y_1\|_2 \leq \rho'$, it follows that

$$\|y_1 - \bar{y}_1\|_2 \leq \operatorname{dist}\big(y_1, S_1(\bar{x})\big) + \varepsilon \leq \hat{dl}_{\rho'}\big(S_1(\bar{x}), T_1(x)\big) + \varepsilon$$
$$\leq \hat{dl}_{\breve{\rho}}\big(S_1(\bar{x}), S_1(x)\big) + \hat{dl}_{\breve{\rho}}\big(S_1(x), T_1(x)\big) + \varepsilon,$$

where the last inequality is a consequence of 7.49; $\breve{\rho}$ is indeed sufficiently large because $\operatorname{dist}(0, T_1(x)) \leq \rho'$, $\operatorname{dist}(0, S_1(x)) \leq \rho'$ and

$$\operatorname{dist}\big(0, S_1(\bar{x})\big) \leq \rho' + \operatorname{exs}\big(S_1(x) \cap \mathbb{B}(0, \breve{\rho}); S_1(\bar{x})\big) \leq \rho' + \hat{dl}_{\breve{\rho}}\big(S_1(x), S_1(\bar{x})\big)$$
$$\leq \rho' + \kappa(\hat{\rho})\|x - \bar{x}\|_2 \leq \rho' + \kappa(\hat{\rho})(\hat{\rho} - \rho).$$

Next, with $\bar{y} = \bar{y}_1 + \bar{y}_2$, one has

$$\|y - \bar{y}\|_2 \leq \|y_1 - \bar{y}_1\|_2 + \|y_2 - \bar{y}_2\|_2$$
$$\leq \hat{dl}_{\breve{\rho}}\big(S_1(\bar{x}), S_1(x)\big) + \hat{dl}_{\breve{\rho}}\big(S_1(x), T_1(x)\big) + \hat{dl}_{\bar{\rho}}(\operatorname{gph} S_2, \operatorname{gph} T_2) + 2\varepsilon$$
$$\leq \kappa(\hat{\rho})\big(\hat{dl}_{\bar{\rho}}(\operatorname{gph} S_2, \operatorname{gph} T_2) + \varepsilon\big) + \sup_{\|x'\|_2 \leq \rho} \hat{dl}_{\breve{\rho}}\big(S_1(x'), T_1(x')\big)$$
$$+ \hat{dl}_{\bar{\rho}}(\operatorname{gph} S_2, \operatorname{gph} T_2) + 2\varepsilon.$$

This establishes that $(\bar{x}, \bar{y}) \in \operatorname{gph}(S_1 + S_2)$ satisfies

$$\max\big\{\|x - \bar{x}\|_2, \|y - \bar{y}\|_2\big\} \leq \max\Big\{\hat{dl}_{\bar{\rho}}(\operatorname{gph} S_2, \operatorname{gph} T_2) + \varepsilon,$$
$$\big(1 + \kappa(\hat{\rho})\big)\hat{dl}_{\bar{\rho}}(\operatorname{gph} S_2, \operatorname{gph} T_2) + \sup_{\|x'\|_2 \leq \rho} \hat{dl}_{\breve{\rho}}\big(S_1(x'), T_1(x')\big) + \varepsilon\kappa(\hat{\rho}) + 2\varepsilon\Big\}.$$

Since $(x, y)$ and $\varepsilon$ are arbitrary, we obtain

$$\operatorname{exs}\big(\operatorname{gph}(T_1 + T_2) \cap \mathbb{B}_*(0, \rho);\ \operatorname{gph}(S_1 + S_2)\big)$$
$$\leq \sup_{\|x'\|_2 \leq \rho} \hat{dl}_{\breve{\rho}}\big(S_1(x'), T_1(x')\big) + \big(1 + \kappa(\hat{\rho})\big)\hat{dl}_{\bar{\rho}}(\operatorname{gph} S_2, \operatorname{gph} T_2).$$

The roles of $(S_1, S_2)$ and $(T_1, T_2)$ can be reversed so the conclusion holds. □

The assumptions on $S_1$ and $T_1$ are much stronger than those on $S_2$ and $T_2$, but this reflects a common situation in practice. For example, let $C, D \subset \mathbb{R}^n$ be nonempty sets and $f, g : \mathbb{R}^n \to \mathbb{R}$ be smooth functions for which there's $\kappa : [0, \infty) \to [0, \infty)$ such that

$$\big\|\nabla f(x) - \nabla f(\bar{x})\big\|_2 \leq \kappa(\rho)\|x - \bar{x}\|_2 \quad \forall x, \bar{x} \in \mathbb{B}(0, \rho), \quad \rho \in [0, \infty),$$

with the same condition holding for $\nabla g$. We can then apply the proposition with

$$S_1(x) = \big\{\nabla f(x)\big\}, \quad T_1(x) = \big\{\nabla g(x)\big\}, \quad S_2(x) = N_C(x), \quad T_2(x) = N_D(x).$$

Now, $\operatorname{dom} S_1 = \operatorname{dom} T_1 = \mathbb{R}^n$ and

$$\hat{dl}_{\breve{\rho}}(S_1(x), S_1(\bar{x})) = \|\nabla f(x) - \nabla f(\bar{x})\|_2$$

when $\breve{\rho} = \infty$, with the same situation for $\hat{dl}_{\breve{\rho}}(T_1(x), T_1(\bar{x}))$. Then, one has

$$\begin{aligned}&\hat{dl}_\rho\big(\operatorname{gph}(\nabla f + N_C), \operatorname{gph}(\nabla g + N_D)\big)\\ &\le \eta = \sup_{\|x\|_2 \le \rho} \|\nabla f(x) - \nabla g(x)\|_2 + \big(1 + \kappa(\hat{\rho})\big)\hat{dl}_{\bar{\rho}}(\operatorname{gph} N_C, \operatorname{gph} N_D)\end{aligned}$$

by the proposition provided that

$$\begin{aligned}\bar{\rho} &\ge \rho + \sup_{\|x\|_2 \le \rho} \max\big\{\|\nabla f(x)\|_2, \|\nabla g(x)\|_2\big\}\\ \hat{\rho} &> \rho + \hat{dl}_{\bar{\rho}}(\operatorname{gph} N_C, \operatorname{gph} N_D).\end{aligned}$$

This fact can then be combined with the theorem of solution error 7.48 to conclude that, with

$$S(x) = \nabla f(x) + N_C(x) \quad \text{and} \quad T(x) = \nabla g(x) + N_D(x),$$

one has

$$\operatorname{exs}\Big(S^{-1}(0) \cap \mathbb{B}(0, \rho);\ T^{-1}\big(\mathbb{B}(0, \delta)\big)\Big) \le \eta$$

if $\delta > \eta$. This means that the distance from a solution in $\mathbb{B}(0, \rho)$ of the generalized equation $0 \in \nabla f(x) + N_C(x)$ to a $\delta$-solution of $0 \in \nabla g(x) + N_D(x)$ is at most $\eta$.

When $C$ and $D$ are convex, the distance between the normal cones can be related to the distance between the sets, which in turn can be bounded using 6.63.

**Proposition 7.51** (approximation of subgradient). *For proper, lsc and convex functions $f, g : \mathbb{R}^n \to \overline{\mathbb{R}}$, $\rho \in [0, \infty)$ and norms (6.36) and (7.29) on $\mathbb{R}^{n+1}$ and $\mathbb{R}^{n+n}$, respectively, suppose that*

$$\rho > \max\big\{\operatorname{dist}_*(0, \operatorname{epi} f), \operatorname{dist}_*(0, \operatorname{epi} g)\big\}.$$

*Then, there are $\alpha, \bar{\rho} \in [0, \infty)$, dependent on $\rho$, such that*

$$\hat{dl}_\rho\big(\operatorname{gph} \partial f, \operatorname{gph} \partial g\big) \le \alpha\sqrt{\hat{dl}_{\bar{\rho}}(\operatorname{epi} f, \operatorname{epi} g)}.$$

*For closed convex $C, D \subset \mathbb{R}^n$, $\rho$ satisfying*

$$\infty > \rho > \max\big\{\operatorname{dist}(0, C), \operatorname{dist}(0, D)\big\}$$

*and norm (7.29) on $\mathbb{R}^{n+n}$, there are $\alpha', \rho' \in [0, \infty)$, dependent on $\rho$, such that*

$$\hat{dl}_\rho(\operatorname{gph} N_C, \operatorname{gph} N_D) \le \alpha'\sqrt{\hat{dl}_{\rho'}(C, D)}.$$

**Proof.** We refer to [5] for a proof, which gives expressions for the constants; see also [108]. □

We end the section with a widely applicable result for composite functions quantifying the graphical convergence in 7.45 and adopt the following terminology: For smooth $f : \mathbb{R}^n \to \mathbb{R}$, we say that $\nabla f$ is *Lipschitz continuous with modulus function* $\kappa$ if $\kappa : [0, \infty) \to [0, \infty)$ and

$$\|\nabla f(x) - \nabla f(\bar{x})\|_2 \le \kappa(\rho)\|x - \bar{x}\|_2 \quad \forall x, \bar{x} \in \mathbb{B}(0, \rho), \quad \rho \in [0, \infty).$$

For smooth $F : \mathbb{R}^n \to \mathbb{R}^m$, we say that $\nabla F$ is *Lipschitz continuous with modulus function* $\mu$ if $\mu : [0, \infty) \to [0, \infty)$ and

$$\|\nabla F(x) - \nabla F(\bar{x})\|_F \le \mu(\rho)\|x - \bar{x}\|_2 \quad \forall x, \bar{x} \in \mathbb{B}(0, \rho), \quad \rho \in [0, \infty).$$

**Theorem 7.52** (approximation error in optimality conditions). *For nonempty closed* $X, \hat{X} \subset \mathbb{R}^n$, *smooth* $f_0, \hat{f}_0 : \mathbb{R}^n \to \mathbb{R}$, *smooth* $F, \hat{F} : \mathbb{R}^n \to \mathbb{R}^m$ *and proper lsc* $h, \hat{h} : \mathbb{R}^m \to \overline{\mathbb{R}}$, *suppose that the gradients* $\nabla f_0$ *and* $\nabla \hat{f}_0$ *are Lipschitz continuous with common modulus function* $\kappa$ *and the Jacobians* $\nabla F$ *and* $\nabla \hat{F}$ *are Lipschitz continuous with common modulus function* $\mu$.

*Consider* $S, \hat{S} : \mathbb{R}^{n+2m} \rightrightarrows \mathbb{R}^{2m+n}$ *given by*

$$S(x, y, z) = \{F(x) - z\} \times \{\partial h(z) - y\} \times \{\nabla f_0(x) + \nabla F(x)^\top y + N_X(x)\}$$
$$\hat{S}(x, y, z) = \{\hat{F}(x) - z\} \times \{\partial \hat{h}(z) - y\} \times \{\nabla \hat{f}_0(x) + \nabla \hat{F}(x)^\top y + N_{\hat{X}}(x)\}.$$

*If* $\rho \in [0, \infty)$, *the norms on* $\mathbb{R}^{2n}$ *and* $\mathbb{R}^{2m}$ *are of the form (7.29) and the norm on* $\mathbb{R}^{n+4m+n}$ *is of the form*

$$\max\{\|x\|_2, \|y\|_2, \|z\|_2, \|u\|_2, \|v\|_2, \|w\|_2\}, \tag{7.32}$$

*then, with* $\eta = \hat{dl}_{\bar{\rho}}(\text{gph}\, N_{\hat{X}}, \text{gph}\, N_X)$ *and* $\hat{\eta} = (1 + \kappa(\hat{\rho}) + \rho\mu(\hat{\rho}))\eta$, *one has*

$$\hat{dl}_\rho(\text{gph}\, S, \text{gph}\, \hat{S}) \le \max\Big\{\eta\breve{\rho} + \sup_{\|x\|_2 \le \rho} \|\hat{F}(x) - F(x)\|_2 + \hat{dl}_{2\rho}(\text{gph}\, \partial\hat{h}, \text{gph}\, \partial h),$$
$$\hat{\eta} + \sup_{\|x\|_2 \le \rho} \|\nabla \hat{f}_0(x) - \nabla f_0(x)\|_2 + \rho \sup_{\|x\|_2 \le \rho} \|\nabla \hat{F}(x) - \nabla F(x)\|_F\Big\}$$

*provided that*

$$\bar{\rho} \ge \rho + \max\Big\{\sup_{\|x\|_2 \le \rho} \|\nabla f_0(x)\|_2, \ \sup_{\|x\|_2 \le \rho} \|\nabla \hat{f}_0(x)\|_2\Big\}$$
$$+ \rho \max\Big\{\sup_{\|x\|_2 \le \rho} \|\nabla F(x)\|_F, \ \sup_{\|x\|_2 \le \rho} \|\nabla \hat{F}(x)\|_F\Big\}$$
$$\hat{\rho} \ge \rho + \hat{dl}_{\bar{\rho}}(\text{gph}\, N_{\hat{X}}, \text{gph}\, N_X)$$
$$\breve{\rho} \ge \max\Big\{\sup_{\|x\|_2 \le \hat{\rho}} \|\nabla F(x)\|_F, \ \sup_{\|x\|_2 \le \hat{\rho}} \|\nabla \hat{F}(x)\|_F\Big\}.$$

**Proof.** Suppose that $(\bar{x}, \bar{y}, \bar{z}, \bar{u}, \bar{v}, \bar{w}) \in \operatorname{gph} S \cap \mathbb{B}_*(0, \rho)$, where $\mathbb{B}_*(0, \rho)$ is the ball centered at the origin of $\mathbb{R}^{n+4m+n}$ with radius $\rho$ measured by (7.32). Then,

$$\bar{u} = F(\bar{x}) - \bar{z}, \quad \bar{v} + \bar{y} \in \partial h(\bar{z}), \quad \bar{w} - \nabla f_0(\bar{x}) - \nabla F(\bar{x})^\top \bar{y} \in N_X(\bar{x}).$$

We construct a point $(x, \bar{y}, z, u, v, w) \in \operatorname{gph} \hat{S}$ that's a certain distance from $(\bar{x}, \bar{y}, \bar{z}, \bar{u}, \bar{v}, \bar{w})$ in the norm (7.32).

Let $\varepsilon \in (0, \infty)$. Since $(\bar{x}, \bar{y}, \bar{z}, \bar{u}, \bar{v}, \bar{w}) \in \mathbb{B}_*(0, \rho)$,

$$\max\{\|\bar{z}\|_2, \|\bar{v} + \bar{y}\|_2\} \leq 2\rho.$$

Moreover, $\operatorname{gph} \partial \hat{h}$ is nonempty, which follows largely from the definition of subgradients; cf. [105, Corollary 8.10]. These facts ensure that there are $z, v \in \mathbb{R}^m$ such that $(z, v + \bar{y}) \in \operatorname{gph} \partial \hat{h}$ and

$$\max\Big\{\|z - \bar{z}\|_2, \big\|(v + \bar{y}) - (\bar{v} + \bar{y})\big\|_2\Big\} \leq \hat{d\!l}_{2\rho}(\operatorname{gph} \partial h, \operatorname{gph} \partial \hat{h}) + \varepsilon.$$

We observe that

$$\big\|\bar{w} - \nabla f_0(\bar{x}) - \nabla F(\bar{x})^\top \bar{y}\big\|_2 \leq \|\bar{w}\|_2 + \big\|\nabla f_0(\bar{x})\big\|_2 + \rho\big\|\nabla F(\bar{x})\big\|_F \leq \bar{\rho},$$

which implies that there's $(x, \hat{w}) \in \operatorname{gph} N_{\hat{X}}$ such that

$$\max\Big\{\|x - \bar{x}\|_2, \big\|\hat{w} - \bar{w} + \nabla f_0(\bar{x}) + \nabla F(\bar{x})^\top \bar{y}\big\|_2\Big\} \leq \hat{d\!l}_{\bar{\rho}}(\operatorname{gph} N_{\hat{X}}, \operatorname{gph} N_X).$$

(Since $\operatorname{gph} N_{\hat{X}}$ is closed due to the closedness of $\hat{X}$ and 7.22, there's no need for $\varepsilon$ on the right-hand side in this case.) We then construct

$$w = \hat{w} + \nabla \hat{f}_0(x) + \nabla \hat{F}(x)^\top \bar{y} \quad \text{and} \quad u = \hat{F}(x) - z.$$

It now follows that $(x, \bar{y}, z, u, v, w) \in \operatorname{gph} \hat{S}$ and we just need to compute the distance between this point and $(\bar{x}, \bar{y}, \bar{z}, \bar{u}, \bar{v}, \bar{w})$ in the norm (7.32). We already have the bounds

$$\|x - \bar{x}\|_2 \leq \hat{d\!l}_{\bar{\rho}}(\operatorname{gph} N_{\hat{X}}, \operatorname{gph} N_X)$$
$$\|z - \bar{z}\|_2 \leq \hat{d\!l}_{2\rho}(\operatorname{gph} \partial h, \operatorname{gph} \partial \hat{h}) + \varepsilon$$
$$\|v - \bar{v}\|_2 \leq \hat{d\!l}_{2\rho}(\operatorname{gph} \partial h, \operatorname{gph} \partial \hat{h}) + \varepsilon.$$

Next, we note that

$$\|x\|_2 \leq \|\bar{x}\|_2 + \|x - \bar{x}\|_2 \leq \rho + \hat{d\!l}_{\bar{\rho}}(\operatorname{gph} N_{\hat{X}}, \operatorname{gph} N_X) \leq \hat{\rho}.$$

The mean value theorem (1.11) establishes that

$$\|\hat{F}(x)-\hat{F}(\bar{x})\|_2 \le \int_0^1 \|\nabla\hat{F}(\bar{x}+\sigma(x-\bar{x}))\|_F \|x-\bar{x}\|_2\, d\sigma \le \sup_{\|x'\|_2\le\hat{\rho}} \|\nabla\hat{F}(x')\|_F \|x-\bar{x}\|_2.$$

Utilizing this fact, we obtain

$$\begin{aligned}\|u-\bar{u}\|_2 &\le \|\hat{F}(x)-\hat{F}(\bar{x})\|_2 + \|\hat{F}(\bar{x})-F(\bar{x})\|_2 + \|z-\bar{z}\|_2\\ &\le \sup_{\|x'\|_2\le\hat{\rho}} \|\nabla\hat{F}(x')\|_F\, \hat{d\!l}_{\bar{\rho}}(\operatorname{gph} N_{\hat{X}}, \operatorname{gph} N_X)\\ &\quad + \sup_{\|x'\|_2\le\rho} \|\hat{F}(x')-F(x')\|_2 + \hat{d\!l}_{2\rho}(\operatorname{gph}\partial h, \operatorname{gph}\partial\hat{h}) + \varepsilon.\end{aligned}$$

Finally, one has

$$\begin{aligned}\|w-\bar{w}\|_2 &\le \|\hat{w}-\bar{w}+\nabla f_0(\bar{x})+\nabla F(\bar{x})^\top\bar{y}\|_2\\ &\quad + \|\nabla\hat{f}_0(x)+\nabla\hat{F}(x)^\top\bar{y}-\nabla f_0(\bar{x})-\nabla F(\bar{x})^\top\bar{y}\|_2\\ &\le \hat{d\!l}_{\bar{\rho}}(\operatorname{gph} N_{\hat{X}}, \operatorname{gph} N_X) + \|\nabla\hat{f}_0(x)-\nabla\hat{f}_0(\bar{x})\|_2 + \|\nabla\hat{f}_0(\bar{x})-\nabla f_0(\bar{x})\|_2\\ &\quad + \|\nabla\hat{F}(x)-\nabla\hat{F}(\bar{x})\|_F\|\bar{y}\|_2 + \|\nabla\hat{F}(\bar{x})-\nabla F(\bar{x})\|_F\|\bar{y}\|_2\\ &\le \hat{d\!l}_{\bar{\rho}}(\operatorname{gph} N_{\hat{X}}, \operatorname{gph} N_X) + \kappa(\hat{\rho})\|x-\bar{x}\|_2 + \sup_{\|x'\|_2\le\rho}\|\nabla\hat{f}_0(x')-\nabla f_0(x')\|_2\\ &\quad + \rho\mu(\hat{\rho})\|x-\bar{x}\|_2 + \rho\sup_{\|x'\|_2\le\rho}\|\nabla\hat{F}(x')-\nabla F(x')\|_F\\ &\le \big(1+\kappa(\hat{\rho})+\rho\mu(\hat{\rho})\big)\hat{d\!l}_{\bar{\rho}}(\operatorname{gph} N_{\hat{X}}, \operatorname{gph} N_X)\\ &\quad + \sup_{\|x'\|_2\le\rho}\|\nabla\hat{f}_0(x')-\nabla f_0(x')\|_2 + \rho\sup_{\|x'\|_2\le\rho}\|\nabla\hat{F}(x')-\nabla F(x')\|_F.\end{aligned}$$

Consequently, the distance between $(x, \bar{y}, z, u, v, w)$ and $(\bar{x}, \bar{y}, \bar{z}, \bar{u}, \bar{v}, \bar{w})$ is at most the larger of the bounds on $\|u-\bar{u}\|_2$ and $\|w-\bar{w}\|_2$. Since $\varepsilon$ is arbitrary, the distance is also bounded by the same expression with $\varepsilon = 0$. This furnishes a bound on $\operatorname{exs}(\operatorname{gph} S \cap \mathbb{B}_*(0,\rho); \operatorname{gph}\hat{S})$. We repeat the argument with the roles of $S$ and $\hat{S}$ reversed and reach the conclusion. □

The theorem breaks down the discrepancy between $\hat{S}$ and $S$ in terms of the differences between $\hat{X}$ and $X$, $\partial\hat{h}$ and $\partial h$, $\nabla\hat{f}_0$ and $\nabla f_0$, $\hat{F}$ and $F$ and also their Jacobians. In applications, only some of these differences may be present resulting in simplifications. If $\hat{X}$ and $X$ and/or $\hat{h}$ and $h$ are convex, then 7.51 can be brought in together with earlier estimates for sets and functions; see §6.J as well as [73]. Alternatively, these sets and functions might be separable and then the following fact is useful.

**Proposition 7.53** (separable functions). *For proper lsc $f_j, g_j : \mathbb{R} \to \overline{\mathbb{R}}$, $j = 1, \dots, n$, consider the functions $f, g : \mathbb{R}^n \to \overline{\mathbb{R}}$ given by*

$$f(x) = \sum\nolimits_{j=1}^n f_j(x_j) \quad \text{and} \quad g(x) = \sum\nolimits_{j=1}^n g_j(x_j).$$

*If the norms on $\mathbb{R}^{2n}$ and $\mathbb{R}^2$ are of the form (7.29), then, for $\rho \in [0,\infty)$, one has*

$$\hat{d\!l}_\rho(\operatorname{gph}\partial f, \operatorname{gph}\partial g) \le \sqrt{n}\max_{j=1,\dots,n} \hat{d\!l}_\rho(\operatorname{gph}\partial f_j, \operatorname{gph}\partial g_j).$$

**Proof.** Let $\varepsilon > 0$ and $(\bar{x}, \bar{v}) \in \operatorname{gph} \partial f \cap \mathbb{B}_*(0, \rho)$, where $\mathbb{B}_*(0, \rho)$ is the ball centered at the origin of $\mathbb{R}^{2n}$ with radius $\rho$ measured by (7.29). Since $(\bar{x}_j, \bar{v}_j) \in \operatorname{gph} \partial f_j$ by 4.63 and $\max\{|\bar{x}_j|, |\bar{v}_j|\} \le \rho$, there's $(x_j, v_j) \in \operatorname{gph} \partial g_j$ with

$$\max\big\{|x_j - \bar{x}_j|, |v_j - \bar{v}_j|\big\} \le \hat{d\!l}_\rho(\operatorname{gph} \partial f_j, \operatorname{gph} \partial g_j) + \varepsilon.$$

We can repeat this construction for all $j = 1, \ldots, n$ and obtain $x = (x_1, \ldots, x_n)$ and $v = (v_1, \ldots, v_n)$. Thus, $v \in \partial g(x)$ by 4.63. With

$$\eta = \max_{j=1,\ldots,n} \hat{d\!l}_\rho(\operatorname{gph} \partial f_j, \operatorname{gph} \partial g_j),$$

one has

$$\max\big\{\|x - \bar{x}\|_2, \|v - \bar{v}\|_2\big\} = \sqrt{n}(\eta + \varepsilon).$$

Thus,

$$\operatorname{exs}\big(\operatorname{gph} \partial f \cap \mathbb{B}_*(0, \rho); \operatorname{gph} \partial g\big) \le \sqrt{n}(\eta + \varepsilon).$$

Repeating the argument with the roles of $f$ and $g$ reversed, we establish that

$$\hat{d\!l}_\rho(\operatorname{gph} \partial f, \operatorname{gph} \partial g) \le \sqrt{n}(\eta + \varepsilon).$$

Since this holds for all $\varepsilon > 0$, the conclusion follows. □

**Exercise 7.54** (separable functions). For $\rho \in [0, \infty)$ and $\theta \in (0, \infty)$, apply 7.53 to bound $\hat{d\!l}_\rho(\operatorname{gph} \partial h, \operatorname{gph} \partial \hat{h})$, where

$$h(u) = \sum\nolimits_{i=1}^m \iota_{(-\infty,0]}(u_i) \quad \text{and} \quad \hat{h}(u) = \sum\nolimits_{i=1}^m \theta \max\{0, u_i\}.$$

**Guide.** Define $\varphi(\alpha) = \iota_{(-\infty,0]}(\alpha)$ and $\hat{\varphi}(\alpha) = \theta \max\{0, \alpha\}$ and draw $\operatorname{gph} \partial \varphi$, $\operatorname{gph} \partial \hat{\varphi}$. □

**Exercise 7.55** (normal cones to boxes). For $\rho \in [0, \infty)$, use 7.53 to bound $\hat{d\!l}_\rho(\operatorname{gph} N_{\hat{X}}, \operatorname{gph} N_X)$, where

$$X = \big\{x \in \mathbb{R}^n \;\big|\; \alpha_j \le x_j \le \beta_j, j = 1, \ldots, n\big\}$$
$$\hat{X} = \big\{x \in \mathbb{R}^n \;\big|\; \hat{\alpha}_j \le x_j \le \hat{\beta}_j, j = 1, \ldots, n\big\},$$

with $-\infty \le \alpha_j \le \beta_j \le \infty$ and $-\infty \le \hat{\alpha}_j \le \hat{\beta}_j \le \infty$.

**Guide.** Write $\iota_X(x) = \sum_{j=1}^n \iota_{X_j}(x_j)$, where $X_j = \{\gamma \in \mathbb{R} \mid \alpha_j \le \gamma \le \beta_j\}$. □

# Chapter 8
# RISK MODELING AND SAMPLE AVERAGES

Expectation functions defined in terms of general distributions give rise to challenges when solving stochastic optimization problems. Even to assess the optimality gap for a candidate decision can become nontrivial. One may pass to sample average approximations of expectation functions, but this leads to the need for error analysis. Regardless of the model of uncertainty and the associated probability distribution, formulations of stochastic optimization problems rely on means to quantify a decision-maker's preference for different outcomes of a random phenomenon. A decision-maker might be much more concerned about poor outcomes than average performance and this requires us to broaden the scope beyond expectation functions to other quantifiers that also need to be computationally approachable. In this chapter, we'll expand on these subjects and provide additional tools for modeling and analyzing problems with uncertainty.

## 8.A Estimation of Optimality Gaps

We know from §3.D that expectation functions often need to be approximated and the resulting solutions may therefore fall short of being minimizers of the actual problem of interest. The optimality gap of a candidate solution furnishes a guarantee about the level of sub-optimality and, if sufficiently small, could allow us to conclude that the candidate solution is satisfactory. We'll now show how one can estimate optimality gaps in the case of expectation functions.

Let's consider the problem

$$\underset{x \in C \subset \mathbb{R}^n}{\text{minimize}}\; Ef(x) = \mathbb{E}\big[f(\boldsymbol{\xi}, x)\big], \tag{8.1}$$

where $\boldsymbol{\xi}$ is a random vector with support $\Xi \subset \mathbb{R}^m$ and $f : \mathbb{R}^m \times \mathbb{R}^n \to \mathbb{R}$. Suppose that we're in possession of a candidate solution $\bar{x} \in C$, which might be the output of an algorithm or even simply guesswork. We would like to estimate its optimality gap



J. O. Royset and R. J-B Wets, *An Optimization Primer*, Springer Series in Operations Research and Financial Engineering, https://doi.org/10.1007/978-3-030-76275-9_8

$$Ef(\bar{x}) - \inf_C Ef,$$

and this may lead to the conclusion that $\bar{x}$ is "good enough." Presumably, neither $Ef(\bar{x})$ nor $\inf_C Ef$ is known in an application, the former because the expectation $\mathbb{E}[f(\boldsymbol{\xi}, \bar{x})]$ involves a nontrivial integral. Our goal is to obtain a conservative bound on the optimality gap.

**Proposition 8.1** (downward bias in stochastic optimization). *For a nonempty closed set $C \subset \mathbb{R}^n$ and a random vector $\boldsymbol{\xi}$ with support $\Xi \subset \mathbb{R}^m$, suppose that $f : \mathbb{R}^m \times \mathbb{R}^n \to \mathbb{R}$ satisfies:*

(a) *$f(\boldsymbol{\xi}, x)$ is integrable for all $x \in \mathbb{R}^n$*
(b) *$f(\xi, \cdot\,)$ is continuous for all $\xi \in \Xi$.*

*If $\boldsymbol{\xi}^1, \boldsymbol{\xi}^2, \dots, \boldsymbol{\xi}^\nu$ are random vectors with the same probability distribution as $\boldsymbol{\xi}$, then*[1]

$$\mathbb{E}\Big[\inf_{x \in C} \frac{1}{\nu} \sum\nolimits_{i=1}^{\nu} f(\boldsymbol{\xi}^i, x)\Big] \leq \inf_C Ef.$$

**Proof.** Since $C$ is nonempty and $Ef$ is real-valued by (a), $\inf_C Ef < \infty$. Let $\varepsilon > 0$. Suppose that $\inf_C Ef \in \mathbb{R}$. Then, there's $\bar{x} \in C$ such that

$$Ef(\bar{x}) \leq \inf_C Ef + \varepsilon.$$

The assumptions ensure that $f(\boldsymbol{\xi}, \bar{x})$ as well as $f(\boldsymbol{\xi}^i, \bar{x}), i = 1, \dots, \nu$, are integrable random variables with common expectation. By the properties of expectations in 3.2, extended by 8.53 to the case when the random variables may not be real-valued, one has

$$\begin{aligned} \inf_C Ef + \varepsilon \geq \mathbb{E}\big[f(\boldsymbol{\xi}, \bar{x})\big] &= \frac{1}{\nu} \sum\nolimits_{i=1}^{\nu} \mathbb{E}\big[f(\boldsymbol{\xi}^i, \bar{x})\big] \\ &= \mathbb{E}\Big[\frac{1}{\nu} \sum\nolimits_{i=1}^{\nu} f(\boldsymbol{\xi}^i, \bar{x})\Big] \geq \mathbb{E}\Big[\inf_{x \in C} \frac{1}{\nu} \sum\nolimits_{i=1}^{\nu} f(\boldsymbol{\xi}^i, x)\Big]. \end{aligned}$$

Next, suppose that $\inf_C Ef = -\infty$. Then, there's $\bar{x} \in C$ such that $Ef(\bar{x}) \leq -1/\varepsilon$. A similar argument yields

$$-\frac{1}{\varepsilon} \geq \mathbb{E}\Big[\inf_{x \in C} \frac{1}{\nu} \sum\nolimits_{i=1}^{\nu} f(\boldsymbol{\xi}^i, x)\Big].$$

Since $\varepsilon$ is arbitrary, the asserted inequality holds. □

The proposition cautions against the flawed statement: "the infimum of an expectation function equals the expectation of the infima." This isn't true in general and the inequality in the proposition typically holds strictly as we discuss further in §8.I. Still, we can use the result to estimate an optimality gap.

Suppose that $\bar{x} \in C$ is a candidate solution for the problem (8.1). Under the additional assumption that

[1] We see from 8.51 and 8.52 that the function $(\xi^1, \dots, \xi^\nu) \mapsto \inf_{x \in C} \nu^{-1} \sum_{i=1}^{\nu} f(\xi^i, x)$ is measurable so its expectation is well defined.

$$\inf_{x \in C} \frac{1}{\nu} \sum_{i=1}^{\nu} f(\xi^i, x)$$

is integrable, we can subtract $Ef(\bar{x})$ from each side of the inequality in 8.1 and invoke 3.2 to obtain that

$$Ef(\bar{x}) - \inf_C Ef \leq \mathbb{E}\Big[\frac{1}{\nu} \sum_{i=1}^{\nu} f(\boldsymbol{\xi}^i, \bar{x}) - \inf_{x \in C} \frac{1}{\nu} \sum_{i=1}^{\nu} f(\boldsymbol{\xi}^i, x)\Big]. \tag{8.2}$$

Consequently, the optimality gap for $\bar{x}$ is bounded from above by an expectation, which in turn can be estimated.

**Estimating Expectations.** Suppose that $\boldsymbol{\eta}$ is an $m$-dimensional random vector and $h : \mathbb{R}^m \to \mathbb{R}$ is integrable. We can obtain an estimate of $\mathbb{E}[h(\boldsymbol{\eta})]$ by Monte-Carlo sampling: Generate $\eta^1, \dots, \eta^\tau$ according to the probability distribution of $\boldsymbol{\eta}$ and compute

$$\frac{1}{\tau} \sum_{j=1}^{\tau} h(\eta^j).$$

Prior to sampling the exact values of $\eta^1, \dots, \eta^\tau$ are unknown. Consequently, we assume that instead of specific values, they're iid random vectors $\boldsymbol{\eta}^1, \dots, \boldsymbol{\eta}^\tau$ representing the *possible* values, each with the same probability distribution as $\boldsymbol{\eta}$. Then, the quantity of interest becomes

$$\frac{1}{\tau} \sum_{j=1}^{\tau} h(\boldsymbol{\eta}^j),$$

which is a random variable. The *central limit theorem* quantifies how close this random variable is to $\mathbb{E}[h(\boldsymbol{\eta})]$ by making an assertion about the probability distribution $P^\tau$ of the random variable

$$\sqrt{\tau}\Big(\frac{1}{\tau} \sum_{j=1}^{\tau} h(\boldsymbol{\eta}^j) - \mathbb{E}\big[h(\boldsymbol{\eta})\big]\Big).$$

If $\mathbb{E}[(h(\boldsymbol{\eta}))^2] < \infty$, then the central limit theorem states that $P^\tau(y) \to \Phi(y)$ as $\tau \to \infty$ for all $y \in \mathbb{R}$, where $\Phi$ is the distribution function of a normal random variable with zero mean and variance $\sigma^2 = \text{var}(h(\boldsymbol{\eta}))$. Thus, the distribution function of the error (scaled with $\sqrt{\tau}$) is getting close to a well-known distribution function for large $\tau$. Using the properties of a normal distribution function, this means that the random variable

$$\frac{1}{\tau} \sum_{j=1}^{\tau} h(\boldsymbol{\eta}^j) \text{ is approximately normal with mean } \mathbb{E}\big[h(\boldsymbol{\eta})\big] \text{ and variance } \frac{\sigma^2}{\tau}.$$

This fact is usually expressed in terms of a *confidence interval* for the unknown $\mathbb{E}[h(\boldsymbol{\eta})]$. Given a sample $\eta^1, \dots, \eta^\tau$, obtained by statistically independent sampling according to the probability distribution of $\boldsymbol{\eta}$, let

$$\mu_\tau = \frac{1}{\tau} \sum_{j=1}^{\tau} h(\eta^j) \quad \text{and} \quad s_\tau^2 = \frac{1}{\tau - 1} \sum_{j=1}^{\tau} \big(h(\eta^j) - \mu_\tau\big)^2$$

be estimates of $\mathbb{E}[h(\boldsymbol{\eta})]$ and $\sigma^2$, respectively. (These are essentially the maximum likelihood estimates obtained in 6.3, but the denominator $\tau$ is replaced by $\tau - 1$ for the variance to ensure an unbiased estimator as is common in practice.) Then,

$$\left[\, \mu_\tau - t_{1-\alpha/2}^{\tau-1} s_\tau/\sqrt{\tau},\;\; \mu_\tau + t_{1-\alpha/2}^{\tau-1} s_\tau/\sqrt{\tau} \,\right]$$

is an approximating *two-sided* $(1-\alpha)100\%$ confidence interval for $\mathbb{E}[h(\boldsymbol{\eta})]$, where $t_\alpha^\tau$ is the $\alpha$-quantile of the Student's $t$-distribution with $\tau$ degrees of freedom[2]. Approximately for large $\tau$, the interval covers the actual expectation $\mathbb{E}[h(\boldsymbol{\eta})]$ at least $(1-\alpha)100\%$ of the time as the interval is recomputed for different samples drawn according to the probability distribution of $\boldsymbol{\eta}$. The approximating *one-sided* $(1-\alpha)100\%$ confidence interval for $\mathbb{E}[h(\boldsymbol{\eta})]$ is

$$\left(-\infty,\;\; \mu_\tau + t_{1-\alpha}^{\tau-1} s_\tau/\sqrt{\tau}\,\right],$$

with a similar interpretation.

Let's now apply this procedure for computing confidence intervals to the right-hand side expectation in (8.2), which in turn bounds the optimality gap. Specifically, we seek to estimate $\mathbb{E}[h(\boldsymbol{\eta})]$, where the random vector $\boldsymbol{\eta} = (\boldsymbol{\xi}^1, \dots, \boldsymbol{\xi}^\nu)$ and

$$h(\boldsymbol{\eta}) = \frac{1}{\nu}\sum\nolimits_{i=1}^{\nu} f(\boldsymbol{\xi}^i, \bar{x}) - \inf_{x\in C} \frac{1}{\nu}\sum\nolimits_{i=1}^{\nu} f(\boldsymbol{\xi}^i, x).$$

This requires a sample $h_1, \dots, h_\tau$ of the random variable $h(\boldsymbol{\eta})$ obtained by statistically independent sampling according to the probability distribution of $h(\boldsymbol{\eta})$. Thus, each $h_j$ is determined by generating a sample $\xi^1, \dots, \xi^\nu$ according to the probability distribution of $\boldsymbol{\xi}$, setting $\eta = (\xi^1, \dots, \xi^\nu)$ and computing $h_j = h(\eta)$.

**Gap Estimation Algorithm.**

Data. $\bar{x} \in C$, $\alpha \in (0,1)$ and $\nu, \tau \in \mathbb{N}$.

Step 0. Set $j = 1$.

Step 1. Generate a sample $\eta = (\xi^1, \dots, \xi^\nu)$ according to the probability distribution of $(\boldsymbol{\xi}^1, \dots, \boldsymbol{\xi}^\nu)$, statistically independent of earlier iterations.

Step 2. Compute

$$h_j = \frac{1}{\nu}\sum\nolimits_{i=1}^{\nu} f(\xi^i, \bar{x}) - \inf_{x\in C} \frac{1}{\nu}\sum\nolimits_{i=1}^{\nu} f(\xi^i, x).$$

Step 3. If $j < \tau$, replace $j$ by $j+1$ and go to Step 1.

Step 4. Compute

$$\hat{\mu} = \frac{1}{\tau}\sum\nolimits_{j=1}^{\tau} h_j, \qquad \hat{\sigma}^2 = \frac{1}{\tau-1}\sum\nolimits_{j=1}^{\tau} (h_j - \hat{\mu})^2, \qquad \left[\,0,\;\; \hat{\mu} + t_{1-\alpha}^{\tau-1}\hat{\sigma}/\sqrt{\tau}\,\right].$$

[2] The Student's $t$-distribution replaces the normal distribution here because the variance also needs to be estimated.

The computed interval is an approximating $(1-\alpha)100\%$ confidence interval for the right-hand side in (8.2), which conservatively bounds the optimality gap $Ef(\bar{x}) - \inf_C Ef$. This means that if the algorithm is repeated, the computed intervals can be expected to cover the optimality gap at least $(1-\alpha)100\%$ of the time. For example, if the procedure is carried out once with $\alpha = 0.05$ and a moderately large $\tau$, then we can assert with high confidence that the optimality gap lies in the computed interval. Typically, $\tau$ is set to around 30 because this seems to be sufficiently large for the probability distribution of a sample average to become quite close to a normal distribution.

The lower bound of zero in the confidence interval is better than the typical $-\infty$ for one-sided confidence intervals; see the discussion prior to the algorithm. However, we know that an optimality gap is nonnegative so this improvement is trivially achieved.

The main challenge in the algorithm is the computation of

$$\inf_{x\in C} \frac{1}{\nu}\sum\nolimits_{i=1}^{\nu} f(\xi^i, x)$$

in Step 2. However, this is a stochastic optimization problem with a finite distribution and could be tractable by the methods of Chap. 3. In fact, it suffices to obtain a lower bound on the infimum as this would still translate into an upper bound on the optimality gap. We can therefore consider lower bounding approximations based on duality and other techniques to reduce the computational burden of Step 2.

If the width of the obtained confidence interval is narrow, then we can say with a certain confidence (depending on $\alpha$) that $\bar{x}$ is nearly as good as a minimizer. If the interval is wide, then $\bar{x}$ *might* be poor but this depends on $\nu$ and $\tau$. A small $\tau$ causes $\hat{\sigma}/\sqrt{\tau}$ to be large and one might consider increasing $\tau$ to improve the results. A small $\nu$ tends to make the inequality (8.2) rather loose and thus widens the obtained confidence interval. The remedy would be to increase $\nu$, but this is especially costly because we then need to solve more difficult optimization problems in Step 2 of the procedure. A refined approach that permits essentially only one or two iterations ($\tau = 1$ or 2) in the gap estimation algorithm is described in [9].

**Example 8.2** (validation of truss design). Let's return to §3.E and the design problem

$$\underset{x\in C,\gamma\in\mathbb{R}}{\text{minimize}}\, \mathbb{E}\Big[\gamma + \frac{1}{1-\alpha}\max\big\{0, f(\boldsymbol{\xi}, x) - \gamma\big\}\Big],$$

where

$$C = \Big\{x \in \mathbb{R}^7 \;\Big|\; 0.5 \le x_j \le 2,\ j = 1,\dots,7,\ \sum\nolimits_{j=1}^{7} x_j \le 8\Big\}.$$

Table 3.4 lists candidate designs obtained by solving approximating problems. We can use the gap estimation algorithm to assess the quality of candidate designs.

**Detail.** Consider the design $\bar{x} = (1.334, 1.469, 0.941, 1.024, 1.079, 1.054, 1.098)$ seen in the first row of Table 3.4 and a corresponding $\bar{\gamma}$ (omitted from the table). With $\nu = 10$, $\tau = 30$ and $\alpha = 0.05$, the gap estimation algorithm returns [0, 39.70]. If $\nu = 10000$, the confidence interval narrows to [0, 15.72]. As discussed prior to the example, this tightening of the confidence interval is caused by reduced slack in (8.2). Still, since the objective function values are around −50 in this problem, these confidence intervals are relatively wide and we can't assert that $(\bar{x}, \bar{y})$ is nearly as good as a minimizer.

Can the wide confidence intervals be caused by too small $\tau$? For $\nu = 10$ and $\nu = 10000$, $\hat{\mu}$ is 35.59 and 15.52, respectively, so the answer is no. The widths of the confidence intervals are dominated by the value $\hat{\mu}$ and not $\hat{\sigma}/\sqrt{\tau}$, which are only 2.41 and 0.12, respectively. There's really no merit to increasing $\tau$.

Although one could increase $\nu$ beyond 10000 in an effort to narrow the confidence interval further, this might become computationally costly and the amount of improvement is unlikely to be large due to the substantial $\nu$ already considered. We're left with the belief that $(\bar{x}, \bar{y})$ is poor.

For the design $\tilde{x} = (1.431, 1.433, 1.028, 1.033, 1.026, 1.026, 1.023)$ seen in the fourth row of Table 3.4 and a corresponding $\tilde{y}$ (omitted from the table), the gap estimation algorithm returns [0, 0.07] when using $\nu = 10000$, $\tau = 30$ and $\alpha = 0.05$. In this case, $\hat{\mu} = 0.05$ and $\hat{\sigma}/\sqrt{\tau} = 0.01$. We can assert that $(\tilde{x}, \tilde{y})$ is at most 0.07 worse than a minimizer with high confidence. □

**Exercise 8.3** (stochastic resources). Consider again the product mix problem in 3.1, but now $\boldsymbol{d}_1$ is uniformly distributed on $[4800, 6200]$, $\boldsymbol{d}_2$ is uniformly distributed between $[3930, 4100]$ and the two random variables are statistically independent. Construct an approximating 95% confidence interval of the optimality gap for the candidate solution $\bar{x} = (1072.6, 0, 251.4, 0)$ using the gap estimation algorithm.

**Guide.** In view of §3.A, the underlying problem is to

$$\underset{x\in[0,\infty)^4}{\text{minimize}}\ \langle c, x\rangle + \mathbb{E}\Big[a_1 \max\Big\{0, \sum\nolimits_{j=1}^{4} t_{1j}x_j - \boldsymbol{d}_1\Big\} + a_2 \max\Big\{0, \sum\nolimits_{j=1}^{4} t_{2j}x_j - \boldsymbol{d}_2\Big\}\Big].$$

One can use approaches from Chap. 3 to accomplish Step 2 of the gap estimation algorithm. □

## 8.B Risk and Regret

In Chap. 3, we formulated problems with uncertainty using recourse cost functions as a means to assess a decision. The specific forms of these functions stem from the cost structure of the corresponding applications and from the need to penalize certain decisions. Regardless of the situation, the recourse cost can be viewed as a random variable for each decision. Thus far, we've compared and optimized such (random) recourse costs, appropriately modified with an initial cost, using their expected values. This provides a rich, versatile and computationally attractive approach to decision-making under uncertainty. The approach can be viewed from another angle too, which brings forth the implied attitude to risk, connects with superquantile risk described in §3.C, points to possibilities beyond expected values when assessing random variables and even gives rise to risk-adaptive approaches to data analytics. This alternative perspective is the subject of the next five sections.

Let's discuss the main idea in the context of the newsvendor problem. We recall from §1.C that a newsvendor faces the problem

$$\underset{x \in \mathbb{R}}{\text{minimize}} \; (\gamma - \delta)x + \mathbb{E}\big[\max\{0, \delta(x - \xi)\}\big],$$

where $x$ is the order quantity, $\xi$ is the random demand, assumed to be nonnegative, and $\gamma < \delta$ are the per-unit order cost and sales price, respectively. In this case, the recourse cost function is $y \mapsto \max\{0, \delta y\}$ and represents the additional cost caused by not meeting the requirement $y = x - \xi \le 0$. Suppose for simplicity that $\gamma = 1$ and let $\alpha = 1 - 1/\delta \in (0, 1)$. Then,

$$(\gamma - \delta)x + \mathbb{E}\big[\max\{0, \delta(x - \xi)\}\big] = x + \frac{1}{1-\alpha}\mathbb{E}\big[\max\{0, \xi - x\}\big] - \delta\mathbb{E}[\xi].$$

The term $\delta\mathbb{E}[\xi]$ is the expected revenue when all the demand is satisfied. The remaining expression on the right-hand side is the sum of the initial cost $x$ (under the assumption that $\gamma = 1$) and the expected cost of unmet demand

$$\frac{1}{1-\alpha}\mathbb{E}\big[\max\{0, \xi - x\}\big].$$

We can think of $\delta\mathbb{E}[\xi]$ as the revenue the newsvendor hopes to obtain on average, but the number is too optimistic as the newsvendor may run out when the demand is high. The cost of unmet demand represents this shortfall. Since $\delta\mathbb{E}[\xi]$ doesn't involve $x$, its presence or absence doesn't change the set of minimizers of the problem. Thus, the newsvendor can just as well consider the problem

$$\underset{x \in \mathbb{R}}{\text{minimize}} \; x + \frac{1}{1-\alpha}\mathbb{E}\big[\max\{0, \xi - x\}\big],$$

which has the $\alpha$-superquantile of $\xi$, i.e., $\text{s-rsk}_\alpha(\xi)$, as the minimum value and the $\alpha$-quantile of $\xi$ as a minimizer by 3.9.

The connection with superquantile risk provides new insight. Faced with an uncertain demand $\xi$, the newsvendor can quantify the level of *risk* in the newspaper-selling business by computing $\text{s-rsk}_\alpha(\xi)$ and comparing it with some threshold $\tau$:

$$\text{If } \text{s-rsk}_\alpha(\xi) \le \tau, \text{ then the risk is deemed acceptable}$$

and the newsvendor remains in the business. Otherwise, the newsvendor quits selling newspapers since the risk is simply too high. The risk quantifies in some sense how high the initial cost plus the random cost of unmet demand can get *even under an optimal order quantity*. The choice of threshold $\tau$ could be guided by the expected revenue $\delta\mathbb{E}[\xi]$ under fully satisfied demand. For example, one might require that the risk is 25% below this expected revenue, i.e., $\tau = 0.75\delta\mathbb{E}[\xi]$.

It's clear that using superquantile risk to assess the prospects associated with selling a product with demand $\xi$ gives deeper insight than simply computing the expected demand: It accounts for the effect of an optimal order quantity and also costly variations in the demand. For example, when faced with the choice of selling either product 1, with demand $\xi_1$, or product 2, with demand $\xi_2$, the corresponding superquantile risks tell us how well we can *mitigate* the respective uncertainties by means of optimizing the order quantity.

**Example 8.4** (risk assessment). Suppose that a firm faces a choice between two uncertain costs in the future described by the random variables $\xi_1$ and $\xi_2$ distributed uniformly between $[-3/2, 1]$ and $[-8, 2]$, respectively. (A negative cost means that the firm receives money.) The comparison between $\mathbb{E}[\xi_1] = -1/4$ and $\mathbb{E}[\xi_2] = -3$ leads to the conclusion that the better choice is to select $\xi_2$; on average it results in a lower cost. However, this may not be the right decision if the firm is concern about the possibility of a high cost.

An alternative assessment that emphasize positive values compares

$$\mathbb{E}\Big[\frac{1}{1-\alpha}\max\{0, \xi_1\}\Big] \quad \text{and} \quad \mathbb{E}\Big[\frac{1}{1-\alpha}\max\{0, \xi_2\}\Big],$$

where $\alpha \in [0, 1)$ provides a scaling factor. These quantities can be thought of as the expected levels of *displeasure* felt by the firm when facing the random costs $\xi_1$ and $\xi_2$, respectively. A negative value of the cost is associated with zero displeasure, while a positive value incurs displeasure proportional to that value. Regardless of $\alpha$, this assessment results in a tie between the two choices with $\mathbb{E}[\max\{0, \xi_1\}] = \mathbb{E}[\max\{0, \xi_2\}] = 1/5$.

Neither of these two approaches for assessing the merit of adopting one cost over the other considers the possibility of mitigating actions by the firm. It turns out that more careful considerations may change the decision.

**Detail.** Let's suppose that the costs $\xi_1$ and $\xi_2$ are given in present money and that displeasure is quantified as above. If the firm is more active and invests $x$ amount of money in a risk-free asset (bonds or bank deposit) now, then the future displeasure, as perceived now, is reduced from

$$\mathbb{E}\Big[\frac{1}{1-\alpha}\max\{0, \xi_i\}\Big] \quad \text{to} \quad \mathbb{E}\Big[\frac{1}{1-\alpha}\max\{0, \xi_i - x\}\Big],$$

as $x$ will be available at the future point in time to offset the cost $\xi_i$. The upfront expense $x$ also needs to be considered and the goal becomes to select the investment $x$ such that

$$x + \frac{1}{1-\alpha}\mathbb{E}\big[\max\{0, \xi_i - x\}\big] \quad \text{is minimized.}$$

The resulting minimum value is the superquantile risk $\text{s-rsk}_\alpha(\xi_i)$ and the comparison becomes between $\text{s-rsk}_\alpha(\xi_1) = 3/4$ and $\text{s-rsk}_\alpha(\xi_2) = 1$, where we adopt $\alpha = 0.8$. Now, the first cost $\xi_1$ is better because its risk is lower. When accounting for the possibility of investing in a risk-free asset, but still assessing displeasure as before, the advantage tilts decisively towards the first cost. The corresponding investments are $1/2$ and $0$, respectively, as computed by 3.9. The trouble with $\xi_2$ is the distinct possibility of high values and that any positive investment is often wasted.

The parameter $\alpha$ scales the displeasure and can be adjusted to reflect the firm's preferences. A value near one indicates a high level of displeasure even with small positive costs and this provides a strong incentive for investing in the risk-free asset. In contrast to the newsvendor problem where $\alpha$ is tied to order cost and sales price, $\alpha$ is here more subjective. A comprehensive treatment of risk management in financial engineering is provided by [41]. □

The example illustrates how a decision-maker can approach choices between random variables. Let's put this into a broad framework for decision-making under uncertainty and adopt a perspective concerned with high values of random variables, which is natural in the case of "costs" and corresponds to our focus on minimizing. (A trivial sign change adjusts the framework to cases where low values, instead of high values, are undesirable.) With this orientation, we adopt the following terminology.

A *measure of regret* $\mathcal{V}$ assigns to a random variable $\xi$ a number $\mathcal{V}(\xi) \in (-\infty, \infty]$, its *regret*, that quantifies the current displeasure with the possible values of $\xi$.

In economics, the *pleasure* experienced from an outcome is often quantified by a utility function. A measure of regret is largely a re-orientation towards displeasure and can be viewed as a disutility function; see [98] for further remarks.

A *measure of risk* $\mathcal{R}$ assigns to a random variable $\xi$ a number $\mathcal{R}(\xi) \in (-\infty, \infty]$ as a quantification of its risk.

For a random variable $\xi$ with support $\Xi$, we denote by $\sup \xi$ the highest point in the support, i.e., $\sup \Xi$, which possibly could be $\infty$. Since the probability that $\xi$ takes on a larger value than $\sup \xi$ is zero, we can think of $\sup \xi$ as the *worst value* of $\xi$. Moreover,

$$\sup \xi = \text{s-rsk}_1(\xi)$$

by 3.6 and the fact that

$$Q(\alpha) \leq \bar{Q}(\alpha) \leq \sup \Xi \quad \forall \alpha \in (0, 1),$$

where $Q(\alpha)$ and $\bar{Q}(\alpha)$ are the $\alpha$-quantile and $\alpha$-superquantile of $\xi$, respectively. The *standard deviation* of $\xi$ is

$$\text{std}(\xi) = \sqrt{\text{var}(\xi)}.$$

**Example 8.5** (measures of regret and risk). We've the following examples of measures of regret $\mathcal{V}$ and measures of risk $\mathcal{R}$.

(a) *Penalty regret* and *superquantile risk* with $\alpha \in (0, 1)$:

$$\mathcal{V}(\xi) = \frac{1}{1-\alpha}\mathbb{E}\big[\max\{0, \xi\}\big] \qquad \mathcal{R}(\xi) = \text{s-rsk}_\alpha(\xi).$$

(b) *Worst-case regret* and *risk*:

$$\mathcal{V}(\xi) = \begin{cases} 0 & \text{if } \sup \xi \leq 0 \\ \infty & \text{otherwise} \end{cases} \qquad \mathcal{R}(\xi) = \sup \xi.$$

(c) *Markowitz regret* and *risk* with $\lambda > 0$:

$$\mathcal{V}(\xi) = \mathbb{E}[\xi] + \lambda\sqrt{\mathbb{E}[\xi^2]} \qquad\qquad \mathcal{R}(\xi) = \mathbb{E}[\xi] + \lambda\,\mathrm{std}(\xi).$$

(d) *Mixed regret* and *risk* with $\lambda_1, \ldots, \lambda_q > 0$ and $\sum_{i=1}^q \lambda_i = 1$:

$$\mathcal{V}(\xi) = \inf_{\gamma_1,\ldots,\gamma_q} \left\{ \sum_{i=1}^q \lambda_i \mathcal{V}_i(\xi - \gamma_i) \,\middle|\, \sum_{i=1}^q \lambda_i \gamma_i = 0 \right\} \qquad \mathcal{R}(\xi) = \sum_{i=1}^q \lambda_i \mathcal{R}_i(\xi).$$

**Detail.** The choices in (a) are utilized by 8.4 and point to a general distinction between measures of regret and measures of risk: the former involves a "passive" assessment of a random variable without mitigating steps, while the latter incorporates the effect of mitigation. In the present case, $\mathcal{V}$ assigns an increasingly large penalty to positive outcomes.

In (b), the regret is $\infty$ if there's a positive outcome in the support $\Xi$ of $\xi$. The risk is the highest value in $\Xi$, which might be $\infty$. (This would be the case if $\xi$ is a normal random variable because then $\Xi = \mathbb{R}$.)

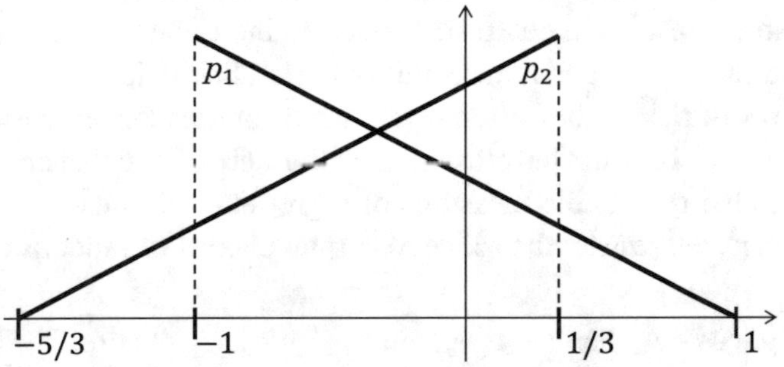


Fig. 8.1: Density functions $p_1$ and $p_2$ with the same Markowitz risk.

While (c) relies on the commonly used mean-squared and standard deviation, the resulting measures of regret and risk fail to distinguish between variation in the values of a random variable on the low side (which might be harmless) and on the high side (which is the main concern under our "cost" orientation). Figure 8.1 shows the density functions $p_1$ and $p_2$ of random variables $\xi_1$ and $\xi_2$, respectively. It turns out that they've the same mean ($-1/3$) and also the same standard deviation ($\sqrt{3}/2$). Thus, a decision-maker that relies on (c) would consider the two random variables equivalent. However, $p_1$ has a troubling upper tail. The concern about $\xi_1$ is revealed by the superquantile risks; $\text{s-rsk}_{0.9}(\xi_1) = 0.58$ and $\text{s-rsk}_{0.9}(\xi_2) = 0.28$.

In (d), a measure of risk $\mathcal{R}$ is constructed from a weighted average of other measures of risk $\mathcal{R}_1, \ldots, \mathcal{R}_q$. For example, one may combine measures of risk with different $\alpha$ from (a). Similarly, but in a more convoluted manner, $\mathcal{V}$ is constructed from other measures of regret $\mathcal{V}_1, \ldots, \mathcal{V}_q$. □

**Exercise 8.6** (exponential distribution). Consider a random variable $\xi$ with distribution function given by $P(\xi) = 1 - \exp(-\xi/\mu)$ for $\xi \geq 0$ and $P(\xi) = 0$ otherwise, where $\mu > 0$. Determine explicit formulae for the measures of regret and risk in 8.5(a,b,c).

**Guide.** For (a), leverage 3.6 to compute superquantiles. □

In applications, we would like to adopt measures of regret and risk that reflect a decision-maker's preferences, avoid embarrassing paradoxes, are computational tractable and facilitate analysis. For example, (c) in 8.5 is problematic in the context of Figure 8.1 as the resulting measures of regret and risk fail to identify the obvious difference between the two alternatives. Moreover, we typically face the choice between many random variables expressed by $f : \mathbb{R}^m \times \mathbb{R}^n \to \mathbb{R}$ in conjunction with an $m$-dimensional random vector $\boldsymbol{\xi}$. We need to assess and compare the random variables $\{f(\boldsymbol{\xi}, x),\ x \in \mathbb{R}^n\}$. Measures of regret and risk are modeling tools that convert these random variables into numbers for the purpose of comparison. Thus, we're led to problems of the form

$$\operatorname*{minimize}_{x \in X \subset \mathbb{R}^n} \mathcal{R}\big(f(\boldsymbol{\xi}, x)\big).$$

The formulation reminds us of those with a composite function $\iota_X + h \circ F$, where $h$ assesses the values of a mapping $F$; see for example (4.2). We can view a measure of risk as playing the role of $h$ as it assesses the potentially infinite number of values of the random variable $f(\boldsymbol{\xi}, x)$. (In contrast, $h$ assesses the finite dimensional vector $F(x)$.) Much of the same can be said about measures of regret, but in decision-making, we prefer to rely on measures of risk as they often reveal opportunities for mitigation; cf. 8.4.

In the following, we'll discuss how to construct measure of risk that are both meaningful and tractable and also reasonable measures of regret that will take on supporting roles. To make this precise, we restrict the scope to certain classes of random variables.

**Probability Spaces.** We can view a random variable as a function from a probability space to the real line. A *probability space* is a triplet $(\Omega, \mathcal{A}, \mathbb{P})$, where $\Omega$ is a set and $\mathcal{A}$ is a collection of subsets of $\Omega$ called *events* for which $\mathbb{P} : \mathcal{A} \to [0, 1]$ computes probabilities. (The collection $\mathcal{A}$ needs to satisfy certain properties such that it forms a so-called sigma-algebra on $\Omega$.) We think of $\omega \in \Omega$ as a future outcome and the probability that $\omega \in A$ for some event $A \in \mathcal{A}$ is then $\mathbb{P}(A)$. Since this probability is known for any event $A \in \mathcal{A}$, the collection $\mathcal{A}$ is a model of the current information about the future outcomes. A *random variable* $\boldsymbol{\xi}$ can then formally be viewed as a measurable function $\boldsymbol{\xi} : \Omega \to \mathbb{R}$, with $\boldsymbol{\xi}(\omega)$ being the value under outcome $\omega \in \Omega$, which we, in the past, expressed simply as $\xi$. Being *measurable* means that

$$\{\boldsymbol{\xi} \leq \alpha\} = \big\{\omega \in \Omega \,\big|\, \boldsymbol{\xi}(\omega) \leq \alpha\big\} \in \mathcal{A} \quad \forall \alpha \in \mathbb{R},$$

i.e., $\mathbb{P}$ can quantify the probability for all sets of the form $\{\boldsymbol{\xi} \leq \alpha\}$. This in turn defines the distribution function $P$ of $\boldsymbol{\xi}$ as

$$P(\alpha) = \operatorname{prob}\{\boldsymbol{\xi} \leq \alpha\} = \mathbb{P}\big(\{\boldsymbol{\xi} \leq \alpha\}\big).$$

We say that the probability space is *finite* if $\Omega$ contains only a finite number of points. Then, every random variable defined on the probability space must be finitely distributed.

In the following, we'll usually assume that there's a given probability space $(\Omega, \mathcal{A}, \mathbb{P})$ and then consider all random variables defined on that space with *finite second moment*:

$$\mathcal{L}^2 = \{\xi : \Omega \to \mathbb{R} \mid \mathbb{E}[\xi^2] < \infty\}.$$

Since the second moment

$$\mathbb{E}[\xi^2] = (\mathbb{E}[\xi])^2 + \text{var}(\xi),$$

$\xi \in \mathcal{L}^2$ entails that the mean and variance of $\xi$ are finite. For $\{\xi, \xi^\nu \in \mathcal{L}^2, \nu \in \mathbb{N}\}$, the random variables $\xi^\nu$ *converge* to the random variable $\xi$, written $\xi^\nu \to \xi$, when

$$\mathbb{E}[(\xi^\nu - \xi)^2] \to 0 \iff \mathbb{E}[\xi^\nu - \xi] \to 0 \text{ and } \text{var}(\xi^\nu - \xi) \to 0.$$

Analogous to the situation in $\mathbb{R}^n$, $C \subset \mathcal{L}^2$ is *closed* if for every $\xi^\nu \in C$ converging to $\xi$, we've $\xi \in C$. Moreover, $C$ is *convex* if

$$(1 - \lambda)\xi_0 + \lambda\xi_1 \in C \quad \forall \lambda \in [0, 1], \quad \xi_0, \xi_1 \in C.$$

A random variable $\xi \in \mathcal{L}^2$ is *constant* if there's $\alpha$ such that

$$\mathbb{P}(\{\omega \in \Omega \mid \xi(\omega) = \alpha\}) = \text{prob}\{\xi = \alpha\} = 1.$$

We denote by $\mathbf{0}$ and $\mathbf{1}$ the constant random variables in $\mathcal{L}^2$ with value zero and one, respectively.

With this terminology, we can now define *functionals* that take as input a random variable from $\mathcal{L}^2$ and return as output a number in $\overline{\mathbb{R}}$. Measures of regret and risk are examples of functionals, but others follow below as well. In the same way as a function can be convex and lsc (cf. §1.D and 4.8), a functional $\mathcal{F} : \mathcal{L}^2 \to \overline{\mathbb{R}}$ may satisfy any of the following properties:

$$\textit{Constancy} : \mathcal{F}(\xi) = \alpha \text{ whenever } \xi \text{ is constant with value } \alpha \in \mathbb{R}.$$

$$\textit{Averseness} : \mathcal{F}(\xi) > \mathbb{E}[\xi] \text{ for all nonconstant } \xi.$$

$$\textit{Convexity} : \mathcal{F}((1 - \lambda)\xi_0 + \lambda\xi_1) \le (1 - \lambda)\mathcal{F}(\xi_0) + \lambda\mathcal{F}(\xi_1) \;\; \forall \xi_0, \xi_1, \; \lambda \in [0, 1].$$

$$\textit{Lsc} : \{\mathcal{F} \le \alpha\} = \{\xi \in \mathcal{L}^2 \mid \mathcal{F}(\xi) \le \alpha\} \text{ is closed } \forall \alpha \in \mathbb{R}.$$

The constancy property is natural for a measure of risk because then the risk of a future cost, known to be equal to $\alpha$, is simply $\alpha$. The averseness condition excludes the possibility $\mathcal{F}(\xi) = \mathbb{E}[\xi]$, which is better treated separately. The convexity and lsc properties are extensions of those for functions on $\mathbb{R}^n$.

**Definition 8.7** (regular measures of regret and risk). A *regular measure of regret* $\mathcal{V}$ is a functional from $\mathcal{L}^2$ to $(-\infty, \infty]$ that's lsc and convex, with

$$\mathcal{V}(\mathbf{0}) = 0, \text{ but } \mathcal{V}(\xi) > \mathbb{E}[\xi] \;\; \forall \xi \ne \mathbf{0}.$$

The quantity $\mathcal{V}(\boldsymbol{\xi})$ is the *regret* of $\boldsymbol{\xi}$.

A *regular measure of risk* $\mathcal{R}$ is a functional from $\mathcal{L}^2$ to $(-\infty, \infty]$ that's lsc, convex and also satisfies the constancy and averseness properties. The quantity $\mathcal{R}(\boldsymbol{\xi})$ is the *risk* of $\boldsymbol{\xi}$.

**Example 8.8** (regular measures of regret and risk). The measures of regret and risk in 8.5(a) are regular.

**Detail.** Since $\mathcal{V}(\boldsymbol{\xi}) = \frac{1}{1-\alpha}\mathbb{E}[\max\{0, \boldsymbol{\xi}\}]$, $\mathcal{V}$ is obviously never $-\infty$. For $\lambda \in (0, 1)$, the convexity of $\xi \mapsto \max\{0, \xi\}$ guarantees that

$$\max\big\{0, (1-\lambda)\xi_0 + \lambda\xi_1\big\} \leq (1-\lambda)\max\{0, \xi_0\} + \lambda\max\{0, \xi_1\},$$

regardless of $\xi_0, \xi_1 \in \mathbb{R}$. By the properties for expectations in 3.2, this implies that

$$\begin{aligned}\mathbb{E}\big[\max\big\{0, (1-\lambda)\boldsymbol{\xi}_0 + \lambda\boldsymbol{\xi}_1\big\}\big] &\leq \mathbb{E}\big[(1-\lambda)\max\{0, \boldsymbol{\xi}_0\} + \lambda\max\{0, \boldsymbol{\xi}_1\}\big]\\ &= (1-\lambda)\mathbb{E}\big[\max\{0, \boldsymbol{\xi}_0\}\big] + \lambda\mathbb{E}\big[\max\{0, \boldsymbol{\xi}_1\}\big] \quad \forall \boldsymbol{\xi}_0, \boldsymbol{\xi}_1 \in \mathcal{L}^2.\end{aligned}$$

Thus, one has

$$\mathcal{V}\big((1-\lambda)\boldsymbol{\xi}_0 + \lambda\boldsymbol{\xi}_1\big) \leq (1-\lambda)\mathcal{V}(\boldsymbol{\xi}_0) + \lambda\mathcal{V}(\boldsymbol{\xi}_1)$$

and we've shown that $\mathcal{V}$ is convex. Next, let $\beta \in \mathbb{R}$ and suppose that $\boldsymbol{\xi}^\nu \to \boldsymbol{\xi}$ with $\mathcal{V}(\boldsymbol{\xi}^\nu) \leq \beta$. Then, $\max\{0, \boldsymbol{\xi}^\nu\} \to \max\{0, \boldsymbol{\xi}\}$ as well, so that

$$\mathbb{E}\big[\max\{0, \boldsymbol{\xi}^\nu\}\big] \to \mathbb{E}\big[\max\{0, \boldsymbol{\xi}\}\big].$$

Consequently, $\mathcal{V}(\boldsymbol{\xi}) \leq \beta$ and $\mathcal{V}$ is lsc. The property $\mathcal{V}(\mathbf{0}) = 0$ holds trivially. It only remains to show that $\mathcal{V}(\boldsymbol{\xi}) > \mathbb{E}[\boldsymbol{\xi}]$ for any random variable $\boldsymbol{\xi}$ except $\mathbf{0}$. Since

$$\frac{1}{1-\alpha}\max\{0, \xi\} - \xi \geq 0 \;\; \forall \xi \in \mathbb{R}$$

we also have

$$\mathbb{E}\Big[\frac{1}{1-\alpha}\max\{0, \boldsymbol{\xi}\} - \boldsymbol{\xi}\Big] \geq 0$$

by 3.2 and this expectation equals zero only when the random variable

$$\frac{1}{1-\alpha}\max\{0, \boldsymbol{\xi}\} - \boldsymbol{\xi}$$

coincides with $\mathbf{0}$, which in turn requires $\boldsymbol{\xi} = \mathbf{0}$. We've shown that the penalty regret in 8.5(a) is regular.

For s-rsk$_\alpha$, we know that a superquantile is no smaller than the corresponding quantile. Consequently, s-rsk$_\alpha(\boldsymbol{\xi}) > -\infty$. In fact, a superquantile coincides with the corresponding quantile when $\boldsymbol{\xi}$ is constant so the constancy property holds. To show convexity, suppose that $\boldsymbol{\xi}_0, \boldsymbol{\xi}_1 \in \mathcal{L}^2$ and $\lambda \in (0, 1)$. Let $Q_0$ and $Q_1$ be the $\alpha$-quantiles of $\boldsymbol{\xi}_0$ and $\boldsymbol{\xi}_1$, respectively. By the convexity of $\mathcal{V}$,

$$\mathcal{V}\big((1-\lambda)(\boldsymbol{\xi}_0 - Q_0) + \lambda(\boldsymbol{\xi}_1 - Q_1)\big) \le (1-\lambda)\mathcal{V}(\boldsymbol{\xi}_0 - Q_0) + \lambda\mathcal{V}(\boldsymbol{\xi}_1 - Q_1).$$

Set $\bar{\gamma} = (1-\lambda)Q_0 + \lambda Q_1$. The prior inequality and the definition of superquantiles in 3.6 imply that

$$\begin{aligned}\bar{\gamma} + \mathcal{V}\big((1-\lambda)\boldsymbol{\xi}_0 + \lambda\boldsymbol{\xi}_1 - \bar{\gamma}\big) &\le \bar{\gamma} + (1-\lambda)\mathcal{V}(\boldsymbol{\xi}_0 - Q_0) + \lambda\mathcal{V}(\boldsymbol{\xi}_1 - Q_1)\\ &= (1-\lambda)\,\text{s-rsk}_\alpha(\boldsymbol{\xi}_0) + \lambda\,\text{s-rsk}_\alpha(\boldsymbol{\xi}_1).\end{aligned}$$

Since the left-hand side is bounded from below by

$$\text{s-rsk}_\alpha\big((1-\lambda)\boldsymbol{\xi}_0 + \lambda\boldsymbol{\xi}_1\big),$$

we've shown that $\text{s-rsk}_\alpha$ is convex. Let $\boldsymbol{\xi}$ be a random variable that isn't constant and $Q$ be its $\alpha$-quantile. Since $\mathcal{V}(\boldsymbol{\xi} - Q) > \mathbb{E}[\boldsymbol{\xi} - Q]$ as just established, one obtains

$$\text{s-rsk}_\alpha(\boldsymbol{\xi}) = Q + \frac{1}{1-\alpha}\mathbb{E}\big[\max\{0, \boldsymbol{\xi} - Q\}\big] > Q + \mathbb{E}[\boldsymbol{\xi} - Q] = \mathbb{E}[\boldsymbol{\xi}]$$

after again bringing in 3.6. We've shown that $\text{s-rsk}_\alpha$ satisfies the averseness property. Finally, we show lsc. Let $\beta \in \mathbb{R}$ and suppose that $\boldsymbol{\xi}^\nu \to \boldsymbol{\xi}$ with $\text{s-rsk}_\alpha(\boldsymbol{\xi}^\nu) \le \beta$. Denote by $Q^\nu$ the $\alpha$-quantile of $\boldsymbol{\xi}^\nu$. It follows that $\{Q^\nu, \nu \in \mathbb{N}\}$ is bounded, which can be established using *Chebyshev's inequality*[3]. Consequently, there are $N \in \mathcal{N}_\infty^\#$ and $Q \in \mathbb{R}$ such that $Q^\nu \to_N Q$ and then it follows using 3.6 that

$$\begin{aligned}\beta \ge \text{s-rsk}_\alpha(\boldsymbol{\xi}^\nu) &= Q^\nu + \frac{1}{1-\alpha}\mathbb{E}\big[\max\{0, \boldsymbol{\xi}^\nu - Q^\nu\}\big]\\ &\to_N Q + \frac{1}{1-\alpha}\mathbb{E}\big[\max\{0, \boldsymbol{\xi} - Q\}\big] \ge \text{s-rsk}_\alpha(\boldsymbol{\xi}).\end{aligned}$$

This establishes that $\text{s-rsk}_\alpha$ is regular.

As a comparison, the measure of risk $\mathcal{R}(\boldsymbol{\xi}) = Q(\alpha)$, where $Q(\alpha)$ is the $\alpha$-quantile of $\boldsymbol{\xi}$, fails the convexity requirement. For example, consider two statistically independent random variables $\boldsymbol{\xi}_0, \boldsymbol{\xi}_1$ with common density function value 0.9 on $[-1, 0]$, value 0.05 on $(0, 2]$ and value zero otherwise. Then, the 0.9-quantile is 0 for both $\boldsymbol{\xi}_0$ and $\boldsymbol{\xi}_1$, but the 0.9-quantile for $\boldsymbol{\xi}_0/2 + \boldsymbol{\xi}_1/2$ is approximately 0.25. A lack of convexity in a measure of risk has serious consequences computationally, but it's also problematic from a modeling point of view as discussed in 3.12. □

A regular measure of risk $\mathcal{R}$ has automatically the property that for $\boldsymbol{\xi} \in \mathcal{L}^2$ and $\alpha \in \mathbb{R}$,

$$\mathcal{R}(\boldsymbol{\xi} + \alpha) = \mathcal{R}(\boldsymbol{\xi}) + \alpha \tag{8.3}$$

as can be seen from [102]. Thus, $\mathcal{R}$ quantifies risk in a manner that's translation invariant; adding a constant amount to an uncertain future cost changes the risk by that amount.

[3] For $\boldsymbol{\xi} \in \mathcal{L}^2$ and $\kappa > 0$, Chebyshev's inequality asserts that $\text{prob}\{|\boldsymbol{\xi} - \mathbb{E}[\boldsymbol{\xi}]| \ge \kappa\} \le \text{var}(\boldsymbol{\xi})/\kappa^2$, which is a consequence of Markov's inequality; see §8.H.

The distinction between regular measures of regret and risk emerges from the assessment of constant random variables. While $\mathcal{V}(\mathbf{0}) = \mathcal{R}(\mathbf{0}) = 0$, a constant random variable $\xi$ with value $\alpha \neq 0$ is treated differently:

$$\mathcal{V}(\xi) > \alpha, \quad \text{but} \quad \mathcal{R}(\xi) = \alpha.$$

This seemingly small difference results in profoundly different roles, with regular measures of regret providing a means to construct regular measures of risk and thereby extending the useful relationship between $\frac{1}{1-\alpha}\mathbb{E}[\max\{0, \xi\}]$ and $\text{s-rsk}_\alpha(\xi)$; see 3.9.

**Theorem 8.9** (regret-risk). *Given a regular measure of regret $\mathcal{V}$, a regular measure of risk $\mathcal{R}$ is obtained by*

$$\mathcal{R}(\xi) = \min_{\gamma \in \mathbb{R}} \left\{\gamma + \mathcal{V}(\xi - \gamma)\right\}.$$

*For every regular measure of risk $\mathcal{R}$, there's a regular measure of regret $\mathcal{V}$, not necessarily unique, that constructs $\mathcal{R}$ through this minimization formula.*

**Proof.** We refer to [99, Theorem 2.2] for a proof of the first claim. The second claim is established by the construction

$$\mathcal{V}(\xi) = \mathcal{R}(\xi) + \alpha\big|\mathbb{E}[\xi]\big|,$$

where $\alpha > 0$. In view of (8.3), the minimization formula then holds because

$$\gamma + \mathcal{V}(\xi - \gamma) = \gamma + \mathcal{R}(\xi - \gamma) + \alpha\big|\mathbb{E}[\xi - \gamma]\big| = \mathcal{R}(\xi) + \alpha\big|\mathbb{E}[\xi] - \gamma\big|$$

and the fact that the minimum value is attained by $\gamma$ being set to $\mathbb{E}[\xi]$. If $\mathcal{R}$ is regular, then $\mathcal{V}$ is lsc and convex because $\alpha|\mathbb{E}[\,\cdot\,]|$ is lsc and convex. Moreover, $\mathcal{V}(\xi) > -\infty$ because $\mathcal{R}(\xi) > -\infty$ and $\alpha|\mathbb{E}[\xi]| \geq 0$. We also have $\mathcal{V}(\mathbf{0}) = 0$. For nonconstant $\xi$,

$$\mathcal{V}(\xi) > \mathbb{E}[\xi] + \alpha\big|\mathbb{E}[\xi]\big| \geq \mathbb{E}[\xi].$$

For constant $\xi \neq \mathbf{0}$,

$$\mathcal{V}(\xi) = \mathbb{E}[\xi] + \alpha\big|\mathbb{E}[\xi]\big| > \mathbb{E}[\xi].$$

We've shown all the properties required for $\mathcal{V}$ to be regular. □

The regret-risk theorem generalizes the situation discussed in 8.4: a regular measure of risk finds the "best" way of reducing displeasure with a mix of uncertain outcomes by optimizing a scalar $\gamma$. Thus, a regular measure of risk provides a deeper assessment of a random variable compared to a regular measure of regret.

The formula in the theorem provides a path to constructing new regular measures of risk from regular measures of regret, which in turn might be motivated by utility functions. Utility functions are widely used in economics and can sometimes be related to decision-makers' preferences; see [42] for a critical discussion. Many utility functions define a regular measure of regret after a sign change and a normalization so that the requirements of 8.7 hold.

**Norm for Random Variables.** The space $\mathcal{L}^2$ of random variables possesses many of the same properties as the space $\mathbb{R}^n$ of $n$-dimensional vectors. It can be associated with a norm that satisfies the three defining properties in 1.15. Specifically,

$$\|\xi\|_{\mathcal{L}^2} = \sqrt{\mathbb{E}[\xi^2]}$$

defines a norm on $\mathcal{L}^2$ and thus satisfies

$$\|\lambda\xi\|_{\mathcal{L}^2} = |\lambda|\|\xi\|_{\mathcal{L}^2} \ \forall\lambda \in \mathbb{R} \qquad \text{(absolute homogeneity)}$$

$$\|\xi_0 + \xi_1\|_{\mathcal{L}^2} \leq \|\xi_0\|_{\mathcal{L}^2} + \|\xi_1\|_{\mathcal{L}^2} \qquad \text{(triangle inequality or subadditivity)}$$

$$\|\xi\|_{\mathcal{L}^2} = 0 \implies \xi = \mathbf{0} \qquad \text{(separation of points).}$$

The last property might be a bit troublesome. For example, consider the probability space $(\Omega, \mathcal{A}, \mathbb{P})$ with $\Omega = [0,1]$ and $\mathbb{P}$ assigning probabilities to subsets of $[0,1]$ according to their length. Then, there are many random variables on this probability space of the form $\xi(\omega) = 0$ for all $\omega \in [0,1]$ except for a finite or countable number of such points. All these random variables have $\|\xi\|_{\mathcal{L}^2} = 0$ because $\text{prob}\{\xi = 0\} = 1$. However, not all of them have $\xi(\omega) = 0$ for every $\omega \in [0,1]$. Do they violate the requirement that $\xi = \mathbf{0}$? No, not under the convention that we consider random variables that differ in value only on a subset $\Omega_0 \in \mathcal{A}$, with $\mathbb{P}(\Omega_0) = 0$, as the *same* random variable. That is, we don't distinguish between $\xi_0$ and $\xi_1$ if there's $\Omega_0 \in \mathcal{A}$ with $\mathbb{P}(\Omega_0) = 0$ such that $\xi_0(\omega) = \xi_1(\omega)$ for all $\omega \in \Omega \setminus \Omega_0$. This is reasonable as $\xi_0$ and $\xi_1$ would then have the same distribution function. Consequently, an equality between random variables such as in $\xi = \mathbf{0}$ means that the two random variables coincide for every $\omega \in \Omega$, possibly except on a set with zero probability.

The properties of $\|\cdot\|_{\mathcal{L}^2}$ implies that the functional defined by

$$\mathcal{F}(\xi) = \|\xi\|_{\mathcal{L}^2}$$

is convex. Moreover, $\xi^\nu \to \xi$ means that $\|\xi^\nu - \xi\|_{\mathcal{L}^2} \to 0$, which in turn implies that $\|\xi^\nu\|_{\mathcal{L}^2} \to \|\xi\|_{\mathcal{L}^2}$ so that $\mathcal{F}$ is certainly lsc.

**Example 8.10** (regular measures of regret and risk; cont.). The measures of regret and risk in 8.5(b,c) are regular and also pair in the sense of the regret-risk theorem 8.9. The measure of regret in 8.5(d) is regular as long as $\mathcal{V}_1, \dots, \mathcal{V}_q$ are regular and the corresponding measure of risk is regular as long as $\mathcal{R}_1, \dots, \mathcal{R}_q$ are regular. Moreover, if $\mathcal{V}_i$ and $\mathcal{R}_i$ pair by satisfying the formula in 8.9, then $\mathcal{V}$ and $\mathcal{R}$ in 8.5(d) also satisfy that formula.

These connections provide alternative formulae for the measures of risk in 8.5(b,c,d) via 8.9.

**Detail.** In view of the regret-risk theorem 8.9, we can prove the regularity of the measures of risk in 8.5(b,c,d) by establishing the regularity of the corresponding measures of regret and then confirming that the formula in 8.9 indeed produces the correct measure of risk.

Let's first examine $\mathcal{V}$ in 8.5(b). Then, $\mathcal{V}(\boldsymbol{\xi}) > -\infty$ and $\mathcal{V}(\mathbf{0}) = 0$ trivially. Let $\Xi$ be the support of $\boldsymbol{\xi}$. For $\boldsymbol{\xi} \neq \mathbf{0}$, we consider two cases: If $\Xi \subset (-\infty, 0]$, then $\mathbb{E}[\boldsymbol{\xi}] < 0$. If this isn't the case, $\mathcal{V}(\boldsymbol{\xi}) = \infty$. In either case, $\mathcal{V}(\boldsymbol{\xi}) > \mathbb{E}[\boldsymbol{\xi}]$. To establish convexity, it suffices to consider $\lambda \in (0, 1)$ as well as $\boldsymbol{\xi}_0$ and $\boldsymbol{\xi}_1$ with supports $\Xi_0$ and $\Xi_1$, respectively, that both are subsets of $(-\infty, 0]$. Then, the random variable $(1 - \lambda)\boldsymbol{\xi}_0 + \lambda\boldsymbol{\xi}_1$ must have support also contained in $(-\infty, 0]$. Consequently,

$$\mathcal{V}\big((1 - \lambda)\boldsymbol{\xi}_0 + \lambda\boldsymbol{\xi}_1\big) \leq (1 - \lambda)\mathcal{V}(\boldsymbol{\xi}_0) + \lambda\mathcal{V}(\boldsymbol{\xi}_1),$$

with zero on both sides of the inequality. The lsc property is established as follows. Let $\boldsymbol{\xi}^\nu \to \boldsymbol{\xi}$ and $\Xi^\nu \subset (-\infty, 0]$ be the support of $\boldsymbol{\xi}^\nu$. Without loss of generality, we assume that $\boldsymbol{\xi}^\nu(\omega) \leq 0$ for all $\omega \in \Omega$. For the sake of contradiction, suppose that $\operatorname{prob}\{\boldsymbol{\xi} > 0\} > 0$. Then, there's $\Omega_0 \in \mathcal{A}$ such that $\mathbb{P}(\Omega_0) > 0$ and $\boldsymbol{\xi}(\omega) > 0$ for all $\omega \in \Omega_0$. Define the random variable $\boldsymbol{\eta} \in \mathcal{L}^2$ by setting $\boldsymbol{\eta}(\omega) = 1$ if $\omega \in \Omega_0$ and $\boldsymbol{\eta}(\omega) = 0$ otherwise. Since

$$\big(\boldsymbol{\xi}^\nu(\omega) - \boldsymbol{\xi}(\omega)\big)^2 \geq \big(\boldsymbol{\xi}(\omega)\big)^2 \;\; \forall \omega \in \Omega_0,$$

it follows by 3.2 that

$$\mathbb{E}\big[(\boldsymbol{\xi}^\nu - \boldsymbol{\xi})^2\big] \geq \mathbb{E}\big[\boldsymbol{\eta}(\boldsymbol{\xi}^\nu - \boldsymbol{\xi})^2\big] \geq \mathbb{E}\big[\boldsymbol{\eta}\boldsymbol{\xi}^2\big] > 0.$$

Since this holds for all $\nu$, it contradicts the fact that $\boldsymbol{\xi}^\nu \to \boldsymbol{\xi}$. Consequently, $\operatorname{prob}\{\boldsymbol{\xi} > 0\} = 0$, which means that the support of $\boldsymbol{\xi}$ is contained in $(-\infty, 0]$. This leads to the conclusion that $\mathcal{V}$ is lsc. We've verified all the requirements for $\mathcal{V}$ to be a regular measure of regret. The relation in the regret-risk theorem 8.9 holds for any $\boldsymbol{\xi}$ because if the support $\Xi$ has $\sup \Xi = \infty$, then

$$\gamma + \mathcal{V}(\boldsymbol{\xi} - \gamma) = \infty \;\; \forall \gamma \in \mathbb{R}.$$

If $\sup \Xi \in \mathbb{R}$, then $\gamma + \mathcal{V}(\boldsymbol{\xi} - \gamma)$ is minimized by $\gamma = \sup \Xi$.

Second, let's consider 8.5(c). We obviously have $\mathcal{V}(\boldsymbol{\xi}) > -\infty$ and $\mathcal{V}(\mathbf{0}) = 0$. If $\boldsymbol{\xi} \neq \mathbf{0}$, then $\mathbb{E}[\boldsymbol{\xi}^2] > 0$ and $\mathcal{V}(\boldsymbol{\xi}) > \mathbb{E}[\boldsymbol{\xi}]$. We recognize that

$$\mathcal{V}(\boldsymbol{\xi}) = \mathbb{E}[\boldsymbol{\xi}] + \lambda \|\boldsymbol{\xi}\|_{\mathcal{L}^2}.$$

For $\boldsymbol{\xi}^\nu \to \boldsymbol{\xi}$, $\mathbb{E}[\boldsymbol{\xi}^\nu] \to \mathbb{E}[\boldsymbol{\xi}]$ and then also $\mathcal{V}(\boldsymbol{\xi}^\nu) \to \mathcal{V}(\boldsymbol{\xi})$ so $\mathcal{V}$ is lsc. The convexity of $\mathcal{V}$ now follow from the convexity of $\|\cdot\|_{\mathcal{L}^2}$ and properties of the expectation; see 3.2. We've shown that $\mathcal{V}$ is regular. The formula in 8.9 holds by recognizing that the minimizer of $\mathbb{E}[(\boldsymbol{\xi} - \gamma)^2]$ is $\mathbb{E}[\boldsymbol{\xi}]$ for any $\boldsymbol{\xi} \in \mathcal{L}^2$; see 3.24.

Third, let's consider 8.5(d). The regularity of $\mathcal{R}$ follows from that of $\mathcal{R}_i$ straightforwardly. For the properties of $\mathcal{V}$ as well as the connection with $\mathcal{R}$ we refer to [102]. □

**Example 8.11** (risk minimization). Suppose that $f : \mathbb{R}^m \times \mathbb{R}^n \to \mathbb{R}$ represents a quantity of interest, $\boldsymbol{\xi}$ is an $m$-dimensional random vector, $X \subset \mathbb{R}^n$, $\mathcal{R}$ is a regular measure of risk and $f(\boldsymbol{\xi}, x) \in \mathcal{L}^2$ for all $x \in \mathbb{R}^n$. Let's consider the problem

$$\underset{x \in X}{\text{minimize}}\ \varphi(x) = \mathcal{R}\big(f(\boldsymbol{\xi}, x)\big).$$

If $X$ is convex and $f(\xi, \cdot)$ is affine for all $\xi \in \Xi$, the support of $\boldsymbol{\xi}$, then $\varphi + \iota_X$ is convex. Thus, a regular measure of risk doesn't disrupt the convexity structure of $f(\xi, \cdot)$.

For any regular measure of regret $\mathcal{V}$ that pairs with $\mathcal{R}$ via the regret-risk theorem 8.9, the problem can be reformulated as

$$\underset{x \in X, \gamma \in \mathbb{R}}{\text{minimize}}\ \gamma + \mathcal{V}\big(f(\boldsymbol{\xi}, x) - \gamma\big).$$

**Detail.** Convexity holds by 1.18(b) provided that $\varphi$ is convex, which we establish using the convexity inequality 1.10. Let $x^0, x^1 \in \mathbb{R}^n$ and $\lambda \in (0, 1)$. Since $f(\xi, \cdot)$ is affine,

$$f\big(\xi, (1-\lambda)x^0 + \lambda x^1\big) = (1-\lambda) f(\xi, x^0) + \lambda f(\xi, x^1).$$

Consequently, the convexity of $\mathcal{R}$ implies that

$$\begin{aligned}\varphi\big((1-\lambda)x^0 + \lambda x^1\big) &= \mathcal{R}\big((1-\lambda) f(\boldsymbol{\xi}, x^0) + \lambda f(\boldsymbol{\xi}, x^1)\big)\\ &\le (1-\lambda)\mathcal{R}\big(f(\boldsymbol{\xi}, x^0)\big) + \lambda \mathcal{R}\big(f(\boldsymbol{\xi}, x^1)\big).\end{aligned}$$

The reformulation is immediate from the regret-risk theorem 8.9 and is most significant when $\mathcal{V}$ is somehow simpler than $\mathcal{R}$. We take advantage of this in §3.C: minimization of $\text{s-rsk}_\alpha(f(\boldsymbol{\xi}, x))$ is replaced by a problem involving

$$\mathcal{V}\big(f(\boldsymbol{\xi}, x)\big) = \frac{1}{1-\alpha}\mathbb{E}\big[\max\{0, f(\boldsymbol{\xi}, x)\}\big].$$

The equivalent problem is convex under the same assumptions as those for the actual problem because $\mathcal{V}$ is also convex.

Risk may also enter as a constraint. For $f_0 : \mathbb{R}^n \to \mathbb{R}$ and $\tau \in \mathbb{R}$, the problem

$$\underset{x \in X}{\text{minimize}}\ f_0(x) \ \text{ subject to }\ \mathcal{R}\big(f(\boldsymbol{\xi}, x)\big) \le \tau$$

is equivalent to

$$\underset{x \in X, \gamma \in \mathbb{R}}{\text{minimize}}\ f_0(x) \ \text{ subject to }\ \gamma + \mathcal{V}\big(f(\boldsymbol{\xi}, x) - \gamma\big) \le \tau.$$

There could be several constraints of this kind and also a measure of risk in the objective function allowing for much flexibility in modeling.

The seemingly complicated situation for mixed risk $\mathcal{R}$ from 8.5(d) simplifies as well. The problem of minimizing $\mathcal{R}(f(\boldsymbol{\xi}, x))$ subject to $x \in X$ is in this case equivalent to

$$\underset{x \in X, \gamma_0, \gamma_1, \ldots, \gamma_q}{\text{minimize}}\ \gamma_0 + \sum\nolimits_{i=1}^{q} \lambda_i \mathcal{V}_i\big(f(\boldsymbol{\xi}, x) - \gamma_0 - \gamma_i\big) \ \text{ subject to }\ \sum\nolimits_{i=1}^{q} \lambda_i \gamma_i = 0.$$

Now, we've $n + q + 1$ variables and one additional linear equation. □

A monotonicity property is also relevant for a functional $\mathcal{F} : \mathcal{L}^2 \to \overline{\mathbb{R}}$:

$$\textit{Monotonicity:} \quad \mathcal{F}(\boldsymbol{\xi}_0) \le \mathcal{F}(\boldsymbol{\xi}_1) \text{ when } \boldsymbol{\xi}_0(\omega) \le \boldsymbol{\xi}_1(\omega)\ \forall \omega \in \Omega.$$

This is a natural requirement because both regret and risk would typically be deemed less for $\boldsymbol{\xi}_0$ than for $\boldsymbol{\xi}_1$ when, for every pair of outcomes $\{\boldsymbol{\xi}_0(\omega), \boldsymbol{\xi}_1(\omega)\}$, the random variable $\boldsymbol{\xi}_0$ never comes out worse than $\boldsymbol{\xi}_1$. In particular, 8.5(a,b) list monotone measures of regret and risk. If $\mathcal{V}_1, \dots, \mathcal{V}_q$ $(\mathcal{R}_1, \dots, \mathcal{R}_q)$ are monotone, then $\mathcal{V}$ $(\mathcal{R})$ in 8.5(d) is monotone as well.

Neither $\mathcal{V}$ nor $\mathcal{R}$ in 8.5(c) is monotone, however. The difficulty is that the standard deviation and mean-squared don't distinguish between variability above and below the mean. Thus, a random variable with a high variability below the mean is deemed "high risk," even though low values represent no "real" risk in our setting focused on cost. For a counterexample, consider the random variables $\boldsymbol{\xi}_0$ and $\boldsymbol{\xi}_1$ and their joint distribution that assigns probability $1/2$ to the outcome $(0, 0)$ and probability $1/2$ to the outcome $(-2, -1)$. Thus, for every outcome, the random variables have either the same value or $\boldsymbol{\xi}_0$ has a value below that of $\boldsymbol{\xi}_1$. Still, with $\lambda = 2$, one has

$$\begin{aligned}\mathcal{R}(\boldsymbol{\xi}_0) &= \mathbb{E}[\boldsymbol{\xi}_0] + \lambda \operatorname{std}(\boldsymbol{\xi}_0) = -1 + 2 \cdot 1 = 1 \\ \mathcal{R}(\boldsymbol{\xi}_1) &= \mathbb{E}[\boldsymbol{\xi}_1] + \lambda \operatorname{std}(\boldsymbol{\xi}_1) = -\tfrac{1}{2} + 2 \cdot \tfrac{1}{2} = \tfrac{1}{2}.\end{aligned}$$

This shows that $\mathcal{R}$ in 8.5(c) lacks monotonicity.

**Proposition 8.12** (monotonicity). *If $\mathcal{V}$ is a monotone regular measure of regret, then the regular measure of risk $\mathcal{R}$ constructed by the regret-risk theorem 8.9 is monotone.*

**Proof.** This follows directly from the formula in 8.9. □

**Example 8.13** (monotonicity in risk minimization). If $\mathcal{V}$ is a monotone regular measure of regret, then the assumption about $f(\xi, \cdot)$ being affine for all $\xi \in \Xi$ in 8.11 can be relaxed to $f(\xi, \cdot)$ being convex for all $\xi \in \Xi$.

**Detail.** In contrast to the arguments in 8.11, we now only have

$$f\big(\xi, (1-\lambda)x^0 + \lambda x^1\big) \le (1-\lambda) f(\xi, x^0) + \lambda f(\xi, x^1).$$

However, the monotonicity of $\mathcal{V}$ and then also of $\mathcal{R}$ by 8.12 imply that

$$\mathcal{R}\Big(f\big(\boldsymbol{\xi}, (1-\lambda)x^0 + \lambda x^1\big)\Big) \le \mathcal{R}\big((1-\lambda) f(\boldsymbol{\xi}, x^0) + \lambda f(\boldsymbol{\xi}, x^1)\big).$$

We can again establish the convexity inequality 1.10 for the function $x \mapsto \mathcal{R}(f(\boldsymbol{\xi}, x))$. A similar argument establishes that $(x, \gamma) \mapsto \gamma + \mathcal{V}(f(\boldsymbol{\xi}, x) - \gamma)$ is also convex. □

## 8.C Risk-Adaptive Data Analytics

The broad approaches to decision-making based on measures of risk as developed in the previous section spill over to data analytics, where the familiar concepts of mean-squared error and standard deviation can be extended. This results in alternative approaches to regression analysis. We start by examining how to quantify the nonzeroness of a random variable.

**Definition 8.14** (regular measure of error and its statistic). A *regular measure of error* $\mathcal{E}$ is a functional from $\mathcal{L}^2$ to $[0, \infty]$ that's lsc and convex, with

$$\mathcal{E}(\mathbf{0}) = 0, \quad \text{but } \mathcal{E}(\boldsymbol{\xi}) > 0 \;\; \forall \boldsymbol{\xi} \neq \mathbf{0}.$$

The quantity $\mathcal{E}(\boldsymbol{\xi})$ is the *error* of $\boldsymbol{\xi}$.

The corresponding *statistic* is $\mathcal{S}(\boldsymbol{\xi}) = \operatorname{argmin}_{\gamma\in\mathbb{R}} \mathcal{E}(\boldsymbol{\xi} - \gamma)$.

Measures of error are important in regression analysis and learning, where the goal is to minimize the nonzeroness of certain random variables representing the "residual" difference between observations and model predictions. The statistic of $\boldsymbol{\xi}$ is the set of scalars, possibly a single number, that best approximate a random variable $\boldsymbol{\xi}$ in the sense of the corresponding regular measure of error.

**Example 8.15** (regular measures of error). We've the following examples of regular measures of error $\mathcal{E}$ and corresponding statistics.

(a) *Koenker-Bassett error* and *quantile statistic* with $\alpha \in (0, 1)$:

$$\mathcal{E}(\boldsymbol{\xi}) = \frac{1}{1-\alpha}\mathbb{E}\big[\max\{0, \boldsymbol{\xi}\}\big] - \mathbb{E}[\boldsymbol{\xi}] \qquad \mathcal{S}(\boldsymbol{\xi}) = \big[Q(\alpha), Q^{+}(\alpha)\big].$$

(b) *Worst-case error* and *statistic*:

$$\mathcal{E}(\boldsymbol{\xi}) = \begin{cases} -\mathbb{E}[\boldsymbol{\xi}] & \text{if } \sup \boldsymbol{\xi} \leq 0 \\ \infty & \text{otherwise} \end{cases} \qquad \mathcal{S}(\boldsymbol{\xi}) = \begin{cases} \{\sup \boldsymbol{\xi}\} & \text{if } \sup \boldsymbol{\xi} < \infty \\ \emptyset & \text{otherwise.} \end{cases}$$

(c) $\mathcal{L}^2$-*error* and *statistic* with $\lambda > 0$:

$$\mathcal{E}(\boldsymbol{\xi}) = \lambda\sqrt{\mathbb{E}[\boldsymbol{\xi}^2]} = \lambda\|\boldsymbol{\xi}\|_{\mathcal{L}^2} \qquad \mathcal{S}(\boldsymbol{\xi}) = \big\{\mathbb{E}[\boldsymbol{\xi}]\big\}.$$

(d) *Mixed error* and *statistic* with $\lambda_1, \ldots, \lambda_q > 0$ and $\sum_{i=1}^{q}\lambda_i = 1$:

$$\mathcal{E}(\boldsymbol{\xi}) = \inf_{\gamma_1,\ldots,\gamma_q}\Big\{\sum\nolimits_{i=1}^{q}\lambda_i\mathcal{E}_i(\boldsymbol{\xi} - \gamma_i)\;\Big|\;\sum\nolimits_{i=1}^{q}\lambda_i\gamma_i = 0\Big\} \qquad \mathcal{S}(\boldsymbol{\xi}) = \sum\nolimits_{i=1}^{q}\lambda_i\mathcal{S}_i(\boldsymbol{\xi}),$$

where $\mathcal{S}_i(\boldsymbol{\xi}) = \operatorname{argmin}_{\gamma\in\mathbb{R}} \mathcal{E}_i(\boldsymbol{\xi} - \gamma)$ and $\mathcal{E}_i$ is a regular measure of error.

**Detail.** For (a), we conclude that the Koenker-Bassett error satisfies the requirements for a regular measure of error by arguing as in 8.8. This functional is well known in statistics because the minimization of $\mathcal{E}(\boldsymbol{\xi} - \gamma)$ produces the $\alpha$-quantile $Q(\alpha)$ of $\boldsymbol{\xi}$; see [58]. The claimed formula for $\mathcal{S}(\boldsymbol{\xi})$ is a consequence of 8.21 below; it coincides with the set of minimizers in the minimization formula for superquantiles as seen in 3.9, where an expression for $Q^+(\alpha)$ is given in the proof.

For (b,c), one can follow the arguments in 8.10 to show that these functionals indeed are regular measures of error. The statistic in (c) is derived at the end of 3.26. We refer to [102] for a proof of (d). □

**Example 8.16** (regression). In many applications, it's useful to be able to predict (or forecast) the value of a random variable $\boldsymbol{\eta}$ from the values of some other random variables $\boldsymbol{\xi}_1, \ldots, \boldsymbol{\xi}_n$. For instance, these other random variables could be input to a system, which we observe or perhaps even control, and $\boldsymbol{\eta}$ is the output of the system, which we hope to predict. Let $\boldsymbol{\xi} = (\boldsymbol{\xi}_1, \ldots, \boldsymbol{\xi}_n)$ and suppose that $\boldsymbol{\eta}, \boldsymbol{\xi}_1, \ldots, \boldsymbol{\xi}_n \in \mathcal{L}^2$.

In *linear regression*, we construct a predictive model of $\boldsymbol{\eta}$ by finding a "best" approximation of the form $\gamma + \langle c, \boldsymbol{\xi}\rangle$. We note that for each $\gamma \in \mathbb{R}$ and $c \in \mathbb{R}^n$, $\gamma + \langle c, \boldsymbol{\xi}\rangle$ is a random variable in $\mathcal{L}^2$. Given a regular measure of error $\mathcal{E}$, this leads to the *regression problem*

$$\underset{(\gamma,c)\in\mathbb{R}^{1+n}}{\text{minimize}}\ \mathcal{E}\big(\boldsymbol{\eta} - \gamma - \langle c, \boldsymbol{\xi}\rangle\big). \tag{8.4}$$

A minimizer $(\gamma^\star, c^\star)$ of the problem defines the random variable $\gamma^\star + \langle c^\star, \boldsymbol{\xi}\rangle$, which then makes the nonzeroness of $\boldsymbol{\eta} - (\gamma + \langle c, \boldsymbol{\xi}\rangle)$ as low as possible in the sense of the selected measure of error. Thus, $\gamma^\star + \langle c^\star, \boldsymbol{\xi}\rangle$ is a best possible approximation of $\boldsymbol{\eta}$ in this sense.

**Detail.** Since $\mathcal{E}$ is convex, it follows by arguments similar to those in 8.11 that the function

$$(\gamma, c) \mapsto \mathcal{E}\big(\boldsymbol{\eta} - \gamma - \langle c, \boldsymbol{\xi}\rangle\big)$$

is convex.

The choice 8.15(c) leads to least-squares regression regardless of $\lambda > 0$. Since the measure of error can be squared without affecting the set of minimizers, in this case one has

$$\text{argmin}_{(\gamma,c)\in\mathbb{R}^{1+n}}\, \mathcal{E}\big(\boldsymbol{\eta} - \gamma - \langle c, \boldsymbol{\xi}\rangle\big) = \text{argmin}_{(\gamma,c)\in\mathbb{R}^{1+n}}\, \mathbb{E}\Big[\big(\boldsymbol{\eta} - \gamma - \langle c, \boldsymbol{\xi}\rangle\big)^2\Big];$$

see 2.37 for the related least-squares problem. In fact, $\boldsymbol{\eta} \mapsto \mathbb{E}[\boldsymbol{\eta}^2]$ is also a regular measure of error and we could just as well have adopted it from the start.

The Koenker-Bassett error in 8.15(a) leads to *quantile regression*. If the underlying probability space $(\Omega, \mathcal{A}, \mathbb{P})$ is finite, then the regression problem in this case becomes

$$\underset{(\gamma,c)\in\mathbb{R}^{1+n}}{\text{minimize}}\ \varphi(\gamma, c) = \frac{1}{1-\alpha}\sum\nolimits_{\omega\in\Omega}\mathbb{P}(\{\omega\})\max\Big\{0, \boldsymbol{\eta}(\omega) - \gamma - \big\langle c, \boldsymbol{\xi}(\omega)\big\rangle\Big\}$$
$$- \sum\nolimits_{\omega\in\Omega}\mathbb{P}(\{\omega\})\Big(\boldsymbol{\eta}(\omega) - \gamma - \big\langle c, \boldsymbol{\xi}(\omega)\big\rangle\Big).$$

This resembles the problem arising in the construction of support vector machines (see §2.H) and can be solved by algorithms from Chap. 2 directly or be reformulated as a large-scale linear optimization problem following the pattern in §2.H.

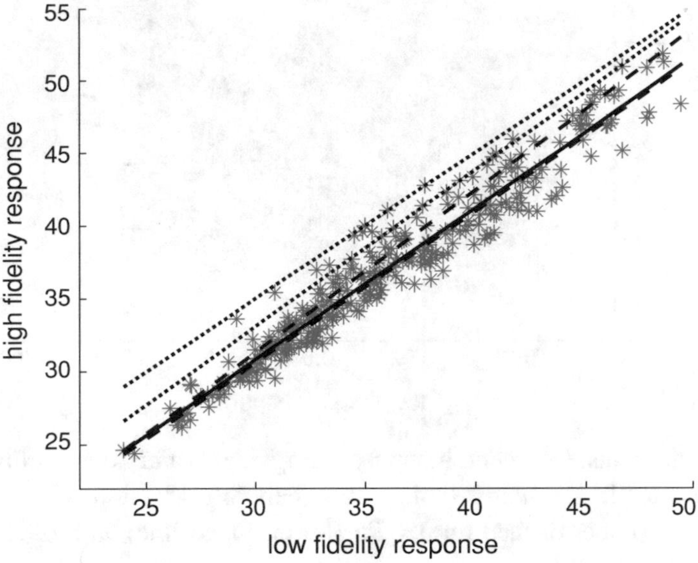


Fig. 8.2: Predicting lift response using high- and low-fidelity simulations. Regression lines for least-squares (solid) and quantile regression with $\alpha = 0.5$ (lower dashed line), 0.75 (upper dashed line), 0.95 (lower dotted line) and 0.995 (upper dotted line).

Figure 8.2 shows 308 data points from the simulation of the lift force of a hydrofoil; see [16] for background information. The first axis gives the uncertain lift force $\boldsymbol{\xi}$ according to inaccurate low-fidelity simulations and the second axis specifies the corresponding uncertain lift force $\boldsymbol{\eta}$ computed using accurate high-fidelity simulations. We obtain predictive models for $\boldsymbol{\eta}$ of the form $\gamma + c\boldsymbol{\xi}$ by least-squares regression and quantile regression with $\alpha = 0.5, 0.75, 0.95$ and $0.995$. A predictive model of this kind can be used to estimate the lift force without having to carry out computationally costly high-fidelity simulations. Figure 8.2 plots the lines given by $\xi \mapsto \gamma + c\xi$ in the five cases. While the lines are quite similar, quantile regression with higher values of $\alpha$ results in more conservative estimates in the sense that $\gamma + c\boldsymbol{\xi}$ tends to exceed $\boldsymbol{\eta}$. Figure 8.3 illustrates 3063 data points from high- and low-fidelity simulations of the displacement of a hydrofoil. Now, the various predictive models are rather different and the benefits from having the opportunity to explore different measures of error become apparent. None of the models are especially accurate, but those based on quantile regression with high $\alpha$ produce mostly conservative predications. Moreover, the slope coefficient $c$ varies with the measure of error, which means that an increase in low-fidelity response has different effects on the predictions. □

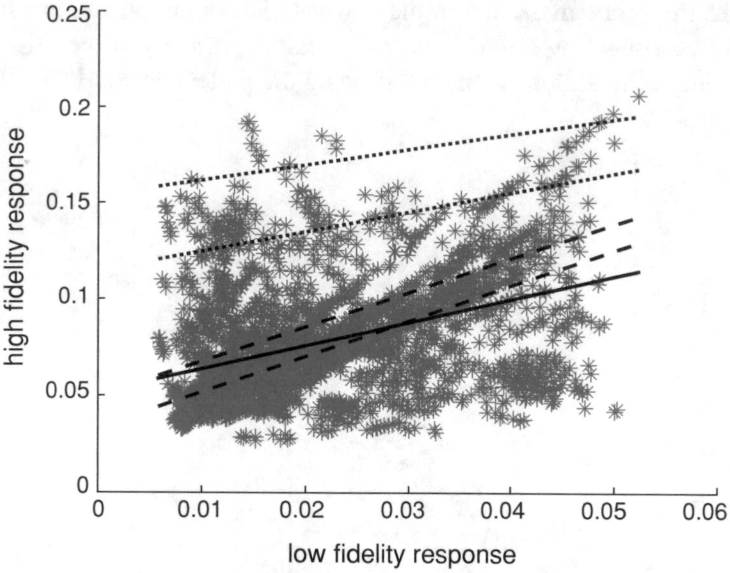


Fig. 8.3: Predicting displacement response using high- and low-fidelity simulations. Regression lines for least-squares (solid) and quantile regression with $\alpha = 0.5$ (lower dashed line), 0.75 (upper dashed line), 0.95 (lower dotted line) and 0.995 (upper dotted line).

**Exercise 8.17** (robustness to outliers). For the data $\{\xi^1, \dots, \xi^{20} \in \mathbb{R}\}$: 0.26, 0.37, 0.62, 0.08, 0.96, 0.48, 0.33, 0.80, 0.16, 0.75, 0.97, 0.94, 0.32, 0.24, 0.72, 0.02, 0.01, 0.61, 0.68, 0.27 and the data $\{\eta^1, \dots, \eta^{20} \in \mathbb{R}\}$: 0.51, 0.69, 1.05, 0.24, 1.49, 1.00, 1.40, 1.31, 0.37, 1.20, 1.61, 1.57, 0.54, 0.40, 1.28, 0.04, 0.15, 1.06, 1.13, 0.53, formulate and solve the regression problems corresponding to quantile regression with $\alpha = 0.5, 0.75$ and 0.9 as well as least-squares regression. Repeat the calculations with the data point $(\xi^7, \eta^7)$ replaced by (0.33, 2.40). Discuss the effect of having such an outlier on the solutions.

**Guide.** For the underlying probability space, simply take $\Omega = \{(\xi^i, \eta^i), i = 1, \dots, 20\}$ and $\mathbb{P}(\{(\xi^i, \eta^i)\}) = 1/20$. Formulations for the regression problems are given in 8.16. Quantile regression can be addressed using a linear optimization problem. □

In addition to nonzeroness, there's also a need for quantifying nonconstancy of a random variable. This can be accomplished using the standard deviation, but there are many other possibilities as well.

**Definition 8.18** (regular measure of deviation). A *regular measure of deviation* $\mathcal{D}$ is a functional from $\mathcal{L}^2$ to $[0, \infty]$ that's lsc and convex, with

$$\mathcal{D}(\boldsymbol{\xi}) = 0 \text{ if } \boldsymbol{\xi} \text{ is constant}; \quad \mathcal{D}(\boldsymbol{\xi}) > 0 \text{ otherwise.}$$

The quantity $\mathcal{D}(\boldsymbol{\xi})$ is the *deviation* of $\boldsymbol{\xi}$.

**Example 8.19** (regular measures of deviation). We've the following examples of regular measures of deviation $\mathcal{D}$.

(a) *Superquantile deviation* with $\alpha \in (0,1)$:

$$\mathcal{D}(\boldsymbol{\xi}) = \text{s-rsk}_\alpha(\boldsymbol{\xi}) - \mathbb{E}[\boldsymbol{\xi}].$$

(b) *Worst-case deviation*:

$$\mathcal{D}(\boldsymbol{\xi}) = \sup \boldsymbol{\xi} - \mathbb{E}[\boldsymbol{\xi}].$$

(c) *Scaled standard deviation* with $\lambda > 0$:

$$\mathcal{D}(\boldsymbol{\xi}) = \lambda \operatorname{std}(\boldsymbol{\xi}).$$

(d) *Mixed deviation* with $\lambda_1, \dots, \lambda_q > 0$ and $\sum_{i=1}^q \lambda_i = 1$:

$$\mathcal{D}(\boldsymbol{\xi}) = \sum_{i=1}^q \lambda_i \mathcal{D}_i(\boldsymbol{\xi}), \quad \text{for regular measures of deviation } \mathcal{D}_1, \dots, \mathcal{D}_q.$$

**Detail.** In (a), we obtain that $\mathcal{D}$ satisfies the requirements of a regular measure of deviation by using the properties of superquantile risk. Interestingly, $\mathcal{D}$ quantifies variability on the "high side." For example, the two random variables $\boldsymbol{\xi}_1$ and $\boldsymbol{\xi}_2$, represented by their density functions in Figure 8.1, have $\mathcal{D}(\boldsymbol{\xi}_1) = 0.91$ and $\mathcal{D}(\boldsymbol{\xi}_2) = 0.61$, when $\alpha = 0.9$. In the sense of superquantile deviation, $\boldsymbol{\xi}_1$ varies more than $\boldsymbol{\xi}_2$ and this is caused by the former's upper tail. In contrast, their standard deviations coincide.

A similar focus on upper tails takes place in (b). The regularity in this case follows using the arguments in 8.10. In that earlier example we essentially also proved the regularity in (c). The regularity of $\mathcal{D}$ in (d) follows directly from the regularity of $\mathcal{D}_i$. □

The regular measures of deviation in 8.19(a,b,c) share the common property that they're obtained from the regular measures of risk in 8.5(a,b,c), respectively, by subtracting $\mathbb{E}[\boldsymbol{\xi}]$. Similarly, the regular measures of error in 8.15(a,b,c) can be constructed from the regular measures of regret in 8.5(a,b,c), respectively, by subtracting $\mathbb{E}[\boldsymbol{\xi}]$. These connections hold in general, as the next theorem asserts, and provide new ways for constructing functionals with desirable properties.

**Theorem 8.20** (expectation translations). *Every regular measure of deviation $\mathcal{D}$ defines a regular measure of risk $\mathcal{R}$ and vice versa through the relations:*

$$\mathcal{R}(\boldsymbol{\xi}) = \mathcal{D}(\boldsymbol{\xi}) + \mathbb{E}[\boldsymbol{\xi}] \quad \textit{and} \quad \mathcal{D}(\boldsymbol{\xi}) = \mathcal{R}(\boldsymbol{\xi}) - \mathbb{E}[\boldsymbol{\xi}].$$

*Similarly, every regular measure of error $\mathcal{E}$ defines a regular measure of regret $\mathcal{V}$ and vice versa through the relations*

$$\mathcal{V}(\boldsymbol{\xi}) = \mathcal{E}(\boldsymbol{\xi}) + \mathbb{E}[\boldsymbol{\xi}] \quad \textit{and} \quad \mathcal{E}(\boldsymbol{\xi}) = \mathcal{V}(\boldsymbol{\xi}) - \mathbb{E}[\boldsymbol{\xi}].$$

**Proof.** These facts follow directly from the definitions. □

We complete the picture by connecting regular measures of error and deviation in a manner that resembles the relation between regular measures of regret and risk; see 8.9.

**Theorem 8.21** (error-deviation). *Given a regular measure of error $\mathcal{E}$, a regular measure of deviation $\mathcal{D}$ is obtained by*

$$\mathcal{D}(\xi) = \min_{\gamma \in \mathbb{R}} \mathcal{E}(\xi - \gamma).$$

*For every regular measure of deviation $\mathcal{D}$ there's a regular measure of error $\mathcal{E}$, not necessarily unique, that constructs $\mathcal{D}$ through this minimization formula.*

*Moreover, if $\mathcal{E}$ is paired with a regular measure of regret $\mathcal{V}$ via 8.20, then*

$$\mathcal{S}(\xi) = \operatorname{argmin}_{\gamma \in \mathbb{R}} \mathcal{E}(\xi - \gamma) = \operatorname{argmin}_{\gamma \in \mathbb{R}} \{\gamma + \mathcal{V}(\xi - \gamma)\} \tag{8.5}$$

*and this set is a nonempty compact interval as long as $\mathcal{V}(\xi - \gamma)$, or equivalently $\mathcal{E}(\xi - \gamma)$, is finite for some $\gamma \in \mathbb{R}$.*

**Proof.** Since $\mathcal{E}(\xi - \gamma) = \gamma + \mathcal{V}(\xi - \gamma) - \mathbb{E}[\xi]$ for the regular measure of regret $\mathcal{V}$ corresponding to $\mathcal{E}$ by 8.20, it follows from 8.9 that

$$\min_{\gamma \in \mathbb{R}} \mathcal{E}(\xi - \gamma) = \min_{\gamma \in \mathbb{R}} \{\gamma + \mathcal{V}(\xi - \gamma)\} - \mathbb{E}[\xi] = \mathcal{R}(\xi) - \mathbb{E}[\xi],$$

where $\mathcal{R}$ is a regular measure of risk. By 8.20, $\mathcal{D}(\xi) = \mathcal{R}(\xi) - \mathbb{E}[\xi]$ defines a regular measure of deviation. This also establishes (8.5). An example of a regular measure of error that produces a regular measure of deviation $\mathcal{D}$ is given by

$$\mathcal{E}(\xi) = \mathcal{D}(\xi) + \alpha \big|\mathbb{E}[\xi]\big|,$$

where $\alpha > 0$; see the arguments in 8.9. The compactness claim about $\mathcal{S}(\xi)$ is proven in [99, Theorem 2.2]. □

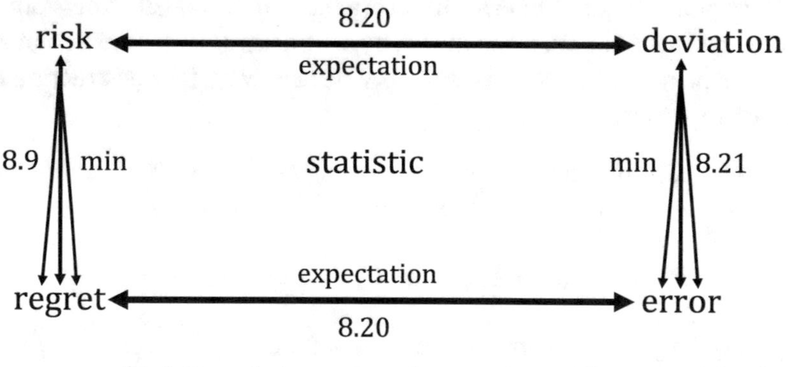


Fig. 8.4: Summary of relations between regular measures of regret, risk, deviation and error that form a risk quadrangle.

Theorems 8.9, 8.20 and 8.21 connect regular measures of regret, risk, deviation and error in a *risk quadrangle* that provides insight as well as computational opportunities. We saw in 8.11 how passing to a measure of regret may simplify risk minimization and, similarly, a regular measure of deviation can be evaluated by first minimizing regret as in

(8.5). One can construct new regular measures of risk and deviation by starting from either a regular measure of regret or a regular measure of error using the relations summarized in Figure 8.4. The "horizontal" connections in the figure are one-to-one as seen in 8.20, while the "vertical" connections aren't unique in the sense that multiple regular measures of regret produce the same regular measure of risk, with a similar situation taking place when passing from error to deviation.

The regular measures of regret, risk, error and deviation labeled (a) in 8.5, 8.15 and 8.19 are connected in a risk quadrangle in the sense of Figure 8.4. Likewise, the regular measures of regret, risk, error and deviation labeled (b,c,d) in 8.5, 8.15 and 8.19 are connected, respectively, and then also form risk quadrangles. Many other risk quadrangles are listed in [102].

**Exercise 8.22** (exponential and normal distributions). First, consider a random variable $\xi$ with distribution function given by $P(\xi) = 1-\exp(-\xi/\mu)$ for $\xi \geq 0$ and $P(\xi) = 0$ otherwise, where $\mu > 0$. Determine explicit formulae for the measures of error in 8.15(a,b,c) and the measures of deviation in 8.19(a,b,c). Verify that 8.21 holds. Second, repeat the task for a normal random variable with mean $\mu$ and standard deviation $\sigma$.

**Guide.** The relations in 8.20, as well as 3.8 and 8.6, furnish several shortcuts. □

**Proposition 8.23** (decomposition). *For a regular measure of error $\mathcal{E}$ and a regular measure of deviation $\mathcal{D}$ paired by 8.21, one has for any $\boldsymbol{\eta}, \boldsymbol{\xi}_1, \dots, \boldsymbol{\xi}_n \in \mathcal{L}^2$:*

$$(\gamma^\star, c^\star) \in \operatorname{argmin}_{(\gamma,c)\in\mathbb{R}^{1+n}} \mathcal{E}\big(\boldsymbol{\eta} - \gamma - \langle c, \boldsymbol{\xi}\rangle\big)$$

$$\Longleftrightarrow c^\star \in \operatorname{argmin}_{c\in\mathbb{R}^n} \mathcal{D}\big(\boldsymbol{\eta} - \langle c, \boldsymbol{\xi}\rangle\big) \quad \text{and} \quad \gamma^\star \in \mathcal{S}\big(\boldsymbol{\eta} - \langle c^\star, \boldsymbol{\xi}\rangle\big),$$

*where $\boldsymbol{\xi} = (\boldsymbol{\xi}_1, \dots, \boldsymbol{\xi}_n)$ and $\mathcal{S}$ is the statistic corresponding to $\mathcal{E}$.*

**Proof.** This equivalence is apparent from the fact that

$$\inf_{(\gamma,c)\in\mathbb{R}^{1+n}} \mathcal{E}\big(\boldsymbol{\eta} - \gamma - \langle c, \boldsymbol{\xi}\rangle\big) = \inf_{c\in\mathbb{R}^n} \inf_{\gamma\in\mathbb{R}} \mathcal{E}\Big(\big(\boldsymbol{\eta} - \langle c, \boldsymbol{\xi}\rangle\big) - \gamma\Big) = \inf_{c\in\mathbb{R}^n} \mathcal{D}\big(\boldsymbol{\eta} - \langle c, \boldsymbol{\xi}\rangle\big)$$

and the definition of $\mathcal{S}$; see 8.14. □

The proposition shows that we can determine the coefficients $(\gamma^\star, c^\star)$ in the regression problem (8.4) using two steps: first, determine $c^\star$ by minimizing deviation and, second, fix $\gamma^\star$ by computing a statistic. From 8.20, we see that minimizing deviation is equivalent to minimizing risk modified by an expectation. Thus, $c^\star$ can often be computed by algorithms for minimizing risk, with minor adjustments.

Separate roles for the "slope" $c$ and the "intercept" $\gamma$ emerge from the proposition. The former is selected to minimize the nonconstancy of $\boldsymbol{\eta} - \langle c, \boldsymbol{\xi}\rangle$, while the latter translates the functional $\boldsymbol{\xi} \mapsto \langle c^\star, \boldsymbol{\xi}\rangle$ up or down to match the correct statistic.

The quality of a predictive model $\gamma^\star + \langle c^\star, \boldsymbol{\xi}\rangle$ obtained by solving a regression problem can be quantified by the nonzeroness of the *residual* $\boldsymbol{\eta} - (\gamma^\star + \langle c^\star, \boldsymbol{\xi}\rangle)$. In fact, the regression problem aims to minimize this nonzeroness. To allow comparison across different applications and circumstances, we normalize the nonzeroness of the residual with the overall variability of $\eta$ as quantified by the corresponding regular measure of deviation. This leads to the following definition.

For a regular measure of error $\mathcal{E}$ paired with a regular measure of deviation $\mathcal{D}$ by 8.21, the *coefficient of determination*, or *R-squared*, of the random vector $\xi = (\xi_1, \dots, \xi_n)$ relative to $\eta$ is given as[4]

$$R^2 = 1 - \frac{\inf_{(\gamma,c)\in\mathbb{R}^{1+n}} \mathcal{E}\big(\eta - \gamma - \langle c, \xi\rangle\big)}{\mathcal{D}(\eta)}. \tag{8.6}$$

It's immediate from the definition that if $(\gamma^\star, c^\star)$ is a minimizer of the regression problem, then

$$\mathcal{E}\big(\eta - \gamma^\star - \langle c^\star, \xi\rangle\big) = 0 \implies R^2 = 1.$$

However, such vanishing error only takes place when the residual

$$\eta - \gamma^\star - \langle c^\star, \xi\rangle = \mathbf{0};$$

the model $\gamma^\star + \langle c^\star, \xi\rangle$ predicts $\eta$ perfectly. Under less ideal circumstances, one would assess the quality of $(\gamma^\star, c^\star)$ by seeing how close $R^2$ is to 1. In general,

$$\inf_{\gamma,c} \mathcal{E}\big(\eta - \gamma - \langle c, \xi\rangle\big) \le \inf_\gamma \mathcal{E}(\eta - \gamma) = \mathcal{D}(\eta).$$

Thus, $R^2 \geq 0$ always and $R^2 = 0$ when $(\gamma, 0)$ minimizes the error. In that case, $\xi$ provides "no information" about $\eta$.

**Example 8.24** (coefficient of determination). In 8.16, we obtained predictive models of the form $\gamma + c\xi$ in a case with a clear trend and little noise (Figure 8.2) and in a more diffuse situation (Figure 8.3). The quality of the models can be assessed using $R^2$ as seen in Table 8.1, which also includes values for the numerator (err) and denominator (dev) in the defining expression for $R^2$ in (8.6). In the case of Figure 8.2, all the predictions are reasonably good with $R^2$ being at least 0.73. One can say that at least 73% of the variability is explained by the predictive models. The predictive models are much less trustworthy in the case of Figure 8.3, where $R^2$ ranges between 0.04 and 0.17.

Table 8.1: Coefficient of determination ($R^2$) for predictive models in 8.16.

| | Figure 8.2 | | | Figure 8.3 | | |
|---|---|---|---|---|---|---|
| $\alpha$ | dev | err | $R^2$ | dev | err | $R^2$ |
| 0.5 | 4.76 | 1.10 | 0.77 | 0.024 | 0.020 | 0.171 |
| 0.75 | 7.97 | 1.83 | 0.77 | 0.044 | 0.038 | 0.136 |
| 0.95 | 12.32 | 3.29 | 0.73 | 0.077 | 0.074 | 0.044 |
| 0.995 | 14.76 | 3.98 | 0.73 | 0.111 | 0.104 | 0.061 |
| least-sq | 33.98 | 1.98 | 0.94 | $9.4 \cdot 10^{-4}$ | $7.8 \cdot 10^{-4}$ | 0.169 |

[4] Here, $\infty/\infty$ is interpreted as 1 and $0/0$ as 0.

**Detail.** As discussed in 8.16, least-squares regression is achieved by utilizing $\mathcal{E}(\eta) = \|\eta\|_{\mathcal{L}^2}$ or, equivalently, the mean-squared error $\mathcal{E}(\eta) = \mathbb{E}[\eta^2]$. The former measure of error has the standard deviation $\mathcal{D}(\eta) = \text{std}(\eta)$ as its corresponding measure of deviation, while the latter corresponds to the variance $\mathcal{D}(\eta) = \text{var}(\eta)$. The coefficient of determination that emerges from the mean-squared error is

$$R^2 = 1 - \frac{\mathbb{E}\big[(\eta - \gamma^\star - \langle c^\star, \xi\rangle)^2\big]}{\text{var}(\eta)}, \text{ with } (\gamma^\star, c^\star) \in \text{argmin}_{(\gamma,c)\in\mathbb{R}^{1+n}} \mathbb{E}\big[(\eta - \gamma - \langle c, \xi\rangle)^2\big].$$

The last row of Table 8.1 reports the corresponding numbers. □

**Exercise 8.25** (range quadrangle). For $\lambda \in (0, \infty)$, show that the following quantities form a risk quadrangle:

$$\mathcal{V}(\xi) = \mathbb{E}[\xi] + \lambda \sup |\xi|$$
$$\mathcal{R}(\xi) = \mathbb{E}[\xi] + \tfrac{\lambda}{2}(\sup \xi - \inf \xi)$$
$$\mathcal{E}(\xi) = \lambda \sup |\xi|$$
$$\mathcal{D}(\xi) = \tfrac{\lambda}{2}(\sup \xi - \inf \xi),$$

where $\inf \xi = \inf \Xi$ and $\Xi$ is the support of $\xi$. Show that the corresponding statistic $\mathcal{S}(\xi) = (\sup \xi + \inf \xi)/2$ if $\sup \xi$ and $\inf \xi$ are finite.

**Guide.** Consult Figure 8.4. Establish that $\mathcal{V}$ is a regular measure of regret using arguments similar to those in 8.10. Then, invoke 8.9, 8.20 and 8.21. □

## 8.D Duality

In Chap. 5, we saw how a rich duality theory provides approximating and sometimes equivalent formulations of optimization problems resulting in alternative computational strategies as well as important insight. The connections are especially tight for convex functions. Since regular measures of regret, risk, error and deviation are also convex, albeit on more general spaces of random variables, we can be hopeful that similar reformulations are available for such functionals as well. In this section, we'll develop a duality theory for convex functionals that results in alternative expressions for measures of risk with significant modeling implications.

Let's start with an extension of conjugates beyond 5.21.

**Definition 8.26** (conjugate functionals). For $\mathcal{F} : \mathcal{L}^2 \to \overline{\mathbb{R}}$, the functional $\mathcal{F}^* : \mathcal{L}^2 \to \overline{\mathbb{R}}$ defined by

$$\mathcal{F}^*(\pi) = \sup\big\{\mathbb{E}[\xi\pi] - \mathcal{F}(\xi) \,\big|\, \xi \in \mathcal{L}^2\big\}$$

is the *conjugate* of $\mathcal{F}$.

The definition extends the earlier notion by replacing the inner product $(u, v) \mapsto \langle u, v\rangle$ by $(\xi, \pi) \mapsto \mathbb{E}[\xi\pi]$, which is an inner product on $\mathcal{L}^2$. A main motivation for restricting the attention to random variables with finite second moment is that a conjugate is then defined on $\mathcal{L}^2$. In contrast, if $\xi$ were only integrable, then $\pi$ would have to be bounded for $\mathbb{E}[\xi\pi]$ to be finite; see [116] for such extensions.

The *domain* of a functional $\mathcal{F} : \mathcal{L}^2 \to \overline{\mathbb{R}}$ is

$$\operatorname{dom}\mathcal{F} = \left\{\xi \in \mathcal{L}^2 \,\middle|\, \mathcal{F}(\xi) < \infty\right\}.$$

The functional is *proper* if $\mathcal{F}(\xi) > -\infty$ for all $\xi \in \mathcal{L}^2$ and $\operatorname{dom}\mathcal{F} \neq \emptyset$.

**Theorem 8.27** (Fenchel-Moreau). *For a proper, lsc and convex functional $\mathcal{F} : \mathcal{L}^2 \to \overline{\mathbb{R}}$, the conjugate $\mathcal{F}^*$ is also proper, lsc and convex and $(\mathcal{F}^*)^* = \mathcal{F}$ so that*

$$\mathcal{F}(\xi) = \sup\left\{\mathbb{E}[\xi\pi] - \mathcal{F}^*(\pi) \,\middle|\, \pi \in \mathcal{L}^2\right\}.$$

**Proof.** This fact is parallel to the Fenchel-Moreau theorem 5.23; see [90, Theorem 5] for a proof. □

Since regular measures of regret, risk, error and deviation are lsc and convex by definition and also proper (because they're finite at some point and never $-\infty$), the Fenchel-Moreau theorem 8.27 applies and furnishes alternative expressions. The usefulness of this fact depends on our ability to characterize the corresponding conjugate functionals. One property of a functional $\mathcal{F} : \mathcal{L}^2 \to \overline{\mathbb{R}}$ is especially salient:

$$\textit{Positive Homogeneity:} \quad \mathcal{F}(\lambda\xi) = \lambda\mathcal{F}(\xi) \quad \forall\lambda \in [0, \infty).$$

It certainly makes sense for a measure of risk to be positively homogeneous as it would imply that rescaling of a random variable, for example from cost in dollars to cost in euros, doesn't fundamentally change the risk associated with the uncertain cost.

**Proposition 8.28** (positive homogeneity). *If $\mathcal{V}$ is a positively homogeneous regular measure of regret, then the regular measure of risk $\mathcal{R}$ constructed by the regret-risk theorem 8.9 is also positively homogenous.*

*If $\mathcal{E}$ is a positively homogeneous regular measure of error, then the regular measure of deviation $\mathcal{D}$ constructed by the error-deviation theorem 8.21 is also positively homogenous.*

**Proof.** Since $\mathcal{R}(\lambda\xi) = \mathcal{R}(\mathbf{0}) = 0 = \lambda\mathcal{R}(\xi)$ when $\lambda = 0$, we can concentrate on $\lambda \in (0, \infty)$. For $f : \mathbb{R}^n \to \overline{\mathbb{R}}$, one has $\lambda \inf f = \inf \lambda f$ generally. Let $\gamma \in \mathbb{R}$ and $\bar{\gamma} = \lambda\gamma$. Then,

$$\lambda\big(\gamma + \mathcal{V}(\xi - \gamma)\big) = \bar{\gamma} + \mathcal{V}\big(\lambda\xi - \bar{\gamma})\big)$$

because $\mathcal{V}$ is positively homogeneous. These facts together with 8.9 establish the first result. The second claim follows by similar arguments via 8.21. □

Since $\xi \mapsto \mathbb{E}[\max\{0, \xi\}]$ is positively homogeneous, $\mathcal{V}$ in 8.5(a) is positively homogeneous and the same holds for s-rsk$_\alpha$ by 8.28. The measures of regret and risk in 8.5(b,c) are also positively homogeneous. For (c), this follows because the functional given by

$$\mathcal{F}(\xi) = \mathrm{std}(\xi) = \|\xi - \mathbb{E}[\xi]\|_{\mathcal{L}^2}$$

is positively homogeneous. In contrast, $\mathcal{F}(\xi) = \mathrm{var}(\xi)$, which surfaces in 8.16, isn't positively homogeneous. For 8.5(d), $\mathcal{V}$ is positively homogeneous when $\mathcal{V}_1, \dots, \mathcal{V}_q$ are positively homogeneous and then $\mathcal{R}$ also has this property. One can see this by following an argument similar to the one in the proof of 8.28.

**Proposition 8.29** (conjugacy under positive homogeneity). *For a positively homogeneous, proper, lsc and convex functional $\mathcal{F} : \mathcal{L}^2 \to \overline{\mathbb{R}}$, one has*

$$\mathcal{F}(\xi) = \sup\{\mathbb{E}[\xi\pi] \mid \pi \in \mathrm{dom}\,\mathcal{F}^*\}.$$

*A nonempty, closed and convex set $C \subset \mathcal{L}^2$ is the domain of $\mathcal{F}^*$ for some positively homogeneous, proper, lsc and convex $\mathcal{F} : \mathcal{L}^2 \to \overline{\mathbb{R}}$ and then*

$$C = \{\pi \in \mathcal{L}^2 \mid \mathbb{E}[\xi\pi] \le \mathcal{F}(\xi)\ \forall \xi \in \mathcal{L}^2\}.$$

**Proof.** These facts follow from the Fenchel-Moreau theorem 8.27. In particular, for a proper, lsc and convex $\mathcal{F}$, one has that $\mathcal{F}$ is positively homogeneous if and only if $\mathcal{F}^*(\pi) = 0$, when $\pi \in \mathrm{dom}\,\mathcal{F}^*$. Since

$$\begin{aligned}\mathrm{dom}\,\mathcal{F}^* &= \{\mathcal{F}^* \le 0\}\\ &= \Big\{\pi \;\Big|\; \sup\{\mathbb{E}[\xi\pi] - \mathcal{F}(\xi) \mid \xi \in \mathcal{L}^2\} \le 0\Big\} = \{\pi \mid \mathbb{E}[\xi\pi] \le \mathcal{F}(\xi)\ \forall \xi \in \mathcal{L}^2\},\end{aligned}$$

we obtain the expression for $C$ in terms of $\mathcal{F}$. □

As a direct application of the proposition, we obtain that for a positively homogeneous regular measure of risk $\mathcal{R}$,

$$\mathcal{R}(\xi) = \sup\{\mathbb{E}[\xi\pi] \mid \pi \in \Pi\}, \tag{8.7}$$

where $\Pi = \mathrm{dom}\,\mathcal{R}^*$ is the *risk envelope* of $\mathcal{R}$.

The proposition asserts that under positive homogeneity, a functional $\mathcal{F}$ is fully characterized by $\mathrm{dom}\,\mathcal{F}^*$. It then becomes important to understand the nature of such domains, especially when they're risk envelopes of regular measures of risk.

**Proposition 8.30** (properties of risk envelopes). *The risk envelope $\Pi$ of a positively homogeneous regular measure of risk $\mathcal{R}$ is closed and convex and, for every $\pi \in \Pi$, one has $\mathbb{E}[\pi] = 1$. Moreover, it has the expression*

$$\Pi = \{\pi \in \mathcal{L}^2 \mid \mathbb{E}[\xi\pi] \le \mathcal{R}(\xi)\ \forall \xi \in \mathcal{L}^2\}.$$

*A nonempty, closed and convex set $C \subset \mathcal{L}^2$ is the risk envelope of some positively homogeneous regular measure of risk on $\mathcal{L}^2$ provided that $C$ also satisfies*

$$\mathbb{E}[\pi] = 1 \quad \forall \pi \in C;$$

*for each nonconstant* $\xi \in \mathcal{L}^2$, *there's* $\pi \in C$ *such that* $\mathbb{E}[\xi\pi] > \mathbb{E}[\xi]$.

**Proof.** Since $\mathcal{R}$ is proper, lsc and convex, the conjugate $\mathcal{R}^*$ is also proper, lsc and convex by the Fenchel-Moreau theorem 8.27. Directly from the definitions, the convexity of $\mathcal{R}^*$ implies the convexity of $\Pi = \operatorname{dom} \mathcal{R}^*$. Since $\operatorname{dom} \mathcal{R}^* = \{\mathcal{R}^* \leq 0\}$ due to positive homogeneity (see 8.29 and its proof) and $\mathcal{R}^*$ is lsc, $\Pi$ is closed.

The expression for $\Pi$ is given in 8.29. The property $\mathbb{E}[\pi] = 1$ for all $\pi \in \Pi$ is established as follows: In $\mathbb{E}[\xi\pi] \leq \mathcal{R}(\xi)$, set $\xi = \mathbf{1}$. Thus, $\mathbb{E}[\pi] \leq \mathcal{R}(\mathbf{1}) = 1$. Similarly, with $\xi = -\mathbf{1}$, we obtain $\mathbb{E}[\pi] \geq 1$.

A nonempty, closed and convex set $C \subset \mathcal{L}^2$ is the domain of $\mathcal{F}^*$ for some positively homogeneous, proper, lsc and convex $\mathcal{F} : \mathcal{L}^2 \to \overline{\mathbb{R}}$ by 8.29. It only remains to show that $\mathcal{F}$ satisfies the constancy and averseness properties. From 8.29,

$$\mathcal{F}(\xi) = \sup\left\{\mathbb{E}[\xi\pi] \,\middle|\, \pi \in C\right\} \quad \forall \xi \in \mathcal{L}^2$$

and, in particular, for the constant random variable $\xi_0$ with value $\alpha$. Thus,

$$\mathcal{F}(\xi_0) = \sup\left\{\mathbb{E}[\xi_0\pi] \,\middle|\, \pi \in C\right\} = \sup\left\{\alpha\mathbb{E}[\pi] \,\middle|\, \pi \in C\right\} = \alpha$$

because $\mathbb{E}[\pi] = 1$ for all $\pi \in C$. We've established the constancy property.

Let $\xi \in \mathcal{L}^2$ be nonconstant. By assumption, there's $\pi_0 \in C$ such that $\mathbb{E}[\xi\pi_0] > \mathbb{E}[\xi]$. Consequently,

$$\mathcal{F}(\xi) = \sup\left\{\mathbb{E}[\xi\pi] \,\middle|\, \pi \in C\right\} \geq \mathbb{E}[\xi\pi_0] > \mathbb{E}[\xi]$$

and the averseness property holds. □

**Example 8.31** (risk envelopes). The risk envelopes corresponding to the regular measures of risk in 8.5 are as follows:

(a) *Superquantile risk envelope*:

$$\Pi = \left\{\pi \,\middle|\, 0 \leq \pi(\omega) \leq 1/(1-\alpha) \ \forall \omega, \quad \mathbb{E}[\pi] = 1\right\}.$$

(b) *Worst-case risk envelope*:

$$\Pi = \left\{\pi \,\middle|\, \pi(\omega) \geq 0 \ \forall \omega, \quad \mathbb{E}[\pi] = 1\right\}.$$

(c) *Markowitz risk envelope*:

$$\Pi = \left\{\mathbf{1} + \lambda\pi \,\middle|\, \mathbb{E}[\pi^2] \leq 1, \ \mathbb{E}[\pi] = 0\right\}.$$

(d) *Mixed risk envelope*:

$$\Pi = \left\{\sum\nolimits_{i=1}^{q} \lambda_i \pi_i \,\middle|\, \pi_i \in \Pi_i\right\},$$

where $\Pi_i$ is the risk envelope of the positively homogeneous regular measure of risk $\mathcal{R}_i$.

**Detail.** All these sets are nonempty, closed, convex and satisfy the additional requirements in the second half of 8.30. Thus, each is the risk envelope for some positively homogeneous regular measure of risk and we only need to show that they produce the "right" one in (8.7).

For (a), let's fix $\xi$ and define

$$\Omega_- = \{\omega \mid \xi(\omega) < Q\}, \quad \Omega_\alpha = \{\omega \mid \xi(\omega) = Q\}, \quad \Omega_+ = \{\omega \mid \xi(\omega) > Q\},$$

where $Q$ is the $\alpha$-quantile of $\xi$. Let $\boldsymbol{I}_+$ be the random variable with $\boldsymbol{I}_+(\omega) = 1$ if $\omega \in \Omega_+$ and $\boldsymbol{I}_+(\omega) = 0$ otherwise. Then, we see from 3.6 that

$$\text{s-rsk}_\alpha(\xi) = Q + \frac{1}{1-\alpha}\mathbb{E}\big[\boldsymbol{I}_+(\xi - Q)\big] = Q + \frac{\mathbb{E}[\boldsymbol{I}_+\xi]}{1-\alpha} - \frac{1-P(Q)}{1-\alpha}Q, \tag{8.8}$$

where $P$ is the distribution function of $\xi$. Let $\boldsymbol{I}_-$ be the random variable with $\boldsymbol{I}_-(\omega) = 1$ if $\omega \in \Omega_-$ and $\boldsymbol{I}_-(\omega) = 0$ otherwise; and let $\boldsymbol{I}_\alpha$ be the random variable with $\boldsymbol{I}_\alpha(\omega) = 1$ if $\omega \in \Omega_\alpha$ and $\boldsymbol{I}_\alpha(\omega) = 0$ otherwise. Then,

$$\mathbb{E}[\xi\pi] = \mathbb{E}[\boldsymbol{I}_-\xi\pi] + \mathbb{E}[\boldsymbol{I}_\alpha\xi\pi] + \mathbb{E}[\boldsymbol{I}_+\xi\pi].$$

To maximize this quantity by selecting $\pi$ in the stipulated set, one needs to associate the higher values of $\xi$ with higher values of $\pi$. Tentatively, let's set $\pi(\omega) = 1/(1-\alpha)$ for $\omega \in \Omega_+$, which then maximizes the third term. Let $\bar{p} = \text{prob}\{\xi = Q\}$. Since $\pi$ is subject to the constraint $\mathbb{E}[\pi] = 1$, we may not be able to set $\pi(\omega)$ to $1/(1-\alpha)$ for $\omega \in \Omega_\alpha$ because

$$\mathbb{E}[\boldsymbol{I}_\alpha\pi] + \mathbb{E}[\boldsymbol{I}_+\pi] = \frac{\bar{p}}{1-\alpha} + \frac{1-P(Q)}{1-\alpha}$$

could exceed 1. This occurs if

$$P(Q) \geq \alpha > P(Q) - \bar{p}.$$

Consequently, the best we can do in maximizing $\mathbb{E}[\xi\pi]$ is to set

$$\pi(\omega) = \frac{P(Q) - \alpha}{(1-\alpha)\bar{p}} \quad \forall \omega \in \Omega_\alpha.$$

This ratio lies between 0 and $1/(1-\alpha)$ because $P(Q) \geq \alpha \geq P(Q) - \bar{p}$; we interpret 0/0 as 0. Moreover, with $\pi(\omega) = 0$ for $\omega \in \Omega_-$, we've

$$\mathbb{E}[\boldsymbol{I}_-\xi\pi] + \mathbb{E}[\boldsymbol{I}_\alpha\xi\pi] + \mathbb{E}[\boldsymbol{I}_+\xi\pi] = 0 + \frac{P(Q)-\alpha}{(1-\alpha)\bar{p}}Q\bar{p} + \frac{\mathbb{E}[\boldsymbol{I}_+\xi]}{1-\alpha}.$$

This expression coincides with (8.8) and we've shown that the risk envelope of $\text{s-rsk}_\alpha$ is given by (a).

The risk envelope in (b) follows almost immediately from the expression

$$\Pi = \{\pi \mid \mathbb{E}[\xi\pi] \le \mathcal{R}(\xi)\ \forall \xi\}.$$

The risk envelope in (c) is seen by setting $\pi = (\xi - \mu)/\sigma$ if $\sigma > 0$, where $\mu = \mathbb{E}[\xi]$ and $\sigma = \text{std}(\xi)$. Then, $\mathbb{E}[\pi^2] = 1$ and $\mathbb{E}[\pi] = 0$ so that $\mathbf{1} + \lambda\pi \in \Pi$. Moreover,

$$\mathbb{E}\big[\xi(\mathbf{1} + \lambda\pi)\big] = \mu + \lambda\mathbb{E}\big[\xi(\xi - \mu)/\sigma\big] = \mu + \frac{\lambda}{\sigma}\big(\mathbb{E}[\xi^2] - \mu^2\big) = \mu + \lambda\sigma.$$

If $\sigma = 0$, then set $\pi = \mathbf{0}$ and again $\mathbb{E}[\xi(\mathbf{1} + \lambda\pi)] = \mu + \lambda\sigma$.

For the risk envelope in (d), one has

$$\sum_{i=1}^{q} \lambda_i \sup\big\{\mathbb{E}[\xi\pi_i] \mid \pi_i \in \Pi_i\big\} = \sup\big\{\mathbb{E}[\xi\pi] \mid \pi \in \Pi\big\}$$

and this establishes the asserted formula. □

The fact that a positively homogeneous regular measure of risk can be expressed as a sup-projection in the sense of (8.7) offers several computational possibilities. In particular, it brings forward an expectation, which subsequently can be approximated using Monte Carlo sampling or other techniques.

**Example 8.32** (dual approach for risk minimization). For $f : \mathbb{R}^m \times \mathbb{R}^n \to \mathbb{R}$, $X \subset \mathbb{R}^n$ and a finitely distributed random vector $\xi$ with support $\Xi \subset \mathbb{R}^m$ and probabilities $\{p_\xi > 0, \xi \in \Xi\}$, let's consider

$$\underset{x \in X}{\text{minimize}}\ \mathcal{R}\big(f(\xi, x)\big),$$

where $\mathcal{R}$ is a positively homogeneous regular measure of risk. In view of (8.7), the problem is equivalently stated as

$$\underset{x \in X}{\text{minimize}}\ \varphi(x) = \sup\Big\{\sum_{\xi \in \Xi} p_\xi f(\xi, x)\pi_\xi \;\Big|\; \pi \in \Pi\Big\},$$

where $\pi = (\pi_\xi \in \mathbb{R}, \xi \in \Xi)$ and $\Pi \subset \mathbb{R}^q$, with $q$ being the number of outcomes in $\Xi$. The reformulation can be solved by the outer approximation algorithm.

**Detail.** In this case, we can set the underlying probability space $(\Omega, \mathcal{A}, \mathbb{P})$ that defines $\mathcal{L}^2$ to simple have $\Omega = \Xi$ and $\mathbb{P}(\{\xi\}) = p_\xi$. Then, all random variables $\pi \in \mathcal{L}^2$ are represented by $q$-dimensional vectors of the form $\pi = (\pi_\xi \in \mathbb{R}, \xi \in \Xi)$. The risk envelope of $\mathcal{R}$ corresponds in this case to a subset of $\mathbb{R}^q$, which we above denote by $\Pi$.

In the case with $\mathcal{R} = \text{s-rsk}_\alpha$ and $\alpha \in (0, 1)$, 8.31(a) implies that

$$\Pi = \Big\{\pi \in \mathbb{R}^q \;\Big|\; 0 \le \pi_\xi \le 1/(1-\alpha)\ \forall \xi \in \Xi,\ \sum_{\xi \in \Xi} p_\xi \pi_\xi = 1\Big\}.$$

Since $\varphi$ is a sup-projection, we can bring in results from §6.E and especially 6.30 as $\Pi$ is compact. Moreover, one can apply the outer approximation algorithm of §6.C. Step 1 in that algorithm would then determine

$$\pi^\nu \in \operatorname{argmax}_{\pi \in \Pi} \sum_{\xi \in \Xi} p_\xi f(\xi, x^{\nu-1}) \pi_\xi,$$

where $x^{\nu-1}$ is the current point. This step is available almost explicitly. From the derivation in 8.31, we see that $\pi^\nu = (\pi^\nu_\xi, \xi \in \Xi)$ has

$$\pi^\nu_\xi = \begin{cases} \frac{1}{1-\alpha} & \text{if } f(\xi, x^{\nu-1}) > Q \\ \frac{P(Q)-\alpha}{(1-\alpha)\bar{p}} & \text{if } f(\xi, x^{\nu-1}) = Q \\ 0 & \text{if } f(\xi, x^{\nu-1}) < Q, \end{cases}$$

where $Q$ is the $\alpha$-quantile of $f(\boldsymbol{\xi}, x^{\nu-1})$,

$$P(Q) = \operatorname{prob}\{f(\boldsymbol{\xi}, x^{\nu-1}) \leq Q\}$$

and $\bar{p} = \sum p_\xi$; the sum is over all $\xi \in \Xi$ with $f(\xi, x^{\nu-1}) = Q$. Thus, $\pi^\nu$ is obtained largely by sorting of the values $\{f(\xi, x^{\nu-1}), \xi \in \Xi\}$.

Step 2 of the outer approximation algorithm amounts to solving

$$\underset{x \in X, \gamma \in \mathbb{R}}{\text{minimize}}\ \gamma \ \text{ subject to } \sum_{\xi \in \Xi} p_\xi f(\xi, x) \pi^k_\xi \leq \gamma, \ \ k = 1, \dots, \nu.$$

If $f(\xi, \cdot)$ is smooth and/or convex and $X$ is relatively simple, this subproblem is approachable by standard algorithms such as those described in earlier chapters. The overall approach reminds us of the active-set strategy in 6.16. In both approaches, one brings in elements of $\xi \in \Xi$ that appear "important" at the present point $x^{\nu-1}$. □

Besides the computational opportunities that emerge from the alternative expression (8.7), we also gain significant insight about the difference between minimizing risk and minimizing expected value. The situation is clearest in the case of monotonicity.

**Proposition 8.33** (conjugacy under monotonicity). *For a monotone, proper, lsc and convex functional $\mathcal{F} : \mathcal{L}^2 \to \overline{\mathbb{R}}$, one has $\boldsymbol{\pi}(\omega) \geq 0$ for all $\omega \in \Omega$ and $\boldsymbol{\pi} \in \operatorname{dom} \mathcal{F}^*$.*

**Proof.** This holds as stated (see [102]), but we only prove it for positively homogeneous $\mathcal{F}$. Fix $\boldsymbol{\pi} \in \operatorname{dom} \mathcal{F}^*$. Then, $\mathbb{E}[\boldsymbol{\xi}\boldsymbol{\pi}] \leq \mathcal{F}(\boldsymbol{\xi})$ for all $\boldsymbol{\xi}$ by 8.29. Construct $\boldsymbol{\xi}_0$ by setting

$$\boldsymbol{\xi}_0(\omega) = \min\{0, \boldsymbol{\pi}(\omega)\} \leq 0 \ \ \forall \omega \in \Omega.$$

Then, $\mathcal{F}(\boldsymbol{\xi}_0) \leq \mathcal{F}(\mathbf{0}) = 0$ by monotonicity and positive homogeneity. Consequently,

$$0 \geq \mathbb{E}[\boldsymbol{\xi}_0 \boldsymbol{\pi}] = \mathbb{E}[\boldsymbol{\xi}_0^2].$$

This means that $\boldsymbol{\xi}_0 = \mathbf{0}$ and $\boldsymbol{\pi}(\omega) \geq 0$ for all $\omega \in \Omega$. □

Superquantile risk and worst-case risk are monotone as mentioned in §8.B and this materializes in 8.31 in the form of nonnegative $\boldsymbol{\pi}$. In contrast, Markowitz risk isn't monotone, which is hinted to by 8.31.

As seen from 8.30 and 8.33, a risk envelope of a positively homogeneous, monotone and regular measure of risk contains exclusively nonnegative random variables with expectation one. This has the consequence that we can interpret the expression $\mathbb{E}[\boldsymbol{\xi}\boldsymbol{\pi}]$ as the expectation of $\boldsymbol{\xi}$ under a different probability distribution. In fact, (8.7) states that $\mathcal{R}(\boldsymbol{\xi})$ is the highest value of this expectation over a set of possible probability distributions defined by $\Pi$. At least under positive homogeneity and monotonicity, this means that an optimization model formulated in terms of a regular measure of risk accounts for the uncertainty in a random vector *and* the uncertainty in the probability distribution of that random vector. While the underlying probability space $(\Omega, \mathcal{A}, \mathbb{P})$ models the uncertainty in the random vector, the choice of measure of risk or, equivalently, the choice of risk envelope, implies a certain ambiguity regarding $\mathbb{P}$: one actually considers a worst-case substitute for $\mathbb{P}$. This is meaningful in practice because $(\Omega, \mathcal{A}, \mathbb{P})$ is often the result of an imperfect modeling process and may not capture reality as accurately as one would like.

**Example 8.34** (uncertainty about probability distribution). In the case of superquantile risk and a finite probability space $(\Omega, \mathcal{A}, \mathbb{P})$, the superquantile risk envelope of 8.31(a) dictates that the nominal expectation of a random variable $\boldsymbol{\xi}$ is replaced by an expected value under an alternative probability distribution that's scaled by numbers between zero and $1/(1-\alpha)$.

**Detail.** Specifically, with $\boldsymbol{\pi} \in \Pi$ from 8.31(a),

$$\mathbb{E}[\boldsymbol{\xi}] = \sum\nolimits_{\omega\in\Omega} \xi(\omega)\mathbb{P}(\omega) \;\text{ is replaced by }\; \mathbb{E}[\boldsymbol{\xi}\boldsymbol{\pi}] = \sum\nolimits_{\omega\in\Omega} \xi(\omega)\pi(\omega)\mathbb{P}(\omega),$$

where $\pi(\omega)\mathbb{P}(\omega)$ can be interpreted as an alternative probability for the outcome $\omega$. Indeed, $\bar{\mathbb{P}}$ given by $\bar{\mathbb{P}}(\omega) = \pi(\omega)\mathbb{P}(\omega)$ is a probability distribution on $\Omega$ because $\bar{\mathbb{P}}(\omega) \geq 0$ and

$$\sum\nolimits_{\omega\in\Omega} \pi(\omega)\mathbb{P}(\omega) = \mathbb{E}[\boldsymbol{\pi}] = 1.$$

The constraints on $\pi(\omega)$ implies that the new probabilities must satisfy

$$0 \leq \bar{\mathbb{P}}(\omega) \leq \frac{\mathbb{P}(\omega)}{1-\alpha}.$$

If $\alpha$ is near zero, then the change in probability is small but it can be made larger by increasing $\alpha$. For example, $\alpha = 0.5$ implies the point of view that $\mathbb{P}$ is suspect to the degree that we plan for events occurring with up to twice as high probability as those nominally stipulated by $\mathbb{P}$. □

## 8.E Subgradients of Functionals

We aim to minimize measures of risk and related quantities and this leads to the need for subgradients of functions of the form

$$x \mapsto \mathcal{R}\big(f(\boldsymbol{\xi}, x)\big).$$

In view of the earlier rules for computing subgradients of composite functions, the path forward is clear: determine the subgradients of $\mathcal{R}$ and $f(\xi,\cdot)$ separately and then combine them using a chain rule. This requires us to extend the meaning of subgradients to functionals defined on $\mathcal{L}^2$ and also develop a new chain rule.

**Definition 8.35** (subgradients of functional). For a convex functional $\mathcal{F}:\mathcal{L}^2\to\overline{\mathbb{R}}$ and a point $\xi_0\in\mathcal{L}^2$ at which $\mathcal{F}$ is finite, $\pi\in\mathcal{L}^2$ is a *subgradient* of $\mathcal{F}$ at $\xi_0$ when

$$\mathcal{F}(\xi)\geq\mathcal{F}(\xi_0)+\mathbb{E}\big[\pi(\xi-\xi_0)\big]\quad\forall\xi,\xi_0\in\mathcal{L}^2.$$

The set of all subgradients of $\mathcal{F}$ at $\xi_0$ is denoted by $\partial\mathcal{F}(\xi_0)$.

The definition mirrors the subgradient inequality 2.17 for functions on $\mathbb{R}^n$ and suffices in the present context as we restrict the scope to convex functionals. It's now apparent that an extension of the Fermat rule holds as well.

For a convex functional $\mathcal{F}:\mathcal{L}^2\to\overline{\mathbb{R}}$ and a point $\xi^\star\in\mathcal{L}^2$ at which $\mathcal{F}$ is finite, one has

$$\text{the Fermat rule:}\qquad \xi^\star\in\operatorname{argmin}\mathcal{F}\iff \mathbf{0}\in\partial\mathcal{F}(\xi^\star),$$

where

$$\operatorname{argmin}\mathcal{F}=\big\{\xi_0\in\operatorname{dom}\mathcal{F}\;\big|\;\mathcal{F}(\xi_0)\leq\mathcal{F}(\xi)\;\forall\xi\in\mathcal{L}^2\big\}.$$

This follows directly from 8.35.

For any $\mathcal{F}:\mathcal{L}^2\to\overline{\mathbb{R}}$, let's define

$$\operatorname{argmax}\mathcal{F}=\operatorname{argmin}(-\mathcal{F}).$$

**Proposition 8.36** (subgradients from conjugate functionals). *For a proper, lsc and convex functional $\mathcal{F}:\mathcal{L}^2\to\overline{\mathbb{R}}$ and a point $\xi_0$ at which $\mathcal{F}$ is finite, one has*

$$\partial\mathcal{F}(\xi_0)=\operatorname{argmax}\big\{\mathbb{E}[\xi_0\pi]-\mathcal{F}^*(\pi)\;\big|\;\pi\in\mathcal{L}^2\big\}.$$

**Proof.** Suppose that

$$\pi_0\in\operatorname{argmax}\big\{\mathbb{E}[\xi_0\pi]-\mathcal{F}^*(\pi)\;\big|\;\pi\in\mathcal{L}^2\big\}.\tag{8.9}$$

By the Fenchel-Moreau theorem 8.27,

$$\mathcal{F}(\xi_0)=\sup\big\{\mathbb{E}[\xi_0\pi]-\mathcal{F}^*(\pi)\;\big|\;\pi\in\mathcal{L}^2\big\}=\mathbb{E}[\xi_0\pi_0]-\mathcal{F}^*(\pi_0)$$

and $\mathcal{F}^*(\pi_0)=\mathbb{E}[\xi_0\pi_0]-\mathcal{F}(\xi_0)$. We see that $\xi_0$ attains the maximum in the definition of $\mathcal{F}^*$, i.e.,

$$\xi_0\in\operatorname{argmax}\big\{\mathbb{E}[\xi\pi_0]-\mathcal{F}(\xi)\;\big|\;\xi\in\mathcal{L}^2\big\}=\operatorname{argmin}\big\{\mathcal{F}(\xi)-\mathbb{E}[\xi\pi_0]\;\big|\;\xi\in\mathcal{L}^2\big\},$$

which in turn implies that $\pi_0\in\partial\mathcal{F}(\xi_0)$ by the Fermat rule. To realize the last claim, note that the functional given by

$$\bar{\mathcal{F}}(\boldsymbol{\xi}) = \mathcal{F}(\boldsymbol{\xi}) - \mathbb{E}[\boldsymbol{\xi}\boldsymbol{\pi}_0]$$

has $\mathbf{0} \in \partial\bar{\mathcal{F}}(\boldsymbol{\xi}_0)$ if and only if $\boldsymbol{\pi}_0 \in \partial\mathcal{F}(\boldsymbol{\xi}_0)$, which is seen from 8.35.

Next, suppose that $\boldsymbol{\pi}_0 \in \partial\mathcal{F}(\boldsymbol{\xi}_0)$. One can then reverse the above arguments to establish that (8.9) again holds. □

The *interior* of $C \subset \mathcal{L}^2$, denoted by $\operatorname{int} C$, consists of every $\boldsymbol{\xi} \in C$ for which there's $\rho > 0$, such that

$$\{\boldsymbol{\xi}_0 \mid \|\boldsymbol{\xi}_0 - \boldsymbol{\xi}\|_{\mathcal{L}^2} \leq \rho\} \subset C.$$

In particular, $\operatorname{int}(\operatorname{dom}\mathcal{F}) = \mathcal{L}^2$ when $\mathcal{F} : \mathcal{L}^2 \to \mathbb{R}$, i.e., $\mathcal{F}$ is a *real-valued functional.*

**Theorem 8.37** (chain rule for functionals). *For a proper, lsc and convex functional $\mathcal{F} : \mathcal{L}^2 \to \overline{\mathbb{R}}$ and $\boldsymbol{\xi} = (\boldsymbol{\xi}_1, \ldots, \boldsymbol{\xi}_n)$, with each $\boldsymbol{\xi}_i \in \mathcal{L}^2$, consider a point $\bar{x} \in \mathbb{R}^n$ such that*

$$\langle \boldsymbol{\xi}, \bar{x}\rangle \in \operatorname{int}(\operatorname{dom}\mathcal{F})$$

*and the function $\varphi : \mathbb{R}^n \to \overline{\mathbb{R}}$ given by*

$$\varphi(x) = \mathcal{F}\big(\langle \boldsymbol{\xi}, x\rangle\big).$$

*Then,*

$$\partial\varphi(\bar{x}) = \Big\{\mathbb{E}[\boldsymbol{\xi}\bar{\boldsymbol{\pi}}] \;\Big|\; \bar{\boldsymbol{\pi}} \in \operatorname{argmax}\big\{\mathbb{E}\big[\langle \boldsymbol{\xi}, \bar{x}\rangle\boldsymbol{\pi}\big] - \mathcal{F}^*(\boldsymbol{\pi}) \;\big|\; \boldsymbol{\pi} \in \mathcal{L}^2\big\}\Big\}.$$

**Proof.** The function $\varphi = \mathcal{F} \circ f(\boldsymbol{\xi}, \cdot)$, where $f(\boldsymbol{\xi}, \cdot) : \mathbb{R}^n \to \mathcal{L}^2$ is given by $f(\boldsymbol{\xi}, x) = \langle \boldsymbol{\xi}, x\rangle$. Since $f(\boldsymbol{\xi}, \cdot)$ is linear, it has an adjoint mapping $g(\boldsymbol{\xi}, \cdot) : \mathcal{L}^2 \to \mathbb{R}^n$ defined by the relation

$$\big\langle x, g(\boldsymbol{\xi}, \boldsymbol{\pi})\big\rangle = \mathbb{E}\big[f(\boldsymbol{\xi}, x)\boldsymbol{\pi}\big] = \mathbb{E}\big[\langle \boldsymbol{\xi}, x\rangle\boldsymbol{\pi}\big] = \big\langle \mathbb{E}[\boldsymbol{\xi}\boldsymbol{\pi}], x\big\rangle \;\; \forall x \in \mathbb{R}^n, \boldsymbol{\pi} \in \mathcal{L}^2.$$

This reveals that $g(\boldsymbol{\xi}, \boldsymbol{\pi}) = \mathbb{E}[\boldsymbol{\xi}\boldsymbol{\pi}]$, which is a vector with components $\mathbb{E}[\boldsymbol{\xi}_i\boldsymbol{\pi}]$, $i = 1, \ldots, n$. We can now invoke [90, Theorem 19], which yields

$$\partial\varphi(\bar{x}) = \Big\{g(\boldsymbol{\xi}, \boldsymbol{\pi}) \;\Big|\; \boldsymbol{\pi} \in \partial\mathcal{F}\big(\langle \boldsymbol{\xi}, \bar{x}\rangle\big)\Big\} \tag{8.10}$$

provided that $\mathcal{F}$ is bounded from above on a neighborhood of $\langle \boldsymbol{\xi}, \bar{x}\rangle$. A fact from functional analysis is that a proper, lsc and convex functional is continuous on $\operatorname{int}(\operatorname{dom}\mathcal{F})$; see [90, Corollary 8B]. Thus, because $\langle \boldsymbol{\xi}, \bar{x}\rangle \in \operatorname{int}(\operatorname{dom}\mathcal{F})$, there's $\delta > 0$, such that

$$\big|\mathcal{F}(\boldsymbol{\eta}) - \mathcal{F}\big(\langle \boldsymbol{\xi}, \bar{x}\rangle\big)\big| \leq 1 \;\; \forall \boldsymbol{\eta} \in \mathcal{L}^2 \;\text{ with }\; \big\|\boldsymbol{\eta} - \langle \boldsymbol{\xi}, \bar{x}\rangle\big\|_{\mathcal{L}^2} \leq \delta.$$

Moreover, one has

$$\mathcal{F}(\boldsymbol{\eta}) \leq \mathcal{F}\big(\langle \boldsymbol{\xi}, \bar{x}\rangle\big) + 1 < \infty$$

for such $\boldsymbol{\eta}$. Consequently, $\mathcal{F}$ is bounded from above in a neighborhood of $\langle \boldsymbol{\xi}, \bar{x}\rangle$ and (8.10) holds. The conclusion then follows by also invoking 8.36. □

We're now well positioned to address several problems arising in risk and deviation minimization. For $f_0 : \mathbb{R}^n \to \mathbb{R}$, a regular measure of risk $\mathcal{R}$ and $\boldsymbol{\eta}, \boldsymbol{\xi}_1, \ldots, \boldsymbol{\xi}_n \in \mathcal{L}^2$, let's consider the problem

$$\operatorname*{minimize}_{x \in X \subset \mathbb{R}^n} f_0(x) + \mathcal{R}\big(\boldsymbol{\eta} - \langle \boldsymbol{\xi}, x \rangle\big), \tag{8.11}$$

where $\boldsymbol{\xi} = (\boldsymbol{\xi}_1, \ldots, \boldsymbol{\xi}_n)$. In light of the discussion in §3.H about stochastic optimization problems with simple recourse, we see that $\mathcal{R}(\boldsymbol{\eta} - \langle \boldsymbol{\xi}, x \rangle)$ can be viewed as quantifying the cost or penalty that a decision $x$ incurs by failing to produce a high amount $\langle \boldsymbol{\xi}, x \rangle$ relative to the "demand" $\boldsymbol{\eta}$. We model this now using a measure of risk instead of an expectation. In the case of multiple products for which there's demand, we could use a customized measure of risk for each one and then sum the resulting risks to produce an objective function consisting of several terms.

In §8.C, we saw that error minimization arising in regression analysis can be accomplished (cf. 8.23) by solving

$$\operatorname*{minimize}_{c \in C \subset \mathbb{R}^n} \mathcal{D}\big(\boldsymbol{\eta} - \langle c, \boldsymbol{\xi} \rangle\big),$$

where $\mathcal{D}$ is the regular measure of deviation that pairs with the regular measure of error of interest. For this problem as well as the risk minimization problem (8.11), the chain rule 8.37 furnishes subgradients for the objective functions. The subgradient method of §2.I and the cutting plane method in 6.18 then emerge as computational possibilities.

The chain rule 8.37 applies to composition with linear mappings and extensions to broader classes of mappings require further technical conditions; see [100, Theorem 4.2]. If the underlying probability space $(\Omega, \mathcal{A}, \mathbb{P})$ is finite, then a functional can be viewed as a function from some finite-dimensional space to $\overline{\mathbb{R}}$ and results can be obtained by the chain rules 4.64 and 6.23.

**Proposition 8.38** (subgradients in risk and deviation minimization). *For a real-valued, positively homogeneous and regular measure of risk $\mathcal{R}$, with risk envelope $\Pi$, and $\boldsymbol{\eta}, \boldsymbol{\xi}_1, \ldots, \boldsymbol{\xi}_n \in \mathcal{L}^2$, consider $\varphi : \mathbb{R}^n \to \mathbb{R}$ given by*

$$\varphi(x) = \mathcal{R}\big(\boldsymbol{\eta} - \langle \boldsymbol{\xi}, x \rangle\big),$$

*where $\boldsymbol{\xi} = (\boldsymbol{\xi}_1, \ldots, \boldsymbol{\xi}_n)$. Then, for any $x \in \mathbb{R}^n$,*

$$\partial\varphi(x) = \Big\{ -\mathbb{E}[\boldsymbol{\xi}\bar{\boldsymbol{\pi}}] \;\Big|\; \bar{\boldsymbol{\pi}} \in \operatorname{argmax}\big\{\mathbb{E}\big[(\boldsymbol{\eta} - \langle \boldsymbol{\xi}, x \rangle)\boldsymbol{\pi}\big] \;\big|\; \boldsymbol{\pi} \in \Pi\big\}\Big\}.$$

*Moreover, with $\mathcal{D}$ being the regular measure of deviation paired with $\mathcal{R}$ in 8.20, the function $\psi : \mathbb{R}^n \to \mathbb{R}$ given by*

$$\psi(c) = \mathcal{D}\big(\boldsymbol{\eta} - \langle c, \boldsymbol{\xi} \rangle\big)$$

*has, for any $c \in \mathbb{R}^n$,*

$$\partial\psi(c) = \Big\{ \mathbb{E}[\boldsymbol{\xi}] - \mathbb{E}[\boldsymbol{\xi}\bar{\boldsymbol{\pi}}] \;\Big|\; \bar{\boldsymbol{\pi}} \in \operatorname{argmax}\big\{\mathbb{E}\big[(\boldsymbol{\eta} - \langle c, \boldsymbol{\xi} \rangle)\boldsymbol{\pi}\big] \;\big|\; \boldsymbol{\pi} \in \Pi\big\}\Big\}.$$

**Proof.** The properties of $\mathcal{R}$ allow us to apply the chain rule 8.37 to the function $\tilde{\varphi} : \mathbb{R}^{1+n} \to \mathbb{R}$ given by

$$\tilde{\varphi}(\gamma, x) = \mathcal{R}\Big(\big\langle(\boldsymbol{\eta}, \boldsymbol{\xi}), (\gamma, x)\big\rangle\Big)$$

and obtain

$$\partial\tilde{\varphi}(\gamma, x) = \Big\{\mathbb{E}\big[(\boldsymbol{\eta}, \boldsymbol{\xi})\bar{\boldsymbol{\pi}}\big] \;\Big|\; \bar{\boldsymbol{\pi}} \in \operatorname{argmax}\big\{\mathbb{E}\big[\big\langle(\boldsymbol{\eta}, \boldsymbol{\xi}), (\gamma, x)\big\rangle\boldsymbol{\pi}\big] \;\big|\; \boldsymbol{\pi} \in \Pi\big\}\Big\},$$

where we also leverage that $\mathcal{R}$ is positively homogeneous so that $\mathcal{R}^*(\boldsymbol{\pi}) = 0$ for $\boldsymbol{\pi} \in \operatorname{dom}\mathcal{R}^* = \Pi$; see 8.29. Since $\varphi(x) = \tilde{\varphi}(F(x))$, where $F : \mathbb{R}^n \to \mathbb{R}^{1+n}$ is given by $F(x) = (1, -x)$, we can apply the chain rule 4.64 to obtain that

$$\partial\varphi(x) = \nabla F(x)^\top \partial\tilde{\varphi}\big(F(x)\big).$$

The qualification (4.14) holds because $\tilde{\varphi}$ is real-valued and convex; cf. 4.65. This leads to the asserted formula for $\partial\varphi(x)$.

For $\partial\psi(c)$, observe that

$$\mathcal{D}\big(\boldsymbol{\eta} - \langle c, \boldsymbol{\xi}\rangle\big) = \mathcal{R}\big(\boldsymbol{\eta} - \langle c, \boldsymbol{\xi}\rangle\big) - \mathbb{E}\big[\boldsymbol{\eta} - \langle c, \boldsymbol{\xi}\rangle\big]$$

by 8.20. Since the gradient of the function

$$c \mapsto -\mathbb{E}\big[\boldsymbol{\eta} - \langle c, \boldsymbol{\xi}\rangle\big] = -\mathbb{E}[\boldsymbol{\eta}] + \big\langle c, \mathbb{E}[\boldsymbol{\xi}]\big\rangle$$

is $\mathbb{E}[\boldsymbol{\xi}]$ at any point $c$, the expression for $\partial\varphi(x)$ and 4.58(c) yield the result. □

The practical use of the subgradient formulae in the proposition requires tractable ways of computing $\boldsymbol{\pi} \in \Pi$ that achieves the maximum in the expression

$$\mathcal{R}(\boldsymbol{\eta}) = \sup\big\{\mathbb{E}[\boldsymbol{\eta}\boldsymbol{\pi}] \;\big|\; \boldsymbol{\pi} \in \Pi\big\},$$

for various random variables $\boldsymbol{\eta} \in \mathcal{L}^2$. Specifically, we say that

$$\bar{\boldsymbol{\pi}} \in \operatorname{argmax}\big\{\mathbb{E}[\boldsymbol{\eta}\boldsymbol{\pi}] \;\big|\; \boldsymbol{\pi} \in \Pi\big\}$$

is a *risk identifier* of $\mathcal{R}$ at $\boldsymbol{\eta}$. For many common measures of risk, there are explicit expressions for risk identifiers; see [100] and the following developments.

**Example 8.39** (risk identifiers for superquantiles). For $\alpha \in (0, 1)$, $x \in \mathbb{R}^n$ and finitely distributed random variables $\boldsymbol{\eta}, \boldsymbol{\xi}_1, \dots, \boldsymbol{\xi}_n$, with values $\{\eta^i, \xi_1^i, \dots, \xi_n^i, i = 1, \dots, q\}$ and probabilities $\{p_i > 0, i = 1, \dots, q\}$, there's an explicit formula for a risk identifier of $\text{s-rsk}_\alpha$ at $\boldsymbol{\eta} - \langle\boldsymbol{\xi}, x\rangle$, where $\boldsymbol{\xi} = (\boldsymbol{\xi}_1, \dots, \boldsymbol{\xi}_n)$.

Let $\varphi : \mathbb{R}^n \to \mathbb{R}$ be the function specified as

$$\varphi(x) = \text{s-rsk}_\alpha\big(\boldsymbol{\eta} - \langle\boldsymbol{\xi}, x\rangle\big).$$

Set $\xi^i = (\xi_1^i, \dots, \xi_n^i)$. A subgradient of $\varphi$ at $x$ is then given by

$$-\sum_{i=1}^{q} p_i \xi^i \bar{\pi}_i, \quad \text{where } \bar{\pi}_i = \begin{cases} \frac{1}{1-\alpha} & \text{if } \eta^i - \langle \xi^i, x \rangle > Q \\ \frac{P(Q)-\alpha}{(1-\alpha)\bar{p}} & \text{if } \eta^i - \langle \xi^i, x \rangle = Q \\ 0 & \text{if } \eta^i - \langle \xi^i, x \rangle < Q, \end{cases}$$

with $Q$ being the $\alpha$-quantile of $\boldsymbol{\eta} - \langle \boldsymbol{\xi}, x \rangle$,

$$P(Q) = \text{prob}\{\boldsymbol{\eta} - \langle \boldsymbol{\xi}, x \rangle \leq Q\}$$

and $\bar{p} = \sum p_i$, where the sum is over all $i$ with $\eta^i - \langle \xi^i, x \rangle = Q$.

**Detail.** Following the pattern in 8.32, we set the underlying probability space $(\Omega, \mathcal{A}, \mathbb{P})$ to have

$$\Omega = \{\omega^i = (\eta^i, \xi_1^i, \ldots, \xi_n^i) \in \mathbb{R}^{1+n} \mid i = 1, \ldots, q\},$$

with $\mathbb{P}(\{\omega^i\}) = p_i$. Then, a random variable $\boldsymbol{\pi} \in \mathcal{L}^2$ can be represented by a $q$-dimensional vector of the form $\pi = (\pi_1, \ldots, \pi_q)$. The risk envelope of $\mathcal{R}$ now corresponds to a subset of $\mathbb{R}^q$:

$$\Pi = \left\{\pi \in \mathbb{R}^q \;\middle|\; 0 \leq \pi_i \leq 1/(1-\alpha),\ i = 1, \ldots, q,\ \sum_{i=1}^{q} p_i \pi_i = 1\right\}.$$

We seek to determine

$$\bar{\pi} \in \text{argmax}_{\pi \in \Pi} \sum_{i=1}^{q} p_i \big(\eta^i - \langle \xi^i, x \rangle\big) \pi_i,$$

but this is essentially accomplished in 8.32; a simple transcription furnishes the risk identifier $\bar{\pi} = (\bar{\pi}_1, \ldots, \bar{\pi}_q)$. The asserted subgradient follows by 8.38.

The computational work required to obtain a subgradient is of order $q \ln q$ as it essentially requires sorting the numbers $\{\eta^i - \langle \xi^i, x \rangle,\ i = 1, \ldots, q\}$. An added benefit from this calculation is that the superquantile risk

$$\text{s-rsk}_\alpha \big(\boldsymbol{\eta} - \langle \boldsymbol{\xi}, x \rangle\big) = \sum_{i=1}^{q} p_i \big(\eta^i - \langle \xi^i, x \rangle\big) \bar{\pi}_i$$

is immediately available. □

**Example 8.40** (superquantile quadrangle). As indicated in 8.5(d), we can construct new regular measures of risk from averaging existing ones. A "continuous" version of this is furnished by the *second-order superquantile risk*

$$\mathcal{R}(\boldsymbol{\xi}) = \frac{1}{1-\alpha} \int_\alpha^1 \text{s-rsk}_\beta(\boldsymbol{\xi})\, d\beta \quad \text{for } \boldsymbol{\xi} \in \mathcal{L}^2,$$

where $\alpha \in [0, 1)$. Certainly, $\text{s-rsk}_\alpha(\boldsymbol{\xi}) \leq \mathcal{R}(\boldsymbol{\xi})$ so a second-order superquantile risk is more conservative than the corresponding superquantile risk. One can show that $\mathcal{R}$ is a regular measure of risk that's also monotone, positively homogeneous and real-valued. It corresponds to the regular measure of regret

$$\mathcal{V}(\boldsymbol{\xi}) = \frac{1}{1-\alpha}\int_0^1 \max\{0, \text{s-rsk}_\beta(\boldsymbol{\xi})\}\, d\beta$$

in the sense of 8.9 and also to the regular measures of deviation and error given by

$$\mathcal{D}(\boldsymbol{\xi}) = \mathcal{R}(\boldsymbol{\xi}) - \mathbb{E}[\boldsymbol{\xi}] \quad \text{and} \quad \mathcal{E}(\boldsymbol{\xi}) = \mathcal{V}(\boldsymbol{\xi}) - \mathbb{E}[\boldsymbol{\xi}]$$

via 8.20. The statistic of $\boldsymbol{\xi}$ is

$$\mathcal{S}(\boldsymbol{\xi}) = \{\text{s-rsk}_\alpha(\boldsymbol{\xi})\},$$

so $\gamma \mapsto \mathcal{E}(\boldsymbol{\xi} - \gamma)$ has the $\alpha$-superquantile of $\boldsymbol{\xi}$ as its unique minimizer.

**Detail.** We refer to [97, 101] for proofs of these assertions. The minimization of composite functions involving second-order superquantile risk is supported by the following expression for risk identifiers.

Let $\boldsymbol{\xi}$ be a finitely distributed random variable with support $\Xi$ and corresponding probabilities $\{p_\xi > 0, \xi \in \Xi\}$. For $\xi \in \Xi$, set

$$P(\xi) = \text{prob}\{\boldsymbol{\xi} \leq \xi\} \quad \text{and} \quad P(\xi_-) = P(\xi) - p_\xi.$$

If the underlying probability space $(\Omega, \mathcal{A}, \mathbb{P})$ has $\Omega = \Xi$ and $\mathbb{P}(\{\xi\}) = p_\xi$ for $\xi \in \Xi$, then $\pi = (\pi_\xi, \xi \in \Xi)$ given by[5]

$$\pi_\xi = \begin{cases} \frac{1}{1-\alpha}\left(\ln\frac{1-\alpha}{1-P(\xi_-)} + 1 + \frac{1-P(\xi)}{p_\xi}\ln\frac{1-P(\xi)}{1-P(\xi_-)}\right) & \text{if } \alpha < P(\xi_-) \\ \frac{1}{1-\alpha}\left(\frac{P(\xi)-\alpha}{p_\xi} + \frac{1-P(\xi)}{p_\xi}\ln\frac{1-P(\xi)}{1-\alpha}\right) & \text{if } P(\xi_-) \leq \alpha \leq P(\xi) \\ 0 & \text{otherwise} \end{cases}$$

is a risk identifier of $\mathcal{R}$ at $\boldsymbol{\xi}$; see [100] for details. □

**Example 8.41** (superquantile regression). Since the regular measure of error $\mathcal{E}$ in 8.40 has a superquantile as its statistic, a procedure that supplements least-squares regression and quantile regression emerges. Suppose that we aim to predict the random variable $\boldsymbol{\eta}$ by means of a model involving the random vector $\boldsymbol{\xi} = (\boldsymbol{\xi}_1, \dots, \boldsymbol{\xi}_n)$. Under the assumption that $\boldsymbol{\eta}, \boldsymbol{\xi}_1, \dots, \boldsymbol{\xi}_n \in \mathcal{L}^2$, we obtain the regression problem

$$\underset{(\gamma,c)\in\mathbb{R}^{1+n}}{\text{minimize}}\ \mathcal{E}\big(\boldsymbol{\eta} - \gamma - \langle c, \boldsymbol{\xi}\rangle\big).$$

The problem depends implicitly on the choice of $\alpha \in [0, 1)$ underpinning $\mathcal{E}$ and the value specifies the superquantile being targeted; see 8.40. A minimizer $(\gamma^\star, c^\star)$ of the problem defines the random variable $\gamma^\star + \langle c^\star, \boldsymbol{\xi}\rangle$, which then makes the nonzeroness of $\boldsymbol{\eta} - (\gamma + \langle c, \boldsymbol{\xi}\rangle)$ as low as possible in the sense of $\mathcal{E}$. We refer to this procedure as *superquantile regression*.

[5] $0 \ln 0$ is interpreted as 0.

**Detail.** By 8.23, we obtain $(\gamma^\star, c^\star)$ in two steps: First, solve

$$c^\star \in \operatorname{argmin}_{c\in\mathbb{R}^n} \mathcal{D}\big(\boldsymbol{\eta} - \langle c, \boldsymbol{\xi}\rangle\big),$$

where $\mathcal{D}$ is given by 8.40. Second, set

$$\gamma^\star = \text{s-rsk}_\alpha\big(\boldsymbol{\eta} - \langle c^\star, \boldsymbol{\xi}\rangle\big).$$

We develop a subgradient formula for the first step and furnish a simple approach to the second step.

Suppose that $\alpha \in [0, 1)$ and $\boldsymbol{\eta}, \boldsymbol{\xi}_1, \ldots, \boldsymbol{\xi}_n$ are finitely distributed random variables, with values $\{\eta^i, \xi_1^i, \ldots, \xi_n^i, i = 1, \ldots, q\}$ and probabilities $\{p_i > 0, i = 1, \ldots, q\}$. We set the underlying probability space $(\Omega, \mathcal{A}, \mathbb{P})$ to have

$$\Omega = \big\{\omega^i = (\eta^i, \xi_1^i, \ldots, \xi_n^i) \in \mathbb{R}^{1+n} \,\big|\, i = 1, \ldots, q\big\},$$

with $\mathbb{P}(\{\omega^i\}) = p_i$. Then, a random variable $\boldsymbol{\pi} \in \mathcal{L}^2$ can be represented by a $q$-dimensional vector of the form $\pi = (\pi_1, \ldots, \pi_q)$. For $c \in \mathbb{R}^n$, the subgradients of the function $\psi : \mathbb{R}^n \to \mathbb{R}$ given by

$$\psi(c) = \mathcal{D}\big(\boldsymbol{\eta} - \langle c, \boldsymbol{\xi}\rangle\big)$$

are furnished by 8.38 in combination with the risk identifiers from 8.40.

Specifically, a subgradient of $\psi$ at $c$ is given by

$$\sum_{i=1}^q p_i \xi^i - \sum_{i=1}^q p_i \xi^i \bar{\pi}_i,$$

where $\xi^i = (\xi_1^i, \ldots, \xi_n^i)$ and[6]

$$\bar{\pi}_i = \begin{cases} \frac{1}{1-\alpha}\Big(\ln \frac{1-\alpha}{1-P_i+\bar{p}_i} + 1 + \frac{1-P_i}{\bar{p}_i} \ln \frac{1-P_i}{1-P_i+\bar{p}_i}\Big) & \text{if } \alpha < P_i - \bar{p}_i \\ \frac{1}{1-\alpha}\Big(\frac{P_i-\alpha}{\bar{p}_i} + \frac{1-P_i}{\bar{p}_i} \ln \frac{1-P_i}{1-\alpha}\Big) & \text{if } P_i - \bar{p}_i \le \alpha \le P_i \\ 0 & \text{otherwise,} \end{cases}$$

with

$$P_i = \text{prob}\big\{\boldsymbol{\eta} - \langle c, \boldsymbol{\xi}\rangle \le \eta^i - \langle c, \xi^i\rangle\big\}$$
$$\bar{p}_i = \text{prob}\big\{\boldsymbol{\eta} - \langle c, \boldsymbol{\xi}\rangle = \eta^i - \langle c, \xi^i\rangle\big\}.$$

To compute this subgradient one would need to sort the numbers $\{\eta^i - \langle c, \xi^i\rangle, i = 1, \ldots, q\}$ to determine $\{P_i, i = 1, \ldots, q\}$. The subgradient method (§2.I) is then immediately available for minimizing $\psi$. The progress of the method can be monitored by computing

$$\psi(c) = \mathcal{D}\big(\boldsymbol{\eta} - \langle c, \boldsymbol{\xi}\rangle\big) = \sum_{i=1}^q p_i\big(\eta^i - \langle c, \xi^i\rangle\big)\bar{\pi}_i - \sum_{i=1}^q p_i\big(\eta^i - \langle c, \xi^i\rangle\big).$$

[6] Again, $0 \ln 0$ is interpreted as 0.

We refer to [101] for approaches to minimizing $\psi$ that rely on solving large-scale linear optimization problems.

With $c^\star$ determined, it only remains to compute

$$\gamma^\star = \text{s-rsk}_\alpha\big(\boldsymbol{\eta} - \langle c^\star, \boldsymbol{\xi}\rangle\big).$$

Since $\boldsymbol{\eta} - \langle c^\star, \boldsymbol{\xi}\rangle$ has a finite distribution, the following expression for superquantiles applies after sorting $\{\eta^i - \langle c^\star, \xi^i\rangle, i = 1, \dots, q\}$, combining duplicate values and updating the corresponding probabilities.

For $\alpha \in [0, 1]$, the $\alpha$-superquantile of a finitely distributed random variable with values $\gamma_1 < \gamma_2 < \cdots < \gamma_r$, which occur with probability $p_1, p_2, \dots, p_r$, respectively, is given by

$$\begin{cases} \sum_{j=1}^r p_j\gamma_j & \text{if } \alpha = 0 \\ \frac{1}{1-\alpha}\Big(\Big(\big(\sum_{j=1}^i p_j\big) - \alpha\Big)\gamma_i + \sum_{j=i+1}^n p_j\gamma_j\Big) & \text{if } \sum_{j=1}^{i-1} p_j < \alpha \le \sum_{j=1}^i p_j < 1 \\ \gamma_r & \text{if } \alpha > 1 - p_r. \end{cases}$$

Note that the values $\gamma_1, \dots, \gamma_r$ are unique and sorted in this formula. □

**Exercise 8.42** (superquantile regression). Repeat 8.17, but now apply superquantile regression instead of quantile regression. In solving the regression problem, proceed as indicated by 8.41 via deviation minimization using the subgradient method to determine the slope coefficients in $c$. Report in detail how the method advances from the starting point $c = 0$ to a minimizer.

**Guide.** Expressions for subgradients are provided by 8.41. □

## 8.F Residual Risk and Surrogates

In its assessment of a random variable, a regular measure of risk considers the possibility of reducing displeasure with a mix of outcomes through mitigation as seen by the regret-risk theorem 8.9. This is made concrete in 8.4, where two random variables are assessed as equally unpleasant by a regular measure of regret. However, one of them emerges as less "risky" because we can offset costly outcomes through a preemptive allocation of resources. This idea can be taken further and leads to the notion of residual risk, with deep connections to data analytics.

**Example 8.43** (hedging firm). Let's consider a firm faced with a cost in the future described in present money by a random variable $\boldsymbol{\eta}$. Suppose that the firm's displeasure with the mix of possible costs is described by a regular measure of regret $\mathcal{V}$. If the firm invests $\gamma \in \mathbb{R}$ in a risk-free asset now, then the future displeasure, as perceived now, is reduced from $\mathcal{V}(\boldsymbol{\eta})$ to $\mathcal{V}(\boldsymbol{\eta} - \gamma)$ because $\gamma$ will be available at the future point in time to offset the cost $\boldsymbol{\eta}$. Though, the upfront cost $\gamma$ needs to be considered and the goal becomes to select the risk-free investment $\gamma$ such that $\gamma + \mathcal{V}(\boldsymbol{\eta} - \gamma)$ is minimized. According to the regret-risk theorem 8.9, the resulting minimum value is the corresponding risk $\mathcal{R}(\boldsymbol{\eta})$ and every $\gamma \in \mathcal{S}(\boldsymbol{\eta})$, the corresponding statistic, furnishes the amount to be invested in the risk-free asset. These are the steps taken in 8.4, but let's now go further.

The firm might consider purchasing $c_j$ shares in stock $j$ with value, in present terms, at the future point in time described by a random variable $\boldsymbol{\xi}_j$. The price of each share is $v_j = \mathbb{E}[\boldsymbol{\xi}_j]$. Let $j = 1, \dots, n$, $c = (c_1, \dots, c_n)$, $v = (v_1, \dots, v_n)$ and $\boldsymbol{\xi} = (\boldsymbol{\xi}_1, \dots, \boldsymbol{\xi}_n)$. Then, because

$$\boldsymbol{\eta} - \big(\gamma + \langle c, \boldsymbol{\xi} \rangle\big)$$

is the future net cost in present terms, the future displeasure, as perceived now, is reduced from

$$\mathcal{V}(\boldsymbol{\eta}) \quad \text{to} \quad \mathcal{V}\big(\boldsymbol{\eta} - \gamma - \langle c, \boldsymbol{\xi} \rangle\big).$$

Though, the upfront cost $\gamma + \langle c, v \rangle$ needs also to be considered and the goal becomes to select a risk-free investment $\gamma$ and risky investments $c \in \mathbb{R}^n$ that solve

$$\underset{(\gamma,c)\in\mathbb{R}^{1+n}}{\text{minimize}}\ \gamma + \langle c, v \rangle + \mathcal{V}\big(\boldsymbol{\eta} - \gamma - \langle c, \boldsymbol{\xi} \rangle\big).$$

According to the regret-risk theorem 8.9, this is equivalent to selecting the risky investments $c \in \mathbb{R}^n$ that solve

$$\underset{c\in\mathbb{R}^{n}}{\text{minimize}}\ \langle c, v \rangle + \mathcal{R}\big(\boldsymbol{\eta} - \langle c, \boldsymbol{\xi} \rangle\big).$$

The (identical) minimum value of the two problems is the *residual risk* of $\boldsymbol{\eta}$ given $\boldsymbol{\xi}$ and is denoted by $\mathcal{R}(\boldsymbol{\eta}|\boldsymbol{\xi})$.

**Detail.** The possibly nonoptimal choices of setting $\gamma = 0$ and/or $c = 0$ correspond to forfeiting moderation of the future cost through risk-free and/or risky investments and give the values $\mathcal{R}(\boldsymbol{\eta})$ and $\mathcal{V}(\boldsymbol{\eta})$. Consequently,

$$\mathcal{R}(\boldsymbol{\eta}|\boldsymbol{\xi}) \le \mathcal{R}(\boldsymbol{\eta}) \le \mathcal{V}(\boldsymbol{\eta}).$$

The differences between these quantities reflect the degree of benefit the firm derives by departing from the passive strategy of $\gamma = 0$ and $c = 0$ to various degrees. The ability to reduce risk by taking positions in stocks, i.e., to *hedge*, is determined by the dependence between $\boldsymbol{\eta}$ and $\boldsymbol{\xi}$.

In a decision-making situation, when comparing two candidate random variables $\boldsymbol{\eta}_1$ and $\boldsymbol{\eta}_2$ representing future costs, a firm's preference for one over the other heavily depends on whether the comparison is carried out at the level of regret, i.e., $\mathcal{V}(\boldsymbol{\eta}_1)$ versus $\mathcal{V}(\boldsymbol{\eta}_2)$, which corresponds to traditional expected utility theory, at the level of risk, i.e., $\mathcal{R}(\boldsymbol{\eta}_1)$ versus $\mathcal{R}(\boldsymbol{\eta}_2)$, or at the level of residual risk, i.e., $\mathcal{R}(\boldsymbol{\eta}_1|\boldsymbol{\xi})$ versus $\mathcal{R}(\boldsymbol{\eta}_2|\boldsymbol{\xi})$. The latter perspective might provide a more comprehensive picture of the situation faced by the firm; it accounts for opportunities to offset a future cost. □

Let's formalize the discussion in the example.

**Definition 8.44** (measure of residual risk). For random variables $\xi_1, \ldots, \xi_n \in \mathcal{L}^2$ and a regular measure of regret $\mathcal{V}$, let $\xi = (\xi_1, \ldots, \xi_n)$ and

$$\mathcal{R}(\eta|\xi) = \inf_{(\gamma,c)\in\mathbb{R}^{1+n}} \mathbb{E}\big[\gamma + \langle c, \xi\rangle\big] + \mathcal{V}\big(\eta - \gamma - \langle c, \xi\rangle\big). \tag{8.12}$$

The functional $\mathcal{R}(\cdot\,|\xi) : \mathcal{L}^2 \to \overline{\mathbb{R}}$ defined by this expression is the *measure of residual risk* given $\xi$ under $\mathcal{V}$. The quantity $\mathcal{R}(\eta|\xi)$ is the *residual risk* of $\eta$ given $\xi$ under $\mathcal{V}$.

The next theorem summarizes key properties of measures of residual risk, some of which are indicated by 8.43. We say that an $n$-dimensional random vector $\xi = (\xi_1, \ldots, \xi_n)$ is *nondegenerate* if

$$\langle c, \xi\rangle \text{ is a constant random variable} \implies c = 0.$$

As an example, if $\xi = (\xi_1, \xi_2)$ has value $(0, 1)$ with probability $1/2$ and value $(1, 0)$ with probability $1/2$, then $\langle (1, 1), \xi\rangle = \xi_1 + \xi_2$ is the constant random variable **1** but $c = (1, 1)$ isn't zero. This means that $\xi$ isn't nondegenerate. Informally, $\xi$ is nondegenerate if there's always some "randomness" in $\langle c, \xi\rangle$ when $c$ is nonzero.

For $\xi = (\xi_1, \ldots, \xi_n)$ with $\xi_j \in \mathcal{L}^2$, $j = 1, \ldots, n$, let's define

$$\mathcal{H}(\xi) = \big\{\gamma + \langle c, \xi\rangle \,\big|\, \gamma \in \mathbb{R}, c \in \mathbb{R}^n\big\} \subset \mathcal{L}^2.$$

This subset specifies the random variables in $\mathcal{L}^2$ that can be reproduced by linear combinations of $\xi_1, \ldots, \xi_n$, possibly also involving a shift $\gamma$. It relates to the regression problem (8.4) as follows: If $\eta \in \mathcal{H}(\xi)$, then there's a minimizer $(\gamma^\star, c^\star)$ of (8.4) such that

$$\mathcal{E}\big(\eta - \gamma^\star - \langle c^\star, \xi\rangle\big) = 0$$

for any regular measure of error $\mathcal{E}$ and, therefore, the coefficient of determination $R^2 = 1$. Thus, $\xi$ fully "explains" the randomness in $\eta$.

**Theorem 8.45** (residual risk). *For random variables $\xi_1, \ldots, \xi_n \in \mathcal{L}^2$ and a risk quadrangle $(\mathcal{V}, \mathcal{R}, \mathcal{E}, \mathcal{D})$ with statistic $\mathcal{S}$ (see Figure 8.4), let $\mathcal{R}(\cdot\,|\xi)$ be the corresponding measure of residual risk given $\xi$ under $\mathcal{V}$, where $\xi = (\xi_1, \ldots, \xi_n)$. Then, for any $\eta \in \mathcal{L}^2$, the following hold:*

(a) $\mathbb{E}[\eta] \le \mathcal{R}(\eta|\xi) \le \mathcal{R}(\eta) \le \mathcal{V}(\eta)$.
(b) *If $\xi$ is constant, then $\mathcal{R}(\eta|\xi) = \mathcal{R}(\eta)$.*
(c) *$\mathcal{R}(\eta|\xi) = \mathbb{E}[\eta]$ if $\eta \in \mathcal{H}(\xi)$, whereas $\mathcal{R}(\eta|\xi) > \mathbb{E}[\eta]$ if $\eta \notin \mathcal{H}(\xi)$ and $\xi$ is nondegenerate.*
(d) *$\mathcal{R}(\cdot\,|\xi)$ is convex and satisfies the constancy property. If $\mathcal{V}$ is positively homogeneous (monotonic), then $\mathcal{R}(\cdot\,|\xi)$ is positively homogeneous (monotonic).*
(e) *If $\xi$ is nondegenerate, then $\mathcal{R}(\cdot\,|\xi)$ is lsc and the infimum in its definition is attained.*
(f) *One has the alternative expressions*

$$\begin{aligned}\mathcal{R}(\boldsymbol{\eta}|\boldsymbol{\xi}) &= \inf_{c\in\mathbb{R}^n} \big\langle c, \mathbb{E}[\boldsymbol{\xi}]\big\rangle + \mathcal{R}\big(\boldsymbol{\eta} - \langle c, \boldsymbol{\xi}\rangle\big)\\ &= \mathbb{E}[\boldsymbol{\eta}] + \inf_{c\in\mathbb{R}^n} \mathcal{D}\big(\boldsymbol{\eta} - \langle c, \boldsymbol{\xi}\rangle\big)\\ &= \mathbb{E}[\boldsymbol{\eta}] + \inf_{(\gamma,c)\in\mathbb{R}^{1+n}} \mathcal{E}\big(\boldsymbol{\eta} - \gamma - \langle c, \boldsymbol{\xi}\rangle\big).\end{aligned}$$

(g) *The point* $(\gamma^\star, c^\star)$ *furnishes the minimum value* $\mathcal{R}(\boldsymbol{\eta}|\boldsymbol{\xi})$ *in (8.12) if and only if*

$$\begin{aligned}&(\gamma^\star, c^\star) \in \operatorname{argmin}_{\gamma,c} \mathcal{E}\big(\boldsymbol{\eta} - \gamma - \langle c, \boldsymbol{\xi}\rangle\big)\\ \Longleftrightarrow\ \ & c^\star \in \operatorname{argmin}_c \big\langle c, \mathbb{E}[\boldsymbol{\xi}]\big\rangle + \mathcal{R}\big(\boldsymbol{\eta} - \langle c, \boldsymbol{\xi}\rangle\big) = \operatorname{argmin}_c \mathcal{D}\big(\boldsymbol{\eta} - \langle c, \boldsymbol{\xi}\rangle\big),\\ & \gamma^\star \in \mathcal{S}\big(\boldsymbol{\eta} - \langle c^\star, \boldsymbol{\xi}\rangle\big).\end{aligned}$$

**Proof.** The first inequality in (a) is a consequence of the fact that the regret of a random variable can't be lower than the corresponding expectation; see definition 8.7 of regular measures of regret. The other inequalities are established by 8.43. In (b), one can take $\tilde{\gamma} = \gamma + \langle c, \mathbb{E}[\boldsymbol{\xi}]\rangle$ and find that

$$\inf_{\tilde{\gamma}\in\mathbb{R}} \tilde{\gamma} + \mathcal{V}(\boldsymbol{\eta} - \tilde{\gamma}) = \inf_{(\gamma,c)\in\mathbb{R}^{1+n}} \mathbb{E}\big[\gamma + \langle c, \boldsymbol{\xi}\rangle\big] + \mathcal{V}\big(\boldsymbol{\eta} - \gamma - \langle c, \boldsymbol{\xi}\rangle\big).$$

Thus, the claim follows. The convexity in (d) holds because the function

$$(\gamma, c, \boldsymbol{\eta}) \mapsto \gamma + \big\langle c, \mathbb{E}[\boldsymbol{\xi}]\big\rangle + \mathcal{V}\big(\boldsymbol{\eta} - \gamma - \langle c, \boldsymbol{\xi}\rangle\big)$$

is convex and, as in 1.21, inf-projections preserve convexity; see [90, Theorem 1]. The constancy property is a consequence of (a) and the fact that $\mathcal{R}$ possesses that property. The positive homogeneity and monotonicity follow easily from the definitions.

We omit the proof of (e); see [99, Theorem 3.4]. Parts (f,g) follow directly from the relations within a risk quadrangle (cf. 8.9, 8.20 and 8.21) and 8.23.

Finally, we consider (c). First, suppose that $\boldsymbol{\eta} \in \mathcal{H}(\boldsymbol{\xi})$. Then, there exists $(\hat{\gamma}, \hat{c}) \in \mathbb{R}^{1+n}$ such that $\boldsymbol{\eta} = \hat{\gamma} + \langle \hat{c}, \boldsymbol{\xi}\rangle$. By (a,f), one has

$$\begin{aligned}\mathbb{E}[\boldsymbol{\eta}] \le \mathcal{R}(\boldsymbol{\eta}|\boldsymbol{\xi}) &= \inf_{c\in\mathbb{R}^n} \big\langle c, \mathbb{E}[\boldsymbol{\xi}]\big\rangle + \mathcal{R}\big(\boldsymbol{\eta} - \langle c, \boldsymbol{\xi}\rangle\big)\\ &\le \big\langle \hat{c}, \mathbb{E}[\boldsymbol{\xi}]\big\rangle + \mathcal{R}\big(\boldsymbol{\eta} - \langle \hat{c}, \boldsymbol{\xi}\rangle\big) = \big\langle \hat{c}, \mathbb{E}[\boldsymbol{\xi}]\big\rangle + \mathcal{R}(\hat{\gamma}) = \hat{\gamma} + \big\langle \hat{c}, \mathbb{E}[\boldsymbol{\xi}]\big\rangle = \mathbb{E}[\boldsymbol{\eta}].\end{aligned}$$

Second, suppose that $\boldsymbol{\eta} \notin \mathcal{H}(\boldsymbol{\xi})$. Then, $\boldsymbol{\eta} - \langle c, \boldsymbol{\xi}\rangle \neq \gamma$ for any $(\gamma, c) \in \mathbb{R}^{1+n}$. Consequently, $\boldsymbol{\eta} - \langle c, \boldsymbol{\xi}\rangle$ isn't constant for any $c \in \mathbb{R}^n$, which by the averseness property of $\mathcal{R}$ implies that

$$\mathcal{R}\big(\boldsymbol{\eta} - \langle c, \boldsymbol{\xi}\rangle\big) > \mathbb{E}\big[\boldsymbol{\eta} - \langle c, \boldsymbol{\xi}\rangle\big].$$

If $\boldsymbol{\xi}$ is nondegenerate, then by (e,f,g) there's $c^\star \in \mathbb{R}^n$ such that

$$\begin{aligned}\mathcal{R}(\boldsymbol{\eta}|\boldsymbol{\xi}) &= \inf_{c\in\mathbb{R}^n} \big\langle c, \mathbb{E}[\boldsymbol{\xi}]\big\rangle + \mathcal{R}\big(\boldsymbol{\eta} - \langle c, \boldsymbol{\xi}\rangle\big)\\ &= \big\langle c^\star, \mathbb{E}[\boldsymbol{\xi}]\big\rangle + \mathcal{R}\big(\boldsymbol{\eta} - \langle c^\star, \boldsymbol{\xi}\rangle\big) > \big\langle c^\star, \mathbb{E}[\boldsymbol{\xi}]\big\rangle + \mathbb{E}\big[\boldsymbol{\eta} - \langle c^\star, \boldsymbol{\xi}\rangle\big] = \mathbb{E}[\boldsymbol{\eta}],\end{aligned}$$

which completes the proof. □

The theorem formalizes properties discussed in 8.43 and establishes that a measure of residual risk has most of the properties of a regular measure of risk. When $\xi$ is nondegenerate, the only deficiency is that it satisfies the averseness property only outside $\mathcal{H}(\xi)$.

**Example 8.46** (normal random variables). For normal random variables $\xi_1$ and $\xi_2$ with mean values $\mu_1$ and $\mu_2$, respectively, and corresponding standard deviations $\sigma_1, \sigma_2 \in (0, \infty)$, let $\rho \in [-1, 1]$ be the correlation coefficient between them, i.e.,

$$\rho = \frac{\mathbb{E}\big[(\xi_1 - \mu_1)(\xi_2 - \mu_2)\big]}{\sigma_1\sigma_2}.$$

The residual risk of $\xi_1$ given $\xi_2$ under the penalty regret in 8.5(a), with $\alpha \in (0, 1)$, is

$$\mathcal{R}(\xi_1|\xi_2) = \mu_1 + \sigma_1\sqrt{1-\rho^2}\,\frac{\varphi(\Phi^{-1}(\alpha))}{1-\alpha},$$

where $\varphi$ and $\Phi$ are the density and distribution functions of a standard normal random variable, respectively. The underlying measure of regret pairs with s-rsk$_\alpha$ in the sense of the regret-risk theorem 8.9 and results in the superquantile-risk formula in 3.8 for normal random variables. The residual risk depends on $\mu_2$ and $\sigma_2$ only via $\rho$.

**Detail.** The minimizer $c^\star$ in 8.45(g) is $\rho\sigma_1/\sigma_2$ in this case and the asserted formula then follows from 8.45(f). For $\rho = \pm 1$, the residual risk is reduced to its minimum possible level of $\mu_1$. The other extreme is attained for $\rho = 0$, when $\mathcal{R}(\xi_1|\xi_2) = \text{s-rsk}_\alpha(\xi_1)$; see 3.8. The sign of the correlation coefficient is immaterial. In terms of investments (cf. the hedging firm in 8.43), this is explained by the fact that both short positions (negative holdings of shares) and long positions (positive holdings of shares) can be utilized to offset risk. □

**Exercise 8.47** (approximation of residual risk). For a regular measure of regret $\mathcal{V} : \mathcal{L}^2 \to \mathbb{R}$ and random variables $\eta, \xi_1, \dots, \xi_n \in \mathcal{L}^2$, with $\xi = (\xi_1, \dots, \xi_n)$, consider the points that furnish the minimum value $\mathcal{R}(\eta|\xi)$ in (8.12), i.e.,

$$C = \operatorname{argmin}_{(\gamma,c)\in\mathbb{R}^{1+n}} \mathbb{E}\big[\gamma + \langle c, \xi\rangle\big] + \mathcal{V}\big(\eta - \gamma - \langle c, \xi\rangle\big).$$

The random variables aren't known, but must be approximated by $\eta^\nu, \xi_1^\nu, \dots, \xi_n^\nu \in \mathcal{L}^2$, with $\xi^\nu = (\xi_1^\nu, \dots, \xi_n^\nu)$. This results in the set of approximating solutions

$$C^\nu = \operatorname{argmin}_{(\gamma,c)\in\mathbb{R}^{1+n}} \mathbb{E}\big[\gamma + \langle c, \xi^\nu\rangle\big] + \mathcal{V}\big(\eta^\nu - \gamma - \langle c, \xi^\nu\rangle\big).$$

Show that LimOut $C^\nu \subset C$ provided that $\eta^\nu \to \eta$ and $\xi_j^\nu \to \xi_j$ for all $j$.

**Guide.** A fact from functional analysis is that a real-valued, lsc and convex functional is continuous; see [90, Corollary 8B]. Combine this with epi-convergence to reach the conclusion. □

The alternative expressions for residual risk in 8.45(f) as well as the minimizers in the corresponding problems establish that determining a residual risk is essentially equivalent to solving the regression problem (8.4). This furnishes a fundamental connection between data analytics and decision-making under uncertainty that enables us to apply algorithms for the former in the solution of the latter and vice versa. In particular, residual risk relates to the risk-minimization problem (8.11) with "initial cost" $f_0(c) = \langle c, \mathbb{E}[\xi]\rangle$; a constraint set $C \subset \mathbb{R}^n$ can also be introduced in the definition of residual risk.

The regression problem (8.4) produces the "best" affine approximation $\gamma^\star + \langle c^\star, \xi\rangle$ of a random variable $\eta$ in the sense of a specific measure of error. Although this is meaningful in many contexts, we can't generally guarantee that the corresponding risk $\mathcal{R}(\eta)$ is bounded from above by $\mathcal{R}(\gamma^\star + \langle c^\star, \xi\rangle)$. However, we can achieve a model $\gamma + \langle c, \xi\rangle$ with this bounding property by making an adjustment to $\gamma^\star$ based on the residual risk $\mathcal{R}(\eta|\xi)$. Although we can carry this out for any regular measure of risk (see [99]), we limit the scope to the positively homogeneous case for simplicity.

Suppose that $\mathcal{D}$ and $\mathcal{R}$ are paired regular measures of deviation and risk in the sense of 8.20 and they are positively homogeneous. The convexity and positive homogeneity of $\mathcal{D}$ imply that for $\eta, \xi_1, \dots, \xi_n \in \mathcal{L}^2$ and $c \in \mathbb{R}^n$, one has

$$\mathcal{D}(\eta) = \mathcal{D}\big(\langle c, \xi\rangle + \eta - \langle c, \xi\rangle\big) \le \mathcal{D}\big(\langle c, \xi\rangle\big) + \mathcal{D}\big(\eta - \langle c, \xi\rangle\big),$$

where $\xi = (\xi_1, \dots, \xi_n)$. After moving a term to the left-hand side and adding $\mathbb{E}[\eta]$ on both sides, this leads to

$$\mathcal{R}(\eta) - \mathcal{D}\big(\langle c, \xi\rangle\big) \le \mathcal{D}\big(\eta - \langle c, \xi\rangle\big) + \mathbb{E}[\eta] = \big\langle c, \mathbb{E}[\xi]\big\rangle + \mathcal{R}\big(\eta - \langle c, \xi\rangle\big),$$

where we also utilize expectation translations in 8.20. If we view $\langle c, \xi\rangle$ as a model that's supposed to estimate $\mathcal{R}(\eta)$ conservatively, then it makes sense to select $c \in \mathbb{R}^n$ so that the right-hand side is minimized. Consequently, $\mathcal{R}(\eta) \le \mathcal{D}(\langle c, \xi\rangle)$ plus a quantity that's made as small as possible. This smallest amount is the residual risk of $\eta$ given $\xi$ under the corresponding regular measure of regret $\mathcal{V}$; see 8.45(f).

**Proposition 8.48** (conservative prediction). *For $\eta, \xi_1, \dots, \xi_n \in \mathcal{L}^2$ and corresponding (in the sense of Figure 8.4) positively homogeneous regular measures of regret, risk and deviation denoted by $\mathcal{V}, \mathcal{R}$ and $\mathcal{D}$, respectively, let $\xi = (\xi_1, \dots, \xi_n)$ and $\mathcal{R}(\cdot\,|\xi)$ be the paired measure of residual risk. Set*

$$c^\star \in \operatorname{argmin}_{c \in \mathbb{R}^n} \mathcal{D}\big(\eta - \langle c, \xi\rangle\big).$$

*Then,*

$$\mathcal{R}(\eta) \le \mathcal{D}\big(\langle c^\star, \xi\rangle\big) + \mathcal{R}(\eta|\xi).$$

*Moreover, the model $\bar{\gamma} + \langle c^\star, \xi\rangle$ satisfies*

$$\mathcal{R}(\eta) \le \mathcal{R}\big(\bar{\gamma} + \langle c^\star, \xi\rangle\big)$$

*provided that*

$$\bar{\gamma} = \mathcal{R}(\boldsymbol{\eta}|\boldsymbol{\xi}) - \langle c^\star, \mathbb{E}[\boldsymbol{\xi}]\rangle = \mathcal{R}(\boldsymbol{\eta} - \langle c^\star, \boldsymbol{\xi}\rangle).$$

**Proof.** This follows by 8.45 and the arguments prior to the proposition. □

While the model $\bar{\gamma} + \langle c^\star, \boldsymbol{\xi}\rangle$ shares the slope coefficients $c^\star$ with the model produced by solving the corresponding regression problem (8.4), which is seen by 8.23, the intercepts are typically different. The conservative approach in the proposition produces $\bar{\gamma} = \mathcal{R}(\boldsymbol{\eta} - \langle c^\star, \boldsymbol{\xi}\rangle)$, whereas (8.4) gives the intercept

$$\gamma^\star \in \mathcal{S}(\boldsymbol{\eta} - \langle c^\star, \boldsymbol{\xi}\rangle)$$

using the corresponding statistic $\mathcal{S}$. For example, if $\mathcal{R} = \text{s-rsk}_\alpha$, then $\gamma^\star$ is the $\alpha$-quantile of $\boldsymbol{\eta} - \langle c^\star, \boldsymbol{\xi}\rangle$ while $\bar{\gamma}$ is the $\alpha$-superquantile of $\boldsymbol{\eta} - \langle c^\star, \boldsymbol{\xi}\rangle$. The model $\bar{\gamma} + \langle c^\star, \boldsymbol{\xi}\rangle$ is thus obtained by translating upward the one produced by the regression problem, just enough to ensure conservative estimates of $\mathcal{R}(\boldsymbol{\eta})$.

**Example 8.49** (surrogate building). Let's consider the $n$-dimensional dynamical system with initial condition $x(0) = \xi \in \mathbb{R}^n$ and differential equation

$$x_j'(t) = -x_j(t)\sin x_{j-1}(t) - \beta x_j(t) + \tau, \quad j = 1, \dots, n,$$

where $x_0(t) = x_n(t)$, $\beta = 2$ and $\tau = 10$. We assume that the initial condition is unknown and model it by the $n$-dimensional random vector $\boldsymbol{\xi}$ with statistically independent normally distributed components having zero mean and standard deviation 2. Suppose that we need to estimate the first state at time 20, i.e., $x_1(20)$, which is a random variable due to the random initial condition $\boldsymbol{\xi}$. Let's denote this random variable by $\boldsymbol{\eta}$. We can use 8.48 to construct a conservative model $\bar{\gamma} + \langle c^\star, \boldsymbol{\xi}\rangle$ of $\boldsymbol{\eta}$ even when the system is high dimensional.

**Detail.** The model, called a *surrogate*, is a normal random variable because any weighted sum of normal random variables is a normal random variable. Thus, the model $\bar{\gamma} + \langle c^\star, \boldsymbol{\xi}\rangle$ can be viewed as the best conservative, normal approximation of $\boldsymbol{\eta}$ in the sense of the chosen regular measures of risk. Let's adopt superquantile risk with $\alpha = 0.8$.

Since the probability distribution of $\boldsymbol{\eta}$ is unknown, we produce an approximation of the probability distribution of $(\boldsymbol{\eta}, \boldsymbol{\xi})$ using sampling: From $\nu$ randomly sampled initial conditions according to the probability distribution of $\boldsymbol{\xi}$, we simulate the dynamical system up to time 20. This produces $\nu$ possible values of $\boldsymbol{\eta}$, which we denote by $\{\eta^1, \dots, \eta^\nu\}$. The corresponding initial conditions are $\{\xi^1, \dots, \xi^\nu\}$. As an approximation, we adopt an underlying probability space with

$$\Omega = \{\omega^i = (\eta^i, \xi^i),\ i = 1, \dots, \nu\}$$

and $\mathbb{P}(\{\omega^i\}) = 1/\nu$. One can now solve the deviation-minimization problem in 8.48 using this approximation and obtain the model $\bar{\gamma} + \langle c^\star, \boldsymbol{\xi}\rangle$. However, we consider one more issue.

Suppose that $n = 1000$ and the simulations of the dynamical system are computationally costly. We would then like $\nu$ to be only moderately large. Moreover, we prefer a sparse model with few nonzero coefficients in $c^\star$ as this points to the important input dimensions.

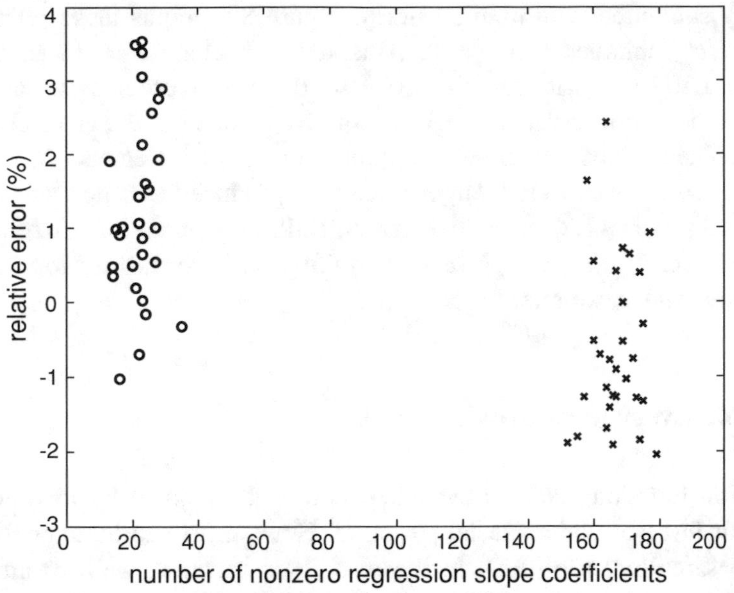


Fig. 8.5: Relative error in sparse normal approximation of response of dynamical system for $\theta = 0.1$ (marked with x) and $\theta = 0.2$ (marked with o), plotted against the number of nonzero coefficients in the obtained $c^\star \in \mathbb{R}^{1000}$.

To achieve this, we follow the pattern in lasso regression (cf. 2.5) and add a regularization term to the deviation-minimization problem so that

$$c^\star \in \operatorname{argmin}_{c \in \mathbb{R}^n} \mathcal{D}\big(\boldsymbol{\eta}^\nu - \langle c, \boldsymbol{\xi}^\nu \rangle\big) + \theta \|c\|_1,$$

where $\theta \in (0, \infty)$, $\mathcal{D}$ is the corresponding superquantile deviation (cf. 8.19(a)) and $(\boldsymbol{\eta}^\nu, \boldsymbol{\xi}^\nu)$ is the approximation of $(\boldsymbol{\eta}, \boldsymbol{\xi})$ obtained by sampling as described above. The intercept $\bar{\gamma}$ is computed as before. The approach produces a sparse, conservative and normal approximation of $\boldsymbol{\eta}$.

The resulting model is guaranteed by 8.48 to be conservative:

$$\text{s-rsk}_\alpha(\boldsymbol{\eta}^\nu) \le \text{s-rsk}_\alpha\big(\bar{\gamma} + \langle c^\star, \boldsymbol{\xi}^\nu \rangle\big).$$

However, the guarantee is only with respect to the approximation $(\boldsymbol{\eta}^\nu, \boldsymbol{\xi}^\nu)$ used in computing $(\bar{\gamma}, c^\star)$. We also would like to know whether $\text{s-rsk}_\alpha(\boldsymbol{\eta}) \le \text{s-rsk}_\alpha(\bar{\gamma} + \langle c^\star, \boldsymbol{\xi} \rangle)$ holds. We can check this empirically by generating a test dataset of 50000 randomly sampled initial conditions and, for each one, simulate the system to obtain 50000 possible values of $x_1(20)$. These values furnish a fairly accurate approximation of the probability distribution of $(\boldsymbol{\eta}, \boldsymbol{\xi})$ from which we can estimate $\text{s-rsk}_\alpha(\boldsymbol{\eta})$ as well as $\text{s-rsk}_\alpha(\bar{\gamma} + \langle c^\star, \boldsymbol{\xi} \rangle)$. The relative error

$$\frac{\text{s-rsk}_\alpha(\boldsymbol{\eta}) - \text{s-rsk}_\alpha\big(\bar{\gamma} + \langle c^\star, \boldsymbol{\xi} \rangle\big)}{\text{s-rsk}_\alpha(\boldsymbol{\eta})}$$

is then easily computed with high accuracy. Figure 8.5 graphs these relative errors in percent for $(\bar{\gamma}, c^\star)$ obtained with $\theta = 0.1$ (marked with x) and $\theta = 0.2$ (marked with o), each replicated 30 times with sample size $\nu = 500$. The figure shows that the resulting 60 models are often conservative (negative relative error), but the error climbs to 2-3% occasionally. The first axis specifies the number of nonzero elements in $c^\star \in \mathbb{R}^{1000}$ for the various models. As expected, larger $\theta$ tends to produce fewer nonzero elements, but this comes at a price of higher relative errors. Still, only about 20 nonzero elements in models of the form $\bar{\gamma} + \langle c^\star, \boldsymbol{\xi} \rangle$ approximate $\boldsymbol{\eta}$ fairly well. For further applications of this approach to surrogates, we refer to [16]. □

## 8.G Sample Average Approximations

An expectation function can't be evaluated unless it's defined by a finite probability distribution or involves some special structure. However, it can be approximated using Monte Carlo sampling; see §3.D. In the next three sections, we'll examine the error associated with such approximations and the ramifications for stochastic optimization problems.

For $f : \Xi \times \mathbb{R}^n \to \overline{\mathbb{R}}$, let $Ef : \mathbb{R}^n \to \overline{\mathbb{R}}$ be the expectation function given by

$$Ef(x) = \mathbb{E}\big[f(\boldsymbol{\xi}, x)\big].$$

We now permit a generalization of the earlier treatment of expectation functions: $\Xi$ doesn't have to be $\mathbb{R}^m$ or a subset thereof, but can be an arbitrary set. Thus, a model of uncertainty as expressed by a random quantity $\boldsymbol{\xi}$ might be more "abstract" and, for example, involve an infinite number of random variables. Moreover, $f$ may not be real-valued, as often assumed earlier, and this enables us to address induced constraints; see the discussion in §3.J.

Given $\xi^1, \dots, \xi^\nu \in \Xi$, the *sample average function*

$$x \mapsto \frac{1}{\nu} \sum\nolimits_{i=1}^{\nu} f(\xi^i, x)$$

furnishes an approximation of $Ef$. It gives rise to a versatile approach for addressing a problem formulated in terms of $Ef$: solve an approximating problem obtained by replacing the expectation function with its sample average function. The approximating problem simply involves a sum and we can draw on a vast array of optimization algorithms for its solution. In some situations, $\{\xi^1, \dots, \xi^\nu\}$ is easily generated by sampling according to the probability distribution of $\boldsymbol{\xi}$. Other times, such as in statistical applications, $\{\xi^1, \dots, \xi^\nu\}$ is data obtained by observing a random phenomenon that's modeled by the random quantity $\boldsymbol{\xi}$. The sample average function is, therefore, readily available in many applications.

To place this approximation-based approach to stochastic optimization on a solid footing, we need to establish a condition under which sample average functions epi-converge to the corresponding expectation function as the sample size $\nu \to \infty$. This can be combined with facts from §4.C, §5.A and §7.I to construct guarantees about the convergence of minimum values and minimizers of the approximating problems to their counterparts for the actual problem. We'll establish the epi-convergence property in 8.56, but only after a deep dive into general integration theory.

Let $(\Xi, \mathcal{B}, \mathbb{P})$ be a probability space. It specifies the model of uncertainty associated with the parameter $\xi$ that enters as the first argument in $f$. We think of this parameter as a random element (variable, vector, matrix, etc.) $\boldsymbol{\xi}$ that takes values according to the probability space by setting

$$\text{prob}\{\boldsymbol{\xi} \in B\} = \mathbb{P}(B) \quad \forall B \in \mathcal{B}.$$

Given $x \in \mathbb{R}^n$, we would like $f(\cdot, x) : \Xi \to \overline{\mathbb{R}}$ to be a random variable with values in $\overline{\mathbb{R}}$ and distribution function $P : \overline{\mathbb{R}} \to [0, 1]$ defined by

$$P(\alpha) = \mathbb{P}\big(\{\xi \in \Xi \mid f(\xi, x) \le \alpha\}\big).$$

Moreover, we need the expectation $\mathbb{E}[f(\boldsymbol{\xi}, x)]$ to be well defined and also have certain properties such as lsc when viewed as a function of $x$. These technical concerns are put to rest by imposing minor assumptions on $f$.

**Measurability.** Relative to a probability space $(\Xi, \mathcal{B}, \mathbb{P})$, a function $h : \Xi \to \overline{\mathbb{R}}$ is *measurable* when

$$\{\xi \in \Xi \mid h(\xi) \le \alpha\} \in \mathcal{B} \quad \forall \alpha \in \overline{\mathbb{R}}.$$

For example, if $\Xi = \mathbb{R}^m$ and $\mathcal{B}$ contains the closed sets in $\mathbb{R}^m$, then $h$ is measurable when $h$ is lsc because the level-sets of $h$ are closed by 4.8. When $h$ is measurable, we say it's a random variable and often denote it by $h(\boldsymbol{\xi})$. Then, in particular, $\xi \mapsto \max\{0, h(\xi)\}$ and $\xi \mapsto \max\{0, -h(\xi)\}$ are also measurable and the expectation

$$\mathbb{E}\big[h(\boldsymbol{\xi})\big] = \mathbb{E}\big[\max\{0, h(\boldsymbol{\xi})\}\big] - \mathbb{E}\big[\max\{0, -h(\boldsymbol{\xi})\}\big]$$

is well defined under the usual convention that $\infty - \infty = \infty$; see the discussion in §3.B. When $\mathbb{E}[h(\boldsymbol{\xi})]$ is finite, which takes place only if both $\mathbb{E}[\max\{0, h(\boldsymbol{\xi})\}]$ and $\mathbb{E}[\max\{0, -h(\boldsymbol{\xi})\}]$ are finite, we say that $h(\boldsymbol{\xi})$ is *integrable*.

A set-valued mapping $S : \Xi \rightrightarrows \mathbb{R}^n$ is *measurable* if the set

$$S^{-1}(O) = \{\xi \in \Xi \mid S(\xi) \cap O \neq \emptyset\} \in \mathcal{B} \quad \text{for all open set } O \subset \mathbb{R}^n.$$

When $S$ is measurable, we say it's a *random set* and denote it by $S(\boldsymbol{\xi})$. This formalizes the preliminary treatment in §3.F.

**Definition 8.50** (random lsc function). A function $f : \Xi \times \mathbb{R}^n \to \overline{\mathbb{R}}$ is *random lsc* if its *epigraphical mapping* $S_f : \Xi \rightrightarrows \mathbb{R}^{n+1}$ given by

$$S_f(\xi) = \text{epi}\, f(\xi, \cdot)$$

is measurable and closed-valued.

As we see in the following, $f$ being random lsc is a basic requirement that addresses any technical concerns about the corresponding expectation function. It's defined relative to a probability space $(\Xi, \mathcal{B}, \mathbb{P})$, which we assume to be given in the following statements.

**Example 8.51** (random lsc function). Either one of the following conditions suffices for $f : \Xi \times \mathbb{R}^n \to \overline{\mathbb{R}}$ to be a random lsc function:

(a) For lsc $g : \mathbb{R}^n \to \overline{\mathbb{R}}$, one has $f(\xi, x) = g(x)$ for all $x \in \mathbb{R}^n, \xi \in \Xi$.
(b) For measurable and closed-valued $C : \Xi \rightrightarrows \mathbb{R}^n$, one has $f(\xi, x) = \iota_{C(\xi)}(x)$ for all $x \in \mathbb{R}^n, \xi \in \Xi$.
(c) For random lsc $f_0 : \Xi \times \mathbb{R}^n \to \overline{\mathbb{R}}$ and closed $C \subset \mathbb{R}^n$, one has $f(\xi, x) = f_0(\xi, x) + \iota_C(x)$ for all $x \in \mathbb{R}^n, \xi \in \Xi$.
(d) $f$ is real-valued, $f(\xi, \cdot\,)$ is continuous for all $\xi \in \Xi$ and $f(\,\cdot\,, x)$ is measurable for all $x \in \mathbb{R}^n$.
(e) $f(\xi, \cdot\,)$ is lsc, convex and $\operatorname{int}(\operatorname{dom} f(\xi, \cdot\,)) \neq \emptyset$ for all $\xi \in \Xi$ and $f(\,\cdot\,, x)$ is measurable for all $x \in \mathbb{R}^n$.
(f) $f$ is lsc, $\Xi = \mathbb{R}^m$ and $\mathcal{B}$ contains the closed subsets of $\mathbb{R}^m$.
(g) For $\lambda_i \in [0, \infty)$ and random lsc $f_i : \Xi \times \mathbb{R}^n \to \overline{\mathbb{R}}$, with $f_i(\xi, \cdot\,)$ proper for all $\xi \in \Xi$, $i = 1, \dots, q$, one has $f(\xi, x) = \sum_{i=1}^q \lambda_i f_i(\xi, x)$ for all $x \in \mathbb{R}^n, \xi \in \Xi$.
(h) For random lsc $f_\alpha : \Xi \times \mathbb{R}^n \to \overline{\mathbb{R}}$, $\alpha \in \mathbb{A}$, and countable $\mathbb{A}$, one has $f(\xi, x) = \sup_{\alpha \in \mathbb{A}} f_\alpha(\xi, x)$ for all $x \in \mathbb{R}^n, \xi \in \Xi$.

**Detail.** These facts are compiled from [105]: (a) follows by 14.30, (b,c) by 14.32 and a slight extension, (d) by 14.29, (e) by 14.39, (f) by 14.31 and (g,h) by 14.44. □

A function $f$ satisfying (d) is called a *Caratheodory function*. This fact together with the composition in (c) already cover many applications. We refer to [105, Chapter 14] for other rules about random lsc functions (where they're called *normal integrands*).

It's immediate from the definition that $f$ being random lsc implies that $f(\xi, \cdot\,)$ is lsc for all $\xi \in \Xi$ (because $S_f$ is closed-valued) and $f(\,\cdot\,, x)$ is measurable for all $x \in \mathbb{R}^n$; see [105, Proposition 14.28]. Thus, $\mathbb{E}[f(\boldsymbol{\xi}, x)]$ is well defined when $f$ is random lsc.

**Proposition 8.52** (measurability of inf-projection). *If $f : \Xi \times \mathbb{R}^n \to \overline{\mathbb{R}}$ is a random lsc function, then the inf-projection $\xi \mapsto \inf_{x \in \mathbb{R}^n} f(\xi, x)$ is measurable.*

**Proof.** This fact follows from [105, Proposition 14.37]. □

We next turn to expectation functions defined by random lsc functions and start by recording basic properties of expectations as a supplement to 3.2.

**Proposition 8.53** (properties of expectations). *For measurable $h, g : \Xi \to \overline{\mathbb{R}}$ and $\alpha \in [0, \infty)$, the following hold:*

(a) $h(\xi) \leq g(\xi)\ \forall \xi \in \Xi \implies \mathbb{E}[h(\boldsymbol{\xi})] \leq \mathbb{E}[g(\boldsymbol{\xi})]$.
(b) $\mathbb{E}[\alpha h(\boldsymbol{\xi})] = \alpha \mathbb{E}[h(\boldsymbol{\xi})]$.
(c) $\mathbb{E}[h(\boldsymbol{\xi}) + g(\boldsymbol{\xi})] \leq \mathbb{E}[h(\boldsymbol{\xi})] + \mathbb{E}[g(\boldsymbol{\xi})]$.

**Proof.** Let's adopt the short-hand notation:

$$h_+(\xi) = \max\{0, h(\xi)\} \text{ and } h_-(\xi) = \max\{0, -h(\xi)\},$$

with similar expressions for other functions besides $h$. It's known from integration theory [40, Section 2.2] that (a) holds for nonnegative functions. Since $h_+(\xi) \leq g_+(\xi)$ for all $\xi \in \Xi$, it follows that

$$\mathbb{E}\big[h_+(\boldsymbol{\xi})\big] \leq \mathbb{E}\big[g_+(\boldsymbol{\xi})\big].$$

Moreover, $h_-(\xi) \geq g_-(\xi)$ for all $\xi \in \Xi$ so that

$$\mathbb{E}\big[h_-(\boldsymbol{\xi})\big] \geq \mathbb{E}\big[g_-(\boldsymbol{\xi})\big].$$

These inequalities put together yield (a).

The property in (b) holds for nonnegative functions [40, Section 2.2] so

$$\begin{aligned}\mathbb{E}\big[\alpha h(\boldsymbol{\xi})\big] &= \mathbb{E}\big[\max\{0, \alpha h(\boldsymbol{\xi})\}\big] - \mathbb{E}\big[\max\{0, -\alpha h(\boldsymbol{\xi})\}\big] \\ &= \mathbb{E}\big[\alpha h_+(\boldsymbol{\xi})\big] - \mathbb{E}\big[\alpha h_-(\boldsymbol{\xi})\big] = \alpha\mathbb{E}\big[h_+(\boldsymbol{\xi})\big] - \alpha\mathbb{E}\big[h_-(\boldsymbol{\xi})\big] = \alpha\mathbb{E}\big[h(\boldsymbol{\xi})\big],\end{aligned}$$

which establishes (b).

For (c), we note that the inequality holds trivially if $\mathbb{E}[h_+(\boldsymbol{\xi})] = \infty$ and/or $\mathbb{E}[g_+(\boldsymbol{\xi})] = \infty$ because then the right-hand side is infinity. Consequently, we concentrate on the case with $\mathbb{E}[h_+(\boldsymbol{\xi})]$ and $\mathbb{E}[g_+(\boldsymbol{\xi})]$ being finite. Since

$$\max\{0, h(\xi) + g(\xi)\} = (h + g)_+(\xi) \leq h_+(\xi) + g_+(\xi) \quad \forall \xi \in \Xi,$$

it follows from (a) that

$$\mathbb{E}\big[(h + g)_+(\boldsymbol{\xi})\big] \leq \mathbb{E}\big[h_+(\boldsymbol{\xi})\big] + \mathbb{E}\big[g_+(\boldsymbol{\xi})\big] < \infty.$$

In view of the finiteness of $\mathbb{E}[h_+(\boldsymbol{\xi})]$ and $\mathbb{E}[g_+(\boldsymbol{\xi})]$, there's $\Xi_0 \in \mathcal{B}$ with $\mathbb{P}(\Xi_0) = 1$ such that $h(\xi) < \infty$ and $g(\xi) < \infty$ for all $\xi \in \Xi_0$. We can therefore continue without loss of generality under the assumption that $h(\xi) < \infty$ and $g(\xi) < \infty$ for all $\xi \in \Xi$.

Next, if there's a set $\Xi_1 \in \mathcal{B}$ with $\mathbb{P}(\Xi_1) > 0$ such that $h(\xi) = -\infty$ for all $\xi \in \Xi_1$, then $\mathbb{E}[h_-(\boldsymbol{\xi})] = \infty$ and the right-hand side in (c) is $-\infty$. Since

$$h(\xi) + g(\xi) = -\infty \quad \forall \xi \in \Xi_1,$$

we've $\mathbb{E}[(h + g)_-(\boldsymbol{\xi})] = \infty$, which then implies that the left-hand side of (c) is also $-\infty$ because $\mathbb{E}[(h + g)_+(\boldsymbol{\xi})]$ is finite.

Finally, we consider the case when $h(\xi)$ and $g(\xi)$ are finite at every $\xi \in \Xi_2$ for some $\Xi_2 \in \mathcal{B}$ with $\mathbb{P}(\Xi_2) = 1$. Then,

$$(h + g)_+(\xi) - (h + g)_-(\xi) = h_+(\xi) - h_-(\xi) + g_+(\xi) - g_-(\xi) \quad \forall \xi \in \Xi_2$$

and because all these numbers are finite, also

$$(h + g)_+(\xi) + h_-(\xi) + g_-(\xi) = (h + g)_-(\xi) + h_+(\xi) + g_+(\xi) \quad \forall \xi \in \Xi_2.$$

The property in (c) holds with equality for nonnegative functions [40, Theorem 2.15]. Applying this fact to the previous equality leads to

$$\mathbb{E}\big[(h+g)_+(\boldsymbol{\xi})\big] + \mathbb{E}\big[h_-(\boldsymbol{\xi})\big] + \mathbb{E}\big[g_-(\boldsymbol{\xi})\big] = \mathbb{E}\big[(h+g)_-(\boldsymbol{\xi})\big] + \mathbb{E}\big[h_+(\boldsymbol{\xi})\big] + \mathbb{E}\big[g_+(\boldsymbol{\xi})\big].$$

Among these six nonnegative terms, the ones with "+" subscripts are finite. If $\mathbb{E}[h_-(\boldsymbol{\xi})]$ and/or $\mathbb{E}[g_-(\boldsymbol{\xi})]$ is infinity, then $\mathbb{E}[(h+g)_-(\boldsymbol{\xi})]$ must also be infinity and both sides in (c) become $-\infty$. If $\mathbb{E}[h_-(\boldsymbol{\xi})]$ and $\mathbb{E}[g_-(\boldsymbol{\xi})]$ are finite, then $\mathbb{E}[(h+g)_-(\boldsymbol{\xi})]$ is finite and one can rearrange terms to obtain

$$\mathbb{E}\big[(h+g)_+(\boldsymbol{\xi})\big] - \mathbb{E}\big[(h+g)_-(\boldsymbol{\xi})\big] = \mathbb{E}\big[h_+(\boldsymbol{\xi})\big] - \mathbb{E}\big[h_-(\boldsymbol{\xi})\big] + \mathbb{E}\big[g_+(\boldsymbol{\xi})\big] - \mathbb{E}\big[g_-(\boldsymbol{\xi})\big],$$

which means that

$$\mathbb{E}\big[h(\boldsymbol{\xi}) + g(\boldsymbol{\xi})\big] = \mathbb{E}\big[h(\boldsymbol{\xi})\big] + \mathbb{E}\big[g(\boldsymbol{\xi})\big].$$

Thus, (c) holds. □

**Proposition 8.54** (Fatou's lemma; extended). *For measurable $h^\nu : \Xi \to \overline{\mathbb{R}}$ and integrable $g : \Xi \to [0,\infty]$ such that $h^\nu(\xi) \geq -g(\xi)$ for all $\xi \in \Xi$ and $\nu \in \mathbb{N}$, suppose that $h : \Xi \to \overline{\mathbb{R}}$ is given by*

$$h(\xi) = \liminf h^\nu(\xi).$$

*Then, $h$ is measurable and*

$$\liminf \mathbb{E}\big[h^\nu(\boldsymbol{\xi})\big] \geq \mathbb{E}\big[h(\boldsymbol{\xi})\big].$$

**Proof.** The measurability of $h$ follows by [40, Proposition 2.7]. The usual Fatou's lemma (see [40, Lemma 2.18]) applies to nonnegative functions and in particular to $f^\nu$ defined by

$$f^\nu(\xi) = h^\nu(\xi) + g(\xi).$$

Thus, one has

$$\liminf \mathbb{E}\big[f^\nu(\boldsymbol{\xi})\big] \geq \mathbb{E}\big[f(\boldsymbol{\xi})\big], \tag{8.13}$$

where $f$ is the nonnegative function with $f(\xi) = \liminf f^\nu(\xi)$. By 8.53,

$$\liminf \mathbb{E}\big[f^\nu(\boldsymbol{\xi})\big] \leq \liminf \Big(\mathbb{E}\big[h^\nu(\boldsymbol{\xi})\big] + \mathbb{E}\big[g(\boldsymbol{\xi})\big]\Big) = \liminf \mathbb{E}\big[h^\nu(\boldsymbol{\xi})\big] + \mathbb{E}\big[g(\boldsymbol{\xi})\big],$$

where the equality holds because $\mathbb{E}[g(\boldsymbol{\xi})]$ is finite.

Next, we consider the right-hand side in (8.13). Since $g$ is integrable, there's $\Xi_0 \in \mathcal{B}$ with $\mathbb{P}(\Xi_0) = 1$ such that $g(\xi) \in \mathbb{R}$ for all $\xi \in \Xi_0$. Thus, we can proceed without loss of generality under the assumption that $g(\xi) \in [0,\infty)$ for all $\xi \in \Xi$. Let $1_+(\boldsymbol{\xi})$ be the random variable with $1_+(\xi) = 1$ if $h(\xi) \geq 0$ and is zero otherwise, and let $1_-(\boldsymbol{\xi})$ be the random variable with $1_-(\xi) = 1$ if $h(\xi) < 0$ and is zero otherwise. Then, for any measurable $k : \Xi \to \overline{\mathbb{R}}$, one has

$$\mathbb{E}\big[k(\xi)\big] = \mathbb{E}\big[1_+(\xi)k(\xi)\big] + \mathbb{E}\big[1_-(\xi)k(\xi)\big],$$

which follows directly from the definition of the expectation. Now,

$$\begin{aligned}\mathbb{E}\big[f(\xi)\big] &= \mathbb{E}\big[1_+(\xi)f(\xi)\big] + \mathbb{E}\big[1_-(\xi)f(\xi)\big]\\ &= \mathbb{E}\big[1_+(\xi)h(\xi) + 1_+(\xi)g(\xi)\big] + \mathbb{E}\big[1_-(\xi)h(\xi) + 1_-(\xi)g(\xi)\big],\end{aligned}$$

where the second equality follows because

$$f(\xi) = \liminf f^\nu(\xi) = \liminf \big(h^\nu(\xi) + g(\xi)\big) = \liminf h^\nu(\xi) + g(\xi) \quad \forall \xi \in \Xi,$$

which in turn is a consequence of the fact that $g$ is real-valued. Since $1_+(\xi)h(\xi)$ and $1_+(\xi)g(\xi)$ are nonnegative, we can invoke the fact that 8.53(c) holds with equality for such random variables (cf. [40, Theorem 2.15]) and obtain

$$\mathbb{E}\big[1_+(\xi)h(\xi) + 1_+(\xi)g(\xi)\big] = \mathbb{E}\big[1_+(\xi)h(\xi)\big] + \mathbb{E}\big[1_+(\xi)g(\xi)\big].$$

Moreover, $h(\xi) \geq -g(\xi)$ because $h^\nu(\xi) \geq -g(\xi)$ for all $\nu$ and then

$$0 \geq 1_-(\xi)h(\xi) \geq -g(\xi) \quad \forall \xi \in \Xi.$$

Two applications of 8.53(a) establish that $1_-(\xi)h(\xi)$ is integrable. It then follows by 3.2 that

$$\mathbb{E}\big[1_-(\xi)h(\xi) + 1_-(\xi)g(\xi)\big] = \mathbb{E}\big[1_-(\xi)h(\xi)\big] + \mathbb{E}\big[1_-(\xi)g(\xi)\big].$$

Combining these facts, we obtain

$$\begin{aligned}\mathbb{E}\big[f(\xi)\big] &= \mathbb{E}\big[1_+(\xi)h(\xi)\big] + \mathbb{E}\big[1_+(\xi)g(\xi)\big] + \mathbb{E}\big[1_-(\xi)h(\xi)\big] + \mathbb{E}\big[1_-(\xi)g(\xi)\big]\\ &= \mathbb{E}\big[h(\xi)\big] + \mathbb{E}\big[g(\xi)\big].\end{aligned}$$

This expression for the right-hand side in (8.13) and the earlier derived expression for the left-hand side lead to the conclusion. □

We're now in a position to return to expectation functions and to establish their basic properties. As a mild surrogate for integrability, we adopt the following terminology:

A random lsc function $f : \Xi \times \mathbb{R}^n \to \overline{\mathbb{R}}$ is *locally inf-integrable* if

$$\forall \bar{x} \in \mathbb{R}^n,\ \exists \delta > 0 \text{ such that } \mathbb{E}\Big[\min\big\{0,\ \inf_{x \in \mathbb{B}(\bar{x},\delta)} f(\xi, x)\big\}\Big] > -\infty.$$

In particular, $f$ is locally inf-integrable if there's an integrable $g : \Xi \to [0, \infty]$ such that $f(\xi, x) \geq -g(\xi)$ for all $\xi \in \Xi$ and $x \in \mathbb{R}^n$ because then

$$\mathbb{E}\Big[\min\big\{0,\ \inf_{x \in \mathbb{B}(\bar{x},\delta)} f(\xi, x)\big\}\Big] \geq \mathbb{E}\big[-g(\xi)\big] > -\infty,$$

where the first inequality follows by 8.53(a).

We note that $\xi \mapsto \inf_{x\in\mathbb{B}(\bar{x},\delta)} f(\xi, x)$ is measurable by 8.52 because

$$(\xi, x) \mapsto f(\xi, x) + \iota_{\mathbb{B}(\bar{x},\delta)}(x)$$

is a random lsc function by 8.51(c), so the expectation in the definition of local inf-integrability is well defined.

Local inf-integrability restricts $f$ from below and thus doesn't interfere with representation of constraints in terms of indictor functions. As an illustration, if $f_0 : \Xi \times \mathbb{R}^n \to \overline{\mathbb{R}}$ is locally inf-integrable random lsc and $C \subset \mathbb{R}^n$ is closed, then

$$(\xi, x) \mapsto f_0(\xi, x) + \iota_C(x)$$

is random lsc by 8.51(c) and also locally inf-integrable as can be seen via 8.53(a). Moreover, any locally inf-integrable random lsc $f$ has $\mathbb{E}[f(\boldsymbol{\xi}, x)] > -\infty$ for all $x \in \mathbb{R}^n$.

**Proposition 8.55** (properties of expectation functions). *For a random lsc function $f : \Xi \times \mathbb{R}^n \to \overline{\mathbb{R}}$, the expectation function $Ef : \mathbb{R}^n \to \overline{\mathbb{R}}$ given by*

$$Ef(x) = \mathbb{E}\big[f(\boldsymbol{\xi}, x)\big]$$

*has the following properties:*

(a) *If $f(\xi, \cdot\,)$ is convex for all $\xi \in \Xi$, then $Ef$ is convex.*
(b) *If $f$ is locally inf-integrable, then $Ef$ is lsc.*

**Proof.** By the convexity inequality 1.10, it suffices to show in (a) that

$$Ef(x^\lambda) \leq (1-\lambda)Ef(x^0) + \lambda Ef(x^1) \quad \forall \lambda \in [0,1] \text{ and } x^0, x^1 \in \mathbb{R}^n,$$

where $x^\lambda = (1-\lambda)x^0 + \lambda x^1$. Since

$$f(\xi, x^\lambda) \leq (1-\lambda)f(\xi, x^0) + \lambda f(\xi, x^1) \quad \forall \xi \in \Xi,$$

one has

$$\mathbb{E}\big[f(\boldsymbol{\xi}, x^\lambda)\big] \leq \mathbb{E}\big[(1-\lambda)f(\boldsymbol{\xi}, x^0) + \lambda f(\boldsymbol{\xi}, x^1)\big]$$

by 8.53(a). Moreover, 8.53(b,c) lead to

$$\mathbb{E}\big[(1-\lambda)f(\boldsymbol{\xi}, x^0) + \lambda f(\boldsymbol{\xi}, x^1)\big] \leq (1-\lambda)\mathbb{E}\big[f(\boldsymbol{\xi}, x^0)\big] + \lambda\mathbb{E}\big[f(\boldsymbol{\xi}, x^1)\big].$$

For (b), let $x^\nu \to \bar{x}$ and define

$$h^\nu(\xi) = f(\xi, x^\nu) \quad \forall \xi \in \Xi.$$

Since $f$ is locally inf-integrable, there's $\delta > 0$ such that

$$\mathbb{E}\Big[\min\big\{0, \inf_{x\in\mathbb{B}(\bar{x},\delta)} f(\boldsymbol{\xi}, x)\big\}\Big] > -\infty.$$

Let $\bar{\nu}$ be such that $x^\nu \in \mathbb{B}(\bar{x}, \delta)$ when $\nu \geq \bar{\nu}$. For all $\xi \in \Xi$, set

$$g(\xi) = -\min\{0, \inf_{x\in\mathbb{B}(\bar{x},\delta)} f(\xi, x)\}.$$

We consider two cases: First, suppose that

$$\mathbb{E}\big[\max\{0, \inf_{x\in\mathbb{B}(\bar{x},\delta)} f(\boldsymbol{\xi}, x)\}\big] = \infty.$$

Then, $\mathbb{E}[\max\{0, f(\boldsymbol{\xi}, x^\nu)\}] = \infty$ by 8.53(a) for $\nu \geq \bar{\nu}$ and also $\mathbb{E}[f(\boldsymbol{\xi}, x^\nu)] = \infty$. Thus,

$$\operatorname{liminf} Ef(x^\nu) \geq Ef(\bar{x})$$

by virtue of the left-hand side being infinity. Second, suppose that

$$\mathbb{E}\big[\max\{0, \inf_{x\in\mathbb{B}(\bar{x},\delta)} f(\boldsymbol{\xi}, x)\}\big] < \infty.$$

Since $\mathbb{E}[\min\{0, \inf_{x\in\mathbb{B}(\bar{x},\delta)} f(\boldsymbol{\xi}, x)\}] > -\infty$, $g(\boldsymbol{\xi})$ is integrable. Moreover, for $\nu \geq \bar{\nu}$,

$$h^\nu(\xi) = f(\xi, x^\nu) \geq \inf_{x\in\mathbb{B}(\bar{x},\delta)} f(\xi, x) \geq \min\{0, \inf_{x\in\mathbb{B}(\bar{x},\delta)} f(\xi, x)\} = -g(\xi)$$

and we can bring in the extended Fatou's lemma 8.54 to establish

$$\operatorname{liminf} \mathbb{E}\big[h^\nu(\boldsymbol{\xi})\big] \geq \mathbb{E}\big[h(\boldsymbol{\xi})\big], \tag{8.14}$$

where $h : \Xi \to \overline{\mathbb{R}}$ is given by

$$h(\xi) = \operatorname{liminf} f(\xi, x^\nu).$$

Since $f(\xi, \cdot)$ is lsc, one has via (5.1) that

$$\operatorname{liminf} f(\xi, x^\nu) \geq f(\xi, \bar{x}).$$

From 8.53(a), we then obtain

$$\mathbb{E}\big[h(\boldsymbol{\xi})\big] \geq \mathbb{E}\big[f(\boldsymbol{\xi}, \bar{x})\big].$$

This inequality combined with (8.14) produce

$$\operatorname{liminf} Ef(x^\nu) \geq Ef(\bar{x}).$$

The conclusion then follows from (5.1). □

One can apply 8.55(b) twice to establish continuity of an expectation function: If $f : \Xi \times \mathbb{R}^n \to \overline{\mathbb{R}}$ is locally inf-integrable random lsc and $-f$ also has these properties, then $Ef$ is both lsc and usc and thus continuous.

We now turn to the main theorem about epi-convergence of sample average functions, but first formalize the setting. Suppose that $\boldsymbol{\xi}^1, \boldsymbol{\xi}^2, \ldots$ are iid random elements (variables, vectors, matrices, etc.), each taking values according to a probability space $(\Xi, \mathcal{B}, \mathbb{P})$. For one set of values $\omega = (\xi^1, \xi^2, \ldots)$, with $\xi^i \in \Xi$, and $\nu \in \mathbb{N}$, the *sample average function* $E^\nu f(\cdot, \omega) : \mathbb{R}^n \to \overline{\mathbb{R}}$ is given by

$$E^\nu f(x, \omega) = \frac{1}{\nu} \sum_{i=1}^{\nu} f(\xi^i, x).$$

Prior to sampling, $\omega$ is unknown and we would like a guarantee that

$$E^\nu f(\cdot, \omega) \xrightarrow{e} Ef$$

for all "possible" $\omega$. Since every $\omega$ is a sequence of sample points from $\Xi$, its possible values lie in the product set

$$\Omega = \Xi \times \Xi \times \cdots.$$

The probability of a subset $\Omega_0 = \Xi_1 \times \Xi_2 \times \cdots$, with $\Xi_i \in \mathcal{B}$, is then

$$\mathbb{P}_\infty(\Omega_0) = \prod_{i=1}^{\infty} \mathbb{P}(\Xi_i).$$

(The validity of this construction is supported by the *Kolmogorov extension theorem*; see for example [32, Section 1.4C].) In the next theorem, we obtain that $E^\nu f(\cdot, \omega) \xrightarrow{e} Ef$ for *almost every* $\omega \in \Omega$, which means that there's $\Omega_0 \subset \Omega$ with $\mathbb{P}_\infty(\Omega_0) = 1$ such that for every $\omega \in \Omega_0$ the epi-convergence takes place. We stress that this provides the needed justification for sample average functions under *independent* sampling. For other cases, we refer to [59].

**Theorem 8.56** (law of large numbers for sample average functions). *For a locally inf-integrable random lsc function $f : \Xi \times \mathbb{R}^n \to \overline{\mathbb{R}}$, one has*

$$E^\nu f(\cdot, \omega) \xrightarrow{e} Ef \ \text{ for almost every } \omega \in \Omega.$$

**Proof.** Let $C_+ \subset \mathbb{R}^{n+1}$ be a countable dense subset[7] of epi $Ef$, $C \subset \mathbb{R}^n$ be a countable dense subset of $\mathbb{R}^n$ that contains the projection of $C_+$ on $\mathbb{R}^n$ and $R_+$ be the nonnegative rational numbers. (We note that $C_+$ would be empty if epi $Ef$ is empty and then $C$ is simply a countable dense subset of $\mathbb{R}^n$.) For $x \in C$ and $\rho \in R_+$, we define $\pi_{x,\rho} : \Xi \to \overline{\mathbb{R}}$ by setting

$$\pi_{x,\rho}(\xi) = \inf_{\bar{x} \in \mathbb{B}^o(x,\rho)} f(\xi, \bar{x}) \ \text{ if } \rho > 0 \quad \text{and} \quad \pi_{x,0}(\xi) = f(\xi, x),$$

where

$$\mathbb{B}^o(x, \rho) = \{\bar{x} \in \mathbb{R}^n \mid \|\bar{x} - x\|_2 < \rho\}.$$

By [59, Theorem 3.4], which is a slight extension of 8.52, every such $\pi_{x,\rho}$ is measurable and we write $\pi_{x,\rho}(\boldsymbol{\xi})$ to indicate this random variable. Since $f$ is locally inf-integrable, it

[7] $C_0$ is a countable dense subset of $C_1 \subset \mathbb{R}^n$ if $C_0$ contains a countable number of points and cl $C_0 = C_1$. For example, the $n$-dimensional vectors formed by rational numbers furnish a countable dense subset of $\mathbb{R}^n$.

follows that for every $x \in C$ there's $\delta_x \in (0, \infty)$ such that

$$-\infty < \mathbb{E}\Big[\min\big\{0, \ \inf_{\bar{x}\in\mathbb{B}(x,\delta_x)} f(\boldsymbol{\xi}, \bar{x})\big\}\Big] \le \mathbb{E}\big[\min\big\{0, \pi_{x,\rho}(\boldsymbol{\xi})\big\}\big] \quad \forall \rho \in [0, \delta_x],$$

where the last inequality holds by 8.53(a). Consequently, for every $x \in C$ and $\rho \in [0, \delta_x] \cap R_+$, the random variable $\pi_{x,\rho}(\boldsymbol{\xi})$ is either integrable or its negative part $\max\{0, -\pi_{x,\rho}(\boldsymbol{\xi})\}$ is integrable with the positive part having

$$\mathbb{E}\big[\max\big\{0, \pi_{x,\rho}(\boldsymbol{\xi})\big\}\big] = \infty.$$

In either case, we can invoke the standard *law of large numbers* (see for example Theorem 22.1 and its corollary in [14]) to establish that there's $\Omega_{x,\rho} \subset \Omega$ with $\mathbb{P}_\infty(\Omega_{x,\rho}) = 1$ such that

$$\frac{1}{\nu}\sum_{i=1}^{\nu} \pi_{x,\rho}(\xi^i) \to \mathbb{E}\big[\pi_{x,\rho}(\boldsymbol{\xi})\big] \quad \forall (\xi^1, \xi^2, \ldots) \in \Omega_{x,\rho}.$$

Since the set

$$K = \big\{(x, \rho) \,\big|\, x \in C, \ \rho \in [0, \delta_x] \cap R_+\big\}$$

is a countable set, $\Omega_0 = \cap_{(x,\rho)\in K} \Omega_{x,\rho}$ has $\mathbb{P}_\infty(\Omega_0) = 1$. Let $\omega = (\xi^1, \xi^2, \ldots) \in \Omega_0$.

We proceed by establishing the two conditions in the characterization 4.15 of epi-convergence. First, suppose that $x^\nu \to x$. There are

$$\big\{\bar{\nu}^k \in \mathbb{N}, \ z^k \in C, \rho^k \in [0, \delta_x] \cap R_+, \ \ k \in \mathbb{N}\big\},$$

with the properties that $z^k \to x$ and $\rho^k \to 0$ as $k \to \infty$ and, for all $\nu \ge \bar{\nu}^k$ and $k \in \mathbb{N}$, one has

$$x \in \mathbb{B}^o(z^{k+1}, \rho^{k+1}) \subset \mathbb{B}^o(z^k, \rho^k) \ \text{ and } \ x^\nu \in \mathbb{B}^o(z^k, \rho^k).$$

Fix $k$ temporarily. Then, for $\nu \ge \bar{\nu}^k$,

$$\begin{aligned}\frac{1}{\nu}\sum_{i=1}^{\nu} f(\xi^i, x^\nu) &\ge \frac{1}{\nu}\sum_{i=1}^{\nu} \inf_{\bar{x}\in\mathbb{B}^o(z^k,\rho^k)} f(\xi^i, \bar{x}) \\ &= \frac{1}{\nu}\sum_{i=1}^{\nu} \pi_{z^k,\rho^k}(\xi^i) \to \mathbb{E}\big[\pi_{z^k,\rho^k}(\boldsymbol{\xi})\big]. \end{aligned} \tag{8.15}$$

Since $f$ is a random lsc function, $f(\xi, \cdot)$ is lsc regardless of $\xi \in \Xi$. This implies that for all $\xi \in \Xi$, one has

$$\pi_{z^k,\rho^k}(\xi) \to \pi_{x,0}(\xi) = f(\xi, x) \ \text{ as } \ k \to \infty;$$

see [105, Definition 1.5; Theorem 1.6].

The nestedness of the balls implies that

$$\pi_{z^k,\rho^k}(\xi) \le \pi_{z^{k+1},\rho^{k+1}}(\xi) \ \ \forall \xi \in \Xi \ \text{ and } \ k.$$

We can now invoke the *monotone convergence theorem* (see [40, Theorem 2.14]) to establish that

$$\mathbb{E}\big[\max\{0, \pi_{z^k,\rho^k}(\boldsymbol{\xi})\}\big] \to \mathbb{E}\big[\max\{0, f(\boldsymbol{\xi}, x)\}\big] \text{ as } k \to \infty.$$

As already noted, $\max\{0, -\pi_{z^k,\rho^k}(\boldsymbol{\xi})\}$ is integrable. Since

$$\max\{0, -\pi_{z^1,\rho^1}(\xi)\} \geq \max\{0, -\pi_{z^k,\rho^k}(\xi)\} \geq 0 \quad \forall \xi \in \Xi \text{ and } k,$$

we can invoke the *dominated convergence theorem* (see [40, Theorem 2.24]) to establish that

$$\mathbb{E}\big[\max\{0, -\pi_{z^k,\rho^k}(\boldsymbol{\xi})\}\big] \to \mathbb{E}\big[\max\{0, -f(\boldsymbol{\xi}, x)\}\big] \text{ as } k \to \infty.$$

Thus,

$$\mathbb{E}\big[\pi_{z^k,\rho^k}(\boldsymbol{\xi})\big] \to \mathbb{E}\big[f(\boldsymbol{\xi}, x)\big] \text{ as } k \to \infty.$$

We can combine this fact with (8.15) and conclude that

$$\operatorname{liminf} \frac{1}{\nu}\sum_{i=1}^{\nu} f(\xi^i, x^\nu) \geq \mathbb{E}\big[f(\boldsymbol{\xi}, x)\big],$$

i.e., $\operatorname{liminf} E^\nu f(x^\nu, \omega) \geq Ef(x)$ and 4.15(a) holds.

Second, let $\bar{x} \in C$. Then,

$$E^\nu f(\bar{x}, \omega) = \frac{1}{\nu}\sum_{i=1}^{\nu} f(\xi^i, \bar{x}) = \frac{1}{\nu}\sum_{i=1}^{\nu} \pi_{\bar{x},0}(\xi^i) \to \mathbb{E}\big[\pi_{\bar{x},0}(\boldsymbol{\xi})\big] = Ef(\bar{x}).$$

This implies that whenever $Ef(\bar{x}) \leq \bar{\alpha}$ for some $\bar{\alpha} \in \mathbb{R}$, one has

$$(\bar{x}, \bar{\alpha}) \in \operatorname{LimInn}\big(\operatorname{epi} E^\nu f(\cdot, \omega)\big). \tag{8.16}$$

To see this, note that with

$$\alpha^\nu = \max\{\bar{\alpha}, E^\nu f(\bar{x}, \omega)\},$$

one has $(\bar{x}, \alpha^\nu) \in \operatorname{epi} E^\nu f(\cdot, \omega)$ for sufficiently large $\nu$ and $(\bar{x}, \alpha^\nu) \to (\bar{x}, \bar{\alpha})$.

Since $\bar{x} \in C$ is arbitrary, we can conclude from (8.16) that

$$\{(x, \alpha) \in \mathbb{R}^{n+1} \mid Ef(x) \leq \alpha,\ x \in C\} \subset \operatorname{LimInn}\big(\operatorname{epi} E^\nu f(\cdot, \omega)\big).$$

One can replace the set on each side of an inclusion by its closure. Applying this here, the right-hand side remains unchanged because inner limits are closed and thus

$$\operatorname{cl}\{(x, \alpha) \in \mathbb{R}^{n+1} \mid Ef(x) \leq \alpha,\ x \in C\} \subset \operatorname{LimInn}\big(\operatorname{epi} E^\nu f(\cdot, \omega)\big). \tag{8.17}$$

Next, let $(\hat{x}, \hat{\alpha}) \in \operatorname{epi} Ef$. By construction of $C_+$, there exists $(x^\nu, \alpha^\nu) \in C_+ \to (\hat{x}, \hat{\alpha})$. Thus, $Ef(x^\nu) \leq \alpha^\nu$ and $x^\nu \in C$ by construction as well, which then implies that

$$(\hat{x}, \hat{\alpha}) \in \text{cl}\,\big\{(x, \alpha) \in \mathbb{R}^{n+1} \,\big|\, Ef(x) \le \alpha,\ x \in C\big\}.$$

In fact, epi $Ef$ is contained in this closure. In combination with (8.17), we've established that

$$\text{epi}\, Ef \subset \text{LimInn}\,\big(\text{epi}\, E^\nu f(\cdot, \omega)\big).$$

This is equivalent to 4.15(b) as seen from the proof of that statement. □

The theorem is a major tool for justifying approximations of expectation functions constructed by Monte Carlo sampling. It can be used in conjunction with, for example, 4.19 and 4.20 to establish epi-convergence for composite functions. If the assumptions of the theorem hold for $f$ as well as $-f$, then one has the stronger conclusion that

$$\text{for almost every } \omega \in \Omega: \quad E^\nu f(x^\nu, \omega) \to Ef(x) \text{ whenever } x^\nu \to x.$$

This can be utilized to show via 4.19(a) that

$$\iota_{X^\nu} + E^\nu f(\cdot, \omega) \xrightarrow{e} \iota_X + Ef(x)$$

provided that $X^\nu \xrightarrow{s} X$.

**Example 8.57** (superquantile-risk minimization). For $f : \Xi \times \mathbb{R}^n \to \mathbb{R}$, $\alpha \in (0, 1)$ and closed $C \subset \mathbb{R}^n$, suppose that $f(\boldsymbol{\xi}, x)$ is integrable for all $x \in \mathbb{R}^n$ and $f(\xi, \cdot)$ is continuous for all $\xi \in \Xi$. As seen in §3.C, the problem of minimizing s-rsk$_\alpha(f(\boldsymbol{\xi}, x))$ over $x \in C$ is equivalent to

$$\underset{(x,\gamma)\in\mathbb{R}^{n+1}}{\text{minimize}}\ \varphi(x, \gamma) = \iota_C(x) + \gamma + \frac{1}{1-\alpha}\mathbb{E}\big[\max\{0, f(\boldsymbol{\xi}, x) - \gamma\}\big].$$

For $\omega = (\xi^1, \xi^2, \ldots)$ obtained by statistically independent sampling according to the probability distribution of $\boldsymbol{\xi}$, one has the approximating problem

$$\underset{(x,\gamma)\in\mathbb{R}^{n+1}}{\text{minimize}}\ \varphi^\nu(x, \gamma, \omega) = \iota_C(x) + \gamma + \frac{1}{1-\alpha}\frac{1}{\nu}\sum\nolimits_{i=1}^{\nu} \max\big\{0, f(\xi^i, x) - \gamma\big\}.$$

We can show that $\varphi^\nu(\cdot, \omega) \xrightarrow{e} \varphi$ for almost every $\omega$.

**Detail.** Let's define

$$g_0(\xi, x, \gamma) = \max\big\{0, f(\xi, x) - \gamma\big\} \text{ and } g(\xi, x, \gamma) = \iota_C(x) + g_0(\xi, x, \gamma).$$

By 8.51(d), $g_0$ is random lsc and the same holds for $g$ via 8.51(c). Moreover, $g$ is locally inf-integrable because it's bounded from below by zero. Let's consider the sample average function $E^\nu g(\cdot, \omega) : \mathbb{R}^{n+1} \to \overline{\mathbb{R}}$ given by

$$E^\nu g(x, \gamma, \omega) = \frac{1}{\nu}\sum\nolimits_{i=1}^{\nu} g(\xi^i, x, \gamma).$$

We can then apply the law of large numbers 8.56 to conclude that $E^\nu g(\cdot, \omega) \xrightarrow{e} Eg$ for almost every $\omega \in \Omega$, where $Eg(x, \gamma) = \mathbb{E}[g(\boldsymbol{\xi}, x, \gamma)]$. After combining this with 4.20(b) and 4.19(a), we realize that

$$(x,\gamma) \mapsto \gamma + \frac{1}{1-\alpha} E^\nu g(x,\gamma,\omega) \text{ epi-converges to } (x,\gamma) \mapsto \gamma + \frac{1}{1-\alpha} Eg(x,\gamma).$$

Since these functions coincide with $\varphi^\nu(\cdot, \omega)$ and $\varphi$, respectively, we've established the claim about epi-convergence. The epigraphical approximation algorithm is justified when applied in this context. The numerical results in §3.C, see Table 3.4, indeed show signs of convergence to a solution as the sample size $\nu$ grows.

While the integrability of $f(\boldsymbol{\xi}, x)$ ensures that $\text{s-rsk}_\alpha(f(\boldsymbol{\xi}, x))$ is finite, it's not needed for the conclusion $\varphi^\nu(\cdot, \omega) \xrightarrow{e} \varphi$; a measurable $f(\cdot, x)$ would still produce a locally inf-integrable random lsc function $g$. □

**Example 8.58** (consistency in regression). For random variables $\boldsymbol{\eta}, \boldsymbol{\xi}_1, \dots, \boldsymbol{\xi}_n \in \mathcal{L}^2$, we recall from 8.16 the regression problem

$$\underset{(\gamma,c)\in\mathbb{R}^{1+n}}{\text{minimize}}\ \varphi(\gamma, c) = \mathcal{E}\big(\boldsymbol{\eta} - \gamma - \langle c, \boldsymbol{\xi}\rangle\big),$$

where $\boldsymbol{\xi} = (\boldsymbol{\xi}_1, \dots, \boldsymbol{\xi}_n)$. Now, we restrict the attention to regular measures of error $\mathcal{E}$ of the form

$$\mathcal{E}(\boldsymbol{\eta}_0) = \mathbb{E}\big[e(\boldsymbol{\eta}_0)\big] \quad \forall \boldsymbol{\eta}_0 \in \mathcal{L}^2,$$

where $e : \mathbb{R} \to \overline{\mathbb{R}}$ is lsc and convex, with $e(0) = 0$ and $e(\gamma) > 0$ for $\gamma \neq 0$.

Suppose that $\{\omega_i = (\eta^i, \xi_1^i, \dots, \xi_n^i),\ i = 1, \dots, \nu\}$ is a dataset generated by statistically independent sampling according to the probability distribution of $(\boldsymbol{\eta}, \boldsymbol{\xi}_1, \dots, \boldsymbol{\xi}_n)$. We view the data as the first $\nu$ vectors of a sequence of vectors $\omega = (\omega_1, \omega_2, \dots)$. The approximating regression problem is

$$\underset{(\gamma,c)\in\mathbb{R}^{1+n}}{\text{minimize}}\ \varphi^\nu(\gamma, c, \omega) = \frac{1}{\nu}\sum\nolimits_{i=1}^{\nu} e\big(\eta^i - \gamma - \langle c, \xi^i\rangle\big),$$

where $\xi^i = (\xi_1^i, \dots, \xi_n^i)$. Then, $\varphi^\nu(\cdot, \omega) \xrightarrow{e} \varphi$ for almost every $\omega$. Let $(\gamma^\nu(\omega), c^\nu(\omega))$ be a minimizer of the approximating problem. As we receive more data and compute $(\gamma^\nu(\omega), c^\nu(\omega))$ for increasing $\nu$, every cluster point $(\gamma^\star(\omega), c^\star(\omega))$ of the generated sequence is a minimizer of the actual regression problem. In the language of statistics, this establishes *consistency* of the estimator defined by the approximating regression problem.

**Detail.** Let's first check that $\mathcal{E}$ is a regular measure of error; cf. 8.14. It follows by 8.53(a) and the nonnegativity of $e$ that $\mathbb{E}[e(\boldsymbol{\eta}_0)] \geq 0$. The convexity of $e$ translates into convexity of $\mathcal{E}$ by an argument similar to that in the proof of 8.55 by using 8.53. The property $\mathcal{E}(\mathbf{0}) = 0$ follows trivially. When $\boldsymbol{\eta}_0 \neq \mathbf{0}$, $e(\boldsymbol{\eta}_0)$ is positive with positive probability, which means that $\mathbb{E}[e(\boldsymbol{\eta}_0)]$ is positive. For a proof of lsc, we refer to [102]. Consequently, $\mathcal{E}$ satisfies all the properties of a regular measure of error.

Among the many possibilities, we've encountered $e$ of the forms:

$$e(\gamma) = \gamma^2 \qquad \text{(least-squares regression)}$$

$$e(\gamma) = \frac{1}{1-\alpha}\max\{0, \gamma\} - \gamma \qquad \text{(quantile regression).}$$

Since $e$ is nonnegative, the claim about epi-convergence holds by the law of large numbers 8.56 as long as the function $f : \mathbb{R}^{1+n} \times \mathbb{R}^{1+n} \to \overline{\mathbb{R}}$ given by

$$f\big((\eta, \xi), (\gamma, c)\big) = e\big(\eta - \gamma - \langle c, \xi\rangle\big)$$

is random lsc. In this case, the probability space $(\Xi, \mathcal{B}, \mathbb{P})$ has $\Xi = \mathbb{R}^{1+n}$, $\mathbb{P}$ is defined by the probability distribution of $(\boldsymbol{\eta}, \boldsymbol{\xi})$ and $\mathcal{B}$ can be taken to include the closed sets. Then, $f$ is lsc because $e$ is lsc and it follows from 8.51(f) that $f$ is random lsc.

Interestingly, these conclusions hold without any concern about the integrability of $e(\boldsymbol{\eta} - \gamma - \langle c, \boldsymbol{\xi}\rangle)$. For example, $e(\gamma) = \gamma^4$ could easily make $\mathbb{E}[e(\boldsymbol{\eta})] = \infty$ even though $\boldsymbol{\eta} \in \mathcal{L}^2$. □

## 8.H Concentration Inequalities

While the previous section establishes that sample average functions epi-converge to expectation functions under mild assumptions, we caution that the approximation error can be sizeable even for seemly large sample sizes. For example, the numerical results in §3.C, see Table 3.4, reveal a substantial discrepancy between results obtained using sample size $\nu = 1000$ and $\nu = 500000$. The central limit theorem (see the discussion in §8.A) indicates that Monte Carlo sampling using sample size $\nu$ has an error of order $\nu^{-1/2}$ in estimating an expectation, which decays at a frustratingly slow rate, and anyhow is only valid in an asymptotic sense as $\nu \to \infty$. In practice, computational considerations and data availability often limit the sample size. This brings forth the need for error analysis given a fixed (finite) sample size.

For a compact set $X \subset \mathbb{R}^n$, let's examine bounds of the form

$$\sup_{x\in X}\left|\frac{1}{\nu}\sum\nolimits_{i=1}^{\nu} f(\xi^i, x) - \mathbb{E}\big[f(\boldsymbol{\xi}, x)\big]\right| \leq \delta$$

because they furnish essential components of error estimates in §6.J, where often $X = \mathbb{B}(0, \rho)$. A complicating factor is that $\xi^1, \dots, \xi^\nu$ isn't known in advance and, in fact, is generated by sampling or obtained through (future) observations of the underlying random phenomenon. Consequently, the sample average can be viewed as a random variable. We need to quantify the probability that this random variable isn't near its mean, which hopefully vanishes at a fast rate as the sample size increases.

We start in the simplified setting with a nonnegative random variable $\boldsymbol{\xi}$, a scalar $\tau \in (0, \infty)$ and the goal of bounding prob$\{\boldsymbol{\xi} \geq \tau\}$ from above. Let $\boldsymbol{I}_\tau$ be the random variable with value 1 if $\boldsymbol{\xi} \geq \tau$ and 0 otherwise. Since the values of $\boldsymbol{\xi}\boldsymbol{I}_\tau$ are no smaller than those of $\tau \boldsymbol{I}_\tau$. It follows from 8.53(a) that

$$\mathbb{E}[\boldsymbol{\xi}] \geq \mathbb{E}[\boldsymbol{\xi}\boldsymbol{I}_\tau] \geq \mathbb{E}[\tau\boldsymbol{I}_\tau] = \tau \text{prob}\{\boldsymbol{\xi} \geq \tau\}.$$

Thus, we obtain *Markov's inequality*:

$$\text{prob}\{\boldsymbol{\xi} \geq \tau\} \leq \frac{\mathbb{E}[\boldsymbol{\xi}]}{\tau}.$$

While Markov's inequality can be applied directly, refinements allow us to go beyond nonnegative random variables and also leverage additional information about $\boldsymbol{\xi}$. In particular, Markov's inequality applied to a nonlinear transformation of the random variable of interest proves to be highly effective.

Suppose that $\boldsymbol{\xi}$ is a real-valued random variable. Then, for $\tau \in (0, \infty)$ and $\lambda \in [0, \infty)$,

$$\text{prob}\{\boldsymbol{\xi} \geq \tau\} \leq \text{prob}\big\{\exp(\lambda\boldsymbol{\xi}) \geq \exp(\lambda\tau)\big\} \leq \exp(-\lambda\tau)\mathbb{E}\big[\exp(\lambda\boldsymbol{\xi})\big], \tag{8.18}$$

where the first inequality holds because the exponential function is nondecreasing and the second one follows by Markov's inequality applied to the random variable $\exp(\lambda\boldsymbol{\xi})$. We would like to obtain the lowest right-hand side by adjusting $\lambda$ and the key quantity to get a handle on is the function $\lambda \mapsto \mathbb{E}[\exp(\lambda\boldsymbol{\xi})]$, which is the *moment-generating function* of $\boldsymbol{\xi}$. In the present context, it's more convenient to work with the logarithm of the moment-generating function: Let $\psi : \mathbb{R} \to \overline{\mathbb{R}}$ be the *log-moment-generating (log-mg) function* of $\boldsymbol{\xi}$, which is given by

$$\psi(\lambda) = \ln \mathbb{E}\big[\exp(\lambda\boldsymbol{\xi})\big], \tag{8.19}$$

with the convention that $\ln \infty = \infty$. Its conjugate function $\psi^* : \mathbb{R} \to \overline{\mathbb{R}}$ has

$$\psi^*(\tau) = \sup_{\lambda \in \mathbb{R}} \big\{\lambda\tau - \psi(\lambda)\big\}.$$

Returning to (8.18) and the goal of finding $\lambda \geq 0$ such that its right-hand side is minimized, we realize that one can just as well minimize the logarithm of the right-hand side. This leads to

$$\inf_{\lambda \geq 0} \big\{-\lambda\tau + \psi(\lambda)\big\} = -\sup_{\lambda \geq 0} \big\{\lambda\tau - \psi(\lambda)\big\}. \tag{8.20}$$

Under the assumption that $\boldsymbol{\xi}$ is integrable, Jensen's inequality 3.5 together with the convexity of the exponential function lead to $\psi(\lambda) \geq \lambda\mathbb{E}[\boldsymbol{\xi}]$ for all $\lambda \in \mathbb{R}$. This means that

$$\lambda\tau - \psi(\lambda) \leq 0 \ \text{ when } \ \lambda < 0, \ \ \tau \geq \mathbb{E}[\boldsymbol{\xi}].$$

Moreover, $\psi(0) = 0$ so $\psi^*(\tau) \geq 0$ for all $\tau \in \mathbb{R}$. We conclude that the restriction to nonnegative $\lambda$ on the right-hand side in (8.20) can be lifted and

$$\inf_{\lambda \geq 0} \big\{-\lambda\tau + \psi(\lambda)\big\} = -\psi^*(\tau)$$

under the assumption that $\tau \geq \mathbb{E}[\xi]$. We've established that a lowest bound on the right-hand side of (8.18) is furnished by the conjugate $\psi^*$.

**Proposition 8.59** (Chernoff's inequality). *For $\tau \in [0, \infty)$ and a real-valued random variable $\xi$ with mean zero, suppose that $\psi : \mathbb{R} \to \overline{\mathbb{R}}$ is the log-mg function of $\xi$ and $\psi^*$ is its conjugate. Then,*

$$\text{prob}\{\xi \geq \tau\} \leq e^{-\psi^*(\tau)}.$$

**Proof.** This is an immediate consequence of the earlier discussion. □

**Example 8.60** (log-mg functions). Let $\psi : \mathbb{R} \to \overline{\mathbb{R}}$ be the log-mg function of a zero-mean random variable $\xi$ and let $\psi^*$ be the corresponding conjugate. Then, the following hold:

(a) If $\xi$ is normal with variance $\sigma^2 \in (0, \infty)$, then

$$\psi(\lambda) = \tfrac{1}{2}\sigma^2\lambda^2 \quad \forall \lambda \in \mathbb{R}$$

$$\psi^*(\tau) = \frac{\tau^2}{2\sigma^2} \quad \forall \tau \in \mathbb{R}.$$

(b) If $\text{prob}\{\xi \in [\alpha, \beta]\} = 1$, with $-\infty < \alpha < \beta < \infty$, then

$$\psi(\lambda) \leq \tfrac{1}{8}(\beta - \alpha)^2\lambda^2 \quad \forall \lambda \in \mathbb{R}$$

$$\psi^*(\tau) \geq \frac{2\tau^2}{(\beta - \alpha)^2} \quad \forall \tau \in \mathbb{R}.$$

**Detail.** For (a), we compute

$$\mathbb{E}\big[\exp(\lambda\xi)\big] = \int_{-\infty}^{\infty} \frac{1}{\sqrt{2\pi}\sigma} \exp\Big(\lambda\xi - \frac{\xi^2}{2\sigma^2}\Big)\, d\xi = \exp\big(\tfrac{1}{2}\sigma^2\lambda^2\big).$$

The expression for the conjugate follows directly from its definition via the optimality condition 2.19.

For (b), we refer to [17, Lemma 2.2] for the upper bound on $\psi$, which in turn leads to

$$\psi^*(\tau) = \sup\nolimits_{\lambda \in \mathbb{R}} \big\{\lambda\tau - \psi(\lambda)\big\} \geq \sup\nolimits_{\lambda \in \mathbb{R}} \big\{\lambda\tau - \tfrac{1}{8}(\beta - \alpha)^2\lambda^2\big\} = \frac{2\tau^2}{(\beta - \alpha)^2},$$

where we again use 2.19. □

A real-valued zero-mean random variable is *subgaussian* with *proxy variance* $\sigma^2 \in (0, \infty)$ if its log-mg function satisfies

$$\psi(\lambda) \leq \tfrac{1}{2}\sigma^2\lambda^2 \quad \forall \lambda \in \mathbb{R},$$

in which case

$$\psi^*(\tau) \geq \frac{\tau^2}{2\sigma^2} \quad \forall \tau \in \mathbb{R}. \tag{8.21}$$

This property can be characterized by the growth of $\mathbb{E}[\xi^k]$ as the power $k$ increases; see [17, Theorem 2.1]. Random variables of this kind has log-mg functions of essentially the same form as those in 8.60 and this leads to productive use of Chernoff's inequality 8.59.

These concepts can now be applied to sample averages. Suppose that $\xi_1, \ldots, \xi_\nu$ are iid real-valued random variables with finite mean $\mu$. Then, the random variable

$$\frac{1}{\nu}\sum_{i=1}^{\nu} \xi_i - \mu$$

has zero mean and we can apply Chernoff's inequality 8.59 provided that we obtain the conjugate of its log-mg function. This can be expressed in terms of the log-mg of $\xi_i - \mu$ by observing that

$$\begin{aligned}\mathbb{E}\Big[\exp\Big(\lambda\Big(\frac{1}{\nu}\sum_{i=1}^{\nu}\xi_i - \mu\Big)\Big)\Big] &= \mathbb{E}\Big[\exp\Big(\sum_{i=1}^{\nu}\lambda\nu^{-1}(\xi_i - \mu)\Big)\Big]\\ &= \mathbb{E}\Big[\prod_{i=1}^{\nu}\exp\big(\lambda\nu^{-1}(\xi_i - \mu)\big)\Big]\\ &= \prod_{i=1}^{\nu}\mathbb{E}\big[\exp\big(\lambda\nu^{-1}(\xi_i - \mu)\big)\big],\end{aligned}$$

where the last equality follows by a fact about statistically independence random variables: the expectation of a product of statistically independent random variables is the product of their expectations. Let $\psi_\nu$ and $\psi$ be the log-mg functions of $\frac{1}{\nu}\sum_{i=1}^{\nu}\xi_i - \mu$ and $\xi_i - \mu$, respectively. The previous equation then implies

$$\begin{aligned}\psi_\nu(\lambda) &= \ln\mathbb{E}\Big[\exp\Big(\lambda\Big(\frac{1}{\nu}\sum_{i=1}^{\nu}\xi_i - \mu\Big)\Big)\Big]\\ &= \sum_{i=1}^{\nu}\ln\mathbb{E}\big[\exp\big(\lambda\nu^{-1}(\xi_i - \mu)\big)\big] = \nu\psi(\lambda\nu^{-1}) \quad \forall\lambda\in\mathbb{R}.\end{aligned}$$

Moreover, a rule for conjugates immediately seen from their definition (cf. [105, equation 11(3)]) gives that

$$\psi_\nu^*(\tau) = \nu\psi^*(\tau) \quad \forall\tau\in\mathbb{R}.$$

**Proposition 8.61** (concentration inequality). *For iid real-valued random variables $\xi_1, \ldots, \xi_\nu$ with finite mean $\mu$, suppose that $\xi_i - \mu$ is subgaussian with proxy variance $\sigma^2 \in (0, \infty)$. If $\tau \geq 0$, then*

$$\mathrm{prob}\Big\{\frac{1}{\nu}\sum_{i=1}^{\nu}\xi_i - \mu \geq \tau\Big\} \leq \exp\Big(-\frac{\nu\tau^2}{2\sigma^2}\Big).$$

**Proof.** If $\psi_\nu^*$ and $\psi^*$ are the conjugates of the log-mg functions of $\frac{1}{\nu}\sum_{i=1}^{\nu}\xi_i - \mu$ and $\xi_i - \mu$, respectively, then Chernoff's inequality 8.59 establishes that the left-hand side in the claimed inequality is bounded by $\exp(-\psi_\nu^*(\tau))$. The discussion prior to the proposition gives that $\psi_\nu^*(\tau) = \nu\psi^*(\tau)$. This fact together with (8.21) yield the conclusion. □

In stochastic optimization with an expectation function $Ef : \mathbb{R}^n \to \overline{\mathbb{R}}$, we're rarely concerned about a single sample average. For each $x \in \mathbb{R}^n$, we approximate

$Ef(x)$ by Monte Carlo sampling. This raises the concerning possibility that while we may approximate one $Ef(x)$ well with high probability, the odds for achieving this simultaneously across many different $x$ might be low. Thus, we need to quantify how much the probability of having low error across many points suffers as the number of points increases.

**Proposition 8.62** (maximal inequality). *For each $j = 1, \ldots, n$, suppose that $\xi_{1j}, \ldots, \xi_{\nu j}$ are iid real-valued random variables with finite mean $\mu_j$ and $\xi_{ij} - \mu_j$ is subgaussian with proxy variance $\sigma_j^2 \in (0, \infty)$. Then, for $\tau \geq 0$,*

$$\text{prob}\Big\{ \max_{j=1,\ldots,n} \Big| \frac{1}{\nu} \sum\nolimits_{i=1}^{\nu} \xi_{ij} - \mu_j \Big| \geq \tau \Big\} \leq 2n \exp\Big( -\frac{\nu \tau^2}{2\sigma^2} \Big),$$

*where $\sigma = \max_{j=1,\ldots,n} \sigma_j$.*

**Proof.** By the concentration inequality 8.61,

$$\text{prob}\Big\{ \frac{1}{\nu} \sum\nolimits_{i=1}^{\nu} \xi_{ij} - \mu_j \geq \tau \Big\} \leq \exp\Big( -\frac{\nu \tau^2}{2\sigma_j^2} \Big).$$

Since $-\xi_{ij} + \mu_j$ is also subgaussian with proxy variance $\sigma_j^2$, we similarly obtain

$$\text{prob}\Big\{ \frac{1}{\nu} \sum\nolimits_{i=1}^{\nu} \xi_{ij} - \mu_j \leq -\tau \Big\} \leq \exp\Big( -\frac{\nu \tau^2}{2\sigma_j^2} \Big).$$

Boole's inequality states that the probability of a union of events is bounded by the sum of the probabilities of the events. We're interested in the events

$$\Big\{ \frac{1}{\nu} \sum\nolimits_{i=1}^{\nu} \xi_{ij} - \mu_j \geq \tau \Big\} \text{ and } \Big\{ \frac{1}{\nu} \sum\nolimits_{i=1}^{\nu} \xi_{ij} - \mu_j \leq -\tau \Big\},\ j = 1, \ldots, n.$$

Since each of these $2n$ events occur with probability at most

$$\exp\Big( -\frac{\nu \tau^2}{2\sigma^2} \Big),$$

the conclusion follows. □

The proposition shows that as the number of expectations $n$ to be estimated grows, the sample size $\nu$ needs to increase as well to make sure all of the estimates are accurate with high probability. This represents a challenge in stochastic optimization with an expectation function $Ef : \mathbb{R}^n \to \overline{\mathbb{R}}$ that may need to be estimated across an uncountable number of points $x \in \mathbb{R}^n$. One can overcome this issue by introducing additional structure so that the expectation function is essentially determined by its values at a finite number of points.

**Theorem 8.63** (error for sample average functions). *For a nonempty compact set $X \subset \mathbb{R}^n$, a probability space $(\Xi, \mathcal{B}, \mathbb{P})$ and a function $f : \Xi \times \mathbb{R}^n \to \mathbb{R}$, suppose that $f(\xi, x)$ is* *integrable for all $x \in X$ and there's an integrable random variable $\kappa : \Xi \to [0, \infty)$ such that*

$$\big|f(\xi, x) - f(\xi, \bar{x})\big| \le \kappa(\xi)\|x - \bar{x}\|_2 \quad \forall \xi \in \Xi,\ x, \bar{x} \in X.$$

*For each $x \in X$, suppose that $f(\boldsymbol{\xi}, x) - \mathbb{E}[f(\boldsymbol{\xi}, x)]$ is subgaussian with proxy variance $\sigma(x)^2 \in (0, \infty)$ and $\kappa(\boldsymbol{\xi}) - \mathbb{E}[\kappa(\boldsymbol{\xi})]$ is subgaussian with proxy variance $\sigma_0^2 \in (0, \infty)$.*

*If $\boldsymbol{\xi}^1, \dots, \boldsymbol{\xi}^\nu$ are iid random elements with common probability distribution $\mathbb{P}$, then for any $\delta \in (0, \infty)$, there are constants $\alpha, \beta \in (0, \infty)$ such that*

$$\operatorname{prob}\Big\{ \sup_{x \in X} \Big| \frac{1}{\nu} \sum_{i=1}^{\nu} f(\boldsymbol{\xi}^i, x) - \mathbb{E}\big[f(\boldsymbol{\xi}, x)\big] \Big| \le \delta \Big\} \ge 1 - \beta e^{-\alpha\nu}.$$

*The constants $\alpha, \beta$ depend on the problem data as follows: For a point $x^0 \in X$, which can be selected arbitrarily, set*

$$\varepsilon = \max\left\{ \frac{1}{2\bar{\kappa} + 1}\left(\delta - \frac{\sigma(x^0)}{\sigma_0}\right),\ \frac{\delta}{4\bar{\kappa} + 2} \right\}, \tag{8.22}$$

*where $\bar{\kappa} = \mathbb{E}[\kappa(\boldsymbol{\xi})]$. Then, there are points $x^1, \dots, x^r \in X$, with $r$ at most*

$$\left(1 + \frac{\sup_{x, \bar{x} \in X} \|x - \bar{x}\|_2}{\varepsilon}\right)^n, \tag{8.23}$$

*such that one can take*

$$\alpha = \frac{\big(\delta - (2\bar{\kappa} + 1)\varepsilon\big)^2}{2\sigma^2} \quad \textit{and} \quad \beta = 2r + 1,$$

*where $\sigma^2 = \max_{j=0,\dots,r} \sigma^2(x^j)$.*

**Proof.** Let $\varepsilon, \gamma, \tau \in (0, \infty)$. Since $X$ is compact, there are points $x^1, \dots, x^r \in \mathbb{R}^n$ such that

$$X \subset \bigcup_{j=1}^{r} \mathbb{B}(x^j, \varepsilon),$$

with $r \in \mathbb{N}$. By the maximal inequality 8.62,

$$\operatorname{prob}\Big\{ \max_{j=1,\dots,r} \Big| \frac{1}{\nu} \sum_{i=1}^{\nu} f(\boldsymbol{\xi}^i, x^j) - \mathbb{E}\big[f(\boldsymbol{\xi}, x^j)\big] \Big| \ge \tau \Big\} \le 2r \exp\Big( - \frac{\nu\tau^2}{2\bar{\sigma}_\varepsilon^2} \Big),$$

where $\bar{\sigma}_\varepsilon = \max_{j=1,\dots,r} \sigma(x^j)$. By 8.61,

$$\operatorname{prob}\Big\{ \frac{1}{\nu} \sum_{i=1}^{\nu} \kappa(\boldsymbol{\xi}^i) - \mathbb{E}\big[\kappa(\boldsymbol{\xi})\big] \ge \gamma \Big\} \le \exp\Big( - \frac{\nu\gamma^2}{2\sigma_0^2} \Big).$$

We can combine these bounds using Boole's Inequality and this leads to

$$\text{prob}\Big\{ \max_{j=1,\dots,r} \Big| \frac{1}{\nu}\sum_{i=1}^{\nu} f(\xi^i, x^j) - \mathbb{E}\big[f(\xi, x^j)\big]\Big| \leq \tau \text{ and}$$

$$\frac{1}{\nu}\sum_{i=1}^{\nu} \kappa(\xi^i) - \bar{\kappa} \leq \gamma\Big\} \geq p = 1 - 2r\exp\Big(-\frac{\nu\tau^2}{2\bar{\sigma}_\varepsilon^2}\Big) - \exp\Big(-\frac{\nu\gamma^2}{2\sigma_0^2}\Big).$$

Let's fix $x \in X$. Then there's $j \in \{1, \dots, r\}$ such that $x \in \mathbb{B}(x^j, \varepsilon)$. Thus, for any $\xi^1, \dots, \xi^\nu$,

$$\Big|\frac{1}{\nu}\sum_{i=1}^{\nu} f(\xi^i, x) - \mathbb{E}\big[f(\xi, x)\big]\Big| \leq \Big|\frac{1}{\nu}\sum_{i=1}^{\nu} f(\xi^i, x) - \frac{1}{\nu}\sum_{i=1}^{\nu} f(\xi^i, x^j)\Big|$$
$$+ \Big|\frac{1}{\nu}\sum_{i=1}^{\nu} f(\xi^i, x^j) - \mathbb{E}\big[f(\xi, x^j)\big]\Big| + \Big|\mathbb{E}\big[f(\xi, x^j)\big) - \mathbb{E}\big[f(\xi, x)\big]\Big|.$$

Two of the three parts in this error bound are further detailed by

$$\Big|\frac{1}{\nu}\sum_{i=1}^{\nu} f(\xi^i, x) - \frac{1}{\nu}\sum_{i=1}^{\nu} f(\xi^i, x^j)\Big| \leq \frac{1}{\nu}\sum_{i=1}^{\nu} \kappa(\xi^i)\varepsilon$$
$$\Big|\mathbb{E}\big[f(\xi, x^j)\big) - \mathbb{E}\big[f(\xi, x)\big]\Big| = \Big|\mathbb{E}\big[f(\xi, x^j) - f(\xi, x)\big]\Big|$$
$$\leq \mathbb{E}\Big[\big|f(\xi, x^j) - f(\xi, x)\big|\Big] \leq \bar{\kappa}\varepsilon,$$

where the derivation involving the expectation relies on 3.2. Collecting these facts, we obtain that with probability at least $p$,

$$\Big|\frac{1}{\nu}\sum_{i=1}^{\nu} f(\xi^i, x) - \mathbb{E}\big[f(\xi, x)\big]\Big| \leq \frac{1}{\nu}\sum_{i=1}^{\nu} \kappa(\xi^i)\varepsilon + \tau + \bar{\kappa}\varepsilon \leq (2\bar{\kappa} + \gamma)\varepsilon + \tau.$$

We've essentially established the claim, but one can fine-tune the values of $\varepsilon, \tau, \gamma$ such that the error bound matches $\delta$ and $p$ is as high as possible. We shy away from actually optimizing $p$ and instead make choices that result in relatively simple expressions. Let $x^0 \in X$, which can be selected arbitrarily, and set $\gamma = 1$, $\varepsilon$ according to (8.22) and $\tau = \delta - (2\bar{\kappa} + 1)\varepsilon$ so that the obtained error bound $(2\bar{\kappa} + \gamma)\varepsilon + \tau = \delta$. This choice is permissible because $\varepsilon, \tau > 0$. It only remains to compute a lower bound on $p$.

Let $\sigma^2 = \max_{j=0,\dots,r} \sigma^2(x^j)$. With these choices of $\varepsilon, \tau, \gamma$, one has

$$2r\exp\Big(-\frac{\nu\tau^2}{2\bar{\sigma}_\varepsilon^2}\Big) + \exp\Big(-\frac{\nu\gamma^2}{2\sigma_0^2}\Big) \leq 2r\exp\Big(-\frac{\nu\tau^2}{2\sigma^2}\Big) + \exp\Big(-\frac{\nu}{2\sigma_0^2}\Big)$$
$$\leq (2r+1)\exp\Big(-\frac{\nu\tau^2}{2\sigma^2}\Big)$$

because $\tau/\sigma \leq 1/\sigma_0$. To see this, note that if $\varepsilon = \delta/(4\bar{\kappa} + 2)$, then

$$\tau = \delta - (2\bar{\kappa} + 1)\varepsilon = \tfrac{1}{2}\delta \leq \frac{\sigma(x^0)}{\sigma_0} \leq \frac{\sigma}{\sigma_0}.$$

If $\varepsilon = (\delta - \sigma(x^0)/\sigma_0)/(2\bar{\kappa} + 1)$, then

$$\tau = \delta - (2\bar{\kappa} + 1)\varepsilon = \frac{\sigma(x^0)}{\sigma_0} \leq \frac{\sigma}{\sigma_0}.$$

A Euclidean ball of radius $\rho$ in $n$ dimensions can be covered by $(1 + 2\rho/\varepsilon)^n$ Euclidean balls of radius $\varepsilon$. Thus, $r$ is bounded by (8.23). □

The error bound in the theorem can be paired with several earlier results. It leads to a bound on the truncated Hausdorff distance between a sample average function and the corresponding expectation function; see 6.59. It can be used to address composite functions in 6.64; the proof of the theorem furnishes expressions for Lipschitz moduli.

While $1 - \beta e^{-\alpha\nu}$ tends to 1 exponentially fast, $\alpha$ can be near zero and $\beta$ can be sizeable making the probability for a small error painfully low even for large $\nu$. The detailed expressions for $\alpha$ and $\beta$ highlight the most influential quantities. Naturally, large $\delta$ as well as small $\bar{\kappa}$ and $\sup_{x,\bar{x}\in X} \|x - \bar{x}\|_2$ are helpful. The dependence on $n$, the dimension of $x$, is especially bothersome and shows that high-dimensional problems can be expected to have substantial sampling error.

## 8.I Diametrical Stochastic Optimization

The potentially substantial error in a sample average function prompts us to look for alternative problem formulations. In this section, we'll examine one approach based on certain sup-projections. We start by getting a better understanding of the nature of the error.

For random variables $\boldsymbol{\xi}_1, \ldots, \boldsymbol{\xi}_\nu$ with common mean $\mu \in \mathbb{R}$, the sample average $\nu^{-1}\sum_{i=1}^{\nu}\boldsymbol{\xi}_i$ is *unbiased*, i.e.,

$$\mathbb{E}\Big[\frac{1}{\nu}\sum\nolimits_{i=1}^{\nu}\boldsymbol{\xi}_i\Big] = \frac{1}{\nu}\sum\nolimits_{i=1}^{\nu}\mathbb{E}[\boldsymbol{\xi}_i] = \mu,$$

which follows by 3.2. Thus, a sample average estimates the mean "correctly on average" even with small $\nu$. In view of 8.1, the situation isn't as favorable for sample average *functions* and their minimum values. For $f : \Xi \times \mathbb{R}^n \to \mathbb{R}$, $C \subset \mathbb{R}^n$ and $\xi^1, \ldots, \xi^\nu \in \Xi$, the minimum value of

$$\varphi^\nu(x, \xi^1, \ldots, \xi^\nu) = \frac{1}{\nu}\sum\nolimits_{i=1}^{\nu} f(\xi^i, x) \tag{8.24}$$

over $x \in C$ is no greater than $\inf_C Ef$ on average by 8.1. The potential downward bias vanishes as $\nu \to \infty$ (as seen in the previous two sections), but could be significant for finite $\nu$. Table 3.4 illustrates the situation with minima of sample average functions being $-80.23$, $-63.21$ and $-54.04$ for sample sizes $\nu = 10, 100, 1000$, respectively, and the minimum of the expectation function being about $-50$. Consequently, it's often beneficial to replace the sample average function with some other approximation that lies *above* it. This would combat the downward bias and potentially improve the accuracy of the solutions obtained.

One possibility is to replace an expectation with a regular measure of risk, which due to its averseness property is more conservative. This is a viable and often productive approach, well motivated by duality theory as seen in the discussion near 8.34. However, let's discuss an alternative approach that's supported by the observation: If a sample average function is low in a neighborhood, then the expectation function tends to be low at least at some point in the neighborhood.

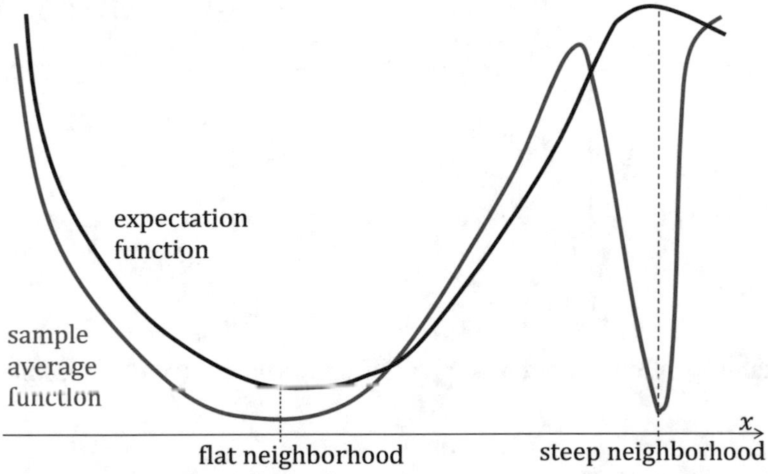


Fig. 8.6: If a sample average function $\varphi^\nu(\cdot, \xi^1, \dots, \xi^\nu)$ is low in a neighborhood, then the expectation function $Ef$ tends to be low at least at some point in the neighborhood.

Figure 8.6 shows a sample average function and the corresponding expectation function. For any fixed $x$, $\varphi^\nu(x, \xi^1, \dots, \xi^\nu)$ is on average equal to $Ef(x)$ due to the unbiased property discussed above, with overestimating and underestimating being roughly equally likely. However, across many $x$, there's likely to be one $x$ at which $\varphi^\nu(x, \xi^1, \dots, \xi^\nu)$ severely underestimates $Ef(x)$ and this tends to be a local minimizer of $\varphi^\nu(\,\cdot\,, \xi^1, \dots, \xi^\nu)$; see the right dip in Figure 8.6. In contrast, if a sample average function is low in a neighborhood, then it's unlikely that, at all the points in the neighborhood, the sample averages come out well below the actual expectations. Thus, the expectation function tends to be low at least at some point in the neighborhood as well; see the left dip in Figure 8.6. In summary, while minimizing a sample average function, there's a distinct possibility that an algorithm finds the "steep" neighborhood on the right in Figure 8.6, which would be unfortunate as any solution in that neighborhood is actually poor with respect to the expectation function. This is indeed the driving force of the downward bias in 8.1 and Table 3.4. We would like to design an algorithm that recognizes a steep neighborhood, avoids it and steers towards a "flat" neighborhood.

Another element enters as well. The Lipschitz modulus of $f(\xi, \cdot)$ emerges in 8.63 via $\bar{\kappa}$. If it's large, then $\beta$ in that theorem is large too. This may significantly increase the sample size required to achieve a certain error $\delta$. Even worse, if $f(\xi,\cdot)$ isn't Lipschitz continuous, then 8.63 doesn't apply.

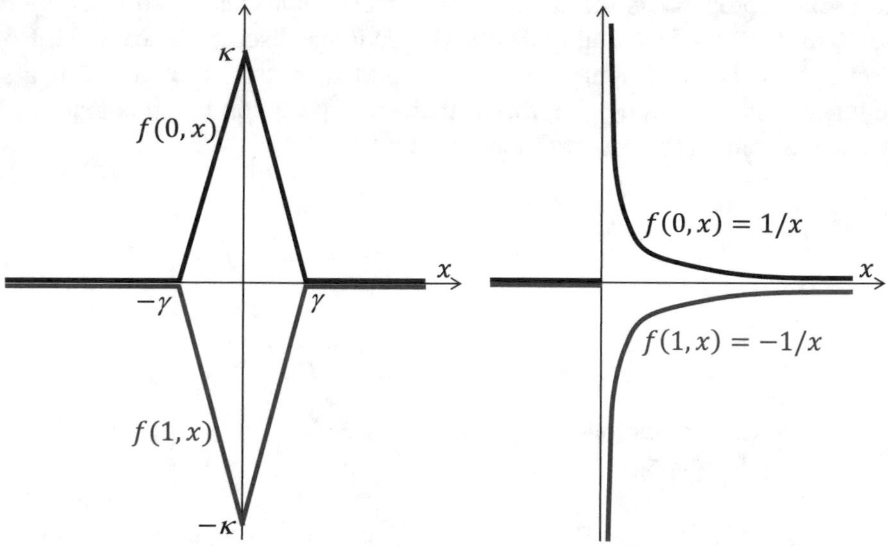


Fig. 8.7: Examples of downward bias with (left) and without (right) Lipschitz continuity.

**Example 8.64** (downward bias). For $\kappa \in (1,\infty)$ and $\gamma \in (0,1)$, let

$$f(\xi,x) = \begin{cases} \kappa x/\gamma + \kappa & \text{if } x \in [-\gamma,0),\ \xi = 0 \\ -\kappa x/\gamma - \kappa & \text{if } x \in [-\gamma,0),\ \xi = 1 \\ -\kappa x/\gamma + \kappa & \text{if } x \in [0,\gamma),\ \xi = 0 \\ \kappa x/\gamma - \kappa & \text{if } x \in [0,\gamma),\ \xi = 1 \\ 0 & \text{otherwise;} \end{cases}$$

see Figure 8.7(left). If $\boldsymbol{\xi}$ takes the values 0 and 1, each with probability 1/2, then $\mathbb{E}[f(\boldsymbol{\xi},x)] = 0$ for all $x \in \mathbb{R}$. In contrast,

$$\varphi^\nu(x,\xi^1,\dots,\xi^\nu) = \frac{1}{\nu}\sum_{i=1}^{\nu} f(\xi^i,x) = \begin{cases} \rho_\nu(\kappa x/\gamma + \kappa)/\nu & \text{if } x \in [-\gamma,0) \\ \rho_\nu(-\kappa x/\gamma + \kappa)/\nu & \text{if } x \in [0,\gamma) \\ 0 & \text{otherwise,} \end{cases}$$

where $\rho_\nu$ is the number of zeros minus the number of ones in the sample $\{\xi^1,\dots,\xi^\nu\}$. Minimizing $\varphi^\nu(\cdot,\xi^1,\dots,\xi^\nu)$ produces a result that's too low on average and the downward bias is proportional to $\kappa$.

**Detail.** Suppose that $\{\xi^1,\dots,\xi^\nu\}$ is obtained by statistically independent sampling according to the probability distribution of $\boldsymbol{\xi}$. Then, with probability nearly 1/2, $\rho_\nu < 0$. Under such circumstances, $x^\nu = 0$ minimizes $\varphi^\nu(\cdot,\xi^1,\dots,\xi^\nu)$ and

$$\varphi^\nu(x^\nu, \xi^1, \dots, \xi^\nu) = \rho_\nu \kappa/\nu.$$

Also with probability near 1/2, $\rho_\nu \ge 0$ and then $x^\nu = 1$ minimizes $\varphi^\nu(\cdot, \xi^1, \dots, \xi^\nu)$ with

$$\varphi^\nu(x^\nu, \xi^1, \dots, \xi^\nu) = \mathbb{E}\big[f(\boldsymbol{\xi}, x^\nu)\big] = 0.$$

Consequently, $\inf \varphi^\nu(\cdot, \xi^1, \dots, \xi^\nu)$ tends to be too low and

$$\mathbb{E}\big[f(\boldsymbol{\xi}, x^\nu)\big] - \inf \varphi^\nu(\cdot, \xi^1, \dots, \xi^\nu) \le \max\{0, -\rho_\nu\}\kappa/\nu, \tag{8.25}$$

with the right-hand side vanishing with increasing $\nu$. However, the error depends on $\kappa$, which is proportional to the Lipschitz modulus $\kappa/\gamma$ of $f(\xi, \cdot)$.

The situation deteriorates further when $f(\xi, \cdot)$ isn't Lipschitz continuous. Let

$$f(\xi, x) = \begin{cases} 1/x & \text{if } x \in (0, \infty),\ \xi = 0 \\ -1/x & \text{if } x \in (0, \infty),\ \xi = 1 \\ 0 & \text{otherwise;} \end{cases}$$

see Figure 8.7(right). With $\boldsymbol{\xi}$ and $\rho_\nu$ as above, $\mathbb{E}[f(\boldsymbol{\xi}, x)] = 0$ for all $x \in \mathbb{R}$ and

$$\varphi^\nu(x, \xi^1, \dots, \xi^\nu) = \begin{cases} \frac{1}{\nu}\rho_\nu/x & \text{if } x \in (0, \infty) \\ 0 & \text{otherwise.} \end{cases}$$

Then, $\inf \varphi^\nu(\cdot, \xi^1, \dots, \xi^\nu) = -\infty$ when $\rho_\nu < 0$, which takes place with probability nearly 1/2. The downward bias is now unbounded. □

As an alternative to minimizing the sample average function $\varphi^\nu(\cdot, \xi^1, \dots, \xi^\nu)$, we consider the sup-projection $\varphi^\nu_\gamma(\cdot, \xi^1, \dots, \xi^\nu) : \mathbb{R}^n \to \overline{\mathbb{R}}$ given by

$$\varphi^\nu_\gamma(x, \xi^1, \dots, \xi^\nu) = \sup_{\|y\| \le \gamma} \varphi^\nu(x + y, \xi^1, \dots, \xi^\nu), \tag{8.26}$$

where $\gamma \in (0, \infty)$, $\|\cdot\|$ is any norm on $\mathbb{R}^n$ and $\varphi^\nu$ is defined in (8.24). We refer to minimizing $\varphi^\nu_\gamma(\cdot, \xi^1, \dots, \xi^\nu)$ as *diametrical stochastic optimization* because the sample average function $\varphi^\nu(\cdot, \xi^1, \dots, \xi^\nu)$ is being "pulled" in diametrically opposite directions by the successive minimization and maximization. Trivially,

$$\varphi^\nu_\gamma(x, \xi^1, \dots, \xi^\nu) \ge \varphi^\nu(x, \xi^1, \dots, \xi^\nu) \quad \forall x \in \mathbb{R}^n$$

and we've achieved a "higher" approximation of $Ef$ than the one furnished by the sample average function $\varphi^\nu(\cdot, \xi^1, \dots, \xi^\nu)$.

Since $\varphi^\nu_\gamma(\bar{x}, \xi^1, \dots, \xi^\nu)$ is low only if $\varphi^\nu(x, \xi^1, \dots, \xi^\nu)$ is low for all points $x$ in a neighborhood of $\bar{x}$, diametrical stochastic optimization tends to avoid situations corresponding to the right dip in Figure 8.6 and instead obtains a significantly better solution more akin to the left dip.

**Example 8.65** (downward bias; cont.). In the two instances from 8.64, we obtain that

$$\mathbb{E}\big[f(\boldsymbol{\xi}, x^\nu_\gamma)\big] - \varphi^\nu_\gamma(x^\nu_\gamma, \xi^1, \ldots, \xi^\nu) = 0$$

regardless of the sample for the first instance, where $x^\nu_\gamma$ is a minimizer of $\varphi^\nu_\gamma(\,\cdot\,, \xi^1, \ldots, \xi^\nu)$. In the second instance,

$$\mathbb{E}\big[f(\boldsymbol{\xi}, x^\nu_\gamma)\big] - \varphi^\nu_\gamma(x^\nu_\gamma, \xi^1, \ldots, \xi^\nu) \leq \max\{0, -\rho_\nu\}\frac{1}{\nu(2\gamma + \varepsilon)}$$

for any $\varepsilon > 0$ and any sample. In either instance, the situation is significantly improved compared to when simply minimizing $\varphi^\nu(\,\cdot\,, \xi^1, \ldots, \xi^\nu)$.

**Proposition 8.66** (error in diametrical stochastic optimization). *For a nonempty compact set $X \subset \mathbb{R}^n$, a random vector $\boldsymbol{\xi}$ with support $\Xi$ and a function $f : \Xi \times \mathbb{R}^n \to \mathbb{R}$, with $-f$ being a locally inf-integrable random lsc function, suppose that $f(\xi, \cdot\,)$ is continuous for all $\xi \in \Xi$ and $f(\boldsymbol{\xi}, x) - \mathbb{E}[f(\boldsymbol{\xi}, x)]$ is subgaussian for all $x \in X$. For fixed $\nu \in \mathbb{N}$, let $\boldsymbol{\xi}^1, \ldots, \boldsymbol{\xi}^\nu$ be iid with the same probability distribution as $\boldsymbol{\xi}$.*

*Then, for any $\alpha \in (0, 1)$ and $\gamma \in (0, \infty)$, there's $\beta > 0$ (independent of $\nu$) such that*

$$\text{prob}\Big\{ \sup_{x\in X} \big\{Ef(x) - \varphi^\nu_\gamma(x, \boldsymbol{\xi}^1, \ldots, \boldsymbol{\xi}^\nu)\big\} \leq \beta\nu^{-1/2}\Big\} \geq 1 - \alpha,$$

*where $\varphi^\nu_\gamma$ is defined by (8.26).*

**Proof.** Let's define the ball

$$\mathbb{B}_*(x, \rho) = \big\{\bar{x} \in \mathbb{R}^n \;\big|\; \|x - \bar{x}\| \leq \rho\big\},$$

where $\|\cdot\|$ is the norm used to define $\varphi^\nu_\gamma$. Since $X$ is compact, there are $x^1, \ldots, x^r \in X$ such that

$$X \subset \bigcup\nolimits_{j=1}^r \mathbb{B}_*\big(x^j, \tfrac{1}{2}\gamma\big),$$

with $r \leq (\eta/\gamma)^n$ for some $\eta \in (0, \infty)$. Since $-f$ is a locally inf-integrable random lsc function, $E(-f)$ is lsc by 8.55. Then, $-Ef$ is lsc by 3.2 and $Ef$ is usc. These functions are also real-valued due to the subgaussian assumption. Let

$$\bar{x}^j \in \operatorname{argmax}_{x\in\mathbb{B}_*(x^j,\gamma/2)\cap X} Ef(x),$$

which exists because $X$ is compact and $Ef$ is real-valued and usc; cf. 4.9.

Since $f(\boldsymbol{\xi}, \bar{x}^j) - \mathbb{E}[f(\boldsymbol{\xi}, \bar{x}^j)]$ is subgaussian, say with proxy variance $\sigma_j^2$, we deduce from the maximal inequality 8.62 that, for any $\tau \geq 0$, one has

$$\text{prob}\bigg\{ \max_{j=1,\ldots,r} \Big\{\mathbb{E}\big[f(\boldsymbol{\xi}, \bar{x}^j)\big] - \frac{1}{\nu}\sum\nolimits_{i=1}^\nu f(\boldsymbol{\xi}^i, \bar{x}^j)\Big\} \geq \tau\bigg\} \leq r\exp\Big(-\frac{\nu\tau^2}{2\sigma^2}\Big),$$

where $\sigma = \max_{j=1,\ldots,r} \sigma_j$. Compared to 8.62, there's no "2" in front of exp on the right-hand side. This is permissible because the left-hand side is the probability of a one-sided event and not a two-sided event; cf. the proof of 8.62. Since $r \le (\eta/\gamma)^n$, $\alpha$ needs to satisfy

$$\left(\frac{\eta}{\gamma}\right)^n \exp\left(-\frac{\nu\tau^2}{2\sigma^2}\right) \le \alpha$$

or, equivalently,

$$\tau \ge \beta\nu^{-1/2} \text{ with } \beta = \sigma\sqrt{2n\ln(\eta/\gamma) - 2\ln\alpha}.$$

Consider an outcome $(\xi^1, \dots, \xi^\nu)$ for which

$$\max_{j=1,\dots,r}\left\{\mathbb{E}\big[f(\boldsymbol{\xi}, \bar{x}^j)\big] - \frac{1}{\nu}\sum\nolimits_{i=1}^{\nu} f(\xi^i, \bar{x}^j)\right\} \le \tau.$$

Such outcomes take place with probability at least $1 - \alpha$ when $\tau \ge \beta\nu^{-1/2}$. Let $\bar{x} \in X$. There's $j \in \{1, \dots, r\}$ such that $\bar{x} \in \mathbb{B}_*(x^j, \gamma/2)$. Since

$$\|\bar{x}^j - \bar{x}\| \le \|\bar{x}^j - x^j\| + \|x^j - \bar{x}\| \le \gamma,$$

one has

$$\varphi_\gamma^\nu(\bar{x}, \xi^1, \dots, \xi^\nu) \ge \frac{1}{\nu}\sum\nolimits_{i=1}^{\nu} f(\xi^i, \bar{x}^j)$$

and then

$$\begin{aligned}\mathbb{E}\big[f(\boldsymbol{\xi}, \bar{x})\big] - \varphi_\gamma^\nu(\bar{x}, \xi^1, \dots, \xi^\nu) &\le \mathbb{E}\big[f(\boldsymbol{\xi}, \bar{x})\big] - \frac{1}{\nu}\sum\nolimits_{i=1}^{\nu} f(\xi^i, \bar{x}^j)\\ \le \mathbb{E}\big[f(\boldsymbol{\xi}, \bar{x})\big] - \mathbb{E}\big[f(\boldsymbol{\xi}, \bar{x}^j)\big] + \tau &\le \mathbb{E}\big[f(\boldsymbol{\xi}, \bar{x}^j)\big] - \mathbb{E}\big[f(\boldsymbol{\xi}, \bar{x}^j)\big] + \tau = \tau,\end{aligned}$$

where the last inequality follows by the construction of $\bar{x}^j$ as the maximizer of $Ef$ over $\mathbb{B}_*(x^j, \gamma/2) \cap X$. We've reached the conclusion.

A final note on measurability: The random vector $\boldsymbol{\xi}$ defines a probability space $(\Xi, \mathcal{B}, \mathbb{P})$ and, under independent sampling, also a $\nu$-fold product probability space $(\Xi^\nu, \mathcal{B}^\nu, \mathbb{P}^\nu)$. From the facts about $f$, the function $\varphi^\nu : \mathbb{R}^n \times \Xi^\nu \to \mathbb{R}$ given by (8.24) has the properties that $-\varphi^\nu(x, \cdot)$ is measurable for all $x$ and $-\varphi^\nu(\cdot, \xi^1, \dots, \xi^\nu)$ is continuous for all $(\xi^1, \dots, \xi^\nu) \in \Xi^\nu$. Thus, by 8.51(d), $-\varphi^\nu$ is a random lsc function. Moreover, $\varphi_\gamma^\nu : \mathbb{R}^n \times \Xi^\nu \to \mathbb{R}$ given by (8.26) is real-valued by virtue of being the maximum value of a continuous function over a compact set; cf. 4.9. For all $x \in \mathbb{R}^n$, $\varphi_\gamma^\nu(x, \cdot)$ is measurable by 8.52. Since $\varphi_\gamma^\nu(\cdot, \xi^1, \dots, \xi^\nu)$ is continuous regardless of the sample by 6.29, we conclude that $\varphi_\gamma^\nu$ is a random lsc function as well by 8.51(d). This further implies that $\varphi_\gamma^\nu - Ef$ is a random lsc function because $Ef$ is usc; cf. 8.51(a,g). Again invoking 8.52, we see that the function

$$(\xi^1, \dots, \xi^\nu) \mapsto \sup_{x\in X}\left\{\mathbb{E}\big[f(\boldsymbol{\xi}, x)\big] - \varphi_\gamma^\nu(x, \xi^1, \dots, \xi^\nu)\right\}$$

is measurable. □

The constant $\beta$ in the proposition is given in the proof and depends on the largest proxy variance, denoted by $\sigma^2$, for $f(\boldsymbol{\xi}, x) - \mathbb{E}[f(\boldsymbol{\xi}, x)]$ at a finite number of different $x$. It also depends on the size of $X$ and thus on the norm used in the definition of $\varphi^{\nu}_{\gamma}$. Under the Euclidean norm $\|\cdot\|_2$, one can leverage the fact that a Euclidean ball of radius $\rho$ in $n$ dimensions can be covered by $(1 + 2\rho/\varepsilon)^n$ Euclidean balls of radius $\varepsilon$. Thus, we can set

$$\beta = \sigma\sqrt{2n\ln(1 + 2\delta/\gamma) - 2\ln\alpha}, \quad \text{where } \delta = \sup_{x,\bar{x}\in X} \|x - \bar{x}\|_2.$$

Since compensating for the downward bias in sample average functions becomes less critical for large sample sizes, one may want to tailor $\gamma$ to $\nu$. If $\gamma$ is proportional to $\nu^{-1}$, then $\beta$ grows at a logarithmic rate as $\nu$ increases and the rate of convergence deteriorates from $\nu^{-1/2}$ to $\sqrt{\nu^{-1}\ln\nu}$, which is insignificant.

In summary, we see that the proposition provides a guaranteed error for diametrical stochastic optimization even for cases where the underlying function $f(\xi, \cdot)$ isn't Lipschitz continuous. This addresses, in part, the concern about the error in 8.63 being dependent on the Lipschitz modulus.

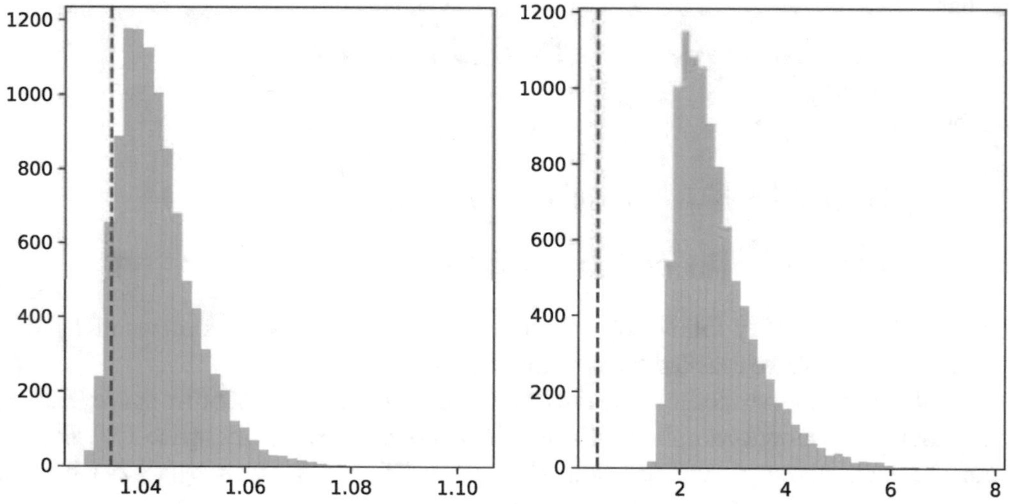


Fig. 8.8: When fitting a neural network in 8.67, one tends to obtain a flat neighborhood when using diametrical stochastic optimization (left) and a steep neighborhood when minimizing a sample average function (right).

**Example 8.67** (learning to recognize handwriting). The task of automatically recognizing handwritten numbers is presently accomplished by constructing a neural network that maps an image of a handwritten number to a label. The parameters in the neural network are fitted such that on a set of images with known labels (the training data), the mapping produces few errors; see 4.1. The fitting amounts to solving

$$\underset{x\in X}{\text{minimize}}\ \frac{1}{\nu}\sum\nolimits_{i=1}^{\nu} f(\xi^i, x), \tag{8.27}$$

where $x$ is the vector of parameters to be fitted, $\xi^1, \ldots, \xi^\nu$ represent a set of $\nu$ images and $f(\xi^i, x)$ quantifies how well the neural network with a parameter vector $x$ predicts the label of the $i$th image. We can think of $\xi^1, \ldots, \xi^\nu$ as a sample drawn according to the probability distribution of a random vector $\boldsymbol{\xi}$ that represents the "true" frequency with which various handwritten numbers occur in the setting where the system will be employed. Thus, the fitting problem is an approximation of the actual problem

$$\operatorname*{minimize}_{x \in X} \ \mathbb{E}\big[f(\boldsymbol{\xi}, x)\big].$$

We're in the setting discussed above and face the possibility that (8.27) produces a solution (a neural network) with low sample average value but high actual expected value; cf. the right dip in Figure 8.6. Diametrical stochastic optimization replaces (8.27) by

$$\operatorname*{minimize}_{x \in X} \ \sup_{\|y\| \le \gamma} \frac{1}{\nu} \sum\nolimits_{i=1}^{\nu} f(\xi^i, x + y) \tag{8.28}$$

and thereby promotes solutions of the kind illustrated by the left dip in Figure 8.6.

**Detail.** The analysis of diametrical stochastic optimization relies on properties of sup-projections (§6.E) and computations can, at least in principle, be carried out using the outer approximation algorithm of §6.C; see [72] for details on using the stochastic gradient descent method from §3.G.

For a class of neural networks characterized by 420000 parameters and the goal of predicting the numbers 0, 1 and 2 using the MNIST dataset with 50% of the labels corrupted, we obtain by solving (8.28) a parameter vector $\bar{x}$ with $\nu^{-1}\sum_{i=1}^{\nu} f(\xi^i, \bar{x}) \approx 1.035$. This value is compared with $\nu^{-1}\sum_{i=1}^{\nu} f(\xi^i, x)$ at 10000 randomly selected $x$ in a neighborhood of $\bar{x}$. The resulting histogram of values is given in Figure 8.8(left), where 1.035 is marked with a dashed line. We see that the values in the neighborhood are also low, typically in the range 1.03 to 1.06. We've identified a "flat neighborhood" of the kind corresponding to the left dip in Figure 8.6. In contrast, a solution of (8.27) results in a parameter vector $\hat{x}$ with $\nu^{-1}\sum_{i=1}^{\nu} f(\xi^i, \hat{x}) \approx 0.45$, which at first might appear better. However, the values in the neighborhood of $\hat{x}$ are quite high, in the range of 1 to 6; see Figure 8.8(right). This means that $\hat{x}$ resembles more the right dip in Figure 8.6. When tested on handwritten numbers not used in the fitting, the neural network with parameter vector $\bar{x}$ makes a mistake only 5% of the time while the one with $\hat{x}$ exhibits 45% errors; see [72]. □

# Chapter 9
# GAMES AND MINSUP PROBLEMS

Multiple agents, each minimizing its own objective function, might be linked through common resources or prices in a market place and this leads to games that can be formulated as generalized equations and then also as optimization problems using the techniques in §7.D. Often the resulting problem is that of minimizing a sup-projection, which we refer to as a minsup problem. This remarkably versatile class of problems is already supported by our knowledge from 6.9, §6.E and 6.60, but further developments are needed to address situations such as those arising when agents interact through their constraints. In this chapter, we'll formulate models of games and minsup problems, consider their approximations and develop computational methods.

## 9.A Nash Games

In a situation with two or more agents, models of their behavior and decision-making usually need to account for the interaction among the agents. The agents could be various firms competing in a market place, countries negotiating international agreements, two sides in an armed conflict and autonomous systems trying to accomplish a task. An agent's decision results in a cost or penalty, but the exact amount depends on the decisions of the others as well. Our goal might be to help one of the agents make a good decision or to understand the overall interaction for the purpose of designing a market place that achieves certain objectives.

In real life, decision-making might be incremental, intertwined with data collection and taking place over multiple stages. Still, even simplified models of a complicated decision process can be useful and provide insight about "best" decisions. We consider only *single-stage games*: each agent makes a decision, all decisions are revealed simultaneously (so no one has the advantage of knowing the decisions of the others while making its own) and the game ends with each agent having to pay a cost that depends on everybody's decisions. A version of this game is the penny-nickel-dime game in 5.39 with two agents, but now we consider $m$ agents. The vector $x_i \in \mathbb{R}^{n_i}$ and function $\varphi_i : \mathbb{R}^{n_i} \times \mathbb{R}^{n-n_i} \to \overline{\mathbb{R}}$

 
J. O. Royset and R. J-B Wets, *An Optimization Primer*, Springer Series in Operations Research and Financial Engineering, https://doi.org/10.1007/978-3-030-76275-9_9

represent the decision and objective function for agent $i$, where $n = \sum_{i=1}^{m} n_i$. Let

$$x_{-i} = (x_1, x_2, \ldots, x_{i-1}, x_{i+1}, \ldots, x_m)$$

be a vector specifying decisions for all the agents except the $i$th one. Thus, $\varphi_i(x_i, x_{-i})$ is the cost for agent $i$ under decision $x_i$ when the other agents have selected $x_{-i}$. This leads to a collection of minimization problems that we refer to as a *Nash game*[1]:

$$\Big\{ \underset{x_i \in \mathbb{R}^{n_i}}{\text{minimize}}\ \varphi_i(x_i, x_{-i}), \quad i = 1, \ldots, m \Big\}$$

The difficulty here is that we can't simply solve one agent's problem because it depends on the other agents' decisions. In fact, it isn't entirely clear what should be considered a solution of such a collection of interdependent optimization problems. An agent might do very well if the others are accommodating, but that isn't realistic if they also optimize their own decisions. We say that $\bar{x} = (\bar{x}_1, \ldots, \bar{x}_m)$ is a *Nash equilibrium* of the game if

$$\bar{x}_i \in \operatorname{argmin}_{x_i \in \mathbb{R}^{n_i}} \varphi_i(x_i, \bar{x}_{-i}), \quad i = 1, \ldots, m$$

This is a natural definition of a *solution* of a Nash game because it identifies a collection of decisions, one for each agent, with the property that no agent can do better by *unilaterally* changing its decision. It coincides with our definition of a solution of a two-player game as a saddle point in §5.E. The task of determining a Nash equilibrium is an example of a *variational problem*.

Of course, a Nash game is an abstraction. In reality, the agents may not be as perfectly focused on optimizing known objective functions and are also driven by other, "irrational" concerns. Still, it remains a useful modeling tool and can provide important insight.

**Example 9.1** (simple Nash game). Let's consider two agents, each deciding a scalar quantity. The problem for agent 1 is

$$\underset{x_1 \in \mathbb{R}}{\text{minimize}}\ \varphi_1(x_1, x_2) = \begin{cases} (x_1 - 1)^2 & \text{if } x_1 + x_2 \le 1 \\ \infty & \text{otherwise} \end{cases}$$

and for agent 2 it's

$$\underset{x_2 \in \mathbb{R}}{\text{minimize}}\ \varphi_2(x_2, x_1) = \begin{cases} (x_2 - \frac{1}{2})^2 & \text{if } x_1 + x_2 \le 1 \\ \infty & \text{otherwise.} \end{cases}$$

For any $\alpha \in [1/2, 1]$, $(\alpha, 1 - \alpha)$ is a Nash equilibrium.

[1] The seminal work by J. Nash was limited to the case $\varphi_i(x_i, x_{-i}) = \varphi_i^0(x_i, x_{-i}) + \iota_{X_i}(x_i)$ for some real-valued $\varphi_i^0$ and $X_i \subset \mathbb{R}^{n_i}$, i.e., the feasible set of one agent isn't influenced by the others. The extension to extended real-valued functions is sometimes called generalized Nash games.

**Detail.** For agent 1, the (unique) minimizing decision is $\bar{x}_1 = 1$ if agent 2 chooses $x_2 \leq 0$, but $\bar{x}_1 = 1 - x_2$ if $x_2 > 0$. For agent 2, the (unique) minimizing decision is $\bar{x}_2 = 1/2$ if agent 1 selects $x_1 \leq 1/2$ and $\bar{x}_2 = 1 - x_1$ if $x_1 > 1/2$. Even though each agent obtains a unique minimizer, there are multiple Nash equilibria. □

In the example, we could identify solutions by inspection, but generally some computational approach needs to be brought in. It's then useful to reformulate the problem. We can view a Nash equilibrium as the solution of the generalized equation $0 \in S(\bar{x})$, where

$$S(\bar{x}) = \prod\nolimits_{i=1}^{m} \big( -\bar{x}_i + \operatorname{argmin}_{x_i \in \mathbb{R}^{n_i}} \varphi_i(x_i, \bar{x}_{-i}) \big)$$

and $\bar{x} = (\bar{x}_1, \ldots, \bar{x}_m)$. If $\varphi_i(\cdot, x_{-i})$ is convex for each $x_{-i} \in \mathbb{R}^{n-n_i}$, then the optimality condition 2.19 ensures that

$$\forall i = 1, \ldots, m: \qquad \bar{x}_i \in \operatorname{argmin}_{x_i \in \mathbb{R}^{n_i}} \varphi_i(x_i, \bar{x}_{-i}) \iff 0 \in \partial_i \varphi_i(\bar{x}_i, \bar{x}_{-i}),$$

where $\partial_i$ indicates that the subgradients are computed with respect to the variables of agent $i$. Further details emerge if

$$\varphi_i(x_i, x_{-i}) = \varphi_i^0(x_i, x_{-i}) + \iota_{C_i(x_{-i})}(x_i)$$

for smooth $\varphi_i^0(\cdot, x_{-i}) : \mathbb{R}^{n_i} \to \mathbb{R}$ and convex $C_i(x_{-i}) \subset \mathbb{R}^{n_i}$. Then, by 2.23 and 4.58, the above equivalence is extended to

$$\begin{aligned} \forall i = 1, \ldots, m: \quad & 0 \in \nabla_i \varphi_i^0(\bar{x}_i, \bar{x}_{-i}) + N_{C_i(\bar{x}_{-i})}(\bar{x}_i) \\ & \iff \bar{x}_i \in C_i(\bar{x}_{-i}), \quad \big\langle \nabla_i \varphi_i^0(\bar{x}_i, \bar{x}_{-i}), x_i - \bar{x}_i \big\rangle \geq 0 \quad \forall x_i \in C_i(\bar{x}_{-i}). \end{aligned}$$

Let's now stack all these gradients in a big vector denoted by $F(\bar{x})$ so that

$$F(\bar{x}) = \big( \nabla_1 \varphi_1^0(\bar{x}_1, \bar{x}_{-1}), \ldots, \nabla_m \varphi_m^0(\bar{x}_m, \bar{x}_{-m}) \big).$$

If there's a convex set $X \subset \mathbb{R}^n$ such that

$$x = (x_1, \ldots, x_m) \in X \iff x_i \in C_i(x_{-i}), \; i = 1, \ldots, m, \tag{9.3}$$

then the optimality conditions can be written as the variational inequality

$$\bar{x} \in X \text{ and } \big\langle F(\bar{x}), x - \bar{x} \big\rangle \geq 0 \quad \forall x \in X.$$

Consequently, under these assumptions, a Nash game can be solved by algorithms for variational inequalities; see Chap. 7.

The condition (9.3) is satisfied if $C_i(x_{-i}) = X_i \subset \mathbb{R}^{n_i}$, i.e., the agents don't interact through their constraints. Then, $X = X_1 \times \cdots \times X_m$. In 9.1, $X$ can be taken to be $\{(x_1, x_2) \in \mathbb{R}^2 \mid x_1 + x_2 \leq 1\}$ and the Nash equilibria are characterized by $(\bar{x}_1, \bar{x}_2) \in X$ and

$$(2\bar{x}_1 - 2)(x_1 - \bar{x}_1) + (2\bar{x}_2 - 1)(x_2 - \bar{x}_2) \geq 0 \quad \forall (x_1, x_2) \in X.$$

We've reduced the game in 9.1 to a variational inequality.

**Example 9.2** (international pollution control). The 1997 Kyoto agreement was an international accord for reducing global emissions of $CO_2$ and other greenhouse gases. One mechanism agreed upon was that countries could offset emissions by investing at home and abroad in pollution reducing projects. The hope was that this would channel resources to developing countries and thus promote their sustainable economic growth. We can study the merit of such a mechanism using a Nash game.

**Detail.** Let $m$ be the number of countries (agents) involved. For country $i$, let $e_i$ be its total emission in a future year, which we determine by solving a Nash game. Suppose that there's a given function $h_i : \mathbb{R} \to \mathbb{R}$ that specifies the country's revenue as a function of emission, i.e., $h_i(e_i)$ is the revenue of country $i$ under emission level $e_i$. The emission needs to stay below an agreed upon target $\tau_i$, but can be abated by investing in various projects at home and abroad. Let $w_{ij}$ be the level of such investments by country $i$ in country $j$. The benefit of this investment is an emission credit $\gamma_{ij} w_{ij}$, where $\gamma_{ij}$ is an agreed upon nonnegative coefficient. Thus, country $i$ is held accountable for its emission level minus its credits:

$$e_i - \sum\nolimits_{j=1}^{m} \gamma_{ij} w_{ij},$$

which can't exceed $\tau_i$. At the same time, no country should get credit from investing in country $i$ unless there's actual emission to offset, i.e.,

$$e_i - \sum\nolimits_{j=1}^{m} \gamma_{ji} w_{ji}$$

must be nonnegative. We also assign a penalty to country $i$ that reflects the detrimental effect of global emissions by means of a function $f_i : \mathbb{R}^m \to \mathbb{R}$. The problem for the $i$th country is then to minimize its penalty plus investment cost minus revenue:

$$\begin{aligned}
&\underset{e_i, w_{i1}, \ldots, w_{im}}{\text{minimize}} \; f_i\Big(e_1 - \sum\nolimits_{j=1}^{m} \gamma_{j1} w_{j1}, \ldots, e_m - \sum\nolimits_{j=1}^{m} \gamma_{jm} w_{jm}\Big) + \sum\nolimits_{j=1}^{m} w_{ij} - h_i(e_i) \\
&\text{subject to } \; e_i - \sum\nolimits_{j=1}^{m} \gamma_{ij} w_{ij} \leq \tau_i, \quad e_i - \sum\nolimits_{j=1}^{m} \gamma_{ji} w_{ji} \geq 0, \quad e_i, w_{i1}, \ldots, w_{im} \geq 0.
\end{aligned}$$

The functions $f_i$ and $h_i$ are presumably increasing so higher emission results in higher penalties and higher revenues. Penalties can be reduced by investments, but they in turn add to the objective function. Country $i$ would like to have no investments and high emissions that are all offset by *other* countries. Of course, this would not be ideal for the other countries! A Nash equilibrium specifies actions for all the countries so that no country would benefit from unilaterally changing their emissions and investments; see [19, 34] for further details. □

**Example 9.3** (power allocation in telecommunication). A Nash game may not always involve actual "agents" that make their own decisions, but could simply be the result of modeling a multi-component system. Suppose that we want to design a system such that the various components are "in balance." For instance, in a telecommunication network wires are usually bundled together but this may cause the signal in one wire to interfere with those in the other wires. Further complication is caused by the common practice of dividing the available frequency band in each wire into a set of parallel subcarriers. The goal then becomes to appropriately control the power associated with each subcarrier and each wire such that the quality of the transmission is acceptable. In this case, we can view each wire as an "agent" that interacts with the other wires.

**Detail.** Suppose that there are $m$ wires, each with $n$ subcarriers. For wire $i$ and subcarrier $j$, we would like to select the power level $x_{ij}$ such that the achievable transmission rate is sufficiently high. For wire $i$, this rate is given by

$$f_i(x_i, x_{-i}) = \sum\nolimits_{j=1}^{n} \ln\Bigl(1 + \frac{\alpha_{ij}x_{ij}}{\beta_{ij} + \sum_{k\neq i}\gamma_{ijk}x_{kj}}\Bigr),$$

where $x_i = (x_{i1}, \dots, x_{in}) \in \mathbb{R}^n$, $x_{-i}$ is the corresponding vector of power levels for the other wires and $\alpha_{ij}, \beta_{ij}, \gamma_{ijk}$ are coefficients. With this rate constrained from below by $\tau_i$ and a goal of minimizing power levels, we obtain the following model for wire $i$:

$$\operatorname*{minimize}_{x_i\in\mathbb{R}^n} \sum\nolimits_{j=1}^{n} x_{ij} \ \text{ subject to } \ f_i(x_i, x_{-i}) \geq \tau_i, \ \ x_{ij} \geq 0, \ \ j = 1, \dots, n.$$

A designer of a telecommunication system may seek to determine a Nash equilibrium for the game consisting of these models and thereby identify suitable power levels; see [75] for extensive developments. □

## 9.B Formulation as Minsup Problems

Nash games and many other problems can be formulated in terms of sup-projections. In fact, minimization of sup-projections emerges as a focal point for a wide variety of variational problems and this enables us to greatly expand the reach of optimization technology. For $X \subset \mathbb{R}^n$, $g : \mathbb{R}^n \times \mathbb{R}^m \to \mathbb{R}$ and $Y : \mathbb{R}^n \rightrightarrows \mathbb{R}^m$, let's consider the *minsup problem*

$$\boxed{\operatorname*{minimize}_{x\in X} \ \sup\nolimits_{y\in Y(x)} g(x, y)}$$

Problems of this form arise in duality theory with a Lagrangian furnishing $g$ (§5.C), dual representations of risk measures (§8.D) and reformulations of generalized equations (§7.D). In contrast to earlier developments (see, in particular, §6.E), we now permit the set over which the maximization is taking place to depend on the outer variable $x$. This section gives three additional examples of minsup problems.

**Example 9.4** (investment planning). Let's recall the setting in 2.57, where $x_j$ is the dollar amount you invest in stock $j$ and $\xi_j$ is the yield of that stock between now and the time of your retirement. With $\xi = (\xi_1, \dots, \xi_n)$ and $x = (x_1, \dots, x_n)$, $\langle \xi, x\rangle$ becomes the total value of your investment portfolio at the time of your retirement, which we would like to maximize or, equivalently, minimize $\langle -\xi, x\rangle$. As an alternative to the model in 2.57, suppose that we account for the uncertainty in $\xi$ using the worst-case measure of risk; see 8.5(b). Specifically, we seek to minimize $\sup_{\xi\in\Xi}\langle -\xi, x\rangle$ over some feasible set, where $\Xi \subset \mathbb{R}^n$ captures the yield values against which we would like to plan. We've reached a minsup problem. The set $\Xi$ may also depend on $x$ to better reflect a decision-maker's preferences.

**Detail.** Suppose that $\xi_j$ has a nominal value $\mu_j$, with a parameter $\bar{\sigma}_j$ quantifying the associated level of uncertainty. For example, $\mu_j$ and $\bar{\sigma}_j$ could be the mean and standard deviation of the yield of the $j$th stock during some earlier time period. This leads to

$$\Xi = \prod\nolimits_{j=1}^{n} [\mu_j - \bar{\sigma}_j,\ \mu_j + \bar{\sigma}_j]. \tag{9.5}$$

In a slightly more sophisticated approach, we may temporarily think of the yield as a normal random variable with mean $\mu_j$ and standard deviation $\sigma_j$. Then, the probability that this random variable takes a value in the interval

$$\big[\mu_j - Q(1-\alpha/2)\sigma_j,\ \mu_j + Q(1-\alpha/2)\sigma_j\big]$$

is $1-\alpha$, where $Q(\beta)$ is the $\beta$-quantile of the standard normal distribution function. Fixing $\alpha$ sufficiently small (for example, $\alpha = 0.05$), the interval can be considered the range of values for $\xi_j$ over which we would like to consider the worst case and we again settle on (9.5), but now with $\bar{\sigma}_j = Q(1-\alpha/2)\sigma_j$. Thus,

$$\xi \in \Xi \iff \Big(\frac{\xi_j - \mu_j}{\sigma_j}\Big)^2 \le \big(Q(1-\alpha/2)\big)^2, \quad j = 1,\dots,n.$$

Instead of accounting for each $j$ individually in this manner, we can aggregate across all $j$ and define

$$\Xi = \Big\{\xi \in \mathbb{R}^n \;\Big|\; \sum\nolimits_{j=1}^{n}\Big(\frac{\xi_j - \mu_j}{\sigma_j}\Big)^2 \le \gamma\Big\},$$

where $\gamma = n(Q(1-\alpha/2))^2$ or some other positive number that reflects the level of conservativeness that's needed. This form of $\Xi$ has a convenient geometric interpretation: $\Xi$ is given by an ellipse centered at the nominal values $(\mu_1,\dots,\mu_n)$ with halfaxes specified by $\sigma_j$ and $\gamma$.

Suppose that $x_j$ represents the *fraction* of money invested in stock $j$. Then, $x_j = 1/n$ for all $j$ is an equal allocation strategy often considered as a benchmark, which one might deviate from only reluctantly. We may decide to increase $\gamma$ for decisions that deviate from the equal allocation strategy. For example, one could replace $\gamma$ by

$$n\big(Q(1-\alpha/2)\big)^2\Big(1 + \sum\nolimits_{j=1}^{n}(x_j - 1/n)^2\Big),$$

which makes $\Xi$ depend on $x$. □

**Example 9.5** (game as minsup problem). For $\varphi_i : \mathbb{R}^{n_i} \times \mathbb{R}^{n-n_i} \to \mathbb{R}$, $C_i(x_{-i}) \subset \mathbb{R}^{n_i}$, $i = 1, \dots, m$, and $n = \sum_{i=1}^m n_i$, the Nash game

$$\Big\{ \operatorname*{minimize}_{x_i \in \mathbb{R}^{n_i}} \varphi_i(x_i, x_{-i}) + \iota_{C_i(x_{-i})}(x_i), \quad i = 1, \dots, m \Big\}$$

can be formulated as a minsup problem. Specifically, $\bar{x} = (\bar{x}_1, \dots, \bar{x}_m) \in \mathbb{R}^n$ is a Nash equilibrium of the game, i.e.,

$$\bar{x}_i \in \operatorname{argmin}_{x_i \in C_i(\bar{x}_{-i})} \varphi_i(x_i, \bar{x}_{-i}), \quad i = 1, \dots, m,$$

if and only if

$$\bar{x} \in \operatorname{argmin}_{x \in X} \sup_{y \in Y(x)} g(x, y) \quad \text{and} \quad \sup_{y \in Y(\bar{x})} g(\bar{x}, y) \le 0, \tag{9.6}$$

where

$$X = \big\{ x \in \mathbb{R}^n \;\big|\; x_i \in C_i(x_{-i}),\ i = 1, \dots, m \big\}, \quad Y(x) = \prod_{i=1}^m C_i(x_{-i})$$

and, for $x = (x_1, \dots, x_m) \in X$ and $y = (y_1, \dots, y_m) \in Y(x)$, one has

$$g(x, y) = \sum_{i=1}^m \big( \varphi_i(x_i, x_{-i}) - \varphi_i(y_i, x_{-i}) \big).$$

The function expressed as $g(x, y)$ for $x \in X$ and $y \in Y(x)$ is referred to as the *Nikaido-Isoda bifunction*. Consequently, if we minimize $\sup_{y \in Y(x)} g(x, y)$ over $x \in X$ and obtain a minimum value of at most zero, then the resulting minimizer is a Nash equilibrium.

**Detail.** While $Y(x)$ generally depends on the value of $x$, the special case with $C_i(x_{-i}) = X_i \subset \mathbb{R}^{n_i}$ results in $Y(x) = X_1 \times \cdots \times X_m$ for all $x$. Thus, if the agents don't interact through their constraints, then the resulting minsup problem involves a sup-projection of the kind analyzed in §6.E.

The following argument establishes the asserted characterization of Nash equilibria. If $\bar{x}$ is a Nash equilibrium, then $\bar{x} \in X$ and

$$\varphi_i(\bar{x}_i, \bar{x}_{-i}) \le \varphi_i(y_i, \bar{x}_{-i}) \ \ \forall y_i \in C_i(\bar{x}_{-i}), \quad i = 1, \dots, m.$$

Thus, $g(\bar{x}, y) \le 0$ for all $y \in Y(\bar{x})$. Given any $x \in X$, one has $x \in Y(x)$ and therefore

$$\sup_{y \in Y(x)} g(x, y) \ge 0.$$

In particular,

$$\sup_{y \in Y(\bar{x})} g(\bar{x}, y) = g(\bar{x}, \bar{x}) = 0.$$

This implies that

$$\bar{x} \in \operatorname{argmin}_{x \in X} \sup_{y \in Y(x)} g(x, y).$$

For the converse, suppose that $\bar{x}$ satisfies (9.6). Then,

$$0 \geq \sup_{y \in Y(\bar{x})} g(\bar{x}, y) = \sum_{i=1}^{m} \Big( \varphi_i(\bar{x}_i, \bar{x}_{-i}) - \inf_{y_i \in C_i(\bar{x}_{-i})} \varphi_i(y_i, \bar{x}_{-i}) \Big).$$

The upper bound of zero and the fact that $\bar{x}_i \in C_i(\bar{x}_{-i})$ for all $i$ imply that each term in the sum must be zero and the asserted equivalence follows. □

**Example 9.6** (Walras barter model as minsup problem). Let's return to the Walras barter model in 7.5, where the holdings of $m$ agents as well as the price of $n$ goods are determined by an optimization problem for each agent and an equilibrium condition that links the agent problems. However, we generalize the problem by allowing the objective function $f_i : \mathbb{R}^n \to \overline{\mathbb{R}}$ of agent $i$ to be nonsmooth and nonconvex. For a price vector $p \in [0, \infty)^n$, agent $i$ determines a holding

$$x_i(p) \in \operatorname{argmin}_{x_i \in C_i(p)} f_i(x_i), \quad \text{with } C_i(p) = \big\{ x_i \in [0, \infty)^n \;\big|\; \langle p, x_i \rangle \leq \langle p, w_i \rangle \big\},$$

by leveraging an initial endowment $w_i \in \mathbb{R}^n$ of the goods. We assume that this set of minimizers consists of a single point for any $p \in [0, \infty)^n$, which means in particular that $\operatorname{dom} f_i \cap [0, \infty)^n \neq \emptyset$. The uniqueness assumption makes sure that there's no ambiguity regarding the agent's decision. For $p \in [0, \infty)^n$, we also define

$$G(p) = \sum_{i-1}^{m} \big( x_i(p) - w_i \big),$$

which quantifies the excess holdings relative to the initial endowments under price vector $p$. In this notation, but consistent with 7.5, $\bar{p}$ is an equilibrium price vector if

$$\bar{p} \geq 0, \quad G(\bar{p}) \leq 0, \quad \big\langle G(\bar{p}), \bar{p} \big\rangle = 0. \tag{9.7}$$

The problem of determining an equilibrium price can be reformulated as that of solving a minsup problem involving the *Walrasian bifunction*

$$g(p, q) = \big\langle G(p), q \big\rangle, \quad \text{with } (p, q) \in [0, \infty)^n \times [0, \infty)^n.$$

**Detail.** If $\bar{p}$ is an equilibrium price vector, i.e., satisfies (9.7), then

$$\bar{p} \in \operatorname{argmin}_{p \in [0,\infty)^n} \sup_{q \in [0,\infty)^n} g(p, q) \quad \text{and} \quad \sup_{q \in [0,\infty)^n} g(\bar{p}, q) \leq 0. \tag{9.8}$$

To see this, simply note that for any $p \in \mathbb{R}^n$,

$$\sup_{q \in [0,\infty)^n} g(p, q) \geq g(p, 0) = 0.$$

Thus, $\bar{p}$ satisfies (9.8) if

$$\sup_{q \in [0,\infty)^n} g(\bar{p}, q) = 0,$$

which is the case because $G(\bar{p}) \leq 0$ and the maximization is over $q \in [0, \infty)^n$.

For the converse, suppose that $\bar{p}$ satisfies (9.8). Thus, $g(\bar{p}, q) \leq 0$ for all $q \in [0, \infty)^n$. In particular, with $q = (1, 0, \ldots, 0)$, this implies that

$$\sum_{i=1}^{m}\big(x_{i1}(\bar{p})-w_{i1}\big)\le 0.$$

We can repeat this argument for $q=(0,1,0,\dots,0)$, etc., and conclude that $G(\bar{p})\le 0$. However, $\bar{p}$ may not satisfy $\langle G(\bar{p}),\bar{p}\rangle = 0$.

We say that a nonnegative price vector $\bar{p}$ satisfying $G(\bar{p})\le 0$ is a *weak equilibrium price vector*. Thus, a price vector $\bar{p}$ satisfying (9.8) is only guaranteed to be a weak equilibrium price vector. The discrepancy with (9.7) is minor as our primary concern might be to find a price vector that ensures holdings don't exceed endowments. This is exactly the condition $G(\bar{p})\le 0$. The additional requirement $\langle G(\bar{p}),\bar{p}\rangle = 0$ asserts that if the holdings of a good are less than the endowments, then the price of that good must be zero. While this is reasonable as the value of an endowment of such an unpopular good is questionable, one might ignore this possibility in an initial analysis.

Under the additional assumption that $f_i$ is decreasing in all arguments[2], $\langle G(\bar{p}),\bar{p}\rangle = 0$ as well. To establish this claim, let's assume for the sake of contradiction that $\langle G(\bar{p}),\bar{p}\rangle < 0$, i.e.,

$$\sum_{j=1}^{n}\sum_{i=1}^{m}\big(x_{ij}(\bar{p})-w_{ij}\big)\bar{p}_j=\sum_{i=1}^{m}\sum_{j=1}^{n}\big(x_{ij}(\bar{p})-w_{ij}\big)\bar{p}_j<0.$$

This means that there's a good $j'$ such that

$$\sum_{i=1}^{m}\big(x_{ij'}(\bar{p})-w_{ij'}\big)\bar{p}_{j'}<0$$

and then also

$$\sum_{i=1}^{m}\big(x_{ij'}(\bar{p})-w_{ij'}\big)<0,$$

i.e., some of good $j'$ remains unclaimed. Moreover, there's an agent $i'$ with

$$\sum_{j=1}^{n}\big(x_{i'j}(\bar{p})-w_{i'j}\big)\bar{p}_j<0,$$

which means that its constraint

$$\langle\bar{p},x_{i'}\rangle\le\langle\bar{p},w_{i'}\rangle$$

isn't active. Since $f_{i'}$ is decreasing in all arguments, the agent can then improve its objective function value by increasing its holding in $j'$. This contradicts the optimality of $x_{i'}(\bar{p})$. □

## 9.C Bifunctions and Solutions

We'll now formalize the treatment of minsup problems and examine conditions under which we can guarantee that they've solutions. Immediately, we face one minor issue: The roles of $-\infty$ and $\infty$ are more delicate. A minimization problem on $\mathbb{R}^n$ can be identified with

[2] A function $f:\mathbb{R}^n\to\overline{\mathbb{R}}$ is decreasing in all arguments when $f(x)>f(\bar{x})$ for $x=(x_1,\dots,x_n)$ and $\bar{x}=(x_1,\dots,x_{j-1},\bar{x}_j,x_{j+1},\dots,x_n)$ with $x_j<\bar{x}_j$.

a function $f : \mathbb{R}^n \to \overline{\mathbb{R}}$ whose value $f(x) = \infty$ signifies that $x$ is infeasible. This allows us to simplify many expressions and reduces the need for checking finiteness, at least initially. Minsup problems are expressed by a mapping with two inputs, one associated with minimization and one with maximization. Since a value of infinity is dramatically different if viewed from the minimization side than from the maximization side, we now need to be more explicit about finiteness.

To eliminate ambiguity and other complications, we express a minsup problem by a mapping $g$ that on a nonempty subset of $\mathbb{R}^n \times \mathbb{R}^m$ returns a *finite value*. This subset is the *domain* of $g$ denoted by $\operatorname{dom} g$. Outside its domain, $g$ isn't defined. We refer to such mappings as *bifunctions*. (In our usual setting, a function $f : \mathbb{R}^n \to \overline{\mathbb{R}}$ is defined everywhere and may have $f(x) = -\infty$ for $x \in \operatorname{dom} f$, but the situation is different for bifunctions.) The family of all these bifunctions is denoted by

$$\operatorname{bfcns}(\mathbb{R}^n, \mathbb{R}^m) = \big\{g : \operatorname{dom} g \to \mathbb{R} \;\big|\; \emptyset \neq \operatorname{dom} g \subset \mathbb{R}^n \times \mathbb{R}^m\big\}.$$

The first input to a bifunction takes a primary role in our development leading us to the following description of the domain of a bifunction. We associate with a bifunction $g$ the set

$$X = \big\{x \in \mathbb{R}^n \mid \exists y \in \mathbb{R}^m \text{ such that } (x, y) \in \operatorname{dom} g\big\}$$

and the set-valued mapping $Y : \mathbb{R}^n \rightrightarrows \mathbb{R}^m$ such that

$$Y(x) = \big\{y \in \mathbb{R}^m \mid (x, y) \in \operatorname{dom} g\big\} \quad \forall x \in \mathbb{R}^n.$$

Thus,

$$\operatorname{dom} g = \operatorname{gph} Y = \big\{(x, y) \in \mathbb{R}^n \times \mathbb{R}^m \;\big|\; x \in X,\; y \in Y(x)\big\},$$

with

$$\operatorname{dom} Y = \big\{x \in \mathbb{R}^n \;\big|\; Y(x) \neq \emptyset\big\} = X.$$

Figure 9.1(left) illustrates the simplest situation when the set-valued mapping $Y : \mathbb{R}^n \rightrightarrows \mathbb{R}^m$ has $Y(x) = D \subset \mathbb{R}^m$ for all $x$. Then, $\operatorname{dom} g$ is the product set $X \times D$. Figure 9.1(right) shows a general case, where admissible values for $y$ are influenced by $x$. A bifunction $g \in \operatorname{bfcns}(\mathbb{R}^n, \mathbb{R}^m)$, with domain described by $(X, Y)$, defines the minsup problem

$$\underset{x \in X}{\text{minimize}} \; \sup\nolimits_{y \in Y(x)} g(x, y).$$

In applications, a bifunction of interest might be defined on a large subset of $\mathbb{R}^n \times \mathbb{R}^m$, possibly everywhere, but the context requires restrictions to some smaller subset, for example, dictated by constraints imposed on the variables. If these constraints are that $x \in X \subset \mathbb{R}^n$ and $y \in Y(x)$ for some $Y : \mathbb{R}^n \rightrightarrows \mathbb{R}^m$, then $g$ in our notation would be the original bifunction restricted to the set

$$\big\{(x, y) \in \mathbb{R}^n \times \mathbb{R}^m \;\big|\; x \in X, y \in Y(x)\big\},$$

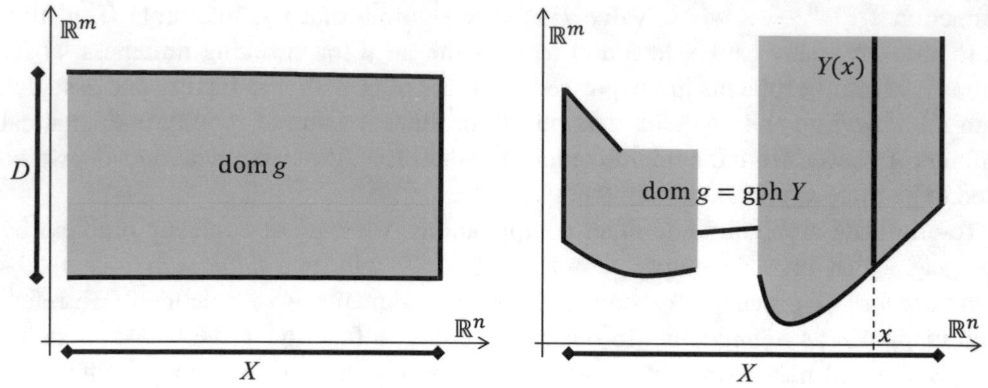


Fig. 9.1: Domain of bifunction: product set (left) and general (right).

which then becomes dom $g$. In other applications, a bifunction $g$ might be the only problem data given. These differences are immaterial to the following development as both are captured by considering bfcns($\mathbb{R}^n, \mathbb{R}^m$).

The *infsup-value* of a bifunction $g \in$ bfcns($\mathbb{R}^n, \mathbb{R}^m$), with domain described by $(X, Y)$, is defined as

$$\operatorname{infsup} g = \inf_{x \in X} \sup_{y \in Y(x)} g(x, y).$$

For $\varepsilon \in [0, \infty)$, the *$\varepsilon$-minsup-points* of $g$ are given by

$$\varepsilon\text{-argminsup}\, g = \left\{x \in X \;\middle|\; \sup_{y \in Y(x)} g(x, y) \leq \operatorname{infsup} g + \varepsilon < \infty\right\}.$$

If $\varepsilon = 0$, then we simply write argminsup $g$ and refer to such points as *minsup-points*.

The *sup-projection* of a bifunction $g \in$ bfcns($\mathbb{R}^n, \mathbb{R}^m$), with domain described by $(X, Y)$, is the function $f : \mathbb{R}^n \to \overline{\mathbb{R}}$ given by

$$f(x) = \begin{cases} \sup_{y \in Y(x)} g(x, y) & \text{if } x \in X \\ \infty & \text{otherwise.} \end{cases}$$

Since dom $g$ is nonempty, $f(x) > -\infty$ for all $x \in \mathbb{R}^n$ but dom $f$ could be a strict subset of $X$, possibly empty, because the supremum expression can be infinity for some or all $x \in X$. Thus, $f$ may not be proper. The minsup problem defined by $g$ is nevertheless equivalently stated as

$$\underset{x \in \mathbb{R}^n}{\text{minimize}}\, f(x).$$

Thus, for $\varepsilon \in [0, \infty)$, one has

$$\inf f = \operatorname{infsup} g \quad \text{and} \quad \varepsilon\text{-argmin}\, f = \varepsilon\text{-argminsup}\, g. \tag{9.9}$$

By 4.9, if $f$ is lsc and level-bounded with nonempty domain, then argmin $f \neq \emptyset$ and this becomes an avenue for establishing the existence of a minsup-point of $g$. Before we make it concrete, we extend the notion of lsc.

A bifunction $g \in \text{bfcns}(\mathbb{R}^n, \mathbb{R}^m)$ is *lsc* when

$$(x^\nu, y^\nu) \in \text{dom}\, g \to (x, y) \implies \begin{cases} \text{liminf } g(x^\nu, y^\nu) \geq g(x, y) & \text{if } (x, y) \in \text{dom}\, g \\ g(x^\nu, y^\nu) \to \infty & \text{otherwise.} \end{cases}$$

The definition is consistent with that for functions defined on $\mathbb{R}^n \times \mathbb{R}^m$ (see (5.1)) in the following sense: If a bifunction $g$ is extended to $\mathbb{R}^n \times \mathbb{R}^m$ by assigning it the value $\infty$ outside dom $g$, then the extension is lsc if and only if $g$ is lsc.

**Proposition 9.7** (lsc sup-projection). *For a bifunction $g \in \text{bfcns}(\mathbb{R}^n, \mathbb{R}^m)$, with domain described by $(X, Y)$, let $f$ be its sup-projection. Then, the following hold:*

(a) *$f$ is lsc if $X$ is closed, $Y$ is isc relative to $X$ and $g$ is lsc.*
(b) *$f$ is lsc if there's $D \subset \mathbb{R}^m$ such that $Y(x) = D$ for all $x \in X$ and, for all $y \in D$, $g(\cdot, y)$ is lsc when extended to all of $\mathbb{R}^n$ by assigning it the value $\infty$ outside $X$.*

**Proof.** Let $x^\nu \to x$ and $\varepsilon \in (0, \infty)$. By (5.1), it suffices to show that

$$\text{liminf } f(x^\nu) \geq f(x). \tag{9.10}$$

Since $f(x^\nu) = \infty$ for $x^\nu \notin \text{dom}\, f$, we can assume without loss of generality that $x^\nu \in \text{dom}\, f$ for all $\nu \in \mathbb{N}$.

For (a), let's consider two cases. First, suppose that $x \in \text{dom}\, f$. Then, $\sup_{y \in Y(x)} g(x, y)$ is finite and there's $y_\varepsilon \in Y(x)$ such that

$$\sup_{y \in Y(x)} g(x, y) \leq g(x, y_\varepsilon) + \varepsilon.$$

Since $Y$ is isc relative to $X$, there exist $y^\nu \in Y(x^\nu) \to y_\varepsilon$ and also $\bar{\nu}$ such that for all $\nu \geq \bar{\nu}$,

$$g(x^\nu, y^\nu) \geq g(x, y_\varepsilon) - \varepsilon$$

because $g$ is lsc. Thus, for $\nu \geq \bar{\nu}$,

$$\sup_{y \in Y(x^\nu)} g(x^\nu, y) \geq g(x^\nu, y^\nu) \geq g(x, y_\varepsilon) - \varepsilon \geq \sup_{y \in Y(x)} g(x, y) - 2\varepsilon.$$

Since $\varepsilon$ is arbitrary, (9.10) holds.

Second, suppose that $x \notin \text{dom}\, f$. Since $x^\nu \in X$ and $X$ is closed, $x \in X$ as well and $\sup_{y \in Y(x)} g(x, y) = \infty$. There's $y_\varepsilon \in Y(x)$ such that $g(x, y_\varepsilon) \geq 1/\varepsilon$. Following a similar argument as earlier, we find that

$$\text{liminf } \big( \sup_{y \in Y(x^\nu)} g(x^\nu, y) \big) \geq 1/\varepsilon - \varepsilon.$$

Since $\varepsilon$ is arbitrary, $f(x^\nu) \to \infty$, (9.10) holds and the proof of (a) is complete.

Essentially, 6.27 furnishes (b), but minor discrepancies lead us to repeat the argument. First, suppose that $x \in \operatorname{dom} f$. Then, $x \in X$ and there's $y_\varepsilon \in D$ such that $f(x) \leq g(x, y_\varepsilon) + \varepsilon$. There's also $\bar{\nu}$ such that

$$g(x^\nu, y_\varepsilon) \geq g(x, y_\varepsilon) - \varepsilon \quad \forall \nu \geq \bar{\nu}$$

because the extension of $g(\cdot, y_\varepsilon)$ is lsc. Thus, for $\nu \geq \bar{\nu}$,

$$f(x^\nu) \geq g(x^\nu, y_\varepsilon) \geq g(x, y_\varepsilon) - \varepsilon \geq f(x) - 2\varepsilon.$$

Since $\varepsilon$ is arbitrary, (9.10) holds.

Second, suppose that $x \notin \operatorname{dom} f$ and $x \notin X$. Select $\bar{y} \in D$. Since the extension of $g(\cdot, \bar{y})$ is lsc and has value $\infty$ at $x$, $g(x^\nu, \bar{y}) \to \infty$ and then $f(x^\nu) \to \infty$. Again, (9.10) is satisfied.

Third, suppose that $x \notin \operatorname{dom} f$ and $x \in X$. Then, $\sup_{y \in D} g(x, y) = \infty$ and there's $y_\varepsilon \in D$ such that $g(x, y_\varepsilon) \geq 1/\varepsilon$. Since the extension of $g(\cdot, y_\varepsilon)$ is lsc, we again find that for sufficiently large $\nu$, one has

$$f(x^\nu) \geq g(x^\nu, y_\varepsilon) \geq g(x, y_\varepsilon) - \varepsilon \geq 1/\varepsilon - \varepsilon.$$

Since $\varepsilon$ is arbitrary, $f(x^\nu) \to \infty$ and (9.10) holds. □

The closedness assumption on $X$ in (a) of the proposition is motivated by the example $g(x, y) = 0$ for $(x, y) \in \operatorname{dom} g$, with the domain being characterized by $X = (0, \infty)$ and $Y(x) = \{1/x\}$ for $x \in X$. Then, the sup-projection of $g$ is given by $f(x) = 0$ for $x \in X$ and $f(x) = \infty$ otherwise. While $Y$ is isc relative to $X$ and $g$ is lsc, $f$ isn't lsc.

**Exercise 9.8** (bifunction). Consider the bifunction given by $g(x, y) = |x|e^y$ when $xy \leq 1$. In the notation just introduced, determine and graph the following quantities: $X, Y, \operatorname{dom} Y$, $f$, $\operatorname{dom} f$, $\operatorname{infsup} g$ and $\varepsilon$-$\operatorname{argminsup} g$. Examine whether $f$ is lsc.

**Guide.** In addition to the visual assessment, use 9.7. □

For $C \subset X \subset \mathbb{R}^n$ and $f : X \to \overline{\mathbb{R}}$, we've the following terminology: The function $f$ is *lsc relative to $C$* if

$$x^\nu \in C \to x \in C \implies \operatorname{liminf} f(x^\nu) \geq f(x).$$

It's *continuous relative to $C$* if

$$x^\nu \in C \to x \in C \implies f(x^\nu) \to f(x).$$

It's *concave relative to $C$* if $C$ is convex, $f$ is real-valued on $C$ and

$$f\big((1-\lambda)x^0 + \lambda x^1\big) \geq (1-\lambda)f(x^0) + \lambda f(x^1) \quad \forall \lambda \in [0,1],\ x^0, x^1 \in C.$$

A certain class of bifunctions, with advantageous properties, is common in applications.

**Definition 9.9** (Ky Fan bifunction). We say that $g \in \text{bfcns}(\mathbb{R}^n, \mathbb{R}^n)$ is a *Ky Fan bifunction* when

(a) for convex $X \subset \mathbb{R}^n$, $\text{dom}\, g = X \times X$
(b) $\forall y \in X$: $g(\cdot, y)$ is lsc relative to $X$
(c) $\forall x \in X$: $g(x, \cdot)$ is concave relative to $X$.

The sup-projection of a Ky Fan bifunction is lsc provided that its domain is defined in terms of a closed set $X$. This follows from 9.7(b) after noting that the extension of $g(\cdot, y)$ to $\mathbb{R}^n$ is lsc when $X$ is closed.

A Ky Fan bifunction $g$ with domain $X \times X$ has the practically important property that the value of its sup-projection at every $x \in X$ is computationally tractable in the sense it only requires the minimization of $-g(x, \cdot)$ over the convex set $X$, which is a convex problem. This opens up the possibility of computing subgradients when $X$ is compact (see 6.30) and applying the outer approximation algorithm from §6.C.

**Theorem 9.10** (Ky Fan inequality). *For a convex compact set $X \subset \mathbb{R}^n$ and a Ky Fan function $g \in \text{bfcns}(\mathbb{R}^n, \mathbb{R}^n)$, with domain $X \times X$, suppose that $g(x, x) \leq 0$ for all $x \in X$. Then,*

$$\text{argminsup}\, g \neq \emptyset \quad \textit{and} \quad g(x^\star, y) \leq 0 \;\; \forall x^\star \in \text{argminsup}\, g, \;\; y \in X.$$

**Proof.** Let $f$ be the sup-projection of $g$. As just argued before the theorem, $f$ is lsc because $X$ is closed. It's also level-bounded due to the compactness assumption. If $f$ is finite at some point, then 4.9 applies and we can conclude that

$$\emptyset \neq \text{argmin}\, f = \text{argminsup}\, g;$$

cf. (9.9). The finiteness of $f$ is trivial if $X$ contains a single point and we can concentrate on the case with $X$ consisting of more than one point. Fix $\bar{x} \in X$ and let $h : \mathbb{R}^n \to \overline{\mathbb{R}}$ be the function defined by

$$h(y) = \begin{cases} -g(\bar{x}, y) & \text{if } y \in X \\ \infty & \text{otherwise.} \end{cases}$$

This is a proper convex function because $X$ is convex and $g(\bar{x}, \cdot)$ is concave relative to $X$. Since $X$ consists of more than one point, we can invoke [89, Theorem 23.4] to establish that there's $\bar{y} \in X$ such that $\partial h(\bar{y}) \neq \emptyset$. (If $\text{int}\, X \neq \emptyset$, then 2.25 guarantees the existence of a subgradient but the claim holds as the reference confirms.) For $v \in \partial h(\bar{y})$, the subgradient inequality 2.17 establishes that

$$-g(\bar{x}, y) = h(y) \geq h(\bar{y}) + \langle v, y - \bar{y} \rangle \;\; \forall y \in X.$$

The right-hand side has a minimum value over all $y \in X$ because $X$ is compact. Thus, there's $\alpha \in \mathbb{R}$ such that $g(\bar{x}, y) \leq \alpha$ for all $y \in X$. This implies that $f(\bar{x}) < \infty$ as needed to conclude that $\text{argminsup}\, g \neq \emptyset$.

For the second claim, there's $\bar{x} \in X$ such that

$$f(\bar{x}) \leq \sup_{x \in X} g(x, x)$$

by [7, Lemma 3, Section 6.3]. Since $g(x, x) \leq 0$ for all $x \in X$ by assumption, this implies that

$$g(x^\star, y) \leq f(x^\star) \leq f(\bar{x}) \leq 0$$

when $x^\star \in \operatorname{argminsup} g$ and $y \in X$. □

The theorem provides a path to showing the existence of solutions for a variety of problems: First, establish that the problem of interest can be formulated as a minsup problem using a particular bifunction. Second, verify the conditions of the theorem for the adopted bifunction.

**Example 9.11** (existence of solution to variational inequality). For $F : \mathbb{R}^n \to \mathbb{R}^n$ and a nonempty, convex and compact set $C \subset \mathbb{R}^n$, the variational inequality

$$x \in C \quad \text{and} \quad \langle F(x), \bar{x} - x \rangle \geq 0 \quad \forall \bar{x} \in C$$

has a solution when $F$ is continuous relative to $C$. We can prove this fact using the Ky Fan inequality 9.10 and the Auslender gap function. This supplements the existence result in 7.11.

**Detail.** The Auslender gap function has

$$f(x) = \sup_{y \in C} \langle F(x), x - y \rangle \quad \forall x \in C.$$

We recall from 7.13 that $x^\star$ solves the variational inequality if and only if

$$x^\star \in \operatorname{argmin}_{x \in C} f(x) \quad \text{and} \quad f(x^\star) = 0.$$

Let's define

$$g(x, y) = \langle F(x), x - y \rangle \quad \forall (x, y) \in C \times C.$$

Thus, $g \in \operatorname{bfcns}(\mathbb{R}^n, \mathbb{R}^n)$ with $\operatorname{dom} g = C \times C$. For all $y \in C$, $g(\cdot, y)$ is continuous relative to $C$ because $F$ is continuous relative to $C$. For all $x \in C$, $g(x, \cdot)$ is concave relative to $C$. Consequently, $g$ is a Ky Fan function. It now follows directly from the Ky Fan inequality 9.10 that

$$\operatorname{argminsup} g \neq \emptyset \quad \text{and} \quad g(x^\star, y) \leq 0 \quad \forall x^\star \in \operatorname{argminsup} g, \quad y \in C$$

because $g(x, x) = 0$ for all $x \in C$ trivially. Thus, via (9.9), we obtain that

$$x^\star \in \operatorname{argmin}_{x \in C} f(x) \quad \text{and} \quad f(x^\star) = 0,$$

which means that $x^\star$ solves the variational inequality. □

**Example 9.12** (existence of fixed point). For $F : \mathbb{R}^n \to \mathbb{R}^n$ and nonempty $C \subset \mathbb{R}^n$, a *constrained fixed-point problem* is that of finding

$$x = F(x) \in C.$$

We associate this problem with the bifunction $g \in \text{bfcns}(\mathbb{R}^n, \mathbb{R}^n)$ defined by

$$g(x, y) = \big\langle x - F(x), x - y \big\rangle \quad \forall (x, y) \in C \times C.$$

The existence of a solution (a fixed point) to a constrained fixed-point problem can be established using the Ky Fan inequality 9.10 and the bifunction $g$ provided that $C$ is convex and compact, $F$ is continuous relative to $C$ and $F(x) \in C$ for all $x \in C$.

**Detail.** First, let's make the connection between the minsup-points of $g$ and fixed points. Let $f$ be the sup-projection of $g$. Then,

$$f(x) = \sup_{x \in C} g(x, y) \geq 0 \quad \forall x \in C$$

because $y = x$ is a possibility in the maximization. If $x^\star$ is a fixed point, then

$$f(x^\star) = g(x^\star, y) = 0 \quad \forall y \in C.$$

Thus, one has

$$x^\star \in \operatorname{argmin} f = \operatorname{argminsup} g;$$

cf. (9.9). If $x^\star \in C$ satisfies $g(x^\star, y) \leq 0$ for all $y \in C$, then

$$\big\langle x^\star - F(x^\star), x^\star - F(x^\star) \big\rangle \leq 0$$

because $F(x^\star) \in C$. Consequently, $\|x^\star - F(x^\star)\|_2^2 \leq 0$ and $x^\star = F(x^\star)$. We've established that

$$x^\star \text{ is a fixed point} \iff x^\star \in \operatorname{argminsup} g \text{ and } g(x^\star, y) \leq 0 \;\; \forall y \in C.$$

Second, $g$ is a Ky Fan function because $g(\cdot, y)$ is continuous relative to $C$ due to the same property for $F$ and $g(x, \cdot)$ is concave relative to $C$. Trivially, $g(x, x) = 0$ for all $x \in C$ so the Ky Fan inequality 9.10 applies and there's $x^\star \in \operatorname{argminsup} g$ such that $g(x^\star, y) \leq 0$ for all $y \in C$. As just established, this means that $x^\star$ is a fixed point. □

**Exercise 9.13** (bilinear form as Ky Fan bifunction). For a positive semidefinite $n \times n$-matrix $A$, consider the bifunction with $g(x, y) = -\langle y, Ax \rangle$ for $(x, y) \in \mathbb{B}(0, 1) \times \mathbb{B}(0, 1)$. Show that $g$ is a Ky Fan bifunction satisfying the assumptions of 9.10. Derive an explicit formula for the sup-projection of $g$.

**Guide.** Use the Cauchy-Schwarz inequality to determine an explicit formula for the sup-projection. □

## 9.D Lopsided Approximation Algorithm

Minsup problems that can't be solved directly require approximations and this leads to the question: Does an alternative bifunction have minsup-points that are close to those of the actual bifunction of interest? Parallel to the central role of epi-convergence for minimization problem, lopsided convergence of bifunctions furnishes the desired guarantees. Leveraging such properties, let's now develop an approximation-based framework for solving the minsup problem

$$\underset{x \in X}{\text{minimize}} \ \sup_{y \in Y(x)} g(x, y),$$

where $g \in \text{bfcns}(\mathbb{R}^n, \mathbb{R}^m)$, with domain described by $(X, Y)$.

**Lopsided Approximation Algorithm.**

Data. Bifunctions $g^\nu \in \text{bfcns}(\mathbb{R}^n, \mathbb{R}^m)$ and tolerances $\varepsilon^\nu \in [0, \infty)$, $\nu \in \mathbb{N}$.
Step 0. Set $\nu = 1$.
Step 1. Compute $x^\nu \in \varepsilon^\nu\text{-argminsup}\, g^\nu$.
Step 2. Replace $\nu$ by $\nu + 1$ and go to Step 1.

The intention is to choose the bifunctions $\{g^\nu, \nu \in \mathbb{N}\}$ so that Step 1 becomes relatively easy to execute compared to solving the actual minsup problem. If $g^\nu$ approximates $g$ in a suitable manner and $\varepsilon^\nu \to 0$, then we can be hopeful that cluster points of the generated sequence $\{x^\nu, \nu \in \mathbb{N}\}$ are minsup-points of $g$. The inner lopsided condition is fundamental in this regard.

**Definition 9.14** (inner lopsided condition). The bifunctions $\{g, g^\nu, \ \nu \in \mathbb{N}\} \subset \text{bfcns}(\mathbb{R}^n, \mathbb{R}^m)$, with domains described by $(X, Y)$ and $(X^\nu, Y^\nu)$, respectively, satisfy the *inner lopsided condition* when

(a) $\forall N \in \mathcal{N}_\infty^\#$, $x^\nu \in X^\nu \underset{N}{\to} x \in X$ and $y \in Y(x)$,
$$\exists y^\nu \in Y^\nu(x^\nu) \underset{N}{\to} y \text{ such that } \liminf_{\nu \in N} g^\nu(x^\nu, y^\nu) \geq g(x, y)$$
(b) $\forall N \in \mathcal{N}_\infty^\#$ and $x^\nu \in X^\nu \underset{N}{\to} x \notin X$,
$$\exists y^\nu \in Y^\nu(x^\nu) \text{ such that } g^\nu(x^\nu, y^\nu) \underset{N}{\to} \infty.$$

While somewhat involved, the inner lopsided condition reminds us of the characterization of epi-convergence in 4.15(a). Just as in that case, the condition requires the approximating quantities to be "high" relative to the actual one. The additional intricacies are caused by the bivariate nature of bifunctions and the need for carefully adhering to their domains. Nevertheless, the inner lopsided condition is largely all what's required to justify the lopsided approximation algorithm.

**Convergence 9.15** (lopsided approximation algorithm). *Suppose that* $g \in \text{bfcns}(\mathbb{R}^n, \mathbb{R}^m)$, *with* $(X, Y)$ *describing* dom $g$, *and* $\{x^\nu, \nu \in \mathbb{N}\}$ *is generated by the lopsided approximation algorithm, with* $(X^\nu, Y^\nu)$ *describing* dom $g^\nu$. *Let*

$$\alpha = \liminf_{\nu \in N} \left( \sup_{y \in Y^\nu(x^\nu)} g^\nu(x^\nu, y) \right).$$

*If the inner lopsided condition 9.14 is satisfied for* $\{g, g^\nu\}$ *and* $x^\nu \underset{N}{\to} \bar{x}$ *for some* $N \in \mathcal{N}_\infty^\#$ *and* $\bar{x} \in \mathbb{R}^n$*, then the following hold:*

(a) $\alpha \geq \operatorname{infsup} g$.
(b) *If* $\alpha < \infty$*, then* $\bar{x} \in X$ *and* $\alpha \geq \sup_{y \in Y(\bar{x})} g(\bar{x}, y)$.

We postpone the proof to the end of the section and instead discuss the consequences of the assertions. The quantity $\sup_{y \in Y^\nu(x^\nu)} g^\nu(x^\nu, y)$ is presumably computed as part of Step 1 of the lopsided approximation algorithm and can be monitored as the algorithm progresses. Thus, $\alpha$ in the convergence statement can be regarded as an output of the algorithm. It furnishes the upper bound in (a), which is important because the usual strategy of obtaining an upper bound on a minimum value by evaluating the objective function at a particular point may not be available. One can expect that computing

$$\sup_{y \in Y(x)} g(x, y)$$

is nontrivial for any $x \in X$ as it involves the actual bifunction. Moreover, a lower bound on $\operatorname{infsup} g$ is often achievable by restricting $Y(x)$ to some (finite) subset so knowledge of an upper bound results in a bracketing of $\operatorname{infsup} g$.

While the convergence statement holds for any tolerances $\{\varepsilon^\nu \in [0, \infty), \nu \in \mathbb{N}\}$, one might let $\varepsilon^\nu \to 0$ with the hope of lowering the left-hand side in (a).

Part (b) of the statement provides the feasibility guarantee $\bar{x} \in X$ under the mild assumption that $\alpha < \infty$. If $\operatorname{infsup} g \geq 0$, which might be known from the problem structure, and $\alpha = 0$, then we can conclude from (b) that actually $\bar{x} \in \operatorname{argminsup} g$ because $\sup_{y \in Y(\bar{x})} g(\bar{x}, y)$ would have to be equal to 0.

**Example 9.16** (mixed complementarity problems). For $F : \mathbb{R}^n \to \mathbb{R}^n$ and

$$C = \{x \in \mathbb{R}^n \mid \alpha_j \leq x_j \leq \beta_j,\ j = 1, \dots, n\},$$

with $-\infty \leq \alpha_j \leq \beta_j \leq \infty$, consider the mixed complementarity problem

$$x \in C \quad \text{and} \quad \langle F(x), \bar{x} - x \rangle \geq 0 \ \ \forall \bar{x} \in C.$$

As discussed in 7.21, the nonsmooth Newton-Raphson algorithm can be brought in to solve such problems if $F$ is smooth and this might be made easier if the box $C$ is small. Thus, a strategy for solving mixed complementarity problems with nonsmooth $F$ and/or large $C$ would be to approximate $F$ by a smooth mapping $F^\nu : \mathbb{R}^n \to \mathbb{R}^n$ and approximate $C$ by a smaller box

$$C^\nu = \{x \in \mathbb{R}^n \mid \alpha_j^\nu \leq x_j \leq \beta_j^\nu,\ j = 1, \dots, n\},$$

where $-\infty \leq \alpha_j^\nu \leq \beta_j^\nu \leq \infty$.

If $\alpha_j^\nu \to \alpha_j$ and $\beta_j^\nu \to \beta_j$ for all $j = 1, \dots, n$, and $F^\nu(x^\nu) \to F(x)$ for all $x^\nu \in C^\nu \to x \in C$, then, via the Auslender gap function, one can confirm the inner lopsided condition 9.14 and thereby justify the lopsided approximation algorithm in light of 9.15.

**Detail.** Let's consider the bifunctions $g, g^\nu \in \text{bfcns}(\mathbb{R}^n, \mathbb{R}^n)$ given by

$$g(x,y) = \langle F(x), x - y \rangle \quad \forall (x,y) \in C \times C$$
$$g^\nu(x,y) = \langle F^\nu(x), x - y \rangle \quad \forall (x,y) \in C^\nu \times C^\nu.$$

The sup-projections of $g$ and $g^\nu$, denoted by $f$ and $f^\nu$, respectively, are essentially the Auslender gap functions for the actual and approximating mixed complementarity problems; see 7.13. There, we also establish that $x^\star$ solves the actual mixed complementarity problem if and only if $x^\star \in \operatorname{argmin} f$ and $f(x^\star) = 0$, with a similar relation for the approximating mixed complementarity problem and $f^\nu$.

The inner lopsided condition 9.14 holds in this case as can be seen as follows. For 9.14(a), let $N \in \mathcal{N}_\infty^\#$, $x^\nu \in C^\nu \xrightarrow[N]{} x \in C$ and $y \in C$. Since $C^\nu \xrightarrow{s} C$, there's $y^\nu \in C^\nu \xrightarrow[N]{} y$. Moreover,

$$\langle F^\nu(x^\nu), x^\nu - y^\nu \rangle \xrightarrow[N]{} \langle F(x), x - y \rangle$$

because $F^\nu(x^\nu) \xrightarrow[N]{} F(x)$. Since $C^\nu \xrightarrow{s} C$, there isn't any sequence $x^\nu \in C^\nu \xrightarrow[N]{} x \notin C$ and 9.14(b) holds automatically.

Next, let's assume that the approximating mixed complementarity problems have solutions. This is a mild assumption. For example, if $C^\nu$ is compact, then the corresponding approximating problem has a solution by 9.11 and the fact that $F^\nu$ is continuous. Moreover, let $\varepsilon^\nu \to 0$.

The lopsided approximation algorithm computes $x^\nu \in \varepsilon^\nu\text{-}\operatorname{argmin} f^\nu$ in Step 1 (cf. (9.9)), which implies that $f^\nu(x^\nu) \le \varepsilon^\nu$ because $\inf f^\nu = 0$ from the assumption about having solutions. Thus, $f^\nu(x^\nu) \to 0$. In this case, Step 1 can be accomplished by the nonsmooth Newton-Raphson algorithm. As long as $\{x^\nu, \nu \in \mathbb{N}\}$ has a cluster point $\bar{x}$, we conclude that

$$\operatorname{infsup} g = \inf f = f(\bar{x}) = 0$$

from (9.9), 9.15(a) and the fact that $f(x) \ge 0$ for all $x \in C$. Thus, $\bar{x}$ solves the actual mixed complementarity problem by 9.15(b). □

**Exercise 9.17** (approximation of sup-projection). For continuous $g_0 : \mathbb{R}^n \times \mathbb{R}^m \to \mathbb{R}$, nonempty closed $D^\nu \subset D \subset \mathbb{R}^m$ and nonempty closed $X \subset \mathbb{R}^n$, define the bifunctions $g, g^\nu \in \text{bfcns}(\mathbb{R}^n, \mathbb{R}^m)$ by

$$g(x,y) = g_0(x,y) \quad \forall (x,y) \in X \times D; \qquad g^\nu(x,y) = g_0(x,y) \quad \forall (x,y) \in X \times D^\nu.$$

Suppose that for each $\bar{y} \in D$ there's $y^\nu \in D^\nu \to \bar{y}$. Show that the inner lopsided condition 9.14 holds in this case. Discuss how the inequality in 9.15(a) can be strict under the present assumptions.

**Guide.** A result under more restrictive assumptions is furnished by 6.33. □

We now turn to the proof of 9.15 and start with an intermediate result, which brings forth the central connection to epi-convergence.

**Proposition 9.18** (inner approximation of sup-projections). *For bifunctions* $\{g, g^\nu, \nu \in \mathbb{N}\} \subset \mathrm{bfcns}(\mathbb{R}^n, \mathbb{R}^m)$ *with corresponding sup-projections* $\{f, f^\nu\}$*, suppose that the inner lopsided condition 9.14 holds. Then,*

$$\mathrm{LimOut}(\mathrm{epi}\, f^\nu) \subset \mathrm{epi}\, f.$$

**Proof.** Let $(X, Y)$ and $(X^\nu, Y^\nu)$ describe the domains of $g$ and $g^\nu$, respectively. Suppose that

$$(x, \alpha) \in \mathrm{LimOut}(\mathrm{epi}\, f^\nu).$$

Then there exist $N \in \mathcal{N}_\infty^\#$ and $\{(x^\nu, \alpha^\nu) \in \mathbb{R}^{n+1}, \nu \in N\}$, with $x^\nu \in X^\nu$,

$$\sup_{y \in Y^\nu(x^\nu)} g^\nu(x^\nu, y) \le \alpha^\nu, \quad x^\nu \to_N x \;\text{ and }\; \alpha^\nu \to_N \alpha.$$

If $x \notin X$, then we can construct a sequence $y^\nu \in Y^\nu(x^\nu)$ such that $g^\nu(x^\nu, y^\nu) \to_N \infty$ as guaranteed by the inner lopsided condition 9.14(b). However,

$$\alpha^\nu \ge \sup_{y \in Y^\nu(x^\nu)} g^\nu(x^\nu, y) \ge g^\nu(x^\nu, y^\nu), \; \nu \in N$$

contradicts $\alpha^\nu \to_N \alpha \in \mathbb{R}$. Thus, $x \in X$. If $\sup_{y \in Y(x)} g(x, y) = \infty$, then there's $\bar{y} \in Y(x)$ such that

$$g(x, \bar{y}) \ge \alpha + 1.$$

Now, the inner lopsided condition 9.14(a) ensures that one can construct $y^\nu \in Y^\nu(x^\nu) \to_N \bar{y}$ such that

$$\mathrm{liminf}_{\nu \in N}\, g^\nu(x^\nu, y^\nu) \ge g(x, \bar{y}).$$

Consequently, one has

$$\begin{aligned} \alpha = \mathrm{liminf}_{\nu \in N}\, \alpha^\nu &\ge \mathrm{liminf}_{\nu \in N} \big( \sup_{y \in Y^\nu(x^\nu)} g^\nu(x^\nu, y) \big) \\ &\ge \mathrm{liminf}_{\nu \in N}\, g^\nu(x^\nu, y^\nu) \ge g(x, \bar{y}) \ge \alpha + 1, \end{aligned}$$

which is a contradiction. Next, let's consider when $\sup_{y \in Y(x)} g(x, y)$ is finite. For $\varepsilon > 0$, pick $y_\varepsilon \in Y(x)$ such that

$$g(x, y_\varepsilon) \ge \sup_{y \in Y(x)} g(x, y) - \varepsilon.$$

Then, the inner lopsided condition again yields $y^\nu \in Y^\nu(x^\nu) \to_N y_\varepsilon$ such that

$$\mathrm{liminf}_{\nu \in N}\, g^\nu(x^\nu, y^\nu) \ge g(x, y_\varepsilon).$$

This leads to

$$\begin{aligned} \mathrm{liminf}_{\nu \in N} \big( \sup_{y \in Y^\nu(x^\nu)} g^\nu(x^\nu, y) \big) &\ge \mathrm{liminf}_{\nu \in N}\, g^\nu(x^\nu, y^\nu) \\ &\ge g(x, y_\varepsilon) \ge \sup_{y \in Y(x)} g(x, y) - \varepsilon \end{aligned}$$

and then also

$$\operatorname{liminf}_{\nu\in N}\big(\sup_{y\in Y^\nu(x^\nu)} g^\nu(x^\nu,y)\big) \ge \sup_{y\in Y(x)} g(x,y)$$

because $\varepsilon$ is arbitrary. Since

$$\alpha = \operatorname{liminf}_{\nu\in N}\alpha^\nu \ge \operatorname{liminf}_{\nu\in N}\big(\sup_{y\in Y^\nu(x^\nu)} g^\nu(x^\nu,y)\big) \ge \sup_{y\in Y(x)} g(x,y),$$

we've established that $(x,\alpha)\in \operatorname{epi} f$ and the conclusion follows. □

We can't proceed and establish LimInn(epi $f^\nu$) $\supset$ epi $f$ as well and then $f^\nu \xrightarrow{e} f$ in the proposition without additional assumptions. For example, consider the bifunctions $g, g^\nu$ with $\operatorname{dom} g = \operatorname{dom} g^\nu = \mathbb{R}\times\mathbb{R}$ and values $g(x,y)=0$ for $(x,y)\in\mathbb{R}\times\mathbb{R}$ and

$$g^\nu(x,y) = \begin{cases} 1 & \text{if } (x,y)\in\mathbb{R}\times\{\nu\}\\ 0 & \text{otherwise.}\end{cases}$$

The inner lopsided condition 9.14 holds. The sup-projection of $g^\nu$ has value 1 for all $x\in\mathbb{R}$ and that of $g$ has value 0 for all $x\in\mathbb{R}$. Thus, the inclusion in 9.18 is strict.

**Proof of 9.15.** Let $f$ and $f^\nu$ be the sup-projections of $g$ and $g^\nu$, respectively. By 9.18, LimOut(epi $f^\nu$) $\subset$ epi $f$ and this is equivalent to satisfying 4.15(a); see the proof of that result. Thus,

$$\alpha = \operatorname{liminf}_{\nu\in N}\big(\sup_{y\in Y^\nu(x^\nu)} g(x^\nu,y)\big) = \operatorname{liminf}_{\nu\in N} f^\nu(x^\nu) \ge f(\bar{x}) \ge \inf f = \operatorname{infsup} g$$

by also invoking (9.9). This establishes (a). If $\alpha<\infty$, then $f(\bar{x})<\infty$ so that

$$\bar{x}\in X \quad\text{and}\quad f(\bar{x}) = \sup_{y\in Y(\bar{x})} g(\bar{x},y)$$

by the definition of $f$. This leads to (b). □

## 9.E Lop-Convergence I

The inner lopsided condition 9.14 is the first element in a series of properties for bifunctions that lead to successively stronger conclusions about solutions generated by the lopsided approximation algorithm and about approximating bifunctions in general. The second element, to be presented next, reminds us of the characterization of epi-convergence in 4.15(b) by insisting on sufficiently "low" approximating bifunctions.

**Definition 9.19** (outer lopsided condition). The bifunctions $\{g, g^\nu, \nu\in\mathbb{N}\}\subset \operatorname{bfcns}(\mathbb{R}^n,\mathbb{R}^m)$, with domains described by $(X,Y)$ and $(X^\nu,Y^\nu)$, respectively, satisfy the *outer lopsided condition* if the following holds for all $x\in X$:

$\exists\, x^\nu \in X^\nu \to x$ such that $\forall N \in \mathcal{N}_\infty^\#$ and $y^\nu \in Y^\nu(x^\nu) \underset{N}{\to} y \in \mathbb{R}^m$, one has

$$\begin{cases} \limsup_{\nu\in N} g^\nu(x^\nu, y^\nu) \le g(x,y) & \text{if } y \in Y(x) \\ g^\nu(x^\nu, y^\nu) \underset{N}{\to} -\infty & \text{otherwise.} \end{cases}$$

**Example 9.20** (mixed complementarity problems; cont.). The bifunctions $\{g, g^\nu\}$ in 9.16 satisfy the outer lopsided condition 9.19.

**Detail.** Let $x \in C$. Since $C^\nu \xrightarrow{s} C$, there exists $x^\nu \in C^\nu \to x$. Then, for arbitrary $N \in \mathcal{N}_\infty^\#$ and $y^\nu \in C^\nu \underset{N}{\to} y$, one has

$$\langle F^\nu(x^\nu), x^\nu - y^\nu \rangle \underset{N}{\to} \langle F(x), x - y \rangle.$$

Since $y \in C$, we've established the outer lopsided condition. □

Jointly, the inner and outer lopsided conditions result in lopsided convergence, which is the counterpart for bifunctions to epi-convergence of functions. In fact, lopsided convergence is equivalent to epi-convergence when applied to bifunctions that don't depend on a second $y$-argument as can be seen directly from the definitions.

**Definition 9.21** (lopsided convergence). If $\{g, g^\nu,\ \nu \in \mathbb{N}\} \subset \mathrm{bfcns}(\mathbb{R}^n, \mathbb{R}^m)$ satisfy the inner and outer lopsided conditions, then $g^\nu$ *converges lopsided*, or *lop-converges*, to $g$, written as $g^\nu \xrightarrow{l} g$.

The lop-convergence is *ancillary-tight* when for every $\varepsilon > 0$ and sequence $x^\nu \to x$ selected while verifying the outer lopsided condition 9.19, there are compact $B_\varepsilon \subset \mathbb{R}^m$ and $\nu_\varepsilon \in \mathbb{N}$ such that

$$\sup_{y \in Y^\nu(x^\nu)\cap B_\varepsilon} g^\nu(x^\nu, y) \ge \sup_{y\in Y^\nu(x^\nu)} g^\nu(x^\nu, y) - \varepsilon \quad \forall \nu \ge \nu_\varepsilon,$$

where $Y^\nu$ describes $\mathrm{dom}\, g^\nu$.

The added requirement for ancillary-tightness is satisfied if for every sequence $x^\nu \to x$ selected while verifying the outer lopsided condition 9.19,

$$\bigcup\nolimits_{\nu\in\mathbb{N}} Y^\nu(x^\nu)$$

is contained in a compact set. Other sufficient conditions exist as well. For example, if $g^\nu \xrightarrow{l} g$, $\inf\sup g > -\infty$ and, for each $\alpha \in \mathbb{R}$ and sequence $x^\nu \to x$ selected while verifying the outer lopsided condition 9.19, there's a compact set $C_\alpha$ such that

$$\bigcup\nolimits_{\nu\in\mathbb{N}} \big\{ y \in Y^\nu(x^\nu) \,\big|\, g^\nu(x^\nu, y) \ge \alpha \big\} \subset C_\alpha,$$

then $g^\nu$ lop-converges to $g$ ancillary-tightly. To see this, let $x^\nu \to x$ be a sequence selected while verifying the outer lopsided condition 9.19. Then,

$$\liminf\big(\sup\nolimits_{y\in Y^\nu(x^\nu)} g^\nu(x^\nu, y)\big) \ge \inf\sup g$$

by the arguments in the proof of 9.15(a). Since $\operatorname{infsup} g > -\infty$ and $g^\nu(x^\nu, y) > -\infty$ for all $y \in Y^\nu(x^\nu)$ and $\nu \in \mathbb{N}$, there's $\bar\alpha \in \mathbb{R}$ such that

$$\sup_{y \in Y^\nu(x^\nu)} g^\nu(x^\nu, y) \geq \bar\alpha + 1 \quad \forall \nu \in \mathbb{N}.$$

The compact set $C_{\bar\alpha}$ guaranteed by the assumption then results in

$$\sup_{y \in Y^\nu(x^\nu) \cap C_{\bar\alpha}} g^\nu(x^\nu, y) = \sup_{y \in Y^\nu(x^\nu)} g^\nu(x^\nu, y) \quad \forall \nu \in \mathbb{N}$$

because the points $y$ excluded on the left-hand side have $g^\nu(x^\nu, y) < \bar\alpha$ and thus don't affect the supremum. For any $\varepsilon > 0$, one has that $B_\varepsilon = C_{\bar\alpha}$ and $\nu_\varepsilon = 1$ verify the ancillary-tightness condition.

**Convergence 9.22** (lopsided approximation algorithm; cont.). *Suppose that* $g \in \operatorname{bfcns}(\mathbb{R}^n, \mathbb{R}^m)$ *has a proper sup-projection and* $\{x^\nu, \nu \in \mathbb{N}\}$ *is generated by the lopsided approximation algorithm, with* $\varepsilon^\nu \to 0$ *and* $\operatorname{dom} g^\nu$ *described by* $(X^\nu, Y^\nu)$*. If* $g^\nu \xrightarrow{l} g$ *ancillary-tightly and* $x^\nu \underset{N}{\to} \bar x$ *for some* $N \in \mathcal{N}_\infty^\#$ *and* $\bar x \in \mathbb{R}^n$*, then*

$$\lim_{\nu \in N} \left( \sup_{y \in Y^\nu(x^\nu)} g^\nu(x^\nu, y) \right) = \operatorname{infsup} g \quad \textit{and} \quad \bar x \in \operatorname{argminsup} g.$$

The convergence statement strengthens the conclusions from 9.15 by confirming that the lopsided approximation algorithm indeed leads to a solution of the actual problem even without any particular knowledge about the value of $\operatorname{infsup} g$. We postpone the proof to the end of the section.

**Example 9.23** (penalty method for minsup problems). For a bifunction $g \in \operatorname{bfcns}(\mathbb{R}^n, \mathbb{R}^m)$, with domain described by $(X, Y)$, suppose that

$$Y(x) = \{ y \in D \mid \varphi(x, y) \leq 0 \}, \quad x \in X,$$

where $D \subset \mathbb{R}^m$ and $\varphi : \mathbb{R}^n \times \mathbb{R}^m \to \mathbb{R}$. The nontrivial dependence on $x$ in $Y$ causes theoretical and computational difficulties. Under such dependence, expressions for subgradients of the corresponding sup-projection (see §6.E) aren't easily available and the outer approximation algorithm (see §6.C) can't be used to solve the resulting minsup problem. Thus, motivated by the penalty method in 4.21, we may consider the approximating bifunction given by

$$g^\nu(x, y) = g(x, y) - \theta^\nu \max\{0, \varphi(x, y)\} \quad \forall (x, y) \in X \times D,$$

with $\theta^\nu \in (0, \infty)$. (A related approach based on augmented Lagrangians is described in 6.9.) Under certain assumptions, $g^\nu$ lop-converges to $g$ ancillary-tightly, which validates an approach based on the lopsided approximation algorithm for solving the minsup problem defined by $g$. In Step 1 of that algorithm, one would need $x^\nu \in \varepsilon^\nu\text{-}\operatorname{argminsup} g^\nu$, but this might be easier to accomplish than solving the actual minsup problem because $\operatorname{dom} g^\nu = X \times D$ and the outer approximation algorithm applies.

**Detail.** Suppose that $\theta^\nu \to \infty$, $X \subset \mathbb{R}^n$ is closed, $D$ is compact, $Y$ is isc relative to $X$, $g$ is continuous relative to $X \times D$ and $\varphi(x, \cdot)$ is lsc for all $x \in X$. To verify the inner lopsided condition 9.14, let $N \in \mathcal{N}_\infty^\#$, $x^\nu \in X \underset{N}{\to} x \in X$ and $y \in Y(x)$. Since $Y$ is isc relative to $X$, there's $y^\nu \in Y(x^\nu) \underset{N}{\to} y$. Using the continuity property of $g$ and the fact that $\varphi(x^\nu, y^\nu) \leq 0$ for all $\nu \in N$, one obtains

$$\operatorname{liminf}_{\nu \in N} \Big(g(x^\nu, y^\nu) - \theta^\nu \max\big\{0, \varphi(x^\nu, y^\nu)\big\}\Big) = g(x, y).$$

Since $X$ is closed, we've established the inner lopsided condition 9.14. For the outer lopsided condition 9.19, let $x \in X$ and set $x^\nu = x$. Suppose that $N \in \mathcal{N}_\infty^\#$ and $y^\nu \in D \underset{N}{\to} y$. If $y \in Y(x)$, then

$$\operatorname{limsup}_{\nu \in N} \Big(g(x^\nu, y^\nu) - \theta^\nu \max\big\{0, \varphi(x^\nu, y^\nu)\big\}\Big) \leq \operatorname{limsup}_{\nu \in N} g(x, y^\nu) = g(x, y).$$

If $y \notin Y(x)$, then $\varphi(x, y) = \delta > 0$. Thus, for $\nu \in N$ sufficiently large, $\varphi(x, y^\nu) \geq \delta/2$ because $\varphi(x, \cdot)$ is lsc and then

$$\theta^\nu \max\big\{0, \varphi(x, y^\nu)\big\} \geq \tfrac{1}{2}\theta^\nu \delta \to \infty.$$

This in turn implies that

$$g(x^\nu, y^\nu) - \theta^\nu \max\big\{0, \varphi(x^\nu, y^\nu)\big\} \underset{N}{\to} -\infty$$

and the outer lopsided condition 9.19 holds. We've established that $g^\nu \xrightarrow{l} g$ and, in fact, this takes place ancillary-tightly because $D$ is compact.

We note that $\varphi(x, y) \leq 0$ could represent multiple constraints by having

$$\varphi(x, y) = \max_{i=1,\dots,q} f_i(x, y),$$

where $f_i : \mathbb{R}^n \times \mathbb{R}^m \to \mathbb{R}$, $i = 1, \dots, q$. Even in such a structured case, the assumption that $Y$ is isc relative to $X$ is delicate and far from automatic even for smooth $f_i$. One sufficient condition is as follows:

$$D = \big\{y \in \mathbb{R}^m \,\big|\, f_0(y) \leq 0\big\}$$

for some convex $f_0 : \mathbb{R}^m \to \mathbb{R}$, $\{f_i, i = 1, \dots, q\}$ are continuous and $\{f_i(x, \cdot), i = 1, \dots, q\}$ are convex for all $x \in \mathbb{R}^n$. Moreover, for each $x \in X$, there's $\bar{y}$ such that

$$f_0(\bar{y}) < 0 \quad \text{and} \quad f_i(x, \bar{y}) < 0,\; i = 1, \dots, q.$$

Then, $Y$ is continuous at every point $x \in X$ by 7.28 and certainly isc relative to $X$. □

**Exercise 9.24** (penalty method). Consider the setting of 9.23 with

$$g(x, y) = 3(x_1 - y)^2 + (2 - y)x_2^2 + 5(x_3 + y)^2 + 2x_1 + 3x_2 - x_3 + \exp(4y^2)$$

for $x \in X$ and $y \in Y(x)$, where $X = \mathbb{R}^3$, $D = [0, 1]$ and

$$\varphi(x, y) = \tfrac{1}{4}\sin(x_1x_2) + y - \tfrac{1}{2}.$$

Solve the problem using the approach of 9.23.

**Guide.** Use the outer approximation algorithm to solve the approximating problem

$$\underset{x\in X}{\text{minimize}}\ \sup_{y\in D} g(x, y) - \theta^\nu \max\{0, \varphi(x, y)\},$$

which in turn requires minimizing $\gamma$ over $(x, \gamma) \in X \times \mathbb{R}$ subject to

$$g(x, y^i) - \theta^\nu \max\{0, \varphi(x, y^i)\} \leq \gamma, \ \ i = 1, \ldots, k,$$

where $y^1, \ldots, y^k \in D$ are points obtained during the iterations of the outer approximation algorithm. Using the fact that

$$\max\{0, \alpha\} = \max_{z\in[0,1]} \alpha z,$$

this subproblem is equivalently stated as

$$\underset{x\in X, \gamma\in\mathbb{R}, z\in[0,1]^k}{\text{minimize}}\ \gamma \ \text{ subject to } \ g(x, y^i) - \theta^\nu z_i\varphi(x, y^i) \leq \gamma, \ \ i = 1, \ldots, k,$$

where $z = (z_1, \ldots, z_k)$. A known "good" solution is $\bar{x} = (-0.3896, -1.1985, -0.2875)$, with $\sup_{y\in Y(\bar{x})} g(\bar{x}, y) = 1.9135$; see [84]. □

**Example 9.25** (existence of minsup-points). The discussion in §9.C establishes conditions under which a minsup-point of a bifunction exists. While these are based on properties of the corresponding sup-projection or the Ky Fan inequality, one can also rely on lop-convergence. Suppose that $g \in \text{bfcns}(\mathbb{R}^n, \mathbb{R}^m)$ is the bifunction of interest, but there are $g^\nu \in \text{bfcns}(\mathbb{R}^n, \mathbb{R}^m)$ lop-converging to $g$ ancillary-tightly. Then, $\text{argminsup}\, g \neq \emptyset$ provided that the sup-projection of $g$ is proper and there's $\{x^\nu \in \varepsilon^\nu\text{-argminsup}\, g^\nu, \nu \in \mathbb{N}\}$ with a cluster point for some $\varepsilon^\nu \searrow 0$.

**Detail.** This fact is an immediate consequence of 9.22. The requirement that $x^\nu \in \varepsilon^\nu$-argminsup $g^\nu$ must exist for positive $\varepsilon^\nu$ is mild because it holds when $\text{infsup}\, g^\nu \in \mathbb{R}$. The existence of a cluster point for $\{x^\nu, \nu \in \mathbb{N}\}$ is more restrictive and would require an additional assumption such as level-boundedness for the sup-projections of $g^\nu$ uniformly across $\nu$. □

**Exercise 9.26** (lack of lopsided convergence). Consider the bifunctions $\{g, g^\nu, \ \nu \in \mathbb{N}\} \subset \text{bfcns}(\mathbb{R}, \mathbb{R})$, with domains described by $X = X^\nu = [0, 1]$, $Y(x) = Y^\nu(x) = \{0\}$ for $x \in [0, 1)$, $Y^\nu(1) = [0, 1 + 1/\nu]$ and $Y(1) = [0, 1]$. Draw $\text{dom}\, g$ and $\text{dom}\, g^\nu$ and show that for each $x \in \mathbb{R}$, one has $Y^\nu(x) \xrightarrow{s} Y(x)$. Establish that regardless of the values the bifunctions take on their respective domains, $g^\nu$ doesn't lop-converge to $g$.

**Guide.** Examine the requirements of the inner lopsided condition 9.14. □

Let's now elaborate on the connections between lopsided convergence and epi-convergence, from which we eventually obtain a proof of 9.22.

**Proposition 9.27** (epi-convergence of slices). *For bifunctions* $\{g, g^\nu, \nu \in \mathbb{N}\} \subset$ bfcns$(\mathbb{R}^n, \mathbb{R}^m)$*, with domains described by* $(X, Y)$ *and* $(X^\nu, Y^\nu)$*, respectively, suppose that* $g^\nu \xrightarrow{l} g$*. Then, for* $x \in X$*, there's* $x^\nu \in X^\nu \to x$ *such that* $h^\nu \xrightarrow{e} h$*, where* $h^\nu, h : \mathbb{R}^m \to \overline{\mathbb{R}}$ *are given by*

$$h^\nu(y) = \begin{cases} -g^\nu(x^\nu, y) & \text{if } y \in Y^\nu(x^\nu) \\ \infty & \text{otherwise} \end{cases} \qquad h(y) = \begin{cases} -g(x, y) & \text{if } y \in Y(x) \\ \infty & \text{otherwise.} \end{cases}$$

**Proof.** We use the characterization of epi-convergence in 4.15. From the outer lopsided condition 9.19, there exists $x^\nu \in X^\nu \to x$ such that the functions $h^\nu, h$ satisfy 4.15(a). Next, let $y \in Y(x)$. From the inner lopsided condition 9.14, one finds $y^\nu \in Y^\nu(x^\nu) \to y$ such that

$$\limsup h^\nu(y^\nu) \le h(y).$$

For $y \notin Y(x)$, $h(y) = \infty$ and the same inequality holds. We've confirmed 4.15(b). □

**Proposition 9.28** (epi-convergence of sup-projections). *For* $\{g, g^\nu, \nu \in \mathbb{N}\} \subset$ bfcns$(\mathbb{R}^n, \mathbb{R}^m)$*, suppose that* $g^\nu \xrightarrow{l} g$ *ancillary-tightly. Then,* $f^\nu \xrightarrow{e} f$*.*

**Proof.** Let $f$ and $f^\nu$ be the sup-projections of $g$ and $g^\nu$, respectively. In view of 9.18 and the characterization 4.15 of epi-convergence as well as its proof, it suffices to show that for all $x \in \mathbb{R}^n$, there's $x^\nu \to x$ such that

$$\limsup f^\nu(x^\nu) \le f(x).$$

In fact, one can restrict the attention to $x \in \text{dom}\, f$ because any other $x$ has $f(x) = \infty$ and the inequality holds trivially.

Let $(X, Y)$ and $(X^\nu, Y^\nu)$ describe the domains of $g$ and $g^\nu$, respectively, and $x \in \text{dom}\, f$. Since $x \in X$, it follows from 9.27 that there's $x^\nu \in X^\nu \to x$ such that $h^\nu \xrightarrow{e} h$, with $h^\nu, h$ being specified in 9.27. In fact, these functions epi-converge tightly, as can be seen as follows. The proof of 9.27 shows that $\{x^\nu, \nu \in \mathbb{N}\}$ is also a sequence that can be used to confirm the outer lopsided condition 9.19. Let $\varepsilon > 0$. Since $g^\nu \xrightarrow{l} g$ ancillary-tightly, there are compact $B_\varepsilon \subset \mathbb{R}^m$ and $\nu_\varepsilon \in \mathbb{N}$ such that

$$\sup_{y \in Y^\nu(x^\nu) \cap B_\varepsilon} g^\nu(x^\nu, y) \ge \sup_{y \in Y^\nu(x^\nu)} g^\nu(x^\nu, y) - \varepsilon \quad \forall \nu \ge \nu_\varepsilon.$$

After multiplying each side with $-1$ and converting sup to inf, we find for all $\nu \ge \nu_\varepsilon$:

$$\inf_{B_\varepsilon} h^\nu = \inf_{y \in Y^\nu(x^\nu) \cap B_\varepsilon} \big\{-g^\nu(x^\nu, y)\big\} \le \inf_{y \in Y^\nu(x^\nu)} \big\{-g^\nu(x^\nu, y)\big\} + \varepsilon = \inf h^\nu + \varepsilon.$$

This confirms that $\{h^\nu, \nu \in \mathbb{N}\}$ is tight; see 5.3. Now, $\inf h^\nu \to \inf h$ by 5.5(d) and then

$$f^\nu(x^\nu) = \sup_{y \in Y^\nu(x^\nu)} g^\nu(x^\nu, y) \to \sup_{y \in Y(x)} g(x, y) = f(x)$$

as required. □

It's clear that ancillary-tight lop-convergence is a stronger condition than epi-convergence of the corresponding sup-projections, i.e., the converse of 9.28 fails. For example, the bifunctions $g, g^\nu$, both with domains $[0, 1] \times [0, 1]$ and given by $g(x, y) = 1$ and $g^\nu(x, y) = y$ for all $(x, y) \in [0, 1] \times [0, 1]$, have proper lsc sup-projections that are identical and this entails epi-convergence. However, the inner lopsided condition 9.14 fails and therefore $g^\nu$ doesn't lop-converge to $g$.

**Proof of 9.22.** Let $f$ and $f^\nu$ be the sup-projections of $g$ and $g^\nu$, respectively. By 9.28, $f^\nu \xrightarrow{e} f$. Since $x^\nu \in \varepsilon^\nu$-argmin $f^\nu$ by (9.9), this has the consequence that

$$\lim_{\nu \in N} f^\nu(x^\nu) \to \inf f$$

by 5.5(c). The convergence to infsup $g$ then follows via (9.9). The fact that $\bar{x} \in$ argminsup $g$ holds in light of 5.5(b) and (9.9). □

## 9.F Lop-Convergence II

We've seen that lopsided convergence is the key property when considering approximations of minsup problems. In view of the rather complicated definition, it's useful to have sufficient conditions for lopsided convergence that are easily verified. In this section, we'll give some conditions and also elaborate on the consequences of lopsided convergence, especially under an additional tightness assumption.

**Proposition 9.29** (sufficiency when $X = X^\nu$). *For bifunctions* $\{g, g^\nu, \nu \in \mathbb{N}\} \subset$ bfcns$(\mathbb{R}^n, \mathbb{R}^m)$, *with domains described by* $(X, Y)$ *and* $(X^\nu, Y^\nu)$, *respectively,* $g^\nu \xrightarrow{l} g$ *provided that*

(a) $\forall \nu \in \mathbb{N}$, *one has* $X^\nu = X$ *and this set is closed*
(b) $\forall x^\nu \in X \to x$, *one has* $Y^\nu(x^\nu) \xrightarrow{s} Y(x)$
(c) $\forall (x^\nu, y^\nu) \in \text{dom}\, g^\nu \to (x, y) \in \text{dom}\, g$, *one has* $g^\nu(x^\nu, y^\nu) \to g(x, y)$.

**Proof.** Let $N \in \mathcal{N}_\infty^\#$, $x^\nu \in X \to x \in X$ and $y \in Y(x)$. By (b), one can find $y^\nu \in Y^\nu(x^\nu) \underset{N}{\to} y$ and then

$$g^\nu(x^\nu, y^\nu) \underset{N}{\to} g(x, y)$$

in view of (c). Thus, the inner lopsided condition 9.14 holds because the closedness of $X$ eliminates the possibility that $\{x^\nu \in X, \nu \in \mathbb{N}\}$ converges to a point outside $X$.

To verify the outer lopsided condition 9.19, let $x \in X$ and choose $\{x^\nu = x, \nu \in \mathbb{N}\}$. Then, $Y^\nu(x) \xrightarrow{s} Y(x)$ by (b). Let $N \in \mathcal{N}_\infty^\#$ and $y^\nu \in Y^\nu(x) \underset{N}{\to} y$. This implies that $y \in Y(x)$ by (b) and

$$g^\nu(x, y^\nu) \underset{N}{\to} g(x, y)$$

as can be seen from (c). □

For example, if $C^\nu = C$ but $F^\nu$ still approximates $F$ in 9.16, then the proposition can be used to quickly confirm lopsided convergence in the context of mixed complementarity problems.

**Proposition 9.30** (sufficiency under product sets). *For* $\{g, g^\nu, \nu \in \mathbb{N}\} \subset \mathrm{bfcns}(\mathbb{R}^n, \mathbb{R}^m)$, *with domain* $X \times D$ *and* $X \times D^\nu$, *respectively, suppose that* $X$ *is closed,* $D^\nu \xrightarrow{s} D$ *and, in terms of some lsc* $\bar{g} : \mathbb{R}^n \times \mathbb{R}^m \to \mathbb{R}$ *for which* $\bar{g}(x, \cdot)$ *is usc for all* $x \in X$*, one has*

$$g(x, y) = \bar{g}(x, y) \;\; \forall (x, y) \in X \times D \quad \text{and} \quad g^\nu(x, y) = \bar{g}(x, y) \;\; \forall (x, y) \in X \times D^\nu.$$

*Then,* $g^\nu \xrightarrow{l} g$.

**Proof.** Let $N \in \mathcal{N}_\infty^\#$, $x^\nu \in X \to x \in X$ and $y \in D$. Since $D^\nu \xrightarrow{s} D$, there's $y^\nu \in D^\nu \xrightarrow[N]{} y$ and then

$$\mathrm{liminf}_{\nu \in N}\, g^\nu(x^\nu, y^\nu) = \mathrm{liminf}_{\nu \in N}\, \bar{g}(x^\nu, y^\nu) \geq \bar{g}(x, y) = g(x, y)$$

because $\bar{g}$ is lsc. Thus, the inner lopsided condition 9.14 holds; the closedness of $X$ eliminates the possibility that $\{x^\nu \in X, \nu \in \mathbb{N}\}$ converges to a point outside $X$.

To verify the outer lopsided condition 9.19, let $x \in X$ and choose $\{x^\nu = x, \nu \in \mathbb{N}\}$. Let $N \in \mathcal{N}_\infty^\#$ and $y^\nu \in D^\nu \xrightarrow[N]{} y$. Since $D^\nu \xrightarrow{s} D$, $y \in D$. Moreover,

$$\mathrm{limsup}_{\nu \in N}\, g(x^\nu, y^\nu) = \mathrm{limsup}_{\nu \in N}\, \bar{g}(x, y^\nu) \leq \bar{g}(x, y) = g(x, y)$$

by the usc of $\bar{g}(x, \cdot)$. □

**Proposition 9.31** (sufficiency when $X \neq X^\nu$). *For bifunctions* $\{g, g^\nu, \nu \in \mathbb{N}\} \subset \mathrm{bfcns}(\mathbb{R}^n, \mathbb{R}^m)$, *with domains described by* $(X, Y)$ *and* $(X^\nu, Y^\nu)$, *respectively, suppose that* $X^\nu \xrightarrow{s} X$ *and, in terms of continuous* $\bar{Y} : \mathbb{R}^n \rightrightarrows \mathbb{R}^m$ *and continuous* $\bar{g} : \mathbb{R}^n \times \mathbb{R}^m \to \mathbb{R}$,

$$Y(x) = \bar{Y}(x) \;\; \forall x \in X \quad \text{and} \quad Y^\nu(x) = \bar{Y}(x) \;\; \forall x \in X^\nu$$

$$g(x, y) = \bar{g}(x, y) \;\; \forall (x, y) \in \mathrm{dom}\, g \quad \text{and} \quad g^\nu(x, y) = \bar{g}(x, y) \;\; \forall (x, y) \in \mathrm{dom}\, g^\nu.$$

*Then,* $g^\nu \xrightarrow{l} g$.

**Proof.** Let $N \in \mathcal{N}_\infty^\#$, $x^\nu \in X^\nu \to x \in X$ and $y \in Y(x)$. Since $\bar{Y}$ is continuous, there's $y^\nu \in \bar{Y}(x^\nu) \xrightarrow[N]{} y$ and then

$$g^\nu(x^\nu, y^\nu) = \bar{g}(x^\nu, y^\nu) \xrightarrow[N]{} \bar{g}(x, y) = g(x, y).$$

This establishes the inner lopsided condition 9.14 because the case $x^\nu \in X^\nu \to x \notin X$ is ruled by $X^\nu \xrightarrow{s} X$. For the outer lopsided condition 9.19, let $x \in X$. Since $X^\nu \xrightarrow{s} X$, there's $x^\nu \in X^\nu \to x$. Let $N \in \mathcal{N}_\infty^\#$ and

$$y^\nu \in Y^\nu(x^\nu) \xrightarrow[N]{} y.$$

The continuity of $\bar{Y}$ implies that $y \in Y(x)$. Moreover, the continuity of $\bar{g}$ ensures that $g^\nu(x^\nu, y^\nu) \xrightarrow[N]{} g(x, y)$ and this suffices for the outer lopsided condition. □

**Example 9.32** (failure of lop-convergence). We caution that certain "natural" conditions aren't sufficient for lopsided convergence. In particular, for bifunctions $\{g, g^\nu, \nu \in \mathbb{N}\} \subset$ bfcns$(\mathbb{R}^n, \mathbb{R}^m)$, with domains described by $(X, Y)$ and $(X, Y^\nu)$, respectively, graphical convergence of $Y^\nu$ to $Y$ isn't sufficient for lopsided convergence even when this is the only source of approximation.

**Detail.** For a counterexample, set $X = \mathbb{R}$ and

$$Y^\nu(x) = Y(x) = \begin{cases} [-1, 1] & \text{if } x \leq 0 \\ \{0\} & \text{otherwise} \end{cases}$$

so that $\operatorname{dom} g = \operatorname{dom} g^\nu$. Let $g^\nu(x, y) = g(x, y)$ for all $(x, y) \in \operatorname{dom} g$. Then, $Y^\nu \xrightarrow{g} Y$ because $\operatorname{gph} Y^\nu = \operatorname{gph} Y$ and this set is closed. However, when considering the inner lopsided condition 9.14 with positive $x^\nu \to 0$ and $y = 1 \in Y(0)$, there's no $y^\nu \in Y^\nu(x^\nu) \to y$ and lop-convergence fails. □

Let's now establish the main consequences of lop-convergence; these supplement 9.15 and 9.22. In part, they rely on an additional tightness assumption.

**Definition 9.33** (tight bifunctions). The bifunctions $\{g^\nu, \ \nu \in \mathbb{N}\} \subset$ bfcns$(\mathbb{R}^n, \mathbb{R}^m)$ are *tight* if for all $\varepsilon > 0$, there are compact $B_\varepsilon \subset \mathbb{R}^n$ and $\nu_\varepsilon \in \mathbb{N}$ such that

$$\inf_{x \in X^\nu \cap B_\varepsilon} \sup\nolimits_{y \in Y^\nu(x)} g^\nu(x, y) \leq \inf_{x \in X^\nu} \sup\nolimits_{y \in Y^\nu(x)} g^\nu(x, y) + \varepsilon \ \ \forall \nu \geq \nu_\varepsilon,$$

where $(X^\nu, Y^\nu)$ describes $\operatorname{dom} g^\nu$.

The bifunctions *lop-converge tightly* if in addition to being tight they also lop-converge ancillary-tightly to some bifunction.

If $\cup_{\nu \in \mathbb{N}} X^\nu$ is contained in a compact set, then tightness holds but other possibilities exist; see the discussion after 5.3.

**Theorem 9.34** (consequences of lop-convergence). *For* $\{g, g^\nu, \nu \in \mathbb{N}\} \subset$ bfcns$(\mathbb{R}^n, \mathbb{R}^m)$, *suppose that* $g^\nu \xrightarrow{l} g$ *ancillary-tightly, the sup-projection of* $g$ *is proper and* $\varepsilon \in [0, \infty)$. *Then, the following hold:*

(a) $\limsup\, (\operatorname{infsup} g^\nu) \leq \operatorname{infsup} g$.
(b) *If* $\varepsilon^\nu \in [0, \infty) \to \varepsilon$, *then*

$$\operatorname{LimOut}(\varepsilon^\nu\text{-argminsup}\, g^\nu) \subset \varepsilon\text{-argminsup}\, g.$$

(c) *If the lop-convergence is tight, then* $\operatorname{infsup} g^\nu \to \operatorname{infsup} g \in \mathbb{R}$ *and there exists* $\varepsilon^\nu \in [0, \infty) \to \varepsilon$ *such that*

$$\operatorname{LimInn}(\varepsilon^\nu\text{-argminsup}\, g^\nu) \supset \varepsilon\text{-argminsup}\, g.$$

**Proof.** Let $f$ and $f^\nu$ be the sup-projections of $g$ and $g^\nu$, respectively. By 9.28, $f^\nu \xrightarrow{e} f$ and then

$$\limsup\,(\inf f^\nu) \le \inf f$$

follows from 5.5(a). Thus, (a) holds by (9.9). Part (b) follows similarly from 5.5(b). For (c), note that the tight lop-convergence implies that $f^\nu \xrightarrow{e} f$ tightly; see 5.3. The conclusion is then a direct application of 5.5(d,e) in conjunction with (9.9). □

In contrast to 9.15 and 9.22, these consequences don't rely on the knowledge of a convergent subsequence of $\varepsilon^\nu$-minsup-points for the approximating bifunctions. Moreover, (c) furnishes the guarantee that *any* minsup-point of $g$ can be approached by $\varepsilon^\nu$-minsup-points of the approximating bifunctions as long as $\varepsilon^\nu$ vanishes sufficiently slowly.

**Example 9.35** (approximation of fixed point). For nonempty, convex and compact sets $C, C^\nu \subset \mathbb{R}^n$ and $F, F^\nu : \mathbb{R}^n \to \mathbb{R}^n$, the constrained fixed-point problems

$$x = F(x) \in C \qquad \text{and} \qquad x = F^\nu(x) \in C^\nu$$

can be expressed in terms of the bifunctions $g, g^\nu \in \text{bfcns}(\mathbb{R}^n, \mathbb{R}^n)$ defined by

$$g(x,y) = \langle x - F(x), x - y\rangle \quad \forall (x,y) \in C \times C$$
$$g^\nu(x,y) = \langle x - F^\nu(x), x - y\rangle \quad \forall (x,y) \in C^\nu \times C^\nu;$$

see the development in 9.12. If $F$ is continuous relative to $C$, $F^\nu$ is continuous relative to $C^\nu$, $F^\nu(x^\nu) \to F(x)$ when $x^\nu \in C^\nu \to x \in C$, $F(x) \in C$ for all $x \in C$, $F^\nu(x) \in C^\nu$ for all $x \in C^\nu$ and $C^\nu \xrightarrow{s} C$, then $g^\nu \xrightarrow{l} g$ tightly. Moreover, the fixed points

$$\{\bar{x}^\nu = F^\nu(\bar{x}^\nu) \in C^\nu,\ \nu \in \mathbb{N}\}$$

of the approximating problems have a cluster point. Any such cluster point $\bar{x}$ satisfies $\bar{x} = F(\bar{x}) \in C$ and, consequently, is a fixed point of the actual problem.

**Detail.** We establish lop-convergence directly from its definition. For the inner lopsided condition 9.14, let $N \in \mathcal{N}_\infty^\#$, $x^\nu \in C^\nu \underset{N}{\to} x \in C$ and $y \in C$. Since $C^\nu \xrightarrow{s} C$, there's $y^\nu \in C^\nu \underset{N}{\to} y$ and then

$$g^\nu(x^\nu, y^\nu) \underset{N}{\to} g(x,y).$$

Part (b) of 9.14 is automatic because $C^\nu \xrightarrow{s} C$. For the outer lopsided condition 9.19, let $x \in C$. Since $C^\nu \xrightarrow{s} C$, there's $x^\nu \in C^\nu \to x$. Now, if $N \in \mathcal{N}_\infty^\#$ and $y^\nu \in C^\nu \underset{N}{\to} y$, then $y \in C$ and again $g^\nu(x^\nu, y^\nu) \underset{N}{\to} g(x,y)$. This completes the proof of lop-convergence.

To establish ancillary-tightness and tightness, we observe that $C^\nu \xrightarrow{s} C$, with $C^\nu, C$ nonempty, convex and compact, implies that there are $\rho \in [0,\infty)$ and $\bar{\nu}$ such that $C^\nu \subset \mathbb{B}(0,\rho)$ for all $\nu \ge \bar{\nu}$. This claim can be established as follows: For the sake of contradiction, suppose that there are $N \in \mathcal{N}_\infty^\#$ and $\{x^\nu \in C^\nu, \nu \in N\}$ escaping to the horizon, i.e., $\|x^\nu\|_2 \underset{N}{\to} \infty$. Let $x \in C$ and $\rho \in (0,\infty)$ be such that $C \subset \mathbb{B}(0,\rho/2)$. Since $C^\nu \xrightarrow{s} C$, there's $y^\nu \in C^\nu \to x$. Let $\bar{\nu}$ be such that $y^\nu \in \mathbb{B}(0,\rho)$ and $x^\nu \notin \mathbb{B}(0,\rho)$ for all $\nu \ge \bar{\nu}, \nu \in N$. Since

$x^\nu, y^\nu \in C^\nu$, the line segment between these two points is contained in $C^\nu$ by convexity. In particular, let $z^\nu$ be the point on this line segment that lies on bdry $\mathbb{B}(0,\rho)$. The sequence $\{z^\nu, \nu \geq \bar\nu, \nu \in N\}$ must have a cluster point $\bar z \in$ bdry $\mathbb{B}(0,\rho)$ because bdry $\mathbb{B}(0,\rho)$ is compact. Since $z^\nu \in C^\nu$ and $C^\nu \xrightarrow{s} C$, $\bar z \in C$. However, this contradicts the fact that bdry $\mathbb{B}(0,\rho)$ doesn't intersect with $C$ and the claim holds.

In light of the claim, both ancillary-tightness and tightness follows immediately as $\cup_{\nu\in\mathbb{N}} C^\nu$ is contained in a compact set. Moreover, the sup-projection of $g$ is proper as seen from 9.12. We can now invoke 9.34.

By the discussion in 9.12, fixed points

$$\left\{\bar x^\nu = F^\nu(\bar x^\nu) \in C^\nu, \nu \in \mathbb{N}\right\}$$

of the approximating problems exist and satisfy $\bar x^\nu \in \operatorname{argminsup} g^\nu$ and $g^\nu(\bar x^\nu, y) \leq 0$ for all $y \in C^\nu$, which means that

$$\sup_{y\in C^\nu} g^\nu(\bar x^\nu, y) = 0.$$

From the claim established above, $\{\bar x^\nu, \nu \in \mathbb{N}\}$ must have a cluster point by virtue of being a bounded sequence. Moreover, any such cluster point $\bar x$ is contained in argminsup $g$ by 9.34(b) and has value

$$\sup_{y\in C} g(\bar x, y) = 0$$

by 9.34(c). This means that $\bar x$ is a fixed point of the actual problem: $\bar x = F(\bar x) \in C$. □

Lopsided convergence addresses minsup problems with a distinct focus on the outer minimization over the inner maximization. In other minsup problems, the minimization and maximization are on more equal footing such as in saddle-point problems based on the theory in Chap. 5. Then, the parallel concept of epi/hypo-convergence emerges as the central tool; see [6].

**Exercise 9.36** (tightness). Consider $\{g, g^\nu, \nu \in \mathbb{N}\} \subset \operatorname{bfcns}(\mathbb{R}, \mathbb{R})$, with $g(x, y) = yx$ and $g^\nu(x, y) = y(x+1/\nu)$ for $(x, y) \in [0, 1]\times[0, \infty)$. Show that $g^\nu \xrightarrow{l} g$, derive explicit formulae for the sup-projections of $g$ and $g^\nu$ and assess whether they're proper functions. Check if tightness and/or ancillary-tightness hold in this case.

**Guide.** Lop-convergence can be established using 9.29. □

## 9.G Approximation of Games

We know from 9.5 that the solutions of a Nash game are fully characterized by the associated Nikaido-Isoda bifunction. Lopsided convergence of such bifunctions can then be used to study approximations in games introduced for computational and/or theoretical reasons. We'll see that such lop-convergence takes place under natural assumptions and this leads to methods for computing Nash equilibria based on the lopsided approximation algorithm.

For $\varphi_i : \mathbb{R}^{n_i} \times \mathbb{R}^{n-n_i} \to \mathbb{R}$ and $C_i : \mathbb{R}^{n-n_i} \rightrightarrows \mathbb{R}^{n_i}$, $i = 1,\ldots,m$, with $n = \sum_{i=1}^m n_i$, let's consider the Nash game

$$\Big\{\underset{x_i \in C_i(x_{-i})}{\text{minimize}}\ \varphi_i(x_i, x_{-i}), \quad i = 1, \dots, m\Big\}. \tag{9.11}$$

As in §9.A, $x_{-i} = (x_1, x_2, \dots, x_{i-1}, x_{i+1}, \dots, x_m) \in \mathbb{R}^{n-n_i}$ captures the decisions of all the agents except the $i$th one. The Nikaido-Isoda bifunction associated with the game has

$$g(x, y) = \sum\nolimits_{i=1}^{m} \big(\varphi_i(x_i, x_{-i}) - \varphi_i(y_i, x_{-i})\big) \ \ \forall x \in X, \ \ y \in Y(x),$$

where

$$X = \big\{x \in \mathbb{R}^n \,\big|\, x_i \in C_i(x_{-i}), \ i = 1, \dots, m\big\}; \qquad Y(x) = \prod\nolimits_{i=1}^{m} C_i(x_{-i}).$$

We assume that $X$ is nonempty because then $\operatorname{dom} g$ is nonempty and $g \in \operatorname{bfcns}(\mathbb{R}^n, \mathbb{R}^n)$. This simply means that there's $x \in \mathbb{R}^n$ furnishing feasible decisions for all the agents.

While (9.11) is the actual game of interest, all its components might be subject to approximations. For $\varphi_i^\nu : \mathbb{R}^{n_i} \times \mathbb{R}^{n-n_i} \to \mathbb{R}$ and $C_i^\nu : \mathbb{R}^{n-n_i} \rightrightarrows \mathbb{R}^{n_i}$, $i = 1, \dots, m$, the approximating games

$$\Big\{\underset{x_i \in C_i^\nu(x_{-i})}{\text{minimize}}\ \varphi_i^\nu(x_i, x_{-i}), \quad i = 1, \dots, m\Big\} \tag{9.12}$$

have their own Nikaido-Isoda bifunctions given by

$$g^\nu(x, y) = \sum\nolimits_{i=1}^{m} \big(\varphi_i^\nu(x_i, x_{-i}) - \varphi_i^\nu(y_i, x_{-i})\big) \ \ \forall x \in X^\nu, \ \ y \in Y^\nu(x),$$

where

$$X^\nu = \big\{x \in \mathbb{R}^n \,\big|\, x_i \in C_i^\nu(x_{-i}), \ i = 1, \dots, m\big\}; \qquad Y^\nu(x) = \prod\nolimits_{i=1}^{m} C_i^\nu(x_{-i}).$$

We assume that $X^\nu$ is nonempty so that $g^\nu \in \operatorname{bfcns}(\mathbb{R}^n, \mathbb{R}^n)$.

**Proposition 9.37** (lop-convergence of Nikaido-Isoda). *For Nikaido-Isoda bifunctions* $\{g, g^\nu, \nu \in \mathbb{N}\} \subset \operatorname{bfcns}(\mathbb{R}^n, \mathbb{R}^n)$ *defined above, suppose that*

$$\forall x^\nu \to x, \ i = 1, \dots, m: \quad C_i^\nu(x_{-i}^\nu) \xrightarrow{s} C_i(x_{-i}) \ \ \textit{and} \ \ \varphi_i^\nu(x_i^\nu, x_{-i}^\nu) \to \varphi_i(x_i, x_{-i}).$$

*Then,* $\{g, g^\nu\}$ *satisfies the inner lopsided condition 9.14. They also satisfy the outer lopsided condition 9.19, which then implies that* $g^\nu \xrightarrow{l} g$, *under the additional assumption that either one of the following two properties holds:*

(a) $C_i^\nu(x_{-i}) \supset C_i(x_{-i})$ *for all* $x_{-i} \in \mathbb{R}^{n-n_i}$, $i = 1, \dots, m$ *and* $\nu \in \mathbb{N}$.
(b) *There are* $D_i^\nu, D_i \subset \mathbb{R}^{n_i}$ *such that* $C_i^\nu(x_{-i}) = D_i^\nu$ *and* $C_i(x_{-i}) = D_i$ *for all* $x_{-i} \in \mathbb{R}^{n-n_i}$, $i = 1, \dots, m$ *and* $\nu \in \mathbb{N}$.

**Proof.** For the inner lopsided condition 9.14, let $N \in \mathcal{N}_\infty^\#$, $x^\nu \in X^\nu \xrightarrow[N]{} x \in X$ and $y \in Y(x)$. By assumption,

$$C_i^\nu(x_{-i}^\nu) \xrightarrow{s} C_i(x_{-i})$$

along the subsequence with $\nu \in N$ for all $i = 1, \dots, m$ so that

$$Y^\nu(x^\nu) \xrightarrow{s} Y(x)$$

as well for such $\nu$; see 4.26. This set-convergence guarantees that there exists $y^\nu \in Y^\nu(x^\nu) \underset{N}{\to} y$ and then $g^\nu(x^\nu, y^\nu) \underset{N}{\to} g(x, y)$ in light of the assumption. Thus, 9.14(a) is satisfied. Next, since

$$X^\nu = \bigcap_{i=1}^{m} \{x \in \mathbb{R}^n \mid x_i \in C_i^\nu(x_{-i})\},$$

it follows by the assumption and 4.29 that

$$\begin{aligned} \text{LimOut } X^\nu &\subset \bigcap_{i=1}^{m} \text{LimOut } \{x \in \mathbb{R}^n \mid x_i \in C_i^\nu(x_{-i})\} \\ &\subset \bigcap_{i=1}^{m} \{x \in \mathbb{R}^n \mid x_i \in C_i(x_{-i})\} = X. \end{aligned}$$

Thus, every cluster point of $\{x^\nu \in X^\nu, \nu \in \mathbb{N}\}$ is in $X$ and 9.14(b) is automatically satisfied.

For the outer lopsided condition 9.19, let $x \in X$. First, suppose that assumption (a) holds. Then, set $x^\nu = x$, which is in $X^\nu$ because $X^\nu \supset X$. Suppose that $N \in \mathcal{N}_\infty^\#$ and

$$y^\nu \in Y^\nu(x) \underset{N}{\to} y \in Y(x).$$

Thus, $g^\nu(x^\nu, y^\nu) \underset{N}{\to} g(x, y)$. This implies the outer lopsided condition as the possibility

$$y^\nu \in Y^\nu(x) \underset{N}{\to} y \notin Y(x)$$

is ruled out by the fact that $Y^\nu(x) \xrightarrow{s} Y(x)$; see 4.26. Second, suppose that assumption (b) holds. Since in this case

$$X = \prod_{i=1}^{m} D_i \quad \text{and} \quad X^\nu = \prod_{i=1}^{m} D_i^\nu,$$

one has $X^\nu \xrightarrow{s} X$ by 4.26 and there's $x^\nu \in X^\nu \to x$. Suppose that $N \in \mathcal{N}_\infty^\#$ and

$$y^\nu \in Y^\nu(x^\nu) \underset{N}{\to} y \in Y(x).$$

Again, $g^\nu(x^\nu, y^\nu) \underset{N}{\to} g(x, y)$. This concludes the proof of the outer lopsided condition as the possibility $y^\nu \in Y^\nu(x) \underset{N}{\to} y \notin Y(x)$ can again be ruled out. □

Although other options exist, it's generally impossible to completely eliminate some qualification of the kind stated as assumptions (a,b) in 9.37. The source of the difficulty is that $X^\nu \xrightarrow{s} X$ may not hold if these assumptions are eliminated. Although LimOut $X^\nu \subset X$ would still hold, as leveraged in the above proof, the inclusion may be strict. As an example, suppose that $m = 2$, $n_1 = n_2 = 1$ and for all $x \in \mathbb{R}^2$, $i = 1, 2$ and $\nu \in \mathbb{N}$, let

$$C_i(x_{-i}) = \{0\} \cup \{x_{-i}\} \quad \text{and} \quad C_i^\nu(x_{-i}) = \{0\} \cup \{x_{-i} + 1/\nu\}.$$

In this case,

$$X = \{x \in \mathbb{R}^2 \mid x_1 = x_2\} \quad \text{and} \quad X^\nu = \{(0, 0), (0, 1/\nu), (1/\nu, 0)\} \ \ \forall \nu.$$

Clearly, LimOut $X^\nu = \{(0,0)\}$ is a strict subset of $X$. This takes place even though

$$C_i^\nu(x_{-i}^\nu) \xrightarrow{s} C_i(x_{-i}) \quad \text{when} \quad x^\nu \to x.$$

The results furnished by 9.15, 9.22 and 9.34 can now be translated into guarantees about cluster points of Nash equilibria of approximating games.

**Proposition 9.38** (convergence of Nash equilibria). *For Nikaido-Isoda bifunctions $\{g, g^\nu, \nu \in \mathbb{N}\} \subset \text{bfcns}(\mathbb{R}^n, \mathbb{R}^n)$ associated with the actual game (9.11) and the approximating games (9.12), suppose that the inner lopsided condition 9.14 holds for $\{g, g^\nu\}$, which in particular would be the case when $g^\nu \xrightarrow{l} g$. If $\{\bar{x}^\nu, \nu \in \mathbb{N}\}$ are Nash equilibria of the approximating games with a cluster point $\bar{x}$, then $\bar{x}$ is a Nash equilibrium of the actual game.*

**Proof.** Let $(X^\nu, Y^\nu)$ and $(X, Y)$ be the descriptions of $\text{dom}\, g^\nu$ and $\text{dom}\, g$, respectively. By 9.5,

$$\bar{x}^\nu \in \text{argminsup}\, g^\nu \quad \text{and} \quad \sup_{y \in Y^\nu(\bar{x}^\nu)} g^\nu(\bar{x}^\nu, y) \leq 0.$$

Suppose that $\bar{x}^\nu \underset{N}{\to} \bar{x}$ for some $N \in \mathcal{N}_\infty^\#$. For $x \in X$, we've $x \in Y(x)$ and $g(x,x) = 0$ so that

$$\sup_{y \in Y(x)} g(x, y) \geq 0.$$

Thus, infsup $g \geq 0$. It follows from 9.15(b) that $\bar{x} \in X$ and $\sup_{y \in Y(\bar{x})} g(\bar{x}, y) = 0$, and then $\bar{x} \in \text{argminsup}\, g$. By 9.5, this implies that $\bar{x}$ is a Nash equilibrium of the actual game. □

One can now combine this proposition with 9.37 to justify the application of the lopsided approximation algorithm to solve a Nash game by means of approximations that satisfy the initial assumptions of 9.37. One setting that falls outside 9.37, however, is that of penalty methods and these are addressed next.

**Example 9.39** (lop-convergence in penalty methods). For computational reasons, an agent's constraints may be removed and replaced with a penalty in the objective function. This natural idea also leads to lop-convergence under mild assumptions and resembles 9.23. Specifically, for $h_i : \mathbb{R}^{n_i} \times \mathbb{R}^{n-n_i} \to \mathbb{R}$, $i = 1, \dots, m$, suppose that the feasible sets in the actual game (9.11) are of the form

$$C_i(x_{-i}) = \{x_i \in \mathbb{R}^{n_i} \mid h_i(x_i, x_{-i}) \leq 0\}.$$

Then, we can construct the approximating games (9.12) by setting

$$C_i^\nu(x_{-i}) = \mathbb{R}^{n_i} \quad \text{and} \quad \varphi_i^\nu(x_i, x_{-i}) = \varphi_i(x_i, x_{-i}) + \theta^\nu \max\{0, h_i(x_i, x_{-i})\},$$

where $\theta^\nu \in (0, \infty)$. In the approximating games, the agents are solving unconstrained problems, but with penalized cost functions. Presumably, it's much simpler to solve the approximating games than the actual one.

The domains of the corresponding Nikaido-Isoda bifunctions $g, g^\nu$ are specified by

$$X = \{x \in \mathbb{R}^n \mid h_i(x_i, x_{-i}) \le 0,\ \ i = 1, \dots, m\}$$

$$Y(x) = \prod\nolimits_{i=1}^m \{y_i \in \mathbb{R}^{n_i} \mid h_i(y_i, x_{-i}) \le 0\},\ \ x \in X,$$

and by $X^\nu = Y^\nu(x) = \mathbb{R}^n$. Let's extend the definition of $Y(x)$ to hold for all $x \in \mathbb{R}^n$. If $\varphi_i$ and $h_i$ are continuous, $Y(x) \neq \emptyset$ for all $x \in \mathbb{R}^n$ and $\theta^\nu \to \infty$, then $g^\nu \xrightarrow{l} g$ provided that the following qualification holds:

$$\forall x, y \in \mathbb{R}^n \text{ with } y \in Y(x),\ \ \exists y^\nu \to y \text{ such that } h_i(y_i^\nu, x_{-i}) < 0\ \forall i, \nu.$$

This example illustrates the situation when $\operatorname{dom} g^\nu$ doesn't approach $\operatorname{dom} g$, but still $g^\nu$ lop-converges to $g$.

**Detail.** We start by establishing the following property: For all $x^\nu \to x \in \mathbb{R}^n$ and $y \in Y(x)$, there's $y^\nu \to y$ such that $y^\nu \in Y(x^\nu)$ for all $\nu$.

To see this, let $x, y \in \mathbb{R}^n$ satisfy $y \in Y(x)$ and $x^\nu \to x$. By the qualification, there's $\bar{y}^k \to y$, as $k \to \infty$, such that $h_i(\bar{y}_i^k, x_{-i}) < 0$ for all $i, k$. Since $h_i$ is continuous, one can find $\{\varepsilon^k > 0,\ k \in \mathbb{N}\}$ such that

$$h_i(\bar{y}_i^k, \bar{x}_{-i}) \le 0\ \ \forall i,\ \bar{x} \in \mathbb{B}(x, \varepsilon^k).$$

Since $x^\nu \to x$, there exists $\nu_1 \in \mathbb{N}$ such that $x^\nu \in \mathbb{B}(x, \varepsilon^1)$ for all $\nu \ge \nu_1$. Moreover, for $k = 2, 3, \dots$, there exists $\nu_k > \nu_{k-1}$ such that $x^\nu \in \mathbb{B}(x, \varepsilon^k)$ for all $\nu \ge \nu_k$. We can then construct the sequence $\{y^\nu, \nu \in \mathbb{N}, \nu \ge \nu_1\}$ by setting

$$y^\nu = \bar{y}^k\ \ \forall \nu_k \le \nu < \nu_{k+1},\ \ k \in \mathbb{N}.$$

Since $\bar{y}^k \to y$, one has $y^\nu \to y$ as well. Thus, for all $k \in \mathbb{N}, i = 1, \dots, m$ and $\nu_k \le \nu < \nu_{k+1}$,

$$h_i(y_i^\nu, x_{-i}^\nu) = h_i(\bar{y}_i^k, x_{-i}^\nu) \le 0$$

because $x^\nu \in \mathbb{B}(x, \varepsilon^k)$. This establishes the claimed property.

We proceed by showing lop-convergence directly from 9.21. Suppose that $N \in \mathcal{N}_\infty^\#$, $x^\nu \xrightarrow{N} x \in X$ and $y \in Y(x)$. By the property just established, there's $y^\nu \xrightarrow{N} y$ such that

$$\max\{0, h_i(y_i^\nu, x_{-i}^\nu)\} = 0,\ \ i = 1, \dots, m.$$

Hence,

$$\begin{aligned}
&\operatorname{liminf}_{\nu \in N} g^\nu(x^\nu, y^\nu)\\
&= \operatorname{liminf}_{\nu \in N} \sum\nolimits_{i=1}^m \Big(\varphi_i(x_i^\nu, x_{-i}^\nu) + \theta^\nu \max\{0, h_i(x_i^\nu, x_{-i}^\nu)\} - \varphi_i(y_i^\nu, x_{-i}^\nu)\Big)\\
&\ge \operatorname{liminf}_{\nu \in N} \sum\nolimits_{i=1}^m \big(\varphi_i(x_i^\nu, x_{-i}^\nu) - \varphi_i(y_i^\nu, x_{-i}^\nu)\big) = g(x, y),
\end{aligned}$$

where the last equality follows from the continuity of $\varphi_i$. Next, suppose that $x^\nu \xrightarrow{N} x \notin X$. Since $Y(x)$ is nonempty, there's $y \in Y(x)$ and then also $y^\nu \xrightarrow{N} y$ with $\max\{0, h_i(y_i^\nu, x_{-i}^\nu)\} = 0$

for all $i$ by the established claim. Thus,

$$g^\nu(x^\nu, y^\nu) = \sum_{i=1}^{m} \Big(\varphi_i(x_i^\nu, x_{-i}^\nu) + \theta^\nu \max\{0, h_i(x_i^\nu, x_{-i}^\nu)\} - \varphi_i(y_i^\nu, x_{-i}^\nu)\Big) \underset{N}{\to} \infty$$

because $\theta^\nu \to \infty$ and

$$\sum_{i=1}^{m} \max\{0, h_i(x_i^\nu, x_{-i}^\nu)\}$$

remains bounded away from 0 as $\nu \to \infty$, $\nu \in N$, by virtue of $x \notin X$ and the continuity of $h_i$. Consequently, the inner lopsided condition 9.14 holds. Next, suppose that $x \in X$ and set $x^\nu = x$, which imply that $\max\{0, h_i(x_i, x_{-i})\} = 0$ for all $i$. Let $N \in \mathcal{N}_\infty^\#$ and $y^\nu \underset{N}{\to} y$. If $y \in Y(x)$, then

$$\begin{aligned}
&\text{limsup}_{\nu \in N}\, g^\nu(x^\nu, y^\nu) \\
&= \text{limsup}_{\nu \in N} \sum_{i=1}^{m} \Big(\varphi_i(x_i, x_{-i}) - \varphi_i(y_i^\nu, x_{-i}) - \theta^\nu \max\{0, h_i(y_i^\nu, x_{-i})\}\Big) \\
&\leq \text{limsup}_{\nu \in N} \sum_{i=1}^{m} \big(\varphi_i(x_i, x_{-i}) - \varphi_i(y_i^\nu, x_{-i})\big) = g(x, y).
\end{aligned}$$

If $y \notin Y(x)$, then $g^\nu(x^\nu, y^\nu) \underset{N}{\to} -\infty$ because $\theta^\nu \to \infty$ and

$$\sum_{i=1}^{m} \max\{0, h_i(y_i^\nu, x_{-i})\}$$

remains bounded away from 0 as $\nu \to \infty$ due to the fact that $y \notin Y(x)$ and $h_i$ is continuous. This establishes the outer lopsided condition 9.19.

The assumption $Y(x) \neq \emptyset$ for all $x \in \mathbb{R}^n$ enables each agent to select a feasible decision regardless of the decisions of the others. The qualification assumes that given decisions by the others and a feasible decision of its own, an agent can unilaterally make its constraint inactive by moving slightly away from its decision. This qualification resembles the Slater condition 5.47. □

## 9.H Walras Barter Model

Let's return to the Walras barter model described in 9.6 and examine the existence of solutions as well as the role of approximations. Using the notation developed there, we observe that the constraint $\langle p, x_i \rangle \leq \langle p, w_i \rangle$ implies an invariance to scaling of the price vector $p$; the agent's decision doesn't change if the trading takes place in euro instead of dollar. Thus, we can just as well "normalize" the price vector and assume it lies in

$$P = \Big\{p \in [0, \infty)^n \,\Big|\, \sum_{j=1}^{n} p_j = 1\Big\}.$$

This results in a slight change in the definition of the Walrasian bifunction:

$$g(p, q) = \big\langle G(p), q \big\rangle \quad \forall (p, q) \in P \times P, \quad \text{where } G(p) = \sum_{i=1}^{m} \big(x_i(p) - w_i\big).$$

We recall that $w_i \in [0, \infty)^n$ is the endowment of agent $i$,

$$C_i(p) = \{x_i \in [0, \infty)^n \mid \langle p, x_i \rangle \le \langle p, w_i \rangle\}$$

and

$$\{x_i(p)\} = \{(x_{i1}(p), \dots, x_{in}(p))\} = \operatorname{argmin}_{x_i \in C_i(p)} f_i(x_i) \tag{9.13}$$

is the decision by agent $i$ under price $p$, which is assumed to be unique; see 9.6. Certainly, $g \in \operatorname{bfcns}(\mathbb{R}^n, \mathbb{R}^n)$.

If $p \mapsto x_{ij}(p)$ is lsc relative to $P$ for all $i$ and $j$, then $g$ is a Ky Fan function because for all $q \in P$, $g(\cdot, q)$ is lsc relative to $P$ and for all $p \in P$, $g(p, \cdot)$ is concave relative to $P$. Moreover, for $p \in P$,

$$g(p, p) = \sum_{j=1}^{n} \sum_{i=1}^{m} (x_{ij}(p) - w_{ij}) p_j = \sum_{i=1}^{m} \sum_{j=1}^{n} (x_{ij}(p) - w_{ij}) p_j$$
$$= \sum_{i=1}^{m} \langle x_i(p) - w_i, p \rangle \le 0$$

because $\langle p, x_i(p) \rangle \le \langle p, w_i \rangle$ for all $i$. Since $P$ is convex and compact, we can apply the Ky Fan inequality 9.10 to establish that $\operatorname{argminsup} g \ne \emptyset$ and $g(\bar{p}, q) \le 0$ for all $\bar{p} \in \operatorname{argminsup} g$ and $q \in P$. From the discussion in 9.6, we conclude that there exists a weak equilibrium price vector provided that $p \mapsto x_{ij}(p)$ is lsc relative to $P$ for all $i, j$. Under the additional assumption that each $f_i$ is decreasing in all arguments, there's also an equilibrium price; see 9.6.

We next examine more closely what it takes for $p \mapsto x_{ij}(p)$ to be lsc relative to $P$. In fact, under moderate assumptions, $p \mapsto x_i(p)$ is continuous relative to $P$, which then establishes the existence of a weak equilibrium price vector by the earlier discussion. Before we state that continuity result, we prove an intermediate fact of independent interest as it supplements 5.5.

**Proposition 9.40** (convergence of minimizers of convex functions). *For lsc convex functions $f, f^\nu : \mathbb{R}^n \to \overline{\mathbb{R}}$, suppose that $f^\nu \xrightarrow{e} f$ and $f$ is proper with* argmin $f$ *consisting of a single point. Then,* argmin $f^\nu \xrightarrow{s}$ argmin $f$.

**Proof.** Let $\rho \in (0, \infty)$, $\{\bar{x}\} = \operatorname{argmin} f$ and $B = \mathbb{B}(\bar{x}, \rho)$. Set $\eta = \inf_{\operatorname{bdry} B} f$. If $\eta < \infty$, then $f(z) < \infty$ for some $z \in \operatorname{bdry} B$ so 4.9 applies,

$$\operatorname{argmin}_{\operatorname{bdry} B} f \ne \emptyset \quad \text{and} \quad \eta = f(\bar{z}) \ \ \forall \bar{z} \in \operatorname{argmin}_{\operatorname{bdry} B} f.$$

Since $\bar{x} \notin \operatorname{bdry} B$, $f(\bar{z}) > \inf f$. Thus, regardless of $\eta$ being finite or not, there's $\delta \in (0, \infty)$ such that

$$f(z) \ge \inf f + \delta \quad \forall z \in \operatorname{bdry} B. \tag{9.14}$$

Since $f^\nu \xrightarrow{e} f$, there's $y^\nu \to \bar{x}$ such that $\limsup f^\nu(y^\nu) \le f(\bar{x})$ by 4.15(b). Thus, there's $\nu_0$ such that

$$f^\nu(y^\nu) \le \inf f + \tfrac{1}{3}\delta \quad \text{and} \quad y^\nu \in B \ \ \forall \nu \ge \nu_0.$$

For the sake of contradiction, suppose that there are $N \in \mathcal{N}_\infty^\#$ and $\{x^\nu \notin B, \nu \in N\}$ such that $f^\nu(x^\nu) \le f^\nu(y^\nu)$ for all $\nu \in N$. For $\nu \in N$ with $\nu \ge \nu_0$, let $z^\nu$ be the unique point in bdry $B$ on the line segment between $x^\nu$ and $y^\nu$, i.e.,

$$z^\nu = (1 - \lambda^\nu)x^\nu + \lambda^\nu y^\nu \quad \text{for some } \lambda^\nu \in [0, 1].$$

The convexity inequality 1.10 applied to $f^\nu$ implies that

$$\begin{aligned} f^\nu(z^\nu) &\le (1 - \lambda^\nu) f^\nu(x^\nu) + \lambda^\nu f^\nu(y^\nu) \\ &\le (1 - \lambda^\nu) f^\nu(y^\nu) + \lambda^\nu f^\nu(y^\nu) = f^\nu(y^\nu) \le \inf f + \tfrac{1}{3}\delta \end{aligned} \tag{9.15}$$

for $\nu \in N$ with $\nu \ge \nu_0$. Since $\{z^\nu, \nu \in N, \nu \ge \nu_0\}$ is contained in the compact set bdry $B$, it has a cluster point $\bar{z} \in$ bdry $B$. After passing to the corresponding subsequence, which we also denote by $N$, we find that

$$\operatorname{liminf}_{\nu \in N} f^\nu(z^\nu) \ge f(\bar{z}) \ge \inf f + \delta,$$

where the first inequality follows by 4.15(a) and the second one by (9.14). Thus, for sufficiently large $\nu \in N$,

$$f^\nu(z^\nu) \ge \inf f + \tfrac{2}{3}\delta.$$

However, this contradicts (9.15) and we conclude that there's $\nu_1 \ge \nu_0$ such that $f^\nu(x^\nu) > f^\nu(y^\nu)$ for all $x^\nu \notin B$ and $\nu \ge \nu_1$. This means that

$$\operatorname{argmin} f^\nu = \operatorname{argmin}_B f^\nu \quad \forall \nu \ge \nu_1.$$

Since $f^\nu + \iota_B$ is lsc, $B$ is compact and $f^\nu(y^\nu) < \infty$ for $\nu \ge \nu_1$, $\operatorname{argmin}_B f^\nu \neq \emptyset$ for such $\nu$ by 4.9. We've shown that argmin $f^\nu$ is nonempty and contained in $\mathbb{B}(\bar{x}, \rho)$ for all $\nu \ge \nu_1$. Since we can repeat this argument with successively smaller $\rho$, $\operatorname{argmin} f^\nu \xrightarrow{S} \{\bar{x}\}$. □

**Proposition 9.41** (continuity of agent decision). *For agent i, suppose that the endowment* $w_i = (w_{i1}, \ldots, w_{in})$ *has* $w_{ij} \in (0, \infty)$ *for all j and* $f_i$ *is level-bounded and of the form*

$$f_i(x_i) = \varphi_i(x_i) + \iota_{X_i}(x_i), \quad \textit{where } X_i = \big\{x_i \in [0, \infty)^n \bigm| h_i(x_i) \le 0\big\},$$

$\varphi_i : \mathbb{R}^n \to \mathbb{R}$ *is strictly convex and* $h_i : \mathbb{R}^n \to \mathbb{R}$ *is convex and satisfies* $h_i(w_i) < 0$. *Then, the decisions* $\{x_i(p), p \in P\}$ *given by* (9.13) *are well-defined and* $p \mapsto x_i(p)$ *is continuous relative to P.*

**Proof.** First, let $\bar{p} \in P$ and $S : \mathbb{R}^n \rightrightarrows \mathbb{R}^n$ be given by

$$S(p) = \big\{x_i \in X_i \bigm| \langle p, x_i \rangle \le \langle p, w_i \rangle\big\}.$$

Since $h_i(w_i) < 0$ and $h_i$ is continuous, there's $\varepsilon > 0$ such that $h_i(x_i) < 0$ for all $x_i \in \mathbb{B}(w_i, \varepsilon)$. Moreover, $\bar{p} \neq 0$ and $w_{ij} > 0$ for all $j$ so $\langle \bar{p}, w_i \rangle > 0$. This implies the existence of $\bar{x}_i \in \mathbb{B}(w_i, \varepsilon)$, with $\bar{x}_{ij} > 0$ for all $j$, such that $\langle \bar{p}, \bar{x}_i \rangle < \langle \bar{p}, w_i \rangle$. We've satisfied the requirements of 7.28 and conclude that $S$ is continuous at $\bar{p}$.

Second, let $p^\nu \in P \to \bar{p}$. Then, $S(p^\nu) \xrightarrow{s} S(\bar{p})$ and, consequently,

$$\iota_{S(p^\nu)} \xrightarrow{e} \iota_{S(\bar{p})};$$

see the discussion after 4.26. Since $\varphi_i$ is continuous, we can invoke the epigraphical sum rule 4.19(a) to establish that

$$\varphi_i + \iota_{S(p^\nu)} \xrightarrow{e} \varphi_i + \iota_{S(\bar{p})}.$$

Third, for $p \in P$, $\varphi_i + \iota_{S(p)}$ is level-bounded, lsc and proper so that its set of minimizers is nonempty by 4.9. Moreover, it contains exactly one point because $\varphi_i$ is strictly convex; cf. 1.16 and 1.18(b). Thus, $\{x_i(p), p \in P\}$ are well-defined. We can then bring in 9.40 to conclude that

$$\operatorname{argmin} \varphi_i + \iota_{S(p^\nu)} \xrightarrow{s} \operatorname{argmin} \varphi_i + \iota_{S(\bar{p})}$$

and we also have $x_i(p^\nu) \to x_i(\bar{p})$. □

The assumption $w_{ij} > 0$ for all $j$ requires that agent $i$ has at least some amount of each good and stems from difficulties when $\langle \bar{p}, w_i \rangle = 0$. For example, consider two goods, $w_i = (1, 0)$, $\bar{p} = (0, 1)$ and no constraints except nonnegativity so that $X_i = [0, \infty)^2$. Then, the feasible set of the agent under $\bar{p}$ is $S(\bar{p}) = [0, \infty) \times \{0\}$. However, $p^\nu = (1/\nu, 1-1/\nu) \in P$ produces the feasible set

$$S(p^\nu) = \big\{x_i \in [0, \infty)^2 \;\big|\; x_{i1} + (\nu - 1)x_{i2} \le 1\big\},$$

which doesn't set-converge to $S(\bar{p})$ even though $p^\nu \to \bar{p}$. Without further assumptions, the unique minimizer $x_i(\bar{p})$ could be $(10, 0)$, which is nowhere near $S(p^\nu)$ for any $\nu$. Thus, $p \mapsto x_i(p)$ isn't continuous at $\bar{p}$.

Next, let's turn to approximations. Suppose that $f_i^\nu : \mathbb{R}^n \to \overline{\mathbb{R}}$ and $w_i^\nu \in [0, \infty)^n$ are substitutes for $f_i$ and $w_i$, $i = 1, \dots, m$. Approximations may arise in the context of a sensitivity analysis that aims to establish conditions under which small changes in the objective functions and the endowments have only minor effect on the agents' decisions and the equilibrium price vector. A model that lacks such stability can easily be misleading. For $p \in P$, the approximations result in the agent problems

$$\forall i = 1, \dots, m: \quad \big\{x_i^\nu(p)\big\} = \operatorname{argmin}_{x_i \in C_i^\nu(p)} f_i^\nu(x_i), \qquad (9.16)$$
$$\text{with } C_i^\nu(p) = \big\{x_i \in [0, \infty)^n \;\big|\; \langle p, x_i\rangle \le \langle p, w_i^\nu\rangle\big\},$$

where we again assume that each set of minimizers is a single point. The corresponding Walrasian bifunction is

$$g^\nu(p, q) = \big\langle G^\nu(p), q\big\rangle \;\; \forall (p, q) \in P \times P, \;\; \text{with } G^\nu(p) = \sum\nolimits_{i=1}^m \big(x_i^\nu(p) - w_i^\nu\big).$$

By 9.29, $g^\nu \xrightarrow{l} g$ provided that $G^\nu(p^\nu) \to G(p)$ for $p^\nu \in P \to p$. Then, in fact, the lop-convergence is both ancillary-tight and tight because $P$ is compact. Several conclusions are now available via 9.15, 9.22 and 9.34. In particular, if $\bar{p}$ is a cluster point of a sequence of equilibrium price vectors $\{p^\nu, \nu \in \mathbb{N}\}$ corresponding to the approximating setting, i.e., $p^\nu \geq 0$, $G^\nu(p^\nu) \leq 0$ and $\langle G^\nu(p^\nu), p^\nu\rangle = 0$, then $\bar{p} \in \operatorname{argminsup} g$ and $g(\bar{p}, q) \leq 0$ for all $q \in P$. Thus, $\bar{p}$ is a weak equilibrium price vector for the actual problem and also an equilibrium price vector under the additional assumption that each $f_i$ is decreasing in all arguments; see 9.6.

The assumption $G^\nu(p^\nu) \to G(p)$ whenever $p^\nu \in P \to p$ can be satisfied by extending the arguments in 9.41.

**Proposition 9.42** (decisions under approximation). *For agent i, suppose that the assumptions of 9.41 hold, $f_i^\nu : \mathbb{R}^n \to \overline{\mathbb{R}}$ has*

$$f_i^\nu(x_i) = \varphi_i^\nu(x_i) + \iota_{X_i}(x_i),$$

*where $\varphi_i^\nu : \mathbb{R}^n \to \mathbb{R}$ is strictly convex, $X_i$ is as in 9.41, $\varphi_i^\nu \xrightarrow{e} \varphi_i$ and $w_i^\nu \to w_i$. Then, the decisions*

$$\{x_i(p), x_i^\nu(p),\ p \in P\}$$

*in (9.13) and (9.16) are well-defined for sufficiently large ν and*

$$x_i^\nu(p^\nu) \to x_i(\bar{p})$$

*provided that $p^\nu \in P \to \bar{p}$. Moreover, if these assumptions hold for all $i = 1, \dots, m$, then*

$$G^\nu(p^\nu) \to G(\bar{p}).$$

**Proof.** Let $p^\nu \in P \to \bar{p}$,

$$C = \{x_i \in X_i \mid \langle \bar{p}, x_i\rangle \leq \langle \bar{p}, w_i\rangle\} \quad \text{and} \quad C^\nu = \{x_i \in X_i \mid \langle p^\nu, x_i\rangle \leq \langle p^\nu, w_i^\nu\rangle\}.$$

We deduce from the proof of 9.41 that $C^\nu \xrightarrow{s} C$ and then $\iota_{C^\nu} \xrightarrow{e} \iota_C$; see the discussion after 4.26. Since $\varphi_i^\nu \xrightarrow{e} \varphi_i$ and $-\varphi_i^\nu \xrightarrow{e} -\varphi_i$ by 4.18, we can invoke the epigraphical sum rule 4.19(a) to confirm that

$$\varphi_i^\nu + \iota_{C^\nu} \xrightarrow{e} \varphi_i + \iota_C.$$

By 9.41, $\{x_i(p), p \in P\}$ is well-defined. We can then bring in 9.40 to conclude that

$$\operatorname{argmin} \varphi_i^\nu + \iota_{C^\nu} \xrightarrow{s} \operatorname{argmin} \varphi_i + \iota_C$$

because $\varphi_i^\nu + \iota_{C^\nu}$ is lsc and convex. Thus, for sufficiently large $\nu$, $\operatorname{argmin} \varphi_i^\nu + \iota_{C^\nu}$ is nonempty and then must contain exactly a single point because $\varphi_i^\nu$ is strictly convex; cf. 1.16 and 1.18(b). Moreover, $\{x_i^\nu(p), p \in P\}$ are well-defined and $x_i^\nu(p^\nu) \to x_i(\bar{p})$. □

**Example 9.43** (two-agent barter model). Suppose that there are two goods traded between two agents and let $x_{ij}$ be the holding of good $j$ by agent $i$. The objective functions for the agents are $f_i = \varphi_i + \iota_{[0,\infty)^2}$, where

$$\varphi_1(x_1) = (x_{11} - 1)^2 + (x_{12} - 1)^2 \quad \text{and} \quad \varphi_2(x_2) = (x_{21} - 2)^2 + (x_{22} - 1)^2.$$

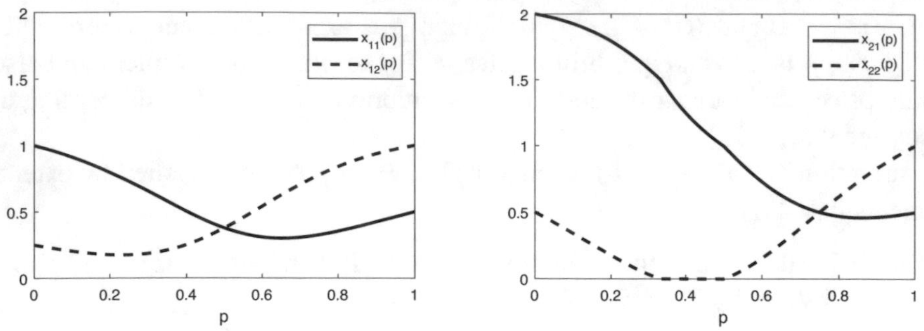


Fig. 9.2: Decisions for agent 1 (left) and agent 2 (right).

Here, $x_1 = (x_{11}, x_{12})$ and $x_2 = (x_{21}, x_{22})$. The endowments are $w_1 = (1/2, 1/4)$ and $w_2 = (1/2, 1/2)$. The objective functions are strictly convex by 1.24 and the other assumptions of 9.41 hold as well. Let $p \in [0, 1]$ be the price for the first good and $1 - p$ the price for the second good, which then result in a price vector $(p, 1 - p) \in P$. Figure 9.2(left) confirms the continuity property in 9.41 of the minimizer $(x_{11}(p), x_{12}(p))$ for agent 1 as a function of $p$. The agent's decision reflects a balance between several concerns and the holding of good 1, $x_{11}(p)$, doesn't necessarily decrease as the price of it increases. The agent also benefits from a higher price as the value of $w_{11}$ increases. Similarly, Figure 9.2(right) presents the minimizer $(x_{21}(p), x_{22}(p))$ for agent 2 as a function of $p$.

The black lines in Figure 9.3 show the excess of holdings over endowments, i.e.,

$$G(p) = \big(x_{11}(p) - w_{11} + x_{21}(p) - w_{21},\ \ x_{12}(p) - w_{12} + x_{22}(p) - w_{22}\big); \tag{9.17}$$

the first component (good 1) is represented by the solid line and the second component (good 2) by the dashed line. An equilibrium takes place where the two lines cross, at $p = 0.615$, because

$$p \geq 0, \quad 1 - p \geq 0, \quad G(p) \leq 0, \quad \big\langle G(p), (p, 1 - p)\big\rangle = 0.$$

**Detail.** For $p \in [0, 1]$, the problem of agent 1 is

$$\operatorname*{minimize}_{x_1=(x_{11},x_{12})} \varphi_1(x_1) \ \text{ subject to } \ p\big(x_{11} - \tfrac{1}{2}\big) + (1 - p)\big(x_{12} - \tfrac{1}{4}\big) \leq 0, \ \ x_{11} \geq 0, \ \ x_{12} \geq 0.$$

The optimality condition 2.45 then produces

$$x_{11}(p) = 1 - \tfrac{1}{2} p y_1; \quad x_{12}(p) = 1 - \tfrac{1}{2}(1 - p) y_1, \qquad \text{where } \ y_1 = \frac{3 - p}{2p^2 + 2(1 - p)^2}$$

is the multiplier associated with the budget constraint. (Here, $p^2$ is indeed $p$ squared.) The problem of agent 2 is

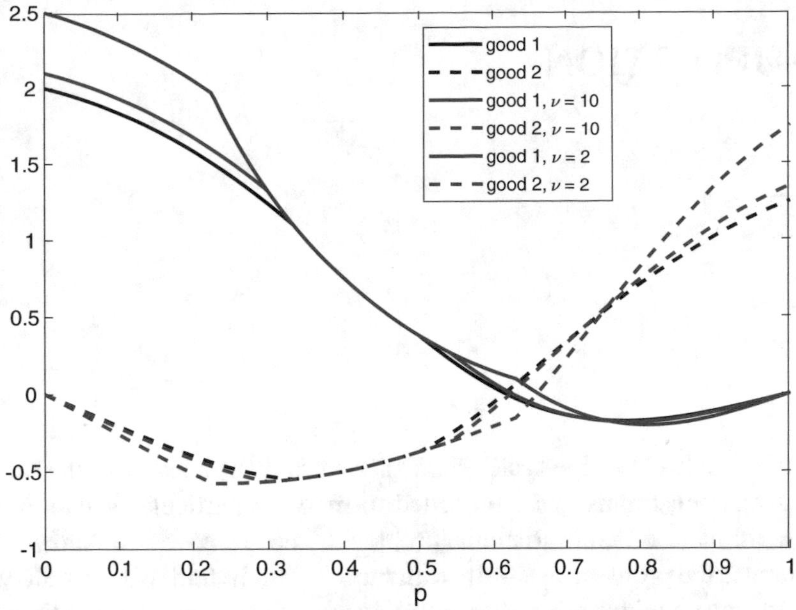


Fig. 9.3: Excess $G(p)$ given by (9.17) in the actual problem (black lines) and in approximations with $\nu = 10$ (blue) and $\nu = 2$ (red).

$$\underset{x_2=(x_{21},x_{22})}{\text{minimize}}\ \varphi_2(x_2) \ \text{ subject to } \ p\big(x_{21} - \tfrac{1}{2}\big) + (1-p)\big(x_{22} - \tfrac{1}{2}\big) \le 0, \ \ x_{21} \ge 0, \ \ x_{22} \ge 0.$$

The optimality condition 2.45 then produces two cases:

$$p \in \big[\tfrac{1}{3}, \tfrac{1}{2}\big]: \qquad x_{21}(p) = \tfrac{1}{2} + \tfrac{1}{2}(1-p)/p; \quad x_{22}(p) = 0$$

$$p \in \big[0, \tfrac{1}{3}\big) \cup \big(\tfrac{1}{2}, 1\big]: \qquad x_{21}(p) = 2 - \tfrac{1}{2}py_2; \quad x_{22}(p) = 1 - \tfrac{1}{2}(1-p)y_2,$$

where

$$y_2 = \frac{1+2p}{p^2 + (1-p)^2}$$

is the multiplier associated with the budget constraint.

Suppose that the objective function of agent 2 is changed to

$$\varphi_2^\nu(x_2) = (1+\varepsilon^\nu)(x_{21} - 2 - \varepsilon^\nu)^2 + (x_{22} - 1 - \varepsilon^\nu)^2,$$

where $\varepsilon^\nu = 1/\nu$, while that of agent 1 remains unchanged. Let $G^\nu(p)$ be the excess of holdings over endowments given by (9.17), but now with $(x_{21}(p), x_{22}(p))$ replaced by the minimizer $(x_{21}^\nu(p), x_{22}^\nu(p))$ under the approximating $\varphi_2^\nu$. As guaranteed by 9.42, $G^\nu(p^\nu) \to G(p)$ when $p^\nu \to p$ because $\varphi_2^\nu \xrightarrow{e} \varphi_2$. Figure 9.3 illustrates this numerically; the blue and red lines represent $G^{10}$ and $G^2$, respectively. We note that the equilibrium shifts from 0.66 to 0.625 and to 0.615 as we pass from $\varphi_2^2$ to $\varphi_2^{10}$ and to $\varphi_2$. Consequently, the price of good 1 decreases (and the price of good 2 increases) as the approximation of the agent's objective function improves. □

# Chapter 10
# DECOMPOSITION

Applications often lead to large-scale optimization problems because of the sheer number of variables and constraints that are needed to model a particular situation and this is also the case for more general variational problems. The presence of unknown parameters tends to increase the problem size further. In fact, we might find that the only viable path forward is to solve a series of smaller subproblems. There are also situations when the data for a problem is distributed among several agents, which are unable or unwilling to share their data despite a common goal of solving the problem. This could be due to privacy rules, common in many healthcare and social media applications, or lack of bandwidth as might be the case for inexpensive autonomous systems. The partitioning of data naturally defines subproblems that are only loosely connected. In this chapter, we'll see how a strategy of problem decomposition can lead to practical algorithms even for enormously large problems.

## 10.A Proximal Alternating Gradient Method

An objective function in a minimization problem may be expressed as the sum of simpler functions, each involving only some of the variables. This is a common situation encountered in lasso regression, where the goal is to minimize $f(x) = \|Ax - b\|_2^2 + \theta\sum_{j=1}^n |x_j|$; see 2.34. Here, only the first of the $1 + n$ terms in this sum involves the full vector $x$. Each of the remaining $n$ terms involves only one variable and the simple absolute-value function. To take advantage of structural properties of this kind, let's extend and combine the ideas of the proximal gradient method (§2.D), which addresses the situation with two convex functions, and those of coordinate descent algorithms from §1.J.

For $x = (x_1, x_2) \in \mathbb{R}^{n_1} \times \mathbb{R}^{n_2}$, with $n = n_1 + n_2$, consider the problem

$$\underset{x\in\mathbb{R}^n}{\text{minimize}}\ f(x) = f_1(x_1) + f_2(x_2) + g(x_1, x_2), \tag{10.1}$$



J. O. Royset and R. J-B Wets, *An Optimization Primer*, Springer Series in Operations Research and Financial Engineering, https://doi.org/10.1007/978-3-030-76275-9_10

where $g : \mathbb{R}^{n_1} \times \mathbb{R}^{n_2} \to \mathbb{R}$ is smooth and $f_i : \mathbb{R}^{n_i} \to \overline{\mathbb{R}}$, $i = 1, 2$, are proper and lsc. In particular, the functions may *not* be convex. We'll concentrate on the partition of $x$ into two subvectors, but will comment on more general partitions at the end of the section.

**Example 10.1** (blind deconvolution). A problem in signal processing is to recover a signal that reaches us only in a corrupted form. Let's assume that the signal is represented by a vector $x \in \mathbb{R}^n$, but we only observe

$$\boldsymbol{y} = Ax + \boldsymbol{\xi},$$

where $A$ is an $m \times n$-matrix that systematically changes the signal and $\boldsymbol{\xi}$ is an $m$-dimensional random vector representing noise that further corrupts the observation $\boldsymbol{y}$, which then also is random. For example, the signal $x$ could represent color values for each pixel in an image that's being blurred and/or compressed in some way. Our goal is to estimate $x$ from an observation of $\boldsymbol{y}$. While we may be willing to make an assumption about the probability distribution of $\boldsymbol{\xi}$, the matrix $A$ could be harder to pin down. Thus, we often need to estimate $A$ and $x$, which is referred to as *blind deconvolution*. The *maximum-a-posteriori* estimate of $(A, x)$ requires us to solve a problem of the form (10.1).

**Detail.** For $g : \mathbb{R}^m \to \mathbb{R}$ and $\alpha \in (0, \infty)$, suppose that the noise $\boldsymbol{\xi}$ has density function $\alpha e^{-g}$. Our prior knowledge (belief) about $x$ is captured by the density function $\beta_1 e^{-f_1}$, where $f_1 : \mathbb{R}^n \to \overline{\mathbb{R}}$ and $\beta_1 \in (0, \infty)$. For example, the density function could be uniform over some subset of $\mathbb{R}^n$, the size of which models the level of uncertainty associated with $x$. Let's denote by $\mathbb{R}^{m \times n}$ the space of $m \times n$-matrices with real values. Then, for $f_2 : \mathbb{R}^{m \times n} \to \overline{\mathbb{R}}$ and $\beta_2 \in (0, \infty)$, we can define a density function $\beta_2 e^{-f_2}$ on the space of such matrices[1] that represents our belief about $A$. According to *Bayes theorem* and under statistical independence between the random quantities, the updated (posterior) density function of our belief regarding $(A, x)$ after an observation $y$ is

$$p(A, x) = \gamma e^{-g(y - Ax)} e^{-f_1(x)} e^{-f_2(A)},$$

where $\gamma$ is a normalization constant. The maximum-a-posteriori estimate of $(A, x)$ is a maximizer of $p$ or, equivalently, a minimizer of its negative logarithm, which leads to the problem

$$\underset{A \in \mathbb{R}^{m \times n}, x \in \mathbb{R}^n}{\text{minimize}} \ f_1(x) + f_2(A) + g(y - Ax)$$

after having dropped constants that don't affect the minimizers. The problem is of the form (10.1), with the unknown quantity $(A, x)$ being partitioned into two parts. □

Faced with (10.1), it appears promising to pursue the idea of alternating between optimizing $x_1$ and optimizing $x_2$. The resulting subproblems are certainly smaller than the actual one and might also have desirable structural properties. Still, the naive iterative scheme

[1] Since an $m \times n$-matrix can always be represent as an $mn$-dimensional vector, a problem expressed in terms of matrices can be reformulated as one with vectors. However, sometimes it's more convenient, computationally and theoretically, to work with matrices. Concepts such as density functions remain essentially unchanged when passing from vectors to matrices.

$$x_1^{\nu+1} \in \operatorname{argmin}_{x_1 \in \mathbb{R}^{n_1}} f_1(x_1) + f_2(x_2^\nu) + g(x_1, x_2^\nu)$$
$$x_2^{\nu+1} \in \operatorname{argmin}_{x_2 \in \mathbb{R}^{n_2}} f_1(x_1^{\nu+1}) + f_2(x_2) + g(x_1^{\nu+1}, x_2)$$

converges to a minimizers of $f$ only under strong assumptions and requires computation of minimizers of potentially nonconvex functions in each iteration. A modified approach adds the proximal terms

$$\tfrac{1}{2\lambda_1^\nu}\|x_1 - x_1^\nu\|_2^2 \quad \text{and} \quad \tfrac{1}{2\lambda_2^\nu}\|x_2 - x_2^\nu\|_2^2$$

to the subproblems and also approximates $g$ by an affine function. This leads to the iterative scheme

$$\begin{aligned} x_1^{\nu+1} \in \operatorname{argmin}_{x_1 \in \mathbb{R}^{n_1}} &\, f_1(x_1) + f_2(x_2^\nu) \\ &+ g(x_1^\nu, x_2^\nu) + \big\langle \nabla_1 g(x_1^\nu, x_2^\nu), x_1 - x_1^\nu \big\rangle + \tfrac{1}{2\lambda_1^\nu}\|x_1 - x_1^\nu\|_2^2 \\ x_2^{\nu+1} \in \operatorname{argmin}_{x_2 \in \mathbb{R}^{n_2}} &\, f_1(x_1^{\nu+1}) + f_2(x_2) \\ &+ g(x_1^{\nu+1}, x_2^\nu) + \big\langle \nabla_2 g(x_1^{\nu+1}, x_2^\nu), x_2 - x_2^\nu \big\rangle + \tfrac{1}{2\lambda_2^\nu}\|x_2 - x_2^\nu\|_2^2, \end{aligned}$$

where $\nabla_i g(x_1, x_2) \in \mathbb{R}^{n_i}$ consists of the partial derivatives with respect to the components of $x_i$, i.e.,

$$\nabla g(x_1, x_2) = \big(\nabla_1 g(x_1, x_2),\ \nabla_2 g(x_1, x_2)\big).$$

The two subproblems defining the iterative scheme can be expressed in terms of the set-valued mappings

$$z \mapsto \operatorname{prx}_{\lambda_1^\nu} f_1(z) \quad \text{and} \quad z \mapsto \operatorname{prx}_{\lambda_2^\nu} f_2(z).$$

To see this, let's strip away terms that don't change the minimizers of the subproblems and obtain

$$x_1^{\nu+1} \in \operatorname{argmin}_{x_1 \in \mathbb{R}^{n_1}} f_1(x_1) + \big\langle \nabla_1 g(x_1^\nu, x_2^\nu), x_1 - x_1^\nu \big\rangle + \tfrac{1}{2\lambda_1^\nu}\|x_1 - x_1^\nu\|_2^2$$
$$x_2^{\nu+1} \in \operatorname{argmin}_{x_2 \in \mathbb{R}^{n_2}} f_2(x_2) + \big\langle \nabla_2 g(x_1^{\nu+1}, x_2^\nu), x_2 - x_2^\nu \big\rangle + \tfrac{1}{2\lambda_2^\nu}\|x_2 - x_2^\nu\|_2^2.$$

For $a, x, y \in \mathbb{R}^n$ and $\lambda \in (0, \infty)$, we generally have

$$\langle a, x - y\rangle + \tfrac{1}{2\lambda}\|x - y\|_2^2 = \tfrac{1}{2\lambda}\big\|x - (y - \lambda a)\big\|_2^2 - \tfrac{1}{2}\lambda\|a\|_2^2.$$

Consequently, we can rewrite the iterative scheme in terms of proximal points and this leads to the following algorithm.

**Proximal Alternating Gradient Method.**

Data. $x_1^0 \in \mathbb{R}^{n_1}$, $x_2^0 \in \mathbb{R}^{n_2}$ and $\{\lambda_1^\nu, \lambda_2^\nu \in (0, \infty),\ \nu = 0, 1, 2, \dots\}$.
Step 0. Set $\nu = 0$.

Step 1. Compute

$$x_1^{\nu+1} \in \operatorname{prx}_{\lambda_1^\nu} f_1\big(x_1^\nu - \lambda_1^\nu \nabla_1 g(x_1^\nu, x_2^\nu)\big).$$

Step 2. Compute

$$x_2^{\nu+1} \in \operatorname{prx}_{\lambda_2^\nu} f_2\big(x_2^\nu - \lambda_2^\nu \nabla_2 g(x_1^{\nu+1}, x_2^\nu)\big).$$

Step 3. Replace $\nu$ by $\nu + 1$ and go to Step 1.

The success of the method relies on an ability to compute the proximal points for $f_1$ and $f_2$. Although this can be difficult in general, important tractable cases exist such as the one in 2.34. If $f_i$ is the indicator function of a polyhedral set, then $\operatorname{prx}_{\lambda_i^\nu} f_i$ corresponds to a projection on the set and is tractable; see §2.D. The proximity operator repository (`proximity-operator.net`) furnishes numerous other examples.

A main strength of the method is its guaranteed performance under the assumption that the objective function $f$ is a *Kurdyka-Łojasiewicz (KL) function*, which holds quite broadly.

**KL functions.** For $\alpha \in (0, \infty]$, let $\mathcal{F}_\alpha$ be the class of concave functions $\varphi : \mathbb{R} \to \overline{\mathbb{R}}$ with the properties that $\varphi(0) = 0$, $\varphi(\gamma) \to 0$ as $\gamma \searrow 0$, $\varphi$ is nonnegative and smooth with a positive derivative $\varphi'(\gamma)$ at every $\gamma \in (0, \alpha)$, and $\varphi(\gamma) = -\infty$ for $\gamma \notin [0, \alpha)$. A proper lsc function $h : \mathbb{R}^n \to \overline{\mathbb{R}}$ satisfies the *KL property* at $\bar{x} \in \operatorname{dom} \partial h$ when there exist $\alpha \in (0, \infty]$, $\varepsilon > 0$ and $\varphi \in \mathcal{F}_\alpha$ such that the following holds:

$$x \in \mathbb{B}(\bar{x}, \varepsilon) \text{ with } h(\bar{x}) < h(x) < h(\bar{x}) + \alpha \implies \varphi'\big(h(x) - h(\bar{x})\big) \operatorname{dist}(0, \partial h(x)) \geq 1.$$

If $h$ satisfies the KL property at every $\bar{x} \in \operatorname{dom} \partial h$, then $h$ is a *KL function*.

The KL property is a growth condition that holds for many proper lsc functions including polynomial functions and indicator functions of sets described by finite collections of equalities and inequalities expressed using polynomial functions; see [15] for more details. For example, if $h(x) = x^2$ and $\bar{x} = 0$, then one can take $\alpha = \infty$ and $\varphi(\gamma) = \sqrt{\gamma}$ so that

$$\varphi'\big(h(x) - h(\bar{x})\big) \operatorname{dist}\big(0, \partial h(x)\big) = \frac{1}{2\sqrt{x^2}} 2|x| = 1.$$

Hence, $h$ satisfies the KL property at 0 in this case and, in fact, at all other points as well.

**Convergence 10.2** (proximal alternating gradient method). *For proper lsc* $f_i : \mathbb{R}^{n_i} \to \overline{\mathbb{R}}$, $i = 1, 2$, *and smooth* $g : \mathbb{R}^{n_1} \times \mathbb{R}^{n_2} \to \mathbb{R}$, *let* $f : \mathbb{R}^n \to \overline{\mathbb{R}}$ *be given by*

$$f(x) = f_1(x_1) + f_2(x_2) + g(x_1, x_2), \quad \text{with } x = (x_1, x_2).$$

*Suppose that* $f$ *is a KL function,* $\inf f, \inf f_1, \inf f_2 > -\infty$ *and the following Lipschitz conditions hold:*

(a) *There's* $\kappa_1 : \mathbb{R}^{n_2} \to [0, \infty)$ *such that for any* $x_2 \in \mathbb{R}^{n_2}$*:*

$$\|\nabla_1 g(x_1, x_2) - \nabla_1 g(\bar{x}_1, x_2)\|_2 \le \kappa_1(x_2)\|x_1 - \bar{x}_1\|_2 \quad \forall x_1, \bar{x}_1 \in \mathbb{R}^{n_1}.$$

(b) *There's* $\kappa_2 : \mathbb{R}^{n_1} \to [0, \infty)$ *such that for any* $x_1 \in \mathbb{R}^{n_1}$*:*

$$\|\nabla_2 g(x_1, x_2) - \nabla_2 g(x_1, \bar{x}_2)\|_2 \le \kappa_2(x_1)\|x_2 - \bar{x}_2\|_2 \quad \forall x_2, \bar{x}_2 \in \mathbb{R}^{n_2}.$$

(c) *For each pair of bounded sets* $B_i \subset \mathbb{R}^{n_i}$*,* $i = 1, 2$*, there's* $\kappa \in [0, \infty)$ *such that*

$$\|\nabla g(x_1, x_2) - \nabla g(\bar{x}_1, \bar{x}_2)\|_2 \le \kappa\|(x_1, x_2) - (\bar{x}_1, \bar{x}_2)\|_2 \quad \forall (x_1, x_2), (\bar{x}_1, \bar{x}_2) \in B_1 \times B_2.$$

*If* $\{x_1^\nu, x_2^\nu,\ \nu \in \mathbb{N}\}$ *is a bounded sequence generated by the proximal alternating gradient method while minimizing* $f$*, with*

$$\lambda_1^\nu \in \big(0, 1/\kappa_1(x_2^\nu)\big) \quad \textit{and} \quad \lambda_2^\nu \in \big(0, 1/\kappa_2(x_1^{\nu+1})\big) \quad \forall \nu \in \mathbb{N},$$

*and* $\kappa_1(x_2^\nu)$ *as well as* $\kappa_2(x_1^\nu)$ *are bounded from above and away from zero uniformly across all* $\nu$*, then* $(x_1^\nu, x_2^\nu) \to x^\star \in \mathbb{R}^n$ *and* $0 \in \partial f(x^\star)$*.*

The mechanism of the algorithm and the step size requirements are remarkably similar to those of the proximal gradient method (§2.D). However, the conclusion is naturally weaker now due to the absence of convexity. Instead of a proof, which can be found in [15], we examine an application where the proximal points in Steps 1 and 2 can be computed explicitly.

**Example 10.3** (text analytics). Suppose that we're interested in summarizing a collection of $n$ documents and we scan the documents for occurrences of $m$ pre-specified words and phrases. This produces an $m \times n$-matrix $A$ with entry $A_{ij}$ being the number of times word (or phrase) $i$ occurs in document $j$. A rank-one approximation $yx^\top$ of $A$ expressed in terms of column vectors

$$x = (x_1, \ldots, x_n) \in \mathbb{R}^n \quad \text{and} \quad y = (y_1, \ldots, y_m) \in \mathbb{R}^m$$

then has the following interpretation: Column $A^j$ of $A$, corresponding to document $j$, is approximated by $yx_j$. Thus, the occurrence of the difference words in document $j$ is given approximately by a scalar coefficient $x_j$ times a vector $y$. This vector can be viewed as corresponding to an artificial "basic document." Since this holds for every $j$, all the $n$ documents are approximated by scaled versions of the same basic document. The components of $y$ with high values then specify the words from the pre-specified list often "used" in the basic document and they're the ones that summarize all the documents. This problem and many of its generalizations can be solved using the proximal alternating gradient method.

**Detail.** One might attempt to compute $(x, y)$ by minimizing the Frobenius norm $\|A - yx^\top\|_F$. However, since $A$ contains nonnegative numbers exclusively, a solution to this problem that involves negative values in $x$ and $y$ becomes difficult to interpret. The vector

$yx_j$ should be nonnegative as it corresponds to our estimate of the number of times the various words from the pre-specified list occur in document $j$. Moreover, $y_i$ is the number of times word $i$ occurs in the basic document and should also be nonnegative. These constraints lead to the problem

$$\underset{x \in \mathbb{R}^n, y \in \mathbb{R}^m}{\text{minimize}} \ \tfrac{1}{2}\|A - yx^\top\|_F^2 \ \text{ subject to } \ x, y \geq 0.$$

A better approximation that results in $q$ basic documents, and not only one, is obtained by replacing $x$ by an $n \times q$-matrix $X$ and $y$ by an $m \times q$-matrix $Y$. Then, we face a *nonnegative matrix factorization problem*

$$\underset{X \in \mathbb{R}^{n \times q}, Y \in \mathbb{R}^{m \times q}}{\text{minimize}} \ \tfrac{1}{2}\|A - YX^\top\|_F^2 \ \text{ subject to } \ X, Y \geq 0,$$

where the constraints are imposed elementwise. A solution $(X, Y)$ has the interpretation that every column of $Y$ can be viewed as a basic document. With $X_j$ being the $j$th row of $X$, the approximation of column $A^j$ (corresponding to document $j$) is then $YX_j^\top$, i.e., a weighted combination of the $q$ basic documents, where $X_j$ furnishes the weights. Again, since this holds for all $j$, we've constructed $q$ basic documents that summarize the $n$ documents.

The nonnegative matrix factorization problem is nonconvex, but the proximal alternating gradient method applies with each step being easily executed. Let's take

$$g(X, Y) = \tfrac{1}{2}\|A - YX^\top\|_F^2, \quad f_1(X) = \begin{cases} 0 & \text{if } X \geq 0 \\ \infty & \text{otherwise,} \end{cases} \quad f_2(Y) = \begin{cases} 0 & \text{if } Y \geq 0 \\ \infty & \text{otherwise.} \end{cases}$$

Thus, the nonnegative matrix factorization problem is of the form (10.1), with the modification that we now optimize over matrices and not vectors. Mimicking the rules for computing gradients of functions on $\mathbb{R}^n$, we obtain the gradient of $g$ with respect to the matrix $X$ as $(YX^\top - A)^\top Y$ and with respect to $Y$ as $(YX^\top - A)X$. The first expression defines the gradient mapping

$$X \mapsto (YX^\top - A)^\top Y,$$

which is Lipschitz continuous with modulus function of the form $\kappa_1(Y) = \|Y^\top Y\|_F$. To see this, note that if $X, \bar{X} \in \mathbb{R}^{n \times q}$, then

$$\big\|(YX^\top - A)^\top Y - (Y\bar{X}^\top - A)^\top Y\big\|_F = \big\|(X - \bar{X})Y^\top Y\big\|_F \leq \|Y^\top Y\|_F \|X - \bar{X}\|_F.$$

Similarly, the second expression defines the gradient mapping

$$Y \mapsto (YX^\top - A)X,$$

which is Lipschitz continuous with modulus function given by $\kappa_2(X) = \|X^\top X\|_F$.

In the case of matrices, the proximal points are defined in terms of the Frobenius norm so that

$$\begin{aligned}\mathrm{prx}_{\lambda} f_1(\bar{X}) &= \mathrm{argmin}_{X\in\mathbb{R}^{n\times q}} \left\{ f_1(X) + \tfrac{1}{2\lambda}\|X-\bar{X}\|_F^2 \right\}\\ &= \mathrm{argmin}_{X\in\mathbb{R}^{n\times q}} \left\{ \tfrac{1}{2}\|X-\bar{X}\|_F^2 \;\middle|\; X \geq 0 \right\},\end{aligned}$$

which is just the projection of $\bar{X}$ to the nearest nonnegative matrix in the sense of the Frobenius norm. This, in fact, is carried out by replacing any negative element in $\bar{X}$ by zero. The situation is similar for $f_2$. Consequently, Step 1 in the proximal alternating gradient method becomes to

$$\text{select } \lambda_1^\nu \in \Big(0,\ 1/\big\|(Y^\nu)^\top Y^\nu\big\|_F\Big), \quad \text{compute } X^\nu - \lambda_1^\nu\big(Y^\nu(X^\nu)^\top - A\big)^\top Y^\nu$$

and replace any negative element in this matrix with zero to form $X^{\nu+1}$. Step 2 reduces to

$$\text{select } \lambda_2^\nu \in \Big(0,\ 1/\big\|(X^{\nu+1})^\top X^{\nu+1}\big\|_F\Big), \quad \text{compute } Y^\nu - \lambda_2^\nu\big(Y^\nu(X^{\nu+1})^\top - A\big)X^{\nu+1}$$

and replace any negative element in this matrix with zero to form $Y^{\nu+1}$. Thus, both steps can be carried out using basic matrix operations, which means that the algorithm is practical even for large instances involving, for instance, millions of documents ($n \sim 10^6$) and thousands of words ($m \sim 10^4$). □

The development in this section deals with (10.1), but can be extended to sums of any finite number of terms, say $k$, and corresponding subvectors; see [15]. One would then have $k$ nontrivial steps in the algorithm, not only Steps 1 and 2. In the extreme, one might partition $x \in \mathbb{R}^n$ into $n$ one-dimensional subvectors so that each iteration of the algorithm involves $n$ one-dimensional problems, which essentially brings us back to coordinate descent algorithms of §1.J, but now also with the separable terms given by $f_i$.

## 10.B Linkage Constraints

While two components of a problem might be connected through a smooth function as seen in the previous section, they may also be linked through a constraint. We'll now examine situations where the difficulty is caused by a linkage constraint that forces the decision vector $x \in \mathbb{R}^n$ to lie in a subspace of $\mathbb{R}^n$. In the absence of this constraint, the problem decomposes into simpler subproblems. Since a subspace constraint can be expressed as the linear equality $Ax = 0$ for some $m \times n$-matrix $A$, this problem class may appear rather specific. However, as we'll see in the next seven sections, it covers a vast number of applications and several situations where a constraint $Ax = 0$ isn't apparent immediately.

**Example 10.4** (linkage across time periods). Let's consider the state of a system at time $t$ represented by a vector $s_t \in \mathbb{R}^m$. If we remain inactive, a given $m \times m$-matrix $B_t$ transforms the state to $B_t s_t$ for the next time period. However, we can intervene with an action $x_t \in \mathbb{R}^n$, which through a known $m \times n$-matrix $A_t$ influences the state as well. The combined effect is that the state of the system at the beginning of time period $t + 1$ is

$$s_{t+1} = A_t x_t + B_t s_t.$$

For a given $f_t : \mathbb{R}^n \times \mathbb{R}^m \to \overline{\mathbb{R}}$, with $f_t(x_t, s_t)$ representing the cost of carrying out action $x_t$ and being in state $s_t$ during time period $t$, this leads to the problem

$$\underset{x_1, s_1, \dots, x_\tau, s_\tau}{\text{minimize}} \sum_{t=1}^{\tau} f_t(x_t, s_t) \text{ subject to } s_{t+1} = A_t x_t + B_t s_t, \quad t = 1, \dots, \tau - 1,$$

where $\tau$ is the number of time periods under consideration. The problem has linear equality constraints linking the various time periods.

**Detail.** The objective function is separable in time, but our decision at time $t$ affects the situation later through the constraints. Let $z = (x_1, s_1, \dots, x_\tau, s_\tau)$. We can organize $A_t$ and $B_t$ for all the time period into a large matrix $D$ such that all the explicit constraints are expressed by $Dz = 0$. Thus, $z$ is restricted to a subspace. An algorithmic strategy for solving the problem could be to minimize $f_t(x_t, s_t)$ for each $t$, which results in $\tau$ simpler problems, but make some adjustments so that the linkage constraint $Dz = 0$ is accounted for indirectly. □

**Subspaces and Orthogonality.** We recall that a subspace of $\mathbb{R}^n$ is a nonempty set $X \subset \mathbb{R}^n$ such that $\alpha x + \beta \bar{x} \in X$ whenever $x, \bar{x} \in X$ and $\alpha, \beta \in \mathbb{R}$. Consequently, $X$ is a subspace of $\mathbb{R}^n$ if and only if $X = \text{null}\, A = \{x \in \mathbb{R}^n \mid Ax = 0\}$ for some $m \times n$-matrix $A$. A vector $\bar{x} \in \mathbb{R}^n$ is *orthogonal* to a subspace $X \subset \mathbb{R}^n$ if $\langle \bar{x}, x \rangle = 0$ for every $x \in X$. The collection of all vectors in $\mathbb{R}^n$ that are orthogonal to $X$ is the *orthogonal complement* of $X$ and is denoted by $X^\perp$. An orthogonal complement is a subspace of $\mathbb{R}^n$ as well; see Figure 10.1.

If $X = \text{null}\, A$ for some $m \times n$-matrix $A$, then $X^\perp = \{A^\top w \mid w \in \mathbb{R}^m\}$. At the same time, the normal cone $N_X(x) = \{A^\top w \mid w \in \mathbb{R}^m\}$ for $x \in X$ by 2.44. Consequently, for any subspace $X$ of $\mathbb{R}^n$, one has

$$y \in N_X(x) \iff x \in X, y \in X^\perp \iff x \in N_{X^\perp}(y), \tag{10.2}$$

which is illustrated by Figure 10.1. This means that optimality conditions for problems constrained to a subspace are relatively simple. In particular, by 2.30, $z \in \text{prj}_X(x)$ if and only if $x - z \in N_X(z)$, which now translates into $z \in X$ and $x - z \in X^\perp$. Similarly, $y \in \text{prj}_{X^\perp}(x)$ if and only if $y \in X^\perp$ and $x - y \in X$. Since both $X$ and $X^\perp$ are convex sets, projections on these subspaces are unique and we can view $\text{prj}_X$ and $\text{prj}_{X^\perp}$ as (point-valued) mappings from $\mathbb{R}^n$ to $\mathbb{R}^n$. They satisfy

$$z \in X, \ x - z \in X^\perp \iff z = \text{prj}_X(x) \iff x - z = \text{prj}_{X^\perp}(x). \tag{10.3}$$

Consequently, for any $x \in \mathbb{R}^n$,

$$x = \text{prj}_X(x) + \text{prj}_{X^\perp}(x). \tag{10.4}$$

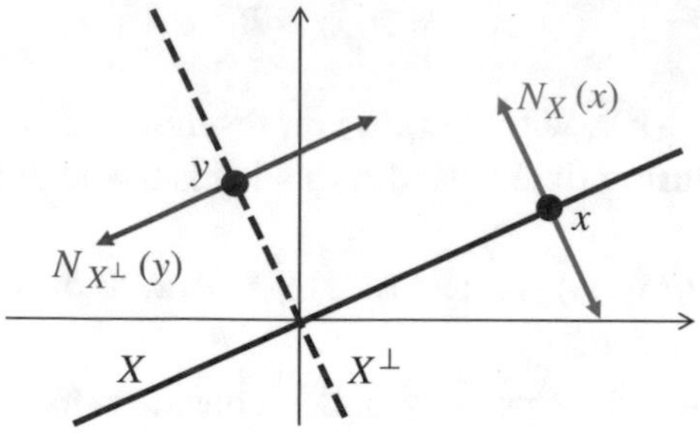


Fig. 10.1: The line $X$ is a subspace of $\mathbb{R}^2$ with orthogonal complement $X^\perp$.

Let's generalize the situation from the previous example. For lsc $f_i : \mathbb{R}^{n_i} \to \overline{\mathbb{R}}$, $i = 1, \dots, m$, with $n = \sum_{i=1}^m n_i$, and a subspace $X$ of $\mathbb{R}^n$, consider the problem

$$\operatorname*{minimize}_{x \in X} \ f(x) = \sum\nolimits_{i=1}^m f_i(x_i), \ \text{ where } x = (x_1, \dots, x_m).$$

While the objective function is *separable*, the problem doesn't separate into $m$ subproblems due to the linkage caused by $X$. Under a qualification (see for example 4.67), the Fermat rule 4.73 produces the optimality condition

$$0 \in \partial f(x) + N_X(x),$$

where

$$\partial f(x) = \partial f_1(x_1) \times \cdots \times \partial f_m(x_m)$$

by 4.63 as long as $f_i(x_i)$ is finite for all $i = 1, \dots, m$. The trouble is that the normal cone to $X$ may not separate along the same lines.

The key principle allowing us to relax the linkage constraint is that of lifting to a higher dimensional space. Instead of viewing the problem as one in terms of $x \in \mathbb{R}^n$, we formulate it using $x$ and $y \in \mathbb{R}^n$, with the latter playing the role of a multiplier vector or dual element. This helps us to appropriately penalize the violation of the linkage constraint after it has been relaxed. Specifically, by (10.2),

$$0 \in \partial f(x) + N_X(x) \iff y \in \partial f(x), \ -y \in N_X(x) \iff y \in \partial f(x), \ x \in X, \ y \in X^\perp,$$

where we use the fact that $-y \in X^\perp$ if and only if $y \in X^\perp$. The last form is of a canonical kind for which we'll develop an algorithm in the next section that repeatedly solves slightly modified versions of the generalized equation $y \in \partial f(x)$ for a fixed $y$ and *without* the linkage constraint on $x$. This appears promising because

$$y \in \partial f(x) \ \text{ separates into } \ y_i \in \partial f_i(x_i), \ i = 1, \dots, m.$$

In the case of 10.4, each of these subproblems would involve only one time period.

The situation is similar when the actual problem of interest is the generalized equation

$$0 \in F(x) + N_{C\cap X}(x), \text{ with } F : \mathbb{R}^n \to \mathbb{R}^n, \text{ closed } C \subset \mathbb{R}^n \text{ and subspace } X \subset \mathbb{R}^n.$$

Under a qualification (see the intersection rule 4.52),

$$N_{C\cap X}(x) = N_C(x) + N_X(x)$$

and then the generalized equation is equivalently stated as

$$0 \in F(x) + N_C(x) + N_X(x) \iff x \in X, \ y \in X^\perp, \ y \in F(x) + N_C(x).$$

The advantage of this reformulation becomes clear when $x = (x_1, \dots, x_m)$, with $x_i \in \mathbb{R}^{n_i}$ and $\sum_{i=1}^m n_i = n$, and

$$F(x) = \big(F_1(x_1), \dots, F_m(x_m)\big) \qquad C = C_1 \times \cdots \times C_m,$$

where $F_i : \mathbb{R}^{n_i} \to \mathbb{R}^{n_i}$ and $C_i \subset \mathbb{R}^{n_i}$ is closed. Then, the generalized equation

$$y \in F(x) + N_C(x) \text{ separates into } y_i \in F_i(x_i) + N_{C_i}(x_i), \ i = 1, \dots, m$$

by 4.44, where $y = (y_1, \dots, y_m)$. These smaller generalized equations can be solved, hopefully, much faster than $0 \in F(x) + N_{C\cap X}(x)$. However, one still needs to account for $x \in X$ and $y \in X^\perp$, but this linkage is relaxed in the algorithm of the next section and then enforced indirectly.

While a problem might be in a separable form from the outset as just seen, the setting can also be constructed through *splitting*. For lsc $f_i : \mathbb{R}^n \to \overline{\mathbb{R}}$, $i = 1, \dots, m$, the problem

$$\operatorname*{minimize}_{x\in\mathbb{R}^n} f(x) = \sum\nolimits_{i=1}^m f_i(x) \tag{10.5}$$

doesn't seem to be of the kind with linkage constraints. However, let's duplicate $m$ times the vector $x$ to obtain $x_1, \dots, x_m \in \mathbb{R}^n$ and then write, equivalently,

$$\operatorname*{minimize}_{z\in\mathbb{R}^{nm}} \sum\nolimits_{i=1}^m f_i(x_i) \text{ subject to } x_1 = \cdots = x_m, \quad \text{with } z = (x_1, \dots, x_m). \tag{10.6}$$

We now have a problem involving a separable objective function $\varphi : \mathbb{R}^{nm} \to \overline{\mathbb{R}}$ given by

$$\varphi(z) = \sum\nolimits_{i=1}^m f_i(x_i)$$

and the linkage constraint

$$z \in X = \big\{(x_1, \dots, x_m) \in \mathbb{R}^{nm} \bigm| x_1 = x_2 = \cdots = x_m\big\}.$$

The set $X$ is indeed a subspace of $\mathbb{R}^{nm}$.

**Proposition 10.5** (orthogonality under splitting). *For the subspace*

$$X = \big\{(x_1, \dots, x_m) \in \mathbb{R}^{nm} \bigm| x_1 = x_2 = \cdots = x_m\big\},$$

*one has the orthogonal complement*

$$X^\perp = \{(y_1, \ldots, y_m) \in \mathbb{R}^{nm} \mid y_1 + \cdots + y_m = 0\}.$$

**Proof.** If $(y_1, \ldots, y_m)$ satisfies $y_1 + \cdots + y_m = 0$, then

$$\langle (w, \ldots, w), (y_1, \ldots, y_m) \rangle = \sum_{i=1}^m \langle w, y_i \rangle = \Big\langle w, \sum_{i=1}^m y_i \Big\rangle = 0$$

for arbitrary $(w, \ldots, w) \in X$ so $(y_1, \ldots, y_m) \in X^\perp$. If $(y_1, \ldots, y_m) \in X^\perp$, then

$$0 = \langle (w, \ldots, w), (y_1, \ldots, y_m) \rangle = \Big\langle w, \sum_{i=1}^m y_i \Big\rangle$$

for all $(w, \ldots, w) \in X$. In particular, $w$ could be the $n$-dimensional vector with one as its $j$th component and zero otherwise. This results in $\sum_{i=1}^m y_{ij} = 0$, where $y_{ij}$ is the $j$th component of $y_i$. After repeating this argument for all $j = 1, \ldots, n$, we conclude that $\sum_{i=1}^m y_i = 0$. Consequently, the formula for $X^\perp$ holds. □

The problem in (10.6) has the optimality condition

$$0 \in \partial\varphi(z) + N_X(z)$$

under a qualification; see 4.67 and 4.73. Parallel to the earlier development, the optimality condition can be expressed using a vector $y \in \mathbb{R}^{nm}$ and this results in

$$y \in \partial\varphi(z), \quad z \in X, \quad y \in X^\perp = \{(y_1, \ldots, y_m) \in \mathbb{R}^{nm} \mid y_1 + \cdots + y_m = 0\},$$

where the first part separates into $y_i \in \partial f_i(x_i)$, $i = 1, \ldots, m$, by 4.63. We've obtained a generalized equation involving subspace constraints.

**Example 10.6** (perturbation). We recall from Chap. 5 that a problem can often be beneficially formulated in terms of a Rockafellian that represents the actual objective function as well as perturbations of model parameters. In fact, this perspective provides the basis for a versatile duality theory. For proper lsc $f : \mathbb{R}^m \times \mathbb{R}^n \to \overline{\mathbb{R}}$, let's consider the problems

$$\Big\{ \underset{x \in \mathbb{R}^n}{\text{minimize}}\, f(u, x), \quad u \in \mathbb{R}^m \Big\},$$

with $u = 0$ furnishing the actual problem of interest. By the Rockafellar condition 5.10,

$$(y, 0) \in \partial f(0, x)$$

is an optimality condition for the actual problem under a qualification. We can write this condition in terms of a subspace constraint.

**Detail.** Let $X = \{0\}^m \times \mathbb{R}^n$, which is a subspace of $\mathbb{R}^{m+n}$ with orthogonal complement

$$X^\perp = \mathbb{R}^m \times \{0\}^n.$$

The optimality condition $(y, 0) \in \partial f(0, x)$ is then equivalently stated as

$$(u, x) \in X, \quad (y, v) \in X^{\perp}, \quad (y, v) \in \partial f(u, x).$$

Numerous possibilities exist within this broad framework. For example, when $u$ represents perturbation of constraints, it might turn out to be relatively simple to solve $(y, v) \in \partial f(u, x)$ given values for $(y, v)$ and relaxed subspace constraint $(u, x) \in X$ because $u$ offers additional flexibility. We return to this setting in §10.H. □

**Exercise 10.7** (search problem). In planning the search for floating debris from an aircraft lost over the ocean, one would divide the area of interest into $n$ squares and then determine how much time and resources should be put towards searching each square. Suppose that the search will be carried out by two aircraft flying at $\eta_1$ and $\eta_2$ miles per hour, respectively, with sensors that effectively can reach out $\omega_1/2$ and $\omega_2/2$ miles, respectively. The debris is located in one square and is stationary. Prior knowledge stipulates the probability that the debris is in square $j$ as $p_j > 0$, with $\sum_{j=1}^{n} p_j = 1$. Let the size of the squares be $\alpha_1, \dots, \alpha_n$ square miles and let the total amount of search time available for the first aircraft be $\tau_1$ and for the second aircraft be $\tau_2$. With $x_{ij}$ being the allocation of time for aircraft $i$ in square $j$, the problem of determining the time allocation $x = (x_{11}, \dots, x_{1n}, x_{21}, \dots, x_{2n})$ that minimizes the probability of not finding the debris can be formulated as

$$\operatorname*{minimize}_{x \in C_1 \times C_2} f(x) = \sum\nolimits_{j=1}^{n} p_j \exp\big(-(\eta_1 \omega_1 x_{1j} + \eta_2 \omega_2 x_{2j})/\alpha_j\big),$$

with

$$C_i = \Big\{(x_{i1}, \dots, x_{in}) \,\Big|\, \sum\nolimits_{j=1}^{n} x_{ij} = \tau_i, \; x_{ij} \geq 0, \; j = 1, \dots, n\Big\}.$$

Show that the problem can be reformulated as one involving the sum of two terms and that it can be reduced to two subproblems via splitting, each associated with just one of the sets $C_i$ and a subspace constraint.

**Guide.** The actual problem can be reformulate as minimizing $f_1(x) + f_2(x)$, where

$$f_1(x) = \tfrac{1}{2} f(x) + \iota_{C_1 \times \mathbb{R}^n}(x) \quad \text{and} \quad f_2(x) = \tfrac{1}{2} f(x) + \iota_{\mathbb{R}^n \times C_2}(x).$$

Then, follow the pattern laid out for (10.5). □

## 10.C Progressive Decoupling Algorithm

For a set-valued mapping $S : \mathbb{R}^n \rightrightarrows \mathbb{R}^n$ and a subspace $X$ of $\mathbb{R}^n$, we'll now develop a broadly applicable algorithm for solving the *linkage problem*

| |
|---|
| find $x \in X$ and $y \in X^{\perp}$ such that $y \in S(x)$ |

The examples of the previous section show that the subspace $X$ could arise organically during the formulation of constraints or be constructed through a process of splitting. We proceed under the assumption that for a given $y \in \mathbb{R}^n$, the generalized equation $y \in S(x)$, or slight modifications thereof, is computationally tractable. For example, $S$ may be separable, resulting in several low-dimensional generalized equations. In essence, the linkage problem is deemed difficult, but simplifies substantially when the subspace constraint $x \in X$ is relaxed and $y$ is fixed.

**Progressive Decoupling Algorithm.**

Data. $x^0 \in X$, $y^0 \in X^\perp$, $\theta \in [0, \infty)$, $\kappa \in (\theta, \infty)$.
Step 0. Set $\nu = 0$.
Step 1. Obtain $\hat{x}^\nu \in \mathbb{R}^n$ by solving

$$y^\nu \in S(x) + \kappa(x - x^\nu).$$

Step 2. Compute $x^{\nu+1} = \mathrm{prj}_X(\hat{x}^\nu)$ and $y^{\nu+1} = y^\nu - (\kappa - \theta)\,\mathrm{prj}_{X^\perp}(\hat{x}^\nu)$.
Step 3. Replace $\nu$ by $\nu + 1$ and go to Step 1.

A direct solution of the linkage problem would need to pay attention to the subspace constraints $x \in X$ and $y \in X^\perp$, but the calculations in Step 1 of the algorithm isn't burdened by such concerns and, hopefully, can be carried out reasonably fast using an algorithm from the earlier chapters as a subroutine. Most significantly, any separable structure in $S$ can now be utilized to break down Step 1 into many smaller subproblems, which may even be solved in parallel.

Step 2 involves the projection on $X$, which defines a point uniquely, and requires us to solve a linear system of equations in less than $2n$ variables; see 2.46. The subspace $X$ is often the nullspace of a sparse matrix, in which case the resulting system of equations is also sparse and solvable in a negligible amount of time.

An explicit expression for $y^{\nu+1}$ is achieved by using (10.4) and the fact that $x^{\nu+1} = \mathrm{prj}_X(\hat{x}^\nu)$. This leads to

$$y^{\nu+1} = y^\nu - (\kappa - \theta)(\hat{x}^\nu - x^{\nu+1}).$$

So the only nontrivial work in Step 2 is associated with the projection on $X$.

Before we state the convergence properties of the progressive decoupling algorithm, let's examine some special cases and introduce some new concepts.

**Example 10.8** (minimization of a sum). For $f : \mathbb{R}^n \to \overline{\mathbb{R}}$, $g : \mathbb{R}^m \to \overline{\mathbb{R}}$ and an $m \times n$-matrix $A$, let's consider the problem

$$\operatorname*{minimize}_{x \in \mathbb{R}^n} f(x) + g(Ax).$$

Under an appropriate qualification (see 4.64 and 4.67), an optimality condition for the problem is

$$0 \in \partial f(x) + A^\top \partial g(Ax). \tag{10.7}$$

This generalized equation can be reformulated as a linkage problem and be solved by the progressive decoupling algorithm.

**Detail.** We rewrite (10.7) as follows:

$$0 \in \partial f(x_1) + A^\top \partial g(Ax_1) \iff y_2 \in \partial g(Ax_1),\ -A^\top y_2 \in \partial f(x_1)$$
$$\iff (x_1, x_2) \in X,\ (y_1, y_2) \in X^\perp,\ (y_1, y_2) \in S(x_1, x_2),$$

where $S : \mathbb{R}^n \times \mathbb{R}^m \rightrightarrows \mathbb{R}^n \times \mathbb{R}^m$, with

$$S(x_1, x_2) = \partial f(x_1) \times \partial g(x_2)$$
$$X = \big\{(x_1, x_2) \in \mathbb{R}^n \times \mathbb{R}^m \bigm| x_2 = Ax_1\big\}$$
$$X^\perp = \big\{(y_1, y_2) \in \mathbb{R}^n \times \mathbb{R}^m \bigm| y_1 = -A^\top y_2\big\}.$$

Since we've reformulated (10.7) as a linkage problem, the progressive decoupling algorithm applies. Step 1 of the algorithm now amounts to solving

$$y_1^\nu \in \partial f(x_1) + \kappa(x_1 - x_1^\nu) \quad \text{and} \quad y_2^\nu \in \partial g(x_2) + \kappa(x_2 - x_2^\nu),$$

with the solutions furnishing $(\hat{x}_1^\nu, \hat{x}_2^\nu)$. If $f$ is convex and proper, then $\hat{x}_1^\nu$ satisfying the first generalized equation is equivalent to

$$\hat{x}_1^\nu \in \operatorname{argmin}_{x_1 \in \mathbb{R}^n} \big\{f(x_1) - \langle y_1^\nu, x_1\rangle + \tfrac{1}{2}\kappa\|x_1 - x_1^\nu\|_2^2\big\}$$

by the optimality condition 2.19. A parallel optimization problem is available if $g$ is convex and proper. Even in the absence of convexity, we might apply one of the many well-developed algorithms (see for example §4.K) to these optimization problems as a means to compute $\hat{x}_1^\nu$ and $\hat{x}_2^\nu$. The algorithms may not produce minimizers in nonconvex cases, but rather points satisfying optimality conditions as needed here. In any case, we consider $f$ and $g$ separately, a significant simplification compared to the actual problem.

Step 2 of the algorithm requires computing

$$(x_1^{\nu+1}, x_2^{\nu+1}) = \operatorname{prj}_X(\hat{x}_1^\nu, \hat{x}_2^\nu),$$

which amounts to having

$$(x_1^{\nu+1}, x_2^{\nu+1}) \in X \quad \text{and} \quad (\hat{x}_1^\nu, \hat{x}_2^\nu) - (x_1^{\nu+1}, x_2^{\nu+1}) \in X^\perp$$

by (10.2). In view of the specific formulae for $X$ and $X^\perp$, this requires us to solve for $(x_1^{\nu+1}, x_2^{\nu+1})$ in

$$x_2^{\nu+1} = Ax_1^{\nu+1} \quad \text{and} \quad \hat{x}_1^\nu - x_1^{\nu+1} = -A^\top(\hat{x}_2^\nu - x_2^{\nu+1}).$$

In the special case $A = I$, the identity matrix, we've the explicit solution

$$x_1^{\nu+1} = x_2^{\nu+1} = \tfrac{1}{2}(\hat{x}_1^\nu + \hat{x}_2^\nu).$$

The final part of Step 2 is always easily accomplished and in this case becomes

$$y_i^{\nu+1} = y_i^\nu - (\kappa - \theta)(\hat{x}_i^\nu - x_i^{\nu+1}), \quad i = 1, 2.$$

The progressive decoupling algorithm provides an alternative to the proximal gradient method (§2.D) and applies, as we see below, also in the nonconvex setting. □

**Example 10.9** (generalized equation with splitting). For $S_i : \mathbb{R}^n \rightrightarrows \mathbb{R}^n$, $i = 1, \dots, m$, let's consider the generalized equation

$$0 \in S_1(x) + \cdots + S_m(x),$$

which could, for example, be an optimality condition for the problem

$$\underset{x \in \mathbb{R}^n}{\text{minimize}} \sum\nolimits_{i=1}^m f_i(x)$$

and then $S_i(x) = \partial f_i(x)$. We can introduce $m$ copies of $x$ and, equivalently, state the problem as a linkage problem. The resulting instance of the progressive decoupling algorithm only needs to consider one of the set-valued mappings $S_i$ at a time.

**Detail.** Let $S : \mathbb{R}^{nm} \rightrightarrows \mathbb{R}^{nm}$ be defined by

$$S(x_1, \dots, x_m) = S_1(x_1) \times \cdots \times S_m(x_m),$$

where $x_i \in \mathbb{R}^n$ for all $i = 1, \dots, m$, and set

$$X = \big\{(x_1, \dots, x_m) \in \mathbb{R}^{nm} \bigm| x_1 = x_2 = \cdots = x_m\big\}.$$

Then,

$$X^\perp = \big\{(y_1, \dots, y_m) \in \mathbb{R}^{nm} \bigm| y_1 + \cdots + y_m = 0\big\}$$

by 10.5. This leads to the reformulations:

$$\begin{aligned}
&0 \in S_1(x) + \cdots + S_m(x)\\
&\iff y_1 + \cdots + y_m = 0, \quad \text{with } y_i \in S_i(x), \quad i = 1, \dots, m\\
&\iff x_1 = \cdots = x_m, \; y_1 + \cdots + y_m = 0, \quad \text{with } y_i \in S_i(x_i), \quad i = 1, \dots, m\\
&\iff (x_1, \dots, x_m) \in X, (y_1, \dots, y_m) \in X^\perp, \quad \text{with } (y_1, \dots, y_m) \in S(x_1, \dots, x_m).
\end{aligned}$$

Thus, the actual generalized equation in $n$ variables corresponds to a linkage problem in $nm$ variables. In this case, Step 1 of the progressive decoupling algorithm amounts to solving the generalized equations

$$y_i^\nu \in S_i(x_i) + \kappa(x_i - x_i^\nu), \quad i = 1, \dots, m$$

and obtaining $(\hat{x}_1, \dots, \hat{x}_m)$. Most significantly, the step is reduced to $m$ sub-steps, each involving only one of the set-valued mappings $S_i$.

Step 2 of the algorithm requires us to compute $(x_1^{\nu+1}, \dots, x_m^{\nu+1})$ as the projection of $(\hat{x}_1, \dots, \hat{x}_m)$ on $X$. By (10.3), this amounts to having

$$(x_1^{\nu+1}, \dots, x_m^{\nu+1}) \in X \quad \text{and} \quad (\hat{x}_1^{\nu}, \dots, \hat{x}_m^{\nu}) - (x_1^{\nu+1}, \dots, x_m^{\nu+1}) \in X^{\perp},$$

which simplify further to

$$x_1^{\nu+1} = \cdots = x_m^{\nu+1} \quad \text{and} \quad 0 = \sum\nolimits_{i=1}^{m} (\hat{x}_i^{\nu} - x_i^{\nu+1}) = \sum\nolimits_{i=1}^{m} \hat{x}_i^{\nu} - m x_1^{\nu+1}.$$

Thus, we need to carry out the averaging

$$x_i^{\nu+1} = (\hat{x}_1^{\nu} + \cdots + \hat{x}_m^{\nu})/m, \quad i = 1, \dots, m.$$

The last part of Step 2 specializes to

$$y_i^{\nu+1} = y_i^{\nu} - (\kappa - \theta)(\hat{x}_i^{\nu} - x_i^{\nu+1}), \quad i = 1, \dots, m.$$

Step 2 is, therefore, easily executed. □

**Example 10.10** (progressive decoupling in Nash game). Let's consider a Nash game with $m$ agents and objective functions $\varphi_i : \mathbb{R}^{n_i} \times \mathbb{R}^{n-n_i} \to \overline{\mathbb{R}}$, $i = 1, \dots, m$, where $n = \sum_{i=1}^{m} n_i$; see §9.A. A solution of the game is a Nash equilibrium, i.e., a vector $\bar{x} = (\bar{x}_1, \dots, \bar{x}_m)$ with

$$\bar{x}_i \in \operatorname{argmin}_{x_i \in \mathbb{R}^{n_i}} \varphi_i(x_i, \bar{x}_{-i}), \quad i = 1, \dots, m,$$

where $\bar{x}_{-i} = (\bar{x}_1, \bar{x}_2, \dots, \bar{x}_{i-1}, \bar{x}_{i+1}, \dots, \bar{x}_m)$. Here, we've a natural decomposition into $m$ subproblems, but their linkage is potentially complicated and not in the form of a subspace constraint. We again pursue the idea of splitting to achieve a reformulation as a linkage problem.

**Detail.** Suppose that $\varphi_i(\cdot, w)$ is convex and proper for any $w \in \mathbb{R}^{n-n_i}$. Then,

$$x_i \in \operatorname{argmin} \varphi_i(\cdot, w) \iff 0 \in \partial_i \varphi_i(x_i, w)$$

by the optimality condition 2.19, where $\partial_i \varphi_i(x_i, w)$ is the set of subgradients of $\varphi_i(\cdot, w)$ at $x_i$. If this holds for all $i$, then the problem of determining a Nash equilibrium $x \in \mathbb{R}^n$ can be expressed as

$$0 \in S_1(x) + \cdots + S_m(x),$$

where the set-valued mapping $S_i : \mathbb{R}^n \rightrightarrows \mathbb{R}^n$ has

$$S_i(x) = \prod\nolimits_{k=1}^{m} S_i^k(x), \quad \text{with} \quad S_i^k(x) = \begin{cases} \partial_i \varphi_i(x_i, x_{-i}) & \text{if } k = i \\ \{0\} \subset \mathbb{R}^{n_k} & \text{otherwise.} \end{cases}$$

Again, $x = (x_1, \ldots, x_m)$, with $x_i \in \mathbb{R}^{n_i}$, and $x_{-i} = (x_1, x_2, \ldots, x_{i-1}, x_{i+1}, \ldots, x_m)$. We're in the setting of 10.9, which furnishes an alternative formulation as the linkage problem

$$(z_1, \ldots, z_m) \in X, \quad (y_1, \ldots, y_m) \in X^\perp, \quad (y_1, \ldots, y_m) \in S(z_1, \ldots, z_m),$$

where $S : \mathbb{R}^{nm} \rightrightarrows \mathbb{R}^{nm}$ has $S(z_1, \ldots, z_m) = S_1(z_1) \times \cdots \times S_m(z_m)$ and

$$X = \{(z_1, \ldots, z_m) \in \mathbb{R}^{nm} \mid z_1 = \cdots = z_m\}$$
$$X^\perp = \{(y_1, \ldots, y_m) \in \mathbb{R}^{nm} \mid y_1 + \cdots + y_m = 0\}.$$

Let's clarify the notation. The vector $x = (x_1, \ldots, x_m) \in \mathbb{R}^{n_1} \times \cdots \times \mathbb{R}^{n_m} = \mathbb{R}^n$ specifies the decisions of all the agents, but is duplicated $m$ times to produce $z_1, \ldots, z_m$. We interpret $z_i \in \mathbb{R}^n$ as the vector containing the $i$th agent's decision as well as its *estimate* of the other agents' decisions. Specifically, we write

$$z_i = (z_{i1}, \ldots, z_{im}) \in \mathbb{R}^{n_1} \times \cdots \times \mathbb{R}^{n_m},$$

where $z_{ii}$ is the $i$th agent's own decision and $z_{ik}$ is its estimate for the $k$th agent. Other vectors such as $y_i$ are partitioned into subvectors similarly.

Given $(z_1^\nu, \ldots, z_m^\nu) \in X$ and $(y_1^\nu, \ldots, y_m^\nu) \in X^\perp$, Step 1 of the progressive decoupling algorithm in this case amounts to solving

$$y_i^\nu \in S_i(z_i) + \kappa(z_i - z_i^\nu), \quad i = 1, \ldots, m.$$

Bringing in the expression for $S_i$, the $i$th subproblem specializes to

$$y_{ii}^\nu \in \partial_i \varphi_i(z_{ii}, z_{i,-i}) + \kappa(z_{ii} - z_{ii}^\nu)$$
$$y_{ik}^\nu = \kappa(z_{ik} - z_{ik}^\nu), \quad k \neq i,$$

where

$$z_{i,-i} = (z_{i1}, z_{i2}, \ldots, z_{ii-1}, z_{ii+1}, \ldots, z_{im}).$$

The equations for $k \neq i$ have the explicit solutions

$$\hat{z}_{ik}^\nu = z_{ik}^\nu + \tfrac{1}{\kappa} y_{ik}^\nu, \quad k \neq i,$$

which then furnish the $i$th agent's estimates of the other agents' decisions. We recall that $z_1^\nu = \cdots = z_m^\nu$ so $z_i^\nu$ can be viewed as a vector of intermediate compromise decisions. The estimates $\hat{z}_{ik}^\nu$ are then (small) modifications of these compromise decisions. This fully specifies $z_{i,-i}$. We can now determine $z_{ii}$ using the generalized equation

$$y_{ii}^\nu \in \partial_i \varphi_i(z_{ii}, \hat{z}_{i,-i}^\nu) + \kappa(z_{ii} - z_{ii}^\nu),$$

where

$$\hat{z}_{i,-i}^\nu = (\hat{z}_{i1}^\nu, \hat{z}_{i2}^\nu, \ldots, \hat{z}_{ii-1}^\nu, \hat{z}_{ii+1}^\nu, \ldots, \hat{z}_{im}^\nu).$$

By the optimality condition 2.19, the resulting solution for $z_{ii}$ is

$$\hat{z}_{ii}^{\nu} \in \operatorname{argmin}_{z_{ii}\in\mathbb{R}^{n_i}} \varphi_i(z_{ii}, \hat{z}_{i,-i}^{\nu}) - \langle y_{ii}^{\nu}, z_{ii}\rangle + \tfrac{1}{2}\kappa\|z_{ii} - z_{ii}^{\nu}\|_2^2.$$

Compared to the actual problem for agent $i$, which is coupled with the other agents, this subproblem utilizes estimates of the other agents' decisions and is thus decoupled. Moreover, it involves two additional terms that attempt to capture the effect of deviation from the compromise decisions and its ramification on the other agents. This, in turn, circles back and influences the $i$th agent. In summary, we've now determined $\hat{z}_i = (\hat{z}_{i1}, \dots, \hat{z}_{im})$.

Parallel to the situation in 10.9, Step 2 of the progressive decoupling algorithm has two parts. First, compute a new compromise decision

$$z_i^{\nu+1} = (\hat{z}_1^{\nu} + \cdots + \hat{z}_m^{\nu})/m, \quad i = 1, \dots, m,$$

which then is the same for all $i$. The resulting subvector $z_{ik}^{\nu+1}$ is the average of what agent $k$ would like to do and the "opinion" of everybody else about what agent $k$ should do. Second, update

$$y_i^{\nu+1} = y_i^{\nu} - (\kappa - \theta)(\hat{z}_i^{\nu} - z_i^{\nu+1}), \quad i = 1, \dots, m.$$

Both parts of the step are easily executed. □

**Exercise 10.11** (solution of Nash game). Apply the progressive decoupling algorithm as laid out in 10.10 to the game in 9.1 starting from

$$z_1^0 = z_2^0 = (0, 0), \quad y_1^0 = y_2^0 = (0, 0)$$

and using $\kappa = 1$ and $\theta = 0$. Compute three iterations.

**Guide.** The first iteration produces

$$\hat{z}_1^0 = \left(\tfrac{2}{3}, 0\right),\ \hat{z}_2^0 = \left(0, \tfrac{1}{3}\right),\ z_1^1 = z_2^1 = \left(\tfrac{1}{3}, \tfrac{1}{6}\right),\ y_1^1 = \left(-\tfrac{1}{3}, \tfrac{1}{6}\right),\ y_2^1 = \left(\tfrac{1}{3}, -\tfrac{1}{6}\right).$$

The second iteration gives

$$\hat{z}_1^1 = \left(\tfrac{2}{3}, \tfrac{1}{3}\right),\ \hat{z}_2^1 = \left(\tfrac{2}{3}, \tfrac{1}{3}\right),\ z_1^2 = z_2^2 = \left(\tfrac{2}{3}, \tfrac{1}{3}\right),$$

which is one of the solutions; cf. 9.1. □

We next turn to the convergence properties of the progressive decoupling algorithm, which rely on the fundamental concept of monotonicity. In the same way as convexity plays a central role in the analysis of minimization problems, the presence of monotonicity in a generalized equation leads to many simplifications and the possibility of developing efficient computational procedures.

**Definition 10.12** (monotonicity). A set-valued mapping $S : \mathbb{R}^n \rightrightarrows \mathbb{R}^n$ is *monotone* when

$$\forall (x, y) \in \operatorname{gph} S,\ (\bar{x}, \bar{y}) \in \operatorname{gph} S : \quad \langle y - \bar{y}, x - \bar{x}\rangle \geq 0.$$

It's *maximal monotone* when, in addition,

$$\forall (x,y) \notin \operatorname{gph} S : \qquad \exists (\bar{x},\bar{y}) \in \operatorname{gph} S \text{ such that } \langle y-\bar{y}, x-\bar{x}\rangle < 0.$$

Equivalently, $S$ is maximal monotone if there's no monotone $T : \mathbb{R}^n \rightrightarrows \mathbb{R}^n$ with $\operatorname{gph} T \supset \operatorname{gph} S$ and $\operatorname{gph} T \neq \operatorname{gph} S$.

The definition of monotonicity reduces to (7.14) when the set-valued mapping always returns a single point. In that case, maximality is less critical because every continuous monotone $F : \mathbb{R}^n \to \mathbb{R}^n$ is automatically maximal monotone when viewed as a set-valued mapping; cf. [105, Example 12.7].

Figure 10.2(left) shows the graph of a set-valued mapping $S$. For any two points in the graph, one is always "north-east" of the other, which means that $S$ is monotone. It's actually maximal monotone because for any point outside the graph, one can find a point in the graph that's neither "north-east" nor "south-west" of the point outside the graph. The arrows indicate that the graph extends east and south, which is essential for maximality. The set-valued mapping $S$ in Figure 10.2(middle) is also monotone. However, the point $(x,y)$ is "north-east" of all the points in $\operatorname{gph} S$ so $S$ isn't maximal monotone. We can also see this by invoking the equivalent definition of maximal monotonicity: extend $\operatorname{gph} S$ in both directions to produce the graph of a monotone set-valued mapping $T$.

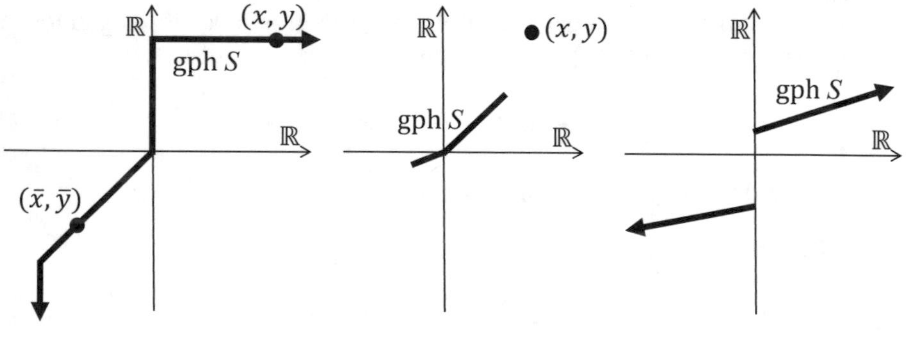


Fig. 10.2: Graphs of three monotone set-valued mappings, but only the left one is maximal.

**Proposition 10.13** (extension to maximality). *For a monotone set-valued mapping* $S : \mathbb{R}^n \rightrightarrows \mathbb{R}^n$, *there's a maximal monotone set-valued mapping* $T : \mathbb{R}^n \rightrightarrows \mathbb{R}^n$ *such that* $\operatorname{gph} T \supset \operatorname{gph} S$.

**Proof.** For a proof based on Zorn's lemma; see [105, Proposition 12.6]. □

An extension of a monotone set-valued mapping guaranteed by the proposition isn't necessarily unique; in Figure 10.2(middle) rays adjoining the right endpoint of $\operatorname{gph} S$ and pointing "north," "east" or anything between will do, with similar flexibility presenting itself at the left endpoint of $\operatorname{gph} S$. Figure 10.2(right) has $S$ monotone, but $\operatorname{gph} S$ "misses" a vertical line connecting the two branches of the graph and thus $S$ isn't maximal.

**Proposition 10.14** (maximal monotonicity). *For $S : \mathbb{R}^n \rightrightarrows \mathbb{R}^n$, one has the facts:*

(a) *$S$ is maximal monotone if and only if $S^{-1}$ is maximal monotone.*
(b) *If $S$ is maximal monotone, then* gph *$S$ is closed so that $S$ is osc.*
(c) *If $S$ is maximal monotone, then $S$ and $S^{-1}$ are both closed-valued and convex-valued.*

**Proof.** These assertions follow straightforwardly from the definition of maximal monotonicity, with the connection between closedness and osc being a consequence of 7.26. □

**Proposition 10.15** (convexity and maximal monotonicity). *For a proper function $f : \mathbb{R}^n \to \overline{\mathbb{R}}$ and its subgradient mapping $\partial f$, the following hold:*

(a) *If $f$ is convex, then $\partial f$ is monotone.*
(b) *If $f$ is lsc and $\partial f$ is monotone, then $f$ is convex and $\partial f$ is maximal monotone.*

**Proof.** For (a), let $f$ be proper and convex, $x, \bar{x} \in \operatorname{dom} f$, $v \in \partial f(x)$ and $\bar{v} \in \partial f(\bar{x})$. Then, the subgradient inequality 2.17 ensures that

$$f(x) \geq f(\bar{x}) + \langle \bar{v}, x - \bar{x} \rangle \quad \text{and} \quad f(\bar{x}) \geq f(x) + \langle v, \bar{x} - x \rangle.$$

Thus,

$$f(x) + f(\bar{x}) \geq f(\bar{x}) + f(x) + \langle v, x - \bar{x} \rangle + \langle v, \bar{x} - x \rangle$$

and then $0 \geq \langle \bar{v}, x - \bar{x} \rangle + \langle v, \bar{x} - x \rangle$. This means that

$$\forall x, \bar{x} \in \operatorname{dom} f,\ v \in \partial f(x),\ \bar{v} \in \partial f(\bar{x}) : \qquad \langle v - \bar{v}, x - \bar{x} \rangle \geq 0.$$

Since $(x, v) \in \operatorname{gph} \partial f$ if and only if $x \in \operatorname{dom} f$ and $v \in \partial f(x)$, we conclude that $\partial f$ is monotone. For (b), we refer to [105, Theorem 12.17]. □

Figure 10.2(left) can be viewed as the graph produced by the set of subgradients for

$$f(x) = \begin{cases} \frac{1}{2}x^2 & \text{if } x \in [-1, 0) \\ x & \text{if } x \in [0, \infty) \\ \infty & \text{otherwise.} \end{cases}$$

This function is certainly proper, lsc and convex. Moreover,

$$\partial f(x) = \begin{cases} (-\infty, -1] & \text{if } x = -1 \\ \{x\} & \text{if } x \in (-1, 0) \\ [0, 1] & \text{if } x = 0 \\ \{1\} & \text{if } x \in (0, \infty) \\ \emptyset & \text{otherwise,} \end{cases}$$

which defines a maximal monotone set-valued mapping.

Figure 10.2(middle) is the graph of the subgradient mapping of the function given by

$$f(x) = \begin{cases} \frac{1}{4}x^2 & \text{if } x \in \big(-\frac{1}{4}, 0\big) \\ \frac{1}{2}x^2 & \text{if } x \in \big[0, \frac{1}{2}\big) \\ 1 & \text{if } x = \frac{1}{2} \\ \infty & \text{otherwise.} \end{cases}$$

This function is proper and convex, with

$$\partial f(x) = \begin{cases} \big\{\frac{1}{2}x\big\} & \text{if } x \in \big(-\frac{1}{4}, 0\big) \\ \big\{x\big\} & \text{if } x \in \big[0, \frac{1}{2}\big) \\ \emptyset & \text{otherwise} \end{cases}$$

so the resulting graph doesn't include the endpoints $(-1/4, -1/8)$ and $(1/2, 1/2)$. The subgradient mapping is monotone, but fails maximality in this case. Since the value $f(1/2)$ can be changed to any number $\alpha \in (1/8, \infty)$ without ruining convexity, $f$ just given isn't the only convex function with subgradients producing this graph.

**Exercise 10.16** (gradient test for convexity). Show that a smooth function $f : \mathbb{R}^n \to \mathbb{R}$ is convex if and only if $\langle \nabla f(x) - \nabla f(\bar{x}), x - \bar{x} \rangle \geq 0$ for all $x, \bar{x} \in \mathbb{R}^n$. Use this fact to establish that the quadratic function given by $f(x) = \frac{1}{2}\langle x, Qx \rangle + \langle c, x \rangle$ is convex if and only if $Q$ is positive semidefinite.

By setting $f = \iota_C$ in 10.15 for some nonempty, closed and convex $C \subset \mathbb{R}^n$, we see that the normal cone mapping $N_C$ is maximal monotone in that case. From this property, one can invoke the following proposition to conclude that the set-valued mapping $F + N_C$, a common form in applications (see §7.A), is maximal monotone when $F : \mathbb{R}^n \to \mathbb{R}^n$ is continuous and monotone.

**Proposition 10.17** (preservation of maximality). *If $F : \mathbb{R}^n \to \mathbb{R}^n$ is continuous and monotone and $S : \mathbb{R}^n \rightrightarrows \mathbb{R}^n$ is maximal monotone, then $F + S$ is maximal monotone.*

**Proof.** By [105, Example 12.7], $F$ is maximal monotone and also nonempty-valued when viewed as a set-valued mapping. We can then invoke [105, Corollary 12.44]. □

The proposition guarantees that if $S$ under consideration by the progressive decoupling algorithm is maximal monotone, then the set-valued mapping given by

$$S^\nu(x) = S(x) + \kappa(x - x^\nu),$$

which we face in Step 1 of the algorithm, is also maximal monotone. In this sense, the change from $S$ to $S^\nu$ isn't detrimental.

**Proposition 10.18** (maximality under separability). *For maximal monotone $S_i : \mathbb{R}^{n_i} \rightrightarrows \mathbb{R}^{n_i}$, $i = 1, \dots, m$, and $S : \mathbb{R}^n \rightrightarrows \mathbb{R}^n$ given by*

$$S(x) = S_1(x_1) \times \cdots \times S_m(x_m),$$

*with $x = (x_1, \dots, x_m)$ and $n = \sum_{i=1}^m n_i$, one has that $S$ is also maximal monotone.*

**Proof.** This follows straightforwardly from the definition. □

We would like to leverage the useful properties of maximal monotone set-valued mappings to establish convergence properties of the progressive decoupling algorithm. However, it would be restrictive to insist on $S$ in the linkage problem being maximal monotone. We temper the severity of such an assumption by recognizing that for any $\theta \in [0, \infty)$,

$$(\bar{x}, \bar{y}) \text{ solves the linkage problem } \iff \bar{x} \in X,\ \bar{y} \in X^{\perp},\ \bar{y} \in S(\bar{x}) + \theta\, \mathrm{prj}_{X^{\perp}}(\bar{x}),$$

which follows immediately from (10.3) because $\bar{x} = \mathrm{prj}_X(\bar{x})$ and then $0 = \mathrm{prj}_{X^{\perp}}(\bar{x})$. This insight shows that we can

$$\text{replace } S \text{ by } S + \theta\, \mathrm{prj}_{X^{\perp}}$$

in the linkage problem, with $\theta \in [0, \infty)$ being selected in a manner that promotes desirable properties such as maximal monotonicity.

**Definition 10.19** (elicitation). For $S : \mathbb{R}^n \rightrightarrows \mathbb{R}^n$ and a subspace $X$ of $\mathbb{R}^n$, we say that $S$ has *elicitable maximal monotonicity at level* $\theta \in [0, \infty)$ if the set-valued mapping

$$\big(S + \theta\, \mathrm{prj}_{X^{\perp}}\big) : \mathbb{R}^n \rightrightarrows \mathbb{R}^n \text{ is maximal monotone.}$$

The property is always relative to a particular subspace $X$, which is understood from the context.

If $S$ is maximal monotone, then it's apparent that $\theta$ can be set to zero in the definition regardless of $X$. In view of 10.15, this would take place when $S$ is the subgradient mapping of a proper, lsc and convex function, but that's just the beginning, however, as the concept of elicitation covers much more territory.

**Example 10.20** (elicitation in optimization). For $f(x) = x_1 x_2$ and a subspace

$$X = \big\{(x_1, x_2) \in \mathbb{R}^2 \;\big|\; x_1 = x_2\big\},$$

the problem of minimizing $f$ over $X$ has the optimality condition

$$-\nabla f(x) \in N_X(x) \iff x \in X,\ y \in X^{\perp},\ y = \nabla f(x);$$

see (10.2) and 2.27. The set-valued mapping $S : \mathbb{R}^2 \rightrightarrows \mathbb{R}^2$ given by $S(x) = \{\nabla f(x)\}$ isn't maximal monotone, but has elicitable maximal monotonicity at level $\theta = 1$.

**Detail.** Since $f$ is proper and lsc but not convex, it follows from 10.15 that $S$ can't be monotone. We also see this from the definition of monotonicity because

$$\big\langle \nabla f(x) - \nabla f(\bar{x}), x - \bar{x} \big\rangle = (x_2 - \bar{x}_2)(x_1 - \bar{x}_1) + (x_1 - \bar{x}_1)(x_2 - \bar{x}_2) < 0$$

for $(x_1, x_2) = (-1, 1)$ and $(\bar{x}_1, \bar{x}_2) = (0, 0)$.

In this case,

$$X^\perp = \{(x_1, x_2) \mid x_1 + x_2 = 0\}.$$

By (10.3), $\bar{x} = \mathrm{prj}_{X^\perp}(x)$ is equivalent to $x - \bar{x} \in X$ and $\bar{x} \in X^\perp$, which simplify to $x_1 - \bar{x}_1 = x_2 - \bar{x}_2$ and $\bar{x}_1 + \bar{x}_2 = 0$. Thus,

$$\bar{x}_1 = \tfrac{1}{2}(x_1 - x_2) \quad \text{and} \quad \bar{x}_2 = \tfrac{1}{2}(x_2 - x_1),$$

which lead to

$$S(x) + \theta\,\mathrm{prj}_{X^\perp}(x) = \left\{ \begin{bmatrix} x_2 + \frac{1}{2}\theta(x_1 - x_2) \\ x_1 + \frac{1}{2}\theta(x_2 - x_1) \end{bmatrix} \right\}.$$

This is the gradient of a function $\varphi$ at $x$, where

$$\varphi(x) = x_1 x_2 + \tfrac{1}{4}\theta(x_1 - x_2)^2.$$

Since $\varphi$ is twice smooth, it's convex if and only if

$$\nabla^2 \varphi(x) = \begin{bmatrix} \frac{1}{2}\theta & 1 - \frac{1}{2}\theta \\ 1 - \frac{1}{2}\theta & \frac{1}{2}\theta \end{bmatrix}$$

is positive semidefinite (cf. 1.23), which takes place for $\theta \in [1, \infty)$. Thus, for such $\theta$, $\varphi$ is convex and then $S + \theta\,\mathrm{prj}_{X^\perp}$ is maximal monotone by 10.15. □

The example hints to a general connection between elicitation for a subgradient mapping of a function $f$ and the convexity of a modified function. In fact, this furnishes a main approach to establish an elicitation level for a set-valued mapping. We start with the following observation, where we recall that $\mathrm{dist}(x, X)$ is the Euclidean distance from $x$ to $X$ so that $\mathrm{dist}^2(x, X)$ is the square of that distance.

**Lemma 10.21** *For a subspace $X$ of $\mathbb{R}^n$, the function $\varphi : \mathbb{R}^n \to \mathbb{R}$ given by*

$$\varphi(x) = \tfrac{1}{2}\,\mathrm{dist}^2(x, X)$$

*is smooth with $\nabla\varphi(x) = \mathrm{prj}_{X^\perp}(x)$.*

**Proof.** For $x \in \mathbb{R}^n$, one has

$$\varphi(x) = \inf_{z \in \mathbb{R}^n} \psi(x, z), \quad \text{with } \psi(x, z) = \tfrac{1}{2}\|z - x\|_2^2 + \iota_X(z).$$

Since $X$ is nonempty, convex and closed, $\psi$ is proper, lsc and convex. Moreover, $\psi(x, z)$ is level-bounded in $z$ locally uniformly in $x$. Thus, 5.13 applies and for every $x \in \mathbb{R}^n$,

$$\partial\varphi(x) = \{v \in \mathbb{R}^n \mid (v, 0) \in \partial\psi(x, \bar{z})\},$$

where $\{\bar{z}\} = \mathrm{argmin}_{z \in \mathbb{R}^n} \psi(x, z) = \mathrm{prj}_X(x)$. By 4.58,

$$\partial\psi(x, z) = \{x - z\} \times \big(z - x + N_X(z)\big).$$

Thus, the condition $(v, 0) \in \partial\psi(x, \bar{z})$ specifies $v = x - \bar{z}$ and $x - \bar{z} \in N_X(\bar{z})$, with the latter holding by (10.2) and (10.3). We've established that

$$\partial\varphi(x) = x - \text{prj}_X(x)$$

and this is a single vector for each $x$. By 7.14, $\text{prj}_X$ is continuous so $\varphi$ is smooth. We then invoke (10.4) to reach the conclusion. □

**Proposition 10.22** (elicitation for subgradient mappings). *For a proper lsc function $f : \mathbb{R}^n \to \overline{\mathbb{R}}$ and a subspace $X$ of $\mathbb{R}^n$, the subgradient mapping $\partial f$ has elicitable maximal monotonicity at level $\theta \in [0, \infty)$ if and only if the function*

$$\big(f + \tfrac{1}{2}\theta\,\text{dist}^2(\cdot, X)\big) : \mathbb{R}^n \to \overline{\mathbb{R}}$$

*is convex.*

**Proof.** By 10.21 and 4.58, one has at any point $x$ for which $f$ is finite,

$$\partial\big(f + \tfrac{1}{2}\theta\,\text{dist}^2(\cdot, X)\big)(x) = \partial f(x) + \theta\,\text{prj}_{X^\perp}(x).$$

Since $f + \frac{1}{2}\theta\,\text{dist}^2(\cdot, X)$ is proper and lsc, it follows from 10.15 that

$$f + \tfrac{1}{2}\theta\,\text{dist}^2(\cdot, X) \text{ is convex} \iff \partial f + \theta\,\text{prj}_{X^\perp} \text{ is monotone},$$

in which case $\partial f + \theta\,\text{prj}_{X^\perp}$ is maximal monotone. □

The proposition establishes that elicitation for subgradient mappings reduces to checking convexity of a regularized version of the corresponding function. The point-to-set distance furnishes the regularization term; a point $x$ is penalized by the square of $\text{dist}(x, X)$. The hope is that this regularization suffices to construct a convex function, which indeed is the case in 10.20. We elaborate on this path to elicitation in §10.G.

We're now ready to state the first of three convergence properties for the progressive decoupling algorithm. Let's quantify the distance to a solution using the norm on $\mathbb{R}^{2n}$ given by

$$\big\|(x, y)\big\|_{\kappa,\theta} = \sqrt{\|x\|_2^2 + \frac{1}{\kappa(\kappa - \theta)}\|y\|_2^2},$$

where $\theta \in [0, \infty)$ and $\kappa \in (\theta, \infty)$ are the parameters of the algorithm. If $x \in X$ and $y \in X^\perp$, then the expression simplifies to

$$\big\|(x, y)\big\|_{\kappa,\theta} = \big\|x + y/\sqrt{\kappa(\kappa - \theta)}\big\|_2 \tag{10.8}$$

because $x$ and $y$ are orthogonal and

$$\|x + \alpha y\|_2^2 = \langle x + \alpha y, x + \alpha y\rangle = \|x\|_2^2 + \alpha^2\|y\|_2^2 \quad \forall \alpha \in (0, \infty).$$

As we see below, this norm reflects the possibility that the algorithm may generated $x^\nu$ that converges faster than $y^\nu$ or vice versa, and these rates are influenced by $\theta, \kappa$.

**Convergence 10.23** (progressive decoupling algorithm). *Suppose that $S$ has elicitable maximal monotonicity at level $\theta \in [0, \infty)$. If the progressive decoupling algorithm with parameters $\theta < \kappa$ has generated $\{x^\nu, y^\nu, \nu \in \mathbb{N}\}$ while solving the linkage problem, which is assumed to have a solution, then $(x^\nu, y^\nu)$ converges to a solution $(\bar{x}, \bar{y})$ of the linkage problem and*

$$\big\|(x^{\nu+1}, y^{\nu+1}) - (\bar{x}, \bar{y})\big\|_{\kappa,\theta} \leq \big\|(x^\nu, y^\nu) - (\bar{x}, \bar{y})\big\|_{\kappa,\theta} \;\; \forall \nu = 0, 1, 2, \dots. \tag{10.9}$$

*Moreover, the point $\hat{x}^\nu$ computed in Step 1 of the algorithm is uniquely defined in each iteration.*

We postpone the proof to the next section and instead examine some implications. As seen from (10.9), the algorithm can't move further away from a solution indicating "steady" progress in each iteration. This holds for any $\theta < \kappa$.

It's clear that a low value of $\theta$ allows for more flexibility in choosing $\kappa$. But, what would be a good value of $\kappa$? We can think of $1/\kappa$ as a step size for the $x$-vector; cf. the discussion of the proximal point method in §1.H and 10.8. A high $\kappa$ implies a low $1/\kappa$ so that $\hat{x}^\nu$ tends to be near $x^\nu$, with the $x$-component of the solution converging fast. In updating the $y$-vector, however, $\kappa$ emerges as a step size with high values of $\kappa$ causing large changes in the $y$-component of the solution and then slower convergence there. A reverse situation takes place when $\kappa$ is low. The guarantee in (10.9) indicates that a balanced approach is achieved, in theory, by setting

$$\kappa = \tfrac{1}{2}\theta + \sqrt{\tfrac{1}{4}\theta^2 + 1}$$

so that $\kappa(\kappa - \theta) = 1$ and $\|(x, y)\|_{\kappa,\theta} = \|(x, y)\|_2$. When $x \in X$ and $y \in X^\perp$, one also has

$$\big\|(x, y)\big\|_{\kappa,\theta} = \|x + y\|_2$$

by (10.8). In practice, some tuning of $\kappa$ can be expected to achieve reasonable computational performance.

The elicitation level $\theta$ may be unknown in a particular application, but this can be overcome by gradually increasing $\kappa$ with the hope that eventually $\kappa$ would exceed $\theta$. This difficulty is further alleviated by the fact that $S + \theta \operatorname{prj}_{X^\perp}$ is maximal monotone whenever $S + \bar{\theta} \operatorname{prj}_{X^\perp}$ is maximal monotone and $\bar{\theta} \leq \theta$. Thus, any $\theta$ greater than a "minimal elicitation level" is also an elicitation level. To see this, note that

$$S + \theta \operatorname{prj}_{X^\perp} = S + \bar{\theta} \operatorname{prj}_{X^\perp} + (\theta - \bar{\theta}) \operatorname{prj}_{X^\perp}$$

with $S + \bar{\theta} \operatorname{prj}_{X^\perp}$ being maximal monotone and $(\theta - \bar{\theta}) \operatorname{prj}_{X^\perp}$ being nonempty-valued and continuous by 7.14. In view of 10.17, the claim follows after establishing the monotonicity of $\operatorname{prj}_{X^\perp}$. By 10.21, the function with

$$\varphi(x) = \tfrac{1}{2} \operatorname{dist}^2(x, X)$$

has $\partial\varphi(x) = \mathrm{prj}_{X^\perp}(x)$. The function $\varphi$ is proper and also convex by virtue of being expressible as an inf-projection; see 1.21. Thus, $\partial\varphi = \mathrm{prj}_{X^\perp}$ is monotone by 10.15.

## 10.D Local Elicitation

While elicitation extends the convergence property of the progressive decoupling algorithm in 10.23 beyond the case of maximal monotone mappings, the bar remains too high for some applications. For example, when $S = \partial f$ for some proper lsc $f : \mathbb{R}^n \to \overline{\mathbb{R}}$, we need $f + \frac{1}{2}\theta\,\mathrm{dist}^2(\cdot, X)$ to be convex by 10.22. A necessity for this to take place is that $f + \iota_X$ is convex, which may not hold in a particular case. It turns out that the progressive decoupling algorithm remains valid for a much broader class of set-valued mappings when initiated sufficiently close to a solution of the linkage problem.

Let's start by considering local notions of monotonicity. For $C \subset \mathbb{R}^n \times \mathbb{R}^n$, a set-valued mapping $S : \mathbb{R}^n \rightrightarrows \mathbb{R}^n$ is *monotone in* $C$ when $\mathrm{gph}\, S \cap C$ is the graph of a monotone set-valued mapping. It's *maximal monotone in* $C$ if there's a maximal monotone set-valued mapping $T : \mathbb{R}^n \rightrightarrows \mathbb{R}^n$ such that $\mathrm{gph}\, T \cap C = \mathrm{gph}\, S \cap C$.

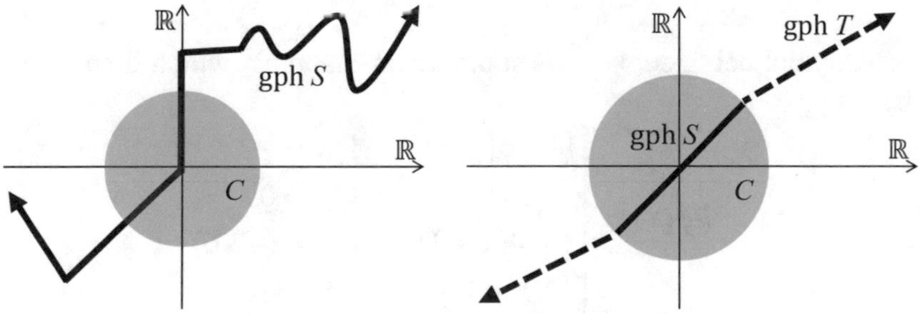


Fig. 10.3: The graph of nonmonotone $S$ (left) defines a monotone set-valued mapping when intersected by the disk $C$ and is thus monotone in $C$. Monotone $S$ (right), which isn't maximal, has a graph that can be extended beyond a disk $C$ to form the graph of maximal monotone $T$ and this makes $S$ maximal monotone in $C$.

Figure 10.3 illustrates how these concepts reduce the considerations to looking at $\mathrm{gph}\, S \cap C$. With these refinements of monotonicity, we can extend the notion of elicitation by invoking it in a local sense.

**Definition 10.24** (local elicitation). For $S : \mathbb{R}^n \rightrightarrows \mathbb{R}^n$ and a subspace $X$ of $\mathbb{R}^n$, we say that $S$ has *elicitable maximal monotonicity at level* $\theta \in [0, \infty)$ *locally near a solution* $(\bar{x}, \bar{y})$ of the linkage problem if the set-valued mapping

$$\big(S + \theta\,\mathrm{prj}_{X^\perp}\big) : \mathbb{R}^n \rightrightarrows \mathbb{R}^n \text{ is maximal monotone in a neighborhood of } (\bar{x}, \bar{y}).$$

**Example 10.25** (local elicitation in composite optimization). For $h : \mathbb{R} \to \overline{\mathbb{R}}$ and $f : \mathbb{R}^n \to \mathbb{R}$, both convex, and a subspace $X$ of $\mathbb{R}^n$, let's consider the problem

$$\operatorname*{minimize}_{x\in X} \varphi(x) = h\big(f(x)\big).$$

Although both $h$ and $f$ are convex, $\varphi$ may not be. The associated generalized equation

$$x \in X, \ \ y \in X^\perp, \ \ y \in \partial\varphi(x),$$

which is a necessary condition for a minimizer under a qualification (see 4.67, (10.2) and the Fermat rule 4.73), involves a potentially nonmonotone subgradient mapping $\partial\varphi$. However, there's a good chance that a restriction of $\operatorname{gph}\partial\varphi$ to an open set containing a solution of the generalized equation produces a monotone mapping. As a consequence, $\partial\varphi$ would have elicitable maximal monotonicity at level $\theta = 0$ locally near that solution.

**Detail.** For a concrete example, let $h(u) = \max\{-2u, u\}$, which may represent penalties for deviating from a target of $f(x) = 0$. If $f(x) = (x+1)^2 - 1$, then

$$\varphi(x) = \begin{cases} -2(x+1)^2 + 2 & \text{if } x \in [-2, 0] \\ (x+1)^2 - 1 & \text{otherwise,} \end{cases}$$

with $\varphi$ certainly not being convex. Its subgradient mapping, which then isn't monotone, is given by

$$\partial\varphi(x) = \begin{cases} [-2, 4] & \text{if } x = -2 \\ [-4, 2] & \text{if } x = 0 \\ \{-4(x+1)\} & \text{if } x \in (-2, 0) \\ \{2(x+1)\} & \text{otherwise.} \end{cases}$$

However, the function $\varphi + \iota_{(0,\infty)}$ is convex, with monotone subgradient mapping

$$\partial(\varphi + \iota_{(0,\infty)})(x) = \begin{cases} \{2(x+1)\} & \text{if } x \in (0, \infty) \\ \emptyset & \text{otherwise.} \end{cases}$$

By 10.13, there's a maximal monotone $T : \mathbb{R} \rightrightarrows \mathbb{R}$ such that

$$\operatorname{gph} T \supset \operatorname{gph} \partial(\varphi + \iota_{(0,\infty)}).$$

At any point $(\bar{x}, 2(\bar{x}+1))$, with $\bar{x} > 0$, there's a neighborhood $C$ of that point such that

$$\operatorname{gph}\partial\varphi \cap C = \operatorname{gph} T \cap C.$$

Thus, $\partial\varphi$ is maximal monotone in $C$. □

When the elicitation is only local, the convergence properties of the progressive decoupling algorithm also localize. Specifically, 10.23 is replaced by the following fact.

**Convergence 10.26** (progressive decoupling algorithm; cont.). *Suppose that $S$ has elicitable maximal monotonicity at level $\theta \in [0,\infty)$ locally near a solution $(\tilde{x},\tilde{y})$ of the linkage problem. Then, there's a neighborhood $B$ of $\tilde{x}+\tilde{y}$ with the following property:*

*If the progressive decoupling algorithm, with parameters $\theta < \kappa$ and $x^0 + y^0 \in B$, is applied to $S$ with the additional condition in Step 1 that $\hat{x}^\nu \in B$, then $\hat{x}^\nu$ is uniquely defined, the generated points $\{x^\nu, y^\nu,\ \nu \in \mathbb{N}\}$ have $x^\nu + y^\nu \in B$ for all $\nu$ and $(x^\nu, y^\nu)$ converges to a solution $(\bar{x},\bar{y})$ of the linkage problem while satisfying (10.9).*

The requirement of starting the algorithm sufficiently close to a solution resembles the situation for Newton's method (§1.G). This brings up the need for a gap function that monitors progress as well as modifications that ensure progress is actually achieved, but we omit further details; see for example §7.F. We find specific bounds on what constitutes "sufficiently close" in [94].

We now turn to the proofs of the asserted convergence properties in 10.23 and 10.26, which rely heavily on the properties of the proximal point method in §1.H. At that early stage, we limited the treatment to smooth functions. However, the proximal point method extends to the solution of the generalized equation.

**Proximal Point Method for Generalized Equations.** As laid out in §1.H for a smooth function $f : \mathbb{R}^n \to \mathbb{R}$, the proximal point method solves the equation $\nabla f(x) = 0$ by computing

$$x^{\nu+1} \in \operatorname{prx}_\lambda f(x^\nu) = \operatorname{argmin}_{x\in\mathbb{R}^n} f(x) + \tfrac{1}{2\lambda}\|x - x^\nu\|_2^2,$$

with parameter $\lambda \in (0,\infty)$. Let's suppose that $f$ is convex. Then, one could equivalently leverage the optimality condition in 2.19 corresponding to this subproblem, which is

$$0 = \nabla f(x^{\nu+1}) + \tfrac{1}{\lambda}(x^{\nu+1} - x^\nu).$$

This is largely the direction of the proof of convergence 1.37. Here, $\nabla f$ is a maximal monotone mapping by 10.15. We can extend this idea to an arbitrary maximal monotone $S : \mathbb{R}^n \rightrightarrows \mathbb{R}^n$. For $\kappa \in (0,\infty)$, one can solve the generalized equation $0 \in S(x)$ by starting at some point $x^0 \in \mathbb{R}^n$ and then computing iteratively $x^{\nu+1}$ such that

$$0 \in S(x^{\nu+1}) + \kappa(x^{\nu+1} - x^\nu). \tag{10.10}$$

In fact, $x^{\nu+1}$ is uniquely defined by the inclusion; see [91]. This is the *proximal point method for generalized equations*. For refinements beyond the maximal monotone case, we refer to [94].

**Convergence 10.27** (proximal point method for generalized equations). *For a maximal monotone set-valued mapping $S : \mathbb{R}^n \rightrightarrows \mathbb{R}^n$, $\kappa \in (0,\infty)$ and $x^0 \in \mathbb{R}^n$, suppose that $\{x^\nu, \nu \in \mathbb{N}\}$ is generated by (10.10) starting from $x^0$. Then, the following hold:*

(a) *$\{x^\nu, \nu \in \mathbb{N}\}$ is bounded if and only if $0 \in S(x)$ has a solution.*
(b) *If $\{x^\nu, \nu \in \mathbb{N}\}$ is bounded, then it converges to a solution of $0 \in S(x)$.*

(c) *For all* $\nu$ *and solution* $x \in S^{-1}(0)$, *one has*

$$\|x^{\nu+1} - x\|_2 \leq \|x^\nu - x\|_2.$$

**Proof.** Theorem 1 and Section 1 of [91] establish these facts. □

Our goal is to leverage this result to obtain the convergence properties of the progressive decoupling algorithm. We proceed through three intermediate steps.

**Proposition 10.28** (localization). *For* $S : \mathbb{R}^n \rightrightarrows \mathbb{R}^n$, $\kappa \in (0, \infty)$ *and* $\tilde{x} \in S^{-1}(0)$, *suppose that S is maximal monotone in a neighborhood of* $(\tilde{x}, 0)$. *Then, there's a neighborhood B of* $\tilde{x}$ *with the following properties:*

(a) *If* $x^\nu \in B$, *then there's a unique point* $x^{\nu+1} \in B$ *satisfying (10.10).*
(b) *If* $\{x^\nu, \nu \in \mathbb{N}\}$ *is generated by (10.10) starting from* $x^0 \in B$ *with* $x^\nu \in B$ *for all* $\nu$, *then the sequence converges to a solution of* $0 \in S(x)$.

**Proof.** Let $C \subset \mathbb{R}^n \times \mathbb{R}^n$ be the neighborhood of $(\tilde{x}, 0)$ referenced in the assumption. Since $S$ is maximal monotone in $C$, there's maximal monotone $T : \mathbb{R}^n \rightrightarrows \mathbb{R}^n$ with

$$\operatorname{gph} S \cap C = \operatorname{gph} T \cap C.$$

By 10.27, the proximal point method for generalized equations when applied to $T$, starting from $x^0 \in \mathbb{R}^n$, converges to some $\bar{x} \in T^{-1}(0)$ with the generated points $\{x^\nu, \nu \in \mathbb{N}\}$, all uniquely defined, satisfying

$$\|x^{\nu+1} - x\|_2 \leq \|x^\nu - x\|_2 \quad \forall x \in T^{-1}(0), \quad \nu = 0, 1, 2, \ldots.$$

Therefore, if the algorithm starts with $x^0 \in \mathbb{B}(\tilde{x}, \delta)$ for some $\delta \in (0, \infty)$, then $x^\nu \in \mathbb{B}(\tilde{x}, \delta)$ for all $\nu$ because $x$ can be taken as $\tilde{x}$ in the previous inequality. From the construction of $\{x^\nu, \nu \in \mathbb{N}\}$, one has

$$\big(x^{\nu+1}, \kappa(x^\nu - x^{\nu+1})\big) \in \operatorname{gph} T,$$

which then converges to $(\bar{x}, 0)$. By choosing $\delta$ sufficiently small, we can thus ensure that

$$\big\{\big(x^{\nu+1}, \kappa(x^\nu - x^{\nu+1})\big), \nu \in \mathbb{N}\big\} \subset C.$$

However, there's no distinction between $T$ and $S$ at points in $C$. Thus, $\{x^\nu, \nu \in \mathbb{N}\}$ is also generated by applying the proximal point method for generalized equations to $S$ under the additional requirement that any solution of $0 \in S(x) + \kappa(x + x^\nu)$ outside of $\mathbb{B}(\tilde{x}, \delta)$ is discarded and $x^{\nu+1}$ is set to a solution in $\mathbb{B}(\tilde{x}, \delta)$, which then exists and is unique. Part (b) follows immediately because the limit point $\bar{x} \in T^{-1}(0)$ also solves $0 \in S(x)$. □

**Proposition 10.29** (rescaling). *If an* $n \times n$*-matrix A is invertible,* $S : \mathbb{R}^n \rightrightarrows \mathbb{R}^n$ *is maximal monotone,* $\kappa \in (0, \infty)$ *and* $x^0 \in \mathbb{R}^n$, *then* $\{x^\nu, \nu \in \mathbb{N}\}$ *generated by setting* $x^{\nu+1}$ *such that*

$$0 \in S(x^{\nu+1}) + \kappa A^\top A(x^{\nu+1} - x^\nu) \tag{10.11}$$

*is uniquely defined, 10.27(a,b) hold and, for all $\nu$ and $x \in S^{-1}(0)$,*

$$\|x^{\nu+1} - x\|_A \leq \|x^\nu - x\|_A, \quad \textit{where } \|x\|_A = \|Ax\|_2.$$

**Proof.** Let $T : \mathbb{R}^n \rightrightarrows \mathbb{R}^n$ be given by

$$T(z) = (A^\top)^{-1} S(A^{-1}z),$$

which is also maximal monotone by [105, Theorem 12.43]. The proximal point method for generalized equations when applied to $T$, starting from $z^0 = Ax^0$, uniquely defines $\{z^\nu, \nu \in \mathbb{N}\}$ and then also $\{x^\nu = A^{-1}z^\nu, \nu \in \mathbb{N}\}$. Under the transformation $z = Ax$, (10.11) can be expressed as

$$0 \in S(A^{-1}z^{\nu+1}) + \kappa A^\top A\big(A^{-1}z^{\nu+1} - A^{-1}z^\nu\big)$$

or, equivalently, as

$$0 \in (A^\top)^{-1} S(A^{-1}z^{\nu+1}) + \kappa A\big(A^{-1}z^{\nu+1} - A^{-1}z^\nu\big) = T(z^{\nu+1}) + \kappa(z^{\nu+1} - z^\nu).$$

Thus, the points $\{x^\nu, \nu \in \mathbb{N}\}$ are also generated by (10.11). Since $\{z^\nu, \nu \in \mathbb{N}\}$ satisfies 10.27(a,b,c) with $S$ replaced by $T$, $\{x^\nu, \nu \in \mathbb{N}\}$ satisfies (a,b) and also

$$\|Ax^{\nu+1} - Ax\|_2 = \|z^{\nu+1} - z\|_2 \leq \|z^\nu - z\|_2 = \|Ax^\nu - Ax\|_2$$

for $z \in T^{-1}(0)$ and $x = A^{-1}z$, which then is a point in $S^{-1}(0)$. In view of the definition of $\|\cdot\|_A$, the final conclusion follows. □

The third and final intermediate proposition leverages heavily the fact that every point $z \in \mathbb{R}^n$ can be written uniquely as $z = x + y$, with $x = \mathrm{prj}_X(z)$ and $y = \mathrm{prj}_{X^\perp}(z)$, for any subspace $X$ of $\mathbb{R}^n$; see (10.4). The proposition shows that the linkage problem can be reformulated as solving $0 \in T(z)$ for a specific $T : \mathbb{R}^n \rightrightarrows \mathbb{R}^n$, with the resulting $\bar{z}$ furnishing a solution $(\mathrm{prj}_X(\bar{z}), \mathrm{prj}_{X^\perp}(\bar{z}))$ of the linkage problem.

**Proposition 10.30** (reformulation). *For $S : \mathbb{R}^n \rightrightarrows \mathbb{R}^n$, a subspace $X$ of $\mathbb{R}^n$, $\theta \in [0, \infty)$ and $\kappa \in (\theta, \infty)$, let $T : \mathbb{R}^n \rightrightarrows \mathbb{R}^n$ be given by*

$$v + u \in T(x + y) \iff v + y \in (S + \theta\, \mathrm{prj}_{X^\perp})(x + u) \quad \textit{when } x, v \in X,\ u, y \in X^\perp.$$

*Then, $T$ is well-defined and*

$$\bar{x} \in X,\ \bar{y} \in X^\perp,\ \bar{y} \in S(\bar{x}) \iff \bar{x} \in X,\ \bar{y} \in X^\perp,\ 0 \in T(\bar{x} + \bar{y}).$$

*Moreover, there's an invertible $n \times n$-matrix $A$ such that*

$$A(x + y) = x + y/\sqrt{\kappa(\kappa - \theta)} \quad \forall x \in X,\ y \in X^\perp$$

*and, for any* $(x^0, y^0) \in X \times X^\perp$*, the points* $\{x^\nu, y^\nu,\ \nu \in \mathbb{N}\}$ *generated by the progressive decoupling algorithm when applied to S, starting from* $(x^0, y^0)$*, are equivalent to the points generated by*

$$0 \in T(z^{\nu+1}) + \kappa A^\top A(z^{\nu+1} - z^\nu), \text{ with } z^0 = x^0 + y^0, \tag{10.12}$$

*in the sense that* $z^\nu = x^\nu + y^\nu$.

**Proof.** For any $z \in \mathbb{R}^n$, the set $T(z)$ is obtained as follows: Compute $x = \mathrm{prj}_X(z)$, $y = \mathrm{prj}_{X^\perp}(z)$ and let

$$T(z) = \big\{ v + u \,\big|\, v \in X,\ u \in X^\perp,\ v + y \in (S + \theta\,\mathrm{prj}_{X^\perp})(x+u) \big\}.$$

Thus, $T$ is well-defined.

The claimed equivalence is a direct consequence of the discussion prior to 10.19 and the definition of $T$; take $u = v = 0$, $x = \bar{x}$, $y = \bar{y}$.

The existence of an invertible matrix $A$ follows because the mapping

$$x + y \mapsto x + y/\sqrt{\kappa(\kappa - \theta)}$$

is bijective. Using the orthogonality between $X$ and $X^\perp$, we see that

$$\big\langle A(x+y), v+u \big\rangle = \big\langle x+y, A(v+u) \big\rangle \quad \forall x, v \in X,\ \ y, u \in X^\perp$$

so $A$ is symmetric, i.e., $A = A^\top$. Thus,

$$A^\top A(x+y) = x + \kappa^{-1}(\kappa - \theta)^{-1} y.$$

With $x^\nu = \mathrm{prj}_X(z^\nu)$ and $y^\nu = \mathrm{prj}_{X^\perp}(z^\nu)$, (10.12) can be written as

$$0 \in T(x^{\nu+1} + y^{\nu+1}) + \kappa(x^{\nu+1} - x^\nu) + (\kappa - \theta)^{-1}(y^{\nu+1} - y^\nu).$$

Let's define

$$u^{\nu+1} = -(\kappa - \theta)^{-1}(y^{\nu+1} - y^\nu),$$

which then is a point in $X^\perp$. Thus,

$$-\kappa(x^{\nu+1} - x^\nu) + u^{\nu+1} \in T(x^{\nu+1} + y^{\nu+1}).$$

By the definition of $T$,

$$-\kappa(x^{\nu+1} - x^\nu) + y^{\nu+1} \in (S + \theta\,\mathrm{prj}_{X^\perp})(x^{\nu+1} + u^{\nu+1}). \tag{10.13}$$

Since $\mathrm{prj}_{X^\perp}(x^{\nu+1} + u^{\nu+1}) = u^{\nu+1}$, this is equivalent to

$$0 \in S(x^{\nu+1} + u^{\nu+1}) + \kappa(x^{\nu+1} - x^\nu) + \theta u^{\nu+1} - y^{\nu+1}.$$

The expression for $u^{\nu+1}$ can be rewritten as

$$y^{\nu+1} = y^{\nu} - (\kappa - \theta)u^{\nu+1}. \tag{10.14}$$

Thus, $\theta u^{\nu+1} - y^{\nu+1} = -y^{\nu} + \kappa u^{\nu+1}$, which leads to

$$0 \in S(x^{\nu+1} + u^{\nu+1}) - y^{\nu} + \kappa(x^{\nu+1} - x^{\nu} + u^{\nu+1}).$$

Now, let $\hat{x}^{\nu} = x^{\nu+1} + u^{\nu+1}$. We then have $\mathrm{prj}_X(\hat{x}^{\nu}) = x^{\nu+1}$ and $\mathrm{prj}_{X^\perp}(\hat{x}^{\nu}) = u^{\nu+1}$, which correspond to Step 2 in the progressive decoupling algorithm, and

$$y^{\nu} \in S(\hat{x}^{\nu}) + \kappa(\hat{x}^{\nu} - x^{\nu}),$$

which corresponds to Step 1. This completes the proof. □

**Proof of 10.23.** The elicitation at level $\theta$ ensures that $S + \theta\,\mathrm{prj}_{X^\perp}$ is maximal monotone and then $T$ defined in 10.30 is also maximal monotone because it's the "partial inverse" of $S + \theta\,\mathrm{prj}_{X^\perp}$; see [121]. By 10.30, the progressive decoupling algorithm is equivalent to carrying out iterations according to (10.12), which results in the convergence properties listed in 10.29 and the claims in 10.23 follow. In particular, the form of $A$ emerging from 10.30 leads to the norm $\|\cdot\|_{\kappa,\theta}$ and the uniqueness of $\hat{x}^{\nu}$ is a consequence of its equivalence with $x^{\nu+1} + u^{\nu+1}$ in the proof of 10.30. □

**Proof of 10.26.** When the elicitation is only local, the arguments in the proof of 10.23 carry over on the basis of the localization result 10.28. □

## 10.E Decoupling in Stochastic Optimization

Stochastic optimization is an area where the progressive decoupling algorithm is especially useful because the models arising in applications are often excessively large, but well structured. In addition to computational concerns, a real-world application may have to contend with data distributed among several agents. The agents could be unable to share their data due to privacy concerns or a lack of connectivity and bandwidth. Consequently, it's not an option to solve the problem directly and decoupling is necessary.

For a finitely distributed random vector $\boldsymbol{\xi}$, with support $\Xi \subset \mathbb{R}^m$ and corresponding probabilities $\{p_\xi > 0, \xi \in \Xi\}$, and a function $f : \Xi \times \mathbb{R}^n \to \overline{\mathbb{R}}$, let's consider the problem

$$\underset{x \in \mathbb{R}^n}{\text{minimize}}\ \mathbb{E}\big[f(\boldsymbol{\xi}, x)\big] = \sum\nolimits_{\xi \in \Xi} p_\xi f(\xi, x), \tag{10.15}$$

which is of the form (10.5) and is thus amenable to splitting. Specifically, we replace $x \in \mathbb{R}^n$ by $x(\xi) \in \mathbb{R}^n$ so that the decision becomes tailored to the outcome $\xi$. Suppose that there are $q$ possible outcomes in $\Xi$. The problem can then be formulated in terms of the enlarged vector

$$z = \big(x(\xi) \in \mathbb{R}^n, \xi \in \Xi\big) \in \mathbb{R}^{nq}$$

and the subspace

$$X = \left\{ z = \big(x(\xi) \in \mathbb{R}^n, \xi \in \Xi\big) \in \mathbb{R}^{nq} \;\middle|\; x(\xi) = x(\bar{\xi}) \;\; \forall \xi, \bar{\xi} \in \Xi \right\}.$$

This leads to the reformulation

$$\underset{z \in \mathbb{R}^{nq}}{\text{minimize}} \sum\nolimits_{\xi \in \Xi} p_\xi f\big(\xi, x(\xi)\big) \;\text{ subject to }\; z = \big(x(\xi), \xi \in \Xi\big) \in X. \tag{10.16}$$

The reformulation is useful because in the absence of the subspace constraint $z \in X$, the problem decouples into the subproblems

$$\left\{ \underset{x(\xi) \in \mathbb{R}^n}{\text{minimize}} \; p_\xi f\big(\xi, x(\xi)\big), \quad \xi \in \Xi \right\},$$

which can be addressed individually. This holds regardless of any additional structure that might be present in $f$. For example, one might have

$$f(\xi, x) = f_0(\xi, x) + \iota_{C(\xi)}(x)$$

for some $f_0 : \Xi \times \mathbb{R}^n \to \mathbb{R}$ and $C(\xi) \subset \mathbb{R}^n$, i.e., the objective function as well as the constraint set depend on the parameter vector $\xi$.

Minimizers of the subproblems furnish decisions that are optimal for specific outcomes and may provide useful insight as part of a what-if study. However, they aren't implementable *before* the value of $\boldsymbol{\xi}$ has been revealed. In particular, they typically don't solve (10.15), where we need to commit to a vector $x$ before the value of $\boldsymbol{\xi}$ is known. The diverse decisions furnished by the subproblems must somehow be combined to produce a single vector. This is the role of the subspace constraint $z \in X$, which is enforced indirectly by the progressive decoupling algorithm.

Let's address the reformulation (10.16) through the condition

$$0 \in S(z) + N_X(z), \;\text{ where }\; S(z) = \prod\nolimits_{\xi \in \Xi} p_\xi \partial_x f\big(\xi, x(\xi)\big), \quad z = \big(x(\xi), \xi \in \Xi\big).$$

Under a qualification, the condition is necessary for a minimizer of (10.16); see 4.67 and the Fermat rule 4.73. Regardless of the circumstances, one can reformulate the condition using (10.2) and this results in the linkage problem

$$z \in X, \quad y \in X^\perp, \quad y \in S(z).$$

We've reached a problem of the form treatable by the progressive decoupling algorithm. In particular, Step 1 of the algorithm decouples into $q$ subproblems.

There's a minor issue with this approach, however. Step 2 of the progressive decoupling algorithm projects the subproblem solutions $(\hat{x}^\nu(\xi), \xi \in \Xi)$ onto $X$. As seen in 10.9, this amounts to the averaging

$$\frac{1}{q} \sum\nolimits_{\xi \in \Xi} \hat{x}^\nu(\xi).$$

While unproblematic if $p_\xi$ is the same for all $\xi \in \Xi$, such averaging may cause slow convergence if some of these probabilities are much larger than the others. It would be more meaningful to give weight to the solutions of the subproblems proportional to their probabilities, i.e., instead in Step 2, compute

$$\sum\nolimits_{\xi\in\Xi} p_\xi \hat{x}^\nu(\xi).$$

A convenient way of handling this is to express the generalized equation

$$0 \in S(z) + N_X(z)$$

in terms of a random vector $\boldsymbol{x} : \Xi \to \mathbb{R}^n$ instead of the vector $z \in \mathbb{R}^{nq}$. We can view the random vector as defined on the probability space $(\Xi, \mathcal{B}, \mathbb{P})$, where $\mathbb{P}(\{\xi\}) = p_\xi$ and $\mathcal{B}$ is the collection of all subsets of $\Xi$. Of course, selecting a point in $\mathbb{R}^{nq}$ can uniquely be associated with selecting such a random vector as the latter involves determining $\{\boldsymbol{x}(\xi) \in \mathbb{R}^n, \xi \in \Xi\}$ and $q$ is the number of outcomes in $\Xi$. The benefit, however, is that the probabilities now emerge as weights in a desirable manner.

Let $\mathcal{H}$ be the space of all random vectors $\boldsymbol{x} : \Xi \to \mathbb{R}^n$ on which we define the inner product

$$\langle \boldsymbol{x}, \bar{\boldsymbol{x}} \rangle_{\mathcal{H}} = \sum\nolimits_{\xi\in\Xi} p_\xi \big\langle \boldsymbol{x}(\xi), \bar{\boldsymbol{x}}(\xi) \big\rangle$$

and the norm

$$\|\boldsymbol{x}\|_{\mathcal{H}} = \sqrt{\langle \boldsymbol{x}, \boldsymbol{x} \rangle_{\mathcal{H}}} = \sqrt{\sum\nolimits_{\xi\in\Xi} p_\xi \|\boldsymbol{x}(\xi)\|_2^2}.$$

With these definitions, $\mathcal{H}$ is a finite-dimensional Hilbert space and thus retains essentially all the properties of a Euclidean space. In particular, we can mimic 10.5 and define the subspace

$$\mathcal{X} = \big\{\boldsymbol{x} \in \mathcal{H} \mid \boldsymbol{x}(\xi) = \boldsymbol{x}(\bar{\xi}) \;\; \forall \xi, \bar{\xi} \in \Xi\big\},$$

which has as its orthogonal complement the subspace

$$\mathcal{X}^\perp = \big\{\boldsymbol{y} \in \mathcal{H} \mid \langle \boldsymbol{x}, \boldsymbol{y} \rangle_{\mathcal{H}} = 0 \;\; \forall \boldsymbol{x} \in \mathcal{X}\big\} = \big\{\boldsymbol{y} \in \mathcal{H} \mid \mathbb{E}[\boldsymbol{y}] = 0\big\},$$

where

$$\mathbb{E}[\boldsymbol{y}] = \sum\nolimits_{\xi\in\Xi} p_\xi \boldsymbol{y}(\xi) \in \mathbb{R}^n.$$

A projection on $\mathcal{X}$, which is defined as

$$\mathrm{prj}_{\mathcal{X}}(\bar{\boldsymbol{x}}) = \mathrm{argmin}\big\{\|\boldsymbol{x} - \bar{\boldsymbol{x}}\|_{\mathcal{H}}^2 \mid \boldsymbol{x} \in \mathcal{X}\big\},$$

then reduces to computing an expectation:

$$\boldsymbol{x} = \mathrm{prj}_{\mathcal{X}}(\bar{\boldsymbol{x}}) \iff \forall \xi \in \Xi : \; \boldsymbol{x}(\xi) = \mathbb{E}[\bar{\boldsymbol{x}}] = \sum\nolimits_{\xi\in\Xi} p_\xi \bar{\boldsymbol{x}}(\xi).$$

Similar to (10.3), one finds that

$$y = \mathrm{prj}_{\mathcal{X}^\perp}(\bar{x}) \iff y = \bar{x} - \mathrm{prj}_{\mathcal{X}}(\bar{x}).$$

With these modifications, we achieve the following version of the progressive decoupling algorithm for solving (10.15) via the conditions

$$x \in \mathcal{X}, \;\; y \in \mathcal{X}^\perp, \;\; y \in \mathcal{S}(x),$$

where $\mathcal{S} : \mathcal{H} \rightrightarrows \mathcal{H}$ is the set-valued mapping defined by

$$\mathcal{S}(x) = \prod\nolimits_{\xi\in\Xi} \partial_x f\big(\xi, x(\xi)\big).$$

**Progressive Hedging Algorithm.**

Data. $x^0 \in \mathcal{X}$, $y^0 \in \mathcal{X}^\perp$, $\theta \in [0, \infty)$, $\kappa \in (\theta, \infty)$.
Step 0. Set $\nu = 0$.
Step 1. For all $\xi \in \Xi$, obtain $\hat{x}^\nu(\xi)$ by solving

$$y^\nu(\xi) \in \partial_x f(\xi, x) + \kappa\big(x - x^\nu(\xi)\big).$$

Step 2. Compute $x^{\nu+1} = \mathrm{prj}_{\mathcal{X}}(\hat{x}^\nu)$ and $y^{\nu+1} = y^\nu - (\kappa - \theta)\,\mathrm{prj}_{\mathcal{X}^\perp}(\hat{x}^\nu)$.
Step 3. Replace $\nu$ by $\nu + 1$ and go to Step 1.

For each $\xi \in \Xi$, Step 1 amounts to finding a solution that satisfies an optimality condition for the subproblem

$$\operatorname*{minimize}_{x\in\mathbb{R}^n} f(\xi, x) - \big\langle y^\nu(\xi), x\big\rangle + \tfrac{1}{2}\kappa\big\|x - x^\nu(\xi)\big\|_2^2.$$

These subproblems may even be solved in parallel to reduce the overall computing time.

In Step 2, $x^{\nu+1}$ is easily determined by averaging the solutions from the subproblems as seen before the algorithm and one can then leverage the alternative expression

$$y^{\nu+1}(\xi) = y^\nu(\xi) - (\kappa - \theta)\big(\hat{x}^\nu(\xi) - x^{\nu+1}(\xi)\big).$$

Since the arguments underpinning 10.23 and 10.26 can just as well be carried out with $(\mathcal{H}, \langle\cdot,\cdot\rangle_{\mathcal{H}})$ in place of $(\mathbb{R}^n, \langle\cdot,\cdot\rangle)$, the progressive hedging algorithm inherits the convergence properties of the progressive decoupling algorithm. The only difference is that progress is now measured in the norm

$$\big\|(x, y)\big\|_{\mathcal{H},\kappa,\theta} = \sqrt{\|x\|_{\mathcal{H}}^2 + \frac{1}{\kappa(\kappa - \theta)}\|y\|_{\mathcal{H}}^2}.$$

The key assumption is that the set-valued mapping $\mathcal{S} : \mathcal{H} \rightrightarrows \mathcal{H}$ has elicitable maximal monotonicity at some level $\theta$, possibly only locally, with elicitation now being defined in terms of monotonicity under the inner product $\langle\cdot,\cdot\rangle_{\mathcal{H}}$; see [93]. For example, if $f(\xi,\cdot)$ is convex and lsc for all $\xi\in\Xi$, then $\mathcal{S}$ has elicitation at level $\theta=0$.

A practical concern emerges in the implementation of the progressive hedging algorithm. Since the number of outcomes in $\Xi$, denoted by $q$, tends to be large, we may end up with many subproblems in Step 1. In fact, one would then have traded one large problem involving $q$ outcomes for $q$ small problems with one outcome each. Practically, it's better to avoid these two extremes and instead adopt a *mini-batch approach* with a moderate number of medium-sized problems. This is trivially accomplished by redefining some of the quantities above. In particular, if the actual problem is

$$\operatorname*{minimize}_{x\in\mathbb{R}^n} \sum\nolimits_{\xi\in\tilde{\Xi}} \tilde{p}_\xi \tilde{f}(\xi,x),$$

with $\tilde{f}:\tilde{\Xi}\times\mathbb{R}^n\to\overline{\mathbb{R}}$ and $\tilde{\Xi}$ having $\tilde{q}$ outcomes, then one can partition $\tilde{\Xi}$ into $q$ disjoint sets $\Xi_i$, $i=1,\dots,q$, so that

$$\tilde{\Xi}=\bigcup\nolimits_{i=1}^{q}\Xi_i.$$

Now, the problem can be expressed in the form (10.15) with

$$\Xi=\{1,\dots,q\},\quad p_i=\sum\nolimits_{\xi\in\Xi_i}\tilde{p}_\xi,\quad f(i,x)=\sum\nolimits_{\xi\in\Xi_i}\tilde{p}_\xi\tilde{f}(\xi,x)/p_i,\quad i\in\Xi.$$

For example, $\tilde{\Xi}$ may involve $\tilde{q}=10^6$ outcomes each occurring with probability $\tilde{p}_\xi=10^{-6}$, but these can be partitioned into 1000 subsets each having 1000 outcomes and then $p_i=10^{-3}$. In theory, any partition is valid, but the practical performance is often better if each $\Xi_i$ contains a diverse set of outcomes. Hence, one may consider maximizing some measure of dissimilarity when constructing $\{\Xi_i, i=1,\dots,q\}$ from $\tilde{\Xi}$.

Sometimes the partition is given by the application due to the fact that the data in $\tilde{\Xi}$ is distributed among several agents. In the above notation, $q$ is then the number of agents and $\Xi_i$ is the data associated with agent $i$. Each subproblem in Step 1 of the progressive hedging algorithm would now involve only the data of one agent and can be carried out by the agent without interaction with the other agents. Thus, the algorithm requires merely a communication of intermediate solutions between a centralized unit (that executes Step 2) and the agents; no data transfer is needed.

**Example 10.31** (truss design). Let's return to the design of a truss structure in §3.E, where the goal is to solve

$$\operatorname*{minimize}_{(x,\gamma)\in C}\ \mathbb{E}\left[\gamma+\frac{1}{1-\alpha}\max\left\{0, f(\boldsymbol{\xi},x)-\gamma\right\}\right].$$

Here, $f(\xi,\cdot)$ is convex for all $\xi$ (see §3.E for the specific formula) and

$$C=\left\{(x,\gamma)\in\mathbb{R}^8 \;\middle|\; 0.5\le x_j\le 2,\ \ j=1,\dots,7,\ \ x_1+\cdots+x_7\le 8\right\}.$$

The eight-dimensional random vector $\boldsymbol{\xi}$ has many possible outcomes because its probability distribution is constructed by Monte Carlo sampling and this makes it hard to solve the problem directly as discussed in §3.E. However, the progressive hedging algorithm applies with elicitation level $\theta = 0$ and allows us to reach a solution by solving a series of smaller problems.

**Detail.** Suppose that the support of $\boldsymbol{\xi}$ has 100000 possible outcomes, all equally likely to occur. While a direct solution to the problem is still possible (see the second-to-last row in Table 3.4), we use this as a test case to see if we can reproduce the results by considering 100 subproblems with only 1000 outcomes each. In a real-world setting, the actual problem may be completely intractable due to a huge number of outcomes and a decomposition approach would be one of the few viable paths to a solution.

Following the steps laid out before the example, the actual problem can be stated as

$$\underset{(x,\gamma)\in C}{\text{minimize}}\ 10^{-2}\sum\nolimits_{i=1}^{100} g(i, x, \gamma),$$

where

$$g(i, x, \gamma) = 10^{-3}\sum\nolimits_{j=1}^{1000}\left(\gamma + \frac{1}{1-\alpha}\max\left\{0, f(\xi^{ij}, x) - \gamma\right\}\right)$$

and $\{\xi^{ij} \in \mathbb{R}^8,\ i = 1, \dots, 100,\ j = 1, \dots, 1000\}$ are the 100000 possible outcomes of $\boldsymbol{\xi}$, partitioned into 100 mini-batches of size 1000.

In this particular case, the progressive hedging algorithm solves for the random vectors $\boldsymbol{x}, \boldsymbol{y} : \Xi \to \mathbb{R}^8$, with $\Xi = \{1, \dots, 100\}$ and $\boldsymbol{x}(i)$ specifying the seven $x$-variables as well as $\gamma$ in the case of the $i$th mini-batch. The algorithm starts from an initial point $\boldsymbol{x}^0(i) = (x^0, \gamma^0)$, with $x^0 = (8/7, \dots, 8/7)$ and $\gamma^0 = -55$ for all $i \in \Xi$. The initial multiplier element $\boldsymbol{y}^0$ has $\boldsymbol{y}^0(i) = 0$ for all $i \in \Xi$. We set the parameters $\theta = 0$ (due to the elicitation level) and $\kappa = 1$. In its optimization form, the subproblems in Step 1 reduce to

$$\left\{\underset{(x,\gamma)\in C}{\text{minimize}}\ g(i, x, \gamma) - \left\langle \boldsymbol{y}^\nu(i), (x, \gamma)\right\rangle + \tfrac{1}{2}\kappa\left\|(x, \gamma) - \boldsymbol{x}^\nu(i)\right\|_2^2, \quad i \in \Xi\right\}.$$

Each subproblem involves only 1000 outcomes (see the definition of $g$) and can be solved quickly as moderately sized quadratic problems. Specifically, adapting the reformulation (3.6) to the present context, we see that the $i$th subproblem can be stated as

$$\begin{aligned}
\underset{(x,\gamma)\in C,\, z\in\mathbb{R}^{1000}}{\text{minimize}}\ & \gamma - \left\langle \boldsymbol{y}^\nu(i), (x, \gamma)\right\rangle + \tfrac{1}{2}\kappa\left\|(x, \gamma) - \boldsymbol{x}^\nu(i)\right\|_2^2 + \frac{1}{1000(1-\alpha)}\sum\nolimits_{j=1}^{1000} z_j \\
\text{subject to}\ & \xi_8^{ij}/\beta_k - \xi_k^{ij} x_k - \gamma \le z_j, \quad j = 1, \dots, 1000,\ k = 1, \dots, 7 \\
& 0 \le z_j, \quad j = 1, \dots, 1000,
\end{aligned}$$

where we also bring in the specific form of $f$; see §3.E. The portion of a minimizer of this subproblem that pertains to $(x, \gamma)$ furnishes $\hat{\boldsymbol{x}}^\nu(i)$.

In Step 2, we first compute $\bar{x} = 10^{-2}\sum_{i\in\Xi}\hat{\boldsymbol{x}}^\nu(i)$. Second, we set

$$\boldsymbol{x}^{\nu+1}(i) = \bar{x} \quad \text{and} \quad \boldsymbol{y}^{\nu+1}(i) = \boldsymbol{y}^\nu(i) - (\kappa - \theta)\left(\hat{\boldsymbol{x}}^\nu(i) - \bar{x}\right) \quad \forall i \in \Xi.$$

Table 10.1 summarizes results produced by the progressive hedging algorithm across 80 iterations. After the first iteration (see the row corresponding to $\nu = 0$), the seven $x$-variables ($\gamma$ isn't reported) have values similar to the poorer designs obtained in §3.E. A reason for this is that the decision variables are here initiated with the same value and the algorithm invokes the quadratic penalty

$$\tfrac{1}{2}\kappa \|(x, \gamma) - \boldsymbol{x}^\nu(i)\|_2^2$$

for any deviation from the current solution. Thus, the progressive hedging algorithm tends to advance in small steps, especially when $\kappa$ is relatively large. The last column of Table 10.1 gives an indication of the variability among the designs produced by the different subproblems; the number reported is the standard deviation of a decision variable obtained across the 100 subproblems, averaged over the seven $x$-variables.

Gradually, the solutions improve and after 80 iterations, we've essentially reproduced the best solution obtained in §3.E; see the last row in Table 3.4. There, we solved a single problem involving 500000 outcomes, but now we solve $80 \cdot 100 = 8000$ problems with 1000 outcomes. The former took 271 seconds, while the latter consumes about $8000 \cdot 0.15 = 1200$ seconds. However, this takes place without using any parallel processing. One could estimate, a bit optimistically, that if the 100 subproblems in each iteration are distributed to 100 processors, the time for the progressive hedging algorithm could be reduced by a factor of nearly 100. The algorithm can also be tuned by adjusting $\kappa, \theta$ as well as the 100-1000 partition. In any case, the true merit of the progressive hedging algorithm and other decomposition methods typically emerges when a direct solution of the actual problem isn't practical and the only viable approach is to solve smaller subproblems. □

Table 10.1: Truss designs during 80 iterations of the progressive hedging algorithm.

| $\nu$ | Design $\boldsymbol{x}^{\nu+1}(1)$ (in mm$^2$) | | | | | | | std. of $\hat{\boldsymbol{x}}^\nu(i)$ |
|---|---|---|---|---|---|---|---|---|
| 0 | 1376 | 1375 | 1053 | 1050 | 1047 | 1050 | 1050 | 22 |
| 9 | 1395 | 1394 | 1044 | 1041 | 1041 | 1042 | 1043 | 18 |
| 19 | 1408 | 1407 | 1039 | 1036 | 1036 | 1037 | 1037 | 17 |
| 39 | 1420 | 1420 | 1034 | 1030 | 1031 | 1033 | 1032 | 14 |
| 59 | 1425 | 1425 | 1031 | 1029 | 1029 | 1031 | 1030 | 13 |
| 79 | 1427 | 1426 | 1031 | 1028 | 1029 | 1030 | 1030 | 11 |

**Exercise 10.32** (federated learning). Consider the situation where two agents collaborate to construct a classifier, which is referred to as *federated learning*. Suppose that the goal is to solve the hinge-loss problem in §2.H, but each agent has only access to half the dataset due to privacy concerns. Solve the problem using the progressive hedging algorithm with the mini-batch approach.

**Guide.** Select a binary classification dataset with at least 500 data points and partition it into three portions: put aside 10% of the data for testing and divide the remaining 90% into two equally-sized mini-batches. First, solve the problem directly using the two mini-batches pooled together and then assess the quality of the resulting classifier using the test data. This furnishes a benchmark against which you'll compare. Second, apply the progressive hedging algorithm using the two mini-batches, i.e., there'll be two subproblems in Step 1 of the algorithm. Report the classifiers produced by the subproblems as well as the "merged" one from Step 2 of the algorithm. Test each one of these using the test data and discuss how they compare with the benchmark classifier as the progressive hedging algorithm iterates. □

In modeling real-world decision processes, we often encounter situations with decisions being taken at different points in time. Later decisions may benefit from the arrival of information not available to earlier ones. In a two-stage decision process, we may first make a strategic decision about the locations of warehouses without fully knowing the customer demand and later make a second decision about routing of delivery trucks after the demand has been revealed; see §3.J for models of this kind. Generally, the process may unfold over $\tau$ stages with decision $x_1$ being followed by observation $\xi_1$ and then decision $x_2$ followed by observation $\xi_2$, and so forth up to decision $x_\tau$ and observation $\xi_\tau$. It now becomes especially useful to view future decisions as random vectors as they indeed depend on the information arriving prior to their execution.

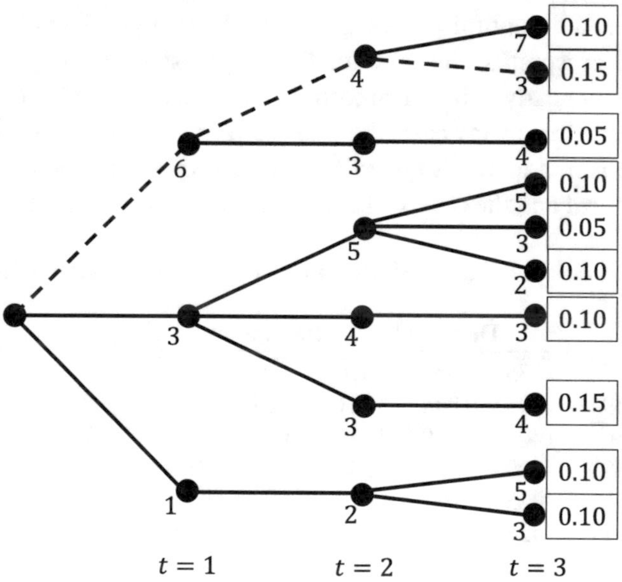


Fig. 10.4: Scenario tree for an information process over three stages.

We start by modeling the uncertainty and assume that in stage $t$ there's a parameter vector $\xi_t \in \mathbb{R}^{m_t}$ which can take a finite number of values. Chaining these together across all the stages, we obtain a *scenario*

$$\xi = (\xi_1, \xi_2, \dots, \xi_\tau) \in \Xi, \text{ with } \Xi \subset \mathbb{R}^{m_1} \times \mathbb{R}^{m_2} \times \cdots \times \mathbb{R}^{m_\tau} \text{ and } m = \sum\nolimits_{t=1}^{\tau} m_t.$$

The set of scenarios $\Xi$ is then also finite. A scenario specifies all the uncertain parameters and is a description of *one* possible series of future events. Typically, we identify a number of possible scenarios against which a decision is assessed. Figure 10.4 illustrates a *scenario tree* that represents the process over three stages with 10 scenarios. One scenario $\xi = (\xi_1, \xi_2, \xi_3)$, marked with a dashed line in the figure, has values (6,4,3), which may represent the demand for a product across three months. Although this could be the scenario that actually unfolds, we don't know that in the beginning of the first month. Then, we're faced with 10 possible scenarios each represented by a path from left to right. The boxed numbers to the right in the figure specify the probability of each of the 10 scenarios. Our decision in the first month should account for these 10 possible future demand sequences.

Suppose that 6 is observed as the demand in the first month. Then, the decision in the second month can reflect this knowledge and the fact that future possibilities have also been narrowed down: Of the 10 scenarios, only the top three ones remain possible. With 6 revealed, the probabilities for scenarios (6,4,7), (6,4,3) and (6,3,4) are

$$0.1/(0.1 + 0.15 + 0.05) = \tfrac{1}{3}, \;\; 0.15/(0.1 + 0.15 + 0.05) = \tfrac{1}{2}, \;\; 0.05/(0.1 + 0.15 + 0.05) = \tfrac{1}{6},$$

respectively. When the demand in the second month is observed as 4, the future has only two possibilities: a demand in the third month of 7 occurring with probability $0.1/(0.1 + 0.15)$ or a demand of 3 occurring with probability $0.15/(0.1 + 0.15)$. The decision in the third month should account for this fact.

As the discussion illustrates, a decision depends on the information available at the time of its execution and is, therefore, viewed as a random vector prior to that time. Suppose that in the third month we need to determine an $n_3$-dimensional vector, but its value should depend on the observations in the two first months. Let $\boldsymbol{x}_3 : \Xi \to \mathbb{R}^{n_3}$ be the random vector that furnishes the decision in the third month, i.e., given a scenario $\xi = (\xi_1, \xi_2, \xi_3)$, $\boldsymbol{x}_3(\xi)$ is the $n_3$-dimensional decision vector for stage 3 in the case of scenario $\xi$. One can think of $\boldsymbol{x}_3$ as specifying contingency plans for each of the scenarios. Similarly, the decision for the second and first months are given by $\boldsymbol{x}_2 : \Xi \to \mathbb{R}^{n_2}$ and $\boldsymbol{x}_1 : \Xi \to \mathbb{R}^{n_1}$, respectively. In total, we need to determine $\boldsymbol{x} = (\boldsymbol{x}_1, \boldsymbol{x}_2, \boldsymbol{x}_3)$ while accounting for the information actually available at the various stages. In the third month, $\boldsymbol{x}_3(\xi)$ is nominally made dependent on the whole scenario $\xi = (\xi_1, \xi_2, \xi_3)$, but in reality, it can only depend on $(\xi_1, \xi_2)$ because they're the quantities observable by that time. Likewise, $\boldsymbol{x}_2(\xi)$ can actually only depend on $\xi_1$ and $\boldsymbol{x}_1(\xi)$ must be the same for all $\xi$ because no additional information about the scenarios is available at the beginning of the first month. In general,

$$\boldsymbol{x}_t(\xi) \text{ depends only on } \xi_1, \xi_2, \ldots, \xi_{t-1} \text{ and not on } \xi_t, \ldots, \xi_\tau. \tag{10.17}$$

This means that we seek to determine contingency plans that furnish courses of action while accounting for the additional information that becomes available in the future. For $f : \Xi \times \mathbb{R}^n \to \overline{\mathbb{R}}$, with $n = \sum_{t=1}^{\tau} n_t$, and probabilities $\{p_\xi > 0, \xi \in \Xi\}$, this leads to the problem

$$\text{minimize} \sum\nolimits_{\xi \in \Xi} p_\xi f\big(\xi, \boldsymbol{x}(\xi)\big) \;\text{ subject to }\; \boldsymbol{x} \in \mathcal{X} \subset \mathcal{H}, \tag{10.18}$$

where $\mathcal{H}$ is the space of all random vectors $\boldsymbol{x} : \Xi \to \mathbb{R}^n$ and

$$\mathcal{X} = \big\{(x_1, \ldots, x_\tau) \in \mathcal{H} \bigm| \forall \xi, \bar{\xi} \in \Xi : \; x_1(\xi) = x_1(\bar{\xi}),$$
$$x_t(\xi) = x_t(\bar{\xi}) \text{ if } \xi_k = \bar{\xi}_k, \;\; k = 1, 2, \ldots, t-1\big\}$$

defines the *nonanticipativity constraints* reflected by (10.17). These constraints specify that the first-stage decision can't depend on the scenario. In the second stage,

$$x_2(\xi) = x_2(\bar{\xi}) \;\text{ when } \xi, \bar{\xi} \in \Xi \text{ match in stage 1.}$$

At the time of the second-stage decision, we don't know whether it's scenario $\xi$ or scenario $\bar{\xi}$ that's unfolding if they coincide in stage 1. Thus, we must have the corresponding decisions $x_2(\xi)$ and $x_2(\bar{\xi})$ being equal. In the third stage,

$$x_3(\xi) = x_3(\bar{\xi}) \;\text{ when } \xi, \bar{\xi} \in \Xi \text{ match in stage 1 and again in stage 2,}$$

with similar effects for the later stages. The nonanticipativity constraints simply amount to linear equality constraints, with zero right-hand side, and thus define a subspace of $\mathcal{H}$.

Following the approach in this section, we can address (10.18) by solving

$$x \in \mathcal{X}, \; y \in \mathcal{X}^\perp, \; y \in \mathcal{S}(x), \;\text{ where } \mathcal{S}(x) = \prod\nolimits_{\xi \in \Xi} \partial_x f\big(\xi, x(\xi)\big).$$

We've reached a problem treatable by the progressive hedging algorithm. Again, Step 1 decomposes into a series of subproblems pertaining to the various scenarios and produces $\hat{x}^\nu$. Step 2 amounts to projecting $\hat{x}^\nu$ onto $\mathcal{X}$. In more detail, for $t \in \{1, 2, \ldots, \tau\}$, let

$$\Xi_{1:t} = \big\{\xi_{1:t} = (\xi_1, \ldots, \xi_t) \bigm| (\xi_1, \ldots, \xi_\tau) \in \Xi\big\}.$$

These are the possible paths up to stage $t$ in a scenario tree. In Figure 10.4, $\Xi_{1:1}$ has three paths, $\Xi_{1:2}$ has six paths, $\Xi_{1:3} = \Xi$ has 10 paths. Moreover, let

$$\Xi(\xi_{1:t}) = \big\{(\bar{\xi}_1, \ldots, \bar{\xi}_\tau) \in \Xi \bigm| (\bar{\xi}_1, \ldots, \bar{\xi}_t) = \xi_{1:t}\big\} \;\text{ when } \xi_{1:t} \in \Xi_{1:t}.$$

These are the scenarios that coincide with $\xi_{1:t}$ up to time $t$. Then, for $t = 1, \ldots, \tau$ and $\xi_{1:t} \in \Xi_{1:t}$, we set

$$\forall \xi \in \Xi(\xi_{1:t}) : \quad x_t^{\nu+1}(\xi) = \bar{x}_t(\xi_{1:t})$$

in Step 2 of the progressive hedging algorithm, where

$$\bar{x}_t(\xi_{1:t}) = \sum\nolimits_{\xi \in \Xi(\xi_{1:t})} p_\xi \hat{x}_t^\nu(\xi) / \alpha_t(\xi_{1:t}) \quad \text{and} \quad \alpha_t(\xi_{1:t}) = \sum\nolimits_{\xi \in \Xi(\xi_{1:t})} p_\xi.$$

In Figure 10.4 with $t = 2$ and $\xi_{1:2} = (6, 4) \in \Xi_{1:2}$, we need to consider the two scenarios $(6, 4, 7)$ and $(6, 4, 3)$ in $\Xi(\xi_{1:2})$ with $\alpha_2(\xi_{1:2}) = 0.10 + 0.15$.

A strength of this approach to multistage decision-making under uncertainty is the ability to handle any number of stages using the same algorithm. The stages can be interconnected in complicated ways, with an objective function that isn't necessarily a sum of costs across the stages. Moreover, the probability for a particular outcome during stage $t$ could very well depend on the outcomes during stages $1, 2, \ldots, t-1$, i.e., we don't need to assume statistical independence of any kind. One can also incorporate mini-batches as discussed before 10.31.

While a real-world decision process may advance over numerous stages, we note that *modeling* of such a process may involve far fewer stages. A reason for this is the common shortage of reliable estimates about events in the distant future, which makes building a representative scenario tree difficult. An aggregated view of decisions in the distant future may then emerge as most meaningful. For example, an inventory control model could involve seven "daily" stages and then an eighth stage representing the next week and beyond. For further details about the progressive hedging algorithm; see [45, 104]. Implementations are addressed in [48] and techniques for constructing scenario trees are given in [78].

## 10.F Strong Monotonicity

It's possible to estimate the rate of convergence of the progressive decoupling algorithm under an enhanced notion of monotonicity. Parallel to the linear rate of convergence guaranteed for the proximal gradient method under strong convexity (cf. 2.33), the rate of the progressive decoupling algorithm is quantifiable when the considered set-valued mapping is strongly monotone.

For $\sigma \in (0, \infty)$, a set-valued mapping $S : \mathbb{R}^n \rightrightarrows \mathbb{R}^n$ is *$\sigma$-strongly monotone* if $S - \sigma I$ is monotone, where $I : \mathbb{R}^n \to \mathbb{R}^n$ is the identity mapping, i.e., $I(x) = x$. Thus, $S$ is $\sigma$-strongly monotone if and only if

$$\forall (x, y) \in \operatorname{gph} S,\ (\bar{x}, \bar{y}) \in \operatorname{gph} S : \qquad \langle y - \bar{y}, x - \bar{x} \rangle \geq \sigma \|x - \bar{x}\|_2^2.$$

This means that $\operatorname{gph} S$ can't have any "flat spots" as then $x \neq \bar{x}$ could imply $y = \bar{y}$ and the inequality can't hold. The set-valued mapping given by $S(x) = \{Ax\}$ is $\sigma_1$-strongly monotone when $A$ is a symmetric positive definite $n \times n$-matrix with smallest eigenvalue $\sigma_1 > 0$ because then

$$\langle Ax - A\bar{x}, x - \bar{x} \rangle = \big\langle x - \bar{x}, A(x - \bar{x}) \big\rangle \geq \sigma_1 \|x - \bar{x}\|_2^2.$$

Informally, we can think of the *strong monotonicity modulus* $\sigma$ of a set-valued mapping $S$ as the smallest rate of increase in the values of $S(x)$ as $x$ moves to higher values.

If a maximal monotone set-valued mapping $S$ is $\sigma$-strongly monotone, then the generalized equation $0 \in S(x)$ has a unique solution. In fact, the inverse mapping $S^{-1}$ is point-valued with the Lipschitz property that

$$\|x - \bar{x}\|_2 \leq \tfrac{1}{\sigma} \|y - \bar{y}\|_2 \quad \forall (x, y), (\bar{x}, \bar{y}) \in \operatorname{gph} S;$$

see [105, Proposition 12.54]. As an application of this fact, the subproblem

$$y^\nu \in S(x) + \kappa(x - x^\nu)$$

in Step 1 of the progressive decoupling algorithm has a unique solution when $S$ is maximal monotone. To see this, note that

$$\begin{aligned}\langle y - \bar{y}, x - \bar{x}\rangle &= \langle v + \kappa x - \bar{v} - \kappa\bar{x}, x - \bar{x}\rangle \\ &= \langle v - \bar{v}, x - \bar{x}\rangle + \kappa\langle x - \bar{x}, x - \bar{x}\rangle \geq \kappa\|x - \bar{x}\|_2^2\end{aligned}$$

for $y = v + \kappa x$ and $\bar{y} = \bar{v} + \kappa\bar{x}$, with $v \in S(x)$ and $\bar{v} \in S(\bar{x})$. Consequently, the set-valued mapping given by

$$T(x) = S(x) + \kappa x$$

is $\kappa$-strongly monotone. It's also maximal monotone by 10.17. Thus, $y^\nu + \kappa x^\nu \in T(x)$ has a unique solution as just claimed. Moreover, the difference in solutions obtained during Step 1 of the progressive decoupling algorithm in two consecutive iterations is bounded by

$$\begin{aligned}\|\hat{x}^{\nu+1} - \hat{x}^\nu\|_2 &\leq \tfrac{1}{\kappa}\big\|y^{\nu+1} + \kappa x^{\nu+1} - y^\nu - \kappa x^\nu\big\|_2 \\ &\leq \tfrac{1}{\kappa}\|y^{\nu+1} - y^\nu\|_2 + \|x^{\nu+1} - x^\nu\|_2.\end{aligned}$$

**Proposition 10.33** (strong convexity and monotonicity). *For a proper lsc function $f : \mathbb{R}^n \to \overline{\mathbb{R}}$ and $\sigma \in (0, \infty)$, one has*

$$f \text{ is } \sigma\text{-strongly convex} \iff \partial f \text{ is } \sigma\text{-strongly monotone.}$$

**Proof.** We recall that $f$ being $\sigma$-strongly convex means that $f - \frac{1}{2}\sigma\|\cdot\|_2^2$ is convex; see the definition in §2.D. In view of the definitions and the subgradient formula from 4.58, the assertion follows directly by 10.15. □

For example, if $f : \mathbb{R}^n \to \mathbb{R}$ is a convex and twice smooth function with Hessian matrices $\{\nabla^2 f(x), x \in \mathbb{R}^n\}$, then its subgradient mapping, which in this case is simply specified by the gradients, is $\sigma$-strongly monotone provided that

$$\sigma = \inf_{x\in\mathbb{R}^n} \sigma_1(x) > 0,$$

where $\sigma_1(x)$ is the smallest eigenvalue of $\nabla^2 f(x)$. This follows from 10.33 because $\nabla^2 f(x) - \sigma I$ is then positive semidefinite for all $x$, where $I$ is the identity matrix, and $f$ is $\sigma$-strongly convex.

With these refinements, we can extend the notion of elicitation.

**Definition 10.34** (elicitation of strong monotonicity). For $S : \mathbb{R}^n \rightrightarrows \mathbb{R}^n$, a subspace $X$ of $\mathbb{R}^n$ and $\sigma \in (0, \infty)$, we say that $S$ has *elicitable $\sigma$-strong maximal monotonicity at level $\theta \in [0, \infty)$* if the set-valued mapping

$$\big(S + \theta\, \mathrm{prj}_{X^\perp}\big) : \mathbb{R}^n \rightrightarrows \mathbb{R}^n \text{ is } \sigma\text{-strongly monotone and maximal monotone.}$$

**Example 10.35** (elicitation in optimization; cont.). Let's return to 10.20 with $S(x) = \{(x_2, x_1)\}$ and $X = \{(x_1, x_2) \mid x_1 = x_2\}$. For $\theta = 2$ and $\sigma = 1$, $S$ has elicitable $\sigma$-strong maximal monotonicity at level $\theta$.

**Detail.** Since $\theta \geq 1$, it follows from 10.20 and the discussion after 10.23 that $S + \theta\,\mathrm{prj}_{X^\perp}$ is maximal monotone. The gradient $\nabla\varphi(x)$, with

$$\varphi(x) = x_1 x_2 + \tfrac{1}{4}\theta(x_1 - x_2)^2,$$

coincides with $(S + \theta\,\mathrm{prj}_{X^\perp})(x)$ as seen in 10.20. By 10.33, the $\sigma$-strong monotonicity of $S + \theta\,\mathrm{prj}_{X^\perp}$ is determined by the $\sigma$-strong convexity of $\varphi$, which, in turn, holds when

$$\nabla^2\varphi(x) - \sigma I = \begin{bmatrix} \frac{1}{2}\theta - \sigma & 1 - \frac{1}{2}\theta \\ 1 - \frac{1}{2}\theta & \frac{1}{2}\theta - \sigma \end{bmatrix}$$

is positive semidefinite. Thus, we see that $\theta = 2$ and $\sigma = 1$ suffice. □

**Convergence 10.36** (progressive decoupling algorithm; cont.). *For $\sigma \in (0, \infty)$, suppose that $S$ has elicitable $\sigma$-strong maximal monotonicity at level $\theta \in [0, \infty)$. If the progressive decoupling algorithm with parameters $\theta < \kappa$ has generated $\{x^\nu, y^\nu,\ \nu \in \mathbb{N}\}$ while solving the linkage problem, then $(x^\nu, y^\nu)$ converges to a solution $(\bar{x}, \bar{y})$ of the linkage problem and*

$$\|x^{\nu+1} - \bar{x}\|_2 \leq \big\|(x^{\nu+1}, y^{\nu+1}) - (\bar{x}, y^\nu)\big\|_{\kappa,\theta} \leq \frac{\kappa}{\kappa + \sigma}\big\|(x^\nu, y^\nu) - (\bar{x}, \bar{y})\big\|_{\kappa,\theta} \quad \forall \nu.$$

**Proof.** The strong monotonicity of $S + \theta\,\mathrm{prj}_{X^\perp}$ ensures that the linkage problem has a solution and then the first convergence statement 10.23 about the progressive decoupling algorithm applies. Thus, it suffices to confirm the claimed rate.

With $u^{\nu+1}$ as in the proof of 10.30, the strong monotonicity of $S + \theta\,\mathrm{prj}_{X^\perp}$ ensures that

$$\big\langle -\kappa(x^{\nu+1} - x^\nu) + y^{\nu+1} - \bar{y},\ x^{\nu+1} + u^{\nu+1} - \bar{x}\big\rangle \geq \sigma\|x^{\nu+1} + u^{\nu+1} - \bar{x}\|_2^2$$

because $\bar{y} \in S(\bar{x})$ and (10.13) holds. Certainly, $x^\nu \in X$ and both $y^\nu, u^\nu \in X^\perp$ for all $\nu$, so after appealing to the orthogonality of such vectors, we find that

$$\|x^{\nu+1} + u^{\nu+1} - \bar{x}\|_2^2 = \|x^{\nu+1} - \bar{x}\|_2^2 + \|u^{\nu+1}\|_2^2.$$

When combined with the expression for $y^{\nu+1}$ in (10.14), the previous inequality can be rewritten as

$$\kappa\big\langle x^\nu - \bar{x} - (x^{\nu+1} - \bar{x}),\ x^{\nu+1} - \bar{x}\big\rangle + \big\langle y^\nu - \bar{y} - (\kappa - \theta)u^{\nu+1},\ u^{\nu+1}\big\rangle \geq \sigma\|x^{\nu+1} - \bar{x}\|_2^2 + \sigma\|u^{\nu+1}\|_2^2.$$

After multiplying out the terms on the left-hand side and moving terms to the right-hand side, we reach

$$\kappa\langle x^\nu - \bar{x},\ x^{\nu+1} - \bar{x}\rangle + \langle y^\nu - \bar{y}, u^{\nu+1}\rangle \geq (\kappa + \sigma)\|x^{\nu+1} - \bar{x}\|_2^2 + (\kappa - \theta + \sigma)\|u^{\nu+1}\|_2^2. \quad (10.19)$$

Let $\gamma = \sqrt{(\kappa - \theta)/\kappa}$. Then,

$$\kappa - \theta + \sigma \ge (\kappa + \sigma)\gamma^2.$$

Incorporating this on the right-hand side of (10.19) leads to

$$\kappa\langle x^\nu - \bar{x},\ x^{\nu+1} - \bar{x}\rangle + \langle y^\nu - \bar{y}, u^{\nu+1}\rangle \ge (\kappa + \sigma)\|x^{\nu+1} - \bar{x} + \gamma u^{\nu+1}\|_2^2 \tag{10.20}$$

after utilizing the orthogonality between $x^{\nu+1} - \bar{x}$ and $u^{\nu+1}$. Similarly, the left-hand side can be expressed as

$$\kappa\big\langle x^\nu - \bar{x} + \kappa^{-1}\gamma^{-1}(y^\nu - \bar{y}),\ x^{\nu+1} - \bar{x} + \gamma u^{\nu+1}\big\rangle$$

and subsequently be bounded from above by

$$\kappa\big\|x^\nu - \bar{x} + \kappa^{-1}\gamma^{-1}(y^\nu - \bar{y})\big\|_2\|x^{\nu+1} - \bar{x} + \gamma u^{\nu+1}\|_2.$$

After incorporating this into (10.20), we achieve

$$(\kappa + \sigma)\|x^{\nu+1} - \bar{x} + \gamma u^{\nu+1}\|_2^2 \le \kappa\big\|x^\nu - \bar{x} + \kappa^{-1}\gamma^{-1}(y^\nu - \bar{y})\big\|_2\|x^{\nu+1} - \bar{x} + \gamma u^{\nu+1}\|_2$$

and then also

$$\|x^{\nu+1} - \bar{x} + \gamma u^{\nu+1}\|_2 \le \frac{\kappa}{\kappa + \sigma}\big\|x^\nu - \bar{x} + \kappa^{-1}\gamma^{-1}(y^\nu - \bar{y})\big\|_2. \tag{10.21}$$

The definition of $u^{\nu+1}$ (see the proof of 10.30) shows that

$$\gamma u^{\nu+1} = -\gamma(\kappa - \theta)^{-1}(y^{\nu+1} - y^\nu).$$

Since

$$\gamma(\kappa - \theta)^{-1} = \kappa^{-1}\gamma^{-1} = 1/\sqrt{\kappa(\kappa - \theta)},$$

we find that (10.21) amounts to

$$\left\|x^{\nu+1} - \bar{x} - \frac{y^{\nu+1} - y^\nu}{\sqrt{\kappa(\kappa - \theta)}}\right\|_2 \le \frac{\kappa}{\kappa + \sigma}\left\|x^\nu - \bar{x} + \frac{y^\nu - \bar{y}}{\sqrt{\kappa(\kappa - \theta)}}\right\|_2.$$

By (10.8), this expression is equivalent to the right-most inequality in the claim. The left-most inequality follows by the definition of $\|\cdot\|_{\kappa,\theta}$. □

The result shows an advantage of having a high modulus of strong monotonicity. In view of the localization proposition 10.28, it's possible to refine the result to the case when $S + \theta\,\mathrm{prj}_{X^\perp}$ is strongly monotone only in a neighborhood of a solution of the linkage problem. Then the same rate is achieved provided that the algorithm is started sufficiently close to that solution and is carried out in a manner that keeps the iterates in the neighborhood as explained in 10.26; see [93, 94], which also account for varying proximal parameter $\kappa$ and inexact solution of the subproblems in Step 1 of the progressive decoupling algorithm.

## 10.G Variational Convexity and Elicitation

Elicitation can be achieved via convexity (see 10.22), but this isn't the only possibility. We'll now examine other properties that lead to elicitation of maximal monotonicity as well as strong monotonicity in a local or global sense. A central concept in this regard is *variational convexity*, which also furnishes a sufficient condition for local minimizers that's of independent interest.

**Proposition 10.37** (elicitation from linearity). *For a nonempty, closed and convex set $C \subset \mathbb{R}^n$, a subspace $X$ of $\mathbb{R}^n$ and an $n \times n$-matrix $A$, suppose that there's $\alpha \in (0, \infty)$ such that $\langle x, Ax \rangle \geq \alpha \|x\|_2^2$ for all $x \in X$. Then, the set-valued mapping $S : \mathbb{R}^n \rightrightarrows \mathbb{R}^n$ given by*

$$S(x) = Ax + N_C(x)$$

*has elicitable maximal monotonicity at level $\theta \in [0, \infty)$ for*[2]

$$\theta \geq \beta^2/\alpha + \delta, \text{ where } \beta = \tfrac{1}{2}\|B\|_o,\ \delta = \|D\|_o, \tag{10.22}$$
$$B \text{ is the } n \times n\text{-matrix with } Bx = \mathrm{prj}_X\big((A + A^\top)\,\mathrm{prj}_{X^\perp}(x)\big) \text{ and}$$
$$D \text{ is the } n \times n\text{-matrix with } Dx = \mathrm{prj}_{X^\perp}\big(A\,\mathrm{prj}_{X^\perp}(x)\big) \quad \forall x \in \mathbb{R}^n.$$

*Moreover, $S$ has elicitable $\sigma$-strong maximal monotonicity at level $\theta$ with $\sigma \in (0, \alpha)$ provided that*

$$\theta \geq \beta^2/(\alpha - \sigma) + \delta + \sigma.$$

**Proof.** In view of 10.17, we obtain maximal monotonicity of $S + \theta\,\mathrm{prj}_{X^\perp}$ from proving monotonicity of $A + \theta\,\mathrm{prj}_{X^\perp}$ because that mapping is continuous by 7.14 and $N_C$ is already maximal monotone; see 10.15 and the discussion following that result.

As can be seen from (10.3), $\mathrm{prj}_{X^\perp}$ is actually a linear mapping so $A + \theta\,\mathrm{prj}_{X^\perp}$ is linear. This means that its monotonicity reduces to showing that

$$\big\langle x, Ax + \theta\,\mathrm{prj}_{X^\perp}(x) \big\rangle \geq 0 \quad \forall x \in \mathbb{R}^n.$$

Moreover, $\mathrm{prj}_X$ is also a linear mapping so

$$x \mapsto \mathrm{prj}_X\big((A + A^\top)\,\mathrm{prj}_{X^\perp}(x)\big) \quad \text{and} \quad x \mapsto \mathrm{prj}_{X^\perp}\big(A\,\mathrm{prj}_{X^\perp}(x)\big)$$

are linear mappings and thus representable by $n \times n$ matrices $B$ and $D$ as asserted.

Let $x \in \mathbb{R}^n$ be arbitrary. In the notation $z = \mathrm{prj}_X(x)$ and $\bar{z} = \mathrm{prj}_{X^\perp}(x)$, and using the orthogonality between $X$ and $X^\perp$, we obtain

[2] $\|\cdot\|_o$ is the operator norm: $\|A\|_o = \sup_{x \in \mathbb{R}^n} \{\|Ax\|_2 \mid \|x\|_2 = 1\}$ for $n \times n$-matrix $A$.

$$\begin{aligned}\big\langle z, (A + A^\top)\bar{z}\big\rangle &= \Big\langle z,\ \mathrm{prj}_X\big((A + A^\top)\,\mathrm{prj}_{X^\perp}(\bar{z})\big)\Big\rangle\\ &\geq -\|z\|_2\Big\|\mathrm{prj}_X\big((A + A^\top)\,\mathrm{prj}_{X^\perp}(\bar{z})\big)\Big\|_2 = -\|z\|_2\|B\bar{z}\|_2 \geq -2\beta\|z\|_2\|\bar{z}\|_2\end{aligned}$$

and, similarly,

$$\begin{aligned}\langle \bar{z}, A\bar{z}\rangle &= \big\langle \bar{z},\ \mathrm{prj}_{X^\perp}\big(A\,\mathrm{prj}_{X^\perp}(\bar{z})\big)\big\rangle\\ &\geq -\|\bar{z}\|_2\big\|\mathrm{prj}_{X^\perp}\big(A\,\mathrm{prj}_{X^\perp}(\bar{z})\big)\big\|_2 = -\|\bar{z}\|_2\|D\bar{z}\|_2 \geq -\delta\|\bar{z}\|_2^2.\end{aligned}$$

These expressions together with $x = z + \bar{z}$ from (10.4) yield

$$\begin{aligned}\big\langle x, Ax + \theta\,\mathrm{prj}_{X^\perp}(x)\big\rangle &= \langle x, Ax\rangle + \theta\langle x, \bar{z}\rangle = \big\langle z + \bar{z}, A(z + \bar{z})\big\rangle + \theta\langle z + \bar{z}, \bar{z}\rangle\\ &= \langle z, Az\rangle + \big\langle z, (A + A^\top)\bar{z}\big\rangle + \langle \bar{z}, A\bar{z}\rangle + \theta\|\bar{z}\|_2^2\\ &\geq \alpha\|z\|_2^2 - 2\beta\|z\|_2\|\bar{z}\|_2 + (\theta - \delta)\|\bar{z}\|_2^2.\end{aligned}$$

The nonnegativity of the last expression for all $x$ is guaranteed when

$$\alpha u_1^2 - 2\beta u_1 u_2 + (\theta - \delta)u_2^2 \geq 0 \quad \forall u = (u_1, u_2) \in \mathbb{R}^2,$$

which can also be expressed as $\langle u, Mu\rangle \geq 0$ for all $u \in \mathbb{R}^2$, where

$$M = \begin{bmatrix} \alpha & -\beta \\ -\beta & \theta - \delta \end{bmatrix}.$$

In turn, this is equivalent to $M$ being positive semidefinite. The determinant test for positive semidefiniteness asserts that we need to have

$$\alpha \geq 0, \quad \theta - \delta \geq 0, \quad \alpha(\theta - \delta) - \beta^2 \geq 0.$$

Since $\alpha$ is positive by assumption, the last inequality amounts to having $\theta \geq \alpha^{-1}\beta^2 + \delta$, in which case $\theta - \delta \geq 0$ is automatic. Thus,

$$\big\langle x, Ax + \theta\,\mathrm{prj}_{X^\perp}(x)\big\rangle \geq 0$$

for such $\theta$ and the claim about elicitation of maximal monotonicity holds.

For strong monotonicity, because of the linearity of $A + \theta\,\mathrm{prj}_{X^\perp}$, we only need to show that

$$\big\langle x, Ax + \theta\,\mathrm{prj}_{X^\perp}(x)\big\rangle \geq \sigma\|x\|_2^2 \quad \forall x \in \mathbb{R}^n.$$

Repeating the above arguments and utilizing the fact that

$$\|z + \bar{z}\|_2^2 = \|z\|_2^2 + \|\bar{z}\|_2^2$$

due to orthogonality, we now seek

$$(\alpha - \sigma)\|z\|_2^2 - 2\beta\|z\|_2\|\bar{z}\|_2 + (\theta - \delta - \sigma)\|\bar{z}\|_2^2 \geq 0.$$

This translates into the requirements

$$\alpha - \sigma \geq 0, \quad \theta - \delta - \sigma \geq 0, \quad (\alpha - \sigma)(\theta - \delta - \sigma) - \beta^2 \geq 0.$$

Thus $\theta \geq (\alpha - \sigma)^{-1}\beta^2 + \delta + \sigma$ as claimed. □

For an $n \times n$-matrix $A$ and $x \in \mathbb{R}^n$, we always have that

$$\langle x, Ax\rangle = \left\langle x, \tfrac{1}{2}(A + A^\top)x\right\rangle,$$

with $(A + A^\top)/2$ being referred to as the *symmetric part* of $A$; the *asymmetric part* is $(A - A^\top)/2$ so that the sum of the symmetric and asymmetric parts recovers $A$. Thus, if the smallest eigenvalue $\sigma_1$ of $(A + A^\top)/2$ is positive, which takes place if and only if $(A + A^\top)/2$ is positive definite, then

$$\langle x, Ax\rangle \geq \sigma_1\|x\|_2^2 \quad \forall x \in \mathbb{R}^n$$

and $\sigma_1$ furnishes a possible value of $\alpha$ in 10.37. The proposition refines this further by allowing us to consider only $x \in X$. For example,

$$A = \begin{bmatrix} 0 & 1 \\ 1 & 0 \end{bmatrix}$$

isn't positive definite, but for

$$x \in X = \left\{(x_1, x_2) \in \mathbb{R}^2 \;\middle|\; x_1 = x_2\right\}$$

one has $\langle x, Ax\rangle = \|x\|_2^2$ so that $\alpha$ can be set to 1 in 10.37.

As seen from the proof of 10.37, one might get slightly better coefficients $\beta$ and $\delta$ if they're set to $\beta = \max\{0, -\beta'/2\}$ and $\delta = \max\{0, -\delta'\}$, where

$$\beta' = \min_{x\in X, y\in X^\perp} \left\{\left\langle x, (A + A^\top)y\right\rangle \;\middle|\; \|x\|_2 = 1, \|y\|_2 = 1\right\}$$
$$\delta' = \min_{y\in X^\perp} \left\{\langle y, Ay\rangle \;\middle|\; \|y\|_2 = 1\right\}.$$

The minimum values of these problems, which are finite by 4.9, can be computed via the smallest eigenvalues of certain matrices; see [46] for details.

**Example 10.38** (elicitation in quadratic problems). For a nonempty, closed and convex set $C$, a symmetric $n \times n$-matrix $A$ and a subspace $X$ of $\mathbb{R}^n$, let's consider

$$\underset{x\in C\cap X}{\text{minimize}} \; f(x) = \tfrac{1}{2}\langle x, Ax\rangle,$$

which can be associated with the condition

$$0 \in \partial f(x) + N_C(x) + N_X(x).$$

In light of 4.52, the condition is necessary for a minimizer under a qualification but that's not a primary concern at the present. Our goal is to find $x \in X$ and $y \in X^{\perp}$ such that $y \in \partial f(x) + N_C(x)$, and thereby satisfy the condition. This lands us squarely in the territory of the progressive decoupling algorithm and elicitation for the set-valued mapping $\partial f + N_C$ becomes important. It turns out that $\partial f + N_C$ has elicitable maximal monotonicity at level $\theta$ provided that $\theta \geq \beta^2/\alpha + \delta$, with $\alpha, \beta$ and $\delta$ as in 10.37.

**Detail.** Since $\partial f(x) = Ax$, the assertion is a direct consequence of 10.37 and one can, therefore, also proceed and elicit $\sigma$-strong maximal monotonicity as laid out there. In view of 10.22,

$$f + \iota_C + \tfrac{1}{2}\theta\, \mathrm{dist}^2(\cdot, X)$$

is a convex function for $\theta \geq \beta^2/\alpha + \delta$ despite the fact that $f$ may not be convex. □

**Proposition 10.39** (elicitation from smoothness). *For a nonempty, closed and convex set $C \subset \mathbb{R}^n$, a subspace $X$ of $\mathbb{R}^n$ and a smooth $F : \mathbb{R}^n \to \mathbb{R}^n$, suppose that there's*

$$\alpha : C \to (0, \infty) \text{ such that } \langle \bar{x}, \nabla F(x)\bar{x} \rangle \geq \alpha(x)\|\bar{x}\|_2^2 \quad \forall \bar{x} \in X, x \in C.$$

*Let $\beta(x)$ and $\delta(x)$ be obtained from* (10.22), *with $A$ replaced by $\nabla F(x)$, and set*

$$\theta_0 = \sup_{x \in C \cap X} \{\beta^2(x)/\alpha(x) + \delta(x)\}.$$

*If $\theta_0 < \infty$, then the set-valued mapping $F + N_C$ has elicitable maximal monotonicity at level $\theta \geq \theta_0$; it has elicitable $\sigma$-strong maximal monotonicity at level $\theta$ for some $\sigma > 0$ when $\theta > \theta_0$.*

**Proof.** In view of (2.5) (see also [105, Proposition 12.3]), the monotonicity of $F$ can be characterized by the properties of $\nabla F$, which then together with the linearity of $\mathrm{prj}_{X^{\perp}}$ allow us to repeat the arguments in 10.37 for each $x \in C$ with $\nabla F(x)$ in the role of $A$. This leads to requirements on $\alpha(x), \beta(x)$ and $\delta(x)$ that are met uniformly across $C$ by the assumptions. There's no need to consider $x \notin C$ because then $N_C(x) = \emptyset$. □

The proposition addresses, for example, the problem of minimizing a twice smooth function $f : \mathbb{R}^n \to \mathbb{R}$ over $C \cap X$, where $C \subset \mathbb{R}^n$ is a nonempty, closed and convex set and $X$ is a subspace of $\mathbb{R}^n$. The problem can be associated with the condition

$$0 \in \nabla f(x) + N_C(x) + N_X(x)$$

and this leads to the Hessian $\nabla^2 f(x)$ playing the role of $\nabla F(x)$. For a local version of 10.39; see [93].

A systematic approach to achieve local elicitation relies on the broad concept of variational convexity, which also brings us to a sufficient condition for local minimizers.

**Definition 10.40** (variational convexity). For a proper lsc function $f : \mathbb{R}^n \to \overline{\mathbb{R}}$, we say that $f$ is *variationally convex* at $\bar{x} \in \mathbb{R}^n$ for $\bar{y} \in \partial f(\bar{x})$ if for some open convex neighborhood $U \times V$ of $(\bar{x}, \bar{y})$ there are convex lsc $g : \mathbb{R}^n \to \overline{\mathbb{R}}$ and $\varepsilon \in (0, \infty]$ such that

(a) $g(x) \leq f(x)$ for all $x \in U$
(b) $g(x) = f(x)$ when $(x, y) \in (U \times V) \cap \operatorname{gph} \partial g$
(c) $(U_\varepsilon \times V) \cap \operatorname{gph} \partial f = (U \times V) \cap \operatorname{gph} \partial g$, with $U_\varepsilon = U \cap \{f \leq f(\bar{x}) + \varepsilon\}$.

The function $f$ is *variationally $\sigma$-strongly convex* at $\bar{x}$ for $\bar{y}$ if $g$ is $\sigma$-strongly convex. In either case, $\varepsilon$ is the *localization threshold*.

In essence, $f$ is variationally convex at $\bar{x}$ for $\bar{y} \in \partial f(\bar{x})$ when there exists a lower-bounding convex lsc function $g$ that coincides with $f$ at $\bar{x}$ and that has the same subgradient mapping as $f$ locally near $(\bar{x}, \bar{y})$.

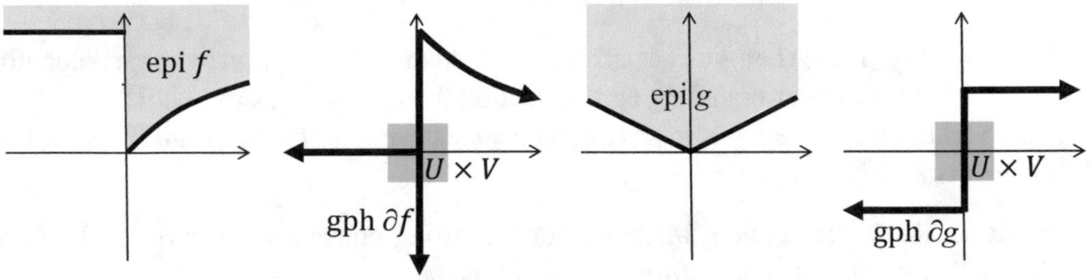


Fig. 10.5: The function $f$ is variationally convex at $\bar{x} = 0$ for $\bar{y} = 0$, with $g$ furnishing a convex function from which this can be established.

Figure 10.5 illustrates the case with

$$f(x) = \begin{cases} 1 & \text{for } x < 0 \\ 1 - e^{-x} & \text{otherwise} \end{cases} \qquad \partial f(x) = \begin{cases} \{0\} & \text{for } x < 0 \\ (-\infty, 1] & \text{for } x = 0 \\ \{e^{-x}\} & \text{otherwise.} \end{cases}$$

Although $f$ isn't convex, it's variationally convex at $\bar{x} = 0$ for $\bar{y} = 0$. To see this, consider $g(x) = |x|/2$ and its subgradient mapping illustrated to the right in the figure. For localization threshold $\varepsilon = 1/2$, $U = (-1/4, 1/4)$ and $V = (-1/4, 1/4)$, we certainly have (a) in the definition satisfied. Since

$$(U \times V) \cap \operatorname{gph} \partial g = \{0\} \times \left(-\tfrac{1}{4}, \tfrac{1}{4}\right)$$

and $g(0) = f(0)$, (b) also holds. The set $U_\varepsilon = [0, 1/4)$ in this case, which makes $(U_\varepsilon \times V) \cap \operatorname{gph} \partial f = \{0\} \times (-1/4, 1/4)$ and we've satisfied (c). The example highlights the role of the localization threshold $\varepsilon$ as a vehicle to handle a discontinuous function $f$ via $U_\varepsilon$ in an effective manner.

**Exercise 10.41** (variational convexity). Examine the function given by $f(x) = \max\{-x - 2, x^3, 4x - 3\}$ for variational convexity at the points 1, 0 and $-1$. At each point, determine all (if any) subgradients for which variational convexity holds.

If $f$ is variationally convex at $\bar{x} \in \mathbb{R}^n$ for $\bar{y} \in \partial f(\bar{x})$, then $f$ satisfies a local version of the subgradient inequality 2.17: With $U_\varepsilon, U, V$ as in the definition, one has

$$(x, y) \in (U_\varepsilon \times V) \cap \operatorname{gph} \partial f \implies f(x') \geq f(x) + \langle y, x' - x \rangle \quad \forall x' \in U. \tag{10.23}$$

To see this, simple note that the convex function $g$ that certifies the variational convexity coincides with $f$ at points $x$ with $(x, y) \in (U_\varepsilon \times V) \cap \operatorname{gph} \partial f$ and then also has $y \in \partial g(x)$; see 10.40. From the subgradient inequality 2.17 applied to $g$, one has

$$g(x') \geq g(x) + \langle y, x' - x \rangle \quad \forall x' \in \mathbb{R}^n.$$

Since $g(x') \leq f(x')$ for $x' \in U$, the claim follows by

$$f(x') \geq g(x') \geq g(x) + \langle y, x' - x \rangle = f(x) + \langle y, x' - x \rangle.$$

The inequality (10.23) helps us visualize what it means for a function to be variationally convex; the familiar lower-bounding approximation from the convex case holds locally.

If the variational convexity is $\sigma$-strong, then we can strengthen the result by using the following general fact.

**Proposition 10.42** (subgradient inequality under strong convexity). *For $\sigma \in (0, \infty)$, if $f : \mathbb{R}^n \to \overline{\mathbb{R}}$ is $\sigma$-strongly convex and $\bar{y} \in \partial f(\bar{x})$, then*

$$f(x) \geq f(\bar{x}) + \langle \bar{y}, x - \bar{x} \rangle + \tfrac{1}{2}\sigma \|x - \bar{x}\|_2^2 \quad \forall x \in \mathbb{R}^n.$$

**Proof.** Since $f - \frac{1}{2}\sigma \|\cdot\|_2^2$ is convex with subgradient $\bar{y} - \sigma\bar{x}$ at $\bar{x}$ (cf. 4.58), the subgradient inequality 2.17 ensures that

$$f(x) - \tfrac{1}{2}\sigma\|x\|_2^2 \geq f(\bar{x}) - \tfrac{1}{2}\sigma\|\bar{x}\|_2^2 + \langle \bar{y} - \sigma\bar{x}, x - \bar{x} \rangle,$$

which simplifies to the asserted expression. □

Using this proposition and arguments similar to those leading to (10.23), we obtain that if $f$ is variationally $\sigma$-strongly convex at $\bar{x} \in \mathbb{R}^n$ for $\bar{y} \in \partial f(\bar{x})$, then

$$(x, y) \in (U_\varepsilon \times V) \cap \operatorname{gph} \partial f \implies \tag{10.24}$$
$$f(x') \geq f(x) + \langle y, x' - x \rangle + \tfrac{1}{2}\sigma\|x' - x\|_2^2 \quad \forall x' \in U.$$

In the absence of convexity, it would generally be difficult to conclude from $0 \in \partial f(\bar{x})$ that $\bar{x}$ is a (local) minimizer, although this is the case in Figure 10.5. However, if $f$ is variationally convex at $\bar{x}$ for $\bar{y} = 0$, then there's a convex function $g$ with $0 \in \partial g(\bar{x})$ so that $\bar{x} \in \operatorname{argmin} g$ by 2.19. Since $f$ coincides with $g$ locally in some sense, this means that $f$ must have a local minimizer at $\bar{x}$. Let's formalize this insight.

**Definition 10.43** (variational second-order sufficient condition). For a proper lsc function $f : \mathbb{R}^n \to \overline{\mathbb{R}}$ and a subspace $X$ of $\mathbb{R}^n$, consider the problem of minimizing $f$ over $X$. We say that the problem satisfies the *variational second-order sufficient condition* at $\bar{x} \in X$ for $\bar{y} \in X^\perp$ if $\bar{y} \in \partial f(\bar{x})$ and

$$\exists \theta \in [0, \infty) \text{ such that } f + \tfrac{1}{2}\theta \operatorname{dist}^2(\cdot, X) \text{ is variationally convex at } \bar{x} \text{ for } \bar{y}.$$

We say that the problem satisfies the *variational strong second-order sufficient condition* with modulus $\sigma \in (0, \infty)$ when the variational convexity is $\sigma$-strong.

These conditions are said to hold with a specific *localization threshold* $\varepsilon \in (0, \infty]$ if the variational convexity holds with localization threshold $\varepsilon$.

At first, there appears to be a mismatch between the definition of variational convexity in 10.40 and how it's invoked here. In 10.43, we're asking for the function

$$f + \tfrac{1}{2}\theta \operatorname{dist}^2(\cdot, X)$$

to be variationally convex at $\bar{x}$ for $\bar{y}$ and then would have expected that $\bar{y}$ should be a subgradient of that function at $\bar{x}$; see 10.40. However, we seemingly only have $\bar{y} \in \partial f(\bar{x})$. It turns out that this is immaterial because $\bar{x} \in X$ and then

$$\partial f(\bar{x}) = \partial f(\bar{x}) + \theta \operatorname{prj}_{X^\perp}(\bar{x}) = \partial\big(f + \tfrac{1}{2}\theta \operatorname{dist}^2(\cdot, X)\big)(\bar{x})$$

as discussion before 10.19; see also the proof of 10.21.

If $f$ itself is variationally convex at $\bar{x}$ for $\bar{y}$ as in Figure 10.5, then the variational second-order sufficient condition is satisfied with $\theta = 0$. However, $\theta$ offers additional possibilities tied to the subspace $X$. The name of 10.43 hints to the following sufficiency condition for a local minimizer, which is distinct from those in §6.A as it doesn't rely on smoothness.

**Theorem 10.44** (sufficiency for local minimizer). *For a proper lsc function $f : \mathbb{R}^n \to \overline{\mathbb{R}}$ and a subspace $X$ of $\mathbb{R}^n$, suppose that the problem of minimizing $f$ over $X$ satisfies the variational second-order sufficient condition at $\bar{x} \in X$ for $\bar{y} \in X^\perp$, then $\bar{x}$ is a local minimizer of the problem.*

*If in addition the variational strong second-order sufficient condition holds with modulus $\sigma$, then there's $\varepsilon > 0$ such that*

$$f(x) \geq f(\bar{x}) + \tfrac{1}{2}\sigma\|x - \bar{x}\|_2^2 \quad \forall x \in X \cap \mathbb{B}(\bar{x}, \varepsilon).$$

**Proof.** Since $\bar{y} \in \partial f(\bar{x})$ and then also

$$\bar{y} \in \partial\big(f + \tfrac{1}{2}\theta \operatorname{dist}^2(\cdot, X)\big)(\bar{x})$$

by the discussion before the theorem, the function $g : \mathbb{R}^n \to \overline{\mathbb{R}}$, with

$$g(x) = f(x) + \tfrac{1}{2}\theta \operatorname{dist}^2(x, X) - \langle \bar{y}, x - \bar{x}\rangle,$$

has $0 \in \partial g(\bar{x})$. Since $f + \frac{1}{2}\theta \operatorname{dist}^2(\cdot, X)$ is variationally convex at $\bar{x}$ for $\bar{y}$, this implies that $g$ is variationally convex at $\bar{x}$ for 0. As argued before 10.43, $\bar{x}$ is then a local minimizer of $g$ over $\mathbb{R}^n$. However, $g(x) = f(x)$ when $x \in X$ and $\bar{y} \in X^\perp$ because $\operatorname{dist}(x, X) = 0$ and the orthogonality between $\bar{y}$ and $x - \bar{x}$ makes $\langle \bar{y}, x - \bar{x}\rangle = 0$. Consequently, $\bar{x}$ is a local minimizer of $f$ over $X$.

Under the variational strong second-order sufficient condition, $g$ is variationally $\sigma$-strongly convex at $\bar{x}$ for 0. By (10.24), this implies that for some open neighborhood $U$ of $\bar{x}$, one has

$$g(x) \geq g(\bar{x}) + \langle 0, x - \bar{x} \rangle + \tfrac{1}{2}\sigma \|x - \bar{x}\|_2^2 \;\; \forall x \in U.$$

In view of the expression for $g$, we then also have

$$f(x) + \tfrac{1}{2}\theta \operatorname{dist}^2(x, X) - \langle \bar{y}, x - \bar{x} \rangle \geq f(\bar{x}) + \tfrac{1}{2}\theta \operatorname{dist}^2(\bar{x}, X) + \tfrac{1}{2}\sigma \|x - \bar{x}\|_2^2 \;\; \forall x \in U.$$

Certainly, $\operatorname{dist}^2(\bar{x}, X) = 0$ and $\operatorname{dist}^2(x, X) = 0$ as well when $x \in X$ and then $\langle \bar{y}, x - \bar{x} \rangle = 0$ by orthogonality. These cancelations produce the required expression. □

As seen in Chap. 5, we can represent an optimization problem using a Rockafellian $f : \mathbb{R}^m \times \mathbb{R}^n \to \overline{\mathbb{R}}$, with minimizing $f(0, x)$ over $x \in \mathbb{R}^n$ being the actual problem of interest. This is closely related to the present perspective. Let

$$X = \big\{(u, x) \in \mathbb{R}^m \times \mathbb{R}^n \;\big|\; u = 0\big\}.$$

Then, the actual problem is equivalently stated as

$$\operatorname*{minimize}_{u \in \mathbb{R}^m, x \in \mathbb{R}^n} f(u, x) \;\text{ subject to }\; (u, x) \in X.$$

If this reformulation satisfies the variational second-order sufficient condition at $(0, \bar{x}) \in X$ for $(\bar{y}, 0) \in X^\perp$ and $f$ is proper and lsc, then 10.44 asserts that $(0, \bar{x})$ is a local minimizer for the reformulated problem and $\bar{x}$ is a local minimizer of the actual problem. In this case, the variational second-order sufficient condition amounts to having

$$(\bar{y}, 0) \in \partial f(0, \bar{x}),$$

which is the Rockafellar condition 5.10, and for some $\theta \in [0, \infty)$ having

$$f + \tfrac{1}{2}\theta \operatorname{dist}^2(\cdot, X)$$

be variationally convex at $(0, \bar{x})$ for $(\bar{y}, 0)$. Since

$$\operatorname{dist}^2\big((u, x), X\big) = \|u\|_2^2,$$

the latter says that

$$(u, x) \mapsto f(u, x) + \tfrac{1}{2}\theta \|u\|_2^2$$

should be variationally convex at $(0, \bar{x})$ for $(\bar{y}, 0)$. Thus, in this broad setting that encapsulates many types of problem structures including constraint and composite functions (see Chap. 5), a local minimizer is guaranteed by the variational convexity property. Most significantly, these claims hold regardless of smoothness that may or may not be present in $f$. Connections with augmented Lagrangians of §6.B are laid out in [95].

Returning to the subject of elicitation, we find that the variational second-order sufficient condition 10.43 is also a key property in that regard.

**Proposition 10.45** (elicitation from second-order condition). *For a proper lsc function $f : \mathbb{R}^n \to \overline{\mathbb{R}}$ and a subspace $X$ of $\mathbb{R}^n$, suppose that the problem of minimizing $f$ over $X$ satisfies the variational second-order sufficient condition at $\bar{x} \in X$ for $\bar{y} \in X^\perp$, with localization threshold $\varepsilon \in (0, \infty]$ and $\theta \in [0, \infty)$ as specified by definition 10.43. Let $S : \mathbb{R}^n \rightrightarrows \mathbb{R}^n$ be defined by*

$$\operatorname{gph} S = \big\{(x, y) \in \operatorname{gph} \partial f \;\big|\; f(x) + \tfrac{1}{2}\theta \operatorname{dist}^2(x, X) \le f(\bar{x}) + \varepsilon\big\}.$$

*Then, $S$ has elicitable maximal monotonicity at level $\theta$ locally near $(\bar{x}, \bar{y})$.*

**Proof.** The assumption of variational second-order sufficient condition implies that $f + \frac{1}{2}\theta \operatorname{dist}^2(\cdot, X)$ is variationally convex at $\bar{x}$ for $\bar{y}$ with localization threshold $\varepsilon$. From the definition of variational convexity in 10.40, there exist an open convex neighborhood $U \times V$ of $(\bar{x}, \bar{y})$ and a convex lsc function $g : \mathbb{R}^n \to \overline{\mathbb{R}}$ such that

$$(U_\varepsilon \times V) \cap \operatorname{gph} \partial\big(f + \tfrac{1}{2}\theta \operatorname{dist}^2(\cdot, X)\big) = (U \times V) \cap \operatorname{gph} \partial g, \tag{10.25}$$

where

$$U_\varepsilon = U \cap \big\{f + \tfrac{1}{2}\theta \operatorname{dist}^2(\cdot, X) \le f(\bar{x}) + \varepsilon\big\}.$$

Since $g$ is convex and lsc and necessarily also proper, $\partial g$ is maximal monotone by 10.15. The conclusion follows if

$$(U \times V) \cap \operatorname{gph}(S + \theta \operatorname{prj}_{X^\perp}) = (U \times V) \cap \operatorname{gph} \partial g$$

because then $S + \theta \operatorname{prj}_{X^\perp}$ is maximal monotone in $U \times V$. In view of (10.25), it only remains to show that

$$\begin{aligned}(U_\varepsilon \times V) \cap \operatorname{gph} \partial\big(f + \tfrac{1}{2}\theta \operatorname{dist}^2(\cdot, X)\big) &= (U_\varepsilon \times V) \cap \operatorname{gph}(\partial f + \theta \operatorname{prj}_{X^\perp})\\ &= (U \times V) \cap \operatorname{gph}(S + \theta \operatorname{prj}_{X^\perp}).\end{aligned}$$

Here, the first equality holds by 10.21 and 4.58(c). To establish the second equality, let

$$(x, y) \in (U \times V) \cap \operatorname{gph}(S + \theta \operatorname{prj}_{X^\perp}).$$

Then, $y - \theta \operatorname{prj}_{X^\perp}(x) \in S(x)$ so that $f(x) + \frac{1}{2}\theta \operatorname{dist}^2(x, X) \le f(\bar{x}) + \varepsilon$ and

$$y - \theta \operatorname{prj}_{X^\perp}(x) \in \partial f(x)$$

by the definition of $S$. Thus,

$$(x, y) \in (U_\varepsilon \times V) \cap \operatorname{gph}(\partial f + \theta \operatorname{prj}_{X^\perp}).$$

The reverse inclusion follows by a similar argument. □

If the localization threshold $\varepsilon = \infty$, then $\partial f$ itself has elicitable maximal monotonicity at level $\theta$ locally near $(\bar{x}, \bar{y})$ by the proposition. The progressive decoupling algorithm applied to the linkage problem of finding $x \in X$ and $y \in X^{\perp}$, with $y \in \partial f(x)$, would then have the local behavior near a solution $(\bar{x}, \bar{y})$ described by 10.26. Thus, we can address the problem of minimizing $f$ over $X$ without insisting on smoothness and convexity; the variational second-order sufficient condition at a solution is enough in some sense.

For example, $\varepsilon = \infty$ emerges by making $f$ in Figure 10.5 symmetric: Consider

$$f(x) = \begin{cases} 1 - e^{x} & \text{for } x < 0 \\ 1 - e^{-x} & \text{otherwise} \end{cases} \qquad \partial f(x) = \begin{cases} \{-e^{x}\} & \text{for } x < 0 \\ [-1, 1] & \text{for } x = 0 \\ \{e^{-x}\} & \text{otherwise.} \end{cases}$$

One can take $g(x) = |x|/2$ as in Figure 10.5, $\varepsilon = \infty$, $U = (-1/4, 1/4)$ and $V = (-1/4, 1/4)$ to establish that $f$ is variationally convex at $\bar{x} = 0$ for $\bar{y} = 0$.

A finite localization threshold would be brought in to handle situations where a portion of gph $\partial f$ near a solution $(\bar{x}, \bar{y})$ isn't relevant for the characterization of a local minimizer due to a discontinuity in $f$; consider negative $x$ in Figure 10.5(left). In such cases, the set-valued mapping $S$ in 10.45 wouldn't coincide with $\partial f$, which fails to be maximal monotone in any neighborhood of $(\bar{x}, \bar{y})$. The horizontal line forming part of gph $\partial f$ in Figure 10.5 exemplifies the lack of maximal monotonicity for $\partial f$ locally near the origin. Still, the local behavior of the progressive decoupling algorithm would be retained when applied to finding $x \in X$ and $y \in X^{\perp}$, with $y \in \partial f(x)$, provided that the localization condition $f(x) + \frac{1}{2}\theta\, \text{dist}^2(x, X) \leq f(\bar{x}) + \varepsilon$ is obeyed during the calculations; see [93].

## 10.H Nonlinear Linkage

As already seen through various applications of splitting, linkage in terms of a subspace constraint can be brought forward in problems without any apparent linear structure. We'll now see an example of this approach for a composite function of the kind encountered in the earlier chapters.

For proper, lsc and convex $h : \mathbb{R}^m \to \overline{\mathbb{R}}$, proper lsc $f_i : \mathbb{R}^{n_i} \to \overline{\mathbb{R}}$, smooth $F_i : \mathbb{R}^{n_i} \to \mathbb{R}^m$ and a subspace $X_0$ of $\mathbb{R}^n$, with $i = 1, \ldots, q$ and $\sum_{i=1}^{q} n_i = n$, consider the problem

$$\underset{(x_1,\ldots,x_q)\in X_0}{\text{minimize}} \sum_{i=1}^{q} f_i(x_i) + h\Big(\sum_{i=1}^{q} F_i(x_i)\Big), \tag{10.26}$$

which has structure that points to the possibility of decomposition. For example, $h$ might be a monitoring function that penalizes certain values of $\sum_{i=1}^{q} F_i(x_i)$ and would represent the constraint

$$\sum_{i=1}^{q} F_i(x_i) \leq 0$$

when $h = \iota_{(-\infty,0]^m}$. The coupling between the variables $x_1, \ldots, x_q$ could then be nonlinear, taking us beyond the range of the approaches in the previous sections. However, this first impression is misleading as a reformulation enables us to pose the problem in a form treatable by the progressive decoupling algorithm.

**Lemma 10.46** *For a convex function* $h : \mathbb{R}^m \to \overline{\mathbb{R}}$, *mappings* $F_i : \mathbb{R}^{n_i} \to \mathbb{R}^m$, $i = 1, \dots, q$, *and* $\alpha \in \mathbb{R}$, *one has*

$$h\Big(\sum_{i=1}^{q} F_i(x_i)\Big) \le \alpha \iff \exists\{u_i \in \mathbb{R}^m, i = 1, \dots, q\} \text{ such that}$$

$$\sum_{i=1}^{q} u_i = 0 \text{ and } q^{-1}\sum_{i=1}^{q} h\Big(q\big(F_i(x_i) + u_i\big)\Big) \le \alpha.$$

**Proof.** Let $z_1, \dots, z_q \in \mathbb{R}^m$. Since $h$ is convex,

$$h\Big(q^{-1}\sum_{i=1}^{q} z_i\Big) \le q^{-1}\sum_{i=1}^{q} h(z_i),$$

which follows essentially from Jensen's inequality 3.5; an extension beyond real-valued functions holds by [105, Theorem 2.2]. Trivially, if $z_1 = \dots = z_q$, then the inequality holds with equality. For any $\bar{z} \in \mathbb{R}^m$, this explains that

$$h(\bar{z}) = \min_{z_1,\dots,z_q} \Big\{q^{-1}\sum_{i=1}^{q} h(z_i) \;\Big|\; q^{-1}\sum_{i=1}^{q} z_i = \bar{z}\Big\}.$$

With $\bar{z} = \sum_{i=1}^{q} F_i(x_i)$ and a change from $z_i$ to $u_i = q^{-1}z_i - F_i(x_i)$, we obtain

$$h\Big(\sum_{i=1}^{q} F_i(x_i)\Big) = \min_{u_1,\dots,u_q} \Big\{q^{-1}\sum_{i=1}^{q} h\Big(q\big(F_i(x_i) + u_i\big)\Big) \;\Big|\; \sum_{i=1}^{q} u_i = 0\Big\},$$

which confirms the claim. □

If $h$ is positively homogeneous, i.e., $\lambda h(z) = h(\lambda z)$ for $\lambda \ge 0$, then $q^{-1}$ cancels the inside $q$ in the lemma so the last expression simplifies to

$$\sum_{i=1}^{q} h\big(F_i(x_i) + u_i\big) \le \alpha.$$

This would be the case if $h = \iota_C$, with $C$ being a cone, or if

$$h(z) = \sum_{i=1}^{m} h_i(z_i) \quad \text{and} \quad h_i(z_i) = \max\{\alpha_i z_i, \beta_i z_i\}$$

for scalars $\alpha_i$ and $\beta_i$, which are common forms in applications.

Using the lemma, we obtain a reformulation of (10.26) as

$$\underset{u\in\mathbb{R}^{mq}, x\in\mathbb{R}^n}{\text{minimize}} \ \varphi(u, x) = \sum_{i=1}^{q} \varphi_i(u_i, x_i) \ \text{subject to} \ (u, x) \in X, \tag{10.27}$$

where $x = (x_1, \dots, x_q)$, $u = (u_1, \dots, u_q)$ and

$$\varphi_i(u_i, x_i) = f_i(x_i) + q^{-1}h\Big(q\big(F_i(x_i) + u_i\big)\Big)$$

$$X = \Big\{(u_1, \dots, u_q, x) \in \mathbb{R}^{mq} \times \mathbb{R}^n \;\Big|\; \sum\nolimits_{i=1}^{q} u_i = 0,\;\; x \in X_0\Big\}.$$

We're now back in a setting amenable to the progressive decoupling algorithm; the objective function is separable with linkage furnished by a subspace constraint. By an argument similar to that leading to 10.5, we obtain

$$X^\perp = \big\{(y_1, \dots, y_q, v) \in \mathbb{R}^{mq} \times \mathbb{R}^n \;\big|\; y_1 = y_2 = \cdots = y_q,\;\; v \in X_0^\perp\big\}.$$

Associated with (10.27), we've the condition

$$0 \in \partial\varphi(u, x) + N_X(u, x),$$

which is necessary for a minimizer in (10.27) under a qualification; see 4.67. Regardless, the condition is equivalently stated as the linkage problem

$$(u, x) \in X,\;\; (y, v) \in X^\perp,\;\; (y, v) \in \partial\varphi(u, x)$$

and can be addressed using the progressive decoupling algorithm.

Step 1 of the algorithm requires the solution of

$$(y^\nu, v^\nu) \in \partial\varphi(u, x) + \kappa\big((u, x) - (u^\nu, x^\nu)\big),$$

which decouples by 4.63 into

$$(y_i^\nu, v_i^\nu) \in \partial\varphi_i(u_i, x_i) + \kappa\big((u_i, x_i) - (u_i^\nu, x_i^\nu)\big), \quad i = 1, \dots, q. \tag{10.28}$$

These generalized equations can be viewed as the optimality conditions for

$$\Big\{ \underset{u_i \in \mathbb{R}^m, x_i \in \mathbb{R}^{n_i}}{\text{minimize}} \;\; f_i(x_i) + q^{-1} h\Big(q\big(F_i(x_i) + u_i\big)\Big) - \langle y_i^\nu, u_i\rangle - \langle v_i^\nu, x_i\rangle$$
$$+ \tfrac{1}{2}\kappa\|u_i - u_i^\nu\|_2^2 + \tfrac{1}{2}\kappa\|x_i - x_i^\nu\|_2^2, \quad i = 1, \dots, q\Big\}. \tag{10.29}$$

These subproblems could be implemented and solved in Step 1 of the progressive decoupling algorithm as there might be efficient subroutines for this purpose. They are much smaller than the actual problem (10.26) when $q$ is moderately large. Let $(\hat{u}_i^\nu, \hat{x}_i^\nu)$ be the solution furnished by the $i$th generalized equation in (10.28), which we combine to form $\hat{u}^\nu = (\hat{u}_1^\nu, \dots, \hat{u}_q^\nu)$ and $\hat{x}^\nu = (\hat{x}_1^\nu, \dots, \hat{x}_q^\nu)$.

For Step 2 of the algorithm, with $w^\nu = q^{-1}\sum_{i=1}^{q} \hat{u}_i^\nu$, we compute

$$\mathrm{prj}_X(\hat{u}^\nu, \hat{x}^\nu) = \big(\hat{u}_1^\nu - w^\nu, \dots, \hat{u}_q^\nu - w^\nu, \mathrm{prj}_{X_0}(\hat{x}^\nu)\big)$$

as can be seen from (10.3). Hence, for $i = 1, \dots, q$,

$$u_i^{\nu+1} = \hat{u}_i^\nu - w^\nu, \qquad x^{\nu+1} = (x_1^{\nu+1}, \dots, x_q^{\nu+1}) = \mathrm{prj}_{X_0}(\hat{x}^\nu), \tag{10.30}$$
$$v_i^{\nu+1} = v_i^\nu - (\kappa - \theta)(\hat{x}_i^\nu - x_i^{\nu+1}), \qquad y_i^{\nu+1} = y_i^\nu - (\kappa - \theta)w^\nu.$$

We note that $y_i^{\nu+1}$ would be the same for all $i$.

As usual, the bottleneck of the progressive decoupling algorithm is the solution of the subproblems (10.29) in Step 1. In the present setting, augmented Lagrangians offer an opportunity to simplify these subproblems significantly. Let

$$F_i^\nu(x_i) = F_i(x_i) + u_i^\nu.$$

We recall from §6.B that the augmented Lagrangian for the Rockafellian $(u_i, x_i) \mapsto f_i(x_i) + q^{-1}h(q(F_i^\nu(x_i) + u_i))$, under augmenting function $\frac{1}{2}\|\cdot\|_2^2$, is $\bar{l}_i^\nu : \mathbb{R}^{n_i} \times \mathbb{R}^m \times (0, \infty) \to \overline{\mathbb{R}}$ given by

$$\bar{l}_i^\nu(x_i, y_i, \kappa) = \inf_{u_i \in \mathbb{R}^m} \Big\{ f_i(x_i) + q^{-1}h\Big(q\big(F_i^\nu(x_i) + u_i\big)\Big) - \langle y_i, u_i\rangle + \tfrac{1}{2}\kappa\|u_i\|_2^2 \Big\}.$$

An augmented Lagrangian often reduces from this definition to simpler expressions. For example, if $h = \iota_{\{0\}^m}$ as in 6.7 and especially (6.9), then

$$\bar{l}_i^\nu(x_i, y_i, \kappa) = f_i(x_i) + \big\langle F_i^\nu(x_i), y_i\big\rangle + \tfrac{1}{2}\kappa\big\|F_i^\nu(x_i)\big\|_2^2. \tag{10.31}$$

The definition of an augmented Lagrangian involves a minimization that resembles the $u$-component of the subproblems (10.29). In fact, using the change of variable $u_i' = u_i - u_i^\nu$,

$$\begin{aligned}
&\inf_{u_i \in \mathbb{R}^m} \Big\{ f_i(x_i) + q^{-1}h\Big(q\big(F_i(x_i) + u_i\big)\Big) - \langle y_i^\nu, u_i\rangle + \tfrac{1}{2}\kappa\|u_i - u_i^\nu\|_2^2 \Big\}\\
&= \inf_{u_i \in \mathbb{R}^m} \Big\{ f_i(x_i) + q^{-1}h\Big(q\big(F_i^\nu(x_i) + u_i - u_i^\nu\big)\Big) - \langle y_i^\nu, u_i\rangle + \tfrac{1}{2}\kappa\|u_i - u_i^\nu\|_2^2 \Big\}\\
&= \inf_{u_i' \in \mathbb{R}^m} \Big\{ f_i(x_i) + q^{-1}h\Big(q\big(F_i^\nu(x_i) + u_i'\big)\Big) - \langle y_i^\nu, u_i' + u_i^\nu\rangle + \tfrac{1}{2}\kappa\|u_i'\|_2^2 \Big\}\\
&= \bar{l}_i^\nu(x_i, y_i^\nu, \kappa) - \langle y_i^\nu, u_i^\nu\rangle.
\end{aligned}$$

This expression allows us to eliminate $u_i$ from (10.29). Consequently, the subproblems in Step 1 of the algorithm reduce to

$$\Big\{ \underset{x_i \in \mathbb{R}^{n_i}}{\text{minimize}}\ \bar{l}_i^\nu(x_i, y_i^\nu, \kappa) - \langle v_i^\nu, x_i\rangle + \tfrac{1}{2}\kappa\|x_i - x_i^\nu\|_2^2, \quad i = 1, \dots, q \Big\}, \tag{10.32}$$

where the term $\langle y_i^\nu, u_i^\nu\rangle$ is dropped as it doesn't affect the solutions. We've successfully reduced the $i$th subproblem from $n_i + m$ to $n_i$ variables.

The reduced subproblems furnish $\{\hat{x}_i^\nu, i = 1, \dots, q\}$, but we also need to update $u_i^\nu$; it enters in the definition of $F_i^\nu$. As seen from (10.30), this requires $\hat{u}_i^\nu$, the $u$-component of a solution from (10.29). Specifically,

$$\{\hat{u}_i^\nu\} = \operatorname{argmin}_{u_i \in \mathbb{R}^m} \Big\{ q^{-1} h\Big(q\big(F_i(\hat{x}_i^\nu) + u_i\big)\Big) - \langle y_i^\nu, u_i \rangle + \tfrac{1}{2}\kappa \|u_i - u_i^\nu\|_2^2 \Big\},$$

with the minimizer being unique because of strict convexity; see 1.16. Again, there's typically an explicit expression available via the augmented Lagrangian. To see this, let

$$g(y) = \sup_{u_i \in \mathbb{R}^m} \Big\{ -f_i(\hat{x}_i^\nu) - q^{-1} h\Big(q\big(F_i^\nu(\hat{x}_i^\nu) + u_i\big)\Big) + \langle y, u_i \rangle - \tfrac{1}{2}\kappa \|u_i\|_2^2 \Big\}$$

so that $g(y) = -\bar{l}_i^\nu(\hat{x}_i^\nu, y, \kappa)$. This makes $g$ convex by 1.18 and lsc by 6.27; the real-valuedness required in the latter result can be circumvented by redefining the set over which the maximization takes place. Since that maximum is always attained, $g$ is also real-valued and thus proper. Moreover, by definition, $g$ is the conjugate of the function

$$u_i \mapsto f_i(\hat{x}_i^\nu) + q^{-1} h\Big(q\big(F_i^\nu(\hat{x}_i^\nu) + u_i\big)\Big) + \tfrac{1}{2}\kappa \|u_i\|_2^2.$$

Thus, the inversion rule 5.37 and the optimality condition 2.19 can be brought in to yield

$$\partial g(y) = \operatorname{argmin}_{u_i \in \mathbb{R}^m} \Big\{ f_i(\hat{x}_i^\nu) + q^{-1} h\Big(q\big(F_i^\nu(\hat{x}_i^\nu) + u_i\big)\Big) - \langle y, u_i \rangle + \tfrac{1}{2}\kappa \|u_i\|_2^2 \Big\}.$$

There's a unique minimizer in this case due to strict convexity, which then is the gradient of $g$ at $y$. By 9.40, $\nabla g$ is continuous so $g$ is smooth and $\bar{l}_i^\nu(\hat{x}_i^\nu, \cdot, \kappa)$ as well. The expression for $\hat{u}_i^\nu$ connects with this development via the change of variable $u_i' = u_i - u_i^\nu$. In detail,

$$\begin{aligned} \{\hat{u}_i^\nu\} &= \operatorname{argmin}_{u_i' \in \mathbb{R}^m} \Big\{ f_i(\hat{x}_i^\nu) + q^{-1} h\Big(q\big(F_i^\nu(\hat{x}_i^\nu) + u_i'\big)\Big) - \langle y_i^\nu, u_i' \rangle + \tfrac{1}{2}\kappa \|u_i'\|_2^2 \Big\} + u_i^\nu \\ &= -\nabla_y \bar{l}_i^\nu(\hat{x}_i^\nu, y_i^\nu, \kappa) + u_i^\nu. \end{aligned}$$

Thus, the computation of $\hat{u}_i^\nu$ reduces to the differentiation of the augmented Lagrangian. In the specific case of (10.31), one obtains

$$\hat{u}_i^\nu = -F_i^\nu(\hat{x}_i^\nu) + u_i^\nu = -F_i(\hat{x}_i^\nu).$$

In summary, we've obtained a version of the progressive decoupling algorithm that addresses (10.26) via a generalized equation, which in turn is a necessary optimality condition under a qualification. Starting from $(u^0, x^0) \in X$ and $(y^0, v^0) \in X^\perp$, one proceeds in Step 1 of the algorithm by solving the augmented Lagrangian problems (10.32) to obtain $\hat{x}_i^\nu$ and then set

$$\hat{u}_i^\nu = -\nabla_y \bar{l}_i^\nu(\hat{x}_i^\nu, y_i^\nu, \kappa) + u_i^\nu.$$

Step 2 amounts to updating according to (10.30) with

$$w^\nu = \frac{1}{q} \sum\nolimits_{i=1}^q \hat{u}_i^\nu.$$

The convergence properties of the algorithm follow from 10.23, 10.26 and 10.36.

# References


1. R.K. Ahuja, T.L. Magnanti, J.B. Orlin, *Network Flows: Theory, Algorithms, and Applications* (Pearson, 1993)
2. F. Alizadeh, D. Goldfarb, Second order cone programming. Math. Program. **95**, 3–51 (2003)
3. Z. Allen-Zhu, Z. Qu, P. Richtárik, Y. Yuan, Even faster accelerated coordinate descent using non-uniform sampling, in *Proceedings of the 33rd International Conference on Machine Learning*, pp. 1110–1119 (2016)
4. S. Asmussen, P.W. Glynn, *Stochastic Simulation: Algorithms and Analysis* (Springer, 2007)
5. H. Attouch, R.J-B Wets, Quantitative stability of variational systems: I. the epigraphical distance. Trans. Am. Math. Soc. **328**(2), 695–729 (1991)
6. H. Attouch, R.J-B Wets, A convergence theory for saddle functions. Trans. Am. Math. Soc. **280**(1), 1–41 (1983)
7. J.-P. Aubin, I. Ekeland, *Applied Nonlinear Analysis* (Wiley Interscience, 1984)
8. H.H. Bauschke, P.L. Combettes, *Convex Analysis and Monotone Operator Theory in Hilbert Spaces* (Springer, 2011)
9. G. Bayraksan, D.P. Morton, Assessing solution quality in stochastic programs. Math. Program. **108**(2–3), 495–514 (2006)
10. A. Beck, *First-Order Methods in Optimization* (SIAM, 2017)
11. A. Ben-Tal, A. Nemirovski, *Lectures on Modern Convex Optimization* (SIAM, 2001)
12. A. Ben-Tal, A. Nemirovski, On polyhedral approximations of the second-order cone. Math. Program. **26**(2), 193–205 (2001)
13. D.P. Bertsekas, *Nonlinear Programming*, 2nd edn. (Athena Scientific, 1999)
14. P. Billingsley, *Probability and Measure*, 3rd edn. (Wiley, 1995)
15. J. Bolte, S. Sabach, M. Teboulle, Proximal alternating linearized minimization for nonconvex and nonsmooth problems. Math. Program. **146**, 459–494 (2014)
16. L. Bonfiglio, J.O. Royset, Multi-disciplinary risk-adaptive set-based design of super-cavitating hydrofoils. AIAA J. **57**(8), 3360–3378 (2019)
17. S. Boucheron, G. Lugosi, P. Massart, *Concentration Inequalities: A Nonasymptotic Theory of Independence* (Oxford University Press, 2013)
18. S.P. Boyd, L. Vandenberghe, *Convex Optimization* (Cambridge University Press, 2004)
19. M. Breton, G. Zaccour, M. Zahaf, A game-theoretic formulation of joint implementation of environmental projects. Eur. J. Oper. Res. **168**(1), 221–239 (2006)
20. G.G. Brown, W.M. Carlyle, J. Salmeron, R.K. Wood, Defending critical infrastructure. Interfaces **36**(6), 530–544 (2006)
21. J.V. Burke, T. Hoheisel, Epi-convergent smoothing with applications to convex composite functions. SIAM J. Optim. **23**(3), 1457–1479 (2013)
22. G.C. Calafiore, L. El. Ghaoui, *Optimization Models* (Cambridge University Press, 2014)
23. W.M. Carlyle, J.O. Royset, R.K. Wood, Lagrangian relaxation and enumeration for solving constrained shortest-path problems. Networks **52**(4), 256–270 (2008)
24. S. Chatterjee, P. Diaconis, Estimating and understanding exponential random graph models. Ann. Stat. **41**(5), 2428–2461 (2013)
25. X. Chen, Smoothing methods for nonsmooth, nonconvex minimization. Math. Program. **134**, 71–99 (2012)




J. O. Royset and R. J-B Wets, *An Optimization Primer*, Springer Series in Operations Research and Financial Engineering, https://doi.org/10.1007/978-3-030-76275-9_2

26. M. Clagett, *Nicole Oresme and the Medieval Geometry of Qualities and Motions: a Treatise on the Uniformity and Difformity of Intensities Known as Tractatus de Configurationibus Qualitatum et Motuum* (University of Wisconsin Press, 1968)
27. R.W. Cottle, J.-S. Pang, R.E. Stone, *The Linear Complementarity Problem* (SIAM, 2009)
28. Y. Cui, J.-S. Pang, B. Sen, Composite difference-max programs for modern statistical estimation problems. SIAM J. Optim. **28**(4), 3344–3374 (2018)
29. S.P. Dirkse, M.C. Ferris, The path solver: a nonmonotone stabilization scheme for mixed complementarity problems. Optim. Methods Softw. **5**(2), 123–156 (1995)
30. O. Ditlevsen, H.O. Madsen, *Strutural Reliability Methods* (Wiley, 1996)
31. S. Dolecki, G. Salinetti, R.J-B Wets, Convergence of functions: Equi-semicontinuity. Trans. Am. Math. Soc. **276**, 409–429 (1983)
32. R.A. Durrett, *Probability: Theory and Examples*, 2nd edn. (Duxbury Press, 1996)
33. Y.M. Ermoliev, V.I. Norkin, R.J-B Wets, The minimization of discontinuous functions: Mollifier subgradients. SIAM J. Control. Optim. **33**(1), 149–167 (1995)
34. F. Facchinei, C. Kanzow, Generalized Nash equilibrium problems. 4OR **5**, 173–210 (2007)
35. F. Facchinei, J.-S. Pang, *Finite-Dimensional Variational Inequalities and Complementarity Problems* (Springer, 2007)
36. F. Facchinei, J. Soares, A new merit function for nonlinear complementarity problems and a related algorithm. SIAM J. Optim. **7**(1), 225–247 (1997)
37. A.R. Ferguson, G.B. Dantzig, The allocation of aircraft to routes. An example of linear programming under uncertain demand. Manag. Sci. **3**(1), 45–73 (1956)
38. M.C. Ferris, J.-S. Pang, Engineering and economic applications of complementarity problems. SIAM Rev. **39**(4), 669–713 (1997)
39. M.C. Ferris, A.B. Philpott, Dynamic risked equilibrium. Oper. Res. (2021)
40. G.B. Folland, *Real Analysis. Modern Techniques and Their Applications*, 2nd edn. (Wiley, 1999)
41. H. Föllmer, A. Schied, *Stochastic Finance: An Introduction in Discrete Time*, 3rd edn. (de Gruyter, 2011)
42. D. Friedman, R.M. Isaac, D. James, S. Sunder, *Risky Curves* (Routledge, 2014)
43. M. Fukushima, Equivalent differentiable optimization problems and descent methods for asymmetric variational inequality problems. Math. Program. **53**, 99–110 (1992)
44. M. Fukushima, Z.Q. Luo, P. Tseng, Smoothing functions for second-order-cone complementarity problems. SIAM J. Optim. **12**(2), 436–460 (2001)
45. D. Gade, G. Hackebeil, S.M. Ryan, J.-P. Watson, R.J-B Wets, D.L. Woodruff, Obtaining lower bounds from the progressive hedging algorithm for stochastic mixed-integer programs. Math. Program. **157**(1), 47–67 (2017)
46. G.H. Golub, Some modified matrix eigenvalue problems. SIAM Rev. **15**(2), 318–334 (1973)
47. C.C. Gonzaga, E. Polak, On constraint dropping schemes and optimality functions for a class of outer approximation algorithms. SIAM J. Control Optim. **17**(4), 477–493 (1979)
48. W.E. Hart, C.D. Laird, J.-P. Watson, D.L. Woodruff, G.A. Hackebeil, B.L. Nicholson, J.D. Siirola, *Pyomo - Optimization Modeling in Python* (Springer, 2017)
49. J.L. Higle, S. Sen, Stochastic decomposition: an algorithm for two-stage stochastic linear programs with recourse. Math. Oper. Res. **16**(3), 650–669 (1991)
50. J.L. Higle, S. Sen, On the convergence of algorithms with implications for stochastic and nondifferentiable optimization. Math. Oper. Res. **17**(1), 112–131 (1992)
51. J.L. Higle, S. Sen, Epigraphical nesting: a unifying theory for the convergence of algorithms. J. Optim. Theory Appl. **84**(2), 339–360 (1995)
52. V. Katz, *A History of Mathematics* (Addison Wesley, 2008)
53. K. Kawaguchi, Deep learning without poor local minima, in *Conference on Neural Information Processing Systems*, pp. 586–594 (2016)
54. K. Kawaguchi, L.P. Kaelbling, Elimination of all bad local minima in deep learning, in *Artificial Intelligence and Statistics (AISTATS)* (2020)
55. E.N. Khobotov, Modification of the extra-gradient method for solving variational inequalities and certain optimization problems. USSR Comput. Math. Math. Phys. **27**(5), 120–127 (1987)
56. A.J. King, S.W. Wallace, *Modeling with Stochastic Programming* (Springer, 2012)
57. D.P. Kingma, L.J. Ba, Adam: a method for stochastic optimization, in *International Conference on Learning Representations*, pp. 1–13 (2015)
58. R. Koenker, *Quantile Regression* (Cambridge University Press, 2005)
59. L.A. Korf, R.J-B Wets, Random lsc functions: an ergodic theorem. Math. Oper. Res. **26**(2), 421–445 (2001)
60. G. Lan, *First-Order and Stochastic Optimization Methods for Machine Learning* (Springer, 2020)
61. A.S. Lewis, S.J. Wright, A proximal method for composite minimization. Math. Program. **158**, 501–546 (2016)

62. A. Mafusalov, S. Uryasev, Buffered probability of exceedance: mathematical properties and optimization. SIAM J. Optim. **28**(2), 1077–1103 (2018)
63. M. Mahsuli, Probabilistic models, methods, and software for evaluating risk to civil infrastructure. Ph.D. thesis, University of British Columbia, 2012
64. P. Marcotte, Application of Khobotov's algorithm to variational inequalities and network equilibrium problems. INFOR **29**(4), 258–270 (1992)
65. R.B. Nelsen, *An Introduction to Copulas*, 2nd edn. (Springer, 2006)
66. A. Nemirovski, A. Juditsky, G. Lan, A. Shapiro, Robust stochastic approximation approach to stochastic programming. SIAM J. Optim. **19**(4), 1574–1609 (2009)
67. Y. Nesterov, *Introductory Lectures on Convex Optimization* (Kluwer Academic Publishers, 2004)
68. Y. Nesterov, Efficiency of coordinate descent methods on huge-scale optimization problems. SIAM J. Optim. **22**(2), 341–362 (2012)
69. J. Nocedal, S. Wright, *Numerical Optimization* (Springer, 2006)
70. M. Norton, V. Khokhlov, S. Uryasev, Calculating CVaR and bPOE for common probability distributions with application to portfolio optimization and density estimation. Ann. Oper. Res. **299**, 1281–1315 (2021)
71. M. Norton, A. Mafusalov, S. Uryasev, Soft margin support vector classification as buffered probability minimization. J. Mach. Learn. Res. **18**, 1–43 (2017)
72. M. Norton, J.O. Royset, Diametrical risk minimization: theory and computations. Mach. Learn. (2020)
73. J.-S. Pang, Error bounds in mathematical programming. Math. Program. B **79**(1–3), 299–332 (1997)
74. J.-S. Pang, D. Chan, Iterative methods for variational and complementarity problems. Math. Program. **24**, 284–313 (1982)
75. J.-S. Pang, G. Scutari, F. Facchinei, C. Wang, Distributed power allocation with rate constraints in Gaussian parallel interference channels. IEEE Trans. Inf. Theory **54**(8), 3471–3489 (2008)
76. N. Parikh, S. Boyd, Proximal algorithms. Found. Trends Optim. **1**(3), 123–231 (2013)
77. E.Y. Pee, J.O. Royset, On solving large-scale finite minimax problems using exponential smoothing. J. Optim. Theory Appl. **148**(2), 390–421 (2011)
78. G. C. Pflug, A. Pichler, Multistage Stochastic Optimization (Springer, 2014)
79. C. Phelps, J.O. Royset, Q. Gong, Optimal control of uncertain systems using sample average approximations. SIAM J. Control Optim. **54**(1), 1–29 (2016)
80. A.B. Philpott, M.C. Ferris, R.J-B Wets, Equilibrium, uncertainty and risk in hydrothermal electricity systems. Math. Program. B **157**(2), 483–513 (2016)
81. J. Pietz, J.O. Royset, Generalized orienteering problem with resource dependent rewards. Nav. Res. Logist. **60**(4), 294–312 (2013)
82. J. Pietz, J.O. Royset, Optimal search and interdiction planning. Mil. Oper. Res. **20**(4), 59–73 (2015)
83. E. Polak, *Optimization: Algorithms and Consistent Approximations* (Springer, 1997)
84. E. Polak, J.O. Royset, On the use of augmented Lagrangians in the solution of generalized semi-infinite min-max problem. Comput. Optim. Appl. **31**(6), 173–192 (2005)
85. E. Polak, J.O. Royset, R.S. Womersley, Algorithms with adaptive smoothing for finite minimax problems. J. Optim. Theory Appl. **119**(3), 459–484 (2003)
86. L. Qi, D. Sun, G. Zhou, A new look at smoothing Newton methods for nonlinear complementarity problems and box constrained variational inequalities. Math. Program. **87**, 1–35 (2000)
87. D. Ralph, Global convergence of damped Newton's method for nonsmooth equations via the path search. Math. Oper. Res. **19**(2), 352–389 (1994)
88. P. Richtárik, M. Takáč, Iteration complexity of randomized block-coordinate descent methods for minimizing composite function. Math. Program. A **144**, 1–38 (2011)
89. R.T. Rockafellar, *Convex Analysis* (Princeton University Press, 1970)
90. R.T. Rockafellar, *Conjugate Duality and Optimization* (SIAM, 1974)
91. R.T. Rockafellar, Monotone operators and the proximal point algorithm. SIAM J. Control Optim. **14**, 877–898 (1976)
92. R.T. Rockafellar, Lagrange multipliers and optimality. SIAM Rev. **35**(2), 183–238 (1993)
93. R.T. Rockafellar, Progressive decoupling of linkage in optimization and variational inequalities with elicitable convexity or monotonicity. Set-Valued Var. Anal. **27**, 863–893 (2019)
94. R.T. Rockafellar, Advances in convergence and scope of the proximal point algorithm. J. Nonlinear Convex Anal. **22**(11), 2347–2375 (2021)
95. R.T. Rockafellar, Augmented Lagrangians and hidden convexity in sufficient conditions for local optimality. Math. Program. (2021)
96. R.T. Rockafellar, J.O. Royset, On buffered failure probability in design and optimization of structures. Reliab. Eng. Syst. Saf. **95**, 499–510 (2010)
97. R.T. Rockafellar, J.O. Royset, Random variables, monotone relations, and convex analysis. Math. Program. B **148**(1), 297–331 (2014)

98. R.T. Rockafellar, J.O. Royset, Engineering decisions under risk-averseness. ASCE-ASME J. Risk Uncertain. Eng. Syst. Part A: Civ. Eng. **1**(2), 04015003 (2015)
99. R.T. Rockafellar, J.O. Royset, Measures of residual risk with connections to regression, risk tracking, surrogate models, and ambiguity. SIAM J. Optim. **25**(2), 1179–1208 (2015)
100. R.T. Rockafellar, J.O. Royset, Superquantile/CVaR risk measures: second-order theory. Ann. Oper. Res. **262**(1), 3–28 (2018)
101. R.T. Rockafellar, J.O. Royset, S.I. Miranda, Superquantile regression with applications to buffered reliability, uncertainty quantification, and conditional value-at-risk. Eur. J. Oper. Res. **234**(1), 140–154 (2014)
102. R.T. Rockafellar, S. Uryasev, The fundamental risk quadrangle in risk management, optimization and statistical estimation. Surv. Oper. Res. Manag. Sci. **18**, 33–53 (2013)
103. R.T. Rockafellar, R.J-B Wets, A Lagrangian finite generation technique for solving linear-quadratic problems in stochastic programming. Math. Program. **28**, 63–93 (1986)
104. R.T. Rockafellar, R.J-B Wets, Scenarios and policy aggregation in optimization under uncertainty. Math. Oper. Res. **16**, 119–147 (1991)
105. R.T. Rockafellar, R.J-B Wets, *Variational Analysis* (Springer, 1998) (2009, 3rd printing)
106. J.O. Royset, Approximations and solution estimates in optimization. Math. Program. **170**(2), 479–506 (2018)
107. J.O. Royset, Approximations of semicontinuous functions with applications to stochastic optimization and statistical estimation. Math. Program. **184**, 289–318 (2020)
108. J.O. Royset, Stability and error analysis for optimization and generalized equations. SIAM J. Optim. **30**(1), 752–780 (2020)
109. J.O. Royset, J.-E. Byun, Gradients and subgradients of buffered failure probability. Oper. Res. Let. **49**(6), 868–873 (2021)
110. J.O. Royset, W.M. Carlyle, R.K. Wood, Routing military aircraft with a constrained shortest-path algorithm. Mil. Oper. Res. **14**(3), 31–52 (2009)
111. J.O. Royset, E.Y. Pee, Rate of convergence analysis of discretization and smoothing algorithms for semi-infinite minimax problems. J. Optim. Theory Appl. **155**(3), 855–882 (2012)
112. J.O. Royset, R.J-B Wets, Multivariate epi-splines and evolving function identification problems. Set-Valued Var. Anal. **24**(4), 517–545 (2016). Erratum, pp. 547–549
113. J.O. Royset, R.J-B Wets, On univariate function identification problems. Math. Program. B **168**(1–2), 449–474 (2018)
114. J.O. Royset, R.K. Wood, Solving the bi-objective maximum-flow network-interdiction problem. INFORMS J. Comput. **19**(2), 175–184 (2007)
115. A. Ruszczynski, *Nonlinear Optimization* (Princeton University Press, 2006)
116. A. Ruszczynski, A. Shapiro, Optimization of convex risk functions. Math. Oper. Res. **31**(3), 433–452 (2006)
117. S. Saha, A. Guntuboyina, On the nonparametric maximum likelihood estimator for Gaussian location mixture densities with application to Gaussian denoising. Ann. Stat. **48**(2), 738–762 (2020)
118. A. Shapiro, D. Dentcheva, A. Ruszczynski, *Lectures on Stochastic Programming: Modeling and Theory* (SIAM, 2009)
119. R.M. Van Slyke, R.J-B Wets, L-shaped linear programs with applications to optimal control and stochastic programming. SIAM J. Appl. Math. **17**(4), 638–663 (1969)
120. M.V. Solodov, B.F. Svaiter, A new projection method for variational inequality problems. SIAM J. Control Optim. **37**(3), 765–776 (1999)
121. J.E. Spingarn, Partial inverse of a monotone operator. Appl. Math. Optim. **10**, 247–265 (1983)
122. D. Sun, J. Han, Newton and quasi-Newton methods for a class of nonsmooth equations and related problems. SIAM J. Optim. **7**(2), 463–480 (1997)
123. M.D. Teter, J.O. Royset, A.M. Newman, Modeling uncertainty of expert elicitation for use in risk-based optimization. Ann. Oper. Res. **280**(1–2), 189–210 (2019)
124. R.J-B Wets, Stochastic programs with fixed recourse: the equivalent deterministic program. SIAM Rev. **16**(3), 309–339 (1974)
125. L.A. Wolsey, G.L. Nemhauser, *Integer and Combinatorial Optimization* (Wiley, 1999)

# Index




J. O. Royset and R. J-B Wets, *An Optimization Primer*, Springer Series in Operations Research and Financial Engineering, https://doi.org/10.1007/978-3-030-76275-9

Printed in the United States
by Baker & Taylor Publisher Services